# 2013 天线与传播国际会议论文集

# Proceedings of the 2013 International Symposium on Antennas and Propagation

## （卷二）

东南大学毫米波国家重点实验室　编

U0396300

东南大学出版社
·南京·

**图书在版编目(CIP)数据**

2013 天线与传播国际会议论文集：Proceedings of the 2013 International Symposium on Antennas and Propagation：英文/东南大学毫米波国家重点实验室编. —南京：东南大学出版社,2013.10

ISBN　978-7-5641-4279-7

Ⅰ.①2… Ⅱ.①东… Ⅲ.①天线-国际学术会议-文集-英文②电波传播-国际学术会议-文集-英文Ⅳ.①TN82-53②TN011-53

中国版本图书馆 CIP 数据核字(2013)第 108576 号

2013 天线与传播国际会议论文集

Proceedings of the 2013 International Symposium on Antennas and Propagation

| | | |
|---|---|---|
| 出版发行 | 东南大学出版社 | |
| 社　　址 | 南京市四牌楼 2 号(邮编:210096) | |
| 出 版 人 | 江建中 | |
| 经　　销 | 全国各地新华书店 | |
| 印　　刷 | 常州市武进第三印刷有限公司 | |
| 开　　本 | 787 mm×1092 mm　1/16 | |
| 印　　张 | 89.75 | |
| 字　　数 | 2304 千字 | |
| 版　　次 | 2013 年 10 月第 1 版 | |
| 印　　次 | 2013 年 10 月第 1 次印刷 | |
| 书　　号 | ISBN　978-7-5641-4279-7 | |
| 定　　价 | 360.00 元(共两卷) | |

本社图书若有印装质量问题,请直接与营销部联系,电话:025-83791830。

# Message from the General Chairmen

The ISAP2013 will be held in Nanjing, China on October 23-25, 2013. On behalf of the conference committees, it is our pleasure to welcome all of you to attend the 2013 International Symposium on Antennas and Propagation (ISAP2013) in Nanjing, one of the most beautiful and ancient cities in China.

The 2013 International Symposium on Antennas and Propagation (ISAP2013) provides an international forum for exchanging information on, and updating progresses of, the most recent research and development in antennas, propagation, electromagnetic wave theory, and other related fields. It is also an important objective of this meeting to promote professional networking among conference participants.

Nanjing was awarded the title of Famous Historic and Culture City because she had been Capitals of China for ten times. Today, Nanjing is the Capital City of Jiangsu Province and one of the important wireless communication hubs in China. The ISAP2013 is sponsored and organized by Southeast University, co-sponsored by University of Electronic Sci. & Tech. of China, and technically co-sponsored by CIE Antenna Society, the IEICE Communications Society, IEEE Antennas and Propagation Society, the European Association on Antennas and Propagation (EurAAP), Laboratory of Science and Technology on Antenna and Microwave, IEEE AP/MTT/EMC Nanjing Joint Chapter, the Jiangsu Institute of Electronics, Journal of Microwaves, etc.

ISAP2013 totally received 420 submissions, and finally 359 papers are accepted after its rigorous peer reviews by TPC members based on their technical merits and interests to the antennas and propagation communities. Among a large number of students' papers, 15 papers were selected by the TPC into the final-list for the best student paper contest.

Finally, please enjoy technical sessions of the conference, and also the Chinese culture in, and the beautiful modern city scenery of, the ancient Nanjing.

Prof. Wei Hong, Prof. Joshua Le-Wei Li, and Prof. Shuxi Gong
General Co-Chairs

October 23, 2013

# ISAP2013 Committee Officers

**General Co-chairs**

*Wei Hong (Southeast University)*

*Le-Wei Li (University of Electronic Science and Technology of China & Monash University)*

*Shu-Xi Gong (STAML，NRIET & XIDIAN)*

**IAC Co-chairs**

*Wen-Xun Zhang (Southeast University)*

*Koichi Ito (Chiba University)*

*Zhi-Ning Chen (National University of Singapore)*

**TPC Co-chairs**

*Xiao-Wei Zhu (Southeast University)*

*Dong-Lin Su (Beijing University of Aeronautics and Astronautics)*

*Yi-Jun Feng (Nanjing University)*

*Zhi-Peng Zhou (STAML，NRIET & XIDIAN)*

**Local Arrangement Co-chairs**

*Ji-Xin Chen (Southeast University)*

*Ya-Ming Bo (Nanjing University of Posts Telecommunications)*

*Bing Liu (Nanjing University of Aeronautics and Astronautics)*

**Exhibition Comm. Co-chairs**

*Ju-Lin He (CIE-AS)*

*Zhe Song (Southeast University)*

*Tie Gao (STAML，NRIET & XIDIAN)*

**Publication Comm. Co-chairs**

*Zhang-Cheng Hao (Southeast University)*

*Wen-Quan Che (Nanjing University of Science and Technology )*

*Zu-Ping Qian (PLA-UTS)*

*Can Lin (Nanjing Research Institute of Electronics Technology)*

**Finance Co-chairs**

*Guang-Qi Yang (Southeast University)*

*Chen Yu (Southeast University)*

*You-Cai Lin (STAML，NRIET & XIDIAN)*

# ISAP2013 IAC Members

Ajay Chakraborty (Indian Institute of Technology Kharagpur)

Chi Hou Chan (City University of Hong Kong)

Dau-Chyrh Chang (Oriental Institute of Technology)

Jin Pan (Univ. of Electronic Sci. and Tech.)

Juan Mosig (EurAAP)

Kam Weng Tam (University of Macau)

Ke Wu (Polytechnique Montreal)

Makoto Ando (Tokyo Institute of Technology)

Mazlina Esa (Universiti Teknologi Malaysia)

Prayoot Akkaraekthalin (King Mongkut's Inst. of Tech. North Bangkok)

W. Ross Stone (Stoneware Limited)

Wen Bin Dou (Southeast University)

Yang Hao (Queen Mary, University of London)

Yilong Lu (Nanyang Technological University)

Yingjie Jay Guo (CISRO)

# ISAP2013 TPC Members

Bingzhong Wang (University of Electronic Science and Technology of China)

Cheng Liao (Southwest Jiaotong University)

Dau-Chyrh Chang (Oriental Institute of Technology)

Derek Gray (The University of Nottingham Ningbo)

Dhaval Pujara ( Nirma University)

Fan Yang (Tsinghua University)

Feng Xu (Nanjing University of Posts Telecommunications)

Guohua Zhai (East China Normal University)

Guoqing Luo (Hangzhou Dianzi University)

Hao Xin (The University of Arizona)

Jian Yang (Chalmers Univ. of Technology)

Jiang Zhu (Apple Co.)

Julien Le Kernec (The University of Nottingham Ningbo)

Jun Hu (University of Electronic Science and Technology of China)

Kam Weng Tam (University of Macau)

Keisuke Konno (Tohoku University)

Kin-Lu Wong (Sun Yat-Sen University)

Kwok Wa Leung (City University of Hong Kong)

Lezhu Zhou (Peking University)

Lixin Guo (Xidian University)

Luyi Liu (Antenova Ltd, Cambridge)

Min Zhang (Shanghai Jiaotong University)

Qingxin Chu (South China University of Technology)

Qun Wu (Harbin Institute of Technology)

Ronghong Jin (Shanghai Jiaotong University)

Ruixin Wu (Nanjing University)

Tiejun Cui (Southeast University)

Weixing Sheng (Nanjing University of Science and Technology)

Wen Bin Dou (Southeast University)

Xiangyu Cao (Air Force Engineering University)

Xiaodong Chen (Queen Mary, University of London)

Xiaoxing Yin (Southeast University)

Xiuping Li (Beijing University of Posts and Telecommunications)

Xueguan Liu (Soochow University)

Xuexia Yang (Shanghai University)

Xun Gong (The University of Central Florida)

Ya-Ming Bo (Nanjing University of Posts Telecommunications)

Yan Zhang (Southeast University)

Yang Hao (Queen Mary, University of London)

Yi Huang (University of Liverpool)

Yi-Jun Feng (Nanjing University )

Yongchang Jiao (STAML，NRIET & XIDIAN)

Yongjun Xie (Beijing University of Aeronautics and Astronautics)

Yueping Zhang (Nanyang Technological University)

Yujian Cheng (University of Electronic Science and Technology of China)

Zhengwei Du (Tsinghua University)

Zhenqi Kuai (Southeast University)

Zhijun Zhang (Tsinghua University)

Zhongxiang Shen (Nanyang Technological Univ. Singapore)

# ISAP2013 Secretariat Staff

Chuan Ge (Southeast University)
Fan Meng (Southeast University)
Wencui Zhu (Nanjing Normal University)
Yinjin Sun (Southeast University)
Mei Jiang (Southeast University)
Jun Chen (Southeast University)
Tao Zhang (Southeast University)
Lina Cao (Southeast University)
Yao Li (Southeast University)
Maomao Xia (Southeast University)
Zhihao Tang (Southeast University)

# ISAP2013 Online Support

Guangqi Yang (Southeast University)
Kaihua Gu (Southeast University)

*2013-10-23 AM*

*Room F, 2ⁿᵈ Floor*

## Keynote Speech

*10:00-10:40*............................................................................... 2

**A teleco's view for better and better customer expectations in multi-band, multi-network, multi-device and multi-demand smart society**
*Dr. Shinichi Nomoto*
*KDDI R&D Laboratories Inc.*

*10:40-11:20*............................................................................... 3

**4G/Multiband Handheld Device Antennas and Their Antenna Systems**
*Kin-Lu Wong, Professor*
*Sun Yat-sen University*

*11:20-12:00*............................................................................... 4

**Rethinking the Wireless Channel for OTA testing and Network Optimization by Including User Statistics: RIMP, Pure-LOS, Throughput and Detection Probability**
*Per-Simon Kildal, Professor*
*Chalmers University of Technology*

---

**ORAL Session: WP-1(A)**
**Adv Ant for Radio-Astr. -1**
Session Chair: Jian Yang, Bo Peng

---

| | | |
|---|---|---|
| 13:30 - 13:50 | Design of Antenna Array for the L-band Phased Array Feed for FAST (Invited Paper)<br>*Yang Wu, Xiaoyi Zhang, Biao Du, Chengjin Jin, Limin Zhang, Kai Zhu* (China) | **11** |
| 13:50 - 14:10 | Progress in SHAO 65m Radio Telescope Antenna (Invited Paper)<br>*Biao Du, Yuanpeng Zheng, Yifan Zhang, Wancai Zhang, Zhiqiang Shen, Qingyuan Fan, Qinghui Liu, Quanbao Ling* (China) | **14** |
| 14:10 - 14:30 | Promoting the Planetary Radio Science in the Lunar and Deep Space Explorations of China<br>*Jinsong Ping* (China) | **17** |
| 14:30 - 14:50 | Initial Considerations of the 5 Meter Dome A Terahertz Explorer (DATE5) for Antarctica<br>*Ji Yang, Zheng Lou, Yingxi Zuo, Jingquan Cheng* (China) | **21** |
| 14:50 - 15:10 | Design Study on Near-Field Radio Holography of the 5-Meter Dome A Terahertz Explorer<br>*Ying-Xi Zuo, Zheng Lou, Ji Yang, Jingquan Cheng* (China) | **25** |

---

**ORAL Session: WP-2(A)**
**Adv Ant for Radio-Astr. -2**
Session Chair: Biao Du, Jian Yang

---

| | | |
|---|---|---|
| 16:00 - 16:20 | Calculation of the Phase Center of an Ultra-wideband Feed for Reflector Antennas (Invited Paper)<br>*Jian Yang* (Sweden) | **30** |
| 16:20 - 16:40 | Dish Verification Antenna China for SKA (Invited Paper)<br>*Xiaoming Chai, Biao Du, Yuanpeng Zheng, Lanchuan Zhou, Xiang Zhang, Bo Peng* (China) | **33** |
| 16:40 - 17:00 | Telescopes for IPS Observations (Invited Paper)<br>*Li-Jia Liu, Lan-Chuan Zhou, Bin Liu, Cheng-Jin Jin, Bo Peng* (China) | **37** |
| 17:00 - 17:20 | Design of the 4.5m Polar Axis Antenna for China Spectral Radio Heliograph Array<br>*Chuanfeng Niu, Jingchao Geng, Yihua Yan, Guodong Yang, Donghe Zhao, Chao Liu, Zhijun Chen, Biao Du, Yang Wu* (China) | **41** |
| 17:20 - 17:40 | The optics of the Five-hundred-meter Aperture Spherical radio Telescope<br>*Chengjin Jin, Kai Zhu, Jin Fan, Hongfei Liu, Yan Zhu, Hengqian Gan, Jinglong Yu, Zhisheng Gao, Yang Cao, Yang Wu* (China) | **44** |

## ORAL Session: WP-1(B)
## New Strategies of CEM-1
Session Chair: Chao-Fu Wang, Haogang Wang

## ORAL Session: WP-2(B)
## New Strategies of CEM-2
Session Chair: Wen-Yan Yin, Lianyou Sun

## ORAL Session: WP-1(C)
## UWB Antennas
Session Chair: Zhongxiang Shen, Qiang Chen

| | | |
|---|---|---|
| 13:30 - 13:50 | Low VSWR High Efficiency Ultra-Wideband Antenna for Wireless Systems Applications<br>*Na Li, Haili Zhang, Yi Jiang* (China) | **85** |
| 13:50 - 14:10 | Printed Slot Antenna for WLAN/WiMAX and UWB Applications<br>*Pichet Moeikham, Prayoot Akkaraekthalin* (Thailand) | **88** |
| 14:10 - 14:30 | A Compact Lower UWB Band PIFA for BAN Applications<br>*XiongYing Liu, Chunping Deng, YuHao Fu* (China) | **92** |
| 14:30 - 14:50 | A TE-shaped Monopole Antenna with Semicircle Etching Technique on ground plane for UWB Applications<br>*Amnoiy Ruengwaree, Watcharaphon Naktong, Apirada Namsang* (Thailand) | **95** |
| 14:50 - 15:10 | Design and Analysis of a Modified Sierpinski Carpet Fractal Antenna for UWB Applications<br>*Bin Shi, Zhiming Long, Jili Wang, Lixia Yang* (China) | **99** |

## ORAL Session: WP-2(C)
## Broadband Antennas
Session Chair: Derek Gray, Jin Shi

| | | |
|---|---|---|
| 16:00 - 16:20 | Broadband Loaded Cylindrical Monopole Antenna<br>*Solenne Boucher, Ala Sharaiha, Patrick Potier* (France) | **104** |
| 16:20 - 16:40 | Wideband frangible monopole for radio monitoring<br>*Derek Gray* (China) | **107** |
| 16:40 - 17:00 | Hybrid Antenna Suitable for Broadband Mobile Phone MIMO System<br>*Minkil Park, Theho Son, Youngmin Jo* (South Korea) | **111** |
| 17:00 - 17:20 | Design of a High-Gain Wideband Microstrip antenna with a Stepped Slot Structure<br>*Weigang Zhu, Tongbing Yu, Weimin Ni* (China) | **115** |
| 17:20 - 17:40 | A Q-Band Dual-Mode Cavity-Backed Wideband Patch Antenna with Independently Controllable Resonances<br>*Tao Zhang, Yan Zhang, Shunhua Yu, Wei Hong, Ke Wu* (China) | **118** |

## ORAL Session: WP-1(D)
## Compact Antennas
Session Chair: Kin-Lu Wong, Yuehe Ge

13:30 - 13:50 Compact Hybrid Dielectric Resonator with Patch Antenna Operating at Ultra-high Frequency Band **123**
*Muhammad Ishak Abdul Sukur, Mohamad Kamal A Rahim, Noor Asniza Murad (Malaysia)*

13:50 - 14:10 A UHF Band Compact Conformal PIFA Array **127**
*Hao Wang, Jing Zhou, Kang Chen, Yong Huang, Jie Wang, Xiaoqi Zhu (China)*

14:10 - 14:30 Small-Size Printed Antenna with Shaped Circuit Board for Slim LTE/WWAN Smartphone Application **131**
*Hsuan-Jui Chang, Kin-Lu Wong, Fang-Hsien Chu, Wei-Yu Li (Taiwan)*

14:30 - 14:50 Compact and Planar Near-field and Far-field Reader Antenna for Handset **133**
*Wenjing Li, Yuan Yao, Junsheng Yu, Xiaodong Chen (China)*

14:50 - 15:10 Printed Loop Antenna with an Inductively Coupled Branch Strip for Small-Size LTE/WWAN Tablet Computer Antenna **136**
*Meng-Ting Chen, Kin-Lu Wong (Taiwan)*

## ORAL Session: WP-2(D)
## Small Antennas
Session Chair: Seong-Ook Park, Wen-Shan Chen

16:00 - 16:20 An Idea for Low-Profile Unidirectional Slot Antennas Based on Its Complementary Dipoles **139**
*Jiang Xiong, Xuelian Li, Bing-Zhong Wang (China)*

16:20 - 16:40 A New Radiation Method for Ground Radiation Antenna **143**
*Hongkoo Lee, Jongin Ryu, Jaeseok Lee, Hyunghoon Kim, Hyeongdong Kim (South Korea)*

16:40 - 17:00 Design of A Novel Quad-band Circularly Polarized Handset Antenna **146**
*Youbo Zhang, Yuan Yao, Junsheng Yu, Xiaodong Chen, Yixing Zeng, Naixiao He (China)*

17:00 - 17:20 Mobile handset antenna with parallel resonance feed structures for wide impedance bandwidth **149**
*Yongjun Jo, Kyungnam Park, Jaeseok Lee, Hyunghoon Kim, Hyeongdong Kim (South Korea)*

17:20 - 17:40 Dual Beam Antenna for 6-Sector Cellular System **151**
*Yoshihiro Kozuki, Hiroyuki Arai, Huiling Jiang, Taisuke Ihara (Japan)*

| 15:10 - 16:00 | **POSTER Session: WP-C** |
| | **Best Student Papers Contest** |

| 15:10 - 16:00 | POSTER Session: WP-P | |
|---|---|---|

## ORAL Session: TA-1(A)
## EurAAP/COST
Session Chair: Per-Simon Kildal

| | | |
|---|---|---|
| 08:00 - 08:20 | A General Technique for THz Modeling of Vertically Aligned CNT Arrays (Invited Paper) | **281** |
| | *Jiefu Zhang, Yang Hao* (United Kingdom) | |
| 08:20 - 08:40 | Efficient numerically-assisted modelling of grounded arrays of printed patches (Invited Paper) | **284** |
| | *María García-Vigueras, Francisco Mesa, Francisco Medina, Raúl Rodríguez-Berral, Juan R. Mosig* (Switzerland) | |
| 08:40 - 09:00 | Correlation Between Far-field Patterns on Both Sides of the Head of Two-port Antenna on Mobile Terminal (Invited Paper) | **288** |
| | *Ahmed Hussain, Per-Simon Kildal, Ulf Carlberg, Jan Carlsson* (Sweden) | |
| 09:00 - 09:20 | Antenna Measurement Intercomparison Campaigns in the framework of the European Association of Antennas and Propagation (Invited Paper) | **290** |
| | *Lucia Scialacqua, F. Mioc, Jiaying Zhang, Lars Foged, M. Sierra-Castañer* (France) | |
| 09:20 - 09:40 | Capacitively-Loaded THz Dipole Antenna Designs with High Directivity and High Aperture Efficiency (Invited Paper) | **294** |
| | *Ning Zhu, Richard WZiolkowski* (United States) | |

## ORAL Session: TA-2(A)
## Wireless Power Transmis.
Session Chair: Le-Wei Li, Qiang Chen

| | | |
|---|---|---|
| 10:30 - 10:50 | Theoretical Analysis, Design and Optimization of Printed Coils for Wireless Power Transmission (Invited Paper) | **298** |
| | *Jia-Qi Liu, Yi-Yao Hu, Yin Li, Le-Wei Li* (China) | |
| 10:50 - 11:10 | Rectifier Conversion Efficiency Increase in Low Power Using Cascade Connection at X-band (Invited Paper) | **302** |
| | *JoonWoo Park, Youngsub Kim, Youngjoong Yoon, Jinwoo Shin, Joonho So* (South Korea) | |
| 11:10 - 11:30 | Interference Reduction Method Using a Directional Coupler in a Duplex Wireless Power Transmission System | **306** |
| | *Kengo Nishimoto, Kenzaburo Hitomi, Takeshi Oshima, Toru Fukasawa, Hiroaki Miyashita, Yoshiyuki Takahashi, Yoshiyuki Akuzawa* (Japan) | |
| 11:30 - 11:50 | A Hybrid Method on the Design of C Band Microwave Rectifiers | **310** |
| | *Chengyang Yu, Biao Zhang, Sheng Sun, Changjun Liu* (China) | |
| 11:50 - 12:10 | Analysis of Near-Field Power Transfer of Multi-Antenna Using Multiport Scattering Parameters | **313** |
| | *Mingda Wu, Qiang Chen, Qiaowei Yuan* (Japan) | |

## ORAL Session: TA-1(B)
## Computational EM
Session Chair: André Barka, Yaming Bo

| | | |
|---|---|---|
| 08:00 - 08:20 | Fast Broad-band Angular Response Sweep Using FEM in Conjunction with Compressed Sensing Technique<br>*Lu Huang, Bi-yi Wu, Xin-qing Sheng* (China) | **318** |
| 08:20 - 08:40 | Finite Macro-Element Method for Two-Dimensional Eigen-Value Problems<br>*Huapeng Zhao, Zhongxiang Shen* (Singapore) | **322** |
| 08:40 - 09:00 | Accelerated Plasma Simulations using the FDTD Method and the CUDA Architecture<br>*Wei Meng, Yufa Sun* (China) | **325** |
| 09:00 - 09:20 | A Near-Surface Interpolation Scheme Based on Radial Basis Function<br>*Canlin Pan, Ming Zhang, Yaming Bo* (China) | **329** |
| 09:20 - 09:40 | 1D Modified Unsplitted PML ABCs for truncating Anisotropic Medium<br>*Zhichao Cai, Shuibo Wang, Haochuan Deng, Lixia Yang, Xiao Wei, Hongcheng Yin* (China) | **333** |

## ORAL Session: TA-2(B)
## EM Scattering
Session Chair: Kiyotoshi Yasumoto, Zhenhai Shao

| | | |
|---|---|---|
| 10:30 - 10:50 | Enhancement of Near Fields Scattered by Metal-Coated Dielectric Nanocylinders<br>*Pei-Wen Meng, Kiyotoshi Yasumoto, Yun-Fei Liu* (China) | **337** |
| 10:50 - 11:10 | Diffraction Components given by MER Line Integrals of Physical Optics across the Singularity on Reflection Shadow Boundary<br>*Pengfei Lu, Makoto Ando* (Japan) | **341** |
| 11:10 - 11:30 | Near-Field Scattering Characters of the Ship<br>*Chonghua Fang, Xuemei Huang, Qiong Huang, Hui Tan, Jing Xiao* (China) | **343** |
| 11:30 - 11:50 | Hybrid SPM to investigate scattered field from rough surface under tapered wave incidence<br>*Qing Wang, Xiao-bang Xu, Zhenya Lei, Yongjun Xie* (China) | **347** |
| 11:50 - 12:10 | Study on the Optical Properties of Nanowires Using FDTD Method<br>*Xiang Huang, Liang Yu, Jin-Yang Chu, Zhi-Xiang Huang, Xian-Liang Wu* (China) | **351** |

---

**ORAL Session: TA-1(C)**
## WLAN Antennas
Session Chair: Jui-Han Lu, Yuan Yao

---

| | |
|---|---|
| 08:00 - 08:20 A Compact Microstrip-Line-Fed Printed Parabolic Slot Antenna for WLAN Applications | **356** |
| *Wanwisa Thaiwirot, Norakamon Wongsin* (Thailand) | |
| 08:20 - 08:40 Dual-band Circularly Polarized Monopole Antenna for WLAN Applications | |
| *Hao-Shiang Huang, Jui-Han Lu* (Taiwan) | **360** |
| 08:40 - 09:00 Tapered Slot Antenna with Squared Cosine Profile for WLAN Applications | |
| *Yosita Chareonsiri, Wanwisa Thaiwirot, Prayoot Akkaraekthalin* (Thailand) | **363** |
| 09:00 - 09:20 A Triple Band Arc-Shaped Slot Patch Antenna for UAV GPS/Wi-Fi Applications | |
| *Jianling Chen, Kin-Fai Tong, Junhong Wang* (China) | **367** |
| 09:20 - 09:40 Dual-Band Printed L-Slot Antenna for 2.4/5 GHz WLAN Operation in the Laptop Computer | **371** |
| *Saran Prasong, Rassamitut Pansomboon, Chuwong Phongcharoenpanich* (Thailand) | |

---

**ORAL Session: TA-2(C)**
## Patch Antennas
Session Chair: Dhaval Pujara, Xinmi Yang

---

| | |
|---|---|
| 10:30 - 10:50 Analysis of L-Probe Fed-Patch Microstrip Antennas in a Multilayered Spherical Media | **376** |
| *Tao Yu, Chengyou Yin* (China) | |
| 10:50 - 11:10 Research on Circularly Polarized Small Disk Coupled Square Ring Microstrip Antenna for GPS Application | **380** |
| *Peng Cheng, Tongbin Yu, Hongbin LI, Wenquan Cao* (China) | |
| 11:10 - 11:30 A 35GHz Stacked Patch Antenna with Dual-Polarized Operations | |
| *Xuexia Yang, Guannan Tan, Yeqing Wang* (China) | **384** |
| 11:30 - 11:50 Design of a Circularly Polarized Elliptical Patch Antenna using Artificial Neural Networks and Adaptive Neuro-Fuzzy Inference System | **388** |
| *Aarti Gehani, Jignesh Ghadiya, Dhaval Pujara* (India) | |
| 11:50 - 12:10 Circularly Polarized Micrstrip Antenna Based on Waveguided Magneto-Dielectrics | |
| *Xinmi Yang, Huiping Guo, Xueguan Liu* (China) | **391** |

## ORAL Session: TA-1(D)
### Measurements
Session Chair: Hiroyoshi Yamada, Ji Yang

08:00 - 08:20 Fast Measurement Technique Using Multicarrier Signal for Transmit Array
Antenna Calibration ................................................................................. **396**
*Kazunari Kihira, Toru Takahashi, Hiroaki Miyashita* (Japan)

08:20 - 08:40 Evaluation of RCS Measurement Environment in Compact Anechoic Chamber
*Naobumi Michishita, Tadashi Chisaka, Yoshihide Yamada* (Japan) ............ **400**

08:40 - 09:00 Stable Parameter Estimation of Compound Wishart Distribution for Polarimetric
SAR Data Modeling ................................................................................. **404**
*Yi Cui, Hiroyoshi Yamada, Yoshio Yamaguchi* (Japan)

09:00 - 09:20 Narrow Pulse Transient Scattering Measurements and Elimination of Multi-path
Interference .............................................................................................. **408**
*Zichang Liang, Wei Gao, Jinpeng Fang* (China)

09:20 - 09:40 A Composite Electromagnetic Absorber for Anechoic Chambers
*Weijia Duan, Han Chen, Mingming Sun, Yi Ding, Xiaohan Sun, Chun Cai,* ... **412**
*Xueming Sun* (China)

## ORAL Session: TA-2(D)
### Radio Propagation
Session Chair: Mazlina Esa, Dongya Shen

10:30 - 10:50 A 3-D FDTD Scheme for the Computation of HPM Propagation in Atmosphere
*Ke Xiao, Shunlian Chai, Haisheng Zhang, Huiying Qi, Ying Liu* (China) ...... **416**

10:50 - 11:10 Analysis of Schumann Resonances based on the International Reference
Ionosphere ............................................................................................... **419**
*Yi Wang, Xiao Yuan, Qunsheng Cao* (China)

11:10 - 11:30 Quantitative Analysis of Rainfall Variability in Tokyo Tech MMW Small-Scale
Model Network ........................................................................................ **422**
*Hung V.Le, Takuichi Hirano, Jiro Hirokawa, Makoto Ando* (Japan)

11:30 - 11:50 A Nyström-Based Esprit Algorithm for DOA Estimation of Coherent Signals
*Yuanming Guo, Wei Li, Yanyan Zuo, Junyuan Shen* (China) ....................... **426**

11:50 - 12:10 Impact of Reconfiguring Inclination Angle of Client's Antenna on Radio Channel
Characteristics of IEEE802.11ac System .................................................. **430**
*Hassan El-Sallabi, Mohamed Abdallah, Khalid Qaraqe* (Qatar)

| 09:40 - 10:30 | **POSTER Session: TA-P** | |
|---|---|---|

## ORAL Session: TP-1(A)
### Body-central Antennas
Session Chair: Koichi ITO, Zhao Wang

| | | |
|---|---|---|
| 13:30 - 13:50 | Multi-Functional Small Antennas for Health Monitoring Systems (Invited Paper) <br> *Chia-Hsien Lin, Koichi Ito, Masaharu Takahashi, Kazuyuki Saito* (Japan) | **569** |
| 13:50 - 14:10 | Performance of An Implanted Tag Antenna in Human Body (Invited Paper) <br> *Hoyu Lin, Masaharu Takahashi, Kazuyuki Saito, Koichi Ito* (Japan) | **573** |
| 14:10 - 14:30 | Design of Low Profile On-body Directional Antenna <br> *Juneseok Lee, Jaehoon Choi* (South Korea) | **577** |
| 14:30 - 14:50 | K-factor Dependent Multipath Characterization for BAN-OTA Testing Using a Fading Emulator <br> *Kun Li, Kazuhiro Honda, Koichi Ogawa* (Japan) | **580** |
| 14:50 - 15:10 | Development of VHF-band Antenna Mounted on the Helmet <br> *Yuma Ono, Yoshinobu Okano* (Japan) | **584** |

## ORAL Session: TP-2(A)
### Body-central Propagation
Session Chair: Yang Hao, Yoshinobu Okano

| | | |
|---|---|---|
| 16:00 - 16:20 | Signal Propagation Analysis for Near-Field Intra-Body Communication Systems (Invited Paper) <br> *Kohei Nagata, Tomonori Nakamura, Mami Nozawa, Yuichi Kado, Hitoshi Shimasaki, Mitsuru Shinagawa* (Japan) | **589** |
| 16:20 - 16:40 | Numerical investigation on a Body-Centric Scenario at W Band (Invited Paper) <br> *Khaleda Ali, Alessio Brizzi, Alice Pellegrini, Yang Hao* (United Kingdom) | **593** |
| 16:40 - 17:00 | A Wearable Repeater Relay System for Interactive Real-time Wireless Capsule Endoscopy (Invited Paper) <br> *Sam Agneessens, Thijs Castel, Patrick Van Torre, Emmeric Tanghe, Günter Vermeeren, Wout Joseph, Hendrik Rogier* (Belgium) | **597** |
| 17:00 - 17:20 | Phase Characterization of 1-200 MHz RF Signal Coupling with Human Body (Invited Paper) <br> *Nannan Zhang, Zedong Nie, Lei Wang* (China) | **601** |
| 17:20 - 17:40 | Electromagnetic Wave Propagation of Wireless Capsule Antennas in the Human Body <br> *Zhao Wang, Enggee Lim, Meng Zhang, Jingchen Wang, Tammam Tillo, Jinhui Chen* (China) | **605** |

## ORAL Session: TP-1(B)
## SIW Antennas & Devices
Session Chair: Jian Yang, Jiro Hirokawa

| | | |
|---|---|---|
| 13:30 - 13:50 | Substrate Integrated Waveguide Antenna Arrays for High-Performance 60 GHz Radar and Radio Systems (Invited Paper)<br>*Ajay Babu Guntupalli, Ke Wu* (Canada) | 610 |
| 13:50 - 14:10 | A New E-plane Bend for SIW Circuits and Antennas Using Gapwave Technology (Invited Paper)<br>*Jian Yang, Ali Razavi Parizi* (Sweden) | 614 |
| 14:10 - 14:30 | Simplified Wavelength Calculations for Fast and Slow Wave Metamaterial Ridged Waveguides and their Application to Array Antenna Design (Invited Paper)<br>*Hideki Kirino, Koichi Ogawa* (Japan) | 617 |
| 14:30 - 14:50 | A Novel SIW Slot Antenna Array Based on Broadband Power Divider<br>*Dongfang Guan, Zuping Qian, Yingsong Zhang, Yang Cai* (China) | 621 |
| 14:50 - 15:10 | Novel Antipodal Linearly Tapered Slot Antenna Using GCPW-to-SIW Transition for Passive Millimeter-Wave Focal Plane Array Imaging<br>*Wen Wang, Xuetian Wang, Wei Wang, Aly E.Fathy* (China) | 625 |

## ORAL Session: TP-2(B)
## Integrated MMW Antennas
Session Chair: Yueping Zhang, Bing Zhang

| | | |
|---|---|---|
| 16:00 - 16:20 | The Substrate and Ground Plane Size Effect on Radiation Pattern of 60-GHz LTCC Patch Antenna Array (Invited Paper)<br>*Lei Wang, Yongxin Guo, Wen Wu* (Singapore) | 630 |
| 16:20 - 16:40 | Ultra-broadband Tapered Slot Terahertz Antennas on Thin Polymeric Substrate (Invited Paper)<br>*Masami Inoue, Masayuki Hodono, Shogo Horiguchi, Masayuki Fujita, Tadao Nagatsuma* (Japan) | 633 |
| 16:40 - 17:00 | A D-Band Packaged Antenna on Low Temperature Co-Fired Ceramics for Wire-Bond Connection with an Indium Phosphide Detector (Invited Paper)<br>*Bing Zhang, Li Wei, Herbert Zirath* (Sweden) | 637 |
| 17:00 - 17:20 | Circuit Model and Analysis of Antenna-in-Package (Invited Paper)<br>*Li Li, Wenmei Zhang* (China) | 641 |
| 17:20 - 17:40 | Design, Simulation and Measurement of a 120GHz On-Chip Antenna in 45 nm CMOS for High-Speed Short-Range Wireless Connectors (Invited Paper)<br>*Noel Deferm, Patrick Reynaert* (Belgium) | 645 |

---

## ORAL Session: TP-1(C)
### Reflector & Air-fed Array
Session Chair: Hisamatsu Nakano, Zhi-Hang Wu

---

13:30 - 13:50  A Wideband Dipole Feed for Big Reflector Antenna
*Jinglong Yu, Chengjin Jin* (China) ... **650**

13:50 - 14:10  Low Sidelobe Compact Reflector Antenna Using Backfire Primary Radiator for Ku-Band Mobile Satellite Communication System on Board Vessel
*Shinichi Yamamoto, Shuji Nuimura, Tomohiro Mizuno, Yoshio Inasawa, Hiroaki Miyashita* (Japan) ... **653**

14:10 - 14:30  Fully Metallic Compound Air-fed Array Antennas for 13 GHz Microwave Radio-link Applications
*Zhi-Hang Wu, Wen-Xun Zhang* (China) ... **657**

14:30 - 14:50  Broadband Circularly Polarized Fabry-Perot Resonator Antenna
*Zhenguo Liu, Yongxin Guo, Na Xie* (China) ... **661**

14:50 - 15:10  Monopulse Fabry-Perot Resonator Antenna
*Zhenguo Liu, Yongxin Guo* (China) ... **664**

---

## ORAL Session: TP-2(C)
### Array for Radar Systems
Session Chair: Cornelis G. van 't Klooster, Fan Yang

---

16:00 - 16:20  Reflect-Array Sub-Reflector in X-Ka Band Antenna
*C. G.van 't Klooster, A. Pacheco, C. Montesano, J. A.Encinar, A. Culebras* (Netherlands) ... **669**

16:20 - 16:40  Design of a 60 GHz Band 3-D Phased Array Antenna Module Using 3-D SiP Structure
*Yuya Suzuki, Satoshi Yoshida, Suguru Kameda, Noriharu Suematsu, Tadashi Takagi, Kazuo Tsubouchi* (Japan) ... **673**

16:40 - 17:00  Design of Low Side Lobe Level Milimeter-Wave Microstrip array antenna for Automotive Radar
*Donghun Shin, Kibeom Kim, Jongguk Kim, Seongook Park* (South Korea) ... **677**

17:00 - 17:20  Realizing Sample Matrix Inversion (SMI) in Digital BeamForming (DBF) System
*Hao Lei, Zaiping Nie, Feng Yang* (China) ... **681**

17:20 - 17:40  Co-aperture dual-band waveguide monopulse antenna
*Yuanwun Liu, Fengwei Yao, Yuanbo Shang* (China) ... **685**

## ORAL Session: TP-1(D)
## Mobile & Indoor Propag.
Session Chair: Zhizhang Chen, Nadir Hakem

| | | |
|---|---|---|
| 13:30 - 13:50 | Propagation Models for Simulation Scenario of ITS V2V Communications<br>*Hisato Iwai, Ryoji Yoshida, Hideichi Sasaoka* (Japan) | **689** |
| 13:50 - 14:10 | Comparison of Small-Scale parameters at 60 GHz for Underground Mining and Indoor Environments<br>*Yacouba Coulibaly, Gilles YDelise, Nadir Hakem* (Canada) | **693** |
| 14:10 - 14:30 | Study on the Effect of Radiation Pattern on the Field Coverage in Rectangular Tunnel by FDTD method and Point Source Array Approximation<br>*Dawei Li, Yuwei Huang, Junhong Wang, Mei-E Chen, Zhan Zhang* (China) | **697** |
| 14:30 - 14:50 | Modelling of Electromagnetic Propagation Characteristics in Indoor Wireless Communication Systems Using the LOD-FDTD method<br>*Meng-Lin Zhai, Wen-Yan Yin, Zhizhang Chen* (China) | **701** |
| 14:50 - 15:10 | Design of Multi-channel Rectifier with High PCE for Ambient RF Energy Harvesting<br>*Zheng Zhong, Hucheng Sun, Yongxin Guo* (Singapore) | **705** |

## ORAL Session: TP-2(D)
## Wire Antennas
Session Chair: Dau-Chyrh Chang, Hiroyuki Arai

| | | |
|---|---|---|
| 16:00 - 16:20 | Loop Antenna Array for IEEE802.11b/g<br>*Dau-Chyrh Chang, Win-Ming Liang* (Taiwan) | **709** |
| 16:20 - 16:40 | Ground Radiation Antenna using Magnetic Coupling Structure.<br>*Hyunwoong Shin, Yang Liu, Jaeseok Lee, Hyunghoon Kim, Hyeongdong Kim* (South Korea) | **713** |
| 16:40 - 17:00 | A Planar Coaxial Collinear Antenna with Rectangular Coaxial Strip<br>*Jiao Wang, Xueguan Liu, Xinmi Yang, Huiping Guo* (China) | **716** |
| 17:00 - 17:20 | Analysis of a Horizontally Polarized Antenna with Omni-Directivity in Horizontal Plane Using the Theory of Characteristic Modes<br>*Shen Wang, Hiroyuki Arai* (Japan) | **720** |
| 17:20 - 17:40 | High Gain Spiral Antenna with Conical Wall<br>*Jaehwan Jeong, Kyeongsik Min, Inhwan Kim, Sungmin Kim* (South Korea) | **723** |
| 17:40 - 18:00 | Asymmetric TEM Horn Antenna for Improved Impulse Radiation Performance<br>*Hyeongsoon Park, JaeSik Kim, Youngjoong Yoon, JiHeon Ryu, JinSoo Choi* (South Korea) | **725** |

| 15:10 - 16:00 | POSTER Session: TP-P | |
|---|---|---|

---

## ORAL Session: FA-1(A)
## A &P for Mobile Comm.
### Session Chair: J. W. Modelski, Eko T Rahardjo

---

08:00 - 08:20 Emerging Antennas for Modern Communication Systems (Invited Paper)
*J. W. Modelski* (Poland) **862**

08:20 - 08:40 Tunable Antenna Impedance Matching for 4G Mobile Communications
*Peng Liu, Andreas Springer* (Austria) **863**

08:40 - 09:00 A dual-band and dual-polarized microstrip antenna subarray design for Ku-band satellite communications
*Yong Fu, Zhiping Yin, Guoqiang Lv* (China) **867**

09:00 - 09:20 Circularly Polarized Microstrip Antenna Array for UAV Application
*Eko TRahardjo, Fitri YZulkifli, Desriansyah YHerwanto, Basari, Josaphat TSri Sumantyo* (Indonesia) **870**

09:20 - 09:40 Interpolation of Communication Distance in Urban and Suburban Areas
*Kazunori Uchida, Masafumi Takematsu, Jun-Hyuck Lee, Keisuke Shigetomi, Junichi Honda* (Japan) **873**

---

## ORAL Session: FA-2(A)
## A &P for MIMO Comm.
### Session Chair: Richard W Ziolkowski, Jiaying Zhang

---

10:30 - 10:50 MIMO 2x2 Reference Antennas – Measurement Analysis Using the Equivalent Current Technique
*Alessandro Scannavini, Lucia Scialacqua, Jiaying Zhang, Lars Foged, Muhammad Zubair, J. L. A. Quijano, G. Vecchi* (France) **878**

10:50 - 11:10 Design of a High Isolation Dual-band MIMO Antenna for LTE Terminal
*Lili Wang, Chongyu Wei, Weichen Wei* (China) **881**

11:10 - 11:30 Slot Ring Triangular Patch Antenna with Stub for MIMO 2x2 Wireless Broadband Application
*Fitri YuliZulkifli, Daryanto , Eko TjiptoRahardjo* (Indonesia) **885**

11:30 - 11:50 Simple Models for Multiplexing Throughputs in Open- and Closed-Loop MIMO Systems with Fixed Modulation and Coding for OTA Applications
*Xiaoming Chen, Per-Simon Kildal, Mattias Gustafsson* (Sweden) **888**

11:50 - 12:10 Channel estimation method using MSK signals for MIMO sensor
*Keita Ushiki, Kentaro Nishimori, Tsutomu Mitsui, Nobuyasu Takemura* (Japan) **892**

## ORAL Session: FA-1(B)
## MMW & THz Antennas
Session Chair: Xiaodong Chen, Makoto Ando

| | | |
|---|---|---|
| 08:00 - 08:20 | Transmission System for Terahertz Pre-amplified Coaxial Digital Holographic Imager (Invited Paper) | 897 |
| | *Wenyan Ji, Haitao Wang, Zejian Lu, Yuan Yao, Junsheng Yu, Xiaodong Chen (China)* | |
| 08:20 - 08:40 | Millimeter Wave Power Divider Based on Frequency Selective Surface (Invited Paper) | 901 |
| | *Wenyan Ji, Haitao Wang, Xiaoming Liu, Yuan Yao, Junsheng Yu, Xiaodong Chen (China)* | |
| 08:40 - 09:00 | Equivalent Radius of Dipole-patch Nanoantenna with Parasitic Nanoparticle at THz band | 905 |
| | *M. K. H. Ismail, M. Esa, N. N. NIk Abd. Malek, S. A. Hamzah, N. A. Murad, M. F. Mohd. Yusoff, M. R. Hamid (Malaysia)* | |
| 09:00 - 09:20 | Design and Implementation of A Filtenna with Wide Beamwidth for Q-Band Millimeter-Wave Short Range Wireless Communications | 909 |
| | *Zonglin Xue, Yan Zhang, Wei Hong (China)* | |
| 09:20 - 09:40 | Design of Terahertz Ultra-wide Band Coupling Circuit Based on Superconducting Hot Electron Bolometer Mixer | 913 |
| | *Chun Li, Lei Qin, Miao Li, Ling Jiang (China)* | |

## ORAL Session: FA-2(B)
## MMW Antennas
Session Chair: Takeshi Manabe, Yan Zhang

| | | |
|---|---|---|
| 10:30 - 10:50 | Design of a Linear Array of Transverse Slots without Cross-polarization to any Directions on a Hollow Rectangular Waveguide | 918 |
| | *Nhu Quyen Duong, Makoto Sano, Jiro Hirokawa, Makoto Ando, Jun Takeuchi, Akihiko Hirata (Japan)* | |
| 10:50 - 11:10 | Design of Package Cover for 60GHz Small Antenna and Effects of Device Box on Radiation Performance | 921 |
| | *Yuanfeng She, Ryosuke Suga, Hiroshi Nakano, Yasutake Hirachi, Jiro Hirokawa, Makoto Ando (Japan)* | |
| 11:10 - 11:30 | A Novel 60 GHz Short Range Gigabit Wireless Access System using a Large Array Antenna | 924 |
| | *Miao Zhang, Jiro Hirokawa, Makoto Ando, Koji Tokosaki, Toru Taniguchi, Makoto Noda (Japan)* | |
| 11:30 - 11:50 | 60 GHz On-Chip Loop Antenna Integrated in a 0.18 μm CMOS Technology | 927 |
| | *Yuki Yao, Takuichi Hirano, Kenichi Okada, Jiro Hirokawa, Makoto Ando (Japan)* | |
| 11:50 - 12:10 | Microstrip Comb-Line Antenna with Inversely Tapered Mode Transition and Slotted Stubs on Liquid Crystal Polymer Substrates | 930 |
| | *Ryohei Hosono, Yusuke Uemichi, Han Xu, Ning Guan, Yusuke Nakatani, Masahiro Iwamura (Japan)* | |

## ORAL Session: FA-1(C)
## Ant. Analysis & Synthesis
Session Chair: Riccardo E Zich, Toru Uno

| | | |
|---|---|---|
| 08:00 - 08:20 | A Modified BBO for Design and Optimization of Electromagnetic Systems (Invited Paper) <br> *Marco Mussetta, Paola Pirinoli, Riccardo EZich* (Italy) | **935** |
| 08:20 - 08:40 | Understanding the Fundamental Radiating Properties of Antennas with Characteristic Mode Analysis <br> *Danie Ludick, Gronum Smith* (South Africa) | **939** |
| 08:40 - 09:00 | Characterization of H2QL Antenna by Simulation <br> *Erwin BDaculan, Elmer PDadios* (Philippines) | **940** |
| 09:00 - 09:20 | FDTD Analysis of Induced Current of PEC Wire Which In Contact with Half Space Lossy Ground by Using Surface Impedance Boundary Condition <br> *Takuji Arima, Toru Uno* (Japan) | **944** |
| 09:20 - 09:40 | Synthesis of Cosecant Array Factor Pattern Using Particle Swarm Optimization <br> *Min-Chi Chang, Wei-Chung Weng* (Taiwan) | **948** |

## ORAL Session: FA-2(C)
## Freq. Selective Surface
Session Chair: Toshikazu Hori, Zhenguo Liu

| | | |
|---|---|---|
| 10:30 - 10:50 | Gain Enhancement for Multiband Fractal Antenna Using Hilbert Slot Frequency Selective Surface Reflector <br> *Chamaiporn Ratnaratorn, Norakamon Wongsin, Prayoot Akkaraekthalin* (Thailand) | **953** |
| 10:50 - 11:10 | Unit Cell Structure of AMC with Multi-Layer Patch Type FSS for Miniaturization <br> *Ying Ming, Kuse Ryuji, Hori Toshikazu, Fujimoto Mitoshi, Seki Takuya, Sato Keisuke, Oshima Ichiro* (Japan) | **957** |
| 11:10 - 11:30 | Scattering Analysis of Active FSS Structures Using Spectral-Element Time-Domain Method <br> *Hao Xu, Jian Xi, Rushan Chen* (China) | **961** |
| 11:30 - 11:50 | A Novel Frequency Selective Surface for Ultra Wideband Antenna Performance Improvement <br> *Huifen Huang, Shaofang Zhang, Yuanhua Hu* (China) | **965** |
| 11:50 - 12:10 | Terahertz Cassegrain Reflector Antenna <br> *Xiaofei Xu, Xudong Zhang, Zhipeng Zhou, Tie Gao, Qiang Zhang, Youcai Lin, Lei Sun* (China) | **969** |

### ORAL Session: FA-1(D)
## EM in Circuits-1
Session Chair: Dhaval Pujara, Qunsheng Cao

| | | |
|---|---|---|
| 08:00 - 08:20 | Simplified Modeling of Ring Resonator (RR) and Thin Wire Using Magnetization and Polarization with Loss Analysis<br>*Dongho Jeon, Bomson Lee* (South Korea) | **973** |
| 08:20 - 08:40 | Transmission Characteristics of Via Holes in High-Speed PCB<br>*He Xiangyang, Lei Zhenya, Wang Qing* (China) | **977** |
| 08:40 - 09:00 | Novel W-slot DGS for Band-stop Filter<br>*Chen Lin, Minquan Li, Wei Wang, Jiaquan He, Wei Huang* (China) | **981** |
| 09:00 - 09:20 | Coupled-Mode Analysis of Two-Parallel Post-Wall Waveguides<br>*Kiyotoshi Yasumoto, Hiroshi Maeda, Vakhtang Jandieri* (Japan) | **984** |
| 09:20 - 09:40 | Systematic Microwave Network Analysis for Arbitrary Shape Printed Circuit Boards With a Large Number of Vias<br>*Xinzhen Hu, Liguo Sun* (China) | **988** |

### ORAL Session: FA-2(D)
## EM in Circuits-2
Session Chair: Trevor S. Bird, Wenmei Zhang

| | | |
|---|---|---|
| 10:30 - 10:50 | A Novel Phase Shifter Based on Reconfigurable Defected Microstrip Structure (RDMS) for Beam-Steering Antennas (Invited Paper)<br>*Can Ding, Jay Y.Guo, Pei-Yuan Qin, Trevor S.Bird, Yintang Yang* (Australia) | **993** |
| 10:50 - 11:10 | Transient Response Analysis of a MESFET Amplifier Illuminated by an Intentional EMI Source<br>*Qifeng Liu, Jingwei Liu, Chonghua Fang* (China) | **997** |
| 11:10 - 11:30 | Crosstalk Analysis of Through Silicon Vias With Low Pitch-to-diameter ratio in 3D-IC<br>*Sheng Liu, Jianping Zhu, Yongrong Shi, Xing Hu, Wanchun Tang* (China) | **1001** |
| 11:30 - 11:50 | Design of a Feed Network for Cosecant Squared Beam based on Suspended Stripline<br>*Huiying Qi, Fei Zhao, Lei Qiu, Ke Xiao, Shunlian Chai* (China) | **1005** |
| 11:50 - 12:10 | The study on Crosstalk of Single Wire and Twisted-Wire Pair<br>*Lijuan Tang, Zhihong Ye, Linglu Chen, Zheng Xiang, Cheng Liao* (China) | **1008** |

| 09:40 - 10:30 | **POSTER Session: FA-P** | |
|---|---|---|

| P.1 | Bow-tie Shaped Meander Slot on-body Antenna | 1013 |
| | *Chen Yang, Guang Hua, Ping Lu, Houxing Zhou* (China) | |
| P.2 | Evaluation Koch Fractal Textile Antenna using Different Iteration toward Human Body | 1017 |
| | *Mohd EzwanJalil, Mohamad KamalRahim, Noor AsmawatiSamsuri, Noor AsnizarMurad, Bashir DBala* (Malaysia) | |
| P.3 | Compact UWB Antenna with Controllable Band Notches Based On Co-directional CSRR | 1021 |
| | *Tong Li, Huiqing Zhai, Guihong Li, Changhong Liang* (China) | |
| P.4 | ULTRA-WIDEBAND DUAL POLARIZED PROBE FOR MEASUREMENT APPLICATION | 1025 |
| | *Yong Li, Meng Su, Yuzhou Sheng, Liang Dong* (China) | |
| P.5 | Conformal Monopulse Antenna Design Based on Microstrip Yagi Antenna | 1029 |
| | *Chen Ding, Wenbin Dou* (China) | |
| P.6 | A Printed Monopole Antenna with Two Coupled Y-Shaped Strips for WLAN/WiMAX Applications | 1032 |
| | *Zhihui Ma, Huiqing Zhai, Zhenhua Li, Bo Yan, Changhong Liang* (China) | |
| P.7 | Planar Circularly Polarized Antenna with Broadband Operation for UHF RFID System | 1036 |
| | *Jui-Han Lu, Hai-Ming Chin, Sang-Fei Wang* (Taiwan) | |
| P.8 | A Frequency Selection Method Based on Fusion Algorithm in Bistatic HFSWR | 1040 |
| | *Weiwei Chen, Changjun Yu, Wentao Chen* (China) | |
| P.9 | Effects of Antenna Polarization on Power and RMS Delay Spread in LOS/OOS Indoor Radio Channel | 1044 |
| | *Zhong-Yu Liu, Li-Xin Guo, Wei Tao, Chang-Long Li* (China) | |
| P.10 | An RF Self-interference Cancellation Circuit for the Full-duplex Wireless Communications | 1048 |
| | *Binqi Yang, Yunyang Dong, Zhiqiang Yu, Jianyi Zhou* (China) | |
| P.11 | Ad Hoc Quantum Network Routing Protocol based on Quantum Teleportation | 1052 |
| | *Xiaofei Cai, Xutao Yu, Xiaoxiang Shi, jin Qian, Lihui Shi, Youxun Cai* (China) | |
| P.12 | The Service Modeling and Scheduling forWireless Access Network Oriented Intelligent Transportation System (ITS) | 1056 |
| | *Xiaojun Wang, Haikuo Dai, Xiaoshu Chen* (China) | |
| P.13 | Radio Channel Modeling and Measurement of a Localization Rescue System | 1060 |
| | *Lunshang Chai, Jiao He, Xingchang Wei* (China) | |
| P.14 | A Weighted OMP Algorithm for Doppler Super-resolution | 1064 |
| | *Xiaochuan Wu, Weibo Deng, Yingning Dong* (China) | |
| P.15 | Developing RSR for Chinese Astronomical Antenna and Deep Space Exploration | 1068 |
| | *Jinsong Ping* (China) | |
| P.16 | Microwave Attenuation and Phase Shift in Sand and Dust Storms | 1069 |
| | *Qunfeng Dong, Yingle Li, Jiadong Xu, Mingjun Wang* (China) | |
| P.17 | Experimental Research on Electromagnetic Wave Attenuation in Plasma | 1072 |
| | *Li Wei, Suo Ying, Qiu Jinghui* (China) | |
| P.18 | Modulation Recognition Based on Constellation Diagram for M-QAM Signals | 1075 |
| | *Zhendong Chou, Weining Jiang, Min Li* (China) | |

## ORAL Session: FP-1(A)
## Antennas for RFID
Session Chair: Kyeong-Sik Min, Wen Wu

| | | |
|---|---|---|
| 13:30 - 13:50 | A Low-Profile Dual-Band RFID Antenna Combined With Silence Element<br>*Yongqiang Chen, Huiping Guo, Xinmi Yang, Xueguan Liu* (China) | **1146** |
| 13:50 - 14:10 | Impedance Matching Design of Small Normal Mode Helical Antennas for RFID Tags<br>*Yi Liao, Yuan Zhang, Kun Cai, Zichang Liang* (China) | **1150** |
| 14:10 - 14:30 | Circularly Polarized Antenna with Circular Shaped Patch and Strip for Worldwide UHF RFID Applications<br>*Yi Liu, Xiong-Ying Liu* (China) | **1154** |
| 14:30 - 14:50 | Material Property of On-metal Magnetic Sheet Attached on NFC/HF-RFID Antenna and Research of Its Proper Pattern and Size On<br>*Naoki Ohmura, Eriko Takase, Satoshi Ogino, Yoshinobu Okano, Shyota Arai* (Japan) | **1158** |
| 14:50 - 15:10 | A Low-Profile Planar Broadband UHF RFID Tag Antenna for Metallic Objects<br>*Zhen-Kun Zhang, Xiong-Ying Liu* (China) | **1162** |

**ORAL Session: FP-1(B)**
**Inversed Scattering**
Session Chair: Yisok Oh, Xincheng Ren

| | | |
|---|---|---|
| 13:30 - 13:50 | Inversion of the dielectric constant from the co-polarized ratio and the co-polarized discrimination ratio of the scattering coefficient | **1167** |
| | *Yuanyuan Zhang, Yaqing Li, Zhensen Wu, Xiaobing Wang* (China) | |
| 13:50 - 14:10 | Microwave Radiation Image Reconstruction Based on Combined TV and Haar Basis | **1171** |
| | *Lu Zhu, Jiangfeng Liu, Yuanyuan Liu* (China) | |
| 14:10 - 14:30 | Imaging of object in the presence of rough surface using scattered electromagnetic field data | **1175** |
| | *Pengju Yang, Lixin Guo, Chungang Jia* (China) | |
| 14:30 - 14:50 | A Simple and Accurate Model for Radar Backscattering from Vegetation-covered Surfaces | **1179** |
| | *Yisok Oh, Soon-Gu Kweon* (South Korea) | |
| 14:50 - 15:10 | The Analysis of Sea Clutter Statistics Characteristics Based On the Observed Sea Clutter of Ku-Band Radar | **1183** |
| | *Zhuo Chen, Xianzu Liu, Zhensen Wu, Xiaobing Wang* (China) | |

## ORAL Session: FP-1(C)
### Slot Antennas
Session Chair: Tsenchieh Chiu, Peng Chen

| | | |
|---|---|---|
| 13:30 - 13:50 | Design of slot antenna loaded with lumped circuit components<br>*Chichang Hung, Tayeh Lin, Hungchen Chen, Tsenchieh Chiu, Dachiang Chang (Taiwan)* | **1188** |
| 13:50 - 14:10 | Narrow-wall confined slotted waveguide structural antennas for small multi-rotor UAV<br>*Derek Gray, Kunio Sakakibara, Yingdan Zhu (China)* | **1192** |
| 14:10 - 14:30 | Pattern Synthesis Method for a Center Holed Waveguide Slot Array Applied to Composite Guidance<br>*Jingjian Huang, Shaoyi Xie, Weiwei Wu, Naichang Yuan (China)* | **1196** |
| 14:30 - 14:50 | Design and Measurement of a Parallel Plate Slot Array Antenna Fed by a Rectangular Coaxial Line<br>*Hajime Nakamichi, Makoto Sano, Jiro Hirokawa, Makoto Ando, Katsumori Sasaki, Ichiro Oshima (Japan)* | **1199** |
| 14:50 - 15:10 | Circularly polarized square slot antenna for navigation system<br>*Yixing Zeng, Yuan Yao, Junsheng Yu, Xiaodong Chen, Youbo Zhang (China)* | **1202** |

## ORAL Session: FP-1(D)
### A&P in Mata-structures
Session Chair: Youngjoong Yoon, Yijun Feng

| | | |
|---|---|---|
| 13:30 - 13:50 | Radiation from a Metahelical Antenna (Invited Paper) | |
| | *Hisamatsu Nakano, Miyu Tanaka, Junji Yamauchi* (Japan) | **1206** |
| 13:50 - 14:10 | Transformation Optical Design for 2D Flattened Maxwell Fish-Eye Lens | |
| | *Guohong Du, Chengyang Yu, Changjun Liu* (China) | **1208** |
| 14:10 - 14:30 | Tunable Electromagnetic Gradient Surface For Beam Steering by Using Varactor Diodes | **1211** |
| | *Jungmi Hong, Youngsub Kim, Youngjoong Yoon* (South Korea) | |
| 14:30 - 14:50 | Asymmetric Electromagnetic Wave Polarization Conversion through Double Spiral Chiral Metamaterial Structure | **1215** |
| | *Linxiao Wu, Bo Zhu, Junming zhao, Yijun Feng* (China) | |
| 14:50 - 15:10 | Metamaterial Absorber with Active Frequency Tuning in X-band | |
| | *Hao Yuan, Bo Zhu, Junming Zhao, Yijun Feng* (China) | **1219** |

| 15:10 - 16:00 | POSTER Session: FP-P |
|---|---|

# *TP-1(C)*

## October 24 (THU) PM

## Room C

## Reflector & Air-fed Array

# A Wideband Dipole Feed for Big Reflector Antenna

Yu Jinglong   Jin Chengjin

National Astronomical Observatories, Chinese Academy of Sciences

Datun Rd. A20, Chaoyang District, Beijing, China, 100012

*Abstract*-A wideband dipole feed with symmetrical E&H plane patterns is presented for use in reflector antennas. In this paper we describe the design, construction and characterization of a wideband dipole feed for FAST antennas covering the frequency range 70 to 140 MHz. Main goals of our design are, 1) covering octave bandwidth, 2) the feed has symmetrical E & H plane patterns, and 3) the physical dimension is suitable for mounting it in the reserved position of the FAST feed cabin . We hope that four frequency range of FAST antennas will be equipped with this wideband dipole feeds. Preliminary simulation results indicate that we have met most of our design goals.

## I. INTRODUCTION

The Five hundred-meter Aperture Spherical Radio Telescope (FAST) is a national facility available for carrying out astronomical and astrophysical studies. It is the largest single-aperture radio telescope in the world with aperture of 500m diameter. Multi-beam and multi-band receivers will be installed covering a frequency range of 70MHz - 3 GHz. The feed system for octave bandwidth below about 1 GHz is currently not available.

It is well known that the most common feeds of the reflector antenna are resonant half wave dipoles and small open-ended waveguides or horn antennas. The half wave dipole can get nearly symmetrical E- and H-planes radiation pattern by using two parallel dipoles with half wavelength spacing [1], or by locating a metal ring of about one wavelength diameter above the dipole [2], [3]. However, it is very difficult to realize octave bandwidth because of limitation of different dipole structure. The feed described in the present paper is kind of two parallel dipole feed using new structure with octave bandwidth.

The horn feeds became very popular during the eighties. In particular, corrugated horns are popular as feeds for dual reflector antennas, and they can have octave bandwidth [4], [5]. The feed used in primary-fed reflectors is often a choke horn. This kind of feed can achieve small variation of the beam-width and the phase center location over a bandwidth of 1.8:1 [6]. But the above mentioned two kinds of horns will be made very large if they used in low frequency bandwidth such as below about 1 GHz.

The main objective of the present work is to research and develop an octave bandwidth feed for the big reflector antennas such as FAST below 1 GHz and to use the feed system to scale in other low frequency range.

## II. CHARACTERIZATION OF FEED

The main goals of the feed system design were to obtain a reasonable aperture efficiency, symmetrical E & H-plane patterns and, a suitable physical dimension for mounting it in the reserved position of the FAST feed cabin. For low frequency, two kinds of feed system have been used in radio telescope. One is the Fat Dipole of ASTRON with 1.56:1 bandwidth [7]. The other one is V-folded Dipole of GMRT with VSWR < 2 over the frequency range 55 to 80 MHz [8], [9]. It is difficult for both of the above mentioned feed to cover octave bandwidth with VSWR < 2. So the desired frequency range of operation (70-140 MHz) for FAST should be researched and kept development.

The design parameters that one can achieve a good performance of the feed are:

a) The distance between the two parallel dipoles is half wavelength spacing.

b) The shape of the dipoles to achieve the required bandwidth.

c) The two parallel dipoles with ground plane.

Some conventional wideband dipole antenna include: the planar wideband dipole (blade, bowtie, diamond, elliptical, etc.), the sleeve volumetric dipole, the droopy-blade dipole and the folded dipole of different variations etc. A modified fat dipole antenna used in ground penetrating radar (GPR) system with a broad bandwidth, between 100MHz and 350MHz is developed by Korea Electro technology Research Institute [10].

A new volumetric ribcage dipole configuration used in wideband UHF for Digital TV system is researched by F. Scappuzzo [11]. In this paper, HFSS is used to model ribcage dipole antenna design with some improvements for this kind of dipole.

a) The coaxial line is used for feed line instead of original design.

b) The shape of the dipoles is modified to achieve the required bandwidth.

## III. FEED DESIGN

The ribcage dipole is similar to the droopy-blade dipole because it occupies the volume beneath the antenna and provides a capacitive coupling to the ground plane; however, it is different because it forms a well-confined metal cavity.

978-7-5641-4279-7

On the other hand, when the sleeves or ribs are missing, the ribcage dipole is reduced to the planar wide-blade dipole.

The ribcage structure with rectangular metal-sheet wings enables a low-cost design. Use HFSS as simulation environment for feed design. At the same time, the coaxial line is used for feed line instead of original design. The geometry for the ribcage dipole, including feed line is shown in Figure 1.

Figure 1. Ribcage dipole above finite ground plane.

Figure 2 presents the simulated return loss for Figure 1 model.

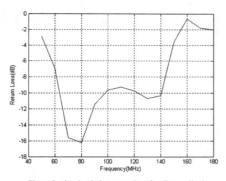

Figure 2. the simulation return loss for ribcage dipole

In this study we focus on the octave bandwidth from 70MHZ to 140MHz. Clearly, the ribcage demonstrates a broadband behavior with good performance.

All simulations are performed in HFSS13, with the PML boundary, with final meshes of about 22,378 tetrahedral, and with a good convergence history. For parametric optimization, the fast frequency sweep is used; all final results have been controlled using discrete frequency sweep.

## IV. IMPROVEMENT OF SHAPE

The shape of the dipoles is the key factor to achieve good performance with wide bandwidth. It is the crux of the matter for the impedance match. So the improvement to the shape of the dipole is necessary.

Figure 3 presents the shape change of the ribcage dipole. The figure is the top view for this dipole. The line changes to curve line from feed point to edge of the dipole. Simulation setup is the same as the above mentioned in section 3.

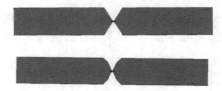

Figure 3. the shape change of the ribcage dipole

The simulation results are shown in figure 4.

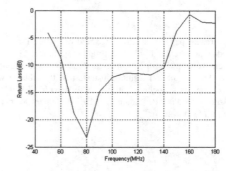

Figure 4. the simulation return loss for shape change of ribcage dipole

We are interested in the feed system consisting of two parallel ribcage dipoles array since it provides symmetric E & H-plane patterns. The configuration of two parallel ribcage dipoles array is shown in figure 5 and the simulation results of radiation pattern is shown in figure 6. The distance between the two parallel dipoles is half wavelength spacing using Figure 1 model.

Figure 5. the simulation model for two parallel ribcage dipole array

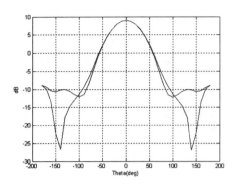

Figure 6. the radiation pattern for two parallel ribcage dipole array

## V. CONCLUSION

The present paper has introduced and discussed the use of the wideband ribcage dipole as a single radiator and in a small array environment with symmetrical E & H plane radiation pattern. The simulation results illustrate that the ribcage with shape change is an octave bandwidth feed with VSWR < 2. This kind of feed is a candidate of radiation element for octave bandwidth feed system.

REFERENCES

[1]   S. Christiansen and Högbom, Radiotelescopes. Cambridge, U.K.: Cambridge Univ. Press, 1969.
[2]   P.-S. Kildal and S. A. Skyttemyr, "Dipole-disk antenna withbeamforming ring," IEEE Trans. Antennas Propag., pp. 529–534,Jul. 1982.
[3]   P.-S. Kildal, S. A. Skyttemyr, and A. A. Kishk, "G/T maximization of a paraboidal reflector fed by a dipole-disk antenna with ring by using the multple-reflection approach and the moment method," IEEE Trans. Antennas Propag., vol. 45, no. 7, pp. 1130–1139, Jul. 1997.
[4]   B. Thomas, G. James, and K. Greene, "Design of wide-band corrugated conical horns for Cassegrain antennas," IEEE Trans. Antennas Propag., vol. 34, no. 6, pp. 750–757, Jun. 1986.
[5]   B. M. Thomas, K. J. Greene, and G. L. James, "A wide-band primefocus horn for low-noise receiver applications," IEEE Trans. Antennas Propag., vol. 38, no. 11, pp. 1898–1900, Nov. 1990.
[6]   Z. Ying, A. A. Kishk, and P.-S. Kildal, "Broadband compact horn feed for prime-focus reflectors," Electron. Lett., vol. 31, no. 14, pp. 1114–1115, Jul. 1995.
[7]   E.E.M. Woestenburg, "LFFE Design Review Report", 27 February 2004.
[8]   N. Udaya Shankar, K.S. Dwarakanath and Shahram Amiri,et al." A 50 MHz System for GMRT",The Low-Frequency Radio Universe ASP Conference Series, Vol. LFRU, 2009
[9]   Shahram Amiri, "A Low Frequency Feed for Big Reflector Antenna", 10th European VLBI Network Symposium and EVN Users Meeting: VLBI and the new generation of radio arrays Manchester, UK September 20-24, 2010
[10]  Korea Electrotechnology Research Institute (KERI) & Microline Co., Ltd., "Buried Small Object Detected by UWB GPR", Proc. of Asia Pacific Microwave Conference 2003 (APMC'03), 2003
[11]  Scappuzzo, F. S. ; Harty, D. D. ; Janice, B. ; Steyskal, H. ; Makarov, S. N , " A Wideband Dipole Array for Directed Energy Applications and Digital TV Reception", 33rd Annual Antenna Applications Symposium September 22-24, 2009

# Low Sidelobe Compact Reflector Antenna Using Backfire Primary Radiator for Ku-Band Mobile Satellite Communication System on Board Vessel

S. Yamamoto, S. Nuimura, T. Mizuno,
Y. Inasawa and H. Miyashita
Mitsubishi Electric Corporation,
5-1-1 Ofuna, Kamakura, Kanagawa, 247-8501 Japan.

*Abstract-* **We developed low side lobe compact reflector antennas of aperture diameter of 60cm and 120cm. They are corresponding to 30 and 60 wavelength. These antennas are axial symmetry antennas using a backfire primary radiator. They are used in Ku-band mobile satellite communication. The requirements of antennas are low side lobes and low cross polarization. The antennas have satisfied ITU-R recommendation S.580-6. This paper introduces the structures and the radiation patterns of the antennas.**

## I. INTRODUCTION

Recent years have seen a growing need for broadband communications services for aircraft, ships, automobiles and other vehicles. Purposes include providing Internet service to passengers, traffic control, and the transmission of high definition video from the sight of disasters and incidents. Introduction of the earth stations on board vessels (ESVs) system using the Ku-band satellite communications systems has been started [1-3]. We propose the two size antennas for the system. They are a 120cm antenna (high gain type) corresponding to global use, and a 60cm antenna (compact type) suitable for a small vessel. It is necessary to conform to international and regional standards and regulations all around the world. For this purpose, antenna needs to be designed to achieve low side lobe with keeping high antenna efficiency. We designed the antenna with which it is satisfied of ITU-R: International Telecommunication Union Radio communications Sector, recommendation S.580-6 radiation pattern mask. In addition, in order to use them at Ku-band, they are operated with linear polarization, and a low cross polarization characteristic is required. In this paper, we report the design and measured results of the compact reflector antennas for the Ku-band satellite communications system on board vessel.

## II. ANTENNA STRUCTURE

In mobile satellite communication, mechanical-drive compact reflector antenna is an attractive candidate. It could achieve the required function and performance in a reasonable cost.

The antenna is scanned to the direction corresponding to the relative location of a satellite and a current position. The antennas are used in radome. In order to make radome size small, it is necessary to make swept volume small. The offset reflector antenna realizes high performance, since there is no blockage, but the swept volume is rather large. The center feed antenna is effective to realize compact reflector antennas.

Dual-reflector antennas have the advantage to optimize aperture distribution by using reflector surface shaping technique. The diameter of the subreflector is several wavelengths at least, in order to achieve the function as a reflector and to realize optimum aperture distribution.

In the single-reflector antenna, making the primary radiator smaller is important in order to reduce blocking, and a backfire primary radiator is effective. A backfire primary radiator is self-supporting structure, so it has the advantage of avoiding the blocking caused by a support structure. However, primary radiation patterns cannot be finely controlled with backfire primary radiators, so there are limits on how far the aperture distribution can be optimized.

Figure 1 shows the antenna structure, which has self-supported backfire primary radiator [4-5] in the centre of a main reflector. A backfire primary radiator consists of a circular feed waveguide and a hat structure supported by a dielectric spacer.

There are three key points of the antenna design for low side lobe level and low cross polarization. They are a backfire primary radiator with a small diameter (small blockage area), the suppression structure of a surface current on the feed

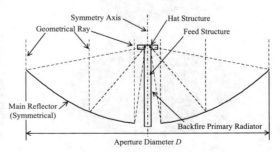

Figure1. Antenna Structure

978-7-5641-4279-7

waveguide, and the hat structure which produces rotational symmetry electromagnetic fields.

For high antenna aperture efficiency, a sharp primary radiation pattern with shallow reflector or a wide primary radiation pattern with deep reflector are chosen in order to suppress the spill-over power from the main reflector. However, for the sharp primary radiation pattern, blockage area by the primary radiator or the subreflector becomes large and side lobe level at the near axis becomes high. We adopt a small backfire primary radiator with a deep reflector.

A surface current on the feed waveguide degrades side lobe patterns especially in near axis of the E-plane. The surface current is suppressed by the corrugation loading in the whole feed waveguide surface and chokes around the aperture of the feed waveguide.

The hat has λ/4 grooves in order to improve the rotational symmetry of the electromagnetic field. The combination of the vertical (parallel to the waveguide axis) and horizontal (normal to the waveguide axis) grooves is effective in stopping the spill-over power from the hat and the main reflector, even if the hat diameter is small. Since reflective conditions differ between E- and H-plane, rotational symmetry is degraded and the cross polarization level is increased. The slope and vertical grooves, instead of the horizontal groove, are suppressed the spill-over, with rotational symmetry maintained.

Figure 2 and 3 show photographs of fabricated antennas for the Ku-band satellite communications system on board vessel. Figure 2 shows 120cm antenna (high gain type antenna). Aperture diameter of the main reflector is 60λ. Figure 2 shows 60cm antenna (compact type antenna). Aperture diameter of the main reflector is 30λ.

## III. RADIATION PATTERN

Figure 3. 60cm antenna (compact type).

An analytical model for the radiation pattern of the antenna is a combined method of electromagnetic analysis and high frequency approximation (physical optics, PO). The electromagnetic fields around the backfire primary radiator are calculated by the finite element method (FEM). Ansoft HFSS is used for the FEM simulation. The currents at the reflector surface are calculated from the electromagnetic fields, and a radiation pattern is calculated by integrating them. This calculation procedure can obtain accurate radiation patterns compared with the conventional calculation method using near or far field radiation patterns of the primary radiator. The ideal reflector surface is used in calculation and the influence of a reflector fabrication error is not include in calculated values. Although the antenna is measured and used with a radome, affects of the radome are not included in calculated values.

Figure 4 shows antenna measurement equipment. These antennas are measured with a compact range measurement system. An inclined fixer is used in order to point the antenna in the direction of 30degree which is the typical elevation angle of satellite station in mid-latitude region.

Figures 5 to 8 show measured and calculated results at transmission frequency band in E-plane (horizontal

Figure 2. 120cm antenna (high gain type).

Figure 4. Antenna measurement equipment.

polarization) and H-plane (vertical polarization), respectively. Since the influence of radome and the influence of a reflector surface error are not taken into calculation, differences are seen between calculated values and measured values. These figures include ITU-R S.580-6 radiation pattern mask. The low side lobe characteristic is achieved over the wide angle range. The radiation patterns are satisfied with the mask.

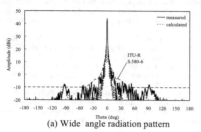

(a) Wide angle radiation pattern

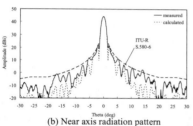

(b) Near axis radiation pattern

Figure 5. 120cm antenna radiation pattern
(E-plane)

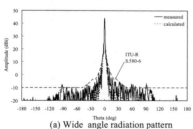

(a) Wide angle radiation pattern

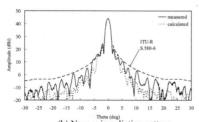

(b) Near axis radiation pattern

Figure 6. 120cm antenna radiation pattern
(H-plane)

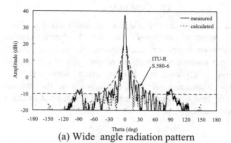

(a) Wide angle radiation pattern

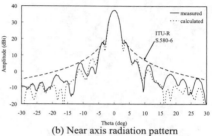

(b) Near axis radiation pattern

Figure 7. 60cm antenna radiation pattern
(E-plane)

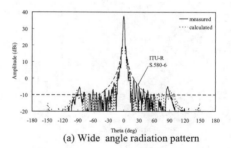

(a) Wide angle radiation pattern

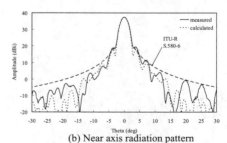

(b) Near axis radiation pattern

Figure 8. 60cm antenna radiation pattern
(H-plane)

## IV. CONCLUSION

We reported the design and measurement results of the compact reflector antennas for the Ku-band satellite communications system on board vessel. The backfire primary radiator, the suppression structure of the surface currents on feed waveguide, and the hat structure with rotational symmetry electromagnetic field are applied to the antenna design. The low side lobe characteristic is achieved over the wide angle range.

## REFERENCES

[1] Y. Konishi, et, al. "Airborne Antenna for Broadband Aeronautical Satellite Communication Systems," IEICE Trans. on Comm., vol.J88-B, no.9, pp.1613-1623, Sept. 2005, in Japanese.

[2] M. Satoh, et al., "Helicopter-Satellite Communication System Developed for Transmission of Disaster and Emergency Information," 21st International Communication Systems Conference and Exhibit, AIAA 2003-2319, 2003.

[3] Y. Inasawa, et al., "Design Method for Low-Profile Dual Shaped Reflector Antenna with an Elliptical Aperture by the Suppression of Undesired Scattering," IEICE Trans. Electron., vol.E91-C, no.4, pp.615-624, April 2008.

[4] P.S. Kildal, et. al, "The Hat Feed: A Dual-Mode Rear-Radiating Waveguide Antenna Having Low Cross Polarization," IEEE Trans. Antenna and Propag., Vol.AP-35, No.9. Sept. 1987.

[5] J. Yang, et. al., "Calculation of Ring Shaped Phase Centers of Feeds for Ring-Focus Parabolids," IEEE Trans. Antenna and Propag., Vol.48, No.4. April 2000.

[6] S. Yamamoto, et. al., "A Ku Band Small Reflector Antenna Using Backfire Primary Radiator for Satellite Communication System on Board Vessel," 2012 International Symposium on Antennas and Propagation, 4A1-3, pp.1173-1176, Nov. 2012.

# Fully Metallic Compound Air-fed Array Antennas for 13 GHz Microwave Radio-link Applications

Zhi-Hang WU[1] and Wen-Xun ZHANG[2]

[1] College of Electronic Science and Engineering, Nanjing University of Posts and Telecommunications, Nanjing, 210003, China.
[2] The State Key Lab. of Millimeter Waves, Southeast University, Nanjing, 210096, China

*Abstract-* **A novel structure of metallic compound air-fed array (CAFA) antennas, as a kind of improved Fabry-Perot resonator (FRP), is proposed in this paper. With stepped metallic ground plate, metallic grating cover and open-ended waveguide feeding, this kind of antenna exhibits features of low profile, low cost and easy integration into microwave out-door unit (ODU). The operating principle, design procedure, simulation analysis and an experimental prototype for application of 13 GHz microwave radio-link are presented. The measured results exhibit: 19.2 dBi peak gain, 3.85% common bandwidth for both 3 dB gain-drop and VSWR≤1.5:1, and good radiation characteristics.**

## I. INTRODUCTION

The concept of CAFA antenna, which could be considered as an improved FPR antenna, has been proposed and studied [1] [2] recently. Compared to traditional FRP antenna, the most essential improvements of CAFA include bandwidth enlargement and/or aperture efficiency enhancement by adopting non-uniform base and/or cover to improve the phase distribution of aperture field [3]-[6]. Most of researches on CAFA so far were based on printed structure, so called as printed CAFA (P-CAFA). P-CAFA owns features of easy fabrication and integration with circuit, but also has some limitations in the out-door applications because of its insufficiency of strength and high cost of microwave dielectric material. In this paper, a metallic CAFA (M-CAFA), which is based on full metallic structure, will be developed to extend its application to radio-link.

Some called metallic EBG resonator antennas, which have similar structure with M-CAFA, had also been studied. They were usually designed for special applications, such as generation of circularly polarized antennas [7] [8] or high performance feed of reflector antenna [9] [10]. But they also had more or less structural encumbrance with additional polarizer or feeding structure, which caused relative high profile and large aperture size and complex feeding. M-CAFA antennas proposed here will exhibit attractive features of compact size, design flexibility, potentiality of wide bandwidth and high aperture efficiency.

Microwave radio-link networks have become the lifeblood of the telecommunications industry, particularly with the rise of high traffic volumes of voice and data for 3G/4G system. One of the main objectives is to provide fast and cost-effective deployment of communications link. Antenna is one of the main components of radio-link system, and usually its size will determine the total size of system to a large extent. Most of the commercial microwave antennas are parabolic antennas, which usually own bulky structure and relative high manufacture cost. The scheme proposed in this paper will provide a very economical solution. Additionally, due to full metallic

structure with low profile and quasi-planar configuration, this kind of antenna has high reliability and can be easily integrated or installed with out-door unit (ODU) or microwave system.

The antenna configuration, design principle and parameter analysis are presented with discussion in detail. A prototype for application of 13 GHz microwave radio-link is designed, fabricated, and measured for verification.

## II. ANALYSIS AND DESIGN

### A. Configuration

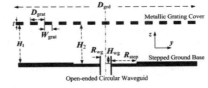

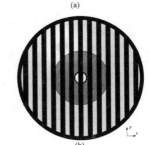

Fig.1. Configuration of proposed M-CAFA antenna. (a) Cutting view with main structural parameters definition. (b) Top view.

The proposed M-CAFA antenna is shown in Fig.1 with definition of most of geometry parameters. It consists of three main parts: a cover of partially reflective surface (PRS) consisting of metallic grating; a base of metal ground with one or more concentric steps; and an open-ended circular waveguide as feed protruded from the centre of base. The outline of antenna shown here is circular with diameter of $D_{grd}$, but it can be also designed into other shape according to required beam shape. The thin grating cover is designed with uniform size: period of grating $D_{grat}$ and width of grating $W_{grat}$. But the ground base is designed with steps $H_{step}=H_2-H_1$, corresponding to different spacing $H_1$ and $H_2$ between cover and base, and inner radius of step $R_{step}$. The circular waveguide is designed to operate in dominant mode of $TE_{11}$ and partly protruded from the base. The radius and insertion length of

978-7-5641-4279-7

waveguide are donated as $R_{wg}$ and $H_{wg}$. The required operating frequency is at $f_0$=13.0 GHz with bandwidth from 12.75 GHz to 13.25 GHz.

### B. Initial Design for Traditional FPR Antenna

As well-known, for a traditional FPR antenna with uniform cover and base, if denote their reflection phase as $\phi_1(f)$ and $\phi_2(f)$ and the reflectivity as $r(f)$ and 1, respectively, then the resonant condition can be presented as:

$$\phi_1(f) + \phi_2(f) = \frac{4\pi H}{c} f - 2N\pi \quad (N=0, 1, 2 \ldots) \quad (1)$$

Where $c$ is the light velocity, $N$ is the order of resonant mode.

On the other side, its directivity gain $D$ is determined by the reflectivity of cover as [11]:

$$D= [1+ r(f)] / [1- r(f)] \quad (2)$$

So, the first step of design is to adopt properly the sizes of grating cover. By considering the operating frequency, realized gain according to (2) and engineering feasibility, the parameters of cover are optimally designed as: $D_{grat}$=10 mm, $W_{grat}$=4mm, with thickness $t$=0.5mm. The reflection magnitude and phase at the cover are shown in Fig.2, in which its reflection phase $\phi_1(f)$=(155~160)$^{\circ}$ and magnitude from 0.96 to 0.97 corresponding to reflectivity $r(f)$=0.92~0.94, covering the frequency range of 12.0~14.0 GHz.

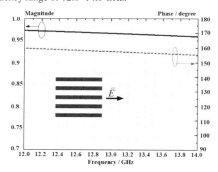

Fig. 2. Reflection coefficients of grating cover. ($D_{grat}$=10 mm, $W_{grat}$=4 mm)

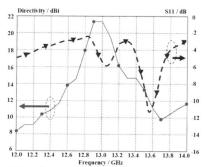

Fig.3. The initial performances of traditional FPR design.

Set the original parameters as: aperture diameter $D_{grd}$=140 mm (6.0 $\lambda_0$), radius of waveguide $R_{wg}$=8.3mm (0.36 $\lambda_0$) and insertion length of waveguide $H_{wg}$=3.5mm (0.15 $\lambda_0$). In the case of ground base without step, the spacing should be calculated from (1) as $H_2$=$H_1$=11.0 mm (0.48 $\lambda_0$). By employing CST-2006 full-wave simulator, the initial performances of traditional FPR design are shown in Fig.3.

Although this design performs 21.7 dBi peak gain and with good radiation pattern at central frequency, but the bandwidth of both gain and impedance matching is obviously insufficient. That is resulted from the nature of a FPR resonator antenna. So, the next design steps should solve these problems.

### C. Improved Design for Gain-Frequency Response

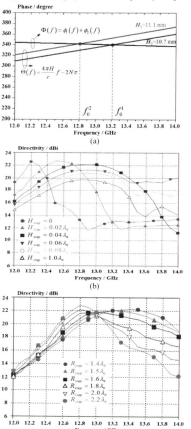

Fig. 4. Gain-frequency response design. (a) Diagram demonstration of resonant condition; (b) Vs. $H_{step}$ ($R_{step}$=1.4 $\lambda_0$, $H_2$=0.49 $\lambda_0$); (c) Vs. $R_{step}$. ($H_{step}$=0.04 $\lambda_0$, $H_2$=0.48 $\lambda_0$).

To enlarge the bandwidth of gain-drop, a key step is to flat the curve of gain-frequency response, based on reducing gradient difference between two sides of equality (1) denoted respectively as:

$$\Phi(f) = \phi_1(f) + \phi_2(f) \qquad (3)$$

$$\Theta(f) = \frac{4\pi H}{c} f - 2N\pi \qquad (4)$$

Where reflection phase $\phi_1(f)$ from the grating-cover and $\phi_2(f) \equiv 180°$ from the metal-base had been given in Fig. 2; and $N=0$ usually for minimal thickness. So the resonant condition (1) can be simplified as

$$\phi_1(f) + \pi = 4\pi H f / c \qquad (5)$$

also be demonstrated in diagram as Fig.4a.

The stepped base makes $H_1 \neq H_2$ ($H_1 < H_2$), corresponding to different resonant sub-frequencies ($f_0{}^1$ and $f_0{}^2$). Their difference $\Delta f = (f_0{}^1 - f_0{}^2)$ can be adjusted by choosing $H_{step} = (H_2 - H_1)$. Compared to the uniform base with only one $f_0$, Fig.4b shows that the gain-frequency response curve of stepped base becomes flatter obviously, and then the gain-drop bandwidth can be significantly enlarged when choosing a proper value of $H_{step}$. Once the stepped base with $H_{step}$ is determined, the bandwidth of gain will also be changed with the radius of inner ground sub-area $R_{step}$, as shown in Fig.4c. A wide and flat gain-frequency response curve can be performed for a proper value of $R_{step} = 1.4\lambda_0$. However, the performance of impedance matching is always dissatisfied without essential improvement, whatever the proper $H_{step}$ and $R_{step}$ are employed, which will be solved in the next design process.

### D. Improved Design for Impedance-matching

The strong reflection returned into the feed seems due to direct reflection from the central part of cover, it may be dispersed by using an additional inverted metal conic structure with radius of $R_{cone}$ and height of $H_{cone}$ as shown in Fig. 5a, by referencing the experience in design a waveguide T-junction. Besides, the insertion length $H_{wg}$ of waveguide plays as an additional adjusted parameter.

Fig. 5b,c,d display the influences of parameters $\{H_{wg}, R_{cone}, H_{cone}\}$ on the frequency response curves of antenna gain and $S_{11}$, respectively, under the case of other proper parameters from the above design procedure as: $D_{grat} = 10$ mm, $W_{grat} = 4$ mm, $D_{grd} = 140$ mm, $R_{wg} = 8.3$ mm, $H_2 = 11.5$ mm, $H_{step} = 0.7$ mm, $R_{step} = 26.2$ mm. Fig. 5b shows that the larger $H_{wg}$ within $(0.05~0.15)\lambda_0$, the wider impedance bandwidth and the slightly lower peak gain.

Fig. 5c shows that $R_{cone}$ affects the bandwidth of both impedance and gain obviously but contradictorily, which means a larger radius of cone is better for impedance bandwidth but results in narrower gain bandwidth and lower peak gain. However, Fig. 5d shows that impedance bandwidth can be improved significantly by tuning $H_{cone}$, while gain deteriorates very slightly. In a word, it is possible to achieve a wholly good performance for both impedance and gain bandwidth by an optimized design procedure.

### III. PROTOTYPE WITH EXPERIMENTAL VERIFICATION

A prototype antenna was fabricated with optimized structural sizes: $D_{grat} = 10$ mm, $W_{grat} = 4$ mm, $D_{grd} = 140$ mm ($6 \lambda_0$),

$R_{wg} = 8.3$ mm ($0.36 \lambda_0$), $H_1 = 10.8$ mm ($0.47 \lambda_0$), $H_2 = 11.5$ mm ($0.5 \lambda_0$), $R_{step} = 26.2$ mm ($1.1 \lambda_0$), $H_{wg} = 3.6$ mm ($0.16 \lambda_0$), $R_{cone} = 4.8$ mm ($0.21 \lambda_0$), $H_{cone} = 6.0$ mm ($0.26 \lambda_0$).

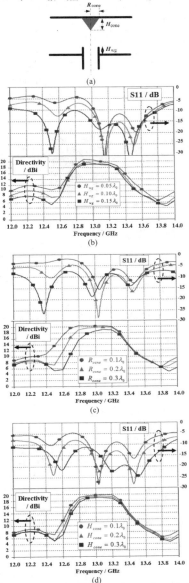

Fig 5. Impedance matching design. (a) Matching configuration; (b) Vs. $H_{wg}$ ($R_{cone} = 0.2 \lambda_0$, $H_{cone} = 0.25 \lambda_0$); (c) Vs. $R_{cone}$ ($H_{wg} = 0.15 \lambda_0$, $H_{cone} = 0.25 \lambda_0$); (d) Vs. $H_{cone}$ ($R_{cone} = 0.2 \lambda_0$, $H_{wg} = 0.15 \lambda_0$).

In Fig. 6a, the measured and simulated curves of gain-frequency response agree with each other very well; the peak gain approaches 19.2 dBi, bandwidth of 3-dB gain-drop covers (12.75~13.37) GHz; its band-pass filter-like curve of gain with distinct features of flat-top and abrupt-edges also benefits the system performance. Fig. 6b shows that measured return loss is better than −14 dB within the required frequency band.

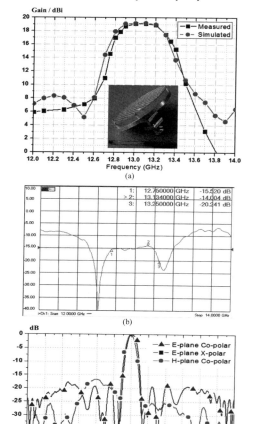

Fig.6. Prototype and performances. (a) Gain-frequency response; (b) Measured return loss; (c) Measured radiation patterns at 13.0 GHz.

Fig.6c presents the measured radiation patterns at designed frequency 13.0 GHz. The simulated ones are not shown for paper brevity. The simulated cross-polarization discrimination (XPD) is less than −40 dB within the main beamwidth in 45°-

plane, and less than even −60 dB within whole angular range in both E- and H-plane. The measured XPD is also so good that the X-polar component in H-plane, which could not be shown in the figure's scale. The measured side-lobe level (SLL) is less than −15 dB in both E- and H-plane, which agrees well with the simulated ones. The radiation performances at the two edge frequencies have not been shown here for paper brevity, however they are quite similar except for a little higher SLL at high frequency.

IV. CONCULSIONS

A novel structure of fully metallic CAFA antenna is proposed. Its design principle and simulation analysis are presented. A prototype for application of 13.0 GHz microwave radio-link has been designed, fabricated and measured for verification. Although the prototype antenna operates in a relative narrow frequency band, it can also be designed for the other wideband systems according to the above method. This kind of antenna exhibits low cost, high reliability, low profile, easy system integration and band-pass filter character. However, only about 20 dBi achievable gain and not consideration of radom are the main insufficiencies in this design, which would be improved in the future design by such as sub-array technique [3].

REFERENCES

[1] W.X. Zhang, D.L. Fu & A.N. Wang, "A compound printed air-fed array antenna," *Proc. Int. Conf. on Electromag. In Advanced Applications*, (Torino, Italy), pp. 1054-1057, Sept. 2007.

[2] W.X. Zhang, "Research progress on printed air-fed antennas", *Proc. 3rd Euro. CAP*, (Berlin, Germany), pp. 1526-1529, Mar. 2009.

[3] Z.H. Wu, W.X. Zhang and D.L. Fu, "A high-efficient compound printed air-fed array based on mushroom-AMC surface and its sub-array technique," *IEEE AP-S*, (San Diego, USA), No. 104.9, July 2008.

[4] Z.H. Wu and W.X. Zhang, "Bandwidth enhancement techniques for printed compound air-fed array antennas", *Int. Symp. on Anten. and Propag.*, (Bangkok, Thailand), pp. 779-782, Oct. 2009.

[5] Zhi-Hang WU and Wen-Xun ZHANG. "Broadband Printed Compound Air-fed Array Antennas with Low Profile". *IEEE Antenna and Wireless Propagation Letters*, Vol. 9, pp. 9-14, 2010.

[6] Z.H. Wu and W.X. Zhang, "On Profile Thickness of Printed Compound Air-fed Array Antenna", *Jour. Electromag. Waves and Appl.*, 24 (2/3), pp. 199-207, 2010.

[7] M. Diblanc, E. Rodes, E. Arnaud, M. Thevenot, T. Monediere, and B. Jecko, "Circularly Polarized Metallic EBG Antenna," *IEEE Microw. and Wireless Components Lett.*, Vol. 15, No. 10, pp.638-640, Oct. 2005.

[8] E. Arnaud, R. Chantalat, Th. Monediere, E. Rodes, and M. Thevenot, "Performance enhancement of self-polarizing metallic EBG antennas," *IEEE Antenna Wireless Propag. Lett.*, Vol. 9, pp.538-541, May 2010.

[9] Andrea Neto, Nuria Llombart, Giampiero Gerini, Magnus D. Bonnedal, and Peter de Maagt, "EBG Enhanced Feeds for the Improvement of the Aperture Efficiency of Reflector Antennas", *IEEE Trans Anten. Propag*, Vol. 55, No. 8, Aug. 2007.

[10] R. Chantalat, C. Menudier, M. Thevenot, T. Monediere, E. Arnaud, and P. Dumon, "Enhanced EBG Resonator Antenna as Feed of a Reflector Antenna in the Ka Band," *IEEE Antenna and Wireless Propagation Letters*, Vol. 7, pp.349-353, 2008.

[11] A.P. Feresidis, G. Goussetis, & S.H. Wang, "Artificial magnetic conductor surfaces and their application to low-profile high-gain planar antennas," *IEEE Trans. AP*-53 (1), pp.209-215, Jan. 2005.

# Broadband Circularly Polarized Fabry-Perot Resonator Antenna

Zhen-Guo LIU[1], Yong-Xin GUO[2], Na XIE[3]

1. State Key Lab. of Millimeter Waves, Southeast University, Nanjing, CHINA, 210096.
2. Department of Electrical and Computer Engineering, National University of Singapore, SINGAPORE, 117576
3. Magnetoeletronic Laboratory, Nanjing Normal University, Nanjing, CHINA, 210097

liuzhenguo@seu.edu.cn, eleguoyx@nus.edu.sg

*Abstract-* The broadband circularly polarized Fabry-Perot resonator (FPR) antenna is presented, which consists of a partially reflective surface (PRS) as cover, a metal-dielectric surface as ground plate and an embedded L-probe coupled rectangular patch as primary radiator. The partially reflective surface covered plate composed of appropriate frequency selective surface (FSS) structure with rectangular patch elements printed on the bottom surface of superstrate. By adjusting the aspect ratio of rectangular patch and the rectangular FSS element, the bandwidth of the circularly polarized (CP) operation can be enhanced. For the proposed antenna, a common frequency bandwidth of 7.7 % for $S_{11} \leq$ -10dB and gain-drop $\leq 3$ dB and axial ratio $\leq 3$ dB is obtained.

## I. INTRODUCTION

Fabry-Perot Resonator (FPR) antennas with attractive features of high-gain, low-profile, and simple feeding have aroused more and more attention for several years [1-4]. It consists of a primary radiator backed with a metal ground plate and consists of a primary radiator backed with a metal ground plate and a partially reflective covered plate [1]. When the spacing between these two plates is about integer times of half wavelength, the energy from the feed is multi-reflected between the cover and ground plate and then the forward radiation can be enhanced remarkably by means of in-phase bouncing. To design a compact and high gain FPR antenna various methods based on different viewpoints and analysis models such as leaky wave model [5], EBG defect model [6], FP cavity model [1, 2], effective material method [3,7] have been attained and proposed.

On the other hand in several wireless communication system and radar applications, where both high gain and circular polarization (CP) are required, FPR antennas can advantageously reduce the complexity of the feeding structure by using the properties of partially reflecting surfaces (PRS). In summary, there are two ways can be used to realize such type of antennas from the viewpoint of primary source (1) a PRS placed directly above a CP feed antenna [8, 9], (2) a PRS arranged directly above a LP feed antenna [10, 11], where PRS acts as a polarizing transform structure to convert linear polarization (LP) of the feed into CP radiation. However, one of the serious disadvantages of this type CP antenna is that its impedance and axial ratio bandwidth are narrow. In this paper, by using the combined techniques incorporating with appropriate aspect ratio of FSS element and feeding patch

design, the broadband circularly polarized FPR antenna can be proposed.

## II. PRINCIPLE AND DESIGN

### A. Principle

In reference [10], the circularly polarized FPR antenna with linear polarization feed is firstly proposed. According to the EM theory, an incident linearly polarized wave $\vec{E}_{in}$, tilting at $45°$ with the plates, can be decomposed into two orthogonal components $E_{in}^{x}$ and $E_{in}^{y}$ of equal amplitude, which can be expressed as:

$$\vec{E}_{in} = \hat{x} \cdot E_{in}^{x} \angle \varphi_x + y \cdot E_{in}^{y} \angle \varphi_y \qquad (1)$$

At the same time, the radiated wave $\vec{E}_{ra}$ from the upper surface of cover plate can also be written as:

$$\vec{E}_{ra} = \hat{x} \cdot E_{ra}^{x} \angle \theta_x + y \cdot E_{ra}^{y} \angle \theta_y \qquad (2)$$

which is relative to the incident wave by

$$\begin{cases} E_{ra}^{x} \angle \theta_x = E_{in}^{x} \angle \varphi_x \cdot T_{ca}^{x} \angle \phi_x \\ E_{ra}^{y} \angle \theta_y = E_{in}^{y} \angle \varphi_y \cdot T_{ca}^{y} \angle \phi_y \end{cases} \qquad (3)$$

where the $T_{ca}^{x}$ and $T_{ca}^{y}$ are the model of transmission functions of FP cavity with respect to x and y components. Due to deployment of rectangular FSS element, the phases of transmission functions of x and y are different. If the condition

$$\begin{cases} \left| E_{ra}^{x} \right| = \left| E_{ra}^{y} \right| \\ \theta_y - \theta_x = \pm 90° \end{cases} \qquad (4)$$

is meet, the circularly polarized radiated wave can be obtained.

### B. Design

Circularly polarized FPR antenna fed by simple dipole with linear polarization is presented in reference [10], but its bandwidth of axial ratio is narrow. In order to improve the property of the bandwidth, an L-probe coupled rectangular patch is proposed as primary radiator, in which the combined techniques incorporating with appropriate aspect ratio of FSS element and coupled rectangular patch are used. Fig.1 shows the geometry of the proposed FPR antenna. The cavity is made of a metal ground plate and a PRS composting of rectangular FSS elements. Both the superstrate of cover and

978-7-5641-4279-7

substrate of base have the same relative permittivity $\varepsilon_r$ =3.2 and thickness $t$. The distance between two parallel planes is $D$. The L-probe coupled patch placed at a distance $h$ from the ground plate. And the L-probe tilts at 45 angle with respect to x direction and the distance between L-probe and coupled patch is $h_1$. The periodicity of FSS in x and y direction is $p_x$ and $p_y$ respectively. The side length of unit cell of FSS in x and y direction is $l_2$ and $w_2$ respectively. As shown in fig.1, the side length of coupled rectangular patch in x and y direction is $l_1$ and $w_1$ respectively. A 45°polarized excitation generated by L-probe can be decomposed into two orthogonal x- and y-components with equal magnitude and phase. But when these two components coupled to the rectangular patch, the different response is occurred due to different side length of rectangular patch, which results in that the radiated wave from coupled rectangular patch is an elliptical polarization. By adjusting the aspect ratio of FSS element and coupled rectangular patch, that is to say, adjusting the transmission function $T_{ca}^x$ and $T_{ca}^y$, finally the broadband circularly polarized FPR antenna can be obtained.

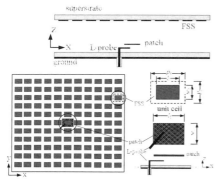

Figure 1. Geometry of proposed antenna

## III. SIMULATION RESULTS

Table 1 show the parameters of geometry of proposed circularly polarized FPR antenna, which has been obtained through the simulation with CST Microwave Studio. The cavity height $D$ is set at 10mm in order to obtain a working frequency around 14 GHz. In simulation, the $w_1/ l_1$ is set to aspect ratio $\tau_1$ and the $w_2/ l_2= p_y/ p_x$ is set to aspect ratio $\tau_2$. Figure 2 presents the frequency responses of axial ratio and gain with respect to the parameters list in Table 1. It is illustrated that the bandwidth of axial ratio lower than 3dB reaches from 13.35 to 14.42GHz. More importantly, the bandwidth of the 3dB gain drop ranges from 12.9 GHz to 14.45GHz, in which the peak value of gain approaches 17.0 dBi at 13.7GHz. The frequency response of reflection coefficient of proposed antenna is shown in Figure 3, where the bandwidth of $S_{11}$ with -10dB threshold is ranged from 13.1 to 15.7GHz. It means that the common or overlapped

bandwidth for axial ratio, gain drop and the impedance matching is about 7.7%.

TABLE I
value of proposed antenna parameters

| Parameter | Dimension (mm) | Parameter | Dimension(mm) |
|-----------|----------------|-----------|---------------|
| $l_1$ | 7.5 | $p_x$ | 6 |
| $w_1$ | 4.8 | $p_y$ | 3.6 |
| $h$ | 3 | $l_2$ | 4.5 |
| $h_1$ | 1 | $w_2$ | 2.7 |
| $D$ | 10 | $t$ | 1.6 |

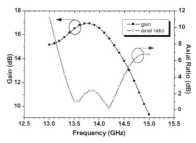

Figure 2. Frequency response of gain and axial ratio

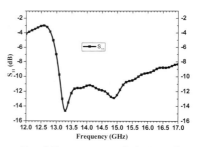

Figure 3. Frequency response of reflection coefficient

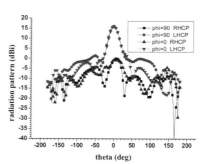

Figure 4. Radiation pattern at $f$=13.4GHz

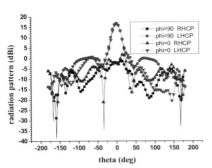

Figure 5. Radiation pattern at $f$=13.7GHz

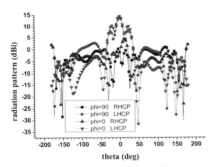

Figure 6. Radiation pattern at $f$=14.4GHz

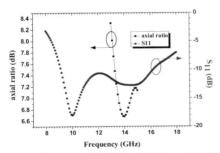

Figure 7. Frequency response of axial ratio and reflection coefficient of L-probe coupled rectangular patch

Figure 4 and Figure 5 show the simulation results of radiation pattern at frequency 13.4GHz and 13.7GHz respectively. At frequency 13.7GHz, the side-lobe level is better as -17.0dB in E and H plane. The -20dB cross-polarization-level (XPL) at broadside and better than 16 dB within half-power beamwidth can be obtained. Figure 6 also shows the radiation pattern at side frequency 14.4GHz, where

the gain approach 15dBi, but the side-lobe-level is only about -5dB.

In order to validate the excited polarization property of the single L-probe coupled rectangular patch, the FPR antenna directly removed the superstrate under the same condition is simulated. Figure 7 shows the simulated result of frequency response of axial ratio and reflection coefficient. It is illustrated that in the common bandwidth the polarization excited by single feed is an elliptical polarization EM wave other than circular polarization, where the minimum axial ratio is 6.65dB. At same time, in the common bandwidth the reflection coefficient is lower than -10dB. From the above statement, it is demonstrated that the combined techniques incorporating with appropriate aspect ratio of the FSS element and the coupled rectangular patch to enhance the bandwidth of circular polarization FPR antenna are in effect.

## IV. CONCLUSION

In this paper, a broadband circularly polarized Fabry-Perot resonator antenna fed by single L-shaped probe coupled rectangular patch and integrated by rectangular FSS element is proposed based on the idea of polarization transform. The bandwidth of axial ratio can be enhanced by adjusting the aspect ratio of rectangular patch and rectangular FSS element, which results in that the excited elliptical polarization EM wave can be transformed into the circular polarization in broad bandwidth.

### ACKNOWLEDGMENT

This work is supported by Open Project of State Key Lab. of Millimeter Waves (Z201203)

### REFERENCES

[1] G.V. Trentini, "Partially reflecting sheet array". *IRE Trans. Antennas Propagat.*,Vol.4, 1956, pp.666-671.

[2] A.P. Feresidis, and J.C. Vardaxoglou, , "High-gain planar antenna using optimized partially reflective surfaces". *IEE Proc Microwave Antennas Propagat* Vol.148, Issue.6, 2001, pp.345–350.

[3] Z.G. Liu, "Fabry-Perot Resonator antenna", *Journal of Infrared Millimeter & Terahertz Waves*, Vol.31,No.4, 2010, pp.391-403.

[4] Z.G. Liu, W.X. Zhang, D.L. Fu, *et al*, "Broadband Fabry-Perot resonator printed antennas using FSS superstrate with dissimilar size", *Microwave & Opt. Tech. Letters*, Vol.50, No.6, 2008. pp.1623-1627.

[5] D. R. Jackson and A. Oliner, "A leaky-wave analysis of the high-gain printed antenna configuration", *IEEE Trans. on Antennas Propag.*, vol. 36, No.7, 1988, pp.905–910.

[6] M. Thevenot, C. Cheype, A. Reineix, and B. Jecko, "Directive photonic-bandgap antennas,"*IEEE Trans. Antennas Propag.*, Vol. 47, No.11, 1999, pp. 2115–2122.

[7] Z.G. Liu, R.Qiang, Z.X.Cao, "A Novel Broadband Fabry-Perot Resonator Antenna with Gradient Index Metamaterial Superstrate", [C], IEEE Int. Symp. Antennas and Propagation, 2010, Toronto, Canada.

[8] Z.G. Liu, Z.X. Cao, "Circularly polarized Fabry-Perot Resonator Antenna", International Conference on Microwave Technology and Computational Elecromagntics 2009, Beijing.

[9] A.R. Weily, K.P. Esselle, T.S. Bird and B.C. Sanders, "High gain circularly polarized 1-D EBG resonator antenna", Electron. Letters, vol. 42, No.18, 2006, pp.1012-1013.

[10] Z.G. Liu, "Fabry-Perot Resonator Antenna with Polarization Transform", IEEE Int. Symp. Antennas and Propagation, 2010, Toronto, Canada.

[11] S. A. Muhammad, R. Sauleau, and H. Legay, "Self generation of circular polarization using compact Fabry–Perot antennas," IEEE Antenna Wireless Propag. Lett., vol. 10, pp. 907–910, 2011.

# Monopulse Fabry-Perot Resonator Antenna

Zhen-Guo LIU[1], Yong-Xin GUO[2]

1. State Key Lab. of Millimeter Waves, Southeast University, Nanjing, CHINA, 210096.
2. Department of Electrical and Computer Engineering, National University of Singapore, SINGAPORE, 117576

liuzhenguo@seu.edu.cn, eleguoyx@nus.edu.sg

*Abstract-* In this paper, a monopulse Fabry-Perot resonator (FPR) antenna is presented, which consists of four sub-array as cells, each one is a Fabry-Perot resonator antenna. An assembly Fabry-Perot resonator cell consists of partially reflective surface (PRS) as cover, a metal-dielectric surface as ground plate and an aperture coupled rectangular patch as primary radiator. Taking advantage of highly directional radiation properties and low complexity with single-feed system of Fabry-Perot resonator antenna, the feeding network of proposed monopulse antenna can be simplified easily, compared with traditional microstrip monopulse antenna. The simulation results show that the maximal gain at operating frequency is 23.7 dBi, and the null-depth of the difference pattern are less than -30 dB in both E-and H-Plane.

## I. INTRODUCTION

The monopulse antennas [1-6] are attractive for high-resolution tracking applications and developed in the variety of technologies, which is a key technique for radar angle estimation. In traditional monopulse radar systems, Cassegrain parabolic antennas [7] or lens antennas [8] are commonly applied. The monopulse comparator in such systems composed of metallic waveguide is usually very complicated and heavy. A lightweight and low cost microstrip array structure has been developed for the monopulse antennas. But the complicated feeding network based on microstrip structure will arise the decrease of antenna radiation efficiency.

On the other hand, a high directive emission based on Fabry-Perot resonator antenna generally consists of a primary radiator backed with a metal ground plate and a partially reflective covered plate [9], has also proposed for several years. When the spacing between these two plates is about integer times of half wavelength, the energy from the feed is multi-reflected between the cover and ground plate and then the forward radiation can be enhanced remarkably by means of in-phase bouncing. Their highly directional radiation properties and low complexity with single-feed system allowing the gain to be increased have aroused more and more attention [10-15]. In view point of effective media, Fabry-Perot resonator antenna can be looked as a kind of lens [11, 16, 17]. In some applications, Fabry-Perot resonator antenna can be acted as a sub-array to make up further highly directive antenna array. However, there is less research on this field.

In this paper, by utilizing the advantage of Fabry-Perot resonator antenna, the monopulse antenna with compact structure and with high efficiency is proposed.

## II. DESIGN AND ANALYSIS OF COMPARATOR NETWORK

### A. Principle

As shown in Fig.1, the comparator network consists of four 3 dB directional coupler and four 90° phase shifter. Excited at different input ports, 1-4, the required amplitude and phase excitations can be generated at output ports which connect to four radiators. Thus a two dimentional monopulse performance can be obtained. As known, the bandwidth of two branch line 3dB directional coupler is narrow. The proposed microstrip-based with 3 branch lines 3dB directional coupler is shown in Fig.2. The parameter of directional coupler is list in table 1. All circuits of comparator are integrated in a single Rogers 5880 substrate with the relative permittivity of 2.2 and the thickness of 0.635mm. Figure 3 shows the simulated results of S parameter of proposed directional coupler.

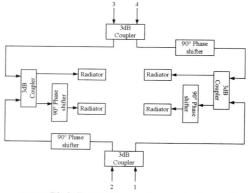

Figure 1. Block diagram of comparator network of monopulse antenna

### B. Design

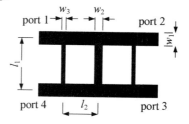

Figure 2. Branch line 3dB directional coupler

Table.1
Parameter of directional coupler (unit: mm)

| $w1$ | $w2$ | $w3$ | $l1$ | $l2$ |
|------|------|------|------|------|
| 1.97 | 1.14 | 0.37 | 5.7 | 4.67 |

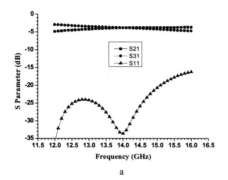

a

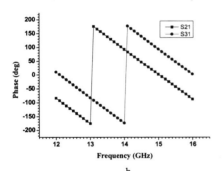

b

Figure 3. S Parameter of Branch line directional coupler
(a) amplitude , and (b) phase

In order to obtain two-dimensional performances of the monopulse antenna, the monopulse comparator or sum–difference feed network is proposed. The configuration of whole comparator is shown in Fig.4, which comprised of four 3 dB hybrid couplers and several 90 ° delay lines. The performance of this comparator was analyzed by HFSS. Excited at different input ports 1-4, the required amplitude and phase excitations can be generated at output ports 5-8. The simulated results of the whole comparator are shown in Fig 5. If the input comes from port 2, the sum beam is then obtained. Meanwhile, if the excitation comes from port 1, the difference radiation pattern will be generated.

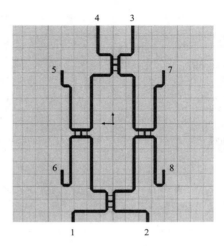

Figure 4. Configuration of whole microstrip comparator

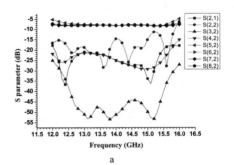

a

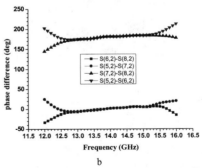

b

Figure 5. S Parameter of proposed comparator (a) amplitude of port 2 (b) phase difference

III. DESIGN AND ANALYSIS OF RADIATOR ELEMENT

The configuration of primary radiator element, aperture coupled microstrip patch antenna, is shown in Fig.6. The radiating microstrip patch element is etched on the top of the antenna substrate, and the microstrip feed line is etched on the bottom of the feed substrate. The fields of microstrip fed line excites the H-shaped aperture etched on the ground plane and then is coupled to the radiating patch. The thicknesses of these two substrates with same dielectric constant are t1 and t2, respectively. The geometry parameters of aperture coupled microstrip patch antenna are listed in table.1. Based on this, a Fabry-Perot resonator antenna fed by aperture coupled microstrip antenna is proposed as shown in figure 7. The squire ring shape FSS is printed on the bottom surface of superstrate, which has the distance hc from the ground plane. The simulation result of S parameter and gain vs frequency of aperture coupled microstrip antenna is shown in figure 8. Figure 9 shows the radiation pattern of E- and H-plane of the proposed Fabry-Perot resonator antenna. The maximum of gain is about 19.5dBi.

TABLE 2
Value of proposed antenna parameters

| Parameter | Dimension (mm) | Parameter | Dimension(mm) |
|---|---|---|---|
| Sx | 13 | $p_x$ | 8.5 |
| Sy | 11 | $p_y$ | 6 |
| sly | 3 | sll | 5 |
| slw | 1 | w1 | 1.97 |
| fl | 2.5 | t1 | 0.635 |
| t2 | 1.575 | | |

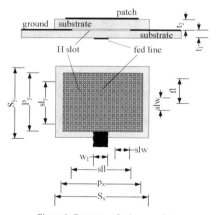

Figure 6. Geometry of primary radiator

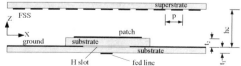

Figure 7. Configuration of Fabry-Perot resonator antenna

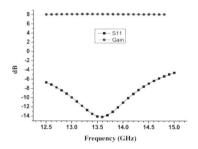

Figure 8. Performance of radiator element aperture coupled patch antenna

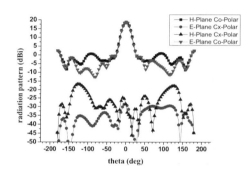

Figure 9. Radiation pattern of Fabry-Perot resonator antenna

## IV. DESIGN AND ANALYSIS OF MONOPULSE ANTENNA

The monopulse comparator network shown above is integrated with an array of four Fabry-Perot resonator antenna fed by aperture coupled patch, as shown in Figure 10, which makes up of Monopulse Fabry-Perot resonator antenna.

Figure 11 and Figure 12 show the E- and H-plane sum and difference beam of monopulse antenna at frequency 14GHz, respectively. It is demonstrated that the gain of sum beam is achieved to 23.7dBi, whereas the gain of the difference beam is less than 2.82 dB. And the null depth of difference beam are less than -30dB in both plane. The difference pattern also shows a good symmetry.

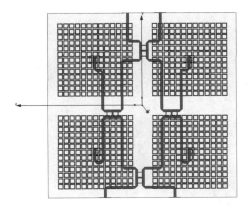

Figure 10. Configuration of proposed Monopulse Fabry-Perot resonator antenna

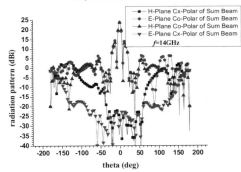

Figure 11. Simulation results of Sum Radiation Pattern @14GHz

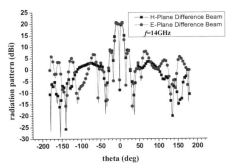

Figure 12. Simulation results of Difference Radiation Pattern @14GHz

V.     CONCLUSION

In this paper, a high gain monopulse antenna consisting of Fabry-Perot resonator antenna acted as sub-array is firstly proposed. Taking advantage of high direction characteristic and single-feed system with low complexity of Fabry-Perot resonator antenna, the feeding network of proposed monopulse antenna can be simplified easily. The simulation results show that the maximal gain at operating frequency is 23.7 dBi, and the null-depth of the difference pattern are less than -30 dB in both E-and H-Plane. This monopulse antenna possesses compact quasi planar architecture, which are particularly suited for airborne applications.

ACKNOWLEDGMENT

This work is supported by Open Project of State Key Lab. of Millimeter Waves (Z201203)

REFERENCES

[1]  S.M. Sherman,Monopulse Principles and Techniques. Dedham,MA: Artech House, 1984.
[2]  C.M. Jackson, "Low cost K-band microstrip patch monopulse antenna," Microwave J., vol. 30, no. 7, pp. 125–126, 1987.
[3]  S. G. Kim and K. Chang, "Low-cost monopulse antenna using Bi-directionally-fed microstrip patch array," Electron. Lett., vol. 39, no. 20, 2003.
[4]  M. Sierra-Castaner,M. Sierra-Perez,M.Vera-Isasa, and J. L. Fernandez-Jambrina, "Low-cost monopulse radial line slot antenna," IEEE Trans. Antennas Propag., vol. 51, no. 2, pp. 256–262, Feb. 2003.
[5]  H. Wang, D.G. Fang, " A Compact Single Layer Monopulse Microstrip Antenna Array", IEEE Trans. Antennas Propag., vol. 54, no. 2, pp. 503–509, Feb. 2006.
[6]  Y.J.Cheng, W.Hong, K.Wu, "94 GHz Substrate Integrated Monopulse Antenna Array", IEEE Trans. Antennas Propag., vol. 60, no. 1, pp. 121–129, Jan. 2012.
[7]  F.R.Connor, "New monopulse tracking radar", Electron. Lett., vol. 19, no. 9, pp. 438-440, 1983
[8]  S.Raman, N.S.Barker, and G.M.Rebeiz, "A W-Band Dielectric-Lens Based Integrated Monopulse Radar Receiver", IEEE Trans. Microwave Theory and Techniques, vol. 46, no. 12, pp. 2308–2316, Dec. 1998.
[9]  G.V. Trentini, "Partially reflecting sheet array". IRE Trans. Antennas Propagat.,Vol.4, 1956, pp.666-671.
[10]  A.P. Feresidis, and J.C. Vardaxoglou, , "High-gain planar antenna using optimized partially reflective surfaces". IEE Proc Microwave Antennas Propagat Vol.148, Issue.6, 2001, pp.345–350.
[11]  Z.G. Liu, "Fabry-Perot Resonator antenna", Journal of Infrared Millimeter & Terahertz Waves, Vol.31,No.4, 2010, pp.391-403.
[12]  Z.G. Liu, W.X. Zhang, D.L. Fu, et al, "Broadband Fabry-Perot resonator printed antennas using FSS superstrate with dissimilar size", Microwave & Opt. Tech. Letters, Vol.50, No.6, 2008. pp.1623-1627.
[13]  D. R. Jackson and A. Oliner, "A leaky-wave analysis of the high-gain printed antenna configuration", IEEE Trans. on Antennas Propag., vol. 36, No.7, 1988, pp.905–910.
[14]  M. Thevenot, C. Cheype, A. Reineix, and B. Jecko, "Directive photonic-bandgap antennas,"IEEE Trans. Antennas Propag., Vol. 47, No.11, 1999, pp. 2115–2122.
[15]  A. Pirhadi, H. Bahrami, and J. Nasri, "Wideband High Directive Aperture Coupled Microstrip Antenna Design by Using a FSS Superstrate Layer", IEEE Trans. Antennas Propag., vol. 60, no. 4, pp. 2101–2105, Apir, 2012.
[16]  Z.G. Liu, R.Qiang, Z.X.Cao, "A Novel Broadband Fabry-Perot Resonator Antenna with Gradient Index Metamaterial Superstrate", [C], IEEE Int. Symp. Antennas and Propagation, 2010, Toronto, Canada.
[17]  Z.G. Liu, "Explanation of broadband Fabry-Perot resonator antenna based on opical viewpoint", [C], NCMMW2013, Chongqing, China.

# *TP-2(C)*

## October 24 (THU) PM

## Room C

## Array for Radar Systems

# Reflect-Array Sub-Reflector in X-Ka Band Antenna

C. G. M. van 't Klooster[1], A. Pacheco[2], C. Montesano[2], J.A. Encinar[3] A. Culebras[4]

[1]ESA, Estec, Noordwijk, The Netherlands
[2]EADS CASA Espacio, Madrid, Spain
[3]University Politecnico Madrid - UPM, Spain
[4]Rymsa Arganda del Rey, Spain

*Abstract-* An X-Ka band high gain antenna configuration is discussed in which the gain is improved by using a reflect-array on the sub-reflector. The implementation is for a dual reflector antenna with a dual frequency feed. Experiences in Spanish institutions have been combined, like reflect-array developments (UPM), feed design and technology (Rymsa) and design and technological manufacturing processes (EADS CASA Espacio). Control of reflection properties (specific design of reflect-array) assists to obtain good performances for an antenna configuration. It allows making efficient use of the antenna aperture for two different bands for high-gain, whilst controlling separate bands.

## I. INTRODUCTION

Interplanetary satellites use frequency bands allocated for Deep Space (S, X or Ka-band) for the telecommunication functions (telemetry, telecommand and data channels). S-band gets overcrowded and interferences with terrestrial services occur more and more frequently, therefore S-band is gradually abandoned for Deep Space. The available bandwidth is only 50 MHz in X-band, leading to 'overpopulation' with more and more interplanetary satellites in service. S/X-band is used on ESA missions like Mars Express and Venus Express. The S/X band antennas have a dichroic sub-reflector transparent for S-band (with a feed in primary focus) and reflective for the X-band feed (secondary focus). These S/X-band antennas use classical surfaces. Some improvement in efficiency is possible by exploiting the feed and dichroic behavior in combination with reflector shaping, but it has appeared to be small.

The mass of such antenna is important. Utilization of novel available materials for reflectors allows reducing the mass. Tri-axial CFRP (Carbon Fiber Reinforced Plastic) reflector technology has matured. A tri-axial CFRP reflector has been used already for a high gain antenna for a Japanese Mars mission (6.5kg [1]) and on various other telecom missions.

One considers today higher frequency bands with 31.8–32.3 GHz for Space-Earth and 34.2–34.7 GHz for Earth to Space. Higher data rates are possible and more bandwidth is available in Ka-band. More users (space-crafts) could be served. Shorter wavelength will impose more stringent requirements on precision for instance for tri-axial or other CFRP reflector technology. Tri-axial technology could be adapted somewhat. Dedicated low-mass technologies are imperative anyway. RF properties, accuracy and dedicated thermal-optical properties are required over wide temperature ranges and need attention. CFRP has somewhat worse reflectivity properties in Ka-band compared to X-band. Solutions for metallization, like thin film with thin metal coating improves RF reflection properties, well-known to CASA.

A high gain antenna optimized for X-band and accurate enough to support Ka-band would allow an improvement of the link budget of several dB's compared to the earlier S/X band antennas. The ESA 35 m Ground Station antennas are prepared to allow Ka-band reception for the Deep-Space bands as well as for Near-Earth applications.

Slightly different frequency bands are also considered in Ka-band for Near-Earth and L-2 missions (Lagrange point). The James Webb Space telescope will use the 26GHz band. In Japan, a mission for space VLBI (VSOP2) considered 37-38 GHz band for data downlink. Such type of missions calls for Ka-band or potentially X-Ka band antennas sometimes with transmit functionality only for a Ka-band data-downlink.

High data rates may require different modulation techniques to be used which has impact on antenna parameters like phase stability and group-delay parameters, to be considered in the design phase. Specifics for high data-rates depend on the mission. Available power on-board the spacecraft is a limiting factor in Deep-Space missions. The utilization of Ka-band has started and so antenna designs for X/Ka band are needed.

The atmospheric loss is higher in Ka-band (compared to X-band) impacting on the link-budget. The loss varies and is strongly dependent on local tropospheric parameters near to the ground station. It will have impact on data transmissions from space. Ka-band data transmissions could be organized as pre-programmed tasks, exploiting storage and re-transmission considerations or other considerations. Carefully planning of (costly) ground stations is an issue with many SC missions.

The paper outlines a few configurations for X/Ka band antennas with a potential to be efficient, with a combination of frequency bands and within which a control of complex boundary conditions could be employed, differently for the different bands. Methods have been investigated, making use of techniques comparable to reflect-array implementations. Reflect-array properties are managed in such a way, that they provide the appropriate EM boundary conditions such as to improve the radiation performances in our case in particular for the Ka-band functionality. But principles are generic and would allow separate control of the two frequency bands. Critical bread-boarding has been carried out for some of the sub-systems, which can be used or are present in an X/Ka band antenna configuration.

978-7-5641-4279-7

## II. ANTENNA CONFIGURATIONS

Different dual-reflector antennas have been investigated, including usual Gregorian, Cassegrain and differently shaped approaches, with a dichroic sub-reflector (inductive or capacitive) or a dual frequency feed to handle two bands from a common aperture within one antenna. Hybrid solutions have been considered also, taking benefit of a combination of other controlling means available. One can coat a reflecting surface (or both reflecting surfaces) with reflect-array like structure to provide control of reflecting properties – or in other words: to provide the appropriate EM boundary conditions. It can be done in a frequency dependent manner or adapted for the two (or more) bands. It allows improving antenna parameters like a higher efficiency. In this way one could for instance use the classical dual reflector geometries or also other reflector geometries (even flat-plate or composed flat-plate like in a limit) and let desirable shaping be provided by the additional effect of a reflect-array for maximum efficiency.

One can also correct discrepancies as caused for instance by the feed. This is what we describe with experiments in support. One can optimise the shaped reflector configuration for X-band and use reflect-array properties for Ka-band optimisation.

The reflect-array could be positioned under constraints on an already shaped sub-reflector. We have the capability to realise desirable EM boundary conditions in a complex manner (full polarisation in principle). Such an approach is an alternative or assists to physically shaping of metal reflectors.

Dual frequency feeds have been developed in the past [2]. A complex three-band feed for the high gain antenna flies now on Cassini spacecraft around Saturn. Optimisation of a dual frequency feed for both X and Ka-band is somewhat complex for the separated spaced transmit and receive frequencies in both X- and Ka-band. The phase centre locations for X- and Ka-band channels can be somewhat different, depending on the design and thus geometrical feed parameters and allowed opening angle for the feed. It requires dedicated and demanding optimisation effort, using good modelling tools.

A dual frequency feed was realised by Rymsa. It was tested and had difference between phase centre for X- and Ka-band. The analysis of the antenna configuration showed the effects which was confirmed by the test results for the latter feed. The X-band result was very good (dual reflector configuration with shaped reflector surfaces). The impact of a displacement of the Ka-band phase centre has been determined and led to a decrease in efficiency.

The team has demonstrated to be able to recover Ka-band performances by inclusion of a Ka-band reflect-array mounted on top of the sub-reflector. The latter reflect-array assisted to introduce the necessary phase response such as to optimise the efficiency for the Ka-band pattern.

Application of a dedicated reflect-array technology allows also to control the two frequency bands separately, with very small influence on X-band, whilst allowing improving the Ka-band performances. In this way the antenna performances can be realised for an X/Ka band antenna with dual-frequency feed.

It is interesting to indicate, that one has other possibilities as well, for instance for shaping or optimisation of shaped surfaces further. One can optimise the dual reflector assembly for X-band and configure a Ka-band sensitive reflect-array on top of the sub-reflector for other purposes, even for some limited beam shaping in Ka-band. It should be even possible to consider a reflect-array on top of the sub-reflector for Ka-band interacting in such a way that it can correct for an off-axis located feed, thus allowing the feeds for X and Ka-band to be placed adjacent in principle. However the needed integer number of phase changes (0°-360°) leads to some discretisation effects with impact on side-lobe performances. For some applications it can reduce the complexity of the dual band feed system in the antenna drastically.

## III. REFLECTOR ASSEMBLY

Within the development activity the shaped dual reflector assembly was determined, using well-known optimization techniques. The reflector configuration has a diameter of 1.3 meter, comparable to the antenna assembly as flying on Venus Express. The shaping has been optimized for the X-band frequency. The precise location for the Ka-band frequency, with its shift along the axis was taken as a new configuration, with the main reflector shaping maintained as in X-band. The optimum shape of the sub-reflector geometry for Ka-band has been determined. The effective difference has been expressed in phase difference and this phase difference has been used to derive a reflect-array configuration which produces such phase difference. The result is an improvement of the Ka-band antenna performances.

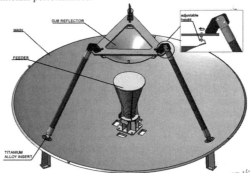

Figure 1. Schematic X/Ka Antenna Configuration (Courtesy CASA)

## IV. FEED ASSEMBLY

A dual frequency feed configuration was investigated and bread-boarded by Rymsa. A feeding part with narrow opening angle was considered for Ka-band part, with a slightly wider opening angle for the X-band part.

The influence of the X-band coupling locations appeared to be very critical to allow good Ka-band performances for both transmit and receive band, requiring good geometrical control. It can be improved in future work. A displacement of the phase centre was measured and was carefully assessed. Fig.1 shows the feed configuration, which for X-band was fed in phase quadrature and for Ka-band by means of a septum polariser. The X-band quadrature network is not shown.

Figure 2. Dual frequency Feed, X/Ka band (Courtesy Rymsa)

Several optimisations and precise complex measurements have been carried out.

V.    REFLECT-ARRAY ASSEMBLY ON A SUB-REFLECTOR

A very wide experience is available at the Universidad Politecnica de Madrid (UPM). A capability for determination of reflect-array parameters for various tasks was demonstrated in [3] where a pencil beam and a contoured beam was designed for different coverage directions. Such type of expertise has been used to derive needed reflect-array parameters for our Ka-band configuration, to be mounted now on the curved sub-reflector. The elements for the reflect-array have a size of about 4 to 5 millimeter, it becomes clear, that manufacturing requirements are very stringent. UPM has derived the reflect-array design for the curved surface that corrects the complex co-polar feed pattern.

EADS CASA Espacio's has proprietary capabilities to design, handle and install thin metallised layer. The mentioned reflect array, was realized with a two-layer design, exploiting laser-assist in the alignment of the layers.

Fig. 3 shows a realized sub-reflector with a Ka-band reflect-array on top.

Figure 3. Sub-reflector with Ka-band reflect-array on top (Courtesy CASA)

The precise testing configuration involved near-field testing of the sub-reflector with a Ka-band feed at the appropriate

distance. The structure to support the feed was analysed with Grasp in order to derive the effect caused on the radiation performances.

The results from the measurements have been imported in Grasp, using spherical wave expansion. In this way the antenna patterns have been derived, using the measured performances (feed with sub-reflector patterns) as input.

A sub-set of results is presented in table 1 below. One column shows the results when the measured results (sub-reflector + feed) are used in the Grasp model. The other column is based on a Grasp model with the feed data injected.

| Band | Freq. (GHz) | Measured í GRASP Model (dB) | GRASP Model (dB) |
|---|---|---|---|
| X-Band | 7.10 | 38.51 | 38.46 |
| | 7.25 | 38.50 | 38.50 |
| | 8.35 | 40.05 | 39.95 |
| | 8.50 | 39.99 | 39.94 |
| Ka-Band | 31.80 | 51.12 | 51.02 |
| | 32.30 | 51.15 | 51.10 |
| | 34.20 | 51.45 | 51.50 |
| | 34.70 | 51.52 | 51.57 |

Table 1. Comparison of directivity values based on Grasp analyses.

As an example the X-band and Ka-band patterns are shown for the case with the measured performances of the feed with sub-reflector (Fig.4).

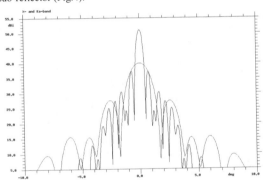

Figure 4. Dual frequency Feed, X/Ka band pattern, using measured data for feed combine with sub-reflector in analysis (Courtesy CASA).

The resulting directivity for X-and Ka-band is respectively 40 dBi and 51.1 dBi (for 8.35 and 32.3 GHz respectively), using the main- and sub-reflector geometry as derived for optimum performances in X-band.

VI.    CONCLUDING REMARKS

The utilization of a reflect-array on top of a sub-reflector has been discussed. A feed pattern with a dislocated Ka-band phase-center has been corrected by means of the latter reflect-array. A sub-reflector with a Ka-band reflect-array has been realized and measurements of the combined feed and sub-reflector (with reflect-array) have been carried out.

## REFERENCES

[1] M. Ichikawa, Y. Kamata, T. Takano, 'Development of On-board Antenna for Mars Spacecraft', *p69 Vol1, ISAP Conf. 1996, Chiba, JP.*

[2] R. Mizzoni, e.a. "The Cassini High Gain Antenna (HGA): a survey on electrical requirements, design and performance" , *IEE Colloquium 9 May 1994,* (in Ieeexplore).

[3] Jose A. Encinar, Leri Sh. Datashvili, J. Agustín Zornoza, Manuel Arrebola, M. Sierra-Castañer, J.L.Besada-Sanmartín, H. Baier, H. Legay "Dual-Polarization Dual-Coverage Reflectarray for Space Applications" *IEEE Trans. Ant and Prop. Vol 54, No 10, Oct. 2006.*

# Design of a 60 GHz Band 3-D Phased Array Antenna Module Using 3-D SiP Structure

Yuya Suzuki, Satoshi Yoshida, Suguru Kameda, Noriharu Suematsu,
Tadashi Takagi and Kazuo Tsubouchi
Research Institute of Electrical Communication, Tohoku University
Katahira 2-1-1, Aoba-ku, Sendai 980-8577, Japan
Email: yysuzuki@riec.tohoku.ac.jp

*Abstract*—A 60 GHz band 3-D phased array antenna module using 3-D system-in-package (SiP) structure is proposed and designed. Two separated $2\times2$ end-fire dipole array antennas on different ($z$-$y$ and $z$-$x$) planes are configured in a multi-stacked substrate module. For $x$-direction, $z$-$y$ plane array antenna is used and for $y$-direction, $z$-$x$ plane array antenna is used. In order to achieve higher antenna gain at intermediate region between $x$ and $y$ directions, hetero-plane beam synthesis using the two separated array antennas is proposed. The simulated result shows 10 dBi beam coverage area of $150°$ in azimuth and $60°$ in elevation.

## I. Introduction

Wireless systems using license-free 60 GHz band have been focused for Gbps ultra-high speed data rate short range communication. The IEEE 802.15.3c STD, published in 2009, defined the specification of beamforming transceiver for 60 GHz wireless personal area network (WPAN) system. Various beamforming array antennas for 60 GHz band WPAN have been developed. Broadside patch array antennas [1], [2] have been widely used due to their easy integration on a substrate surface (uni-plane). 2-dimensional array can be realized, but the coverage is limited to within $\pm45°$ from the perpendicular axes. In order to obtain wider coverage, end-fire antenna composed of two $1\times4$ quasi-Yagi arrays oriented orthogonally has been proposed [3]. Since two arrays are oriented orthogonally, coverage of the azimuth is over $\pm45°$.

In this paper, we propose a 3-D phased array antenna module having two separated $2\times2$ end-fire dipole array antenna. These antennas are orthogonally integrated in a stacked substrate module fabricated in 3-D SiP technology [4]-[7]. A $2\times2$ array antenna allows the beamforming functionality in both azimuth and elevation directions [6], [7]. By combining the beams emitted from two separated array antennas (hetero-plane beam synthesis), wider coverage in azimuth direction can be obtained which exceeds twice of each array antenna.

## II. Configuration of Proposed 3-D Phased Array Antenna Module

The overview of the proposed 60 GHz band 3-D phased array antenna module is shown in Fig. 1. Five multilayered organic substrates are vertically stacked using 3-D SiP technology. In order to realize wide coverage in the azimuth, $2\times2$ dipole phased array-A and B are populated on the $+x$ and $+y$ direction, respectively. A-plane and B-plane are defined as a plane perpendicular to the $+x$ and $+y$-axis, respectively.

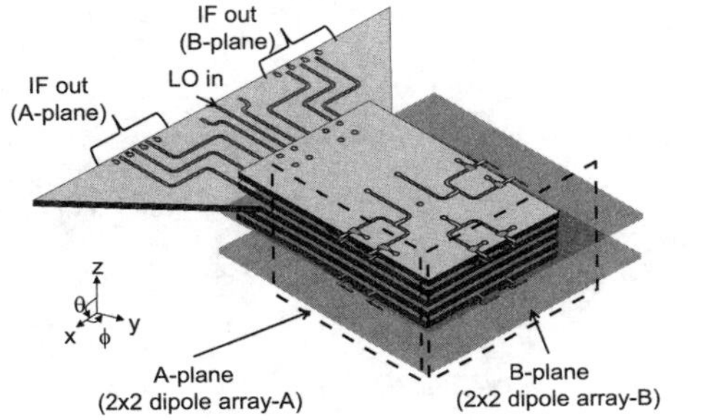

Fig. 1. Overview of the 60 GHz band 3-D phased array antenna module using 3-D SiP structure.

A block diagram of the proposed 3-D phased array antenna module is shown in Fig. 2. Eight passive 60 GHz band double balanced MMIC diode mixers (HMC-MDB169, Hittite Microwave Corporation) were used as downconverter. A conventional T-junction was used to split the LO signal. To evaluate 3-D radiation patterns, the amplitude and phase of the received signal in each direction are measured for each element antennas [8]. Since 60 GHz multichannel measurement of the amplitude and phase is quite difficult, eight passive mixers are used for downconverting the received RF signal to an IF signal at each element antenna. The amplitude and phase of the downconverted IF signals are measured by multichannel oscilloscopes. Frequency of the IF signal is 1 MHz to allow observation of both amplitude and phase using a multichannel oscilloscope. By controlling the phase of each IF signal, beamforming is realized.

The structure of the 60 GHz band 3-D phased array antenna module using 3-D SiP structure is shown in Fig. 3. A side view of the module is shown in Fig. 3(a). Five multilayered organic substrates are vertically stacked using 3-D SiP technology. MEGTRON6 substrate (Panasonic Electric Industry Co.) is used as multilayered organic substrates. Datasheet values of 3.5 for relative permittivity and 0.002 for dielectric loss tangent at 2 GHz are used in the EM simulation. Copper ball interconnections are used in the 3-D SiP structure for the 60 GHz band LO signal transmission [9] and the 1 MHz IF signal transmission. The copper balls are soldered and fixed, and serve to bond and support the various substrates. The planar dipole antennas are located on the top and bottom

978-7-5641-4279-7

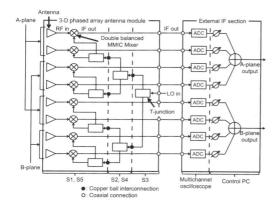

Fig. 2.   Block diagram of the proposed 3-D phased array antenna module.

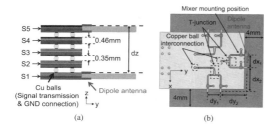

Fig. 3.   Structure of the 3-D phased array antenna module using 3-D SiP structure: (a) side view and (b) front view of top substrate S5.

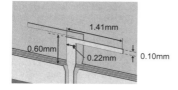

Fig. 4.   Structure and dimension of the element antenna.

substrate. Vertical element spacing $dz$ is $0.65\,\lambda_0$. $\lambda_0$ is defined as free-space wavelength at 60 GHz. S2 and S4 substrates are used to provide a wide $dz$ spacing (about one-half wavelength). A front view of top substrate S5 is shown in Fig. 3(b). The pattern of bottom substrate S1 is same as S5 substrate, but S1 substrate is inverted. Therefore, the signal line and mixer mounting position of S5 substrate are placed on its top layer, but those of S1 substrate are placed its bottom layer. The 60 GHz band LO signal is equally divided for eight passive mixers using T-junctions in the substrates. The substrate length (the length as measured from the top edge of the grounded coplanar waveguide (GCPW)) is 4 mm. This length results in a $-3$ dB beamwidth of 90° in both the $\phi$ and $\theta$ directions [7]. $dx_1$ and $dy_1$ are defined as horizontal element spacings of dipole array-B and A, respectively. And their values are both $0.50\,\lambda_0$. $dx_2$ and $dy_2$ are defined as the distance in the $x$- and $y$-axis direction between dipole array-A and B, respectively.

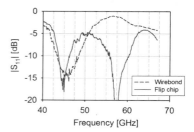

Fig. 5.   Comparison of measured return loss of mounted MMIC mixer.

In the next section, these are set as a parameter to analyze the impact on the gain and beamwidth when hetero-plane beam synthesis is performed. A structure and dimension of the element antenna is shown in Fig. 4. The dipole antenna is placed on the substrate's top and bottom surfaces. The antenna is fed by a 0.60 mm pair line, which is directly connected to the GCPW. The one-sided antenna length is 1.41 mm [10].

When the proposed phased array antenna module is measured as a receiver, a part of received signal of the element antenna is reflected from the RF port of mixer. These phenomenon results in fluctuation of phase and amplitude of received signal. Therefore, it is necessary to reduce reflection at input port of mixer. The measurement results of return loss of the mixer mounted by wire bonding and flip-chip mounting is shown in Fig. 5. It is confirmed that flip-chip mounting has better return loss characteristic than wire bonding at 60 GHz band. Therefore, flip-chip mounting is selected in this work.

### III.   ARRAY DESIGN FOR HETERO-PLANE BEAM SYNTHESIS

Proposed 3-D phased array antenna module can realize wider coverage in the azimuth direction by switching the dipole array-A and B. However, the gain drops as the scanning angle becomes large. Therefore, it is difficult to cover the intermediate region (around $\phi = 45°$) by only switching between dipole array-A and B. To solve this problem, hetero-plane beam synthesis using the two separated array antennas is proposed. The beam synthesis is realized by combining the beams emitted from two separated phased arrays (dipole array-A and B). The dipole array-A and B are oriented orthogonally but the polarization is parallel to the azimuth plane. Therefore, hetero-plane beam synthesis is available in the intermediate region.

An example of EM simulation results of hetero-plane beam synthesis is shown in Fig. 6. EM simulation software Microwave Studio (Computer Simulation Technology Co.) was used. A 3-D radiation pattern in which only dipole array-A is excited at $(py, pz) = (135°, 180°)$ is shown in Fig. 6(a). Similarly, only dipole array-B is excited at $(px, pz) = (135°, 180°)$ is shown in Fig. 6(b). $px$, $py$ and $pz$ are defined as the phase difference between the adjacent elements in the $x$, $y$ and $z$ directions, respectively. These phase differences are selected assuming use of 3-bit phase shifter. Since S1 substrate pattern is reversed S5 pattern, the element antennas

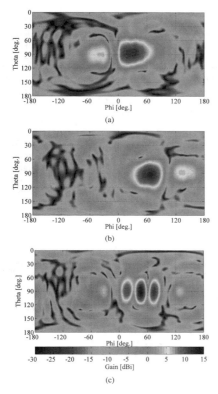

Fig. 6. Simulation results of a 3-D radiation pattern at phase difference $px = 135°$, $py = 135°$ and $pz = 180°$ at 60 GHz ($dx_1$, $dy_1 = 0.5\lambda_0$, $dx_2$, $dy_2 = 1.0\lambda_0$) : (a) only dipole array-A is excited, (b) only dipole array-B is excited and (c) hetero-plane beam synthesis pattern by simultaneous excitation both dipole array-A and B.

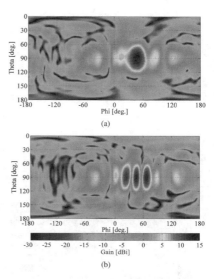

Fig. 7. Simulation results of hetero-plane beam synthesis patterns of dipole array-A and B at phase difference $px = 135°$, $py = 135°$ and $pz = 180°$ at 60 GHz ($dx_1$, $dy_1 = 0.5\lambda_0$) : (a) $dx_2$, $dy_2 = 0.5\lambda_0$ and (b) $dx_2$, $dy_2 = 1.5\lambda_0$.

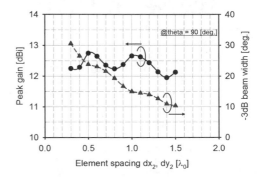

Fig. 8. Simulation results of the impact of array spacing on the gain and beamwidth when hetero-plane beam synthesis is performed ($dx_1$, $dy_1 = 0.5\lambda_0$).

on S1 substrate have opposite phase to element antennas on S5 substrate. Hence, $pz = 180°$ realizes the main beam direction to $\theta = 90°$. Array gain in the intermediate region is less than 10 dBi when only dipole array-A or B is excited as shown in Fig. 6(a) and 6(b). Hetero-plane beam synthesis pattern by simultaneous excitation both dipole array-A and B ($dx_2$, $dy_2 = 1.0\lambda_0$) at ($px$, $py$, $pz$) = (135°, 135°, 180°) is shown in Fig. 6(c). Beam synthesis at intermediate region is confirmed and peak gain is 12.7 dBi at ($\theta$, $\phi$) = (90°, 46°). Since the array spacing is relatively wide ($1.0\lambda_0$), sidelobe level is relatively large. However, this is acceptable because during beamforming, the optimal beam is selected. The positions of side lobes will therefore be covered by the main lobes of other beams eventually.

In order to determine optimal array spacing, parametric study using EM simulator is performed. $dx_2$ and $dy_2$ are set as a parameter to analyze the impact on the gain and beamwidth when hetero-plane beam synthesis is performed. Synthesized patterns with $dx_2$, $dy_2 = 0.5\lambda_0$ and $1.5\lambda_0$ are shown in Fig. 7(a) and 7(b), respectively. It is confirmed that beamwidth and sidelobe level become narrower and higher

respectively, as the spacing is extended. Simulation results of the impact of array spacing on the gain and beamwidth are summarized as shown in Fig. 8. Gain reaches a peak at $dx_2$, $dy_2 = 0.5\lambda_0$ and $1.0\lambda_0$. $x$-$y$ plane beamwidth is more than 10° throughout simulated spacing. This value is enough for beamforming under the assumption of the use of 3-bit phase shifter. The narrow spacing results in increase of density of the feeding network. Hence, considering the actual implementation, the complexity of the implementation is reduced as the array spacing is extended. Therefore, optimal array spacing of $dx_2$, $dy_2 = 1.0\lambda_0$ is selected.

Simulation results of beamforming coverage area where

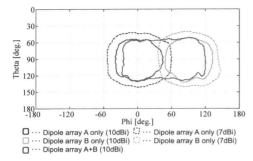

□ ··· Dipole array A only (10dBi)　⊡ ··· Dipole array A only (7dBi)
□ ··· Dipole array B only (10dBi)　⊡ ··· Dipole array B only (7dBi)
□ ··· Dipole array A+B (10dBi)

Fig. 9. Simulation results of beamforming coverage area where gain exceeds 7 dBi and 10 dBi at 60 GHz ($dx_1$, $dy_1$ = 0.5 $\lambda_0$, $dx_2$, $dy_2$ = 1.0 $\lambda_0$).

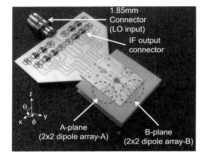

Fig. 10. Fabricated 60 GHz band 3-D phased array antenna module using 3-D SiP structure.

gain exceeds 7 dBi and 10 dBi at 60 GHz are shown in Fig. 9. All the simulations are conducted under the condition of $dx_2$, $dy_2$ = 1.0 $\lambda_0$. 64 phase states are simulated to each array, assuming use of a 3-bit phase shifter. 60° beam scanning coverage area (gain of over 10 dBi) of both the azimuth and elevation are obtained at both 2×2 dipole array-A and B (blue and orange solid lines). However, the intermediate region ($\phi$ = 30°–60°) is not covered with 10 dBi. But 7 dBi coverage areas are overlapped (blue and orange dashed lines). Therefore, beam synthesis is available in the intermediate region. Using hetero-plane beam synthesis, the intermediate region is covered with 10 dBi gain (red solid line). These results show that the 10 dBi beam coverage area which covers 150° in azimuth and 60° in elevation is achieved by switching and hetero-plane beam synthesis of two separated array antennas.

## IV. Fabrication

A fabricated 60 GHz band 3-D phased array antenna module using 3-D SiP structure is shown in Fig. 10. Five multilayered organic substrates are vertically stacked by 3-D SiP technology using 350 $\mu$m $\phi$ copper balls. A passive 60 GHz MMIC mixer is flip-chip mounted for each antenna element, and it downconverts RF signal into 1 MHz IF frequency. The phase of each element antenna is controlled using an external IF section. The IF output ports and LO input port are placed on the 3rd substrate. The intrinsic module size excluding I/O area is 18 mm×17 mm×3.7 mm.

## V. Conclusion

A 60 GHz band 3-D phased array antenna module using 3-D SiP structure is proposed and designed. 2×2 dipole phased arrays configured in the multi-stacked substrate module are placed orthogonally. In order to overcome a drop of antenna gain in the intermediate direction between the orthogonal array antennas, hetero-plane beam synthesis is employed. By the EM simulation, 1.0 $\lambda_0$ is selected as a optimal array spacing between the orthogonal array and the edge of module. This array spacing brings higher gain and wider beamwidth for beamforming under the assumption of the use of 3-bit phase shifter with easier implementation. EM simulation results show that the proposed phased array antenna module can achieve 150° and 60° beam scanning coverage area (gain of over 10 dBi) in azimuth and elevation, respectively. Based on this study, three orthogonal end-fire phased arrays can be integrated in a module and it will show a wider beam coverage of 240° in the azimuth plane.

## Acknowledgment

The authors would like to thank Core Research for Evolutional Science and Technology program (CREST) of the Japan Science and Technology Agency (JST) for fund support.

## References

[1] S. Lin, K. B. Ng, H. Wong, K. M. Luk, S. S. Wong, and A. S. Y. Poon, "A 60GHz digitally controlled RF beamforming array in 65nm CMOS with off-chip antennas," *Radio Frequency Integrated Circuits Symposium*, June 2011.

[2] M. Fakharzadeh, M.-R. Nezhad-Ahmadi, B. Biglarbegian, J. Ahmadi-Ahokouh, and S. Safavi-Naeini, "CMOS phased array transceiver technology for 60 GHz wireless application," *IEEE Trans. Antennas Propag.*, vol. 58, no. 4, pp. 1093–1104, April 2010.

[3] A. L. Amadjikpe, D. Choudhury, G. E. Ponchak, and J. Papapolymerou, "Location specific coverage with wireless platform integrated 60-GHz antenna systems," *IEEE Trans. Antennas Propag.*, vol. 59, no. 7, pp. 2661–2671, July 2011.

[4] K. Tsubouchi, M. Yokoyama, and H. Nakase, "A new concept of 3-dimentional multilayer-stacked system-in-package for software-defined-radio," *IEICE Trans. on Electron.*, vol. E84-C, no. 12, pp. 1730–1734, Dec. 2001.

[5] N. Suematsu, S. Yoshida, S. Tanifuji, S. Kameda, T. Takagi, and K. Tsubouchi, "A 60 GHz-band 3-dimensional system-in-package transmitter module with integrated antenna," *IEICE Trans. on Electron.*, vol. E95-C, no. 7, pp. 1141–1146, July 2012.

[6] Y. Suzuki, S. Yoshida, S. Tanifuji, S. Kameda, N. Suematsu, T. Takagi, and K. Tsubouchi, "60 GHz band 2×4 dipole array antenna using multi stacked organic substrates structure," in *Proc. Int. Symp. on Antennas and Propag.*, Oct. 2012.

[7] S. Yoshida, Y. Suzuki, T. T. Ta, S. Kameda, N. Suematsu, T. Takagi, and K. Tsubouchi, "A 60-GHz band planar dipole array antenna using 3-D SiP structure in small wireless terminals for beamforming applications," *IEEE Trans. Antennas Propag.*, vol. 61, no. 7, pp. 3502–3510, July 2013.

[8] T. T. Ta, S. Yoshida, Y. Suzuki, S. Tanifuji, S. Kameda, N. Suematsu, T. Takagi, and K. Tsubouchi, "A 3-D radiation pattern measurement method for a 60-GHz-band WPAN phased array antenna," in *Proc. Asia Pacific Microwave Conf.*, Dec. 2012.

[9] S. Yoshida, S. Tanifuji, S. Kameda, N. Suematsu, T. Takagi, and K. Tsubouchi, "60-GHz band copper ball vertical interconnection for MMW 3-D system-in-package front-end modules," *IEICE Trans. on Electron.*, vol. E95-C, no. 7, pp. 1276–1284, July 2012.

[10] S. Yoshida, S. Tanifuji, S. Kameda, N. Suematsu, T. Takagi, and K. Tsubouchi, "A high-gain planar dipole antenna for 60 GHz band 3-D system-in-package modules," in *Proc. IEEE AP-S Int. Symp.*, July 2011.

# Design of Low Side Lobe Level Milimeter-Wave Microstrip array antenna for Automotive Radar

Dong-hun Shin[1], Ki-beom Kim[2], Jong-guk Kim[3] Seong-ook Park[4]

Dept. of Electrical Engineering,KAIST[1,2,4]., LG Innotek Components R&D Center[3]

335 Gwahak-ro, Yuseong-gu, Daejeon, 305-701, Korea[1,2,4] , 1271 Sa-3 dong Sangrok-gu Ansan-si Gyeonggi-do, Korea[3]

husido@kaist.ac.kr[1], smiledawn@kaist.ac.kr[2], jgkimd@lginnotek.com[3], sopark@ee.kaist.ac.kr[4]

*Abstract*-A low profile array antenna with a low sidelobe was designed using Taylor array pattern synthesis and null filling concept is adopted using genetic algorithm. The antenna column is consists of 18 element antennas, gap coupled antenna is proposed in this paper to obtain low radiation coefficient and direct coupled antenna is used for high radiation coefficient. 45° polarization is added to reduce the interference.

The antenna has 20.8dBi gain and sidelobe level of elevation direction is under -20dB and azimuth direction is under -25dB.

## I. INTRODUCTION

Requirement for a multitude of different automotive radar based comfort and safety functions are compiled[1]. Typical antenna requirements are high gain and moderate to low sidelobes.

While for the long range ACC applications an angular range of ±4° to ±8° is usually sufficient, short or medium range applications have different requirements in distance and angular range.

Planar antennas are common components in sensing applications due to their low cost, low profile and simple integration with systems. The planar antennas are widely used despite the loss that caused by dielectric loss and conductor loss. One special advantage of planar elements is that they can easily adopt array structures combining very simple elements. Due to all these characteristics, planar antennas are very good candidates for mm-wave applications in radar, sensing or communications.

Polarization is also an important factor in automotive radar. Some systems use 45° polarization, so that radar signals from cars travelling in opposite directions and facing each other will be orthogonal polarization to reduce interference[2]. Alternatively, vertical polarization is less prone to multipath propagation between a car's bottom and the road surface.

Pattern synthesis is the process of choosing the antenna parameters to obtain desired radiation characteristics, such as the specific position of the nulls, the desired sidelobe[3] and beam width of antenna pattern. In literature there are many works concerned with the synthesis of antenna array like as binominal distribution, Dolph-Chbyshev ampltitude distri-bution, Taylor distribution, etc. [4]-[8].

This paper presents straight forward design procedure to synthesize radiation patterns of series fed fish bone array antenna at 76 ~ 77 GHz. The designed antenna which has low

sidelobe level is designed using the concept of Taylor pattern synthesis and genetic algorithm. Gap coupled array concept is proposed for low radiation coefficient.

## II. ANALYSIS OF ANTENNA DESIGN.

### A. TAYLOR DISTRIBUTION

The antenna for automotive radar system typically requirements low sidelobes level to reducing the unwanted reflection. The sidelobes may be included in the angle determination scheme, so that the sidelobe level requirements are rather relaxed.

Taylor synthesis is one of widely used for reduced sidelobes array beam pattern. This is more suitable for series fed from one side due to the array weights for the end elements usually do not increase as much as compared with Chebyshev arrays, although this technique is not optimum in that the relation between beam width and sidelobe level.

Consider a linear array of N-Element, the array weights $\{w_i, i = 1, 2, \cdots, N\}$ are given by

$$w_i = \frac{1}{N}\left\{1 + 2\sum_{n=1}^{\overline{n}-1} f(n, A, \overline{n})\cos\left(\frac{2\pi z_i}{N}\right)\right\} \quad (1.1)$$

Where the parameter $\overline{n}$ is the number used to decide the number of close-in sidelobes to be set with a constant sidelobe level. Other parts of the preceding expression are defined as

$$z_i = i - \left(\frac{N}{2} + 1\right) \quad (1.2)$$

$$f(n, A, \overline{n}) = \begin{cases} \dfrac{[(\overline{n}-1)!]^2}{(\overline{n}-1+n)!(\overline{n}-1-n)!}\prod_{m=1}^{\overline{n}-1}\left[1-\left(\dfrac{n}{u_m}\right)^2\right] & |n| < \overline{n} \\ 0 & |n| \geq \overline{n} \end{cases} \quad (1.3)$$

$$u_m = \begin{cases} \pm a\sqrt{A^2 + \left(m - \dfrac{1}{2}\right)^2} & 1 \leq m < \overline{n} \\ \pm m & \overline{n} \leq m < \infty \end{cases} \quad (1.4)$$

$$a = \frac{\overline{n}}{\sqrt{A^2 + (\overline{n}-0.5)^2}} \quad (1.5)$$

$$A = \frac{1}{\pi}\cosh^{-1} R \quad (1.6)$$

978-7-5641-4279-7

## TALOR ARRAY PATTERN SYNTHESIS

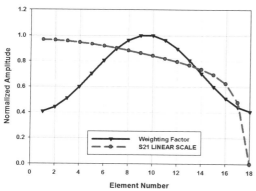

Figure 1. Taylor array pattern synthesis.

$$R = 10^{SLL/20} \qquad (1.7)$$

The beam width of the Taylor array in degree is approximated as

$$\Theta_{Taylor} \doteq 2\sin^{-1}\left\{ \frac{a}{L\pi}\sqrt{\left(\cosh^{-1}R\right)^2 - \left(\cosh^{-1}\frac{R}{\sqrt{2}}\right)^2} \right\} \quad (1.8)$$

The weight factor of each antenna element and the transmission parameter, $S_{21}$ are shown in Fig.1. in which the element number increases from bottom to top, and its sidelobe level target is less than -20dB.

### B. ANTENNA ELEMENT DESIGN.

Series fed microstrip antennas are widely used for many applications. This paper design procedure to synthesize radiation patterns of series fed fishbone type microstrip antennas. 45° polarization is added to this concept by employing inclined antenna elements, which are connected to the feed line on a corner. In order to design an antenna based on the radiation coefficient, the antenna elements which can be tapered without affecting their radiation behavior are designed.

Direct coupled antenna is widely used in microstrip array antenna. However, Taylor array synthesis needs the radiation coefficient from very low to very high. In this paper the gap coupled antenna is proposed for very low radiation coefficient. The gap coupled antenna with 150um coupling gap has the radiation coefficient much smaller than direct coupled antenna

with 150um coupling width.

A Single element with feed line is here designed and simulated, in order to obtain the optimum parameters when connected in array. I designed the low side lobe level (Elevation: -20dB) antenna which is consists of 4 gap coupled antennas for low radiation coefficient and 14 direct coupled antennas for high radiation coefficient. Each element's antenna width, antenna length and coupling width are tapered to achieve better sidelobe suppression.

### C. WAVEGUIDE TO MICROSTRIP LINE TRANSITION.

The millimeter wave above 60GHz, SMA connector is not suitable because of its small size and effected electromagnetic field by soldering. Alternatively, waveguide still used in many applications such as millimeter wave system, and in some precision test application. Therefore, waveguide to microstrip line transitions are required to measure the antenna pattern.

A conventional type of transition needs a metal short block with a quarter-wavelength on the substrate[9-10]. In recently the proximity coupling type[11] has been developed. It can be composed of a single dielectric substrate attached to the waveguide without metal short block..

The upper ground pattern with a notch and the microstrip line are located on the upper plane of the dielectric substrate. A coupling patch element which used for coupling the single is patterned on the lower plane of the dielectric substrate and it is overlaps on a microstrip line on the upper plane. Via holes are surrounding the aperture of the waveguide.

The parameters of designed transition are described in Table 1, and S-parameters of the reflection $S_{11}$ and the transmission $S_{21}$ are calculated as shown in Figure 4.

From the results, reflection is under -20dB in desired frequency and the insertion loss in -0.5dB with 6.7mm microstrip line.

Table 1 Parameter of transition.

| Parameter | Value | Parameter | Value |
|---|---|---|---|
| $\varepsilon_r$ | 2.2 | WG_X | 2.54 mm |
| Line Gap | 0.1 mm | WG_Y | 1 mm |
| VIA Gap | 0.7 mm | CP_X | 1.716 mm |
| VIA Diameter | 0.3 mm | CP_Y | 0.957 mm |

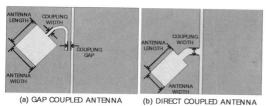

(a) GAP COUPLED ANTENNA    (b) DIRECT COUPLED ANTENNA

Figure 2. Two types of single element antenna.

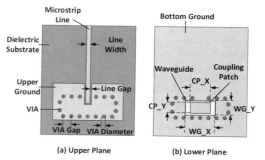

(a) Upper Plane    (b) Lower Plane

Figure 3. The configurations of transition.

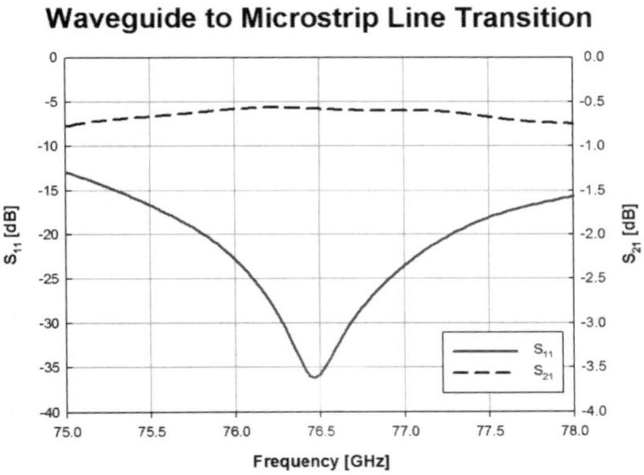

## Waveguide to Microstrip Line Transition

Figure 4. Reflection and insertion loss of transition.

## III.    ANTENNA DESIGN.

### A.   Single Column Design

The antenna is designed using the dielectric substrate Taconic TLY-5 (thickness : 5mil , dielectric constant : 2.2), and the thickness of the copper layer is 1/2oz. The model consists of the fishbone type microstrip antenna with its initial dimensions. The antenna is connected from both sides with the high impedance microstrip line.

Taylor array pattern synthesis is adopted in this design to obtain the low side lobe level. The radiation coefficient of each antenna is optimized by deducing the suitable coupling widths, antenna width and antenna length of each antenna.

Two types of coupling technique are used for array, one is gap coupled type for low radiation coefficient at first 4 antennas. Other is direct coupled type for large radiation coefficient at last 14 antennas.

The radiation coefficient is determined by adjusting parameter like as coupling gap, coupling width, antenna width and antenna length.

### B.   Array Antenna Design.

The long range radar(LRR) system is sensitively sensing the obstacle that is located at front area by using pencil beam antenna. However, due to null point of radiation, it is vulnerable for sensing the pedestrian or obstacle on the side of road. Some systems add to sensing product for this problem, but it is not good solution because of cost and complexity increasing. So that, I proposed the null filling antenna for complement the weakness of LRR using genetic algorithm for pattern synthesis.

The genetic algorithm proposed to determine the amplitude weight in same phase.  First the initial population of chromosomes that are a binary string representing all perturbed weight of each port is randomly generated. Second, the radiation pattern of each chromosome is calculated, and the

Figure 5. Fabricated antenna.

maximum side lobe level and the null point are founded. Then the fitness of each chromosome is evaluated. Base on the evaluated fitness value, a pair of chromosomes is selected and mated. The selection rule is based on a roulette wheel and the crossover is achieved at a single random point. Then the mated chromosomes can be mutated according to the mutation probability. In this manner a new offspring is generated. The iterations is stopped after the evaluation of fitness.

The calculating conditions are under  -40dB sidelobe level, no null point from -40° to 40° and under 10 ratio of weights.

The calculated weight from MATLAB and designed weight are presented in Table 2 in here the port is numbering from left to right.

Figure 6 shows the radiation pattern in azimuth plane. The solid line means the pattern using null filling weight and dash line means the pattern using same weight.

Table 2. Amplitude Weight for Null filling.

| Port | Ideal Weight | Designed Weight |
|------|------|------|
| P4,P5 | 1 | 1 |
| P3,P6 | 0.2985 | 0.2871 |
| P2,P7 | 0.1316 | 0.1406 |
| P1,P8 | 0.1286 | 0.1298 |

## Null Filling in Azimuth Plane

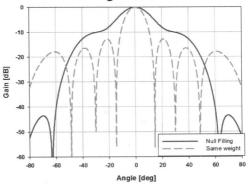

Figure 6. The Calculated Directivity in Azimuth Plane.

## IV. RESULT

Fig. 7 and Fig. 8 show the simulated radiation pattern of overall antenna. It can be seen that a high gain of 20.8dBi is obtained at the design frequency of 76.5 GHz.

Fig. 7 shows the measured radiation pattern in elevation-plane of the developed antenna at the design frequency of from 76 GHz to 77 GHz. The half-power beam width is 4.8 degrees and the sidelobe level is reduced to lower than –20 dB.

Fig. 8 shows the measured radiation pattern in azimuth-plane at the design frequency of from 76 GHz to 77GHz. The half-power beam width is 18.3° and the sidelobe level is –25.7 dB. It shows that the null point is not exists in -40° to 40°.

## Radiation Pattern in Elevation Plane

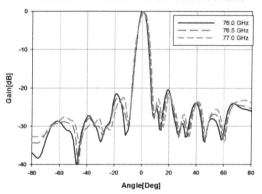

Figure 7. Elevation Radiation Pattern.

## Radiation Pattern in Azimuth Plane

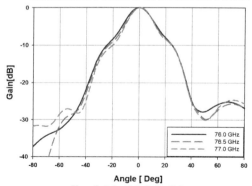

Figure 8. Azimuth Radiation Pattern.

## V. CONCLUSTIONS

The automotive antenna is designed in this paper. Taylor pattern synthesis and genetic algorithm are used for sidelobe level reduction. Two types of coupling technique are used for array, one is gap coupled type for low radiation coefficient, another is direct coupled type for large radiation coefficient.

These antenna elements are modeled taken into account the feed line and all other parasitic effects and optimized for the operating frequency as well as for a given radiation coefficient.

As a result of the design, the antenna has a high gain of 20.8dBi and a low sidelobe level. The array antenna designed here will be utilized as not only a fixed beam antenna but also beamforming antenna for automotive antenna.

ACKNOWLEDGMENT

This work was supported by the LG Innotek through the 77GHz automotive radar project under contact G01130026.

REFERENCES

[1] Mende, R.; Rohling, H, "New Automotive Applications for Smart Radar Systems," . German Radar Symposium GRS 2002, pp. 35-40, Sept. 2002
[2] Schneider, R.; Wenger, J, "System aspects for future automotive radar ", IEEE MTT-S International Microwave Symposium 1999, pp.293-296, June 1999
[3] Aniruddha Basak. Et al, "A modified invasive weed optimization algorithm for time modulated linear antenna array synthesis", IEEE Congress on Evolutionary Computation(CEC) 2010 pp. 1-8.
[4] C.L.Dolph, "A current distribution for broadside arrays which optimizes the relationship between beam width and side-lobe level," Proc IRE 34 pp.335-348, June. 1946
[5] T.T Taylor, "Design of line source antennas for narrow beamwidth and low side lobes", IRE AP Trans 4 pp 16-28 Jan 1955.
[6] A.T.Villeneuve, Taylor, "Patterns for discrete pattern arrays", IEEE AP-S Trans Vol.32 pp 1089-1094 October 1984.
[7] W.W.Hansen and J.R.Woodyard, "A new principle in directional antenna design", Proc. IRE 26 pp 333-345 March 1938.
[8] E.T.Bayliss, "Design of Monopulse Antenna difference Patterns with low sidelobes", Bell Syst. Tech.J.47 pp623-650 May-June 1968.
[9] Ho, T. Q, Shih, Y. C. "Spectral-domain analysis of E-Plane waveguide to microstrip transitions", IEEE Trans. Microw. Theory Tech., Feb., Vol. 37, 388-392. 1989.
[10] Leong, Y, & Weinreb, S., "Full band waveguide to microstrip probe transitions," IEEE MTT-S 1999 pp. 1435-1438.May. 1999.
[11] Iizuka, H, Watanabe, T, Sato, K, & Nisikawa, K. Millimeter-wave microstrip line to waveguide transition fabricated on a single layer dielectric substrate," IEICE Trans. Commun.,2002. , pp. 1169-1177. Jun. 2002

# Realizing Sample Matrix Inversion (SMI) in Digital BeamForming (DBF) System

Hao Lei, Zaiping Nie and Feng Yang
University of Electronic Science and Technology of China
NO.4, Section2, North Jianshe Road
Chengdu, Sichuan 610054 China

*Abstract*-**Digital beamforming is a kind signal processing in adaptive array antennas, based on the disposal of data produced by sampling signal. DBF can adjust beam pattern based on the arrival direction of signal, and produce pattern nulling in the direction of disturb signal. Sample matrix inversion(SMI). Algorithm is one kind method in DBF. This paper will introduce my work on simulating and verifying QRD-SMI algorithm based on Virtex 5, an popular FPGA chip produced by XILINX company. QR matrix decomposition is used to transfer matrix to a upper triangle matrix which is more easily to solve the inverse matrix.**

*Keywords: Digital beamforming, QR-SMI, matrix inversion, FPGA.*

## I. INTRODUCTION

SMI Algorithm is based on the standard that make signal to interference-plus-noise ratio(SINR)[1] maximum, which the desired array signal and the unwanted signal have largest proportion. Linearly constrained minimum variance (LCMV) optimal weight vector satisfies the following linear equation

$$R_{XX} \mathrm{w}_{opt} = \mu s^* \qquad (1)$$

Where $R_{xx}$ is the sampling covariance matrix, $\mu$ is any proportionality constant, $s$ is operation vector of required signal, that is static weight which control the beam in the target direction[2]. Based on the maximum likelihood criterion, N signal sampled data may constitute the best estimate $R_{xx}$. In practical engineering, inversing $R_{xx}$ will consume a lot of time and hardware resources, the method to reduce the computational and hardware resources is using the Givens rotation to realize matrix QR decomposition. Changing the problem that solving weight vector W into a problem solving triangular linear equation. Since the covariance matrix $R_{xx}$ can be expressed as $R_{xx} = X_n^H X_n$, Xn is a $n \times M$ signal sampled matrix which n and M present the number of sample data and the number of array antennas respectively. We can get the equation $X_n^H X_n w = S$. If there is a unitary matrix Q will triangulate matrix $X_n$,

$$QX_n = \begin{bmatrix} A_n \\ 0 \end{bmatrix} \qquad (2)$$

where An is M×M upper triangular matrix, an important equation can be deduced as follow:

$$X_n^H X_n = X_n^H Q^H Q X_n = (QX_n)^H QX_n = [A_n^H, 0^H] \begin{bmatrix} A_n \\ 0 \end{bmatrix} = A_n^H A_n \quad (3)$$

The advantage of these processes is that we have lots of methods to get weight vector more easily. Technical difficulty is how to change the sampled signal matrix $X_n$ into a triangular matrix $A_n$ in FPGA. Givens rotation is an appropriate method. Givens rotation is particularly suitable for adaptive array applications, because it better to be applied in FPGA.

## II. GIVENS ROTATION

Complex Givens rotation can be presented as the following elementary transformations:

$$\begin{bmatrix} C & S^* \\ -S & C \end{bmatrix} \begin{bmatrix} 0 & .. & x_i & x_{i+1} & ... & x_k \\ 0 & .. & y_i & y_{i+1} & ... & y_k \end{bmatrix} = \begin{bmatrix} 0 & .. & x_i' & x_{i+1}' & ... & x_k' \\ 0 & . & 0 & y_{i+1}' & ... & y_k' \end{bmatrix} \quad (4)$$

Where $G = \begin{bmatrix} C & S^* \\ -S & C \end{bmatrix}$ mustbe a unitary matrix simultaneously,

so matrix G must meet following equation

$$\begin{cases} -Sx_i + Cy_i = 0 \\ S^*S + C^*C = 1 \\ C^* = C \end{cases} \qquad (5)$$

Assuming X (n-1) have already achieved triangulation. When $\mathrm{x}^T = [x_1(n),...,x_M(n)]$ is being inputed, $n \times M$ sampled data matrix can be written as follow:

$$X(n) = \begin{bmatrix} X(n-1) \\ x^T(n) \end{bmatrix} \qquad (6)$$

Rotating matrix X(n) by a new n×n unitary matrix

$$\overline{Q}(n-1) = \begin{bmatrix} Q(n-1) & 0_{n-1} \\ 0_{n-1}^T & 1 \end{bmatrix} \qquad (7)$$

An new equation can be presented as follows:

$$\overline{Q}(n-1)X(n) = \begin{bmatrix} Q(n-1)X(n-1) \\ x^T(n) \end{bmatrix} = \begin{bmatrix} R(n-1) \\ 0 \\ x^T(n) \end{bmatrix}$$

$$= \begin{bmatrix} z_{11} & z_{12} & \cdots & z_{1M} \\ & z_{22} & \cdots & z_{2M} \\ & & \cdots & \cdots \\ & & & z_{MM} \\ 0 & \cdots & \cdots & \cdots \\ x_1(n) & x_2(n) & \cdots & x_M(n) \end{bmatrix} \qquad (8)$$

Defining the first givens rotation matrix G1(n) that is eliminating the $x1(n)$, so we get equation (9)

978-7-5641-4279-7

$$G_1(n)\overline{Q}(n-1)X(n)=\begin{bmatrix} z_{11}{}' & z_{12}{}' & \cdots & z_{1M}{}' \\ & z_{22} & \cdots & z_{2M} \\ & & \cdots & \cdots \\ & & & z_{MM} \\ 0 & \cdots & \cdots & \cdots \\ 0 & x_2^{(1)}(n) & \cdots & x_M^{(1)}(n) \end{bmatrix} \quad (9)$$

Defining the second givens rotation[3] matrix $G2(n)$, $G2(n)$ is used to eliminate $x_2^{(1)}(n)$, a new equation can be written as follow:

$$G_2(n)G_1(n)\overline{Q}(n-1)X(n)=\begin{bmatrix} z_{11}{}' & z_{12}{}' & \cdots & \cdots & z_{1M}{}' \\ & z_{22}{}' & \cdots & \cdots & z_{2M}{}' \\ & & & \cdots & \\ 0 & & & & z_{MM} \\ 0 & 0 & x_3^{(1)}(n) & \cdots & x_M^{(2)}(n) \end{bmatrix} \quad (10)$$

After M times rotations, a new upper triangular matrix[4] can be presented as follow:

$$G_M(n)...G_2(n)G_1(n)\overline{Q}(n-1)X(n)=G(n)\begin{bmatrix} R(n-1) \\ 0 \\ x^T(n) \end{bmatrix}=\begin{bmatrix} R(n) \\ 0 \end{bmatrix} \quad (11)$$

The above describes a recursive method to get the upper triangular matrix. But these are based on the formula derivation, how to achieve Givens rotation in circuit or in FPGA chip? Achieving givens rotation in FPGA is the focus of this paper. As we know, it is difficult to do some matrix processing in circuit or in FPGA. FPGA is based on parallel thought, FPGA has no integrated math library to help us to do some math processes. Compared to floating-point arithmetic, fixed-point arithmetic is more complex. Fixed-point arithmetic needs to be considered data overflow, fractional process. Next I will introduce Systolic array to implement givens rotation in FPGA.

III. SYSTOLIC ARRAY AND CORDIC

Systolic array[5] is a parallel pipelined and high-speed signal processing algorithm. Systolic array have lots of advantages, including modular and locality. Systolic array is based on multiprocessor architecture.

All processors have rhythm synchronization. Systolic achieve a high degree of parallelism and pipelining and local communication between processors. Now I will discuss how to achieve givens rotation by systolic array.

Realizing upper triangulation by systolic including two important modules, boundary element and internal element. A 4x4 systolic array is presented in Fig1

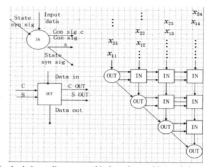

Fig. 1. 4x4 systolic arrays and its boundary and internal element

Boundary element is core unit of systolic array, located on the diagonal of systolic array. Boundary element have two main core functions, firstly it must finishing processing the input data to zero, secondly it must provide control signals(Con sig c, Con sig s) to internal element produced by zero processing. Internal element is located in the upper right diagonal of systolic array, its function is receiving control signal generated from the boundary element and controlled by Con sig c and Con sig s. 4x4 systolic arrays can process the input data sampled from 4 antennas, but we need control the order data enters shown in Fig1.

How to design boundary element and internal element. They must finish function and cost minimal resources. I introduce CORDIC algorithm to achieve this goal. Complex operation such as multiplication and division, can be achieved by shifting processing and addition and subtraction merely. This greatly reduces the implementation complexity required by lots of processing units of the systolic array.

The full name of CORDIC is Coordinate Rotation Digital Computer, be invented by Jack. E. Volder in 1959.Now CORDIC has become a modern numerical model in hardware acceleration. Now introducing the basic principles on cordic[6].

Let the vector $(x_0, y_0)$ clockwise rotation $\theta$, the new coordinates$(x_1, y_1)$ can be written as follows:

$$\begin{cases} x = x_0\cos\theta + y_0\sin\theta \\ y = -x_0\sin\theta + y_0\cos\theta \end{cases} \quad (12)$$

Where above formula can be changed to the follow equation

$$\begin{cases} x = K(x_0 + y_0\tan\theta) \\ y = K(-x_0\tan\theta + y_0) \\ z_1 = z_0 - \theta \\ K = (1+\tan^2\theta)^{-1/2} \end{cases} \quad (13)$$

The original rotation can be instead of rotating a series of multiple rotations

$$\theta = \delta_1 a_1 + \delta_2 a_2 + ... + \delta_n a_n \qquad \delta_i = \pm 1 \quad (14)$$

$\delta_i$ presents the direction of the base angle rotation, if $\tan a_i = 2^{-i}$, multiplying $2^{-i}$ can be get by shifting right i bit, this operation is extremely simple for digital circuit, so a important equation can be deduced according to (13):

$$\begin{cases} x_{i+1} = (x_i + y_i \delta_i 2^{-i}) \\ y_{i+1} = (-x_i \delta_i 2^{-i} + y_i) \\ z_{i+1} = z_i - \delta_i \tan^{-1}(2^{-i}) \\ K_i = \prod_{i=0}^{n}(1+2^{-2i})^{-1/2} \end{cases} \quad (15)$$

When n is large enough, equation15 can make $z_i$ to be close to zero, so the angle can be written as follow:

$$\theta = \sum_{i=1}^{n} d_i \tan^{-1}(2^{-i}) = -z \quad (16)$$

When we rotate this vector to x-axis, the value of x-axis is the amplitude of original vector. Simultaneously Z presents the angel the original vector rotated to the x-axis. Now the recursive process is not a pure rotation, finally we need to make compensate to the amplitude of vector. When n is large enough, K is close to 0.6072, a constant[7]. When all rotation finished, we can process compensation by making shifting and addition and subtraction. Because the sampled data is complex number, we must rotate the complex number to a real number firstly. So the Givens rotation of a complex number can be presented by equation (17)

$$G(n,i) = \begin{pmatrix} c_i & s_i^* \\ -s_i & c_i \end{pmatrix} = \begin{pmatrix} 1 & \\ & \exp(j\varphi_i) \end{pmatrix}\begin{pmatrix} \cos\theta_i & \sin\theta_i \\ -\sin\theta_i & \cos\theta_i \end{pmatrix}\begin{pmatrix} 1 & \\ & \exp(-j\varphi_i) \end{pmatrix}$$

$$\varphi_i = \tan^{-1}\frac{\mathrm{Im}(x_i)}{\mathrm{Re}(x_i)}, \theta_i = \tan^{-1}\frac{|x_i|}{r_i} \quad (17)$$

## IV. SIMULATION IN FPGA

FPGA has strong parallel and pipeline properties which are extremely ideal for systolic array. Simultaneously, systolic requires globe clock which FPGA can provide. The basic structure of QRD-SMI is presented in Fig2.

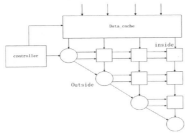

Fig. 2. Basic structure of 4x4 QRD-SMI

The implement of QRD-SMI including four basic module, control module, data buffer module, outside module and inside module. The program is described by VERILOG which is specialized hardware description language. The block diagram of outside module is presented in Fig.3.

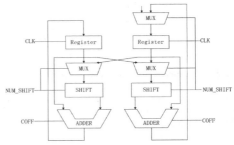

Fig. 3. The block diagram of outside module.

Next I will describe the operation of outside module. When NUM_SHIFT=0, the input data is registered in input register. In the next 24 system clock, the module does not accept data. Before 20 cycles, multiplexer and shift register design shift bits, simultaneously adder or subtraction is selected by the COFF signal. In the last 4 cycles, module must finish amplitude correction. In the 24th cycle, NUM_SHIFT will be zero and module will receive external data again.

In the outside module, the COFF signal is designed by iteration results. If the data is greater than zero, COFF is high, the vector will clockwise rotate. On the contrary, COFF is low, the vector will anticlockwise. At the same time, the signal of NUM-SHIFT and COFF will be conveyed to inside module. These signals will control shift operation and add operation in inside module. Multiple processors will work together through this mode. For complex rotation, the module must rotate the imaginary part to zero. It takes 23 cycles again. So the operation for complex data takes 47 cycles in total.

Testing this program in ISE13.1 platform. In Fig.4, x1, x2, x3, x4 constitute 4x4 input matrix, the high 16 bits are real part of signal, the low 16 bits are imaginary part. y1—y10 are outputs of upper triangular matrix. We can compare matrix RR(result of our algorithm) with matrix R (result of QR function in MATLAB), we can find deviation is in acceptable range.

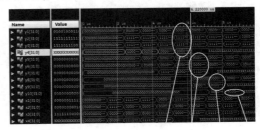

$$RR = \begin{bmatrix} 200 & -3+200i & -201+1i & 200i \\ 0 & 4 & 1+5i & 1i \\ 0 & 0 & 1 & 1+2i \\ 0 & 0 & 0 & 1 \end{bmatrix}$$

$$R = \begin{bmatrix} 200.62 & -3.23+200.96i & -201.31+0.71i & 0.18+200.12i \\ 0 & 3.84 & 1.23+5.14i & -0.41+0.77i \\ 0 & 0 & 1.40 & 1+2.02i \\ 0 & 0 & 0 & 0.70 \end{bmatrix}$$

Fig. 4. The simulated results in ISE platform.

Assuming a simulation model, signal frequency is 2GHz, the direction of desired signal is 0°, SNR is 0dB. The direction of disturb signal is -20°, SNR is 40dB , the number of input data are 64. ISE simulation data is exported to MATLAB to produce image of pattern as Fig.5, we can get a similar result.

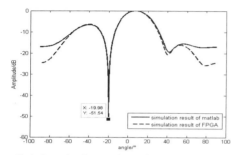

Fig.5. Comparing givens rotation result with matlab result

Downloading the program into Virtex5 chip, observing the signal in chip by internal logic analyzer called CHIPSCOPE. Simulating input signal through VIO, ILA, ICON IPCORE. Observing the output in simulation and the output in Virtex5, We can verify the implementation of the program downloaded in the chip. Two results are shown in Fig6. By comparison, two results are identical. This will verify the correctness of the program. But the input data is limit, it is difficult for us to verify the large amount of input data. The frequency of system clock is 50MHz.

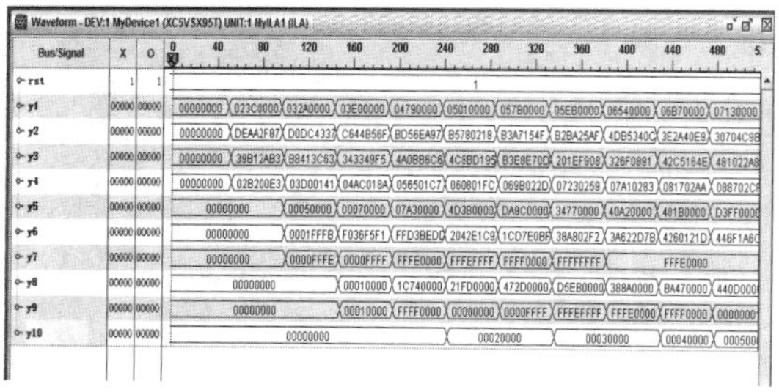

Fig.6. Output in chip and the output of simulation status

## V. CONCLUSION

The process which QRD-SMI is realized in FPGA is introduced. We use the CORDIC algorithm to complete the Givens rotation. However, one operation cost 47 cycles, system's ability to handle high data rate is limited. The program need to optimized to meet faster system clock. If the program can run in 100MHz, two times the data rate can be handle. In addition, reducing cycles of one process by optimizing the frame of program.

## REFERENCES

[1] Yu Lei, Wei Liu, Langley R., "SINR Analysis of the Subtraction-based SMI Beamform," *IEEE Trans. on Signal Processing* vol. 58, pp. 5926-5932. 2010.

[2] Wang, A. K, and Leary, J, "SMI based beamforming algorithms for TDMA signals," *the Thirty-First Asilomar Conference on*, vol. 2, pp.1326-1330. 1997.

[3] Min-Woo Lee and Ji-Hwan Yoon, "High-speed tournament givens rotation-based QR Decompositon Architecture for MIMO Reciever," *IEEE International Symposium on*., pp.21-24. Feb. 2012.

[4] Bienati, N. Spagnolini, U. Zecca, M "An adaptive blind signal separation based on the joint optimization of Givens rotations," acoustics, Speech and signal processing.2001 *IEEE International Conference on*, vol.5, pp.2809-2812. Aug. 2001.

[5] Aslan, S, Niu, S., Saniie, J., "FPGA implementation of fast QR decomposition based on givens rotations., vol.6, pp.470-473. 2012.

[6] Chih-Hsiu Lin.An-Yeu Wu, "Mixed-scaling-rotation CORDIC algorithm and architecture for high performance vector rotational DSP application," *Circuits and system I: regular Papers, IEEE transaction on*., vol.55, pp. 2385-2396. Apr. 2005.

[7] Cheng-Shing Wu, An-Yeu,, "Modified vector rotational cordic algorithm and architecture,"*Circuits and systems : Analog and digital signal processing, IEEE transaction on*., vol.48, pp.546-561. Dec. 2001.

# Co-aperture dual-band waveguide monopulse antenna

Yuan-Yun Liu, Feng-wei Yao, Yuan-Bo Shang

Shanghai Key Laboratory of Electromagnetic Effect for Aerospace Vehicles , Shanghai, 200438,China

jojoyao@163.com

*Abstract-* This paper describe a novel design of dual band monopulse antenna array ,which is made of a X-band and a K-band in the same planar radiating surface. The K-band array is the traditional waveguide slot antenna ,which is interfaced to the X-band antenna. The simulated and measured radiation patterns at center frequency are both presented. The side-lobe level for the sum pattern of two frequency are all less than -22dB , which have been achieved in the experiment .

## I.  INTRODUCTION

Dual mode compound guidance system has many advantages, such as long distance, high-precision and high hit probability, which has wide application foreground and becomes the precision-guided weapon development orientation in future.

Compound monopulse dual-frequency antenna array is the key component of dual mode guidance system, which may provide the information of the elevation, azimuth and distance through one pulse. This information can be used to realize the precise bearing and tracking. In Traditional monopulse radar system, the common type of the dual band antenna are Cassegrain parabolic antenna with double feeds [1], which is acceptable when the available aperture dimensions are enough wide with respect to the wavelength. In paper[2],the dual-band feed of the lower band with four horns and the higher band feed with a monopluse multimode horn is presented , while the efficiency of lower band antenna is only 15% due to the space between two adjacent horn in two-dimensions

In order to reduce the profile of parabolic antenna, a dual band antenna have been made also as reflect array [3], with the advantage of employing planar instead of parabolic re-radiating surface. This implies cost reduction, but less efficiency and power gain, which is similar to parabolic antenna.

In this paper, a novel planar dual band monopulse antenna is presented, the K-band array is the traditional waveguide slot antenna to satisfy the need of lower sidelobe levels and greater efficiency. On the same planar radiating layer ,the X-band antenna array is interlaced to K-band slot array. The lower levels include the feeding waveguides and the monopulse comparators of X-band and K-band respectively .The simulated and measured radiation patterns at center frequency are both presented. Its structure and experimental results are presented as follow.

## II.  DESCRIPTION OF THE ARRAY

The proposed dual-band monopulse antenna array as shown in Fig.1, K-band array is the traditional waveguide slot antenna ,which can accurate control the aperture excitation amplitude through changing the offset of cutting slots into the broad wall of waveguide .The X-band radiator is rectangular waveguide ,which can be fed from end. The X-band antenna array is interlaced to K-band slot array on the same aperture. This planar array is made of 8×8 radiating elements in X-band and 18×18 slots in K-band, both divided in four sub-arrays and arranged in a circle .

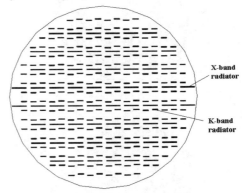

Figure 1. Schematic of  Dual-band Antenna

The feeding networks and comparators are all on the lower levels ,which use waveguide structure . The comparators of different frequency are all formed by four 3dB hybrid junctions. The four inputs of the comparator are connected with four sub-arrays ,four outputs of the comparator are connected to the sum ,H-plane difference, E-plane difference and matching load respectively.

## III.  SIMULATED RESULTS

The simulated 3-dimentional radiation patterns both for sum and for difference beams at X-band and K-band are shown in Fig.2 and Fig.3,and it is seen that the first side lobe of sum are all below -22dB at different frequency. The gain of X-band and K-band is 23.0dB and 30.0dB respectively.

978-7-5641-4279-7

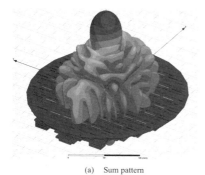

(a)    Sum pattern

(b) E-plane difference pattern

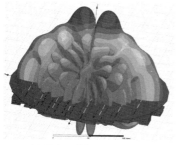

(c)  H-plane difference pattern
Figure 2  simulated 3-dimensional patterns at X-band

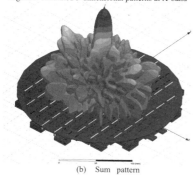

(b)   Sum  pattern

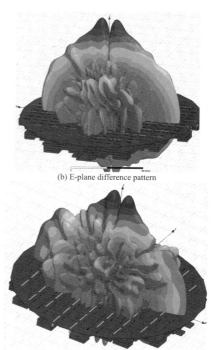

(b) E-plane difference pattern

(c)  H-plane difference pattern
Figure 3  simulated 3-dimensional patterns at K-band

## IV.    MEASURED RESULTS

A test antenna was fabricated and the radiation patterns were measured in an anechoic chamber. The figures highlight a good agreement between theory and experiment.

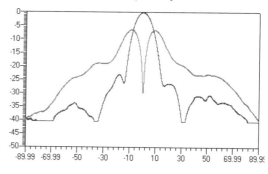

Figure 4  measured sum and difference patterns at X-band

In detail, the first side lobe of X-band and K-band for sum pattern are -22dB and -27dB respectively, the null depth of different frequency for difference pattern are -30dB and -28dB respectively. At dual band ,the efficiency of this compound antenna array is above 40 percent ,which is similar to one frequency antenna.

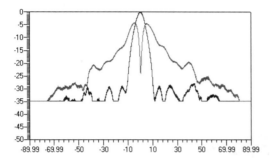

Figure5  measured sum and difference patterns at K-band

## V.    CONCLUSION

A novel dual band monopulse planar antenna has been introduced in this paper, which is made of a X-band array and a K-band array in the same radiating surface. The -22dB side-lobe level for the X-band sum pattern and -27dB side-lobe level for the K-band sum pattern have achieved in the experiment at center frequency and the efficiency is above 40 percent, which is similar to the efficiency of single frequency monopulse antenna.

## REFERENCES

[1]  John D K. Antenna. 3rd ed. Beijing:Publishing House of Electronics Industry, 2006.

[2]  wu xiang "Development of Monopulse Common Reflector Dual Frequency Antenna", ModernRadar, 2011, 33(11):56—62.

[3]  Mark Zawadzki and John Huang, "A dual-band reflectarray for X- and Ka- bands", PIERS2003,Honolulu,2003

[4]  Giuseppe colangelo, "Shared aperture dual band printed antenna" ,Electromagnetic in advanced        application 2011 international conference,1092-1095.

# *TP-1(D)*

## October 24 (THU) PM

## Room D

## Mobile & Indoor Propag.

# Propagation Models for Simulation Scenario of ITS V2V Communications

H. Iwai, R. Yoshida, H. Sasaoka

Doshisha University

Tatara-Miyakodani 1-3, Kyotanabe, Kyoto, 610-0321 Japan

E-mail: {iwai@mail, dum0366@mail4, hsasaoka@mail}.doshisha.ac.jp

*Abstract-* **Unified computer simulation scenario of ITS vehicle to vehicle (V2V) communication has been proposed. In the scenario, several aspects of V2V communications, such as physical layer systems, communication protocols, road traffic models, environment models and so on are defined. In this paper we present propagation models adopted in the scenario. We first introduce a propagation loss estimation model that has been adopted as the basic model and also present two specific models. We show the calculated results by using the ray tracing method for the two models.**

## I. INTRODUCTION

In recent years, research activities aiming at improving safety performance of automobiles by applying the wireless communication have been vigorously carried out all over the world [1]. For instance, in order to prevent collisions of vehicles at intersections with poor visibility, a method where position information is exchanged between vehicles by using vehicle-to-vehicle communication (V2V) is attracting attention. As a new way of the development of such ITS applications, computer simulations are often used to assess the effect of the developed systems and to evaluate the system design of a variety of applications [2, 3]. So far, software simulations for the development of the ITS applications have been done "layer by layer". For example, communication systems for ITS and traffic flow have been considered in simulations separately. In order to appropriately simulate and evaluate the ITS applications where behaviors of several component layers are mutually and closely related, unified operation of all layers must be done in the simulations. These movements of the integration of the simulators for ITS applications have already begun in Japan and worldwide. In Japan, ITS Simulator Promotion Committee was formed in Japan Automobile Research Institute (JARI) and the unified ITS simulator was discussed by experts in various aspects of ITS research. One of the purposes of the committee is to develop a common simulation and evaluation methodology for ITS applications. As the results of the energetic work of the committee, "Standard Simulation and Evaluation Scenario for ITS Applications (VER1.0)" was established in 2010. Also in the following year (2011), the scenario was revised in order to present more detailed information assuming a concrete ITS application. In the revised scenario three actual applications, safe driving assistance application, energy saving application and telematic application, are presented, and the additional exploration and review of the scenario are performed. It includes the detailed description of the traffic flow simulation model, the radio wave propagation model

and the road model. The results of such activities are compiled as VER. 1.1 [4].

In this paper, we introduce the radio wave propagation models defined in the scenario.

## II. STANDARD SIMULATION AND EVALUATION SCENARIO FOR ITS APPLICATIONS

"Standard Simulation and Evaluation Scenario for ITS Applications" includes three applications as mentioned above. For each application, road model, traffic flow simulation model and communication protocol model are defined. For the wireless communication system the IEEE802.11p system is commonly assumed for the all applications.

The details of the simulation procedures for the three applications are presented and the evaluation methodologies are also described in order to make the comparison of the simulated results easy and fair. Since it is important for such scenario to be used commonly and widely, the document describing the scenario is freely available at the website of JARI [4].

## III. REQUIRED PROPERTIES FOR RADIO PROPAGATION MODEL

As a radio propagation model for simulations of general wireless communications, the following propagation characteristics should be modeled and simulated.

 A. *Propagation loss*
 B. *Fading*
 C. *Reflections*
 D. *Diffractions*
 E. *Shadowing*

In the standardized scenario, the above characteristics are considered as follows.

*A. Propagation loss*

The most important environment for simulations of ITS applications is thought as a populous city area since it is traffic accident-prone. In such environment, the propagation becomes so-called a "street canyon" propagation, where radio waves propagate along the road. The street canyon propagation model in ITU-R Recommendation P.1411-5 (hereinafter, P.1411 model) is adopted as the basic model of the scenario, since it is a typical model of such environments and is used worldwide. However, for the safe driving assistance application, the environment is difficult to be simulated by P.1411 model and it is necessary to simulate the individual environments more precisely. Therefore individual propagation models are constructed for the applications.

978-7-5641-4279-7

*B. Fading*

In order to make the implementation of simulation easier, signal level fluctuation due to the fading is not considered in the standard scenario. However, for simulations with small time granularity, the fading should be considered. In such a case, Rayleigh fading model is recommended in the scenario.

*C. Reflections, D. Diffractions, E. Shadowing*

Since P.1411 model is supposed to predict the propagation loss in actual street canyon environments, such effects as reflection, diffraction and shadowing by the surrounding buildings and vehicles are included in the propagation loss calculation formula. However, if obstacles are very close to the transmitter (Tx) or the receiver (Rx), the influence cannot be considered by the formula. Since the current scenario only focuses on the basic part of ITS simulations, the effect is not taken into consideration. It requires further study.

In the models of the safe driving assistance applications, these effects are individually taken into account by the ray tracing calculation.

IV. RADIO PROPAGATION MODEL IN STANDARD SCENARIO

*A. P.1411 Model*

ITU-R recommendation P.1411-5 [5] is a recommendation giving various propagation models of point-to-point transmission in short-range propagation environments. In the recommendation, the propagation model for a street canyon environment is included. In the standard scenario, the model is used as the basic model. Some modifications are made in order to adjust the model to be used appropriately in ITS simulations. Note that the model is originally created for the prediction of the propagation loss in a micro-cell communication environment, not for a V2V environment.

The propagation loss of a LOS environment is calculated by the street canyon LOS model in P.1411 where the equivalent antenna height is set to zero. The equivalent antenna height is omitted in order to simplify the simulation. The propagation loss in the LOS environment $L_{LOS}$ in (dB) is given by the following equation where $d$ (m) is the distance between Tx and Rx.

$$L_{LOS} = L_{bp} + 6 + \begin{cases} 20\log_{10}\left(\dfrac{d}{R_{bp}}\right) & \text{for } d \le R_{bp} \\ 40\log_{10}\left(\dfrac{d}{R_{bp}}\right) & \text{for } d > R_{bp} \end{cases}. \quad (1)$$

$R_{bp}$ and $L_{bp}$ are the distance from Tx to the breakpoint in (m) and the loss at the breakpoint in (dB), respectively. They are given by the following equations where $h_b$ and $h_m$ are the Tx and the Rx antenna heights in (m) and $\lambda$ is the wavelength in (m) at the carrier frequency.

$$R_{bp} \approx \frac{4 h_b h_m}{\lambda}, \quad L_{bp} = \left| 20\log_{10}\left(\frac{\lambda^2}{8\pi h_b h_m}\right) \right|. \quad (2)$$

The loss in the NLOS environment, $L_{NLOS}$ (dB), is given by the following formula.

$$L_{NLOS} = -10\log_{10}\left(10^{-L_r/10} + 10^{-L_d/10}\right), \quad (3)$$

where $L_r$ and $L_d$ are propagation losses in dB of the reflected waves and the diffracted waves, respectively. They are given by the following formulae.

$$L_r = 20\log_{10}(x_1 + x_2) + x_1 x_2 \frac{f(\alpha)}{w_1 w_2} + 20\log_{10}\left(\frac{4\pi}{\lambda}\right), \quad (4)$$

$$L_d = 10\log_{10}[x_1 x_2 (x_1 + x_2)] + 2D_a$$
$$- 0.1\left(90 - \alpha\frac{180}{\pi}\right) + 20\log_{10}\left(\frac{4\pi}{\lambda}\right). \quad (5)$$

The distances $x_1$ and $x_2$ and the road widths $w_1$ and $w_2$ are shown in Fig. 1 and the units are in (m). $\alpha$ is the angle of the two crossing roads in (rad). $f(\alpha)$ and $D_a$ are given by the following expressions.

$$f(\alpha) = \frac{3.86}{\alpha^{3.5}}, \quad D_a = \left(\frac{40}{2\pi}\right)\left[\arctan\left(\frac{x_2}{w_2}\right) + \arctan\left(\frac{x_1}{w_1}\right) - \frac{\pi}{2}\right]. \quad (6)$$

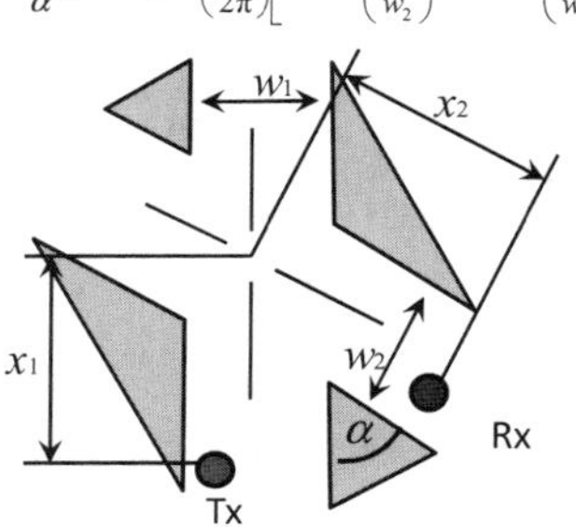

Fig. 1 Distances and road widths of NLOS environment.

P.1411 generally gives realistic values of propagation loss observed in the actual environment. However, in a viewpoint of the implementation to ITS simulations it has some defects. One is the relation of the magnitudes of the predicted propagation losses by LOS and NLOS formulae at areas close to intersections. In such areas, there are some cases where the propagation loss predicted by the NLOS formula is less than that by the LOS formula. It is not realistic, hence we make a modification to the original P.1411 model.

In the modified method, the propagation loss in the vicinity of intersections is given by the larger value of the propagation losses obtained by the following two methods : (See Fig. 2)

- Propagation loss calculated by the LOS formula, where $d$ is the distance of the straight path between Tx and Rx.

- Propagation loss calculated by the NLOS formula with the distances $x_1$ and $x_2$ in Fig. 2.

Further, in P.1411 model, a propagation loss calculation method for turning at intersections twice or more times is not defined. It is calculated below in the modification. Using the distances $x_1'$, $x_2'$, and $x_3'$ shown in Fig. 3, the larger propagation loss among the following two losses is selected.

- Propagation loss calculated by the NLOS formula where $x_1 = x_1'$ and $x_2 = x_2' + x_3'$.

- Propagation loss calculated by the NLOS formula where $x_1 = x_1' + x_2'$ and $x_2 = x_3'$.

The above method is considered based on the measured result [6].

P.1411 model is used as the propagation loss model for the energy saving application, the telematics application, and a part of the safe driving assistance application.

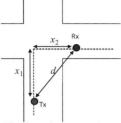

Fig. 2 Modification where Tx and Rx are close to intersection.

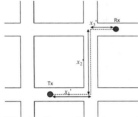

Fig. 3 Modification for turning at intersections twice.

B. *Propagation models for safe driving assistance applications*

P.1411 model gives the propagation loss in a typical propagation environment, but it does not consider the particular circumstances, such as individual road shape, presence or absence of blocking vehicles, curving road, etc. Such environmental influences are required to be taken into consideration for ITS simulations especially for the safe driving assistance application. Therefore, in the standard scenario specific propagation models are defined for the application. The following three specific applications are presented in the safe driving application.

- Collision avoidance application at T-shape junctions.
- Right-turn collision avoidance application at cross-shape intersections (right-turn application).
- Rear-end collision avoidance application at a curve (curve application).

T-junction environment is assumed for the first application. Since this environment is close to the cross-shape intersection like the assumed environment in P.1411 model, P.1411 model is used as the propagation model for the application.

We use P.1411 model as the basic model assuming street canyon propagation environment. However, in the environments of right-turn and curve applications, the propagation loss cannot be well predicted by P.1411 model since the blocking by vehicles and walls generates the essential factor of the propagation. However, a well-established propagation loss estimation formula like P.1411 model does not exist for such environments. Therefore, we calculate the propagation loss in the applications by the ray tracing method. In the scenario document, the propagation loss in these applications is presented by tables of the propagation loss values at discrete locations of Tx and Rx. Users of the scenario are supposed to interpolate the discrete values to obtain the loss at arbitrary positions of Tx and Rx.

It is necessary to identify the carrier frequency for the ray tracing calculation. We select two frequencies 760MHz and 5.8GHz considering the frequencies used for ITS in Japan and the world.

Figure 4 shows the assumed environment of the right-turn application, where a collision of two vehicles moving in the opposite lanes is anticipated. The right-turn waiting vehicle at the intersection is between the two moving vehicles. Since the direct path (both in vision and in radio) is blocked by the right-turn waiting vehicle, if one of the vehicles (Tx) turns right at the intersection, the possibility of the collision is high. It is a typical situation of traffic accident, and such situation is often assumed for the evaluation of the collision avoidance application by ITS.

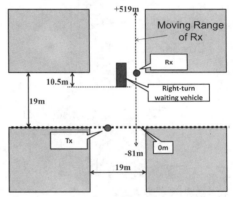

Fig. 4 Assumed road model for right-turn application.

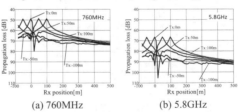

(a) 760MHz        (b) 5.8GHz

Fig. 5 Example of propagation loss of right-turn application.

Figure 5 shows some results of the ray-tracing calculation for the model of Fig. 4. In the figure, a large propagation loss is predicted for Tx at 0m and Rx at 20m. This shows the case where the direct path is blocked by the closely-located blocking vehicle.

Figure 6 shows the environment model for the curve application. In the application, an accident, where a vehicle entering into the curve rear-ends the stopping vehicle beyond the curve, is assumed. The situation often occurs in traffic jam. Assuming a metal wall inside the curve, the situation is NLOS, where the direct path between Tx and Rx is blocked by the wall. The case is assumed for the worst case, since in usual actual roads, such as highways, another metal wall is also installed outside of the curve. It generates reflected paths and the propagation loss becomes smaller.

Similarly to the right-turn application, since well-established path loss estimation model for the situation does not exist, we calculate the path loss of this model by the ray tracing and give two-dimensional tables of the propagation loss with 10m interval of Tx and Rx positions.

In the calculation of the propagation loss of this application, we divide the propagation model into two cases, LOS and NLOS, according to the positions of Tx and Rx.

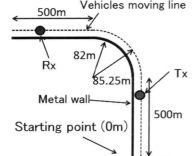

Fig. 6 Assumed road model for curve application.

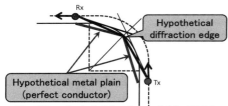

Fig. 7 Hypothetical plain edge model for NLOS.

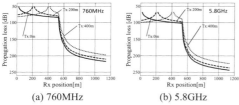

Fig. 8 Example of propagation loss of curve application.

For the LOS situation, we use the free space path loss. For the NLOS situation over the curve as shown in Fig. 7 the propagation loss is calculated by Uniform Theory of Diffraction (UTD). We assume hypothetical metal plains to model the long curve and calculate the diffraction loss by the hypothetical edge by UTD.

In a viewpoint of the precise estimation of the propagation loss, there remains a room for discussion of the usage of this model. We adopt the model for simple calculation of propagation loss, since the environment model is hypothetical and therefore it is not a purpose of the model to calculate an exact value of the loss.

Figure 8 shows examples of the propagation loss of the curve application. The Tx and Rx positions are expressed by the length of the ways from the starting point shown in Fig. 6. In the LOS situation, where the Rx position is less than around 500m the loss is considerably small, while the loss increases very rapidly when Rx is beyond the curve.

## V. SUMMARY

In this paper, we introduce the propagation model defined in "Standard Simulation and Evaluation Scenario for ITS Applications." We adopt P.1411 model to calculate the propagation loss as the basic formula. Since it has some defects to be used in the ITS simulations, we make two modifications to the model. For the safe driving assistance application, two specific environment models are presented and the propagation loss in the environments is calculated by the ray-tracing method.

### ACKNOWLEDGMENT

This paper is concerned with the propagation model in the "Standard Simulation and Evaluation Scenario for ITS Applications" discussed in ITS Simulator Promotion Committee established at Japan Automobile Research Institute (JARI). The authors obtained many valuable advices through the discussion on the propagation models at the committee. The authors wish to express their sincere appreciation to the members of the committee.

### REFERENCES

[1] W. Chen, et al., "Ad hoc peer-to-peer network architecture for vehicle safety communications," IEEE Communications Magazine, vol.43, no.4, pp.100-107, April 2005.

[2] T. Umedu, et al., "An intervehicular-communication protocol for distributed detection of dangerous vehicles", IEEE Transactions on Vehicular Technology, vol. 59, no. 2, pp.627-637, Feb. 2010.

[3] Y. Yamao, et al., "Vehicle-roadside-vehicle relay communication network employing multiple frequencies and routing function," Proceedings of ICWCS 2009, pp.413-417, Sept. 2009.

[4] ITS Simulator Promotion Committee, "Standard simulation and evaluation scenario for ITS applications," http://www.jari.or.jp/research-department/its/H23_simyu/ (in Japanese)

[5] ITU-R P.1411-5, "Propagation data and prediction methods for the planning of short-range outdoor radio communication systems and radio local area networks in the frequency range 300MHz to 100GHz," Oct. 2009.

[6] H. Urayama, et al., "700MHz band propagation loss model for roadside-to-vehicle communications in urban area," IEICE Technical Report, AP2011-146, pp.1-6, Jan. 2012. (in Japanese)

# Comparison of Small-Scale parameters at 60 GHz for Underground Mining and Indoor Environments

Yacouba Coulibaly MIEE, Gilles Y. Delisle LFIEEE, Nadir Hakem MIEEE

Université du Québec en Abitibi-Témiscamingue (LRTCS)
Val d'Or, Qc, Canada, J9P 1Y3
coulibaly@ieee.org, gilles.delisle@uqat.ca, nadir.hakem@uqat.ca

*Abstract*—This paper describes the small scale performances of a 60 GHz channel in an underground mine and in an indoor laboratory using the same channel sounder. The frequency domain channel measurements have been carried out in the 59 GHz to 61 GHz frequency band. Based on the measurements, the two environments are characterized and compared in terms of delay spread, coherence bandwidth. The values of the delay spread are less than 3 ns for both environments. Finally a relationship between the delay spread and the coherence bandwidth is found for both scenarios.

*Keywords-component; 60 Ghz; Channel; underground; Measurement; Characterization*

## I. INTRODUCTION

There are currently three main ways of deploying communication network in an underground mine: leaky feeders, fiber optics and wireless communication. Wireless technology is advantageous compared to the others because it can be deployed anywhere, it is generally a low cost option and it is more resilient than wired technology [1]. Wireless technology can be applied to different applications in the mining industry. It can improve the production through automation, increase safety, reduce maintenance costs and allow the transfer of voice, data and video in remote locations of the mine. Additionally, in the context of an underground mine disaster, it is important to have a wireless network in order to facilitate and improve rescue operations. However, in order for wireless communications to be effective, the specific channels must be characterized with the help of measurements; this is sometimes a painstaking work in an underground mining environment. Furthermore, the design of communication systems in this specific industry is a challenge for any engineer as such systems have to operate in confined environments, which are complex and support a variety of applications.

These different applications can require high data rates. As the lower frequency bands used in wireless standard such as WiFi (Wireless Fidelity) and UWB (Ultra Wide Band) have become congested, the 60 GHz ISM frequency band techniques has been proposed as an alternative communication scheme for high data rates [2]. This frequency band is of great interest because of the massive universal unlicensed spectrum available for communication systems (5-7 GHz). The absorption of the electromagnetic energy is more important at 60 GHz than at lower frequencies used for wireless communications. The signal energy reduces approximately by one half every 200 meters. Therefore, it cannot travel far beyond the intended recipient. Due to high oxygen absorption, the 60 GHz frequency band has a high spatial frequency reuse and it is ideal for high secure short-range wireless communications. The 60 GHz technology can be deployed in indoor and underground environments. Furthermore, it can have a broad range of applications in the underground mining environment. It can be used for the deployment of sensor networks and the transmission of video, the use of high data rate multimedia systems.

Over the last few years, different experimental studies have been carried out to characterize the 60 GHz channel in indoor environments [3-5], cars [6], tunnels [7] and hospitals [8]. However, limited work has been done to characterize and understand the properties of the underground mine radio propagation channel at millimeter frequency and especially at the 60 GHz ISM band for wireless communication systems [9].

An underground mine gallery presents some structural similarities to indoor environments. They both have long corridors, wide halls and intersections. But an underground mine is distinct because it is a confined environment with a curved roof and many other obstacles, such as the walls, the electric wires, the telecommunications cables, the ventilation system and the pipes. Thus, there are slight differences in an underground mine from other indoor environments. Large scale results of an indoor laboratory and an indoor mine have already been compared [9]. The objective of this paper is to characterize a 60 GHz underground mine channel and compare its small scale parameter to those of an indoor laboratory. This paper is organized as follows. Section II gives a brief description of the underground mine environment, the indoor laboratory environment and the measurements procedure. Section III presents the comparison of experimental data in terms of delay spread and coherence bandwidth. Section IV concludes the results of this study.

978-7-5641-4279-7

## II. MEASUREMENT TECHNIQUES

### A. Measurements environments

Figure 1. Underground mine environment..

The underground measurements were conducted inside a gallery of a real mine, named CANMET (Canadian Center for Minerals and Energy Technology). This mine is located at Val d'Or, which is 520 km north of Montreal, Canada. The gallery is at a level of 70 m underground. Its height, width and length are approximately 2 m, 3 m, and 75 m, respectively. Fig 1 shows the measurement setup inside the underground mine. Indoor measurements were carried out in the CANMET laboratory (Fig.2). It has a width of 4 m, a length 8 m and a height 4m, respectively.

Figure 2. Indoor environment.

### B. Measurements setup

For the characterization of the two environments, the same frequency sounder based on a Vector Network Analyzer (VNA), a power divider, a local oscillator, a power amplifier (PA), a low noise amplifier (LNA), frequency multipliers, filters, mixers, cables and antennas have been used. A power divider is used to divide a 2.25 GHz synthetized signal into two channels. The signal of both channel are then multiplied twenty four times to obtain the 54 GHz signals at the local oscillator (LO) output of both mixers. The baseband signal, which is between 5 and 7 GHz, is generated by port 1 of the VNA and then upconverted to frequencies between 59 GHz and 61 GHz. The resultant signal is then passed through a band pass filter

followed by a power amplifier before it is radiated in free space through a directional antenna.

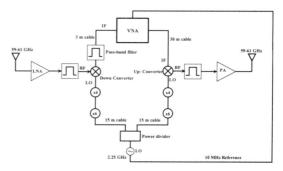

Figure 3. Measurement set-up

.The signal received by the same kind of directional antenna, is amplified by a 60 GHz low noise amplifier and then filtered by a 60 GHz bandpass filter. This radio frequency (RF) signal is downconverted to a baseband signal between 5 GHz and 7 GHz. It is also filtered, and connected to the second port of the VNA The LO at the frequency of 2.25 GHz and the VNA are synchronized through a 10 MHz reference clock.

Both antennas used at the transmission and the reception were vertically polarized directional horn antennas (CERNEX CRA15507520). Their operating frequency bands range from 50-GHz to 75 GHz with a maximum gain of 20 dB. Both half power beamwidths in the azimuth and elevation planes are 12 degrees.

The Agilent E8363 VNA measures the frequency transfer function with 2048 stepped frequency points in the range of 5 GHz to 7 GHz. This gives frequency resolution of 0.97 MHz, a maximum delay of 1.0309 us and the maximum path length of 309 m which is enough for 5m, the maximum distance that separate both the transmitter (TX) and the receiver (RX). A Through Reflect Line (TRL) calibration was done before the measurements. A reference measurement was then performed with the transmitting antenna and receiving antenna 1m apart. The calibration therefore removes the effects of the LNA, the PA, the cables, connectors and antennas from the measured frequency responses.

In order to characterize the propagation channel in small scale, the receiver is moved on a grid square with 9 points (3 X 3) where the distance between each adjacent point is equal to 2.50 cm, which is half of the wavelength in free space at the frequency of the 60 GHz. The antenna is accurately moved on this virtual array by using a VELMEX positioning system. A laser beam has been used to fix the height of the antennas from the ground at 1.5 meters. The measurements were done in light of sight (LOS) for a transmitter-receiver distance of 1 m to 4 m in the two environments. All measurements were done in the middle of the gallery and the laboratory. All measurements were performed with minimal human movement and activity. In all cases, ten consecutive sweeps were averaged at each distance to obtain an important statistical data of the channel and to also reduce the effects of random noise

### III. RESULTS

#### A. Delay spread

In an underground environment, the presence of multipath components is considerably important due to the reflection and scattering from the ground and surrounding rough surfaces. Multiple paths, with various delays, will cause intersymbol interference (ISI). This ISI is a limiting factor of the maximum data rates of a communication system. The RMS delay spread can give the maximum data rate of the channel without using other measures such as channel equalization.

$$\tau_{rms} = \sqrt{\frac{\sum_k a_k^2 \cdot \tau_k^2}{\sum_k a_k^2} - \left(\frac{\sum_k a_k^2 \cdot \tau_k}{\sum_k a_k^2}\right)^2} \qquad (1)$$

The delay spreads are calculated according to equation (1) by taking the multipath components excess delay $\tau_k$ with amplitude $a_k$ within 30 dB threshold of the peak value of the power delay profile (PDP). This threshold can be selected because most of the energy of the different PDP is taken in consideration. The comparison of the delay spread of the underground mine and the indoor laboratory is plotted on Fig 4. It can be seen that, for the underground mine and the indoor laboratory, that the delay spread varies from 2.49 ns to 6.75 ns and from 0.17 ns to 2.48 ns, respectively. The average values for the mine and the laboratory are 3.61 ns and 0.78 ns, respectively.

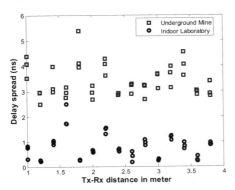

Figure 4. Delay spread in both environments.

These low values of the RMS delay spread are due to the confinement of the electromagnetic waves in the room and the laboratory. At 60 GHz, the diffraction, the scattering and the penetration of the electromagnetic waves around the obstacles are considerably reduced. It can also notice that the

delay spread is not correlated to the transmitter-receiver distance.

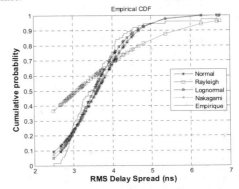

Figure 5. Cumulative distribution of the RMS delay spread in the underground mine and different distributions.

The cumulative distribution function of the delay spread for the underground mine with a threshold of 30 dB has been plotted in Fig. 5. This measured data is fitted to a normal, lognormal, Raleigh, and Nakagami distributions in order to find the one which represents the best the measured data. The Aikake Information Criterion (AIC), which is a method applied to evaluate the goodness of a statistical fit, was used. Fig 5 also shows the CDF of the measured with the fitted distribution functions. The lognormal distribution, which has the lowest AIC, is the best fit for the measured data.

#### B. Coherence bandwidth

The coherence bandwidth (Bc) is one metric which can be obtained from from the frequency domain characterization. It is the the frequency range over which the channel amplitudes are correlated. It is obtained from the the autocorrelation function of the measured frequency response.

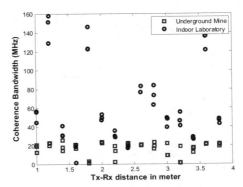

Figure 6. Coherence bandwidth spread in both environments.

Fig.6 shows the Bc for the two environments when the autocorrelation has dropped to 90% of the peak value. This difference is due to the fact there are less diffraction and scattering in the indoor environment. The coherence bandwidth can be related to the delay spread in a heuristic way. The mathematical form of this equation is :

$$B_{coh,0.9} \ (GHz) = \frac{A}{\tau_{rms}(ns)} \qquad (2)$$

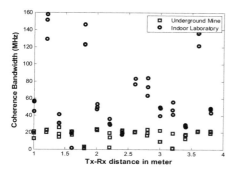

Figure 7.  Coherence bandwidth spread in both environments.

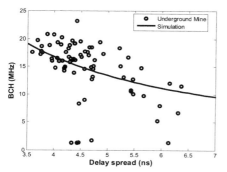

Figure 8.  Coherence bandwidth as a function of the delay spread for the underground mine.

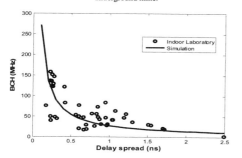

Figure 9.  Coherence bandwidth as a function of the delay spread for the indoor laboratory.

This can be used to deduce the coherence bandwidth from the RMS delay spread and vice versa when one set of values are known. The values of 0.06 [10], 0.063 [3] and 0.285 [2,8] have all been found in the literature. Fig. 7 and Fig. 8 show the coherence bandwidth as a function of the corresponding delay spread for the underground mine and the laboratory, respectively. The values of the constant A are found to be 0.054, and 0.027, for the underground mine and the laboratory, respectively.

## IV. CONCLUSION

This paper has presented the small scale results for the 60 GHz propagation for an underground mining environment and an indoor environment. Measurements were taken in a line of sight case over 4 m using directional horn antennas. The confinement of the electromagnetic wave is visible in both environments. This results in low values of the delay as the diffraction, the scattering and the penetration of the wave around the obstacles are reduced. The delay spread varies between 2.49 ns and 6.75 ns with a mean value of 3.24 ns for the underground mine with a threshold of 30 dB. The lognormal distribution represents best the RMS delay spread in this situation. For both scenarios, a relationship is found between the delay spread and coherence bandwidth

REFERENCES

[1]  S. Outalha, R. Le, P.M. Tardif, "Toward a unified and digital communication system for underground mines", Revue of Canadian Institute of Mining, Mettalurgy and Petrolium, Vol. 93, No. 1044, pp. 100-105, 2000.

[2]  Su-Khiong Yong, Pengfei Xia, Alberto Valdes-Garcia, 60 GHz technology for GBPS WLAN and WPAN. From theory to practice, John Wiley & Sons Ltd, 2011, pp.3–61.

[3]  P. F.M. Smulders, "'Statistical Characterization of 60 GHz Indoor Radio," IEEE Transactions on Antennas and Propagation, Vol. 57, No. 10, October 2009.

[4]  N. Moraitis, P. Constantinou, "Indoor channel Measurements and Characterization at 60GHz for wireless local area network," IEEE transactions on Antennas and Propagation, Vol. 52, No. 12, December 2004.

[5]  S. Geng, J. Kivinen, X. Zhao, P. Vainikainen, "Millimeter-wave propagation channel characterization for short-range wireless communications," IEEE Transactions on Vehicular Technology, Vol. 58, No. 1, January 2009, pp.1–7.

[6]  M. Schack, M. Jacob, and T.Kurner, "Comparison of in-car UWB and 60 GHz channel measurements," Third European Conference on Antennas and Propagation, pp.640–644, 2009.

[7]  N. Prediger and A. Plattner, " Propagation measurements at 60 GHz in railroad tunnels," in Proc. IEEE MTT-S International Symposium Propagation, pp.1085–1087, 1994.

[8]  M. Kyro, K. Haneda, J, Simola ; K, Nakai ; K. Takiszawa ; H. Hagiwara and P. Vainikainen, " Measurement Based Path Loss and Delay Spread Modeling in Hospital Environments at 60 GHz ", Wireless Communications, IEEE Transactions, Vol 10, No 8, pp. 2423-2427, Aug 2011.

[9]  C. Lounis ; N. Hakem ; G. Y. Delisle and Y. Coulibaly " Large-scale characterization of an underground mining environment for the 60 GHz frequency band," in Proc. IEEE ICWCUCA,Clermont Ferrand, France, August 2012, pp.1–4, 2012.

[10]  T. Zwick, T. J. Beukema, and H. Nam, "Wideband channel sounderwith measurementsand model for the 60 GHz indoor channel radio," IEEE Trans. Veh. Technol., vol. 54, no. 4, pp. 1266–1276, 2005

[11]  N. Moraitis, " Measurements and Characterization at 60GHz of wideband indoor radio channel at 60 GHz," IEEE Trans. Wireless Commun., vol. 5, no. 4, pp. 880–889, Apr. 2006

# Study on the Effect of Radiation Pattern on the Field Coverage in Rectangular Tunnel by FDTD method and Point Source Array Approximation

Da-Wei Li[1,2,*], Yu-Wei Huang[1,2], Jun-Hong Wang[1,2], Mei-E Chen[1,2], and Zhan Zhang[1,2]

[1]Key Laboratory of All Optical Network & Advanced Telecommunication Network of MOE,
Beijing Jiaotong University, Beijing 100044, China,
[2]Institute of Lightwave Technology,
Beijing Jiaotong University, Beijing 100044, China,
*11111001@bjtu.edu.cn

*Abstract*-In this paper, the effect of radiation performance of antenna array in confined space on the field coverage is analyzed. In order to find the suitable radiation pattern that can give smoother wave coverage in confined space, a simple but efficient numerical method based on the parallel FDTD method and point source array approximation is presented. The current on each array element is obtained from the desired radiation pattern using synthesis method. This current distribution is then set to the FDTD mesh as the excitation. By this method, field coverage of three kinds of antennas with different radiation patterns in a rectangular tunnel is analyzed. The results show that the antenna with narrower beam width can give smoother field coverage in the tunnel.

## I. INTRODUCTION

With the rapid development of mobile communication, the prediction of the wave propagation and field coverage of the wireless link becomes an important work, especially for the case of confined space. Confined space is defined as one kind of regions with known physical boundaries, such as tunnels, mines, etc. Usually, there are blind zones of wireless communication in these regions and the overall communication quality is not satisfying, because the wave propagation in tunnel environment is very complicate, so antennas with simple radiation patterns can not ensure the overall quality of the wireless communication in confined area. Therefore, antennas with special radiation pattern that is 'suitable' for field coverage in tunnel-like confined space are preferable.

It is a common way to obtain the actual field coverage of a wireless communication system by measurement. But measurement is an expensive and time consuming way, and it still cannot give the overall property of field coverage of some real systems [1]-[2]. Unlike measurement, full wave methods, such as the finite-difference time-domain (FDTD) technique, can not only provide the accurate result of field coverage but also provide the effect insight of environments on the field distribution, especially for the case of complicated environment in confined area. Based on FDTD, an Integrative Modeling Method (IMM) was proposed in [3], in which the whole RF link including the antennas and their environments is integratively modeled and simulated. Of course, the

increasing capacity of storage and computing resources also make it possible for FDTD to simulate large-scale problems.

However, if practical antenna structures are considered in the FDTD mesh together with the complicated tunnel environment, much higher spatial resolution is needed to accurately simulate the small geometrical features of the antenna. As a result, the computation time and the memory requirements will be increased significant if uniform mesh scheme is still used. Certainly, it is possible to combine the FDTD with other numerical schemes, such as the frequency-domain method of moments (FD-MoM) to increase the numerical efficiency [4]. Also hybrid technique based on the surface equivalence theorem can also be used to solve the multi-scale problem [5]. These methods although reduce the computation load in some extension, it introduces other difficulties in combining of different methods. The objective of this paper is looking for the 'suitable' radiation pattern (the first step of actual antenna design) which can give smoother field coverage in tunnels. A technique combined FDTD with antenna array synthesis is presented in this paper, in which the antenna array is composed of a series of current elements with freedom parameters of current amplitude and phase, and the element length occupies only one grid of FDTD mesh discretized in terms of the environment scale. After the 'suitable' pattern is obtained, the real antenna or antenna array structure can be optimized to get an actual radiation pattern that closes to the 'suitable' pattern.

## II. DESCRIPTION OF THE NUMERICAL METHOD

In the field coverage evaluation of confined space, FDTD method is widely used [6]-[7]. However, as already mentioned, these approaches have to face the challenges of density mesh and long iterative time, because fine FDTD mesh should be used in modeling the practical antenna structures. On the other hand, in order to obtain the 'suitable' pattern for confined space application, different patterns should be introduced into FDTD computing domain for numerical experiment. But how to introduce these patterns is also a problem to solve.

To overcome these problems, method for simulating the propagation environment and practical antenna has been proposed [5]. But additional implementation cost is required to simulate the antenna and propagation environment separately. This method can only reflect the effect of a specific antenna but is not suitable for simulation the effect of different radiation patterns on the field coverage. In order to reduce the complexity of the calculation, instead of using real antennas, infinitesimal dipoles (point sources) are used as transmitting and receiving antennas in [8]. However, the utilization of point sources is not easy to study the effect of different radiation patterns on the field coverage. In this paper, point source array is used, so not only the computation complexity is reduced but also the desired radiation patterns are easily obtained by adjusting the amplitude and phase of the feeding current on each array element (current element). Two steps should be done in this method: 1) the amplitudes and phases of array elements are obtained according to desired pattern by synthesizing method; 2) take this antenna array as the excitation of FDTD to get the actual field distribution in confined space.

### A. Source design

As mentioned, in order to reduce the number of FDTD mesh grids, a point source array is employed to obtain the desired far-field pattern. This array is composed of a series of current elements with freedom parameters of current amplitude and phase, and the element length occupies only one grid of FDTD mesh discretized in terms of the environment scale. By synthesizing the phases and amplitudes of the current elements, the point source array can generate the required far-field pattern. It's worth to mention that although the impedances of point source array may have big difference with the realistic antenna array, the radiation pattern which is mainly relates to current distribution of array is coinciding to that of the realistic antenna array. Therefore, this point source approximation can meet the accuracy requirements of the radiation pattern simulation if the amplitudes and phases are set reasonably.

### B. Parallel FDTD technique

In order to shorten the computation time, parallel FDTD technique is used in this paper. MPI (Message Passing Interface), as an important standard of message transmission, is widely used in many fields of science computation. Combining MPI with FDTD to realize parallel computation has been proven to be an efficient way in improving the computing speed. The parallel algorithm of FDTD utilizes a one-cell overlap region to exchange the information between adjacent sub-domains, as shown in Fig. 1. The iteration formulas are given by (1) and (2). In the algorithm, only the tangential magnetic fields are exchanged at each time step as shown in Fig. 1 [9].

$$E_z^{n+1,processor1}(i,j,k+\frac{1}{2}) = \frac{\varepsilon_z - 0.5\sigma_z\Delta t}{\varepsilon_z + 0.5\sigma_z\Delta t}E_z^n(i,j,k+\frac{1}{2}) +$$

$$\frac{1}{\varepsilon_z + 0.5\sigma_z\Delta t}[\frac{H_y^{n+\frac{1}{2}}(i+\frac{1}{2},j,k+\frac{1}{2}) - H_{y1}^{n+\frac{1}{2},processor2}}{0.5[\Delta x(i)+\Delta x(i-1)]}] -$$

$$\frac{1}{\varepsilon_z + 0.5\sigma_z\Delta t}[\frac{H_x^{n+\frac{1}{2}}(i,j+\frac{1}{2},k+\frac{1}{2}) - H_x^{n+\frac{1}{2}}(i,j-\frac{1}{2},k+\frac{1}{2})}{0.5[\Delta y(j)+\Delta y(j-1)]}]$$

(1)

$$E_z^{n+1,processor2}(i,j,k+\frac{1}{2}) = \frac{\varepsilon_z - 0.5\sigma_z\Delta t}{\varepsilon_z + 0.5\sigma_z\Delta t}E_z^n(i,j,k+\frac{1}{2}) +$$

$$\frac{1}{\varepsilon_z + 0.5\sigma_z\Delta t}[\frac{H_{y2}^{n+\frac{1}{2},processor1} - H_y^{n+\frac{1}{2}}(i-\frac{1}{2},j,k+\frac{1}{2})}{0.5[\Delta x(i)+\Delta x(i-1)]}] -$$

$$\frac{1}{\varepsilon_z + 0.5\sigma_z\Delta t}[\frac{H_x^{n+\frac{1}{2}}(i,j+\frac{1}{2},k+\frac{1}{2}) - H_x^{n+\frac{1}{2}}(i,j-\frac{1}{2},k+\frac{1}{2})}{0.5[\Delta y(j)+\Delta y(j-1)]}]$$

(2)

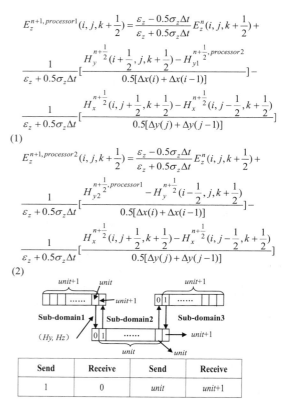

Figure 1. Diagram of parallel FDTD for 1D case in x-direction

### C. Path loss calculation

Assuming that the wave is in linear polarization, the equations for the power calculation from field of one point source are expressed by (3) and (4), where $Z_0$ is the wave impedance in vacuum. During the FDTD simulation, the $E_z$ component for vertical polarization case is recorded. However, when there is more than one single source, the mutual coupling between the sources is significant, and the transmitter power of the array cannot be calculated use this simply equation again. We have to compute the integral of flux density on the surface surrounding the source array to get the complete transmitting power, as expressed by (5), where $E_h$ and $H_h$ are the tangential electric and magnetic fields

$$P_t = \frac{2\pi}{3}Z_0 I^2 \left(\frac{L}{\lambda}\right)^2 \tag{3}$$

$$P_r = \frac{3\lambda^2}{8\pi Z_0}E^2 \sin^2\theta \tag{4}$$

$$P_t = \oint_s \vec{E}_h \times \vec{H}_h \cdot \hat{n}ds \tag{5}$$

Using (4) and (5), the path loss can be calculated by

$$Path\ Loss = P_r / P_t \qquad (6)$$

## III. RESULTS AND ANALYSIS

The radiation patterns of the point source arrays are given in Fig. 2 and Fig. 3. The radiation pattern marked with blue line is generated by a planar array with 5×6 elements (5 in z-direction and 6 in y-direction) denoted by array 1. The array elements are excited in phase and with the same amplitude, a distance of half wave length is set between elements. The originations of the elements are in z-direction. This array is a broadside array and its main beam is pointing to the x-axis. In order to focus the energy to the positive direction of x-axis, a reflector plate is placed behind the array. Array 2 is a 8×4 planar array (8 in x-direction and 4 in y-direction) with element spaces of a quarter wave length in x-direction and half wave length in y-direction. The originations of the elements are in z-direction. The phase difference of current between elements in x-direction is $\pi/2$, and the amplitudes are the same. So the array is an end-fire array and its main beam is wider.

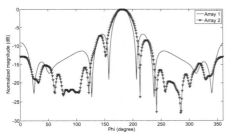

Figure 2. Radiation pattern of the transmitting antennas in H-plane (theta=90)

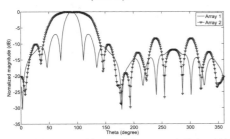

Figure 3. Radiation pattern of the transmitting antennas in E-plane (phi=0)

FDTD simulation is carried out in 3D for antenna arrays radiating in a rectangular tunnel with of 6×6 m² cross section and 200 m in length. The operating frequency is 900 MHz. Cell sizes of $dx = dy = dz = 33$ mm and $dx = dy = dz = 27.75$ mm are used in the simulations for array 1 and array 2 respectively. An eight-cell PML is used to truncate the simulation domain. The relative permittivity and conductivity of the tunnel wall is 6.8 and $3.4\times10^{-2}$ S/m respectively, which is similar to the value of actual tunnel. The tunnel wall is set to

0.5 m in depth. The parallel computing program utilizes 96 processes.

The path loss of the wave in the tunnel generated respectively by the two arrays comparing with that from a dipole is shown in Fig. 4. From Fig. 4 we can see that the broadside antenna array with narrower beam has better radio transmission characteristic in tunnel, it is superior to that of omni-directional antenna. This is because that the radiation energy of directional antenna is more concentrated on the tunnel axis compared to the omni-directional antenna. So the reflected wave from the nearby tunnel walls (the nearby region of the antenna) is weak. If the radiation beam is wider, then most of the energy is reflected by the nearby tunnel walls, and significant multi-path effect occurs, which further results in a significant path loss.

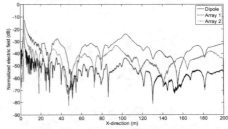

Figure 4. Path loss in the rectangular tunnel for different transmitting antennas

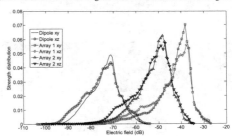

Figure 5. Field strength distribution

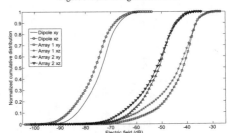

Figure 6. Field strength cumulative distribution

Fig. 7 gives the intensity distribution of electric field in the tunnel. It can be clearly seen the formation of strength and weak regions in the propagation of electromagnetic wave. To get a better communication quality, radio coverage in tunnel should be as uniform as possible. In order to evaluate the uniformity of field coverage, concept of field deviation is proposed, which gives the difference between the field value of each test point and the average of field values of all the test points. Fig. 5 and 6 show the field distribution and field cumulative distribution respectively. From Fig. 5, we can conclude that antenna with narrower beam can give narrower intensity distribution curve and steeper slope of the cumulative distribution curve, which indicates that a better uniformity of the field strength can be get.

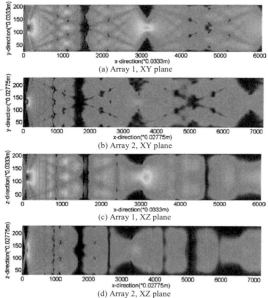

(a) Array 1, XY plane

(b) Array 2, XY plane

(c) Array 1, XZ plane

(d) Array 2, XZ plane

Figure 7. Electric field distributions in the cross sections of the tunnel for different antenna array

CONCLUSION

In this paper, point source array is used to approximately but effectively study the influence of different radiation patterns on the field coverage characteristics in a rectangular tunnel. The field intensity distributions in a tunnel excited by arrays with different radiation patterns are analyzed. From the results, it is concluded that directional antenna can give better field uniformity in a long straight tunnel.

ACKNOWLEDGMENT

This work was supported in part by NSFC under grant no. 61271048 and in part by the National Key Basic Research Project under grant no. 2013CB328903.

REFERENCES

[1]  R. S. He, Z. D. Zhong, B. Ai, J. W. Ding, Y.Q. Yang and A. F. Molisch, "Short-Term Fading Behavior in High-Speed Railway Cutting Scenario: Measurements, Analysis, and Statistical Models," *IEEE Trans. Antennas and Propag.*, vol. 61, no. 4, pp. 2209-2222, April 2013.

[2]  K. Haneda, A. Khatun, M. Dashti, T. A. Laitinen, V.-M. Kolmonen, J.-I. Takada and P. Vainikainen "Measurement-Based Analysis of Spatial Degrees of Freedom in Multipath Propagation Channels," *IEEE Trans. Antennas and Propag.*, vol. 61, no. 2, pp. 890-900, Feb. 2013.

[3]  S. Pu, J. H. Wang, and Z. Zhang, "Estimation for Small-scale Fading Characteristics of RF Wireless Link Under Railway Communication Environment Using Integrative Modeling Technique," *Progress In Electromagnetics Research*. vol.106, 395–417 , 2010.

[4]  Z. Huang, K. R. Demarest, and R. G. Plumb, "An FDTD/MoM hybrid technique for modelling complex antennas in the presence of heterogeneous grounds," *IEEE Trans. Geosci. Remote Sensing*, vol. 37, no. 6,pp. 2692–2698, Nov. 1999.

[5]  A. Sani, Y. Zhao, Y. Hao, A. Alomainy, and C. Parini, "An Efficient FDTD Algorithm Based on the Equivalence Principle for Analyzing Onbody Antenna Performance," *IEEE Trans. Antennas and Propag.*, vol. 57, no.4, pp. 1006-1014, April 2009.

[6]  A. C. Austin, M. M. J. Neve, G. B. Rowe, and R. J. Pirkl , "Modeling the Effects of Nearby Buildings on Inter-Floor Radio-Wave Propagation," *IEEE Trans. Antenn. Propag.* vol.57, no.7, 2155–2161, 2009.

[7]  A. Alighanbari and C. D. Sarris, "Rigorous and efficient time-domain modeling of electromagnetic wave propagation and fading statistics in indoor wireless channels," *IEEE Trans. Antennas Propag.*, vol. 55, no.8, pp. 2373–2381, 2007.

[8]  J. Naganawa, M. Kim and J. Takada, "On Point Source and Observation Modeling for Path Loss Calculation Using FDTD method," in *Proc. Int. Symp. Antennas and Propagation*, Nagoya, Japan, Oct. 2012-Nov. 2012, pp. 745–748.

[9]  Yu, W. H., Y. J. Liu, T. Su, N.-T. Hunag, and R. Mittra, "A Robust Parallelized Conformal Finite Difference Time Domain Field Solver Package Using the MPI Library," *IEEE Antennas Propag. Magazine*. Vol.47, No.3, 39–59, 2005.

# Modelling of Electromagnetic Propagation Characteristics in Indoor Wireless Communication Systems Using the LOD-FDTD method

Meng-Lin Zhai[1], Wen-Yan Yin[1,2], and Zhizhang (David) Chen[3]

[1] Centre for Microwave and RF Technologies, Key Lab of Ministry of Education of Design and EMC of High-Speed Electronic Systems, Shanghai Jiao Tong University, Shanghai 200240, CHINA

[2] Centre for Optical and EM Research (COER), State Key Lab of MOI, Zhejiang University Hangzhou 310058, CHINA; Tel: 0086-571-88206526

[3] Department of Electrical and Computer Engineering, Dalhousie University, Halifax, Nova Scotia, Canada
E-mail: emilychn@sjtu.edu.cn, wyyin@sjtu.edu.cn and z.chen@dal.ca

*Abstract*—Wireless technologies has been attracting much attention due to its wide application. Thus it is important to predict the electromagnetic propagation characterization precisely. Full-wave time-domain electromagnetic methods are usually effective in rigorously modelling and evaluating wireless channels. However, their computational expenditures are expensive, when dealing with electrically large size problems consisting of fine structures. Thus, in this paper, a rigorous full-wave numerical solution, via the locally one-direction finite difference time domain (LOD-FDTD) method for lossy media, which reduces computational time by removing the Courant-Friedrich-Levy (CFL) stability condition, is applied to characterize the wireless channel. By comparing the simulation results with the conventional FDTD, the proposed method demonstrates both good simulation efficiency and high accuracy. Post-processing of the simulation results lead to effective channel characterization with path loss exponent and probability distribution of path loss.

## I. INTRODUCTION

There have been many demands for studying electromagnetic propagation characterization of wireless communication systems. To ensure accurate propagation prediction and optimal realization of a wireless system, it is very important to find an efficient and accurate way to model the propagation environments which may include buildings walls, furniture, human body and *etc*. Although experimental approaches are reliable and close to reality, they can be very expensive due to the requirement for special equipment and setup, and in many cases, it is impossible to pre-set the needed testing environment.

Fortunately, recent progress in the Finite-Difference-Time-Domain (FDTD) method [1] has provided an alternate choice other than experiments for accurately simulating wireless channels [2-4]. However, due to the Courant-Friedrich-Levy (CFL) stability condition [1], which imposes an upper limit on the time step size, it often takes long computational time in simulating electrically large size structures containing fine structures.

In this paper, the locally one-direction finite-difference time-domain (LOD-FDTD) method for lossy media, which removes the Courant Friedrich Levy (CFL) condition and allows the use of large time step in comparison with the conventional FDTD method [5-6], is successfully implemented for studying an indoor wireless channel. As demonstrated, the proposed equation maintains both good simulation efficiency and accuracy. After post-processing the simulation results, pulse wave propagation characteristics are captured including path loss exponent and probability distribution of path loss.

## II. FORMULATION

Consider an isotropic and lossy medium with permittivity of $\varepsilon$, permeability of $\mu$ and conductivity of $\sigma$; the time-dependent Maxwell's equation can be written in the following matrix form:

$$\frac{\partial V}{\partial t} + \sigma V = c(A+B)V , \qquad (1)$$

where $V^n = [E_x^n, E_y^n, E_z^n, Z_0 H_x^n, Z_0 H_y^n, Z_0 H_z^n]$,

$$A = \begin{bmatrix} 0 & 0 & 0 & 0 & 0 & \partial y \\ 0 & 0 & 0 & \partial z & 0 & 0 \\ 0 & 0 & 0 & 0 & \partial x & 0 \\ 0 & \partial z & 0 & 0 & 0 & 0 \\ 0 & 0 & \partial x & 0 & 0 & 0 \\ \partial y & 0 & 0 & 0 & 0 & 0 \end{bmatrix},$$

and

$$B = \begin{bmatrix} 0 & 0 & 0 & 0 & -\partial z & 0 \\ 0 & 0 & 0 & 0 & 0 & -\partial x \\ 0 & 0 & 0 & -\partial y & 0 & 0 \\ 0 & 0 & -\partial y & 0 & 0 & 0 \\ -\partial z & 0 & 0 & 0 & 0 & 0 \\ 0 & -\partial x & 0 & 0 & 0 & 0 \end{bmatrix}. \qquad (2)$$

By applying the Crank-Nicolson scheme to (1), we have

$$(I + \frac{\sigma\Delta t}{2} - \frac{c\Delta t}{2}A - \frac{c\Delta t}{2}B)V^{n+1} = (I - \frac{\sigma\Delta t}{2} + \frac{c\Delta t}{2}A + \frac{c\Delta t}{2}B)V^n \qquad (3)$$

978-7-5641-4279-7

Equation (3) can be approximated by the following factorization:

$$(I+\frac{\sigma\Delta t}{4}-\frac{c\Delta t}{2}A)(I+\frac{\sigma\Delta t}{4}-\frac{c\Delta t}{2}B)V^{n+1}$$
$$=(I-\frac{\sigma\Delta t}{4}+\frac{c\Delta t}{2}A)(I-\frac{\sigma\Delta t}{4}+\frac{c\Delta t}{2}B)V^{n} \quad (4)$$

In the LOD-FDTD method [6], (4) is solved in two steps

$$(I+\frac{\sigma\Delta t}{4}-\frac{c\Delta t}{2}A)V^{n+1/2}=(I-\frac{\sigma\Delta t}{4}+\frac{c\Delta t}{2}A)V^{n} \quad (5)$$

$$(I+\frac{\sigma\Delta t}{4}-\frac{c\Delta t}{2}B)V^{n+1}=(I-\frac{\sigma\Delta t}{4}+\frac{c\Delta t}{2}B)V^{n+1/2}. \quad (6)$$

Take $E_x^{n+1/2}$ for example, from (5), we have

$$(1+\frac{\sigma\Delta t}{4\varepsilon})E_x^{n+1/2}-\frac{\Delta t}{2\varepsilon}\partial yH_z^{n+1/2}=(1-\frac{\sigma\Delta t}{4\varepsilon})E_x^{n}+\frac{\Delta t}{2\varepsilon}\partial yH_z^{n} \quad (7)$$

$$H_z^{n+1/2}-\frac{\Delta t}{2\mu}\partial yE_x^{n+1/2}=H_z^{n}+\frac{\Delta t}{2\mu}\partial yE_x^{n} \quad (8)$$

Here we only consider about electric conductivity.

Substituting (8) into (7) yields the updating equations of $E_x^{n+1/2}$

$$(1+\frac{\sigma\Delta t}{4\varepsilon}-\frac{\Delta t^2}{4\mu\varepsilon}\partial_y^2)E_x^{n+1/2}=(1-\frac{\sigma\Delta t}{4\varepsilon}+\frac{\Delta t^2}{4\mu\varepsilon}\partial_y^2)E_x^{n}+\frac{\Delta t}{\varepsilon}\partial yH_z^{n} \quad (9)$$

As described in [7-8], in order to improve the computational efficiency, we can introduce some auxiliary quantities. For example, $e_x^{n+1/2}$ can be introduced into (9), and

$$(\frac{1}{2}+\frac{\sigma\Delta t}{8\varepsilon}-\frac{\Delta t^2}{8\mu\varepsilon}\partial_y^2)e_x^{n+1/2}=E_x^{n}+\frac{\Delta t}{2\varepsilon}\partial yH_z^{n} \quad (10)$$

where

$$e_x^{n+1/2}=E_x^{n+1/2}+E_x^{n} \quad (11)$$

For the other field components, their updating equations can be obtained in a similar manner.

To check on the accuracy of the proposed formula, a homogeneous cavity, filled with a lossy material with $\varepsilon_r = 4$, $\mu_r = 1$ and $\sigma = 0.05$ S/m, is chosen as the test case. A uniform mesh of $50 \times 30 \times 9$ with a cell size of $1mm$ is employed, and a current line source $Jz$ with a Gaussian pulse waveform of $tw = 150$ ps and $tc = 450$ ps is placed from the bottom to the top at the cavity center. The observation point is located at the center of the computational domain.

For comparison, the cavity is computed with the conventional FDTD method with $CFLN=1$ and the proposed method with $CFLN=1$ and 5, respectively. Here, $CFLN$ (Courant-Friedrich-Levy number) is the ratio of the time step to the CFL time step limit, i.e. $CFLN = \Delta t / \Delta t_{FDTD\_Max}$. To characterize the difference between the proposed method and the FDTD which is considered as the reference, the relative error is calculated by

$$e=\frac{\left|E_z(t)-E_z^{ref}(t)\right|}{Max\left|E_z^{ref}(t)\right|}\times100\% \quad (12)$$

where $E_z(t)$ is the electric field value obtained with the proposed LOD-FDTD method, and $E_z^{ref}(t)$ is the electric field obtained with the FDTD method.

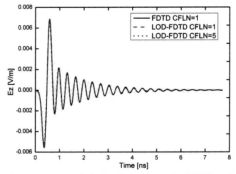

Fig.1.The Ez-component obtained with the conventional FDTD and the proposed method with *CFLN*=1 and 5.

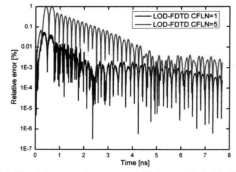

Fig.2. The relative errors of the proposed LOD-FDTD method with different *CFLN*s in reference to the FDTD results.

Fig.1 shows the computed electric field at the center of the computational domain. Fig.2 shows the relative errors of the proposed method. As can be seen, the proposed equations demonstrate high accuracy.

### III. NUMERICAL RESULTS AND DISCUSSION

Simulations are carried out in a 3-D office as shown in Fig.3. The overall computational dimensions are $6 \times 5 \times 3$ $m^3$. The human body model is assumed to be made of homogenous human tissue liquid with $\varepsilon_r = 53.43$ and $\sigma = 0.76$ S/m [9]. The size of the body model is chose to be $1.61 \times 0.30 \times 1.78$ $m^3$. In the simulation, one sine-modulated Gaussian pulse $p(t)=\sin(2\pi ft)\exp(-((t-t_c)/t_w)^2)$, with $f = 50$ MHz, $t_w = 15$ ns and $t_c = 3t_w$, is applied in $E_z$ at the transmitter point $T_x$, and the receiver position marked as $R_x$ as denoted in Fig. 3. The space outside the room is considered as free space, which was then terminated with nine convolutional perfectly matched layers (CPML) [10].

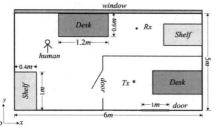

Fig.3. The floor plan of an office. The height is set to be $3m$ in the z-direction.

To save computational memory and reduce computational time, non-uniform meshes are employed. For the regions close to human body model, fine meshes $\Delta x = \Delta y = \Delta z = 1$ $cm$ are used, while for the rest regions, coarse meshes $\Delta x = \Delta y = \Delta z = 4$ $cm$ are applied. The slowly graded non-uniform meshes are employed between fine and coarse ones so as to reduce numerical reflections. As a result, the overall numerical mesh cells are $239 \times 174 \times 253$. The simulation is carried on an Intel Pentium PC with 32 GB RAM. Fig.4 shows the time-dependent electric field $E_z$ at the observation point $R_x$ obtained with conventional FDTD method and the proposed method with $CFLN=4$. As shown in Table I, although the proposed formula uses more memory, it does consumes about 33.1% less CPU time with $CFLN = 4$ than the FDTD method.

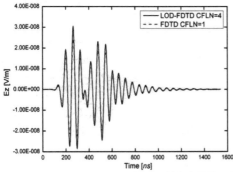

Fig.4 Time-dependent $E_z$ recorded at $R_x$ obtained with both FDTD and our proposed method with $CFLN=4$.

TABLE I

COMPARISON OF THE COMPUTATIONAL EXPENSES OF THE CONVENTIONAL FDTD AND THE PROPOSED LOD-FDTD METHOD

|  | FDTD | LOD-FDTD |
|---|---|---|
| CFLN | 1 | 4 |
| Memory (MB) | 660 | 915 |
| Number of steps | 20,000 | 5,000 |
| Total CPU time (seconds) | 118,617 | 79,310 |

A simplified wireless channel model relating to the path loss

with the transmitter-receiver separation distance is based on the following formula [11]:

$$PL(d)=F+10n\log_{10}d+G \qquad (13)$$

where $F$ is a constant, $d$ is the distance between transmitter and receiver. $n$, known as the path loss exponent (PLE), is a measure of how fast the signal energy decreases with $d$. The PLE plays an important role in the channel model and defines the signal fading of a given channel. $G$ is a Gaussian variable in dB (a lognormal variable in linear scale).

The path loss is computed at 5602 sample grid points of a uniform mesh across the floor plan and shown in Fig. 5 (a) as a function of $d$. The mean values of path loss are also calculated for each fixed $d$ value as plotted in Fig. 5 (b). The simulation results are used to extract the PLE, which is the slope of the best-fit lines.

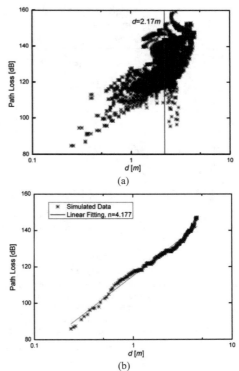

Fig.5. (a) Path loss computed at 5602 sample points, taken uniformly across the floor plan of Fig.3.The solid line represents a fixed $d = 2.17m$. (b) The mean values of path loss and path loss exponent best-fit line.

For a fixed $d$, for example, $d = 2.17m$ as indicated by the solid line in Fig.5 (a), the probability density function (PDF) of path loss follows a Gaussian distribution. Figure 6 shows the results along with the best-fit Gaussian curves solved by the least mean-square method; there $\alpha_1$ and $\alpha_2$ represent the mean and the standard deviation of the associated normal random variable, respectively. As observed, the Gaussian distributions offer a good approximation to the distribution of path loss.

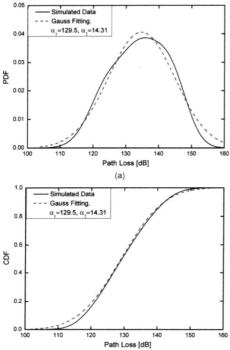

Fig.6. PDF and CDF for path loss (the transmitter and receivers are 2.17m apart). The best fit Gaussian models are also shown here: (a) for PDF and (b) for CDF.

## IV. CONCLUSION

In this paper, a modified LOD-FDTD formulation for lossy media has been presented for modelling the wireless channel. By comparing the simulation results with the conventional FDTD, the proposed method is shown to have good simulation efficiency and accuracy. After post-processing the simulation results, propagation characteristics are captured such as path loss exponent and probability distribution of path loss.

### ACKNOWLEDGMENT

This work is supported the National Science Foundation under Grant 60831002 of China and the Sate Key Lab of Science and Technology of EMC, the Centre for Ship Development and Design of China.

The third author acknowledges the finical support by the Natural Science and Engineering Research Council of Canada under Grant STPGP #396417 and its Discovery Grant #155230-2012.

### REFERENCES

[1] A. Taflove and S. C. Hagness, *Computational Electrodynamics: the Finite-difference Time-domain Method*, 2nd Ed., Artech House, Boston, MA, 2000.

[2] M. F. Iskander and Z. Yun, "Propagation prediction models for wireless communication systems," *IEEE Trans. Microwave Theory Tech.*, vol. 50, pp. 662–673, Mar. 2002.

[3] Z. Yun, M. Iskander, and Z. Zhang, "Complex-wall effect on propagation characteristics and MIMO capacities for an indoor wireless communication environment," *IEEE Trans. Antennas Propag.*, vol. 52, no. 4, pp. 914–922, 2004.

[4] A. Alighanbari and C. D. Sarris, "Rigorous and efficient time-domain modeling of electromagnetic wave propagation and fading statistics in indoor wireless channels," *IEEE Trans. Antennas Propag.*, vol. 55, no. 8, pp. 2373–2381, Aug. 2007.

[5] W. Fu and E. L. Tan, "Development of split-step FDTD method with higher-order spatial accuracy," *Electron. Lett.*, vol. 40, no. 20, pp. 1252-1254, Sep. 2004.

[6] E. L. Tan, "Unconditionally stable LOD-FDTD method for 3-D Maxwell's equations," *IEEE Microw. Wireless Comp. Lett.*, vol. 17, no. 2, pp. 85-87, Feb. 2007.

[7] E. L. Tan, "Fundamental schemes for efficient unconditionally stable implicit finite-difference time-domain methods," *IEEE Trans .Antennas Propagat.*, vol. 56, no. 1, pp. 170-177, Jan. 2008.

[8] T.H. Gan and E. L. Tan, "Unconditionally Stable Fundamental LOD-FDTD Method with Second-Order Temporal Accuracy and Complying Divergence," *IEEE Trans .Antennas Propagat.*, vol. 61, no. 5, pp. 2630-2638, May. 2013.

[9] CENELEC, EN 50383, "Basic standard for the calculation and measurement of electromagnetic field strength and SAR related to human exposure from radio base stations and fixed terminal stations for wireless telecommunication systems(110 MHz{40 GHz)," 2002.

[10] I. Ahmed, E. H. Khoo, and E. P. Li, "Development of the CPML for three-dimensional unconditionally sable LOD-FDTD method," *IEEE Trans. Antennas Propag.*, vol. 58, no. 3, pp. 832–837, Mar. 2010.

[11] T. S. Rappaport, *Wireless Communications: Principles and Practice.* Upper Saddle River, NJ: Prentice Hall PTR, 2001.

# Design of Multi-channel Rectifier with High PCE for Ambient RF Energy Harvesting

Zheng Zhong[1,2], Hucheng Sun[1], Yong-Xin Guo[1,2]

[1]National University of Singapore, Singapore 117583

[2]National University of Singapore (Suzhou) Research Institute, Suzhou, Jiangsu Province, China, 215123

Email; eleguoyx@nus.edu.sg

*Abstract-* **An efficient and successful multi-channel rectifier for ambient RF energy harvesting has been designed and evaluated at GSM-1800 and UTMS-2100 bands. This design could sufficiently enhance the RF-to-DC power conversion efficiency (PCE) in weak signal reception environment. The validity of the proposed design is verified by detailed experimental results, which indicate that a maximum PEC of 38.8% and an output DC voltage of 280 mV have been observed over an optimized 6400Ω resistive load by collecting relatively low ambient RF power.**

## I. INTRODUCTION

Recently, there is a growing interest in wireless power transmission (WPT) technology [1] which is in the forefront of electronic development. The main function of WPT is to allow electrical devices to be continuously charged without the constraint of a power cord. Meanwhile, with the rapid development of commercial communication services, the environment of our modern society is full of radio energy that can be recycled. Therefore, radio frequency (RF) energy harvesting, which is considered as a green and efficient power solution in the future, has rapidly become a hot spot of research. One of its most attractive applications is the self-powered devices and systems, which are of immediate significance to solve the battery recharging and replacement issues. There is no doubt that the ambient RF energy harvesting is provided with wide space and opportunities for development, but at the same time, it will also face two practical difficulties. Firstly, the power level of the ambient RF energy available through public telecommunication services is relatively low [1]. Secondly, due to the character of rectifier circuit using physical diode components, the less input RF power, the less RF-to-dc power conversion efficiency (PCE) will be achieved [2]. Therefore, such a weak ambient RF energy will lead to a low RF-to-dc PCE, which makes the RF energy harvesting become dispensable. Unfortunately, currently, various methods presented under large RF input power condition [3, 4] are not suitable to deal with these two problems.

In this paper, we propose a novel multi-band rectifier to solve these problems. This multi-band rectifier can make the best use of the ambient RF energy of each signal channel so as to accumulate the RF power into a sufficiently higher power level, where the power conversion efficiency could also reach a considerable value. The detailed design and experimental results have been presented in the following sections.

## II. AMBIENT RF POWER MEASUREMENT

As mentioned above, the environment is full of radio energy that can be used for daily recycled. Therefore, in order to make the best use of these ambient RF energy in each channel, the power densities of ambient RF energy available through public telecommunication services have been investigated by a wide-band horn antenna (ANT-DR18S) and spectrum analyzer (Agilent 8565EC). Measurements are made on the roof of engineering building of national university of Singapore (NUS), which simply represents the residential environment in urban area of Singapore. The related power densities are calculated then from Friis transmission equation, as shown in Table I. From the measurement results, it is clear that only down-link channels in three bands, GSM-900, GSM-1800 and UTMS-2100, dominant the ambient RF energy since powers from base-stations are more persistent and stable than powers from portable devices.

TABLE I
TYPE SIZES FOR PAPERSAMBIENT POWER DENSITY OF EACH SINGLE CHANNEL OF DIFFERENT PUBIC TELECOMMUNICATION BANDS (MEASURED BY HORN ANTENNA)

| Band | Downlink Frequency (MHz) | Received Power (dBm) | Antenna Gain (dBi) | Power Density ($\mu W / m^2$) |
|---|---|---|---|---|
| GSM-900 | 925-960 | -35 ~ -25 | 2 | 23.8 ~ 256.7 |
| GSM-1800 | 1805-1880 | -25 ~ -15 | 10 | 143.9~ 1560.6 |
| UTMS-2100 | 2110-2170 | -25 ~ -15 | 10 | 196.6~ 2079.2 |

Meanwhile, the total RF power of all ambient communication bands that the horn antenna received is measured by power meter and it varies from -20 to -15 dBm (10 ~ 31.6 μW). It also could be found in Table I that the power densities of GSM-1800 and UTMS-2100 bands are obviously much larger than that of GSM-900 band. Therefore, with the consideration of the physical size of the final RF energy harvesting system, the following multi-channel rectifier in the paper has been designed to cover only GSM-1800 and UTMS-2100 bands.

978-7-5641-4279-7

## III. MULTI-CHANNEL RECTIFIER DESIGN

As the input RF power is relatively low (< -10dBm), the series-diode rectifying topology has been adopted in this design as it could achieve a higher efficiency under such a condition [2]. Figure 1 shows the maximum RF-to-dc power conversion efficiency that this series-diode topology structure can achieve. This figure is simulated by with practical components (HSMS-2852 and PCB RO4003C) in ADS. It can be seen from Figure 1, when power level is low (< -20dBm), efficiency increases significantly as power increases as mentioned previously.

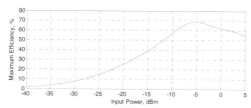

Figure 1.Maximum RF-to-dc power conversion efficiency in series diode topology (Diode: HSMS-2852, PCB: RO4003C)

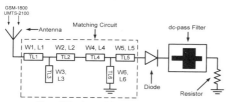

Figure 2.Topology of the proposed rectifier.

The entire topology of the proposed rectifier is illustrated in Figure 2. The substrate used is 32-mil-thick RO4003C with dielectric constant of 3.38. As mentioned before, in this design, the series-mounted diode topology is adopted since it tends to have a higher efficiency under low input RF power conditions. From [5], dual-frequency impedance matching can be achieved by a signal T-shape transmission line structure. However, this method can match at only two single frequency points. Therefore, to increase the bandwidth of our wide-channel rectifier, two T-shape microstrip-line structures have been adopted in this circuit, which are shown as TL1 to TL6. To provide a dc-patch in the circuit, the microstrip line TL6 is grounded by via-holes. A Schottky diode Avago HSMS-2852 ($V_{th}$ = 150 mV, $C_j$ = 0.18 pF, $R_s$ = 25 Ω) is inserted between the matching circuit and the dc-pass filter to convert the microwave power into dc power. The dc-pass filter is realized by a simple stepped-impedance microstrip line low-pass filer, followed by a resistive load to extract the dc power. To further reducing the parasitic effect of lumped components, a cross-shape microstrip-line structure has been introduced into this circuit design to replace the traditional lumped capacitor components,

The parameters of the matching circuit, dc-pass filter and the resistive load can be initially calculated and then optimized in the software Advanced Design System (ADS) by setting appropriate goals. At the first stage, a preliminary rectifier containing only a diode, a dc-pass filter, and a resistor, was optimized under low input power with the goal of high efficiency at both 1.84 GHz and 2.14 GHz. In this way, the optimal dc-pass filter and the load resistance was obtained. After that, the matching circuit was optimized separately to match the input impedance of the rectifier to 50 Ω at both 1.84 GHz and 2.14 GHz. The optimized parameters are (unit: millimeter): W1 = W2 = W3 = 2.2, L1 = 7.8, L2 = 19.9, L3 = 10.4, W4 = W5 = W6 =2.7, L4 = 12.5, L5 = 9.3, L6 = 10.8. The load resistance is optimized as 6.4kΩ. Figure 3 shows the photograph of the fabricated rectifier.

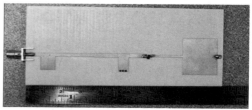

Figure 3.Photograph of the fabricated dual-band rectifier.

## IV. MEASUREMENT RESULTS

To verify the frequency-power related performance of our multi-channel rectifier, its single-tone RF-to-dc power conversion efficiency performance has been tested first. A signal generator (Agilent E8257D) is used to generate this single-tone input signal. Figure 4 shows the measured PCE against frequency at different power levels (-30dBm to -18dBm). The measured conversion efficiency can be obtained by equation (1),

$$\eta(\%) = \frac{V_L^2}{R_L} \times \frac{1}{P_{in}} \times 100 \qquad (1)$$

where $V_L$ is the output dc voltage on the resistor, $R_L$ is the resistance value, and $P_{in}$ is the power level of the single-tone input signal generated by a signal generator. It can be seen from Figure 4 that higher efficiencies can be achieved in the frequency ranges of 1.81-1.87 GHz and 2.11-2.17 GHz, which demonstrates a good agreement with the initial objective, revealing the rectifier's capability to harvest the RF power in the GSM-1900 and UMTS-2100 bands. Meanwhile, it can be seen from the figure that when several channels in either band are activated, the power of these channels can be conversed at the same time at a high efficiency. In fact, the input power to the rectifier can be regarded as multi-tone when measuring in the ambience since the ambient RF power is distributed over the two bands.

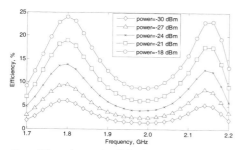

Figure 4.Measured conversion efficiencies at different frequencies

To further validate the efficiency enhancement of this multi-channel rectifier, its two-tone RF-to-DC PCE performance has also been explored. With the help of two Agilent E8257D signal generators and a power splitter (ZFRSC-42), the measured PCE against each single and dual-tone input power has been measured and plotted in Figure 4. The input signals frequencies are 1.80 GHz and 2.14 GHz, respectively, which are two of the major frequencies of the two bands. In this experiment, the input power of these two frequencies keeps the same for simplicity. As the ambient RF power is relatively low, -30dBm input has been taken for example to discuss the efficiency enhancement. It can be seen from Figure 5, when input power is -30dBm, the RF-to-Dc PEC is only 5.29% and 4.79% for 1.83GHz and 2.14GHz, respectively. However, the same two-tone input RF signal can lead to a higher PCE of this multi-channel rectifier to 11.65%, which is more than twice of each single-tone input situation. It is clear to prove that an addition of input RF signals at different frequencies will help generate a high conversion dc power and a higher efficiency. Hence, this fact implies that through this multi-channel rectifier, dispersed RF powers that are separated in different channels can be harvested at the same time with a considerable higher RF-to-dc PCE.

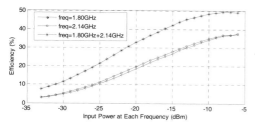

Figure 5.Measured conversion efficiency at different input power levels

Last but not least, this multi-channel rectifier has been connected to the same horn antenna (ANT-DR18S) to test its ambient RF energy harvesting capability. In this experiment, the maximum output dc voltage of the rectifier is above 280 mV when the received RF power varies from -20 to -15 dBm. The PCE of this multi-channel rectenna can be derived

according to the maximum input power (-15dBm, 0.032mW), which is above 38.8% for ambient RF energy harvesting. For a clear comparison, the PCE enhancement of this multi-channel rectifier has been listed in Table II.

TABLE II
COMPARISON OF RF-TO-DC PCE FOR SINGLE-TONE, DUAL-TONE AND MULTI-CHANNEL RF ENERGY HARVESTING

| Frequency Band | Received Power (dBm) | PCE (%) |
|---|---|---|
| Single tone (1.80 GHz) | -30 | 5.6 |
| Dual tone (1.80 & 2.14) | -30 | 12.3 |
| GSM-1800 & UTMS-2100 | -20 ~ -15 | >38.7 |

## V. CONCLUSION

In this paper, a multi-channel rectifier suitable for urban area telecommunication RF power harvesting is proposed. While the RF power at each single channel is relatively low but there are several channels active at the same time, this rectifier can harvest these dispensable RF powers that are separated in different channels together with a considerable higher RF-to-dc PCE above 38.8% when the total ambient RF energy is less than -15dBm. This multi-channel power collection method is suitable for urban area telecommunication RF power harvesting. While the RF power at each single channel is relatively low but there are several channels active at the same time, this multi-channel system can harvest these dispensable RF powers that are separated in different channels together with a considerable higher RF-to-dc PCE compared with these traditional RF energy harvesting system. Since the ambient RF energy harvesting is emerged as an urgent and challengeable issue in the creation of wireless sensor networks which depend on truly autonomous devices, this study can be treated as a very pointed attempt in this respect.

## ACKNOWLEDGMENT

This work was supported in part by Singapore Ministry of Education Academic Research Fund Tier 1 project R-263-000-667-112, in part by the National University of Singapore (Suzhou) Research Institute under the grant number NUSRI-R-2012-N-010.

## REFERENCES

[1] H.J. Visser, A.C.F. Reniers, and J.A.C. Theeuwes, "Ambient RF energy scavenging: GSM and WLAN power density measurements", European Microwave Conf., pp. 721-724, Oct. 2008

[2] V. Marian, C. Vollaire, J.Verdier, and B. Allard, "Potentials of an adaptive rectenna circuit", IEEE Antennas Wireless Propag. Lett., vol. 10, pp. 1393-1396, 2011

[3] J.O. McSpadden, L. Fan, and K. Chang, "Design and experiments of a high-conversion-efficiency 5.8-GHz rectenna", IEEE Trans. Microwave Theory Tech., vol. 46, issue12, pp. 2053-2060, Dec. 1998

[4] Y.H. Suh, and K. Chang, "A high-efficiency dual-frequency rectenna for 2.45- and 5.8-GHz wireless power transmission", IEEE Trans. Microwave Theory Tech., vol. 50, pp. 1784-1789, July 2002

[5] M.A. Nikravan, and Z. Atlasbaf, "T-section dual-band impedance transformer for frequency-dependent complex impedance loads", Electron. Lett., vol. 47, pp. 551-553, April 2011

# *TP-2(D)*

## October 24 (THU) PM

## Room D

## Wire Antennas

# Loop Antenna Array for IEEE802.11b/g

Dau-Chyrh Chang[1], Win-Ming Liang[2]

[1]Oriental Institute of Technology, Taiwan

[2]Yuan Ze University, Taiwan

*Abstract*-Two elements antenna array with high efficiency is developed for IEEE802.11b/g in this paper. The antenna element is one wavelength loop over ground reflector. In order to matching the impedance for the desired band, the parasitic small loop is used with the main loop. The measured results of reflection coefficient are smaller than 15 dB and maximum gain 10.5 dBi with efficiency 85% for the desired band. The measurement results are agreed with that of simulations.

## I. INTRODUCTION

In nowadays, the WiFi IEEE802.11b/g is popular as part of personal communication systems. The band for IEEE802.11b/g is within 2.414 GHz to 2.4835 GHz. The maximum data transmission rate for 802.11g is 54 Mbps (Mega bit per second). In order to maintain the data throughput, the signal to noise ratio S/N should be large. The bandwidth and efficiency of traditional patch antenna element is narrow and low to support the high transmission data rate. In order to overcome the defect of traditional patch antenna element, high efficiency loop antenna is considered and parasitic element for wideband matching is implemented. In order to increase the antenna gain of the loop antenna element, antenna array with two elements and the ground reflector is used to enhance the directivity and gain.

## II. RESULTS OF SIMULATION

In order to design this antenna array, commercial available simulation tool is used to design the antenna array. The size of the two elements antenna array is 40 mm x 100 mm x 8 mm with FR4 0.4 mm thickness substrate. In order to test the antenna array, the size of ground reflector is 100 mm by 120 mm. The spacing between substrate and ground reflector is about 12 mm. The length of loop is around one wavelength at 2.45 GHz. The purpose of small parasitic loop is for impedance matching. Figure 1a is the top view of the antenna array and figure 1b is the side view of the array. Figure 2 is the simulation result of reflection coefficient. The bandwidth of reflection coefficient at -10 dB is 87 MHz. Figure 3 show the simulation current distribution at 2.45 GHz. The maximum current distribution is in y direction on the loop. That means the E-plane pattern is yz-plane. Figure 4 is the simulation power gain pattern at E-/H-plane at 2.45 GHz. The gain of the array is about 10.8 dBi. The half power beamwidth are 58 degrees and 52 degrees at E-plane and H-plane respectively. The null beamwidth is about 180 degrees in H-plane. Figure 5 is the simulated power gain and efficiency versus frequency. The power gain are 10.6 dBi,

10.8 dBi, and 10.9 dBi at 2.4 GHz, 2.45 GHz, and 2.5 GH respectively. The overall efficiency inside the band is over 95%.

## III. RESULTS OF MEASUREMENET

Figure 6 is the hardware implementation of the antenna array. Figure 7 is the measured reflection coefficient. The value reflection coefficient is smaller than 14 dB in the band. Figure 8 is the measured power pattern in both E-/H-planes at 2.4 GHz. The power gain is about 11 dBi. The beamwidth are 58 degrees and 52 degrees at E-/H-planes respectively. Figure 9 is the measured power gain and efficiency with respect to frequency. The measured gain at 2.45 GHz is about 11 dBi. For the desired band the gain is over than 10.5 dBi. The maximum efficiency is about 90% at 2.4 GHz. The overall efficiency is over than 85% for the desired band.

## IV. CONCLUSION

The two elements antenna array for IEEE802.11b/g is developed. The element is loop antenna above ground plane. The simulation gain is 10.9 dBi and the measured gain is 11 dBi at 2.45 GHz. The simulated efficiency is about 95% and the measured efficiency is 85% at 2.45 GHz. The measurement results of gain, efficiency, and beamwidth are quite close to that of simulation. Future four elements antenna array based on this construction with higher antenna gain and dual bands for IEEE802.11/a/b/g is also under developed.

REFERENCES

[1] US 6525694 B2. High gain printed loop antenna. 25, Feb. 2003.
[2] Cai, M. and M. Ito, New Type of Printed Polygonal Loop Antenna, IEE proceedings -H. vol. 138, No. 5, Oct. 1991, pp. 389-396.
[3] Johnson, R.C. and H. Jasik, Antenna Engineering Hand Book, McGraw Hill (2d, Ed. 1984), Chapter 5.

978-7-5641-4279-7

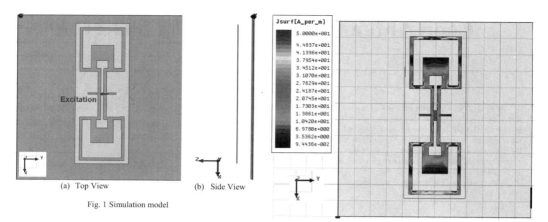

(a) Top View        (b) Side View

Fig. 1 Simulation model

Fig. 3 Current Distribution at 2.45 GHz

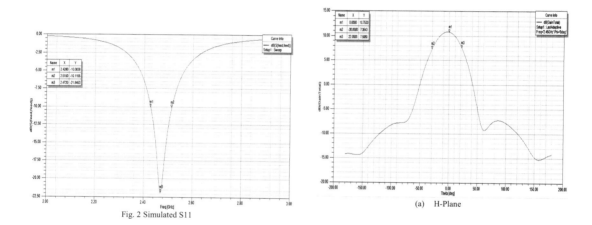

Fig. 2 Simulated S11

(a)    H-Plane

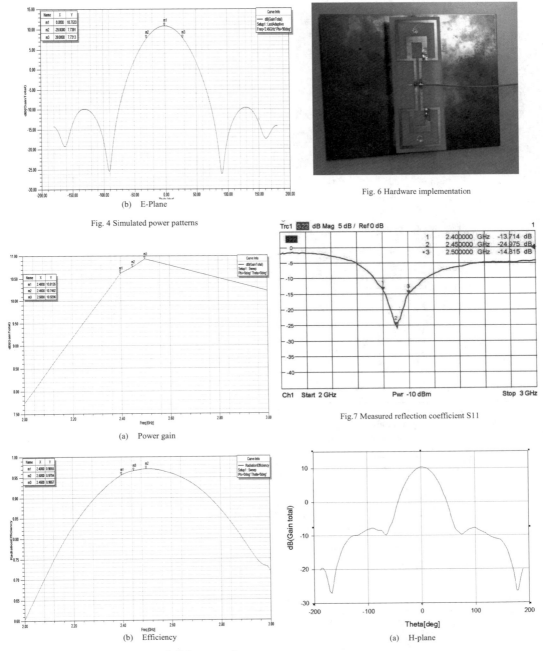

(b)   E-Plane

Fig. 4 Simulated power patterns

Fig. 6 Hardware implementation

(a)   Power gain

Fig.7 Measured reflection coefficient S11

(b)   Efficiency

(a)   H-plane

Fig. 5 Simulated power gain and overall efficiency versus frequency

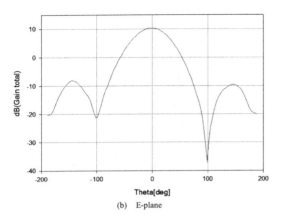

(b)   E-plane

Fig.8 Measured antenna power patter

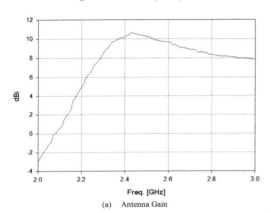

(a)   Antenna Gain

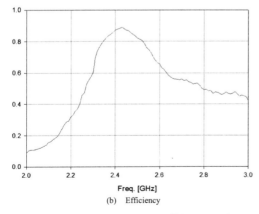

(b)   Efficiency

Fig.9 Measured antenna gain and overall efficiency versus frequency

# Ground Radiation Antenna using Magnetic Coupling Structure.

Hyun woong Shin, Yang Liu, Jaeseok Lee, *Hyung-hoon Kim, Hyeong-dong Kim

Hanyang University

*Kwangju Women's University

*Abstract* - In this paper, radiation performance of a ground radiation antenna using magnetic coupling structure is compared with a conventional planar inverted-F antenna (PIFA) in terms of the impedance bandwidth and radiation efficiency. The size of the proposed ground radiation antenna is 5 × 10 mm², reduced up to approximately 50% while the impedance bandwidth is obtained as 560 MHz (2174~2734MHz), improved by approximately 300% compared to the PIFA. The size of the ground plane is 50 × 20 mm², intended for USB dongle and headset applications.

## I. INTRODUCTION

Recent mobile antennas are being required to be small because the space allocated for antenna in modern mobile devices is becoming smaller. It is difficult to design an antenna with a wide impedance bandwidth and high radiation efficiency using small internal antennas [1], [2]. It was observed in [3], [4] that a wide impedance bandwidth can be achieved by enhancing the coupling between an antenna and the ground plane, even when the antenna is a non-radiating coupling element. Recently, a novel design was proposed [5], [6] in which elements such as an antenna or coupler are not employed, but rather a capacitor is inserted into a small non-ground area of a mobile device. In this work, performance of ground radiation antenna was compared to PIFA, demonstrating that ground radiation antenna can have a better performance than PIFA that has an additional antenna element. PIFA is frequently used as a small and multi-band antenna [7] and the comparison presented here was not conducted in the previous work [5]. In addition, it is shown that the proposed concept can be used in the case of a small ground (50 × 20 mm²), intended for USB dongle and headset applications. Here, simulation data were obtained in HFSS and experimental data were obtained using a network analyzer and a three-dimensional anechoic chamber.

## II. ANTENNA DESIGN AND ANALYSIS

As shown in Figure 1. (a), the size of the ground plane is 50 × 20 mm² and printed on a 1mm thick FR-4 substrate ($\varepsilon_r$=4.4 $\delta$=0.02). In the proposed antenna, the capacitance controls the resonance frequency [5], [6]. The PIFA uses a 20.5mm copper line for radiator and the proposed antenna fully utilizes the ground plane for radiation with the use of two chip capacitors.

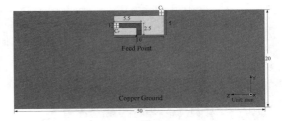

(a)

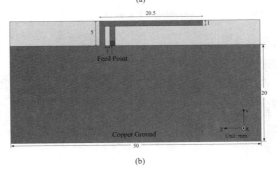

(b)

Figure 1. Geometries of (a) ground antenna (b) PIFA.

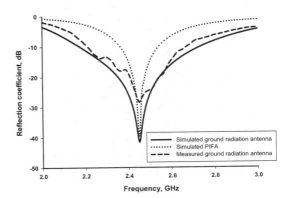

Figure 2. Simulated and measured return loss.

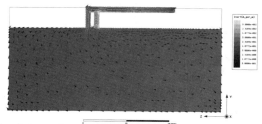

Figure 3. Computed normalized surface current density of (a) ground radiation antenna (b) PIFA at 2.45GHz.

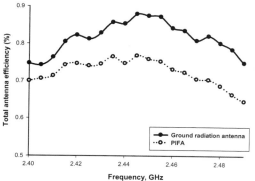

Figure 4. Total antenna efficiency of the Ground radiation antenna and PIFA.

Capacitor $C_L$ is located at the end of the radiator line controlling the resonance frequency and $C_F$ is located in the feed structure controlling the input impedance. The simulated and measured return loss characteristics are shown in Figure 2. The proposed antenna and conventional PIFA are simulated at resonance frequency of 2.45GHz. The surface current density of both antennas at resonance frequency is shown in Figure 3. Loop-type current is formed around the antenna and dipole-type current induced by the loop-type ground can be seen.

Compared to the reference antenna, the ground has a wider current distribution on the ground. Which indicates that the ground antenna can provide a higher ground radiation resistance.

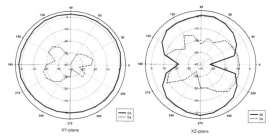

Figure 5. Measured radiation patterns produced by ground radiation antenna at 2.45GHz.

## III. EXPERIMENTAL RESULT AND DISCUSSION

Simulation and measured return loss as shown in Figure 2. The ground radiation antennas impedance bandwidth with VSWR= 2 are 310MHz and PIFA is 170MHz operating at 2.4GHz. Ground antenna has wide impedance bandwidth, wider bandwidth than PIFA. Radiation efficiency as shown in Figure 4. The ground radiation antenna has good radiation efficiency than PIFA. The average efficiency of ground radiation antenna is 81.12% and PIFA is 71.87%. As shown in Figure 4. The total antenna efficiency of the ground antenna is higher than that of the reference antenna. These results indicate that the ground radiation antenna provides better radiation than PIFA. Figure 5. shows the measured radiation patterns of the ground radiation antenna. The measured radiation patterns are in very good agreement with the simulated radiation patterns.

## IV. CONCLUSION

Simulated and measured return loss is shown in Figure 2. The ground radiation antenna and PIFA has impedance bandwidth under VSWR=2:1 of 560MHz and 170MHz. respectively at operating frequency of 2.4GHz. Ground radiator antenna has a wider impedance bandwidth than PIFA. Radiation efficiency is shown in Figure 4. Ground radiation Antenna has better radiation efficiency than PIFA. The average radiation efficiency of ground radiation antenna is 81.12% and PIFA is 71.87%. By using this method, the size of the antenna has become smaller than conventional PIFA.

## ACKNOWLEDGMENT

This research was funded by the MSIP (Ministry of Science, ICT & Future Planning), Korea in the ICT R&D Program 2013.

## REFERENCES

[1] Hansen, R.C.: 'Fundamental limitations in antennas', Proc. IEEE, vol.69, Feb. 1981

[2] Volakis, J.L: 'Antenna engineering handbook'. 4th ed. McGraw-Hill, 2007

[3] Vainikainen, P., Olikainen, J., Kivekas, O., and Kelander, L.:'Resonator-based analysis of the combination of mobile handset antenna and chassis', IEEE Trans. Antenna Propag., vol. 50, no. 10, Oct. 2002

[4] Famdie, C.T., Schroeder, W.L., and Solbach, K.: 'Numerical analysis of characteristic modes on the chassis of mobile phones', Proc. EuCAP, Oct. 2006

[5]  O. Cho, H. Choi, and H. Kim, "Loop-type ground antenna using capacitor," Electron. Lett., vol.47, no.1, pp.11-12, Jan. 2011.

[6]  H. Choi, o. Cho, and J. Lee, "Embedded antenna using ground of terminal device," Patent PCT/KR2010/002314, to RadiNa Inc. Ltd., Korean Intellectual Property Office, Daejeon, Korea, 2010.

[7]  K.L. Wong, Planar Antennas for Wireless Communications, Wiley, New York, 2003

# A Planar Coaxial Collinear Antenna with Rectangular Coaxial Strip

Jiao Wang, Xueguan Liu, Xinmi Yang, Huiping Guo

School of Electronics and Information Engineering, Soochow University, SuZhou, China 215006

*Abstract*—A novel planar Coaxial Collinear (COCO) antenna with rectangular coaxial strip is presented. The planar COCO antenna is fed by microstrip and comprises several radiation sections of rectangular coaxial strips which are cross-linked one by one. A prototype has been designed and fabricated to cover the band of 2.4-2.48 GHz. The simulated and measured results show that the fractional -10 dB bandwidth of the prototype is 24.5%. With this band, the prototype possesses an omni-directional radiation pattern and achieves a gain of over 4 dBi in the direction of $\theta = 50°$. The proposed antenna has characteristics of planar shape, compact size, wideband and easy fabrication. It is suitable for base station application of WiFi, RFID et al.

*Index Terms*—planar COCO antenna, cross-linked rectangular coaxial strips, planar antenna.

## I. INTRODUCTION

Modern wireless communication is requiring antennas with low profile, light weight, compact size, easy integration with circuits, good consistency and accurate fabrication with mature process. Many traditional bulky antennas can't meet all these demands with respect to their heavy weight, large size or fabrication difficult. However, planar antennas can overcome all these shortcomings. Therefore, many researchers have made contribution to planarizing these bulky antennas in order to overcome the above shortcomings. For example, planar Archimedean spiral antenna [1], which demonstrates excellent axial ratio and gain-bandwidth performance in 2-18 GHz, is miniature and easy to fabricate. Hatem Rmili et al. [2] designed a compact and light printed dipole antenna, which can work in dual-band. A compact planar Yagi antenna used in portable RFID devices is described in [3]. Moreover, the famous Vivaldi antenna can also be regarded as the planarization of horn antenna [4]. All the above attempts to planarize traditional bulky antennas were successful and the related traditional antennas maintain satisfactory performance after planarization. And yet, planarization scheme of some other bulky antennas need to be further investigated. One example is coaxial collinear (COCO) antenna.

The classical COCO antenna was first proposed by B. B. Balsleyh and W. L. Ecklund [5]. It is composed of several segments of half-wavelength coaxial cables. The inner and outer conductors of the two adjacent cables are stagger connected. So that it makes the phase of one unit and the next unit same, and theoretically their amplitude is approximately same. The COCO antenna is widely used in radar and communication systems due to its low cost and structural simplicity. But it is time consuming to tune because of the complicated processing. What's more, the COCO antenna will encounter efficiency degradation when its operation frequency exceeds 1 GHz and display a poor uniformity in electrical parameters. For this reason, it is meaningful to research on planar COCO antenna.

The conventional planar COCO antenna [6] consists of serially fed microstrip metallic patches that are alternately printed on the top and bottom surfaces of the substrate. In Ref. [7], a printed COCO antenna with balanced microstrip as its feed line and 8 back-to-back dipole arrays is presented. The bandwidth of both the antennas is limited. In this paper, a novel scheme for planarizing the traditional coaxial collinear (COCO) antenna, which can realize wideband, is proposed. In this scheme, crossed-linked rectangluar coaxial sections are utilized as radiation elements and microstrip is used as feed line. A prototype of the novel planar COCO antenna has been designed and fabricated. The simulated and measured results of the prototype validate the proposed planarization scheme.

## II. ANTENNA STRUCTURE

Fig. 1 depicts the geometry of the proposed antenna. It has four sections of rectangular coaxial strip, three connections and a feeding microstrip. Each section of rectangular coaxial strip is half wavelength ($\lambda_g/2$) long at central frequency 2.45 GHz. In consideration of fabrication, the antenna is composed of two layers of FR4 epoxy substrate, as shown in Fig. 1 (a) (b). The separated pieces of upper/lower ground attach to the upper/lower surface of the top/bottom substrate layer. The separated inner strips attach to the upper surface of the top substrate layer or the lower surface of the bottom substrate layer alternately. The antenna is fed by microstrip which occupies a small part of the bottom substrate layer. So the bottom substrate layer is longer than the upper one (Fig.1 (c)).

The adjacent sections of rectangular coaxial strip are stagger connected to each other as shown by Fig. 1 (c). As an example, Fig. 2 illustrates how the first two rectangular coaxial sections close to the microstrip feeding line are connected to each other. That is, the inner strip of the first section is connected to the end of the lower ground of the second section and the inner strip of the second section is connected to the end of the upper ground of the first section. Both connetions are through via holes and the ground ends touching the via holes are cut by two symmetric triangles and hence exhibit gradual width. Fig. 1(a) (b) shows the geometries of the connection.

## III. DESIGN PROCEDURE

In this section, we will describe the procedure for designing the planar COCO antenna in detail. Firstly, the planar elements

978-7-5641-4279-7

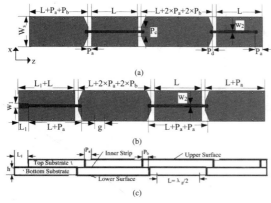

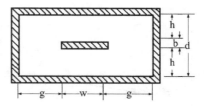

Fig. 3. Cross section of a rectangular coaxial strip.

Fig. 1. Geometry of the proposed antenna. (a) Top View. (b) Bottom View. (c) Side view.

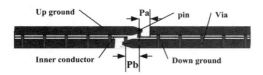

Fig. 2. Cross-linked rectangular coaxial sections.

of the proposed antenna are presented. Secondly, the cross-linked rectangular coaxial sections are designed and optimized. Finally, the dimensions of the antenna are optimized to obtain a good performance.

### A. Design of Rectangular Coaxial Strip

The proposed planar element is rectangular coaxial strip. Rectangular coaxial strip can be regarded as the transformation of coaxial cable: just flatten the outer conductor of the coaxial cable, and make the inner conductor into flat strip line. By this way, coaxial can be planarized with rectangular coaxial strip.

Rectangular coaxial strip is shown in Fig. 3. According to the research in [8], when the width ($w$) of the flat inner conductor is less than one-quarter of the outer conductor's width, the expression for characteristic impedance should be the following:

where, $\varepsilon_0$ is permittivity of free space , $\varepsilon_r$ is relative permittivity of substrate and c is velocity of propagation in free space.

In this paper, the 50 $\Omega$ rectangular coaxial strip, the side walls of which are replaced by via holes in Fig. 2, is fabricated on two-layer FR4 epoxy substrate with relative permittivity of 4.4. Here we set the spacing (g) between the adjacent via holes as 6.5mm. According to (1), the dimensions $b$, $h$ and $w$ of the rectangular coaxial strip are set to 0.018 mm, 1.6 mm and 1.4 mm, respectively.

### B. Design of Cross-Linked Rectangular Coaxial Sections

A parametric study was carried out to find out the influence of certain geometries on antenna's impedance matching. The involving geometries include the spacing between a via hole and its neighboring isolated ground piece reside in the same substrate layer $P_a$, the spacing between the adjacent via holes $P_b$, the width and the length of the triangle cut from the ground end ($P_c$ & $P_d$).

As observed from Fig. 4 (a) (b), neither $P_b$ nor $P_c$ has significant influence on the return loss of the proposed antenna. On contrast, $P_d$ has remarkable influence on the return loss of the proposed antenna, as shown by Fig. 4 (c). An increase of $P_d$ will results in a decrease of return loss. In our final design, $P_b$, $P_c$ and $P_d$ are optimized to 3 mm, 2 mm and 7 mm, respectively.

### C. Antenna Discussion

As illustrated in Fig. 5, $P_a$ has influence greatly over antenna performance. Firstly, the resonance frequency decreases as $P_a$ increases (Fig. 5 (a)). Secondly, the antenna gain increases as $P_a$ increases (Fig. 5 (b)). The final choice of $P_a$ in our design is 3.8 mm.

The prototype antenna was optimized with full wave simulation solver HFSS, and then fabricated with two-layer FR4 epoxy substrate with dielectric constant of 4.4 and loss tangent of 0.02. Both layers are 1.6 mm high. Fig. 6 demonstrates the top and bottom view of the manufactured planar COCO antenna. The design parameters are L=30 mm, $W_s$=20 mm, $W_2$=1.4 mm, $P_a$=3.8 mm, $P_b$=3 mm, $P_c$=2 mm, $P_d$=7 mm, g=6.5 mm. The feeding microstrip which is designed to have a characteristic impedance of 50 $\Omega$ is a 6 mm long and 2.8 mm wide strip. The radius of the pin is 0.5 mm. The overall size of the planar COCO antenna is $20 \times 157 \times 3.3 mm^3$.

### IV. RESULTS AND DISCUSSION

The prototype antenna was measured using Agilent vector network analyzer E5071B. Fig. 7 shows the simulated and measured reflection coefficients of the prototype. The measured resonance frequency is 2.48 GHz, while the simulated value is 2.43 GHz. The measured and simulated fractional -10 dB bandwidth is 24.5% and 12.4%, respectively. The discrepancy between the measured and the simulated results is mainly due to fabrication error and measurement error.

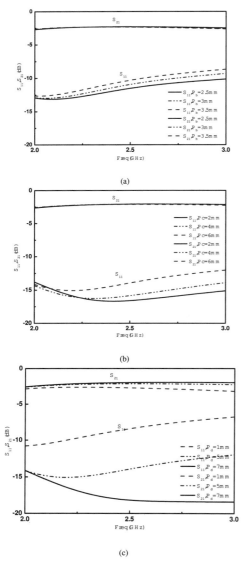

Fig. 4. Effects of varying the (a)$P_b$, (b)$P_c$, (c)$P_d$ on the impedance matching.

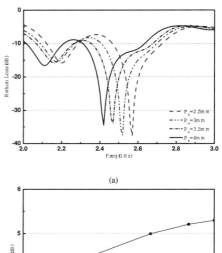

(a)

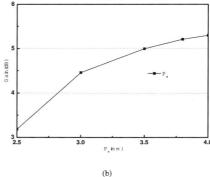

(b)

Fig. 5. (a) Effects of varying $P_a$ on the antenna return loss. (b) Simulated gain of the prototype antenna.

Fig. 6. Photograph of the proposed antenna.

Fig. 8 gives the simulated and measured radiation patterns of the planar COCO antenna. As shown by Fig. 8 (a), the radiation pattern is nearly symmetric with respect to z axis and the prototype antenna radiates nearly omni-directional in the directions with fixed angle. It is also found that the maximum radiation of the prototype antenna is about 50° off the z axis and hence the antenna is suitable for base station application.

A plot of the measured gain in the directions of $\theta = 50°$ over the operational frequency band is illustrated in Fig. 9. It is found from this figure that the maximum measured gain of 5.06 dB is obtained at 2.44 GHz. For comparison, the simulated antenna gain at 2.45GHz is 5.29 dB. The simulated results are in good agreement with the measured results. As shown in Fig. 9 (b) (c), the radiaton pattern in the direction of $\theta = 50°$ at 2.4 GHz and 2.48 GHz are also omni-directional.

V. CONCLUSION

A novel compact planar COCO antenna with rectangular coaxial strip has been proposed. The antenna is composed of four cross-linked rectangular coaxial sections and a microstrip

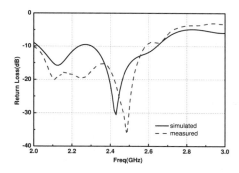

Fig. 7.  Simulated and measured return losses of the prototype antenna.

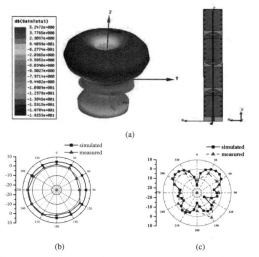

(a)

(b)                                          (c)

Fig. 8.  (a) Simulated 3D radiation pattern at 2.45 GHz. (b) Simulated and measured radiation pattern with $\theta = 50°$ at 2.45 GHz. (c) Simulated and measured radiation pattern in yoz plane at 2.45 GHz.

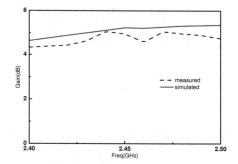

(a)

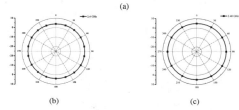

(b)                          (c)

Fig. 9.  (a)Simulated and measured gain of the prototype antenna. (b) Simulated radiation pattern with $\theta = 50°$ at 2.4 GHz. (c) Simulated radiation pattern with $\theta = 50°$ at 2.48 GHz.

feed line. A prototype of the proposed antenna is designed and fabricated. The measured results show that the planar COCO antenna operates in the frequency band of 2.0-2.58 GHz and the gain in the band of 2.4GHz-2.48GHz ranges from 4.33 to 5.06 dB. The proposed antenna has the features of planar shape, compact size and easy fabrication. It is suitable for the base station application of WiFi, RFID et al.

## ACKNOWLEDGMENT

This work was supported in Part by Suzhou Key Laboratory of RF and Microwave Millimeter Wave Technology, and in part by the Natural Science Foundation of the Higher Education Institutions of Jiangsu Province under Grant No. 12KJB510030. The authors would like to thank the staffs of the Jointed Radiation Test Center of Soochow University.

## REFERENCES

[1]  D. Berry, R. Malech, and W. Kennedy, "The reflectarray antenna," *Antennas and Propagation, IEEE Transactions on*, vol. 11, no. 6, pp. 645–651, 1963.

[2]  H. Rmili, J. Floc'h, P. Besnier, and M. Drissi, "A dual-band printed dipole antenna for imt-2000 and 5-ghz wlan applications," in *Wireless Technology, 2006. The 9th European Conference on*, 2006, pp. 4–7.

[3]  P. Nikitin and K. V. S. Rao, "Compact yagi antenna for handheld uhf rfid reader," in *Antennas and Propagation Society International Symposium (APSURSI), 2010 IEEE*, 2010, pp. 1–4.

[4]  P. J. Gibson, "The vivaldi aerial," in *Microwave Conference, 1979. 9th European*, 1979, pp. 101–105.

[5]  T. Judasz and B. Balsley, "Improved theoretical and experimental models for the coaxial colinear antenna," *Antennas and Propagation, IEEE Transactions on*, vol. 37, no. 3, pp. 289–296, 1989.

[6]  R. Hill, "A twin line omni-directional aerial configuration," in *Microwave Conference, 1978. 8th European*, 1978, pp. 307–311.

[7]  Y. Xiaole, N. Daning, and W. Wutu, "An omnidirectional high-gain antenna element for td-scdma base station," in *Antennas, Propagation EM Theory, 2006. ISAPE '06. 7th International Symposium on*, 2006, pp. 1–4.

[8]  T.-S. Chen, "Determination of the capacitance, inductance, and characteristic impedance of rectangular lines," *Microwave Theory and Techniques, IRE Transactions on*, vol. 8, no. 5, pp. 510–519, 1960.

# Analysis of a Horizontally Polarized Antenna with Omni-Directivity in Horizontal Plane Using the Theory of Characteristic Modes

\# Shen Wang and Hiroyuki Arai

Graduate School of Engineering, Yokohama National University

79-5, Tokiwadai, Hodogaya, Yokohama, Kanagawa, 240-8501, Japan

\# ws.augustan@gmail.com

*Abstract*-In this paper we propose a horizontally polarized antenna with omni-directivity in horizontal plane. Ripple coefficient of the radiation pattern in horizontal plane improve to ±0.1 dB. The theory of characteristic modes is used to analyze why the omni-directivity property of the proposed antenna increased comparing with that before optimization.

## I. INTRODUCTION

Recent years, orthogonally polarized composite antennas are proposed for wireless communications because of their excellent characteristics to increase channel capacity and keep compact antenna size. However, a horizontally polarized antenna with dipole-like omni-directivity is not easy to realize. [1][2] presented notch array antennas as horizontally polarized antenna for practical applications. As an important index to evaluate circular degree, ripple coefficient of radiation patterns in horizontal plane of the antennas proposed in [1] are ±1~2.5 dB. It indicates more than 40% energy weaken at somewhere in the horizontal plane. In [2], increased quantity of notch cause the ripple coefficient reduced to about ±0.65~0.8 dB.

To seek for other approaches except adding array element to improve directivity property, we use the theory of characteristic modes which is one of the best methods to help us penetratingly understand antenna operating principles. This paper is devoted to apply the theory of characteristic modes to provide an in-depth physical insight into the behavior of notch array antenna and subsequently improve its directivity property. In this paper, the proposed antennas are simulated by FEKO based on method of moment (MoM).

## II. BRIEF REVIEW OF THE THEORY OF CHARACTERISTIC MODES

The theory of characteristic modes was first developed by Garbacz [3] and was later refined by Harrington and Mautz [4], [5] in 1971. Characteristic modes are current modes numerically obtained for discretionarily shaped conducting bodies, and provide a physical explanation of the radiation phenomena taking place on the antenna. The characteristic modes can be obtained from the following particular weighted eigenvalue equation:

$$X(J_n) = \lambda_n R(J_n), \tag{1}$$

where the $\lambda_n$ are the eigenvalues which are real, the $J_n$ are the eigencurrents. $R(x)$ and $X(x)$ are the real and imaginary parts of the impedance operator. A mode is at resonance when its eigenvalue $|\lambda_n| = 0$, and is inferred that the smaller the magnitude of the eigenvalue, the more efficiently the mode radiates when it is excited. Besides, there is another more visualized representation of the eigenvalues, which is based on the use of characteristic angles and is defined as:

$$\alpha_n = 180° - tan^{-1}(\lambda_n). \tag{2}$$

The characteristic angles physically characterizes the phase difference between the characteristic current $J_n$ and the associated characteristic field $E_n$. Hence, a mode is at resonance when its characteristic angle is or close to 180°. Additionally, when the characteristic angle is near 90° or 270°, the mode is thought mainly storing energy. In addition, it should be noticed that characteristic modes are independent of any kind of excitation but only depend on the shape and size of the conducting object.

## III. ANALYSIS OF PROPOSED ANTENNAS

In order to obtain horizontally polarized radiation and omni-directivity in horizontal plane, a notch array antenna with four array elements is originally proposed at 1.5 GHz and its geometry is shown in Fig. 1(a). There are four notches cut out from the bottom conductor layer, and four microstrip lines feed them, respectively, on the upper layer. A power divider is constructed and fed from the center. The relative dielectric coefficient of dielectric-layer ε = 2.6, while the thickness of the substrate is 0.8 mm. Horizontally polarized waves with semi-omni-directivity in horizontal plane (the plane of antennas' surface) radiated by each notch element, compose an dipole-like omni-directional composite radiation pattern as expected.

In order to further investigate the operating principle of this antenna, we use the theory of characteristic modes to analyze the antenna. Fig. 2 shows current schematics on top layer associated with characteristic current modes $J_1$~$J_4$. The

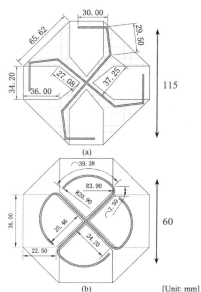

(a)

(b) [Unit: mm]

Fig. 1 Geometry of the proposed notch array antenna, original one (a) and final one (b). Area in gray denotes microstrip lines arranged on top layer of substrate, area with gray panes denotes ground conductor arranged on bottom layer.

978-7-5641-4279-7

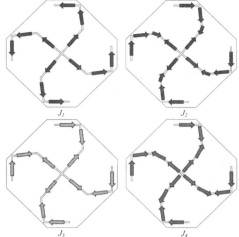

Fig. 2 Current schematics on top layer associated with characteristic current modes $J_1 \sim J_4$.

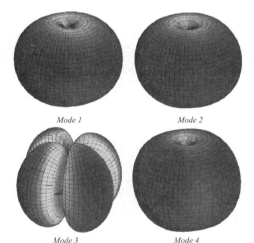

Fig. 3 Normalized radiation patterns of the originally proposed antenna associated with *Mode 1~Mode 4* at their resonant frequencies.

length of each branch of microstrip lines is 96.8 mm, that approximately equals $\frac{1}{2}\lambda_{g[0.96 \text{ GHz}]}$, $(\frac{1}{4}+\frac{1}{2})\lambda_{g[1.44 \text{ GHz}]}$ and $\lambda_{g[1.92 \text{ GHz}]}$. This length causes about 1 standing wave in *mode 1*, 1.5 standing wave in *mode 2* and *mode 3*, 2 standing waves in *mode 4*, respectively. Fig. 3 shows normalized radiation patterns of *mode 1 ~ mode 4*. It can be verified that $J_1$, $J_2$ and $J_4$ contribute to exciting the notches located on bottom layer for the omni-directional radiation pattern is thought composed by semi-omni-directional patterns radiated by the four notches. $J_3$ contributes to exciting the tail-end of microstrip lines to make them operating as monopoles. Due to the in-phase current on the opposite tail-end of microstrip lines and the approximate half-wave distance between them, double 8-shaped radiation patterns are orthogonally formed and can be confirmed in Fig. 3. It should be noticed that $J_1 \sim J_4$ are not the all current modes occurring at concerned frequency spectrum, but the modes possible to be

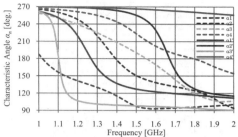

Fig. 4 Variation with frequency of the characteristic angles associated with current modes of the originally proposed antenna (dotted line) and the finally proposed antenna (solid line).

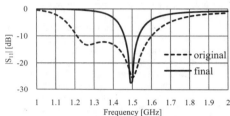

Fig. 5 $|S_{11}|$ characteristics of the originally proposed antenna (dotted line) and the finally proposed antenna (solid line).

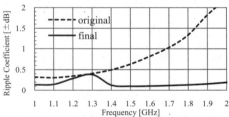

Fig. 6 Ripple coefficient characteristics of the originally proposed antenna (dotted line) and the finally proposed antenna (solid line).

excited by the central feeding. Discussion about other current modes impossible to be excited is ignored in this paper.

Dotted lines in Fig. 4 present the variation with frequency of the characteristic angles $\alpha_1 \sim \alpha_4$ associated with current modes $J_1 \sim J_4$ of the proposed antenna. It is found the characteristic *mode 1*, *mode 2*, *mode 3* and *mode 4* are in resonant state at 1.02 GHz, 1.4 GHz, 1.55 GHz and 1.8 GHz, respectively.

Dotted line in Fig. 5 shows input characteristics $|S_{11}|$ of the originally proposed antenna. Dotted line in Fig. 6 presents the ripple coefficient characteristics. In the range of bandwidth 1.2 ~ 1.6 GHz, the ripple coefficient varies in $\pm 0.35$ ~ 0.8 dB and increases along with operating frequency. By observing and comparing the curves of $|S_{11}|$ and characteristic angles, we may affirm that the double resonances of the antenna are generated by the cooperating of $J_1$, $J_2$ and $J_2$, $J_4$. As one of the most direct reason, radiation pattern of the actually excited antenna performed as that of *mode 1*, *mode 2* and *mode 4*. Additionally, it is considered that $J_3$ is not adequately excited by the actual feeding, since the actual current distribution and radiation pattern do not perform as that associated with *mode 3*. However, we believe that the existence of *mode 3* which radiates orthogonally double 8-

shaped patterns within antenna's operating spectrum does have effect on actual radiation characteristics so that the ripple coefficient rapidly rises when the frequency is above 1.4 GHz.

In order to improve the ripple coefficient characteristic, an optimized notch array antenna is proposed and its geometry is shown in Fig. 1(b). The entire area reduces by 43 % of the original one. Shape of notches becomes a little shorter and wider. Microstrip lines are 26 % shorter and their tail-ends are designed to circular-arc.

Solid lines in Fig.4 present the variation with frequency of the characteristic angles $\alpha_1 \sim \alpha_4$ associated with current modes $J_1 \sim J_4$ of the finally proposed antenna. It can be found that resonant frequency of *mode 1* and *mode 2* rises to 1.27 GHz and 1.68 GHz, respectively, due to the shortening of microstrip lines. Resonant frequency of *mode 4* rises over 2 GHz. What merits special attention is that *mode 3* resonates at 1.1 GHz, which is outside the range between the resonant frequencies of *mode 1*, *mode 2* and *mode 4*, and outside the actual operating band 1.46 GHz ～ 1.54 GHz which can be confirmed by $|S_{11}|$ curve shown in Fig. 5 by solid line. Profit from this change, ripple coefficient reduced to ±0.1 dB in the concerned operating band, shown by solid line of Fig. 6. The variation of *mode 3*'s resonant frequency is thought due to the lengthening of microstrip lines' tail-ends which operate as monopoles. Besides, slopes of the curves near 180° exhibit bigger. This variation indicates narrower bandwidth of the antenna. The actual resonance at 1.5 GHz is regarded as the cooperating of $J_1$ and $J_2$. However, existence of *mode 3*'s resonance does not make the antenna resonating at 1.1 GHz. But its influence to antenna can be verified by ripple coefficient curve. We find the values become much lower except the spectrum around 1.2 ～ 1.3 GHz. The peak-like variation at that band is considered as a cooperating of $J_1$ and $J_3$. However, the influence by *mode 3* is not very remarkable because the distance between the opposite tail-end of microstrip lines is only quarter wave that the directive gain of 8-shaped patterns decreases relatively.

## IV. CONCLUSIONS

A horizontally polarized notch array antenna is proposed in this paper. We used the theory of characteristic modes to analyze the potential current modes of the antenna so that factors affect radiation pattern's ripple coefficient can be found out. In this case, we can confirm the wherefore as the existence of *mode 3*, in which the tail-end of microstrip lines operate as monopoles to make the radiation pattern unsmooth. To solve the problem, we optimized and proposed another design in smaller size. The redesign of antenna's structure especially the microstrip lines makes *mode 3* resonate at lower frequency that is outside antenna's operating band. Profit from this change, the ripple coefficient reduced to ±0.1 dB in operating band.

## REFERENCES

[1] S. Wang, H. Arai, H. Jiang and K. Cho, "A compact orthogonal dual-polarization combined antenna for indoor MIMO base station," Antenna Technology and Applied Electromagnetics (ANTEM), 2012 15th International Symposium, pp. 1–3, Jun. 2012.

[2] S.Wang, H. Arai, H. Jiang, K. Cho and S. Li, "Bandwidth enhancement of a compact dual-polarized indoor base station antenna," 2013 International Workshop on Antenna Technology (iWAT), pp. 59–62, May. 2013.

[3] R. J. Garbacz and R. H. Turpin, "A generalized expansion for radiated and scattered fields," IEEE Trans. Antennas Propag., vol. AP-19, pp. 348–358, May 1971.

[4] R. F. Harrington and J. R. Mautz, "Theory of characteristic modes for conducting bodies," IEEE Trans. Antennas Propag., vol. AP-19, no. 5, pp. 622–628, Sept. 1971.

[5] R. F. Harrington and J. R. Mautz, "Computation of characteristic modes for conducting bodies," IEEE Trans. Antennas Propag., vol. AP-19, no. 5, pp. 629–639, Sep. 1971.

# High Gain Spiral Antenna with Conical Wall

Jae-Hwan Jeong[1], Kyeong-Sik Min[2], In-Hwan Kim[3] and Sung-Min Kim[4]

Department of Radio Communication Engineering
Korea Maritime University,
Dongsam-Dong, Youngdo-Ku, Busan, 606-791, Korea
jjhfifi@nate.com[1]   ksmin@hhu.ac.kr[2]   ehsdlsp2@nate.com[3]   min7947@naver.com[4]

*Abstract*-This paper presents design for a spiral antenna with conical wall to realize the high gain. To improve the axial ratio and the gain of spiral antenna, the conical wall and the optimized Archimedean slit on ground plane are novel designed for the conventional antenna with the circular cavity wall and with the 4.5-turn slit. The good axial ratio of 1.9 dB below and the improved gain of 9.5 dBi above are measured by the added conical wall and the novel designed slit on ground plane, respectively. The measured E-field radiation patterns and main beam directivity toward +z axis direction are agreed well with the simulated results. The proposed antenna will be applied for the NLJD system.

## I. INTRODUCTION

In recent years, the electronic semi-conductor device industry has rapidly developed and becomes minimization. The super minimal semi-conductor with the high performance is frequently used for memory chips with big capacity. To detect a tiny chip made by the semi-conductor or the false junction material which is composed of a semi-conductor and a metal, a non-linear junction detector (NLJD) system has been developed [1]. In this paper, authors designed the high gain spiral antenna with novel Archimedean spiral slit on ground plane to realize the circular polarization and designed the novel cavity added conical wall to realize the high gain.

## II. ANTENNA DESIGN

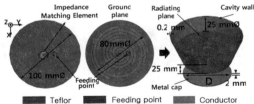

Figure 1. Spiral antenna composed of conical wall and optimized spiral slit ground structure.

Figure 1 shows spiral antenna composed of conical wall and optimized spiral slit ground structure. A diameter of the spiral antenna is 80 mmØ, and Archimedean slit is located on ground plane. The substrate of antenna is used for the teflon dielectric material having relative permittivity of 2.1 and height of 0.6 mm. The cavity wall thickness of 0.2 mm with FR-4_epoxy and metal cap thickness of 2 mm are considered in design. The required antenna bandwidth including transmitting frequency

and receiving frequency is from 2.4 GHz to 7.36 GHz. The Tx band is from 2.4 to 2.48 GHz, and the Rx band is from 4.84 to 4.92 GHz for 2nd harmonic frequency and from 7.28 to 7.36 GHz for 3rd harmonic frequency.

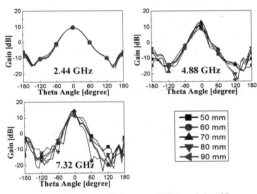

Figure 2. Simulated radiation pattern(X-Z) by variation of D.

Figure 2 shows the simulated radiation pattern characteristics by variation of the metal cap diameter D. Gain shows the maximum value at the 2nd and the 3rd frequency, when D equals 70 mmØ. The simulated gain of spiral antenna added conical wall at 2.44 GHz, 4.88 GHz and 7.32 GHz of center frequency of interested bands appears 9.74 dBi, 12.67 dBi and 14.05 dBi. Gain at Tx frequency, 2nd and 3rd harmonic frequency shows higher about 2.5 dBi, 5.4 dBi and 3.5 dBi than conventional one of refernce [2], respectively.

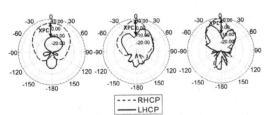

Figure 3. Simulated radiation patterns of LHCP and RHCP for antenna structure in figure 1.

978-7-5641-4279-7

Figure 3 shows the simulated radiation patterns of LHCP and RHCP for antenna structure in Figure 1. The XPD (cross polarization discrimination) is about 19.225 dB, 20.9 dB and 21.142 dB at 2.44 GHz, 4.88 GHz, and 7.32 GHz, respectively, where theta = 0° and phi = 0°. Therefore, the good axial ratio with 3 dB below as shown in Figure 5(b) is obtained at the interested bands.

### III. MEASUREMENT

Figure 4. Photograph of a fabricated antenna.

In order to verify the propriety of a proposed antenna, the novel antenna with the optimized slit on ground plane and with the added conical wall was fabricated as shown in Figure 4. Figure 5 (a) shows the comparison between the simulated and the measured return loss of designed and fabricated antenna, respectively. The measured return loss shows reasonable agreement with the simulated one, even it is slightly different. Figure 5 (b) shows the comparison of the simulated and the measured axial ratio. The simulated axial ratio and the measured axial ratio show good agreement and it keeps 3 dB below at the interested band.

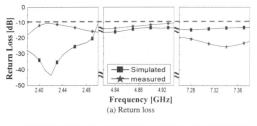

(a) Return loss

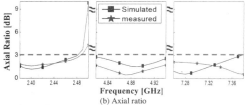

(b) Axial ratio

Figure 5. Comparison between the simulated result and the measured result of a proposed novel antenna.

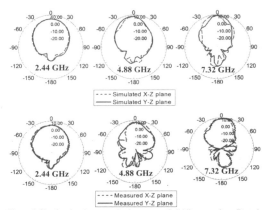

Figure 6. Simulated and measured gain patterns (Solid line: y-z plane, Dotted line: x-z plane).

Figure 6 shows the comparison between the simulated and the measured 2-D gain patterns at 2.44 GHz, 4.88 GHz and 7.32 GHz. Solid line and dotted line indicate the main E-field polarization of the x-z plane and of the y-z plane, respectively. The measured E-field gain patterns are showed very good agreement with the simulation results as shown in Figure 6.

### IV. CONCLUSION

This paper proposed a design for the spiral antenna with conical wall to obtain the high gain. An application of the proposed antenna is the NLJD system. To improve the gain and the axial ratio of spiral antenna, the conical wall and the newly designed Archimedean slit on ground plane are considered for the conventional antenna with the circular cavity wall and with the conventional 4.5-turn slit. The improved gain of 9.74 dBi, 12.67 dBi and 14.05 dBi at 2.44 GHz, 4.88 GHz and 7.32 GHz and the good axial ratio of 1.9 dB below at the interested band are realized by the added conical wall and the newly designed slit from current distribution control on ground plane, respectively. The measured return loss, axial ratio and E-field patterns are agreed well with simulation results.

### ACKNOWLEDGEMENT

This research was supported by Basic Science Research Program through the National Research Foundation of Korea(NRF) funded by the Ministry of Education, Science and Technology(2013)

### REFERENCES

[1] Audiotel International Limited, "Non-linear junction detector', U. S. 01360667, 11.12, 2003.

[2] Jeong-won Kim, Kyeong-sik Min, In-hwan Kim, and Chan-jin Park, "Triple Band Spiral Antenna for Non-Linear Junction Detector" Proceedings of ISAP2012, Nagoya, Japan, 3B1-4, pp. 802-805, Nov. 201

# Asymmetric TEM Horn Antenna for Improved Impulse Radiation Performance

#Hyeong Soon Park[1], Jae Sik Kim[1], Young Joong Yoon[1] and Ji Heon Ryu[2], Jin Soo Choi[2]
[1]The Electrical and Electronic Engineering Department, Yonsei University
Sinchon-dong, Seodaemun-gu, Seoul, South Korea, phsmicro@yonsei.ac.kr
[2]Agency for Defense Development, Daejeon, South Korea, rjh@add.re.kr

*Abstract* – A modified TEM horn antenna which has an asymmetric plates, two plates with different heights, has improved pulse radiation performance. This paper presents an optimized asymmetric TEM (AsyTEM) horn antenna and a same size of conventional exponentially tapered TEM (ETEM) horn antenna to verify the improvement of the impulse radiation performance. The AsyTEM horn antenna is shown to yield an enhanced reflection coefficient at the low frequency of the operation bandwidth. The results show that the AsyTEM horn antenna has reflection coefficient less than -10 dB in the frequencies from 2.6 GHz to over 20 GHz while the ETEM has 2.9 GHz to over 20 GHz. In the case of being compared with the ETEM horn antenna, it also had an improved impulse radiation gain which increased over 0.6.

## I. INTRODUCTION

The ultrawideband (UWB) antennas are recently attracting attention to satisfy the demand for a variety of applications. The UWB antennas provide many advantages, such as improved detection ability, adaptive ranging performance, the higher target resolution, and so on. Thus, the UWB antennas are used for the electromagnetic compatibility (EMC) measurement, the ground penetrating radar (GPR) system, and the broadband communication systems. Also the UWB antennas are applied to the transient radar cross section (RCS) measurements, the synthetic aperture radar (SAR) system for which the radiation of transient waveform is interesting through bandwidths exceeding one decade [1]. A lot of studies on the UWB antennas have worked and many types of the antennas which satisfy UWB characteristics are well established [2]-[4].

Specifically, the TEM horn antennas are widely used as the UWB antenna for having the merit of wideband, low dispersion, unidirectional pattern and easy construction [5]. Several researches have been preceded to improve the performance of the TEM horn antenna.

In this paper, the modified TEM horn antenna with the exponentially tapered asymmetric plates, two plates with different heights, has been designed.

When the TEM horn plates have the different heights, the TEM horn antenna can achieve an improved low frequency radiation ability. For a short pulse excitation, more pulse energy is distributed in the relatively low frequency range than that in the high frequency range. Improving the low frequency radiation means the improvement of the radiation efficiency [6].

To verify compatibility, the same size of the conventional exponentially tapered TEM (ETEM) horn antenna and the

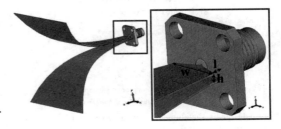

Figure 1. Feed transition of the TEM horn antenna

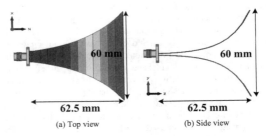

(a) Top view      (b) Side view

Figure 2. the ETEM horn antenna geometry

AsyTEM horn antenna are simulated. The simulated results show that the AsyTEM horn antenna exhibits the better reflection coefficient at the low frequency and the improved impulse radiation gain compared with the ETEM horn antenna.

## II. ANTENNA DESIGN

### A. Feed Transition

The TEM horn antenna is usually equipped with a balun structure at the feed position for guide a travelling wave without any impedance discontinuities. As shown in Figure 1, the smooth transition has been applied to pass from the inner conductor of the SMA connector to the upper plate and the lower plate is extruded to the SMA connector. The height (h), the length (l) and the width (w) of the transition structure is optimized to operating the 50 Ω line [7].

### B. Horn Profiles

Figure 2 shows the geometry of the ETEM horn antenna including the transition structure. The main issue of the TEM

978-7-5641-4279-7

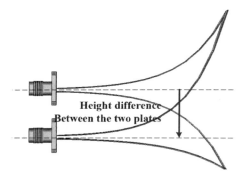

Figure 3. The size and geometry comparison of the AsyTEM and the ETEM horn antenna

horn design is matching the characteristic impedance of the guide from 50 Ω at feeding part to 120π Ω at the aperture. To smoothly varying characteristic impedance from feed to aperture, the tapered plates form of TEM antenna can be linear, exponential and Klopfenstein, etc. Linear tapered antennas can be built easily as compared with an exponentially taped antenna and Klopfenstein tapered antenna yield the smallest minor ripples characteristic in the reflection coefficient [8]. However, exponentially tapered plates have the advantage of the smooth impedance variations than the linear tapered and can be achieve more wideband characteristics in same size. Meanwhile, Klopfenstein tapered antenna has the disadvantage of the complicated design [9]. Thus, the exponentially tapered form is chosen to the basic structure shape. The TEM plates width is calculated by the means of equations for the parallel plate waveguide, as proposed in [9]. The designed ETEM horn antenna is 62.5 mm long, and the aperture size is 60X60 mm$^2$ respectively.

*C. Effect of the Asymmetric Plates*

Usually the TEM horn antenna is a symmetrical structure to guarantee the end-fire radiation characteristic on main direction.

In [10], the asymmetrical anti-podal taper slot antenna (TSA) is introduced to further enhance the bandwidth of the conventional taper slot antenna. However, the asymmetrical TSA has inclined peak direction of the radiation pattern. To obtain the similar characteristic in [10], the three dimensions of the asymmetric plates are applied to the ETEM as shown in Figure 3. As illustrated in this Figure, the AsyTEM horn antenna has the same aperture size and antenna length as the ETEM horn antenna.

The AsyTEM horn antenna can be considered as a combination of the two different profiles of TEM horn antennas. To maintain the dimensions of the AsyTEM horn antenna as same as the ETEM horn antenna, the heights of the upper plate and the lower plate are complementary.

Figure 4 indicates the E-field distribution on near two antennas at 2.6 GHz in the y-z plane. This figure shows that the contour

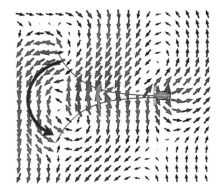

(a) E-field at 2.6 GHz of the ETEM horn antenna

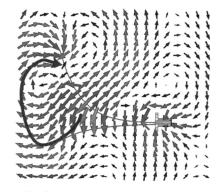

(b) E-field at 2.6 GHz of the AsyTEM horn antenna

Figure 4.Simulated contour path of the E-field vectors

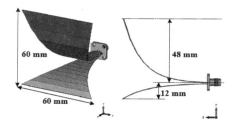

Figure 5. Final geometry of the AsyTEM horn antenna

path of the E-field vectors differs from each other.

The E-field vectors in the AsyTEM are not perpendicular to the edge of the each plate, which means that the equivalent aperture has been extended. Thus, the lower boundary of the resonant band can be downsized [10]. As expected, this result is the same as the analyzed phenomenon of the asymmetrical planar type TSA.

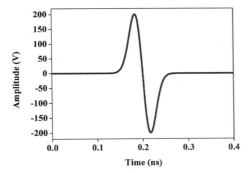

Figure 6. Input bipolar pulse in the time domain

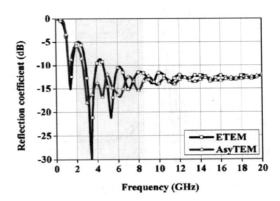

Figure 8. The reflection coefficient of the two antennas

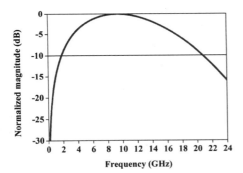

Figure 7. Power spectral density (PSD) of the input pulse

For optimum performance of the AsyTEM horn antenna, finding the optimized values of the different heights is preceded. The final values of the different heights have been founded with CST Microwave studio software using parameter sweep function.

The final geometry of the AsyTEM horn antenna is shown in figure 5. As shown in this figure, the horn length and the aperture size are the same as the ETEM horn antenna, but the height of the upper plate is 48 mm, the lower plate is 12 mm respectively.

### III. RADIATION CHARACTERISTICS

First of all, the input voltage pulse is presented to simulate the radiated field of the two antennas. Figure 6 shows that pulse form is bipolar pulse with a null mean value.

Compared with the monopolar pulse, when received antenna receives pulse, the bipolar pulse can reduce pulse distortion. It also has several other advantages [11]. The basic Gaussian pulse is described analytically as [12]

$$v(t) = \frac{A}{\sqrt{2\pi}\sigma} \exp(-\frac{t^2}{2\sigma^2}) \qquad (1)$$

The bipolar pulse is given by the first derivative of the Gaussian pulse

$$v'(t) = \frac{At}{\sqrt{2\pi}\sigma^3} \exp(-\frac{t^2}{2\sigma^2}) \qquad (2)$$

Where $A$ is the amplitude of the Gaussian pulse and $\sigma$ defines the width of the pulse.

The peak value of the pulse is 200 V, the width of the input pulse is 150 ps and the rise time is about 24 ps respectively. The normalized power spectral density (PSD) of the input pulse is plotted in Figure 7. PSD spectrum can be achieved by Fourier transform of the input pulse. As shown in this figure, the bandwidth of the signal is about from 2 GHz to over 20 GHz and centered around 8 GHz. It is clear that the PSD of the input pulse meet the condition of the operating bandwidth of the two antennas.

Figure 8 shows the computed reflection coefficient as function of the frequency for the two antennas. Both antennas can cover the UWB characteristic. The bandwidth (S11<-10 dB) of the ETEM horn antenna is 2.6 over 20 GHz and the AsyTEM horn antenna is 2.9 to over 20 GHz respectively.

The AsyTEM horn antenna has the enhanced bandwidth range at the low frequency. This result is already explained in the section II, figure 4. The AsyTEM horn antenna also has the better reflection coefficient compared with the ETEM horn antenna from about 2 to 8 GHz band. Mentioned in [6], improved low frequency characteristic makes the radiation efficiency better for a short pulse excitation. To obtain the transient radiated far-field result, E-field probe is 5 m away from the antenna in the simulation.

However, the AsyTEM horn antenna has difference length of the upper and the lower plates that cause the inclination of the peak direction in the radiation pattern. The farfield peak direction of the AsyTEM horn antenna rises up to about 13 degree in the elevation angle from the bore sight direction.

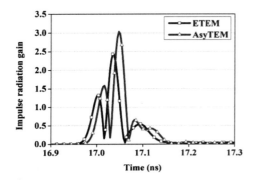

Figure 9. Impulse radiation gain of the two antennas

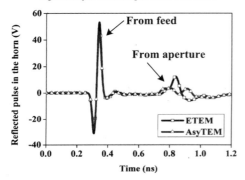

Figure 10. The reflected voltage in the horn

Thus, E-field probe to the AsyTEM horn antenna is located in the same distance and rises up to the peak direction for the correct comparison with the ETEM horn antenna.

Figure 9 indicates the impulse radiation gain of two antennas. The impulse radiation gain of the peak direction is defined as the ratio of the absolute level of a transient radiated far- field ($E_r$, unit : V/m) times the received distance (d, unit : m), and divided by the peak level of the voltage accepted by the antenna ($V_{peak}$, unit : V):

$$impulse\ radiation\ gain = \frac{|E_r \times d|}{V_{peak}} \quad (3)$$

As a result of this UWB improvement, it is obviously seen at the impulse radiation gain graph in figure 9 that both of the antennas show the low pulse dispersion. The AsyTEM horn antenna has the higher peak gain and the narrower received beam width than that of the ETEM horn antenna with the same physical size.

The maximum impulse radiation gain value of the ETEM horn antenna is 2.43 while the AsyTEM horn antenna is 3.03. Therefore, the AsyTEM horn antenna has improved impulse

radiation gain which increased over 0.6 compared with the ETEM horn antenna.

The pulses reflected from the horn are given in Figure 10. In this figure, there is seen to be similar reflections between the two antennas. The biggest difference in the reflection occurs at the aperture. The ETEM horn antenna has the higher reflection voltage pulse from the aperture than the AsyTEM horn antenna. This result influences the radiated pulse from antenna that makes impulse radiation gain worse.

## IV. CONCLUSION

The modified TEM horn antenna with the exponentially tapered asymmetric plates, two plates with different heights, is proposed to improve the impulse radiation performance compared with the same size of the exponentially tapered TEM horn antenna. To verify compatibility, the ETEM horn antenna and the proposed TEM horn antenna are simulated. The proposed antenna can improve the reflection coefficient at the low frequency range and low reflected pulse in the horn aperture. Therefore the impulse radiation gain of the UWB antenna system can be enhanced compared with the system with the conventional TEM horn antenna.

## ACKNOWLEDGMENT

This paper was supported by Agency for Defense Development (ADD) under the contract number UD130008GD.

## REFERENCES

[1] Desrumaux L., Godard A., Lalande M., Bertrand V., Andrieu J., Jecko B., "An Original Antenna for Transient High Power UWB Arrays: The Shark Antenna," *IEEE Trans. Antennas and Propagation*, vol. 58, pp. 2515-2522, August 2010.
[2] Constantine A. Balanis, *Antenna theory - analysis and design*, 3rd ed., John Wiley & Sons, 2005, pp. 549-602.
[3] John DS. Kraus, Ronald J. Marhefka, *Antennas - For all applications*, 3rd ed., McGraw-hill, 2002, pp. 378-400.
[4] Sabath F., Giri D.V., Rachidi F., Kaelin A., *Ultra-wideband, Short Pulse Electromagnetics 9*, New york: Springer, 2010, pp. 189-294.
[5] Sara Banou Bassam, Jalil-Agha Rashed-Mohassel, "A Cheby-shev tapered Tem horn antenna," *Progress In Electromagnetics Research (PIER)*, vol. 2, pp. 706-709, 2006.
[6] Xiaolong Liu, Gang Wang, Wenbing Wang, "Design and performance of TEM horn antenna with low-frequency compensation," *Asia-Pacific Conf. of Environmental Electromagnetics*, pp. 306-309, November 2003.
[7] Godard,et al., "Size reduction and radiation optimization on UWB antenna," *IEEE Radar Conf.*, pp.1-5, May 2008.
[8] Klopfenstein R.W., "A transmission line taper of improved design," *Proceedings of the IRE*, vol. 44, pp. 31-35, January 1956.
[9] Kyungho Chung, Pyun S., Jaehoon Choi, "Design of an ultrawide-band TEM horn antenna with a microstrip-type balun," *IEEE Trans. Antennas and Propagation*, vol. 53, pp. 3410-3413, October 2005.
[10] Wenhua Chen, Yuan Yao, Zhijun Zhang, Zhenghe Feng, Yaqin Chen, "Design of unsymmetrical anti-podal taper slot element for array antenna," *IEEE Antennas and Propagation Society (AP-S)*, pp.1-4, July 2008.
[11] Diot J- C. et al., "Optoelectronic ultra-wide band radar system: RUGBI," *Radar Conf. EURAD*, pp.81-84, October 2005.
[12] Li Li, Pei Wang, Wu Xiao-dong, Zhang Jiakai, "Improved UWB pulse shaping method based on Gaussian derivatives," *IET international conf. CCWMC*, pp.438-442, November 2011.

# *TP-P*

## October 24 (THU) PM

## Room E

# An UWB Rotated Cross Monopole Antenna

Jian Ren, Xueshi Ren, Yingzeng Yin
National Laboratory of Science and Technology on Antennas and Microwaves
Xidian University
2# Taibai South Road
Xi'an, Shaanxi 710071 China
renjianroy@gmail.com, xsren@mail.xidian.edu.cn, yzyin@mail.xidian.edu.cn

*Abstract*-**Configurations of planar rotated cross monopole antennas have been investigated. The rotated cross patch comprises a vertical microstrip and three rectangular patches (area A, B, and C). By rotating the horizontal patches (area B and C), the bandwidth of the antenna can be significantly enhanced. The effect of the rotated angle of B and C on the bandwidth has been studied. The measured results show that The bandwidth of impedance bandwidth (10-dB reflection coefficient) is as wide as 6.97 GHz (2.29–9.26 GHz) or about 120.6% . which is about two times that of the corresponding conventional cross monopole antenna. The proposed antenna has a ultra-wide bandwidth which can cover DECT/IMT-2000/3G/UMTS/2.45-GHz/5.2-GHz/5.8-GHz ISM band (WLAN, IEEE 802.11b and g)/ Bluetooth/ZigBee 2.4 GHz/2.5-GHz WiMAX/3.5-GHz WiMAX bands.**

**Index Terms—rotated monopole antenna, ultra-wideband.**

## I. INTRODUCTION

With the rapidly development of wireless communication, the multiband antenna has become a hot area. Covering multiband using only one antenna becomes a challenging issue. One of the solutions is using wideband antenna. Planar monopole antennas usually have these advantages at the price of relatively narrow bandwidth, which might not be wide enough to support modern digital wireless communication systems. To get a wideband monopole antenna, different methods have been proposed, such as multilayer structures [1] or parasitic elements [2]. In addition, beveling the rectangle of the monopole also can improve the impedance bandwidth to 6:1 for VSWR=2 [3]. In [4], a planar cross monopole antenna was investigated. The cross-shaped patch comprises vertical microstrip and two rectangular patches. The antenna exhibits a wide impedance bandwidth of over 70%. In[5], a printed antenna composing of a trapezoid ground plane and an elliptical monopole patch was proposed. The antenna has a measured impedance bandwidth from 1.02 GHz to 24.1 GHz with a VSWR = 2. To get wide operation band, the author In [6] proposed a printed wide-slot antenna fed by a microstrip line with a rotated slot for bandwidth enhancement, the measured impedance bandwidth, defined by 10 dB return loss, can reach an operating bandwidth of 2.2 GHz at operating frequencies around 4.5 GHz.

In this paper, an ultra-wideband rotated cross monopole antenna was presented. The antenna has wide operation and width of 6.97 GHz (2.29–9.26 GHz) or about 120.6% for return loss=10dB, which can cover DECT/IMT-2000/3G/U-MTS/2.45-GHz/5.2-GHz/5.8-GHz ISM band(WLAN, IEEE-

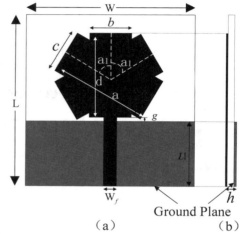

Figure 1  Geometry of the proposed antenna: (a) top view, (b) side view

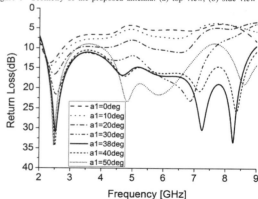

Figure 2 Simulated return loss against frequency for the proposed antenna with various a1, W=L=50mm, L1=15mm, a=9mm, b=9mm, c=13mm, d=28mm, g=0.5mm

802.11b and g)/Bluetooth/ZigBee/2.4-GHz/2.5-GHz WiMAX-/3.5-GHz WiMAX bands. By rotating the two patch of the antenna (area A and B), the impedance bandwidth of proposed cross monopole antenna can be significantly enhanced.

978-7-5641-4279-7

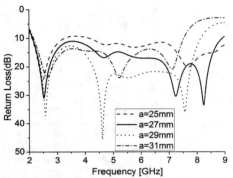

Figure 3 Simulated return loss against frequency for the proposed antenna with various a, W=L=50mm, L1= 15mm, , b=9mm, c=13mm, d=28mm, g=0.5mm, a1= 38°

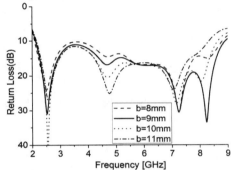

Figure 4 Simulated return loss against frequency for the proposed antenna with various b, W=L=50mm, L1=15mm, a=9mm,, c=13mm, d=28mm, g=0.5mm, a1= 38°

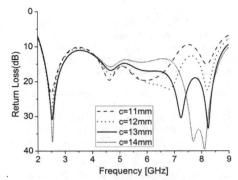

Figure 5 Simulated return loss against frequency for the proposed antenna with various c, W=L=50mm, L1 = 15mm, a=9mm, b=9mm, d=28mm, g=0.5mm, a1= 38°

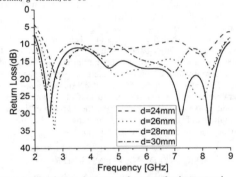

Figure 6 Simulated return loss against frequency for the proposed antenna with various d, W=L=50mm, L1=15mm, a=9mm, b=9mm, c=13mm, g=0.5mm, a1= 38°

## II. ANTENNA DESIGN AND EXPERIMENT RESULTS

### A. ANTENNA DESIGN

The geometry of the proposed rotated cross monopole antenna is shown in Fig.1. The rotated cross monopole antenna can be easily printed on dielectric substrate. It is constituted of rotated cross-shaped planar monopole fed by 50- microstrip line and a ground plane. The rotated cross patch is consisting of three parts-the vertical rectangular patch (A) and two rotated rectangular patches (B and C) , which has a rotated angle of a1. The vertical rectangular patch (A) acts as a radiator as well as feed for the cross patch antenna. The rotated monopole radiator and $50\,\Omega$ -microstrip feed line are printed on the same side of the dielectric substrate while the ground plane locates on the other side. To simplify the design, the part B and C have the same dimensions of $a \times c$ . The size of the patch A is $b \times d$ and the whole antenna is fed by a 50-$\Omega$ microstrip line with a width of $W_f$ . The proposed antenna is printed on an FR4 substrate with a thickness of 1.6mm and a relative permittivity of 4.4, which has a dimension of W× L.

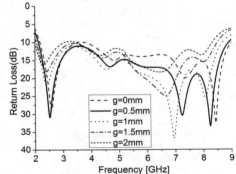

Figure 7 Simulated return loss against frequency for the proposed antenna with various g, W=L=50mm, L1=15mm, a=9mm, b=9mm, c=13mm, d=28mm, a1= 38°

The dimension of ground plane for the proposed antenna is W×L1. The feed gap width between the ground plane and the feed point is $g$ .

*B. PARAMETER STUDY*

In order to understand the effects of various parameters and to optimize the performance of the final design, a parameter study is carried out in this section. The whole Simulation is carried out using CST, a commercial electromagnetic simulator based on Finite Difference Time Domain (FDTD).

In this antenna, the most critical parameter is the rotated angle of the horizontal patch. It has a significant effect on the return loss. Fig.2 shows the simulated reflection coefficient for the reference antenna with different a1. It shows that that the return loss varied significant as different a1. The return loss decreases to 10dB gradually as a1 increases. When a1 is larger than 40 °, the return loss become large. To get a wide operation band, we chose the a1 as 38 °. The bandwidth of impedance bandwidth (10-dB reflection coefficient) is as large as 6.97 GHz (2.29–9.26 GHz). Fig.3-6 show the return loss with different a, b, c, and d. It is from the simulated results shown that a and d have significant effect on the bandwidth of the rotated monopole antenna. The width of the rotated rectangle patches b and c only has effect on the impedance of the antenna at the high band. At the low band operation, the antenna has a stable return loss which varies little as b and c have different value. The width of the gap between the ground plane and the feed point ,g, also has effect on the bandwidth of the antenna. Fig. 7 shows the return loss of the antenna with different g. As the g become larger, the impedance match at the high band becomes worse, meanwhile, the resonant frequency shifts to a little lower frequency.

## III. SIMULATED AND MEASURED RESULTS

The prototype of the proposed antenna with optimal geometrical parameters as shown in Fig. 1 is constructed and measured. The antenna proposed here is fabricated on commercially available FR4 substrate with h=1.6mm, $\varepsilon$ =4.4. The geometric dimensions of the proposed antenna are as follows: W=L=50mm, L1=15mm, a=9mm, b=9mm, c=13mm,d=28mm, g=0.5mm, a1=38°. The antenna was fabricated and the photograph of it is shown in Figure 8. The return-loss performance was measured using Agilent Vector Network Analyzer. Fig.12 shows the simulated and measured return loss of the proposed antenna. The measured results show that the bandwidth of impedance bandwidth (10-dB reflection coefficient) is as large as 6.97 GHz (2.29–9.26 GHz) or about 120.6%. A little difference between the simulated and the measured results can be observed, the measured resonant frequency of the antenna shifts to the high frequency There may be because the substrate that we used is commercial available and the permittivity is not strictly accurate.

Fig.9-11 give the simulated radiation patterns at 2.4GHz, 3.5GHz and 7GHz. It seen that at the low band ,the proposed antenna has a monopole-like patterns. A good omnidirectional radiation pattern is obtained. It is also noted that radiation patterns are in symmetry with respect to the antenna axis since the proposed antenna's structure is symmetrical and a low cross polarization level can be observed at the whole band.

Figure 8 The prototype of the proposed antenna

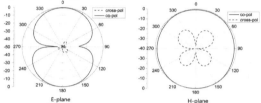
Figure 9 simulated E-plane and H-plane radiation for the antenna with a1=38° at $f$ =2.4GHz

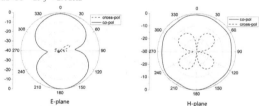
Figure 10 simulated E-plane and H-plane radiation for the antenna with a1=38° at $f$ =3.5GHz

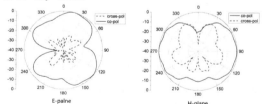
Figure 11 simulated E-plane and H-plane radiation for the antenna with a1=38° at $f$ =7GHz

## IV. CONCLUSION

A printed UWB monopole-like antenna fed by a 50-micro-strip line with rotated patches for band width enhancement has been demonstrated. By rotating the horizontal patch of the cross monopole, the bandwidth of the antenna can be significantly enhanced. Experimental results show that the impedance band width determined by 10dB return loss can reach nearly 6.97-GHz which is about two times that of the

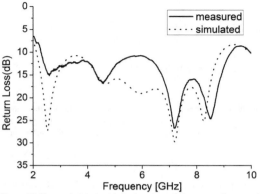

Figure 12 Measured and simulated return loss of the proposed monopole antenna

corresponding conventional microstrip-line-fed cross monopole. The antenna can cover DECT/IMT-2000/3G/UMT-/2.45-GHz/5.2GHz/-5.8GHz ISM band (WLAN, IEEE 802.11b and g)/ Blue-tooth/ZigBee 2.4 GHz/2.5-GHz WiMAX/3.5-GHz WiMAX bands . Moreover, the proposed antenna possesses

nearly omni-directional radiation characteristics at the low band and  low cross polarization level can be observed at the whole band

REFERENCES

[1]  H. Smith and P. Mayes, "Stacking resonators to increase the bandwidth of low-profile antennas," Antennas and Propagation, IEEE Transactions on, vol. 35, pp. 1473-1476, 1987.

[2]  C. Wood, "Improved bandwidth of microstrip antennas using parasitic elements," in IEE Proceedings H (Microwaves, Optics and Antennas), 1980, pp. 231-234.

[3]  M. J. Ammann and Z. N. Chen, "Wideband monopole antennas for multi-band wireless systems," Antennas and Propagation Magazine, IEEE, vol. 45, pp. 146-150, 2003.

[4]  C. Tseng and C. Huang, "A wideband cross monopole antenna," Antennas and Propagation, IEEE Transactions on, vol. 57, pp. 2464-2468, 2009.

[5]  S. Zhong, X. Liang and W. Wang, "Compact elliptical monopole antenna with impedance bandwidth in excess of 21: 1," Antennas and Propagation, IEEE Transactions on, vol. 55, pp. 3082-3085, 2007.

[6]  J. Jan and J. Su, "Bandwidth enhancement of a printed wide-slot antenna with a rotated slot," Antennas and Propagation, IEEE Transactions on, vol. 53, pp. 2111-2114, 2005

# A Design of Miniaturized Dual-band Antenna

Lu Wang[1], Jingping Liu[1], Qian Wei[1], Safieddin Safavi-Naeini[2]

[1]School of Electronic and Optical Engineering, Nanjing University of Science and Technology

Nanjing 210094, Jiangsu, China

[2]Electrical and Computer Engineering, University of Waterloo

N2L 3G1, ON, Canada

*Abstract*: **Based on the principle and cavity mode theory of microstrip antenna, this article introduces the theoretical approaches of slotting in the non-radiative side to get the dual characteristics. We have designed a dual-band microstrip antenna with a U-shaped slot, which can make the high-frequency be 5.25GHz and the low-frequency be 1.75GHz by changing the length and width of the slot.**

## I. INTRODUCTION

In recent years, with the development of communication system and radio system, the technology for antenna's miniaturization and multi-frequency becomes more and more important. Because of its low profile, small volume and light weight, mcirostrip antennas have been widely used in many fields [1].

To make the antenna miniaturization, we can increase the dielectric constant of the dielectric substrate[2], slot on the surface and short loading[3]. We can also make antenna achieve multi-band by using a variety of different resonant modes ( such as radiation patch $TM_{10}$ or $TM_{01}$ mode ), the antenna operates in the mode, which can be achieved in dual or multi- frequency band antenna; patch is loaded by a single method or open grooves, changing the patch field distribution of natural modes, and thus interfere with the mode excitation of the resonant frequency, so that the antenna can operate in dual-band or multi –band; single layer substrate used in a structure covering the plurality of the radiation patch .

According to modes theory, the first three modes of the microstrip patch have the same polarization plane, they are $TM_{10}$ $TM_{20}$, $TM_{30}$, which apply the theory for the realization of the dual frequency and multiple frequency. In these models, $TM_{10}$ is a typical pattern in practical application, $TM_{20}$ mode resonant frequency is twice as high as the $TM_{10}$ mode, and $TM_{30}$ mode resonant frequency is three times higher than the TM10 mode. But in the actual application, the last two resonant modes are not often used. This is because on the surface of the patch, the current distribution is not uniform in $TM_{20}$ mode, which the antenna radiation pattern can produce zero [4]. The antenna radiation direction plan will produce larger side lobe in $TM_{30}$ mode. To take advantage of the above three models to implement dual-band characteristics, usually adopted method is slotted on the patch. Loading cracks in the patch surface can change the path current movements on the surface of the patch, which can change the current distribution. Through the gap on the patch loading position is different, can be roughly divided into the slot method: slot at the edge of the radiation and the radiation in slot.

## II. ANTENNA DESIGN

Slotting at the radiation patch edge is used widely to achieve dual-band antenna [5]. This method is simple and convenient. It can not only make the antenna gain double frequency characteristic, but also the process is simple. Compared with slotting at the edge of the radiation, slot near the radiation side, which will make the current distribution small. The related antenna radiation characteristics, such as antenna pattern, cross polarization performance will be better. This design loads the u-shaped slot in the radiation boundary, by changing the slot length and width can easily adjust the antenna of high frequency resonance point and low frequency resonance points, meet the antenna need.

### 2.1 U-shape slot antenna

The U-shape antenna structure is shown in Fig.1. According to the figure, the size of the patch is L×W, the width of the slot is w, length of the slot is Ls and the slot is 1mm far away from the edge of patch. The thickness of the dielectric substrate is h. The dielectric constant is $\varepsilon_r$. The coaxial feed point from the center axis of the patch is m.

This design of miniaturized multi-band antenna should make the high frequency be 5.25GHz and the low frequency is 1.75GHz.

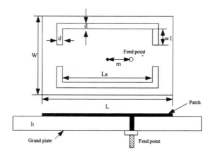

Fig.1. U-shape slot microstrip antenna structure.

978-7-5641-4279-7

According to the above theory, choosing high dielectric constant and thick dielectric substrate can get miniaturization antenna, so this design uses rogers4350 as the substrate and is thickness is 1.524mm. to get a good frequency impedance matching, we uses coaxial feed.

After calculation, the size of the antenna is: L=32mm, W=28mm, m=5mm. Changing the length Ls、width w can affect the ratio of high frequency and low frequency. Detailed discussions of the analysis are as follows.

*(1) Length* Ls

Change the length Ls while the other parameters maintain the same size, S11 parameters of the antenna is shown in Fig. 2.

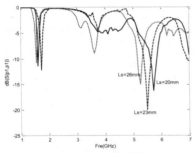

Fig.2. S11 parameters with different Ls.

As can be seen from the figure 2, changing Ls affects the S11 parameters of the antenna. The high frequency changes a lot but the low frequency variation is not obvious. The current of the patch surface will not produce high frequency response if Ls is too short, which cannot form dual-frequency.

*(2) Width* d

Change the width d while the other parameters maintain the same size, S11 parameters of the antenna is shown in Fig.3.

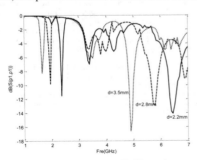

Fig.3. S11 parameters with different d.

As can be seen from the Fig.3, the high frequency and low frequency vary a lot by changing w. With the increasing d, the high and low frequency becomes small, which is because d affects the Antenna impedance characteristics.

Considering the effect of the slot length and width to the antenna, we can optimize the design parameters of the antenna during simulation, which can obtain the desired high and low bands.

After simulation and optimization in the HFSS, the size of the antenna is: L=32mm，W=26mm，m=5mm，L=24，w=3mm，w1=5.5mm。

During the HFSS simulation environment, S11 simulation results of the antenna is shown in Fig.4.

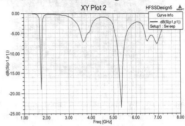

Fig.4. S11 parameters of the antenna.

Antenna radiation pattern is shown in Fig.5.

It can be seen from the radiation pattern that the high-frequency 5.25GHz gain can reach about 3dB while the low frequency 1.75GHz gain is 6dB.It also can be seen that the antenna at high frequency 5.25GHz exits secondary lobes.

*2.2 Antenna measured results*

Based on the simulation, we use the CAD drawing to draw the antenna size and work out this U -shaped slot antenna microstrip antenna.

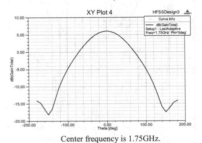

Center frequency is 1.75GHz.

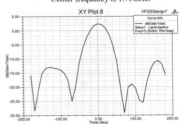

Center frequency is 5.25GHz.
Fig.5. Antenna radiation pattern.

This antenna is printed on a substrate rogers4350, the relative dielectric constant of the board was 3.55 , the thickness is 1.524mm, scale is 40mm * 50mm, at the same time, using the probe of 1mm diameter of SMA coaxial feed，its Physical antenna is shown in Fig.6.

Fig.6. Physical Antenna.

After analyzing by the network analyzer, the physical dual-band antenna S-parameters is shown in Fig.7.

As can be seen from Fig.7, the antenna reflection parameters of the test curve and the curve obtained software simulation is consistent. After testing, the direction of the antenna 5.25GHz diagram is shown in Fig.8.

Fig.8 shows that the measured results and simulation results are basically consistent. It can be seen from the figure, only a slight deviation of the measured and simulation, the actual measured gain is slightly lower than the simulation results.

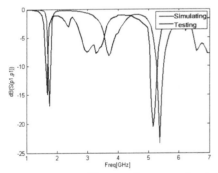

Fig.7. S-parameters of the antenna diagram comparison chart.

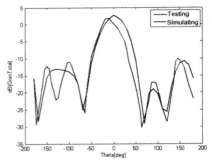

Fig.8. Measured antenna pattern.

The paper designed to achieve the dual-band antenna though, but from the simulated and measured results can be seen very narrow bandwidth of the antenna, with the current requirements of ultra- wideband characteristics compared, there is a large gap between the need for further optimization and improvement. In addition, the size of the antenna could be improved, so that the antenna can truly achieve miniaturization, ease of integration in the circuit board.

## III. CONCLUSIONS

This paper introduces some theoretical basis and technical methods to figure out the miniaturization and dual-frequency of the antenna in detail. Based on this theory, a dual-band structure with a slot along the non-radiation side has been investigated. According to the theory, we have designed a dual-band microstrip antenna with a U-shaped slot, which can make the high-frequency be 5.25GHz and the low-frequency be 1.75GHz by changing the length and width of the slot.

### REFERENCES

[1] Lin Changlu. *Modern Antenna Design* [M]. People Post Press,1990.
[2] Lo. T Ketal. Miniature aperture-coupled microstrip antenna of very high permittivity[J]. *Electron, Letts*, 1997, Vol.33, pp. 9-10.
[3] Chow Y. L, Wail K. L. Miniaturizing patch antenna by adding a shorting pin near the feed probe a folded monopole equivalent[J].*IEEE Antennas and Propagation Society International Symposium*, 2002,4, pp.6-9.
[4] Chen W. H. Dual-band reconfigurable antenna for wireless communication [J]. *Microwave and Optical Technology Letters*. 2004.3,40(6), pp.50-55.
[5] Zi Dong Liu, Hall P S. Wake，Dual-fequency planar inverrted-F antenna[J]. *IEEE Antennas and Propagation*, 1997, 45(10), pp.1451-1458.

# Design of Single-Layer Single-Feed Patch Antenna for GPS and WLAN Applications

Zeheng Lai, Jiade Yuan
College of Physics and Information Engineering
FuZhou University, FuZhou, P.R.China
E-mail:lzh_fzu@sina.cn

*Abstract*-A design of single-layer single-feed dual-frequency patch antenna with different polarizations is presented for the first time. The structure of the proposed antenna is a single layer patch, which mutually nested inside and outside patches and connected together with four conducting strips. Two operating frequencies of the proposed antenna can be observed in the simulated and measured results. The inside patch operates at the high frequency and outside square patch operates at the low frequency. The center frequency of the antenna prototype for Global Positioning System (GPS) is 1.575 GHz with circular polarization and 2.4 GHz for Wireless Local Area Networks (WLAN) with linear polarization respectively. The impedance bandwidth and axial ratio of the proposed antenna is studied through electromagnetic simulations and measurements.

*Index Terms*-circular polarization, single-layer, single-feed, dual-frequency, patch antenna, GPS, WLAN.

## I. INTRODUCTION

With the rapid development of wireless communication, Miniaturization dual-frequency antenna is one of the hot spots in these researches in recent years. Microstrip antenna has been applied widely owing to their attractive features of low profile, easy to manufacture, light weight and low cost. They are often used in multifunctional wireless products to reduce the number of the required antennas [1]. Several designs related to the dual-frequency, single-layer or single-feed antennas have been reported [2-5], and these designs can provide the antenna to be with specific radiation characteristic at each operating frequency to comply with system requirements.

The antenna in ref.[2] consists of two radiating elements which are arranged in a stacked structure, in which the top is a square patch and the bottom is a corner truncated square-ring patch [3], and the two patches are connected together with four conducting strips. Two operating frequencies can be observed in the simulated and measured results. One element is designed for GPS operation and it has circular polarization (CP) broadside radiation. The other element operates at the WLAN band and its radiation is linear polarization (LP) conical pattern. Moreover, the two elements can be simultaneously excited by a single feed [4].

The design in ref.[5] realizes dual-frequency by using a half-ring structure and a half-circular patch structure, It consists of a half-width circular microstrip antenna and a narrow half-ring patch, which are connected by a microstrip feeding line[6], and has a revised formula to calculate the resonant frequency of the proposed narrow half-ring structure[7]. The similar antenna performances can also be carried out by a single-layer circular patch antenna surrounded by two concentric annular rings [8].

In this paper, a new type of single-layer single-feed and dual-frequency microstrip antennas for GPS and WLAN application is proposed. A special slot is used to make the patch become the nested patches, in which the inside and outside patches correspond to high and low resonance frequency respectively. The circular polarization characteristics in outside patch are realized by truncating equal corners in the patch, while inside patch operates at the WLAN band. Finally, some structure parameters of the proposed antenna are discussed and analyzed for the effection to the reflected coefficient and the axial ratio.

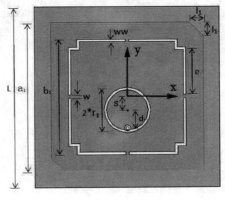

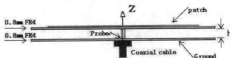

Fig.1. Geometry of the single-layer single-feed dual-frequency patch antenna

978-7-5641-4279-7

## II. ANTENNA STRUCTURE

The proposed single-layer single-feed dual-frequency patch antenna is shown in Fig.1. The antenna is designed by nested inside and outside square patches, having an outside length of $a_1$ and an inside length of $b_1$. The antenna patch is printed on a FR4 substrate_2 with the thickness $h_2$=0.8mm and the relative permittivity $\varepsilon_r$=4.4. The height of air layer is $h$ which is placed above the substrate_1 with the thickness $h_1$=0.8mm and the relative permittivity $\varepsilon_r$=4.4. The inside patch and the outside patch are connected by four conducting strips, which have the same width $w$ and are attached to four edges of the square patch respectively. The truncated corners of outside patches have equal side length $l_1$ for right hand circular polarization operation. The center of circular patch with radius $r_1$ in inside patch is along y-axis, and it is at a distance of $s$ away from the center of the inside patch. Four slots with equal size are cut between inside and outside patch and the side length and the width are marked as $e$ and $ww$, as shown in Fig.1. A coaxial probe with radius $r$=0.7mm is used to excite the antenna, and the distance between the feed point and the center of the square-ring slot is $d$. The proposed antenna is fabricated and its photograph is shown in Fig.2.

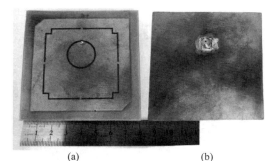

(a)                           (b)

Fig. 2. Photograph of the proposed antenna: (a) front view (b) ground plane

## III. MEASURED AND SIMULATED RESULTS

The reflected coefficient ($S_{11}$) and the axial ratio are obtained by the electromagnetic simulation software HFSS. The optimized structure sizes of the proposed antenna are shown in Table I. The simulated results and measured results of the $S_{11}$ are exhibited in Fig.3. It is clearly observed that good agreements between the simulated results and the measured results are obtained, and some slight differences occur at the frequency could be due to the dimension error during the manufacture. It has two center operating frequencies of 1.575GHz and due to the square patch antenna

TABLE I

| Parameters | $a_1$ | $b_1$ | $e$ | $h_1$ | $h_2$ | $h$ | $l_1$ |
|---|---|---|---|---|---|---|---|
| Values/mm | 66 | 51 | 20 | 0.8 | 0.8 | 4 | 6 |
| Parameters | $w$ | $ww$ | $s$ | $d$ | $r$ | $r_1$ | $L$ |
| Values/mm | 1.5 | 1 | 6 | 8 | 0.7 | 9 | 80 |

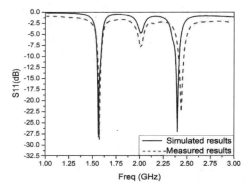

Fig. 3. Measured and simulated reflected coefficient of the proposed antenna

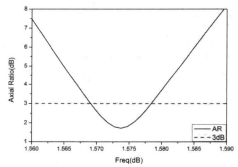

Fig.4. Simulated axial ratios of the proposed antenna

operating at the $TM_{11}$ mode, and the perturbations of the truncated corners allow the fundamental mode to split into two orthogonal degenerated modes. The $S_{11}$ of the GPS band referred to -10 dB is from 1.546 to 1.886 GHz. The $S_{11}$ of the WLAN band referred to -10 dB is from 2.37 to 2.42 GHz.

Fig.4 shows axial ratio characteristics at the GPS band with the center frequency of 1.575GHz. The obtained results in Fig.4 clearly indicate that the antenna has good CP performance and its CP bandwidth referred to 3 dB.

## IV. PARAMETER ANALYSES AND DISCUSSION

The key structure parameters of the proposed antenna, such as $b_1$, $l$ and $w$, have significant effects on the $S_{11}$ and the axial ratio. When other dimensions are fixed, as for the effects of $b_1$, the simulated results are given in Fig.5. When $b_1$ is increased, the resonant frequency of the inside patch, defined as the frequency with minimum return loss, is moved to a lower frequency but the resonant frequency of outside patch is almost constant. The variations of $b_1$ directly affect the performance of the proposed antenna at the operating frequency. The optimization about the position of coaxial probe is shown in Fig.6. As $d$ is increased, the reflected

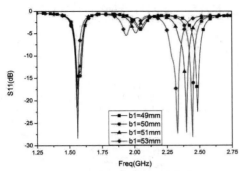

Fig. 5. Simulated return loss of various $b_1$ for the antenna

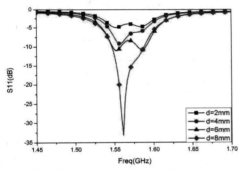

Fig. 6. Simulated return loss of various $d$ for the antenna

Coefficient ($S_{11}$) of GPS is obviously moved to a lower point. However, the position $d$ cannot be set too small or large, or it will causes resonance frequency point to mismatch. A proper parameter $d$ is required to take into account when achieve good impedance bandwidth characteristics of the proposed antenna.

## V.  CONCLUSIONS

In this paper, a new antenna with single-layer single-feed for GPS and WLAN has been presented. The antenna is nested by two patches within each other, where a circle patch is located in the inner nested patch. The proposed antenna has some advantages, such as low profile, easy to manufacture and light weight. Simulation and test results show that the proposed antenna exhibits a good performance.

### ACKNOWLEDGMENT

This work was supported by the Natural Science Foundation of Fujian Province of China (2011J01348) and the China Post doctoral Science Foundation (2013M531540).

## REFERENCES

[1] Ming Chen, "A Compact Dual-Band GPS Antenna Dedign," IEEE Trans. Antennas Propag. ,vol.12, pp. 245-248,2013.

[2] Shun-Lai Ma and Jeen-Sheen Row, "Design of Single-Feed Dual-Frequency Patch Antenna for GPS and WLAN Applicationss," IEEE Trans. Antennas Propag. ,vol.59, pp. 3433-3436, Sep.2011.

[3] C. M. Su and K. L. Wong, "A dual-band GPS microstrip antenna,"Microwave Opt. Technol. Lett., vol. 33, pp. 238–240, May 20, 2002.

[4] S. H. Chen, J. S. Row, and C. Y. D. Sim, "Single-feed square-ring patch antenna with dual-frequency operation," Microwave Opt. Technol. Lett., vol. 49, pp. 991–994, Apr. 2007.

[5] Xin Hu and Yuanxin Li, "Novel Dual-Frequency Microstrip Antenna With Narrow Half-Ring and Half-Circular Patch," IEEE Trans. Antennas Propag. ,vol. 12, pp.3-6,2013.

[6] R. Li, B. Pan, J. Laskar, and M. M Tentzeris, "A compact broadband planar antennas for GPS, DCS-1800, IMT-2000 and WLAN applications," IEEE Antennas and Wireless Propog. Letter,vol. 6, pp. 25-27,2007.

[7] C.Y.Pan ,T.S.Horng  and  W.S.Chen, "Dual wideband printed monopole antenna for WLAN/WiMAX applications,"IEEE Antennas and Wireless Propog. Letter,vol. 6, pp. 149-151, 2007.

[8] X. L. Bao and M. J. Ammann, "Dual-frequency circularly-polarized patch antenna with compact size and small frequency ratio," IEEE Trans. Antennas Propag. , vol. AP-55, pp. 2104–2107, Jul. 2007.

# The design of an ultra-wide band spiral antenna

Li Zhi  Zhang Xi-yu  Sun Quan-guo
Southwest Institute of Electronic Equipment
Chengdu 610036 P. R. China

*Abstract-* **A new kind of spiral antenna is designed which combined the planer spiral antenna with the helical antenna, Ultra-wide band (bandwidth 25:1), miniaturized size, good radiation pattern and circular polarization can be realized by this method.**

## I. INTRODUCTION

As the development of electronic warfare, the research work on miniaturized wide-band circular polarization antenna techniques has become the hot research point.

The traditional spiral antenna is composed of planer spiral antenna and helical antenna, the planer spiral antenna has the characteristic of low profile, wideband and circular polarization. The helical antenna has the characteristic of high gain. We combined them together. Then we can realize miniaturized, wide band, high gain and circular polarization antenna.

## II. THE DESIGN OF THE ANTENNA

### A. The Design of Spiral

In this paper we use Archimedes spiral to realize the radiation of high frequency. According to antenna theory, the size of spiral is decided by the low frequency. Usually the arm of the outermost circle will be almost a wavelength. So we can figure out the size of the planer spiral antenna, if the antenna worked at the lowest frequency with wavelength $\lambda_{max}$, use the formula (1).

$$D \approx \frac{\lambda_{max}}{\pi} \tag{1}$$

We can easily figure out the aperture of the antenna. if we use rippled spiral. We can minimize the size to 2/3 or more. But it was still too large for the application.

The antenna contains two parts, the Archimedes spiral and the helical antenna.

Firstly we designed the Archimedes spiral. According to the simulation and optimization, the diameter of the spiral is 60mm. the planer Archimedes spiral is shown in figure 1.It worked in high frequency.

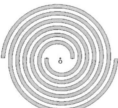

Figure 1. The planer Archimedes spiral

Secondly we designed the part which worked in low frequency. The use of helical [5] can reduce the aperture of the antenna. Figure 2 shows the helical antenna.

Figure 2. The helical antenna

Then we combined planer spiral with helical. The aperture of the antenna is reduced to 3/5 in compare with the planer spiral antenna of the same size. The wide band and miniaturized size is realized.

### B. The Design of Balun

To match the balance structure and the wideband character of the antenna [1], We were used to choose the traditional Marchand balancer to be the feed circuit in traditional planer spiral antenna design, but it is hard to achieve ultra-wide band (bandwidth 25:1).the structure of Marchand balancer is complex and consist a lot of different part, which make it hard to machining. The microstrip line-parallel wire balancer has the character of wideband and the structure of the balancer is simple, So excellent coherence and low cost can be easily realized. In this paper we choose microstrip line-parallel wire balancer to be the feed circuit.

The antenna is fed by 50ohm coaxial-cable. We need to figure out the output impedance of the balancer, if the output impedance of the balancer is equal to the input impedance of the Archimedes spiral or almost equal. Then we realized the matching circuit [3].

Figure 3 shows the model of microstrip line-parallel wire balancer [3].

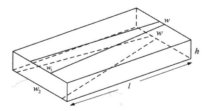

Figure 3. The microstrip line-parallel wire balancer

978-7-5641-4279-7

## C. The Design of Back Cavity and Absorbing Material

In order to achieve unidirectional beam, a back cavity must be mounted to the spiral antenna, the cavity may be a hollow metal cylinder. The depth is equal to the quarter wavelength of the central frequency of the operating band. But its bandwidth is restricted [2], so we need to add absorbing material to the cavity to realize broad bandwidth and unidirectional radiation, in this paper the helical part can be used as the cavity and underprop of the planer spiral, as the design and simulation, the depth of the cavity is about 60mm. Then absorbing materials is added to the cavity to absorb the reflected energy which can destruct the forward beam. A multilayer absorbing material is designed to realize ultra-wide band wave absorbing [4], so tune the parameter of the planer spiral, helical and the absorbing material, we can adjust the resonance frequency and the bandwidth easily.

Figure 4 shows the back cavity and absorbing material of the antenna.

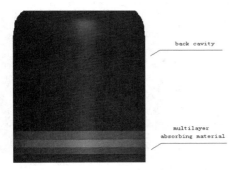

Figure 4. The back cavity and absorbing material of the antenna

### III. THE SIMULATION AND TEST RESULT OF THE ANTENNA

#### A. The Simulation of The Antenna

The performance of the antenna can be simulated in the software HFSS. We put the frequency changing parameter into HFSS which make the simulation more accuracy. At last we compare the simulation result to the test result, they are almost the same.

#### B. The Radiation Pattern of The Antenna

In figure 5 "h" represent normalized radiation pattern for Horizontal Polarization, and "v" represent the normalized radiation pattern for vertical Polarization, "BW" represent the bandwidth of the antenna .

From the figure we can see that the radiation patterns for Horizontal Polarization and vertical Polarization are basically the same.

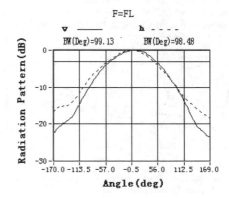

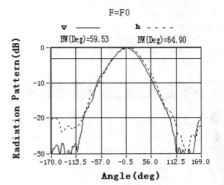

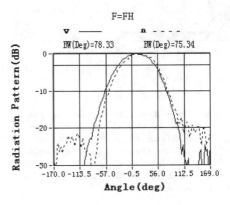

Figure 5. Measured radiation pattern for Horizontal and vertical Polarization at the low, central and high frequencies.

*C. The Measured Axial Ratio of The Antenna*

Figure 6 shows that both the simulation and test axial ratio of the antenna is less than 3 in wide band (bandwidth 25:1).

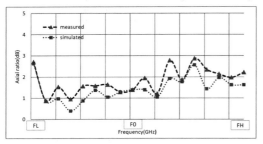

Figure 6. Measured and Simulated axial ratio of circular polarization

*D. The Measured Gain of The Antenna*

The gain of the antenna is in the range of -6dB to 5dB in wide band (bandwidth 25:1).according to antenna theory the antenna realized higher gain than the traditional planer spiral antenna of the same size.

*E. The Measured VSWR of The Antenna*

Figure 7 shows that the VSWR of the antenna is less than 3 in wide band (bandwidth 25:1).

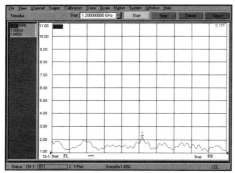

Figure 7. Measured VSWR of the antenna

According to the test result, we can see the antenna has good radiation pattern, low VSWR and axial ratio.

## IV. CONCLUSION

In order to meet the demand of ultra-wide band property of miniaturized antenna, an ultra-wide band spiral antenna is designed and manufactured which achieved 38%-0ff aperture in comparison with conventional Archimedean spiral antenna. The subtle approach of combing the planer spiral with the helical antenna make the antenna presents good radiation property and ultra-wide band character.

## ACKNOWLEDGMENT

Thanks Gong Xiao-dong for his good advices during the researching days. Thanks Zuo Le, Luo Xu and Kang Ze-ling for the helpful suggestions during the simulation and test days.

## REFERENCES

[1] SONG Zhao-hui QIU Jing-hui ZHANG Sheng-hui LIU Zhi-hui and YANG Cai-tian, "Study of A Equiangular Spiral Antenna and the Relevant Wideband Balun" Guidance and Fuze, vol. 24, no.2, pp. 36-39, Jun. 2003.
[2] Baixiao Wang, Aixin Chen "Design of an Archimedean spiral antenna" 2008 8th International Symposium on Antennas, Propagation and EM Theory, pp.348-351 Nov. 2008.
[3] Zhenhua Chen,Qunsheng Cao"Study of a two-arm sinuous antenna and t he relevant wideband balun"2008 International Conference on Microwav e and Millimeter Wave Technology, pp.1837-1840, April.2008.
[4] Nahid Rahman, Anjali Sharma, Mahmut Obol, Mohammed Afsar, Sande ep Palreddy and Rudolf Cheung,"Broadband absorbing material design a nd optimization of cavity-backed, two-arm Archimedean spiral antennas" 2010 IEEE Antennas and Propagation Society International Symposium, pp.1-4, July.2010.
[5] Hao Fang and Yaxian Liu, "Design of normal axis helical antennas", 2011 2nd International Conference on Artificial Intelligence, Management Science and Electronic Commerce (AIMSEC),pp.1054-1056, Aug.8.

# A Wide Bandwidth Circularly Polarized Microstrip Antenna Array Using Sequentially Rotated Feeding Technique

YU Tongbin[1], LI Hongbin[2], ZHONG Xinjian[1], YANG Tao[1], ZHU Weigang[1]

1. Institute of Communications Engineering, PLA University of Science and Technology, Nanjing 210007, China;
2. The First Engineers Research Institute of the General Armaments Department, Wuxi 214035, China

Abstract-A wide bandwidth circularly polarized microstrip antenna array using sequentially rotated feeding technique is presented in this paper. First a stacked microstrip antenna with tuning stub is designed, which meets wide bandwidth application. Then sequentially rotated feeding technique is introduced, basing on which a 4 element stacked microstrip antenna array with preferable axial ratio performance is achieved. The corresponding antenna array is fabricated and measured, the results of which shows that the antenna has the advantages of wide bandwidth, high gain and low axial radio, which has good application foreground.

## I. INTRODUCTION

Due to light weight and low profile, Circularly polarized microstrip antenna has been widely used in the satellite communications. However, traditional microstrip antenna usually has a narrow bandwidth that is less than 3% [1-3], which limits the development and application of the microstrip antenna. Recently, stacked microstrip antennas have been widely investigated and used, because these antennas usually have a wide bandwidth, high gain and simple structure [4-6]. What's more, to improve the axial radio performance of the antenna array, sequentially rotated feeding technique is researched in this paper, which can improve the polarization purity without introducing other complex elements [7-9].

Basing on the stacked microstrip antenna element and the sequentially rotated feeding technique, a novel 4 elements microstrip antenna array is designed and fabricated. By the help of HFSS10 software, the array has been optimized, which impedance bandwidth achieves 54.5% from 1.6 GHz to 2.8GHz, and 3dB gain bandwidth achieves 19% from 1.9 to 2.3GHz, while its maximal gain is 13.5dB. Across the 3dB gain bandwidth, axial ratio of the array is less than 1.5dB. The array has been measured in the anechoic chamber, and good agreement is obtained between the simulated and measured results. The wide bandwidth and the simple structure make the designed antenna meet many applications in the modern satellite communication.

## II. ELEMENT DEGIGN

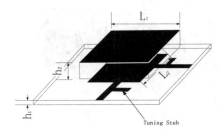

Figure 1. The configuration of element

The stacked microstrip antenna has been used as the array's element, as shown in Figure1. the element has two radiating patches, which are the upper and the lower patch. The lower patch whose side length is denoted $L_1$ works as inspiring element, while the upper patch whose side length is denoted $L_2$ works as parasitical element. The height between the two patches is $h_2$. Two side microstrip line feeds along orthogonal directions with 90° phase difference are used to imply circular polarization. As shown in Figure1, the feeding lines and the inspiring patch are placed at the same surface of the substrate, which relative permittivity is 2.2 and height denoted to be h1 is 0.8mm. The tuning stub is a short open 50 Ω micrsrip transmission line, which length and distance from the lower patch can be used to optimize the impedance, as a result, the height between the upper and lower patches can maintain as a small value. With the help of HFSS10 software, the parameters have been optimized and validated, as L1=58, L2=46, h2=6mm. To obtain good impedance, the length of the tuning stub is tuned to be 8mm, while the distance between the lower patch and the tuning stun is 2.5mm.

## III. SEQUENTIAL FEEDING DESIGN

Sequential rotation of circularly polarized array feeding involves applying both a physical rotation to the element feed point and an appropriate phase offset to the element. And it can be used to both the odd and even arrays[7].Basing on the stacked circularly polarized antenna designed above and the sequentially rotated feeding technique, 4 element circularly polarized microstrip antenna array is designed, which feeding network is shown in Figure 2. The Wilkinson power divider has been used here to provide two feeding ports that have equal magnitude and 90° phase difference. The outside length of the ground plane denoted to be Lg is 20cm and the space

978-7-5641-4279-7

between the patches denoted to be d is 90mm(about 0.6λ, λ is the free space wavelength). As seen in figure 2, the elements of the array is rotated by 90° one by one, while the feeding phase in turn is also changed 90°. So the Sequential rotation of array is achieved.

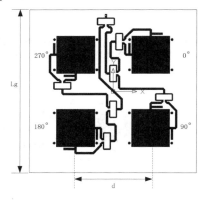

Figure 2 The structure of the feeding network

## IV. MEASUREMENTS

To verify the technique used above, the sequentially rotated feeding mirostrip antenna array has been fabricated in this paper. First, the return loss ($S_{11}$) plots of array is measured, as shown in Figure3. It can be seen that, the measured and simulated results have some difference, which probably owns to the SMA interface and loss of the substrate. Still the measured bandwidth defined by $S_{11}$ less than -10dB is about 25% from 1.6 to 3GHz . The measured result is a little better than the simulated ones.

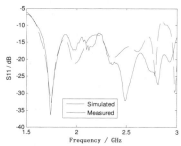

Figure 3 The simulated and measured $S_{11}$

Then the gain, the axial radio (AR) and the radiation pattern of the array has been measured in the anechoic chamber. Figure 4 shows the simulated plots and measured points of the gain and AR against frequency. It can be seen that, the antenna's gain around 2.1GHz is a little smaller the simulated ones, and the axial radio around 2GHz is greater. The reason is likely the tolerance of the fabrication, which is similar to the simulated result. From figure 4, the measured gain band width

defined by 1 dB is 14.3% from 1.95GHz to 2.25GHz while its maximum gain is 13.2dB at 2.2GHz. Within the 1 dB gain bandwidth, the measure AR is below 2 dB.

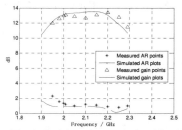

Figure 4 Tne simulated plots and measured points of the gain and AR against frequency

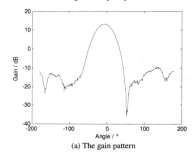

(a) The gain pattern

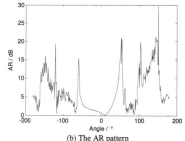

(b) The AR pattern
Figure 5 The measured far-field radiation patterns at 2GHz

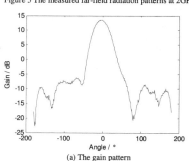

(a) The gain pattern

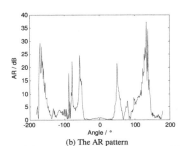

(b) The AR pattern

Figure 6 The measured far-field radiation patterns at 2.2GHz

Figure 5 and Figure 6 gives the measured far-field radiation patterns at the frequency 2GHz and 2.2GHz. It can be seen the radiation patterns have good symmetric characteristic and the axial radio is small, which are profited from the sequentially rotated feeding technique.

The measured results given above make out that the performance of the microstrip antenna array can be notably improved by using the stacked antenna element and sequentially rotated feeding technique. The broadband impedance bandwidth is achieved by making the lower patch and the upper patch work at near frequency. The sequentially rotated feeding technique improves the polarization purity and the symmetric characteristic of the radiation pattern. The corresponding microstrip antenna array is shown in figure 7.

Figure 7 The fabricated antenna

## V. CONCLUSION

A 4 elements stacked microstrip antenna array using sequentially rotated feeding technique is presented in this paper. The measured results approve that by using stacked microstrip antenna, the broadband impedance bandwidth and high gain have been obtained. And the sequentially rotated feeding technique can improve the polarization purity for sure.

References

[1] H.Iwasaki.'A circularly polarized small-size microstrip antenna with a cross slot', IEEE Trans. Antennas Propagat., vol. 44,no.10, Oct. 1996: 1399-1401 .

[2] Chia-Luan Tang, Jui-Han Lu, Kin-Lu Wong.' Circularly polarised equilateral-triangular microstrip antenna with truncated tip',Electron.Lett., vol. 34,no.13, June.1998: 1277-1278 .

[3] K.L.Wong,Y.F.Lin.' Circularly polarised microstrip antenna with a tuning stub', Electron.Lett., vol. 34,no.9, April.1998: 831-832.

[4] Nishiyama, E.; Aikawa, M.; Egashira, S.' Stacked microstrip antenna for wideband and high gain'. IEE Proc.-Microw. Antennas Propag., Vol. 151, No. 2, April 2004:143-148.

[5] S. Egashira, E.Nishiyama, 'Stacked microstrip antenna with wide bandwidth and high gain'. IEEE Trans. Antennas Propagat., vol. 44, no.11.pp.605–609, nov.1996:1533-1534.

[6] Nasimuddin, K. P. Esselle, and A. K. Verma.'Wideband Circularly Polarized Stacked Microstrip Antennas'. IEEE Antennas and Wireless Propagation Letters.vol.6.no.A.2007: 21-24.

[7] P.S.Hall,'Application of sequential feeding to wide bandwidth, circularly polarised microstrip patch arrays'. IEE Proc.H. vol.136.no.5.Oct. 1989 :390-398

[8] W.K.Lo, C.H.Chan, K.M.Luk,' Circularly polarised microstrip antenna array using proximity coupled feed'. Electronics Letters.vol.34.no.23. Nov. 1998: 2190-2191

[9] W.K.Lo, C.H.Chan, K.M. Luk, 'Circularly polarised patch antenna array using proximity-coupled L-strip line feed', Electron.Lett., vol. 36,no.14, July 2000: 1174-1175 .

# Equivalent Radius Analytic Formulas of Substrate Integrated Cylindrical Cavity

Luan Xiu-zhen , Tan Ke-jun
School of Information Science & Technology
Dalian Maritime University, Dalian, China

*Abstract*—Equivalent radius of a Substrate Integrated Cylindrical Cavity (SICC) is an important parameter for calculating the resonant frequency and designing a SICC device. In this paper, based on two sets of equivalent width analytic formulas of Substrate Integrated Waveguide (SIW), the equivalent radius analytic formulas of SICC were derived by the conformal transformation method for the first time. In order to verify the validity of these formulas, using the formulas given in this paper calculated the equivalent radii and the corresponding resonant frequencies of $TM_{010}$ mode SICCs , and the resonant frequencies of $TM_{010}$ mode SICCs were also calculated by the method introduced in [12] and simulated by electromagnetic simulation software. The results show that the formulas given in this paper have higher precision, and are convenient, suitable to more application fields, so they will play important role in the analysis and design of SICC devices.

*Index Terms—substrate integrated cylindrical cavity; resonant frequency; conformal transformation method; equivalent radius.*

## I. INTRODUCTION

In the last years, the Substrate Integrated Waveguide (SIW) was studied widely. SIW is a type of dielectric-filled rectangular waveguide that is synthesized in a planar substrate with two linear arrays of metallic vias. By comparison with conventional metallic rectangular waveguide, SIW has the characteristics of compact bulk, planar structure, easy to fabricate, etc., is very suitable to integrated circuits. Presently, a large number of SIW-based devices have been implemented, and many of them, like filters, oscillator and power divider, are based on the substrate integrated cavities [1]-[8] . In conventional metallic waveguide cavity, resonant frequency can be determined in closed form for canonical shape. In [1]-[2], the cavities are made of substrate integrated rectangular waveguides, their resonant frequencies can be determined by using the similar methods of conventional metallic rectangular cavities, because the equivalent width analytic formulas of SIW have been given [9]-[10] . In [3]-[8], the oscillator, filters and power divider are made on the substrate integrated cylindrical cavities. Substrate Integrated Cylindrical Cavity (SICC) is a cylindrical cavity with sidewall of metallic vias instead of solid metallic sidewall connecting the solid metallic top and bottom plates，

as shown in Fig.1. For SICC, the analytic calculation of the resonant frequency can be cumbersome, because the equivalent radius analytic formula of SICC has not been given at present. In [11], nonlinear eigenvalues method was adopted to analyze arbitrarily shaped substrate integrated waveguide resonators. In [12], the resonant frequency determination methods of $TM_{010}$ mode and $TM_{110}$ mode circular substrate integrated resonators are given. In this paper, based on two sets of equivalent width analytic formulas of SIW given in [9] and [10], two sets of analytic formulas for calculating the equivalent radius of SICC are derived by using the conformal transformation method, respectively. Using the equivalent radius analytic formulas given in this paper, the resonant frequencies of SICC can be determined easily. As an example, the equivalent radii and the corresponding resonant frequencies of $TM_{010}$ mode SICCs are calculated using the formulas given in this paper, and the resonant frequencies are compared with the results calculated by the method given in [12] and the simulated results of electromagnetic simulation software.

## II. ANALYTIC FORMULAS FOR CALCULATING THE EQUIVALENT RADIUS OF SUBSTRATE INTEGRATED CYLINDRICAL CAVITY

The geometry of SICC is shown in Fig.1. Similar to SIW can be equivalent to solid metallic sidewall rectangular waveguide filled with dielectric, the SICC can be equivalent to a solid metallic sidewall cylindrical cavity if the radius $R$ of SICC is replaced by the equivalent radius $R_{eq}$ .Conformal transformation method was adopted to derive the equivalent radius analytic formula of SICC in this paper. Fig.2 shows the transformation process in detail.

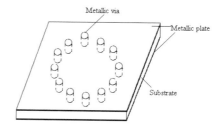

Fig.1 Geometry of the substrate integrated cylindrical cavity

978-7-5641-4279-7

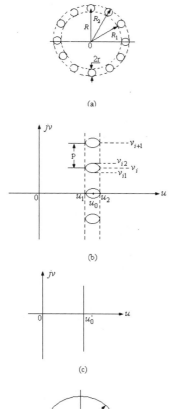

(a)

(b)

(c)

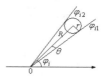

(d)

Fig.2 Transformation process of

substrate integrated cylindrical cavity

In Fig.2 (a), the number of metallic vias is $n$, the radius of metallic via is $r$, the radius of SICC is $R$ , so the inner circle radius $R_1 = R - r$ , and the exterior circle radius $R_2 = R + r$ . Suppose Fig.2 (a) is in $Z$-plane, and $Z = |Z|e^{j\varphi}$ , then the transformation $w = \ln(Z) = \ln|Z| + j\varphi = u + jv$ transforms the $Z$-plane into $w$-plane, as show in Fig.2 （b）. In the $w$-plane , $u = \ln|Z|, v = \varphi$. So the circles which radii are $R_1$ and $R_2$ in $Z$-plane are transformed into beelines $u_1 = \ln R_1$ and $u_2 = \ln R_2$

in $w$-plane, respectively. The circular array of metallic vias in $Z$-plane is transformed into linear array between $u_1$ and $u_2$ in $w$-plane.

In Fig.2 （b）

$$u_0 = (u_1 + u_2)/2 = \ln(R_1 R_2)/2$$
$$= \ln[(R-r)(R+r)]/2 \tag{1}$$

$$v_i = \varphi_i, \quad v_{i1} = \varphi_{i1}, \quad v_{i2} = \varphi_{i2} \tag{2}$$

Where, $\varphi_i = i\dfrac{2\pi}{n}(i = 1,2,\cdots,n)$ , $\varphi_{i1} = \varphi_i - \theta$ , $\varphi_{i2} = \varphi_i + \theta$ , and

$$\theta = \sin^{-1}(r/R) \tag{3}$$

The meanings of $\varphi_i, \varphi_{i1}, \varphi_{i2}$ and $\theta$ are shown in Fig.3.

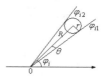

Fig.3 Geometry parameters of metallic via

In Fig.2 (b), the separation between adjacent metallic vias is

$$p = v_{i+1} - v_i = \varphi_{i+1} - \varphi_i = \dfrac{2\pi}{n} \tag{4}$$

There are three equivalent methods for metallic via equivalent radius, that is

$$r_{eq} = \dfrac{1}{2}(u_2 - u_1) = \dfrac{1}{2}(\ln R_2 - \ln R_1) = \dfrac{1}{2}\ln(\dfrac{R+r}{R-r}) \tag{5a}$$

$$r_{eq} = \dfrac{1}{2}(v_{i2} - v_{i1}) = \theta = \sin^{-1}(\dfrac{r}{R}) \tag{5b}$$

$$r_{eq} = \dfrac{1}{2}[\dfrac{1}{2}\ln(\dfrac{R+r}{R-r}) + \sin^{-1}(\dfrac{r}{R})] \tag{5c}$$

It can be proven that the results calculated from the three formulas are nearly equal. It indicates that metallic via is also cylindrical in $w$-plane.

The metallic vias array in Fig.2 (b) can be equivalent into solid metallic plane, as shown in Fig.2(c). In this paper, two sets of equivalent width analytic formulas of SIW given in [9] and [10] are used to derive the position of the equivalent solid metallic plane, respectively.

The inner area of circular metallic vias array in Fig.2 (a) is transformed into the area of $u < u_1$ in Fig.2 (b). Suppose another similar metallic vias array is at $u = -u_0$, then the two metallic vias arrays compose a SIW. So in Fig.2 (c), according to [9], the equivalent solid metallic plane position $u_0'$ can be determined by the following formulas

$$u_0' = a_{eq}/2 \tag{6}$$

Where $a_{eq}$ is the equivalent width of the corresponding SIW, it can be determined by

$$a_{eq} = 2u_0 \cdot \overline{a} = \ln[(R-r)\cdot(R+r)]\cdot\overline{a} \tag{7}$$

In which, $\overline{a}$ is the normalized equivalent width, it is determined by [9]

$$\bar{a} = \xi_1 + \frac{\xi_2}{\dfrac{p}{2r_{eq}} + \dfrac{\xi_1 + \xi_2 - \xi_3}{\xi_3 - \xi_1}}$$

$$= \xi_1 + \frac{\xi_2}{\dfrac{2\pi}{n\ln(\dfrac{R+r}{R-r})} + \dfrac{\xi_1 + \xi_2 - \xi_3}{\xi_3 - \xi_1}} \qquad (8)$$

Where

$$\xi_1 = 1.0198 + \frac{0.3465}{\dfrac{2u_0}{p} - 1.0684}$$

$$\xi_2 = -0.1183 - \frac{1.2729}{\dfrac{2u_0}{p} - 1.201} \qquad (9a)$$

$$\xi_3 = 1.0082 - \frac{0.9163}{\dfrac{2u_0}{p} + 0.2152}$$

That is

$$\xi_1 = 1.0198 + \frac{0.3465}{\dfrac{n\ln[(R-r)\cdot(R+r)]}{2\pi} - 1.0684}$$

$$\xi_2 = -0.1183 - \frac{1.2729}{\dfrac{n\ln[(R-r)\cdot(R+r)]}{2\pi} - 1.201} \qquad (9b)$$

$$\xi_3 = 1.0082 - \frac{0.9163}{\dfrac{n\ln[(R-r)\cdot(R+r)]}{2\pi} + 0.2152}$$

According to [10], the equivalent solid metallic plane position $u_0'$ can also be determined by

$$u_0' \approx u_0 + \frac{p}{4}\ln(\frac{p}{4r_{eq}})$$

$$= \ln\left\{ \sqrt{(R-r)\cdot(R+r)} \cdot [\frac{\pi}{n\ln(\dfrac{R+r}{R-r})}]^{\frac{\pi}{2n}} \right\} \qquad (10)$$

Through the transformation $Z' = e^w$, then the solid metallic plane $w = u_0'$ in Fig.2(c) is transformed into the solid metallic cylindrical sidewall, its radius (that is the equivalent radius of SICC) is

$$R_{eq} = e^{u_0'} \qquad (11)$$

As shown in Fig.2 (d).

The equivalent radius of SICC can be calculated from (6)~(9) and (11), these formulas are derived from [9].

From (10) and (11), we can get another formula of the equivalent radius of SICC derived from [10], that is

$$R_{eq} = \sqrt{(R-r)\cdot(R+r)} \cdot [\frac{\pi}{n\ln(\dfrac{R+r}{R-r})}]^{\frac{\pi}{2n}} \qquad (12)$$

In the design of SIW, the via diameter is chosen to be equal or smaller than a tenth of the wavelength of the maximum operation frequency and the separation between adjacent vias is equal to or smaller than twice the diameter of the metallic via, so in the design of SICC, if corresponding to Fig.2(b) then

$$2r_{eq} = \ln(\frac{R+r}{R-r}) \le \frac{\lambda_{min}}{10}$$

$$p = \frac{2\pi}{n} \le 4r_{eq} = 2\ln(\frac{R+r}{R-r})$$

Where $\lambda_{min}$ is the minimum wavelength corresponding to the maximum operation frequency.

If corresponding to Fig.2 (a), then

$$r \le \frac{\lambda_{min}}{20}$$

$$R \cdot \frac{2\pi}{n} \le 4r$$

That is

$$\begin{cases} \dfrac{\pi}{n} \le \ln(\dfrac{R+r}{R-r}) \le \dfrac{\lambda_{min}}{10} \\ \dfrac{\pi R}{2n} \le r \le \dfrac{\lambda_{min}}{20} \end{cases} \qquad (13)$$

### III. CALCULATION OF SUBSTRATE INTEGRATED CYLINDRICAL CAVITY RESONANT FREQUENCY AND EQUIVALENT RADIUS

Conventional solid metallic sidewall cylindrical cavities usually work at $TE_{11p}$, $TM_{01p}$ and $TE_{01p}$ modes. $TM_{010}$ mode SICC can be realized in substrate easily, and can be excited by SIW, so in this paper, the $TM_{010}$ mode SICC is studied.

In the design process of cavity, the resonant frequency is a fundamental parameter needed to be determined accurately. The resonant frequency $f_0$ of $TM_{010}$ mode solid metallic sidewall cylindrical cavity can be determined by the cavity radius $R$, that is

$$f_0 = \frac{c}{2.62R\sqrt{\varepsilon_r}} \qquad (14)$$

Where, $c$ is the light velocity in air, $\varepsilon_r$ is the relative permittivity constant of dielectric substrate. Formula (14) can also be used to calculate the resonant frequency of $TM_{010}$ mode SICC if the cavity radius $R$ is replaced by the equivalent radius $R_{eq}$ of SICC. $R_{eq}$ can be calculated by the formulas (6)~(9), (11) or (12) derived in the above section.

In order to comparison, with the help of electromagnetic simulation software HFSS, the resonant frequencies of $TM_{010}$ mode SICC are determined by the calculation of eigenmode. Otherwise, the resonant frequencies are also calculated by the method given in [12] and by the formula (14) with the actual radius $R$ of SICC, as shown in Fig.2(a).

Fig.4 shows the curves of resonant frequency $f_0$ versus the via radius $r$ of SICC when the SICC radius $R$=8mm. In which, "**" denote the simulated results by HFSS; "---"denote the calculated results from the formulas (6)~(9), (11) and (14) ; "oo" denote the calculated results from the formulas (12) and (14) ; "—" denote the calculated results using the method given in [12]; "++" denote the calculated results from the formula (14) and the actual radius $R$ of SICC.

It can be seen from Fig.4 that the resonant frequencies calculated from the formulas (6)~(9), (11) , (14) and calculated by the method given in [12] are close to the simulated resonant frequencies, they are relatively accurate, and they have almost the coequal precision ; the resonant frequencies calculated from the formula (14) and the actual SICC radius $R$ have the maximal errors compared to the simulated results, are inaccurate, so the actual radius $R$ of SICC cannot be simply used to calculate the resonant frequencies of SICC, should be replaced by the equivalent radius $R_{eq}$ of SICC.

Fig.5 shows the equivalent radius curves versus actual radius of SICC. The equivalent radii are calculated from formulas (6~9) and (11). It can be seen from Fig.5 that the equivalent radius increases with the decrease of via radius. This can be indicated that when the via radius decreases, the gap between adjacent vias increases when the number of vias and radius of SICC are unchangeable, then the electromagnetic field distribution space in SICC is extended, so the equivalent radius increases.

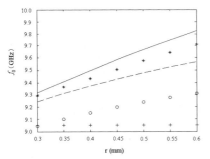

Fig. 4  Resonant frequency curve of SICC versus via radius

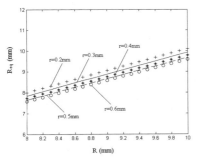

Fig. 5  Equivalent radius curve versus actual radius of SICC

## IV. CONCLUSION

In this paper, two sets of analytic formulas for calculating the equivalent radius of SICC are derived by the conformal transformation method based on two sets of equivalent width analytic formulas of SIW, respectively. In order to comparison, the resonant frequencies of $TM_{010}$ mode SICCs are calculated from the two sets equivalent radius , actual radius and by the

method given in [12], these results are compared with the simulated results obtained from electromagnetic simulation software HFSS. The results show that the formulas (6)~(9) , (11) given in this paper and (14) are more precision, similar to the method given in [12]. However, the formulas given in [12] are only for calculating the resonant frequencies of $TM_{010}$ mode and $TM_{110}$ mode SICCs. The formulas given in this paper are for calculating the equivalent radius of SICC. Equivalent radius is a more basic parameter compared with resonant frequency. Resonant frequency of any mode SICC can be calculated from the equivalent radius of SICC, and the other parameters of SICC can also be calculated from its equivalent radius, so the formulas given in this paper are suitable to more application fields, and they are handy, so they will play important role in the analysis and design of SICC devices.

## REFERENCES

[1] Xiao-Ping Chen and Ke Wu, "Substrate Integrated Waveguide Cross-Coupled Filter With Negative Coupling Structure," IEEE TRANSACTIONS ON MICROWAVE THEORY AND TECHNIQUES, VOL.56, NO.1, JANUARY 2008:142-149.

[2] F. Xu, X. Jiang and K. Wu. "Efficient and accurate design of substrate-integrated waveguide circuits synthesized with metallic via-slot arrays," IET Microw. Antennas Propag., 2008,2,(2): 188-193.

[3] Qiang Liu; Yang Yang; Kama Huang, "Design of X-band oscillator based on substrate integrated circular cavity," 2012 International Conference on Microwave and Millimeter Wave Technology (ICMMT), Volume: 1, 2012: 1 – 3.

[4] Jian Gu; Yong Fan; Dakui Wu , "A LTCC band pass filter based on half-mode SICC structure," 2010 International Conference on Microwave and Millimeter Wave Technology (ICMMT), 2010: 1245 - 1247

[5] Zhigang Zhang; Yong Fan; Yujian Cheng; Yonghong Zhang , "A compact multilayer dual-mode substrate integrated circular cavity (SICC) filter," 2011 China-Japan Joint Microwave Conference Proceedings (CJMW), 2011: 1 – 4.

[6] Shaolun Jiang, Qike Chen and Yong Fan, "A Compact Ku-Band filter based on Substrate Integrated Circular Cavity," International Symposium on Intelligent Signal Processing and Communication System, December, 2010:1-3.

[7] Jian Gu, Yong Fan, Yonghong Zhang, "A Low-Loss SICC Filter Using LTCC Technology for X-Band Application," 2009 IEEE International Conference on Applied Superconductivity and electromagnetic Devices, September 25-27, 2009: 152 - 154.

[8] Dakui Wu, Yong Fan, and zongrui He. "Vertical Transition and Power Divider Using substrate Integrated Circular Cavity," IEEE MICROWAVE AND WIRELESS COMPONETS LETTERS, VOL.19, NO.6, JUNE 2009:371-373.

[9] L. Yan, W. Hong, "Investigations on the propagation characteristics of the Substrate Integrated Waveguide based on the Method of Lines," IEE Proceedings-H: Microwaves, Antennas and Propagation, vol.152, NO.1,2005:35-42.

[10] W. Che, K. Deng, D. Wang and Y.L. Chow. "Analytical equivalence between substrate-integrated waveguide and rectangular waveguide," IET Microw., Antennas Propag., 2 ,(1), 2008:35-41.

[11] Giovanni Angiulli, Emilio Arnieri, Domenico De Carlo, and Giandomenico Amendola. "Fast Nonlinear Eigenvalues Analysis of Arbitrarily Shaped Substrate Integrated Waveguide (SIW) Resonators," IEEE TRANSACTIONS ON MAGNETICS, VOL. 45, NO.3, MARCH 2009:1412-1415.

[12] Giandomenico Amendola, Giovanni Angiulli, Emilio Arnieri, and Luigi Boccia, "Resonant Frequencies of Circular Substrate Integrated Resonators," IEEE MICROWAVE AND WIRELESS COMPONENTS LETTERS, VOL. 18, NO.4, APRIL 2008: 239-241.

# Analysis about the Influence of Terrain on the Fair-weather Atmospheric Electric Field Measurements

Jun Liu,  Jia-qing Chen,  Li-zhi Yang,  Xiang-yu Liu

PLA Univ. of Sci. & Tech. Nanjing Jiangsu 211101, China

**Abstract:** When the field mill is located in complex terrain, such as depression and mountain, the measuring error caused by terrain is always difficult to correct, which will lead the deficiency of observation data about comparability and consistency. Aiming at this problem, this paper conducts a simulation with Maxwell 3D software, thereby discusses the distribution of ground atmospheric electric field on the depression and mountain, as well as the influence area of both terrains. The results have a good reference value to the installation and error correction of the field mill.

## I  INTRODUCTION

As the common equipment for atmospheric electric field detection, field mill can be used in short-term thunderstorm warning by measuring intensity and polarity of the electric field, and the measured date also has great significance in the study of fair weather electricity, thunderstorm electricity and flashing lightning [1, 2]. During the observation, distinct differences always exist in the measuring results of different observation sites and field mills, owing to the influence of instrument and installation environment [3]. If the field mill is located close to tall objects (i.e., buildings, big trees, towers), the field mill reading will be lower due to the shadowing effect of these objects .On the other hand, If the field mill is mounted above the ground or installed at elevated objects, it will suffer from field enhancement [4, 5]. Nowadays, parallel plate calibration [6] and improved field mill component's performance are the preferred methods to improve the veracity of the detective data, but there are no effective correction methods for the error correcting caused by the installation environment, especially when the field mill located in depression and mountain.

Former researchers have thought that, an enhancement factor of 2.75 due to mountain itself with reference to the mountains surrounding terrain [7]. And the influence can be ignored if the distance between observation point and depression (mountain) was 5 times more than the vertical scale of depression (mountain) [8]. However, influenced by various factors, the above conditions are difficult to meet during actual installation [9]. Therefore, in order to provide the reference for the installation and error correction, this paper would analyze the distribution of ground electric field,

and the influence area of both terrains with Maxwell 3D software.

## II  CALCULATION MODEL

As the theory of parallel plate electric field calibration, during the simulation calculation, we have firstly used two pieces of parallel-plate to produce the vertical downward uniform electric field $E_0$, and the upper plate is set with high potential, the bottom plate and model are set the potential of 0 V. Then, simplified the depression (mountain) into symmetrical ideal model, which is located in the center of the bottom plate. Half length of the plate is 10 times as long as the radius of model, in order to reduce the error caused by the plate's edge, seen as figure 1, the two plates space are 10 times as the height of depression (mountain). The model and the bottom plate are both taken the soil, and relative dielectric constant of soil is set at 10, conductivity at 0.05 S/m, medium between the plates and upper plate are both set as air [10, 11]. Finally, the value of background electric field is set at 200

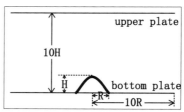

Figure 1.Simplified model for the mountain

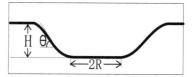

Figure 2.The vertical section of the depression's Simplified model

V/m, considering the influence of aerosols on the atmospheric electric field. Dirichlet boundary condition is adopted and the iteration process would stop while the energy error comes to

978-7-5641-4279-7

0.2%. Finally we choose the electric field above the ground at 1m to analysis and discuss.

In addition, we have assumed that the depression is made up of two parts: the circular flat bottom and the surrounding slope. Its structure is shown in *figure 2*. The radius of the bottom is $R$, the vertical distance between bottom and level ground is H, and the gradient of slope is θ.

## III SIMULATION RESULT ANALYSIS

*A. Distribution of ground atmospheric electric field on the mountain*

Nowadays, the widely used field mill generally detect only on the vertical direction of the atmospheric electric field [10]. However, the fine atmosphere electric field has not only the vertical component and also the horizontal one, due to the uneven isopotential surface causing by the fluctuation with the terrain. So we would decompose the simulation field data into vertical component ($E_v$) and horizontal component ($E_h$).

As shown in the figure 3, Electric field intensity (E) at the foot of the mountain is smaller than $E_0$. Along with the mountainside to a certain height, E begins to be greater than $E_0$, and the maximum value is located at the top, about 2.5 times of $E_0$. That would give an illustration that, due to the existence of mountain, shadowing effect of electric field obviously existed at the foot of the mountain, while the top has been significantly enhanced. In addition, influence is relatively smaller in the middle to lower part.

In order to make a contrastive analysis , the paper sets up three mountain models with the same height and gradient of 14°(H:R=2:8)、21.8°(H:R=2:5)、33.7°(H:R=2:3). It can be found from *figure 4* that, the distribution of electric field corresponding to the mountains with different gradient are also different, and the horizontal component continues to increase with the augment of the gradient. But the vertical electric field component at the foot of the mountain showed a trend of decrease with the augment of the gradient. Demonstrates that the greater the gradient, the stronger the

Figure 3.Atmospheric electric field on the mountain and around the mountain

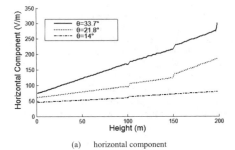

(a)    horizontal component

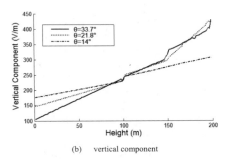

(b)    vertical component

Figure 4.Atmospheric electric field on the mountain with different gradient

shadowing effect on the foot electric field. However, the vertical component on the top has no obvious rule. The increase of the gradient may do not necessarily contribute to the increase of vertical component, while the distortion of atmospheric electric field becoming more serious. The increase of electric field on the top may be mainly caused by the horizontal component. As a matter of fact, the distribution of the ground electric field is more complex, due to the uneven mountain's surface.

*B. Distribution of ground atmospheric electric field in the depression*

In order to study the distribution of ground atmospheric electric field in the depression, three models of depression have been established , as the slope height H=60m with the slope gradient θ=26.7° (depression A), H=40m with θ=26.7° (depression B), and H=40m with θ=18.4° (depression C) , all the models' bottom radius R=100m.

As shown in *figure 5*, abscissa stands for the distance to the center of bottom, and the bottom is mainly composed of $E_v$, approximately thinking that E is equal to $E_v$. However, E is always smaller than $E_0$. Along the center to the slope, E firstly increases, and then decreases, the minimum is at the junction between the bottom and the slope, after that, the $E_v$ and $E_h$ of the slope are both increasing. Directly reflect that, the existence of depression has weakened the surface electric field of the bottom and some areas lower on the slope.

At the same time, through the comparison of electric field distribution curves between depression A and B we have found that, the horizontal and vertical component of A are both smaller than that of B, and the bottom also appears the similar phenomenon, except for the part near the junction area, whose electric field is basically the same. Thus we can conclude that, the higher the height, the stronger the shadowing effect for the slopes with the same gradient.

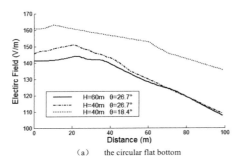

(a)　the circular flat bottom

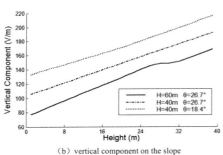

(b)　vertical component on the slope

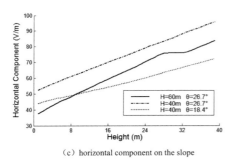

(c)　horizontal component on the slope

Figure 5. Atmospheric electric field in the depression

The same method is used to analyze the differences between depression B and C, which are with the same height

and different gradients, the distributions of electric field on them do exist difference, too. The electric field at the bottom of B is obvious smaller than that of C. As for the electric field on the slope , B is generally smaller than C, as the responding vertical component is smaller than C , while the horizontal is bigger than C. As a result, the increase of slope's gradient can also enhance the shadowing effect, and mainly cause the decrease of the vertical component.

Moreover, with the increase of R, some unaffected areas begin to appear in the center of the bottom, under the condition of keeping the H and θ invariable. If E is equal to95% $E_0$, we could assume that the atmospheric electric field would not be affected by terrain. Therefore, as shown in *Figure 6*, unaffected areas began to appear when the R increases to 300 m, and with the further increase of R, the unaffected area is also enlarging. The electric field in bottom may obtain the most influence, in case that R equals 0 m, and other conditions keep unchanged.

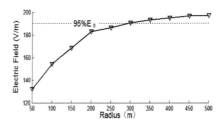

Figure 6. Atmospheric electric field in the center of the bottom along with the change of radius

## IV　INFLUENCE DISTANCE

The measuring accuracy of field mill generally depends on several factors, such as: instrument measurement precision, influence of terrain environment, influence of heavy weather and so on. Combined with the electric field instrument accuracy and the error of the simulation process itself, we could assume that the atmospheric electric field is not affected by terrain when E is equal to95% $E_0$. Thus, we can define the influence distance (D) as the distance between the level edge of the mountain (depression) and unaffected point (E=95% $E_0$), and the corresponding influence coefficient is λ (λ=D/H).

Table 1 shows that λ is almost smaller than 3 with the increase of gradient and the same height of the mountain. Thus, when the distance between observation point and level edge of mountain is more than 3 times compared the vertical height of mountain, the influence on the atmosphere electric field measure of the mountain can be ignored, that was different with the previous conclusion. Compared with the distortion of the electric field of depression, the effect of depression on the field around is not obvious. As it shown in table 2, with L being the distance between the observation point and level edge of depression, the effect on the electric field around the depression has not obvious deviated from $E_0$, besides a small area near the edge , and it is the same by    changing the height of the depression's slope.

Table 1.Influence distance of the mountains with different gradient

| θ (°) | 14 | 16 | 18 | 22 | 17 | 30 | 34 | 39 | 45 | 53 | 63 |
|-------|----|----|----|----|----|----|----|----|----|----|----|
| λ | 2.2 | 2.9 | 2.0 | 2.3 | 2.6 | 3.1 | 2.9 | 2.1 | 2.3 | 2.5 | 2.9 |

Table 2.Atmospheric electric field around the depression (H=100m、θ=45°)

| L (m) | 0 | 0.5H | 1.0H | 1.5H | 2.0H | 2.5H | 3.0H | 3.5H | 4.0H | 4.5H |
|-------|---|------|------|------|------|------|------|------|------|------|
| E (V/m) | 250.1 | 205.6 | 204.7 | 204.2 | 203.6 | 203.2 | 202.7 | 202.2 | 201.7 | 201.2 |

## V  CONCLUSION

In summary, the existence of mountain has an obvious shadowing effect on the electric field at the foot of the mountain, and the increase of the gradient of mountain can enhance the shadowing effect. And the existence of depression has weakened the surface electric field of the bottom and some areas lower on the slope; as for the slopes with the same gradient, the increase of the height would lead to the enhance of the shadowing effect; when comes to the slopes with the same height, the increase of the gradient would also lead to the enhance of the shadowing effect, while mainly cause the decrease of the vertical component.

When the distance between observation point and mountain's level edge is more than 3 times compared the vertical height of mountain, the influence on the atmosphere electric field measurements of mountain terrain can be ignored; as for the installation of field mill around the depression, getting away from the area adjacent to it is the only advice.

## REFERENCES

[1] C. T. R. Wilson, "The electric field of a thundercloud and some of its effects," *J.* Proc. Phys. Soc. London, vol.37(32), pp.1088-1478.

[2] J. Montanya, J. Bergas, B. Hermoso, "Electric field measurements at ground level as a basis for lightning hazard warning," *J.* Journal of Electrostatics, 2004, vol. 60, pp.241-246.

[3] Yang Zhong-jiang, Zhu Hao, et al. "Research on source of error and analytical processing of atmospheric electric field data," *J.* Trans Atmos Sci, 2010, vol.33(6), pp.751-756.

[4] A. Mosaddeghi, D. Pavanello, et al. "Effect of nearby buildings on electromagnetic fields from lightning," Journal of Lightning Research, 2009, vol.1, pp.52-60.

[5] Yoshihiro Baba, Vladimir A. Rakov, "Electric fields at the top of tall building associated with nearby lightning return strokes," Journals and Magazines, 2007, vol.49 (3), pp. 632-643.

[6] Fushan Luo, Yuhui He, et al, "Calibration method of electric field," *Chin. J. Space Sci.*, 2007, vol.27(3), pp.223-226.

[7] Helin Zhou, Gerhard Diendorfer, et al. "Fair-weather atmospheric electric field measurements at the Gaisberg Mountain in Austria," *J.* Piers Online, 2011, vol.7(2), pp.181-185.

[8] Weimin Chen," Lightning Principles," Bei Jing: China Meteorological Press, 2003.

[9] EFM-100 Atmospheric Electric Field Monitor Installation/Operators Guide.

[10]Guoqiang Liu, Lingzhi Zhao, Xuya Jiang, "Ansoft Engineering Electromagnetic Field Finite- element Analysis," Bei Jing: Electronic Industry Press, 2005.

[11] Bihua Zhou, Hui Jiang, et al, "Influence of surface features on atmospheric electric field near ground," *J.* Chinese Jouranal of Radio Science, 2010, vol.25 (5) pp.839-843.

# A Novel Reconfigurable Bandpass Filter Using Varactor-Tuned Stepped-Impedance-Stubs

Min Ou[#1] Yuhang He[#2] ,Liguo Sun[#3]

#Department of Electronic Engineering and Information Science,University of Science and Technology of China

Address: 443 Huangshan Rd, Hefei, Anhui, China, 230027

[1]oumin@mail.ustc.edu.cn
[2]vivahyh@mail.ustc.edu.cn
[3]liguos@ustc.edu.cn

*Abstract*-A novel reconfigurable bandpass filter using varactor-tuned stepped impedance resonator (SIR) is introduced to achieve tunable central frequency and bandwidth. The varactor-tuned stepped impedance resonator structure consists of a stepped impedance with open or short-circuited stub and a varactor between the stubs. It's convenient to change the transmission zeros with the appropriate DC voltage controlling the values of varactor capacitors. In addition, the ratios between the two sections of SIRs can be easily changed to reformulate the resonant frequencies and then more design freedom of filter is obtained. The experimental results validate the design.

## I. INTRODUCTION

With the rapid development of multiple frequency bands wireless communication system, the reconfigurable bandpass filters which can offer tunable center frequencies or bandwidths become necessary. Recently, many kinds of reconfigurable filters were reported. A lot of design approaches have been developed. For example，a new switching structure or resonator is proposed to realize reconfigurable center frequency[1]. However the insertion loss suffers from deterioration caused by complex structures. In [2], the bandwidth is adjusted with the assistance of varactors, and the position of transmission zeros is tuned by changing the DC voltage upon varactors. Nowadays, filters based on PIN diodes could switch the operating frequencies band between UWB and wireless LAN readily [3][4]. However, the designs discussed above provide a single method to achieve reconfigure center frequencies or bandwidths, independently.

In this paper a new reconfigurable filter using SIR structures and varactors is proposed. Center frequency and bandwidth are reconfigurable with similar structures in the design. The paper is organized as follows. Section II describes the theory of the proposed SIR structures and introduces the design of the reconfigurable SIR filter. And in Section III the experimental data are shown and compared with the simulation results. Finally, conclusions are given in Section IV.

## II. RECONFIGURABLE FILTER DESIGNS

*A. short-circuited SIR with varactors*

Fig.1.gives the general structure of a short-circuited

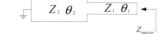

Fig. 1 The general structure of a short-circuited SIR

SIR, the input impedance of Fig. 1 can be derived from the following equation:

$$Z_i = jZ_1 \frac{Z_1 \tan\theta_1 + Z_2 \tan\theta_2}{Z_1 - Z_2 \tan\theta_1 \tan\theta_2} \quad (1)$$

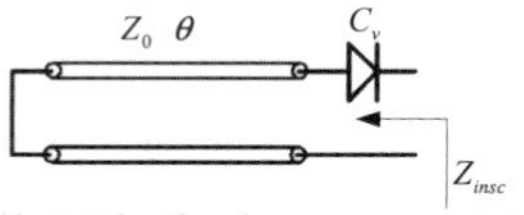

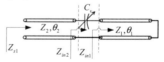

Fig. 2 Short-stubs of a microwave resonant structure

The input impedance corresponding to the structure displayed in Fig.2, is obtained from (2), and the transmission zero can be expressed as (3)

$$Z_{insc} = \frac{1 - \omega C_v Z_0 \tan\theta}{j\omega C_v} \quad (2)$$

$$1 - \omega C_v Z_0 \tan\theta = 0 \quad (3)$$

Where $C_v$ , $\omega$ , $Z_0$ and $\theta$ are the variable capacitance attributed to the varactor, the angular resonant frequency, the characteristic impedance, and the electrical length of the stub, respectively. Equation (3) indicates that the angular resonant frequency $\omega_0$ varies as $C_v$ changes, when $Z_0$ and $\theta$ are fixed.

As demonstrated above, the resonant structures shown in Figs. 2 induce tunable transmission zeros. However, regarding the realization of a reconfigurable filter, a relatively complicated tunable resonator configuration (Fig.3) is employed.

Fig. 3 Short-circuited SIR with a varactor

The input impedance $Z_{si}$ of the whole circuit showed in Fig. 3 is given by Equation (4). Next, the corresponding S-parameter of this microwave network can be expressed as Equation (5).

$$Z_{S1}(\omega) = \frac{S(\omega)}{M(\omega)} = jZ_2 \frac{(Z_1 \tan\theta_1 + Z_2 \tan\theta_2)\omega C_v - 1}{(Z_2 - Z_1 \tan\theta_1 \tan\theta_2)\omega C_v + \tan\theta_2} \quad (4)$$

$$S_{21}(\omega) = \frac{1}{1 + Z_{S1}/2Z_0} = \frac{1}{1 + S(\omega)/2Y_0 M(\omega)} \quad (5)$$

And $Z_0$ is the characteristic impedance of the transmission line in shunt with the two varator-tuned resonant structures. Here, the transmission zeros are obtained by imposing $S(\omega)=0$, whereas the transmission poles are obtained by $M(\omega)=0$. It is shown that the capacitance of the varactor and the tunable transmission zero location are expressed as (6) and (7). Specifically, Figure 4 shows the variation of the zero/pole location with respect to the varactor capacitance.

$$C_v(\omega) = \frac{1}{Z_2\omega + Z_1\omega\cot\theta_1} \tag{6}$$

$$C_v(\omega) = \frac{\tan\theta_2}{Z_2\omega - Z_2\omega\cot\theta_1\tan\theta_2} \tag{7}$$

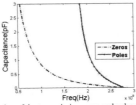

Fig. 4 The location of the transmission zeros and poles with respect to the capacitance of varactor

Based on the discussion above, a simple filter model is achieved in Fig.5, and the simulation results are displayed in Fig.6.Table I gives the data of the tunable bandwidth changed with the variable capacitance.

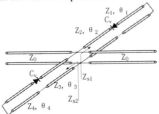

Fig. 5 Filter structure based on short-circuited SIRs with varactors

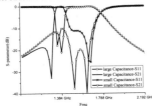

Fig. 6 Simulation results on filter structure showed in Fig.5.

TABLE.1 Simulation data on filter structure showed in Fig.5

| Capacitance | bandwidth | IL | RL |
|---|---|---|---|
| 3.6pF | 340MHz | $\leq 0.1dB$ | $\geq 10dB$ |
| 8pF | 440MHz | $\leq 0.4dB$ | $\geq 10dB$ |

It can be concluded that The FBW of filter structure based on short-circuited SIRs with varactors has a relative reconfigurability.

B. *open-circuited SIR with varactors*

Fig.7.gives the general structure of an open-circuited SIR,

Fig. 7 The general structure of a short-circuited SIR

the input impedance of Fig. 7 can be derived from the following equation:

$$Z_{inoc} = jZ_1\frac{Z_1\tan\theta_1\tan\theta_2 - Z_2}{Z_1\tan\theta_1 + Z_2\tan\theta_2} \tag{8}$$

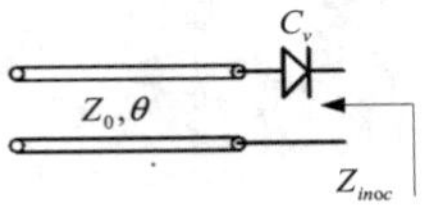

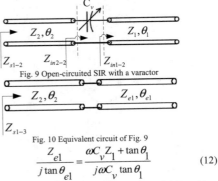

Fig. 8 Open-circuited SIR with a varactor

The input impedance of the structure shown in Fig.8, can be obtained from Equation (9), and then transmission zeros, can be evolved in (10), finally, the capacitance $C_v$ is expressed as Equation (11).

$$Z_{inoc} = \frac{\omega C_v Z_0 + \tan\theta}{j\omega C_v\tan\theta} \tag{9}$$

$$\omega C_v Z_0 + \tan\theta = 0 \tag{10}$$

$$C_v(\omega) = -\frac{\tan\theta}{\omega Z_0} \tag{11}$$

Obviously, the capacitance is negative in the case of $\theta \leq 90°$.Under this condition, the capacitance is an inductor. the structure of Fig. 7 is reconstructed as a part of Dual Behavior Resonator (DBR) [6].

Fig. 9 Open-circuited SIR with a varactor

Fig. 10 Equivalent circuit of Fig. 9

$$\frac{Z_{e1}}{j\tan\theta_{e1}} = \frac{\omega C_v Z_1 + \tan\theta_1}{j\omega C_v\tan\theta_1} \tag{12}$$

With the aid of Eq. (12), an open-circuited SIR with a varactor shown in Fig.9 can be transformed to Fig 10, which is based on DBR.

III. EXPERIMENTAL VERIFICATION

Fig. 11 and Fig.12 shows the transmission-line circuit of the BPF realized by the aforementioned design approach. The filter is simulated with Ansoft HFSS. $\theta = 36°$, $\theta = 54°$,at the frequency 3.7GHz, The varactor SMV1405 manufactured by Skyworks[8] is used. Filters are fabricated on Rogers4003C and measured by Agilent network analyzer. All the results are presented in the following page.

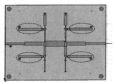

Fig. 11 Open-circuit structure simulation model

Fig. 12 Photograph of the fabricated reconfigurable BPF based on open-circuit structure

As a result, the following measured data are in highly agreement with the simulated data and Table II listed the center frequency and the maximum insertion loss (IL) and return loss (RL) at different DC bias.

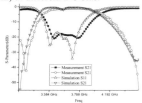

Fig. 13(a) Simulated and measured S-parameter the fabricated reconfigurable BPF

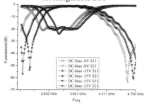

Fig. 14 (b) Measured S-parameter the fabricated reconfigurable BPF with DC-bias vary

TABLE II Measured S-parameter data the fabricated reconfigurable BPF with DC-bias vary

| DC bias | Center Frequency | IL | RL |
|---------|-----------------|------|------|
| 0V | 3.6GHz | ≤1.5dB | ≥18dB |
| 15V | 3.8GHz | ≤1.5dB | ≥18dB |
| 30V | 3.9GHz | ≤1.5dB | ≥18dB |

Additionally, another filter based on short-circuit structure discussed in section II, is fabricated. The measured S-parameters are in highly agreement with the simulation. Also, the final results reveal that the FBW can be adjusted from 3.1GHz to 3.7GHz.

Fig. 15 Photograph of the fabricated reconfigurable BPF based on short-circuit structure

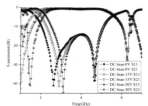

Fig. 16 Measured S-parameter the fabricated reconfigurable BPF with DC-bias vary

TABLE III Measured S-parameter data the fabricated reconfigurable BPF with DC-bias vary

| DC bias | bandwidth | IL | RL |
|---------|-----------|------|------|
| 0V | 3.7GHz | ≤0.5dB | ≥14dB |
| 15V | 3.4GHz | ≤0.5dB | ≥14dB |
| 30V | 3.1GHz | ≤0.5dB | ≥14dB |

The Measurement of fabricated filters indicates that the structures discussed above can realize tunable center frequency and FBW.

## IV. CONCLUSION

Filters with similar structures are proposed and can achieve tunable bandwidths or center frequencies. The presented approach aims for the bandwidth and center frequencies reconfigurable realized with simple open and short stubs. The demonstrated BPFs have a relatively low insertion loss, relatively wider FBWs and center frequencies tuning range.

## ACKNOWLEDGMENT

Thanks to Micro-/Nano-Electronic System Integration Center of University of Science and Technology of China., and the community of 2013 International Symposium on Antennas and Propagation.

## REFERENCES

[1] Peng Wen Wong, Hunter IC. Electronically Reconfigurable Microwave Bandpass Filter. IEEE Transactions on Microwave Theory Techniques. 2009;57(12): 3070–3079.

[2] Jyun-Yu Chen, Hsuan-Ju Tsai, Nan-Wei Chen. Bandwidth Reconfigurable Microwave Bandpass Filter, Microwave Symposium Digest (MTT), 2011 IEEE MTT-S International. Baltimore.2011:1-4.

[3] Karim M.F, Yong-Xin Guo, Chen Z.N, Ong L.C. Miniaturized reconfigurable and switchable filter from UWB to 2.4 GHz WLAN using PIN diodes. Microwave Symposium Digest, 2009. MTT '09. IEEE MTT-S International. Boston. 2009:509-512

[4] Karim M.F, Yong-Xin Guo, Chen Z.N, Ong L.C. Miniaturized Reconfigurable Filter Using PIN Diode For UWB Applications. Microwave Symposium Digest, 2008 IEEE MTT-S International. Atlanta. 2008:1031-1034.

[5] Quendo C, Eric Rius, Person C. Narrow bandpass filters using dual behavior resonators. IEEE Transactions on Microwave Theory Techniques.2003;51(3):734–743.

[6] Quendo C, Eric Rius, Person C. Narrow Bandpass Filters Using Dual-Behavior Resonators Based on Stepped-Impedance Stubs and Different-Length Stubs. IEEE Transactions on Microwave Theory Techniques.2003;52(3):1034-1044.

[7] Chang-Zhou Hua, Chen Miao, Wen Wu. A novel dual-band bandpass filter based on DBR. Microwave Conference, 2009. APMC 2009. Asia Pacific. Singapore. 2009:1383-1386

[8] Skyworks Solutions, Inc. silicon abrupt junction varactors datasheet.

# A New Class of Multi-Band Filters Based on The Same Phase Extension Scheme

Xumin Yu *Member*, Xiaohong Tang *Member, IEEE*, Fei Xiao, *Member, IEEE,* and Xinyang He

*Abstract*—This paper presents a new class of Multi-band filters based on same phase extension scheme, in which each channel is dedicated to selected band and center frequency by using the scheme of same phase extension. The equivalent circuit is the combination of multi-inline networks connected with two manifold waveguides, the structure of which has the same phase extension. The advantage of this method is that each in line network represents an individual channel. A dual-mode tri-band filter is measured and described.

*Index Terms*—Band-pass filter, manifold waveguide, multi-band, resonator filter.

## I. INTRODUCTION

Microwave multiband filters have been attracting considerable attention [1]-[6]. Recent designs of these types of filters, which generate transmission zeros between the bands, involve cross-coupled resonators, as in the single-band case. The shortage of coupling schemes is that the sensitivity of the individual pass-bands to manufacturing errors is connected with the whole order of the filters, but not with the individual bands. A dual-mode dual-band filter was introduced in [7]. In that filter, each individual band is controlled by a dedicated polarization of the dual-mode resonators, and a transmission zero is increased to improve the rejection of two bands. However, such an approach can not present more pass-bands filters.

In this paper, we propose a new design of multi-band filter, in which each individual band is controlled by a dedicated channel. The manifold-coupled approach, which is a kind of same phase extension scheme, is used to connect each channel's input and output ports. This kind of multiband filter requires the presence of all channel filters at the same time so that the effect of channel interactions can be compensated in the designed process. The manifold itself is a transmission line, such as a coaxial line, a rectangular waveguide, or some other low-loss structure. It is possible to achieve a channel performance in the manifold-coupled configuration, which is close to the one that can be obtained from a channel filter by itself. As an example, a dual-mode tri-band filter is designed and presented.

## II. THEORY AND EQUIVALENT CIRCUIT

The equivalent circuits of multiband filters are parallel circuits. In the microwave field, the ideal parallel circuits must meet two requirments. One is the common point of parallel circuits has a common pass power for every frequency in the pass-band; the other is that the point also has a common phase for every frequency in the pass-band. In order to design the ideal common point in parallel circuits, a piece of metal, the area of which is as small as possible, will be used generally. In other transmission structures, the more compact the structure of common part is, the better the achieved performance of common point is. However, when the parallel circuits increase, the design of the common point becomes more difficult.

A manifold-coupled multiplexer, which is capable of realizing optimum performance for absolute insertion loss、amplitude and group delay response, has been known for decades [8]-[13]. The manifold-coupled approach is a scheme of the common point for more parallel circuits. Using this method, the common port of each parallel circuit can share the same phase at pass-band frequencies by carefully designing. The power differences of common port, which may be due to different transmission ways, are too small to be ignored. The same technology with careful arrangements can be used to design robust manifold-coupled multi-band filters.

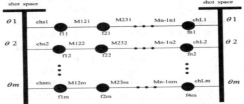

Fig.1. The equivalent circuit of m-band n-order filters

The equivalent circuit of this class of filters, in terms of inverters and resonators, consists of *m* non-interacting paths connected with two manifold transmission lines, as shown in Fig.1. If n cavities are used, each path contains n resonators that are directly coupled by inverters. The *m* paths are connected to the input and output nodes with manifold. Each channel can be designed separately. Naturally, this equivalent circuit is valid for the pass-bands just like a manifold multiplexer. A specific response yielded by the extraction of the parameters of the equivalent circuit in Fig.1 can be carried out by optimization, which is the same as a manifold multiplexer in [11].

As an example, a six-cavity three-band filter with the following specifications, 40 MHz, 60 MHz and 100 MHz centered at 14 GHz, 14.11 GHz and 14.28 GHz, is presented. The in-band return loss in three pass-bands is 20 dB. The three

978-7-5641-4279-7

bands are separated by two transmission-zero at 14.047 GHz and 14.2455 GHz. The initial coupling matrix of channel filter is obtained:

$$M = \begin{bmatrix} 0 & 1.2 & 0 & 0 & 0 & 0 \\ 1.2 & 0 & 0.98 & 0 & 0 & 0 \\ 0 & 0.98 & 0 & 0.745 & 0 & 0 \\ 0 & 0 & 0.745 & 0 & 0.98 & 0 \\ 0 & 0 & 0 & 0.98 & 0 & 1.2 \\ 0 & 0 & 0 & 0 & 1.2 & 0 \end{bmatrix} \quad (1)$$

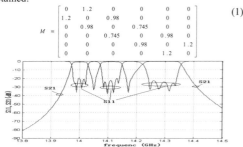

Fig.2. Response of the whole filter circuit (Q=8000)

After connecting the whole channels with manifold by cascading these net [ABCD] matrices of each section in the structure, the response of this filter is shown in Fig.2.The unloading Q 8000 is assumed. All the specifications are met.

### III. FILTER STRUCTURES

Compared with a manifold-coupled multiplexer shown in Fig.3, a manifold-coupled multiband filter's approach shown in Fig.4 is viewed as one more manifold to connect each channel's output.

The most important requirement in this configuration is to preserve realizability in the whole structure. The two designed manifold transmission lines can be connected with each channel's input and output smoothly. The channel filter in this paper (Fig.5) can be designed based on the dual-mode or single-mode channels, which have the same length L. Every channel has two cylindrical cavities and two sub rectangular waveguides, which have been carefully designed.

For the structure in Fig.6, the resonators are uniform sections of a cylindrical waveguide in which a coupling screw and two frequency tuning screws are placed in the middle of the cavity. There is a coupling aperture between two cavities. This cylindrical dual-mode configuration channel was used to design output multiplexer in [12]. From this type of channel filter, it is able to gain high Q, low insertion loss, small size and cross-coupled resonators.

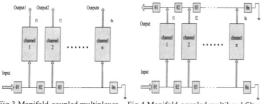

Fig.3.Manifold-coupled multiplexer    Fig.4.Manifold-coupled multiband filter

Fig.5. One channel of the proposed manifold-coupled dual-mode multi-band filter

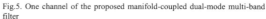
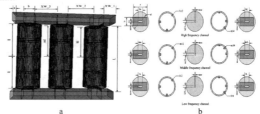

Fig.6 The size of tri-bands filter (a. Layout of the proposed manifold-coupled dual-mode multiband filter; b. Coupling structures of three channels in the proposed manifold-coupled dual-mode multiband filter)

### IV. DESIGN

The design approach used in this paper is perfectly illustrated by the structure in Fig.6 as an example. All the channel filters are electrically connected to each other through the near-loss-less manifold waveguide. The design consists in implementing the equivalent circuit in Fig.1. The manifold multi-band filters need to be considered as a whole, and thus optimization method has been utilized to achieve the final design. To speed up the overall optimization, it is efficient to analyze each channel's input/output to common port transfer characteristic individually, and the common port return loss.

#### A. Common Port Return Loss

In the multi-band filters, the input and output are common port of all channels. The manifold in this paper is the waveguide, and the junctions are E-plane. The junctions are best characterized with three-port S parameters, and symmetry of junctions about the vertical axis in Fig.7, $S11 = S22$ and $S32 = -S31$. The S-parameter matrices are calculated by assuming a matched termination at each port. When the termination at one port is arbitrary, the two-port S parameters between the others can be defined using the following formula.

*1) Admittance* $Y_{L2}$ *($\neq 1$) at port 2:*

$$\begin{bmatrix} S'_{11} & S'_{31} \\ S'_{31} & S'_{33} \end{bmatrix} = \begin{bmatrix} S_{11} & S_{13} \\ S_{31} & S_{33} \end{bmatrix} + \frac{\Gamma_2}{1-\Gamma_2 S_{11}} \begin{bmatrix} S_{21}^2 & kS_{21}S_{31} \\ kS_{21}S_{31} & S_{31}^2 \end{bmatrix}$$

(2)

, where $\Gamma_2 = (1-Y_{L2})/(1-Y_{L2})$ and $Y_{L2}$ is the admittance at port 2 of the junction and $k = -1$ for E-plane junctions. This modified S matrix can also be applied, if $Y_{L1}$ is terminating port 1.

*2) Admittance* $Y_{L3}$ *($\neq 1$) at port 3:*

$$\begin{bmatrix} S'_{11} & S'_{12} \\ S'_{21} & S_{22} \end{bmatrix} = \begin{bmatrix} S_{11} & S_{12} \\ S_{21} & S_{22} \end{bmatrix} + \frac{\Gamma_3 S_{21}^2}{1-\Gamma_3 S_{22}} \begin{bmatrix} 1 & k \\ k & 1 \end{bmatrix} \quad (3)$$

, where $\Gamma_3 = (1-Y_{L3})/(1-Y_{L3})$ and $Y_{L3}$ is the admittance at port 3 of the junction and $k = -1$.

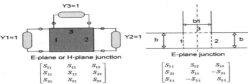

Fig.7. E-plane waveguide junction and S-parameter matrix representation

The channel filter input admittances $Y_{F1}$, $Y_{F2}$ and $Y_{F3}$ are determined at the frequency points in the band rang. The new S parameters of each junction are calculated using (3). These new S parameters are converted to [ABCD] and cascaded in manifold structure with $\theta_{M1}$, $\theta_{M2}$ and $\theta_{M3}$. The along-manifold admittances $Y_{M1}$, $Y_{M2}$ and $Y_{M3}$ are obtained as shown in Fig.8.

$$Y_{Mi} = \frac{1 + S_{11i}}{1 - S_{11i}} \quad \dots \dots \text{(4)}$$

, where $i = 1, 2 \dots, n+1$, and $n$ is the number of channels on the manifold. The CPRL (Common port return loss) can be obtained from the final admittance $Y_{M4}$ as shown in Fig.8.

$$RL_{CP} = -20 \log_{10} \frac{1 + Y_{Mn+1}}{1 - Y_{Mn+1}} (dB) \quad \dots \dots \text{(5)}$$

When the $Y_{F2}$, and $Y_{F3}$ are determined, $\theta_{M1}$, $\theta_{M2}$ and $\theta_{M3}$ are optimized to generate better CPRL. In this scheme, it implies that the symmetry structure of each channel may be more easily designed. However, in order to obtain the transmission zero between bands, asymmetry structure needs to be used in dual-mode cavities.

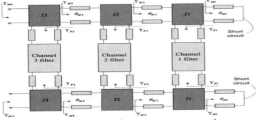

Fig.8. The design of common ports return loss.

### B. Channel Filter Design

Generally, each band of multi-band filters is widely spaced, although they react with each other through the manifold directly. The channel interactions are not high. The doubly terminated networks are used to design the channel filter. In this paper, three doubly terminated filters are designed for three channels before optimization.

As shown in Fig.6, two coupling screws in each channel are used to realize M12 and M34 in the coupling matrix, the inter-cavity rectangle coupling iris realizes M23, and the rectangle coupling irises between cylindrical and rectangle waveguide realize the M01 and M45, respectively. The high and low frequency channels have the same corresponding coupling-routing diagram. Only one coupling screw of the middle frequency channel turns 90° to obtain the two transmission zeros.

Fig.9. The circuit representing a closed cavity is used to design the input and output coupling coefficient.

The design of input/output coupling irises for each channel is done by simulating a single closed cavity coupled to the input/output manifold waveguide. The coupling coefficients, which are the result of optimization, are related to the maximum of the derivative of the phase of S11 with respect to frequency [9]. The relationship between the derivative of the phase of the reflection coefficient and the input/output coupling coefficients can be established by using the circuit in Fig.9. The coupling coefficient is given by [9].

The electromagnetic simulation structure and equivalent circuit of inter-cavity coupling irises are shown in Fig.10. The equivalent circuit parameters can be obtained directly from the S-parameters though (6). The normalized coupling coefficients and the angle $\phi$ are given in (7) as follows:

$$jX_s = \frac{1 - S_{21} + S_{11}}{1 - S_{11} + S_{21}} \qquad jX_p = \frac{2S_{21}}{(1 - S_{11})^2 - S_{21}^2} \quad \text{(6)}$$

$$\phi = -\tan^{-1}(2X_p + X_s) - \tan^{-1}(X_s)$$

$$K = \left| \tan\left( \phi / 2 + \tan^{-1}(X_s) \right) \right| \quad \text{(7)}$$

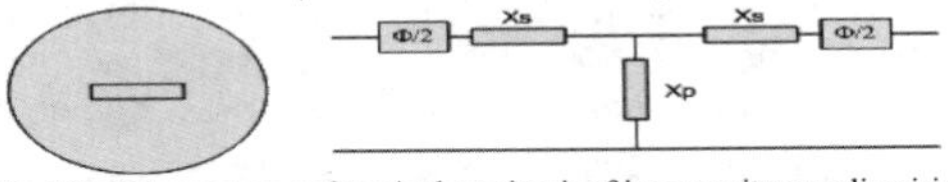

Fig.10. EM structure and equivalent circuit of inter-cavity coupling irises.

The coupling screws can be designed in the same way. The length of resonators is half a guided wavelength at the designed frequency and they are corrected by a phase term to account for the loading by the irises and screws. The actual lengths are related to $\phi_i$ of irises and screws, which is given in (8) as follows.

$$L_i = \frac{\lambda_g}{2\pi} \left( \pi - \frac{1}{2}(\phi_i + \phi_{i+1}) \right) \quad \text{(8)}$$

### V. APPLICATIONS AND RESULTS

To illustrate the design procedure, a four-cavity dual-mode three-band filter was designed and optimized. The filter has three bands: 13.98 GHz-14.02 GHz, 14.08 GHz-14.14 GHz and 14.24 GHz-14.34 GHz. The in-band return loss is approximate 20 dB.

The initial coupling matrix of one channel is given in (1) and the final response of the whole filter is given in Fig.2. The initial design was optimized by using Mician's commercial software package μ Wave Wizard. The response of the optimized structure is shown in Fig.11. The transmission zero outside of the pass-band is generated by cross coupling structure in the channel filter itself. The dimensions of the optimized filter are a=19.05 mm, b=9.525 mm, d=22.5 mm, mw=2 mm, L=72 mm, XW_1=20.9847 mm, XW_2=23.6211 mm, XW_3=11.0616 mm, ll=17.9165 mm, ml=17.683 mm, hl=17.2155 mm, ls=6.6346 mm, l23=3.3238 mm, ll=6.5667 mm, ms=7.0559 mm, m23=4.12 mm, ml=6.942 mm, hs=7.5396 mm, h23=5.0122 mm, and hl=7.459 mm. The iris thickness is t=0.5 mm.

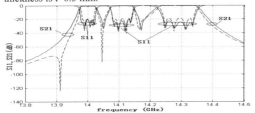

Fig.11. Response of optimized four-cavity three-band dual-mode filter. Dashed lines: EM simulation from Mician's µ Wave Wizard. Solid lines: ideal response.

## VI. SENSITIVITY ANALYSIS

The investigation of the design's sensitivity to errors in the dimensions of the structure is an important step for designing microwave filters, prior to fabrication. The filters' dimensions in Fig.6. are randomly changed by ± 0.025mm, and the resulting responses. are plotted together in Fig.14. For each channel with its own structure, the sensitivity of each band is very similar to a single filter. Since the manifold waveguide connects the input and output of each channel, the dimensions of manifold waveguide is more sensitive to affecting each channel. The low narrow band shows more sensitivity of frequency than the other bands do, just like a narrow band filter. The more width the designed pass-band is, the less sensitivity it shows.

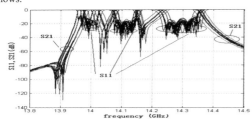

Fig.12. Sensitivity analysis of the tri-band dual-mode filter

## VII. MEASUREMENT

The measured result of a dual-mode tri-bands filter has been shown in fig.13. The filter has three bands: 13.98GHz-14.01 GHz, 14.09 GHz-14.15 GHz and 14.25 GHz-14.35 GHz. Two transmission zeros are generated along the three bands. The photo of the dual-mode tri-bands filter has been shown in fig.14. The presented filter is a litter different from the simulated one, which is caused by error of processing and debugging. However, the measured result, that confirms the same phase extension scheme, is useful for multi-band filter implementing.

## VIII. CONCLUSION

This paper presents a new class of Multi-band filters, in which each channel is dedicated to selected band and center frequency by using the scheme of same phase extension. Two common manifold waveguides are used to connect several separate channels. This implies that each channel models an individual filter, which can present a straight-forward initial design. The advantage of same phase extension is that it is a kind of method to implement common point in parallel circuits. In the filter, a transmission zero between two pass-bands is generated by adjusting a coupling screw of one channel. Simulation result and sensitivity analysis have been elucidated. The lowest band of measured result is narrow than that of simulated result due to error of processing and debugging. However, the measured result, that confirms the same phase extension scheme, is useful for multiband filter implementing, either. Designs, in which the structure of resonator and manifold transmission line are different, are also possible.

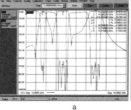

a             b

Fig.13. Measured response and Photo dual-mode tri-bands filter(a.Measured response; b. Photo )

REFERENCES

[1] G. Macchiarella and S. Tamiazzo, "Design techniques for dual-pass-band filters," *IEEE Trans. Microw. Theory Tech.*, vol. 53, no.11, pp. 3265-3271, Nov. 2005.

[2] M. Mokhtaari, J. Bornemann, K. Rambabu, and S. Amari, "coupling matrix design of dual and triple passband filters," *IEEE Trans. Microw. Theory Tech.*, vol.55, no. 11, pp. 3940-3946, Nov. 2006.

[3] V. Lunot, F. Seyfert, S. Bila, and A. Nasser, "Certified computation of optimal multiband filtering functions," *IEEE Trans. Microw. Theory Tech.*, vol. 56, no. 1, pp. 105-112, Jan. 2008.

[4] P. Lenoir, S. Bila, F. Seyfert, D. Baillargeat, and S. Verdeyme, "Synthesis and design of asymmetrical dual-band bandpass filters based on equivalent network simplification," *IEEE Trans. Microw. Theory Tech.*, vol. 54, no. 7, pp. 3090-3097, Jul. 2006.

[5] G. Macchiarella and S. Tamiazzo," Dual-band filters for base station multi-band combiners," in *IEEE MTT-S Int. Microw. Symp. Dig.*, Jun. 2007, pp. 1289-1292.

[6] Y. Zhang, K. A. Zaki, J. A. Ruiz-Cruz, and A. E. Atia, "Analytical synthesis of generalized multi-band microwave filters," in *IEEE MTT-S Int. Microw. Symp. Dig.*, Jun. 2007, pp. 1273-1276.

[7] S. Amari, and M. Bekheit, "A new class of dual-mode dual-band waveguide filters," *IEEE Trans. Microw. Theory Tech.*, vol. 56, no. 8, pp. 1938-1944, Nov. 2008.

[8] J. D. Rhodes and R. Levy, "Design of general manifold multiplexers," *IEEE Trans. Microwave Theory Tech.* MTT-27, 111-123 (1979).

[9] J. D. Rhodes and R. Levy, "A generalized multiplexer theory," *IEEE Trans. Microwave Theory Tech.* MTT-27, 99-110 (1979).

[10] C. Kudsia, R. Cameron, and W. C. Tang, "Innovation in microwave fiters and multiplexing network for communication satellite systems, *IEEE Trans. Microwave Theory Tech.* MTT-40, Jun. 1992, pp. 1133-1149.

[11] J. Bandler, R. Biernacki, S. Chen, P. Grobelny, and R. Hemmers, Space mapping technique for electromagnetic optimization, *IEEE Trans. Microwave Theory Tech.* MTT-42, Dec. 1994, pp. 2536-2544.

[12] R. J. Cameron, C. M. Kudsia, and R. R. Mansour, *Microwave filters for communication systems*. Chapter 18 New Jersey, U.S.: WILEY Press, 2007.

[13] J. Bandler, S. Daijavad, and Q.-J. Zhang, "Exact simulation and sensitivity analysis of multiplexing networks," *IEEE Trans. Microwave Theory Tech.* MTT-34, Jan. 1986, pp. 102-111.

# A Novel C-band Frequency Selective Surface Based on Complementary Structures

Jing Wang, Ming Bai

School of Electronic and Information Engineering, Beihang University

No.37 Xueyuan Road

Beijing, 100191 China

*Abstract*-A new design method for band-pass frequency selective surfaces (FSSs) using complementary loading structure is introduced, which is applied to achieve C pass-band performance in 60 degree angle of incidence. Each FSS element consists of three square patches and two square apertures as complementary patterns etched in the middle tier, which is backed by dielectric slabs. The effect to the transmission characteristics of the square apertures screens as the load of the square patches screens is investigated. This novel design method provides a new design thought for tailoring the response of the FSS and has proved efficacy in designing high performance FSS. An equivalent circuit model is given for predicting the characteristics of the designed FSS, and a good match between the simulated and required transmission coefficients is obtained. Furthermore, the cases of different angles of incidence waves and cascading FSSs are also measured and examined.

## I. INTRODUCTION

The technology of frequency selective surfaces (FSSs) has been developed rapidly, which can be widely applied to space filters, reflectors, radar absorbers and frequency windows of radomes [1]. FSSs have been developed since many years ago (50's). Classical books on this topic were written by B. A. Munk [2]. FSSs are usually constructed of periodic metallic patches or aperture elements within one or more metallic screens backed by dielectric slabs. The first type behaves like a band-stop filter while the second type like a band-pass filter. In recent years this topic has received new breaths coming from the new concepts of complementary structure. It stems from Babinet's principle, whereby a hybrid of two closely coupled FSS, a layer of conducting elements and a layer of aperture elements are etched either side of a dielectric substrate. Babinet's principle, relating to optics, states: 'when the field behind a screen with an opening is added to the field of a complementary structure, the sum is equal to the field when there is no screen' [3]. In 1994 H. Wakabayashi proved the reflection properties of the inverse and normal type FSS are complementary to each other [4]. Soon after, researches on the electromagnetic model for complementary FSSs of linear dipoles and single rings were developed. The dipole complementary structure was used to design band-pass filter by D. S. Lockyer [3]. Then, Shunli Li and Liguo Liu investigated the effect to the transmission characteristics of the Jerusalem complementary FSS [5].

In this paper, a multi-layer FSS structure based on complementary structure is designed which is aimed at achieving low-loss pass-band in 4-8GHz. These metal screens are tightly coupled, producing stable resonant frequency and wide band.

## II. FSS DESIGN

### A. *Basic Structure*

According to design requirements for C-band transmission characteristics, the basic thought is using multilayer FSSs to produce multiple resonance points to support a wide band and complementary structure to adjust transmission performance.

Under the condition of large incident angle, resonance points of different polarization modes will change a lot; in addition, ripple amplitude in the pass-band will increase obviously so that transmission characteristics may be damaged. Aiming at these problems, some ideas are concluded as follows:

(1) Too much layers for FSSs coupling design are not recommended, and the number of resonance points used to support pass-band should be considered.

With less resonance points supporting wide frequency band, low Q-value resonant units are needed to balance the low-frequency cut-off characteristic and pass-band flatness. Using more resonant points, the low-frequency cut-off characteristic seems better, but ripple amplitude will increase obviously under the condition of large angle. A third-order design with five-layer FSSs are discussed here.

(2) Medium layers should be used reasonably to improve the impedance matching characteristics of structure and air under the condition of large incident angle.

The use of the dielectric layers can help improve the TE and TM polarization stability, and reduce the ripple in the pass-band. The typical multilayer dielectric layers matching design for structure with wide band in large incident angle is arrange dielectric constant decrease progressively from inside to outside.

(3) Complementary units are chosen, because their opposite trend of resonant frequency on transmission characteristics.

The configuration of the novel structure is shown in Figure 1. It's symmetrical. There are two types of elements in this design, namely the square aperture and square patch element which are complementary in geometry [3]. We choose the square element due to its simplicity and better bandwidth. The difference is the dimension of the patch element is smaller than the aperture one. The structure constitutes coupled

978-7-5641-4279-7

inductive and capacitive surfaces, resulting in a band-pass response. It is composed of three dielectric substrates and five metal layers including two square apertures and three square patches etched in the middle substrate.

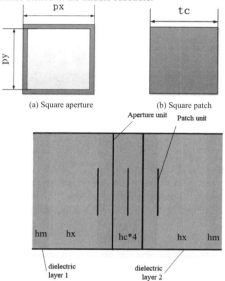

(a) Square aperture      (b) Square patch

(c) Array of designed structure

Figure 1. The proposed FSS and its periodic cells

### B. Analysis of Equivalent Circuit Model

Each aperture screen behaves like a series LC circuit and the patch one behaves like a parallel LC circuit [2]. With the coupling effect between the metallic layers represented by a relatively small mutual inductance, a unit cell of the proposed FSS can be modeled by a simplified equivalent circuit model as shown in Figure 2.

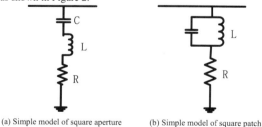

(a) Simple model of square aperture      (b) Simple model of square patch

(c) Simplified equivalent circuit

Figure 2. Model of designed structure

The substrates between metal screens can be considered as a short transmission line [6].The equivalent circuit model of the FSS can be simplified into a lumped-element model. The value of mutual inductance can be ignored because it's relatively small, and its impedance is derived as

$$Z = \frac{jwL_2 + 1/(jwC_2)}{-w^2 L_2 C_1 - 1/(w^2 L_1 C_2) + L_2/L_1 + C_1/C_2 + 1} \quad (1)$$

### III. NUMERICAL SIMULATIONS

In order to demonstrate the properties of this structure, we have numerically simulated the incidence plane waves at 60°.

Study shows that changes of some parameters have great influence on model's performance. The laws of key parameters' influence are summarized in table 1(take the value gets bigger as an example).

TABLE I
LAWS OF KEY PARAMETERS' INFLUENCE

| parameters | TE polarization | TM polarization |
|---|---|---|
| Thickness of outermost layer(hm) | Pass-band become narrower | Pass-band become narrower |
| Thickness of second layer(hx) | Effective bandwidth drops | Little effect |
| Distance between two metal screens(hc) | pass-band moves to lower frequency | Pass-band become wider |
| Length of square patch(tc) | pass-band moves to lower frequency | pass-band moves to lower frequency |
| Length of square aperture(py) | Pass-band become wider | pass-band moves to lower frequency |
| Period of unit(px) | pass-band moves to higher frequency | Pass-band become narrower |

By optimizing, a set of curves which meet the requirements of application are completed. The transmission character of TE and TM polarization is shown in Figure 3.

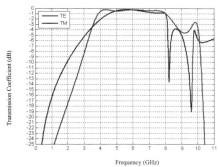

Figure 3.Curves of transmission coefficients at 60°

This design can realize a stable pass-band and gain desired transmission coefficients of TE polarization and TM polarization which are limited between 0 and -2 dB in C-band. At the same time, angle stability is analyzed (as shown in figure 4).

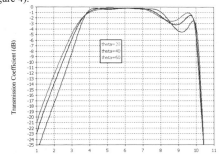

(a) TE polarization

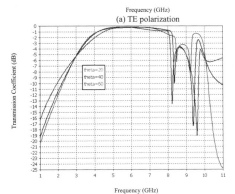

(b) TM polarization

Fig.4. angle stability

According to these figures, the transmission curves for TE and TM waves remain quite well in C-band.

The Q-value of designed FSS combination unit is low so that a wide pass-band can be realized. The bandwidth of TE polarization is close to 4.6GHz, and that of TM is close to 4.1GHz.The transmission coefficients in pass-band are all lower than -2dB. Its polarization stability in C band is quite well in 60 degree angle of incidence, and the resonant frequency is stable when angles change from 20° to 60°。

## IV. CONCLUSION

In this paper, a FSS working at C-band is designed based on a new design method for band-pass characteristic using complementary loading structure. The designed FSS meets the requirements even under the condition of large incidence angle, and realizes stable resonant frequency .This novel design provides a new way to realize band-pass performance and has practical value for the development of FSS in the future.

REFERENCES

[1] R. A. Hill and B. A. Munk. "The effect of perturbating a frequency selective surface and its relation to the design of a dual- band surface", *IEEE Transactions on Antennas and Propagation,* 44(3):368- 374, 1996.
[2] B. A. Munk, *Frequency-Selective Surfaces: Theory and Design.* New York: Wiley-Interscience, 2000.
[3] D. S. Lockyer, J. C .Vardaxoglou and R. A. Simpkin, "Complementary frequency selective surfaces", *IEE Proc –Microw.Antennas Propag ,* Vol. 147, No 6, Decernber 2000
[4] H. Wakabayashi, M. Kominami ,H. Kusaka and H. Nakashima, "Numerical simulations for frequency-selective screens with complementary elements", *IEE Proc.-Microw. Antennas Propag., Vol. 141, No. 6, December 1994.*
[5] Shunli Li and Liguo Liu, "A novel design methodology for bandpass Frequency Selective Surfaces Using Complementary Loading Structure", *IEEE International Symposium on Microwave, Antenna, Propagation and EMC Technologies for Wireless Communications,* 2009, pp. 831-833.
[6] K. Sarabandi and N. Behdad, "A frequency selective surface with miniaturized elements," *IEEE Trans. Antennas Propag.,* vol. 55, no. 5,

# Three-dimensional Random Modeling of Particle Packing Through Growth Algorithm and the Microwave Propagation in the Model

Zhi Xian Xia, Hao chi Zhang, Yu Jian Cheng*, Yong Fan

EHF Key Laboratory of Fundamental Science

School of Electronic Engineering

University of Electronic Science and Technology of China

Chengdu, 611731, P. R. China

E-mail:chengyujian@uestc.edu.cn

*Abstract*-A three-dimensional random modeling method based on growth algorithm is proposed in this paper. The models can be employed to investigate practical problem of coal particle packing with different size range. Full-wave simulations based on the models are able to recognize microwave attenuation and propagation characteristics. Some useful conclusions are obtained for microwave power application.

## I. INTRODUCTION

Material will absorb some microwave energy when it was radiated by microwave. Microwave energy will cause some effects on the material such as temperature rising or promoting other chemical reactions. On the contrary, the existence of the material will change the field distribution of the microwave. Field interaction between material and microwave makes kinds of special process possible. It is a vast area of science, and has appealed an increasing interest from kinds of researchers [1]. An important factor in microwave field interaction is the complex permittivity of the material, especially the image part in other words the loss tangent. The loss tangent stands for the ability to absorb microwave energy [2]. Different materials have different loss tangents so as to different reaction to microwave. On the other hand, the absorption will also vary with the microwave frequency.

In practice, material is directly exposed to microwave for radiation. If the material is uniform field distribution in the material can be analyzed by classical electromagnetic theory. Then, the attenuation and propagation characteristics can be analyzed. However, many materials in practice are not uniform. Some of them are composed of thousands of small particles. The macro object used in practice is piled up by micro particles. There exist air interspaces everywhere in the object which means the object is a mixture of air and material. The microwave propagation in such models is worth investigation.

There is special modeling software for arbitrary shaped particle packed together in civil engineering [3]. Models can perfectly simulate the actual accumulation of stones in the concrete. However, such modeling method in electromagnetic field and analysis of the microwave attenuation and propagation in such models were scarcely reported. Some papers about mixtures were published, but they concentrated on equivalent permittivity of the mixture. They aimed to forecast the discipline between equivalent permittivity and the variation of the mixing ratio [4]-[6], paid little attention to modeling and microwave attenuation and propagation. Investigation of microwave attenuation and propagation in such models will help to improve the application of microwave power in material processing, such as microwave coal desulphurization, food heating, etc.

This paper proposes a random modeling method to investigate object piled up by small particles. Full-wave simulation based on such models is suggested to study microwave attenuation and propagation. The structure of this paper is as follows. Firstly, three-dimensional random modeling method is proposed. Models of the same size are built up to demonstrate the feasibility of the modeling method. Next, the measured complex permittivity of coal is employed to construct the model. With the full-wave simulation software Ansoft HFSS, the microwave attenuation and propagation in mixture of lossy medium coal and air is studied. Models composed of small particles of different sizes are simulated at different frequencies. Then, the influences of the operating frequency and the size of particle on the performance of the attenuation and propagation can be recognized. The propagation characteristic of the gapless objects is displayed for comparison as well.

## II. MODELING OF PARTICLES PACKING

978-7-5641-4279-7

*A. Model Process*

Take the simplicity into consideration particles are all supposed to be spheres. Absolutely, sphere can be changed into polyhedron easily with further program. Polyhedron can constitute by several intersected planes which are picked randomly in the sphere, or linked by dots which are picked randomly in the sphere.

The modeling program starts with a region with specific sizes, the maximum and minimum diameter of spheres. Those parameters can be set according to practical requirement. A modeling flowchart is shown in Fig. 1 to present details of the modeling method.

In this modeling program, it is supposed that one unit of the incremental of sphere diameter is 0.05mm. When the distance between two spheres is less than 0.01mm they are assumed to be intersected.

The models obtained each time are different considering the randomness of the modeling process. Three models of the same size are shown in Table I. The length $l$, width $s$ and height $h$ of the investigated region are 50mm, 50mm and 20mm, respectively. The diameter range is 3mm~6mm. Another three models of different diameters and heights are displayed in Table II. The height is 10mm, 20mm, and 25mm, respectively.

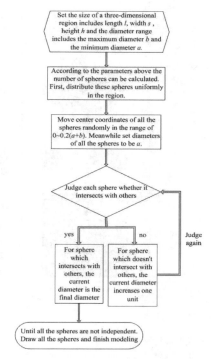

Figure.1 Flowchart of the modeling process

TABLE I

Three Random Models with the Same Size

| | | | |
|---|---|---|---|
| Southeast view | | | |
| Left view | | | |
| Top view | | | |

TABLE II

Three Random Models with the Different Diameters

| | Diameter 2-3mm | Diameter 3-6mm | Diameter 6-10mm |
|---|---|---|---|
| Southeast view | | | |
| Left view | | | |
| Top view | | | |

*B. Model Analysis*

Random function was applied in the modeling program to investigate the practical situation of particle packing. As is well known, for objects with size that is far less than the microwave wavelength the inner structure has little effect on microwave field distribution. Generally, when $c/d$ is less than 0.01, $c$ is assumed to be far less than $d$. 3GHz is the highest frequency we investigated in this paper. The wavelength of 3GHz is 100mm. Sphere with radius less than 1mm can be considered negligible since 1/100 equals to 0.01. Thus, for diameters 2mm is set as a boundary. Actually, simulating models with very small spheres are both time-consuming and meaningless. Therefore, models will be analyzed all have diameters which are greater than 2mm.

III. SIMULATED RESULTS

The external shape of the model built by the above method is cuboids. Thus, it is convenient to put the model into perfect rectangular waveguide to analyze the characteristics of attenuation and propagation. The dielectric permittivity of the model is set as 4.5, while the loss tangent is 0.1, which are measured results of coal powder in our lab. When diameters of the spheres range from 3~6mm and microwave frequency at 2.45GHz field distribution is shown in Fig. 2.

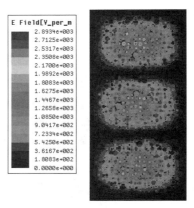

Figure.2 Field distribution at frequency 2.45GHz

As shown in Fig. 2, the field distribution is affected by the spheres. A number of simulations are conducted to recognize the effects of different sizes on the attenuation and propagation characteristics. Simulation results are listed in Table III. To ensure accuracy, several models of the same size are built and simulated. Transmission coefficients listed in Table III are transformed from γ of TE modes in perfect rectangular waveguide to γ of TEM modes in free space [7].

TABLE III

Transmission Coefficient $\gamma = \alpha + j\beta$ with Different Sizes under Different

Frequencies

| $\gamma$ $d$ | $f$ 0.915 GHz | 1.5 GHz | 2 GHz | 2.45 GHz | 3 GHz |
|---|---|---|---|---|---|
| 0 mm | 2.02+ j40.6 | 3.36+ j65.9 | 4.52+ j87.1 | 5.56+ j105.9 | 6.85+ j128.7 |
| 2~3 mm | 2.02+ j40.9 | 3.13+ j66.1 | 4.52+ j86.9 | 5.56+ j105.9 | 6.84+ j128.8 |
| 3~6 mm | 2.02+ j40.9 | 3.02+ j66.1 | 4.48+ j86.6 | 5.56+ j106.0 | 5.68+ j129.6 |
| 6~10 mm | 2.02+ j40.9 | 3.01+ j66.1 | 5.68+ j86.5 | 5.56+ j106.0 | 5.68+ j129.5 |
| 10~15 mm | 2.01+ j40.8 | 3.13+ j66.07 | 5.45+ j86.6 | 5.56+ j105.9 | 5.44+ j129.8 |
| 15~20 mm | 2.01+ j40.9 | 3.36+ j65.9 | 4.52+ j87.0 | 5.56+ j106.0 | 6.85+ j128.8 |
| 30~35 mm | 2.00+ j41.1 | 3.36+ j65.9 | 4.51+ j87.0 | 5.55+ j105.9 | 5.30+ j128.1 |

Higher attenuation coefficient indicates more microwave power is absorbed, and thus leads to a higher efficiency of microwave energy application. The overall trend in Table III is non-monotonic. For the frequency of microwave from 915MHz to 3000MHz the wavelength ranges from 327mm to 100mm. Results of the first group vary little from the model without air. The results verified the fact that spheres with small size have little effect on microwave propagation. Generally, objects are ground to powder in order to be radiated completely. However, according to the results above there is no need to grind objects when low-frequency microwave is used for material processing. With crushed objects of small size, the same performance of processing and efficiency can be achieved. This will greatly decrease physical pre-treatment processes.

Take the convenience of physical pre-treatment processes and penetration depth into consideration, the best size for different frequencies can be obtained. This will help to simplify grinding treatment and increase utilization efficiency of microwave power application significantly.

## IV. CONCLUSION

A three-dimensional random modeling method is proposed in this paper. The model is able to investigate the problem of practical particle packing. Simulations based on the model are conducted to investigate the microwave attenuation and propagation characteristics. Through several simulations, some useful conclusions are obtained to guide the microwave power application, especially for application like coal desulphurization.

ACKNOWLEDGMENT

This work is supported in part by the National Basic Research Program of China under Grant No. 2012CB214900 and in part by the Program for Changjiang Scholars and Innovation Team in University under Grant No. IRT1113.

REFERENCES

[1]  W. Andrew, *Microwave RF applicators and probes for material heating, sensing, and plasma generation*, Elsevier Inc, USA, 2010
[2]  L. F. Chen, C. K. Ong, C. P. Neo, V. V. Varadan and V. K. Varadan, *Microwave electronics-Measurement and material characterization*, John Wiley & Sons Ltd, 2004
[3]  X. Y. Gu, "Preliminary study on characteristics of structure of voids in particle assemblies generated by PFC3D", Master degree thesis of Tsinghua University, December 2008
[4]  A. H. Silwola, "Effective permittivity of dielectric mixtures", *IEEE Trans. Geos. R. S.*, Vol.26, Issue.4, pp. 420-429, July 1988
[5]  A. H. Silwola, *Electromagnetic Mixing Formulas and Applications*, The Institution of Electrical Engineers, London, 1999
[6]  A. H. Sihvola, E. Alanen, "Studies of mixing formulae in the complex plane", *IEEE Trans. Geos. R. S.*, Vol.29, No.4, pp. 679-687, July 1991
[7]  S. W. L, *Microwave engineering basis*, Press for University of Aeronautics and Astronautics of Beijing, Beijing, 1995
[8]  Z. X. Xia, Y. J. Cheng and Y. Fan, "Frequency-reconfigurable $TM_{010}$-mode reentrant cylindrical cavity for microwave material processing", *Journal of E.M. W. A.*, Vol.27, issue.5, pp. 605-614, January 2013
[9]  Q. H. Jin, S. S. Dai and K. M. Huang, *Microwave Chemistry*, China Science Press, Beijing, 1999

# Simulation study of a waveguide power combining network

Z.X. Wang  B. Xiang  M.M. He  W.B. Dou
State Key Lab of Millimeter Waves, Southeast University
Nanjing, 210096, P.R. China

*Abstract*- A waveguide based four way millimeter-wave power combining network is studied using HFSS in this paper. The power combining network is composed of a power divider sub-network, four metal PCB cavities and a power combiner sub-network. A compact waveguide directional coupler is designed to act as the power combiner and power divider. Parameters and shapes of the PCB cavities are carefully studied to avoid the resonant modes of the cavities, and a rhombus-like PCB cavity which has good transmission coefficient is presented.

## I. INTRODUCTION

The demand for high millimeter-wave output power has greatly increased in the military and civil field in Recent years. However, the output power from an individual solid-state device is rather modest at millimeter-wave frequencies, and the effective way to obtain higher output power is to combine the output powers from a number of solid-state devices. Conventional hybrid-type power-combining techniques based on printed circuits, such as the Wilson power divider, Lange coupler, and branch-line coupler, suffer from heavy power loss and narrow bandwidth at millimeter frequencies [1]. To avoid the drawbacks of circuit-type power combiners, quasi-optical [2-6] and waveguide-based [7-11] power-combining approaches have been proposed because of their low insertion loss, and their high power-combining efficiency. Compared to quasi-optical power combining system, the waveguide-based power combining network takes less space, which is attractive in many practical applications.

A compact four-way waveguide-based power combining network is proposed for high power combining application at frequency band 75~77GHz in this paper. The designed power network is composed of a power divider sub-network, a power combiner sub-network and four PCB cavities. The power divider/combiner sub-network consisting of three waveguide directional couplers are carefully designed in a compact size using standard WR12 waveguide and the shape and parameters of the PCB cavities are also studied.

## II. DESIGN OF THE POWER COMBINING NETWORK

### A. Logic structure of the power combining network

In this paper, we present a design of four way waveguide power combining network at V-band, the logic diagram is shown in Fig.1, where the input electromagnetic wave is divided into four way waves by the three dividers, and at the output side the four way waves are combined by three combiners. Between the divider sub-network and the combiner sub-network, there are four PCBs which are used

to install the amplifier chips in the future and set in their respective metal cavities.

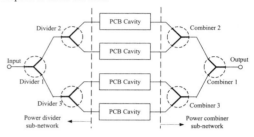

Fig. 1 Logic diagram of the power combining network

### B. Design of the directional coupler

Waveguide directional couplers are used as power divider and power combiner in this design, the structure of an usual broad-wall waveguide directional coupler is shown in Fig.2, the two waveguide are coupled through several slots in the broad wall. When the directional coupler is used as power combiner, the two input ports say port 1 and port 2 are well isolated. However, in a Tee junction power combiner the isolation between the two input ports is good only when the incidences are of same amplitude in phase ( H plane Tee junction) or same amplitude out of phase( E plane Tee junction), and this condition is easy to be damaged in a complex network of long transmission lines, which will finally degrade the performance of the power combining network.

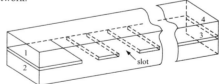

Fig.2 Structure of an usual waveguide broad-wall directional coupler

To reduce the size of the power combining network, the waveguide (WR12) directional coupler is designed in a compact form with only one coupling slot and circular arc structure for matching, see Fig.3, the simulated scattering parameters of the directional coupler are shown in Fig.4, the reflection and isolation are good than -20dB at 75~77GHz.

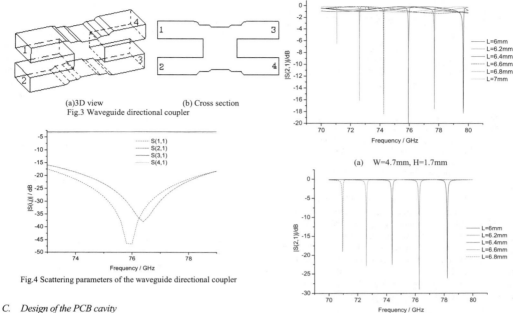

(a)3D view          (b) Cross section

Fig.3 Waveguide directional coupler

Fig.4 Scattering parameters of the waveguide directional coupler

(a)   W=4.7mm, H=1.7mm

(b)   W=4.2mm, H=1.7mm

Fig.6 Transmission coefficient of the rectangular cavity

## C.  Design of the PCB cavity

The transmission performance of the PCB circuit is greatly affected by the shape and size of the metal cavity in which the PCB circuit is installed, and thereby affect the performance of the power combining network. At first, a microstrip line printed on Rogers 5880 dielectric board (thickness 0.127mm) and then installed in a rectangular metal cavity is studied, see Fig.5, where the two ends of the cavity are connected to the waveguide through microstrip-waveguide transitions, making it easy to connect with the power divider network and the power combiner network. The cavity height H is set to 1.7mm, and then calculate the S-parameter of the PCB cavity with different cavity length L and cavity width W using HFSS12. The simulated transmission results of the rectangular cavity are shown in Fig.6, it is clear that for different values of L and W there exist bad performance point at certain frequencies, the reason is supposed to be that there inspired resonant modes in the rectangular cavity, which do damage to the transmission characteristics of the metal cavity.

It is possible to make the bad performance point locating outside of the working frequency band by carefully selecting the sizes of the cavity, for example by selecting W=4.2mm, H=1.7mm and L>6.6mm(or L<6.2mm) there will be no bad performance point in the frequency band 75GHz~77GHz (see Fig.6 (b) ). To further eliminate the bad point, rhombus-like cavity which is helpful to avoid the resonant modes is proposed, see Fig.7, there also added a small rectangular cavity at the side of the rhombus-like cavity for setting auxiliary circuit, and the simulation results of different sizes of L are shown in Fig.8, it is evident that transmission parameters are better than that of the rectangular cavity.

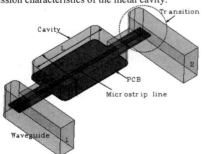

Fig.5 Rectangular PCB cavity

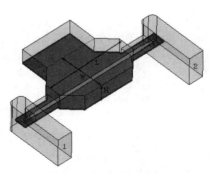

Fig.7 Rhombus-like cavity (W=4.2mm, H=1.7mm)

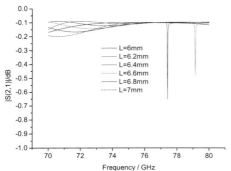

Fig.8 Transmission coefficient of the rhombus-like cavity

*D. Structure of power combining network and simulated results*

Finally, according to the logic diagram in Fig.1, using the waveguide directional coupler in Fig.3 and the Rhombus-like cavity in Fig.7, the whole structure of the power combining network is obtained and shown in Fig.9, the connections between the waveguide directional couplers are carefully designed to reduce the size of the whole structure. Simulated results are shown in Fig.10, and the scattering parameters are good at the design frequency band.

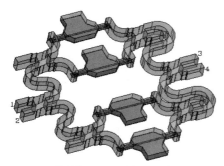

Fig.9 The power combing network

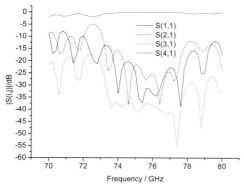

Fig.10 Simulated results of the power combing network

## III. CONCLUSION

A waveguide-based four-way power combining network is designed in this paper, the power combining network contains six waveguide directional couplers and four rhombus-like PCB cavities. The structure of the power combining network is simulated using HFSS 12, the transmission and reflection coefficients are satisfactory at the design frequency band. The required work frequency band is not so wide in this design, so the waveguide directional coupler can be designed in compact size with only one coupling slot, which surely confines the frequency bandwidth of the coupler and accordingly confines the bandwidth of the power combining network. In wide band system, the multi-coupling-slot waveguide directional coupler should be used to broaden the frequency bandwidth of the power combing network.

## ACKNOWLEDGMENT

This paper is supported by NSFC 61071046, NSFC 61101020.

Joint supported by Aeronautical Science Foundation and RF simulation Key Lab of Avionics System.

## REFERENCES

[1] Kai Chang and Cheng Sun, "Millimeter-wave power-combing techniques," *IEEE Transaction on Microwave Theory and Techniques*, Vol.31, No.2, pp.91-107, 1983.

[2] Ge J. X., Li S. F. and Chen Y. Y., "Millimeter wave quasi-optical power combiner," *Electronics Letters*, Vol.27, No.10, pp.880-882, 1991.

[3] M. Kim, E. A. Sovero, J. B. Hacker, et al., "A 100-element HBT grid amplifier," *IEEE Transaction on Microwave Theory and Techniques*, Vol.41, No.10, pp.1762-1771,1993.

[4] Michael P. DeLisio and Robert A. York, "Quasi-Optical and Spatial Power Combining," *IEEE Transaction on Microwave Theory and Techniques*, Vol.50, No.3, pp.929-936, 2002.

[5] M.F. Durkin, R. J. Eckstein, M. D. Mills, et al., "35-GHz active aperture," *IEEE MTT-S Int. Microwave Symp. Dig.*, Los Angeles, CA, pp.425-427, 1 98 l.

[6] J. W. Mink, "Quasi-optical power combining of solid-state millimeter-wave sources," *IEEE Transaction on Microwave Theory and Techniques*，Vol.34, No.2, pp.273-279, 1 986.

[7] Kaijun Song, Yong Fan, and Zongrui He, "Broadband Radial Waveguide Spatial Combiner," *IEEE Microwave and Wireless Components Letters*, Vol.18, No.2, pp.73-75, 2008.

[8] Xiaoqiang Xie, Ruimin Xu, Rui Diao and Weigan Lin, "A New Millimeter-wave Multi-way Power Dividing/Combining Network Based on Waveguide-Microstrip E-plan Dual-Probe Structure," *Global Symposium on Millimeter Waves*, pp.127-159, 2008.

[9] Danyu Wu, Xiaojuan Chen, Gaopeng Chen, Xinyu Liu, "Novel high efficiency broadband Ku band power combiner," *International Conference on Microwave and Millimeter Wave Technology*, pp. 258-261, 2010.

[10] E.G. Wintucky, R.N. Simons, J.C. Freeman, C.T. Chevalier, A.J. Abraham, "High-efficiency three-way Ka-band waveguide unequal power combiner," *IET Microwaves, Antennas and Propagation*, Vol.6, No.11, pp. 1195-1199, 2012.

[11] Kang Zhiyong, Chu Qingxin, Wu Qiongsen, "A compact Ka-band broadband waveguide-based traveling-wave spatial power combiner with low loss symmetric coupling structure," *Progress in Electromagnetics Research Letters*, Vol.36, pp. 181-190, 2013.

# Closed-Form Design Equations for Four-Port Crossover with Arbitrary Phase Delay

Ge Tian[1], Chen Miao[1], Jin-Ping Yang[2,3], Sheng-Cai Shi[2,3] and Wen Wu[1]

*1 Ministerial Key Laboratory of JGMT, Nanjing University of Science and Technology, Nanjing, 210094, China*
*2 Purple Mountain Observatory, CAS, NanJing, 210008, China*
*3 Key Lab of Radio Astronomy, CAS, NanJing, 210008, China*

*Abstract-* **A novel procedure to analyse and design four-port crossover using admittance matrix is proposed. The closed-form design equations for crossover with arbitrary phase delay are obtained. For verification, both simulated and measured results of a fabricated crossover are given. The 15-dB return loss bandwidth is 35.4 %. And the return loss and isolation between adjacent ports are both below -24.7 dB with -0.7 dB insertion loss between the input and output.**

*Index Terms* —**Four-port crossover, admittance matrix, arbitrary phase delay, closed-form design equations.**

## I. INTRODUCTION

With the increasing complexity of microwave integrated circuits, designers are faced with the challenge of layout and routing. When two lines cross over each other, the traditional way to isolate signals on the same intersection area is to use via holes, air bridges, or bond wires. However, these structures have many shortcomings, one of which is that the characteristics of matching and phase delay are no good in higher frequency. Four-port planar crossover, the special case of couplers is a good candidate to solve these problems [1-7]. It has been widely used in Butler matrix for phased array systems [8, 9]. The even-odd-mode method has been used to analyse these structures [10]. But it is difficult to compute and simplify the *S*-parameter expressions manually if the crossover is complicated or cascaded. The admittance matrix has more direct and fundamental corresponding relationship with not only scattering properties but also topology structure. Since every admittance-parameter has a definite meaning, which is the input admittance or the transfer admittance when all other ports are short-circuited. Thus, for a given circuit model, its admittance matrix can be built more easily.

In this paper, a design method for a crossover with arbitrary phase delay based on the admittance matrix is introduced and closed-form design equations are obtained. To verify the design concept, a microstrip crossover worked at 6 GHz is designed. Simulations and measurements are presented to confirm that this approach is easy, simple and efficient.

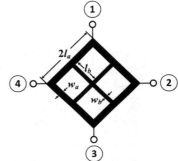

Figure 1. Layout of the crossover.

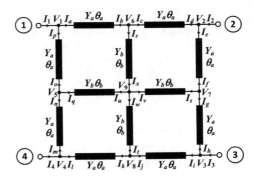

Figure 2. Equivalent circuit model of the crossover.

## II. ADMITTANCE MATRIX APPROACH FOR CROSSOVER

The crossover is a special case of couplers, which allows a pair of signals to cross each other while maintaining isolation between the two signal paths. It also has the matching properties at all ports and a given phase delay $\theta$ between the ports in signal path. Thus, the scattering matrix of the crossover is as follows:

978-7-5641-4279-7

$$S = e^{-j\theta} \begin{bmatrix} 0 & 0 & 1 & 0 \\ 0 & 0 & 0 & 1 \\ 1 & 0 & 0 & 0 \\ 0 & 1 & 0 & 0 \end{bmatrix} \quad (1)$$

By adapting the standard scattering matrix to admittance matrix conversion operation with port admittance $Y_0$, its admittance matrix is obtained immediately:

$$Y = Y_0 (U - S)(U + S)^{-1}$$

$$= jY_0 \begin{bmatrix} -\cot\theta & 0 & \csc\theta & 0 \\ 0 & -\cot\theta & 0 & \csc\theta \\ \csc\theta & 0 & -\cot\theta & 0 \\ 0 & \csc\theta & 0 & -\cot\theta \end{bmatrix} \quad (2)$$

where $U$ is the unit matrix.

It is noted that all of the admittance parameters are zero or pure imaginary number, which means the network is lossless. It is also observed that the admittance matrix is rotational symmetric, which means the network is also rotational symmetric. And then, the crossover can be designed and fabricated using ordinary passive components, such as microstrip lines or lumped components.

Fig. 1 shows the layout of the microstrip crossover. It is composed of a square ring and inner crossed lines. The corresponding equivalent circuit model is depicted in Fig. 2, in which the characteristic admittance and the electric length of the square ring section are $Y_a$ and $\theta_a$. The counterparts of the inner crossed lines are $Y_b$ and $\theta_b$. In order to calculate its admittance matrix, the relationship between branch currents and node voltages is shown as follows:

$$\begin{cases} I_a = Y_{11}^a V_1 + Y_{12}^a V_6 \\ I_b = Y_{12}^a V_1 + Y_{11}^a V_6 \\ I_c = Y_{11}^a V_6 + Y_{12}^a V_2 \\ I_d = Y_{12}^a V_6 + Y_{11}^a V_2 \\ I_e = Y_{11}^a V_2 + Y_{12}^a V_7 \\ I_f = Y_{12}^a V_2 + Y_{11}^a V_7 \\ I_g = Y_{11}^a V_7 + Y_{12}^a V_3 \\ I_h = Y_{12}^a V_7 + Y_{11}^a V_3 \end{cases} \quad (3)$$

$$\begin{cases} I_i = Y_{11}^a V_3 + Y_{12}^a V_8 \\ I_j = Y_{12}^a V_3 + Y_{11}^a V_8 \\ I_k = Y_{11}^a V_8 + Y_{12}^a V_4 \\ I_l = Y_{12}^a V_8 + Y_{11}^a V_4 \\ I_m = Y_{11}^a V_4 + Y_{12}^a V_5 \\ I_n = Y_{12}^a V_4 + Y_{11}^a V_5 \\ I_o = Y_{11}^a V_5 + Y_{12}^a V_1 \\ I_p = Y_{12}^a V_5 + Y_{11}^a V_1 \end{cases} \quad (4)$$

$$\begin{cases} I_q = Y_{11}^b V_5 + Y_{12}^b V_9 \\ I_u = Y_{12}^b V_5 + Y_{11}^b V_9 \\ I_r = Y_{11}^b V_6 + Y_{12}^b V_9 \\ I_x = Y_{12}^b V_6 + Y_{11}^b V_9 \\ I_s = Y_{11}^b V_7 + Y_{12}^b V_9 \\ I_v = Y_{12}^b V_7 + Y_{11}^b V_9 \\ I_t = Y_{11}^b V_8 + Y_{12}^b V_9 \\ I_w = Y_{12}^b V_8 + Y_{11}^b V_9 \end{cases} \quad (5)$$

where
$$Y_{11}^a = -jY_a \cot\theta_a$$
$$Y_{12}^a = jY_a \csc\theta_a$$
$$Y_{11}^b = -jY_b \cot\theta_b$$
$$Y_{12}^b = jY_b \csc\theta_b$$

According to Kirchhoff's current law, the following equations can be obtained:

$$\begin{cases} I_b + I_c + I_r = 0 \\ I_f + I_g + I_s = 0 \\ I_k + I_j + I_t = 0 \\ I_o + I_n + I_q = 0 \\ I_x + I_v + I_w + I_u = 0 \end{cases} \quad (6)$$

$$\begin{cases} I_1 = I_a + I_p \\ I_2 = I_d + I_e \\ I_3 = I_i + I_h \\ I_4 = I_l + I_m \end{cases} \quad (7)$$

Based on the definition of admittance matrix, the admittance parameters corresponding to Fig. 2 can be calculated from equations (3-7):

$$Y' = \begin{bmatrix} Y_{11} & Y_{12} & Y_{13} & Y_{12} \\ Y_{12} & Y_{11} & Y_{12} & Y_{13} \\ Y_{13} & Y_{12} & Y_{11} & Y_{12} \\ Y_{12} & Y_{13} & Y_{12} & Y_{11} \end{bmatrix} \qquad (8)$$

where

$$Y_{11} = 2Y_{11}^a + 2Y_{12} - Y_{13}$$

$$Y_{12} = -\frac{(Y_{12}^a)^2 Y_{11}^b}{2Y_{11}^a Y_{11}^b + (Y_{11}^b)^2 - (Y_{12}^b)^2}$$

$$Y_{13} = -\frac{(Y_{12}^a)^2 (Y_{12}^b)^2}{(2Y_{11}^a + Y_{11}^b)(2Y_{11}^a Y_{11}^b + (Y_{11}^b)^2 - (Y_{12}^b)^2)}$$

Letting the admittance matrix expression (2) be equal to admittance matrix expression (8), three equations with four unknown parameters ($Y_a$, $\theta_a$, $Y_b$ and $\theta_b$) are obtained. Subsequently, the parameters can be derived in term of the given phase delay $\theta$:

$$\tan\theta_a = \pm\sqrt{\frac{3+\cos\theta}{1-\cos\theta}}$$

$$Y_a = \frac{Y_0}{2}\sqrt{\frac{3+\cos\theta}{1+\cos\theta}}$$

$$\theta_b = 90^o \qquad (9)$$

while $Y_b$ can be arbitrary value.

In this paper, we impose $\theta = 40^o$, $Y_0 = 1/50 = 0.02$ S, and $Y_b = 0.0088$ S to give an example of designing microsrtip crossover. So, the parameters of the crossover can be calculated as $\theta_a = 76^o$, $Y_a = 0.0146$ S according to equation (9).

## III. Results and Discussions

The designed crossover is fabricated on Rogers RO4003 substrates with dielectric constant 3.55 and thickness 0.813 mm. It has center frequency of 6GHz. Simulations are carried out with CST Microwave Studio and the measurements are performed using a vector network analyzer (Agilent 8722ES). Following the above approach, 0.3 mm is chosen for the width of inner crossed lines, with $w_a = 1$ mm, $l_a = 7.5$ mm and $l_b = 9.25$ mm.

Simulated and measured frequency responses of the crossover are shown in Fig. 3, together with phase delay between the diagonal ports (port 1 and 3) in Fig. 4. The matching ($S_{11}$) and isolation ($S_{12}$ and $S_{14}$) characteristics are below -24.7 dB and -27 dB at 6 GHz. The insertion loss of transmission characteristics show $S_{13} = -0.7$ dB and the 15-dB return loss bandwidth is 35.4 %. The measured phase delay between the ports in signal path is 42.4°. Fig. 5 shows the photograph of the fabricated crossover.

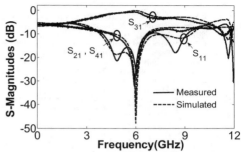

Figure 3. Simulated and measured responses of the crossover.

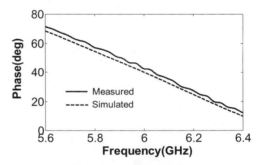

Figure 4. Phase delay between the ports in signal path.

Figure 5. Photograph of the fabricated crossover.

## IV. Conclusion

In this paper, a design method for crossover with arbitrary phase delay is proposed. The method derives closed-form design equations based on the admittance matrix. Simulations and measurements show that the procedure is easy, simple and efficient. One crossover is fabricated and measured for validity confirmation. It shows good insertion loss between diagonal ports and excellent isolation between adjacent ports. In addition, the crossover exhibits an accurate required phase delay between the ports in signal path.

## REFERENCES

[1]  J. J. Yao, C. Lee, and S. P. Yeo, "Microstrip branch-line couplers for crossover application," *IEEE Trans. Microw. Theory Tech.*, vol. 59, no. 1, pp. 87-92, Jan. 2011.

[2]  J. J. Yao, "Nonstandard hybrid and crossover design with branch-line structures," *IEEE Trans. Microw. Theory Tech.*, vol. 58, no. 12, pp. 3801-3808, Dec. 2010.

[3]  Y. C. Chiou, J. T. Kuo, and H. R. Lee, "Design of compact symmetric four-port crossover junction," *IEEE Microw. Wirel. Compon. Lett.*, vol. 19, no. 9, pp. 545-547, Sep. 2009.

[4]  Y. Chen, and S. P. Yeo, "A symmetrical four-port microstrip coupler for crossover application," *IEEE Trans. Microw. Theory Tech.*, vol. 55, no. 11, pp. 2434-2438, Nov. 2007.

[5]  J. Shao, H. Ren, B. Arigong, C. Z. Li, and H. L. Zhang, "A fully symmetrical crossover and its dual-frequency application," *IEEE Trans. Microw. Theory Tech.*, vol. 60, no. 8, pp. 2410-2416, Aug. 2012.

[6]  F. L. Wong and K. K. M. Cheng, "A novel, planar, and compact crossover design for dual-band applications," *IEEE Trans. Microw. Theory Tech.*, vol. 59, no. 3, pp. 568-573, Mar. 2011.

[7]  Z. W. Lee and Y. H. Pang, "Compact planar dual-band crossover using two-section branch-line coupler," *Electron. Lett.*, vol. 48, no. 21, pp. 1348-1349, Oct. 2012.

[8]  Y. S. Jeong and T. W. Kim, "Design and analysis of swapped port coupler and its application in a miniaturized Butler matrix," *IEEE Trans. Microw. Theory Tech.*, vol. 58, no. 4, pp. 764-770, Apr. 2010.

[9]  W. L. Chen, G. M. Wang, and C. X. Zhang, "Fractal-shaped switched-beam antenna with reduced size and broadside beam," *Electron. Lett.*, vol. 44, no. 19, pp. 1110-1111, Sep. 2008.

[10] D. M. Pozar,: "*Microwave Engineering*" (Wiley, New York, 2005)

# Super-resolution and Frequency Spectrum Characteristics of Micro-structured Array Based on Time Reversal Electromagnetic Wave

Huilin Tu, Shaoqiu Xiao, Jiang Xiong and Bingzhong Wang

Institute of Applied Physics, University of Electronic Science and Technology of China

Chengdu, 610054, China

*Abstract*-**This paper has a research on super-resolution and frequency spectrum characteristics of micro-structured array. It proposes a model with different distributions of thin metal wires placed around coaxial probes to verify the filter effect of loaded metal wires on signal frequency spectrum. With time reversal technique, it has been proved that under random loaded metal wires, single-frequency signals have the same super-resolution focusing properties with the broadband signals. Additionally, this paper analyzes and concludes the variation of signal spectrums as well as the effects on super-resolution performance for different uniform distributions of wires. The analyzed and simulated results have important guiding significance to the modeling, quantifiable design and analysis of novel micro-structured array used in multi-antenna wireless communication system with super-resolution characteristics.**

## I. INTRODUCTION

In order to further improve the capacity and rate of wireless communication, compact antenna array adapted in mobile station has received wide attention in recent years, which owns independent channels and spacing much smaller than wavelength. Constructing the technique of super-resolution transmission of electromagnetic waves is important to achieve multi-independent channels in compact space of mobile station. The earliest study of super-resolution phenomenon began with sub-wavelength optical imaging [1-2], which got super-resolution characteristics through probing the high frequency components corresponding with the fine structures of objects in near field. In 2002, Rosny found super-resolution phenomenon in near field by using time reversal (TR) acoustic [3]. TR technique was introduced in electromagnetic fields in 2004 [4]. In 2007, the super-resolution focusing characteristics of far field can be achieved by TR electromagnetic wave in a rich multi-path environment introduced in [5]. Ref. [5] used a type of micro-structured antenna array with /30 spacing which converted evanescent wave to propagating wave to build multi-channel and high-speed wireless communication system. Although the research shows it is possible to construct the compact antenna array in compact space, there is no related design theory of micro-structured antenna array. In 2009, G. D. Ge had done some preliminary explorations about factors which had influences on super-resolution transmission characteristics of TR electromagnetic wave [6]. It is given that the multi-path effect caused by micro-structures is the critical factor to realize

super-resolution characteristics. According to the conclusion, kinds of sub-wavelength antenna arrays etched with different micro-structures were analyzed and designed in [7]-[9], which could be used for high-speed multi-antenna communication systems under TR technique. However, these researches did not find and build general design method of related micro-structures which were designed randomly. So it is necessary to deeply study the interaction principles between micro-structured antenna and electromagnetic field for modeled and quantifiable design of compact multi-antenna system.

This paper takes coaxial probe antennas loaded with thin metal wires as the research object to explore the interaction principles between micro-structured antennas and electromagnetic field. The influence of thin metal wires on frequency response characteristics of radiated signal has been analyzed by comparing the radiation results of different distributions of loaded wires with that of no wires. Based on the antenna model loaded with random thin metal wires, focusing transmission characteristics of single and wide frequency signals has been studied as well as the influences of micro-structures loaded with different distributions of uniform thin metal wires on radiated signals and super-resolution transmission performance. According to these studies, this paper achieves many principles about the interaction between sub-wavelength micro-structured antenna and electromagnetic wave, which have important guiding significance to modeling analysis and quantifiable design of micro-structured antenna array.

## II. SIMULATION AND ANALYSIS OF THIN METAL WIRES STRUCTURES

The frequency response characteristics of the interaction between thin metal wires structures and electromagnetic wave is the basis to understand the property of micro-structured antennas. This paper firstly analyzes the problem with or without thin metal wires placed around the probe antennas as the research object.

As shown in Fig. 1(a), the simulated prototype is placed in open space. The used frequency spectrum band of Gaussian modulated pulse signal is from 2 to 6GHz. The details of the probe array are sketched in Fig. 1(b) which consists of five coaxial probes λ/15 apart from one other numbered 1 to 5, where λ is the wavelength of the central frequency 4GHz. The

978-7-5641-4279-7

length of coaxial probe is $\lambda/4$. The probe array is $3\lambda$ away from the time reversal mirror (TRM), which consists of three bowtie antennas $\lambda/2$ apart from one other numbered 6 to 8 [7]. Here, CST Microwave Studio commercial software is used to simulate the prototype.

To discuss the influences on signal frequency spectrums which TRM receiced under differrent ways of loading thin metal wires, the simulation content contains as follows:

(1) no wires distributed around probes, as shown in Fig. 1(b);

(2) random distribution of thin metal wires around probes, as shown in Fig. 2(a), the length and radius of wires are $0.6\lambda$ and $\lambda/500$, respectively;

(3) uniform distribution of wires as shown in Fig. 2(b), the spacing of wires is $\lambda/25$, the features of wires are the same to 2th content. There are twelve wires around every wire.

The simulation procedure can be described as follows: a pulse signal $i(t)$ which frequency band is from 2 to 6GHz is transmitted from the 3th probe antenna and the signal received at $m$ th bowtie antenna of the TRM is denoted as $r_m(t)$ m=(6,7,8). The procedure is applied to all three contents. To further investigate the filter effect of the thin metal wires, we use the normalized frequency spectrum amplitude of $r_m(t)$ to compare with the normalized frequency spectrum of the input signal, as shown in Fig. 3(a)-(c).

According to the results shown in Fig.3 (a)-(c), the frequency spectrums received at TRM are relatively uniform and smooth in Fig. 3(a). By comparing Fig. 3(b) with Fig. 3(c), it shows that the selected frequency performance at specific frequency points under random distribution is more obvious than that of uniform distribution. This phenomenon can be explained that period structures formed by uniform loaded metal wires show band-gap characteristics at specific frequency points, while non-period structures formed by random loaded wires have wider stop-band characteristics which show the property of narrower selected frequency.

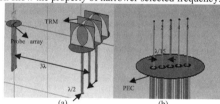

(a)        (b)

Fig. 1 The simulated protype. (a) Probe array, TRM. (b) Probe array

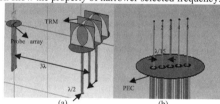

(a)        (b)

Fig. 2. Micro-structured array. (a) Random distribution. (b) Uniform distribution

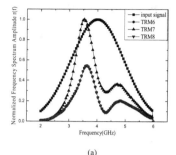

(a)

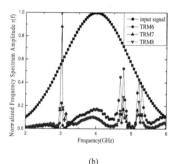

(b)

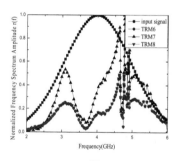

(c)

Fig. 3. Signal frequency spectrum distribution. (a) No thin metal wires. (b) Non-uniform distribution. (c) Uniform distribution

## III. SIMULAITION AND ANALYSIS OF SUPER-RESOLUTION FOCUSING

Under TR technique, in order to determine the influence of the single frequency point corresponding to the peak of frequency spectrum received at 3th antenna based on the micro-structured array shown in Fig. 3(b) on super-resolution property, the simulation content contains as follows: the frequency bandwidth of input signal which 3th antenna transmits are 3.02GHz and 2-6GHz, respectively.

The simulation procedure can be described as follows:

firstly, a pulse signal $i(t)$ is transmitted from the 3th probe antenna and the signal received at $m$ th bowtie antenna of the TRM is denoted as $r_m(t)$ (m=6,7,8). Secondly, $r_m(t)$ is reversed by first-in-last-out way to get $r_m(-t)$. Thirdly, each signal is retransmitted back from the $m$ th bowtie antenna to the micro-structured array at the same time. The same procedure is also applied to single frequency signal, frequency of which is at 3.02GHz. We denote the signals received at the $n$ th micro-structured antenna as $r_n^{TR}(t)$ (n=1, 2, 3, 4, 5). To observe super resolution characteristics after TR clearly, we use the Parserval theorem to express the signal energy in frequency domain:

$$P_n = \sum_k \left| r_n^{TR}(f_k) \right|^2 \times \Delta f \qquad (k = 1, 2, \cdots) \qquad (1)$$

where n is the number of micro-structured array, and $\left| r_n^{TR}(f_k) \right|$ is the frequency spectrum amplitude at frequency $f_k$, which corresponds to the uniform sampling frequency points between 2GHz and 6GHz. $\Delta f$ is the interval of uniform sampling. $P_n$ is the energy of the signal received at the $n$ th micro-structure antenna in the specified frequency bandwidth. To analyze frequency spectrum and super resolution characteristics, we draw out the normalized frequency spectrum amplitude of $r_n^{TR}(t)$ (n=1, 2, 3, 4, 5) and $P_n$ (n=1, 2, 3, 4, 5) of two simulation results, as shown in Fig.4 (a)-(b) and Fig. 5.

According to the results shown in Fig. 4(a), it is clear that the signal frequency spectrums received at probe array are still narrow-band when the model uses random distribution

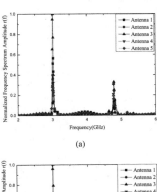

(a)

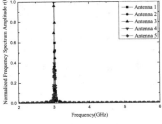

(b)

Fig. 4. Signal frequency spectrum distribution. (a) Bandwidth is from 2GHz to 6GHz. (b) Single frequency at 3.02GHz

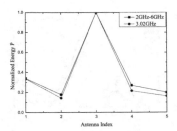

(c)

Fig. 5. Results for different input signals

combined with TR technique. As shown in Fig. 5, energy of the $n$ th (n=1, 2, 4, 5) antenna is 33% lower than the energy of 3th antenna. It could be concluded that the micro-structured array shows super-resolution characteristics clearly and the single frequency point contributes the majority of energy to make the model show super-resolution performance. So under non-period random loaded thin metal wires, the narrow frequency spectrum resulted from the narrow-band characteristics of structures can achieve good super-resolution focusing property. And current literatures show that related researches focus on super-resolution focusing field of ultra-wideband time reversal electromagnetic signal.

## IV. SIMULATION AND ANALYSIS OF DIFFERENT MICRO-STRUCTURES

Under uniform distributions, we denote the length of thin metal wires as $l$ and the spacing of wires as $d$. To discuss influences of the two variables on signal frequency spectrums and super-resolution characteristics which received at the $n$ th (n=1, 2, 3, 4, 5) antenna, this part simulates different metal wires structures with different variable values. The probe array spacing in Fig. 1(b) turns to be $\lambda/4$ to make the variable $d$ to take more values.

Based on the model shown in Fig. 1, we simulate six models as follows: when $l$ is fixed to $0.6\lambda$, $d$ is changed to $\lambda/75$, $2\lambda/75$ and $\lambda/25$, respectively; when $d$ is fixed to $\lambda/25$, $l$ is changed to $3\lambda/5$, $2\lambda/5$ and $\lambda/5$, respectively. The simulation procedure is the same with part III.

Because of the limited space, we only list the normalized frequency spectrum amplitude of two models which have the best super-resolution characteristics, as shown in Fig. 6(a)-(b). And the normalized $P$ by the energy of the 3th antenna of the six models are showed in Fig. 7(a)-(b), respectively.

According to Fig. 5(a), it shows that the signal frequency spectrum received at 3th antenna is clearly different from others. So it is in Fig. 6(a). But when Fig. 6(a) is compared to Fig. 5(a), it has the property of narrow band. Fig. 5(b) and 6(b) show that models with $l = 3\lambda/5$, $d = \lambda/25$ and $l = \lambda/5$, $d = \lambda/25$ have the property of super-resolution at the target antenna which is numbered 3, because the energies of other antennas are all nearly below 50%. When $d$ inceases or $l$ reduces, the coupling effects among probe antennas will decease, thus the

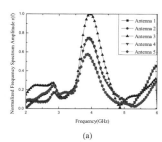

(a)

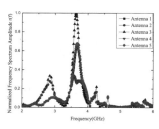

(b)

Fig. 6. signal frequency spectrum distribution. (a) $l = 3/5\lambda$, $d = \lambda/25$. (b) $l = \lambda/5$, $d = \lambda/25$.

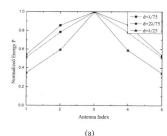

(a)

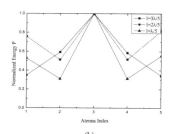

(b)

Fig. 7. results for different varible. (a) $l$ is fixed to $3\lambda/5$, $d$ is changed to $\lambda/75$, $2\lambda/75$ and $\lambda/25$, respectively; (b) $d$ is fixed to $\lambda/25$, $l$ is changed to $3\lambda/5$, $2\lambda/5$ and $\lambda/5$, respectively.

super-resolution property gets more clearly. As the length and spacing of thin metal wires vary, the frequency point corresponding to the peak of signal frequency spectrum received at 1-5 th antenna will take some offsets. In a word, the variables have effects on wideband and super-resolution characteristics of the models. So by analyzing the two variables, we can find out a model to meet the design requirements.

## V. CONCLSION

This paper has studied signal frequency spectrums under different distributions of metal wires based on traditional coaxial probe structures. It shows different filter performances of different distributions of metal wires. With TR technique, the super-resolution focusing performance is determined by the frequency point at the peak of signal frequency spectrum under random distribution of metal wires. The length and spacing of metal wires have important effects on the focusing performance and bandwidths. These results show that we can realize super-resolution characteristics applied in multi-antenna system by quantitative design of the micro-structures. Using the principles concluded from this paper to design super-resolution and ultra-wideband antenna will be the next direction in our research. Other micro-structured general design method different from thin metal wires also needs deep exploration.

## REFERENCES

[1] R. Merlin, "Radiationless Elecrtonmagnetic Interference: Evanescent-Field Lenses and Perfect Focusing,' *Science*, vol. 317, pp. 927-929, August 2007.

[2] A. Grbic A, J. Lei J and R. Merlin, "Near-Field Plates: Subdiffraction Focusing with Patterned Surfaces," *Science*, vol. 320, pp. 511-513, April 2008.

[3] J. D. Rosny and M. Fink, "Overcoming the Diffraction Limit inWave Physics Using a Time-Reversal Mirror and a Novel Acoustic Sink," *Phys.Rev.Lett.*, vol. 89, pp. 124301, September 2002.

[4] G. Lerosey, et al. "Time Reversal of ElectromagneticWaves," *Phys. Rev. Lett.* Vol. 92, pp. 193904, May 2004.

[5] G. Lerosey, J. D. Rosny, A. Tourin and M. Fink, "Focusing Beyond the Diffraction Limit with Far-Field Time Reversal," *Science*, vol. 315, pp. 1120-1122, February 2007.

[6] G.-D. Ge, B.-Z. Wang, H.-Y. Huang and G. Zheng, "Super-resolution characteristics of time-reversed electromagnetic wave," *Acta Phys. Sin.* Vol. 58, 2009.

[7] G.-D. Ge, B.-Z. Wang, D. Wang, D.-S. Zhao, and S. Ding, "Subwavelength Array of Planar Monopoles With Complementary Split Rings Based on Far-Field Time Reversal," *IEEE Trans. Antennas Propag.* 9 vol. 59, pp.4345-4350, November 2011.

[8] G.-D. Ge, R. Zang, D. Wang, S. Ding and B.-Z. Wang, "Subwavelengtharray of planar antennas with defect oval rings based on far-field time reversal," *IET Electronics Letters*, vol. 47, pp. 16-17, August 2011.

[9] Z.-M. Zhang, B.-Z. Wang and G. -D. Ge, "A subwavelength antenna array design for time reversal communication," *Acta Phys. Sin.* vol. 61, pp.058402,2012.

# A Miniaturized Tunable Bandpass Filter Using Asymmetric Coupled Lines and Varactors

Yuhang He, Min Ou, Liguo Sun

Department of Information Science and Technology, University of Science and Technology of China

Email: vivahyh@mail.ustc.edu.com

*Abstract*—Nowadays, it becomes more and more important to miniaturize bandpass filters(BPFs) in RF front-ends because of size requirement of communication systems. As to achieve multifunction of RF front-end, BPFs with tunable passband, on the other hand, is another popular direction of BPF design. In this paper, a new three-order chebyshev BPF using asymmetry coupled lines and varactors, whose passband is widely tunable while the size is extremely reduced at the same time, is presented. Measurement results show that the center frequency of the proposed BPF has an amazing large tuning range from about 521MHz to 1123MHz for different bias voltage of varactors.

*Index Terms*—bandpass filter, miniaturization, tunable

## I. INTRODUCTION

With the development of wireless communication systems, the size of the RF front-end is required smaller and smaller. Since the band pass filter (BPF) is quite an important part of the RF front-end, realizing BPF by new methods that ensure the filter not only reach excellent performance but also meet the requirement of size miniaturization, has become important for the development of microwave filters. Various methods have been reported to achieve miniaturization of BPFs[1]-[4]. Among them, the process using microstrip coupled lines is one of the most popular because of its light weight, easy and low cost of processing. In [3] and [4], a structure of coupled lines with shunt capacitors is used to reduce the size of BPFs, which is very attractive owing to its simplicity.

Meanwhile, tunable filters are also required by the expansion of new communication systems. A number of different design approaches for tunable filters have been proposed[5]-[7]. A tunable filter using p-i-n diodes is given in [5]. Varactors and micro-electromechanical systems (MEMS) are used to make BPFs tunable in [6] and [7], respectively. However, none of the above methods concentrate on the size of the tunable BPFs.

In this paper, we propose a new method for the BPF design based on microstrip asymmetric coupled lines and shunt varactors, by which the center frequency of the BPF can be tuned by varactors, while the size is miniaturized at the same time. In section II, design details about how the size is miniaturized and how the center frequency of the BPF is tunable by the proposed method is given. A third-order miniaturized tunable Chebyshev BPF using the proposed method for verifications is represented in Section III. EM simulation and measurement results of the designed BPF using different capacitors and SMV1148 varactors are shown in Section IV. Finally, a conclusion is drawn in Section V.

## II. PROPOSED NEW METHOD

The principle of the proposed new method is given by two steps. The structure of asymmetric coupled lines and shunt capacitors to realize miniaturization of BPFs is demonstrated in part A. Thereafter, we pay attention to the relation between the center frequency of the BPF and the capacitance. The complete block diagram of miniaturized tunable BPFs is given in part B.

### A. Miniaturization of BPFs Using Coupled Lines and Shunt Capacitors

The structure of asymmetric coupled lines with shunt capacitors is employed for the design of miniaturized tunable BPF. The schematic of an asymmetric coupled-line, whose electrical length is θ, is shown in Figure1.(a). The equivalent circuit for it, in which an transmission line is loaded with two shunt different stubs, is given in Figure1.(b)[8]. In Figure1.(c), the two shunt stubs can be separated into two parts respectively, so that the center part(shadowed), which is a π shaped structure, can be equal to a J-inverter. Finally, the asymmetric coupled lines are expressed by a J-inverter loaded with two new shunt stubs, as Figure1.(d) shows. The characteristic impedances for two stubs are $Y_{s1}$ and $Y_{s2}$, respectively. The value of J-inverter, $Y_{s1}$ and $Y_{s2}$ are calculated as following [4]:

$$J = \frac{(Y_{0\pi}^a - Y_{0C}^a)}{2\sin\theta} \tag{1}$$

$$Y_{S1} = \frac{(Y_{0\pi}^a + Y_{0C}^a)}{2} \tag{2}$$

$$Y_{S2} = Y_{0C}^b + \frac{(Y_{0\pi}^a - Y_{0C}^a)}{2} \tag{3}$$

where $Y_{0C}^a$, $Y_{0\pi}^a$, $Y_{0C}^b$ and $Y_{0\pi}^b$ are the C- and π- mode admittance of strip a and b.

978-7-5641-4279-7

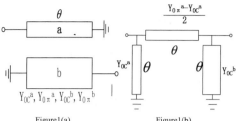

Figure1(a).                    Figure1(b).

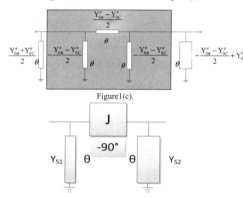

Figure1(c).

Figure1(d).

Figure1. (a) The schematic of asymmetric coupled lines (b), (c) and (d) are the three equivalent circuit of asymmetric coupled lines[4]

When N-1 asymmetric coupled lines are cascaded together with shunt capacitors before and after each, the block diagram is then expressed in Figure2.(a), while the equivalent circuit can be described as a Nth-order BPF containing N shunt resonators($B_1, B_2...B_N$) and N-1 J-inverters, as shown in Figure2.(b). Each shunt resonator is made up of owe or two shunt stubs and one capacitor. If the capacitance of each shunt resonator is the same, then we have [4]:

$$B_1(\omega) = \omega C - Y_{s1,1} \cot \theta \qquad (4)$$

$$B_i(\omega) = \omega C - (Y_{S2,i-1} + Y_{S1,i}) \cot \theta \qquad (5)$$

$$B_N(\omega) = \omega C - Y_{s2,N-1} \cot \theta \qquad (6)$$

where $i = 2,3,...,N-1$, C is the capacitance, $\theta$ is the electrical length, and $B_i$ represents the susceptance of each shunt resonator. As is known to all, the shunt resonators have to resonate at center frequency $\omega_0$, so that we can get

$$B_1(\omega_0) = \omega_0 C - Y_{S1,1} \cot \theta_0 = 0 \qquad (7)$$

then the capacitance is got from

$$C = \frac{Y_{S1,1}}{\omega_0} \cot \theta_0 \qquad (8)$$

$\theta_0$ represents the electrical length of center frequency. What

should be emphasized is $\theta_0$ is determined by designers rather than 90° in a conventional structure without capacitors, so that theoretically the size of the BPF can be miniaturized to arbitrary degree as long as value of C is reasonable..

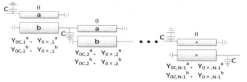

Figure2(a).    The block diagram of miniaturized BPF

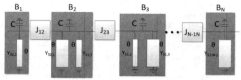

Figure2(b).    The equivalent of the miniaturized BPF

The admittance slope of each section is then

$$b_i = \frac{\omega_0}{2} \frac{dB_i}{d\omega}\Big|_{\omega=\omega_0} = \frac{\omega_0}{2} \frac{dB_1}{d\omega}\Big|_{\omega=\omega_0} = \frac{1}{2} Y_{S1,1} (\cot\theta_0 + \theta_0 \csc^2\theta_0) \qquad (9)$$

where $i = 1,2,...,N$. Thus, the capacitance can be rewritten by $b_i$

$$C = \frac{2b_i}{\cot\theta_0 + \theta_0 \csc^2\theta_0} \frac{\cot\theta_0}{\omega_0} \qquad (10)$$

From (10) it is not difficult to find out that $b_i$ of each section is the same because of the invariable capacitance. On the other hand, the admittance slope of the first section can be determined as follows[8]:

$$b1 = \frac{g_0 g_1}{\Delta Z_0} \qquad (11)$$

$$J_{ii+1} = \Delta \sqrt{\frac{b_i b_{i+1}}{g_i g_{i+1}}} = \frac{\Delta b_1}{\sqrt{g_i g_{i+1}}} \qquad (12)$$

where $\Delta$ is the fractional bandwidth, $Z_0$ and $g_i$ are the characteristic impedance and the i-th section's Chybeshev low-pass prototype value respectively. It is clear that once $\omega_0$, $\Delta$, and the prototype values of Chebyshev low-pass filter are determined, the capacitance and the J-inverter of each section are decided by (10)-(12). Thus, the C- and $\pi$-mode admittance of strip a and b of each section can be calculated by (1) -(3).

However, the physical dimensions of asymmetric lines still need to be solved. Many techniques have been proposed such as Green's function integral equation method[9], full-wave method[10], the approximate method[11] and so on. In this paper, the approximate method in [11] is developed by using numerical computation relation between coupling coefficients and the coupled lines' width of each strip/gap between strips [12].

*B. Further Design to make the Miniaturized BPF Tunable*

Notice that the electrical length can be describe (13)

$$\theta = \beta l = \frac{\omega}{v_p} l$$

where l is the physical length of the coupled-line and $v_p$ is a constant. So (10) can finally be rewritten as

$$C = \frac{2b_1}{\cot(\frac{\omega_0 l}{v_p}) + (\frac{\omega_0 l}{v_p})\csc^2(\frac{\omega_0 l}{v_p})} \frac{\cot(\frac{\omega_0 l}{v_p})}{\omega_0} \quad (14)$$

As is shown in (14), the relation between C and $\omega_0$ notice us that the center frequency can be tuned by changing the value of shunt capacitors. Figure3 gives the relation between $\omega_0$ and C, from which we can observe distinctly that $\omega_0$ increases rapidly as the capacitance becomes smaller. Therefore, a miniaturized tunable BPF, whose block diagram is shown in Figure 4, center frequency tunable by capacitors in series with varactors, instead of capacitors only, is demonstrated. $C_{CAP}$ and $C_{VRA}$ are the capacitance of the capacitor and the varactor, respectively.

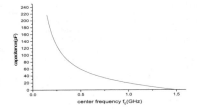

Figure3. Relation between center frequency and capacitance (l=9.35mm)

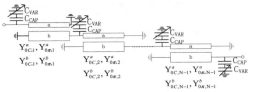

Figure4. The block diagram of the tunable miniaturized BPF

## III. DESIGN OF A THIRD-ORDER MINIATURIZED TUNABLE BPF

A third-order Chebyshev miniaturized tunable BPF with a 0.1-dB passband ripple level and 10% bandwidth is designed by the proposed method for verifications. The third-order filter has two sections of asymmetric coupled lines, at which frequency $\theta_0$ is miniaturized to 30° at $\omega_0$=500MHz to define physical dimensions. According to (1)-(12), we can get parameters such as the capacitance C, and $Y^a_{0C,i}$, $Y^a_{0\pi,i}$, $Y^b_{0C,i}$, $Y^b_{0\pi,i}$(i=1,2), the C- and $\pi$- mode admittance of strip a and b easily. The BPF is fabricated on a ROGERS RT6010

substrate with a relative dielectric constant of 10.2 and a thickness of 0.635mm. The physical dimensions are listed in Table 1, where $W_a$ is the width of strip a, $W_b$ is the width of strip b, S is the Gap spacing between strip a and b. Besides, the photograph of the BPF is given in Figure5. The whole size of the BPF is only 20.8mm*5.7mm.

TABLE I
PHYSICAL DIMENSIONS OF THE FILTER

|  | $W_a$ | $W_b$ | S | Length |
|---|---|---|---|---|
| Section1 | 1.7 mm | 0.6 mm | 0.5 mm | 9.35 mm |
| Section2 | 0.6 mm | 1.7 mm | 0.5 mm | 9.35 mm |

Figure5. The photograph of the tunable miniaturized BPF

## IV. EM SIMULATION AND MESUREMENT RESULTS

Different values of capacitors and varactors in series with capacitors are both tested for the designed third-order BPF to make comparisons. Figure6 gives simulation and measurement results when different Murata GQM type capacitors are shunted. The capacitance is 27pF, 15pF, 10pF and 5pF, respectively, while the center frequency is 550MHz, 717MHz, 837MHz and 1168MHz, respectively.

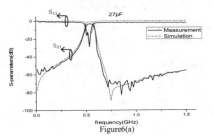

Figure6(a)

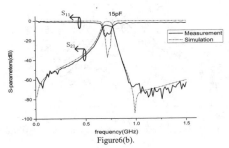

Figure6(b).

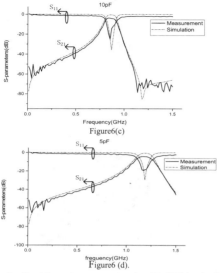

Figure6(c)

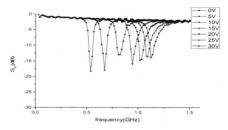

Figure7(b). Measured S$_{11}$ parameters of the BPF with different bias voltage

REFERENCES

[1] J.-S. Hong and M. J. Lancaster, *"Development of new microstrip pseu-do-interdigital bandpass filters,"* IEEE Trans. Microw. Theory Tech., vol. 5, no. 8, pp. 261–263, Aug. 1995.

[2]C.-H. Liang and C.-Y. Chang, *"Compact wideband bandpass filters using stepp-line resonators and interdigital coupling structures,"* IEEE Mcrowave Wireless Compon. Lett., vol 19, no. 9, pp.551-553, Sep 2009

[3]S. Lee and Y. Lee, *"Generalized miniaturization method for coupled-line bandpass filters by reactive loading,"* IEEE Microw. Theory Tech.,vol . 58, no. 9, pp. 2383–2391, Sep. 2010.

[4]J.H. Park, S. LEE and Y. lee, *"Generalized miniaturization method for couled-line bandpass filters by reactive loading,"* Microwave Theory and Techniques IEEE Transactions on 60.2 (2012): 261-269.

[5]Shu, Y-H., Julio A. Navarro, and Kai Chang,"*Electronically switchalb_ e and tunable coplanar waveguide-slotline band-pass filters,"* Microwave Theory and Techniques, IEEE Transactions on 39.3 (1991): 548-554

[6]Kim, Byung-Wook, and Sang-Won Yun, *"Varactor-tuned combline band-pass filter using step-impedance microstrip lines,"* Microwave The_ ory and Techniques, IEEE Transactions on 52.4 (2004): 1279-1283.

[7]Islam, Md Fokhrul, M. Ali, and B. Yeop Majlis. *"Tunable bandpass fi_ lter using RF MEMS variable capacitors,"* Microwave Conference, 2008. APMC 2008. Asia-Pacific. IEEE, 2008.

[8]Matthaei, George L., Leo Young, and E. M. T. Jones, *"Microwave filt_ ers, impedance-matching networks, and coupling structures,"* New York: McGraw-Hill, 1964.

[9]V. K. Tripathi and C. L. Chang, *"Quasi-TEM parameters of non-sysm_ metrical coupled microstrip lines,"* Int. J. Electron., vol. 45, no. 2, pp.251-223, Aug. 1978.

[10]R. K. Mongia, I. J. Bahl, P. Bhartia, and J. Hong, *"RF and Microwave Coupled-Line Circuits,"* Dedham, MA: Artech House, 2007.

[11]P.-K. Ikalainen and G.-L. Matthaei, *"Wide-band, forward-coupling microstrip hybrids with high directivity,"* IEEE Trans. Microw. Theory Tech., vol. 35, no. 8, pp. 719–725, Aug. 1987.

[12] S. Kal, D. Bhattacharya and N.-B. Chakraborti, *"Normal-mode para_ meters of microstrip and coupled lines of unequal width,"* IEEE Trans. Microwave Theory & Tech., vol. 32, no. 2, pp. 198–200, Aug. 1987.

Figure6 (d).

Figure6. Simulation and measurement results of the BPF shunting (a)27pF, (b)15pF, (c) 10pF, (d)5pF capacitors

Then we use capacitors in series with SMV 1148 varactors to replace capacitors so as to make center frequency tunable. The measurement results of S-parameters with different bias voltage on varactors are given in Figure7. The center frequency changes from 521MHz to 1123MHz and the miniaturization ratio varies from 65% to 25%.

## V. CONCLUSION

A new method for miniaturized tunable BPF is proposed by using asymmetric coupled lines with shunt capacitors cascaded by varactors in this paper. Design detail of the method is represented. A third-order bandpass filter, whose center frequency can be tuned from 521MHz to 1123MHz is designed, simulated and fabricated. Experiment results show good agreement with simulation ones, although there is slightly difference in passband flatness and fractional bandwidth.

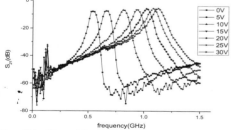

Figure7(a). Measured S$_{21}$ parameters of the BPF with different bias voltage

# Analysis of Terahertz Smith-Purcell Radiation Generated from Tapered Grating by PIC Simulation

Wenxin Liu, Zhao Chao and Yong Wang

Key Laboratory of High Power Microwave Sources and Technologies, Chinese Academy Science, Beijing,100190, China

Institute of Electronics, Chinese Academy Science, Beijing,100190, China

*Abstract-* A novel method for generating Terahertz (THz) Smith-Purcell (SP) radiation from a tapered gating is presented in this paper. For analyzing the characteristics of this kind of grating, the three-dimensional (3D) particle-in-cell simulations are employed, the time of start oscillator, saturated power, the electric field near the tapered grating are analyzed with the help of three-dimensional PIC simulation. On the other hand, the radiation power generated from the tapered grating and normal one is compared. The results of PIC simulations show that the radiation power can be remarkably enhanced and the time of start oscillation can be reduced by the tapered grating.

## I. INTRODUCTION

As we all know, SP radiation is emitted when an electron passes near the surface of a periodic metallic grating [1]. The radiation wavelength λ observed at the angle θ measured from a direction of surface grating is deter-mined by

$$\lambda = \frac{D}{|n|}(\frac{1}{\beta} - \cos\theta),\tag{1}$$

where $D$ is the grating period, $\beta c$ the electron velocity, $c$ the speed of light, and the integer $n$ the spectral order. In recent years, there is a substantial interest in the super-radiant SP radiation, since it is a promising alternative in the development of THz sources [2-4]. The THz sources, a currently active research area, are importance of in a variety of applications to nanostructures, medical and industrial imaging, and material science[5,6]. In order to improve the performance of such kind of device, it is necessary to find an efficient mechanism for the beam-wave interaction. Many methods and theories have been studied. The THz Smith-Purcell radiation is enhanced by making use of two-stream instability, Kim et al use the prebunched beam for the improvements of THz radiation, Li[7] et al use the grating with a sidewall to confine with the electron beam for the improvement of Smith-Purcell Radiation, etc. As far as we know, many methods about improvement the THz radiation characteristics are in terms of electron beam.

In present work, a tapered grating for generating THz Smith-Purcell radiation is presented. The radiation characteristics are analyzed with the help of 3D PIC simulation, on the other hand, the saturated power, the saturated time and electric field are studied. The results of PIC simulations show that the radiation power can be remarkably enhanced and the saturated time can be reduced by the tapered grating.

The organization of present work is organized as follows: the descriptions of PIC simulation are depicted in section 2, the results and discussions are in the section 3, and some conclusions are presented in Section 4.

## II. PIC SIMULATION

### 2.1 Descriptions of Simulation Geometry

The basic simulation geometry is shown in Fig. 1. A grating with rectangular form is set at the center of the bottom of simulation box and a Cartesian coordinate system is adopted with the origin at the center of the grating. The surface of the grating is assumed to consist of a perfect conductor whose grooves are parallel and uniform in the z-direction. For the x-directions, the depth of groove is identified called normal grating, shown in Fig. 2. For the improvement of Smith-Purcell radiation, a tapered grating is adopted. The depth of grating is from shallow to deep, which is called tapered grating, shown in fig.3, the depth of grating taper is 0.01mm in present work.

The beam-wave interaction and radiation propagation occur in the vacuum area, which is enclosed by a special region (called *free-space* in CHIPIC[8] language), where the incident electromagnetic waves and electrons can be absorbed. The whole simulation area is divided into a mesh with rectangular cells of small size in the region of beam propagation and grating, and large size in the rest of region.

The code uses the finite-difference time-domain (FDTD) method [8] to solve Maxwell's equations for the electro-magnetic fields and the motion of charged particles is found by numerically integrating the relativistic Lorentz force equations. The continuity equation is solved, yielding current and charge densities needed for a consistent solution. Various sorts of boundary conditions may be enforced and the properties of the materials forming the boundaries may be specified.

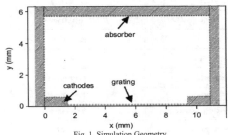

Fig. 1 Simulation Geometry

978-7-5641-4279-7

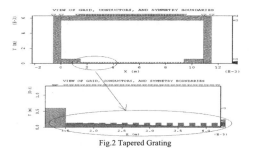

Fig.2 Tapered Grating

Fig.3 Normal Grating

## 2.2 Description of Simulation Parameters

The main parameters of the grating and electron beams are summarized in Table 1. The periodic of grating is 0.2mm, the width and depth of groove are equal to 0.1mm. Since it is a two-dimensional simulation, it assumes that all fields and currents are independent of the z-direction. The current and voltage of electron beam system is 1000A per meter and 40kV, respectively. We point out the choice of such a large value of current to speed up the long computations, and note that the

Table I

MAIN PARAMETER FOR SIMULATIONS

| Parameters | Single beam |
|---|---|
| Grating Period | D=0.2mm |
| Groove width | d=0.1mm |
| Groove depth | h=0.1mm |
| Depth taper | Ht=0.01mm |
| Beam voltage | V=40.0kV |
| Beam current | I=1000A/m |
| Beam thickness | 2b=0.04mm |
| Beam-grating distance | R=0.04mm |
| Guiding magnetic | Bx=2T |

current value mentioned in this paper represents the current per meter in the z-direction. The external magnetic field with 2 Tesla is used in order to ensure stable beam propagation above the grating. As to the diagnostics, CHIPIC allows us to observe a variety of physical quantities such as electromagnetic fields as functions of time and space, power outflow, and electron phase-space trajectories, etc. We can set the relevant detectors anywhere in the simulation area.

### III. RESULTS OF PIC SIMULATION

Recently, Li and coworkers [2,7], Donohue and Gardelle[9] have addressed the superradiant SP radiation with the help of 2D or 3D PIC simulations. They studied the evanescent wave, electron beam bunching and radiation gain, the loss of grating, etc. Shin[10] reported that evanescent tunneling transmission of effective surface plasmon polaritons between two counterstreaming beams noticeably increased the SP radiation. Here, we focus on the enhancement of terahertz SP radiation by the tapered grating.

### 3.1 Dispersion Characteristics

According to refs.11, the dispersion curves along with the beam line for 40kV are given in Fig.4. The solid line is theoretically calculated and the square dots are obtained from the PIC simulations. From the Fig.4, the dispersion curve of simulation is kept in agreement with the theoretical calculation. The frequency of operating point (solid dot) is about 391.5GHz, it shows the SP operates at backward wave.

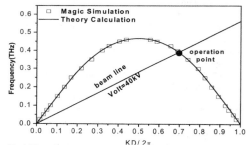

Fig.4. Dispersion curves, parameters: period D=0.2mm, slot width d=0.1mm and groove depth h=0.1mm.

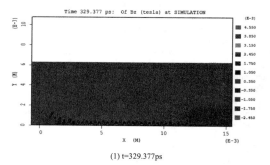

(1) t=329.377ps

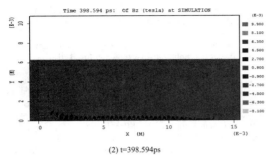

(2) t=398.594ps

Fig.5 The Bz magnetic field

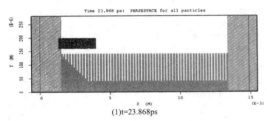

(1)t=23.868ps

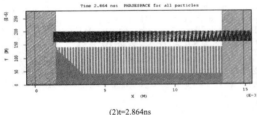

(2)t=2.864ns

Fig. 6 Bunching beam in phasespace at different time

## 3.2 Radiation Characteristics of Tapered Grating

To know the radiation characteristics of tapered grating, the radiation magnetic field Bz and bunched beam are studied with the help of 3D PIC simulations.

Figure 5 is the *z*-direction of magnetic field (Bz) distribution. Fig.5(1) is Bz field at the moment of 329.377ps, and Fig.5(2) is that of 398.594ps. From the amplitude of Bz, we can find that the amplitude of Bz is increased at the negative x-direction. This shows that the THz SP device operates at the backward wave. This result is consistent with that dispersion curve.

Figure 6 is the bunching beam at the different time at the THz SP tapered device. From the electron beam bunching beam in the phase-space, the electron beam can be bunched by the beam runs a distance, which is the least length of grating, otherwise, the electron beam will not be bunched. On the other hand, the distance between the bunched beam is equal to grating period. It shows the electron beam is bunched due to the periodical magnetic and electric field near the grating.

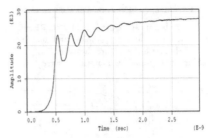

Fig.7 Output Power

Fig.8 Frequency of output Power

## 3.3 Radiation Power

The radiation power of THz SP radiation generating from the tapered grating is shown in Fig.7. This figure shows the time of start oscillator is about 0.55ns, and it reaches the saturated power is about 2.5ns, the amplitude is about 28kW.

The frequency of radiation signal is shown in Fig.8, which is obtained by the DB scale of fig.7, and the operating frequency is about 0.38THz. The operation frequency is lower than the theoretical calculation because of space charge and finite boundary condition during the PIC simulation, however, this case in the theoretical calculation is not considered.

On the other hand, the comparisons of radiation power generated from the tapered grating with normal grating are shown in Fig.9.From this figure, the time of start oscillator is shorter than the normal grating, the former time is about 0.55ns, and the latter time is about 0.8ns. Moreover, the saturated time of tapered grating is shorter than that of normal grating, the former is about 2.5ns and the latter time is about 3.0ns, the radiation power of tapered grating is about 28kW, and that of normal grating is about 22kW. The enhancement of radiation power, shorten saturated time and time for start oscillation are because of the tapered grating, which leads to the increasing impedance.

Moreover, to know the enhancement of radiation power for tapered grating, the longitudinal electric field is analyzed by the PIC simulation, the results are shown in fig.10. The longitudinal electric field is stronger than that of normal grating , which can result to the stronger bunching beam in the tapered grating system, leading to the enhancement of radiation power. The saturated time can be shortened by the stronger electric field.

## IV. EXPERIMENT SETUP FOR TERAHERTZ RADIATION

The electron beam can be produced by the RF photocathode driven by femosecond laser system, and the experimental setups for THz Smith-Purcell are shown in Fig.11. The charge of bunching can be tested by intergration current transformer.

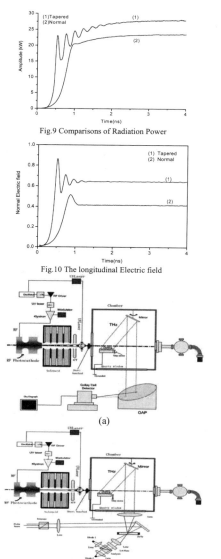

Fig.9 Comparisons of Radiation Power

Fig.10 The longitudinal Electric field

Fig.11. The experimental setup: (a) power measurement (b)frequency measurement;

When the THz wave escapes from the vacuum chamber, its frequency can be measured by the improved Martin-Purplett interferometer, shown in Fig. 11(a). And the intensity of THz CTR can be detected by Gollay cell detector combinations with off-axis parabolic mirrors, shown in Fig. 11(b).

## V. CONCLUSIONS

THz SP radiation producing from a tapered grating is presented in this paper, and the radiation characteristics are investigated with the help of 3D PIC simulations, moreover, the experiment setup of THz SP radiation is displayed. The enhancement of radiation power, the shorten time of start oscillation and saturated time are observed through the PIC simulations. On the other hand, the longitudinal electric field because of the tapered grating, which leads to the increase of coupling impedance, is observed in this work. The results of PIC simulation show that the radiation characteristics can be improved by the using of a tapered grating, which has some unique prospects in the THz radiation power.

### ACKNOWLEDGMENT

This work was supported in part by National Natural Science Foundation of China (Grants No: 10905032, 11275004), and in part by Knowledge Innovation Project of The Chinese Academy of Sciences (Grants No:YYYJ-1123-5), and in part by Science Foundation of The Chinese Academy of Sciences(Grants No:CXJJ-11-M33), and in part by National High Technology Research and Development Program of China(Grants No: 2011AA8122007A).

### REFERENCES

[1]  S.J.Smith, E.M.Purcell, Phys.Rev. 92, (1953)1069
[2]  W.Liu, Z.Yang, et al, Nuclear Instruments and Methods in Physics Research A 570 (2007) 171
[3]  J.Urata,et al,Phys.Rev.Lett.80(1998) 51
[4]  S.E.Korbly,A.S.Kesar, etal, Phys. Rev.Lett.,94 (2005), 054803
[5]  J.Faist,et al, Science, 264 (1994) 5
[6]  X.-C. Zhang, B. Hu, J. Darrow, and D. Auston, Appl.Phys. Lett. 56(1990)1011.
[7]  D. Li, , et al, Appl. Phys. Lett. 91, 221506(2007)
[8]  D. Jun, et al., Electro-magnetic Field Algorithm of the CHIPIC Code, 2005
[9]  J.T. Donohue, J. Gardelle, Phys. Rev. ST Accel. Beams 9 (2006) 060701.
[10] Y.-M. Shin, J.-K. So, K.-H. Jang, J.-H. Won, A. Srivastava, G.-S.Park, Appl. Phys. Lett. 90 (2007) 031502
[11] H.L. Andrews, C.A. Brau, Phys. Rev. ST Accel. Beams 7 (2004)070701

# An Efficient Modal Series Representation of Green's Function of Planar Layered Media for All Ranges of Distances from Source Using CGF-PML-RFFM

A. Torabi, A. A. Shishegar

Department of Electrical Engineering, Sharif University of Technology, Tehran, Iran

*Abstract-* **A closed-form series representation for spatial Green's function of planar layered media for all distances from source, is presented. By terminating the structure by perfectly matched layer (PML) that is backed by perfect electric conductor (PEC) in semi-infinite layer at the top and/or bottom, the discreet set of surface wave (SW) poles is complemented by eigenmodes of the closed structure by PML which construct the continuous spectrum contribution of the original structure. Then applying characteristic Green's function (CGF) technique, a closed-form representation of spatial Green's function is derived. Very close to the source, where the large number of modes must be considered, the method is become inefficient. By combining CGF technique and rational function fitting method (RFFM), Green's function of very near field would be efficiently constructed with few number of poles extracted in modified VECTFIT algorithm in similar form of CGF-PML result. In this way, an efficient modal series representation is derived by using CGF-PML and CGF-RFFM for far from and close to the source respectively. The main advantage of this representation is that for desired accuracy the number of required modes is controllable. Excellent agreements with direct numerical integration of the spectral integral are shown in several examples.**

## I. INTRODUCTION

One of the most extensively studied topics in electromagnetics is the analysis of printed circuits embedded in planar layered media. To have excellent accuracy and also fast computation, integral equation based techniques could be used. In these methods such as electric field integral equation (EFIE) potential form of the Green's functions are required. The spatial domain Green's function of vector and scalar potential are represented as oscillatory Sommerfeld integrals and generally do not have analytical solution. Due to highly oscillating integrands and slow decaying, numerical integration is expensive.

In the literature, different forms of series representation have been proposed for layered media. It is showed that by combining discrete complex images technique (DCIT) with characteristic Green's function (CGF) method and using well-known Weyl's identity, a closed form expression for spatial Green's function can be obtained [1], [2]. CGF-CI is fast since it needs no numerical integration attempt. In DCIT [3], the most important and cumbersome steps is the analytic extraction of the surface wave poles and also quasi-static part of the spectrum before the complex exponential approximation via the generalized pencil of function (GPOF) method [4]. Although by choosing the sampling path

attentively, extraction of surface waves part can be removed but for long distances from the source, extraction of surface wave poles is usually necessary for high precision. Numerical technique like finite difference (FD) [5] can be convenient to implement but it has not sufficient accuracy specially for near field regions. Another interesting technique to evaluate the Sommerfeld integrals resulting in series presented in [6] which is based on a reduced order modeling of differential equation of spectral domain. By applying VECTFIT algorithm for spectral Green's function in rational function fitting method (RFFM), a uniform series expression for spatial Green's function can be obtained [7], [8]. The main challenge of RFFM is that the generated complex poles behave exponentially increasing in semi-infinite layer at the top and/or bottom like leaky wave poles. Perfectly matched layer (PML) was introduced by Berenger has been used to analyze open waveguide problems involving microstrip substrate. In this technique, by adding a metal backed PML to an open layered media, the structure becomes closed without changing its electromagnetic behavior. Then, an efficient continuous spectrum contribution in terms of a set of discreet modes of the closed waveguide is possible [8].

In this paper, combination of characteristic Green's function method and PML technique is used for series representation of spatial domain Green's function of infinite planar dielectric substrate. The main advantage of CGF-PML result is the analytically knowledge of source and observation points dependencies. Very close to the source, to have sufficient accuracy, a large number of PML modes must be considered in derived series expression which makes the method inefficient due to slowly convergent. For very near field regions combination of CGF method and RFFM is utilized. In this way field distribution could be expressed with few numbers of poles whereas the final form of CGF-RFFM is similar to CGF-PML. Therefore, with finite number of poles, a uniform series representation of spatial Green's function is derived for all distances from the source.

This paper is organized as follows. CGF-PML is briefly described in section II. CGF-RFFM is studied in section III. Numerical results and validation of the method are presented in section IV. Eventually, conclusion is provided in section V. Meanwhile $e^{j\omega t}$ time dependency is used.

## II. CGF-PML FORMULATION

For Green's function of magnetic vector potential, $A_z$, the

978-7-5641-4279-7

following Helmholtz's equation

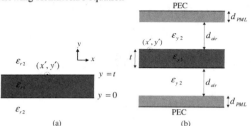

(a)

Figure 1. (a) A line source on a infinite dielectric substrate, (b) A line source on an infinite dielectric substrate closed structure by PML.

$$[\nabla^2 + k_0^2 \varepsilon_r(x,y)]A_z(x,y;x',y') = -\delta(x-x')\delta(y-y'), \quad (1)$$

should be solved along with proper boundary conditions. In the CGF formulation [1], [2], the studied structure is separated into two 1-D layered media (called $N_x$ (read normal to x) and $N_y$ (read normal to y)) i.e.

$$\varepsilon_r(x,y) = \varepsilon_x(x) + \varepsilon_y(y). \quad (2)$$

In this way, $G_x$ and $G_y$ are solutions of the 1-D Helmholtz's equation in the $N_x$ and $N_y$ layered media respectively. That is

$$\frac{d^2 G_\gamma}{d\gamma^2} + \left(\varepsilon_r(\gamma)k_0^2 + \lambda_x\right)G_\gamma = -\delta(\gamma-\gamma'), \quad (3)$$

where $\gamma = x$, $y$ and $\lambda_x + \lambda_y = 0$. For an infinite dielectric substrate shown in Fig. 1(a), it can be shown that this separation is rigorously possible [1], [2]. One choice for the $N_x$ and $N_y$ structures' parameters is: $\varepsilon_{x1} = \varepsilon_{x2} = 0$, $\varepsilon_{y1} = \varepsilon_{r1}$ and $\varepsilon_{y2} = \varepsilon_{r2}$. Therefore, using CGF technique formulation, $A_z$ of the original 2-D Helmholtz's equation is represented exactly in terms of its Characteristic Green's function $G_x$ and $G_y$ as [1]

$$A_z(x,y;x',y')$$
$$= \left(\frac{-1}{2\pi j}\right)\oint_{C_{\lambda_y}} G_x(x,x',-\lambda_y)G_y(y,y',\lambda_y)d\lambda_y, \quad (4)$$

$$G_y(y,y',\lambda_y) =$$
$$\frac{\left(1+R_y e^{-j2\beta_{y1}y_<}\right)\left(1+R_y e^{-j2\beta_{y1}(t-y_>)}\right)e^{-j\beta_{y1}(y_>-y_<)}}{2j\beta_{y1}\left(1-R_y^2 e^{-j2\beta_{y1}t}\right)}, \quad (5)$$

$$R_y = \frac{\beta_{y1}-\beta_{y2}}{\beta_{y1}+\beta_{y2}}, \quad \beta_{y1,2} = \sqrt{\varepsilon_{y1,2}k_0^2 + \lambda_y},$$

$$G_x(x,x',\lambda_x) = \frac{e^{-j\sqrt{\lambda_x}|x-x'|}}{2j\sqrt{\lambda_x}}, \quad (6)$$

and integration contour $C_{\lambda_y}$ should enclose only the singularities of $G_y$ characteristic Green's functions, in a counterclockwise sense as shown in Fig. 2(a) [1]. Numerical integration of the integral in (4) is very expensive and time-consuming. To circumvent this integration we can use PML technique. Consider again an infinite substrate like

$N_y$ structure which is shown in Fig. 1(a) with thickness of $t$. Then, above and below the substrate, an air regions are considered with thickness of $d_{air}$ terminated by a PMLs that are backed by PEC shown in Fig. 1(b). For PMLs, thickness of $d_{PML}$ and material parameters $\kappa_0$ and $\sigma_0$ are considered. It is shown that by stretching the coordinates, the air region can be combined with the PML to form a single layer with complex thickness $\tilde{d}_{air} = d_{air} + d_{PML}(\kappa_0 - j\sigma_0/\omega\varepsilon_0)$ [10]. In this way, modal analysis of the closed structure would be relatively simple.

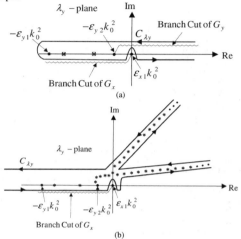

(a)

(b)

Figure 2. Integration path in $\lambda_y$ plane, (a) ⊠: SW poles of $G_y$ (b) ✳: SW poles of $G_y^{(c)}$.

If PMLs properly work, then CGF $G_y$ of open structure in Fig. 1(a) would be the same as the CGF of the closed structure of Fig. 1(b) which is named by $G_y^{(c)}$ ((c) denotes 'closed') which can be achieved simply by using usual spectral techniques. CGF $G_y^{(c)}$ has no branch cut and branch point singularities. Whereas, $G_y^{(c)}$ will have infinite discreet poles singularities which can be computed by argument principle method (APM) [11]. The poles of $G_y^{(c)}$ are the eigenmodes of the closed structure and can be categorized into odd and even TE modes. Then, closed path $C_{\lambda_y}$ must include infinite number of discreet poles of $G_y^{(c)}$, which is demonstrated in Fig. 2(b). Now by using well-known Cauchy-Riemann theorem for closed path $C_{\lambda_y}$ in Fig. 2(b), a series representation for integral of (4) can be obtained as (7) which can be truncated by definite number of PML poles efficiently. Except real poles, two types of eigenmodes can be regarded which are called *Berenger* and *quasi-leaky* wave poles. It must be noted that independent of source and observation location, extraction of PML poles must be done for once and therefore for multiple $\mathbf{r}$ and $\mathbf{r}'$, $A_z$ is simply computed by (7).

$$A_z(x,y;x',y') = \sum_{p=1}^{\infty} -\text{Res}_p \frac{e^{-j\beta_{xp}|x-x'|}}{2j\beta_{xp}}, \quad (7)$$

where $\beta_{xp} = \sqrt{-\lambda_{yp}}$ and $\text{Res}_p = \lim\limits_{\lambda_y \to \lambda_{yp}} (\lambda_y - \lambda_{yp}) G_y^{(c)}$.

### III. CGF-RFFM FORMULATION

For large distances from the source (large $|x - x'|$), series of equation (7) may be truncated by finite number of poles $N_{PML}$ poles) with good accuracy. This is because of that the higher order modes have large absolute value of imaginary part. But for very close to the source considerable number of modes must be considered. In order to have efficient series representation CGF-RFFM is utilized for small $|x - x'|$. Using modified VECTFIT (vector fitting) algorithm [8] for CGF $G_y$,(5) will give

$$G_y(y, y', \lambda_y) \approx \sum_{p=1}^{N_{RFFM}} \frac{\text{Res}_p}{\beta_x^2 - \beta_{xp}^2} = \sum_{p=1}^{N_{RFFM}} \frac{-\text{Res}_p}{\lambda_y - \lambda_{yp}}. \quad (8)$$

$\text{Res}_p$ is the p*th* pole and residue respectively and $N_{RFFM}$ is the total number of poles (including SW poles) for approximating $G_y$. It is assumed that the poles are extracted on the proper Riemann sheet. For (8) uniformly sampled points are used i.e. $\beta_x \in [-t_{max}k_0 - j\delta, t_{max}k_0 + j\delta]$, where $\delta$ is a small number or even zero and $t_{max}$ must be greater than $\sqrt{\varepsilon_{r,max}}$ ($\varepsilon_{r,max}$ is the maximum value of relative dielectric constant), in order to ensure that all the SW poles are included. To obtain more accurate approximation larger $t_{max}$ with more sample points can be used. Modified VECTFIT algorithm is so fast and will converge almost with arbitrary initial guesses for poles and residues. By substituting (9) in (4) for integration path $C_{\lambda_y}$ shown in Fig. 3, we will have

$$A_z(x, y; x', y') \approx \sum_{p=1}^{N_{RFFM}} -\text{Res}_p \frac{e^{-j\beta_{xp}|x - x'|}}{2j\beta_{xp}}. \quad (9)$$

Except surface wave poles of the original structure, some other complex poles which are extracted in VECTFIT algorithm in rational function fitting are responsible to construct the field comes from the continuous spectrum contribution. In fact in the CGF-RFFM against CGF-PML, the obtained poles are dependent to source and observation point location. Therefore, for specific source and observation point location, the best complex poles which result in exact $A_z$ are produced in CGF-RFFM. This is the main disadvantage of CGF-RFFM which makes this method inefficient in large scale problem. But for small distances from the source, to have result similar to CGF-PML, it is useful to utilized CGF-RFFM. The main reason is that by few number of complex poles, very near field can be expressed in a series form, although for each $y$ and $y'$ configuration, a separate VECTFIT algorithm must be run. VECTFIT algorithm is fast enough and is more inexpensive than CGF-PML series computation with large number of poles for small $|x - x'|$.

### IV. NUMERICAL RESULTS

In this section some numerical examples are shown to prove the efficiency and versatility of the proposed method. The codes have been carried out on a 2.26 GHz personal computer.

Let us first define a parameter $d_s$ which denotes the distance from the source where for $|x - x'| < d_s$ and $|x - x'| > d_s$,

CGF-RFFM and CGF-PML are used respectively.

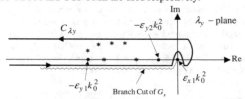

Figure 3. Integration path in $\lambda_y$ − plane, ∗: poles extracted by VECTFIT.

The final result can be written as

$$A_z(x, y; x', y') = \prod_{d_s}(x, x') \sum_{q=1}^{N_{RFFM}} -\text{Res}_q^r \frac{e^{-j\beta_{xq}^r |x - x'|}}{2j\beta_{xq}^r}$$
$$+ \left(1 - \prod_{d_s}(x, x')\right) \sum_{q=1}^{N_{PML}} -\text{Res}_q^p \frac{e^{-j\beta_{xq}^p |x - x'|}}{2j\beta_{xq}^p} \quad (10)$$

where $\prod_{d_s}(x, x')$ is 1 for $|x - x'| < d_s$ and 0 for $|x - x'| > d_s$. $N_{RFFM}$ is the number of poles (with $\beta_{xq}^r$, $\text{Res}_q^r$) extracted for $y$ and $y'$ at hand in CGF-RFFM and $N_{PML}$ is the number of poles (with $\beta_{xq}^p$, $\text{Res}_q^p$) in CGF-PML method. $d_s$ can control the $N_{PML}$ in the way that for smaller $d_s$, bigger $N_{PML}$ must be considered to CGF-PML's result be accurate and vice versa. The important point is that the extracted poles in VECTFIT algorithm in CGF-RFFM are not essentially orthogonal as PML modes of close structure in CGF-PML method. Therefore, to have efficient usage of Green's function in mode-matching type problems, small $d_s$ s may be regarded. To validate the accuracy of the proposed method numerical integration of (4) along the path $C_{\lambda_y}$ is evaluated.

In Fig. 4 the result of $A_z$ for dielectric slab shown in Fig. 1(a) with $t = 0.2\lambda$, $\varepsilon_{r1} = 10$ and $\varepsilon_{r2} = 1$ is shown. The excitation source is located at $x' = 0$ and $y' = t$ and distribution of field is considered on the upper surface of the slab, $y = t$. For $d_s = 0.2\lambda$, CGF-RFFM is used with $N_{RFFM} = 8$ poles and CGF-PML with 50 poles. By considering the smaller $d_s$ ($d_s = 0.1\lambda$) with the same CGF-RFFM results, 70 PML poles must be taken to have excellent match with exact numerical integration. For CGF-PML, $d_{air} = 0.1\lambda$ and also we choose $d_{PML} = 0.05\lambda$ and $\kappa_0 - j\sigma_0 / \omega\varepsilon_0 = 6 - j6$ for PMLs. To search the efficiency of CGF-RFFM-PML, let us have a dielectric substrate with $t = 0.2\lambda$, $\varepsilon_{r1} = 15$ and $\varepsilon_{r2} = 1$. $A_z$ for near field is shown in Fig. 5. For $d_s = 0.5\lambda$, CGF-RFFM-PML is implemented with 12 poles for CGF-RFFM and 30 poles for CGF-PML. It can be seen that accurate modal series representation by 42 poles is obtained in comparison with the numerical integration. By decreasing $d_s$ to $0.2\lambda$, 62 poles are required due to more 20 poles considered in CGF-PML method. To compare the rapidity of the method, let us consider a comparison shown in Fig. 6 for dielectric slab with $t = 0.3\lambda$, $\varepsilon_{r1} = 20$ and $\varepsilon_{r2} = 1$ for line source located at $x' = 2\lambda$ and $y' = 0$. CGF-PML result which was also developed in, is illustrated for $d_s = 0.1\lambda$ (without using CGF-RFFM). With 50 PML poles some deviations can be seen near the $x' + d_s$ which is due to insufficient number of PML modes. $A_z$ for $d_s = 0.02\lambda$ in CGF-PML with 350 poles is also shown which it's required time is at least 9 minutes. This time is mainly related to poles extraction and computation of truncated series in (7). Even so, by using just 40 poles for $d_s = 0.2\lambda$ in CGF-PML and 12 poles in CGF-RFFM, efficient series expression of $A_z$ can be obtained with 52 poles in less than 1 minute.

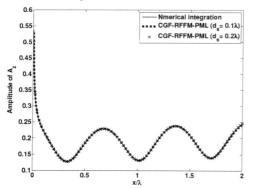

Figure 4. Amplitude of $A_z$ for numerical integration and the CGF-RFFM-PML for dielectric slab with $t = 0.2\lambda$, $\varepsilon_{r1} = 10$ and $\varepsilon_{r2} = 1$, where $y = y' = t$ and $x' = 0$ for $d_s = 0.2\lambda, 0.1\lambda$, $N_{RFFM} = 8$ and $N_{PML} = 50, 70$.

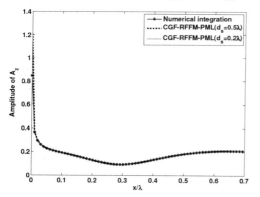

Figure 5. Amplitude of $A_z$ for numerical integration and the CGF-RFFM-PML in the near field for dielectric slab with $t = 0.2\lambda$, $\varepsilon_{r1} = 15$ and $\varepsilon_{r2} = 1$, where $y = y' = t$ and $x' = 0$ for $d_s = 0.5\lambda, 0.2\lambda$, $N_{RFFM} = 12$ and $N_{PML} = 30, 50$.

## V. CONCLUSION

A closed-form and efficient series representation for spatial Green's function of a dielectric slab for all distances from source is presented by using perfectly matched layer method and CGF technique. By combining CGF technique and rational function fitting method, efficient modal series expression of Green's function is achieved in very near field,

with few poles extracted in modified VECTFIT algorithm. The main advantage of this representation is that the number of required modes is controllable.

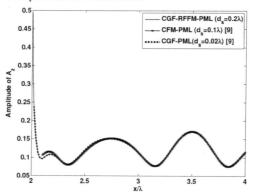

Figure 6. Amplitude of $A_z$ for CGF-PML and the CGF-RFFM-PML for dielectric slab with $t = 0.3\lambda$, $\varepsilon_{r1} = 20$ and $\varepsilon_{r2} = 1$, where $y = y' = t$ and $x' = 2\lambda$ for different $d_s$ and $N_{PML}$.

ACKNOWLEDGMENT

The authors wish to thank Iran Telecommunication Research Center (ITRC) for supporting this research.

REFERENCES

[1] R. Faraji-Dana, An Efficient and Accurate Green's Function Analysis of Packaged Microwave Integrated Circuits. Thesis (Ph.D.)–University of Waterloo, 1993.
[2] A. Shishegar and R. Faraji-Dana, "A closed-form spatial Green's function for finite dielectric structures," Electromagnetics, vol. 23, no. 7, pp. 579–594, 2003.
[3] J. Yang, Y. Chow, G. Howard, and D. Fang, "Complex images of an electric dipole in homogeneous and layered dielectrics between two ground planes," IEEE Trans. Microwave Theory Tech., vol. 40, no. 3, pp. 595–600, Mar 1992.
[4] Y. Hua and T. Sarkar, "Generalized pencil-of-function method for extracting poles of an EM system from its transient response," IEEE Trans. Antennas Propagat., vol. 37, no. 2, pp. 229–234, Feb 1989.
[5] V. Okhmatovski and A. Cangellaris, "A new technique for the derivation of closed-form electromagnetic Green's functions for unbounded planar layered media," IEEE Trans. Antennas Propagat., vol. 50, no. 7, pp. 1005–1016, 2002.
[6] A. Cangellaris and V. Okhmatovski, "Novel closed-form Green's function in shielded planar layered media," IEEE Trans. Microwave Theory Tech., vol. 48, no. 12, pp. 2225–2232, 2000.
[7] V. Okhmatovski and A. Cangellaris, "Evaluation of layered media Green's functions via rational function fitting," IEEE Microw. Wireless Compon. Lett., vol. 14, no. 1, pp. 22–24, Jan 2004.
[8] V. Kourkoulos and A. Cangellaris, "Accurate approximation of Green's functions in planar stratified media in terms of a finite sum of spherical and cylindrical waves," IEEE Trans. Antennas Propagat., vol. 54, no. 5, pp. 1568–1576, May 2006.
[9] F. Olyslager and H. Derudder, "Series representation of Green dyadic for layered media using PMLs," IEEE Trans. Antennas Propagat., vol. 51, no. 9, pp. 2319–2326, Sep 2003.
[10] W. C. Chew and W. H. Weedon, "A 3-D perfectly matched medium from modified maxwell's equations with stretched coordinates," Microwave Opt. Tech. Lett, vol. 7, pp. 599–604, 1994.
[11] E. Anemogiannis and E. Glytsis, "Multilayer waveguides: efficient numerical analysis of general structures," Lightwave Technology, Journal of, vol. 10, no. 10, pp. 1344–1351, Oct 1992.

# Parallel Computation of Complex Antennas Around the Coated Object Using Hybrid Higher-Order MoM and PO Technique

Ying Yan, Xunwang Zhao, Yu Zhang, Changhong Liang
Science and Technology on Antenna and Microwave Laboratory
Xidian University, Xi'an, Shaanxi, 710071, People's Republic of China

Jingyan Mo
School of Communication and Information Engineering
Shanghai University, Shanghai, 200021, People's Republic of China

Zhewang Ma
Graduate School of Science and Engineering, Saitama University
255 Shimo-Okubo, Sakura-ku, Saitama-shi, Saitama 338-8570, Japan

*Abstract-* This paper presents a novel hybrid technique for analyzing complex antennas around the coated object, which is termed the iterative vector fields with Physical Optics (PO). A closed box is used to enclose the antenna and then the complex field vector components on the box' surfaces can be obtained using Huygens principle. The equivalent electromagnetic currents on Huygens surface are computed by Higher-order Method of Moments (HOB-MoM) and the fields scattered from the coated platform are calculated by PO method. Moreover, the parallel technique based on Message Passing Interface (MPI) and Scalable Linear Algebra Package (ScaLAPACK) is employed to accelerate the computation, and good load balance and parallel efficiency are obtained under the proposed parallel scheme. Finally, some numerical examples are presented to validate and to show the effectiveness of the proposed method on solving the practical engineering problems.

## I. INTRODUCTION

In recent years, many researches on hybrid methods emerged. Hybrid FDTD-UTD, MoM-UTD was applied to analyze phased array antennas mounted on airborne platform in [1, 2, 3], but most of them focus on the PEC problem. The iterative MOM-NURBS PO technique was reported to analyze the antenna around NURBS surface [3], while the antenna in this report is limited to the type of line, and the model in PO region is of PEC material. As we know, the platform, such as aircraft, ship or car rarely has pure PEC surface in real-life. Generally, there are several kinds of dielectric materials coated on their surface to achieve different goals, especially for military application. Coating has become a more and more significant method on reducing RCS to make vehicles hidden. In this case, the performance of antenna near the platform will be deteriorated. Thus, it is necessary to study the characteristics of the antenna near an object with dielectric materials. Furthermore, various kinds of complex antennas should be considered in order to make the simulation more realistic. Commercial software *HOBBIES* [4] (Higher-order Basis

Functions Based Integral Equation Solver), is a general purpose frequency domain electromagnetic integral equation solver which provides the reasonable solutions based on Higher-order MoM. The employment of Higher-order basis can significantly reduce the number of unknowns and the memory required as well [5]. Even if the efficiency has been enhanced, the hardware requirements are also unbearable when the model simulated has electronically large dimensions. Thus, PO technique is utilized to calculate the scattered field from the platform.

In addition, parallel technique based on MPI [6] and ScaLAPACK [5] libraries is employed to further speed-up the computation and to improve the computational efficiency. The block-partitioned scheme for the large dense MoM matrix combined with the process-cyclic scheme for the PO discretized triangles is designed to achieve excellent load balance and high parallel efficiency.

## II. MODIFYING THE COMPLEX VECTOR FIELD OVER HUYGENS SURFACE

The whole model simulated has to be divided into two parts that antennas are considered as the MoM region; while the scatterer as the PO region. As shown in Figure 1, the antennas are enclosed by a cube box. There will be a short-lived interaction between these two regions until the electromagnetic currents on the box' surface become stable. The effect of the scatterer on the antenna can be equivalent to the action on electric and magnetic currents on the box. The iteration process can be summarized as follows:

(1) Equivalence theorem tells us the whole computational region of electromagnetic problems can be decomposed into two separate parts-inner and outer. We could set antenna simulated in the inner area and only consider the outer part outside the antennas. In this case, the electromagnetic fields in the inner area should be set to zero. Moreover, there should be

978-7-5641-4279-7

electric and magnetic currents $\vec{J}_s, \vec{M}_s$ induced on the interface surface of inner and outer regions to maintain the radiation fields from antennas to the outside region.

The equivalent currents can be determined by the electric and magnetic fields $\vec{E}$ and $\vec{H}$, which can be expressed as

$$\vec{J}_s = \hat{n} \times \vec{H}, \vec{M}_s = \vec{E} \times \hat{n} \tag{1}$$

Where $\hat{n}$ is the outward unit normal vector of the equivalent surface, and $\vec{E}$, $\vec{H}$ are the radiation fields of antennas on the equivalent surfaces without platform in PO region. The interaction between MoM region and PO region can be transformed into the one between the equivalent surface and the PO region, and the antennas in the box can be considered as non-existed.

In this paper, a cube shaped box is selected as the equivalent surfaces, whose six surfaces are all uniformly divided into some small square patches. The side length of a square is selected as $\lambda/10$ to guarantee the computational precision. To each square patch, its center point is regarded as the sampling point, at which the radiation fields of antennas without platform $\vec{E}^i, \vec{H}^i$ can be obtained by the EM solver *HOBBIES* so that the initial electromagnetic currents $\vec{J}_s(0)$ and $\vec{M}_s(0)$ can be calculated as

$$\vec{J}_s(0) = \hat{n} \times \vec{H}^i, \vec{M}_s(0) = \vec{E}^i \times \hat{n} \tag{2}$$

(2) According to the dual theorem, the electric field produced by the electric and magnetic currents can be expressed as

$$\vec{E}(\vec{r}) = (je^{-jkR}\Delta s/4\pi\omega\varepsilon)\left\{\frac{k^2R^2 - 3jkR - 3}{R^5}\left\{\vec{R}\times\left[\vec{J}_s \times \vec{R}\right]\right\} - \frac{1+jkR}{R^3}\cdot 2\vec{J}_s\right\} - \frac{\Delta s}{4\pi}\left\{\frac{1+jkR}{R^3}e^{-jkR}\cdot\left[\vec{M}_s \times \vec{R}\right]\right\} \tag{3}$$

In equation (3), $\vec{R}$ is the vector from source point $\vec{r}'$ to the observation point $\vec{r}$. The square patch is considered small enough that the currents on it can be uniform and are equal to those at the center. The notation $\Delta s$ stands for the area of each square patch.

(3) The electromagnetic currents on the equivalent surface are considered as source points; they will radiate electric and magnetic fields to the PO model, and then all the illuminated PO triangles radiate fields back to the equivalent source points so as to modify the electromagnetic currents as follows:

$$\Delta \vec{J}_s = (-\hat{n}) \times \vec{H}, \Delta \vec{M}_s = \vec{E} \times (-\hat{n}) \tag{4}$$

Let $\vec{J}_s(k)$ and $\vec{M}_s(k)$ to be the currents after the $k$th iteration, the ones after $k+1$th can be expressed as

$$\vec{J}_s(k+1) = \vec{J}_s(0) + \Delta\vec{J}_s$$
$$\vec{M}_s(k+1) = \vec{M}_s(0) + \Delta\vec{M}_s \tag{5}$$

This process should be repeated until the convergence condition is satisfied

$$\frac{\left\|\vec{J}_s(k+1) - \vec{J}_s(k)\right\|}{\left\|\vec{J}_s(k)\right\|} \leq \varepsilon, \frac{\left\|\vec{M}_s(k+1) - \vec{M}_s(k)\right\|}{\left\|\vec{M}_s(k)\right\|} \leq \varepsilon \tag{6}$$

Where $\varepsilon$ is the threshold of the iteration and is set to be $10^{-5}$ in this paper, $\|\cdot\|$ denotes the 2-norm of complex vector.

(4) After the currents on equivalent surface become stable, it will radiate fields directly to the observation points in free space, adding the scattered fields from PO region we can obtain the total fields.

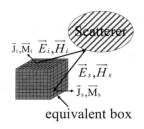

$\vec{J}_i, \vec{M}_i \quad \vec{E}_i, \vec{H}_i$

$\vec{E}_s, \vec{H}_s$

$\vec{J}_s, \vec{M}_s$

equivalent box

Figure 1. Model illustrating

## III. PARALLEL STRATEGY BASED ON MPI AND SCALAPACK

### A. PO part

Distributing to processes the computation of PO currents at all points over the illuminated triangles is the main task to parallelize the PO method. We use the same number of points over each triangle; hence we can partition triangles rather than points among processes. What is necessary is to ensure that the triangles in the lit PO region are partitioned equally to each process. This can be realized by distributing the triangles in neighboring regions, either lit or shadowed, to different processes.

### B. MoM part

To parallelize the solution of the large dense matrix in a MoM problem, typically one needs to divide the matrix between processes in such a way that two important conditions are fulfilled: each process should store approximately the same amount of data, and the computational load should be equally distributed among the processes that run on different nodes.

The parallel LU decomposition based on the ScaLAPACK library package [5] is employed as the solver and the block storage scheme is designed accordingly. For explanation purpose, the MoM matrix equation is rewritten in a general form as

$$AX = B \tag{7}$$

where $A$ denotes the complex dense matrix, $X$ is the unknown vector to be determined and $B$ denotes the given source vector.

For example, we assume that the matrix $A$ is divided into 6×6 blocks, which are distributed to 6 processes in a 2×3 process grid. Figure 2 (a) and (b) show to which process the blocks of $A$ are distributed using the ScaLAPACK's distribution methodology [5].

| Process unit index | 0 | 1 | 2 | 0 | 1 | 2 |
|---|---|---|---|---|---|---|
| 0 | 11 | 12 | 13 | 14 | 15 | 16 |
| 1 | 21 | 22 | 23 | 24 | 25 | 26 |
| 0 | 31 | 32 | 33 | 34 | 35 | 36 |
| 1 | 41 | 42 | 43 | 44 | 45 | 46 |
| 0 | 51 | 52 | 53 | 54 | 55 | 56 |
| 1 | 61 | 62 | 63 | 64 | 65 | 66 |

| 0 (0,0) | 2 (0,1) | 4 (0,2) | 0 (0,0) | 2 (0,1) | 4 (0,2) |
|---|---|---|---|---|---|
| 1 (1,0) | 3 (1,1) | 5 (1,2) | 1 (1,0) | 3 (1,1) | 5 (1,2) |
| 0 (0,0) | 2 (0,1) | 4 (0,2) | 0 (0,0) | 2 (0,1) | 4 (0,2) |
| 1 (1,0) | 3 (1,1) | 5 (1,2) | 1 (1,0) | 3 (1,1) | 5 (1,2) |
| 0 (0,0) | 2 (0,1) | 4 (0,2) | 0 (0,0) | 2 (0,1) | 4 (0,2) |
| 1 (1,0) | 3 (1,1) | 5 (1,2) | 1 (1,0) | 3 (1,1) | 5 (1,2) |

(a) a matrix consisting of 6×6 blocks (b) rank and coordinates of each process owning the corresponding blocks in (a)

Figure 2. Block-cyclic distribution of a matrix

By varying the dimensions of the blocks of $A$ and those of the process grid, different mappings can be obtained. This scheme can be referred to as a "block-cyclic" distribution procedure.

Note that the storage required for the vectors $X$ and $B$ is negligible compared with that of the large dense matrix $A$. Therefore, the entire vectors can be stored in each process.

Load balance is critical to obtain an efficient operation of a parallel code. This parallel scheme of matrix filling is able to achieve the good load balance. Little communication between processes is necessary during the matrix filling [5].

## IV. NUMERICAL EXAMPLES

*1. Simulation of a patch antenna near a coated cube*

In this example, the model of a patch antenna near a coated cube has been simulated. Besides the comparison of field results with Higher-order MoM, the times consumed are also compared and the parallel efficiency is drawn as well.

The simulation frequency is 440 MHz. A closed box centered at (0, 0, 8) encloses the antenna, whose edge length is 0.55m. Through the discretization, all the six surfaces of the box have been meshed into 8×8 squares, at whose center are the equivalent electromagnetic sources. The model in PO region is a cube with edge length 4.0m, whose surfaces are coated with one layer of dielectric material. The parameters of the coating can be referenced from the 1st material in Table I. The cube has been discretized into 47204 small triangles.

This simulation employs the "magic cube" supercomputer in SSC [7]. Let us take 4 CPUs as a base, 8, 16, 32, 64, 128 CPUs are tested respectively to draw the parallel efficiency curve, as shown in Figure 4. From which we can find the good parallel effect of this paper's method, even though the efficiency drops at 128 CPUs, which is due to the reason that the problem simulated is not big enough, the more CPUs used, the longer communication times required. Moreover, the computational time comparison between this method and Higher-order MoM has been presented in Table II and we can easily find this method's superiority in time requirement.

Finally, the field results' comparison is presented in Figure 5. The pattern on $XOY$ is of omnidirectional characteristic; while there are some disagreements near $\theta = 0°, 180°$ on $YOZ$, which is due to the edge diffraction effect from the cube.

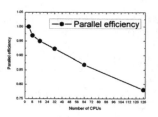

Figure 3. A patch antenna near a coated cube

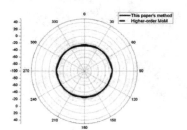

Figure 4. Parallel efficiency

TABLE I
INFORMATION OF THE COATING

| | Relative Electric Permittivity | Loss Tangent | Relative Magnetic Permeability | Thickness (mm) |
|---|---|---|---|---|
| 1st material | 16 | 0.275 | 1.0 | 5 |
| 2nd material | 2.5 | 0.00045 | 1.0 | 5 |

TABLE II
TIME COMPARISON

| | Number of CPUs | Total time (Second) |
|---|---|---|
| This paper's method | 128 | 84 |
| Higher-order MoM | 128 | 1007 |

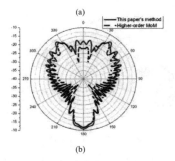

(a)

(b)

Figure 5. Results comparison. (a) XOY plane; (b) YOZ plane.

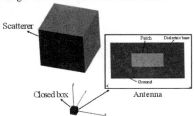

## 2. Phased antenna array near a coated aircraft

In this example, the disturbed radiation pattern of a phased patch antenna array located near a coated aircraft has been analyzed. The simulation frequency is 440MHz. The aircraft is of the dimensions 26.4 $\lambda$ ×32.3 $\lambda$ ×7.0 $\lambda$ and is coated with two layers of dielectric materials, whose parameters can be referenced in Table I. The patch array, whose 11×11 elements radiates beam maximum in the direction of $\theta = 30°, \phi = 90°$, is excited by a Taylor shaped drive distribution and located along XOY coordinate plane. It is worth describing here the angle $\theta$ is started from XOY plane, while $\phi$ is still from X-axis. A cube whose edge length is 3.14m is chosen as the closed box to surround the antenna array, the center of the box is located at (1.434, 1.28, 0.0175) and all of the surfaces has been partitioned into 46×46 small squares.

This simulation employs 256 CPUs in SSC and requires 3 iterations, 21164 seconds to finish the calculation, in which 10229 seconds for sheltering and 9465 seconds for iteration. The disturbed pattern in 3-D is illustrated in Figure 7. We can easily find the effect of the aircraft on the characteristic of the antenna. Another main beam nearly symmetrical to the original one about XOZ coordinate plane appears due to the reflection of the aircraft. The side lobe has been raised about maximum 15dB, while the back lobe has been raised about 20dB.

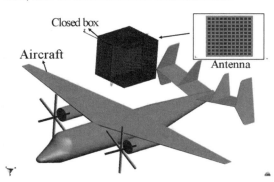

Figure 6. Phased antenna array near a coated aircraft

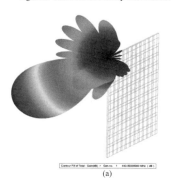

(a)

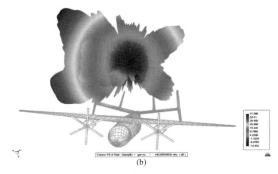

(b)

Figure 7. (a) The radiation pattern (dB) of the antenna array; (b) The disturbed radiation pattern (dB) after the antenna array being located near a coated aircraft.

From the examples above, we can find this paper's method needs very few iteration times to obtain the convergent solution. It is capable of solving the practical engineering problems including complex antennas and electrically large platform.

ACKNOWLEDGMENT

This work is supported by National High Technology Research and Development Program ("863"Program) of China (2012AA01A308); the National Natural Science Foundation of China (61072019); the Fundamental Research Funds for the Central Universities of China (JY10000902002) and the Foundation of Science and Technology on Antenna and Microwave Laboratory (9140C070502110C0702). The computational resources utilized in this research were provided by Shanghai Supercomputer Center.

REFERENCES

[1] Lei, J.-Z., C.-H. Liang, W. Ding, and Y. Zhang, "Analysis of airborne phased-array antennas using hybrid method of parallel FDTD and UTD", Chinese Journal of Radio Science, Vol. 24, No. 1, 2009.
[2] Z. L. He, K. Huang, and C. H. Liang, "ANALYSIS OF COMPLEX ANTENNA AROUND ELECTRICALLY LARGE PLATFORM USING ITERATIV VECTOR FIELDS AND UTD METHOD", Progress In Electromagnetics Research M, Vol. 10, 103-117, 2009.
[3] Ming Chen, Yu Zhang, Xun-Wang Zhao, Chang-Hong Liang, "Analysis of Antenna Around NURBS Surface With Hybrid MoM-PO Technique", IEEE Transactions on Antennas and Propagation, Vol. 55, No. 2, 2007, pp.407-413.
[4] Y. Zhang, T.K. Sarkar, X.W.Zhao, D. García-Doñoro, W.X.Zhao, M. Salazar-Palma and S.W. Ting, Higher Order Basis Based Integral Equation Solver (HOBBIES). Hoboken, NJ: Wiley, 2012.
[5] Y. Zhang and T.K. Sarkar, Parallel solution of Integral Equation Based EM Problems in the Frequency Domain. Hoboken, NJ: Wiley, 2009.
[6] Du Z.H, Parallel Programming technology of high performance computation: MPI programming, Qinghua University Press, Beijing, 2001.
[7] http://www.ssc.net.cn/

# Combination of Ultra-Wide Band Characteristic Basis Function Method and Asymptotic Waveform Evaluation Method in MoM Solution

A.-M. Yao[1], W. Wu[1], J. Hu[1,2] and D.-G. Fang[1]

[1] School of Electronic and Optical Engineering, Nanjing University of Science and Technology, China, 210094

[2] State Key Laboratory of Millimeter Waves, China, 210096

*Abstract*-A novel technique which combines the ultra-wide band characteristic basis function method (UCBFM) and the adaptive multi-point expansion algorithm for asymptotic waveform evaluation (AWE) method is proposed. The UCBFM and the AWE method are two kinds of model order reduction (MOR) methods, and their proposed combination can be applied for fast evaluation of wide band scattering problems. In the proposed approach, ultra-wide band characteristic basis functions (UCBFs) is solved according to the highest frequency in the range of interest, and the adaptive multi-point expansion algorithm for AWE is based on a simple binary search algorithm. Provided numerical results validate the proposed method, and suggest that it has a high efficiency.

*Index Term*-model order reduction methods; ultra-wide band characteristic basis function method; asymptotic waveform evaluation; method of moments; scattering problem

## I. Introduction

One of the most popular methods for radar cross sections (RCS) prediction is the frequency domain integral equation solved using method of moments (MoM) [1], but it places a heavy burden on the CPU time as well as memory requirements when electrically large structures are analyzed. Moreover, they require the impedance matrix to be generated and solved for each frequency sample; hence, if the response over a wide frequency band is of interest, the MoM is computationally intensive. This problem is especially serious when the RCS is highly frequency dependant and fine frequency steps are required to get an accurate representation of the frequency response.

Several techniques have been proposed to alleviate this problem. In [2, 3], the characteristic basis function method (CBFM), is able to reduce the size of the MoM matrix. In the CBFM, the object is divided into a number of blocks, and high-level basis functions called characteristic basis functions (CBFs) are derived for these blocks, which are discretized by using the conventional triangular patch segmentation and Rao-Wilton-Glisson (RWG) basis functions [4]. In [5], the asymptotic waveform evaluation (AWE), is proposed for predicting the RCS over a band of frequencies. Since it needs the MoM matrix inversion at central frequency, the AWE technique can

hardly deal with wideband electromagnetic scattering problems from electrically large object or multi-objects. So in [6], AWE based on CBFM, is proposed to analyze wideband electromagnetic scattering problems. This method uses the mutual coupling method for generating CBFs, that is time consuming and memory demanding, in CBFM and applies single-point expansion for AWE. In [7], A simple binary search algorithm is described to apply AWE at multiple frequency points to generate an accurate solution over a specified frequency band. Since the CBFs depend upon the frequency, they need to be generated repeatedly for each frequency. Hence, in [8], the ultra-wide band characteristic basis function method (UCBFM) is used on a wide band, without having the generation of CBFs for each frequency repeatedly. The CBFs calculated at the highest, termed UCBFs, entail the electromagnetic behavior at lower frequency range; thus, it follows that they can also be employed at lower frequencies without going through the time consuming step of generating them again. However, in the UCBFM, it is still time consuming for the computation of wideband RCS since it requires repeated solving of the reduced matrix equations at each frequency.

In this paper, The combination of the UCBFM with the adaptive multi-point expansion algorithm for AWE is introduced for fast evaluation of wide band scattering problems. In the following sections, the principles of the UCBFM and the AWE based on a simple binary search algorithm are outlined firstly, and then the combination scheme of the UCBFM/AWE method is developed. Finally, two classical scattering problems are analyzed and the comparisons of the proposed method and traditional methods are provided.

## II. Formulation

### A. UCBF Method [8]

Let us consider a complex 3-D object illuminated by a plane wave. In a conventional MoM, the whole surface is divided into triangles with size ranging from $\lambda/10$ to $\lambda/20$. Applying this to the electric field integral equation, one can obtain a dense and complex system of the form

$$Z(k)I(k) = V(k) \tag{1}$$

[2.] This work is supported by State Key Laboratory of Millimeter Waves (K201306).

978-7-5641-4279-7

In (1), $Z$ is the MoM matrix of dimension $N \times N$, $I$ and $V$ are vectors of dimension $N \times I$, where $N$ is the number of unknown current coefficients and $k$ is the wave number of the free space. For large and complex problem, the matrix filling and matrix equation solving are quite time consuming.

The CBFM begins by dividing the object to be analyzed into blocks. For the best division scheme and the number of blocks M one may refer to [9]. These blocks are characterized through a set of CBFs, constructed by exciting each block with multiple plane waves (MPW), incident from $N_{PW}$ uniformly spaced $\theta$ and $\varphi$-angles. To calculate the CBFs on the generic $i$th block , one must solve the following system

$$Z_{ii}(k) J_i^{CBF} = V_i^{MPW} \quad (2)$$

In (2), $Z_{ii}$ is an $N_i \times N_i$ sub-matrix corresponding to the $i$th block, $J_i^{CBF}$ is a $N_i \times N_{PW}$ matrix containing original CBFs, and $V_i^{MPW}$ is a $N_i \times N_{PW}$ matrix containing excitation vectors, where $N_i$ is the number of unknowns relative to $i$th block. In order to extract $Z_{ii}$ from the original MoM matrix, a matrix segmentation procedure can be used. Next, a new set of orthogonal basis functions, which are linear combinations of the original CBFs, are constructed via the singular value decomposition (SVD) approach. Thus, the redundant information because of the overestimation is eliminated. For simplicity, one can assume that the average number of CBFs after SVD is $K$. Consequently, the solution to the entire problem is expressed as a linear combination of the $M \times K$ CBFs, as follows

$$I(k) = \sum_{m=1}^{M} \sum_{k=1}^{K} \alpha_m^k (k) J_m^{CBF_k}(k) \quad (3)$$

where $J_m^{CBF_n}$ is the $n$th CBF of the $m$th lock. By using the above CBFs, the original large MoM matrix can be reduced, and unknowns are changed to weight coefficient vector $\alpha$ whose order is much smaller than original current coefficient vector $I$. Finally, after solving the reduced system and substituting solution back to (3), one can obtain the solution of single frequency point. For the multi frequency point, although the above procedures of the CBFM can be repeated, the UCBFM is more efficient.

The ultra-wide band characteristic basis functions (UCBFs) is the CBFs generated at the highest frequency. Since the UCBFs can adequately represent the solution in the entire band of interest, they are used for lower frequencies without going through the time consuming step of generating them again. Fig.1 shows the flowchart of the UCBFM.

### B. AWE Method [7]

For the given frequency $f_0$ (corresponding to the wave number $k_0$ ) in a wide band, current coefficient vector $I(k_0)$ can be obtained one by one by solving (1) by using the above UCBF. It is still time consuming for wide-band problems. For such problems, some interpolation/extrapolation schemes like the AWE method are always powerful. The basic procedure of the AWE is to first expand the $I(k)$ into Taylor series, then to use

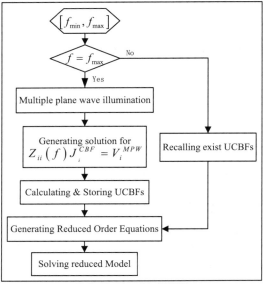

Fig 1. Flowchart of the UCBFM

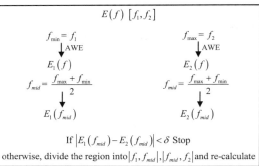

Fig 2. Flowchart of the adaptive multi-point expansion algorithm.

the Padé approximation to generate the rational function as

$$I(k) = \sum_{n=0}^{L+M} m_n (k - k_0)^n = \frac{\sum_{i=0}^{L} a_i (k - k_0)^i}{1 + \sum_{j=1}^{M} b_j (k - k_0)^j} \quad (4)$$

where the moment vectors $m_n$ can be calculated by the following recursive formula

$$m_0 = Z^{-1}(k_0) V(k_0) \quad (5)$$

$$m_n = Z^{-1}(k_0) \left[ \frac{V^{(n)}(k_0)}{n!} - \sum_{i=1}^{n} \frac{Z^{(i)}(k_0) m_{n-i}}{i!} \right] \quad (6)$$

where $Z^{(i)}(k_0)$ is the $i$th derivative with respect to $k$ of $Z(k)$ and evaluated at $k_0$. Similarly, $V^{(n)}(k_0)$ is the $n$th derivative with respect to $k$ of $V(k)$. The coefficients $a_i$ and $b_j$ are obtained by solving the following linear equation

$$\sum_{j=1}^{M} m_{L+i-j} b_j = -m_{L+i} \quad i = 1, 2, ..., M \quad (7)$$

$$a_i - \sum_{j=0}^{i-1} m_j b_{i-1} = m_i \quad i = 1, 2, ..., M \quad (8)$$

where $b_0 = \{1, 1, ..., 1\}$.

The adaptive algorithm for the AWE is shown in Fig.2, where $E(f)$ is the desired quantity. When single-point expansion does not satisfy the requirement over the whole frequency band, the multi-point expansion becomes necessary. The adaptive multi-point expansion algorithm given in Fig.2 is available for this purpose.

### C. UCBFM/AWE Method

It can be seen from (5)–(6) that the computation of the derivatives of $Z(k)$ and $V(k)$ is the key step in the AWE technique. In order to avoid solving the inversion of large matrix in (5) and (6), UCBFM is combined with the adaptive multi-point expansion algorithm for AWE technique to calculate the wide band scattering characteristics from electrically large or multiple objects. The flowchart of UCBFM/AWE is shown in Fig.3. The coefficients $m_n$ in (5) and (6) are calculated by UCBF method. In UCBFM/AWE algorithm, the current distribution at each frequency within the band is expressed as a linear combination of UCBFs.

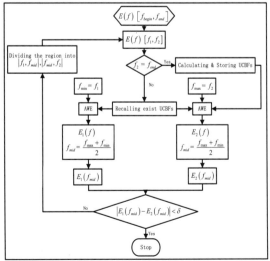

Fig 3. Flowchart of UCBFM/AWE method.

### III. NUMERICAL RESULTS

To demonstrate the efficiency and accuracy of the UCBFM/AWE method, two numerical examples are investigated. The objects of the numerical simulations are illuminated by a normally incident theta-polarized plane wave; a conventional triangular patch segmentation and RWG basis function are employed [4]. All the simulations were run on a notebook equipped with 2 Dual Core at 2.3GHz (only one core was used) and 8GB of RAM.

TABLE I
COMPUTATIONAL TIMES FOR THE DIFFERENT METHODS OF A SPHERE

|  | Conventional MoM | CBFM/AWE | UCBFM/AWE |
|---|---|---|---|
| Total time | 203.227 m | 32.908 m | 15.453 m |
| Saving | − | 83.81% | 92.39% |

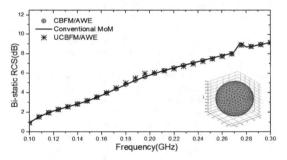

Fig 4. Bi-static RCS of the PEC sphere from 0.1GHz to 0.3GHz using different methods.

TABLE II
COMPUTATIONAL TIMES FOR THE TWO METHODS OF A PLATE

|  | Conventional MoM | UCBFM/AWE |
|---|---|---|
| Total time | 30.127 h | 6.814 h |
| Saving | − | 77.38% |

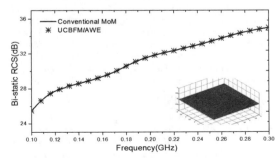

Fig 5. Bi-static RCS response of the PEC plate by UCBFM/AWE and conventional MoM method.

The first example is the RCS by a PEC sphere with diameter of 2.0m from 0.1GHz to 0.3GHz. The geometry is automatically divided into two blocks, as shown in Fig.4. The discretisation in triangular patches is carried out at the highest frequency, with a mean edge length of 0.1λ, which involves almost 1293 unknowns. After SVD procedure we totally obtain 228 UCBFs. We set $L = M = 4$, $\delta = 0.1$ for AWE expansion and chose $N_{PW} = 800$ for UCBFM. The results are compared with those derived by using CBFM/AWE and conventional MoM, as shown in Fig.4. The RCS response of the PEC sphere is for the observation angle at $\theta = 180°$, $\varphi = 0°$. In UCBFM/AWE method, the expansion points are 0.1GHz and 0.3GHz, with a frequency increment of 1MHz. While in CBFM/AWE technique the expansion points are: 0.1GHz, 0.15GHz, 0.2GHz and 0.3GHz, with a same frequency increment. For conventional MoM method the step of frequency sweep is 5MHz, as shown in Fig.4. The bi-static RCS solved by UCBFM/AWE and CBFM/AWE coincide very well with the conventional MoM. Hence, the presented method is accurate in wide band electromagnetic scattering analysis.

To show the efficiency of the UCBFM/AWE, the total simulation time which include matrix filling time and solving time is shown in Table I. With a frequency increment of 1MHz, the conventional MoM method requires about 203.227 minutes to obtain the solution. However, for UCBFM/AWE method, to obtain the same accuracy, only 15.453 minutes are needed, which is 13.15 times faster than the conventional MoM method.

The next example is a $4m \times 4m$ PEC plate. The frequency range starts from 0.1GHz and terminates at 0.3GHz; a length of $4m$ is equal to $4\lambda$ at 0.3GHz. The plate is divided into sixteen blocks, as shown in Fig.5. The discretisation in triangular patches is carried out at highest frequency, with a mean edge length of 0.1λ; this leads to total unknowns of 5299. After SVD procedure we obtain 971 UCBFs. These UCBFs calculated at 0.3GHz can be used for any frequency in the range. The bi-static RCS observed at $\theta = 180°$, $\varphi = 0°$ is shown in Fig.4, which calculated by UCBFM/AWE and conventional MoM method, respectively.

The bi-static RCS shows an excellent match for the two methods. In this numerical simulation we set $L = 4$, $M = 4$, $\delta = 0.1$ for AWE expansion and chose $N_{PW} = 800$ for UCBFM. Over the whole frequency band, we obtain six expansion points, with a frequency increment of 1MHz for Padé approximation, while the step of frequency sweep is 5MHz for conventional MoM method. Table II shows the efficiency of the presented method. With a frequency increment of 1MHz, we obtain a total simulation time of about 6.814 hours with UCBFM/AWE, whereas the use of conventional MoM leads to 30.127 hours.

## IV. Conclusion

In this paper, a new approach that combines the UCBFM with the adaptive multi-point expansion algorithm for AWE is successfully implemented to fast and efficiently analyze wide band scattering problems. Numerical results given demonstrate the high accuracy and efficiency of the proposed method. The scattering problem of very electrically large objects and multi-ple objects can be handled, since large matrix inverse computation is not necessary in the hybrid method and UCBFM allows the decrease of time in CBFs compared to that obtained by CBFM. The AWE method in the UCBFM was used in order to further reduce the computational time required for wide band analysis.

## References

[1]  R.H. Harrington, *Field Computation by Moment Methods*, New York: Macmillan, 1968, pp.49-70.

[2]  V.V.S. Prakash and R. Mittra, "Characteristic Basis Function Method: A new technique for efficient solution of method of moments matrix equation," *Microwave Opt. Technol. Lett.*, vol.36, no.2, pp.94-100, Feb. 2003.

[3]  E. Lucente, A. Monorchio and R. Mittra, "An iteration free MoM approach based on excitation independent characteristic basis functions for solving large multiscale electromagnetic scattering problems," *IEEE Trans. Antennas Propag.*, vol.58, no.7, pp.999-1007, Apr. 2008.

[4]  S.M. Rao, D.R. Wilton and A.W. Glisson, "Electromagnetic scattering by surfaces of arbitrary shape," *IEEE Trans. Antennas Propag.*, vol.30, no.3, pp.409-418, Feb. 1982.

[5]  C.J. Reddy, M.D. Deshpande and C.R. Cockrell, "Fast RCS computation over a frequency band using method of moments in conjunction with asymptotic waveform evaluation technique," *IEEE Trans. Antennas Propag.*, vol.46, no.8, pp.1229-1233, Aug. 1988.

[6]  Y.F. Sun, Y. Du and Y. Shao, "Fast computation of wideband RCS using characteristic basis function method and asymptotic waveform evaluation technique," *Journal of Electronics(CHINA)*, vol.27, no.4, pp.453-457, Apr. 2010.

[7]  J.P. Zhang and J.M. Jin, "Preliminary study of AWE method for FEM analysis of scattering problems," *Microwave Opt. Technol. Lett.*, vol.17, no.1, pp.7-12, Jan. 1998.

[8]  M.D. Gregorio, G. Tiberi, A. Monorchio and R. Mittra, "Solution of wide band scattering problems using the characteristic basis function method," *IET Microw. Antennas Propag.*, vol.6, no.1, pp.60-66, Jan. 2012.

[9]  K. Konno, Q. Chen, K. Sawaya and T. Sezai, "Optimization of block size for CBFM in MoM," *IEEE Trans. Antennas Propag.*, vol.60, no.10, pp.4719-4724, Jan. 2012.

# A Mesh-Tearing Sub-Entire Domain Basis Function Method for Improved Electromagnetic Analysis of Strong-Coupled Cube Array

Lin LIU, Xiaoxiang HE, Chen LIU, Yang YANG

IEEE Conference Publishing

College of Electronic and Information Engineering, Nanjing University of Aeronautics and Astronautics

29 Yudao ST., Baixia Dist., Nanjing 210016, China

*Abstract-* A mesh-tearing sub-entire domain (MTSED) basis function method for improved electromagnetic (EM) scattering analysis of strong-coupled finite periodic structures is proposed in this paper. By tearing the coarse mesh of classic SED basis function into several smaller ones, the modeling precision for strong-coupled arrays can be improved more than 13dB maximally with less than four times computational time increasing compared to classic SED method. The mesh-tearing technique based on both simplified SED (MTSSED) basis function and accurate SED (MTASED) basis function are also investigated in detail. The detailed algorithm is presented and the numerical results demonstrate that the proposed MTSED method is an accurate method for electromagnetic scattering analysis of strong-coupled periodic structures.

## I. INTRODUCTION

Nowadays, finite periodic structures such as photonic band-gap (PBG) crystals [1] and phased-array antennas [2], frequency selective surfaces (FSS) [3] are applied widely in electromagnetic engineering. And accurate and efficient techniques for analysis of periodic structures are always based on periodic boundary condition (PBC), which is either applied in the frequency domain in the context of the method of moments (MoM) [4], and the finite element method (FEM) [5], or implemented by using an equivalent delay condition in the time domain analyses, such as the finite difference time domain (FDTD) [6].

Among these full-wave analysis techniques, the MoM is a robust approach to deal with electromagnetic scattering of periodic structures. However, the conventional MoM requires $O(N^2)$ memory and $O(N^3)$ computational complexity, which are unaffordable for large-scale strong-coupled problems with desired accuracy. Several physical-based entire-domain basis functions have been developed to reduce the number of unknowns. For example, a macro basis function (MBF) was employed to analyze finite printed antenna arrays [7]. The synthetic basis function (SBF) is similar to MBF, which was applied in analyzing large-scale non-periodic arrays [8]. Both MBF and SBF are time consuming because they consider the mutual coupling effects in an iterative way. The characteristic basis function (CBF) is another kind of physical-based entire-domain basis function [9]. The CBF method uses a new type of high-level (secondary) basis function to calculate the mutual

coupling, so the CBF can be obtained directly. For each single cell, $N^2$ CBFs should be considered, where $N$ represents the number of cells. Some other techniques have also been presented to analyze the planar circuits and antenna arrays [10].

Recently, to decrease the memory requirement, an accurate sub-entire domain (ASED) basis function method [11] is proposed by Cui et al. Different from MBF and CBF, the mutual coupling effects are considered by using dummy cells. And all the elements are divided into 9 kinds according to the ASED. Consider one periodic structure with $N$ elements and $M$ discrete edges on each unit cell and the large-scale problem involves $N_0=NM$ unknowns. The ASED basis function decomposed the original problem into two smaller-size problems, one part contains $9M$ unknowns and the other part contains only $N$ unknowns. To further simplify the calculation procedure, a simplified sub-entire domain (SSED) basis function method is proposed [12] which can reduce the number of unknowns. When computing the basis function, the mutual coupling among the elements is neglected directly so that it is shared by all the elements. We compared and discussed the efficiency and accuracy of the ASED and SSED methods in [13].

In this paper, a mesh-tearing sub-entire domain basis function method (MTSED) is proposed to analyze the scattering of strong-coupled finite periodic structures. And with slight computational complexity increasing, the modeling accuracy can be improved significantly.

## II. THEORY

*a. MTSED basis function method*

Consider a two-dimensional periodic structure with $N$ perfectly electric conducting (PEC) cells in free space is illuminated by a plane wave, the magnetic field integral equation (MFIE) can be written as:

$$\frac{1}{2}\mathbf{J}(\mathbf{r}) - \frac{1}{4\pi}\mathbf{n}(\mathbf{r}) \times \nabla \times \int_S G(\mathbf{r}, \mathbf{r}')\mathbf{J}(\mathbf{r}')d\mathbf{r}' = \mathbf{n}(\mathbf{r}) \times \mathbf{H}^{inc}(\mathbf{r}) \quad (1)$$

Where $\mathbf{H}^{inc}(\mathbf{r})$ represents the incident wave, $\mathbf{n}(\mathbf{r})$ is the unit normal vector, $G(\mathbf{r}, \mathbf{r}')$ is the Green's function in free space,

978-7-5641-4279-7

$\mathbf{J}(\mathbf{r})$ represents the electric current distribution of the whole periodic structure, which can be written as:

$$\mathbf{J}(\mathbf{r}) = \sum_{n=1}^{N} \beta_n \mathbf{f}_n(\mathbf{r}) \qquad (2)$$

where $\mathbf{f}_n(\mathbf{r})$ represents the SED basis function on the $n$th cell which can be computed by conventional MoM with dense mesh (usually about $\lambda/10$), $\beta_n$ denotes the corresponding coefficient. After using the Galerkin's procedure and the SED basis function, (1) can be written in a matrix form:

$$[\mathbf{Z}] \cdot [\beta] = [\mathbf{V}] \qquad (3)$$

Where $[\beta] = (\beta_1, \beta_2, \beta_3 \cdots \beta_N)^T$ is the expansion coefficient vector and the elements of $[\mathbf{V}]$ and $[\mathbf{Z}]$ can be written as:

$$V_m = \iint_{f_m} \mathbf{f}_m(\mathbf{r}) \cdot [\mathbf{n}(\mathbf{r})] \times \mathbf{H}^{inc}(\mathbf{r}) d\mathbf{r} \qquad (4)$$

$$Z_{mn} = \frac{1}{2} \int_{f_m = f_n} \mathbf{f}_m(\mathbf{r}) \cdot \mathbf{f}_n(\mathbf{r}) d\mathbf{r}' +$$
$$\frac{1}{4\pi} \int_{f_m} \mathbf{f}_m(\mathbf{r}) \cdot \mathbf{n}(\mathbf{r}) \times \int_{f_n} \mathbf{f}_n(\mathbf{r}) \nabla' G(\mathbf{r}, \mathbf{r}') d\mathbf{r}' d\mathbf{r} \qquad (5)$$

In the SED method, the current distribution (SED basis function) on each cell that is very important to the final solution is computed with approximate (ASED) or even no (SSED) mutual coupling consideration. As a result, the error caused by the approximation is inevitable. Furthermore, the size of the coarse mesh for the whole array analysis is chosen the same as the unit cell for the whole array analysis, which is much larger than $\lambda/10$. For weakly-coupled arrays, SED method in [11], [12] can approach acceptable accuracy. However, for strong-coupled arrays, SED method even can not make the matrix solver converge to the real solution.

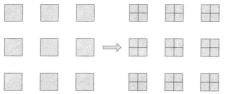

Figure 1. Periodic structures applied with MTSED method when k is 4.

In order to improve the computational accuracy, a MTSED method is proposed in this paper. According to the main idea of MTSED method, the coarse mesh on each cell is to be torn into $k$ parts, as shown in Fig.1. With the increase of mesh density, the accuracy is improved. Theoretically, when the coarse mesh is refined to be about $\lambda/10$, the MTSED method will appear to be the conventional MoM and the accuracy can be ensured. Of course, the CPU time will be unaffordable in the case of large-scale arrays.

In MTSED method, the overall electric current $\mathbf{J}(\mathbf{r})$ in equation (2) is modified as:

$$\mathbf{J}(\mathbf{r}) = \sum_{n=1}^{kN} \beta_n \mathbf{f}_n^T(\mathbf{r}) \qquad (6)$$

Here $\mathbf{f}_n^T(\mathbf{r})$ represent the basis functions on the torn parts of each cell. Thus, (3) can be rewritten as:

$$[\mathbf{Z}^T] \cdot [\beta^T] = [\mathbf{V}^T] \qquad (7)$$

Here, $[\beta^T]$ with a size of $kN$ denotes the tearing expansion coefficient vector, the elements of $[\mathbf{Z}^T]$ and $[\mathbf{V}^T]$ are corresponding to that of $[\mathbf{Z}]$ and $[\mathbf{V}]$ respectively, except that the integration is over the torn parts of each cell. According to MTSED method, the memory consumption is expanded $k^2$ times for matrix storage. The CPU time for matrix filling is the same as classic SED which is the most time consuming procedure and the CPU time for matrix solving is increased because of the expansion of matrix scale. As a result, the increasing of CPU time is slight. The MTSED method is a compromise consideration between MoM full wave analysis and classic SED method.

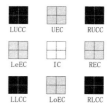

Figure 2. Periodic structures with MTSED method when k is 4.

*b. MTSSED basis function method*

The SSED method is a simplified solution while the electric current distribution on each element is computed by conventional MoM without considering the coupling from neighboring cells. With the coarse mesh of SSED method torn into $k$ parts, the unknowns are increased from $N$ to $kN$ and the CPU time and memory consumption are also increased for accurate array analysis. Not only the current magnitude and phase are amended as it works in classic SSED, but also the current distribution on the whole array is modified. And the precision can be improved consequently as we presented in [14].

*c. MTASED basis function method*

The mutual coupling of neighbouring elements is considered approximately in ASED method for basis function calculation. The whole elements are catalogued into 9 kinds of basis functions as illustrated in Fig.2. The relatively accurate electric current distributions on the interior cell (IC), the left edge cell (LeEC), the right edge cell (REC), the upper edge cell (UEC), the lower edge cell (LoEC), the left upper corner cell (LUCC), the right upper corner cell (RUCC), the left lower corner cell (LLCC), and the right lower corner cell (RLCC) are obtained respectively with conventional MoM. Then the coarse mesh on each cell is torn into $k$ parts according to MTASED and they are substituted as basis functions in equation (6). With the coupling from neighbouring cells taken into account, the accuracy is further improved.

III. NUMERICAL RESULTS AND DISCUSSIONS

*Metal cube array*

Firstly, we give the bistatic radar cross section (RCS) of a strong-coupled 6×6 metal cube array. And the element dimension is $0.5\lambda \times 0.5\lambda \times 0.5\lambda$ and the gap between neighbouring element is $0.5\lambda$ in normal incidence case. The number of torn meshes ($k$) set to be 40. As a result, the coarse mesh that is set to be $0.4\lambda \times 0.4\lambda \times 6$ in classic SED method has been refined to be about $0.13\lambda \times 0.13\lambda$ in MTSED method and the memory consumption is expanded to 1600 times for matrix storage. From Fig.3 (a) we can observe that the RCS curve obtained by MTSSED method is more accurate than that of SSED method compared to FEKO with $0.1\lambda$ mesh size. There is about 5.39dB error at 0° and 11.99dB error at 90° with SSED method which is mainly caused by the ignorance of mutual coupling from neighbouring elements in the simplified sub-entire domain basis function computation. With the application of MTSSED method, the accuracy has been improved significantly for the current distribution on each cell is modified with relatively dense mesh. The error at 0° and 90° has reduced to 1.91dB and 3.13dB, respectively. The average error for all angles has decreased from 2.68dB to 0.89dB. Fig.3 (b) shows the RCSs calculated by ASED method, MTASED method and FEKO with the same mesh. It is obvious that the RCS calculated by MTASED method is more consistent with FEKO full wave analysis. The error at 0° has reduced from 2.77dB to 0.02dB and the error at 90° has reduced from 4.65dB to 1.23dB. Consequently, the MTASED method is more accurate than MTSSED method for the more accurate basis function.

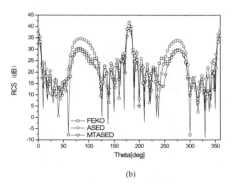

(b)

Figure.3. RCSs of the 6×6 cube array in normal incidence case. (a) MTSSED method. (b) MTASED method.

When the angle of incident wave is oblique, the increase of multiple interactions between array elements will be enhanced. Therefore, the accuracy of SED method will usually be declined. Fig.4 shows the bistatic RCSs of a 6×6 cube array consists of associated elements calculated by FEKO with the same mesh, SED method and MTSED method in oblique incidence case ($\theta$=45°). The torn mesh on each single cell is about $0.15\lambda \times 0.15\lambda$ ($k$=42). From Fig.4 (a) we can observe that the error at 0° has decreased from 8.81dB to 1.54dB and the error at 90° has reduced from 5.06dB to 3.99dB with MTSSED method and there is significant improvement of accuracy at 74° with the error decreased from 11.45dB to 0.66 dB. When MTASED method is employed, as shown in Fig.4 (b), the error at 0° has decreased from 5.73dB to 2.34dB and the error at 90° has reduced from 5.74dB to 2.34dB and there is significant improvement of accuracy at 36° with the error decreased from 15.22dB to 1.86dB. Compared to SED method, the computational accuracy has been improved obviously with MTSED method in oblique incidence case. The accuracy is expected to be further improved with denser torn meshes or with basis function computed with more layers of dummy elements.

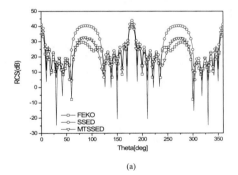

(a)

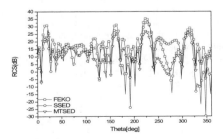

(a)

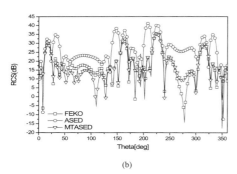

(b)

Figure.4. RCSs of the 6×6 cube array in normal incidence case (θ=45°). (a) MTSSED method. (b) MTASED method.

To test the efficiency of the proposed method, we record the computation time when computing the RCSs of different arrays with SED method and MTSED method, as shown in Table I. Our codes are implemented by a personal computer with Intel(R) Core(TM) 2 Duo CPU T6400-2.00GHz and 2GB RAM. We can conclude that there is about 2 times and 5 times computation time increasing compared to classic SED method with the application of MTSSED and MTASED method, respectively. Although the efficiency of MTASED method is lower than that of MTSSED method with the same torn meshes, the MTASED method can usually achieve a better accuracy. To improve the efficiency of our proposed method, parallel programming such as message passing interface (MPI) [15], shared address space (OpenMP) [16] and meta process model (MpC) [17] can be incorporated.

TABLE I

COMPARISON OF COMPUTATION TIME OF 6×6 CUBE ARRAY

| Theta | Time[s] | | | |
|---|---|---|---|---|
| | SSED | MTSSED | ASED | MTASED |
| 0° | 224.57 | 343.85 | 205.75 | 1071.64 |
| 45° | 223.56 | 589.35 | 211.68 | 972.51 |

## IV. CONCLUSION

In this paper, the MTSED method is proposed by tearing the coarse mesh of SED basis function on each unit cell into several parts. The numerical results show that the modelling precision for strong-coupled arrays can be improved more than 13 dB maximally with less than four times computational time increasing compared to classic SED method which demonstrate that the proposed MTSED method is an accurate method for electromagnetic scattering analysis of strong-coupled finite periodic structures.

## REFERENCES

[1] Jaewon Choi, Seoul, Chulhun Seo, "Power amplifier design using the novel PBG structure for linearity improvement and size reduction," *Microwave Conference, 2007. APMC 2007. Asia-Pacific.*

[2] Shang Jun-ping, Xu Ping, Jiang Shuai, Fu De-min, "Study on a fast measurement and identifying malfunction elements in phased array antennas," *Synthetic Aperture Radar, 2009. APSAR 2009. 2nd Asian-Pacific Conference on.*

[3] Dorsey W. M., McDermitt C. S., Bucholtz F., Parent M. G., "Design and performance of frequency selective surface with integrated photodiodes for photonic calibration of phased array antennas," *IEEE Trans. Antennas Propag.*, vol. 58, no. 8, pp. 2588-2593, 2010.

[4] Jianxun Su, Xiaowen Xu, Mang He, "Hybrid PMM-MOM method for analyzing the RCS of finite array," *Microwave Technology & Computational Electromagnetics (ICMTCE), 2011 IEEE International Conference on.*

[5] Y. Cai, C. Mias, "Faster 3D finite element time domain-floquet absorbing boundary condition modeling using recursive convolution and vector fitting." *IET Microwave, Antenna & Propag.*, vol. 3, no. 2, pp. 310-324, March 2009.

[6] D. Y. Li and C. D. Sarris, "A new approalch for the FDTD modeling of antenna over periodic structures." *IEEE Trans. Antennas Propag.*, vol. 59, no. 1, pp. 310-314, Nov. 2011.

[7] Gonzalez-Ovejero, D., Mesa, F., Craeye, C., "Accelerated macro basis functions analysis of finite printed antenna arrays through 2D and 3D multipole expansions." *Antennas and Propagation, IEEE Transactions on*, vol. 61, no. 2, pp. 707-717, Nov. 2013.

[8] Bo Zhang, Gaobiao Xiao, Junfa Mao, Yan Wang, "Analyzing large-scale non-periodic arrays with synthetic basis functions." *Antennas and Propagation, IEEE Transactions on* , Vol.58.

[9] J. Yeo and R. Mittra, "Numerically efficient analysis of microstrip antennas using the characteristic basis function method (CBFM)," *Antennas and Propagation Society International Symposium*, vol. 4, pp. 85-88, Nov.2003.

[10] V. V. S. Prakash and R. Mittra, "Characteristic basis function method: A new technique for efficient solution of method of moments matrix equations," *Microwave Opt. Technol. Lett.*, vol. 36, pp. 95-100, Jan. 2003.

[11] W.B. Lu, T.J. Cui, Z.G. Qian, X.X. Yin, and W. Hong, "Accurate analysis of large-scale periodic structures using an efficient sub-entire-domain basis function method," *IEEE Trans. Antennas Propag.*, vol. 52, no. 11, pp. 3078-3085, Nov. 2004.

[12] W.B. Lu, T.J. Cui, X.Y. Yin, Z.G. Qian and W. Hong, "Fast algorithms for large-scale periodic structures using subentire domain basis functions," *IEEE Trans. Antennas Propag.*, vol. 53, no. 3, pp. 1154-1162, Mar. 2005.

[13] W.B. Lu, T.J. Cui, and H. Zhao, "Acceleration of fast multipole method for large-scale periodic structures with finite sizes using sub-entire-domain basis functions," *IEEE Trans. Antennas Propag.*, vol. 55, no. 2, pp. 414-421, Feb. 2007.

[14] C. Liu and X. X. He, "A mesh-tearing simplified sub-entire domain basis function method for finite periodic structures." *2012 International Conference on Microwave and Millimeter Wave Technology*, May 2012.

[15] Y. Zhang, X.W. Zhao, Donoro D.G., S.W. Ting and Sarkar T.K., "Parallelized hybrid method with higher-order MoM and PO for analysis of phased array antennas on electrically large platforms," *IEEE Trans. Antennas Propag.*, vol. 58, no. 12, pp. 4110-4115, Dec. 2010.

[16] B. Sun, L.L. Ping and X.X. He, "Acceleration of time-domain finite-element method in electromagnetic analysis with OpenMP." *2010 International Conference on Microwave and Millimeter Wave Technology*, pp. 845-8, May 2010.

[17] Midorikawa H., "The performance analysis of portable parallel programming interface MpC for SDSM and pthread," *IEEE International Symposium on Cluster Computing and the Grid*, vol. 2, pp. 889-896, May 2005.

# Apply Complex Source Beam Technique for Effective NF-FF Transformations

Shih-Chung Tuan [1], Hsi-Tseng Chou[2] and Prabhakar H.Pathak[3]

1. Dept. of Communication. Eng., Oriental Institute of Technology, Taiwan
2. Dept. of Communication Eng., Yuan Ze University, Taiwan
3. Dept. of Electrical and Computer Eng., Ohio State University, USA

*Abstract*- **This paper presents a general technique for an effective near-field to far-field transformation based on the complex source beam (CSB) method. In this technique, a set of CSBs launched from the center of antenna under test is used to represent the antenna's radiation. The weightings of CSBs are found by matching the near-fields measured on a pre-described surface in the conventional measurement systems. All three conventional measurement techniques including planar, cylindrical and spherical measurement systems are examined to validate the CSB method with accuracy comparisons in the near- and far-field patterns.**

## I. INTRODUCTION

Near-field (NF) measurement systems are widely used to measure the antenna's radiation characteristics. Depending on the natures of antennas in terms of their radiation directivities, various system configurations have been developed to provide effective measurements of antenna radiations at the development stages. Three typical and most widely used ones are the planar, cylindrical and spherical near-field measurement systems, which measure the radiation fields on the planar, cylindrical and spherical surfaces, respectively enclosing the antennas under test (AUT). In order to obtain the far-field(FF) radiation characteristics, which are most commonly used to justify the antenna characteristics in the antenna community, the near-field to far-field(NF-FF) transformations need to be performed using the measured near-field information on the respective surfaces. In the past, NF-FF transformation techniques are developed and performed according to the natures of system configurations. For example, a fast Fourier transformation (FFT) scheme can be used to efficiently obtain the far-field radiations of highly directive antennas, such as reflector antennas, in the planar near-field measurement system. On the other hand, cylindrical and spherical wave type expansions have also been used in the literatures for the cylindrical and spherical system. There are several drawbacks in those transformation techniques. First, those transformation techniques are relatively independent in their natures, and do not share common characteristics in the expansions. Second, those techniques experience different degrees of truncation diffraction effects in the transformation procedures. Several studies and techniques have been reported in the literatures to discuss the truncation errors and reduce these effects based on the numerical method of moment (MoM), which increases the computational complexity. Third,

the utilization of measured data has been limited in these conventional transformation techniques. Those drawbacks have significantly limited the applications of NF measurement systems.

This paper presents an effective approach to alternatively resolve the above mentioned shortcomings based on the applications of complex source beam (CSB) techniques[1,2,4]. In particular, a set of CSBs are selected to serve as field basis functions launched radially from the center of AUT. The measured NF data on the selected surface boundaries is represented in terms of this CSB set. The advantages of this CSB technique, responding to each of the above mentioned shortcoming, are apparent due to a fact that the CSB is a solution of Maxwell's equations. First, this CSB technique can be applied universally to a NF measurement system with a relatively arbitrary scan surface including the conventional three NF systems mentioned in the previous paragraphs, which are examined in this paper to validate the technique. Secondly, each CSB function is continuous in the spatial extent. The superposition of CSBs after the NF expansion provides a continuous field variation along the selected surface of measurement system. The truncation errors due to the existence of measurement surface boundaries can be reduced without the need to impose additional taper functions. Third, after the expansion each of the CSB functions can be treated independently, and can be used in conjunction with other simulation techniques to justify AUT's characteristics when it is used in a realistic practice. For example, in [3] CSB techniques have been developed to efficiently analyze the reflector antenna when it is illuminated by the radiation of a feed antenna.

## II. CHARACTERISTIC OF CSB FIELDS

The CSB is an electromagnetic field radiated from a current source located in a complex space. This section reviews the near- and far-field formulations to exhibit the characteristics of field variations from a near-zone to far zone, which makes the NF to FF transformation in a nature of wave propagation. the CSB can be formulated as

$$\bar{E}_{m,e}(\bar{r}) = \frac{jk}{4\pi} \left\{ \begin{matrix} Z_0 \hat{R} \times \hat{R} \times d\bar{p}_e \\ \hat{R} \times d\bar{p}_m \end{matrix} \right\} \frac{e^{kb}e^{-jkz\left(1+\frac{x^2+y^2}{2(z^2+b^2)}\right)}e^{-\frac{kb}{2}\left(\frac{x^2+y^2}{z^2+b^2}\right)}}{z+jb}. \tag{1}$$

978-7-5641-4279-7

the formulation in (1) clearly shows a Gaussian field distribution in the direction (x-y plane) transverse to the beam axis ($\hat{z}$). The CSB field in the far zone of beam waist can be formulated by using the following approximation:

$$\overline{E}_{m,e}(\overline{r}) = \frac{jk}{4\pi}\left\{\frac{Z_0\hat{r}\times\hat{r}\times d\overline{p}_e}{\hat{r}\times d\overline{p}_m}\right\}\frac{1}{r}e^{-jkr}e^{jk\overline{r}_o\cdot\hat{r}}e^{kb\hat{b}\cdot\hat{r}} \qquad (2)$$

In this section, the strategies to represent the radiations of antenna under test (AUT) by a set of CSBs are described. The procedure to obtain the CSBs' weightings by matching its field values with the measured data on the measurement surface is also developed. Once the weightings are found, the CSBs can be ready and used to find the far field radiation. Figure1. shows the discretized grids of CSBs' axes on the virtual spherical surface, $S_{dis}$, enclosing the AUT. Figure2., where a measurement surface, $S_m$, is assumed to measure the NF data. It is noted that $S_m$ enclosing AUT can be relatively arbitrary, and are planar, cylindrical and spherical for the three conventional NF measurement systems. Assuming electrical current sources, the CSB representation of AUT's radiation can be generally expressed as

$$\overline{E}_{AUT}(r) = \sum_{n=1}^{N}\left(A_n\overline{G}_n^1(\overline{r}) + B_n\overline{G}_n^2(\overline{r})\right) \qquad (3)$$

Where

$$\begin{bmatrix}\overline{G}_n^1(\overline{r})\\\overline{G}_n^2(\overline{r})\end{bmatrix} = e^{-kb}\frac{jk}{4\pi}Z_0\hat{R}_n\times\hat{R}_n\times\begin{bmatrix}\hat{u}_{1,n}\\\hat{u}_{2,n}\end{bmatrix}\frac{e^{-jkR_n}}{R_n} \qquad (4)$$

$$\begin{cases}E_{1,p}^{mea} = \sum_{n=1}^{N}\left[A_n\left(\overline{G}_n^1(\overline{r}_p^m)\cdot\hat{u}_{1,p}\right) + B_n\left(\overline{G}_n^2(\overline{r}_p^m)\cdot\hat{u}_{1,p}\right)\right]\\E_{2,p}^{mea} = \sum_{n=1}^{N}\left[A_n\left(\overline{G}_n^1(\overline{r}_p^m)\cdot\hat{u}_{2,p}\right) + B_n\left(\overline{G}_n^2(\overline{r}_p^m)\cdot\hat{u}_{2,p}\right)\right]\end{cases} \qquad (5)$$

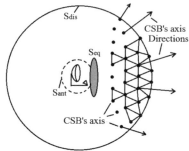

Figure 1: Illustration of CSB discretization grids to launch the axes of CSBs. The relation to AUT's size to justify the resolution of CSB grids is also shown.

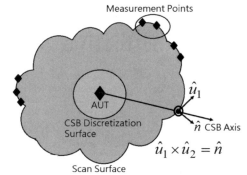

Figure 2: Applications of CSB technique to represent AUT's radiation and the relationship to the measurement surface.

. A minimum least square error scheme can be used to solve (5) for $(A_n, B_n)$ $n = 1 \sim N$, which can be achieved via a simple MATLAB command by

$$[D] = [G]\backslash[E], \qquad (6)$$

where

$$[D] = \begin{bmatrix}A_n\\\vdots\\B_n\\\vdots\end{bmatrix}_{2N\times1} \quad ; \quad [E] = \begin{bmatrix}E_{1,p}^{mea}\\\vdots\\E_{2,p}^{mea}\\\vdots\end{bmatrix}_{2M\times1} \qquad (7a);(7b)$$

and

$$[G] = \begin{bmatrix}\overline{G}_n^1(\overline{r}_p^m)\cdot\hat{u}_{1,p} & \cdots & \overline{G}_n^2(\overline{r}_p^m)\cdot\hat{u}_{1,p} & \cdots\\\vdots & \ddots & \vdots & \\\overline{G}_n^1(\overline{r}_p^m)\cdot\hat{u}_{2,p} & \cdots & \overline{G}_n^2(\overline{r}_p^m)\cdot\hat{u}_{2,p} & \cdots\\\vdots & \ddots & \vdots & \end{bmatrix}. \qquad (8)$$

In the practical implementation, most of the computation time is used to compute (8), which however needs to be computed only once regardless of any AUT, This matrix can be stored in advance once the system setup has been established. Once these coefficients are found, (2) can be used to find the far-field of AUT.

## III. NUMERICAL EXAMPLES

The CSB technique is validated by its treatments of NF-FF transformations for the three conventional measurement systems, which include planar, cylindrical and spherical near-field measurement systems. In particular, the NSI systems (modeled by NSI 300V series Planar/cylindrical NF Antenna Measurement System and NSI-700S-90 Spherical NF Antenna Measurement System, respectively for the three systems) are used to measure the radiations of a standard gain horn-type antenna, SGH430 at a frequency of 2.45GHz, which has been widely used as a reference of antenna's gain comparison. The validations on NF representations and far-field patterns are examined and shown in the following.

*A. PLANAR NEAR-FIELD MEASUREMENT*

Figure3. It shows Implementation of CSB technique in the planar near-field measurement system. Figure4. It shows the comparison of radiation patterns between CSB synthesized and measured data, respectively. On Figure 4(a) and (b), the FF data was taken on the two principal planes ($\phi = 0^o$ and $\phi = 90^o$), respectively.

*B. CYLINDRICAL NEAR-FIELD MEASUREMENT*

Figure5. It shows Implementation of CSB technique in the cylindrical near-field measurement system. On Figure 6 (a) and (b) the FF data was taken on the two principal planes ($\phi = 0^o$ and $\phi = 90^o$), respectively.

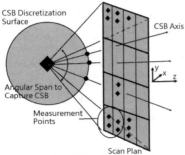

Figure 3: Implementation of CSB technique in the planar near-field measurement system.

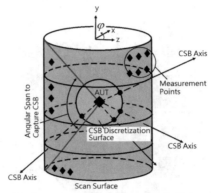

Figure 5: Implementation of CSB technique in the cylindrical near-field measurement system.

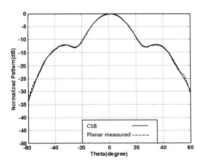

Figure 4(a) The FF data was taken on the E-planes

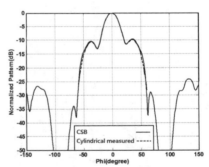

Figure 6(a) The FF data was taken on the E-planes

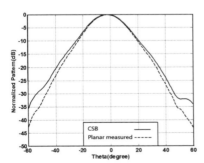

Figure 4(b) The FF data was taken on the H-planes

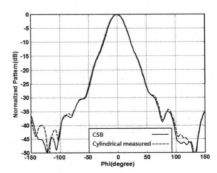

Figure 6(b) The FF data was taken on the H-planes

## C. SPHERICAL NEAR-FIELD MEASUREMENT

Figure7. It shows Implementation of CSB technique in the spherical near-field measurement system. On Figure 8 (a) and (b) the FF data was taken on the two principal planes ($\phi = 0^o$ and $\phi = 90^o$), respectively.

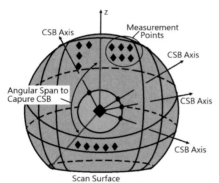

Figure 7: Implementation of CSB technique in the spherical near-field measurement system.

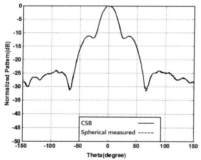

Figure 8(a) The FF data was taken on the E-planes

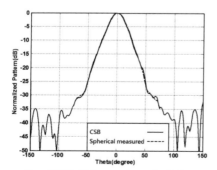

Figure 8(b) The FF data was taken on the H-planes

## IV. CONCLUSIONS

This paper presents an effective approach to alternatively resolve the above mentioned shortcomings based on the applications of complex source beam (CSB) techniques. In this technique, a set of CSBs launched from the center of antenna under test is used to represent the antenna's radiation. The weightings of CSBs are found by matching the near-fields measured on a pre-described surface in the conventional measurement systems.

### REFERENCES

[1] J.B Keller and W. Streifer, "Complex Rays with an Application to Gaussian Beam" J. Opt. Soc. Amer., vol.61,pp.40-43,1971.
[2] G.A Deschamps, "Gaussian Beam as a Bundle of Complex Rays" Electronics Letters, vol.7, pp.684-685,Nov, 1971.
[3] H-T Chou, Shih-Chung Tuan, H-H Chou " Transient Analysis of Scattering From a Perfectly Conducting Parabolic Reflector Illuminated by a Gaussian Beam Electromagnetic Field" *IEEE Transactions on Antennas and Propagation, Vol. 58, No.5, pp.1711~1719, May 2010.*
[4] George C. Zogbi " Reflection and Diffraction of General Astigmatic Gaussian Beams From Curved Surfaces and Edges," Ph.D. Dissertation, The Ohio State University, June 1994.

# A Compressed Best Uniform Approximation for Fast Computation of RCS over Wide Angular-band

Zhiwei Liu[1,2], Shan He[1], Yueyuan Zhang[1], Xiaoyan Zhang[1,2], Yingting Liu[1], Jun Xu[1]

1 East China Jiaotong University, Nanchang, 330013, China

2 The State Key Laboratory of Millimeter Wave, Nanjing, 210096, China

*Abstract*–Using the best uniform approximation hybridized with Singular Value Decomposition (BUA-SVD) is proposed to reduce the time requirement for computation of monostatic radar cross-section (RCS). In contrast to our previous work, the best uniform approximation technique is applied to compute the key excitation vectors instead of electric current vectors. Reduction of the number of multiple excitation vectors can lead to significantly reduced computation time. Moreover, with low-rank property, the excitation vectors may be further compressed by SVD, resulting in a more efficient method. SVD can lead to reduce the computational complexity further with the best uniform approximation. Numerical results demonstrate that BUA-SVD is efficient for monostatic RCS calculation with high accuracy.

## I. INTRODUCTION

Electromagnetic (EM) wave scattering problems address the physical issue of detecting the diffraction pattern of the EM radiation scattered from a large and complex body when illuminated by an incident incoming wave. A good understanding of these phenomena is crucial to RCS calculation, antenna design, EM compatibility, and so on. All these simulations are very demanding in terms of computer resources, and require efficient numerical methods to compute an approximate solution of Maxwell's equations. Using the equivalence principle, Maxwell's equations can be recast in the form of integral equations that relate the electric and magnetic fields to the equivalent electric and magnetic currents on the surface of the object. To solve integral equations by applying traditional method of moment in [1], the computation complexity for the iterative solver is $O(kN^2)$ and the memory requirement is $O(N^2)$, where $N$ refers to the number of unknowns and $k$ refers to the number of iterative steps. Obviously, it is impractical to use personal computer to solve equations with more than 10,000 unknowns. This difficulty can be overcome by using the multi-level fast multi-pole algorithm (MLFMA) [2, 3]. The use of MLFMA reduces the memory requirement to $O(NlogN)$ and computation complexity to $O(kNlogN)$.

Although MLFMA results in an efficient solution of the integral equation, it is still time-consuming for the calculation of monostatic RCS since it requires repeated solution of the integral equation at each incident angle. As is well known, many interpolation methods have been proposed to circumvent this difficulty. Conventional interpolation methods, such as asymptotic waveform evaluation (AWE) and the cubic-spline (CS) interpolation method, can easily approximate the monostatic RCS. AWE which is introduced in [4] and [5] is a kind of classical method which is widely used in computational EM. It utilizes the high-order derivatives of the incident current vector at the interval center to extrapolate the value of nearby points, and is considered to be the AWE extrapolation method in [6] has been introduced by Wei. In the AWE technique which is introduced in [6] and [7], the induced current is expanded in the Taylor series around an angle, and the Padé approximation is used to improve the accuracy. As a result, the induced current vector is expected to be accurate near this sample, but imprecise when the incident angle moves away from the angle.

As an alternative technique, the best uniform approximation in [8] has been introduced by Chebyshev. This approximation is important in approximation theory because the roots of the Chebyshev polynomials of the first kind, which are also called Chebyshev nodes, are used as nodes in polynomial interpolation. The resulting interpolation polynomial minimizes the problem of Runge's phenomenon and provides an approximation that is close to the polynomial of best approximation to a continuous function under the maximum norm. This approximation leads directly to the method of Clenshaw-Curtis quadrature. Therefore, the best uniform approximation can improve efficiency and save much time.

It is noteworthy that selection of sampling points is crucial for interpolation and extrapolation methods. The non-uniform sampling method is more flexible and efficient than the uniform sampling method. As a result, optimally selecting those angles that would be most informative will reduce the number of repeated solutions which one must consider for monostatic scattering computations. In [8], the best uniform approximation is proposed to optimally select the most informative angles in monostatic RCS curve, resulting in an efficient computation of monostatic scattering. In [9], it is reported that multiple excitation vectors or right hand side vectors can be compressed by use of the low-rank property. Inspired by [10], the multiple right hand sides can be approximately described by a low-rank form. In linear algebra, the singular value decomposition (SVD) is a factorization of a real or complex matrix, with many useful applications in signal processing and statistics. Applications which employ the SVD include computing the pseudoinverse, least squares fitting of data, matrix approximation, and determining the rank, range and null space of a matrix. In this paper, the

978-7-5641-4279-7

combination of the best uniform approximation and singular value decomposition (BUA-SVD) is applied to efficient computation of monostatic RCS. The numerical simulations demonstrate that this framework can reduce the computation time significantly.

## II.    THEORY OF BUA-SVD

### A.  the Best Uniform Approximation (BUA)

The best uniform approximation in [8] has been introduced by Chebyshev as an interpolation technique. In order to get the RCS of the target more quickly, the algorithm must be applied multilevel fast multipole method (MLFMA) point by point calculation within a given frequency band. To combine it with the best uniform approximation, the specific process is as follows:

For a given frequency band $f \in [f_m, f_n]$, corresponding to the wave number $k \in [k_m, k_n]$, do the coordinate transformation first. Let

$$\tilde{k} = \frac{2k - (k_m + k_n)}{k_n - k_m}, \qquad (4)$$

so The surface current can be written as:

$$I(k) = I\left(\frac{\tilde{k}(k_n - k_m) + (k_m + k_n)}{2}\right), \qquad (5)$$

where $\tilde{k}_i \in [-1,1]$.

Assume that $T_l(\tilde{k})$ ($l=1,2,\ldots,n$ ) as the $l$-order Chebyshev polynomials, and it is defined as :

$$T_0(\tilde{k}) = 1,\ T_1(\tilde{k}) = \tilde{k},\ T_{l+1} = 2\tilde{k}T_l(\tilde{k}) - T_{l-1}(\tilde{k}),\ 2 \le l \le n. \qquad (6)$$

So the Chebyshev Approximation of $I(k)$ can be expressed as:

$$I(k) = I\left(\frac{\tilde{k}(k_n - k_m) + (k_m + k_n)}{2}\right) \approx \sum_{l=0}^{n-1} c_l T_l(\tilde{k}) - \frac{c_0}{2}. \qquad (7)$$

We suggest $\tilde{k}_i$ ($i=1,2,\ldots,n$) as the $n$-th zreo point of $T_n(\tilde{k})$ ($\tilde{k} \in [-1,1]$ ), so

$$\tilde{k}_i = \cos(\frac{i - 0.5}{n}\pi), i = 1, 2, \cdots, n \qquad (8)$$

$$c_l = \frac{2}{n}\sum_{i=1}^{n} I(k_i) T_l(\tilde{k}_i) \cdot \qquad (9)$$

Here, $k_i$ is called the Chebyshev nodes in $[k_m, k_n]$. Where

$$k_i = \frac{\tilde{k}_i(k_n - k_m) + (k_m + k_n)}{2} (i = 1,2,\cdots,n) \qquad (10)$$

Above all, we calculate the approximate current throughout the whole frequency band to analyze the EM scattering characteristics of the target quickly.

### B.  Singular Value Decomposition (SVD)

Theoretically, the combination of MoM and MLFMA is able to accurately analyze the scattering of any geometry. Improved by the best uniform approximation, the computation of a monostatic RCS can be accelerated greatedly. However, in some cases, the number of coefficients of the interpolation polynomials is so large as to compromise the efficient calculation of monostatic scattering. This process can be computationally prohibitive for electrically large objects. In order to alleviate this difficulty, a singular value decomposition based method is proposed and discussed in this section.

Firstly, a brief review of compression of right hand sides is given. The computation of monostatic RCS can be considered as linear equations with multiple right hand sides

$$\mathbf{A} \cdot \mathbf{X} = \mathbf{B} \qquad (11)$$

where $\mathbf{A}$ is the impedance matrix, $\mathbf{X}$ is the multiple complex coefficient vector of RWG basis and $\mathbf{B}$ is the multiple right hand side generated by the incident wave. In addition

$$\mathbf{X} = [\mathbf{x}(\theta_1), \mathbf{x}(\theta_2), \ldots, \mathbf{x}(\theta_n)] \qquad (12)$$

$$\mathbf{B} = [\mathbf{b}(\theta_1), \mathbf{b}(\theta_2), \ldots, \mathbf{b}(\theta_n)] \qquad (13)$$

where $\theta_i$ is the $i^{th}$ incident angle. Using traditional singular value decomposition, the matrix $\mathbf{B}$ can be described in the form of an eigenvalue and eigenvector.

$$\mathbf{B} = \mathbf{U} \cdot \mathbf{\Sigma} \cdot \mathbf{V}^H \qquad (14)$$

The superscript 'H' denotes the conjugate transpose. If the dimension of $\mathbf{B}$ is $N \times M$, the dimension of matrices $\mathbf{U}$、 $\mathbf{\Sigma}$ and $\mathbf{V}$ are $N \times M$ , $M \times M$, $M \times M$, respectively. $N$ is the number of unknowns. $\mathbf{\Sigma}$ is a diagonal matrix including all the eigenvalues of $\mathbf{B}$ while $\mathbf{U}$ and $\mathbf{V}$ contain all the eigenvectors of $\mathbf{B}$. When $\mathbf{B}$ is the multiple right hand sides in the linear system connecting with the SIE used for monostatic RCS, the matrix $\mathbf{B}$ is low-rank and can be approximately described as a low-rank SVD form.

$$\mathbf{B} = \mathbf{U}_k \cdot \mathbf{\Sigma}_k \cdot \mathbf{V}_k^H \qquad (15)$$

where the dimension of matrices $\mathbf{U}_k$、 $\mathbf{\Sigma}_k$ and $\mathbf{V}_k$ are $N \times k$, $k \times k$, $M \times k$, respectively. Only the $k$ largest eigenvalues and corresponding eigenvectors are reserved. Substituting (15) into (11), the linear equations can be rewritten as

$$\mathbf{X} \approx (\mathbf{A}^{-1} \cdot \mathbf{U}_k) \cdot \mathbf{\Sigma}_k \cdot \mathbf{V}_k^H \qquad (16)$$

Here, $\mathbf{A}^{-1} \cdot \mathbf{U}_k$ can be computed by any iterative solver. If using a direct solver to compute the inversion of matrix $\mathbf{A}$, the proposed method will become useless. Therefore, the number of repeated solutions of $\mathbf{Ax} = \mathbf{b}$ is $k$ for SVD method. Using traditional method, the number is $M$. Generally, $k$ is much smaller than $M$ which leads to an efficient method for computation of monostatic RCS over a wide angular band.

Using the best uniform approximation strategy, we can write the induced current into the sum of the samples shown in (15). We rewrite this formulation in matrix form

$$\mathbf{X} = \mathbf{A}^{-1} \cdot \mathbf{B}_s \cdot \mathbf{C} \qquad (17)$$

where $\mathbf{X}$ is a matrix contains all induced currents over the whole angular band. $\mathbf{C}$ is the coefficient matrix of the non-uniform interpolation method with the dimension of $s \times M$. $\mathbf{B}_s$ contains all the key samples of the excitation vectors. According to formulation (17), the required number of repeated solution of $\mathbf{Ax} = \mathbf{b}$ is $s$. Using singular value decomposition for matrix $\mathbf{B}_s$

$$\mathbf{B}_s = \mathbf{U}_{sk} \cdot \mathbf{\Sigma}_{sk} \cdot \mathbf{V}_{sk}^H \qquad (18)$$

Then

$$\mathbf{X} = (\mathbf{A}^{-1}\cdot\mathbf{U}_{sk})\cdot\mathbf{\Sigma}_{sk}\cdot\mathbf{V}_{sk}{}^{H}\cdot\mathbf{C} \qquad (19)$$

As a result, the required number of repeated solution is reduced to $sk$.

### C. Combination of BUA with SVD

The key problem for SVD is to obtain the decomposition form of multiple right hand sides. The traditional SVD method is a good analytical solution for this problem. However, SVD requires the computation of the matrix including all right hand sides and the complexity of the computation time of SVD is $O(nm^2 + mn^2)$, where $m$ and $n$ are with respect to the number of rows and columns. When the number of unknowns or right hand vectors is large, this analytical solution is not practical. In order to alleviate this difficulty, the best uniform approximation algorithm may be applied and this performs more efficiently than the traditional SVD method.

### III. RESULTS AND DISCUSSION

In this section, a number of numerical results are presented to demonstrate the accuracy and efficiency of the proposed method for fast calculation of monostatic RCS over a wide angular band. All experiments are conducted on a Quad-Core AMD Opteron (tm) with 4.00 GB local memory and run at 2.31 GHz in single precision. Two geometries are applied to illustrate the performance of our proposed method. They consist of an Almond with 1210 unknowns and a cube with 49260 unknowns. In our numerical experiments, the two geometries are illuminated by a plane wave with the incident pitch angles range from 0 to 180 deg. The frequency is 3.0 GHz for the Almond and 300 MHz for Cube. For all cases the azimuth angle is 0 deg.

The algorithm produces a sequence of decompositions of a matrix into a sum of low-rank matrix and error matrix. Neither the original matrix nor the error matrix will be computed completely. Choosing the Convergence Error (CE) is the most important consideration in SVD. In order to avoid the numerical error, CE is required to be sufficiently small. However, large CE is needed for efficiency. In Fig.1, the results of the monostatic RCS of Almond by three different CE are compared with the reference result (the Direct Method). The reference result is the RCS curve computed with repeated solution at each angle. Other curves are computed by our proposed method. We select a part of the curve where the difference is much bigger with the incident pitch angles range from 20 to 100 deg. From this figure, when CE is set to be 0.1, the RCS curve is not accurate enough. When CE is set to be 0.01, the proposed method will perform a good result. A larger number of vectors leads to larger computation time. If using the proposed method without compression (CE=0.0), it would Consume more time because of the high-rank. As a result, despite of the number of excitation vectors, the value of CE is set to be 0.01 in this paper to keep the RCS curve accurate enough.

Since the number of right hand vectors is small for 1-D angular sweep in this paper, it is feasible to apply the proposed

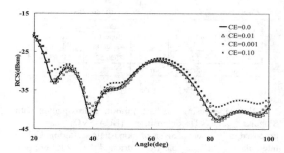

Fig.1 Comparison of the results by Proposed Method with different CE

method for computing the eigenspace of multiple right hand sides. For monostatic RCS simultaneous theta and phi sweep, the number of right hand sides is 721×721(4 point per deg). As is shown in Fig.2-3, the monostatic RCS curve of Almond and Cube which computed by the proposed method is compared with the curve computed by the reference result repeatedly. It is obvious that the proposed method is accurate since there is no significant difference between the RCS result obtained by the reference result and the proposed method. As is shown in TABLE I, when compared with the reference result, the proposed method provides little advantage on total computation time since the number of right hand side is small.

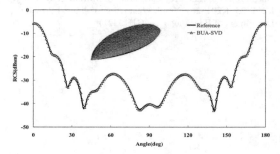

Fig.2 The monostatic RCS results for Almond

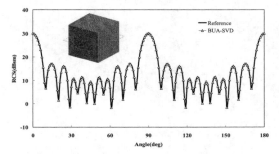

Fig.3 The monostatic RCS results for Cube

TABLE I
COMPUTATION TIME OF MONOSTATIC RCS WITH DIFFERENT METHODS

| Geometry | Unknown | f(GHz) | Computation Time (s) | |
|---|---|---|---|---|
| | | | Repeated Solution | Proposed Method |
| Almond | 1210 | 3.0 | 3520.578 | 231.9844 |
| Cube | 49260 | 0.3 | 53497.78 | 1672.125 |

## IV.  CONCLUSION

In this paper, using the best uniform approximation with Singular Value Decomposition (BUA-SVD) has been proposed for efficient analysis of monostatic scattering. Unlike interpolation of the electric current, the best uniform approximation algorithm is used to approximate the multiple right hand sides on a set of non-uniform sampling angles and SVD is employed to reduce the consumption time automatically. The most informative angles may be selected by this procedure. Moreover, applying SVD to compute the eigenvectors of the selected vectors leads to reduced times for the iterative solutions of linear systems. Numerical experiments demonstrate that the proposed method is more efficient when compared with the traditional direct method.

## ACKNOWLEDGMENT

The authors would like to thank the support of National Science Foundation of China (No: 61061002, 61261005); Open Project of State Key Laboratory of Millimeter Wave (No: K201325, K201326); Youth Foundation of Jiangxi Provincial Department of Education (No: GJJ13321).

## REFERENCES

[1] R. F. Harrington. "Field Computation by Moment Methods," Malabar, Fla.: R. E. Krieger, 1968.

[2] W. C. Chew, J. M. Jin, E. Michielssen, and J. M. Song, "Fast and Efficient Algorithms in Computational Electro-magnetics," Boston, MA: Artech House, 2001.

[3] J. M. Song, C. C. Lu and W. C. Chew, "Multilevel Fast Multipole Algorithm for Electromagnetic Scattering by Large Complex Objects," IEEE Trans. Antennas Propagat, vol. 45, no. 10, pp. 1488-1493, October 1997.

[4] Y. E. Erdemli, J. Gong, C. J. Reddy, and J. L. Volakis. "Fast RCS Pattern Fill Using AWE Techunique," IEEE Trans. Antennas Propagat, vol. 46, no.11, pp. 1752-1753, November 1998.

[5] R. D. Slong, R. Lee, and J. F. Lee. "Multipoint Galerkin Asymptotic Waveform Evaluation for Model Order Reduction of Frequency Domain FEM Electromagnetic Radiation Problems," IEEE Trans. Antennas Propagat, vol. 49, no. 10, pp. 1504-1513, October 2001.

[6] X. C. Wei, Y. J. Zhang, and E. P. Li. "The Hybridization of Fast Multipole Method with Asymptotic Waveform Evaluation for the Fast Monostatic RCS Computation," IEEE Trans. Antennas Propagat, vol. 52, no. 2, pp. 605-607, February 2004.

[7] Y. E. Erdemli, J. Gong, C. J. Reddy, and J. L. Volakis. "Fast RCS Pattern Fill Using AWE Technique," IEEE Trans. Antennas Propagat, vol. 46, no. 11, pp. 1752-1753, November 1998.

[8] C. Tian, Y. J. Xie, Y. Y. Wang, and Y. H. Jiang,"Fast Solutions of Wide-Band RCS Pattern of Objects Using MLFMA with the Best Uniform Approximation" Journal of Electronics & Information Technology. Vol.31,no.11, pp.2772-2775, November 2009.

[9] Z. Peng, M. B. Stephanson, J. F. Lee, "Fast computation of angular responses of large-scale three-dimensional electromagnetic wave scattering," IEEE Trans. Antennas Propagat, vol. 58, no. 9, pp. 3004-3012, September 2010.

[10] M. D. Pocock and S. P. Walker, "Iterative Solution of Multiple Right Hand Side Matrix Equations for Frequency Domain Monostatic Radar Cross Section Calulation," Applied Computational Electromagnetic Society(ACES) Journal, vol. 13, no. 1, pp.4-13,1998.

# Study on the Thin-Film Solar Cell with Periodic Structure Using FDFD Method

Yuan Wei[a], Bo Wu[a], Zhixiang Huang[a,*], Xianliang Wu[a,b,*]

[a]Key Laboratory of Intelligent Computing and Signal Processing, Anhui University, Hefei 230039,China
[b]School of Electronic and Information Engineering, Hefei Normal University, Hefei 230061,China
*Email: zxhuang@ahu.edu.cn,xlwu@ahu.edu.cn

*Abstract*-The optical absorption properties of the thin-film solar cell (SC) is an important manifestation of its performance. Finite-difference frequency-domain (FDFD) method is employed to discretize the Maxwell's equations for the designed structure. The relationship between the absorbed power density with the design structure and the incident angle were deeply explored. Our numerical results can provide a reference for the design and optimization of thin film solar cells.

*Keywords*-Finite-difference Frequency-Domain Method, Thin-Film Solar Cell, Periodic Structure

## I. INTRODUCTION

Since solar energy is renewable, clean energy. In recent years, the research of the solar cell [1] (SC) is also increasing attention. However, thin-film SC is one important part of it; while its energy conversion efficiency is low. So there are a lot of work to do with energy conversion efficiency and reducing costs. We design suitable structure of the SC, and then use FDFD method for its numerical processing. Also, the relationship between the absorption efficiency with structure and incident angle is analyzed.

In order to improve the absorption efficiency of SC, the surface plasmon effect is introduced. Surface plasmon resonances (SPRs) are collective oscillations of the free electrons that are confined to surfaces and interact strongly with light resulting in a polarization. SPRs usually occur at the interface between a dielectric with the positive dielectric constant $\varepsilon_r^d$ and a metal with the negative dielectric constant $\varepsilon_r^m$, and SPRs only exist when they satisfy $\mathrm{Re}(-\varepsilon_r^m) > \varepsilon_r^d$ [2]. The light absorption of SC is strengthened by SPRs, by comparing the field distribution of the two design structures.

In order to describe the propagation and scattering of sunlight within the SC，Maxwell's equation should be strictly solved. The finite-difference time-domain (FDTD) [3] and the FDFD methods are usually used to discretize the Maxwell's equation. For noble metals (such as Ag and Au), the complex dielectric constants [4] have to be described by a large number of summation terms in Lorentz-Drude model [5] in FDTD method. While FDFD method [6,7] can use an experimentally tabulated dielectric constants of the dispersive materials directly. In addition, FDFD method has a unique advantage in processing periodic structure and oblique incidence. FDFD method is adopted here, and the hybrid absorbing boundary condition (ABC) is used to reduce the spurious numerical reflections due to the perfectly matched layer (PML) can't work very well under periodic boundary condition [8].

## II. THEORETICAL AND NUMERICAL CALCULATION MODEL

### A. Model and FDFD Algorithm

Two two-dimensional plasmonic thin-film amorphous silicon (A-Si) SC structures with periodic structure are depicted in Fig.1. Since the s-polarized incident light cannot excite the SPRs[5], we mainly consider the p-polarized light with the electromagnetic components of $H_z$, $E_x$, and $E_y$. And all the materials are assumed non-magnetic ( $\mu_r$=1).

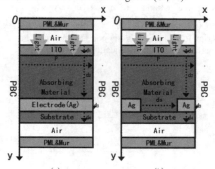

Figure 1. The unit cell of thin-film SC. (a) structure 1; and (b) structure 2,

As shown in Fig.1, the two structures include indium tin oxide (ITO), absorbing materials (A-Si), electrodes (Ag) and substrate with thickness $d_1$, $d_2$, $d_3$, $d_4$, respectively. The distance between two adjacent electrodes is $d_s$ in the structure 2. The structure 2 is given to enhance the surface plasmon effect. The ABC is used with the PML and Mur absorbing boundary conditions, which is employed at the top and the bottom of the SC structure. The periodic boundary conditions (PBC) are imposed at the left and right sides of the unit cell.

For the isotropic and inhomogeneous media with the complex dielectric constant of $\varepsilon_r(x, y)$, the wave equation for the total field is given by [9]

$$\frac{\partial}{\partial x}\left(\frac{1}{\varepsilon_r(x, y)}\frac{\partial H_z^t}{\partial x}\right) + \frac{\partial}{\partial y}\left(\frac{1}{\varepsilon_r(x, y)}\frac{\partial H_z^t}{\partial y}\right) + k_0^2 H_z^t = 0 \quad (1)$$

with $k_0$ is the wave number of free space. Using the second-order central differences, we have

978-7-5641-4279-7

$$\frac{\partial}{\partial x}(\frac{1}{\varepsilon_r(x,y)}\frac{\partial H_z^t}{\partial x})=\frac{1}{\Delta x}\left(\frac{H_z^t(i+1,j)-H_z^t(i,j)}{\varepsilon_r(i+1/2,j)\Delta x}-\right.$$
$$\left.\frac{H_z^t(i,j)-H_z^t(i-1,j)}{\varepsilon_r(i-1/2,j)\Delta x}\right)+O(\Delta x^2) \qquad (2)$$

and $\Delta x$ is the spatial step in $x$ direction. The general treatment of the inhomogeneous material is shown in Fig. 2.

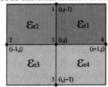

Figure 2. The inhomogeneous material treatment.

For the $p$-polarized incident light, the following averaging techniques is used,

$$\frac{1}{\varepsilon_r(i+1/2,j)}=\frac{1}{2}(\frac{1}{\varepsilon_{r1}}+\frac{1}{\varepsilon_{r4}}) \qquad (3)$$

$$\frac{1}{\varepsilon_r(i-1/2,j)}=\frac{1}{2}(\frac{1}{\varepsilon_{r2}}+\frac{1}{\varepsilon_{r3}}) \qquad (4)$$

Using the notation of $\Phi_1=H_z^t(i,j-1)$, $\Phi_2=H_z^t(i-1,j)$, $\Phi_3=H_z^t(i,j)$, $\Phi_4=H_z^t(i+1,j)$, and $\Phi_5=H_z^t(i,j+1)$, where the subscript 1,2,3,4 and 5 is the node as shown in Fig.2. (1) can be discretized into FDFD equation as

$$\sum_{m=1}^{5}c_m\Phi_m=0 \qquad (5)$$

with

$$c_1=\frac{1}{2}(\frac{1}{\varepsilon_{r1}}+\frac{1}{\varepsilon_{r2}})\cdot\frac{1}{\Delta y^2} \qquad (6)$$

$$c_2=\frac{1}{2}(\frac{1}{\varepsilon_{r2}}+\frac{1}{\varepsilon_{r3}})\cdot\frac{1}{\Delta x^2} \qquad (7)$$

$$c_3=-\frac{1}{2}(\frac{1}{\varepsilon_{r1}}+\frac{1}{\varepsilon_{r2}}+\frac{1}{\varepsilon_{r3}}+\frac{1}{\varepsilon_{r4}})\cdot(\frac{1}{\Delta x^2}+\frac{1}{\Delta y^2})+k_0^2 \qquad (8)$$

$$c_4=\frac{1}{2}(\frac{1}{\varepsilon_{r4}}+\frac{1}{\varepsilon_{r1}})\cdot\frac{1}{\Delta x^2} \qquad (9)$$

$$c_5=\frac{1}{2}(\frac{1}{\varepsilon_{r3}}+\frac{1}{\varepsilon_{r4}})\cdot\frac{1}{\Delta y^2} \qquad (10)$$

The total-field $H_z^t$ can be calculated by solving (5), and the incident-field $H_z^{inc}$ is known, so the scattered-field $H_z^s$ is obtained by $H_z^t=H_z^{inc}+H_z^s$.

### B. Boundary Conditions

As shown in Fig.1, the ABCs are used at the top and bottom respectively in order to reduce reflections. The complex coordinate PML [10] is applied as

$$\frac{\partial^2 H_z^s}{\partial x^2}+\frac{1}{s_y}\frac{\partial}{\partial y}(\frac{1}{s_y}\frac{\partial H_z^s}{\partial y})+k_0^2 H_z^s=0 \qquad (11)$$

with

$$s_y=\begin{cases}1-j_0\dfrac{\sigma(y)}{\omega\varepsilon_0}, \text{in } PML\\[2mm] 1,\quad \text{else}\end{cases} \qquad (12)$$

where $\varepsilon_0$ is the permittivity of free space, $\omega$ is the angular frequency of the incident light, and conductivity $\sigma$ is employed by

$$\sigma(j)=\frac{C}{\Delta y}(\frac{j-1/2}{L})^Q, j=1,2,...,8 \qquad (13)$$

$$\sigma(j+1/2)=\frac{C}{\Delta y}(\frac{j}{L})^Q, j=0,1,...,8 \qquad (14)$$

where $L$ is the layer number of PML, $Q$ is the order of the polynomial, and $C$ is a constant. The optimized parameters are chosen as $L=8$, $Q=3.7$, $C=0.02$. Using the second-order central differences, we have

$$\frac{1}{s_y}\frac{\partial}{\partial y}(\frac{1}{s_y}\frac{\partial H_z^s}{\partial y})=\frac{1}{s_y(j)\cdot\Delta y}\left[\frac{H_z^s(i,j+1)-H_z^s(i,j)}{s_y(j+1/2)\cdot\Delta y}\right.$$
$$\left.-\frac{H_z^s(i,j)-H_z^s(i,j-1)}{s_y(j-1/2)\cdot\Delta y}\right] \qquad (15)$$

Taking the topmost lay of the structure $y=0$ as an example, the second-order Mur absorbing boundary conditions [11] can be employed by

$$\left[\frac{\partial}{\partial y}-j_0(k_0+\frac{1}{2k_0}\frac{\partial^2}{\partial x^2})\right]H_z^s\Bigg|_{y=0}=0 \qquad (16)$$

and its discretized form is given by

$$f_1 H_z^s(i,j)+f_2 H_z^s(i-1,j)+$$
$$f_3 H_z^s(i+1,j)+f_4 H_z^s(i,j+1)=0 \qquad (17)$$

with

$$f_1=2\exp(j_0 k_0\Delta y)-2k_0^2\Delta x^2\exp(j_0 k_0\Delta y)-2 \qquad (18)$$

$$f_2=f_3=1-\exp(j_0 k_0\Delta y) \qquad (19)$$

$$f_4=2k_0^2\Delta x^2 \qquad (20)$$

The periodic boundary conditions along the $x$ direction can be described as

$$H_z^s(x+P,y)=H_z^s(x,y)\exp(-j_0 k_0\cos\theta\cdot P) \qquad (21)$$

$$H_z^s(x,y)=H_z^s(x+P,y)\exp(j_0 k_0\cos\theta\cdot P) \qquad (22)$$

with $P$ is the periodic, and $\theta$ is the incident angle along $x$ direction.

### III. NUMERICAL RESULTS AND ANALYSIS

For the two-dimensional plasmonic thin-film SC with periodic structure as shown in Fig.1, the absorbing material is A-Si, the electrode is Ag, and the substrate is SiO2. The geometric parameters of the left structure are set as $d_1=25nm$, $d_2=140nm$, $d_3=40nm$, $d_4=30nm$, and $P=200nm$. Taking into account the consistency of the structure, the geometric parameters of the right structure are set as $d_1$=25nm、 $d_2$=120nm、 $d_3$=40nm、 $d_4$=30nm, $P$=200nm, and $d_s$=100nm. The $y$-directed incident field is the $p$-polarized plane wave with $H_z^{inc}(x,y)=\exp(-jk_0(x\cdot\cos\theta+y\cdot\sin\theta))$, the spatial step is set as $\Delta x=\Delta y$=0.5nm, $\theta$ is the incident angle. Here we chose $\theta$ equals 30°, 60°, and 90°, respectively.

Fig.3 shows the absorbed power density $\eta$ for the two structures of A-Si when the incident angle is 90°. The formula is given by

$$\eta=\frac{\int_{Sa}\sigma_a|\mathbf{E}|^2\,ds}{\Delta Sa} \tag{23}$$

where $\Delta Sa$ is the area of the A-Si, and $\sigma_a$=-$\omega\varepsilon_0\mathrm{Im}(\varepsilon_{ra})$ is the conductivity of the A-Si. As shown in Fig.3, the absorbed power density of structure 2 is bigger than structure 1 at low-frequency portion (i.e. $\lambda$>660nm). The $|\mathbf{E}|^2$ of (23) will enhance since the generated of SPR, So the structure 2 shows stronger absorption than structure 1.

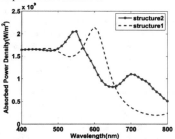

Figure 3. The absorbed power density by the A-Si for the two structures

The generalized reflection coefficient and transmission coefficients of the two structures are showed in Fig.4, the formula is employed as following

$$\begin{cases} R=\dfrac{\left|\dfrac{1}{P}\int_0^P H_z^s(x,y_r)\exp(jk_0x\cos\theta)dx\right|^2}{b^2} \\[20pt] T=\dfrac{\left|\dfrac{1}{P}\int_0^P H_z^t(x,y_t)\exp(jk_0x\cos\theta)dx\right|^2}{b^2} \end{cases} \tag{24}$$

where $b$ is amplitude of the incident light. The absorption $A(\lambda)$ is introduced for comparison with the absorbed power density. It is defined as

$$A(\lambda)=1-R(\lambda)-T(\lambda) \tag{25}$$

Fig.4 (b) shows the absorption $A(\lambda)$, the trends of the absorption coincide with the trends of the absorbed power density, they all reflect the fact that the absorptions of structure 2 are strengthened at low-frequency portion due to the SPRs.

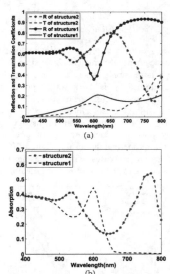

(a)

(b)

Figure 4. Spectrum results for the two structures. (a) The reflection and transmission coefficients; and (b) the absorption coefficients.

Fig.5 shows the total magnetic field $H_z^t$ at $\lambda$=600nm, $\lambda$=650nm, $\lambda$=710nm, and $\lambda$=780nm, respectively. We can clearly see the enhancement of the total magnetic field at the interface of Ag layer and A-Si layer due to the SPRs.

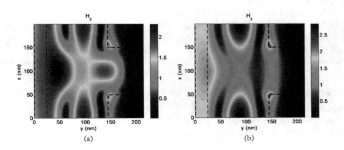

(a)                    (b)

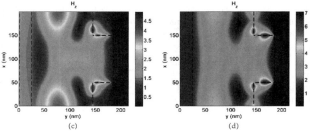

Fig.5 The $H_z^r$ field distribution for the structure 2 at different wavelengths. (a)$\lambda$=600nm; (b) $\lambda$=650nm; (c) $\lambda$=710nm; and (d) $\lambda$=780nm;

Enhancement factor $\kappa=\eta_2/\eta_1$ is employed to explanation why the enhancement of structure 2 is stronger than the enhancement of structure 1, where $\eta_2$ is the absorbed power density of structure 2, $\eta_1$ is the absorbed power density of structure 1. Next, we discuss the relationship between the incident angles with the enhancement factor as show in Fig.6. We can clearly see the enhancement factor is very large at low-frequency portion, which shows that structure 2 has better absorption properties due to the excited SPRs, and the enhancement factor increases with increasing of incident angle. It can increase to about 4.4 for the vertical incidence.

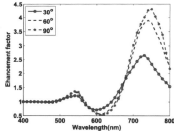

Figure 6. The relationship between enhancement factors with different incident angle.

## IV. CONCLUSION

FDFD algorithm is used to numerical simulation of thin-film solar cell with periodic structure; the results show that the absorption of thin-film solar cell is related to the structure and the incident angle. Major impact on the structure is due to the SPRs which are excited at the interface of absorbing material and metal electrodes. In addition, the absorption in different incident angle is given due to FDFD algorithm can easily deal with the problem of oblique incidence. The results of our work can provide a theoretical basis and technical support for the design and optimization of the actual organic thin-film solar cells.

## ACKNOWLEDGMENT

This work was supported by the National Natural Science Foundation of China under Grant (Nos. 60931002, 61101064, 51277001, 61201122), DFMEC (No.20123401110009) and NCET (NCET-12-0596) of China, Distinguished Natural Science Foundation (No.1108085J01), and Universities Natural Science Foundation of Anhui Province (No. KJ2011A002), and the 211 Project of Anhui University.

## REFERENCES

[1] J. Nelson, *The Physics of Solar Cells, London: Imperial College Press*, 2003.

[2] R. Zia, M. D. Selker, P. B. Catrysse, and M. L. Brongersma "Geometries and materials for subwavelength surface plasmon modes," *J. Opt. Soc. Am*, vol. 21, pp. 2442-2446, Dec. 2004.

[3] SUN Chen, LI Chuan hao, SHI Rui ying, SU Kai, GAO Hong tao, and DU Chun lei. "A Study of Influences of Metal Nanoparticles on Absorbing Efficiency of Organic Solar Cells," *Acta Photonica Sinica*, vol. 41, pp. 1335-1341, Nov. 2012.

[4] M. Qiu, and S. L. He, "A nonorthogonal finite-difference time-domain method for computing the band structure of a two-dimensional photonic crystal with dielectric and metallic inclusions," *J. Appl. Phys*, vol. 87, pp. 8268-8275, Jun. 2000.

[5] G. Veronis, and S. Fan, "Overview of Simulation Techniques for Plasmonic Devices," in Surface Plasmon Nanophotonics, M. L. Brongersma, and P. G. Kik, eds. *Springer, Dordrecht, The Netherlands*, 2007.

[6] Z. M. Zhu, and T. G. Brown, "Full-vectorial finite-difference analysis of microstructured optical fibers," *Opt. Express*, vol. 10, pp. 853-864, Aug. 2002.

[7] C. P. Yu, and H. C. Chang, "Compact finite-difference frequency-domain method for the analysis of twodimensional photonic crystals," *Opt. Express*, vol. 12, pp. 1397-1408, Apr. 2004.

[8] A. F. Oskooi, L. Zhang, Y. Avniel, and S. G. Johnson, "The failure of perfectly matched layers and towards their redemption by adiabatic absorbers," *Opt. Express*, vol. 16, pp. 11376-11392, Jul. 2008.

[9] W. C. Chew, *Waves and Fields in Inhomogenous Media*, New York, 1990.

[10] W. C. Chew, W. H. Weedon. "A 3-D perfectly matched medium from modified Maxwell's equations with stretched coordinates," *Microw. Opt. Technol. Lett*, vol. 7, pp. 599-604. Sep. 1994.

[11] G. Mur, "Absorbing boundary-conditions for the finite-difference approximation of the time-domain electromagnetic-field equations," *IEEE Trans. Electromagn. Compat.* Vol. 23, pp. 377-382, Nov.1981.

# Accurate Evaluation of RCS on the Structure of Aircraft Inlets

Jingyan Mo[1,2], Weidong Fang[2], Haigao Xue[2]
1 Shanghai Aerospace System Engineering Institute
2 School of Communication and Information Engineering, Shanghai University
Shanghai, 200021, China, mojingyan@163.com

Ying Yan
School of Electronic Engineering, Xidian University
Xi'an, 710071, Shaanxi, China, yymqn1008@163.com

Zhewang Ma
Graduate School of Science and Engineering, Saitama University
255 Shimo-Okubo, Sakura-ku, Saitama-shi, Saitama 338-8570, Japan

*Abstract*- A parallel version of Method of Moments (MoM) with Higher-Order Basis functions (HOB-MoM) with the Out-of-Core technique (OOC) is proposed to compute the Radar Cross Section (RCS) of an aircraft with engine inlets in this paper. The block-partitioned parallel scheme for the large dense MoM matrix is designed to achieve excellent load balancing and high parallel efficiency. The OOC technology is employed to break through the random access memory (RAM) limitation. Some numerical results demonstrate that the higher-order basis used in this paper is superior to the conventional RWG basis and is suitable to solve various electrically large problems such as the computation of RCS.

## I. INTRODUCTION

Radar Cross Section (RCS) computation of complex large structures has been paid more and more attention in vehicle stealth design. Recently, so many numerical methods, such as Finite Element Method (FEM), Finite Difference Time Domain (FDTD), Finite Integral Time Domain Method (FITD), Fast Multi-pole Method (FMM), MoM [1] are used to compute RCS. However, as the simulation frequency becomes higher, the MoM method based on Rao-Wilton-Glisson basis functions (RWGs) [3, 5] produces a large number of matrix unknowns for electrically large complex structures. To reduce the number of unknowns and to accelerate the computation, the fast multi-pole method (FMM) is a feasible approach. But when the complex cavity structure such as aircraft's engine is considered, the convergence of FMM may be a problem. For such structures, both the techniques of FEM and FDTD [7] have the huge working amount to finish the discretization, which will also produce large number of unknowns. To handle this problem, another choice is to use MoM with Higher-Order basis (HOB-MoM), which can significantly reduce the number of unknowns. But unfortunately, HOB-MoM still requires much memory for the simulation of large complex structures. Thus, OOC technology is of great value.

In stealth aircraft design, the engine inlet is a main scattering source to be considered. Typical engine inlet is a one-end open waveguide cavity with large size for radar frequency and radar waves entering in it will have severe

978-7-5641-4279-7

effect of resonation and multi-reflect. For this kind of structures, the FMM will spend more simulation time and iteration steps to obtain a stable and accurate solution. HOB-MoM is a good choice to deal with this type of problems.

Apart from the section of introduction, the second section of this paper presents the basic theory used in this paper; Section three presents some numerical examples to validate the capacity and application of this paper's method; Finally, Section four gives the conclusion.

## II. BASIC THEORY

### 1. Higher Order Basis Functions

Flexible geometric modelling can be achieved by using truncated cones for wires and bilinear patches to characterize surfaces [2]. The surface current over a bilinear surface is decomposed into its p and s-components, as shown in Fig.1 (a). The p-current component can be treated as the s-current component defined over the same bilinear surface with an interchange of the p and s coordinates. The approximations for the s-components of the electric and magnetic currents over a bilinear surface are typically defined by:

$$\vec{J}_s(p,s) = \sum_{i=0}^{N_p} [c_{i1}\vec{E}_i(p,s) + c_{i2}\vec{E}_i(p,-s) + \sum_{j=2}^{N_s} a_{ij}\vec{P}_{ij}(p,s)] \quad (1)$$

Where $c_{i1}, c_{i2}, (i=0,1,...,N_p)$ are defined as

$$c_{i1} = \sum_{j=0}^{N_s} a_{ij}(-1)^j, c_{i2} = \sum_{j=0}^{N_s} a_{ij} \quad (2)$$

The edge basis functions $\vec{E}_i(p,s)$ and the patch basis functions $\vec{P}_{ij}(p,s)(i=0,...N_p, j=2,...,N_s)$ are expressed by (3) and (4) respectively

$$\vec{E}_i(p,s) = \frac{\vec{\alpha}_s}{|\vec{\alpha}_p \times \vec{\alpha}_s|} p^i \vec{N}(s) \quad (3)$$

$$\vec{P}_{ij}(p,s) = \frac{\vec{\alpha}_s}{|\vec{\alpha}_p \times \vec{\alpha}_s|} p^i \vec{S}_j(s) \quad (4)$$

where $\vec{\alpha}_p, \vec{\alpha}_s$ are the unitary vectors defined as

$$\vec{\alpha}_p = \frac{\partial \vec{r}(p,s)}{\partial p}, \vec{\alpha}_s = \frac{\partial \vec{r}(p,s)}{\partial s} \tag{5}$$

The parametric equation of such an isoparametric element can be written in the following form as

$$\vec{r}(p,s) = \vec{r}_{11}\frac{(1-p)(1-s)}{4} + \vec{r}_{12}\frac{(1-p)(1+s)}{4} + \vec{r}_{21}\frac{(1+p)(1-s)}{4} + \vec{r}_{22}\frac{(1+p)(1+s)}{4} \tag{6}$$
$$-1 \le p \le 1, -1 \le s \le 1$$

where $\vec{r}_{11}, \vec{r}_{12}, \vec{r}_{21}, \vec{r}_{22}$ are the position vectors of its vertices, and the $p$ and $s$ are the local coordinates.

A right-truncated cone is determined by the position vectors and the radii of its beginning and its end, $\vec{r}_1, a_1, \vec{r}_2, a_2$, respectively, as shown in Fig. 1 (b). Generalized wires (i.e., wires that have a curvilinear axis and a variable radius) can be approximated by right-truncated cones.

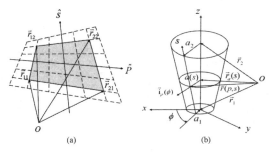

Figure 1. Geometric model: (a) A bilinear surface defined by four vertices

### 2. Out-of-Core factorization and solving

When the MoM matrix is too large to the computer at hand, the space of hard disk should be used to extend the storage capability. When the matrix is too large, they can be stored on the hard disk. As the matrix starts, one portion of the matrix is computed at a time and then written to the hard disk; another matrix portion will be computed and written, which will be repeated until the complete matrix filling has been finished. It is necessary to mention that, for the in-core matrix filling algorithm, the entire matrix is filled at one step, but need not be written out. The main idea of designing the out-of-core algorithm is to improve the performance of in-core algorithm. When performing the out-of-core LU factorization, each portion of the matrix is read into RAM and performs the LU decomposition respectively. Then the results after being factorized are written back to the hard disk. The code then precedes the next portion of the matrix until the whole matrix has been LU factored.

The decomposition in this paper is oriented along the column; each in-core filling is a slab of the matrix, as shown in Fig.2.

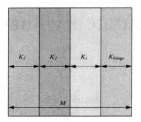

Figure 2. Data decomposition for storing an out-of-core matrix

The width of the $i$th out-of-core slab $K_i$, as illustrated in Figure 2, is bounded by

$$M = \sum_{i=1}^{I_{slab}} K_i \tag{7}$$

At the last out-of-core fill ($i=I_{slab}$), the number of unfilled columns is

$$M_{fringe} = M - \sum_{i=1}^{I_{slab}-1} K_i \tag{8}$$

When only one process is used, the out-of-core matrix filling can easily be done with a slight modification to the serial code. However, when more processes are used, the assignment to fill each slab is distributed over $p$ processes by using the block distribution scheme of ScaLAPACK [2]. Thus, the matrix filling schemes need to be designed in such a way so as to avoid redundant calculation for better parallel efficiency.

### III. NUMERICAL EXAMPLES

#### 1. Comparison with the measurement results

To validate the accuracy and efficiency of the proposed parallel Higher-order basis MoM methodology, a benchmark of a truncated cone is simulated to compare its RCS results with those from the measurement.

The model simulated is an end-capped truncated cone oriented along the z-axis and centred in the plane $z = 0$ (illustrated in Fig. 3). The elevation angle ($\theta$) is taken from the positive z-axis and the azimuth angle ($\phi$) from the positive x-axis. There are several interesting points about this target. First, it shows the RCS response of targets with single curvature (common in structural parts of an aircraft, such as the fuselage). It is also important to know the diffraction mechanism at curved edges. Reflection from planar surfaces with curved edges can also be observed. Therefore, this target is especially suitable for the validation of the prediction of objects with flat surfaces delimited by curved edges and for evaluation of curved edge contributions.

The model is simulated at 7 GHz. The RCS pattern for HH polarizations and corresponding to $\phi = 0$ and $\theta = 0° \sim 180°$ is simulated. The incident direction is perpendicular to the generatrix.

Figure 3. Truncated cone description: size (in mm). The height of the target is 200 mm. The major diameter is 200mm and the minor one is 100mm.

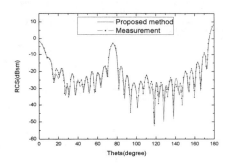

Figure 4. RCS pattern of the truncated cone.Frequency:7GHz. Polarization: HH. Comparison with Measurement.

Fig. 4 shows the RCS results we wish to observe, from which we can find there are three main lobes. Two of them correspond to the specular reflection from the two bases of the cone. The minor one corresponds to $\theta = 0°$ and the major one to $\theta = 180°$, they are of the different levels resulting from the different areas of the corresponding bases. The other main lobe corresponds to the angle at which the generatrix is perpendicular to the incident direction. Diffraction from the curved edges becomes important in the intermediate region between the main lobes. Finally, the simulated RCS results are compared with those from the measurement and good agreement can be seen.

*2. Simulation of aircraft with inlets*

In this section, the RCS of an aircraft's head with two engine inlets has been computed using HOB-MoM with the Out-of-Core algorithm. The models are shown in Fig.5. The canopy and radome are built as metal surface. The dimensions of the whole aircraft model are 20m×15m×4.2m, and those of the aircraft's head with inlets are 9.5m×4.6m×1.8m. Monostatic RCS in horizontal plane is computed at 1.6 GHz. Accordingly, the electrical sizes of the head model are 51 $\lambda$ ×24.5 $\lambda$ ×9.6 $\lambda$ 。

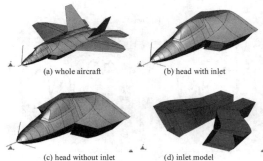

(a) whole aircraft    (b) head with inlet

(c) head without inlet    (d) inlet model

Figure 5. Aircraft and inlet model

Two different configurations of the aircraft's head are compared: the first is the model with inlets, and another is the model without inlets (the inlet mouth is closed by two metal plates). For the computational configuration, monostatic RCS on three planes of theta-cut (-10°,0°,10°) are computed, there are 121 phi angles(phi=-60°~60°) for each theta-cut plane and the direction of phi=0° is along the nose of the aircraft. In addition, mesh results and simulation setting are presented in Fig.6. Finally, the simulation results are shown in Fig.7. The blue line in which represents the results for the model with inlets, while the red one for the model without inlets.

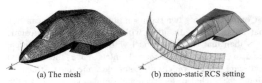

(a) The mesh    (b) mono-static RCS setting

Figure 6. Mesh and computation setting

The number of unknowns by using HOB-MoM is 135315 and RAM required is about 273 GB. On the contrary, the number of unknowns produced by RWG MoM may reach the level of 2.7 million. The problem is simulated on a Lenovo workstation with 12 i7 CPU cores, totally 64 GB RAM and 4TB hard disk. 273GB hard disk is used to store the matrix; the RAM is used as the 'buffer' and it takes about 20 hours for simulating all the 363 angles.

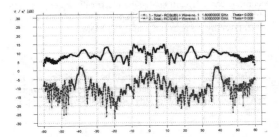

(a) Theta=0°

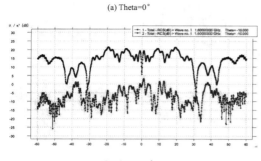

(b) Theta=-10°

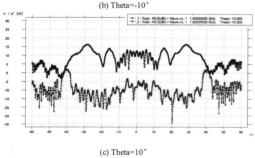

(c) Theta=10°

Figure 7. H polarization incidence mono-RCS (E-total)

Comparing those two results in upper pictures, we can find that the aircraft with inlets will produce larger monostatic RCS than that from the model without inlets. For different incident angles, the S-shaped inlets have the different scattering effect, which indicates that the inlet structures should be included for accurate simulation of the stealth aircraft.

## IV. CONCLUSION

In this paper, RCS computation of electrically large complex structures by using HOB-MoM with Out-of-Core technique is presented, which can successfully reduce the RAM required to store the MoM impedance matrix. Through the analysis of some numerical examples, the conclusion can be drown that combining HOB-MoM and Out-of-Core technique can efficiently solve the electrically large problems which have strong resonance phenomenon, and this can not be achieved by the conventional RWG MoM.

## ACKNOWLEDGMENT

This work is partly supported by the National High Technology Research and Development Program ("863"Program) of China (2012AA01A308) and the NSFC (61072019).

## REFERENCES

[1] R. F. Harrington, *Field Computation by Moment Methods*, IEEE Series on Electromagnetic Waves. New York: IEEE, 1993.

[2] Y. Zhang and T.K. Sarkar, *Parallel solution of Integral Equation Based EM Problems in the Frequency Domain*. Hoboken, NJ: Wiley, 2009.

[3] S.M. Rao, D.R. Wilton, and A.W. Glisson, "Electromagnetic scattering by surfaces of arbitrary shape", *IEEE Trans. Antennas Propagation.*, vol. P-30, no.3: 409-418, 1982.

[4] Zhang, Y., M. Taylor, T. K. Sarkar, H. Moon, and M.-T. Yuan, "Solving large complex problems using a higher-order basis: Parallel in-core and out-of-core integral-equation solvers", *IEEE Antennas and Propagation. Mag.*, vol. 50, No. 4: 13-30, 2008.

[5] Y. Zhang, M. Taylor, T. K. Sarkar, A. De, M.-T. Yuan, H. Moon, and C.-H. Liang, "Parallel in-core and out-of-core solution of electrically large problems using the RWG basis functions," *IEEE Antennas and Propagation. Mag.*, vol. 50, no. 5: 84–94, 2008.

[6] Y. Zhang, T. K. Sarkar, M. Taylor, and H. Moon, "Solving MoM problems with million level unknowns using a parallel out-of-core solver on a high performance cluster," *IEEE Antennas and Propagation Soc. Int. Symp.*, Charleston, SC, USA, 2009.

[7] Ramesh Garg, *Analytical and Computational Methods in Electromagnetics*, Artech House, 2008.

# Error Analysis of a Novel Absorbing Boundary Condition for the 3-Step LOD-FDTD Method

Lina Cao, Jianyi Zhou

State Key Laboratory of Millimeter Waves, Southeast University

Nanjing, Jiangsu, 211111, China

*Abstract*-In this paper, the estimation of the numerical error of a novel absorbing boundary condition (ABC) proposed for the split-step finite-difference time-domain (FDTD) method and the incident angle of the plane wave is investigated. Compared with two other previously reported Mur first-order ABCs, the proposed ABC is simpler and requires less additional computation. Besides, it also has much better absorption performance.

## I. INTRODUCTION

The finite-difference time-domain (FDTD) method is widely used for solving electromagnetic-wave problems [1]. The split-step (SS) FDTD method and the locally one-dimensional (LOD) FDTD [2]-[4] have attracted much attention recently thanks to their unconditional stability, simplicity and flexibility. In [2], the 3-D LOD-FDTD is split into a two-step approach but is complicated in its formulations. And recently it is split into a three-step approach with reduced computational complexity [3]. In [4], the LOD-FDTD method has the second-order temporal accuracy and involves simpler updating procedures.

When the LOD-FDTD methods are applied to open structures, truncated boundary conditions are required to convert the unlimited physical space to the limited calculating space. Generally, there are two kinds of truncated boundary conditions. One is the employment of nonphysical absorbing media, such as the perfect matched layer (PML) [5]-[7]. The other one is the ABC [8]-[12] based on travelling wave equations. Compared with the PMLs, the ABCs require much less computation and memory at the cost of introducing greater reflection.

In [10], a consistent implementation of Mur ABC into three-step 3-D LOD-FDTD method is introduced, which exhibits better wave-absorbing capability than the conventional schemes. In [11], a novel ABC is proposed according to the intrinsic updating procedures in the SS-FDTD method. The proposed ABC requires less additional computation and causes less reflection than the traditional Mur first-order ABCs.

In this paper, the estimation of the numerical error of the novel ABC [11] for the 3-Step LOD-FDTD method is analyzed. The proposed ABC is simpler and requires less additional computation. In addition, it has much better absorption performance compared with two other Mur first-order ABCs.

## II. FORMULATIONS

Maxwell's equations in isotropic and lossless media are given as

$$\nabla \times H = \varepsilon \frac{\partial E}{\partial t} \quad \nabla \times E = -\mu \frac{\partial H}{\partial t}. \tag{1}$$

For simplicity, only the 2D TM($y$) waves are considered. Therefore, these equations can be expressed in the Cartesian coordinates as

$$\frac{\partial U}{\partial t} = \mathbf{A}U + \mathbf{B}U \tag{2}$$

where

$$\mathbf{A} = \begin{bmatrix} 0 & \frac{1}{\varepsilon}\frac{\partial}{\partial z} & 0 \\ \frac{1}{\mu}\frac{\partial}{\partial z} & 0 & 0 \\ 0 & 0 & 0 \end{bmatrix} \mathbf{B} = \begin{bmatrix} 0 & 0 & -\frac{1}{\varepsilon}\frac{\partial}{\partial x} \\ 0 & 0 & 0 \\ -\frac{1}{\mu}\frac{\partial}{\partial x} & 0 & 0 \end{bmatrix}$$

and $[U] = [E_y, H_x, H_z]^T$.

It should be noted that $\mathbf{A}$ and $\mathbf{B}$ are sparse matrices whose elements are related to the spatial derivatives along the $x$- and $z$-directions, respectively. The updating procedure of the three-step LOD-FDTD can be written as follows [2]:

$$\left(\mathbf{I} - \frac{\Delta t}{4}\mathbf{A}\right)U^{n+\frac{1}{4}} = \left(\mathbf{I} + \frac{\Delta t}{4}\mathbf{A}\right)U^n \tag{3a}$$

$$\left(\mathbf{I} - \frac{\Delta t}{2}\mathbf{B}\right)U^{n+\frac{3}{4}} = \left(\mathbf{I} + \frac{\Delta t}{2}\mathbf{B}\right)U^{n+\frac{1}{4}} \tag{3b}$$

$$\left(\mathbf{I} - \frac{\Delta t}{4}\mathbf{A}\right)U^{n+1} = \left(\mathbf{I} + \frac{\Delta t}{4}\mathbf{A}\right)U^{n+\frac{3}{4}} \tag{3c}$$

According to [2] and [10], (3) can be rearranged as

$$\frac{\partial}{\partial t}U^{n+\frac{1}{8}} = 2\mathbf{A}U^{n+\frac{1}{8}} \tag{4a}$$

$$\frac{\partial}{\partial t}U^{n+\frac{1}{2}} = 2\mathbf{B}U^{n+\frac{1}{2}} \tag{4b}$$

$$\frac{\partial}{\partial t}U^{n+\frac{7}{8}} = 2\mathbf{A}U^{n+\frac{7}{8}} \tag{4c}$$

For conciseness, the first updating step of the LOD-FDTD method is presented in detail. The second and third updating step can be derived in the same way.

978-7-5641-4279-7

More specifically, substituting field components into (3a) reads

$$E_y^{n+\frac{1}{4}} = E_y^n + \frac{\Delta t}{4\varepsilon}\left[\frac{\partial H_x^{n+\frac{1}{4}}}{\partial z} + \frac{\partial H_x^n}{\partial z}\right] \quad (5a)$$

$$H_x^{n+\frac{1}{4}} = H_x^n + \frac{\Delta t}{4\mu}\left[\frac{\partial E_y^{n+\frac{1}{4}}}{\partial z} + \frac{\partial E_y^n}{\partial z}\right] \quad (5b)$$

$$H_z^{n+\frac{1}{4}} = H_z^n \quad (5c)$$

This scheme at the $z=1$ grid is illustrated. To simplify the calculating process, some approximations are made.

$$E_y^{n+\frac{1}{4}}(i,1) - E_y^n(i,1) \approx \frac{\Delta t}{2\varepsilon\Delta z}[H_x^n(i,1) - H_x^n(i,0)] \quad (6a)$$

$$H_x^{n+\frac{1}{4}}(i,1) - H_x^n(i,1) \approx \frac{\Delta t}{2\mu\Delta z}[E_y^{n+\frac{1}{4}}(i,1) - E_y^{n+\frac{1}{4}}(i,0)] \quad (6b)$$

$$H_z^{n+\frac{1}{4}}(i,1) = H_z^n(i,1) \quad (6c)$$

Note that in the above formulations (6a)-(6c), all equations are explicit, therefore reducing computational complexity.

In the subsections, the ABC implementations for $E_y$ variable at the boundary $z=0$ are devised. The field components at other boundaries can be derived similarly.

*A. The proposed ABC*

In the first updating step, for an outgoing wave at boundary $z=0$, according to equation (4a) one can obtain that

$$\frac{\partial E_y^{n+\frac{1}{8}}}{\partial t} = -2c\frac{\partial E_y^{n+\frac{1}{8}}}{\partial z} \quad (7)$$

where $c = 1/\sqrt{\mu\varepsilon}$ is the wave velocity.

Equation (7) can be discretized as

$$E_y^{n+\frac{1}{4}}(i,0) + \xi_1 E_y^{n+\frac{1}{4}}(i,1) = \xi_1 E_y^n(i,0) + E_y^n(i,1) \quad (8)$$

where $\xi_1 = (2\Delta z - c\Delta t)/(2\Delta z + c\Delta t)$.

In the second updating step, according to the method in [11], at the boundary $z=0$, $E_y$ does not need to be updated.

$$E_y^{n+\frac{3}{4}}(i,0) = E_y^{n+\frac{1}{4}}(i,0) \quad (9)$$

The third updating step is similar to the first updating step.

$$E_y^{n+1}(i,0) + \xi_1 E_y^{n+1}(i,1) = \xi_1 E_y^{n+\frac{3}{4}}(i,0) + E_y^{n+\frac{3}{4}}(i,1) \quad (10)$$

*B. The ABC in [10]*

Follow the method in [10], one can obtain the ABC in each updating step as

$$E_y^{n+\frac{1}{4}}(i,0) + \xi_1 E_y^{n+\frac{1}{4}}(i,1) = \xi_1 E_y^n(i,0) + E_y^n(i,1) \quad (11)$$

$$E_y^{n+\frac{3}{4}}(i,0) + E_y^{n+\frac{3}{4}}(i,1) = E_y^{n+\frac{1}{4}}(i,0) + E_y^{n+\frac{1}{4}}(i,1) \quad (12)$$

$$E_y^{n+1}(i,0) + \xi_1 E_y^{n+1}(i,1) = \xi_1 E_y^{n+\frac{3}{4}}(i,0) + E_y^{n+\frac{3}{4}}(i,1) \quad (13)$$

*C. Conventional Mur ABC*

At boundary $z=0$, a one-way outgoing wave equation is as follow:

$$\frac{\partial E_y^{n+\frac{1}{8}}}{\partial t} = -c\frac{\partial E_y^{n+\frac{1}{8}}}{\partial z} \quad (14)$$

Discretization of (14) leads to

$$E_y^{n+\frac{1}{4}}(i,0) + aE_y^{n+\frac{1}{4}}(i,1) = aE_y^n(i,0) + E_y^n(i,1) \quad (15)$$

Similarly, in the second and third updating step one can obtain

$$E_y^{n+\frac{3}{4}}(i,0) + bE_y^{n+\frac{3}{4}}(i,1) = bE_y^{n+\frac{1}{4}}(i,0) + E_y^{n+\frac{1}{4}}(i,1) \quad (16)$$

$$E_y^{n+1}(i,0) + aE_y^{n+1}(i,1) = aE_y^{n+\frac{3}{4}}(i,0) + E_y^{n+\frac{3}{4}}(i,1) \quad (17)$$

where $a = (4\Delta z - c\Delta t)/(4\Delta z + c\Delta t)$
$b = (2\Delta x - c\Delta t)/(2\Delta x + c\Delta t)$.

### III. NUMERICAL RESULTS

The numerical error of the ABC, which includes the error introduced by the approximation of the equation and the error introduced by truncation of the expression of the field, can be derived by virtue of a homochromous plane wave as:

$$E_y^n(i,k) = E_0 e^{j(k_x i\Delta x + k_z \Delta z + jwn\Delta t)}$$

All propagation directions are considered to study the numerical error of the ABC. Let $k_x = k\cos\theta$, $k_z = k\sin\theta$ and $\theta$ is incident angle with respective to $x$-axis, as shown in Fig. 1.

A 2-D computation domain is studied and the cell size in each direction is 1mm. Note that CFLN is defined as the ratio between the time step in the SS-FDTD and the maximum CFL limit in the standard FDTD. The numerical error is calculated as $|E_{observed} - E_{ref}|/\max(|E_{ref}|)$, where $E_{ref}$ is obtained at the same position with an extended time step that virtually generates no reflection.

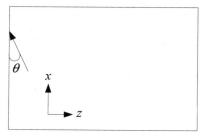

Figure 1. Wave propagates upon the ABC at $z=0$

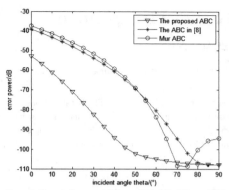

Figure 2. Numerical errors of different ABCs with different incident angles.

The effects of wave propagation direction on the numerical error of the proposed ABC, the ABC in [10] and the conventional Mur ABC are investigated.

Fig. 2 shows the relationship between numerical errors and the incident angle of the plane wave under the conditions f=10GHz and CFLN=1.

According to Fig. 2, the proposed ABC has the best absorption performance compared with the ABC in [10] and the conventional Mur ABC.

## IV. CONCLUSIONS

The estimation of the numerical error of a novel ABC [11] proposed for the split-step finite-difference time-domain (FDTD) method and the incident angle of the plane wave is investigated. The proposed ABC is simpler and has much better absorption performance than two other traditional Mur first order ABCs. Therefore, the proposed ABC can be used conveniently in the split-step FDTD method to solve open structure EM problems.

## ACKNOWLEDGMENT

This work is supported in part by the National Natural Science Foundation of China under Grant 60921063, in part by the National Science and Technology Major Project of China under Grant 2010ZX03007-002-01, and 2011ZX03004-003.

## REFERENCES

[1] Yee, K.S., "Numerical solution of initial boundary value problems involving Maxwell's equations in isotropic media," *IEEE Trans. Antennas Propag.*, 1966, 14, pp. 302 – 307

[2] J. Lee and B. Fornberg, "A split step approach for the 3-D Maxwell's equations," *J. Comput. Appl. Math.*, vol.158, pp. 485-505, 2003.

[3] I. Ahmed, E. K. Chua, E. P. Li, and Z. Chen, "Development of the three-dimensional unconditionally stable LOD-FDTD method," *IEEE Trans. Antennas Propag.*, vol. 56, no. 11, pp. 3596–3600, Nov. 2008.

[4] E. L. Tan, "Unconditionally stable LOD-FDTD method for 3-D Maxwell's equations," *IEEE Microw. Wireless Compon. Lett.*, vol. 17,pp. 85–87, Feb. 2007.

[5] V. Nascimento, B. Borges, et. al., "Split-field PML implementations for the unconditionally stable LOD-FDTD method," *IEEE Microw. Wireless Compon. Lett.*, vol. 16, no. 7, pp. 398–400, Jul. 2006.

[6] I. Ahmed, E. Li, and K. Krohne, "C onvolutional perfectly matched layer for an unconditionally stable LOD-FDTD method," *IEEE Microw. Wireless Compon. Lett.*, vol. 17, no. 12, pp. 816–819, Dec. 2007.

[7] I. Ahmed, E. H. Khoo, and E. Li, "Development of CPML for three dimensional unconditionally stable LOD-FDTD method," *IEEE Trans. Antennas Propag.*, vol. 58, no. 3, pp. 832–837, Mar. 2010.

[8] Jianyi Zhou and Jianing Zhao, "Efficient high-order absorbing boundary condition for the ADI-FDTD method," *IEEE Microw. Wireless Compon. Lett.*, vol. 19, no. 1, pp. 6–9, Jul. 2009.

[9] Jianing Zhao and Jianyi Zhou, "An improved quadratic interpolation ABC for ADI-FDTD method," in 2009 Asia Pacific Microwave Conference,pp60-63,Dec.2009.

[10] Feng Liang, Gaofeng Wang, Hai Lin, and Bing-Zhong Wang, "Mur absorbing boundary condition for thr ee-step 3-D LOD-FDTD method, " *IEEE Microw. Wireless Compon. Lett.*, vol. 20, pp. 589–591, Nov. 2010

[11] Jianyi Zhou,and Jiangning Zhao., "A novel absorbing boundary condition for the 3-D split-step FDTD method," *IEEE Microw. Wireless Compon. Lett.*, vol. 22, no. 4, pp. 167-169,April. 2012.

[12] Zhenhai Shao, Wei Hong, and Jianyi Zhou, "Generalized Z-domain absorbing boundary conditions for the analysis of electromagnetic problems with finite-difference time-domain method," *IEEE Transactions on Microwave Theory and Techniques.*, vol. 51, no. 1, pp. 82-90, Jan. 2003.

# Application of Liao's ABC in 2-D FDFD Algorithm

YU Tongbin, ZHANG lei, ZHAO yanhui

Institute of Communications Engineering, PLA University of Science and Technology, Nanjing 210007, China

njtyytb@sina.cn

*Abstract*-Liao's absorbing boundary condition (ABC) has been widely applied in the early finite difference time domain (FDTD) method, due to its advantages of simple form, good absorption effect and easy programming. In this paper, the frequency domain form of Liao's ABC is derived and successfully applied in 2-D frequency domain finite difference (FDFD) algorithm. The numerical experiments validate that Liao's ABC still shows excellent absorption effect in 2-D FDFD algorithm as it does in the 2-D FDTD algorithm.

## I. INTRODUCTION

Due to the limitation of the computer capacity, appropriate boundary condition should be assigned to accurately simulate the electromagnetic process of free space at the boundaries of the domain when solving the open region electromagnetic problems. To solve this problem, many ABC were proposed by scholars, such as Mur ABC [1], Liao's ABC [2], PML [3,4], UPML [5] and CPML [6], and so on. PML is so far the most effective method which is widely used in many numerical calculations of electromagnetic such as EM compatibility problems [7-9], EM scattering problems [10-12]. In order to obtain perfect absorption effect, much more boundary cells must be set for PML, which greatly increases the computational consuming, especially in FDFD calculations.

Liao's ABC was derived by extrapolating wave function in time and space domain based on the Newton's backward differential polynomial. It owns advantages of simple form, easy implementation and better absorption effect. It can be directly used near the corners of a computational domain without any modification and can also meet the accuracy requirements for most of numerical calculations of engineering. However, lack of stability has restricted its application, especially for the higher-order Liao's ABC. Thus W. C. Chew and other scholars analyzed the reason of the instability of Liao's ABC, and stabilized Liao's ABC by adding small damping factor [13-15]. Lei Zhang and Tong-Bin Yu improved the stability of Liao's ABC remarkably by order-weighting method [16].

FDFD is one of the earliest methods for numerical calculation of electromagnetic, whose theory is simple. But it needs to solve large matrix equation in calculating electromagnetic problem. Therefore, it is particularly important for FDFD algorithm to save the computing space.

We usually apply the Mur ABC in FDFD calculations, though its absorbing effect is poor. In this paper, Liao's ABC is introduced into the FDFD algorithm, and improves the accuracy of FDFD by easy implementation.

## II. THEORY

The FDTD method is well known for the solution of the wave equation:

$$\left( \frac{\partial^2}{\partial x^2} + \frac{\partial^2}{\partial y^2} + \frac{1}{c^2}\frac{\partial^2}{\partial t^2} \right)\phi(x,y;t) = 0 \qquad (1)$$

Where $\phi$ is scalar field, which is computed on a rectangular finite-difference grid with grid spacing $\Delta s$ and time step $\Delta t$: $\phi(m\Delta s, n\Delta s; l\Delta t) = \phi_{m,n}^l$ .Using central differencing in time and space domain, the wave equation is approximated as:

$$\frac{\phi_{m+1,n}^l - 2\phi_{m,n}^l + \phi_{m-1,n}^l}{\Delta s^2} + \frac{\phi_{m,n+1}^l - 2\phi_{m,n}^l + \phi_{m,n-1}^l}{\Delta s^2}$$
$$- \frac{1}{c^2}\frac{\phi_{m,n}^{l+1} - 2\phi_{m,n}^l + \phi_{m,n}^{l-1}}{\Delta t^2} = 0 \qquad (2)$$

At the boundary of the finite computational domain, an ABC must be used to model radiation in a free space.

For a "right" boundary at $x=x_{max}$, Liao's ABC gives the updated boundary field $\phi(x_{max}, y_j, t+\Delta t)$ in terms of field values at previous times lying along a straight line perpendicular to the boundary:

$$\phi(x_{max}, y_j, t+\Delta t) = \sum_{i=1}^{N} (-1)^{i+1} C_i^N$$
$$\phi(x_{max} - i\alpha c\Delta t, y_j, t-(i-1)\Delta t) \qquad (3)$$

Where $C_i^N$ is the binomial coefficient $N!/[i!(N-i)!]$, $N$ is the order of the boundary condition, $i$ is the space step, $\alpha c\Delta t$ is the space sample interval. If we set $\alpha c\Delta t$ equal to $\Delta s$, the sample interval just coincides with the grid, so the form of Liao's ABC is very simple, the updated boundary field of Liao's ABC is given as:

$$\phi(x_{max}, y_j, t+\Delta t) = \sum_{i=1}^{N}(-1)^{i+1}C_i^N\phi(x_{max-i}, y_j, t-(i-1)\Delta t) \quad (4)$$

The Liao's ABC in FDTD algorithm can be seen as a discrete time system. This feature meets the demands of time discrete transform in signal and system course. Therefore it can be taken Z-transforming, and be transformed into frequency domain. We present the Liao's ABC in frequency domain as following:

$$\phi(x_{max}, y_j) = \sum_{i=1}^{N}(-1)^{i+1}C_i^N\phi(x_{max-i}, y_j)*e^{(-ij\omega\Delta t)} \quad (5)$$

In the application of Liao's ABC in FDFD algorithm, it should be paid attention to the selection of time interval $\Delta t$, which is not as the one in FDTD algorithm, for it needn't take the stability into account. In the simulation, we can get better results when we set the time interval $\Delta t = \Delta s / c$. This problem will be mentioned again in the following analysis.

### III. NUMERICAL VALIDATION

In order to validate the calculation accuracy of Liao's ABC in FDFD, we compare the field error computed by Liao's ABC with the one computed by Mur ABC in the cases of single frequency. The global error computed by Liao's ABC is compared with the one computed by Mur ABC in the band ranging from 0.1GHz to 30GHz, as it is done in [17]. In the numerical experiments, we take the TM wave as the example, and calculate the radiation field of infinite line source in free space. The size of the calculating region is (50*50) cells. The line source is set at the center of the calculating region (25*25). The time interval is set as dt = dx/c. The analytical solution of the radiation field of line source in free space is as following:

$$E_z(r, \omega) = \frac{\omega\mu}{4}IH_0^{(2)}(kr) \quad (6)$$

Where, $H_0^{(2)}(\cdot)$ is the second class zero-order Hankel function.

Figure.1 shows the relative field error of the numerical solution calculated by Liao's third-order ABC, Mur second-order ABC on contrast with the analytical solution. It can be seen that the maximum calculation error of Liao's third-order ABC at the boundaries is 6%, and the one of Mur second-order ABC is 10%. In the computing region, the maximum calculation error of Liao's third-order ABC is 3%, and the one of Mur ABC is 5%. Though the field calculated by Mur ABC is more accurate at local region, it is not as accurate as the ones by Liao's third-order ABC at most of the region, especially at the boundaries.

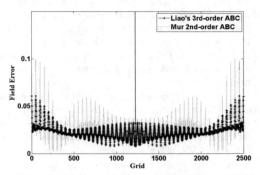

Figrue1. The radiation field error of the line source in free space calculated by Liao's third-order ABC and the Mur ABC (f=300MHz, δ=λ/20, δ is the grid size)

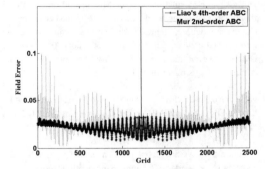

Figure.2 The radiation field error of the line source in free space calculated by Liao's fourth-order ABC and the Mur ABC (f=300MHz, δ=λ/20)

Figure 2 shows the relative field error of the numerical solution calculated by Liao's fourth-order ABC, Mur second-order ABC on contrast with the analytical solution. It can be seen that the maximum error of Liao's fourth-order ABC is 2.5% in the entire computational domain, which declines remarkably at the boundaries compared with the one of Liao's third-order ABC. The maximum error of Mur second-order ABC is 10% at the boundaries and 5% in the computed region. The field values error of Mur ABC is slightly better in local region, but is larger than Liao's fourth-order ABC in most of region.

In order to analyze the impact of the grid size to the absorption effect of Liao's ABC and Mur ABC, we calculate the field error in small grid of λ/40.

Figure 3 shows the relative field error of the numerical solution calculated by Liao's third-order ABC, Mur second-order ABC on contrast with the analytical solution in small grid. It can be seen that the absorption effect of Liao's third-order ABC is very good, which is better than the one of Mur second-order ABC. The calculated error of Liao's ABC is no more than 6% in the entire computing region and is less than 1% at the central region. However, the calculated error of Mur ABC is more than 12% at the boundaries and 3% in the center region, which is poor compared with the ones calculated in large grid.

Figure 4 shows the relative error field of the numerical solution calculated by Liao's fourth-order ABC and Mur second-order ABC on contrast with the analytical solution in small grid. It can be seen that the calculation error of Liao's fourth-order ABC is very small, which has obvious advantage compared with Mur second-order ABC, The calculation error of Liao's ABC is no more than 3% in the entire calculating region and less than 1% at most of region, which can meet the accuracy requirements in majority of engineering calculations. However, the calculation error of Mur second-order ABC is more than 12% at the boundaries and 3% in the center region. Comparing figure1, figure2 with figure 3, figure 4, we can find out that the calculation accuracy of Liao's ABC becomes better. However, the one of Mur ABC becomes poor when calculating by using small grid.

What is given above is the case of single-frequency field values error, the following gives the global error calculated by Liao's ABC and Mur ABC in the band ranging from 0.1GHz to 30GHz, which is used to analyze the absorption effect of Liao's ABC.

Figure 5 shows the global error calculated by Liao's ABC and Mur second-order ABC in the band ranging from 0.1GHz-30GHz. The region size is (50*50) cells. It can be seen that the global error calculated by Liao's second, third, fourth-order ABC decrease by 4%, 8%, 12% comparing with the one calculated by Mur second-order ABC. The Liao's ABC shows good absorption effect.

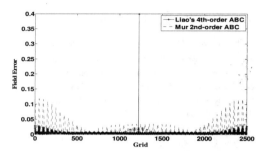

Figrue4. The radiation field error of the line source in free space calculated by Liao's fourth-order ABC and the Mur ABC (f=300MHz, δ=λ/40)

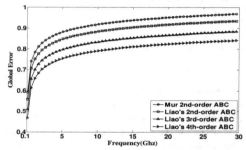

Figure 5 The global error of Liao's ABC and Mur ABC (f=300MHz,δ=λ/20)

Liao's second-order ABC and Liao's third-order ABC is smaller, but the Liao's fourth-order ABC shows better absorption effect. The global error calculated by Liao's second, third, fourth-order ABC in small grid decrease by 27%, 30%, 57% compared with the one calculated by Mur second-order ABC, which is better than the one calculated in large grid. So the advantage of Liao's ABC in absorption effect is more obvious compared with the one in large grid.

## IV.  CONCLUTION

In this paper, the form of Liao's ABC in frequency domain is deduced, and applied in the 2-D FDFD algorithm. Liao's ABC shows better absorption effect compared with Mur ABC in 2-D FDFD algorithm, without increasing the amount of storage and the complication of implementation..

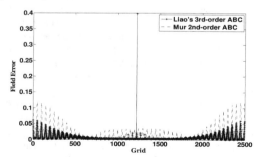

Figrue3. The radiation field error of the line source in free space calculated by Liao's third-order ABC and the Mur ABC (f=300MHz, δ=λ/40)

## REFERENCES

[1]  G. Mur. "Absorbing boundary conditions for the finite-difference approximation of time-domain electromagnetic field equations"[J]. IEEE Trans Electromang Compat, Nov. 1981, EMC-23(4):377-382.

[2] Z. P. Liao, H. L. wong, B. P. Yang et al. "A transmitting boundary for transient wave analysis" [J]. Scientia Sinica (seriesA), Oct.1984, 1063-1076.

[3] J. P. Berenger. "A perfectly matched layer for the absorption of electromagnetic waves" [J]. Comput. Phys.1994, 114(2):185-200.

[4] J. P. Berenger. "Three-dimensional perfectly matched layer for the absorption of electromagnetic waves" [J]. Comput. Phys. Sept.1996, 127(2):363-379.

[5] S. D. Gedney. "An anisotropic perfectly matched layer absorbing media for the truncation of FDTD lattices"[J]. IEEE Trans On Antennas and Propag, Dec. 1996, 44:1630-1639.

[6] J. A. Roden, Gedney. S. D. "Convolutional PML(CPML)and efficient FDTD implementation of the CFS-PML for arbitrarymedia"[J]. Micro. Opt. Tech. Lett, Dec 2000, 27:334-339.

[7] Reineix A, Gazave J, Guiffaut C. "Some Improvements in FDTD Method for EMC Applications"[C]// Electromagnetics in Advanced Applications, 2007. ICEAA 2007.

[8] Yu Zhang; Ming Chen; Wei Ding; Changhong Liang. "EMC Analysis of Antennas Involving Dielectric Bodies With MoM-FDTD Algorithm and Network Theory"[J]. Microwave, Antenna, Propagation and EMC Technologies for Wireless Communications, 2005. 1:768-771

[9] Bagc V, Yilmaz A E, Michielssen E. "EMC/EMI analysis of electrically large and multiscale structures loaded with coaxial cables by a hybrid TDIE-FDTD-MNA approach"[J]. IEEE Antennas and Propagation Society International Symposium, 2005.2B:14-17.

[10] He Y, Kojima T. "Three-dimensional analysis of light-beam scattering from magneto-optical disk structure by FDTD method"[J].IEEE Antennas and Propagation Society International Symposium Digest . 1997. 4:2148-2151.

[11] Mochizuki S. Watanabe S Taki, M Yamanaka, Y Shirai H. "A new iterative MoM/FDTD analysis for EM scattering by a loop antenna"[J]. IEEE Antennas and Propagation Society Symposium, 2004. 2:1495-1498.

[12] Dogaru T, Lam Nguyen. "FDTD Models of Electromagnetic Scattering by the Human Body"[J]. Antennas and Propagation Society International Symposium 2006. 1995-1998.

[13] Moghaddam M, Chew W C. "Stabilizing Liao's absorbing boundary conditions using single-precisionarithmetic"[C]//IEEE AP-S/URSI Int. Symp. Dig, London, Ontario, Canada, June 1991 430-433.

[14] W. C. Chew, R. L. Wagner. "A modified form of Liao's absorbing boundary condition"[C]//IEEE AP Int. Symp. Dig, Chicago, IL, July 1992, 535-539.

[15] Wagner R. L, Chew W. C. "An analysis of Liao's absorbing boundary condition"[J]. Electromagn. Waves Applicat, 1995, 9:993-1009.

[16] Lei Zhang, Tong-Bin Yu. "A Method of Improving the Stability of Liao's Higher-Order Absorbing Boundary Condition", PIERM,27, 2012,167-178

[17] Katz. D. S, Thiele E T, Taflove A. "Validation and extension to three dimensional of the Berenger PML absorbing boundary condition for FDTD meshes"[J]. IEEE Microwave Guided Wave Lett. 1994, 4:268-270.

# A hybrid FE-BI-DDM for electromagnetic scattering by multiple 3-D holes

Zhiwei Cui,* Yiping Han, and Meiping Yu

School of Science, Xidian University, Xi'an, Shaanxi 710071, China

*zwcui@mail.xidian.edu.cn

*Abstract*-**A hybrid finite element-boundary integral-domain decomposition method (FE-BI-DDM) is introduced to efficiently solve the problem of electromagnetic scattering by multiple three dimensional (3-D) holes. Specifically, each hole is modeled by the edge-based finite element method. The holes are coupled to each other through the boundary integral equation based on Green's function. The resultant coupling system of equations is solved by an iterative domain decomposition method. Some numerical results are included to illustrate the validity and capability of the proposed method.**

## I. INTRODUCTION

Conducting plates with multiple holes are widely used in engineering structures, e.g. missiles, aircraft, optic experiment platform, grating couplers etc. The accurate and efficient simulation of the scattering by such a structure is of vital importance in many areas of optical and electrical engineering, such as grating manufacture, surface defect detection, slot antenna design and radar cross section control, but is also a very challenging task.

For the problem of scattering by multiple two-dimensional (2-D) holes, there have been some numerical investigations. Dejoie *et al.* [1] applied an integral equation method to solve the problem of an elastic half-plane containing a large number of randomly distributed, non-overlapping, circular holes. Lee and Chen [2] adopted a semi-analytical approach that is based on null-field integral equation to solve the scattering problem of flexural waves in an infinite thin plate with multiple circular holes. Later in [3], they employed a multipole Trefftz method to analyze the scattering of flexural wave by multiple circular holes in an infinite thin plate. Recently, Alavikia and Ramahi [4] studied the problem of scattering from multiple 2-D holes in infinite metallic walls by means of a finite element method (FEM) that uses the surface integral equation with the free-space Green's function as the boundary constraint.

While some work has been done in the area of solving the problems of scattering by multiple 2-D holes, little work on the solution of three dimensional (3-D) problems have been reported in the previous literature. In the present work, we introduce a domain decomposition of the hybrid finite element -boundary integral (FE-BI) method described in [5] to solve the problem of electromagnetic scattering by multiple 3-D holes embedded in conducting plates. In the implementation of the method, the edge-based FEM is applied inside each hole to derive a linear system of equations associated with the unknown fields. The boundary integral equation (BIE) is then applied on the apertures of all the holes to truncate the computational domain and to connect the matrix subsystem generated from each hole. To reduce computational burdens, an iterative domain decomposition method (DDM) based on the substructuring method [6] and the multilevel fast multipole algorithm (MLFMA) [7] is developed to efficiently solve the coupling system of equations. In the following sections, the theoretical and implementation of the proposed method are firstly given. Some numerical results are then presented.

## II. MATHEMATICAL FORMULATION

### A. FE-BI Formulation for Scattering from a Single Hole

We first consider the problem of electromagnetic scattering by a single 3-D hole in a conducting plate, as shown in Fig. 1. The volume of the hole is denoted as $V$, and the planar surface area of the lower and upper aperture are denoted as $S_1$ and $S_2$, respectively. Also, the hole is assumed to be filled with an inhomogeneous material having a relative permittivity $\varepsilon_r$ and relative permeability $\mu_r$, and the plane wave is assumed to be incident to the upper aperture. In accordance with the FE-BI method described in [5], the electric field inside the hole and at the apertures of the hole can be obtained by seeking the stationary point of the functional

$$
\begin{aligned}
F(\mathbf{E}) = \iiint_V & \left[ \frac{1}{\mu_r}(\nabla\times\mathbf{E})\cdot(\nabla\times\mathbf{E}) - k_0^2\varepsilon_r\mathbf{E}\cdot\mathbf{E} \right]dV \\
& - 2k_0^2 \iint_{S_1} \mathbf{M}_1(\mathbf{r})\cdot\left[ \iint_{S_1} \mathbf{M}_1(\mathbf{r}')\cdot\overline{\overline{G}}_0(\mathbf{r},\mathbf{r}')dS' \right]dS \\
& - 2k_0^2 \iint_{S_2} \mathbf{M}_2(\mathbf{r})\cdot\left[ \iint_{S_2} \mathbf{M}_2(\mathbf{r}')\cdot\overline{\overline{G}}_0(\mathbf{r},\mathbf{r}')dS' \right]dS \\
& + 2jk_0Z_0 \iint_{S_2} \mathbf{M}_2(\mathbf{r})\cdot\mathbf{H}^{inc}(\mathbf{r})dS
\end{aligned}
\tag{1}
$$

where $\mathbf{M}_1$ and $\mathbf{M}_2$ are the equivalent magnetic currents, which

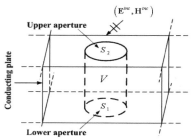

Fig. 1. Geometry of a 3-D hole in a conducting plate.

978-7-5641-4279-7

are related to the electric field on $S_1$ and $S_2$ by $\mathbf{M}_1 = \mathbf{E}_1 \times \hat{n}_1$ and $\mathbf{M}_2 = \mathbf{E}_2 \times \hat{n}_2$. $\hat{n}_1$ and $\hat{n}_2$ denote the outward unit vector normal to $S_1$ and $S_2$, respectively. Also, $\mathbf{H}^{inc}$ is the incident magnetic field, $k_0$ and $Z_0$ are the free-space wave number and impedance, and $\overline{\overline{G}}_0$ is the free-space dyadic Green's function.

Using the edge-based FEM with tetrahedral elements [8], the functional can be converted into a matrix equation

$$
\begin{bmatrix}
K_{II} & K_{IL} & K_{IU} \\
K_{LI} & K_{LL}+P & 0 \\
K_{UI} & 0 & K_{UU}+Q
\end{bmatrix}
\begin{Bmatrix}
E_I \\
E_L \\
E_U
\end{Bmatrix}
=
\begin{Bmatrix}
0 \\
0 \\
b
\end{Bmatrix}
\tag{2}
$$

where $\{E_I\}$ is a vector containing the discrete electric fields inside $V$, while $\{E_L\}$ and $\{E_U\}$ are the vectors containing the discrete electric fields on the lower and upper aperture of the hole. The matrix $[K]$ is contributed by the volume integral in (1), whereas matrices $[P]$ and $[Q]$ are contributed by the dual surface integrals. For the sake of clear description, we define two integral operators

$$
L(\mathbf{g}_j) = -2k_0^2 \iint_{S_1} \mathbf{g}_j(\mathbf{r}') \cdot \overline{\overline{G}}_0(\mathbf{r},\mathbf{r}') dS' \tag{3}
$$

$$
U(\mathbf{g}_j) = -2k_0^2 \iint_{S_2} \mathbf{g}_j(\mathbf{r}') \cdot \overline{\overline{G}}_0(\mathbf{r},\mathbf{r}') dS' \tag{4}
$$

where $\mathbf{g}_j$ are the RWG vector basis functions[9] defined on triangular elements. With the integral operator, the elements of matrices $[P]$ and $[Q]$ can be rewritten as

$$
P_{ij} = \iint_{S_1} \mathbf{g}_i \cdot L(\mathbf{g}_j) dS \tag{5}
$$

$$
Q_{ij} = \iint_{S_2} \mathbf{g}_i \cdot U(\mathbf{g}_j) dS \tag{6}
$$

where $\mathbf{g}_i$ and $\mathbf{g}_j$ denote the testing and expansion basis functions, respectively. In this notation, the testing and expansion locations are both on the apertures of the same hole.

### B. Extension to Multiple Holes

To extend the method described above to case of multiple holes, we propose an expanded system of equations with individual $[P]$ and $[Q]$ sub-matrices [10]. The new $[P]$ and $[Q]$ sub-matrices take on the form

$$
\left[P_{ij}\right]^{mn} = \iint_{S_L} \mathbf{g}_i^m \cdot L_{mn}(\mathbf{g}_j^n) dS \tag{7}
$$

$$
\left[Q_{ij}\right]^{mn} = \iint_{S_U} \mathbf{g}_i^m \cdot U_{mn}(\mathbf{g}_j^n) dS \tag{8}
$$

The new integral operators $L_{mn}$ and $U_{mn}$ are defined as

$$
L_{mn}(\mathbf{g}_j^n) = -2k_0^2 \iint_{S_L} \mathbf{g}_j^n(\mathbf{r}') \cdot \overline{\overline{G}}_0(\mathbf{r},\mathbf{r}') dS' \tag{9}
$$

$$
U_{mn}(\mathbf{g}_j^n) = -2k_0^2 \iint_{S_U} \mathbf{g}_j^n(\mathbf{r}') \cdot \overline{\overline{G}}_0(\mathbf{r},\mathbf{r}') dS' \tag{10}
$$

where the subscript "$S_L$" represents the surface integration in (7) and (9) is performed over the lower apertures of all the holes. Similarly, the subscript "$S_U$" represents the surface integration in (8) and (10) is performed over the upper apertures of all the holes. Also, the indices $[m,n]$ represent the testing and expansion holes, respectively.

According to the principle of edge-based FEM, there is no direct coupling between the unknowns of any two independent computational domains. Therefore, the FEM matrix $[K]$ is only defined when $m = n$. In other words, the holes are coupled to each other only through the BIE based on Green's function. As a result, the coupling system of equations for $m$ holes in a conducting plate would be expressed in the matrix notation

$$
\begin{bmatrix}
\widetilde{K}_{II} & \widetilde{K}_{IL} & \widetilde{K}_{IU} \\
\widetilde{K}_{LI} & \widetilde{K}_{LL}+\widetilde{P} & 0 \\
\widetilde{K}_{UI} & 0 & \widetilde{K}_{UU}+\widetilde{Q}
\end{bmatrix}
\begin{Bmatrix}
\widetilde{E}_I \\
\widetilde{E}_L \\
\widetilde{E}_U
\end{Bmatrix}
=
\begin{Bmatrix}
0 \\
0 \\
\widetilde{b}
\end{Bmatrix}
\tag{11}
$$

where

$$
[\widetilde{P}] =
\begin{bmatrix}
P^{11} & P^{12} & \cdots & P^{1m} \\
P^{21} & P^{22} & \cdots & P^{2m} \\
\vdots & \vdots & \ddots & \vdots \\
P^{m1} & P^{m2} & \cdots & P^{mm}
\end{bmatrix}
\tag{12}
$$

$$
[\widetilde{Q}] =
\begin{bmatrix}
Q^{11} & Q^{12} & \cdots & Q^{1m} \\
Q^{21} & Q^{22} & \cdots & Q^{2m} \\
\vdots & \vdots & \ddots & \vdots \\
Q^{m1} & Q^{m2} & \cdots & Q^{mm}
\end{bmatrix}
\tag{13}
$$

With the aid of the following notation

$$
diag.\sum_{i=1}^{m}[K_i] =
\begin{bmatrix}
K_1 & & & \\
& K_1 & & \\
& & \ddots & \\
& & & K_m
\end{bmatrix}
\tag{14}
$$

The FEM associated matrix sub-blocks are of the form

$$
[\widetilde{K}_{II}] = diag.\sum_{i=1}^{m}[K_{II}^i], \quad [\widetilde{K}_{LL}] = diag.\sum_{i=1}^{m}[K_{LL}^i],
$$

$$
[\widetilde{K}_{IL}] = diag.\sum_{i=1}^{m}[K_{IL}^i], \quad [\widetilde{K}_{LI}] = diag.\sum_{i=1}^{m}[K_{LI}^i], \tag{15}
$$

$$
[\widetilde{K}_{IU}] = diag.\sum_{i=1}^{m}[K_{IU}^i], \quad [\widetilde{K}_{UI}] = diag.\sum_{i=1}^{m}[K_{UI}^i],
$$

$$
[\widetilde{K}_{UU}] = diag.\sum_{i=1}^{m}[K_{UU}^i]
$$

### C. Iterative Domain Decomposition Solver

To efficiently solve the coupling system of equations (11), we develop an iterative domain decomposition solver based on the substructuring method [6] and MLFMA [7]. Specifically, we first eliminate the unknowns in the interior of each hole according to the following equation

$$\{E_I^i\} = -[K_{II}^i]^{-1}[K_{IL}^i]\{E_L^i\} - [K_{II}^i]^{-1}[K_{IU}^i]\{E_U^i\} \quad (16)$$

where $i = 1, 2, \cdots, m$. Once all the interior unknowns are eliminated, the original coupling system of equations can be reduced to a small one which only includes the unknowns on the apertures of all the holes, as follows

$$\begin{bmatrix} A + \tilde{P} & B \\ C & D + \tilde{Q} \end{bmatrix} \begin{Bmatrix} \tilde{E}_L \\ \tilde{E}_U \end{Bmatrix} = \begin{Bmatrix} 0 \\ \tilde{b} \end{Bmatrix} \quad (17)$$

in which

$$[A] = diag.\sum_{i=1}^{m}[A^i], [B] = diag.\sum_{i=1}^{m}[B^i],$$
$$[C] = diag.\sum_{i=1}^{m}[C^i], [D] = diag.\sum_{i=1}^{m}[D^i] \quad (18)$$

where

$$[A^i] = [K_{LL}^i] - [K_{LI}^i][K_{II}^i]^{-1}[K_{IL}^i] \quad (19)$$

$$[B^i] = -[K_{LI}^i][K_{II}^i]^{-1}[K_{IU}^i] \quad (20)$$

$$[C^i] = -[K_{UI}^i][K_{II}^i]^{-1}[K_{IL}^i] \quad (21)$$

$$[D^i] = [K_{UU}^i] - [K_{UI}^i][K_{II}^i]^{-1}[K_{IU}^i] \quad (22)$$

The solution to the reduced system (17) can be obtained by an iterative solver such as the generalized minimum residual (GMRES) method. However, traditional GMRES iteration method incurs very high computational cost and memory requirements with the increasing of the unknowns on the apertures of all the holes. In addition, the conventional approaches to computing the BIE matrix elements consume a considerable portion of the total solution time, and this, in turn, can place an inordinately heavy burden on the CPU regarding memory and time. To overcome these difficulties, we employ the MLFMA to the BIE matrix to significantly reduce the memory requirement and computational complexity. The detailed description of the MLFMA is given in [7] and is not repeated here.

### III. NUMERICAL RESULTS

To illustrate the validity of the proposed method for scattering by multiple holes, we consider two identical circular holes in a conducting plate. Both holes have a diameter of $1.0\lambda$ and depth of $0.5\lambda$, and are separated by $1.0\lambda$ in $x$-direction, $\lambda$ being the operating wavelength. For numerical solution, each hole is discretized independently into 2855 tetrahedral elements. As a result, a total of 6644 FEM unknowns and 514 BIE unknowns are generated. By virtue of the iterative domain decomposition solver described above, only one half of the FEM unknowns need to be deal with. Since the number of BIE unknowns is not so much, the reduced system can be easily solved. The monostatic RCS computed by the proposed FE-BI-DDM are plotted in Fig. 2, and compared with those from the method of moments (MOM)

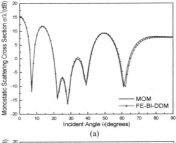

(a)

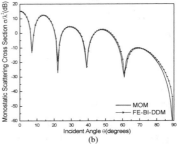

(b)

Fig. 2. Comparison of the monostatic scattering cross sections for two identical circular holes in a conducting plate obtained from FE-BI-DDM and MOM. (a) $\theta\theta$ polarization; (b) $\varphi\varphi$ polarization.

based on the surface boundary integral equation. As is evident from the figure, good agreement is obtained between the two methods.

Then, we investigate the scattering behavior of four identical circular holes in a conducting plate. Each hole has a diameter of $2.0\lambda$ and depth of $3.0\lambda$. These holes are arranged in three ways and are separated by distances of $0.5\lambda$. In the first way, these holes are placed along $y$-axis. In the second way, these holes are placed along $x$-axis. In the last way, these holes are arranged periodically in both the $x$ and $y$ directions, i.e. a $2\times 2$ array. Fig.3 depicts the computed scattering cross section as a function of the angle of incidence in the $xoz$ plane. As can be seen, when these holes are placed along $y$-axis, the scattering cross section curve in the $xoz$ plane is relatively smooth, whereas the curves corresponding to the other two ways have several peaks. This indicates that the coupling effects among these holes are stronger in the last two ways than that in the first one.

Finally, to illustrate the capability of the proposed method to solve large problems, we consider a $10\times 10$ array of circular holes in a conducting plate. Each hole of the array has a diameter of $1.0\lambda$ and depth of $2.0\lambda$. The periodicity is $2.0\lambda$ in both $x$ and $y$ directions. A plane wave with incidence angles $\theta^{inc} = 45°$, $\varphi^{inc} = 0°$ and polarization angle $\alpha = 0°$ is obliquely incident to the he upper apertures of the array. We calculate the bistatic scattering cross section of the array, and the results are shown in Fig.4. In this problem, a total of 1142000 tetrahedral elements are used to discretize the array holes,

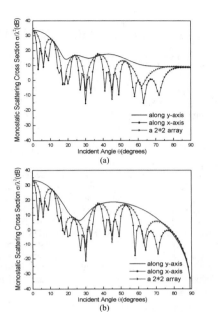

(a)

(b)

Fig. 3. Monostatic scattering cross section of four identical circular holes in a conducting plate: (a) $\theta\theta$ polarization; (b) $\varphi\varphi$ polarization.

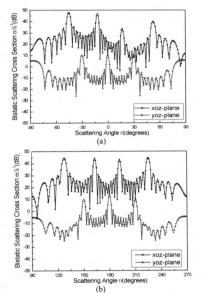

(a)

(b)

Fig. 4. Bistatic scattering cross section for a $10 \times 10$ array of circular holes in a conducting plate: (a) above the plate; (b) below the plate.

which result in 1200300 FEM unknowns and 51400 BIE unknowns. Although the number of FEM unknowns is very large, only one percent of those need to be dealt with. Furthermore, the reduced full system can be converted into a sparse one by virtue of the MLFMA. In this problem, 5-level MLFMA is used, and the memory required is about 630Mb.

## IV. CONCLUSION

This paper presents a domain decomposition of the FE-BI method for the analysis of electromagnetic scattering from multiple 3-D holes in conducting plates. In the implementation of the method, the edge-based FEM is used to obtain the solution of vector wave equation inside each hole. The BIE using the free-space Green's function is applied on the apertures of the holes as a global boundary condition. To reduce the computational burdens of the method, an iterative substructuring method was employed to solve the coupling system of equations. As a result, the original problem was reduced to a small one which only includes the unknowns on the apertures. Furthermore, the reduced system was solved by the CGM, where MLFMA is employed to reduce the memory requirement and computational complexity. Some numerical results are presented to demonstrate the validity and capability of the proposed method.

## ACKNOWLEDGMENT

This work was supported by the he Fundamental Research Funds for the Central Universities (Grant No. k5051307003).

## REFERENCES

[1] A. Dejoie, S.G. Mogilevskaya, and S.L. Crouch, "A boundary integral method for multiple circular holes in an elastic half-plane," *Eng. Anal. Boundary Elem.*, vol. 30, no. 6, pp. 450-464, June 2006.
[2] W.M. Lee and J.T. Chen, "Scattering of flexural wave in thin plate with multiple holes by using the null-field integral equation approach," *CMES Comput. Model. Eng. Sci.*, vol. 37, no.3, pp. 243-273, 2009.
[3] W.M. Lee and J.T. Chen, "Scattering of flexural wave in a thin plate with multiple circular holes by using the multipole Trefftz method," *Int. J. Solids Struct.*, vol. 47, no. 9, pp. 1118-1129, May, 2010.
[4] B. Alavikia and O.M. Ramahi, "Electromagnetic scattering from multiple sub-wavelength apertures in metallic screens using the surface integral equation method," *J. Opt. Soc. Am. A*, vol. 27, no.4, pp. 815-826, April 2010.
[5] J.M. Jin and J.L. Volakis, "Electromagnetic scattering by and transmission through a three-dimensional slot in a thick conducting plane," *IEEE Trans. Antennas Propag.*, vol. 39, no.4, pp. 543-550, April 1991.
[6] A. Toselli and O. Widlund, *Domain Decomposition Methods-Algorithms and Theory*, Springer: Berlin, 2005.
[7] J.M. Song, C.C. Lu, and W.C. Chew, "Multilevel fast multipole algorithm for electromagnetic scattering by large complex objects," *IEEE Trans. Antennas Propag.*, vol. 45, no.10, pp. 1488-1493, October 1997.
[8] J.M. Jin, *The Finite Element Method in Electromagnetics*. Wiley: New York, 1993.
[9] S.M. Rao, D.R. Wilton, and A.W. Glisson, "Electromagnetic scattering by surfaces of arbitrary shape," *IEEE Trans. Antennas Propag.*, vol. 30, no.3, pp. 409-418, May, 1982.
[10] R. Kindt, K. Sertel, E. Topsakal, and J. Volakis, "An extension of the array decomposition method for large finite-array analysis," *Microw. Opt. Technol. Lett.*, vol. 38, no.4, pp. 323-328, August, 2003.

# Numerical Simulation of Complex Inhomogeneous Bodies of Revolution

Y. B. Zhai[1],   J. F. Zhang[2], T. J. Cui[3]

1. No. 38 Research Institute, Cetc, Hefei 230031, China;

2, 3. Department of Radio Engineering, Southeast University, Nanjing 210096, China

*Abstract*-A hybrid technique is presented that combines the finite element and boundary integral methods for simulating electromagnetic scattering from complex inhomogeneous bodies of revolution. In the proposed method, higher order elements are used in the interior FEM region, first-order elements are used in the surface MOM region. The method has been successfully applied to investigate scattering by penetrable axisymmetric devices with inhomogeneous and anisotropic permittivity and permeability, which is important and helpful to analyze the scattering characteristics of metamaterials.

## I. INTRODUCTION

Electromagnetic scattering from bodies of revolution (BOR) has been studied extensively. The rotational symmetry of the scatterer allows the problem to be solved efficiently using a two-dimensional (2D) computational technique.

A very popular method is the method of moments (MoM) based on integral equation formulations. For homogeneous scatterers [1]-[7], such as perfect conductors, homogeneous dielectrics, integral formulation only requires a surface discretization and automatically take into account the radiation condition. Moreover, the generation of the elements of the method of moments' matrix occupies a major portion of the computational time and  many method is presented for accurately and efficiently evaluating the oscillating integrals [2]-[3]. However, MoM formulations yield a dense full matrix, and computational complexity becomes very large for inhomogeneous general materials that require volume MoM [4].

The finite-element method (FEM), which yields a sparse banded matrix, is characterized by very flexible material handling capabilities and is often preferred over the MoM for problems involving complex structures and inhomogeneous materials. When applied to the open-region problems, the FEM mesh must be truncated by an artificial boundary to obtain a bounded computational domain, and to render the problem manageable. Perfectly Matched Layers (PML) is one useful way for mesh truncation without altering the sparsity of the FEM matrix [5]. In order reduce the significant artificial reflection of PMLs at near grazing angles, PML need to be placed at a distance away from the BOR surface. Another widely used method is the hybrid finite element method, which uses a boundary integral to truncate the FEM mesh [6]-[7]. The method has the advantages of both the integral equation methods and the FEM. The integral equation mesh used to truncate the interior finite element region can be made very

close to the body, while the FEM mesh is capable of modeling complex inhomogeneous materials.

In this paper, we present a hybrid technique that combines the finite element and boundary integral methods for simulating electromagnetic scattering from complex inhomogeneous bodies of revolution. In the proposed method, higher order elements are used in the interior FEM region, first-order elements are used in the surface MOM region. The method has been successfully applied to investigate scattering by penetrable axisymmetric devices with inhomogeneous and anisotropic permittivity and permeability, which is important and helpful to analyze the scattering characteristics of metamaterials.

## II. FORMULATIONS

The object and surrounding space are broken into interior regions and the freespace regions out to a defined surface. In the interior region, a finite-element discretization of a weak form of the wave equation is used to model the geometry and the electric fields [8]:

$$\iiint_V \left[ (\nabla \times E^*) \cdot \overline{\overline{\mu}}_r^{-1} \cdot (\nabla \times E) - k_0^2 E^* \cdot \overline{\overline{\varepsilon}}_r^{-1} \cdot E \right] - jk_0\eta \iint_S E^* \cdot (\hat{n} \times E) dS = 0 \tag{1}$$

in which $k_0$ is the free space wave number, $\eta_0$ is the intrinsic impedance of free space, $\overline{\overline{\mu}}_r$ and $\overline{\overline{\varepsilon}}_r$ are the relative permittivity and permeability of the medium, which have the following symmetry form:

$$\overline{\overline{\mu}}_r = \begin{pmatrix} \mu_{\rho\rho} & 0 & \mu_{\rho z} \\ 0 & \mu_{\phi\phi} & 0 \\ \mu_{\rho z} & 0 & \mu_{zz} \end{pmatrix}, \quad \overline{\overline{\varepsilon}}_r = \begin{pmatrix} \varepsilon_{\rho\rho} & 0 & \varepsilon_{\rho z} \\ 0 & \varepsilon_{\phi\phi} & 0 \\ \varepsilon_{\rho z} & 0 & \varepsilon_{zz} \end{pmatrix}. \tag{2}$$

$S$ is the surface of the penetrable object $V$ , and $\hat{n}$ in the normal vector on $S$ , pointing from object into the free space region. $E^*$ is an appropriately chosen set of testing functions, which is known. The entire electric field $E$ inside the body and $\hat{n} \times H$ on the boundary  are the unknown quantities to be determined.

On the boundary $S$ , Fictitious electric $J = \hat{n} \times H$ and magnetic $M = -\hat{n} \times E$ surface currents, equivalent to the tangential magnetic and electric fields just on the exterior of the boundary surface, are defined. These currents produce the true field outside $V$ and zero field inside $V$ . $J = \hat{n} \times H$ could be substituted directly into (1) thereby enforcing continuity of the tangential magnetic field between interior and exterior regions. Following the hybridization procedure of FEM and

MoM described in [6], the other two formulations for the relationship between $\boldsymbol{E}$, $\boldsymbol{J}$ and $\boldsymbol{M}$ are:

$$\iint_S [\boldsymbol{J}^* \cdot \boldsymbol{E} - \boldsymbol{J}^* \cdot K(\boldsymbol{M}) - \boldsymbol{J}^* \cdot \frac{1}{2} \hat{n} \times \boldsymbol{M} + \boldsymbol{J}^* \cdot L(\boldsymbol{J})] dS$$
$$= \iint_S \boldsymbol{J}^* \cdot \boldsymbol{E}^i dS \tag{3}$$

$$\iint_S [\boldsymbol{J}^* \cdot K(\boldsymbol{J}) - \boldsymbol{J}^* \cdot \frac{1}{2} \hat{n} \times \boldsymbol{J} + \frac{1}{\eta_0^2} \boldsymbol{J}^* \cdot L(\boldsymbol{M})] dS$$
$$= \iint_S \boldsymbol{J}^* \cdot \boldsymbol{H}^i dS \tag{4}$$

In these equations, $\boldsymbol{E}^i$ and $\boldsymbol{H}^i$ are incident fields, $L$ are $K$ operators involving the free-space Green's function as defined in.

To take advantage of the rotational symmetry of the problem, the fields and the surface currents are expanded in the Fourier modes as :

$$\boldsymbol{E} = \sum_{m=-\infty}^{+\infty} \left[ E_{t,m}(\rho, z) - \hat{\phi} E_{\phi,m}(\rho, z) \right] e^{jm\phi} \tag{5}$$

$$\boldsymbol{J} = \sum_{m=-\infty}^{+\infty} \left[ \sum_{i=1}^{N_s - 1} (a_{m,i}^t \frac{T_i}{\rho} \hat{t} - a_{m,i}^t \frac{T_i}{\rho} \hat{\phi}) \right] e^{jm\phi} \tag{6}$$

$$\boldsymbol{M} = \sum_{m=-\infty}^{+\infty} \left[ \sum_{i=1}^{N_s - 1} (b_{m,i}^t \frac{T_i}{\rho} \hat{t} - b_{m,i}^t \frac{T_i}{\rho} \hat{\phi}) \right] e^{jm\phi} \tag{7}$$

where $N_s$ is the number of segments along angular cross section of the boundary $S$, $T_i$ is a standard triangular basis function, the tangential direction $\hat{t}$ is defined by $\hat{t} = \hat{n} \times \hat{\phi}$. The unknown field $E_{t,m}$ and $E_{\phi,m}$ is expanded as:

$$E_{t,0} = \sum_{i=1}^{8} e_{t,i}^e N_i^e , \quad E_{\phi,0} = \sum_{i=1}^{8} e_{\phi,0}^e N_i^e , \tag{8}$$

for $m = 0$ ;

$$E_{t,\pm 1} = \mp j \hat{\rho} E_{\phi,\pm 1} + \sum_{i=1}^{8} e_{t,i}^e \rho N_i^e , \quad E_{\phi,0} = \sum_{i=1}^{8} e_{\phi,0}^e N_i^e , \tag{9}$$

for $m = \pm 1$ ;

$$E_{t,0} = \sum_{i=1}^{8} e_{t,i}^e \rho N_i^e , \quad E_{\phi,0} = \sum_{i=1}^{8} e_{\phi,i}^e N_i^e , \tag{10}$$

for $|m| > 1$,

where $N_i^e$ and $N_i^e$ are the second-order hierarchical scalar base functions [9] and vector base functions [10] for triangular element, respectively.

The expansions (5), (6) and (7) are substituted into (1), (3) and (4), and the Rayleigh-Ritz approach is used, where the weight functions are chosen to be equal to the basis function. This process yields a symmetric matrix equation:

$$\begin{pmatrix} Z_m^{FE} & B_m^{EJ} & 0 \\ B_m^{EJ} & G_m^{JJ} & G_m^{JM} \\ 0 & G_m^{MJ} & G_m^{MM} \end{pmatrix} \begin{pmatrix} e_m \\ a_m \\ b_m \end{pmatrix} = \begin{pmatrix} 0 \\ V_m^J \\ V_m^M \end{pmatrix}, \tag{11}$$

$Z_m^{FE}$, $B_m^{EJ}$ and $B_m^{EJ}$ are sparse finite element matrix, and $G_m^{JJ}$, $G_m^{JM}$, $G_m^{MJ}$ and $G_m^{MM}$ are dense moment matrix. The oscillating integrals is evaluated accurately and efficiently to accelerate the generation of the elements of the method of moments' matrix [2].

The above procedure should in principle be carried out for each of the Fourier modes $m = 0, \pm 1, \pm 2, \cdots$ . In practice, however, a rule of truncating the infinite Fourier modes for plane wave incidence is $M_{max} = k_0 \rho_{max} \sin(\theta) + 6$ [5], where $\rho_{max}$ is the largest cylindrical radius of the body. Such a rule is valid for $k_0 \rho_{max} \sin(\theta) > 3$. Furthermore, the solution for each negative modes ($m < 0$) is simply related to that of the corresponding positive modes ($m > 0$). Hence the solutions need to be computed for the nonnegative modes only.

III. NUMERICAL EXAMPLES

In this section, numerical results are presented to show the validity and accuracy of the proposed technique. In all examples, the interior FEM mesh length is chosen as about $\lambda / 10$, and the boundary mesh length is chosen as about $\lambda / 15$, where $\lambda$ is the free space wavelength.

First, we consider an anisotropic spherical cloak with free space inside. The cloak is characterized in spherical coordinates by a relative permittivity of (12), which is transformed into cylindrical coordinates shown in [11]. The sphere has outer radius $R_2 = 1$ m and inner radius $R_1 = 0.5$, and is characterized by $\varepsilon_R = R_2(r - R_1)^2 / (R_2 - R_1)r^2$ and $\varepsilon_\theta = R_2 / (R_2 - R_1)$. It should be note that at the inner boundary $r = R_1$, the material properties of the cloak behaves an unavoidable singular with $\varepsilon_R = 0$, which lead to theoretically outside the domain of application of the finite-element method. However, the numerical results shows that the cloak can be simulated by the presented method. The interior FEM mesh length is chosen as about $\lambda / 15$ and the boundary is set $1\lambda$ away from the cloak. Fig. 1 illustrates the real parts of $E_y$ in the $xz$ plane when the plane wave is incident at 15GHz with the angle of incidence $\theta^i = 180^0$ or $\theta^i = 30^0$. From the figure, we observe that the cloaking effect is very clear and the plane wave is almost unaltered outside the cloaking shell, which is nearly the same as the analytical result. Also, the electric field inside the cloaked region is nearly zero with a maximum leakage about 0.3% of the radiating field into the cloaked region, which is more accurate than the result by DDA formalism [12] and FEM with PML [13].

$$\overline{\overline{\mu}}_r = \overline{\overline{\varepsilon}}_r = \varepsilon_R \hat{r}\hat{r} + \varepsilon_\theta \hat{\theta}\hat{\theta} + \varepsilon_\theta \hat{z}\hat{z} \tag{12}$$

$$\varepsilon_{\rho\rho} = \varepsilon_R \sin^2 \theta + \varepsilon_\theta \cos^2 \theta$$
$$\varepsilon_{\phi\phi} = (\varepsilon_R - \varepsilon_\theta) \sin \theta \cos \theta , \tag{13}$$
$$\varepsilon_{zz} = \varepsilon_R \cos^2 \theta + \varepsilon_\theta \sin^2 \theta$$

Then, the presented method is applied to simulate a large 3-D cylindrical gradient-index lens [14], which can transform spherical waves to plane waves. The lens is cylindrically symmetric around the z axis. The centre of the lens was located at the origin of the cylindrical coordinate, the radius (R) and the thickness (H) are 0.8 m and 0.4 m respectively, shown in Fig. 2(a). A 3-D electric dipole directing in the y-direction is placed at the point $(0, 0, -0.6m)$ in the Cartesian coordinates to excite spherical waves. The current dipole moment is $Il = 0.025$ Am, and the working frequency is f = 1.2 GHz. The field of the dipole can be easily expanded to Fourier modes.

The parameter of the lens are described by: $\varepsilon_r = \mu_r = 1.83 - (\sqrt{0.4^2 + \rho^2} - 0.4)/0.46$, which can be realized using metamaterials. The far radiation of the gradient-index lens are shown in Fig. 2(b). The results by 3D CG-FFT method are also given in the figure for comparison. Obviously, good agreement is achieved.

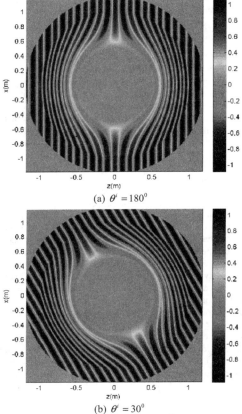

(a) $\theta^i = 180^0$

(b) $\theta^i = 30^0$

Figure 1. The real part of the electronic field $E_y$ in the vicinity of the cloaked sphere

## IV. CONCLUSIONS

In this paper, we have presented the finite element and boundary integral methods for simulating electromagnetic scattering from complex inhomogeneous bodies of revolution. Numerical examples were presented to demonstrate the accuracy and capability of the method.

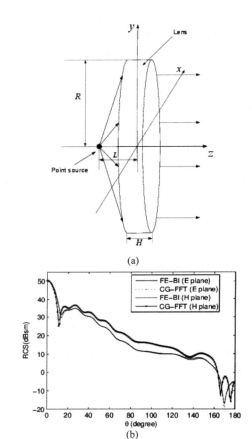

(a)

(b)

Figure 2. 3D gradient-index lens and its far field. (a) 3D gradient-index lens excited by point source. (b) far field RCS.

REFERENCES

[1] M. G. Andreasen, "Scattering from bodies of revolution," *IEEE Trans. Antennas Propgat.,* vol. 13, pp. 303-310, Mar. 1965.
[2] W. M. Yu, D. G. Fang and T. J. Cui, "Closed Form Modal Green's Functions for Accelerated Computation of Bodies of Revolution," *IEEE Trans. Antennas Propgat.,* vol. 56, pp. 3452-3461, 2008.
[3] X. Rui, J. Hu and Q. H. Liu.S, "Fast Inhomogeneous Plane Wave Algorithm for Homogeneous Dielectric Body of Revolution," *Microw. Opt. Tech. Lett.,* vol. 52, pp. 1915-1922, 2010.
[4] A. A. Kucharski, "A method of moments solution for electromagnetic scattering by inhomogeneous dielectric bodies of revolution," *IEEE Trans. Antennas Propgat..* vol. 48, pp. 1202-1210, 2000.
[5] A. D. Greenwood and J. M. Jin, "A novel efficient algorithm for scattering from a complex BOR using mixed finite elements and cylindrical PML," *IEEE Trans. Antennas Propgat.,* vol. 47, pp. 620-629, 1999.
[6] D. J. Hoppe and L. W. Epp and J. Lee, "A hybrid symmetric FEM /MOM formulation applied to scattering by inhomogeneous bodies of revolution," *IEEE Trans. Antennas Propgat.,* vol. 42, pp. 798-805, Jun. 1994.
[7] E. A. Dunn and J.K. Byun and E.D. Branch and J.M. Jin, "Numerical simulation of BOR scattering and radiation using a higher order FEM," *IEEE Trans. Antennas Propgat.,* vol. 54, pp. 945-952, Mar. 2006 .

[8]  J. M. Jin, "The Finite Element Method in Electromagnetics," New York: Wiley, 1993.

[9]  J. P. Webb and S. McFee, "The use of hierarchical triangles in finite-element analysis of microwave and optical devical," *IEEE Trans. Magn.*, vol. 27, pp. 4040-4043, Mar. 1991.

[10]  L. S. Andersen and J. L. Volakis, "Development and application of a novel class of hierarchical tangential vector finite elements for electromagnetics," *IEEE Trans. Antennas Propgat.*, vol. 47, pp. 112-120, Jan. 1999.

[11]  H. Chen, B. I. Wu, B. Zhang and J. A Kong, "Electromagnetic wave interactions with a metamaterial cloak," *Phys. Rev. Lett.* vol. 99, 063903, 2007.

[12]  Y. You, G. W. Kattawar, P. W. Zhai, and P. Yang, "Zero back scatter cloak for aspherical particles using a generalized DDA formalism finite elements for electromagnetics," *Opt. Express.*, vol. 16, pp. 2068-2079, 2008.

[13]  Y. B. Zhai, X. W. Ping, W. X. Jiang, and T. J. Cui, "Finite element analysis of three dimensional axisymmetrical invisibility cloaks and other metamaterial devices," *Commun. Comput. Phys.*, vol. 8, pp. 823-834, 2010.

[14]  H. F. Ma, J. F. Zhang, X. Chen, Q. Cheng and T. J. Cui, "CG-FFT algorithm for three dimensional inhomogeneous and biaxia metamaterials," *Opt. Express.*, vol. 19, pp. 49-64, 2009.

# Full Wave Simulation of the Transfer Response of the TX and RX Antennas in the Full-Duplex Wireless Communication Systems

Yunyang Dong, Jianyi Zhou, Binqi Yang, Jianing Zhao
State Key Lab. of Millimeter Waves
School of Information Science and Engineering
Southeast University
Nanjing, China

*Abstract*-In this paper, the full wave EM investigation on the self-interference channel characteristic of the full duplex wireless communication systems is presented. The FDTD method is used to obtain the channel response with different antenna configurations. The simulated results agree well with the measurement.

## I. INTRODUCTION

The spectrum resource is very important for modern wireless communication systems. The full duplex technique can double the spectrum efficiency by using the same frequency for both the transmitting and receiving simultaneously. The full duplex technique is very attractive, and may become a focus in the wireless communication field.

The largest challenge of a full-duplex system is to reduce the self-interference sufficiently [1], because the self-interference can be millions to billions of times stronger (60dB-90dB) than a received signal.

Well known digital and analog techniques, even combined them together [1], are not able to cancel self-interference sufficiently for full duplex. Motivated by these limitations, recent work has explored antenna placement as an additional cancelation technique. Antenna Separation is the simplest passive self-interference cancellation mechanism and it consists in loss in interference power due to propagation losses caused by separating the transmitting and receiving antennas at a node [6]. However, the reduction of the interference obtained with the antenna separation is limited especially for the mobile terminals.

To further cancel self-interference, J. I. Choi et al. propose a new technique, called antenna cancelation [1]. The system consists of two transmit antennas and a receive antenna. The distances between one TX and the RX is $\lambda/2$ longer than another. Although promising, antenna cancellation-based designs have some limitations. One of the limitations is the bandwidth constraint, a theoretical limit which prevents supporting wideband signals, which is crucial for a communication system.

It is found that the reduction of the self-interference is mainly limited by the transmitting response between the TX antennas and the RX antennas. Although the response can be measured, it is inconvenient during the tuning and optimization. On the other hand, the computation EM method can calculated the response steadily, especially for the optimization. Therefore, the FDTD method is used to simulate the channel characteristic of several cases.

In order to solve open spaces efficiently, absorbing boundary conditions (ABC) must be employed to truncate the computation domain. The first-order Mur ABC is a good choice, but the reflected wave is still larger than wanted. The second-order Mur ABC is often divergent. A new second-order absorbing boundary condition was used to FDTD method and the effect is similar to second-order Mur ABC.

## II. MODEL FOR SELF-INTERFERENCE CANCELATION

A simple model is shown in Figure 1, where two nodes communicate with each other in the full-duplex mode. Each node has two antennas, one antenna is used for transmission and the other antenna is used for reception. $e_a(t)$ is the signal transmitted from Node a, $r_a(t)$ is the signal received at Node a, $h_{aa}(t)$ is the wireless channel from TX to RX of Node a, $e_b(t)$ is the signal transmitted from Node b, and $h_{ba}(t)$ is the wireless channel from Node b to Node a. Therefore, the received signal of the Node a is

$$r_a(t) = e_a(t) * h_{aa}(t) + e_b(t) * h_{ba}(t) \qquad (1)$$

Where "$*$" is the convolution operator. If only Node a is used then the received signal at Rx of Node a is equal to $e_a(t) * h_{aa}(t)$. The estimate of $h_{aa}(t)$ is very useful to cancel the self-interference. Full Wave Simulation of the Transfer Response of the TX and RX Antennas in the full-duplex system can be used to get $h_{aa}(t)$.

As the communication from TX and RX is achieved through electromagnetic waves and such a macroscopic phenomenon is governed by Maxwell's equations. So the electromagnetic simulation can be used to find the self-interference channel $h_{aa}(t)$ and/or $H_{aa}(j\omega)$.

Figure 1. A simple model of full-duplex system

978-7-5641-4279-7

### III. ANALYSIS OF NUMERICAL EXPERIMENTS

The FDTD method [10] is a very powerful tool to simulate the EM problems in both the time domain and the frequency domain. It is very suitable for the channel simulation of the full duplex systems.

#### A. Traditional explicit FDTD method

The iterative formula of explicit FDTD method are

$$E^{n+1} = \frac{\frac{1}{\Delta t} - \frac{\sigma}{2\varepsilon}}{\frac{1}{\Delta t} + \frac{\sigma}{2\varepsilon}} E^n + \frac{\frac{1}{\varepsilon}}{\frac{1}{\Delta t} + \frac{\sigma}{2\varepsilon}} (P - Q) H^{n+\frac{1}{2}} \quad (2)$$

$$H^{n+\frac{1}{2}} = H^{n-\frac{1}{2}} + \frac{\Delta t}{\mu} (P - Q) E^n \quad (3)$$

Where $P$ and $Q$ are discrete differential operator matrixes, and $\delta_x$, $\delta_y$ and $\delta_z$ are central difference operators:

$$P = \begin{bmatrix} 0 & 0 & \delta_y \\ \delta_z & 0 & 0 \\ 0 & \delta_x & 0 \end{bmatrix} \quad (4)$$

$$Q = \begin{bmatrix} 0 & \delta_z & 0 \\ 0 & 0 & \delta_x \\ \delta_y & 0 & 0 \end{bmatrix} \quad (5)$$

$$\delta_x f_{i,j,k} = \frac{f_{i+\frac{1}{2},j,k} - f_{i-\frac{1}{2},j,k}}{\Delta x} \quad (6)$$

$$\delta_y f_{i,j,k} = \frac{f_{i,j+\frac{1}{2},k} - f_{i,j-\frac{1}{2},k}}{\Delta y} \quad (7)$$

$$\delta_z f_{i,j,k} = \frac{f_{i,j,k+\frac{1}{2}} - f_{i,j,k-\frac{1}{2}}}{\Delta z} \quad (8)$$

And the iterative formula are used on Yee cell.

A new absorbing boundary condition was used to this traditional FDTD method and the effect is similar to second-order Mur ABC.

Figure 2. A scheme of antenna cancellation

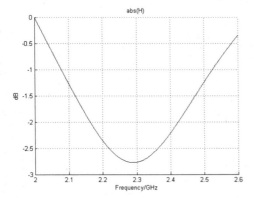

Figure 3. The channel of Scheme 1

Figure 4. Another scheme of antenna cancellation

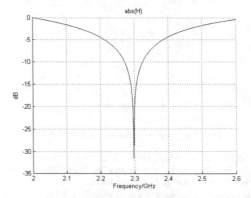

Figure 5. Channel without antenna cancellation

#### B. Antenna cancellation simulation

Antenna cancelation consists of two transmit antennas and a receive antenna. There are two schemes to achieve cancellation and one of them are shown in Figure 2. The left and right antennas are used to transmit signals, whereas the middle one to receive signals. The distances between one TX and the RX is $\lambda/2$ longer than another. So the two signal send by transmit antennas will arrive at the receive antenna with a phase difference of $\pi$ which just implements a cancellation.

Using traditional explicit FDTD method, an ordinary case is calculated. All the antennas are placed in the central of the calculation space. The left antenna is 150mm away from the central antenna, whereas the right antenna 215mm away from the central antenna. The excitation is sinusoidal modulated Gaussian pulse with a carrier frequency of 2.3GHz.

In order to highlight the shape, we normalize the absolute value to the maximum and obtain the channel character $H_{aa}(j\omega)$ is shown in Figure 3. In the figure, the amplitude of $H_{aa}(j\omega)$ is attenuated of about 3dB at the center frequency. We can see the scheme of antenna cancellation can have a suppression on self-interference.

Another scheme is shown in Figure 4. The distances between the TX and the RX are the same, but one TX is excited $T/2$ later than another. So the two signal send by transmit antennas will arrive at the receive antenna with a phase difference of $\pi$ which just implements a cancellation.

In the space one can calculate, all the antennas are placed at the central part of the calculation space. The transmit antennas are 150mm away from the central antenna. The right antenna is excited $T/2$ later than the left. The excitation is sinusoidal modulated Gaussian pulse with a carrier frequency of 2.3GHz.

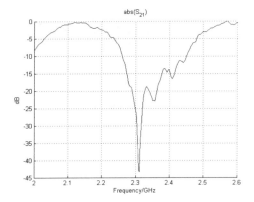

Figure 6. The channel tested with vector network analyze

The transmission characteristic is shown in Figure 5. The amplitude of $H_{aa}(j\omega)$ is attenuated of about 30dB at the center frequency more than the first scheme.

To prove this, an experiment is done like what has been shown in Figure 4. With the use of vector network analyzer, a similar result can be got shown in Figure 6.

We can see this scheme of antenna cancellation have a better effect on self-interference. However, the slope near the recess tends to infinite, in another words, the bandwidth of this scheme is extremely narrow, which prevents supporting wideband signals. Furthermore, the second scheme is particularly sensitive to the changes in environment. A small change that will make the operating point shift to an undesirable position.

*C.  Self-interference channel simulation*

At first, in order to get the channel characteristics between the transmit antenna and receive antenna in free space, both the antennas are placed at the central part of the calculation space. The transmit antennas are 130mm away from the

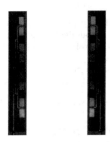

Figure 8. Antennas used for experiment

central antenna. The excitation is sinusoidal modulated Gaussian pulse with a carrier frequency of 3GHz.

Then an iron box is set behind the antennas with a distance of 300mm acted as a vector network analyzer. At the same time, at the bottom of the antennas, a wooden table was added covered the entire horizontal space. Comparing these two conditions, the channel between transmit and receive antennas shows different amplitude fluctuations at different frequencies.

Furthermore, the concrete walls are introduced in around the calculation space. The channel transfer function shown in Figure 7 becomes more ups and downs. As this trend, when the surrounding scatters become extremely rich, the transfer function will be quite similar with the actual.

*D.  Comparison of the results of experiment, commercial software and our program*

To explore the effect of our program, a test is made that showed in Figure 8. Around the antennas some scatters are introduced at the same time.

With the use of vector network analyzer, one can clearly know the transmission characteristics $S_{21}$ between TX and RX. Export data to MATLAB and the result is shown in Figure 9 with green solid line. The figure shows the center frequency of the antenna is 2.3GHz, and the bandwidth is about 1GHz.

Next, using a commercial software, a simulation model is created and solved and the result is shown in Figure 9 with

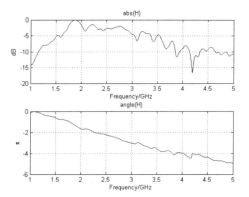

Figure 7. Channel with the presence of table, box and cement walls

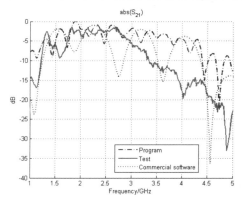

Figure 9. $|S_{21}|$ between TX and RX in scattering space

blue short dash line. From the figure, one can see the simulation result of this commercial software is different with the actual result and the central frequency are not the same.

Finally, using the program, set medium according to the real position. The excitation is sinusoidal modulated Gaussian pulse with a carrier frequency of 2.3GHz. The transmission characteristic is shown in Figure 9 with red dotted line. The figure graphs the simulation result of our program, which is the same with the actual result especially near the central frequency.

In conclusion, the program has a better simulation result near the central frequency. Using the program, the response between transmit and receive antennas is more similar to measurement result. With the assistance of the adaptive algorithm, a better self-interference suppression will be got.

## IV. CONCLUSION

In order to promote the development of full-duplex systems it is important to have signal models. We try to contribute this area by providing a simplest model. We have showed the self-interference channel $h_{aa}(t)$ can be got in analog domain. Assistance with the adaptive algorithm, a better self-interference suppression can be got. With the combine of other techniques, a better full-duplex system will be used in the future doubling the resource utilization.

## REFERENCES

[1] J. I. Choi, M. Jain, K. Srinivasan, P. Levis, and S. Katti. Achieving single channel, full duplex wireless communication. In Proceedings of the sixteenth annual international conference on Mobile computing and networking, MobiCom'10, pages 1-12, New York, NY, USA, 2010. ACM.

[2] B. Radunovic, D. Gunawardena, P. Key, A. Proutiere, N. Singh, V. Balan, and G. Dejean. Rethinking indoor wireless mesh design: Low power, low frequency, full-duplex. In Wireless Mesh Networks (WIMESH 2010), 2010 Fifth IEEE Workshop on, pages 1 -6, 2010.

[3] S. Sen, R. R. Choudhury, and S. Nelakuditi. CSMA/CN: carrier sense multiple access with collision notification. In Proceedings of the sixteenth annual international conference on Mobile computing and networking, MobiCom '10, pages 25-36, New York, NY, USA, 2010. ACM.

[4] I. L. Gheorma and G. K. Gopalakrishnan. Rf photonic techniques for same frequency simultaneous duplex antenna operation. IEEE Photonics Letters, 19(13), July 2007.

[5] D. W. Bliss, P. A. Parker, and A. R. Margetts. Simultaneous transmission and reception for improved wireless network performance. In Proceedings of the 2007 IEEE Workshop on Statistical Signal Processing, 2007.

[6] M. Duarte and A. Sabharwal. Full-duplex wireless communications using off-the-shelf radios: Feasibility and first results. In Forty-Fourth Asilomar Conference on Signals, Systems, and Components, 2010.

[7] M. Jain, J. I. Choi, T. Kim, D. Bharadia, S. Seth, K. Srinivasan, P. Levis, S. Katti, and P. Sinha. Practical, real-time, full duplex wireless, in Proc. 2011 ACM MobiCom, pp. 301-312.

[8] M. Duarte, C. Dick, and A. Sabharwal. Experiment-Driven Characterization of Full-Duplex Wireless Systems. IEEE Transactions on Wireless Communications, 2013.

[9] F. Zheng, Z. Chen and J. Zhang, Toward the development of a three dimensional unconditionally stable finite-difference time domain method, IEEE Trans. Microwave Theory and Tech., vol.48, no. 9, pp. 1550-1558, September 2000.

[10] Yee K. S. Numerical solution of initial boundary value problems involving Maxwell's equations in isotropic media. IEEE Transactions on Antennas and Propagation, 1966, 14(3):302-307.

[11] Zhao Jianing, Improved absorbing boundary condition based on linear interpolation for ADI—FDTD method. Journal of Southeast University(English Edition), Vol.25, No.3, PP.289—293, Sept. 2009.

[12] J. Shibayama, M. Muraki, J. Yamauchi, and H. Nakano, "Efficient implicit FDTD algorithm based on locally one-dimensional scheme," Electron. Lett., vol. 41, pp. 1046–1047, Sep. 2005.

[13] J. Shibayama et al., "Efficient implicit FDTD algorithm based on locally one-dimensional scheme," Electron. Lett., vol. 41, pp. 1046–1047, Sep. 2005.

# Research of radio wave propagation in forest based on Non-uniform mesh Parabolic Equation

Qinghong Zhang   Cheng Liao   Nan Sheng   Linglu Chen

(*Institute of Electromagnetics, Southwest Jiaotong University, Chengdu 610031, China*)

*Abstract*- Non-uniform mesh technology of parabolic equation method is introduced to solve the radio wave propagation problems in forest more efficiently. The simulation errors of radio wave propagation in forest with different uniform grids are analyzed first. The results indicate that fine grid in forest area is necessary to ensure the simulation accuracy. As typical scenes for simulation are usually forest region within a large area, using uniform fine grids requires large computation memory. Hence, the non-uniform mesh technology is employed. The performance of non-uniform mesh technology in the application of local forest scene is analyzed. The result obtained by non-uniform meshes is in good agreement with that obtained by uniform fine grids. Besides, it makes nearly three times reduction of computation time compared with the one with uniform fine grids in this example. This demonstrates that the employment of non-uniform mesh parabolic equation method for forest scenes simulation can improve the computational efficiency greatly.

*Key words*- Parabolic equation, Forest, Non-uniform mesh, Radio wave propagation

## I.   INTRODUCTION

The research of radio wave propagation in complex electromagnetic environment is an important project in modern electromagnetic field [1, 2]. In the rural and suburban areas, forest is an important factor of affecting the radio wave propagation over a long distance [3, 4]. For the scattering and absorption of trees, the attenuation and phase shift of signal will happen, which affects the target identification, remote sensing and wireless communication greatly. Therefore, it has very important practical application value to study the forest effect on wave propagation.

The parabolic equation (PE) method, introduced by Leontovich and Fock in 1946 [5], is applied to the underwater acoustics problems at the earliest and has been used extensively in electromagnetic field since the mid-1980s. PE can deal with the inhomogeneous medium and complex boundary conditions and has been widely used to simulate the radio wave propagation characteristics in complex

electromagnetic environment [6, 7].

Comparing with experimental results, Tamir obtained that forest can be viewed as a dissipative dielectric slab in solving the radio wave propagation problems in range 2-200 MHz, and the effective dielectric constant approached 1 [8]. As effective dielectric constant satisfies the PE approximation conditions, the radio wave propagation characteristics can be easily solved through PE.

In recent years, scholars at home and abroad have done some research about the radio wave propagation problems in forest using PE. The radio wave propagation characteristics are simulated by Palud [9], including the source located inside and outside forest. Through the finite-difference algorithms of PE, the radio wave propagation problems over Irregular Terrain partly covered by forest are solved by Holo and the results are compared with experiments [10]. Besides, the radio wave propagation characteristics in areas covered by forest entirely and partly respectively are analyzed by Jianyan Guo based on the split-step Fourier transform algorithm [4, 11]. At present, the uniform grid is used in all PE simulations for forest. The computing resources and calculating time will increase when using the uniform fine mesh, and the computational accuracy will reduce when using the uniform coarse grid. To this end, the non-uniform mesh technology of PE is introduced in this paper and from the simulation result it is found that the computational efficiency is improved greatly for wave propagation problems in large scale forest environment.

## II.   OVERVIEW OF PARABOLIC EQUATION

In rectangular coordinate system, we assume that $e^{-i\omega t}$ is the time-dependence of fields, where $\omega$ is the angular frequency. The field component $\psi$ satisfies the two-dimensional scalar wave equation [12]

$$\frac{\partial^2 \psi(x,z)}{\partial x^2} + \frac{\partial^2 \psi(x,z)}{\partial z^2} + k_0^2 n^2 \psi(x,z) = 0 \qquad (1)$$

Where $\psi$ is the electric field for horizontal polarization or magnetic field for vertical polarization. $n$ is the refraction index. $k_0$ is the wave number in vacuum. Introduce the reduced function associated with the paraxial direction $x$

$$u(x,z) = e^{-ik_0 x}\psi(x,z) \qquad (2)$$

Now plugging (2) into (1) and factoring(1), we can get the

978-7-5641-4279-7

forward parabolic equation

$$\frac{\partial u}{\partial x} = -ik_0(1-Q)u$$

(3)

Where $Q$ is the pseudo-differential operator and is defined by

$$Q = \sqrt{\frac{1}{k_0^2}\frac{\partial^2}{\partial z^2} + n^2}$$

(4)

By using the Feit-Fleck approximation [13] of $Q$, the wide angle parabolic equation(WAPE) is obtained

$$\frac{\partial u(x,z)}{\partial x} = ik_0\left[\sqrt{1+\frac{1}{k_0^2}\frac{\partial^2}{\partial z^2}}-1\right]u(x,z)+ \qquad (5)$$
$$ik_0(n-1)u(x,z)$$

The WAPE can be solved using the split-step Fourier transform algorithm(SSFT), which is represented as

$$u(x_0+\Delta x,z) = e^{ik_0\Delta x(n-1)}F^{-1}\left\{e^{i\Delta x\left(\sqrt{k_0^2-p^2}-k_0\right)}F[u(x_0,z)]\right\} \qquad (6)$$

Where $u(x_0,z)$ is the initial field. $\Delta x$ is the range step. $F$ and $F^{-1}$ indicate the Fourier transformation and inverse Fourier transformation respectively. $p=k_0\sin\alpha$ is the variant of z, $\alpha$ is the grazing angle. SSFT is a algorithm marching over range steps and its range step is almost not limited by wavelength, so it can be easily applied to solve the radio wave problem in large scale complex environment.

### III. NON-UNIFORM MESH TECHNOLOGY FOR PARABOLIC EQUATION

#### A. Analysis of range step in forest

The influence of range step on simulation results is analyzed for 100 MHz and 200 MHz. We assume that the antenna height is 16 m, the receiving height is 10 m, atmosphere is ideal uniform air, the relative dielectric constant and the conductivity of the earth surface are 20 and 0.01 S/m respectively. the forest condition: height 19 m, equivalent complex permittivity 1.064+0.037i.

Compared with the Tamir model, Propagation loss(PF) on the receiving height for 100MHz and 200MHz are simulated, which are showed in figure 1 and figure 2.

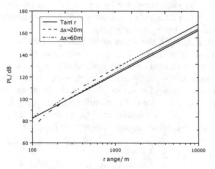

Figure 1 Propagation Loss for frequency 100MHz

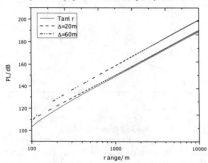

Figure 2 Propagation Loss for frequency 200MHz

Figure 1 and Figure 2 indicate that the error of PE increases with the increasing of range step. Compared with the Tamir model, the max error is 6dB and 12dB respectively for 100 MHz and 200MHz when the range step is 60m. Thus, we should reduce the range step to improve the simulation efficiency for forest scene.

#### B. Non-uniform mesh technology

From the previous analysis, we get that the range step should be reduced to improve the computational efficiency for solving the radio wave propagation problems in forest. In large scale areas, if the space is divided by the uniform fine mesh, the computing time will be added greatly, which is bad for rapidly solving the radio wave propagation problems. Therefore, the paper introduces the non-uniform mesh technology of PE. The general idea: the uniform fine mesh is adopted in areas covered by forest where the electromagnetic field changes quickly and the uniform coarse grid is used in the open areas where the electromagnetic field changes slowly to reach the equilibrium between calculation accuracy and computation time.

### IV. NUMERICAL EXAMPLES

The radio wave propagation characteristics in partly

forested area are simulated based on the non-uniform mesh technology and the results are compared with the uniform fine mesh and the uniform coarse grid.

In this example, the forest exists in range from 10 km to 20 km, the max range is 32 km, source frequency is 0.1 GHz. Other parameters are set the same as the previous example. The range step of uniform fine mesh is set to 10 m and the uniform coarse grid is set to 100 m and 150 m respectively. the minimal range step of non-uniform mesh is 10 m and the maximal range step is 200 m. The contour of propagation factor on the receiving height is plotted in Figure 3.

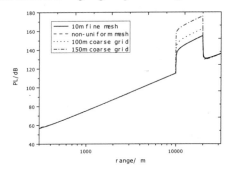

Figure 3 Propagation Loss

The figure 3 shows that there is excellent agreement for different grids in the area without forest. The uniform coarse grid will bring large error in the forest area and the error increases with the increasing of grid size. The good agreement between the non-uniform mesh and the uniform fine mesh validates the correctness of non-uniform mesh technology in forest.

The calculating time of uniform fine mesh and non-uniform mesh is showed in Table I.

TABLE I
CALCULATING TIME

| mesh type | time / s |
| --- | --- |
| uniform fine mesh | 29.028 |
| non-uniform mesh | 9.656 |

Table 1 shows us that the calculating time of uniform fine mesh and non-uniform mesh are 9.656 s and 29.028 s respectively. Compared with the uniform fine mesh, the computing speed of PE is improved by nearly three times with the uniform fine mesh, which is very important in simulating the electromagnetic characteristics rapidly in large scale areas, especially in the three-dimensional space. From the previous analysis we can see that PE with non-uniform mesh has good precision, this demonstrates that the non-uniform mesh technology can improve the computational efficiency greatly when solving the wave propagation problems in forest.

V. CONCLUSION

In this paper, the non-uniform mesh technology is employed to improve the computation efficiency when the simulation scenes are local forest within a large area. The numerical results show the advantages of non-uniform mesh technology. Comparing with the result using uniform fine grids, the one using non-uniform meshes can reduce the computation time while keeping the simulation accuracy.

ACKNOWLEDGMENT

This work is supported by the Fundamental Research Funds for The Central Universities (SWJTU12ZT08) and the Research Fund of Key Laboratory of CEMC Science & Technology of CAEP (FZ2012-2-O1).

REFERENCES

[1] Xuan Shao, Xiaoliang Chu, Jian Wang and Jinju Xu, "Study on effect of wind waves on radar echoes in atmosphere duct oversea," *Acta Phys. Sin.* vol. 61, pp. 159203, 2012.

[2] Hui Du, Gang Wei, Yuanming Zhang and Xiaohui Xu, "Experimental investigations on the propagation characteristics of internal solitary waves over a gentle slope," *Acta Phys. Sin.* vol. 62, pp. 064704, 2013.

[3] Zhirong Cai, Yonghua Liu and Xiulu Zhang, "Investigation on Radiowave Propagation in Forest Environments in Central China," *JOURNAL OF CHINA INSTITUTE OF COMMUNICATIONS.* vol. 18, pp. 87-92, 1997.

[4] Jianyan Guo, Jianying Wang, Yunliang Long and Zhuqian Gong, "Analysis of radio propagation in partly forested terrain environment using parabolic equation approach," *CHINESE JOURNAL OF RADIO SCIENCE.* vol. 23, pp. 1045-1050, 2008.

[5] LEONTOVICH M A and FOCK V A, "Solution of propagation of electromagnetic waves along the earth's surface by the method of parabolic equation," *JPhys USSR.* vol. 10, pp. 13-23, 1946.

[6] Huibin Hu, Junjie Mao and Shunlian Chai, "Application of atmosphere refractivity profile in wide-angle parabolic equation," *Journal of Microwaves.* vol. 22, pp. 5-8, 2006.

[7] Guangcheng Li, Lixin Guo, Zhensen Wu and Jinhai Liu, "Influence of obstacle towards electromagnetic wave propagation in the evaporation duct," *Chinese Journal of Radio Science.* vol. 26, pp. 621-627, 2011.

[8] Tamir, "On radio-wave propagation in forest environments," *IEEE Transactions on Antennas and Propagation.* vol. 15, pp. 806-817, 1967.

[9] M Le Palud, "Propagation modeling of VHF radio channel in forest environments," *IEEE Military Communications Conference.* vol. 2, pp. 609-614, October 2004.

[10] HOLM P, ERIKSSON G, KRANS P, Lundborg B, Lafsved E, Sterner U and Waern A, "Wave propagation over a forest edge-parabolic

equation modelling vs . measurements,". *IEEE Symp. PIMRC.* Lisboa Portugal, vol. 1, pp. 140-145, September 2002.

[11]  Jianyan Guo, Jianying Wang and Yunliang Long, "Parabolic equation model for wave propagation in forest environments," *Chinese Journal of Radio Science.* vol. 22, pp. 1042- 1046, 2008.

[12]  LEVY M, "Parabolic Equation Methods for Electromagnetic Wave Propagation," London: *IEEE*, 2000.

[13]  FEIT M D and FLECK J A, "Light propagation in graded-index fibers,"

[14]  *Applied Optics.* vol. 17, pp. 3990-3998, 1978.

# Fast Calculation Method of Target's Monostatic RCS Based on BCGM

X. F. Xu    J. Li

Key Laboratory of Astronautic Dynamics, Xi'an Satellite Control Center

Xi'an 710043, China

*Abstract*-When the target's monostatic RCS (Radar Cross Section) is calculated by the traditional method sweeping with frequency and degree, the matrix equation of each interval point must be solved, which cost a lot of time. To solve the problem, the calculation method of monostatic RCS using method of moments is analyzed. Based on the BCGM (bistatic conjugate gradient method), one solving method using the last sweeping point result as the sweeping point initial iterative values is proposed. The numerical results indicate that the iterative times can be effectively reduced by the method, and the efficiency will be higher if the sweeping points are denser.

## I. INTRODUCTION

The RCS of the target is relative to the frequency and the angle, so the matrix equation of the electric current distribution of the target is needed to be solved in frequency and angle domain. Using traditional frequency sweeping and angle sweeping calculating method, the matrix equation must be solved repetively at a series of points in the hope frequency and angle domain. For accurate solution, we must reduce frequency or angle interval, which means that the repetition of matrix equation solving increases greatly. It inevitably makes significantly increased amount of calculation and spending a lot of calculation time and computer memory.

Matrix equation solving speed is a important factor that affects RCS computing time overall scanning range. Multilevel fast multipole technique can reduce the computing amount from $O(N^2)$ to $O(N \log N)$ [1-4]. Adopting conditional processing method can effectively reduce the number of iterations, which is very effective for the single scanning point [5,6]. But for the calculation of multiple scanning points, especially when scanning points are very dense, the iteration times of each scanning point won't decrease due to the increased scanning points. The total times of iteration is still more. In this paper, the method using the last point iterative value as the current scanning point initial iterative value is presented. The method can effectively reduce the times of iterations. There are more populous scanning points (the smaller scanning interval), the iterations times of single scanning point will be less, and there will be higher efficiency.

## II. BASIC THEORY

For 3d scattering problem, according to the ideal conductor surface boundary conditions, the electric field integral equation can be obtained.

$$E_{inc}(r) = ik\eta \int_S G(r,r') \cdot J_s(r')dS' \qquad (1)$$

$E_{inc}$ represents the incident field, $G(r,r')$ is the green's function. The surface current of the scatterer is expanded with the RWG basis function [7], and the current basis function of the nth edge is

$$f_n(r) = \begin{cases} \dfrac{l_n}{2A_n^+}\rho_n^+ & r \in T_n^+ \\[2mm] \dfrac{l_n}{2A_n^-}\rho_n^- & r \in T_n^- \\[2mm] 0 & other \end{cases} \qquad (2)$$

Scatterer surface current density can be approximated as

$$J_s(r) \approx \sum_{n=1}^{N} I_n f_n(r) \qquad (3)$$

Galerkin's method is adopted, which selects the basis functions as a test function. And the matrix equation was derived.

$$Z_{mn} = ik\eta \int_{\Delta_m} f_m(r) \cdot \int_{\Delta_n} G(r,r') \cdot f_n(r')dS'dS \qquad (4)$$

$$V_m = \int_{\Delta_m} f_m(r) \cdot E_{inc}(r)dS \qquad (5)$$

The scatterer surface currents can be got by solving the matrix equation, and the scattering field of the any point in the space will be calculated. The calculation formula of RCS is

$$\sigma = \lim_{r\to\infty} 4\pi r^2 \frac{\left|E_s^2\right|}{\left|E_{inc}^2\right|} \qquad (6)$$

Observing from equation 1, we can get $Z_{mn} \approx Z_{nm}$, so the impedance matrix is a symmetrical complex matrix. Compared with the conjugate gradient method (CGM), one time matrix vector multiplication is only needed by using bistatic conjugate gradient method (BCGM). Under the same accuracy, BCGM needs a third of the iteration times of CGM. So, the computing can be speeded up 5 ~ 6 times by using BCGM. But due to the calculation error, the generated impedance matrix is only a approximate symmetric matrix. In order to make the impedance matrix symmetry, after filling impedance matrix, a standard plural symmetric matrix is generated using the following method, and half of storage can be saved.

$$Z_{mn} = \frac{Z_{mn} + Z_{nm}}{2}, \quad Z_{nm} = Z_{mn} \qquad (7)$$

978-7-5641-4279-7

For the monostatic RCS calculation of the complex structure scatterer, if want to get accurate results, scanning points require relatively close, so scanning interval will be very small. Imagining that, if the angle or frequency interval are close to zero ($\Delta\theta \to 0, \Delta f \to 0$), then every changing of a scanning point, scatterer surface current distribution almost unchange, so the two adjacent scanning points' results will be slightly different. Based on this idea, this paper puts forward a method of that using the last scanning point iteration value as the current scanning point initial iteration. It is easy to judge, the method will reduce the times of iterations and speed up the convergence. When $\Delta\theta \to 0, \Delta f \to 0$, it may not require iteration or a few times of iteration, the computation will quickly converge and meet the accuracy requirements.

### III. NUMERICAL CALCULATION AND ANALYSIS

In order to verify the validity of the method, firstly, the monostatic RCS of a combination model of a sphere and a cone is calculated with angle sweeping. The sphere diameter and the bottom diameter and the height of the cone are all $\lambda$. The sphere center is located in the apex of the cone. For the incident wave frequency, the unknown number is 3717, and $\varphi = 0°$. With changing the angle of the incident wave, the monostatic RCS with different $\Delta\theta$ are shown in figure 1.

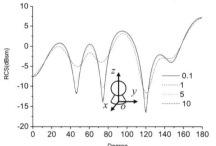
Figure 1. The monostatic RCS of the combination.

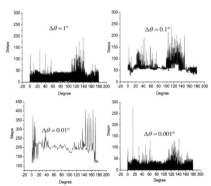
Figure 2. The iterative times of each scanning point.

As can be seen from figure 1, the monostatic RCS will be more precise with the increase of the scanning points ($\Delta\theta$ reduced). For the irregular surface structure targets, more scanning points are needed to obtain accurate numerical results. The iterative times of each scanning point with different $\Delta\theta$ are shown in figure 2. As can be seen from figure 2, $\Delta\theta$ is smaller, the iteration times of each scanning point is less. The average iteration times of each scanning point with different $\Delta\theta$ are shown in table1.

TABLE I
THE AVERAGE ITERATION TIMES OF EACH SCANNING POINT

| $\Delta\theta$ | 1.0° | 0.1° | 0.01° | 0.001° | 0.0001° |
|---|---|---|---|---|---|
| The number of scanning points | 181 | 1801 | 18001 | 180001 | 1800001 |
| The total iteration times | 37477 | 123586 | 261929 | 285842 | 289286 |
| The average iteration times | 207.06 | 68.62 | 14.55 | 1.59 | 0.16 |

As can be seen from figure 2, $\Delta\theta$ is smaller, the average iteration times is less.

Secondly, the affection to monostatic RCS with different scanning intervals of frequency is studied. The monostatic RCS of a metal sphere is calculated. $\varphi = 0°$ and $\theta = 0°$, by sweeping the frequency of incident wave, the structure and the monostatic RCS are shown in figure 3.

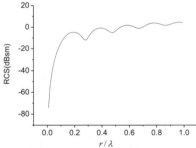
Figure 3. The monostatic RCS of a metal sphere.

The average iteration times of each scanning point with different $\Delta f$ (frequency interval) are shown in figure 4. $\Delta f$ is smaller, the average iteration times is less, by which the effection of the method is also verified.

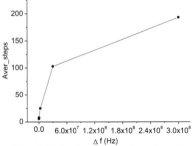
Figure 4. The iterative times of each scanning point.

## IV. CONCLUSION

For the calculation of the monostatic RCS of the targets with irregular surface structure, the scanning points are denser, and the results are more accurate. The numerical calculation indicates that, the method presented in this paper is very suitable for solving the dense scanning points problem. The scanning points populated more densely (the scanning interval are smaller), the less iteration times is needed, so the monostatic RCS calculating speed of the whole scanning range can be effectively improved.

## REFERENCES

[1]    Z. H. Fan, J. L. Liu, Y. Q. Hu, etc., "Multilevel Fast Multipole Algorithm for EM Scattering of Conducting Objects Above Lossy Half-space," Chinese Journal of Computational Physics, vol. 27(1), pp. 95-100, 2010.

[2]    J. Hu, Z. P. Nie, J. Wang, etc., "Multilevel fast multipole algorithm for solving scattering from 3-D electrically large object," Chinese Journal of Radio Science, vol. 19(5), pp. 509-514, 2004.

[3]    M. S. Tong, W. C. Chew, "Multlevel fast multipole algorithm for elastic wave scattering by large three-dimensional objects," Journal of Computational Physics, vol. 228, pp. 921-932, 2009.

[4]    X. F. Xu, X. Y. Cao, J. J. Ma, "Efficient calculation of vehicular antennas' radiation patterns," Progress In Electromagnetics Research Symposium. pp. 255-257, 2009.

[5]    J. L. Guo, J. Y. Li, Y. L. Zhou, etc., "Incomplete LU pre-conditioner technology for fast analysis of conducting objects," Chinese Journal of Radio Science, vol. 23(1), pp. 141-145, 2008.

[6]    S. G. Wang, X. P. Guan, D. W. Wang, etc., "A Preconditioner in Iterative Solution of Higher-Order MoMs," Joournal of Microwaves, vol. 24(4), pp. 20-23, 2008.

[7]    S. M. Rao, D. R. Wilton, A. W. Glisson, "Electromagnetic scattering by surfaces of arbitrary shape," IEEE Trans. Antennas Propagat, vol. 30 (5), pp. 409-418, 1982.

# New Discretized Schemes of 1st Mur's ABC and Its Applications to Scattering Problems

Quan Yuan, Lianyou Sun and Wei Hong

STATE KEY LABORATORY OF MILLIMETER WAVES, SOUTHEAST UNIVERSITY

Nanjing, 210096, China

*Abstract- The new discretized schemes of Mur's absorbing boundary condition (ABC) based on finite difference frequency domain method has been presented in this paper. The analysis figures out that their numerical accuracy are second-order. Combining the new schemes with five-point format of Helmholtz equation to solve scattering problems, the numerical results show that the accuracy of numerical solution with the new schemes is higher than that with ordinary discretized scheme of Mur's absorbing boundary condition.*

## I. INTRODUCTION

Finite Difference Method (FD) is one of the most prevailing methods on computational electromagnetics due to its simplicity and low cost. In this paper finite difference frequency domain (FDFD) is used to analyze cylinder scattering problem. Generally, the scattering region is infinite, so it will be truncated. At the truncation boundary, 1st Mur's absorbing boundary condition (ABC) will be set. The ABC is a Robin's condition. Its numerical discretized equations are of first-order accuracy. For Helmholtz equation, its numerical accuracy of discretized equations is second-order. So the numerical accuracies of Helmholtz equation and ABC condition are mismatched. In order to improve the numerical accuracy of ABC, the new numerical discretized equations of ABC are presented here, and they have second-order numerical accuracy. Then, two kinds of numerical discretized equations of ABC with different numerical accuracies are employed to solve the scattering problems. The numerical results show that the numerical solution with second-order numerical precision equations of ABC is better.

## II. NEW NUMERICAL SCHEME OF MUR BOUNDARY CONDITION

Mur boundary condition, which proposed by Gerrit Mur [1] in 1981 based on outgoing wave theory [3], is very easy to implement and has quite good results if properly applied, so until nowadays this type of boundary conditions still widely used in both FDTD and FDFD. Due to including first-order derivative, its discretized numerical equations are only first-order accuracy. In order to enhance its numerical accuracy, some researches use second-order Mur condition. However, it is not only very complicate, but also it is still first-order numerical accuracy. In fact, boundary condition including second-order derivative is not reasonable for Helmholtz equation. Here, the new discretized numerical equations for Mur's ABC will be presented and they are of second-order numerical accuracy.

Considering the scattering problem of a PEC square cylinder with infinite length and with $TM_z$ plane wave incidence, it can be expressed as follow 2-D problem

$$\begin{cases} \dfrac{\partial^2 \phi}{\partial x^2} + \dfrac{\partial^2 \phi}{\partial y^2} + k^2 \phi = 0, (x,y) \in \Omega & (1) \\[2mm] \phi = 0, \text{on } \Gamma_1 & (2) \\[2mm] \dfrac{\partial \phi}{\partial n} + jk\phi = 0, \text{on } \Gamma_2 & (3) \end{cases}$$

Where $\phi = E_z$ here. As Fig. 1 shows, $\Omega$ is the cross section of scattering area, $\Gamma_1$ is the PEC boundary, $\Gamma_2$ is the truncated boundary, $n$ is outgoing unit normal vector of $\Gamma_2$. Equation (3) is the well-known 1st Mur's absorbing boundary condition.

Just for simplicity, only the left boundary (x=0) of $\Gamma_2$ will be considered here. Equation (3) is usually discretized as:

$$\frac{\partial \phi_0}{\partial x} - jk\phi_0 = \frac{\phi_2 - \phi_0}{h} - jk\phi_0 + O(h) \qquad (4)$$

Here, $\phi_0 = \phi(x_0, y_0)$ is the left boundary point shown in Fig.2, and $\phi_2 = \phi(x_0 + h, y_0)$, $h$ is mesh step. Apparently, equation (4) has only first-order numerical accuracy, while the traditional 5 nodes format of Helmholtz equation has second-order numerical accuracy, their precisions are mismatched.

Considering the condition that quantity $\phi$ satisfies equation (1) at all points in computational domain, so do the points at boundary (not include the corner points). As Fig. 2 (a) shows, use Taylor expansion of point 0 $(x_0, y_0)$ at 1, 2, 3:

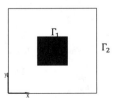

Figure 1. The cross sections of the scattering area for example 2

978-7-5641-4279-7

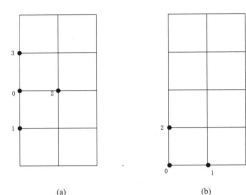

(a)                    (b)

Figure 2. Mesh and nodes for 1st Mur ABC

$$\phi_1 = \phi(x_0, y_0) - \frac{\partial \phi(x_0, y_0)}{\partial y} h + \frac{1}{2!} \frac{\partial^2 \phi(x_0, y_0)}{\partial y^2} h^2 + O(h^3) \qquad (5)$$

$$\phi_2 = \phi(x_0, y_0) + \frac{\partial \phi(x_0, y_0)}{\partial x} h + \frac{1}{2!} \frac{\partial^2 \phi(x_0, y_0)}{\partial x^2} h^2 + O(h^3) \qquad (6)$$

$$\phi_3 = \phi(x_0, y_0) + \frac{\partial \phi(x_0, y_0)}{\partial y} h + \frac{1}{2!} \frac{\partial^2 \phi(x_0, y_0)}{\partial y^2} h^2 + O(h^3) \qquad (7)$$

Such that:

$$\phi_1 + \phi_3 + 2\phi_2 - 4\phi_0 = \frac{\partial^2 \phi(x_0, y_0)}{\partial y^2} h^2 + \frac{\partial^2 \phi(x_0, y_0)}{\partial x^2} h^2 + 2 \frac{\partial \phi(x_0, y_0)}{\partial x} h + O(h^3) \qquad (8)$$

$\phi$ also satisfies Helmholtz equation at $(x_0, y_0)$, so

$$\frac{\partial^2 \phi(x_0, y_0)}{\partial y^2} h^2 + \frac{\partial^2 \phi(x_0, y_0)}{\partial x^2} h^2 = -k^2 h^2 \phi(x_0, y_0) \qquad (9)$$

Substitute formula (9) into formula (8), obtain the formula below:

$$\phi_1 + \phi_3 + 2\phi_2 - 4\phi_0 + k^2 h^2 \phi_0 - 2 \frac{\partial \phi(x_0, y_0)}{\partial x} h = O(h^3) \qquad (10)$$

Therefore

$$\frac{\partial \phi(x_0, y_0)}{\partial x} = \frac{1}{2h}(\phi_1 + \phi_3 + 2\phi_2 - 4\phi_0 + k^2 h^2 \phi_0) + O(h^2) \qquad (11)$$

Substitute formula (11) into formula (3) (noticed $n = -x$), obtain:

$$\frac{\partial \phi(x_0, y_0)}{\partial x} - jk\phi(x_0, y_0)$$
$$= \frac{1}{2h}(\phi_1 + \phi_3 + 2\phi_2 - 4\phi_0 + k^2 h^2 \phi_0)$$
$$- jk\phi(x_0, y_0) + O(h^2) \qquad (12)$$

So the new scheme of 1st Mur left boundary points except corner points is:

$$\phi_1 + \phi_3 + 2\phi_2 - 4\phi_0 + k^2 h^2 \phi_0 - 2jkh\phi_0 = 0 \qquad (13)$$

Comparing to formula (4), the formula (13) has second-order numerical accuracy obviously. So if using this scheme to discrete 1st Mur boundary, it can obtain better accuracy than that of formula (4) theoretically. Same procedure can be used to derivate the scheme of 1st Mur boundary condition at other nodes of the boundary, and the same result like (13) can be obtained.

At the corner point of computation domain in Fig.2 (b), the above procedure can be used similarly. Considering that the corner point has two different normal direction, a weighted absorbing boundary condition, which combine 1st Mur boundary at both normal direction is constructed here:

$$\alpha \left( \frac{\partial \phi_0}{\partial x} - jkE_{z0} \right) + (1 - \alpha) \left( \frac{\partial \phi_0}{\partial y} - jk\phi_0 \right) = 0 \qquad (14)$$

$$\phi_1 = \phi(x_0, y_0) + \frac{\partial \phi(x_0, y_0)}{\partial x} h + \frac{1}{2!} \frac{\partial^2 \phi(x_0, y_0)}{\partial x^2} h^2 + O(h^3) \qquad (15)$$

$$\frac{\partial \phi(x_0, y_0)}{\partial x} = \frac{1}{h}(\phi_1 - \phi(x_0, y_0) - \frac{1}{2!} \frac{\partial^2 \phi(x_0, y_0)}{\partial x^2} h^2) + O(h^2) \qquad (16)$$

$$\phi_2 = \phi(x_0, y_0) + \frac{\partial \phi(x_0, y_0)}{\partial y} h + \frac{1}{2!} \frac{\partial^2 \phi(x_0, y_0)}{\partial y^2} h^2 + O(h^3) \qquad (17)$$

$$\frac{\partial \phi(x_0, y_0)}{\partial y} = \frac{1}{h}(\phi_2 - \phi(x_0, y_0) - \frac{1}{2!} \frac{\partial^2 \phi(x_0, y_0)}{\partial y^2} h^2) + O(h^2) \qquad (18)$$

Substitute formula (16), (18) into formula (14), gets:

$$\alpha \left( \frac{\partial \phi_0}{\partial x} - jk\phi_0 \right) + (1 - \alpha) \left( \frac{\partial \phi_0}{\partial y} - jk\phi_0 \right)$$
$$= \frac{\alpha}{h} \left( \phi_2 - \phi_0 - \frac{1}{2!} \frac{\partial^2 \phi_0}{\partial x^2} h^2 - jkh\phi_0 \right)$$
$$+ \frac{1-\alpha}{h} \left( \phi_1 - \phi_0 - \frac{1}{2!} \frac{\partial^2 \phi_0}{\partial y^2} h^2 - jkh\phi_0 \right)$$
$$+ O(h^2) \qquad (19)$$

In order to eliminate the second order terms by using (2), select $\alpha = 0.5$, then formula (19) becomes:

$$\alpha \left( \frac{\partial \phi_0}{\partial x} - jk\phi_0 \right) + (1 - \alpha) \left( \frac{\partial \phi_0}{\partial y} - jk\phi_0 \right)$$
$$= \frac{1}{2h} (\phi_1 + \phi_2 - 2\phi_0 + 0.5h^2 k\phi_0$$
$$- 2jkh\phi_0)$$
$$+ O(h^2) \qquad (20)$$

So the new scheme at this corner points is:

$$\phi_1 + \phi_2 - 2\phi_0 + 0.5k^2h^2\phi_0 - 2jk\phi_0 h = 0 \qquad (21)$$

This formula holds at both four corner points.

The formula (13) and (21) are the new discretized scheme of 1st Mur's absorbing boundary condition. Comparing to formula (4), they have second-order numerical accuracy.

### III. NUMERICAL RESULTS

#### A. Example 1: square cylinder scattering problem

A scattering problem of PEC square cylinder with infinite length (Fig. 1) which discussed above is computed here to verify the efficiency of the new scheme of Mur boundary condition derived above. Let $\lambda$ be the wavelength and the mesh step h is $1/30\lambda$, the side length of square cylinder is $2/5\ \lambda$. The width of square domain along x-axis is $6/5\ \lambda$. Considering $TM_z$ plane wave incidence along the x-axis, so only z direction electric field existed, and this problem can be expressed just like equations (1), (2), (3). At the truncated boundary, the traditional 1st Mur boundary discretized by formula (4) and (13), (21) respectively, noticed that when using (4), the corner is handled with mean absorption field conditions [6]. The numerical result of absolute value of total field with the new scheme of 1st Mur ABC is shown in Fig. 3.

In order to verify the effect of the new scheme, a finer mesh with step $h_F = 0.5h$ is used to calculate the same problem with the traditional 1stMur boundary condition, and its result is considered as reference solution. Comparing with the reference solution, the relative errors of numerical solutions computed by tradition scheme (formula (4)) and new scheme (formula (13) and (21)) of 1st Mur ABC respectively are presented in Fig.4.

Table I lists the mean relative error, maximum relative error, mean square error and iteration numbers of both scheme.

According to numerical results of Figure 4 and table I, formula (13) and (21), the new scheme of discretized equations for Mur ABC, can obtained more accurate numerical solution.

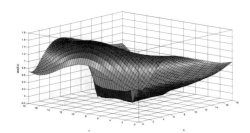

Figure 3 Numerical solution of total field for example 1

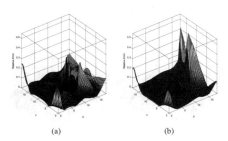

(a)                    (b)

Figure 4 Relative errors comparing to the reference solution. (a) New scheme of 1st Mur ABC. (b) Traditional 1st Mur ABC

TABLE I

RESULT OF EXMAPLE 1

| | Mean Relative Error | Maximum Relative Error | Mean Square Error | Iteration Number |
|---|---|---|---|---|
| Traditional Mur | 0.0576 | 0.5506 | 0.1175 | 1106 |
| New Scheme | 0.0480 | 0.3093 | 0.0749 | 965 |

#### B. Example 2: Two circle cylinders scattering problem

Another scattering problem of two PEC circle cylinders with infinite length is analyzed here. Fig. 7 is the cross section of the scattering area contains the cylinders. Such as the example I above, let mesh step $h = 1/30\ \lambda$, and the diameter of two cylinders is $10h$. The distance between both cylinders is $18h$. Two schemes of 1st Mur's ABC, i.e. formula (4), (13) and (21) are used to solve the scattering problem. The reference solution is solved with finer mesh $h_F = 1/60\lambda$. Here, the Compatible Sub-gridding Method [6] is used to approximate the circular boundary. The total field numerical solutions shown in Fig. 5, Comparing with the reference solution, the relative errors are shown in Fig.6. The mean relative errors, the maximum errors, the mean square errors and the iteration numbers for both scheme are presented in Table II.

The numerical results of Figure 5 and table II also validate that using the formula (13), (21) at the boundary can obtained more accurate numerical solution than traditional scheme.

## IV. CONCLUSION

In order to improve the accuracy of numerical solution for scattering problem, the new discretized equations of Mur's ABC, i.e. formula (13) and (21), are presented in this paper. Theoretically, the new discretized equations are of second-order numerical accuracy. Combining the new equations with the normal 5 points format of Helmholtz equation, two numerical examples of scattering problems are solved. Comparing to traditional discretized equations of Mur's ABC which is first-order numerical accuracy, the results show that the new scheme can significantly improve the accuracy of numerical solution.

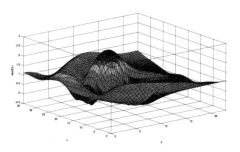

Figure 5. The numerical solution of total field for example 2

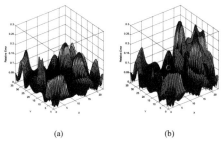

(a)                             (b)

Figure 6. .Relative errors comparing to the reference solution. (a) New scheme of 1st Mur. (b) Traditional 1st Mur

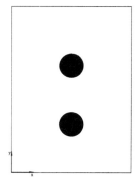

Figure 7. The cross sections of the scattering area for example 2

TABLE II

RESULT OF EXMAPLE 2

|  | Mean Relative Error | Maximum Relative Error | Mean Square Error | Iteration Numbers |
|---|---|---|---|---|
| Traditional Mur | 0.0675 | 0.2838 | 0.0867 | 23697 |
| New Scheme | 0.0440 | 0.2012 | 0.0571 | 21663 |

## REFERENCES

[1] G.Mur, "Absorbing Boundary Conditions for the Finite-Difference Approximation of the Time-Domain Electromagnetic-Field Equations," IEEE Trans. Electromagn. Compat. 23,377 (1981)1).

[2] K. S. Yee "Numerical solution of initial boundary value problems involving Maxwell's equations in isotropic media," IEEE Trans. Antennas Propagat., vol. AP-14, pp.302 -307 1966

[3] B.Engquist a d A. Majda, "Absorbing boundary conditions for the numerical simulation of waves," Math. Comp., vol. 31, pp.629 -651 1977

[4] Rui Fan, Lianyou Sun and Wei Hong, "Compatible sub-gridding and domain decomposition method applied to the slotted waveguide,"ICMMT, 2012

[5] J. W. Nehrbass and R. Lee, "Optimal Finite-Difference Sub-Gridding Techniques Applied to the Helmholtz Equation," IEEE Trans. ON Microwave Theory and Techniques, vol. 48, pp. 976-984, June 2000.

[6] Y. B. Li, L. Y. Sun and W. Hong, "Helmholtz equation sub-grid method for multi-transition region," National Microwave Meeting, Qingdao China, pp. 1150-1152, Jul 2011.

# Full-Wave Analysis of Multiport Microstrip Circuits by Efficient Evaluation of Multilayered Green's Functions in Spatial Domain

Zhe Song [#] and Yan Zhang

[#] *State Key Lab of Millimeter Waves, Southeast University*
*Nanjing, China*
[#]zhesong@emfield.org

*Abstract* — In this paper, a full wave analysis of multiport microstrip circuits is presented, which is based on the efficient computation of spatial domain dyadic Green's functions and accurate modeling of method of moments (MoM) with perfectly matched ports. With fast and accurate extracting all the surface wave and adequate leaky wave modes, the spatial domain dyadic Green's functions of layered structures can be evaluated by combining the discrete complex image method (DCIM) and the all modes method in near and non-near region, respectively. The Delta-Gap model is adopted to model the exciting port, while the others are perfectly matched by pre-determining the current exponentially decaying. The matrix pencil method (MPM) are adopted to extract the scattering parameters of a 4-port microstrip branch-line hybrid. Very good agreement between simulation and measurement results have been found from the provided example.

## I. INTRODUCTION

The fast development in microwave and millimeter wave integrated circuit design has been playing an important role in the modern communication technologies. Therefore, more and more attention has been paid to the rigorous, accurate and fast modeling and simulation methods of multilayered circuits [1]. The multilayered dyadic Green's functions in spatial domain and the mixed potential integral equation (MPIE) have been proved to be one of the most efficient method in full wave analysis of complicated planar multilayered structures with the MoM [2]-[5]. As is well-known, the multilayered Green's functions can be expressed in closed form in spectral domain, and then inversed to spatial domain by calculating the Sommerfeld integrals (SI) [2], [3]. In our recent work [6], [7], [10], the spatial domain Green's functions have been fast and accurately evaluated by means of the combination of the DCIM [8] and the all modes method [9] in near and non-near region, respectively, where the surface and leaky wave modes are extracted by means of a modified dichotomy method and the consecutive perturbation algorithm [10]. With the obtained dyadic Green's functions, the RWG-based MoM can be well constructed with adopting the Delta-Gap voltage excitation model [12].

In this paper, a full wave analysis of multiport microstrip circuits is presented, which is based on the efficient computation of spatial domain dyadic Green's functions and accurate modeling of MoM with perfectly matched ports. The multilayered Green's functions involved in this paper are calculated by means of the combination of the DCIM and the all modes method that have been developed recently. To validate the correctness and efficiency of the presented method, a 4-port branch-line hybrid coupled line filter is simulated, fabricated and measured. Very good agreement between simulation and measurement results have been found from the provided example.

## II. EVALUATION OF GREEN'S FUNCTIONS

With equivalent expression of the transmission line Green's functions (TLGF) in fractional forms, all the components of the dyadic Green's functions can be rewritten in a convenient way in spectral domain, as is shown in [6]. By using the polynomial denominator of the Green's functions, all the surface wave modes for lossless case can be accurately extracted by a modified dichotomy method [6]. Furthermore, for the layered microstrip structures with moderate thickness and loss, not only all the surface wave but also adequate leaky wave modes have to be extracted to ensure the smoothness of the kernel in the SI. By applying the consecutive perturbation algorithm proposed in [6] and [10], all the necessary modes can be traced and their corresponding residues can be analytically calculated simultaneously. For convenience, the computation method of the spatial domain Green's functions of layered medium is summarized as follows [10]:

a) In near field region, $\rho \leq 0.05\lambda$ DCIM [8], [14]:

$$G(\rho) = \frac{A}{4\pi}\left(G_{Q-S} + G_{SWP} + G_{CIM}\right) \tag{1}$$

$$G_{Q-S} = G\big|_{\rho \to \infty} \cdot \frac{e^{-jk_0\rho}}{\rho} \tag{2}$$

$$G_{SWP} = -2\pi j \sum_{i}^{N_{SWP}} \text{Res}_i H_0^{(2)}\left(k_{\rho i}\rho\right)k_{\rho i} \tag{3}$$

$$G_{CIM} = \sum_{i=1}^{N_c} a_i \frac{e^{-jk_0 r_i}}{r_i} \tag{4}$$

where $\rho$ is the projection of the distance between the source and field point on *x-o-y* plane, $A$ is a coefficient depending on the medium, $G_{Q-S}$, $G_{SWP}$ and $G_{CIM}$ stand for the quasi-static term, the surface wave term and the complex image term, respectively. $\text{Res}_i$ and $H_0^{(2)}(\bullet)$ stand for the residue of the *i-th* surface wave mode and the zeroth-order Hankel function of the second kind. $a_i$ and $r_i$ are the amplitudes and complex distances of the image terms.

978-7-5641-4279-7

b) In non-near field region, $\rho > 0.05\lambda$ All modes method:

$$\int_{SIP}(\cdot)dk_\rho = \int_\Gamma(\cdot)dk_\rho - 2\pi j \cdot (\sum_i^{N_{SWP}} R_{k_{\rho i}} + \sum_j^{N_{LWP}} R_{k_{\rho j}}) \quad (5)$$

$$\int_\Gamma(\cdot)dk_\rho = \int_\Gamma[\tilde{G}^+(k_\rho) + \tilde{G}^-(k_\rho)] \cdot H_n^{(2)}(k_\rho\rho)k_\rho dk_\rho \quad (6)$$

where $\Gamma$ stands for the deformed branch cut as depicted in [5]. $\tilde{G}^+$ and $\tilde{G}^-$ stand for the spectral-domain Green's functions on the top and bottom Riemann sheet, respectively. $N_{SWP}$ and $N_{LWP}$ are the number of surface and leaky wave modes, respectively. All the extracted modes involved in (5) can be fast and accurately extracted by the consecutive perturbation algorithm as is proposed in [6] and [10].

## III. MPIE FORMULATION AND MoM

The EFIE governing the total current density can be established by enforcing the boundary condition that revokes the vanishing of the total tangential electric field on the conductor surface [4]. However, to avoid the two-dimensional infinite Sommerfeld integrals with highly oscillating, slowly decaying and hyper-singular kernel involved in the EFIE, the MPIE has been preferred and widely used in layered structures, which is composed of vector and scalar potentials with weakly singular kernels [13], [17]. The MPIE can be formulated as below:

$$\bar{E}_t^{imp} = \left[ \begin{array}{c} j\omega\mu_0 \langle \bar{\bar{G}}^A(\bar{r}\,|\,\bar{r}');\bar{J}(\bar{r}') \rangle \\ -\dfrac{1}{j\omega\varepsilon_0} \nabla \left( \langle G^\Phi(\bar{r}\,|\,\bar{r}'), \nabla_S' \cdot \bar{J}(\bar{r}') \rangle \right) \end{array} \right]_t \quad (7)$$

where $\bar{E}_t^{imp} = V_p\delta(\bar{r} - \bar{r}_p)$ stands for the impressed electric field, $\bar{r}_p$ is the location of the port. $\bar{\bar{G}}^A(\cdot)$ and $G^\Phi(\cdot)$ are the dyadic and scalar Green's functions of the vector and scalar potential, respectively.

After obtaining the Green's functions in spatial domain, the MoM can convert the MPIE into a matrix equation. The triangular patches and RWG basis functions are adopted in this paper for the sake of modeling the arbitrarily shaped geometries. With the Galerkin's procedure applying to (7), the integral equation can be derived as [17]:

$$-j\omega\varepsilon_0 \int_{T_m} \bar{E}_t^{imp}(\bar{r}) \cdot \bar{f}_m(\bar{r})ds$$

$$= k_0^2 \sum_{n=1}^N I_n \int_{T_m} \int_{T_n} \bar{f}_n(\bar{r}') \cdot \bar{\bar{G}}^A(\bar{r}\,|\,\bar{r}') \cdot \bar{f}_m(\bar{r})ds'ds \quad (8)$$

$$+ \sum_{n=1}^N I_n \int_{T_m} \nabla \left( \int_{T_n} G^\Phi(\bar{r}\,|\,\bar{r}')\left(\nabla_S' \cdot \bar{f}_n(\bar{r}')\right)ds' \right) \cdot \bar{f}_m(\bar{r})ds$$

where $\bar{f}_n$ and $\bar{f}_m$ stand for the RWG basis and weighting functions, respectively. $T_n$ and $T_m$ are the triangular pairs containing the source ($\bar{r}'$) and field ($\bar{r}$) point, respectively. By using the Green's identity and the numerical Gaussian integral over triangular meshes, the integral equation (8) can be deduced as a matrix equation. As is proposed in [12], the delta-gap voltage model is adopted to excite each physical

port, which introduces half-RWG subsections to approximate the induced current from the mathematical point of view. Therefore, the MoM matrix equation can be expressed as [12]:

$$\begin{bmatrix} Z^{ff} & Z^{fh} \\ Z^{hf} & Z^{hh} \end{bmatrix} \cdot \begin{bmatrix} I^f \\ I^h \end{bmatrix} = \begin{bmatrix} 0 \\ V^{inc} \end{bmatrix} \quad (9)$$

where the superscripts $f$ and $h$ refer to full- and half-RWG subsections, respectively. In detail, the matrix element involved in (9) can be calculated as [17]:

$$Z_{mn} = k_0^2 \int_{T_m} \int_{T_n} \bar{f}_n(\bar{r}') \cdot \underline{G}^A(\bar{r}\,|\,\bar{r}') \cdot \bar{f}_m(\bar{r})ds'ds$$

$$- \int_{T_m} \int_{T_n} \left(\nabla_S' \cdot \bar{f}_n(\bar{r}')\right) G^\Phi(\bar{r}\,|\,\bar{r}')\left(\nabla_S \cdot \bar{f}_m(\bar{r})\right)ds'ds \quad (10)$$

Meanwhile, the right hand vector in (9) can be calculated by integration of the impressed excitation electric field over the half-RWG meshes, as is shown below:

$$V^{inc} = -j\omega\varepsilon_0 \int_{T_p^-} V_p \delta(\bar{r} - \bar{r}_p) \cdot \hat{n}_p \cdot \frac{l_p}{2A_p^-} \bar{\rho}_p^- ds \quad (11)$$

$$= j\omega\varepsilon_0 \cdot V_p$$

where $l_p$ is the length of the triangular edge on the port, $A_p^-$ is area of triangle $T_p^-$, $\bar{\rho}_p^-$ is a position vector towards the free vertex of $T_p^-$.

For multiport circuits, all the non-exciting ports should be perfectly matched by enforcing the surface current decaying from the reference plane to the port end exponentially. In this paper, a modified method based on [18] is used to realize the perfectly matching. In detail, the propagation constant on microstrip feed line is firstly determined from the surface current distribution, and then, the reference planes can be selected at least 1 "wavelength" away from the port ends. In this section of the feed line, the normalized current can be pre-determined by artificially enforced as an exponential decaying form. It is not difficult to solve the modified MoM matrix equation when these current coefficients on the matching feedline are known in advance.

As is well-known, the calculation of matrix element will encounter singularities when the field and source points are very close to each other. In this paper, the singular parts of the matrix element are treated with the method derived in [15]. To observe the recognizable standing-wave feature on the feed line, the reference planes should be selected away from not only the discontinuities but also the exciting ports [12], [14], [17]. The matrix pencil method (MPM) is adopted in this paper. After the fitting operation, the current distribution on feed lines can be written as [16]:

$$I(z) \approx \sum_{i=1}^N p_i \exp(\gamma_i z)$$

$$= \sum_{i=1}^N p_i \exp\left[(\alpha_i + j\beta_i)z\right], \quad z > 0 \quad (12)$$

where $p_i$ is the amplitude of the $i$th mode. $\alpha_i$ and $\beta_i$ stand for the propagation constant of the $i$th mode. The reference plane is selected at $z = 0$. From the physical point of view, the first two terms, $(p_1\ \alpha_1\ \beta_1)$ and $(p_2\ \alpha_2\ \beta_2)$, are just the incident and reflected wave of the dominant mode [14]. The $S_{11}$ can be easily obtained from them. By applying this fitting algorithm to each port, the scattering matrix can be finally obtained.

## IV. NUMERICAL EXAMPLES

To verify the method involved in this paper, a microstrip branch-line hybrid is simulated, fabricated and measured, as is shown in Fig. 1. The relative permittivity of the substrate is 2.2 and the thickness is 0.508mm (Rogers 5880). The detailed configuration is provided in Fig.1(a), while the photos of the hybrid is shown in Fig.1(b). In order to facilitate the meshing procedure with triangular patches, the oblique cutting is used on the feed lines.

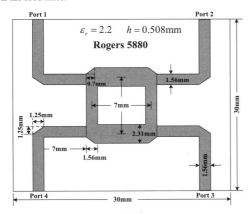

(a)

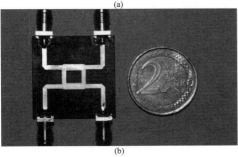

(b)

Fig 1 Microstrip branch-line hybrid.

(a) configuration and material. (b) photo of the filter

As is mentioned above, all the others should be perfectly matched except the exciting port. Both the simulation and measurement results are plotted in Fig. 2. From this figure, very good agreement can be found in the waveform of the S-parameters, however, a frequency offset can also be observed. This can be explained by the unstable constitutive parameters of the substrate in this high frequency band.

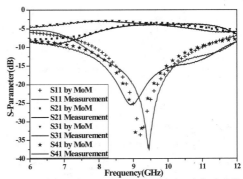

Fig 2 Simulation and Measurement results of the branch-line hybrid

## V. CONCLUSION

In this paper, a full wave analysis of microstrip multiport circuits is presented, which is based on the efficient computation of Green's functions and the accurate modeling of method of moments. With extraction of all the surface wave modes and adequate leaky wave modes, the spatial domain Green's functions of layered medium structure can be obtained by combining the DCIM and the all modes method in near and non-near regions, respectively. The Delta-Gap model is adopted to describe the excitation ports, while all the other non-exciting ports are perfectly matched by artificially enforcing the current distribution exponentially decaying on the feed line. To calculate the S-parameters, the MPM is used to extract the incident and reflected wave of the dominant mode. A 4-port branch-line hybrid is simulated, fabricated and measured to verify the method involved in this paper. Very good agreement between simulation and measurement results have been observed from the example.

## VI. ACKNOWLEDGEMENT

This work was supported in part by the National Basic Research Program of China (No. 2009CB320203, and No. 2010CB327400) and in part by the National Nonprofit Industry Specific Research Program of China (No. 200910041-2).

## REFERENCES

[1] T. Itoh, Ed., *Numerical Techniques for Microwave and Millimeter-Wave Passive Structures*, New York: Wiley, 1989, ch,3.

[2] R. E. Collin, *Field Theory of Guided Waves*, New York: McGraw-Hill, 1960.

[3] W. C. Chew, *Waves and Fields in Inhomogeneous Media*, ser. Electromagn. Waves Piscataway, NJ: IEEE Press, 1995.

[4] R. F. Harrington, *Field computation by Moment Methods*. Melbourne, FL: Krieger, 1983.

[5] D. G. Fang, *Antenna Theory and Microstrip Antennas*. Beijing: Science Press, 2006.

[6] Z. Song, H. -X. Zhou, J. Hu, W. –D. Li and W. Hong, "Accurate Location of All Surface Wave Modes for Green's Functions of a Layered Medium by Consecutive Perturbations," *SCIENCE CHINA Information Sciences*. 2010;53(11):2363-2376.

[7] Z. Song, et al., "A method of locating leaky wave poles of spectral Green's functions for a layered medium by consecutive frequency perturbation," in *Electrical Design of Advanced Packaging & Systems Symposium*, (EDAPS). *IEEE*, 2009

[8] D. G. Fang, J. J. Yang, and G. Y. Delisle. "Discrete image theory for horizontal electric dipoles in a multilayered medium above a conducting ground plane," *IEE Proc. H*, pp.135 (5): 297-303, 1988.

[9] B. Wu and L. Tsang, "Fast computation of layered medium of Green's functions of multilayers and lossy media using fast all-modes method and numerical modified steepest descent path method," *IEEE Trans. Microwave Theory Tech.*, vol. 56, pp. 1446-1454, 2008.

[10] Song Z, Zhou H-X, Zheng K-L et. al, Accurate Evaluation of Green's Functions for Lossy Layered Medium by Fast Locating Surface and Leaky Wave Modes [J]. *IEEE Antennas and Propagation Magazine*, vol.55, no.1, pp.92-102.

[11] S. M. Rao, D. R. Wilton, A. W. Glisson., "Electromagnetic scattering by surfaces of arbitrary shape," *IEEE Trans. Antennas Propagat.*, vol. 30, pp. 409-418, 1982.

[12] G. V. Eleftheriades and J. R. Mosig, "On the network characterization of planar passive circuits using the method of moments," *IEEE. Trans. Microwave Theory Tech.*, vol. 44, pp. 438-445, 1996.

[13] K. A. Michalski and J. R. Mosig, "Multilayered media Greens functions in integral equation formulations," *IEEE. Trans. Microwave Theory Tech.*, vol. 45, pp. 508-519, 1997.

[14] F. Ling. Fast electromagnetic modeling of multilayer microstrip antennas and circuits. Ph. D. Thesis in Elect. Eng., Illinois Univ. at Urbana-Champaign, 2000.

[15] D. R. Wilton, S. M. Rao, A. W. Glisson, D. H. Schaubert, et al. "Potential integrals for uniform and linear source distributions on polygonal and polyhedral domains," *IEEE Trans. Antennas Propagat.*, vol. 32, pp. 276-281, 1984.

[16] Y. Hua and T. K. Sarkar, "Generalized pencil-of-function method for extracting poles of an EM system from its transient response," *IEEE Trans. Antennas Propagat.*, vol. 37, pp. 229-234, 1989.

[17] M.-J. Tsai, F. D. Flaviis, O. Fordham N. G. Alexopoulos, "Modeling planar arbitrarily shaped microstrip elements in multilayered media," *IEEE Trans. Microwave Theory Tech.*, vol. 45, pp. 330-337, 1997.

[18] E. K. L. Yeung, J. C. Beal and Y. M. M. Antar, "Multilayer microstrip structure analysis with matched load simulation, " *IEEE Trans. Microwave Theory Tech.*, vol. 43, pp. 143-149, 1995.

# Analysis of the Antenna in Proximity of Human Body Base on the Dual-Grid FDTD Method

H. S. Zhang, K. Xiao, L. Qiu, H. Y. Qi, L. F. Ye, S. L. Chai

College of Electronic Science and Engineering
National University of Defense Technology
Changsha, 410073, China

*Abstract*- Dual-grid finite-difference time-domain(DG-FDTD) method is suitable for electromagnetic simulation of multi-scale problems, this paper validate the accuracy of DG-FDTD through a simple simulation. An accurate human model was built based on Zubal human digital phantom, which was used to analyze the electrical performance of the antenna on human body. Simulation results show that human body had a significant influence on the antenna due to the high relative permittivity of the body, the closer from human body, the greater impact on antenna. Because of the shielding of the human body, the antenna radiation power on the backside of human body is 5dB smaller than which in front of human body.

## I. INTRODUCTION

With the development of body-centric wireless communications, more and more communications systems contain antennas in proximity human body, such as mobile phone, Google Glasses and wearable antennas. Body-centric wireless communications will become one of the main ways of communication for end-user in the future [1-3]. It has abundant applications in smart home, personal healthcare, space exploration and military. In these communication systems, wearable antenna's performance will play a key role, but since the antenna is located on the body surface or inside the body, it will be affected by human body inevitably. From the electromagnetic point of view, the human body can be consider as a highly inhomogeneous, and lossy dielectric object, so the effect of human body must be taken into account in antenna design [4].

There are several difficulties in the co-simulation of body with antenna. The first is to establish an accurate model of the human body. up to now, there are several human digital models can be used for electromagnetic simulation[5,6], but some of them are integrated in electromagnetic software, and some of them are not free to get. This paper adopts Zubal Phantom data to establish the human model [7]. The second is to select the appropriate simulation method. Finite-difference time-domain (FDTD) method is a direct time-domain method to solve Maxwell's Equations[8,9], it is suitable to deal with the complex and highly inhomogeneous human body. However, a high spatial resolution is often required for the description of antenna's small geometrical features. Therefore, considering the size of human body and a uniform meshing for the whole FDTD domain, the human body will be a large oversampling area, and the computation time and memory requirements will increase greatly.

For such a multi-scale problems, there are some solutions based on FDTD. Such as sub-gridding FDTD scheme and FDTD combined with other numerical techniques. Sub-gridding FDTD consists of using different resolutions for the description of different areas, finer mesh for antenna area and sparser mesh for body area. Although these method reduces the computation time and memory requirements, they have some drawbacks such as late time instability and reflection from interface between the grids [10].

To overcome these disadvantages, the Dual-grid finite-difference time-domain (DG-FDTD) method is proposed by Raphael Gillard, which has been successfully used to analyze transmission channel between on-body devices [11-13]. This technique divides the original problem into two sub problems each characterized by a different mesh. Then the equivalent principle is used on the surface of two sub problems to reduce the reflection from interface. Hence , for the first, this paper adopts DG-FDTD method to analyze the wearable antenna with accurate human digital model.

## II. SIMULATION METHOD

Consider the problem presented in Fig. 1. As it is shown in this figure, the antenna is located in proximity of a dielectric block which represents its environment. Given the proximity of the environment, we must simulate the overall problem to take into account the coupling effects that may generate disturbances in the radiation patterns and input impedance. The simulation is divided into two FDTD simulations steps.

First step of the DG-FDTD. The first step's aims are characterizing the antenna accurately and recording its primary radiation by the use of an excitation recording surface placed around it. Thus the antenna is described using a fine mesh to describe its geometrical features. Absorbing boundary condition is adopted to simulate the infinite open space. The field components on the excitation recording surface are

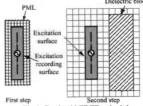

Fig.1. Dual-grid FDTD principle

978-7-5641-4279-7

recorded in a data file for every time step. These records will be used as the excitation of the next step of DG-FDTD. This step starts at $t_0$ and ends at $t_1$ when all the electromagnetic energy radiates outside the excitation recording surface.

Second step of the DG-FDTD. The second step's calculation region contains the whole problem in which both the antenna and the dielectric block locate. It aims at calculating the antenna radiation while taking into account the coupling effect between the antenna and the dielectric block. The whole region is described using a coarse mesh to reduce memory requirement. Absorbing boundary condition located outside the whole region is adopted to calculate the antenna radiation. The field components recorded in first step are used as excitation source in this step, which is realized by using the total-field/scattered-field boundary condition on the excitation surface to ensure that only scattered-field exist inside the excitation surface. This step starts at $t_0$ and ends at $t_2$ generally longer than $t_1$ to allow the electromagnetic energy radiates outside the whole region. The Mesh density of the second step is chosen in order to correctly represent the dielectric block, but it only provides a coarse description of the antennas. More specifically, it is not fine enough to simulate the antenna radiation in first step, but it is just enough to simulate the reflected signal received by the antenna in second step.

Since grid density and time step are inconsistent in the two steps, an interpolation process of the field component recorded in first step is need before it is used as excitation source in second step. It is a fine-to-coarse interpolation process, the spatial interpolation can be realized by weighted averaging the electromagnetic fields of neighboring grid, and the temporal interpolation can be realized by weighted averaging the electromagnetic fields of adjacent time.

Post processing. After the above two steps, we can calculate the reflection coefficient $S_{11}$ and the antenna radiation pattern. S11 is calculated as follows:

$$S_{11} = \frac{V_{tot}(f) - Z_0 I_{tot}(f)}{V_{tot}(f) + Z_0 I_{tot}(f)} \tag{1}$$

$$V_{tot}(t) = V_{step1}(t) + V_{step2}(t) \tag{2}$$

$$I_{tot}(t) = I_{step1}(t) + I_{step2}(t) \tag{3}$$

Where $V_{tot}(f)$ and $I_{tot}(f)$ are the Fourier transform of the corresponding voltage $V_{tot}(t)$ and current $I_{tot}(t)$ of the feeding point. The $V_{tot}(t)$ and $I_{tot}(t)$ are the sum of the simulation results of the two steps. Radiation pattern calculation is completed in the second step, which is the same as the conventional FDTD method.

Before adopting the DG-FDTD principle to analyze the antenna in proximity of human body, we validate its accuracy through a simple problem. As it is shown in Fig. 1, a dipole antenna is located in proximity of a dielectric block, the size of the dielectric block is 160mm×24mm×200mm, and the relative permittivity is 50. The length of dipole antenna is 136mm. The distance between antenna and dielectric block is 24mm.

Fig.2. Model for validating the DG-FDTD principle

This problem is simulated using two different FDTD principles. In DG-FDTD principle, the first step's volume size is 32×32×168, spatial step is dx=dy=dz=1mm, time step is dt=1.7ps, iterations is 10000. The second step's volume size is 50×22×60, spatial step is dx=dy=dz=4mm, time step is dt=6.8ps, iterations is 10000. A fine meshing FDTD simulation is used as the reference, it's fine spatial step is dx=dy=dz=1mm for the whole calculate domain, time step is dt=1.7ps, iterations is 30000.

Fig. 3 shows the reflection coefficients of the antenna in free space and place near the dielectric block two different FDTD principles. It is seen from Fig. 3 that, the resonant frequency of the dipole antenna is at 1.04GHz. Since the presence of the dielectric block, the resonant frequency

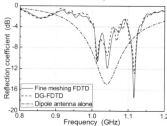

Fig.3. S11 parameters of Fine meshing FDTD and DG-FDTD

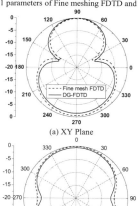

(a) XY Plane

(b) XZ Plane

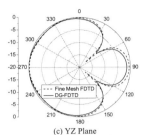

(c) YZ Plane

Fig.4. Normalized radiation patterns of FDTD and DG-FDTD at 1.1GHz

increases to 1.12GHz and the operation band becomes narrow. The reflection coefficients calculated by DG-FDTD shows excellent agreement with the reflection coefficients calculated by fine meshing FDTD principle. Fig. 4 shows the radiation patterns of the antenna calculated by the two FDTD principles agree with each other very well.

By the comparison of reflection coefficient and radiation patterns of two different FDTD principles, we can conclude that the DG-FDTD algorithm is accurate enough to analyze the antenna in proximity of human body.

### III. ANALYSIS OF THE ANTENNA NEARBY HUMAN BODY

In order to accurately predict the effect of the human body on antenna, the human model should be as accurate as possible. This paper adopts Zubal Phantom data [7], which is established base on computer tomography(CT) and magnetic resonance imaging (MRI) data of two living human males. It is segmented into 88 different organ types, the resolution of Zubal Phantom data is 3.6×3.6×3.6mm. Three-dimensional human model is shown in Fig.5. In order to show clearly, it draws only brain, eyes, bones, and stomach. The permittivity and conductivity [14] for the 31 tissues at 1GHz are shown in Table□.

| TABLE□ PERMITTIVITY AND CONDUCTIVITY OF HUMAN TISSUES AT 1GHZ | | |
|---|---|---|
| Tissue | $\varepsilon_r$ | $\sigma$ [S/m] |
| Skin Dry | 40.936 | 0.89977 |
| Brain White Matter | 38.577 | 0.6219 |
| spinal cord | 32.252 | 0.59997 |
| Bone Cancellous | 20.584 | 0.36395 |
| Muscle | 54.811 | 0.97819 |
| Lung Inflated | 21.825 | 0.47406 |
| Heart | 59.29 | 1.2836 |
| Liver | 46.401 | 0.89708 |
| Gall Bladder | 58.997 | 1.2883 |
| Kidney | 57.939 | 1.4495 |
| Cartilage | 42.317 | 0.82886 |
| Oesophagus | 64.797 | 1.2316 |
| Stomach | 64.797 | 1.2316 |
| Small Intestine | 58.872 | 2.2179 |
| Colon | 57.482 | 1.1274 |
| Pancreas | 59.47 | 1.0788 |
| Fat | 5.447 | 0.053502 |
| Blood | 61.065 | 1.5829 |
| Bone Marrow | 5.4854 | 0.042803 |
| Lymph | 59.47 | 1.0788 |
| Thyroid | 59.47 | 1.0788 |
| Trachea | 41.779 | 0.80232 |
| Spleen | 56.611 | 1.3227 |
| Testis | 60.259 | 1.2527 |
| Prostate | 60.259 | 1.2527 |
| Duodenum | 64.797 | 1.2316 |
| Mucous Membrane | 45.711 | 0.88181 |
| Cerebellum | 48.858 | 1.308 |
| Tongue | 55.017 | 0.97508 |
| Lens | 46.399 | 0.82431 |
| Tooth | 12.363 | 0.15566 |

The generally forms of wearable antenna are dipole antenna or microstrip patch antenna. In order to compare conveniently, we will use dipole antenna to analyze the effect of human body. As it is shown in Fig. 5(b), the antenna is located at human chest, 25mm or 36mm away from human body. In DG-FDTD algorithm, the first step's volume size is 20×20×96, spatial step is dx=dy=dz=1.8mm, time step is dt=3.1ps, iterations is 10000. The second step's volume size is 96×48×249, spatial step is dx=dy=dz=7.2mm, time step is dt=12.4ps, iterations is 15000.

Fig. 6 shows the reflection coefficients of the antenna. S11_36mm and S11_25mm denote the reflection coefficients

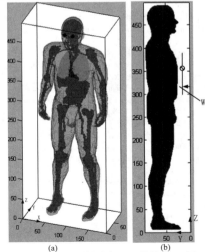

(a)        (b)

Figure. 5. (a)Zubal human digital phantom (b) Position of the antenna

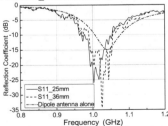

Figure.6. S11 of dipole antenna on human body

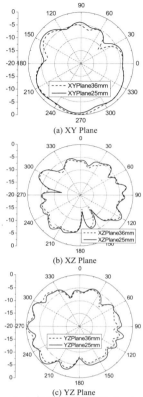

(a) XY Plane

(b) XZ Plane

(c) YZ Plane

Fig.7. Normalized radiation patterns of dipole antenna at 1GHz

with distances 36mm and 25mm away from human chest. It is seen from Fig. 6 that, the resonant frequency decreases and the operation band becomes broad. The closer to human body, the more resonance frequency shift. At a distance of 36mm and 25mm from the body, the reflection coefficients are both good. Antenna can radiate effectively.

Fig. 7 shows the radiation patterns of the antenna at 1GHz. It is seen from Fig. 7 that, there is little difference between the radiation patterns with different distances. But they are both impacted by the human body. It is seen from Fig. 7(a) and Fig. 7(c) that, The main direction of the radiation is in front of human chest. The radiated power behind the human body is 5dB smaller than which in front of chest. Omni-directional property is usually required in wearable antenna design, so it is possible to place more antennas on body surface. It is seen from Fig. 7(b) that, the radiation pattern is not symmetric about the YZ plane. This is because the human organs are not symmetric about the YZ plane in Zubal human model. This also means that the human body effects the antenna's radiation greatly. Accurate human model is essential in wearable antenna design.

## IV. CONCLUSION

This paper tests the accuracy of the DG-FDTD method, and establishes an accurate human model for FDTD simulation. The simulation results of antenna in proximity of human chest show that, human body has significant effect on the reflection coefficient and radiation pattern of antenna. The conclusions may be useful for wearable antenna design.

## REFERENCES

[1]  S.H. Peter,  H. Yang. Antenna and propagation for body-centric wireless communications[M]. Boston: Artech House，2006.

[2]  S. Zhu, L. Richard. Dual-Band Wearable Textile Antenna on an EBG Substrate [J]. IEEE  Transactions on Antennas and Propagation, 2009, 57(4), 926-935.

[3]  K. Timothy, F. Patrick, W. Andrew. Body-Worn E-Textile Antennas: The Good, the Low-Mass, and the Conformal[J]. IEEE Transactions on Antennas and Propagation,2009,57(4), 910-918.

[4]  S. Andrea, Z. Yan, H. Yang, et al. An Efficient FDTD Algorithm Based on the Equivalence Principle for Analyzing Onbody Antenna Performance[J]. IEEE Transactions on Antennas and Propagation, 2009, 57(4), 1006-1014.

[5]  Nat. Inst. Health Nat. Library Med. Board of Regents, "Electronic imaging: board of regents," Bethesda, MD, Tech. Rep. NIH 90-2197, 1990  http://www.brooks.af.mil/AFRL/HED/hedr/

[6]  K. Michael, B. Maurice, S. Sheldon, et al. Formulation and Characterization of Tissue Equivalent Liquids Used for RF Densitometry and Dosimetry Measurements [J]. IEEE Transactions on Microwave Theory and Techniques, 2004, 52(8), 2046-2056.

[7]  G. Zubal, C. Harrell, E. Smith, et al. Yale university school of medicine, New Haven, CT USA. http://noodle.med.yale.edu/

[8]  D.B. Ge, Y.B. Yan. The Finite-difference Time-domain Method for Electromagnetics [M]. Xi'an: Xi'an University of Electronic Science and Technology Publishing House, 2005.

[9]  A. Taflove, S. Hagness, Computetional electrody namics finite-difference time-domain method[M] Third edition. Boston: Artech House，2005.

[10]  M. Okniewski, Three-dimensional Sub-gridding Algorithm for FDTD[J]. IEEE Transactions on Antennas and Propagation, 1997, 45(3), 422-427

[11]  M. Celine, L. Renaud, G. Raphael. An Efficient Bilateral Dual-Grid FDTD Approach Applied to On-Body Transmission Analysis and Specific Absorption Rate Computation [J]. IEEE Transactions on Microwave Theory and Techniques, 2010,58(9), 2375-2382.

[12]  P. Romain, G. Raphael, L. Renaud, et al. Dual-Grid Finite-Difference Time Domain Method[J]. IEEE Microwave and wireless Components Letters, 2008, 18(10), 656-658.

[13]  M. Celine, A. Thierry, G. Raphael, et al. Analysis of the Transmission Between On-Body Devices Using the Bilateral Dual-Grid FDTD Technique [J]. IEEE Antennas and wireless Propagation Letters, 2010,9, 1073-1075.

[14]  G. Camelia, G. Sami. Compilation of the dielectric properties of body tissue at RF and microwave frequencies. http://niremf.ifac.cnr.it/tissprop/htmlclie/htmlclie.htm

# Design of Millimeter Wave Waveguide-Fed Omnidirectional Slotted Array Antenna

Mohsen Chaharmahali[1] & Narges Noori[2]

[1]Shahre Rey Branch of Islamic Azad University, Iran

[2]Department of Communication Technology, Cyberspace Research Institute, Tehran, Iran

*Abstract*-In this paper the design of a linear array of slots is presented at 38 GHz. The array is fabricated with a total number of 24 slots, 12 on each broad side wall of a WR22 rectangular waveguide, to form an omnidirectional radiation pattern. The offsets of the individual slots are calculated for two cases: when all slots are uniformly excited and when Dolph-Chebyshev synthesis method is used to achieve a radiation pattern with low sidelobe level.

## I. INTRODUCTION

Millimeter wave technology occupies the frequency spectrum from 30 GHz to 300 GHz with wavelength between one and ten millimeters. This technology has attracted a great deal of interest due to its unique features. The millimeter wave frequencies were primarily used in military applications and radio astronomy [1]-[3]. Now, this part of the spectrum is used in a wider range of applications such as medical imaging, collision avoidance radars, inter-vehicle communication, indoor cellular systems, personal area network, high-definition video streaming, point-to-multipoint communication links, multi-gigabit file transmission, surveillance systems and security [4]-[5].

Antennas occupy an important place in millimeter wave systems. Different types of planar, horn, dielectric and lens antennas are used in single antenna and array configurations to achieve the desired features such as gain, beamwidth and sidelobe level at millimeter wave frequencies [10].

Slotted array are popular antennas because of their simple geometry, mechanical strength, high efficiency, high gain and polarization purity. They are used in radar systems, communications and navigation applications. These antennas can provide excellent pattern control at millimetre wave frequencies. In [11], the design and development of a series fed 1 x 12 resonant slotted array antenna is presented. The slot dimensions, their spacing and positions are calculated for broadside radiation pattern.

In this paper, the design of a linear array of slots is presented at 38 GHz. Since an array of slots in one side of a waveguide does not radiate uniformly on both sides, two identical rows of 12 slots are fabricated on the broad side walls of a WR22 waveguide to form an omnidirectional radiation pattern. The offsets of the individual slots are calculated for two cases: when all slots are uniformly excited

and when Dolph-Chebyshev synthesis method is used to achieve a radiation pattern with low sidelobe level.

## II. THEORY

A waveguide-fed slotted array is usually fabricated by milling a set of rectangular shaped slots in a common wall of a rectangular waveguide. One advantage of a waveguide slotted array is that the radiating aperture and the feed network can be made from waveguides [12]. This makes the design easier since matching networks are not required. The transverse dimensions of the waveguide are chosen so that only the $TE_{10}$ mode can propagate.

The modal fields within a waveguide must necessarily be known to understand how to properly place the slots. Figure 1 shows a longitudinal slot displaced from the center line of an air-filled rectangular waveguide. This slot interrupts $x$ - directed current. The modal fields analysis shows that more displacement causes greater radiation into outer space. The lengths and offsets of the slots must be chosen so that the desired electric field intensity in amplitude and phase is achived in each slot. This will insure the specific radiation pattern. A linear array consists of $N$ slots is shown in Fig. 2. The slots are made by spacing their centers at electrical half wavelength intervals along the waveguide, where the successive slots are placed in the opposite sides of the centerline of the waveguide. The center of the last slot is located at a quarter-wavelength from the closed end of the waveguide. The electrical wavelength in the waveguide, $\lambda_g$, is obtained by:

$$\lambda_g = \left[ \left( \frac{1}{\lambda_0} \right)^2 - \left( \frac{1}{\lambda_c} \right)^2 \right]^{-\frac{1}{2}} \tag{1}$$

where $\lambda_0$ is the wavelength in free space and $\lambda_c$ is the cutoff wavelength. In the linear array of Fig. 2, the following formulas are used to calculate the slots length and displacement [12]:

$$\frac{Y_n^a}{G_0} = K_1 f_n \sin k l_n \frac{V_n^s}{V_n} \tag{2}$$

$$\frac{Y_n^a}{G_0} = \frac{K_2 f_n^2}{Z_n^a} \tag{3}$$

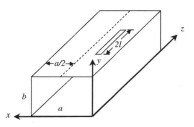

Figure 1. A longitudinal slot displaced from the center line of a rectangular waveguide.

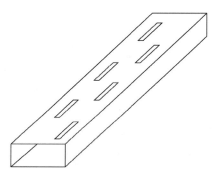

Figure 2. A linear array consists of $N$ slots.

where $V_n$, $V_n^s$, $Y_n^a$ and $l_n$ are the mode voltage, peak slot voltage, active admittance and length of the $n$ th slot, respectively, $G_0$ is the characteristic admitance of the equivalent transmission line and $k$ is the propagation constant of the air. Also,

$$K_1 = -j\left[\frac{8}{\pi^2 \eta G_0} \frac{(a/b)}{(\beta/k)}\right]^{\frac{1}{2}} \qquad (4)$$

$$K_2 = \frac{292(a/b)}{0.61\pi(\beta/k)} \qquad (5)$$

and

$$f_n = \frac{\cos\beta l_n - \cos k l_n}{\sin k l_n}\sin\frac{\pi x_n}{a} \qquad (6)$$

$$Z_n^a = Z_{nn} + Z_n^b \qquad (7)$$

in which

$$Z_n^b = \sum_{m=1}^{N}\frac{V_m^s \sin k l_m}{V_n^s \sin k l_n}Z_{nm} \quad \text{for } m \neq n \qquad (8)$$

$$Z_{nn} = \frac{K_2 f_n^2}{Y_n/G_0} \qquad (9)$$

where $\beta$ is the propagation constant in the waveguide, $Z_n^a$, $Z_{nn}$ and $Z_n^b$ are the active impedance, self-impedance and mutual impedance of the $n$ th equivalent loaded dipole, respectively, $Y_n$ is the isolated self-admittance of the $n$ th slot and $Z_{nm}$ (with $m \neq n$) is the mutual impedance between the $n$ th and $m$ th equivalent dipoles.

The linear array of $N$ slots, shown in Fig. 2, does not radiate uniformly on both sides. In order to have an omnidirectional radiation pattern, identical row of $N$ slots must be fabricated on the opposite broad side wall of the waveguide. The total number of slots, $2N$, is selected to satisfy desired gain and beamwidth.

### III. NUMERICAL SIMULATIONS AND RESULTS

Based on the theory presented in the previous section, the design of an omnidirectional array is presented at 38 GHz. The length of each slot is set to $\lambda_0/2$. The array is fabricated on a WR22 waveguide. A total number of 24 slots, 12 on each broad side wall, are positioned on the waveguide to form an omnidirectional radiation pattern. Two cases are considered to calculate offsets of the individual slots. In the first case, all slots are uniformly excited. In the second case, Dolph-Chebyshev synthesis method is used to achieve a radiation pattern with low sidelobe level. In the first case, all slots have an offset equal to 0.5 mm. In the second case, the resulted slot offsets are:

$$x_1 = x_{12} = x_{13} = x_{24} = 0.2976 \text{ mm}$$

$$x_2 = x_{11} = x_{14} = x_{23} = 0.3222 \text{ mm}$$

$$x_3 = x_{10} = x_{15} = x_{22} = 0.4515 \text{ mm}$$

$$x_4 = x_9 = x_{16} = x_{21} = 0.5725 \text{ mm}$$

$$x_5 = x_8 = x_{17} = x_{20} = 0.6676 \text{ mm}$$

$$x_6 = x_7 = x_{18} = x_{19} = 0.7201 \text{ mm}$$

In Fig. 3, we present the three-dimensional radiation pattern of the arrays to compare the omnidirectional radiation property of the 24-slot arrays with the case that only 12 slots are placed on one broad side wall of the waveguide. Furthermore, the azimuth radiation patterns of these arrays are shown in Fig. 4. The azimuth radiation pattern of the 12-slot array shows that this array does not uniformly radiate on both sides of the waveguide. However, both 24-slot arrays provide pretty omnidirectional radiation patterns. Also, Dolph-Chebyshev excitation leads to smaller gain variations over the full 360° azimuth.

The elevation radiation patterns of the 24-slot arrays are shown in Fig. 5. It can be seen that Dolph-Chebyshev synthesis method results in a 5 dB better sidelobe level than the uniform excitation. Figure 6 shows the VSWR of these arrays.

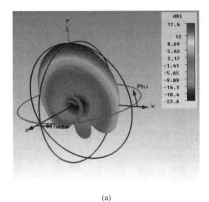

(a)

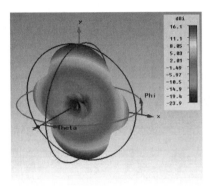

(b)

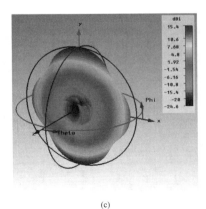

(c)

Figure 3. Three-dimensional radiation pattern of the arrays. (a) 12-slot array, (b) 24-slot array with uniform excitation, (c) 24-slot array with Dolph-Chebyshev excitation.

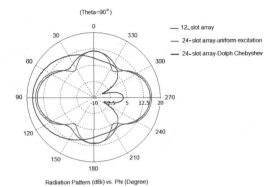

Figure 4. Azimuth radiation patterns of the arrays.

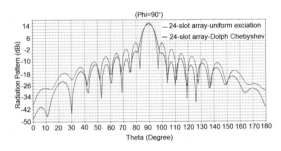

Figure 5. Azimuth radiation patterns of the 24-slot arrays.

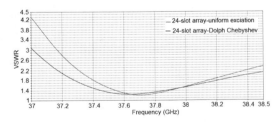

Figure 6. VSWR of the 24-slot arrays.

## IV. CONCLUSIONS

A 24-slot waveguide-fed omnidirectional array antenna has been designed at 38 GHz. Two different sets of slot offsets have been calculated for uniform and Dolph-Chebyshev excitations. The last set leads to a radiation pattern with a 5 dB better sidelobe level. It also results in smaller gain variations over the full 360° azimuth.

## REFERENCES

[1]  C. W. Tolbert, A. W. and C. O. Britt, "Phantom radar targets at millimeter radio wavelengths," *IRE Transactions on antennas and Propagation,* vol. 6, no. 4, pp. 380 – 384, 1958.

[2]  J. W. Meyer, "Radar astronomy at millimeter and submillimeter wavelengths," *Proceedings of the IEEE,* vol. 54, no. 4, pp. 484–492, 1966.

[3]  F. B. and E. K. Reedy, "Millimeter RADAR at Georgia Tech," in Proc. *S-MTT International Microwave Symposium Digest,* pp. 152, June 1974.

[4]  S. S. Ahmed, A. Schiessl, F. Gumbmann, M. Tiebout, S. Methfessel, L. Schmidt, "Advanced Microwave Imaging," *IEEE Microwave Magazine,* vol. 13, no. 6, pp. 26–43, 2012.

[5]  S. Futatsumori, A. Kohmura, N. Yonemoto, K. Kobayashi, Y. Okuno, "Small transmitting power and high sensitivity 76GHz millimeter-wave radar for obstacle detection and collision avoidance of civil helicopters," in Proc. *9th European Radar Conference,* pp. 178–181, Oct. 31-Nov. 2, 2012.

[6]  D. Gunton, "Millimeter Wave Technology in ITS," in Proc. *6th International Conference on ITS Telecommunications,* June 2006.

[7]  H. Ogawa, "Millimeter-wave wireless personal area network systems," in Proc. *IEEE Radio Frequency Integrated Circuits Symposium,* June 2006.

[8]  R. Appleby and R. N. Anderton, "Millimeter-Wave and Submillimeter-Wave Imaging for Security and Surveillance," *Proceedings of the IEEE,* vol. 95, no. 8, pp. 1683–1690, 2007.

[9]  D. Lockie, D. Peck, "High-data-rate millimeter-wave radios," *IEEE Microwave Magazine,* vol. 10, no. 5, pp. 75–83, 2009.

[10]  K. C. Huang and D. J. Edwards, *Millimetre Wave Antennas for Gigabit Wireless Communications,* John Wiley and Sons, 2008.

[11]  R. Kumar, P. K. Verma and M. Singh, "Design and Development of a 1 x 12 Series Fed Linear Slotted Array Antenna at 38 GHz," in Proc. *Applied Electromagnetics Conference,* pp. 1–3, Sep. 2009.

[12]  R. S. Elliott, *Antenna Theory and Design,* John Wiley and Sons, 2003.

# *FA-1(A)*

## October 25 (FRI) AM

## Room A

## A &P for Mobile Comm.

# Emerging Antennas for Modern Communication Systems

**J. W. Modelski**

Warsaw University of Technology, Institute of Radioelectronics

Nowowiejska str., 15/19,00-665 Warsaw, Poland

j.modelski@ire.pw.edu.pl;

This paper presents three types of antennas, which have been created and tested at the Warsaw University of Technology – antennas with photonic, ferroelectric and semiconductor elements.

Nowadays, a continuous progress in new generation services offered to the users of communication systems led to the need for very rapid development of both backbone and access networks. The strongest growth is visible for wireless systems, which offer mobility, flexibility, ease of installation and modernization of the network. Modern wireless devices: mobile phones, laptops, tablets etc., stimulate the development of services that combine simultaneous transmission of voice, data and multimedia content, resulting in the need to ensure the requirements for the transmission parameters for each of them. These requirements include: high capacity, high data rate, the security of transmitted information, system reliability, scalability and flexibility. To use the advantages of fiber-optic communication, while leaving the benefits of wireless transmission, such as flexibility and the ability to develop and modify the structure of the network, an idea of Wireless over Fiber (WoF) systems has been widely introduced. Our idea was designing and investigation of photonic antenna stations for bidirectional transmission in the last mile. Three different photonic antenna stations in the last mile WoF links have been tested. For construction of antenna stations commercially available optoelectronic devices have been used. The solutions differ from each other by the mean of ensuring separation between transmitting and receiving modes of operation. The separation is realized on the microwave side. All antenna stations have been designed, realized and measured in various kinds of measurements. The results prove that all propositions can be successfully applied to uplink and downlink IEEE 802.11b/g wireless LAN systems, employing WoF technique.

An smart or adaptive antennas are the most suitable for wireless communication, especially for 3G and 4G systems. The key property of the intelligent technology is the ability to respond automatically by changing an appropriate radiation pattern. The best solution will be the possibility for dynamic reconfiguring of the antenna aperture. Many solutions of the reconfigurable antennas have been described in the literature. The key elements of the antenna, which has been presented in this paper, are individually controlled SPIN diodes. This antenna shows the extensive functionalities. First of all, it can be used as the conventional frequency scanning antenna. It is easy to see, that the reconfigurable antenna can direct radiation beam to desire direction. The first additional possibility in comparison with the conventional waveguide slot antenna, is that the reconfigurable antenna can be used for operating at one frequency, but with generating two or more different radiation patterns at different moments. Changing the configuration of the reconfigurable elements between e.g. two configurations, there is a possibility to achieve two different radiation patterns for the same frequency. The second extending possibility of the presented reconfigurable antenna is that the antenna can operate at different frequencies with supporting radiation in the same direction.

Advances in several areas of materials science have led to a variety of new materials with strong potential applications to microwave and millimeter-wave components. The high tunability and low dielectric losses are only the desired properties of material which can be applied in the tunable micro- and mm-wave devices. A number of the device configurations are a promising solution to inexpensive steering. A new low-cost scan antenna concept (without phase shifters) has been presented. The substrate of the presented microstrip antenna has been made using a ceramic–polymer composite with modified ferroelectric powder Ba Sr TiO and an appropriate polymer (grains of the powder were sprayed into polymer with the use of a specific method). The ceramic–polymer composite was designed to change permittivity in response to an applied electric control field for antenna utilization. It allows changing the electrically phase constant of the propagation wave and in result - changes of the main beam direction. Currently, different compositions of the ceramic polymer with modified ferroelectric powder have been investigating.

978-7-5641-4279-7

# Tunable Antenna Impedance Matching for 4 G Mobile Communications

Peng Liu, Student Member, IEEE, and Andreas Springer, Member, IEEE

Institute for Communications Engineering and RF-Systems, Johannes Kepler University Linz
Altenberger Strasse 69, 4040 Linz, Austria
E-mail: {p.liu,a.springer}@nthfs.jku.at

*Abstract*—We analyze a Π-network for tunable antenna impedance matching in 4G mobile communications. Fundamental limits on designability and practical limits on implementation of the network are presented. The performance of the network is evaluated in terms of coverage (the area on Smith chart that can be matched) and the maximum achievable tuning bandwidth. We also derive the required component values and component $Q$ from coverage, bandwidth, as well as the efficiency of the network. We conclude that the Π-network (synthesized for the maximum achievable bandwidth) is able to cover E-UTRA Band 4 and 10 for VSWR up to 8 and all other bands for VSWR up to 10. To achieve that, in a 50 Ω RF environment, it requires a maximum capacitance within the range 13.64–2.7 pF, and a maximum inductance in the range 34.1–6.8 nH in frequency range 700–3500 MHz. If the power loss is limited to 0.5–0.6 dB, it requires $Q_L$ of 70 and $Q_C$ of 70–100.

*Index Terms*—Antenna tuning, 4G mobile communication, impedance matching, $Q$ factor, Π-network.

## I. INTRODUCTION

Mobile communication systems are evolving into the 4th generation (4G) which is targeting a downlink peak data rate of 1 Gbps (100 Mbit/s for high and 1 Gbit/s for low mobility) and an uplink peak data rate of 500 Mbps [1], [2]. To reliably achieve such high-speed communication, efficient transmission and reception of signals are required. One of the main components influencing transmission and reception is the mobile antenna. Because mobile devices operate in proximity to the human body, the antenna impedance is affected by the body and the hand that holds the device [4], [5]. A change in antenna impedance creates mismatch between the antenna and the RF front-end which significantly degrades the power efficiency of the radio link [7]. We have presented in [8] that, in receive antenna diversity systems, antenna mismatch can cause severe degradation of the system performance in circumstances of multiple diversity antennas being simultaneously mismatched.

To maintain link quality, an antenna tuning unit (ATU) is therefore used to dynamically match the antenna impedance to the RF front-end. ATUs typically use lumped Π- or L-networks, which are composed of tunable capacitors and fixed inductors, to achieve tunable antenna impedance matching. The topology, performance and tuning method have been studied in a variety of publications, e.g., [7], [9], [10]. However, none of these papers is focused on the 4G LTE application. This gives rise to the motivation of our work.

In this paper, we start from the 4G LTE application requirements. Then we analyze one of the most widely used matching networks — the low-pass Π-network, and present

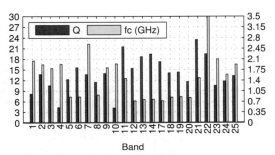

Fig. 1. Quality factor of E-UTRA FDD frequency bands. $f_c$ represents the center frequency of each band.

its performance in terms of coverage (the area on Smith chart that can be matched), the achievable tuning bandwidth and power efficiency. From these performance indicators, the required component values and component $Q$ (quality factor) are derived.

Different from papers which show the performance of matching networks at specific frequencies, we study the matching network in a normalized way and present results that can be de-normalized for different frequencies, bandwidths and application requirements. The results achieved will show how does the Π-network fit 4G LTE applications and also give guidelines on how to determine what tuning components are required.

## II. THE E-UTRA FREQUENCY BANDS

The spectrum allocation for 4G LTE (also known as E-UTRA, the Evolved Universal Terrestrial Radio Access) is defined in the 3GPP technical specification [3]. The E-UTRA band definition contains FDD (frequency division duplexing) and TDD (time division duplexing) bands.

From the matching network design point of view, we are concerned with the $Q$ of the bands, which is defined as

$$Q = \frac{\text{Center frequency of the band}}{\text{Bandwidth}}. \tag{1}$$

The band $Q$ is a suitable measure to assess the tunability requirements of a matching network.

The $Q$ of the FDD bands (Band 1 to 25) are shown in Fig. 1. We see that $Q$ varies from 4 on Band 4 and 10 both at around 1.9 GHz to 23 on Band 21 at around 1.5 GHz. It is

978-7-5641-4279-7

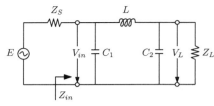

Fig. 2. Π-network

also interesting to notice that, except for Band 4 and 10, all the other FDD bands have a $Q$ no less than 8.

The $Q$ for TDD bands varies from 13 on Band 41 at around 2.6 GHz to 134 on Band 34, which has a very small bandwidth of 15 MHz, at around 2 GHz.

Despite the fact that some bands have an extremely large $Q$, we do not have to design extremely narrow band matching networks accordingly, because the primary goal of designing a matching network is to cover the band with sufficiently low return loss, rather than to suppress out-of-band transmission. On the contrary, we will see it is the small $Q$, which means large bandwidth, that makes designing of the matching network a challenging task. Therefore, we put more emphasis on small $Q$ and will show how $Q$, together with the antenna impedance, is related to the designability of the matching network.

III. The Π Impedance Matching Network

A. Topology

The objective of antenna matching is to match the antenna impedance to the RF front-end, which typically has an impedance of 50 Ω, to minimize reflection, i.e., to minimize VSWR (voltage standing wave ratio) which can go up to 10 : 1 [6].

The low-pass Π-network, shown in Fig. 2, is one of the most simple and widely used networks for antenna impedance matching [7], [10] due to its harmonic rejection capability and wide coverage [12]. Unlike L-networks, whose $Q$ is uniquely determined by the load, the Π-network has one more degree of freedom for the user to design the bandwidth of matching.

In theory, the Π-network can provide complete Smith chart coverage [10]. In practice, its coverage is limited by available component values, or tunability of the components. Furthermore, the achievable bandwidth of the matching network is determined by the load impedance. Now questions arise: For a given area, can the network match all the load impedances falling in it? Does it meet the bandwidth requirements? What components are needed? What efficiency can we expect? The following sections will answer these.

B. Fundamental Limitations

It is customary to analyze a matching network by terminating it with a real load. In this subsection we assume that in Fig. 2 $Z_S = R_S$ and $Z_L = R_L$.

The loaded $Q$ of the Π-network can be found to be

$$Q_L = \frac{1}{2} \left( \omega C_1 R_S + \omega C_2 R_L \right). \tag{2}$$

The necessary and sufficient conditions for designability of the Π-network are given in [12] as

$$Q_L \geq \begin{cases} \frac{1}{2} \sqrt{\dfrac{R_L}{R_S} - 1} & \text{for } R_L \geq R_S \\ \frac{1}{2} \sqrt{\dfrac{R_S}{R_L} - 1} & \text{for } R_S > R_L \end{cases}. \tag{3}$$

The fundamental limitations on the designability given in (3) indicate that the degree of mismatch, measured by $R_L/R_S$, determines the minimum achievable $Q$, or equivalently, the maximum achievable bandwidth. Qualitatively, the larger the degree of mismatch, the lower is the achievable bandwidth. From another viewpoint, once $Q$ is given, the maximum allowed mismatch is determined and thus the coverage of the network is determined. For bands with a small $Q$, the coverage realizable by the Π-network is correspondingly small.

C. Practical Limitations

In addition to the fundamental limitations, there are also practical limitations on the feasibility of the matching network. We highlight these aspects: (1) feasibility of the component values, (2) tunability of the tuning components, and (3) availability of the component $Q$ which are derived from the power efficiency of the network.

For tunable capacitors, the tunability is defined as $(C_{\max} : C_{\min})$, and for tunable inductors as $(L_{\max} : L_{\min})$. The maximum and minimum component values are determined by the goal of matching as well as the architecture and technology of tuning devices. [10] introduces some of the architectures for equivalently tunable inductance and capacitance.

The coverage of the Π-network (synthesized for the maximum achievable bandwidth) with respect to $b_C = \omega C Z_0$ ($C = \max(C_1, C_2)$) and $x_L = \omega L/Z_0$ is illustrated in Fig. 3. It is shown in Fig. 3a that to match the impedances located in the outer left part of the Smith chart larger $b_C$, thus larger $C$, is required and in Fig. 3b that to match the impedances located in the outer right part larger $x_L$, thus larger $L$, is needed. To cover the area defined by VSWR $\leq 10$, the required $b_{c,\max}$ and $x_{L,\max}$ are both 3, which correspond to $C_{\max}$ of 13.64–2.7 pF and $L_{\max}$ of 34.1–6.8 nH in the frequency range 700–3500 MHz accordingly.

D. Bandwidth

We measure the 6 dB bandwidth $BW_{6\,dB}$ of the complex-loaded matching network as defined by $|S_{11}| \leq -6$ dB. To allow direct comparison between the maximum achievable bandwidths and the E-UTRA bandwidth requirements, we compute the corresponding quality factor from

$$Q_{L,6dB} = \frac{f_C}{BW_{6dB}}, \tag{4}$$

where $f_C$ is center frequency of the 6 dB band.

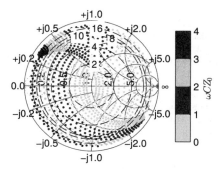

(a) Coverage vs. susceptance of capacitor

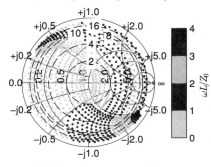

(b) Coverage vs. reactance of inductor

Fig. 3. Coverage of the Π-network (synthesized for maximum bandwidth) versus component values. The characteristic impedance $Z_0$ is used for normalization. The labeled concentric circles indicate different VSWR values.

As shown in Fig. 4, there is $Q_{L,6dB} < 4$ within the area defined by VSWR $\leq 8$ (except for a reasonably small area beyond this threshold) and $Q_{L,6dB} < 8$ within the VSWR $\leq 10$ area. Recall that Band 4 and 10 have $Q = 4$ and all the other bands have $Q \geq 8$.

### E. Power Efficiency

All practical components exhibit a certain amount of loss that is usually modeled with an equivalent series resistance $R$. The degree of loss is measured by the components $Q$, which is defined for a capacitor as $Q_C = 1/(\omega RC)$, and for an inductor as $Q_L = \omega L/R$.

For a given component $Q$, we can express the equivalent component value in complex form which takes into account the loss of a capacitor as $C_e \approx C(1 - j/Q_C)$ and the loss of an inductor as $L_e = L(1 - j/Q_L)$.

To evaluate the power efficiency of the matching network, we design the network neglecting loss first and then simply replace the ideal components with practical components by substituting their equivalent complex values. Although this

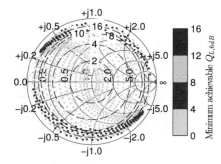

Fig. 4. Minimum achievable $Q_{L,6dB}$ (corresponding to the maximum achievable bandwidth) of the Π-network. The bandwidth is defined by $|S_{11}| \leq -6$ dB. The labeled concentric circles indicate different VSWR values.

creates some mismatch, it won't affect the evaluation of the power efficiency of the matching network itself. The power efficiency of the Π-network is defined as

$$\eta = \frac{P_L}{P_{in}} = \frac{|V_L|^2 G_L}{|V_{in}|^2 G_{in}}, \tag{5}$$

where $P_L$ and $P_{in}$ denote the power delivered to the load (the antenna) and the input power which is the power supplied by the source to the matching network. $V_L$ and $V_{in}$ are the load and input voltages. $G_L$ and $G_{in}$ are the load and input admittances, respectively.

The power efficiency of the Π-network with respect to component $Q$ within the area VSWR $\leq 10$ is shown in Fig. 5b. If the power loss is limited to 0.5–0.6 dB, it requires $Q_L$ of 70 and $Q_C$ of 70–100. Our result shows good agreement with [7] which gives an efficiency of 86.3%($-0.64$ dB), 91.7%($-0.38$ dB) and 95.8%($-0.19$ dB) for $Q_L$ of 60, 100 and 200 respectively.

By contrast, Fig. 5a and Fig. 5c, which show the power efficiency within the areas VSWR $\leq 8$ and VSWR $\leq 16$ respectively, give very different results. Assume $Q_C \gg Q_L$, to achieve insertion loss below 0.5 dB, covering the area VSWR $\leq 8$ requires a $Q_L$ of 50, but covering the area VSWR $\leq 16$ requires a $Q_L$ of 100. This is due to the fact, as shown in Fig. 3, that at the edge of the Smith chart the contour lines of VSWR, $Q_L$ and $Q_C$ are getting increasingly dense.

Apart from the numbers, we see that when $Q_C < Q_L$, the loss of the capacitors is dominant, and when $Q_L < Q_C$, the loss of the inductor dominates. Another interesting point is that, due to the flatness of the power efficiency curves at high $Q$ area, the larger the component $Q$ the more difficult it is to further improve the efficiency of the matching network by increasing it.

These give us practical guidelines on selecting component $Q$: (1) the component $Q$ of the capacitors and the inductor should not differ too largely such that the poor one covers up

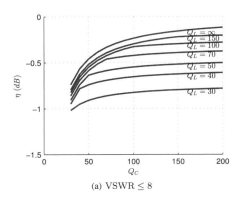

(a) VSWR $\leq 8$

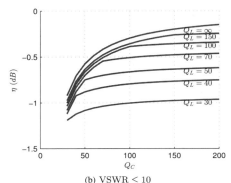

(b) VSWR $\leq 10$

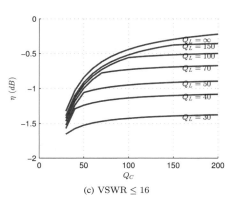

(c) VSWR $\leq 16$

Fig. 5. Power efficiency $\eta$ of the $\Pi$-network versus component $Q$. $\eta$ represents the worst case power efficiency in Smith chart within the area defined by VSWR.

the performance of the better one, (2) it is less efficient and more costly to further improve a high component $Q$ to further improve the power efficiency of the network, and (3) trade-offs must be taken between the coverage of the matching network and the component $Q$, which is very sensitive to large VSWR.

## IV. CONCLUSION

A $\Pi$ impedance matching network has been studied from the 4G mobile antenna impedance matching perspective, and analyzed in terms of coverage, achievable bandwidth, the required component values and power efficiency.

We conclude that the $\Pi$-network (synthesized for the maximum achievable bandwidth) is able to cover Band 4 and 10 for VSWR up to 8 and all other bands for VSWR up to 10. To achieve that, in a typical $50\,\Omega$ RF environment, it requires $C_{\max}$ of 13.64–2.7 pF, and $L_{\max}$ of 34.1–6.8 nH in the frequency range 700–3500 MHz. If the power loss is limited to 0.5–0.6 dB, it requires $Q_L$ of 70 and $Q_C$ of 70–100.

## ACKNOWLEDGMENT

This work has been performed in the ENIAC project ARTEMOS (Project No. 270683-2), in which the partner JKU is funded by the Austrian Research Promotion Agency (FFG) and the ENIAC Joint Undertaking.

## REFERENCES

[1] 3GPP TR 36.913, "Technical Report, 3rd Generation Partnership Project; Technical Specification Group Radio Access Network; Requirements for Further Advancements for E-UTRA (LTE Advanced)," V11.0.0, 2012-09.
[2] Recommendations ITU-R M.1645, "Framework and Overall Objectives of the Future Deployment of IMT-2000 and Systems beyond IMT-2000," 2003.
[3] 3GPP TS 36.101, "Technical Specification, 3rd Generation Partnership Project; Technical Specification Group Radio Access Network; Evolved Universal Terrestrial Radio Access (E-UTRA); User Equipment (UE) radio transmission and reception," V10.8.0, 2012-09.
[4] J. Toftgard, S.N. Hornsleth, J.B. Andersen, "Effects on portable antennas of the presence of a person," *IEEE Transactions on Antennas and Propagation*, vol.41, no.6, pp.739–746, Jun 1993.
[5] K. Ogawa, T. Matsuyoshi, "An analysis of the performance of a handset diversity antenna influenced by head, hand, and shoulder effects at 900 MHz: Part I — Effective gain characteristics," *IEEE Transactions on Vehicular Technology*, vol.50, no.3, pp.830–844, May 2001.
[6] D. J. Qiao, D. Choi, Y. Zhao, D. Kelly, T. Hung, D. Kimball, M. Y. Li, P. Asbeck, "Antenna impedance mismatch measurement and correction for adaptive CDMA transceivers," *IEEE MTT-S International Microwave Symposium Digest*, June 2005.
[7] F. C. W. Po, E. de Foucauld, D. Morche, P. Vincent, E. Kerherve, "A novel method for synthesizing an automatic matching Network and its control unit," *IEEE Transactions on Circuits and Systems I: Regular Papers*, vol.58, no.9, pp.2225–2236, Sept. 2011.
[8] P. Liu, A. Springer, "Impact of mobile antenna mismatch on receive antenna diversity in frequency-flat Rayleigh fading channels", *Vehicular Technology Conference (VTC Fall), 2013 IEEE*, 2–5 September, 2013.
[9] H. Song, S.-H. Oh, J.T. Aberle, B. Bakkaloglu, C. Chakrabarti, "Automatic antenna tuning unit for software-defined and cognitive radio," *2007 IEEE Antennas and Propagation Society International Symposium*, pp.85–88.
[10] Q. Z. Gu, J.R. De Luis, A.S. Morris, J. Hilbert, "An analytical algorithm for Pi-network impedance tuners,"*IEEE Transactions on Circuits and Systems I: Regular Papers*, vol.58, no.12, pp.2894–2905, Dec. 2011.
[11] C. Bowick, J. Blyler, and C. J. Ajluni, *RF circuit design*. Amsterdam; Boston: Newnes/Elsevier, 2008.
[12] Y. C. Sun, J.K. Fidler, "Design of $\Pi$ impedance matching networks," *1994 IEEE International Symposium on Circuits and Systems*, vol.5, pp.5–8, 30 May–2 Jun 1994.

# A dual-band and dual-polarized microstrip antenna subarray design for Ku-band satellite communications

Fu Yong[1][2], Yin Zhi-ping[1], Lv Guo-qiang[1]

[1]Academe of Opto Electronic Technology , Hefei University of Technology , Hefei Anhui 230009, P.R.China

[2]School of Instrument Science and Opto-electronics Engineering, HFUT, Hefei Anhui 230009, P.R.China

E-mail: fuyong0807@163.com,  zpyin@hfut.edu.cn

*Abstract*-In this paper, a dual-band and dual-polarized microstrip subarray antenna is designed for Ku-band phased array antenna system used in satellite mobile communication. The antenna has a vertical polarization for receiving band and a horizontal polarization for transmitting band. Aperture coupled feed, symmetry feed and parallel feed techniques are used to achieve high gain and high isolation level. The simulated results show that the proposed 4×4 subarray antenna obtains good performance. The return loss values of both ports are lower than -15dB. The isolation between two ports is higher than 30dB in receiving band. The proposed array antenna obtains the gain of 16.7dB in receiving band, and 18.3dB in transmitting band.

## I. INTRODUCTION

In recent years, the satellite mobile communication is rapidly increasing due to its unique advantages of wide area coverage, fast service initiation and high data rate. The low profile system using microstrip antenna have rapidly emerged and may become the main trends and evolution goals of the future mobile communications [1,2]. Sharing the same aperture in the transmitting (Tx) and receiving (Rx) bands is a proper way to reduce the size and weight of the antenna. But it may also cause some design challenges. The most important one is that the two bands signal may be coupled to each other to an undesired level, although the orthogonal polarization technique is normally adopted. In this paper, a dual-band and dual-polarized microstrip subarray antenna is designed for the Ku-band phased array antenna system used in satellite mobile communication. In order to obtain high isolation and good cross-polarization level, a medial ground layer with feeding slot is used to separate the feeding network of two band. A 4×4 microstrip antenna subarray with good performance is constructed by using electromagnetic simulation software optimizing.

## II. ANTENNA STRUCTURE AND DESIGN

The basic design of multilayered dual-polarized antenna is shown in Figure 1. This structure can flexibly get the best bandwidth and gain by adjusting the dielectric constant and thickness of each layers [3,4]. The upper substrate layer adopts RO5870 material with permittivity 2.2 and thickness 0.787 mm. In order to reduce the insertion loss in the feed network and increase the strength of the antenna plate, the lower substrate layer use RO4350 material with permittivity 3.48 and thickness 0.508 mm.

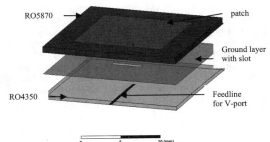

Figure 1. The dual-band dual-polarized unit

The driven patch and the horizontally polarized feed network are both formed on the layer 1 while the vertically polarized feed network is printed under the layer 3. Two polarization feed networks are separated by a ground plane which locates between layer 1 and layer 3. The coupling slots on the ground plane are used as exciting aperture. Since the array antenna is designed for two bands, 12.25~12.75GHz for receiving and 14.00~14.50GHz for transmitting, where the higher band operates in the horizontal polarization while the lower band operates in the vertical polarization, in order to improve cross-polarization level, the microstrip feedline of the horizontally polarized port (H-port) on layer 1 must be vertical to the aperture-coupled feedline of the vertical polarization port (V-port) on layer 3 [5]. Surface current distributions of the two bands are shown in Figure 2. It can be seen that the current vector of the two bands are orthogonal, then the orthogonal isolation between the two port is basically guaranteed.

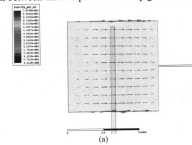

(a)

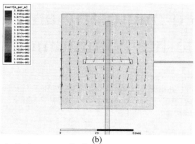

(b)

Figure 2. Surface current distribution at a)14.25GHz and b)12.5GHz

Figure 3 shows a 4×4 subarray antenna constructed by parallel feed network which is easy to satisfy amplitude requirement by adjusting power distribution ratio, and phase requirement by adjusting the electrical length of the feedline [6]. Meanwhile, by using the co-phase feeding technique, the design of parallel feed network is simple and effective for the absence of inverter production. To improve the array antenna performances in terms of gain, efficiency and cross-polarization as much as possible, we adopt symmetrical feeding network [7,8].

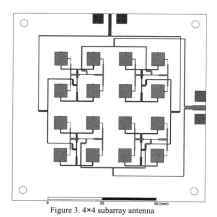

Figure 3. 4×4 subarray antenna

III. SUTIMULATION AND RESULTS

Figure 4 shows return loss curves of both H-port and V-port. At the operating frequency, both ports have return loss values lower than -15dB. The isolation between two ports is higher than 40dB in operateing band. The results show that the proposed array antenna obtains enough bandwidth at H-port, but the bandwidth at V-port is inferior to H-port in order to ensure the high isolation performance.

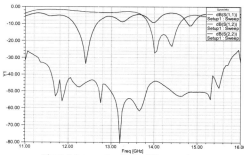

Figure 4. S-parameters for the 4×4 array antenna

Figure 5(a)-(d) plot the main polarization and cross-polarization curves of the two ports in the $\varphi = 0°$ and $\varphi = 90°$ plane. From figure 5(a) and (d), in the $\varphi = 0°$ plane, the peak gain levels are 18dB for H-port and 16.7dB for V-port respectively. However, cross polarization levels are nearly -10db because of feedlines for V-port also inspired ground layer. From figure 5(b) and (c), in the $\varphi = 90°$ plane, cross polarization levels for both ports are below -30dB in the beamwidth.

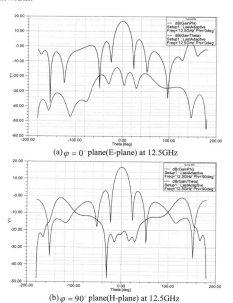

(a) $\varphi = 0°$ plane(E-plane) at 12.5GHz

(b) $\varphi = 90°$ plane(H-plane) at 12.5GHz

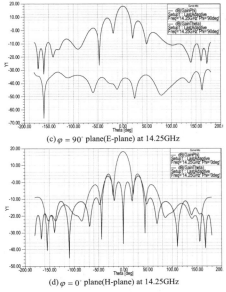

(c) $\varphi = 90°$ plane(E-plane) at 14.25GHz

(d) $\varphi = 0°$ plane(H-plane) at 14.25GHz

Figure 5. Antenna radiation patterns for two ports

## IV. CONCLUSION

In this paper, aperture coupled feed, symmetry feed and parallel feed techniques are used to achieve a high gain, high isolation of the dual-band and dual-polarized microstrip antenna. The antenna structure is simple, easy to fabricate. Two kinds of polarized within the working frequency band achieve maximum gain of 18.6 dB and 16.7 dB respectively, cross-polarization below -30 dB. The above results show that the antenna subarray has a good performance, can be the basis for the design to meet the needs of phased array antenna for satellite communication transceiver.

### REFERENCES

[1] P. Mousavi, M. Fakharzadeh,ect, "1K Element Antenna System for Mobile Direct Broadcasting Satellite Reception," *IEEE Trans. Broadcasting*, vol. 56, pp. 340-349, Sept. 2010.

[2] S.H.Son,U.H.Park ect. "Mobile antenna system for Ku-band satellite Internet service."*IEEE Vehicular Technology Conference, 2005. VTC 2005-Spring.*, Vol. 1, pp.234-327 , 2005.

[3] Hamed Hasani and Custódio Peixeiro,"Dual-band, dual-polarized microstrip reflectarray antenna in Ku band," *Antennas and Propagation Conference (LAPC), 2012 Loughborough* , pp.1-3,Nov. 2012.

[4] Martynyuk, S., F. Dubrovka, and P. Edenhofer. "A novel dual-polarized Ku-band antenna subarray." *Mathematical Methods in Electromagnetic Theory, 2002. MMET'02.2002 International Conference on.* Vol.1, pp.195-197,Spet.2002.

[5] Nanbo Jin and Yahya Rahmat-Samii. "A dual-polarized Ku-band microstrip antenna array: a compact feeding network and arefinement strategy using microwave holographic imaging."*Antennas and Propagation Society International Symposium*, pp.2120-2123, Jun. 2007.

[6] Weily,Andrew R., and Nasiha Nikolic. "Dual-Polarized Planar Feed for Low-Profile Hemispherical Luneburg Lens Antennas."*Antennas and Propagation, IEEE Trans.*, Vol.60, pp. 402-407. 2012

[7] Li,Wenjing,ect,"Wide bandwidth dual-frequency dual-polarized microstrip array antenna for Ku-band applications." *Antennas and Propagation Society International Symposium (APSURSI), 2012 IEEE.* pp.1-2, 2012.

[8] Ohtsuka, M., ect. "A dual-polarized planar array antenna for Ku-band satellite communications." *Antennas and Propagation Society International Symposium, 1998. IEEE.* Vol. 1,pp.16-19, 1998.

# Circularly Polarized Microstrip Antenna Array for UAV Application

Eko Tjipto Rahardjo[1], Fitri Yuli Zulkifli[1], Basari[1], Desriansyah Yudha Herwanto[1],
and Josaphat Tetuko Sri Sumantyo[2]

[1]Department of Electrical Engineering, Universitas Indonesia
Kampus Baru UI Depok
Depok 16424, INDONESIA
[2]Microwave Remote Sensing Laboratory, Chiba University
Chiba, JAPAN

*Abstract*-This paper present a circularly polarized microstrip antenna array for Unmanned Aerial Vehicle (UAV). The antenna element is designed using conventional rectangular microstrip antenna which is electromagnetically coupled fed by a microstrip line. The 1x4 element microstrip antenna array is then assembled on the surface of UAV cylindrical fuselage. The measurement result showed for 1x4 element antenna array that it operates at frequency 5.6 GHz with return loss characteristic of -22.24 dB,. The impedance bandwidth at VSWR < 2 is 720 MHz. Moreover, the radiation pattern show omnidirectional. Both simulation and experimental results show the proposed antenna radiation characteristics can be achieved as specified.

## I. INTRODUCTION

Recently UAV (Unmanned Aerial Vehicle) is becoming common in use for both civil authorities and military use. For military, the UAV can be used for spionase as well as for missile launch attack, while for civilian, UAV is widely used for exploring, mapping, monitoring and surveillance [1]. The UAV's mission usually required various link types such as telemetry, telecommand and payload communication link. Payload communication link usually needs larger bandwidth; therefore the UAV needs antenna with wideband characteristic.

In Indonesia, the UAV research has also been carried out e.g. at the Agency for Development of Research and Technology (BPPT) that has developed various types of UAV. These UAV are for civilian purposes [2].

Microwave Remote Sensing Laboratory of Chiba University has developed UAV that use linearly polarized microstrip antenna for its payload communication link [3]. However, due to the aircraft manuver, the linearly polarized antenna used will result in the disruption of data link communication. To keep data link continuiously available therefore a circularly polarized antenna is needed.

In this paper a circularly polarized microstrip array antenna is proposed in order to have communication from UAV to Ground Control Station and vice versa from any direction.

## II. ANTENNA DESIGN

In the proposed antenna system, a single element circularly microstrip patch antenna is used. The geometry of antenna is shown in Fig. 1 which is a rectangular patch antenna with corner truncation to achieve circular polarization [4]. The antenna is designed to be working at ISM band 5.6 GHz. The antenna is electromagnetically coupled with microstrip line feeding system [5]. This single element antenna is first designed to perform it basic radiation characteristics.

Furthermore a 1x4 element antenna array is assembled on the surface of cylindrical fuselage as shown in Fig. 2. Each element is located at the top, bottom and the 2 sides of the fuselage to obtain omnidirectional pattern. Each antenna element then connected through 1 to 4 power divider.

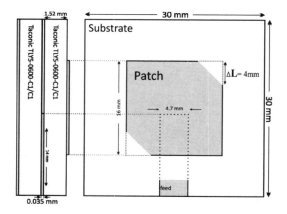

Figure 1. A single element circularly polarized microstrip patch antenna

978-7-5641-4279-7

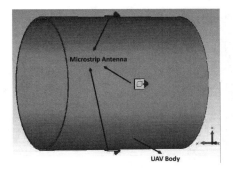

Fig. 2   Antenna array configuration on the surface of cylindrical fuselage

## III.  RESULTS AND DISCUSSION

The proposed single element antenna was simulated using CST Microwave Studio software with excellent radiation characteristics. Furthermore the antenna element is arrayed on the surface of cylindrical fuselage. The simulation results showed the return loss characteristic and the radiation pattern characteristics. The simulated return loss characteristic exhibit value of − 14.25 dB at the design frequency. The result also showed the antenna array impedance bandwidth is about 952 MHz at VSWR < 2 as depicted in Fig. 3. Radiation patterns showed almost omnidirectional for both phi = 0 and phi = 90 as expected as shown in Fig. 5 and Fig 6, respectively.

The proposed antenna array was then fabricated and measured. The antenna was fabricated using Taconic dielectric substrate with relative permittivity of 2.2, thickness of 1.52 mm and tanδ = 0.0009. The measurement of return loss characteristic was performed using a HP 8753D network analyzer. The result is shown in Fig. 2 where exhibit return loss value of -22.24 dB and VSWR < 2 impedance bandwidth of 720 MHz. The measured radiation patterns showed omnidirectional for both phi = 0 and phi = 90.

The simulated and measured radiation characteristics are displayed in Fig. 2 to Fig. 4. It can be seen that the results showed very good agreement in the desire frequency range. Both radiation pattern at phi = 0 and phi = 90 were measured at 5.6 GHz.

Furthermore, the antenna array gain was also measured which show 5 dB. The circularly polarized radiation pattern shows in Fig. 7.

Fig. 3. The photograph of fabricated assembled antenna on the surface cylindrical fuselage

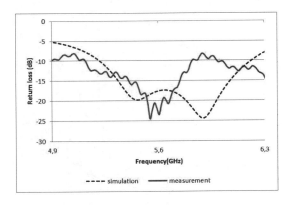

Fig. 4. Return loss characteristic comparison of 1x4 microstrip antenna array.

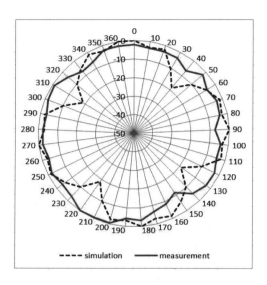

Fig. 5. Radiation pattern comparison between simulation and experiment result at phi = 0.

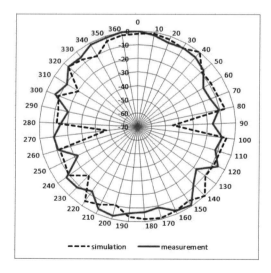

Fig. 6. Radiation pattern comparison between simulation and experiment result at phi = 90

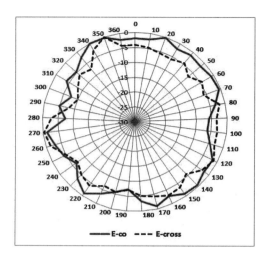

Fig. 7. Circularly polarized radiation pattern measurement for microstrip antenna array on UAV fuselage

## IV. CONCLUSION

This paper proposed a circularly polarized 1x4 microstrip antenna array for small UAV. The proposed antenna has been designed, simulated and measured using Taconic dielectric substrate with dielectric permitivitty 2.2. Both simulation and experiment results of the proposed antenna radiation characteristic using cylindrical model fuselage show a good agreement. Further study will be carried out to study for original size of UAV.

ACKNOWLEDGMENT

The authors would like to thanks to the Universitas Indonesia for granting the research grant under scheme RUUI 2012.

REFERENCES

[1] http://www.antaranews.com/berita/371972/pesawat-tanpa-awak-wulungmiliki-banyak-fungsi
[2] Berita Teknlogi Hankam,Transportasi & Manufakturing. Available: http://http://www.bppt.go.id/
[3] http://indonesiaproud.wordpress.com/2012/06/20/josaphat-tetuko-srisumantyo-kembangkan-uav-terbesar-di-asia-jx-1/
[4] J.R. James & P.S. Hall, "Handbook of Microstrip Antennas," *Peter Peregrinus Ltd.*, London, UK, 1989, ch. 4: Circular Polarisation and Bandwidth by M. Haneishi & Y. Suzuki.
[5] E.T. Rahardjo, S. Kitao, and M. Haneishi, "Planar Antenna Excited by Electromagnetically Coupled Coplanar Waveguide," *Electronics Letters*, vol. 29 , no. 10, pp. 870 - 872, May 1997

# Interpolation of Communication Distance in Urban and Suburban Areas

K. Uchida[1], M. Takematsu[1], J.H. Lee[1], K. Shigetomi[1] and J. Honda[2]

[1]Fukuoka Institute of Technology
Fukuoka 811-0295 Japan

[2]Electronic Navigation Research Institute
Chofu 182-0012 Japan

*Abstract- This paper is concerned with a numerical method to interpolate the distribution of communication distance function in urban and suburban areas. According to the 1-ray model, it is shown that the propagation order of distance $\beta$ depends only on the height of transmitting antenna and the amplitude modification factor $\alpha$ can be derived from the path loss obtained theoretically or experimentally. We propose a numerical method for estimating communication distance functions in urban and suburban areas by employing the path loss computed by the Okumura-Hata model. Numerical examples are shown to demonstrate the effectiveness of the proposed method.*

## I. INTRODUCTION

Recently, we have proposed the 1-ray model characterized by the amplitude modification factor $\alpha$ and the propagation order of distance $\beta$ as well as the field matching factor $\gamma$ in order to numerically simulate electromagnetic (EM) wave propagation in complicated EM environments such as urban and suburban areas or along random rough surfaces (RRSs) [1]. Originally the parameter $\beta$ was proposed by the Okumura-Hata model when introducing the empirical equations for the path loss in urban, suburban and open areas in Japan [2].

According to the Okumura-Hata model [2], $\beta$ depends only on the antenna height of base station (BS), and the path loss of EM wave propagation is given only in the urban, suburban and open areas. As a result, the 1-ray model combined with the Okumura-Hata model enables us to evaluate propagation characteristics only in these three areas. In order to deal with the EM propagation in other regions, however, it is necessary to propose a new method by use of a numerical interpolation.

It is evident that the path loss is closely associated with the complexity of propagation environments especially in the urban areas with high-rise buildings. In this paper, we propose a new algorithm to construct the communication distance function in urban and suburban areas which plays an important role for allocating base stations optimally in a complicated EM environment.

Section 1 is the introduction of the present paper. Section 2 reviews the propagation characteristics of the 1-ray model. Section 3 deals with the path loss of the 1-ray model. Section 4 shows numerical examples for the gross floor area ratio in Fukuoka city and the amplitude modification factor as well as the communication distance function. Section 5 concludes the present investigation.

## II. 1-RAY MODEL

The 1-ray model in far zone ($r \gg \lambda$) yields the electric field expression as follows [3],[4]:

$$\mathbf{E}(r) = 10^{\alpha/20} 10^{(\beta-1)\gamma/20} r^{(1-\beta)} \mathbf{E_i}(r) \qquad (1)$$

where $r$ is the distance from source to receiver and the field matching factor $\gamma$ [dB] as well as the field matching distance $\Gamma$ [m] are defined by

$$\gamma = 20\log_{10}\Gamma, \quad \Gamma = r\,|\mathbf{E_t}(r)|\,/\,|\mathbf{E_i}(r)| \quad . \qquad (2)$$

The incident electric field in the free space is given by [3],[4]

$$\mathbf{E_i}(r) = \sqrt{30 G_s P_s}\,\sin\theta_s \mathbf{\Theta^V}(\mathbf{r},\mathbf{p_s}) e^{-j\kappa_0 r}/r \qquad (3)$$

and the total electric field above a uniform ground plane is given by [3],[4]

$$\mathbf{E_t}(r) = \mathbf{E_i}(r) + \sqrt{30 G_s P_s}\,\sin\theta_0 \mathbf{e_r}(r) e^{-j\kappa_0 r_0}/r_0 \qquad (4)$$

where the time dependence $e^{j\omega t}$ is assumed. Moreover, $\kappa_0$ is the wave number in the free space, and $r_0$ is the total distance of the reflection ray above the uniform ground plane.

The transmitting and receiving antennas are small dipoles with gain $G_s = G_r = 1.5$ and direction vectors $\mathbf{p_s}$ and $\mathbf{p_r}$. Other detailed discussions for notations can be seen elsewhere [5]. Then the received power is given by

$$P_r = \lambda^2 G_r |\mathbf{E}(r) \cdot \mathbf{p_r}| / 4\pi Z_0 \;, \quad Z_0 = \sqrt{\mu_0/\varepsilon_0} \qquad (5)$$

where $\lambda$ is the wavelength and $Z_0$ is the intrinsic impedance of the free space.

Fig.1 shows the received powers for $\alpha = 0$ with parameter $\beta = 1.0$, 1.5 and 2.0 where we have assumed that the base

978-7-5641-4279-7

station (BS) antenna height is $h_b$ =30 [m] and the mobile station (MS) antenna height is $h_m$ =1.5 [m].

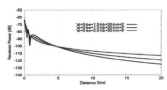

Figure 1 Received power computed using $\gamma$ versus distance with a parameter $\beta$ .

It is interesting that the three curves coincide at $r=\Gamma$ defined in Eq.(2) which can be approximated in far zone as follows;

$$\gamma=\gamma_0=20\log_{10}\Gamma_0 \quad , \quad \Gamma=\Gamma_0=2\kappa_0 h_b h_m \ . \qquad (6)$$

The accuracy of Eq.(6) can be numerically confirmed as shown in Fig.2 depicting the normalized field matching length $\Gamma/\Gamma_0$ versus the normalized distance $r/\Gamma_0$. It is demonstrated that the three curves for different BS antenna heights, $h_b$= 30, 100 and 200 [m], are almost coincident with each other, and 94.4% accuracy at $r/\Gamma_0$=1 together with 99.6% accuracy at $r/\Gamma_0$=5 has been achieved.

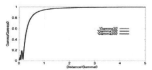

Figure 2 $\Gamma/\Gamma_0$ vs. $r/\Gamma_0$ for three different BS antenna heights.

Thus, we can conclude that $\gamma$ in Eq.(1) can be replaced by $\gamma_0$ defined by Eq.(6) for $r > \Gamma_0$, and the electric field can be rewritten as follows: .

$$\mathbf{E}(r) = 10^{\alpha/20} 10^{(\beta-1)\gamma_0/20} \, r^{(1-\beta)} \mathbf{E_i}(r) \ . \qquad (7)$$

Fig.3 shows received powers computed by Eq.(7) with the same parameters as Fig.1. It is evident that Fig.3 is in good agreement with Fig.1 for $r>\Gamma_0$, and a monotonic property is well depicted for $r<\Gamma_0$ .

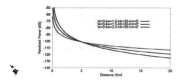

Figure 3 Received powers computed by using $\gamma_0$ .

III. PATH LOSS

The path loss expressed in [dB] is given by the difference between the transmitted power $P_s$ and the received power $P_r$ . This definition of the path loss in [dB] is summarized as follows [2];

$$L_p \ [dB] = P_s \ [dBW] - P_r \ [dBW] = P_s \ [dBW] - |\mathbf{E}| \ [dBV/m]$$
$$+10\log_{10} Z_0 - 10\log_{10}(\lambda^2/4\pi) \ . \qquad (8)$$

Eq.(8) indicates that received electric field intensity can be obtained from the path loss as long as the input power $P_s$ is specified. Substituting Eq.(7) into Eq.(8) leads to

$$L_p \ [dB] = -\alpha - (\beta-1)\gamma_0 \ + 20\beta\log_{10} R + 20\log_{10} f_c$$
$$+60\beta + 20\log_{10}(4\pi/300) \qquad (9)$$

where conversions of units have been made in accordance with the Okumura-Hata model; that is $f$ [Hz] $\rightarrow$ $f_c$ [MHz] and $r$ [m] $\rightarrow$ $R$ [Km] [2]. It should be noted that Eq.(9) indicates that $\alpha$ can be obtained analytically as far as the path loss $L_p$ and $\beta$ are known theoretically or experimentally.

A. Evaluation of $\beta$

One of the interesting features of the Okumura-Hata model is that $\beta$ depends only on BS antenna height $h_b$ as follows:

$$\beta=2.25-0.33\log_{10} h_b \qquad (10)$$

Now we consider a theoretical derivation of Eq.(10) from the 1-ray model discussed so far. According to Eq.(6), an enhancement of field intensity of the 1-ray model occurs when the BS and/or MS antenna heights are increased. This is why its amplitude is proportional to $\Gamma_0$ given in Eq.(6).

Now we assume that the field enhancement caused by the increment of BS antenna height from $h_b^{min}$ to $h_b$ may be compensated by the propagation order of distance $\beta$ at the smallest distance denoted by

$$\Gamma_0^{min} = 2\kappa_0 h_b^{min} h_m^{min} \qquad (11)$$

where $h_m^{min}$ is the minimum MS antenna height. Then we have the following relation

$$h_b h_m /(h_b^{min} h_m^{min}) = (\Gamma_0^{min})^{2-\beta} \ . \qquad (12)$$

Taking the logarithm of Eq.(12) leads to

$$\beta=2+[\log_{10}(h_b^{min}) - \log_{10}(K)]/\log_{10}(\Gamma_0^{min})$$
$$-\log_{10}(h_b)/\log_{10}(\Gamma_0^{min}) \qquad (13)$$

where $K=h_m/h_m^{min}$ .

Fig.4 depicts an enhancement of the field intensity due to the increase of BS antenna height from 30 to 100 [m]. According to the Okumura-Hata model [2], $h_b$ ranges from 30 to 200 [m] and so we could choose $h_b^{min} = 30$ [m]. On the other hand, $h_m$ varies from 1 to 10 [m] and so we could choose $K \approx 5$ since the median of MS antenna height is 5 [m]. Moreover, since Eq.(6) provides $\Gamma_0^{min} \approx 1.005$ [Km] for $f_c = 800$ [MHz], Eq.(13) can be re-written as follows:

$$\beta \approx 2 + 0.26 - 0.33 \log_{10}(h_b) . \qquad (14)$$

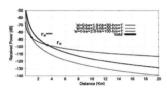

Figure 4.Received powers with different BS antenna heights. .

Fig.5 shows an example of $\beta$ calculated by Eq. (14) in comparison with that computed by Eq.(10). It is found that the two results are in good agreement and the 1-ray model could provide us an physical insight of the Okumura-Hata model..

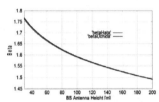

Figure 5 Proposed $\beta$ compared with the Okumura-Hata model.

### B. Evaluation of $\alpha$

Since $\beta$ is given by Eq.(10), Eq.(9) provides us an explicit expression for $\alpha$ as follows:

$$\alpha = -L_p \text{ [dB]} - (\beta-1)\gamma_0 + 20\beta\log_{10}R$$
$$+ 20\log_{10}f_c + 60\beta + 20\log_{10}(4\pi / 300) . \qquad (15)$$

The most important feature of Eq.(15) is described as $\alpha$ can be evaluated if and only if the value of path loss $L_p$ is given theoretically or experimentally [8], [9].

Based on the Okumura-Hata model, for examples, when $f_c = 800$ [MHz], $h_b = 30$ [m] and $h_m = 1.5$ [m], Eq.(15) provides us $\alpha_u = -37.3$ [dB], $\alpha_s = -27.7$ [dB] and $\alpha_o = -9.3$ [dB] in urban, suburban and open areas, respectively. However, it should be noted that Eq.(15) is restricted only to the urban areas of a large city or a medium-small city together with suburban and open areas, since the Okumura-Hata model provides the path loss only in these three areas. In other regions, we have to interpolate the value of $\alpha$ numerically [10].

### C. Communication Distance

Now we assume that the antenna orientation is arranged so that the maximum received power can be obtained. Let $E_{min}$ be the minimum detectable electric field intensity and $D_c$ the maximum communication distance. Then Combining Eq.(3) and Eq.(7) leads to the following expression

$$D_c = 10^{\alpha/20\beta} \times 10^{(\beta-1)\gamma_0/20\beta} \times (30G_sP_s)^{1/2\beta} \times (E_{min})^{-1/\beta} \quad (16)$$

Consequently, we can evaluate the communication distance by Eq.(16), when the amplitude modification factor $\alpha$ is known theoretically or experimentally. One method to estimate communication distance is to employ the known path loss as described in Eq.(15).

### IV. NUMERICAL EXAMPLES

The statistics of buildings and houses in a city of Japan are described in terms of two parameters; one is the building coverage ratio denoted by $DK$ indicating the total covered area on all floors of all buildings on a certain site area, and the other is the floor area ratio denoted by $DY$ indicating the ratio of the total floor area of buildings to the site area. Moreover, these statistical quantities are classified into two categories, net and gross values. The term "net" means that only building site is considered, and the term "gross" means that not only building site areas but also road and park areas are included for the two statistical parameters. We have obtained the gross floor area ratio $GDY$ from the Fukuoka local government as shown in Fig.6.

Figure 6.Gross floor ration in Fukuoka city.

Now we assume that the amplitude modification factor $\alpha$ could be strongly correlated with the gross floor area ratio $GDY$. Based on this assumption, we can estimate $\alpha$ in terms of the interpolation with respect to $GDY$ as follows [11]:

$$\alpha = \frac{(y-y_s)(y-y_o)}{(y_u-y_s)(y_u-y_o)}\alpha_u + \frac{(y-y_u)(y-y_o)}{(y_s-y_u)(y_s-y_o)}\alpha_s + \frac{(y-y_u)(y-y_s)}{(y_o-y_u)(y_o-y_s)}\alpha_o \quad (17)$$

where y is given by a function of GDY data x as follows:

$$y_p = x_p^{0.4}, \qquad (p=u,s,o). \qquad (18)$$

Fig.3 shows an example of interpolated $\alpha$ in Fukuoka city where we have selected "Tenjin" as an urban area with the amplitude modification factor $\alpha_u = -37.3$ [dB] and the gross floor area ratio $x_u$ =435.8 , "Hakozaki" as a suburban area with $\alpha_S$ =-27.7 [dB] and $x_S$ =87.2 and "Heiwa" as an open area with $\alpha_O$ =-9.3 [dB] and $x_O$ =5.2 . It is worth noting that "Tenjin" is a commercial district and "Heiwa" is a graveyard area.

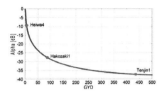

Figure 7 Interpolated $\alpha$ in Fukuoka city.

Figure 8.Distribution of $\alpha$ in Fukuoka city.

Fig.8 shows an example of distribution of the amplitude modification factor $\alpha$ interpolated from the gross floor area ratio in Fukuoka. It should be noted that $\alpha = 0$ [dB] is assumed in the sea region and the maximum attenuation $\alpha$ =-37.3 [dB] is observed in the urban area.

Figure 9.Distribution of communication distance in Fukuoka.

Fig.9 shows an example of distribution of interpolated communication distance in Fukuoka in a 2D-way. It should be noted that the unit of communication distance is expressed as $D_c$ [Km] which is computed by use of $\alpha$ distribution shown in Fig.8. The distribution of communication distance as shown in this figure can be considered to be a candidate for the communication distance function which is useful for an optimal allocation of base stations in urban areas [12].

## V. CONCLUSION

First, we have reviewed the 1-ray model in order to estimate propagation characteristics in complicated EM environments such as urban and suburban areas or random rough surfaces. Second, we have discussed the path loss in conjunction with the amplitude modification factor $\alpha$ and the propagation order of distance $\beta$ , and also we have proposed a numerical method to evaluate these two parameters. Third, we have shown the gross floor area ratios in Fukuoka city by using the vector interpolation, and also we have proposed a method to relate the amplitude modification factor $\alpha$ and the gross floor area ratio GDY. Numerical results have been shown in a 2D way to demonstrate the example of a distribution of communication distance or a communication distance function in Fukuoka city.

More detailed discussions are required on the statistical relationship between ( $\alpha$ , $\beta$ ) and (GDK, GDY). It deserves as a future investigation.

### ACKNOWLEDGMENT

The work was supported in part by a Grand-in Aid for Scientific Research (C) (24560487) from Japan Society for the Promotion of Science. The authors are indebted to the Fukuoka city hall for providing building and housing data in Fukuoka.

### REFERENCES

[1] K. Uchida and J. Honda, "An Algorithm for Allocation of Base Stations in Inhomogeneous Cellular Environment", Proceedings of 2011 International Conference on Network-Based Information Systems (NBiS-2011), pp.507-512, Sept. 2011.

[2] M. Hata, "Empirical Formula for Propagation Loss in Land Mobile Radio Services", IEEE Trans. Veh. Technol., vol. VT-29, no.3, pp.317-325, Aug. 1980.

[3] K. Uchida, J. Honda, T. Tamaki and M. Takematsu, "Two-Rays Model and Propagation Characteristics in View of Hata's Empirical Equations", IEICE Technical Report, AP2011-14, pp.49-54, May 2011.

[4] K. Uchida, J. Honda, Jun-Hyuck Lee, "A Study of Propagation Characteristics and Allocation of Mobile Stations", IEICE Technical Report, IN2011-98, pp.31-36, MoMuC2011-32, pp.31-36, Nov. 2011.

[5] K. Uchida, K. Shigetomi，M. Takematsu and J. Honda, "An Estimation Method for Amplitude Modification Factor Using Floor Area Ratio in Urban Areas", Proceedings of ITCS 2013, July. 2013. (to appear)

[6] Yasuto Mushiake, "Antennas and Radio Propagation", Corona Publishing Co., LTD. Tokyo, (1985).

[7] Robert E. Collin, "Antennas and Radiowave Propagation", McGraw-Hill Book Company, New York, (1985).

[8] K. Uchida, M. Takematsu, J. Honda, "An Algorithm to Estimate Propagation Parameters Based on 2-Ray Model", Proceedings of NBiS-2012, Melbourne, pp.556-561 (2012-09).

[9] K. Uchida, M. Takematsu, Jun-Hyuck Lee, J. Honda, "Field Distributions of 1-Ray Model Using Estimated Propagation Parameters in Comparison with DRTM", Proceedings of BWCCA-2012, Victoria, pp.488-493 (2012-11).

[10] K. Shigetomi, J. Honda and K. Uchida:" On Estimation of Amplitude Modification Factor and Propagation Order of 1-Ray and 2-Ray Models in Comparison with Hata's Equations", Proceedings of KJJC-2012, Seoul, Korea, pp.145-148, May 2012.

[11] W. H. Press, B. P. Flannery, S. A. Teukolsky, and W. T. Vetterling, "Numerical Recipes in FORTRAN: The Art of Scientific Computing," 2nd ed. Cambridge, England: Cambridge University Press, Chapter 3., pp. 99-122, 1992.

[12] K. Uchida, M. Takematsu, J.H. Lee and J. Honda :"An Adaptive Algorithm Based on PSO for Generating Inhomogeneous Triangular Cells", Proceedings of CISIS-2012, Palermo, Italy, pp.654-659, July 2012.

# *FA-2(A)*

## October 30 (FRI) AM

## Room A

## A &P for MIMO Comm.

# MIMO 2x2 Reference Antennas – Measurement Analysis Using the Equivalent Current Technique

A. Scannavini, L. Scialacqua, J. Zhang, L. J. Foged
SATIMO Italian Office
Via Castelli Romani, 59
00040 – Pomezia, Italy

Muhammad Zubair, J. L. A. Quijano, G. Vecchi
Antenna and EMC Lab, Politecnico di Torino
Turin, Italy

*Abstract-* **The reference antenna concept has been created to eliminate the uncertainties linked to the unknown antenna performances of the LTE 2x2 MIMO reference devices [1]. The wireless industry through the CTIA (The Wireless Association) and 3GPP (3G Partnership Project) standardization bodies has been using such antennas for characterizing the methodologies being proposed for the MIMO OTA tests [2]. The developments on the antenna concept and report on the measured performances at uniform incoming power distribution, figures and correlation between different labs have been presented in [3]. In this paper we present analysis of the measurements by using the equivalent radiating current technique (EQC). This technique is based on an integral equation formulation of the inverse source problem upon rigorous application of the equivalence principle [4]–[9]. The application of EQC enables to investigate the radiating details of the device and measurement setup. The aim of this paper is to show how the measurement set up can impacting the current distributions of the reference antennas and hence the performances.**

## I. INTRODUCTION

Long term Evolution (LTE) adopts multi-antenna mechanisms to increase coverage and physical layer capacity. Therefore, Multi-input Multi-Output (MIMO) antenna systems are required on LTE systems.

The MIMO 2x2 Reference Antenna concept was created based on the need of eliminate the unknown antenna performance of the available LTE MIMO 2x2 devices for radiated data throughput measurements. The adoption of the reference antennas, eliminate part of the measurement uncertainty, and increase repeatability among different MIMO OTA test methodologies and test facilities.

The reference antennas were designed to cover three LTE bands (2, 7 and 13) respectively 1.9GHz, 2.7GHz, and 750MHz. Conceptually for each band three antennas were designed to emulate a "good" MIMO antenna system FoM, i.e. low correlation coefficient ($\rho$<0.1), high system efficiency (SE>90%) and low gain imbalance (GI$\cong$0dB). Respectively the "nominal" MIMO antenna system has moderate correlation coefficient ($\rho\leqslant$0.5), moderate system efficiency (SE$\geqslant$50%) and low gain imbalance (GI$\cong$0dB). And finally the "bad" MIMO antenna system, having poor correlation

coefficient ($\rho\geqslant$0.9), moderate-to-poor system efficiency (SE $\leqslant$50%) and low gain imbalance (GI$\cong$0dB).

Measured data on these antennas have been presented in [3]. This paper will be presenting the analysis of the measurements results by using the equivalent current formulation [4]. The current distributions on the reference antennas will be used for understanding the impact of the measurement set up on the reference antennas performances such as radiation patterns, efficiency, gain imbalance and correlation coefficient.

## II. REFERENCE ANTENNA CONCEPT

### A. Reference antennas design

The reference antennas prototype is shown in figure 1 for the LTE Band 13 Good case.

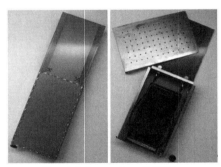

Figure 1. Reference antennas Prototype

The reference antenna needs to solve the potential problem with connecting any external antenna into a portable device (right picture). The connection between the portable device RF port and the external antenna, normally a coaxial cable, can potentially carry current in the outer conductor. The associated radiation perturbs the antenna system radiation and influence system parameters like correlation coefficient, absolute gain and gain imbalance. For this reason the antenna was conceived attaching MIMO 2x2 external antennas to a RF enclosure, where the DUT and its RF connections are located.

*B. Measured Performances*

In Table 1 the reference antennas measured performances are reported:

TABLE I: MEASURED PERFORMANCES OF THE MIMO ANTENNAS

| Band | Configuration | Antenna 1 | | | Antenna 2 | | | Gain Imbalance (dB) | | | Mag Complex Cor. Coef. | | |
|---|---|---|---|---|---|---|---|---|---|---|---|---|---|
| | | 1930MHz | 1960MHz | 1990MHz | 1930MHz | 1960MHz | 1990MHz | 1930MHz | 1960MHz | 1990MHz | 1930MHz | 1960MHz | 1990MHz |
| | | SE (%) | SE (%) | SE (%) | SE (%) | SE (%) | SE (%) | | | | | | |
| 2 | "Good" 2x2 MIMO | | | | | | | | | | | | |
| | "Nominal" 2X2 MIMO | 81 | 82.7 | 84.2 | 83.9 | 84.2 | 83.5 | -0.28 | -0.25 | -0.19 | 0.18 | 0.18 | 0.19 |
| | | 60.8 | 59.1 | 56.3 | 59.9 | 57.3 | 54 | 0.11 | 0.19 | 0.17 | 0.52 | 0.48 | 0.45 |
| Band | Configuration | 2620MHz | 2655MHz | 2690MHz | 2620MHz | 2655MHz | 2690MHz | 2620MHz | 2655MHz | 2690MHz | 2620MHz | 2655MHz | 2690MHz |
| 7 | "Good" 2x2 MIMO | | | | | | | | | | | | |
| | "Nominal" 2X2 MIMO | 81.5 | 87.8 | 87.3 | 80.1 | 85.6 | 84.8 | 0.24 | 0.26 | 0.27 | 0.1 | 0.07 | 0.05 |
| | | 55.7 | 61.8 | 63.3 | 55.0 | 61.2 | 63 | -0.03 | -0.01 | 0.02 | 0.32 | 0.29 | 0.27 |
| Band | Configuration | 746MHz | 751MHz | 756MHz | 746MHz | 751MHz | 756MHz | 746MHz | 751MHz | 756MHz | 746MHz | 751MHz | 756MHz |
| 13 | MIMO | 76.1 | 78.1 | 77.2 | 76 | 78.1 | 77.6 | 0.14 | 0.07 | 0.04 | 0.00 | 0.01 | 0.01 |
| | "Nominal" 2X2 | 50.5 | 50.9 | 49.6 | 50.9 | 51.5 | 50.1 | 0.25 | 0.22 | 0.3 | 0.58 | 0.53 | 0.49 |

The above is a special case of uniform distribution of incoming wave.

It must also be noted that the reference antennas have been tested in a SATIMO StarLab 15 by using coax cable for feeding the antenna under test (AUT). The coax cable influences on the radiation patterns has been studied in the following sections by using the equivalent current technique.

III. EQUIVALENT CURRENT TECHNIQUE

Measurement of small passive antennas often requires a coax cable to be connected to the antenna port. If the AUT is electrically small, as for the case of the reference antennas comparing current flowing back from the radiator to the outer surface of the cable will result in a second radiation and cause the measured radiation pattern to be inaccurate.

The measurement diagnostic and filtering capabilities of the equivalent current technique is used here to detect and then spatial filtering the interactions between the coax cable and the antenna itself. The method is based on the application of the Equivalence Theorem. The electric and magnetic current can be computed on an arbitrary surface by using either the Near Field (NF) or Far Field (FF) measured data. The field is then re-evaluated on the whole sphere. Enforcing explicitly Love's equivalence by a field boundary integral identity, the reconstructed currents are proportional to the actual field on the equivalent surface. The method becomes important for diagnostic and filtering purposes. Figure 2 shows the diagnostic tool when applying to the reference antennas case.

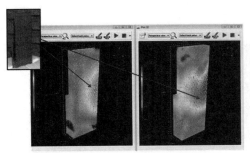

Figure 2. Current distributions on the reference antennas.

IV. MEASUREMENTS ANALYSIS

The LTE BAND 13 (751MHz) Good, and LTE BAND 7 (2.655MHz). Good reference antennas were considered for the analysis. Figure 3 shows the current distributions (J,M) for the LTE BAND 13 Good antenna case.

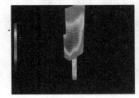

Figure 3. Measured current distributions – J (left), M (right).

It can be seen that there is some energy on the coax cable which could be re-radiated and cause the issue in the radiation pattern. This can be seen by looking at the 2D directivity plot in figure 4 when comparing it with the simulated (no coax cable effect). As anticipated, the equivalent current technique will allow us to spatially filter out the energy on the cable and re-evaluate the field on the whole sphere. Figure 5 shows the comparison of the 2D directivity pattern between simulated and filtered.

It can be seen that the 2D directivity pattern for the filtered case is closer to the simulated pattern than the measured one.

For the LTE BAND 7 Good antenna case, the effect of connecting a balun in between the coax cable and the antenna feeding point is observed. The expected effect is that the currents are choked due to presence of the balun itself. Figure 6 shows a comparison of the J current on the coax cable with and without the balun.

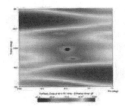

Figure 4. 2D Directivity comparison between simulated (left) and measured(right).

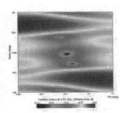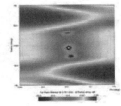

Figure 5. 2D Directivity comparison between simulated (left) and filtered (right).

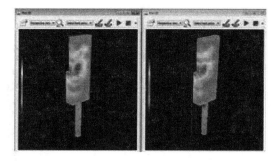

Figure 6. Current distributions comparison – only coax cable (left), with balun (right)

It can be seen that in both cases there is no energy on the cable hence the 2D directivity patterns would expect to be similar. Figure 7 shows the 2D directivity patterns comparison.

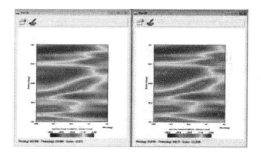

Figure 7. 2D directivity pattern comparison – only coax cable (left), with balun (right)

As expected, there are no substantial differences between the two patterns.

## VI. CONCLUSIONS

In this paper the usefulness of an antenna diagnostic tool was shown when testing electrically small antennas. The powerfulness of spatially filtering the currents on the coax cable was also reported by comparing the simulated vs measured and filtered 2D directivity patterns.

Due to the fact that the effect of the balun was not seen for the LTE BAND 7 case, the next step would be to measure the LTE BAND 13 antenna by using a balun in order to see how the current distributions on the coax cable and hence the 2D pattern will be impacted.

### ACKNOWLEDGMENT

SATIMO would like to thank Motorola Mobility LLC for sharing the antenna design and simulation data results.

### REFERENCES

[1] I. Szini, G.F. Pedersen, J. Estrada, A. Scannavini, L.J. Foged, "Design and Verification of MIMO 2x2 Reference Antennas" *APS 2012 – Chicago (IL)*

[2] 3GPP TR 37.977 v 0.6.0, "Verification of radiated multi-antenna reception performance of User Equipment (UE)", *May 2013*

[3] I. Szini, G.F. Pedersen, A. Scannavini, L.J. Foged, "MIMO 2x2 Reference Antennas Concept", *EuCap2012*

[4] J. Araque and G. Vecchi, "Improved-accuracy source reconstruction on arbitrary 3-D surfaces," IEEE Antennas Wireless Propag. Lett., vol. 8, pp. 1046–1049, 2009.

[5] J. A. Quijano and G.Vecchi, "Field and source equivalence in source reconstruction on 3D surfaces," Prog. Electromagn. Res., no. PIER 103, pp. 67–100, 2010.

[6] J. L. Araque Quijano, L. Scialacqua, J. Zackrisson, L. J. Foged, M. Sabbadini, G. Vecchi "Suppression of undesired radiated fields based on equivalent currents reconstruction from measured data", IEEE Antenna and wireless propagation letters, vol. 10, 2011 p314-317.

[7] L. A. Quijano, G. Vecchi, L. Li, M. Sabbadini, L. Scialacqua, B. Bencivenga, F. Mioc, L. J. Foged "3D spatial filtering applications in spherical near field antenna measurements", 32rd Annual Symposium of the Antenna Measurement Techniques Association, AMTA, October 2010, Atlanta, Georgia, USA.

[8] J. Foged, L. Scialacqua, F. Saccardi, J. L. Araque Quijano, G. Vecchi, M. Sabbadini, "Practical Application of the Equivalent Source Method as an Antenna Diagnostics Tool", 33rd Annual Symposium of the Antenna Measurement Techniques Association, AMTA, October 2011, Englewood, Colorado, USA.

[9] www.satimo.com/software/insight

# Design of a High Isolation Dual-band MIMO Antenna for LTE Terminal

Lili Wang[1], Chongyu Wei[2], Weichen Wei[3]

[1,2]Qingdao University of Science and Technology, Qingdao, 266061, P.R.China

[3]Melbourne University, VIC, Australia

Email:youranjinsh@163.com

*Abstract*-This paper presents a dual-band MIMO array antenna working in LTE frequency band. A dual-band PIFA antenna is used for every single antenna element. The PIFA antenna is achieved by using the slotted technology. Using coupling feed and adding a capacitive patch to the short wall to reduce the coupling between two antenna elements can improve the isolation of the MIMO system. The designed antenna can work on the LTE 2.3GHz and LTE 2.6GHz bands synchronously, corresponding band widths are 2300MHz-2400MHz and 2570MHz-2620MHz respectively. Simulation results show that within the operating frequency bands, isolation can be less than -25dB without significantly affecting the radiation characteristics, which demonstrates the feasibility of this design.

*Keywords*-LTE, MIMO, Isolation, Dual-band, PIFA

## I. INTRODUCTION

LTE (Long Term Evolution) is a key technology for the evolution of the wireless communication system from 3G to 4G. In order to meet requirements of high data rate, LTE system uses the multi-antenna technology. Multiple-input multiple-output (MIMO) wireless communication has become one of the most critical technologies in LTE and future 4G system. By using multiple antennas for receiving and transmitting, it can greatly improve the capacity of the communication system. However, there is often mutual-coupling between the antenna elements, which reduces the independence of the antenna elements, thus severely limiting the improvement of the capacity of the communication system. Therefore, the multi-antenna decoupling technology has become one of the key issues in mobile terminals.

This paper only discusses MIMO antenna for LTE terminal that consists of dual antenna elements. Reference [1] increases the isolation between antennas by changing the antenna polarization characteristics and slotting on the ground plane. Reference [2] uses the DGS (Defected Ground Structure), which means reducing the permittivity of the dielectric plate or digging up part of the media below the antenna. Or add PBG (Photonic Band Gap) material or EBG (Electromagnetic Band Gap) material between the antennas such as Reference [3]. Reference [4] uses DMN (Decoupling Matching Networks) structure to achieve decoupling. Reference [5] applies microstrip line to link the two antenna elements, to achieve the purpose of offsetting the coupling effects as some of the energy will transfers to another antenna.

Antenna elements that are array compactly often have stronger coupling. One of the reasons is that there is strong common currents distribution on the ground plane. Thus, low coupling between antenna elements can be achieved by reducing the currents or changing the currents distribution on the ground plane. When introduce DGS structure, currents on the ground plane will be weaken due to the action of the stop band and thereby increasing the isolation Reference [6]. This technology to some extent is undesirable for practical application, since the ground plane structure of the actual mobile terminals may not allow such a change. However, lower the coupling between antennas elements by reducing currents on the ground plane is a desirable manner. The method proposed in Reference [7] can significantly improve the isolation between antenna elements in MIMO system. In this paper, both the capacitive feed and capacitive load structures can be used to prevent currents on the radiation patch flowing directly to the ground plane through short wall and feed point, taking a step forward from an antenna port flows into another antenna port. Thereby reducing the coupling between antenna elements and lowering the correlation between antennas. Its essence is achieving the low coupling of MIMO multi-antenna system by reducing the currents on the ground plane.

## II. ANTENNA MODELING AND ANALYSIS

PIFA antenna has simple structure, small size and can achieve a wide bandwidth as well as achieve multi-band. The single antenna element of the MIMO system proposed in this paper also uses PIFA antenna. In the literatures that introduce how to reduce the mutual coupling, to a great extent, single-frequency antennas are researched. This is because the structure of a single-frequency antenna is simple and easy to decouple. Here we achieve the dual-band characteristics by using meander technology based on PIFA antenna. The operating frequency of the antenna is from 2300MHz to 2400MHz and from 2570MHz to 2620MHz, both of them are LTE bands. Resonant frequencies are about 2.35GHz and 2.6GHz respectively. In order to reduce the coupling between the two antenna elements, capacitive coupling feed and capacitive load technology are used at the same time.

### A Antenna Design

Figure 1(a) is the side view of a single antenna element .The two antenna elements are placed at the same top edge of the ground plane, keeping 2mm and 6mm respectively from the long side and the short side of the ground plane. The radiation patch dimensions are (L, W) = (22mm, 9mm). The antenna height from the ground is 6mm. Instead of coaxial feed, the currents will be fed to the metal microstrip through coaxial, then the metal microstrip feed line will feed the currents to the

978-7-5641-4279-7

radiation patch. The feed line is not directly connected to the radiation patch, but a capacitive feed patch is used at the top of the feed line to achieve currents feed using coupling feed method. The capacitive feed patch has length of 4mm along the X-axis and 3mm along the Y-axis direction. The distance between the feed patch and the radiation patch is 0.5mm. The metal microstrip feed line with a width of 2mm places in the center of the feed patch, and parallels to the short wall. The width of the short wall is 1mm, and a capacitive load patch which has length of 4mm along the X-axis and 9mm along the Y-axis direction is connected to the short wall. The distance between the load patch and the ground plane is 0.5mm. By using slotted structure, the dual-band characteristics will be realized. Slot gap structure and relative parameters are shown in Figure 1 (b). The gap width Slot = 1mm, and the gap keep distance of W0 = 2mm with each edge of the patch, L1_Slot = 18mm, L2_Slot = 14mm.

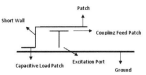

(a)    Side view of the antenna element

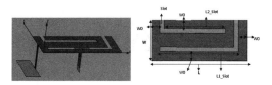

(b) Antenna element structure and structural parameters
Fig.1 Geometry of the antenna structure

B  Antenna Structure Analysis

The so-called coupling feed is to retain an adjustable gap between the feed patch and the radiation patch. Capacitive coupling feed can offset the sensitivity brought by the feed line and increase the degree of freedom in debugging and matching, and this can also be used as a mean of increasing bandwidth. Here, it also can reduce the currents that flow directly to the ground plane through the feed line. As a result, the isolation between the antenna elements to some extent will be improved.

Currents flowing to the ground plane through the short wall is one of the important reasons resulting in strong coupling between antenna elements. Add a capacitive load patch to the short wall can reduce currents that flows directly through the short wall to the ground plane obviously, which can play a good role in improving the isolation.

The coupling feed patch and the capacitive load patch constitute an adjustable capacitor with the radiation patch and the ground plane respectively. To the feed patch, the capacitor value can be adjusted by changing the distance between the feed patch and the radiation patch and the feed patch area. And to the load patch, there are the same methods to change

the capacitor value. The change of the capacitor value will affect the characteristics of a single antenna and the isolation between the antenna elements.

To achieve a dual-band PIFA antenna and taking into account the coupling feed structure at the same time, design a gap structure on the radiation patch shown in figure1(b). The capacitive coupling feed patch placed on the middle of the short side of the radiation patch connected to the microstrip line. Adjust the position of the feed point and the short wall and the capacitance of the two parallel plate capacitors can be used to achieve impedance matching. As a result, a good antenna property can be implemented.

III.  SIMULATION RESULTS AND ANALYSIS

HFSS electromagnetic simulation software is used in this paper for simulating and analyzing the designed MIMO antenna. The related simulation results are shown in figure 2, 3 and 4.

Figure 2 is the simulation result of the return loss of the MIMO antenna. As shown in figure 2, the resonance characteristics of the two antennas change little and there is only small difference at the low frequency resonant point. Put the reflection coefficient S11 ≤ -6 dB as a standard, we can measure that the return loss over the frequency range from 2.28GH-2.46GHz is less than -6dB which covering LTE 2300GHz-2400GHz band with a bandwidth of 180MHz. And the return loss of 2.54GHz—2.63GHz also covers LTE 2570GHz—2620GHz band with a bandwidth of 90MHz. Their center frequencies are approximately 2.35GH and 2.6GHz. The return loss is up to -25dB or so.

Figure 3 is the simulation result of the isolation between the two ports of the antenna system. To the MIMO system consists of the traditional FIFA antenna elements, isolation is far from the requirement by MIMO. However, to the new designed dual-band MIMO antenna system, the isolation within working bands can up to -25dB or less. Within 2.3GHz band, the minimum isolation is -26.6dB and the maximum isolation can up to -35.9dB. The isolation of the 2.6GH band ranges from a minimum isolation of -28.4dB to a maximum isolation of -30.8dB. In the case of affecting the radiation characteristics little, the isolation of the MIMO antenna system is effectively improved.

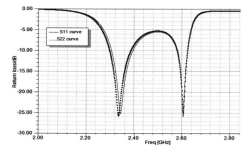

Fig. 2   Simulated return loss curve of MIMO antenna

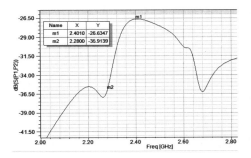

Fig. 3   Simulated isolation curve of MIMO antenna

Figure 4 is the current distribution on the ground plane. Figure (a) is the situation without capacitive feed patch and capacitive load patch. Figure (b) is the antenna structure that this paper proposed.

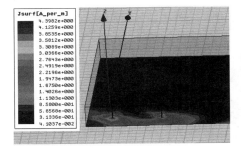

(a)   Current distribution on the ground plane of the traditional PIFA antenna

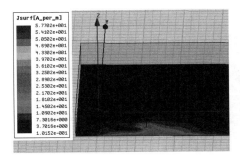

(b)   Current distribution on the ground plane of the PIFA antenna proposed in this paper

Fig. 4   Current distribution on the ground plane

It can be clearly seen that when capacitive load patch is added, the current density on the nearby region of the short wall on the ground plane is significantly reduced. The current density on the nearby region under the feed patch to some extent is also decreased. This result proves that short wall has

played a significant role in guiding the current flowing. And when the capacitive load patch is added, it can significantly prevent the currents flows to the ground plane through the short wall. Current on the ground plane that is located under the radiation patch significantly larger than other positions of the ground which proves that the ground plane as an important part of the antenna plays a key role in guiding the flow of currents and involved in the electromagnetic radiation.

## IV.   MEASUREMENT RESULTS

The designed antenna is fabricated and measured (As shown in Figure 5). The measurement results of the dual-band MIMO antenna are shown in Figure 6. As can be seen from the measurement curves, the resonant points are basically consistent with the simulation results. However, the return loss values, in particular for the low frequency resonant point, have slight difference compared with the simulation results. Overall, the bandwidth can meet the requirements. It can also be seen from Figure 6, the measurement result of the isolation can not be fully consistent with the simulation curve. However, the trend of the two curves is identical and the measurement result of the isolation can achieve the range of values of the simulation. The measurement results show that the antenna has a good practicability.

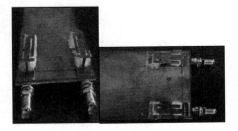

Fig. 5   Photograph of the dual-band MIMO antenna

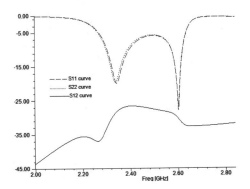

Fig. 6   Return loss curves and isolation curve of the measurement results

## V.   CONCLUSIONS

This paper presents a dual-band MIMO array antenna working in LTE frequency band. Coupling feed patch and capacitive load patch are used to reduce the current density on the ground plane, so that improving the isolation between the antenna elements. Focus on the design of a high isolation and dual-band MIMO antenna system, a PIFA antenna structure is used for every single antenna element. The slotted technology is used to achieve the dual-band. The simulation and measurement results proof that within the operating frequency bands, less than-25dB isolation can be obtained in this MIMO antenna system without significantly affecting the radiation characteristics. Therefore, this antenna structure has a good reference value to the design and application of the MIMO antenna for the terminal in the future.

REFERENCES

[1] Qinxin Chu, Jianfeng Li, "A Compact Wider Dual-Band MIMO Antenna Array for Mobile Phone," School of Electronic and Information Engineering,, South China University of Technology, Guangzhou, China.

[2] Zhaohui Song, Zhiyong Ding, "High Isolation and Low Correlation Small Size Multi-Antenna for MIMO Mobile Terminals," China Communications, 2012.5, pp. 100-107

[3] F．Yang and Y．R．Samii, "Microstrip antennas integrated with electromagnetic band—gap EBG structures：A low mutual coupling design for array applications," IEEE Trans．Antennas Propag. V01．5l, no．10，pp. 2936—2946，Oct. 2003．

[4] Fei Zhao, Shouzheng Zhu, "One kind decoupling analysis of a 710MHz LTE antenna," .Modern Electronics Technique. Jul. 2011, VO l. 34 No .13

[5] Guoqing Hu, "Research for coupling of the terminal antenna used in the MIMO system," Xi'an University of Electronic Science and Technology, 2011.3.11

[6] Sha Cui, "Research and design of a miniaturization MIMO antenna applicable for the mobile terminal" Xi'an University of Electronic Science and Technology2012.3.21

[7] Siyan Lou，Xiaolin Li，Zufan Zhang, "High isolation dual-element modified PIFA array for MIMO application" .The Journal of China Universities of Posts and Telecommunications. June 2012, 19(3): 1–6

# Slot Ring Triangular Patch Antenna with Stub for MIMO 2x2 Wireless Broadband Application

Fitri Yuli Zulkifli, Daryanto, Eko Tjipto Rahardjo

Antenna Propagation and Microwave Research Group (AMRG)
Department of Electrical Engineering, Faculty of Engineering, Universitas Indonesia
Kampus Baru UI Depok, Indonesia
Tel: 021-7270078, Fax: 021-7270077
Email: yuli@eng.ui.ac.id, anto.eui@gmail.com, eko@eng.ui.ac.id

*Abstract*-A compact slot ring triangular patch antenna with stub has been designed for MIMO 2x2 wireless broadband application. This compact antenna consists of only one layer substrate; however has broadband characteristic due to the slot ring inserted to the triangular patch antenna. The MIMO 2x2 elements achieved mutual coupling suppression to lower than 25 dB by placing the single element perpendicular to each other. In addition, the maximum antenna gain obtained is 5.9 dBi.

## I. INTRODUCTION

In Indonesia, the government has regulated for wireless broadband application the frequency band from 2.3 GHz to 2.39 GHz [1] with impedance bandwidth of 90 MHz. This wireless broadband application includes Worldwide Interoperability Microwave Access (WIMAX) application which needs Multiple Input Multiple Output (MIMO) characteristic due to high capacity and reliability requirements of wireless communication systems without increasing transmitted power or bandwidth [2].

To achieve broadband characteristic, several studies have been conducted by using aperture coupled antenna [3], air filled substrate [4] and slot antenna [5].

For the mutual coupling suppression between elements in the MIMO system, [3] used defected ground structure, while [6] added slit on the ground plane and [7] added spacing of λ/4 between elements. Theses paper achieved around 20 dB mutual coupling suppression.

In this paper, a compact slot ring triangluar patch antenna with stub is designed to achieve broadband characteristic and mutual coupling suppression of more than 20 dB between elements of the MIMO antenna.

## II. ANTENNA DESIGN

The proposed antenna design is depicted in Fig. 1. The antenna shape is triangular patch with insertion of slot ring to obtain broadband characteristic. Microstrip line feeding technique is used and good impedance matching is achieved by adding stub to the microstrip line. The antenna is printed on FR4-Epoxy substrate of 50 mm x 50 mm with permittivity 4.4 and height 1.6 mm.

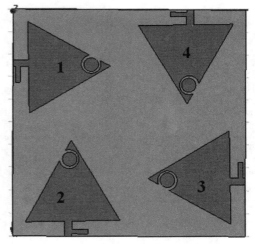

Fig.1 The proposed antenna design

The antenna consists of four elements perpendicular to each other to form the 2 x 2 MIMO configuration. Two antenna elements are for the transmitters (port 2 and port 4) and two elements for the receivers (port 1 and port 3). The antenna design is very compact because it consists of only one substrate and the elements are placed directly near to each other without any spacing.

The antenna design excites linear polarization. The polarization of port 2 is linear in the x direction, while port 1 in the y direction. Therefore, this can reduce the mutual coupling between antenna port 1 and port 2. This also occurs for the other pair of ports of the MIMO antenna.

## III. SIMULATION AND MEASUREMENT RESULTS

The simulation result of the proposed antenna using HFSS software is shown in Fig. 2. All four ports show that the antenna works at the frequncy center 2.35 GHz.

978-7-5641-4279-7

Observed at return loss of -10 dB, port 1, port 2, port 3 and port 4 has impedance bandwidth of 112 MHz, 112 MHz, 113 MHz and 107 MHz, respectively.

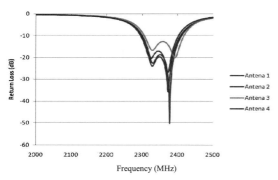

Fig.2. Return loss simulation result

The perpendicular position of the antenna plays an important role towards the mutual coupling reduction of the antenna. Focusing on the port as the tranmitter, the mutual coupling of the port 2 and port 4 towards the nearest port is shown in Fig. 3. The simulated mutual coupling suppression of S21, S41, S23 and S43 at center frequency 2.35 GHz is -33.1 dB, -40.1 dB, -32.9 dB and -34.9 dB, respectively. This mutual coupling result is suppressed well below -20 dB.

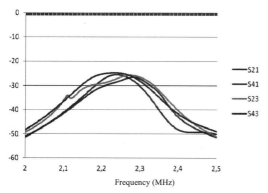

Fig.3. Mutual coupling simulation result

The simulation results show that the antenna design fulfills the aforementioned parameters, therefore the antenna design is fabricated and measured in an anechoic chamber.

The measurement result of the return loss is depicted in Fig. 4 for all of the antenna elements. At return loss of -10 dB, port 1, port 2, port 3 and port 4 has impedance bandwidth of 105 MHz, 108 MHz, 110 MHz and 120 MHz, respectively.

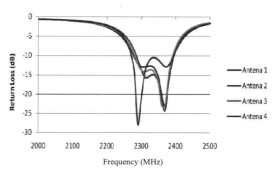

Fig.4. Return loss measurement result

The measured mutual coupling was also measured at the center frequency 2.35 GHz and shows the S21, S23, S41 and S43 is -25.31 dB, -25.22 dB, -25.17 dB and -25.6 dB, respectively.

In addition, the maximum antenna gain obtained at center frequency 2.35 GHz is 5.9 dB.

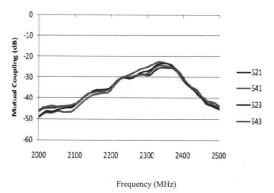

Fig.5. Mutual coupling measurement result

Moreover, the input impedance of the antenna was also measured and the results is shown in Fig. 6(a) for the port 1 Fig. 6(b) for port 3.

The input impedance for port 2 and port 4 was also measured at 2.35 GHz, therefore the impedance for port 1, port 2, port 3 and port 4 is 40.35-j11.07 Ω; 37.59-j7.37 Ω; 40.27-j9.6 Ω and 37.94-j21.23 Ω, respectively.

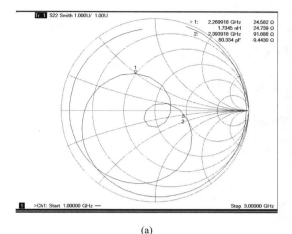

(a)

(b)

Fig.6. Input impedance measurement result
(a) port 1 (b) port 2

The measured and simulated results show good agreement with similar results. A slight discrepancy between simulation and measurement results are due to imperfect fabrication of the antenna. The slot ring inserted in the triangular patch is very small and the connection of the antenna with sma connector was not designed with the HFSS software, therefore these causes are mostly possible to be the cause of the discrepancy.

A slight difference of dimension between the design and fabrication can cause a slight shift of frequency when the antenna works at high frequencies.

## IV. CONCLUSION

The compact slot ring triangular patch antenna with stub has been designed, fabricated and measured for MIMO 2x2 wireless broadband applications. The measured impedance bandwidth of the antenna for all ports is more than 100 MHz while the mutual coupling effect is suppressed to lower than -25 dB. In addition, the maximum antenna gain obtained is 5.9 dBi.

REFERENCES

[1] www.depkominfo.go.id
[2] H. Zhang, Z. Wang, J. Yu, and J. Huang, "A compact MIMO antenna for wireless communication," *IEEE Antennas and Propagation Magazine*, Vol. 50, No. 6, December 2008.
[3] F.Y. Zulkifli and E.T. Rahardjo, "Compact MIMO microstrip antenna with defected ground for mutual coupling suppression" Progress in Electromagnetics Research Symp.(PIERS), Marrakesh, Morocco, 20-23 Mar. 2011, pp. 89-92
[4] M.N. Shakiv, M.T. Islam and N. Misran, "Design of a Broadband low cross-polarization W-shape microstrip patch antenna for MIMO system", RF and Microwave conference (RFM) 2008, Kuala Lumpur, Malaysia, 2-4 Dec. 2008, pp.313-314.
[5] R. Karimian, H. Oraizi, S. Fakhte and M. Farahani, "Novel F-Shaped Quad-Band Printed Slot Antenna for WLAN and WiMAX MIMO Systems", IEEE Antennas and Wireless Propagat. Lett., Vol. 12, 2013, pp. 405-408
[6] H. Li, J. Xiong and S. He, "A compact planar MIMO antenna system of four elements with similar radiation characteristics and isolation system", IEEE Antennas and Wireless Propagat. Lett. Vol. 8, 2009, pp. 1107 – 1110.
[7] Fakhr, R. S., "Compact size and dual band semicircle shaped antenna for MIMO applications", Progress In Electromagnetics Research C, Vol. 11, pp.147-154, 2009.

# Simple Models for Multiplexing Throughputs in Open- and Closed-Loop MIMO Systems with Fixed Modulation and Coding for OTA Applications

Xiaoming Chen[*], Per-Simon Kildal[*], Mattias Gustafsson[+]

[*]Chalmers University of Technology, Gothenburg, Sweden
[+] Huawei Technologies Sweden AB, Gothenburg, Sweden

*Abstract*-In this paper, we study the multiplexing throughputs of open- and closed-loop multiple-input multiple-output (MIMO) systems with fixed modulation and coding (i.e., fixed maximum data rate). For the open-loop case, we assume the simplest linear decoder, i.e., zero forcing (ZF). For the closed-loop case, we assume singular value decomposition (SVD). Using the threshold receiver model (accounting for the advanced coding), the throughput for either case can be readily obtained via simulations. MIMO antenna degradations such as correlation and imbalanced antenna efficiencies can be readily included in the developed throughput model. In addition to the usual presentation of the average throughput in a fading channel, we introduce the user-distributed detection probability of the (data) streams.

## I. Introduction

The long term evolution (LTE) of the universal mobile telecommunications system (UMTS) is attracting more and more attentions due to its throughput enhancement capability. In order to characterize (and improve) the LTE system, currently there is a growing interest in developing efficient over-the-air (OTA) system for testing LTE devices [1]-[5]. To complement the OTA measurements, throughput models have been presented in [3], [5]. Specifically, a single-input multiple-output (SIMO) throughput model based on the concept of the threshold receiver was presented in [3], a useful approximation also being studied in [6]. The SIMO throughput model was extended to multiple-input multiple-output (MIMO) systems with full spatial multiplexing in [5]. Due to the unavailability of closed-loop MIMO testing instruments, all these previous studies are for open-loop MIMO systems, where the channel state information (CSI) is only available at the receive side.

In this paper, we extend the MIMO throughput model further to include the closed-loop MIMO system (where the CSI is known at both MIMO sides). For simplicity, we assume fixed modulation and coding (i.e., fixed maximum data rate) as in [3], [5], and beamforming based on singular-value-decomposition (SVD) at both MIMO sides. Using the developed throughput model, the effects on the MIMO system throughput of correlation and power imbalance due to different antenna efficiencies are studied. In addition to the usual perception of the average throughput in a fading channel, we introduce a user-distributed throughput presentation, which

can be interpreted as the probability of the supported streams under certain signal to noise ratio (SNR) values.

## II. Multiplexing Throughput

### A. SISO Throughput Model

The single-input single-output (SISO) throughput model serves as a building block for developing the MIMO throughput model later on. It has been presented in [3]. Nevertheless, for the sake of completeness, we present it here briefly.

In order to model the advance coding in the modern telecommunications system with lowest possible complexity, we resort to the threshold receiver, whose block-error-rate (BLER) in an additive white Gaussian noise (AWGN) channel can be expressed as

$$P_e(\gamma) = \begin{cases} 1, & \gamma < \gamma_{th} \\ 0, & \gamma > \gamma_{th} \end{cases} \quad (1)$$

where $\gamma$ is the received SNR in the AWGN channel and $\gamma_{th}$ is the threshold value. In a fading channel, the average BLER is

$$\overline{P_e}(\overline{\gamma}) = \int_0^{\gamma_{th}} f(\gamma/\overline{\gamma})d\gamma = F\left(\gamma_{th}/\overline{\gamma}\right) \quad (2)$$

where $\overline{\gamma}$ represents the average $\gamma$ and $F$ denotes the cumulative distribution function (CDF) of $\gamma$. Interestingly, (2) agrees with the outage theorems [7], which states that with powerful coding the average BLER can be well approximated by the outage probability of the fading channel.

The throughput of a SISO system can be easily modeled as

$$T_{put}\left(\overline{\gamma}\right) = T_{put,max}\left(1 - F(\gamma_{th}/\overline{\gamma})\right) \quad (3)$$

where $T_{put,max}$ denotes the maximum data rate. Note that $1-F$ in (3) is the complementary CDF (CCDF) of the fading channel.

### B. MIMO Channel

In a flat fading channel, the MIMO system can be modeled as

$$\mathbf{y} = \mathbf{Hx} + \mathbf{n} \quad (4)$$

where $\mathbf{H}$ is the MIMO channel matrix, $\mathbf{x}$ and $\mathbf{y}$ are the transmitted and received signal vectors, respectively, and $\mathbf{n}$ is

978-7-5641-4279-7

the noise vector with independent identically distributed (i.i.d.) Gaussian variables. Note that, for an LTE MIMO system working in a frequency-selective and quasi-static fading channel [8], the model (4) can be regarded as one subcarrier of the orthogonal frequency division multiplexing (OFDM) system, which can be easily extended to model the frequency-selective fading channel [3], [5].

In MIMO OTA tests, the device under test (DUT) is usually measured as a receiver. In order to characterize the DUT alone, and the transmit antennas (whose efficiencies are calibrated out) are usually uncorrelated. The MIMO channel including the overall antenna effect (at the receive side) then can be expressed as [9]

$$\mathbf{H} = \mathbf{R}^{1/2}\mathbf{H}_w \qquad (5)$$

where $\mathbf{H}_w$ denotes the spatially white MIMO channel with i.i.d. complex Gaussian elements, $\mathbf{R}^{1/2}$ is the Hermitian square root of $\mathbf{R}$, which is

$$\mathbf{R} = \mathbf{\Xi} \circ \mathbf{\Phi} \qquad (6)$$

where $\mathbf{\Xi} = \sqrt{\mathbf{e}}\sqrt{\mathbf{e}}^T$ with $\mathbf{e}$ denoting a vector consisting the embedded radiation efficiencies at each port of the MIMO antenna, the superscript $^T$ is the transpose operator, $\sqrt{\ }$ is the element-wise square root, $\circ$ denotes element-wise product, and the correlation matrix $\mathbf{\Phi}$ consists of the complex correlation coefficients of the MIMO antenna.

### C.  Open-Loop MIMO System

In an open-loop MIMO system with full spatial multiplexing (i.e., the number of independent data streams equals the number of transmit antennas $N_T$), we assume a ZF receiver [10] for simplicity.

Let $\mathbf{h}_i$ be the $i$th column of $\mathbf{H}$ and $x_i$ be the $i$th element of $\mathbf{x}$, (4) can be rewritten as

$$\mathbf{y} = \mathbf{h}_i x_i + \sum_{j \neq i} \mathbf{h}_j x_j + \mathbf{n}. \qquad (7)$$

The first term in the right side of (7) stands for the $i$th stream, and the second term represents the interferences from all the other streams with respect to (w.r.t.) the $i$th stream.

The ZF equalizer for the $i$th stream corresponds to the $i$th row of the pseudo-inverse of $\mathbf{H}$, $(\mathbf{H}^H\mathbf{H})^{-1}\mathbf{H}^H$, where the superscript $^H$ stands for conjugate transpose. It projects the $i$th data stream into the subspace orthogonal to the one spanned by $\mathbf{h}_1,\ldots, \mathbf{h}_{i-1}, \mathbf{h}_{i+1},\ldots, \mathbf{h}_{N_T}$. Assuming equal allocation of transmit power and left multiplying the pseudo-inverse of $\mathbf{H}$ to (4), the SNR of the $i$th stream can be easily derived as

$$\gamma_i = \frac{1}{\left[\left(\mathbf{H}^H\mathbf{H}\right)^{-1}\right]_{i,i}} \qquad (8)$$

where $[\mathbf{X}]_{i,i}$ denotes the $i$th diagonal element of the matrix $\mathbf{X}$. Note that, without loss of generality, (8) assumes $\mathrm{E}[|x_i|^2]/\mathrm{E}[|n_i|^2] = 1$. The MIMO throughput is the sum of the throughputs of all the streams:

$$T_{\mathrm{put}}\left(\overline{\gamma}\right) = \sum_i T_{\mathrm{put,max},i}\left(1 - F\left(\gamma_{th}/\gamma_i\right)\right) \qquad (9)$$

where $T_{\mathrm{put,max},i}$ denotes the maximum data rate of the $i$th stream. Since the CSI is unknown at the transmitter, $T_{\mathrm{put,max},i} = T_{\mathrm{put,max}}/N_T$ and $\overline{\gamma}_i = \overline{\gamma}/N_T$ .

### D.  Closed-Loop MIMO System

When the CSI is known at the transmitter, it is natural to use SVD-based MIMO configuration [10]. However, due to limited feedback in frequency-division duplex (FDD) systems, codebook-based precoding [11] is used instead. The codebook-based precoding uses, in essence, quantized CSI (i.e., partial CSI), which incurs performance degradation. In this paper, for simplicity, we assume full CSI at the transmitter (e.g., SVD-based MIMO). The corresponding results can be regarded as upper bounds for the codebook-based precoding counterparts.

Let the SVD of $\mathbf{H}$ be $\mathbf{H} = \mathbf{U}\mathbf{\Lambda}\mathbf{V}^H$, where $\mathbf{U}$ and $\mathbf{V}$ are the unitary matrices, and $\mathbf{\Lambda}$ is a diagonal matrix consisting the singular values of $\mathbf{H}$. The precoding and power allocation is done by multiplying the signal vector $\mathbf{s}$ by $\mathbf{VP}$, $\mathbf{x} = \mathbf{VPs}$, where $\mathbf{P}$ is a diagonal matrix whose elements correspond to the allocated power for each stream. The decoding is done by multiplying $\mathbf{y}$ by $\mathbf{U}^H$, $\mathbf{r} = \mathbf{U}^H \mathbf{y}$. The resulting (interference-free) parallel MIMO channel is

$$\mathbf{r} = \mathbf{\Lambda}\mathbf{Ps} + \mathbf{z} \qquad (10)$$

where $\mathbf{z} = \mathbf{U}^H \mathbf{n}$. Note that $\mathbf{s}$ and $\mathbf{z}$ have the same statistics as $\mathbf{x}$ and $\mathbf{n}$, respectively, in that $\mathbf{V}$ and $\mathbf{U}$ are unitary.

Note that the well known water-filling algorithm for optimally allocating power [10] (i.e., determining optimal $\mathbf{P}$) is derived from Shannon's extended capacity formula, not for the practical throughput case. Thus it is not suitable for practical LTE system [12]. Actually, equal power allocation is implemented in current LTE systems [11]. As a result, this paper assumes equal power allocation (at the transmitter), i.e., $\mathbf{P} = \mathbf{I}$. (Quantized precoding with limited feedback will be studied in future work.) With the assumption of $\mathrm{E}[|x_i|^2]/\mathrm{E}[|n_i|^2] = 1$, the SNR of the $i$th stream equals the square of the $i$th element of $\mathbf{\Lambda}$,

$$\gamma_i = \lambda_i^2 . \qquad (11)$$

The SVD-based MIMO throughput is the sum throughputs of all the parallel streams as expressed in (9).

Next, we show that combining precoding $\mathbf{W} = \mathbf{V}$ (that is also a unitary matrix) with ZF can achieve the same performance of a SVD-based MIMO system.

*Proof*: A 2×2 MIMO system with a precoder $\mathbf{W}$ (in a flat-fading channel) can be modeled as

$$\mathbf{y} = \mathbf{HWs} + \mathbf{n} = \mathbf{U}\mathbf{\Lambda}\mathbf{V}^H\mathbf{Ws} + \mathbf{n} . \qquad (12)$$

Since $\mathbf{V}$ and $\mathbf{W}$ are unitary matrices, one can choose $\mathbf{W}$ such that

$$\mathbf{V}_{eq} = \mathbf{V}^H\mathbf{W} = \begin{bmatrix} \cos\theta & -\sin\theta \\ \sin\theta & \cos\theta \end{bmatrix} \qquad (13)$$

where $\theta \in [0, \pi/2]$. The SNR of the 2×2 MIMO system with precoding $\mathbf{W}$ and ZF receiver is

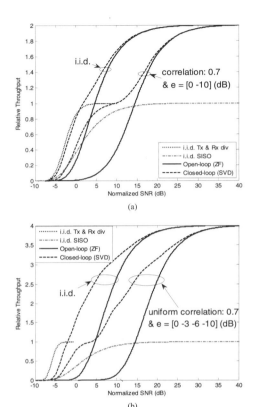

(a)

(b)

Figure 1. MIMO Throughput: (a) 2×2 MIMO systems; (b) 4×4 MIMO systems. As references, the throughputs of SISO and transmit (Tx) and receive (Rx) diversity systems in i.i.d. channels are plotted in the same figure. Note that the maximum throughput of a single stream is normalized to one in both graphs. Thus, the maximum throughput 2×2 and 4×4 MIMO systems are 2 and 4, respectively.

$$\gamma_i = \frac{1}{\left[\left(\left(\mathbf{HV}_{eq}\right)^H \mathbf{HV}_{eq}\right)^{-1}\right]_{i,i}} = \frac{1}{\dfrac{\cos^2\theta}{\lambda_i} + \dfrac{\sin^2\theta}{\lambda_l}} \quad (14)$$

where $i, l \in \{1, 2\}$ and $i \neq l$. Let $\theta = 0$ (i.e., $\mathbf{W} = \mathbf{V}$), (14) reduces to (11).                                                             □

This observation is rather intuitive. When $\theta = 0$, $\mathbf{V}^H \mathbf{W}$ becomes identity matrix, i.e., the precoder $\mathbf{W}$ diagonalizes the transmit singular vector matrix; and the receive singular vector matrix can be diagonalized similarly by a linear ZF decoder. Thus, parallel channels identical to the SVD case are obtained.

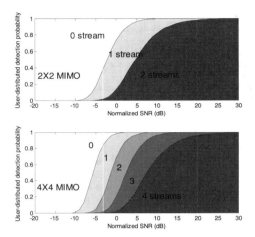

Figure 2. User-distributed detection probability of 2×2 (upper) and 4×4 (lower) open-loop MIMO systems. The different color regions represent the probability of the maximum number of streams supported.

## III. SIMULATIONS AND RESULTS

Using the MIMO channel and throughput models developed in the previous section, we can study the effects of correlations and power imbalance between antenna ports on the MIMO throughput in a fading channel. We generate 2×2 and 4×4 i.i.d. Gaussian matrices with 20000 realizations and introduce power imbalance and correlation to the MIMO channel via (5) and (6).

Fig. 1 shows the throughputs of the 2×2 and 4×4 MIMO systems with both open- and closed-loop configurations. The throughputs of SISO and transmit (Tx) and receive (Rx) diversity systems in i.i.d. channels are plotted in the same figure as references. Note that the relative throughput is defined as the throughput normalized to its maximum value, and that the normalized SNR is the average received SNR divided by the threshold value. Also note that the maximum ratio combining (MRC) is assumed in simulations. The uniform correlation of 0.7 (for the 4×4 MIMO case) means that the correlation coefficient between any two antenna ports is 0.7. As expected, the closed-loop MIMO configuration offers better throughput performance compared with its open-loop counterpart, and that both correlation and power imbalances (due to different antenna efficiencies) affect the throughput performance adversely. Note that the maximum throughput of the 4×4 MIMO system is double that of the 2×2 MIMO system (at the expense of more SNR required to achieve the full spatial multiplexing). Another interesting observation is that the throughput of the 2×2 closed-loop MIMO system at low SNR is slightly smaller than that of the transmit and receive diversity system. This can be explained by the fact that the received SNR of the (MRC) diversity system corresponds to the summation of all the eigenvalues (11), whereas throughput performance at low SNR is given by the largest eigenvalue for the closed-loop (or SVD-based) MIMO system. The inferior throughput of the closed-loop MIMO with respect to the diversity system becomes more noticeable in that the diversity system uses all the four eigenvalues, whereas the closed-loop MIMO system still relies on the largest eigenvalue in the low SNR regime.

Fig. 1 corresponds to the average throughput of a single user in a fading environment. Another useful presentation of the simulated throughput is to plot the throughput as the user-distributed detection probability. For illustration purpose, we focus on the open-loop MIMO systems with ZF receivers in an i.i.d. flat-fading channel. Fig. 2 shows the user-distributed detection probability for the 2×2 and 4×4 open-loop MIMO system, where different color regions represent the probability of the maximum number of streams supported. For instance, when the average SNR equals the threshold value (i.e., at 0-dB normalized SNR), the probability of detecting one, two, three, and four streams by using the 4×4 open-loop MIMO system are 92%, 65%, 20%, and 2%, respectively.

## IV. CONCLUSION

In this paper, based on the threshold receiver [3], [6], we have developed throughput models for open- and closed-loop MIMO systems. Specifically for the closed-loop MIMO system, we show that by diagonalizing the transmit eigenvector matrix with a unitary precoder matrix, for which CSI must be known on Tx side, a ZF decoder offers the same system performance as the SVD-based MIMO. We compared the open- and closed-loop MIMO configurations (corresponding to CSI not know and known, respectively, on Tx side) under different correlations and power imbalances. As expected, the closed-loop MIMO configuration has better performance than the open-loop MMIO configuration, especially at low SNR. Nevertheless, it is clearly seen in Fig. 1 that, at low SNR, the diversity scheme offers the highest throughput; spatial multiplexing can only be achieved at high SNR. In addition to the usual presentation of the average throughput, we present the user-distributed detection probability, based on which, the probability of the number of supported streams is readily shown. The detection probability of two bit streams can be easily obtained from the ZF algorithm, and it is then almost the same as obtained by SVD for large SNR. For low SNR the detection probability of one bit stream is better for ZF than for SVD. Thus, when characterizing quality of wireless devices we can use the detection probability of bit streams as a quality metric, with

the multiple bit stream cases having the i.i.d. result obtained by a ZF algorithm as the maximum reference cases, and we can determine quality in terms of degradation in dBiid relative to this reference at a certain probability level such as 95%.

## ACKNOWLEDGMENT

This work has been supported in part by the cooperation project "uBTS HO-MIMO antenna" funded by Huawei Technologies Sweden AB. The first author would also like to acknowledge the Alice and Lars Erik Landahl Grant fund.

## REFERENCES

[1] A. A. Glazunov, V. M. Kolmonen, and T. Laitinen, MIMO Over-The-Air Testing. *LTE-Advanced and Next Generation Wireless Networks: Channel Modelling and Propagation*, John Wiley & Sons, 2012.

[2] N. Arsalane, M. Mouhamadou, C. Decroze, D. Carsenat, M. A. Garcia-Fernandez, and T. Monediere, "3GPP Channel Model Emulation with Analysis of MIMO-LTE Performances in Reverberation Chamber," *Int. J. Antennas Propaga.*, vol. 2012, Article ID 239420, 8 pages, 2012.

[3] P.-S. Kildal, C. Orlenius, and J. Carlsson, "OTA testing in multipath of antennas and wireless devices with MIMO and OFDM," *Proc. IEEE*, vol. 100, no. 7, pp. 2145-2157, July 2012.

[4] X. Chen, P.-S. Kildal, and J. Carlsson, "Fast converging measurement of MRC diversity gain in reverberation chamber using covariance-eigenvalue approach," *IEICE Trans. Electronics*, vol. E94-C, no.10, pp.1657-1660, Oct. 2011.

[5] X. Chen, P.-S. Kildal, and M. Gustafsson, "Characterization of Implemented Algorithm for MIMO Spatial Multiplexing in Reverberation Chamber," *IEEE Trans. Antennas Propag.,* vol. 61, no. 8, Aug. 2013.

[6] A. Toyserkani, E. Ström, and A. Svensson, "An analytical approximation to the block error rate in Nakagami-m non-selective block fading channels," *IEEE Trans. Wireless Commun.,* vol. 9, pp. 1543-1546, 2010.

[7] N. Prasad and M. K. Varanasi, "Outage theorems for MIMO block-fading channels," *IEEE Trans. Inf. Theory*, vol. 58, no. 7, pp. 2159-2168, July 2006.

[8] X. Chen and P.-S Kildal, "Theoretical derivation and measurements of the relationship between coherence bandwidth and RMS delay spread in reverberation chamber", *Proceedings of the 3rd European Conference of Antenna and propagation* (EuCAP), 23-27 March, 2009.

[9] X. Chen, P.-S. Kildal, J. Carlsson, and J. Yang, "MRC diversity and MIMO capacity evaluations of multi-port antennas using reverberation chamber and anechoic chamber," *IEEE Trans. Antennas Propag*, vol. 61, no. 2, pp. 917-926, Feb. 2013.

[10] A. Paulraj, R. Nabar and D. Gore, *Introduction to space-time wireless communication*, Cambridge University Press, 2003.

[11] *LTE Physical Layer – General Description*, 3rd Generation Partnership Project, Tech. Rep. TS 36201, V8.1.0, Nov. 2007.

[12] G. Berardinelli, L. Temiño, S. Frattasi, et al., "On the feasibility of precoded single user MIMO for LET-A uplink," *J. Commun.*, vol. 4, no. 3, pp. 155-163, April 2009.

# Channel estimation method using MSK signals for MIMO sensor

Keita Ushiki, Kentaro Nishimori, Tsutomu Mitsui* and Nobuyasu Takemura

Graduate School of Science and Technology, Niigata University, Samsung Yokohama Laboratories

Ikarashi 2-nocho 8050, Nishi-ku Niigata-shi, 950-2181 Japan

Email : ushiki@gis.ie.niigata-u.ac.jp, nishimori@ie.niigata-u.ac.jp

*Abstract*—We have proposed an intruder detection method by using Multiple Input Multiple Output (MIMO) channel. Although the channel capacity on the MIMO transmission is severely degraded in time variant channels, we can take advantage of this feature *MIMO Sensor* applications. We have already demonstrated the effectiveness of MIMO sensor using wireless LAN signals at 2.4GHz band. On the other hand, the transceiver should be simplified from a point of view on power saving. In this paper, we deal a narrowband Minimum Shift Keying (MSK) which is used in RF-ID and so on, and propose a signal synchronization method for the channel estimation using the narrowband MSK signal. Moreover, the basic performance by the proposed channel estimation method is verified when considering the intruder detection by MIMO sensor.

*Index Terms*—MIMO sensor, minimum shift keying, frequency offset, time correlation

## I. INTRODUCTION

Reliable security systems have been recently attracted much attention. Microwave sensors using existing signals, such as Frequency Modulation (FM), Television (TV) broad-casts signals and so on, have been studied, because the microwave sensors can detect the signal even in Non Line of Sight (NLOS) environment unlike conventional infrared light and camera for security usage. The microwave sensor by using received signal strength indicator (RSSI) of wireless LAN based detection method etc, are proposed [1]. This method is relatively simple but there is an issue for the detection accuracy [2]. In order to solve this problem, an intruder detection using array signal processing proposed [2][3]. In this method, Single Input Multiple Output (SIMO) channel is assumed and the variation of 1st eigenvector, which is obtained by the correlation matrix of the received signal, is utilized as a cost function of intruder detection.

We proposed an intruder detection method which utilizes channel matrix in Multiple Input Multiple Output (MIMO) channels [4], in order to enhance detection performance in [2]. We call this method *MIMO Sensor* [5]. Although the channel capacity on the MIMO transmission is severely degraded in time variant channels [6], we can take advantage of this feature in MIMO Sensor applications. Since not only receiving but also transmitting diversity effects are obtained by using MIMO transmission, higher reliability for the intruder detection is expected by using the MIMO Sensor compared to the SIMO sensor [5].

IEEE802.11n based Wireless LAN signals are used for the evaluation on the MIMO sensor [5], because MIMO-OFDM system has been incorporated into IEEE802.11n standard and channel state information (CSI) for the MIMO channel can be easily obtained when considering MIMO-OFDM system. On the other hand, the transceiver should be simplified from a point of view on power saving. Hence, we introduce narrowband signals into the MIMO sensor. Generally speaking, differential detection is employed for the narrowband signals because the transmission rate is low and large carrier and timing offsets between the transmitter and receiver exist. Hence, the performance of CSI estimation is severely degraded if an actuate carrier and timing offset compensation is not employed. In this paper, we propose an synchronization method of carrier and timing offsets for the CSI estimation by narrowband minimum shift keying (MSK) signals, which are used in RF-ID systems.

The remainder of this paper is organized as follows. Section III shows the principle of MIMO sensor. The new synchronization method of carrier and timing offsets using MSK signals is explained in Section III. Section V shows effectiveness of the new method by the measurement for the intruder detection.

## II. PRINCIPLE OF MIMO SENSOR

Fig. 1 shows a concept of MIMO sensor [5]. Fig. 1(a) and (b) compare the variation of channel matrix in the MIMO channel due to a person. Although the channel capacity of MIMO transmission is severely degraded in the time variant channels [6], we utilize the variation of channel matrix in MIMO channel as an input of *Sensor*. When $M$ and $N$ are the numbers of transmitting and receiving antennas, the channel matrix $\boldsymbol{H} \in \mathbb{C}^{N \times M}$ is change to $\boldsymbol{H}' \in \mathbb{C}^{N \times M}$ due to the intrusion by the person. We realize the intrusion detection by checking the variation of the channel matrix, $\boldsymbol{H}$. The variation of the channel matrix can be expressed by time correlation function. Let us assume that $h_{no,ij}$ ($i = 1 \sim N, j = 1 \sim M$) is a component at the channel matrix without people in the room. When $h_{ij}(t)$ is a component of the channel matrix at

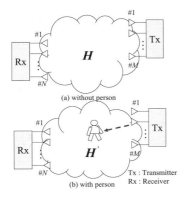

Fig. 1. Concept of MIMO sensor (Tx:Transmitter, Rx:Receiver).

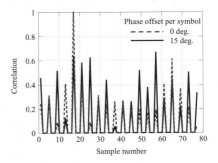

Fig. 2. Example of sliding correlation with carrier offset.

time $t$, the time correlation, $\rho_H(t)$ is represented by

$$\rho_{\mathrm{H}}(t) = \frac{\left| \sum_{i=1}^{N} \sum_{j=1}^{M} h_{no,ij}^{*} h_{ij}(t) \right|}{\sqrt{\sum_{i=1}^{N} \sum_{j=1}^{M} |h_{no,ij}|^2} \sqrt{\sum_{i=1}^{N} \sum_{j=1}^{M} |h_{ij}(t)|^2}}. \quad (1)$$

## III. CHANNEL ESTIMATION METHOD BY USING MSK SIGNALS

Although the transceiver is simplified by using narrowband signal, large frequency and timing offsets arise due to local oscillators, A/D and D/A convertors between transmitters and receivers. Fig. 2 shows an example of sliding correlation when considering the frequency offset. In order to detect the initial timing of data packet, the sliding correlation between *known* transmit and receive signals is employed in IEEE802.11n based MIMO-OFDM system [7]. Since the transmission rate is very high compared to the frequency offset in MIMO-OFDM system, the sliding correlation is effective for detecting the initial timing of data packet [7]. We measured that 14.1° per symbol is observed when the transceiver for RF-ID system is used where the frequency is 440 MHz and symbol rate is 2.4KHz [8]. The peak should be observed when the sample number is 17 in Fig. 2. The peak of sliding correlation cannot be obtained at the sample number of 17 when considering the phase offset per symbol is 15° while the sliding correlation is properly employed without the frequency offset. Hence, the accurate signal synchronization method for frequency and timing offset required when using the narrowband signals.

In this paper, the feature of MSK signals is utilized for the accurate signal synchronization. Fig. 3 shows the signal format for channel estimation and its phase variation using the MSK signals. In order to discriminate the signals from transmitter 1 (Tx1) and 2 (Tx2), different unique words (UWs,

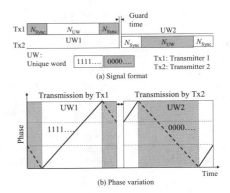

Fig. 3. Signal format for channel estimation and its phase variation.

UW1 and UW2) are assigned for Tx1 and Tx2, respectively. Since the phase variation is $\pi/2$ $(-\pi/2)$ per symbol for a bit "1" ("0"), the phase variations are plus and minus for UW1 and UW2, respectively, and the signals from Tx1 and 2 can be discriminated at the receivers.

Since it is very difficult to obtain the initial timing of data packet as shown in Fig. 2, the phase variations on the previous and next data bits for UW1 and UW2 are reversed as shown in Fig. 3. By preparing the data format in Fig. 3, the initial timing and phase variation of UW1 and UW2 can be accurately obtained. $N_{\mathrm{UW}}$ and $N_{\mathrm{Sync}}$ denote the number of bits for UW and signals of reversed phase variation.

After the detection on initial timing of UW1 and UW2, the frequency offset is employed by using the phase variations on UW1 and UW2. Fig. 4 shows the phase difference between transmit and receive signals for UW1. Since the accurate phase variation on UW1 is $\pi/2 \cdot N_{\mathrm{UW}}$, the phase difference per symbol, $\Delta\theta_1$ is $\Delta\theta_1 = (\theta_1 - (\pi/2) \cdot N_{\mathrm{UW}})/N_{\mathrm{UW}}$ when the actual phase variation is $\theta_1$. In the proposed method, the

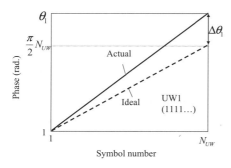

Fig. 4.   Phase difference between transmit and receive signals (UW1).

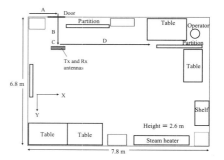

Fig. 5.   Measurement environment.

following equation is used instead of this equation:

$$\Delta\theta_1 = \frac{1}{N_{UW} - 1} \sum_{t=1}^{N_{UW}-1} (\theta_1(t+1) - \theta_1(t)) - \frac{\pi}{2}, \qquad (2)$$

where $\theta_1(t)$ denotes the phase at the time $t$. After the calculation of Eq.(4), the received signal, $r(t)$ at the time $t$ is compensated as

$$r'(t) = r(t) \exp(-j \cdot \Delta\theta_1 \cdot t) \qquad (3)$$

where $r'(t)$ denotes a compensated signal at the time $t$. Regarding UW2, the phase difference, $\Delta\theta_2$ due to the frequency offset is obtained by using $\pi/2$ instead of $-\pi/2$ on the second term of right side in Eq.(2).

The timing offset arises as the phase difference on the initial timing of UW1 and UW2 after the compensation on frequency offset. When the phase due to timing offset is $T_{s_j}$, the compensated signal, $r''(t)$ is denoted as

$$r''(t) = r'(t) \exp(-j \cdot \Delta T_{s_j}) \quad (j = 1, 2). \qquad (4)$$

After the compensation of the frequency and timing offsets, the channel estimation is employed. When the signals for UW1 and UW2 are $s_{UW1}(t)$ and $s_{UW2}(t)$, the estimated channel, $\tilde{h}_{ij}$ ($i = 1 \sim 2, j = 1 \sim 2$) are denoted as

$$\tilde{h}_{i1} = \sum_{t=1}^{N_{UW}} \frac{r(t) \exp(-j \cdot \Delta\theta_1 \cdot t) \exp(-j \cdot \Delta T_{s_1})}{N_{UW} \cdot s_{UW1}(t)}, \quad (5)$$

$$\tilde{h}_{i2} = \sum_{t=1}^{N_{UW}} \frac{r(t) \exp(-j \cdot \Delta\theta_2 \cdot t) \exp(-j \cdot \Delta T_{s_2})}{N_{UW} \cdot s_{UW2}(t)}. \quad (6)$$

## IV. INTRUDER DETECTION PERFORMANCE BY PROPOSED METHOD

In order to confirm the effectiveness of proposed channel estimation method, we conducted the measurement for the intruder detection. The measurement environment is shown in Fig. 5. A person moved the courses A to D in Fig. 5. The total measurement time was 14 sec. and the features on courses A to D are as follows:

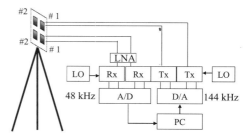

Fig. 6.   Setup of antennas and transceiver for measurement.

(A) The person moves across corridor and opens door with 4 sec.

(B) After opening door, the person moves to the antennas across Y-axis with 3 sec.

(C) The person changed the direction from Y-axis to X-axis with 2 sec.

(D) The person moves to the operation across X-axis with 5 sec.

Fig. 6 shows the setup of antennas and transceiver for the measurement. The measurement parameters are shown in Table I. In order to confirm the effectiveness of proposed method, local oscillator and clock by Tx is independently employed with those by Rx. As shown in Fig. 6, the transmit and receive antennas for Tx and Rx are set up and down. As the antenna element, micro strip antenna is adopted and its 3-dB beam width is approximately 80°. The element spacing of the array antenna is one wavelength. $N_{UW}$ and $N_{Sync}$ in Fig. 3 are 8 and 2, respectively.

In order to clarify the basic performance of proposed method, the time correlation characteristics without people is plotted in Fig. 7. The results without and with signal synchronization (compensation of signal) are shown in Fig. 7. As can be seen in Fig. 7, the time correlation is changed even in the static environment. On the other hand, the time correlation is approximately one regardless of the measurement time by

TABLE I
PARAMETERS FOR MEASUREMENT.

| Frequency | 2.4 GHz band |
|---|---|
| Bandwidth | 12kHz |
| Sampling rate (D/A) | 144 kHz |
| Sampling rate (A/D) | 48 kHz |
| Transmit power | -10dBm |
| Modulation scheme | MSK |

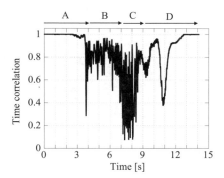

Fig. 9. Time correlation characteristics with an intruder (w/ compensation).

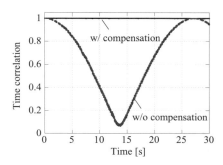

Fig. 7. Time correlation characteristics without people.

applying the proposed method.

Figs.8 and 9 shows the time correlation versus the measurement time without and with signal synchronization (compensation of signals). Eq.(1) is used to obtain the time correlation. As can be seen in Fig.8, the time correlation becomes low according to the time. Since the antenna is directed toward the end-fire direction when the person walks around the final point of D, it is difficult to detect the change on the propagation channel due to the person. Hence, the correlation should be approximately one at the final point of D, and the correlation is approximately one as shown in Fig.9. Moreover, the time correlation is changed according to the situation on the courses

A to B when the proposed method is employed. Therefore, it is shown that the proposed method can accurately estimate the change on the propagation channel due to the person.

## V. CONCLUSION

This paper has proposed the signal synchronization method for the channel estimation using the narrowband MSK signals. The proposed method realizes the signal synchronization by giving the reversal of phase variation on front and back of unique word even in a severe condition on frequency and timing offsets between transmitters and receivers. Moreover, it is shown that the proposed channel estimation method is effective when considering the intruder detection by MIMO sensor.

### ACKNOWLEGEMENT

The part of this work was supported by S1 Corporation.

### REFERENCES

[1] M. Nishi, S. Takahashi, and T. Yoshida, "Indoor human detection systems using VHF-FM and UHF-TV broadcasting waves," Proc. of IEEE International Symposium on Personal Indoor and Mobile Radio Communication 2006 (PIMRC2006), Session TJ7.3, Sept. 2006.
[2] S. Ikeda, H. Tsuji, and T. Ohtsuki, "Indoor Event Detection with Eigenvector Spanning Signal Subspace for Home or Office Security," IEICE Trans. Commun., vol.E92-B, no.7,pp.2406–2412, July 2009.
[3] S. Ikeda, T. Ohtsuki, and H. Tsuji, "Signal-Subspace-Partition Event Filtering for Eigenvector-Based Security System Using Radio Waves," IEEE International Symposium on Personal Indoor and Mobile Radio Communications (PIMRC2009), Tokyo, Japan, Sep. 2009.
[4] D Gesbert et al., "From theory to practice: an overview of MIMO space-time coded wireless systems," IEEE Journal on Selected Areas in Commun., vol.21, no.3, pp.281–302, April, 2003.
[5] K. Nishimori, Y.Koide, D. Kuwahara, N. Honma H. Yamada and H. Makino, "MIMO Sensor –Evaluation on Antenna Arrangement–," Proc. of EuCAP2011, pp. 2924-2928, April, 2011.
[6] J. W. Wallace and M. A. Jensen, "Time-Varying MIMO Channels: Measurement, Analysis, and Modeling," IEEE Trans. Antenna & Propagation, vol. 54, no.11, Nov. 2006.
[7] IEEE 802.11n, $http://www.ieee802.org/11n/$
[8] K. Nishimori, T. Mitsui, K. Ushiki and N. Takemura, "Channel estimation method using narrowband FSK signals for MIMO sensor," IEICE Trans. Commun. (in Japanese Edition), Vol.J-96B, No.9, 2013 (to be published).

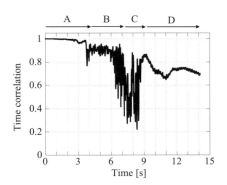

Fig. 8. Time correlation characteristics with an intruder (w/o compensation).

# *FA-1(B)*

## October 26 (FRI) AM

## Room B

## MMW & THz Antennas

# Transmission System for Terahertz Pre-amplified Coaxial Digital Holographic Imager

Wenyan Ji[#], Haitao Wang[#], Zejian Lu[#], Yuan Yao[#], Junsehng Yu[#], Xiaodong Chen[*]

[#]School of Electronic Engineering, Beijing University of Posts and Telecommunications, 279 Box, No.10 Xitucheng Road, Haidian District, 100876, Beijing, China

[*] School of Electronic Engineering and Computer Science, Queen Mary, University of London, UK

wenyanji_1989@126.com

*Abstract-* **The application of terahertz wave holographic imaging technique in the study of the disease detection has attracted great interest. In this paper, a high-resolution terahertz coaxial digital holographic imaging transmission system operating at 3THz is presented. This design is used in the THz imaging system for the study of the disease detection. The physical optical propagation analysis method (POP) and an advanced three dimensional numerical calculation software MATLAB are used to analyze the results. The simulation results indicate that the design meet the requirement of the proposed system.**

*Index Terms* — **THz imaging, Digital holography, pre-amplification, spatial resolution, THz propagation.**

## I. INTRODUCTION

Currently, THz imaging has received worldwide attention due to its extensive application potentials. Owning to the nondestructive and non ionizing capability, THz imaging shows great application prospects in biomedicine, security, and quality control [1]-[3]. Especially for the occasions where the visible light can't pass through and the contrast radio of the X-ray imaging system is not enough, THz imaging system is almost the only choice. With the development of suitable sources and detectors, there has been much research interest in THz imaging. The first THz imaging system based on electro-optic THz time-domain spectroscopy (THz-TDS) techniques was introduced by Hu in 1995 [4]. Years later, in order to improve the resolution to $\lambda/4$, S. Hunsche put forward the THz near-field imaging [5]. Then Mitrofanov reported the improved THz near-field imaging based on photoconductive antenna mechanism [6]. However, due to the long wavelength, the spatial resolution of these THz imaging systems is constrained and the obtained results are blurred. Therefore, improving the imaging resolution and image quality of THz imaging systems is always a research focus.

Digital holography is a new imaging technique, which numerically records the hologram and reconstructs the target from the hologram. Compared with the traditional THz-TDS imaging system, the greatest advantage of holography is that it can perform not only structural imaging, but also functional imaging, which can improve the image quality. Recently, digital holography has become a fast-growing research field with increasing attention [7]-[8]. Mahon carried out off-axis

digital holography experiment using a 100 GHz Gunn diode oscillator [9]. However, the resolution of the result is limited by the long wavelength. Usually, the resolution of the system is determined by the numerical aperture of the detector. Nevertheless, the detector applied in THz is still in the earlier stage of commercialization. To achieve high spatial resolution, the recording distance between the object and the detector must be very small, which makes it difficult to realize in practical engineering. Generally, the optical structure can be classified as the coaxial Fresnel digital holographic system without pre-amplified, the lensless Fourier transform digital holographic system, and the pre-amplified off-axis digital holographic system [10].

In this paper, considering the high-resolution, the coherence length and the image reconstruction, the transmission system of the pre-amplified coaxial digital holographic system is presented and numerically evaluated.

## II. PRINCIPLE OF THE DESIGN

This article reports on the use of holography and THz techniques to design a transmission system with target specifications of < 100um spatial resolution, a maximum imaging area of $\Phi 10mm \times 50mm$, and $11^0$ divergence angle.

The source spot of the system is QCL which resembles a Gaussian distribution. Based on the assumption of the paraxial approximation, the Gaussian beam propagation can be induced. In a homogeneous medium, the propagation of electromagnetic waves can be written by the Helmholtz equation

$$\nabla^2\psi + k^2\psi = 0 \tag{1}$$

Where $\psi$ is the component of E or H. Beam radius along the optical path can be deduced as flow

$$w = w_0[1+(\frac{\lambda z}{\pi w_0^2})^2]^{0.5} \tag{2}$$

The divergence of a Gaussian beam in the far-field region is given by the beam divergence half angle:

$$\theta \approx \frac{\omega(z)}{z} \approx \frac{\omega_0}{z_0} = \frac{\lambda}{\pi\omega_0} \tag{3}$$

In order to give full play to the role of the lens, usually the resolution capability of the detector is chosen to be higher than

978-7-5641-4279-7

the lens resolution capability. Therefore, the spatial resolution of the pre-amplified coaxial holographic system is determined by the numerical aperture of the lens. Fig.1 shows the diagram of the pre-amplified coaxial digital holographic system.

The ultimate resolution of the lens and the resolution of the detector are shown as follows

$$\triangle\delta=0.61\frac{\lambda}{\sin\theta}=0.61\frac{\lambda}{NA} \tag{3}$$

$$\triangle\delta=\frac{1}{M}\frac{\lambda d}{L_{CCD}} \tag{4}$$

Where $\triangle\delta$ is the spatial resolution, NA is the numerical aperture of the lens, M is the magnification, $L_{CCD}$ is the size of the effective detection area of the detector, d is the record distance.

For the pre-amplified coaxial digital holographic system, the minimum recording distance can be estimated by

$$\frac{1}{2\Delta N}=\frac{2}{\lambda}\sin\frac{\alpha_{max}}{2} \tag{5}$$

$$d_{min}=\frac{MX-L_{CCD}}{2\tan\alpha_{max}} \tag{6}$$

Where $\Delta N$ is the element size of the plane array detector, X is the size of the sample, d is the distance between the object and the CCD. In order to make the best of the imaging area of the detector, the spot in which all the object light can be received is chosen as the actual recording spot of the whole system, which is shown in Fig. 1.

The actual recording distance and the optical path difference can be estimated by the following equations:

$$d=\frac{MX-L_{CCD}}{MX+D}d_i \tag{7}$$

$$\Delta L=\sqrt{(\frac{D-X}{2})^2+d_0^2}+\sqrt{(\frac{D+L_{CCD}}{2})^2+d_\phi^2}-d_0-d_\phi \tag{8}$$

In this system, the Off-Axis-Parabolic (OAPs) are used to change the direction of the optical and make the beam parallel. Fig.2 shows the parameters of the 90° off-axis parabolic mirror. Parabolic mirrors are the most common type of aspherical mirrors used in optical instruments. They are free

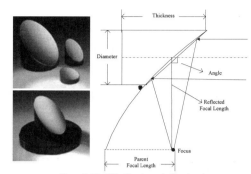

Figure 2. The 90° off-axis parabolic mirror

from spherical aberration and thus focus the parallel beam to a point or point source to infinity. In some applications the central part of the mirror obscures the beam path and therefore distorts the beam. For such systems, off-axis mirrors offer many advantages versus the traditional paraboloids, including minimized system size and cost.

Following the previously introduced processes, The structure of the pre-amplified coaxial digital holographic system is shown in Fig. 3.The key parameters of the pre-amplified coaxial digital holographic system can be determined. And the parameters are presented in Table I.

### III. SIMULATION VERIFICATION

Following the process previously introduced, the whole transmission system can be established, which is shown in Fig. 4. A POP method is used to analysis the results of the system. Fig.5-Fig.9 shows the irradiance and beam size in each surface.

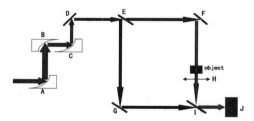

Figure 3.The optical structure diagram of the pre-amplified coaxial digital holographic system. (A,B,C: OAP, D,F,G: mirror, E:splitter, H: lens, I: combiner, J:detector.)

TABLE I
PARAMETERS OF THE TRANSMISSION SYSTEM

| λ | X | M | NA | $L_{CCD}$ |
|---|---|---|---|---|
| 100um | 10mm | 6 | 0.8 | 10mm |
| •δ | D | $d_0$ | d | •L |
| 76um | 40mm | 15mm | 45mm | 12.7mm |

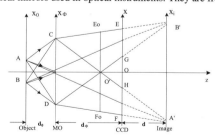

Figure 1.The diagram of the pre-amplified coaxial digital holographic system

Figure 4. The diagram of the transmission part of the pre-amplified coaxial digital

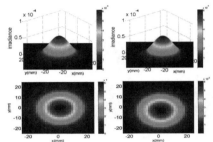

Figure 5. The irradiance and beam size on the surface A and B.

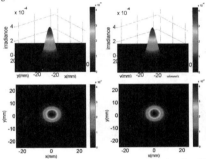

Figure 6. The irradiance and beam size on the surface C and D.

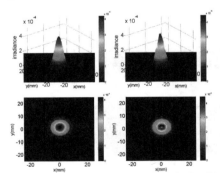

Figure 7. The irradiance and beam size on the surface E and F.

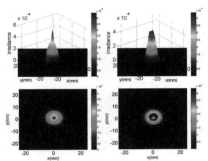

Figure 8. The irradiance and beam size on the surface G and H.

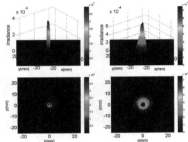

Figure 9. The irradiance and beam size on the surface I and J.

From the figures we can see that the THz beam transmits through the whole system successfully with no large scattering, and the beam size reaches the requirements.

## IV. CONCLUSION

This paper presents the design of the transmission system for the THz pre-amplified coaxial digital holographic imager. The system is simulated by a three dimensional visual software, and an advanced numerical calculation software MATLAB is used to analyze the results. The simulated results are presented, and the required specifications are acquired.

### ACKNOWLEDGMENT

This research is supported by the special-funded program on national key scientific instruments and equipment development with number 2011YQ13001802 and the Fundamental Research and Funds for the Central Universities.

### REFERENCES

[1] E. Pickwell and V. P. Wallace, "Biomedical application of terahertz technology," J. Phys. D 39, R301-R310 (2006).
[2] C. Jansen, S. Wietzke, O. Peters, M. Scheller, N. Vieweg, M. Salhi, N. Krumbholz, C. Jordens, T. Hochrein, and M. Koch, "Terahertz imaging: applications and perspectives," Appl. Opt. 49, E48-E57 (2010).
[3] H. B. Wallace, "Analysis of RF imaging applications at frequency over 100GHz," Appl. Opt. 49, E38-E47 (2010).
[4] B. B. Hu and M. C. Nuss, "Imaging with terahertz waves," Opt. Lett. 20, 1716-1718 (1995).
[5] S. Hunsche, M. Koch, I. Brener, "THz imaging in the near-field," May. Lasers and Electro-Optics, Vol11, E64-E65 (1997).

[6] O. Mitrofanov, M. Lee, L.N, K.W. West, J.D. Wynn, J. Federici, "THz transmission near-field imaging with very high spatial; resolution," Lasers and Electro-Optics, E515-E516 (2001).

[7] U. Schnars and W. Juptner, "Digital recording and numerical recording struction of holograms," Meas. Sci. Technol, vol. 13,no. 9.pp.85-101, 2002.

[8] A.F. Doval, "A systematic approach to tv holography," Meas. Sci. Technol, vol. 11,p. 36,Jan. 2000.

[9] R.J. Mahon, J.A. Murphy, W. Lanigan, Opt. Commun. 260, 469 (2006)

[10] B. A. knyazev, A. L. Balandin, V. S. Cherkassky, Y. Y. Chjoporova, "Classic holography, tomography and speckle metrology using a high-power terahertz free electron laser and real-time image detectors," Sept. IRMMW. 2010.

# Millimeter Wave Power Divider Based on Frequency Selective Surface

Wenyan Ji[#], Haitao Wang[#], Xiaoming Liu[#], Yuan Yao[#], Junsehng Yu[#], Xiaodong Chen*

[#]School of Electronic Engineering, Beijing University of Posts and Telecommunications, 279 Box, No.10 Xitucheng Road, Haidian District, 100876, Beijing, China

* School of Electronic Engineering and Computer Science, Queen Mary, University of London, UK

wenyanji_1989@126.com

*Abstract*- **Investigations into a device that separates one input signal into two signals of equal power, namely 3dB power divider, were carried out. This paper presents 3dB power divider using double layer metal slotted structure without any dielectric plate. The proposed structure with low coupling effect operates at 54GHz and has a 2GHz bandwidth. All the operating principles and simulation characteristics of the device are discussed in this paper.**

## I. INTRODUCTION

Power dividers are widely used in various microwave applications such as antenna feeds, balanced mixers, balanced amplifiers, and phase shifters due to its compactness, small weight and high reliability. For power divider, the properties such as insertion loss, isolation, phase imbalance, amplitude imbalance and other indicators have an important impact on the performance of the whole system. The most widely known power dividers are the Wilkinson, hybrid ring, and T-junction [1]-[2]. At millimeter frequency band, the miniaturized broadband power divider design is trapped by the following problems, including the smaller distance and the mutual coupling effect. Much efforts has been devoted to the design of the power diving devices [3]-[6] with lower loss in nature in the past two decades. Soroka et al[3], Bozzi et al[4] and Dittloff et al[5] individually reported power dividers based on waveguide structures. Taking isolation into consideration, Song et al[6] reported power dividers based on integrated waveguide in 2007. Although certain isolation between output ports has been realized, the insertion loss is considerably higher than those waveguide dividers.

Quasi-optical technology is an advanced technique in which various kinds of quasi-optical devices are used to adjust the electromagnetic signals. Quasi optics involves beams of radiation propagating in free space which are limited in lateral extent when measured in terms of wavelengths and for which diffraction is of major importance. In the area of quasi-optical techniques, the quasi-optical components can be used as polarization processing device, filters and diplexers, quasi-optical ferrite devices, resonators, quasi-optical power combining components and other quasi-optical components[7].

After compared with other design methods, the characteristics such as strong ability in dealing with high-power radiation, lower path and insertion loss, no isolation problem and powerful in multi-polarization processing, make

using the quasi-optical technique to design the power divider be the most suitable scheme.

There are many ways in the design of quasi-optical devices, such as dielectric plate, frequency selective surface (FSS), polarization grating and so on. In this paper, two types of FSS structure were proposed.

## II. PRINCIPLE OF THE METHOD

Due to the high frequency of millimeter-wave, many factors need to be taken into consideration when choosing the design scheme. Using dielectric plate to allocate the power has various advantages, such as easy to design, more parameters can be optimized.

$$\begin{cases} \Gamma = \dfrac{\sqrt{\varepsilon_{r2}} - \sqrt{\varepsilon_{r1}}}{\sqrt{\varepsilon_{r2}} + \sqrt{\varepsilon_{r1}}} \\[3mm] T = \dfrac{2\sqrt{\varepsilon_{r1}}}{\sqrt{\varepsilon_{r2}} + \sqrt{\varepsilon_{r1}}} \end{cases} \tag{1}$$

Where εr1 and εr2 represent the dielectric constant of the two medium respectively. Then the ratio of the reflected energy and the incident energy can be expressed as $\Gamma^2$. In order to design the 3dB power divider, the reflection coefficient should satisfy the condition (2). After deduced from (1) and (2), the final design parameters of the dielectric plate can be summarized as (3)□

$$\Gamma^2 = 0.5 \tag{2}$$

$$\varepsilon_{r2} = 9\varepsilon_{r1} \tag{3}$$

However, this design method also brings out some disadvantages, for instance, the operating bandwidth, the material loss and the power affordability of the material. Thus, this approach has the possibility in practical application, but the material is a critical problem.

FSS is a periodic assembly of one- or two-dimensional resonant structures, either as apertures in a thin conducting sheet or as metallic patches on a substrate, which may have a band-pass or band-stop function respectively. The FSS structure has a phenomenon with high impedance surface that reflects the plane wave in-phase and suppresses surface wave [8]. FSS are used in various applications, such as reflector

978-7-5641-4279-7

antenna system of a communication satellite, deep space exploration vehicle for multi frequency operations, band pass radomes for missiles etc [9]-[11]. Frequency Selective Surface is one of the most widely used device in quasi-optical system, with the main function of selectively reflecting and transmitting the electromagnetic waves of different frequencies. After compared with dielectric plate, easy to realize and simple structure make FSS become the most suitable design scheme.

In this paper, considering the simple structure, the enough bandwidth and the coupling effect in improving the in-band characteristics, the double-layer metal slotted structure was chosen. And two double-layer slotted shapes were chosen for the simulation.

### A. Rectangular slotted structure

In the process of FSS design, all the characteristics of FSS are not only related with the geometric shape, the size, the arrangement, the cycle, the thickness of the dielectric substrate, the electromagnetic properties, the number of layers of the unit cell, but also with the incident direction and polarization direction of the incident wave. As stated in the Introduction, the design started with the double-layer metal slotted FSS unit cell with the following parameters. All the structure diagram of this slotted FSS is showed in Fig. 1(a), the incident angle and incident way is shown in Fig. 1(b) and Fig. 1(c) presents the three-dimensional simulation model.

The detailed parameters of the FSS structure are listed in Table I and the specific definitions of them are marked in Fig. 1(a).

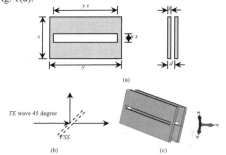

(a)

(b)                    (c)

Figure 1. (a) Structure diagram of the slotted FSS (b) The incident angle and incident way (c) Three-dimensional simulation model

Table I
Fixed parameters

| parameter | The definition of parameter | value |
|---|---|---|
| d | The distance between the two layers | 0.65mm |
| t | The thickness of the metal layer | 0.071mm |
| x | The periodic length in x direction | 3.7mm |
| xs | The slot width in x direction | 0.3mm |
| y | The periodic length in y direction | 6.2mm |
| ys | The slot width in y direction | 5.88mm |
| theta | The spherical angle of incident direction | 45deg |
| phi | The azimuth of the incident plane | 90deg |

### B. Hexagonal structure

All the structure diagram of this slotted FSS is showed in Fig. 2(a), the incident angle and incident way is shown in Fig. 2(b) and Fig. 2(c) presents the three-dimensional simulation model.

The detailed parameters of the FSS structure are listed in Table II and the specific definitions of them are marked in Fig. 2(a).

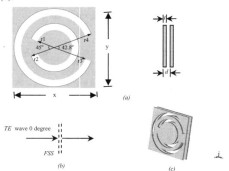

(a)

(b)                    (c)

Figure 2. (a) Structure diagram of the slotted FSS (b) The incident angle and incident way (c) Three-dimensional simulation model

Table II
Fixed parameters

| parameter | The definition of parameters | value |
|---|---|---|
| d | The distance between the two layers | 0.5mm |
| t | The thickness of the metal layer | 0.15mm |
| x | The periodic length in x direction | 4.4mm |
| y | The periodic length in y direction | 4.4mm |
| r1 | The radius of circle | 1.14mm |
| r2 | The radius of circle | 1.46mm |
| r3 | The radius of circle | 1.78mm |
| r4 | The radius of circle | 2.10mm |
| theta | The spherical angle of incident direction | 0deg |
| phi | The azimuth of the incident plane | 0deg |

### III. SIMULATION RESULTS

Following the process previously introduced, the two types of frequency selective surface unit cell can be established, which are shown in Fig. 1[c] and Fig. 2[c] respectively. And Fig. 3- Fig. 7 display the simulation results of the rectangular slotted structure, while Fig. 8-Fig. 12 display the simulation results of the hexagonal structure.

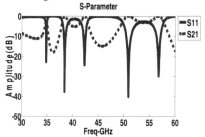

Figure 3.   TE wave reflection and transmission s-parameters

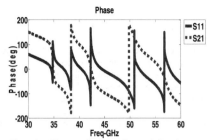

Figure 4.　TE wave reflection and transmission phase

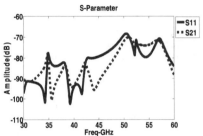

Figure 5.　S-Parameters of TE coupling to TM

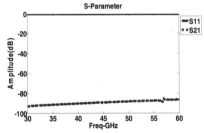

Figure 6.　TM wave reflection and transmission S-Parameter

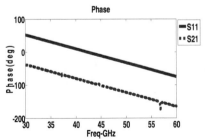

Figure 7. TM wave reflection and transmission phase

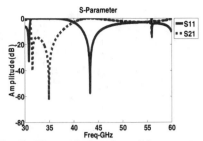

Figure 8.　TE wave reflection and transmission s-parameters

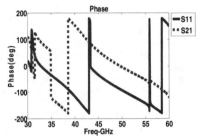

Figure 9.　TE wave reflection and transmission phase

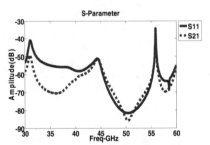

Figure 10. S-Parameters of TE coupling to TM

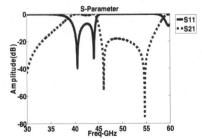

Figure 11. TM wave reflection and transmission S-Parameter

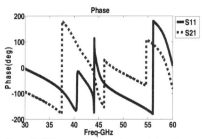

Figure 12. TM wave reflection and transmission phase

## IV. CONCLUSION

In this paper, two types of double-layer metal slotted frequency selective surface used for the 3dB power divider application were introduced and realized. After analyzing and calculating from the simulation results, it is clear that the rectangular slotted structure power divider operates at 54 GHz, with the bandwidth of 2GHz, and meets 90 degree phase difference requirements, while the hexagonal structure failed to satisfy the phase difference requirements. This type of FSS named as 3dB power divider can be applied in the aerospace situation where we need the power in two directions be equal.

ACKNOWLEDGMENT

This research is partly supported by the Fundamental Research Funds for the Central Universities of China and the Civil Aerospace Science and Technology Pre-research Program of China.

For A structure, Fig. 3 and Fig. 4 are the simulation results of S-Parameter and phase respectively when the incident wave is TE wave, also the detailed values of S-Parameter in different frequency points are shown in Table III. Fig. 4 shows that the phase difference in 53GHz-55GHz is about $90^0$. Both the figure and data prove that the proposed FSS structure can realize the design of 3dB power divider. Fig. 5 represents the transmission and reflection S-Parameters of TE wave couples to the TM wave, both of them are less than -70dB in 53GHz to 55GHz band, which means that the coupling component is very small. The simulation results are shown in Fig. 6 and Fig. 7 when choose the TM wave as the incident wave.

For B structure, Fig. 8 and Fig. 9 are the simulation results of S-Parameter and phase respectively when the incident wave is TE wave, also the detailed values of S-Parameter in different frequency points are shown in Table IV. Fig. 9 shows that the phase difference in 53GHz-55GHz is about $90^0$.But the incident angle is $0^0$, so the phase difference can't meet the design requirement, it should be $180^0$. Therefore, the B structure can't be used to design the 3dB power divider. And Fig. 10, Fig. 11 and Fig. 12 represent the parametric curves just as introduced in A structure.

Table III
TE INCIDENT WAVE S-PARAMETERS FOR A STRUCTURE

| Freq(GHz) Amp(dB) | 53 | 54 | 55 |
|---|---|---|---|
| Reflection | -3.1911 | -2.7641 | -3.3418 |
| Transmission | -2.8372 | -3.2714 | -2.7024 |

Table IV
TM INCIDENT WAVE S-PARAMETERS FOR B STRUCTURE

| Freq(GHz) Amp(dB) | 53 | 54 | 55 |
|---|---|---|---|
| Reflection | -3.1246 | -3.0186 | -2.8998 |
| Transmission | -2.8991 | -3.0021 | -3.1237 |

REFERENCES

[1] E. Wilkinson, "An N-way hybrid power divider", IEEE Transactions on Microwave Theory and Techniques, vol. 8, no. 1, pp. 116–118, January 1960.
[2] David M. Pozar Microwave Engineering Third Edition[M]. Beijing Electronic Industry Press, 2006; 274-277
[3] A.S.Soroka, A.O.Silin,"Simulation of Multichannel Waveguide Power Dividers",MSMW'98 Symposium Proceedings. Kharkov, Ukraine, vol 2,pp.634-635, Sept.1998
[4] Maurizio Bozzi', Luca Perregrinil,"A Compact, Wideband, Phase Equalized Waveguide DividerKombiner for Power Amplification",33rd European Microwave Conference - Munich pp.155-158, 2003.
[5] J. Dittloff, J. Bornemann," Computer Aided Design of Optimum E-or H-Plane N-Furcated Waveguide Power Divider,"17th European Microwave Conference. pp.181-186,Otc 2003.
[6] K. Song, Y. Fan and Y. Zhang, "Investigation of a power divider using a coaxial probe array in a coaxial waveguide," Microwave, Antennas & Propagation, IET, vol.1, Issue 4.pp.900-903, Aug. 2007
[7] Goldsmith, P.F. "Quasi-Optical Techniques" Proceedings of the IEEE Volume: 80 Issue:11
[8] Hsing-Yi Chen and Yu Tao, —Bandwidth Enhancement of a U-Slot Patch Antenna Using Dual-Band Frequency-Selective Surface With Double Rectangular Ring Elements□, MOTL Vol. 53 No. 7, pp 1547-1553, (July 2011).
[9] Sung, G.H.-h, Sowerby, K.W.Neve , M.J.Williamson A.G , A Frequency selective Wall for Interface Reduction In Wireless Indoor Environments□ Antennas and Propagation Magazine, IEEE ,Vol 48,Issue 5, pp 29-37,Oct 2006.
[10] D.H. Werner and D. Lee, —A Design Approach for Dual-Polarised Multiband Frequency Selective Surface using Fractal Elements□ IEEE International Symposium on Antennas and Propagation digest vol. 3,Salt Lake City Utah, pp. 1692-1695, July 2000.
[11] T.K. Wu, —Frequency Selective Surface and grid array□, A Wiley Interscience publication, pp. 5-7, 1995.

# Equivalent Radius of Dipole-patch Nanoantennas with Parasitic Nanoparticle at THz band

M. K. H. Ismail[1], M. Esa[1], N. A. Murad[2], N. N. Nik Abd. Malik[2], M. F. Mohd. Yusoff[2], M. R. Hamid[1]

[1]UTM-MIMOS COE for Telecommunication Technology
[2]Department of Communication Engineering
Faculty of Electrical Engineering
Universiti Teknologi Malaysia
81310 UTM Johor Bahru, Malaysia

*Abstract*-Recent developments in nano-technology have jumped from well established radiowave concept and analogy. Many designs have been proposed to investigate the capabilities of this high frequency regime. In this article, the dipole-patch nanoantenna has been designed based on effective wavelength scaling and equivalent radius concept. The square cross-sectional dimension has been adapted from circular radius equivalent geometry. The concept that has been used could be a prime knowledge for designing more complex patch nanoantennas structures. In addition, the nanoantennas have a unique interest because of its capability in focusing light into a small gap region and it can be tuned at various frequency bands. Due to this, a parasitic nanoparticle is introduced at the center of the feed gap to provide a double gap region. This has enabled the field radiation to increase by approximately 22%. The enhancement factor of the field radiation will increase the efficiency of energy collector and light-emitting applications.

## I. INTRODUCTION

Terahertz (THz) technology has received voluminous attention worldwide. Devices exploiting this waveband are set to become increasingly important in a very diverse range of applications. Despite such a great potential, the analysis to describe the properties of THz devices is still lacking and more investigation are needed to be done. Numerous efforts have been published in employing radio frequency analogy into THz regime. Those works are now being realized and it is interesting to see how radiowave design analogy is being redefined at THz region.

A straightforward design rules for radiowave is not valid at THz. The design parameters are not proportional to wavelength, $\lambda$, because the penetration of wave cannot be ignored anymore [1]. The different concept between radiowave and THz antennas has been discovered through plasmonic nanowire antennas where the volume current is considered for THz analysis [2]. Furthermore, the frequency dispersion of the effective permittivity must be taken into account for THz region [3].

Taking into account of the design rules, the performance of nanoantennas need to be further considered. The field enhancement is one vital property in THz. The bowtie antenna is studied with dipole, where the bowtie is capable in enhancing the field better than dipole [4]. Numerous articles have been reported, to demonstrate the field enhancement in order to increase their performances [5], [6], [7].

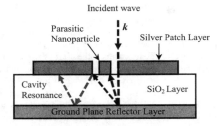

Figure 4. Cross sectional geometry of a dipole-patch with parasitic nanoparticle

## II. DIPOLE PATCH NANOANTENNA DESIGN

The dipole-patch nanoantenna was designed based on effective wavelength, $\lambda_{eff}$. The direct scaling law for effective wavelength, $\lambda_{eff}$ has been derived from [8] as

$$\lambda_{eff} = n_1 + n_2[\lambda/\lambda_p] \qquad (1)$$

where $\lambda_p$ is the plasma wavelength and $n_1 + n_2$ are dependent coefficients of dimensions with regard to the antenna geometry.

A single antenna segment (rod wire) has been used to derive the effective wavelength for a half-wave dipole. The antenna radius, $R$ is assumed to be smaller than the wavelength, $\lambda$ ($R \ll \lambda$). The effective wavelength can be written as:

$$\lambda_{eff} = \frac{\lambda}{\sqrt{\varepsilon_s}} \sqrt{\frac{4\pi^2 \varepsilon_s \left(R^2/\lambda^2\right)\tilde{z}(\lambda^2)}{1 + 4\pi^2 \varepsilon_s \left(R^2/\lambda^2\right)\tilde{z}(\lambda^2)} - 4R} \qquad (2)$$

where $\varepsilon_s$ is dielectric constants for vacuum medium [8].

Later, the Drude theory has been generalised to define the frequency dependency of silver permittivity at THz band. According to [7],[9], the silver permittivity is given by $\varepsilon_{Ag} = \varepsilon_\infty - \omega_p^2/\omega(\omega-i\delta)$. The properties of silver, Ag used are $\varepsilon_\infty \approx 3.57$, $\lambda_p \approx 135$ nm and $R = 5$ nm. The symbol of $\varepsilon_\infty$ is defined

978-7-5641-4279-7

as a contribution of dielectric function due to interband transitions. The silicon oxide, $SiO_2$ has been applied to be dielectric layer with bulk dielectric constant of 2.13. The correlation of effective wavelength based on Equation (2) is shown in Fig. 1.

At this point, an equivalent radius concept is used by substituting the circular rod radius, $R$ with noncircular cross section, $l \times l$ which is the equivalent radius of the circular rod as depicted in Fig. 2a. The equivalent radius correlation can be referred to [10] as

$$R = 0.59l \qquad (3)$$

By replacing $R = 5$ nm, $l \cong 8.5$ nm. The proposed dipole-patch cross section will then become $l \times l = 8.5$ nm x 8.5 nm as depicted in Fig. 2b. In this article, two cases has been studied, which are $L_1 = 300$ nm and $L_2 = 200$ nm. The feed gap distance between dipole-patch arms is set at 2 nm.

The parasitic nanoparticle is then introduced at the center of feed gap. The distance between nanoparticle and dipole arm is 2 nm. The objective is to enhance the field strength of the dipole-patch nanoantenna. Higher field strength is predicted as incident waves having two feed gap to trap and collect the field. The presence of a ground plane reflector will enable a directive lobe. This will furthermore resulted in the incident wave to be collected more effectively through the beam angle. The dipole-patch with parasitic nanoparticle is illustrated in Fig. 3.

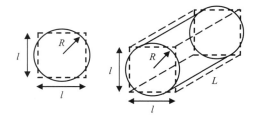

(a)        (b)

Figure 2: Circular cylinder geometrical shape and equivalent noncircular cross section.

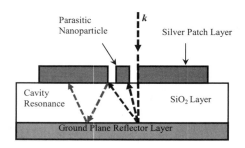

Figure 3. Cross sectional geometry of a dipole-patch with parasitic nanoparticle

## III. RESULTS AND DISCUSSION

In order to verify the theory of equivalent radius concept at THz band, half-wave dipole-patches is simulated and numerically analysed. A dipole-patch of $L_1 = \lambda_1/2 = 300$ nm has been designed. The dominant peak is observed at $\lambda_1 = 2068$ nm (or 145 THz) as shown in Fig. 4. On the other hand, an effective wavelength is calculated as $\lambda_{eff1} = 2L_1 = 600$ nm which agrees well with an incident wavelength of $\lambda = 2080$ nm as portrayed in Fig. 1. A second dipole-patch of $L_2 = \lambda_2/2 = 200$ nm is designed to further investigate the theory. The simulated dominant peak is observed at $\lambda_2 = 1580$ nm (or 190 THz) and the calculated $\lambda_{eff2} = 2L_2 = 400$ nm is shown at $\lambda = 1520$ nm. All results are tabulated in Table 1. It has been demonstrated that the effective wavelength described in (2) with noncircular equivalent geometrical is closely equivalent to the simulated result.

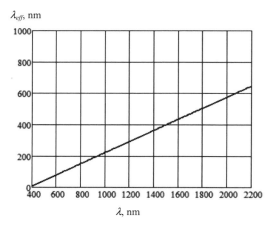

$\lambda_{eff}$, nm

$\lambda$, nm

Figure 1. Cross sectional geometry of a dipole-patch with parasitic nanoparticle

Table 1: Simulations and calculations of dipole-patch resonance with respect to effective wavelength

| Half-wave dipole, $L$ (nm) | Effective Wavelength, $\lambda_{eff}$ (nm) | Resonances | |
|---|---|---|---|
| | | Simulations (nm) | Calculations (nm) |
| $L_1 = 300$ | 600 | 2068 | 2080 |
| $L_2 = 200$ | 400 | 1580 | 1520 |

Next, the local electrical field strength is probed at the location of the centre feed gap. The plane wave is used to excite the dipole-patch antenna with a value of 1 V/m. Fig. 3 shows the local electrical field for $L_1$ and $L_2$ where the values are recorded at 392 V/m and 380 V/m, respectively.

Then, the field increment with parasitic nanoparticle is studied. The nanoantennas are denoted as $L_{1p}$ and $L_{2p}$. The field distributions for both nanoantennas are shown in Fig. 5. The corresponding dominant resonances peaked at 145 THz and 190 THz, respectively. The peak resonance remains the same with the results shown in the dipole-patch without parasitic nanopartical. It is that the dominant resonance has not been controlled by the parasitic element. On other hand, higher field strength is observed as expected. The corresponding field radiations have been increased approximately at 440 V/m and 418 V/m, respectively. The increments for both cases are $\cong 11\%$ at each gap. If two feed gaps are considered, the total field collection can be achieved up to 22%.

The local electrical field distribution for $L_1$ is observed at the reference plane through y-axis. It can be seen that the field distribution is concentrated higher at the feed gap as depicted in Fig. 6. Moreover, for $L_{1n}$, two peaks of electrical field concentrated at the two feed gaps in order to double the field strength as depicted in Fig. 7. The model has been generated using electromagnetic modeling software, CST Studio illustrated that the field flows and concentrated at the feed gap. This evidence signified a correct location to amass the energy and transfer it for conversion process. Therefore, an efficiency of nanoantennas can be enhanced for energy collector and light emitting applications.

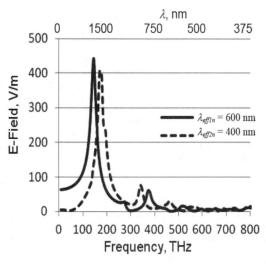

Figure 5: Spectral resonances of dipole-patch nanoantenna with parasitic element.

Figure 6: Local electrical field amplitude distributions of dipole-patch nanoantenna.

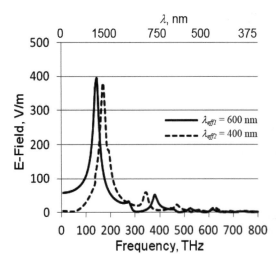

Figure 4: Spectral resonances of half-wave dipole-patch nanoantenna.

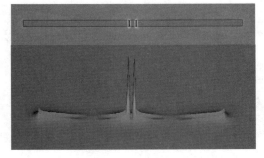

Figure 7: Local electrical field amplitude distributions of dipole-patch nanoantenna with parasitic nanoparticle.

## IV. CONCLUSION

This article presents an analytical model to describe the dipole-patch nanoantennas with parasitic nanoparticles at THz band. The effective wavelength analogy of circular wire has been successfully transformed to noncircular cross-section of dipole-patch. Having obtained an equivalent radius concept, it is simple to expand the radiowave theory into THz region. The agreement is not limited to the equivalent radius of cross-section only. It can also enhance the electrical field radiation with the introduction of parasitic nanopartical at the feed gap. The increment of 11% has been achieved through the proposed design. This field enhancement is a crucial property for energy collector and light-emitting applications.

## ACKNOWLEDGMENT

The work is supported by Universiti Teknologi Malaysia (UTM), Research University Grant votes 08J55, 08J51 and Malaysia Ministry of Education (formerly Ministry of Higher Education), Fundamental Research Grant Scheme vote 4F039. The authors would like to acknowledge Faculty of Electrical Engineering, UTM for conference support. The authors would also like to thank Agensi Angkasa Negara (ANGKASA), Ministry of Science, Technology & Innovation, and Public Service Department of Malaysia for supporting PhD studies of the first author.

## REFERENCES

[1] Lukas Novotny and Niek van Hulst. 2011. Antennas for light. *Nature Photonic.* Vol. 5. pp. 83-90

[2] Jens Dorfmuller, Ralf Vogelgesang, Worawut Khunsin, Carsten Rockstuhl, Cristoph Etrich and Klaus Kern. 2010. Plasmonic Nanowire Antennas: Experiment, Simulation and Theory. *Nano Letter.* Vol. 10. pp. 3596-3603

[3] F. J. Gonzalez, G. Almpanis, B. A. Lail, and G. D. Boreman. 2004. Wave propagation in planar antennas at THz frequencies. *Antennas and Propagation Society International Symposium.* Vol. 1. pp. 113-116

[4] Sakhno M. V and Gumenjuk-Sichevska J. V. 2010. Electric Field Enhancement Computation for Intergrated Detector of THz range. *International Kharkov Symposium on Physic and Engineering of Microwave, Millimeter & Submillimeter Wave (MSMW)*

[5] Hideaki Tanaka, Yusuke Sugitani, Jiro Kitagawa, Yutaka Kadoya, Francois Blanchard, Hideki Hirori, Atushi Doi, Masaya Nagai and Koichiro Tanaka. 2010. Enhancement of THz field in a gap of dipole antenna. *35th International Conferences on Infrared, Millimeter and Terahertz Wave (IRMMW- THz)*

[6] Matthias D. Wissert, Andreas W. Schell, Konstantin S. Ilin, Michael Siegel and Hans-Jurgen Eisler. 2009. Nanoengineering and characterization of gold dipole nanoantennas with enhanced integrated scattering properties. *Nanotechnology.* Vol. 20. 7pp

[7] Bhuwan P. Joshi and Qi-Huo Wei. 2008. Cavity resonance of metal-dieletric-metal nanoantennas. *Optics Express.* Vol. 16, No. 14. pp. 10315-10322

[8] Lukas Novotny. 2007. Effective Wavelength Scaling for Optical Antennas. *Physical Review Letters.* Vol. 98. No. Issue 26. 4pp.

[9] Paul R. West, Satoshi Ishii, Gururaj V. Naik, Naresh K. Emani, Vladimir M. Shalaev and Alexandra Boltasseva. 2010. Searching for better plasmonic materials. *Laser & Photonics Rev.* Vol. 4. Issue 6. pp. 795-808

[10] Constantine A. Balanis. 2005. Antenna Theory: Analysis and Design. *3rd Edition, New York, Chichester, Brisbane, Toronto, Singapore, John Wiley & Son.*

# Design and Implementation of A Filtenna with Wide Beamwidth for Q-Band Millimeter-Wave Short Range Wireless Communications

Zonglin XUE, Yan ZHANG, *Member IEEE*, Wei HONG, *Fellow IEEE*
State Key Laboratory of Millimeter Waves, School of Information Science and Engineering
Southeast University, Nanjing, 210096, P. R. China
zlxue@emfield.org, yanzhang@seu.edu.cn, weihong@seu.edu.cn

*Abstract*-An integrated millimeter wave filtenna is proposed for Q-band indoor short range wireless communications (IEEE 802.11aj). Comprising of a substrate integrated waveguide (SIW) filter and a printed angled dipole with a reflector, the proposed filtenna performs a bandpass response over 43.5-47GHz and the wide beamwidth larger than 120° in main cuts. The proposed filtenna and its counterpart without the filter are simulated, fabricated, and measured. By comparing the measured return loss and radiation patterns between the filtenna and the original antenna, it is concluded that the performance of the filtenna is kept nearly no change within the operating frequency band while the out of band signal is effectively filtered.

*Keywords*-Angled dipole, Filter, Wide Beam-Width, Filtenna, Substrate integrated waveguide(SIW).

## I. INTRODUCTION

With the rapid development of communication technology, demand for high speed data transmission with limitations of allowed power and available frequency spectrum leads to the development of millimeter-wave (mmW) short range wireless communications. In China, the millimeter wave high data rate transmission standard called as Q-LINKPAN was proposed in 2010 and get adopted as 802.11aj (45GHz) by IEEE802.11 Working Group in 2012[1]-[3]. The spectrum of 43.5-47GHz in Q-band is proposed for the next generation WPAN application, supporting the ultra high speed access and interconnection among a variety of wireless terminals in the indoor environment.

Filtennas working at mmW band have been reported in literatures[4]-[6]. Usually, these filtennas exhibit sharp transmission characteristics due to the SIW filter contribution, which is expected for mmW communication systems.

According to the demand of indoor short-range wireless communication, the corresponding access devices need to provide complete indoor coverage. We propose to use a structure of printed angled dipole with reflector to realize a type of access-point antenna with beamwidths of 120° in two main cuts for the indoor application demand of IEEE 802.11aj (45GHz) standard[7].

In this paper, Q-band filtennas consisting of a five-order SIW inductive window filter and a printed angled dipole with reflector are designed, fabricated and measured. The measured results agree well with simulations.

## II. DESIGN OF THE SIW FILTER

SIW filters possess the advantages of low loss, high power capability, high efficiency, high Q-factor and low cost[8]-[12].

Considering the selectivity requirement, a five-order SIW inductive window filter with five inline SIW cavities for 43.5~47GHz is designed. The geometry of the five-order SIW inductive window filter is shown in Fig. 1. The corresponding parameters are shown in Table 1. Fig. 2 shows the simulated S-parameters of five-order SIW inductive window filter.

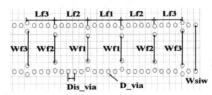

Figure 1. The geometry of five-order SIW inductive window filter

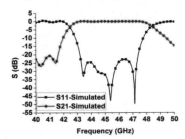

Figure 2. The simulated S-parameters of the five-order SIW inductive window bandpass filter

Table 1 The parameters of the five-order SIW filter(in mm)

| Parameters | Values | Parameters | Values |
|---|---|---|---|
| Lf1 | 2.81 | Wf1 | 1.78 |
| Lf2 | 2.71 | Wf2 | 1.95 |
| Lf3 | 2.3 | Wf3 | 2.46 |
| D_via | 0.3 | Wsiw | 3.3 |
| Dis_via | 0.6 | | |

978-7-5641-4279-7

### III. DESIGN OF THE FILTENNA

Based on the design of the printed angled dipole with reflector fed by SIW-CPW transition and SIW-RWG transition, Q-band filtennas consisting of a five-order SIW filter and the original angled dipole antenna are designed.SIW is used as the feed line, which provides differential feed and a tight integration with the reflector.

The SIW-CPW transition can achieve fine impedance matching by adjusting the slot length and the bending angle of CPW into SIW[13]. Meanwhile, the SIW-RWG transition can achieve fine impedance matching by adjusting the joint cavity size and the air window size connecting SIW and RWG[14]. The RWG uses the standard rectangular waveguide: WR19, whose size is 4.775mm*2.388mm, and the test flange is the UG-383.

In the design the substrate of Rogers 5880 with permittivity of 2.2 and the thickness of 0.508mm is used for designing and implementing the filtennas.Fig.3 shows the geometry of filtennas fed by SIW-CPW transition. Fig. 4 shows the geometry of filtennas fed by SIW-RWG transition.

The length of SIW filter is about 15mm, and the whole size of the filtennas are 20mm*34.2mm fed by SIW-CPW and 30mm*48.4mm fed by SIW-RWG. The optimized physical parameters of filtennas fed by SIW-CPW transition are shown in Table 2. The optimized physical parameters of filtennas fed by SIW-RWG transition are shown in Table 3. The parameters of the five-order SIW inductive window filter are adjusted slightly to make the fabrication available. All optimized parameters are got from CST software[15].

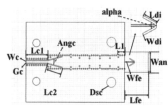

Figure 3. The geometry of filtennas fed by SIW-CPW transition

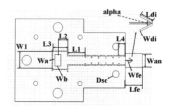

Figure 4. The geometry of filtennas fed by SIW-RWG transition

### IV. ANTENNA MEASUREMENTS

The photos of the filtennas and the original antennas without filters prototypes are shown in Fig. 5 and Fig. 6, which are fabricated with standard PCB process.

Table 2 The optimized parameters of filtenna fed by SIW-CPW transition(mm)

| Parameters | Values | Parameters | Values |
| --- | --- | --- | --- |
| alpha(°) | 55 | Dsc | 2.5 |
| Ldi | 1.35 | Angc(°) | 28 |
| Wdi | 0.2 | Wc | 0.6 |
| Wan | 5 | Gc | 0.2 |
| Wfe | 0.17 | Lc1 | 6 |
| Lfe | 2.4 | Lc2 | 3.9 |
| L1 | 1.9 | | |

Table 3 The optimized parameters of filtenna fed by SIW-RWG transition(mm)

| Parameters | Values | Parameters | Values |
| --- | --- | --- | --- |
| alpha | 55 | Wa | 4.2 |
| Ldi | 1.35 | Wb | 2.3 |
| Wdi | 0.2 | W1 | 6.64 |
| Wan | 5 | L1 | 6.7 |
| Wfe | 0.17 | L2 | 3.75 |
| Lfe | 2.4 | L3 | 5.86 |
| Dsc | 2.5 | L4 | 2.4 |

|S11| of four antennas are simulated and measured with vector network analyzer, and as shown in Fig. 7 and Fig. 8, well agreement between the simulated and measured results is achieved.

Fig. 7 shows the simulated and measured |S11| of printed angled dipole with reflector and the corresponding filtenna both fed by SIW-CPW transition. The measured impedance bandwidth of printed angled dipole with reflector fed by SIW-CPW transition for |S11| ≤ -10dB is 43.25-48.5GHz. The measured impedance bandwidth of the corresponding filtenna fed by SIW-CPW transition for |S11|≤-10dB is 43.15-47.4GHz. It can be seen that the impedance bandwidth of two antennas covers the band 43.5GHz-47GHz. And integrating filter with the original antenna steepens the |S11| curve due to the signals out of the band being suppressed effectively.

Fig.8 shows the simulated and measured |S11| of printed angled dipole with reflector and the corresponding filtenna both fed by SIW-RWG transition. The measured impedance bandwidth of the printed angled dipole with reflector fed by SIW-RWG transition for |S11|≤-10dB is 43.3-47.2GHz. The measured impedance bandwidth of the corresponding filtenna fed by SIW-RWG transition for |S11| ≤ -10dB is 43.35-47.1GHz. As pointed out above, it can be seen that the impedance bandwidth of two antennas covers the band 43.5GHz-47GHz, and integrating filter with the original antenna steepens the |S11| curve due to the signals out of the band being suppressed effectively.

Apparently the seamless integration of five-order SIW filter and the printed angled dipole with reflector will reduce the impact on antennas from the signals out of the band.

Figure 5. The photos of the filtennas and the original antennas without filters prototypes fed by SIW-CPW transition

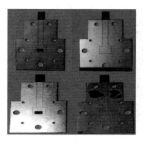

Figure 6. The photos of the filtennas and the original antennas without filters prototypes fed by SIW-RWG transition

The radiation patterns of four antennas are measured in the anechoic chamber. Fig. 9 shows the measured and the simulated radiation patterns of printed angled dipole with reflector and the measured radiation patterns of the corresponding filtenna both fed by SIW-CPW transition at E-plane (x-y plane) and H-plane (y-z plane) at 45.25GHz. Fig. 10 shows the measured and the simulated radiation patterns of printed angled dipole with reflector and the measured radiation patterns of the corresponding filtenna both fed by SIW-RWG transition at E-plane (x-y plane) and H-plane (y-z plane) at 45.25GHz.

Some metallic screws for fixing antennas in measurement and feeding transitions may have some little effect on radiation patterns. It can be seen that integrating filter just affects the radiation pattern of the radiation element slightly. The simulated 3-dB beamwidth of printed angled dipole with reflector fed by SIW-CPW transition in E-plane and H-plane is 135.8 degree and 126.1 degree. The simulated 3-dB beamwidth of corresponding filtenna fed by SIW-CPW transition in E-plane and H-plane is 138.5 degree and 120.2 degree. The simulated 3-dB beamwidth of printed angled dipole with reflector fed by SIW-RWG transition in E-plane and H-plane is 144.3 degree and 130 degree. The simulated 3-dB beamwidth of corresponding filtenna fed by SIW-RWG transition in E-plane and H-plane is 143.8 degree and 127.1 degree.

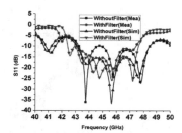

Figure 7. The simulated and measured |S11| of printed angled dipole with reflector and the corresponding filtenna both fed by SIW-CPW transition

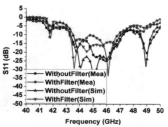

Figure 8. The simulated and measured |S11| of printed angled dipole with reflector and the corresponding filtenna both fed by SIW-RWG transition

Table 4 shows the measured gains and the simulated gains of the four antennas of 43.5, 45.25 and 47GHz.

The measured gains of the printed angled dipole with reflector fed by SIW-CPW transition varies from 5.34 to 5.76 dBi. The measured gains of the corresponding filtenna fed by SIW-CPW transition varies from 4.1 to 5.28 dBi. The measured gains of the printed angled dipole with reflector fed by SIW-RWG transition varies from 4.4 to 6.06 dBi. The measured gains of the corresponding filtenna fed by SIW-RWG transition varies from 4.04 to 5.34 dBi.

Table 4 The measured gains and the simulated gains of the four antennas

| Gain(dBi) | Fed by SIW-CPW | | | | Fed by SIW-RWG | | | |
|---|---|---|---|---|---|---|---|---|
| | Original | | Filtenna | | Original | | Filtenna | |
| Freq (GHz) | Mea | Sim | Mea | Sim | Mea | Sim | Mea | Sim |
| 43.5 | 5.76 | 6.1 | 4.1 | 5.4 | 4.4 | 5.61 | 4.04 | 5.04 |
| 45.25 | 5.34 | 6.07 | 5.27 | 5.7 | 6.06 | 6.38 | 5.34 | 5.35 |
| 47 | 5.49 | 6.37 | 5.28 | 6.1 | 5.73 | 5.83 | 5.13 | 5.44 |

V.    CONCLUSION

In this paper, the filtennas operating at 43.5GHz-47GHz band for Q-Band Millimeter-Wave short range wireless communication are designed, simulated and measured. Compared with the angled dipole with reflector, the performance of the filtenna is kept nearly no change within the operating frequency band while the out of band signal is effectively filtered. It can be seen that the designs, based on SIW technology, features of lightweight, compact size, low cost, and easy integration. It is believed that the filtennas presented in the article are promising for mmW ultra-throughput access and interaction requirements.

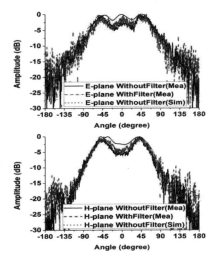

Figure 9. The measured and the simulated radiation patterns of printed angled dipole with reflector and the measured radiation patterns of the corresponding filtenna fed by SIW-CPW transition

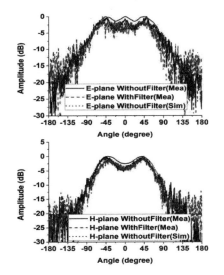

Figure 10. The measured and the simulated radiation patterns of printed angled dipole with reflector and the measured radiation patterns of the corresponding filtenna fed by SIW-RWG transition

ACKNOWLEDGMENT

This work was supported in part by National 973 project 2010CB327400 and in part by NSF of Jiangsu province under Grant SBK201241785.

REFERENCES

[1] W. Hong, et al (Invited), "Millimeter wave and THz communications in China," *IEEE IMS'2011*, Baltimore, USA, June 5-10, 2011.

[2] W. Hong, et al (Invited), "CMOS ICs for the Proposed Chinese Millimeter Wave Communication Standard Q-LINKPAN,"*APMC'2012*, Kaohsiung, Taiwan, Dec. 4-7, 2012.

[3] W. Hong, et al (Invited), "Recent Advances in Q-LINKPAN/IEEE 802.11aj (45GHz) Millimeter Wave Communication Technologies," APMC'2013, Seoul, Korea, Nov. 5-8, 2013.

[4] C. Yu, and W. Hong, "37-38GHz substrate integrated filtenna for wireless communication application," Microwave and optical technology letters, vol.54, No.2 , February 2012, pp. 346-351.

[5] S. H. Yu, W. Hong, C. Yu, H. J. Tang, J. X. Chen, and Z. Q.Kuai, "Integrated Millimerer Wave Filtenna for Q-LINKPAN Application,"6th European Conference on Antennas and Propagation (EuCAP'2012), Prague, March 26-30, 2012, pp. 1333-1335.

[6] C. Yu, W. Hong, Z. Q. Kuai, and H. M. Wang, "Ku-Band Linearly Polarized Omnidirectional Planar Filtenna," IEEE Antennas and Wireless Propagation Letters, vol. 11, 2012, pp. 310-313.

[7] Z. L. Xue, Y. Zhang, and W. Hong, "Design of Antenna with Wide Beamwidth for Q-Band Millimeter-Wave Short Range Wireless Communication," National Conference on Microwave and Millimeter waves (NCMMW2013), Chong Qing, China, May, 2013, pp.594-597.

[8] Z. C. Hao, W. Hong, X. P. Chen, J. X. Chen, K. Wu, and T. J. Cui, "Multilayered substrate integrated waveguide (MSIW) elliptic filter," IEEEMWCL, vol.15, no.2, 2005, pp.95-97.

[9] Y. L. Zhang, W. Hong, K. Wu, J. X. Chen and H. J. Tang, "Novel substrate integrated waveguide cavity filter with defected ground structure," IEEE Trans.on MTT, vol.53, no.4, April, 2005, pp.1280-1287.

[10] Z. C. Hao, W. Hong, J. X. Chen, X. P. Chen, and K. Wu, "Compact Super-Wide Bandpass Substrate Integrated Waveguide (SIW) Filters," IEEE Trans. on MTT, vol.53, no.9, 2005, pp.2968-2977.

[11] H. J. Tang, W. Hong, Z. C. Hao, J. X. Chen, and K. Wu, "Optimal design of compact millimeter wave SIW circular cavity filters," Electron. Lett., vol.41,no.19, 2005, pp.1068-1069.

[12] H. J. Tang, W. Hong, J. X. Chen, G. Q. Luo, and K. Wu, "Development of Millimeter-Wave Planar Diplexers Based on Complementary Characters of Dual-Mode Substrate Integrated Waveguide Filters With Circular and Elliptic Cavities," IEEE Trans. on MTT, vol.55, no.4, pp.776-782, 2007.

[13] Z. B.Wang, S. Adhikari, D. Dousset, C. W. Park, and K. Wu, "Substrate Integrated Waveguide (SIW) Power Amplifier Using CBCPW-to-SIW Transition for Matching Network," IEEE MTT-S International Microwave Symposium Digest, Montreal, Canada, June 17-22, 2012, pp. 1-3.

[14] K. D. Wang, W. Hong, and Ke Wu, "Broadband Transition between Substrate Integrated Waveguide (SIW) and Rectangular Waveguide for Millimeter-Wave Applications," Applied Mechanics and Materials Vols, 2012(130-134), pp. 1990-1993.

[15] CST Microwave Studio, User Manual Version 5, CST GmbH, Darmstadt, Germany (2004).

# Design of Terahertz Ultra-wide Band Coupling Circuit Based on Superconducting Hot Electron Bolometer Mixer

Chun Li, Lei Qin, Miao Li, Ling Jiang

Information Science and Technology College, Nanjing Forestry University, Nanjing, 210037

E-mail: 419726258@qq.com

**Abstract: Superconducting hot-electron-bolometer (HEB) mixer is the most sensitive detector in the terahertz (THz) frequencies. The design of RF coupling circuit is very crucial to the mixer's sensitivity. In this paper, we present a theoretical model for the calculation of the feed point matching impedance of the coupling circuit, and demonstrate the simulation results with the aid of 3D electromagnetic simulation software (HFSS) and a lumped-gap source method similar to quasi optical antennas in the frequency range of 0.9-1.3 THz. The influences of the slice tolerance of substrate thickness, the high-order modes, and the feed-point size on the embedding impedance at 0.9-1.3 THz have been studied in detail. The mixer's embedding impedance is simulated to be around 35-j10□ in the whole working band whose bandwidth can reach 36%. The simulation results are in good agreement with the theoretical results, which provide helpful instructions for the future developments of ultra-wide band and highly sensitive superconducting hot-electron-bolometer (HEB) mixer.**

*Key words : superconducting hot-electron-bolometer (HEB) mixer; embedding impedance; wide band matching; high-order mode; substrate thickness*

## I. INTRODUCTION

Superconducting hot-electron-bolometer (HEB) mixer is the most sensitive detector in the terahertz frequency band, based on the fixed-tuned waveguide coupling circuit and the good performance of the beam. It has been widely used in radio telescopes on the ground and space[1-3].

Embedding impedance is an important parameter to characterize the coupling efficiency between the mixer and the coupling circuit, which furthermore affects the mixer's sensitivity[4,5]. In order to achieve the high sensitivity and wide frequency bandwidth suitable for astronomical observation in terahertz frequency band, a fixed-tuned waveguide coupling circuit with ultra-wide band is proposed from 0.9 to 1.3 THz. We present the equivalent circuit model of the RF coupling circuit according to transmission line theory. The model is demonstrated to be in good agreement with the calculated results by using HFSS software. In addition, the effects of substrate thickness, the low-pass microstrip filter, and the feedpoint size on the embedding impedance have been investigated. This paper aims to optimize the structure of the transition of waveguide to microstrip and low-pass microstrip filter, further improves the embedding impedance matching to the mixer and increases the working frequency bandwidth of the mixer.

The mixer coupling circuit working at 0.9-1.3 THz, which presented in this paper will be used in Atacama Submillimeter Telescope Experiment Telescope in Japan, to achieve the observation of astrochemistry molecular spectral line.

## II. THEORETICAL CALCULATION

Figure 1 shows the schematic of the waveguide coupling circuit of the superconducting HEB mixer. RF signal input into the waveguide port, then pass through the transition of waveguide to microstrip, finally enter the HEB mixer in the feed point. The two symmetric low-pass microstrip filters (i.e., choke filter) are used to choke the RF signal and transmit the low frequency signal after mixing.

The feed point located at the superconducting HEB mixer (of 50-Ω normal-state resistance), which has a 2-μm width and 0.2-μm height, as shown in Fig. 1 (a). The matching impedance at lumped-gap source port is made up of several parts: the main mode transfer impedance $Z_{g10}$, backshort impedance under the back part of short circuit $Z_{bs}$, symmetrical impedance $Z_{chk1}$ and $Z_{chk2}$ of low-pass microstrip filter, as shown in Fig. 1 (b). The transition

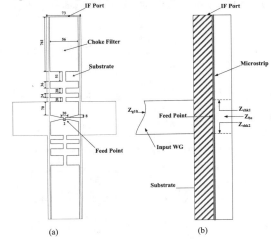

Fig.1 (a) Front view and (b) side view of a waveguide HEB mixer

978-7-5641-4279-7

probe from waveguide to microstrip (shadows in Fig. 2) are symmetrical according to the location of feed point.

The high-order modes probably exsit at the feed point due to the discontinuity of the transition of waveguide to microstrip. These non-propagating mode changes the E-field distribution of the feed point. We present the equivalent circuit model of the propagation of the high-order modes, as shown in Fig. 3.

The impedance of the waveguide main mode $TE_{10}$ and high-order modes are given by equation (1) and (2).

$$Z_{g10} = 2 \frac{b}{a} \frac{\eta}{\sqrt{1 - (\frac{f_c}{f})^2}} \quad (1)$$

here $\eta = \sqrt{\frac{\mu_0}{\varepsilon_0}} \approx 377\Omega$

$$Z_{mn} = i\eta \frac{b}{a} \frac{c}{2\pi f} ((\frac{2\pi f}{c})^2 - (\frac{n\pi}{b})^2)((2 - \delta_0)\gamma_{mn})^{-1} \quad (2)$$

$$\delta_0 = \begin{cases} 1 & n = 0 \\ 0 & n \neq 0 \end{cases}$$

$$\gamma_{mn} = \alpha + i\beta_{mn} \quad (3)$$

$$\alpha = \frac{8.686}{\eta\sigma\delta\, b\sqrt{1 - (\frac{f_c}{f})^2}} (2\frac{b}{a}(\frac{f_c}{f})^2 + 1) \quad (4)$$

$$\beta = \sqrt{(\frac{2\pi f}{c})^2 - (\frac{m\pi}{a})^2 - (\frac{n\pi}{b})^2} \quad (5)$$

In the equations, $c$ is the speed of light in a vacuum, under the temperature of 300 K, $\delta$ equal to $42.5 * 10^6 (\Omega)^{-1}$, about 65 nm, $Z_{mn}$ is the impedance of the $TE_{mn}$ and $TM_{mn}$ modes.

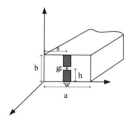

Fig.2 Size of coupling probe at the feed point is marked by cross

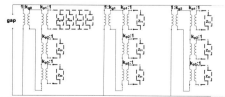

Fig.3 Equivalent circuit of the waveguide impedance at the feed point for different waveguide modes

The position of the metal and geometry dimension in waveguide is shown in Fig. 2, the coupling coefficients are shown in equation (6, 7).

$$k_{pm} = \sin(m\pi\frac{s}{a})\sin(m\pi\frac{w}{2a})(m\pi\frac{w}{2a})^2 \quad (6)$$

$$k_{gn} = \cos(n\pi\frac{h}{b})\sin(n\pi\frac{g}{2b})(n\pi\frac{g}{2b})^2 \quad (7)$$

Waveguide back is a short circuit, the resistance under the main mode is a function of frequency, as shown in equation(8).

$$Z_{bs} = Z_{g10} \frac{Z_t + Z_{g10} \tanh(\gamma_{10}l_{bs})}{Z_{g10} + Z_t \tanh(\gamma_{10}l_{bs})} \quad (8)$$

$Z_t$ as terminal short circuit impedance, approximate to zero under ideal conditions. The distance $l_{bs}$ from feed point to the back part short circuit is 38 μm.

In superconducting HEB mixer, due to the symmetrical structure of two low pass filter, $Z_{chk1}$ and $Z_{chk2}$ are equivalent. Considering each part's impedance contributions, the feeding point impedance can be calculated by equivalent circuit of Fig. 4.

Concerning of the capacitance effect of medium plate substrate, a capacitance constant $Y_{cex} = 12.3$ fF [6-7] should be added into the circuit. In Fig. 4, $X_L$ and $Y_R$ represent the parasitic impedance of high-order modes, as shown in the following expressions.

$$X_L = \sum_{m=2}^{2MI} k_{pm}^2 Z_{m0} \quad (9)$$

$$Y_R = \sum_{n=1}^{NI} [\sum_{m=1}^{MI} Z_{mn} (\frac{k_{pm}}{k_{gn}})^2]^{-1} \quad (10)$$

For symmetric structure of probe (s = a / 2, h = b / 2), obtained from equation (6, 7), there is no coupling coefficients. In frequency range of 0.9-1.3 THz, the influence of high-order modes are very small, thus $X_L$ and $Y_R$ approximate to zero.

III. SCALE-MODEL OPTIMIZATION

The transmission of multiple modes will lead to dispersion phenomenons and signal distortions, and reduce the transmission power into the HEB mixer. To ensure that only dominant mode of waveguide propagate in the range of 0.9-1.3 THz, the wavelength must satisfy the following conditions:

$$\lambda_{c01} < \lambda < \lambda_{c10} \quad (11)$$

$$\lambda_{c20} < \lambda < \lambda_{c10} \quad (12)$$

For single-mode propagation, the most suitable size ranges below:

$$a < \lambda < 2a \quad and \quad \frac{b}{a} \leq 2 \quad (13)$$

In conclusion, within the scope of 0.9-1.3 THz, the width and height of the waveguide is chose to be 200 μm and 100 μm

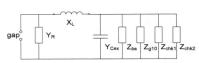

Fig.4 Equivalent network of impedance matching calculation

respectively. Fig. 5 is the model used in the simulation of the HFSS, colored areas are microstrip lines, to choke RF signals and transmit the low frequency signal.

Microstrip low-pass filter[8] in mixer is designed on the single crystal substrates ($\varepsilon_r$= 4.65), placed in a rectangular cavity with the size of 73 μm by 73 μm. The upper and lower microstrip filter is symmetrically arranged. High-impedance line works similar to series inductance, low-impedance line similar to parallel capacitor. Characteristic impedance of the filter ranges as shown follows:

$$Z_{0C} < Z_o < Z_{0L} \qquad (14)$$

$Z_{OL}$, $Z_{OC}$ corresponding to high- and low-impedance line characteristic impedances. Low-impedance line works as short circuit. The high-impedance line is equivalent to open circuit, and the width is limited for fabrication process and influence line of current-carrying capacity. In order to facilitate the optimization of the model, one part of the microstrip filters will be investigated for simulation, as shown in Fig. 6.

Initial size of the microstrip line is showed in Tab. I. We optimize the width and the length of impedance lines such as $a_1$, $a_2$, $a_3$, $b_1$, $b_2$, $b_3$, as shown in Fig. 6.

As seen from the reflection coefficient, with the increase of frequency, low pass filter has a bad performance at the high frequency band. With the increase of length of $b_1$, the reflection coefficient at higher frequency approach to 1. It means the RF signal is choked from IF port and enter the mixer, thus the size of the $b_1$ is fixed 70 μm, as shown in Fig. 7. Considering the entire optimizations, the simulation reflection coefficient is shown in Fig. 8.

## IV. OPTIMIZATION OF SUBSTRATE THICKNESS

We calculated the embedding impedance for different substrate thicknesses of the HEB mixer. The embedding impedance is calculated by the equation of $Z_{emb}=Z_0*(1+\Gamma)/(1-\Gamma)$, here $\Gamma$ is complex reflection coefficient at feed point. As exhibited in Fig. 9, the embedding impedance change is most flat in the whole frequency band when the substrate thickness is 35 μm. We consider the effect of higher-order mode in the HFSS simulation, and set the feed point port to be multi-mode. The simulation results indicate that the higher-order modes have slight impact on the embedding impedance. The simulated embedding impedance $Z_{emb}$ is approximately 35-j10 Ω, which is close to the impedance of the mixer(~ 35 Ω), it can achieve broad-band impedance matching within the scope of 0.9-1.3 THz.

We compare the calculation of the equivalent circuit theory model in Fig. 4 with the simulation results by using HFSS, and found that the two methods show basically identical result. The tolerance of the resistance and reactance is within 5 Ω, as shown in Fig. 10.

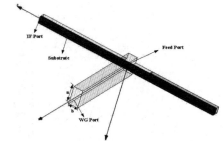

Fig.5 3D view of simulated model by HFSS

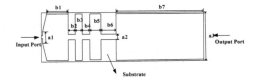

Fig.6 Planar graph of rectangular waveguide microstrip filter

TABLE. I

DIMENSIONS OF MICROSTRIP LINE BEFORE AND AFTER OPTIMIZATIONS（μm）

|  | $a_1$ | $a_2$ | $a_3$ | $b_1$ | $b_2$ | $b_3$ | $b_4$ | $b_5$ | $b_6$ | $b_7$ |
|---|---|---|---|---|---|---|---|---|---|---|
| before | 18 | 5 | 43 | 48 | 34 | 44 | 28 | 36 | 51 | 749 |
| after | 20 | 3 | 56 | 70 | 20 | 24 | 28 | 36 | 51 | 761 |

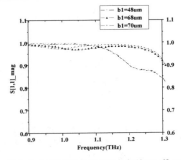

Fig.7 Reflection coefficient of microstrip filter when $b_1$ is 48 μm, 68 μm, 70 μm.

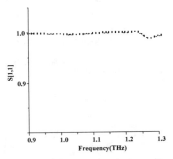

Fig.8 Reflection coefficient of RF signal propagates in the microstrip filter.

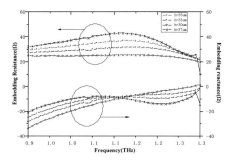

Fig.9 Embedding impedance at feed point when the substrate thickness is 33μm，35μm，37μm

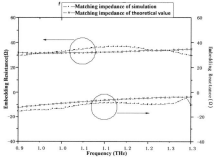

Fig.10 Comparison of the embedding impedance between the theoretical calculation and HFSS simulation.

## V. CONCLUSION

We have successfully designed a broadband terahertz coupling circuit of superconducting HEB mixer. We have optimized the coupling structure of the HEB mixer such as the dimensions of the rectangular waveguide, the microstrip filter, and the substrate thickness. The simulation result shows that the embedding impedance of the mixer is about 35-j10 in the frequency range of 0.9-1.3 THz. We present the equivalent coupling circuit with taking account into high-order modes. The theoretical calculation results demonstrate the simulation results by HFSS with the error below 5 Ω. The relative working bandwidth reaches 36 %. This paper is of great significance to implement the ultra-wide band terahertz mixer in the future.

## REFERENCE

[1] S. Cherednichenko, P. Khosropanah, E. Kollberg, M. Kroug and H. Merkel, "Terahertz superconducting hot-electron bolometer mixers," Physica C 372-376 407-415 (2002).
[2] W. Zhang et al, "Quantum noise in a terahertz hot electron bolometer mixer," Appl. Phys. Lett. 96(11), pp 111113-1-111113-3, (2010).
[3] W. Zhang et al, "Noise temperature and beam pattern of an NbN hot electron bolometer mixer at 5.25 THz,"Journal of Applied Physics, 108(9) (2010)
[4] L. Jiang and S. C. Shi, "Investigation on the resonance observed in the embedding-impedance response of an 850-GHz waveguide HEB mixer," IEEE Microw. Wireless Compon. Lett., vol.15, no. 4, April 2005.
[5] W Zhang et al, "Scaled Model Measurement of the Embedding Impedance of a 660-GHz Waveguide SIS Mixer With a 3-Standard Deembedding Method," IEEE Microw. Wireless Compon. Lett., vol.13, no. 9, September 2003.
[6] T.H. Büttgenbach, T.D. Groesbeck, and B.N. Ellison. A scale mixer model for SIS waveguide receivers. Int. J. Infrared Millimeter Waves, vol. 17, 1685, 11(1):1–20, Jan 1990.
[7] C.E. Honingh. A Quantum Mixer at 350 GHz based on Superconductor-Insulator-Superconductor SIS-Junctions. PhD thesis, Rijksuniversiteit Groningen, 1993.
[8] Cao Liang zu, Fu li, Design and Manufacture of Microwave Low pass Filter with Semi-lumped Components. Journal of Microwaves, 2007, 21(8)

# *FA-2(B)*

## October 31 (FRI) AM

## Room B

## MMW Antennas

# Design of a Linear Array of Transverse Slots without Cross-polarization to any Directions on a Hollow Rectangular Waveguide

Duong Nhu Quyen[1], Makoto Sano[1], Jiro Hirokawa[1], Makoto Ando[1], Jun Takeuchi[1,2], and Akihiko Hirata[2]
[1]Dept. of Electrical and Electronic Eng., Tokyo Institute of Technology
S3-19, 2-12-1 O-okayama, Meguro-ku, Tokyo 152-8552, JAPAN
E-mail: quyen@antenna.ee.titech.ac.jp
[2]Microsystem Integration Laboratories, NTT Corporation, Atsugi, 243-0198, JAPAN

*Abstract*-In this paper, we propose a linear array of transverse slots spaced with the half guided wavelength, which are fed by two hollow rectangular waveguides. No cross-polarization is radiated in any directions. The reflection is suppressed to below −14 dB in a wide frequency range and almost uniform illumination of all the slots is achieved in an eight-element array designed at 125 GHz.

## I. INTRODUCTION

The linear array of inclined slots on the narrow-wall is commonly used to radiate the polarization parallel to the waveguide axis [1]. As can be seen in Fig. 1, since slots with opposite inclination angles are alternatively placed with the separation of the half guided wavelength, it could radiate high cross-polarization at wide angles. In a transverse slot array on the broad-wall of a waveguide, grating lobes should be suppressed by filling dielectric material [2] or introducing transverse corrugations as slow-wave structures [3], because the slot spacing is one guided wavelength. In this paper, we propose a linear array of transverse slots with the half guided wavelength spacing fed by two rectangular waveguides. No cross-polarization is radiated in any directions.

## II. ANTENNA CONFIGURATION

Fig. 2 shows the antenna configuration. The antenna consists of an E-plane T-junction, two feeding waveguides, and radiating slots. The two feeding waveguides are alternatively fed through the E-plane T-junction. The radiating slots are placed with the interval of the half guided wavelength. The odd-numbered and even-numbered slots are fed through the upper and lower waveguides, respectively, in order to excite all the slots in phase. All the slots are perpendicular to the direction of the array arrangement, so that no cross-polarization is radiated in any directions. The antenna will be fabricated by diffusion bonding of laminated thin metal plates [4] [5].

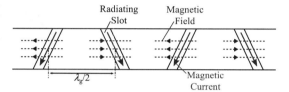

Fig. 1 A linear array of inclined slots on the narrow wall of a rectangular waveguide

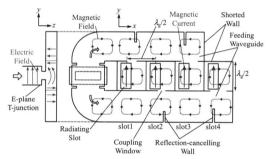

Fig. 2 Overall structure

## III. DESIGN OF COMPONENTS

### A. Radiating elements

Fig. 3 shows the design model of the radiating element. The radiation slot is placed on a cavity. The distance between the center of the slot and the shorted wall is the quarter guided wavelength.

Firstly, the resonant slot length is determined to have the reflection phase of 0 degree by analyzing it on a shorted waveguide. The same slot length is used for all the radiating elements. The coupling of the slots is controlled by the width of the coupling window. The reflection is suppressed by adjusting the length and the position of the reflection-canceling wall for the travelling wave excitation.

978-7-5641-4279-7

An array of eight slots is designed at 125 GHz. Four radiating elements are fed with each feeding waveguide. The radiating elements with the couplings of 25%, 33.3%, 50%, and 100% are designed for the uniform excitation. The radiating element with 100% coupling is realized by shorting the end of the feeding waveguide and its window width is same as that with 50% coupling.

Fig. 4 shows the reflection of the radiating element with 50% coupling for various slot widths. As can be seen in Fig. 4, the radiating element with 0.6mm slot width has the widest bandwidth. When the slot width is more than 0.6mm, the resonant condition is not satisfied. We adopt the slot width of 0.6mm to realize wideband operation.

## B. Feeding part

Fig. 5 shows the design model of the feeding part. The feeding part consists of an E-plane T-junction and H-bends. The E-plane T-junction is fed by a standard rectangular waveguide in the 120-GHz band (WR-8). The dimensions of the feeding window, the length and the position of reflection-canceling walls are determined to minimize the reflection at 125 GHz. The E-plane T-junction is connected to the H-bends, and the radius of the rounded-edge of the H-bends is tuned to suppress the reflection.

The reflection of the overall feeding part is shown in Fig. 6. The reflection is below −20 dB in a broad bandwidth.

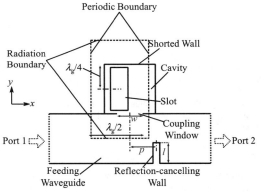

Fig. 3 Design model of the radiating element

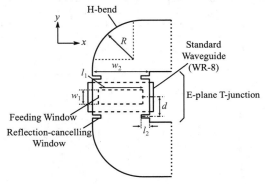

Fig. 5 Design model of the feeding part

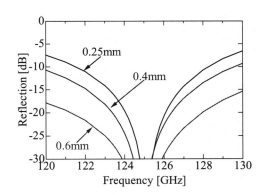

Fig. 4 The reflection of the radiating element with 50% coupling for various slot widths

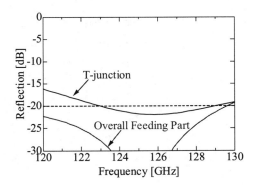

Fig. 6 Reflection of the feeding part

## IV. OVERALL STRUCTURE

The radiating part is connected to the feeding part to form the 8-slot array antenna. The spacings between the slots are determined using the transmission phase and radiation phase to excite all the radiating elements in phase [6]. The overall antenna structure is analyzed using Ansoft HFSS.

The radiation patterns in the E-plane at 125GHz is shown in Fig. 7. The HPBW is about 9 degrees and the first sidelobe level is −13 dB. The cross-polarization is suppressed to below −35 dB even though the wide slots are used.

The amplitude and phase on each slot are shown in Fig. 8. The deviations of the amplitudes and phases are less than 0.5 dB and 4 degrees. Almost uniform illumination is achieved.

Fig. 9 shows the frequency characteristic of the reflection. The reflection is below −14 dB in 120-130 GHz. However, the center frequency is shifted to 124 GHz.

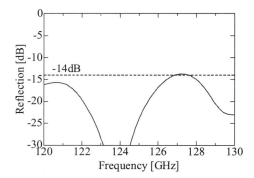

Fig. 9 Reflection of the overall structure

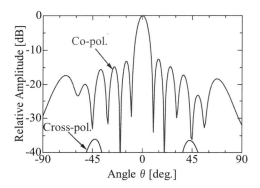

Fig. 7 Radiation patterns in the E-plane

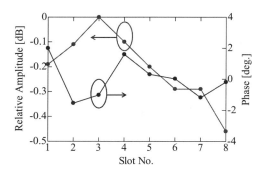

Fig. 8 Amplitude and phase on the slots

## V. CONCLUSION

The authors have designed and analyzed a linear array of transverse slots spacing with the half guided wavelength, which are fed by two rectangular waveguides. The authors have designed an eight-element array and confirmed the uniform aperture illumination in the simulation. The measured results will be shown in the conference.

## ACKNOWLEDGEMENT

This work is conducted in part by the Strategic Information and Communications R&D Promotion Programme, the Ministry of Internal Affairs and Communications.

## REFERENCES

[1] M. G. Chemin, "Slot admittance data at Kaband," IRE Trans. Antennas Propagat., vol. AP-4, pp. 632-636, Oct. 1956.
[2] L. Josefsson, "A waveguide transverse slot for array applications," IEEE Trans. Antenn. Propagat., vol. AP-41, no. 7, pp. 845-850, July 1993.
[3] D. M. Pozar, Microwave Engineering, 4th ed. New York, NY, USA: Wiley, 2011, ch.8.
[4] R. W. Haas, D. Brest, H. Mueggenburg, L. Lang, and D. Heimlich, "Fabrication and performance of MMW and SMMW platelet horn arrays," Int. J. Infrared Millimeter Waves, vol. 14, no. 11, pp. 2289–2294, Sep. 1993.
[5] M. Zhang, J. Hirokawa, and M. Ando, "Design of a partially-corporate feed double-layer slotted waveguide array antenna in 39 GHz band and fabrication by diffusion bonding of laminated thin metal plates," IEICE Trans. Commun., vol. E93-B, no. 10, pp. 2538–2544, Oct. 2010.
[6] K. Sakakibara, J. Hirokawa, M. Ando, and N. Goto, "A linearly-polarized slotted waveguide array using reflection-cancelling slot pairs," IEICE Trans. Commun., vol. E77-B, pp. 511-518, Apr. 1994.

# Design of Package Cover for 60 GHz Small Antenna and Effects of Device Box on Radiation Performance

Y. She, R. Suga *, H. Nakano, Y. Hirachi, J. Hirokawa and M. Ando
Tokyo Institute of Technology
Tokyo 152-8552, Japan
*Aoyama Gakuin University
Kanagawa 252-5258, Japan

*Abstract-* **This paper shows the simulated performance of the small antenna with the effects of the package cover and the device box at 60 GHz band. The total thickness of the package has been reduced from 2.6 mm to 1.4 mm by using the designed cover. The gain is enhanced and the reflection is suppressed lower than -15 dB by using different ways of packaging. On the other hand, the small antenna is put in a 120 mm x 75 mm x 30 mm polycarbonate box in simulation. The radiation performance due to the dielectric box case in practical device has also been simulated and discussed.**

## I. INTRODUCTION

In millimeter wave applications, the performance of the small antennas for the mobile terminals is sensitive in package surroundings. For instance, the ways of setting the antenna redome and the effects of the cover or the device box will both affect the antenna performance. In order to meet the requirements of the practical commercial systems, it is important to optimize the effects of the surroundings in antenna design.

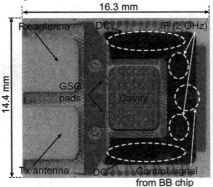

Figure 1. Photo of unpackaged antenna.

The authors take the example of the 60GHz small package with an end-fire radiation antenna [1], [2] to discuss the affects of the antenna cover and the device box in practice. Figure1 shows the photo and the size of the antenna. The side view of the configuration of the small antenna is shown in Fig.2. The 60GHz CMOS chip was mounted in the package made of low-cost, multilayered substrate and connected to the antenna by bonding wires or by the flip chip. The feeding circuit of the antenna was made of a microstrip line (MSL) layer and a post-wall waveguide layer. The power is radiated from the side face of the substrate with the end-fire radiation direction.

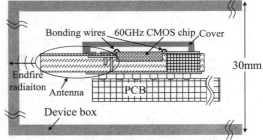

Figure 2. Side view of antenna package in device box

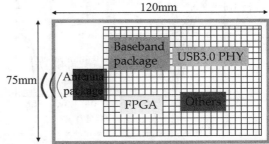

Figure 3. Antenna package mounted on PCB with other circuits in device box
(Approximate size of the device box: 120mm x 75mm x 30mm)

In the practical device, this antenna should be covered and mounted on PCB with other circuits as Fig.3.[3] The performance of the antenna will be affected by these surroundings. These influences should be estimated in advanced. On the other hand, it can also be utilized to enhance the performance of the antenna. In this paper, the gain enhancement and the reflection suppression affected by the cover using four different ways of packaging have been shown. The affect of the thickness of the cover and the cover position has been discussed. The radiation performance due to the dielectric box in practical mounting has also been simulated and discussed.

978-7-5641-4279-7

## II.    GAIN PERFORMANCE WITH PACKAGE COVER

Usually the package cover is used to protect the chip, the printed circuits and the bonding wires in the package. However, the performance of the small antenna is affected by the package cover. In other words, it can be utilized to facilitate fine adjustments on the antenna performance in packaging. Generally, the cover is the thinner the better and it is better to keep a stable performance in the antenna gain at certain frequency band.

TABLE I
DIFFERENT COVERS

| Case No. | Schematic diagram | Total thickness |
|---|---|---|
| ① w/o cover | | 1mm |
| ②1.6mm cover | | 2.6mm (cover 1.6mm) |
| ③0.4mm cover | | 1.4mm (cover 0.4mm) |
| ④two-step cover | | 1.4mm (cover 0.4mm) |

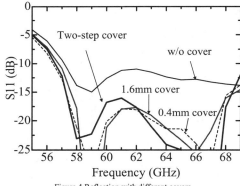

Figure 4 Reflection with different covers

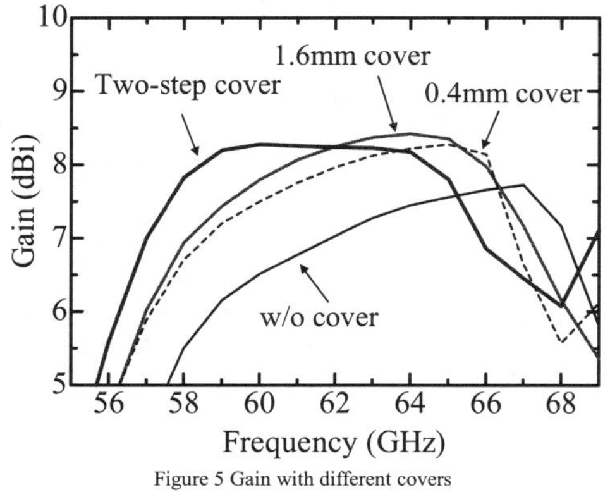

Figure 5 Gain with different covers

In table I, Case① to Case ④ show the schematic diagrams and estimated total thickness of the antennas respectively. The location of the antenna part in the package is shown in Fig.2. Case ① is the 1mm-thick unpackaged antenna without cover. Case ② is the 2.6mm-thick conventional antenna package with 1.6mm cover. Case ③ and Case ④ reduce the cover thickness to 0.4mm. Case ③ is the 1.4mm-thick antenna package with 0.4mm cover. Case④ is the antenna package with a two-step cover. The input impedance is set as 50 Ω in the input GSG pad in the simulation. The covers are set between the postwall waveguide and the slab waveguide in order to suppress the higher mode in transition.

Figure 4 and 5 show the estimated reflection and gain. The package covers increase the effect radiation area in thickness and enhance the gain of the antenna. The reflection S11 has been suppressed from -10 dB to -15 dB since the higher order mode is suppressed. According to the S11 and gain of Case② and case③, a thinner cover reduces the total thickness of the antenna while the antenna performance of the main beam does not change so much. On the other hand, a two-step cover as case④ is introduced to enhance the gain in lower frequency band and get a small variation in gain characteristic. However, it results the impact by the higher order mode in higher frequencies.

Assuming 1W excitation, Fig. 6 shows the simulated far field radiation pattern at 60 GHz in the linear scale. The antennas are holding an end-fire radiation in the main beam while the side lobes and the backward radiation are affected by different covers.

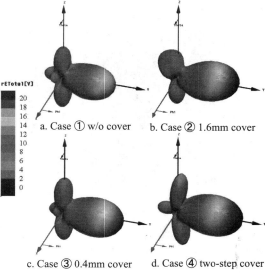

a. Case ① w/o cover    b. Case ② 1.6mm cover

c. Case ③ 0.4mm cover    d. Case ④ two-step cover

Figure.6 Far field radiation pattern of antenna package at 60GHz

## III. EFFECTS OF DEVICE BOX

In the practical device, the antenna performance is also largely affected by the device box. For instance, in the demonstration system of Sony[3], this antenna package is put at the edge of a rectangular polycarbonate box with 120mm long, 75mm wide and 30mm high. Figure 7 and 8 show the estimated reflection and gain of the antenna packages with the consideration of this condition. Figure 9 shows far field radiation pattern at 60 GHz. Compare with the w/o box cases in Sect.II, it affects little to the reflection while largely to gain and radiation pattern. The antenna with the two-step cover has a higher gain and can hold a flat gain performance in two channels of 60 GHz system.

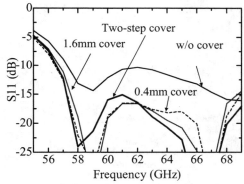

Figure 7.Reflection with different covers in polycarbonate box

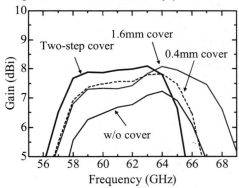

Figure 8.Gain with different covers in polycarbonate box

## IV. CONCLUSIONS

Effects of the package cover and the device box on the radiation performance both should be estimated and it can also be utilized to enhance the performance of the antenna. By optimizing the cover, the authors designed the thin cover to reduce the total thickness of the antenna from 2.6 mm to 1.4 mm with at most 2 dBi gain enhancement and the reflection kept suppressed lower than -15 dB. The radiation performance

is also affected by the dielectric box in the practical device. The small antenna is put in a 120 mm x 75 mm x 30 mm polycarbonate box in simulation. The overall optimization as well as the experimental verification is left for the future study.

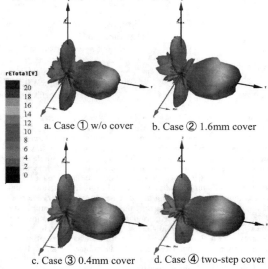

a. Case ① w/o cover    b. Case ② 1.6mm cover

c. Case ③ 0.4mm cover    d. Case ④ two-step cover

Figure.9 Far field radiation pattern of antenna at 60GHz in polycarbonate box

### ACKNOWLEDGMENT

Authors thank Mr. Makoto Noda of Sony Corporation for collaboration in developing handsets. This work was conducted in part as "the Research and Development for Expansion of Radio Wave Resources" under the contract of the Ministry of Internal Affairs and Communications.

### REFERENCES

[1]     R. Suga, H. Nakano, Y. Hirachi, J. Hirokawa and M. Ando, "A Small Package With 46-dB Isolation Between Tx and Rx Antennas Suitable for 60-GHz WPAN Module," IEEE Trans. Microw. Theory Tech., vol.60, no.3, pp.640-646, Mar. 2012.
[2]     R. Suga, H. Nakano, Y. Hirachi, J. Hirokawa and M. Ando, "Cost-Effective 60-GHz Antenna Package With End-Fire Radiation for Wire less Fire-Transfer System," IEEE Trans. Microw. Theory Tech., vol.58, no.12, pp.3989-3995, Dec. 2010.
[3]     Y. Asakura, K. Kondou, M. Shinagawa, S. Tamonoki, M. Noda, "Prototype of 3-Gb/s 60-GHz Millimeter-wave-based Wireless File-transfer System," 2013 URSI Commission B International Symposium on Electromagnetic Theory (EMTS), 21PM1E, Hiroshima, Japan, May. 2013.

# A Novel 60 GHz Short Range Gigabit Wireless Access System using a Large Array Antenna

Miao Zhang[#1], Jiro Hirokawa[#2], Makoto Ando[#3], Koji Toyosaki[#4], Toru Taniguchi[*1] and Makoto Noda[+1]

\# Department of Electrical and Electronics, Tokyo Institute of Technology

2-12-1-S3-19, O-okayama, Meguro-ku, Tokyo, 152-8552, Japan

\* Laboratory, Japan Radio Co., Ltd.

5-1-1 Shimorenjaku, Mitaka, Tokyo 181-8510, Japan

\+ Device Solution Business Group, Sony Corporation

2-3-10 Shimomeguro, Meguro-ku, Tokyo, 153-0064 Japan

*Abstract-* **A novel concept of short range wireless access system is proposed for multi-Gb/s data transfer in the 60 GHz band. An extremely large array antenna (64x64 elements) generates a quasi-plane wave and forms communication zone proportional to the antenna size. The key is that the stable signal coverage up to 10 m with uniform illumination still stays in the Fresnel region of the array antenna. The link budget based upon Friis Transmission Equation is not applicable in this short range communication system. The mobile user standing everywhere in the coverage, receives a constant signal almost free of multipath. For demonstrations, the circularly-polarized waveguide slot arrays are designed and fabricated first. They are installed in the prototype of a short range Gigabit wireless access system as the transmitting antenna. The system performances of not only the RF receiving level but also the bit error rate et al. will be evaluated to verify our proposal.**

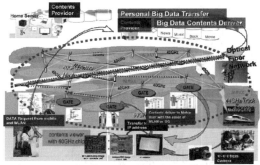

Figure 1. Image of a millimeter-wave Gigabit Network for realizing the mobile cloud service.

## I. INTRODUCTION

The high-performance mobile wireless terminals such as tablet PC, smartphone et al. are fast becoming widespread in recent years. In addition, a cloud computing system is widely expected to realize a free high-capacity data transfer with the server. However, the present mobile communication systems are facing the spectrum congestion worldwide, and their traffics are exploding and facing their limitation as the weak points. The "wireless-fiber project" to realize millimeter-wave wireless communication systems has been developed by the authors for years [1]. One of the two outputs from this project is a 60 GHz short range file transfer system with a maximum throughput of 3.5 Gb/s [2]. It has the potential of speed-up to 6.3 Gb/s [3], and a burst type download of high capacity files becomes possible. The other output is a 38 GHz-band 1Gbps FWA (Fixed Wireless Access) system for the outdoor backhauls [4]. These indoor and outdoor millimeter wave technologies open up the utilization of radio frequency resources and solve the frequency congestion in lower frequencies.

This paper presents a new concept of "short range communication" as the high-speed multi-Gb/s short range wireless access system in millimeter-wave band. The GATE (Gigabit Access Transponder Equipment) will provide the mobile terminals with an access to the cloud service. An extremely large array antenna with a high gain of 45dBi is realized in 60GHz band. It will be installed in the GATE as the wireless access point antenna, and will provide a stable and large signal-reception zone, which has an area proportional to the antenna aperture and a communication distance up to 10 m. In addition, the peak electric field density is favorably suppressed by introducing the large aperture antenna instead of the small one, and it is safe in terms of SAR (Specific Absorption Rate) for human body [5]

## II. THE MMW GIGABIT NETWORK AND 60 GHz GATE

Fig.1 gives the image of the millimeter-wave Gigabit network to realize the mobile cloud service. The key is to make full use of the two outputs from our "wireless-fiber project". That is the mobile terminals are connected with GATE by 60 GHz compact range communication while GATEs are the access points to the mesh network consisting of the 40 GHz FWA systems. Fig. 2 illustrates the 60 GHz GATE to be equipped in stations and other public areas. It provides the Gigabit access to the mobile users when they are staying in the coverage area of GATE.

The communication distance between the mobile wireless terminal and the GATE is relative short from tens of centimeters to several meters. It is necessary to specify the reception zone first which is different from the conventional far-field wireless systems in terms of user-friendliness and access performance. Those far-field evaluation indices as represented by Friis Transmission Equation as well as the

978-7-5641-4279-7

antenna gain and HPBW cannot be easily applied as before. For example, the design of link budget based upon Friis Transmission Equation is not applicable for the GATE.

### III. COMMUNICATION ZONE FORMATION BY APPLYING LARGE APERTURE ANTENNA

In this paper, we propose a new concept of "Compact range communication" where the user standing in the clearly defined coverage is illuminated as if it was receiving the plane wave. To realize this ultimate radio environment, the GATE antenna must have the antenna aperture large enough so that the user antenna may be well in the near-field or Fresnel region. The communication zone with uniform signal intensity has been investigated by varying the aperture sizes of the GATE antennas [5]. Our findings are briefly summarized as follows:

1) For the large aperture antenna (251 mm square, 45 dBi), the electric field intensity keeps almost unchanged in a short range up to 10 m. A larger reception area for the wireless terminal can be achieved by adopting a large aperture antenna in the access point compared with the small one as illustrated in Fig. 3;

2) In terms of safety issue for the human body, the peak electric field density associated with the large aperture (high gain) antenna is much weaker than those of the standard small antennas, which is also demonstrated in Fig. 3.

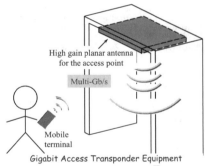

High gain planar antenna for the access point

Multi-Gb/s

Mobile terminal

Gigabit Access Transponder Equipment

Figure 2. The concept of GATE for short range high-speed data transfer.

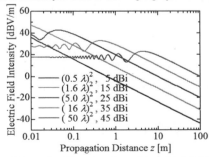

Figure 3. Propagation distance dependence of the electric field intensity generated by the antennas with various aperture sizes

3) Reflection from walls and obstacles outside of the coverage is negligible since the field strength is rapidly decreasing away from the coverage, and multipath radio environment may be neglected in principle.

### IV. SYSTEM EVALUATION INCLUDING BB AND RF CIRCUITS

To verify our proposal of applying an extremely large aperture antenna to form the clearly defined communication zone for GATE, the circularly-polarized waveguide slot arrays [6] with different numbers of elements are designed. Fig. 4 shows the photograph of the 2×2, 4×4, 8×8, 16×16, 32×32 and 64×64 elements antennas fabricated by diffusion bonding of thin copper plates. The antenna characteristics of reflection, aperture illumination and antenna gain in far field region are evaluated first. The aperture efficiencies of more than 70% are achieved for all antennas and are acceptable to be applied in the system evaluation.

At first, the electric field intensity is simply evaluated by connecting a RF circuit on the backside of antennas. The 16×16, 32×32 and 64×64-element arrays are used as the transmission antenna, while a V-band waveguide probe is used as the receiving antenna. A continuous wave without modulation at 60 GHz is adopted. The receiving signal intensity is measured when changing the communication distance between the Tx and Rx antennas. As illustrated in Fig. 5, relatively good agreement between predicted and measured result is observed.

Secondly, the prototype of a short range Gigabit wireless access system including a BB module [2] and RF front end is established. This system only uses the second channel of 60 GHz band WPAN system [7], covering the frequency range of 59.40 ~ 61.56 GHz. A maximum throughput of 3.5 Gb/s is expected by adopting the modulation of QPSK. The longitudinal distance between the Tx and Rx terminals is fixed at 10 m. The position of the Rx terminal is shifted transversely within the range of ± 20 cm. The bit error rate (BER) as summarized in Fig. 6 is measured to evaluate the system performances. By adopting a rate-14/15 low-density parity-check (LDPC) code, a communication zone with a BER below 1e-11 proportional to the array size is realized as expected.

Figure 4. Circularly-polarized waveguide slot arrays fabricated by diffusion bonding of copper plates in the 60 GHz band.

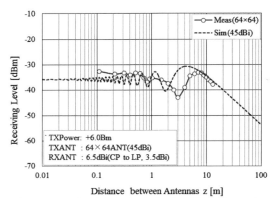

Figure 5. Measured receiving level as the function of distance between the transmitting and receiving antenna.

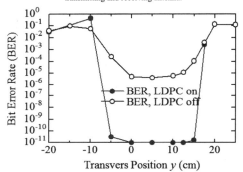

Figure 6. Measured bit error rate as the function of transverse position of receiving antenna.

## V. CONCLUSION

A new concept of 60GHz Gigabit 'Access Transponder Equipment (GATE) for short range and short-time file transfer is proposed. A larger reception area in short range can be achieved by using the large aperture antenna compared with the small one, even though the high gain antenna has narrower HPBW in the far-field. In addition, the large aperture antenna provides wide and stable reception area regardless of the reflected waves. The circularly-polarized corporate-fed waveguide slot array antennas with 2×2, 4×4, 8×8, 16×16, 32×32 and 64×64 elements are designed and fabricated in the 60 GHz band. The system evaluation including the BB module and RF front end has been conducted. Relatively good agreement between predicted and measured receiving level as the function of antennas distance is observed. An error-free communication zone proportional to the array size is also realized as expected.

## ACKNOWLEDGMENT

This work was conducted in part as "the Research and Development for Expansion of Radio Wave Resources" under the contract of the Ministry of Internal Affairs and Communications, Japan.

## REFERENCES

[1] M Ando, "Tokyo Tech Wireless-Fiber Project for Millimeter Wave Systems", The International Workshop on Millimeter Wave Wireless Technology and Applications -Tokyo Institute of Tech., pp.8-13, December 6, 2010.

[2] Y. Asakura, K. Kondou, M. Shinagawa, S. Tamonoki and M. Noda, "Prototype of 3-Gb/s 60-GHz Millimeter-wave-based Wireless File Transfer System," 2013 URSI Commission B International Symposium on Electromagnetic Theory (EMTS), Session: 21PM1E-4, Hiroshima, Japan, May 20-24, 2013.

[3] K. Okada et al., "A Full 4-Channel 6.3Gb/s 60GHz Direct-Conversion Transceiver with Low-Power Analog and Digital Baseband Circuitry," IEEE International Solid-State Circuits Conference (ISSCC), pp.218-219, San Francisco, CA, USA, Feb. 2012.

[4] J. Sato et al., "A 38 GHz-band 1 Gbps TDD FWA System Using Co-polarization Dual Antenna with High Spatial Isolation" IEEE Radio & Wireless Sym., Jan. 2012.

[5] M. Zhang, J. Hirokawa and M. Ando, "Receptionable Area Enlargement in MMW Short Range Communication Using Waveguide Slot Antennas with Large Number of Elements," IEEE AP-S International Symposium (USNC/URSI National Radio Science Meeting), Session: 454.6, Chicago, IL, USA, July 8-14, 2012

[6] Y. Miura, J. Hirokawa, M. Ando, K. Igarashi, and G. Yoshida, "A high-efficiency circularly-polarized aperture array antenna with a corporate-feed circuit in the 60 GHz band," IEICE Trans. Electron., vol.E94-C, no.10, pp.1618-1625, Oct. 2011

[7] [Online]. Available: http://ieee802.org/15/index.html

# 60 GHz On-Chip Loop Antenna Integrated in a 0.18 μm CMOS Technology

[#]Yuki Yao [1], Takuichi Hirano [2], Kenichi Okada [3]
Jiro Hirokawa [1], Makoto Ando[1]
[1]Dept. of Electrical and Electronic Eng., Tokyo Institute of Technology
S3-19, 2-12-1 O-okayama, Meguro-ku, Tokyo, 152-8552 Japan
E-mail : {yao, jiro, mando}@antenna.ee.titech.ac.jp
[2]Dept. of International Development Eng., E-mail : hira@antenna.ee.titech.ac.jp
[3]Dept. of Physical Electronics, E-mail : okada@ssc.pe.titech.ac.jp

*Abstract-* **This paper describes the electromagnetic (EM) field analysis of a 60 GHz on-chip loop antenna integrated in a 0.18 μm CMOS Technology. The simulation was compared with the measurement. The reflection coefficient showed good agreement between simulation and measurement by assuming a conductive layer of about 1 μm in the simulation. The radiation efficiency is calculated and it was found that the radiation efficiency can be improved by reducing conductivity of the silicon substrate. The radiation efficiency of 82.6% can be achieved if conductivity is less than 0.1 S/m.**

*Index terms-* **On-chip loop antenna, Millimeter-wave, CMOS substrate, Conductive layer, Electromagnetic simulation**

## I. INTRODUCTION

The CMOS technology can be applied to chips for wireless systems in the 60-GHz band [1]. Recently, in order to integrate the antenna with low cost single chip and CMOS RF front-end circuitry, on-chip antennas have been studied by many researchers [2]. It is important to reduce a connection loss between the CMOS chip and the antenna. The authors have simulated a dipole antenna on the Si CMOS substrate and confirmed good agreement between simulation and measurement by assuming a conductive layer on the surface of the Si CMOS substrate [3].

In this paper, an on-chip loop antenna is simulated considering the conductive layer. Simulation result is compared with measurement. The effect of the conductivity of the Si CMOS substrate on the radiation efficiency is investigated.

## II. STRUCTURE OF LOOP ANTENNA

Figure 1 shows the structure of the on-chip loop antenna. The size of a silicon chip is 5 mm square and the thickness of silicon substrate is 320 μm. The relative permittivity and conductivity of the silicon substrate are 11.9 and 6300 S/m, respectively. The metal is aluminum and space is filled with SiO₂ insulator. The conductivity of the conductive layer is $10^5$ S/m with the thickness of about 1 μm [3]. A passivation layer is on the top surface of the chip. Figure 2 (a) shows the

micrograph of the fabricated on-chip loop antenna. The loop antenna consists of the top metal layer and first metal layer which are connected by vias at the upper left corner in the figure. The loop antenna is fed by the microstrip line and the microstrip line is connected to a GSG pad for 100 μm-pitch GSG probe.

Figure 2 (b) shows the analysis model. Figure 3 shows the 3-D model and the radiation boundary in HFSS. The finite element method (FEM) based EM simulator, Ansoft HFSS, is used for simulation. A vacuum radiation box size is 10mm×10mm×5mm. The two lumped ports between signal and ground pad in the GSG pad with 100Ω internal impedance are used for excitation modeling. The reflection coefficient is obtained by converting the 2×2 S-matrix of the two lumped ports [4].

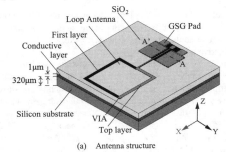

(a)   Antenna structure

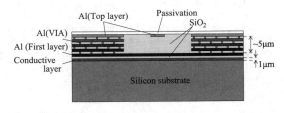

(b)   Cross-sectional view (A-A' in (a))

Figure 1: Structure of the on-chip loop antenna

978-7-5641-4279-7

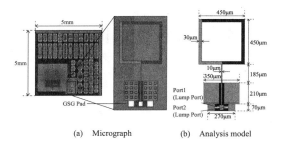

(a) Micrograph     (b) Analysis model

Figure 2: On-chip loop antenna on a silicon CMOS substrate

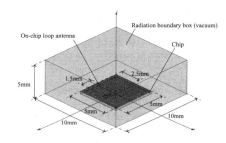

Figure 3: The 3-D model with the radiation boundary box in HFSS

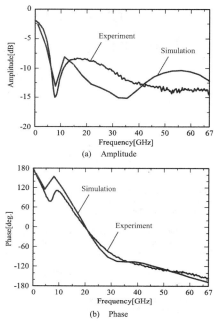

(a) Amplitude

(b) Phase

Figure 4: Frequency characteristic of reflection coefficient

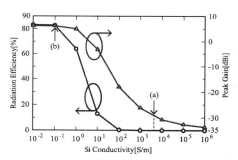

Figure 5: Radiation pattern and gain

## III. RESULTS

Figure 4 shows the frequency characteristics of the reflection coefficient. The reflection coefficient is in good agreement between simulation and measurement by assuming a conductive layer.

The conductivity of Si CMOS substrate is changed from $10^{-2}$ S/m to $10^{6}$ S/m at 60 GHz, and the peak gain and the radiation efficiency of the on-chip loop antenna are shown in Figure 5. The radiation efficiency and the peak gain increase as the conductivity of the Si CMOS substrate decreases. The radiation efficiency is 0.021% and the peak gain is -31.2 dBi when the conductivity is 6300 S/m.(a) And the radiation efficiency and the peak gain are saturated to be 82.6% and 6.11 dBi when the conductivity is 0.1 S/m.(b) The residual loss is due to the finite conductivity of metal. Figure 6 shows the 3D radiation patterns when the conductivity is 6300 S/m and 0.1 S/m. They have almost the same patterns, though the gain difference is more than 35dBi.

## IV. CONCLUSION

The on-chip loop antenna has been simulated by the EM simulator. The reflection coefficient showed good agreement between simulation and measurement by assuming a conductive layer with the high conductivity on the Si CMOS

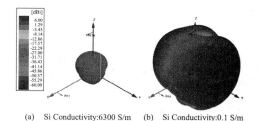

(a) Si Conductivity:6300 S/m    (b) Si Conductivity:0.1 S/m

Figure 6: Gain pattern (@60 GHz)

substrate surface. The gain and the radiation efficiency can be improved by reducing the conductivity of Si CMOS substrate.

## ACKNOWLEDGMENT

This work was partly supported by Semiconductor Technology Academic Research Center (STARC), MEXT/JSPS KAKENHI Grant Number 24760291 and VLSI Design and Education Center (VDEC), The University of Tokyo in collaboration with Cadence Design Systems Inc. and ROHM Semiconductor.

## REFERENCES

[1]   C. H. Doan, S. Emami, D. A. Sobel, A. M. Niknejad, and R. W. Brodersen, "Design considerations for 60 GHz CMOS radios," *IEEE Communications Magazine*, vol. 42, no. 12, pp. 132-140, 2004.

[2]   J. Grzyb et al., "Packaging Effects of a Broadband 60GHz Cavity-Backed Folded Dipole Superstrate Antenna," *IEEE AP-S International Symposium 2007*, Session: 331.3, pp.4365-4368, June 2007.

[3]   J. Mihira, T. Hirano, K. Okada, J. Hirokawa, and Makoto Ando, "Characteristic of On-Chip Dipole Antennas Considering Conductive layer on the Surface of Silicon Substrate," *Proceedings of International Symposium on Antennas and Propagation (ISAP)*, Paper ID: FrD2-2, Oct. 2011.

[4]   T. Hirano et al., Numerical Simulation - From Theory to Indus-try, Mykhaylo Andriychuk (Ed.), InTech, ISBN: 978-953-51-0749-1,pp.233-258, Sept. 2012.

# Microstrip Comb-Line Antenna with Inversely Tapered Mode Transition and Slotted Stubs on Liquid Crystal Polymer Substrates

Ryohei Hosono[1], Yusuke Uemichi[1], Han Xu[1], Ning Guan[1], Yusuke Nakatani[2], and Masahiro Iwamura[2]

[1]Optics and Electronics Laboratory, Fujikura Ltd., 1440, Mutsuzaki, Sakura, Chiba, JAPAN

[2]Printed Circuit Development Division of Fujikura Ltd., 1440, Mutsuzaki, Sakura, Chiba, JAPAN

*Abstract*-In recent years, millimeter-wave technology has attracted much attention in high-speed wireless communications, imaging and radar applications. Low-cost devices are very important for prevalence of the technology in consumer applications. Antennas for such applications are required to have high gain and low loss due to the high absorption in air-propagation and the high operation frequency. In this paper, a microstrip comb-line antenna constructed on liquid crystal polymer substrates is proposed. The antenna consists of 10 radiation elements and a termination element lined up like a comb and a waveguide-to-microstrip mode transition. Each radiation element has a slot and the mode transition has an inversely tapered section. The antenna realized a return loss less than -15 dB and a radiation with a maximum gain of 13 dBi and a side lobe level less than -10 dB at 60GHz. Radiation efficiency reached to about 90%.

## I. INTRODUCTION

In recent years, microwave wireless network is almost reaching its limitation of capability for quickly increased network traffics. Millimeter-wave communication is one of effective solution for solving this problem because of its large spectrum. Millimeter-wave technology has attracted much attention not only on communications but also on imaging and radar applications. Low-cost devices are very important for prevalence of the technology in consumer applications.

Antennas for such applications are required to have high gain and low loss due to the high absorption in air-propagation and the high operation frequency. Several types of antennas with high gain and low side-lobe level have been proposed [1]. Among them, microstrip-line antenna has low profile and simple structure and can be cost-effectively fabricated.

In this paper, microstrip comb line antenna is fabricated on Liquid Crystal Polymer (LCP) substrate and new configurations for realizing reflection cancellation and high gain in spite of keeping low side lob level are proposed. LCP substrate has advantages of its lower permittivity rather than LTCC (Low Temperature Co-fired Ceramic) substrate and loss tangent, similarity to thin filmed polyimide (PI) substrate with respect to the fabrication process easier and cheaper than teflon substrate such as polytetrafluoroethylene (PTFE). As the substrate thickness less than one tenth of a wavelength is preferable for suppressing dispersion depicted in [2], it equals to the thickness less than 0.5 mm at 60 GHz, thin film such as

flexible printed circuit (FPC) technology is applicable. By using this technology, proposed antenna which has slotted stubs for reflection cancelation and high gain and mode transition between rectangular waveguide and microstrip line with inverse tapered shape is fabricated precisely. Good agreement is obtained between electromagnetic simulation results and those of measurement. Low reflection less than 15 dB and high gain greater than 13 dBi with maximum side-lobe level around 10 dB are obtained at 60 GHz in measured results so it is demonstrated that proposed antenna is consistent with high gain and low side-lobe level keeping with low reflection. In addition, radiation efficiency greater than 90% at 60 GHz is also realized.

## II. DESIGN CONCEPTS

Figure 1 shows a configuration of proposed antenna. Microstrip antenna has comb shape whose radiation elements are partially slotted and with waveguide-to-microstrip mode transition. Antenna is terminated by the stub which is also looks like a tooth of comb. The mode transition is located in base of comb line antenna and it has a conductor with metalized via connected to ground plane to suppress needless radiation and reflection between waveguide and microstrip line. There is gap between mode transition and microstrip line for excitation of mode of microstirp line. The gap width is changed at longitudinal direction and a curve of transition is inversely tapered. Inversely tapered mode transition structure contributes to suppress needless radiation compatible with high gain and low reflection between waveguide and microstrip line. The reason why this structure is effective is because currents which are caused useless reflection and radiation are concentrated on bottom side of microstrip line and the gap between microstrip line and mode transition.

978-7-5641-4279-7

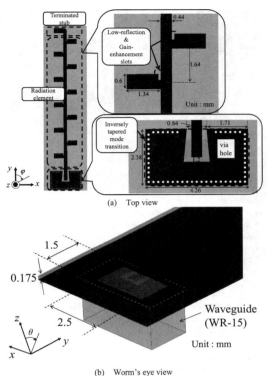

(a) Top view

(b) Worm's eye view
Figure 1 Configuration of proposed antenna.

To show the validity of proposed antenna, parametric studies are demonstrated below by using electromagnetic simulation. Simulation is carried out with commercial software HFSS$^{TM}$. It is assumed that relative permittivity is 3 and loss tangent is 0.003. The thickness of substrate is set to be 0.175 mm for design. First of all, several types of mode transition structures are compared, as shown Fig. 2. Exponential-function-taper is applied for configuration and width of larger and smaller gaps are same in models (b) and (c). Other geometries are optimized for obtaining resonance around 60 GHz. Structure (a) is similar to that of previous reports [3]-[5] so it is regarded as conventional structure in this investigation. As far as I knew previous report, the shape of mode transition between waveguide and microstrip similar to the type of (a) is mainly reported. Figure 3 (a) shows simulated input characteristics and (b) shows simulated radiation pattern in $yz$-plane at 60 GHz for antennas with these structures. $|S_{11}|<$-10 dB around 60 GHz can be achieved in all structures. The result of model (c) has especially $|S_{11}|<$-15 dB around 60 GHz. A bandwidth of $|S_{11}|<$-10 dB which has 1.1 % is obtained and it is wider than that of other structures which have 0.5 % bandwidth of model (a) and 0.8 % bandwidth of model (b), respectively. Maximum gain of models (a)-(c) are 12.8, 13.1 and 13.0 dBi so slight difference is appeared in radiation characteristics in

$yz$-plane. Contribution for low reflectance and broadening bandwidth by inverse tapered mode transition between rectangular waveguide and microstrip is successfully demonstrated.

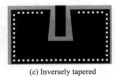

(a) Straight

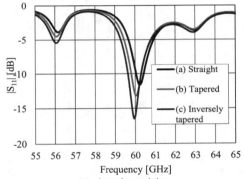

(b) Tapered                    (c) Inversely tapered

Figure 2 Mode transitions for comparison.

(a) input characteristics

(b) radiation characteristics in $yz$-plane
Figure 3 Simulated input and radiation characteristics for several transitions.

Figure 4 shows the simulated input and radiation characteristics for antennas with and without slots to confirm the effects of low reflection and gain enhancement. The geometry of mode transition is set to be model (c). From input characteristics, reflectance is reduced from -7 dB to -16 dB at 60 GHz by adding slots. From radiation characteristics, maximum gain is increased from 12.3 dB to 13.0 dB at 60 GHz by slots. Additionally, a beam shape of radiation pattern is straightened so radiation to desired direction is achieved by the slots. To explain how it works effectively, Figure 5 shows the current vectors of stubs for two types of antennas. Past papers are reported the effect of slots for microstirp comb line antenna [3]-[4] though they set slots into different position form this investigation. In these past papers, physical phenomena are not sufficiently denoted and parametric studies are mainly reported. It can be explained that slots of stubs generate current pair whose directions are opposite each other. Electric fields by current pair are canceled due to their vicinity and useless radiation and reflection finally reduced.

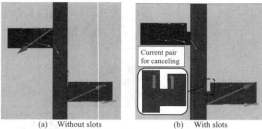

(a) Without slots          (b) With slots

Figure 5 Current vectors of stubs for antennas with and without slots.

III. EXPERIMENT RESULTS

Figure 6 (a) shows the fabricated antenna on LCP substrate for measurement and (b) shows the configuration for radiation measurement. In figure 6 (b), glass-epoxy spacer contributes to avoid useless deformation of antenna under test (AUT) due to its thickness and flexibility and PTFE screw. Nut for alignment can be expected to make AUT free from influence in radiation measurement rather than the configuration composed of nylon screw and fitting pins such as [6]. This configuration is also new proposal in this investigation and it is simpler and expected to be lower losses than previous one which is included in test jig and long microstrip lines for avoiding feeding structure [7]. By using this configuration, measured input and radiation characteristics are shown in Fig. 7 and Fig. 8 comparing with simulated ones. According to these figures, good agreements are obtained in both results. Figure 9 shows radiation efficiency and maximum gain for measured antenna. Maximum value of radiation efficiency is 90 % at 60 GHz and relative bandwidth of efficiency greater than 60 % is 2.4 %. From result of fabricated antenna, low reflection and high gain and efficiency are finally obtained at design frequency so this structure can be applicable at millimeter wave frequency.

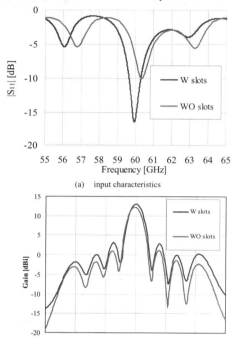

(a) input characteristics

(b) radiation characteristics

Figure 4 Simulated input and radiation characteristics for antennas with and without slots.

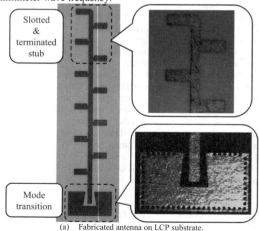

(a) Fabricated antenna on LCP substrate.

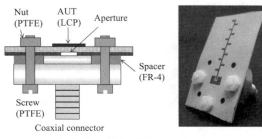

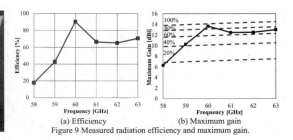

(a) Efficiency      (b) Maximum gain

Figure 9 Measured radiation efficiency and maximum gain.

(b) Configuration for radiation measurement.

Figure 6 Fabricated antenna and configuration for measurement.

## IV. CONCLUSION

In this paper, microstrip comb line antenna on liquid crystal polymer substrate is proposed. Mode transition shape and slotted stubs of radiation element play a role for low reflectance, small reflectance bandwidth and gain enhancement keeping low side-lobe levels. Proposed antenna is fabricated by flexible printed circuit technology and it is measured with new measurement configuration suitable for flexible device and avoidance of influence by fixture. Good agreement between simulated and measured results and good operation at millimeter wave frequency is finally confirmed and it is suitable for low cost and mass-product.

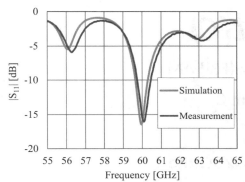

Figure 7 Simulated and measured input characteristics.

## REFERENCES

[1] D. Liu, B. Gaucher, U. Pfeiffer, and J.Grzyb, "Advanced Millimeter-wave Technologies : Antennas, Packaging and Circuits," , John Wiley & Sons, Chichester, U.K., 2009.

[2] L. Devlin, "Designing Cost Competitive E-band Radio Front-ends," Automated RF & Microwave Measurement Society (ARRMS) Conference, Oxfordshire, U.K., April 2013.

[3] Y. Hayashi, Y. Kashino, K. Sakakibara, N. Kikuma, and H. Hirayama, "Measured Performance of Millimeter-Wave Microstrip Comb-Line Antenna using Reflection-Canceling Slit Structure," Proc. Int. Symp. Ant. and Propag. 2007, pp. 1118-1121, Niigata, Japan, Aug. 2007.

[4] Y. Hayashi, K. Sakakibara, M. Nanjo, S. Sugawa, N. Kikuma, and H. Hirayama, "Millimeter-Wave Microstrip Comb-Line Antenna Using Reflection-Canceling Slit Structure," IEEE Trans. Ant. and Propagat., vol. 59, no. 2, pp. 398-406, Feb. 2011.

[5] S. Sugawa, K. Sakakibara, N. Kikuma, and H. Hirayama, "Design of Microstrip Comb-Line Antenna Array Composed of Elements with Matching Circuit," Proc. Int. Symp. Ant. And Propag. 2009, pp. 652-655, Bangkok, Thailand, Oct. 2009.

[6] C. Oikonomopoulos-Zachos, D. Titz, M. Martínez-Vázquez, F. Ferrero, C. Luxey, and G. Jacquemod, "Accurate Characterisation of a 60 GHz Antenna on LTCC Substrate," Proc. 5th European Conf. Ant. and Propag. pp. 3117-3121, Roma, Italy, April, 2011.

[7] A.E.I. Lammimen, J. Saily, and A.R. Vimpari, "60-GHz Patch Antennas and Arrays on LTCC With Embedded-Cavity Substrates," IEEE Trans. Ant. and Propagat., vol. 56, no. 9, pp. 2865-2874, Sept. 2008.

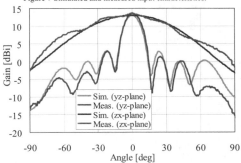

Figure 8 Simulated and measured radiation characteristics at 60 GHz.

# *FA-1(C)*

## October 27 (FRI) AM

## Room C

## Ant. Analysis & Synthesis

# A Modified BBO for Design and Optimization of Electromagnetic Systems

Marco Mussetta[1], Paola Pirinoli[2], Riccardo E. Zich[1]

[1] Politecnico di Milano, Dipartimento di Energia, Via La Masa 34, Milano, Italy - marco.mussetta@polimi.it
[2] Dipartimento di Elettronica e Telecomunicazioni, Politecnico di Torino, Torino, Italy - paola.pirinoli@polito.it

*Abstract*—**Several improvements of the Biogeography Based Optimization (BBO), have been recently introduced, in order to increase the optimization performances of the standard BBO algorithm, namely $M_mC_nBBO$. In this paper we compare the different proposed variations and apply them to benchmark functions and standard electromagnetic problems.**

*Index Terms*—*global optimization, Biogeography Based Optimization, Band-pass filter*

## I. INTRODUCTION

Evolutionary global optimization is nowadays largely applied to different types of engineering problems [1]. Some of these approaches, *i.e.* Genetic Algorithm (GA) [2], Particle Swarm Optimization (PSO) [3] and their hybrids [4], are well-assessed but they require high computational time to optimize complex problems, as those involving electromagnetic aspects, thus the scientific community is still developing new techniques.

In the last recent years, among the newly introduced optimization algorithms, there is the Biogeography Based Optimization [5], based on the science of Biogeography, *i.e.* the study of the geographical distribution of biological organisms. BBO shows very good features when applied to benchmark functions, but it is less performing when used in some real-world problems (see e.g. [6]). For this reason several improved versions of the original algorithm has been recently proposed [7]-[10].

In particular, the authors have recently implemented a different migration model and the concept of cataclysms [11,12], with the aim of avoiding the algorithm stagnation. If the possibility of using different migration model has already been explored in [10], even if here they are used in a different way, the idea of cataclysm has been more recently introduced [11,12]. The here-presented results of the application of the proposed methods, named in the following $M_mC_n$-BBO, to several benchmark functions and real-world electromagnetic problems show their effectiveness.

## II. BIOGEOGRAPHY BASED OPTIMIZATION

As mentioned, BBO is based on the study of the geographical distribution of biological organisms. Even if BBO shares some features with other evolutionary optimization methods, it has also some unique characteristics. In fact, in BBO the problem possible solutions are identified as islands or habitats, and its operators are based on the concept of migration, to share information between the problem solutions. In particular, the BBO algorithm introduces four new parameters: *suitability index variable* (SIV) represents a variable that characterize habitability in an island, i.e. in a solution; *habitat suitability index* (HSI), represents the goodness of the solution, similarly to the fitness score concept in GA; *immigration rate* ($\lambda$) indicates how likely a solution is to accept features from other solutions; *emigration rate* ($\mu$) indicates how likely a solution is to share its features with other solutions.

A low performing habitat has a low emigration rate and high immigration rate (in fact, the maximum possible immigration rate occurs when there are zero species in the habitat), while a high performing solution has a high emigration rate and low immigration rate, in fact, when HSI increases, the number of species grows, the habitat becomes more crowded, and more species are able to leave the island to explore other possible habitats, thus increasing the emigration rate.

### A. Modified Migration Model

Different modification regarding the migration model were proposed in [10], affecting the way $\lambda$ and $\mu$ are updated during iterations. In particular, while in the standard BBO the migration model is linear:

$$\lambda = 1 - f(y)$$

$$\mu = f(y)$$

in order to improve the share of information between high performing solution, similarly to [10] two other model are here introduced, a quadratic one:

$$\lambda = \left(1 - f(y)\right)^2$$

$$\mu = f^2(y)$$

and a cosine migration:

$$\lambda = \frac{1}{2}\left(\cos(f(y) \cdot \pi) + 1\right)$$

$$\mu = \frac{1}{2}\left(-\cos(f(y) \cdot \pi) + 1\right)$$

### B. Cataclysm

In order to avoid premature stagnation, a novel implicit restart procedure, named "cataclysm" was introduced in [11]: when the best HSI among all habitats did not improve in the last $C_n$ iterations, all the habitats are destroyed (cataclysm) and new ones are randomly generated. In order to preserve the best habitat elitism applies and no cataclysm occurs again before at least $5n$ generations have passed.

Therefore, by changing the migration model and $n$ it is possible to obtain an (infinite) set of schemes, that are identified by a corresponding name; the first part of the algorithm name codified the type of migration model used: "$M_L$", "$M_Q$" and "$M_C$" indicates a scheme in which the migration model is respectively the linear, the quadratic or the cosine one, while $n$ directly appears as the subscript of "$C$".

## III. PRELIMINARY ANALYSIS

The recently proposed $M_m C_n$-BBO was tested against the standard BBO in order to to assess the best configuration, *i.e.* to define which is the best migration model and the proper separation between two following cataclysms, considering different values of $n$.

The standard BBO together with different variations of the $M_m C_n$-BBO have been applied first to several benchmark functions. Table I (from [12]) reports the final mean ($\mu$) and standard deviation ($\sigma$) values for the optimization of the *step* and Griewank function (see eg. [5]). These data have been obtained as the average over 50 independent trials, and using a population of 20 individuals for 200 iterations; the first row of the table corresponds to the standard BBO ($M_L$, $C_0$). By analyzing these preliminary results the authors found that:

- the different migration models seem to have a small influence on the performances of the optimizer, even if the quadratic scheme seems to work almost always slightly better;

- the presence of the cataclysm plays a very important role in increasing the performances of the BBO, in conjunction with any of the migration models;

- it is more effective when the value of $n$ is small, i.e. when cataclysms are quite frequent.

## IV. NUMERICAL RESULTS

In view of these promising results, the proposed modified version of BBO has been checked against common electromagnetic problems, namely: the optimization of a microstrip filter (as shown in [12]) and the design of a planar array. In particular, an array of 9×9 elements is considered, with the element excitation (amplitude and phase) and position as free parameters. Here only the results relative to the optimization of a planar array are shown, but others will be presented at the conference.

The curves of convergence obtained as the average value over 10 independent trials with the standard BBO, the $M_Q$-BBO, the $M_C$-BBO and the $M_Q C_{10}$-BBO are reported in Table II and plotted in Fig. 1. In this case the $M_Q C_{50}$-BBO outperforms the other schemes, even if also other $M_Q$ schemes gives good results.

Additional results, relative to other electromagnetic optimization problems, will be presented at the conference, showing that usually the best convergences are obtained with $n$ equal to 5 or 10, generally in conjunction with the quadratic or linear migration scheme.

## ACKNOWLEDGEMENT

This work is part of the ASPRI educational program of Politecnico di Milano; the students involved in this project are: Baldassarre Luca, Bargiacchi Edoardo, Bosetti Luca, Ceci Alessandro, Cioppa Gregorio, Cocci Edoardo, Gastaldello Niccolò, Massarweh Lotfi, Mulassano Loris, Niccolai Alessandro, Papetti Viola, Passoni Davide, Pelamatti Julien, Purpura Giovanni, Rossi Giorgio, Rossi Pietro, Rozza Eleonora, Sacramone Simone, Saetta Alessandro, Secondi Alessandro, Spinelli Marco.

The here considered standard BBO algorithm was taken from [13].

TABLE I.   FINAL VALUES OF BENCHMARK FUNCTIONS OPTIMIZATION, WITH DIFFERENT PARAMETERS (AVERAGE OVER 50 TRIALS)

| Migration model: $M_m$ | Cataclysm: $C_n$ | Step Function | | Griewank function | |
|---|---|---|---|---|---|
| | | $\mu$ | $\sigma$ | $\mu$ | $\sigma$ |
| L | 0 | 94.3 | 37.32 | 1.84 | 0.3 |
| Q | 0 | 59.4 | 21.48 | 1.42 | 0.19 |
| C | 0 | 75.9 | 36.44 | 1.77 | 0.4 |
| L | 5 | 0 | 0 | 0.024 | 0.014 |
| Q | 5 | 0 | 0 | 0.021 | 0.008 |
| C | 5 | 0 | 0 | 0.02 | 0.007 |
| L | 10 | 0.3 | 0.8 | 0.23 | 0.53 |
| Q | 10 | 0.98 | 6.36 | 0.135 | 0.32 |
| C | 10 | 1.36 | 9.33 | 0.26 | 0.63 |
| L | 20 | 65.66 | 63.5 | 1.51 | 0.88 |
| Q | 20 | 25.3 | 31.8 | 1.035 | 0.65 |
| C | 20 | 40.68 | 39.55 | 1.06 | 0.8 |

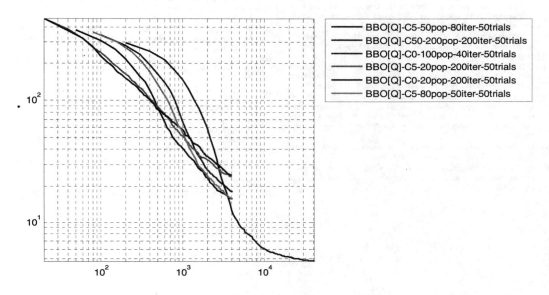

Figure 1.  Convergences curves of the different considered schemes applied to the optimization of a planar array.

TABLE II.  COMPARISON OF $M_MC_NBBO$ FOR PLANAR ARRAY OPTIMIZATION

| Migration model | Cataclysm $n$ | Population | Iteration | Final value | Std. dev. |
|---|---|---|---|---|---|
| Q | 50 | 200 | 200 | 4,7904 | 5,9645 |
| Q | 0 | 200 | 200 | 8,1484 | 7,7736 |
| Q | 0 | 200 | 20 | 14,6529 | 9,5654 |
| Q | 5 | 200 | 20 | 15,3996 | 11,4604 |
| Q | 5 | 100 | 40 | 15,4252 | 12,5037 |
| Q | 5 | 50 | 80 | 15,7631 | 13,6975 |
| Q | 5 | 80 | 50 | 15,9084 | 11,5191 |
| Q | 0 | 100 | 40 | 17,9288 | 12,8635 |
| Q | 5 | 20 | 200 | 19,3865 | 16,5959 |
| Q | 0 | 20 | 200 | 24,5947 | 19,8467 |
| Q | 50 | 20 | 200 | 25,144 | 18,2325 |
| C | 5 | 20 | 200 | 25,75 | 21,8699 |
| L | 5 | 20 | 200 | 26,4743 | 17,7922 |

## REFERENCES

[1] P. Antonio, D. Caputo, A. Gandelli, F. Grimaccia, M. Mussetta, "Architecture and methods for UAV-based heterogeneous sensor network applications", *Proc. SPIE*, Vol. 8532, 2012.

[2] R. E. Zich, M. Mussetta, M. Tovaglieri, P. Pirinoli, M. Orefice, "Genetic optimization of microstrip reflectarrays", *IEEE Int. Symp. of Antennas Propagat.*, vol.2,pp. 128-131, 2002.

[3] L. Matekovits, M. Mussetta, P. Pirinoli, S. Selleri, R.E. Zich, "Improved PSO algorithms for electromagnetic optimization", *2005 IEEE Antennas and Propagation Society International Symposium*, Vol. 2, pp. 33-36.

[4] E.A. Grimaldi, F. Grimaccia, M. Mussetta, P. Pirinoli, R.E. Zich, "A new hybrid genetical-swarm algorithm for electromagnetic optimization", *Proc. of ICCEA 2004, the 3rd IEEE International Conference on Computational Electromagnetics and Its Applications*, pp. 157-160.

[5] D. Simon "Biogeography-Based Optimization", *IEEE Trans. Evolutionary Comp.*, Vol. 12, pp. 702-713, 2009.

[6] L. Teagno, D. Tonella, P. Pirinoli, "Some Investigations on New Optimization Techniques for EM Problems", *Proc. EuCAP 2012*, Prague, March 2012.

[7] M.R. Lohokare, S.S. Pattnaik, S. Devi, B.K. Panigrahi, K.M. Bakwad, J.G. Joshi, "Modified BBO and calculation of resonant frequency of circular microstrip antenna", *Proc. 2009 Conf. on Nature & Biol. Inspired Comp.*, Dec. 2009, pp. 487 – 492.

[8] Weiyin Gong, Zhihua Cai, C.X. Ling, "DE/BBO: a hybrid differential evolution with biogeography-based optimization for global numerical optimization", *Soft Computing*, Vol. 15, pp. 645-665, 2010.

[9] H. Ma, D. Simon, "Blended Biogeography-based optimization for constrained optimization", *Evolutionary Comp.*, Vol. 24, pp. 517-525, 2011.

[10] H. Ma M. Fei, Z. Ding, J. Jin, "Biogeography-based optimization with ensemble of migration models for global numerical optimization", *Proc. IEEE Congress on Evolutionary Computation*, June 2012.

[11] M. Mussetta, P. Pirinoli, "MmCn-BBO Schemes for Electromagnetic Problem Optimization", *Proc. EuCAP 2013, the 7th European Conference on Antennas and Propagation*, Gothenburg, Sweden, 8-12 April 2013.

[12] M. Mussetta, P. Pirinoli, R.E. Zich, "Application of Modified BBO to Microstrip Filter Optimization", *Proc. 2013 IEEE AP-S/USNC-URSI Symposium*, Orlando, FL, July 2013.

[13] Prof. Dan Simon's homepage: http://academic.csuohio.edu/simond/

# Understanding the Fundamental Radiating Properties of Antennas with Characteristic Mode Analysis

Danie Ludick, Gronum Smith

EM Software & Systems - (Pty.) Ltd.

32 Techno Avenue, Techno Park, Stellenbosch, South Africa

## SUMMARY

Characteristic mode analysis (CMA) is a useful design tool that equips antenna designers with a non-brute force way of systematically extracting the radiating properties of a structure. These properties are presented by CMA in the form of eigenvalues and eigenvectors, i.e. the solution of a generalised eigenvalue equation that is formulated from the MoM impedance matrix. An important aspect to note at this stage, is that the eigenvalue equation can be solved without taking into account the effect of sources. Only the geometrical properties of the structure are considered.

One can deduce valuable information form the eigenvalues that are calculated. Reactive power is proportional to the magnitude of the eigenvalue, with modes radiating more efficiently as the eigenvalue approaches zero. In addition, the sign of the eigenvalue provides information on the type of energy being stored, i.e., in the form electric and magnetic energy. Derived quantities such as modal significance, modal weighting coefficients and modal input power also provide additional information on how external sources couple with, and excite, certain modes. The information that is obtained in this manner, aids the developer in choosing and exciting a desired behavior of the structure, something that is particularly useful in the mobile phone industry.

One particular challenge to CMA, however, is the manner in which quantities such as eigenvalues are presented as a function of frequency. At each discrete frequency sample, the eigenvalues are sorted according to modal significance, i.e., based on the efficiency with which they radiated. At higher frequencies, the ordering of modes may differ to that obtained at lower frequencies, as these higher order modes may contribute more towards the radiated or scattered power of the structure. To mitigate this problem, we have developed a tracking algorithm to keep the ordering of modes as consistent as possible over frequency.

The talk will first discuss the fundamental concepts of CMA, where the quantities such as eigenvectors, eigenvalues, modal significance, modal excitation coefficients and the modal weighting coefficients will be introduced. Thereafter, a step-by-step overview will be provided for the mode-tracking algorithm that is implemented in our solver. The efficiency of our technique will be illustrated at the hand of various practical examples.

Danie Ludick was born in the Free State, South Africa, on July 30, 1985. He received the B.Eng. degree in electrical and electronic engineering with computer science (*cum laude*) and the M.Sc.Eng. degree in electronic engineering (*cum laude*) from the University of Stellenbosch, South Africa, in 2007 and 2009, respectively. His masters thesis was based on the efficient analysis of focal plane arrays for the Square Kilometre Array (SKA) radio telescope using electromagnetic simulation techniques. He is currently part of the development team for the computational electromagnetic software package, FEKO. He is also presently pursuing a Ph.D. degree in electronic engineering at the University of Stellenbosch. His main research interests include computational electromagnetic simulation, domain decomposition techniques and antenna design.

Dr. Gronum Smith obtained his PhD from the University of Stellenbosch in 1993 and established EM Software and Systems with Dr Frans Meyer in 1994. EMSS started as a consulting company with first projects on characterisation of EM fields in the Naval environment, Bio-Electromagnetics, radiation hazards in naval and mobile phone industry, radome development and antenna placement on aircraft. These projects required good EM simulation and EMSS started cooperation with, and further development, of FEKO. From 2000 Gronum focussed on global marketing and distribution of FEKO

978-7-5641-4279-7

# Characterization of H2QL Antenna by Simulation

Erwin B. Daculan
ECE Department
De La Salle University – Manila
2401 Taft Ave., Malate, Metro Manila, Philippines
ebdaculan@gmail.com

Elmer P. Dadios
ECE Department
De La Salle University – Manila
2401 Taft Ave., Malate, Metro Manila, Philippines
elmer.dadios@dlsu.edu.ph

*Abstract*— **A novel wire antenna was named hybrid dual quad loop (H2QL) antenna. This paper attempted to characterize the H2QL antenna by simulating the antenna impedance, SWR and antenna pattern. EZNEC+ v5.0 was used to simulate the antenna parameters. The simulation was done at both UHF (431.35 MHz) and VHF (145.22 MHz) bands using three wire diameters (3.175, 6.35 and 9.525 mm). Simulation results and preliminary analysis were presented. Variant configuration was also simulated and the result was presented. Further study on other variants and performance of experiments were recommended.**

*Index Terms*—**computational EM, hybrid antenna, EZNEC**

## I. Introduction

The first antennas were made of wires. Michael Faraday experimented on electromagnetic radiation and used loop antenna to receive it. Wire antennas were as old as man-made radio wave propagation. As time passed, they became less popular. Novel wire antennas still blotted some professional publications now and then, but it rarely stirred up the general interest.

An antenna may be extended to three basic components : the driven element, the reflector, and the director. The set-up was especially true for wire antennas – dipole, loop, etc. That limitation (i.e., maximum of three basic components) had been the case for decades. This paper asked two different questions: *If there can be a fourth component, where will it be found? How will it influence the antenna parameters?*

## II. Related Works

With reference to the direction of propagation, the reflector was placed at the back of the driven element and the director was placed at the front. There was one place where a fourth element could be located – at the same plane as the driven element. That possibility was investigated using the hybrid dual quad loop (H2QL) antenna shown in Fig. 1.

The H2QL antenna was made up of two coplanar quad loops – segments 1, 2, 3 and 4 for first loop, and segments 5, 6, 7 and 8 for second loop – and a single parasitic element called *pinoy* (marked as segment 9). All segments were of the same diameter and material. The *pinoy* was a novel parasitic element that is neither a reflector nor a director. This element was on the same plane as the two quad loops and was placed at the midpoint between the two quad loops parallel to segments 1 and 5.

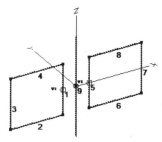

Fig. 1. Configuration of H2QL Antenna

Loop antennas were classified in [1] as either electrically small (total length $\leq 0.1\lambda$) or electrically large (total length $\approx \lambda$). The antenna under test measured exactly $1 \lambda$ as length of its perimeter. Consequently, the direction of propagation was normal to the plane of the loop.

A prior experimental study [2] served as impetus to this work. The experiments only measured the received signal strength and SWR values of the antenna using HP 8920A RF Communications Test Set equipment and Diamond SX-600 SWR meter, respectively. The current work dealt with antenna modeling using EZNEC+ v5.0 simulation software based on numerical electromagnetic code (NEC).

Gerald Burke was credited to be the main contributor for the development of the numerical electromagnetic code (NEC) that came out of Lawrence Livermore National Laboratory (LLNL). Since its introduction in the late 70s, the code had been adapted into several other antenna modeling software. Two of the most popular variants were the EZNEC by Roy Lewallen, a radio amateur, and the GNEC by Nittany-Scientific. Both variants had been used for the simulation needs of graduate theses and professional publications such as in [3] and [4] for EZNEC and in [5] and [6] for GNEC.

NEC presumed round wire antenna elements that are thin relative to the element length. This presumption limited what types of antenna structures or configuration can be replicated. This presumption, though, worked well with the antenna under test. Unfortunately, no working rule of thumb had been found in what possibly is the limit of the expression "*thin relative to the antenna length*". Nevertheless, rules had been established on the ratio between the smallest division of a single straight wire element (known as wire segment in the software) and the diameter of the wire element. Much more could be learned from [7] and [8].

978-7-5641-4279-7

An antenna manufactured by BAZ Spezialantennen [9] sounded similar to the H2QL antenna introduced here. It was named *hybrid double quad antenna*. Figure 2 showed the actual configuration of the hybrid double quad loop. It was composed of two quad antennas oriented like a diamond. One vertex from each quad was joined together and the feed point was located at that joint. Unfortunately, BAZ Spezialantennen website did not show clearly how the feed point was actually connected – whether it was shorted out or not.

Fig. 2. Configuration of Hybrid Double Quad Antenna

### III. PROBLEM SETTINGS

The parameters of the H2QL antenna remained largely unexplored. The objectives of this work were to design the configuration of this antenna for simulation, simulate its performance, and correlate the output data for approximate relationship of the antenna parameters being considered. Simulation was done through the use of EZNEC+ v.5.0 simulation software based on numerical electromagnetic code – version 2 (NEC2) with calculating engine NEC-2D at ground type of free space.

*Scope and Limitations*

Aluminum alloy 6063-T5 was assumed in the simulation. It had a resistivity of $3.2 \times 10^8$ Ω·m and a relative permeability of 1.000022. This was assumed as the characteristic of the wire since it was the best fit for the type of aluminum bars available locally in the Philippines. The test wire diameters used were 3.175, 6.35 and 9.525 mm. The test frequencies used were 431.35 MHz at UHF and 145.22 MHZ at VHF.

*Hypotheses*

Two hypotheses were forwarded : (1) the addition of a new parasitic element – called *pinoy* – placed in between two co-planar quad loops will significantly change the values of the standing wave ratio, impedance and gain of the antenna under test, and (2) the length of the *pinoy* element will also prove significant in determining the above performance parameters.

The name *pinoy* was chosen by the author for personal reason. It had no etymological connection to the function of that particular parasitic element. As such, it should be treated simply as a label.

### IV. RESULTS AND DISCUSSIONS

Figure 3 showed that standing wave ratio is dependent on the size of the wire. This relationship was more pronounced at higher frequencies. The results, though, was based on the assumption that the device using the antenna had a 50-Ω resistance. This was the case for other parameters under test.

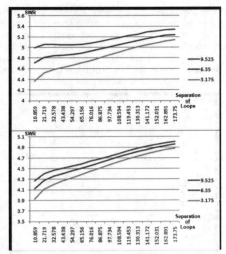

Fig. 3. SWR of the Antenna without Pinoy at UHF (top) and VHF (bottom)

Figure 4 showed that impedance is dependent on the size of the wire. Again, it was more pronounced at higher frequencies. At certain distance of the separation of loops, the diameter of the wires became insignificant.

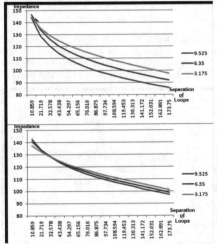

Fig. 4. Impedance of the Antenna without Pinoy at UHF (top) and VHF (bottom)

Figure 5 showed that the wire diameter was insignificant in the determination of antenna gain. Nevertheless, the graph of each wire diameter seemed to be quite linear.

In addition, all three parameters started off at lower values in the VHF band compared to the measurements at the UHF band. The perimeter of the loop antenna at both bands is equivalent to one wavelength of each designated frequency.

The VHF set-up allowed for better accuracy since the ratio between length of wire segment and diameter of the wire is larger due to longer wavelength at that band – a condition for better accuracy when using numerical electromagnetic code (NEC) as stated in [10] and [11]. Throughout all the simulation runs, the ratio between the length of the wire segment and the diameter of the wire is kept relatively constant.

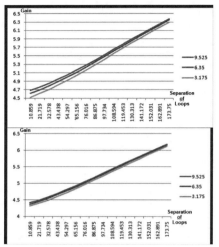

Fig. 5. Gain of the Antenna without Pinoy at UHF (top) and VHF (bottom)

Figures 6, 7, and 8 showed the SWR, impedance, and gain – respectively – of the antenna at different *pinoy* lengths and separations of loops. The *pinoy* was incrementally increased by 0.03125λ. The length started at 0.03125λ and ended at 0.5λ.

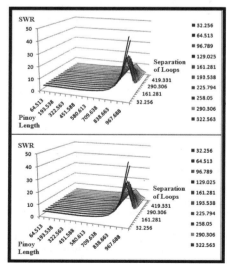

Fig. 6. SWR of the Antenna with Pinoy at UHF (top) and VHF (bottom)

Smaller SWR was obtained with wider separation of the loops and longer *pinoy* length. Highest SWR is found at *pinoy* length of about 0.4375λ with the narrowest separation of the loops. Lowest SWR is found at the longest *pinoy* length with the widest separation of loops. The maximum *pinoy* length is set at 0.5λ.

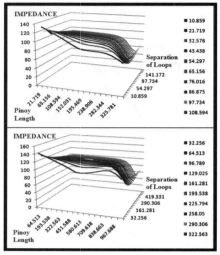

Fig. 7. Impedance of the Antenna with Pinoy at UHF (top) and VHF (bottom)

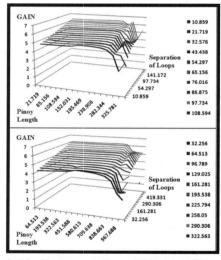

Fig. 8. Gain of the Antenna with Pinoy at UHF (top) and VHF (bottom)

The lowest impedance was attained at *pinoy* length of about 0.4375λ at the widest separation of loops. Unfortunately, the phase shift due to the presence of reactive components was not part of the considered results shown in Fig. 7. It would have given a better understanding on how to design the impedance-matching network and how much effective power was transmitted by the antenna and how much was lost along the transmission line.

If taken separately from the other performance parameters, higher gain was supposed to define a better antenna. The received signal was better amplified at higher gains – resulting to better signal reception. Gain was used here synonymous with directivity. More rapid gain change happened when separation of loops was gradually increased than when *pinoy* length was gradually increased as illustrated in Fig. 8. From the results shown, the best gain was obtained from *pinoy* length at around $0.3125\lambda$ and the separation of the loops at around $0.25\lambda$. Gain increased almost linearly as separation of loops and/or *pinoy* length increased – until the *pinoy* length reached about $0.375\lambda$. The variation of the gain from this length up to $0.5\lambda$ was yet to be described mathematically.

The seemingly rapid change of the parameters at *pinoy* lengths between $0.375\lambda$ and $0.5\lambda$ stirred the interest of the researcher to look elsewhere for other clues. Interesting results were observed when looking at the propagation pattern of the antenna. Figure 9 showed a sample of the propagation patterns as *pinoy* length was gradually increased while the separation of loops remained constant. As can be gleaned from the last row (especially the first two patterns), the direction of propagation had shifted.

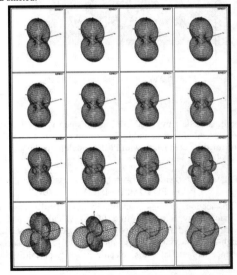

Fig. 9. Propagation Patterns at Different Pinoy Lengths

When a single-wire reflector was added to the antenna, an interesting propagation pattern resulted. Figure 10 showed the resultant pattern when the separation of loops was at $0.0625\lambda$, *pinoy* length is at $0.4375\lambda$, reflector length is at $0.5\lambda$, and the distance between the reflector and *pinoy* is at $0.09375\lambda$. The antenna was operated at the identified UHF frequency.

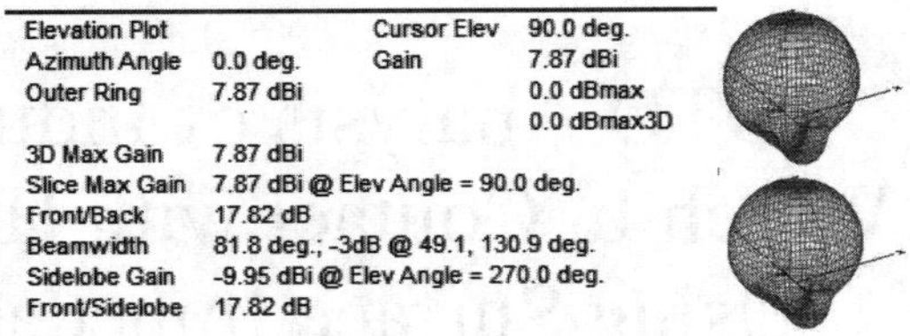

| Elevation Plot | | Cursor Elev | 90.0 deg. |
|---|---|---|---|
| Azimuth Angle | 0.0 deg. | Gain | 7.87 dBi |
| Outer Ring | 7.87 dBi | | 0.0 dBmax |
| | | | 0.0 dBmax3D |
| 3D Max Gain | 7.87 dBi | | |
| Slice Max Gain | 7.87 dBi @ Elev Angle = 90.0 deg. | | |
| Front/Back | 17.82 dB | | |
| Beamwidth | 81.8 deg.; -3dB @ 49.1, 130.9 deg. | | |
| Sidelobe Gain | -9.95 dBi @ Elev Angle = 270.0 deg. | | |
| Front/Sidelobe | 17.82 dB | | |

Fig. 10. H2QL Antenna with Wire Reflector

## V. CONCLUSION AND RECOMMENDATIONS

The H2QL antenna showed some promise as a directional antenna and some probability of being reconfigurable. The *pinoy* exhibited an influence to all antenna parameters under test. The most significant influence happened when the *pinoy* length is between $0.375\lambda$ and $0.5\lambda$.

With the addition of a wire reflector and/or a wire director, the antenna under test may exhibit other interesting variations of the parameters. It is therefore recommended that a study on H2QL antenna with a wire reflector and/or director be done at different lengths of the reflector/director, *pinoy* and separation of the loops. Finally, it is recommended that the antenna under test is used in an array.

## REFERENCES

[1] Glen S. Smith. "Loop Antennas" in Antenna Engineering Handbook, 4th ed., John L. Volakis (ed.), New York : McGraw-Hill, 2007, pp. 5-1 – 5-25.

[2] Erwin B. Daculan, "Short Experimental Study of a Hybrid Dual Quad Loop Antenna for VHF/UHF Commercial TV Broadcast," 25th SPP National Congress, Laguna, Philippines, 2007.

[3] Thomas Moses. "A Survey of Antennas for Wireless Communication Systems." M.S. Thesis, Florida State University, United States, 2008.

[4] M. Salmi. "Phasing of a HF Antenna Array." M.S. Thesis, University of Oulu, Finland, 2007.

[5] Timothy L. Pitzer and James A. Fellows. "Linear Ensemble Antennas Resulting from the Optimization of Log Periodic Dipole Arrays Using Genetic Algorithms." IEEE Congress on Evolutionary Computations, Sheraton Vancouver Wall Centre Hotel, Vancouver, BC, Canada, 2006.

[6] Kho Swee Jin, John C. McEachen and Gurminder Singh. "RF Characteristics of Mica-Z Wireless Sensor Network Motes." Internet : IEEE Xplore, 2006 [18 December 2012].

[7] Kho Swee Jin, John C. McEachen and Gurminder Singh. "RF Characteristics of Mica-Z Wireless Sensor Network Motes." Internet : IEEE Xplore, 2006 [18 December 2012].

[8] Steve Stearns, "Antenna Modelling for Radio Amateurs," ARRL Pacificon Antenna Seminar, San Ramon, CA, 17-19 October 2008, Internet : FARS [18 December 2012].

[9] BAZ Spezialantennen, www.amateur-radio-antenna.com/amateurradio-antenna/double-quad-antenna/index.php

[10] G.J. Burke and A.J. Poggio, "Numerical Electromagnetics Code (NEC) : Method of Moments – Part II : Program Description – Code," Lawrence Livermore National Laboratory, USA, Internet : LLNL [18 December 2012].

[11] G.J. Burke and A.J. Poggio, "Numerical Electromagnetics Code (NEC) : Method of Moments – Part III : User's Guide," *ibid*.

# FDTD Analysis of Induced Current of PEC Wire Which In Contact with Half Space Lossy Ground by Using Surface Impedance Boundary Condition

Takuji Arima [1], Toru Uno

*Tokyo University of Agruculture and Technology*
*2-24-16, Naka-cho, Koganei-shi, Tokyo JAPAN*
[1] t-arima@cc.tuat.ac.jp

*Abstract—* In recent years, electromagnetic waves are widely used. On the other hand, a fraction of electromagnetic waves are absorbed into the human body. In the case of far filed exposure, whole body resonance phenomena are observed at VHF frequency band when the component of E-field is parallel to the human body's height. In this phenomenon intensive induced current is observed in the human body. The induced current maybe depends on earth condition. In this paper, in order to estimate an effect of lossy flat earth for induced current, the induced current of object which is in contact with earth is analyzed by the FDTD method. The flat earth is modeld by using surface impedance boundary conditions to reduce calculation resources.

## I. Introduction

In recent years, electromagnetic waves are widely used. On the other hand, a fraction of electromagnetic waves are absorbed into the human body. In the case of far filed exposure, whole body resonance phenomena are observed at VHF [1] frequency band when the component of E-field is parallel to the human body's height as shown in Fig.1. In this phenomenon intensive induced current is observed in the human body. The induced current maybe depends on earth condition. In international guidelines, the limitations of the exposures are provided [2]. In order to analyze this phenomenon, an FDTD method[3] is widely used, because, the numerical human data including the electric property of tissue is provided as the Voxel model, the Voxel data can be included to FDTD analysis easily.

In the FDTD analysis, Perfect Matching Layer absorbing boundary conditions (PML) is effective to model half space flat earth. However, PML is required relatively large computation resources. Furthermore, electric property of earth is a lossy media. An implementation of the PML for lossy media is complicated, because the PML for lossy media should be implemented as dispersive materials.

On the other hand, in [4], our group has proposed efficient modeling method to model half space flat earth by using Surface Impedance Boundary Condition (SIBC).

In this paper, in order to estimate an effect of lossy flat earth for induced current, the induced current of object which is in contact with earth is analyzed by the FDTD method. The

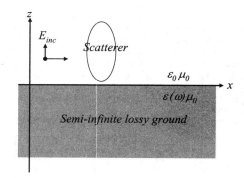

Fig.1 Scatterer over the semi-infinite ground

flat earth is modeld by using SIBC. In this paper, the PEC wire is used instead of human body. The effectiveness of the method is confirmed by comparing with PML modeling method.

## II. The Modeling Method of Half Space Lossy Flat Earth in the FDTD Method by using SIBC

In this section, an implementation method of SIBC to model half space lossy flat earth into FDTD calculation and its estimation method are explained briefly. A derivation of SIBC for high-loss material has been indicated in [5]. The surface impedance of material whose electric property permittivity $\sigma_s$ conductivity $\sigma_s$ permeability $\varepsilon_0$ is obtained as

$$Z_s(\omega) = \sqrt{\frac{j\omega\mu_0}{\sigma_s + j\omega\varepsilon_s}} \tag{1}$$

The purpose of this paper to introduce SIBC of lossy soil that instead of real lossy flat earth. The FDTD method is time domain analysis method. Unfortunately, time domain expression of eq.1 may include Bessel functions. The Bessel functions are not easy to introduce the FDTD method, because special treatment is required.

On the other hand, electric properties of soil in VHF frequency band are indicated in table.1

Table 1: electric property of soil

| Frequency[MHz] | σ [S/m] | εr |
|---|---|---|
| 19.9 | 0.001 | 15.0 |
| 31.6 | 0.001 | 15.0 |
| 39.8 | 0.00105 | 15.0 |
| 50.1 | 0.0011 | 15.0 |
| 63.0 | 0.0012 | 15.0 |
| 79.4 | 0.0016 | 15.0 |

From Table.1, the surface impedance can be approximated by using binomial approximation as

$$Z_s(\omega) = \sqrt{\frac{\mu/\varepsilon_s}{1 + \frac{\sigma_s}{j\omega\varepsilon_s}}} \simeq Z\left(\frac{1}{1 + \frac{\sigma_s/2}{j\omega\varepsilon_s}}\right) \quad (2)$$

where $Z = (\mu_0/\varepsilon_s)^{1/2}$. Eq.(2) can be transform easily to time domain. The time domain expression is simple expression, so eq.(2) is useful for the FDTD calculation. The tangential component of electric field at the flat earth surface can be expressed by using surface impedance and magnetic field. Therefore the computation resources can be reduced by using the surface impedance, because, the electric and magnetic field which is inside of lossy flat earth electric filed which is unnecessary.

Next, we will explain briefly how to formulate the obtained surface impedance into the FDTD method. The FDTD geometry is shown in Fig.2. The electric field on the SIBC is calculated by using surface impedance. Next, we will indicate briefly how to formulate obtained surface impedance into the FDTD method. The FDTD geometry is shown in Fig.3. The electric field on the SIBC is calculated by using surface impedance as

$$E_x(t) = \int_0^t Z_s(\tau)H_y(t-\tau)d\tau \quad (3)$$

Descritizing eq.(3) and using the FDTD notation, $H_y^{n+\frac{1}{2}}$ is expressed as

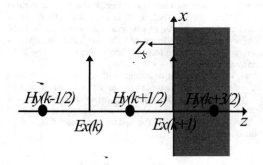

Fig.2 FDTD cell model

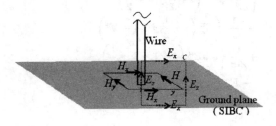

Fig.3 SIBC modeling

$$H_y^{n+\frac{1}{2}}\left(i+\frac{1}{2},j,k+\frac{1}{2}\right) = \frac{1}{\frac{\mu}{\Delta t} + \frac{Z+\chi^0}{\Delta z}}$$
$$\left\{\frac{\mu}{\Delta t}H_y^{n-\frac{1}{2}}\left(i+\frac{1}{2},j,k+\frac{1}{2}\right)\right.$$
$$-\frac{1}{\Delta z}\left(E_z^n\left(i+\frac{1}{2},j,k\right) - \Phi^n\right)$$
$$\left.+\frac{1}{\Delta z}\left\{E_y^n\left(i+1,j,k+\frac{1}{2}\right) - E_y^n\left(i,j,k+\frac{1}{2}\right)\right\}\right\} \quad (4)$$

where

$$\chi^0 = -\frac{\sigma_s}{2\varepsilon_s}\int_0^{\Delta t} e^{-\frac{\sigma_s}{2\varepsilon_s}t}dt$$

$$\Phi^{n-1} = -\frac{\sigma_s}{2\varepsilon_s}H_y^{n-\frac{1}{2}}\int_{\Delta t}^{2\Delta t} e^{-\frac{\sigma_s}{2\varepsilon_s}t}dt \quad - \quad e^{-\frac{\sigma_s}{2\varepsilon_s}\Delta t}\Phi^{n-2}$$

respectively.

Next, we expand SIBC for PEC wire in contact with half space ground. Fig. 3 shows FDTD cell on the ground plane. In this figure $H_y$ and $H_x$ are obtained by using SIBC as eq.(4). In order to model PEC wire which in contact with lossy flat earth, $E_z$ component which just under the wire should be $E_z=0$. However, the calculated induced current is not good agreement with a PML result. The PML result is using few cells thickness soil and calculation reason of soil area is truncated by the PML which is matched with lossy soil. In this case, the PML is dispersive material. Therefore, the implementation of this PML is relatively complicated.

Next, we will explain treatment method for in contact with SIBC cells. A special treatment is needed to obtain $H_y$ and $H_x$ vicinity of the wire. The integral form of Faraday's law is effective to model this special condition. Applying Faraday's law to contour $C$ in Fig.3, the update equation of $H_y$ can be obtained as

$$H_y^{n+\frac{1}{2}}\left(i+\frac{1}{2},j,k+\frac{1}{2}\right) = \frac{2}{\frac{\mu}{\Delta t}+\frac{z+\chi^0}{\Delta z}}$$
$$\left\{\frac{\mu}{\Delta t}H_y^{n-\frac{1}{2}}\left(i+\frac{1}{2},j,k+\frac{1}{2}\right)\right.$$
$$-\frac{1}{\Delta z}\left(E_a^n\left(i+\frac{1}{2},j,k\right)-\Phi^n\right)$$
$$+\frac{1}{\Delta z}\left\{0.5E_z^n\left(i+1,j,k+\frac{1}{2}\right)\right.$$
$$\left.\left.- E_z^n\left(i,j,k+\frac{1}{2}\right)\right\}\right\}_{(5)}$$

### III. IMPROVING CALCULATION ACCURACY

In this section, improving calculation accuracy technique is proposed. In order to improve the calculation accuracy, we introduce quasi-static approximation[6]. In the original FDTD method, the electric field distribution is assumed as constant. This may be on reason of accuracy deterioration. Static fields are dominant near the antenna. Fig.4 shows analysis model to obtain static field distribution near the antenna on the flat earth. In this model, the antenna is uniformly charged. The electro static potential $\phi(\rho,z)$ is obtained as

$$\phi(\rho,z) = \frac{\sigma}{4\pi\varepsilon}\left\{\frac{\log\left|z+\sqrt{z^2+\rho^2}\right|}{\log\left|z-l+\sqrt{(z-l)^2+\rho^2}\right|}\right.$$
$$\left.+R\frac{\log\left|z+l+\sqrt{(z+l)^2+\rho^2}\right|}{\log\left|z+\sqrt{(z)^2+\rho^2}\right|}\right\}$$

Where.

$$R=(1-\varepsilon_r)/(1+\varepsilon_r)$$

The obtained electric field can be included into FDTD method by using integral form of Faraday's law[6].

### IV. RESULTS

In order to confirm the effectiveness of this paper method, induced current by plane wave incident is calculated. The calculated induced current in PEC wire which in contact with lossy flat earth plane is shown in Fig. 5. The length of PEC wire is 175cm. The imputed pulse is plane wave and frequency is 79.4MHz. The ground is semi-dry condition, the electric properties are $\varepsilon_s=15.0$ $\sigma_s=0.0016$. In Fig.5, solid line is calculated by using exact ground plane modeling which uses few cells thickness soil and calculation reason of soil area is truncated by the PML which is matched with lossy soil. The calculated induced current is good agreement with PML result. Therefore the proposed method is effective to model PEC wire which touched semi-infinite ground plane.

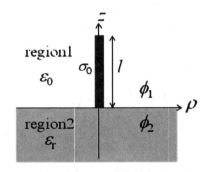

Fig.4 SIBC modeling

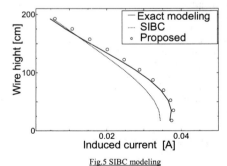

Fig.5 SIBC modeling

## V. CONCLUSION

In this paper, the induced current in PEC wire which in contact with lossy flat earth plane was calculated. In the calculation, SIBC was used to model lossy flat earth plane The result is good agreement with PML result. Therefore the proposed method is effective to model PEC wire which in contact with lossy flat earth plane.

## REFERENCES

[1]  ]OM P. Gandhi, "State of the Knowledge for Electromagnetic Absorbed Dose in Man and Animals", Proceedings of the IEEE, vol. 68, no.1, pp.24-32, January, 1980.

[2]  International Commission on Non-Ionizing Radiation Protection，"Guidelines for Limiting Exposure to Time-varying Electric, Magnetic, and Electromagnetic Fields (up to 300 GHz)"，1998.

[3]  T.Uno, Y.He and S.Adachi, "Perfectly Matched Layer Absorbing Boundary Condition for Dispersive Medium", IEEE  Microwave and Guided Wave Lett., vol. 7, no. 9, pp.264-266, sept. 1997.

[4]  Takuji Arima, Soichi Watanabe, Kanako Wake and Toru Uno,"A Numerical Analysis of Induced Current in Human Standing Over Low-Loss Ground Plane by FDTD Method",Proc. 2009 Internatinal Symposium on Anennas and Propagations, pp253-256, Bangkok, Thailand, 2009.10

[5]  R. J. Luebbers, F. P. Hunsberger, K. S. Kunz, R. B. Standler, M. Schneider; A frequency-dependent finite-difference time-domain formulation for dispersive materials, IEEE Trans. Electromagn. Compat., vol. 32, pp. 222 - 227, Aug. 1990.

[6]  T.Arima and T.Uno  IEICE Transaction B, vol.J85-B(2), pp.200-206,2002(in Japanese)

# Synthesis of Cosecant Array Factor Pattern Using Particle Swarm Optimization

Min-Chi Chang and Wei-Chung Weng
Department of Electrical Engineering
Chi Nan University
301 University Rd., Puli 54561, Taiwan
Email: wcweng@ncnu.edu.tw

*Abstract*-A 24-element symmetrically, equally spaced linear array was synthesize by particle swarm optimization to obtain the cosecant beam pattern in this study. Detailed settings of the array and PSO were presented. Obtained results show that the cosecant squared beam is successfully achieved. Compared the results obtained in this study with those of the published literature, the comparison shows that few numbers of array elements with less number of required iterations by proposed method can achieve the same desired goal.

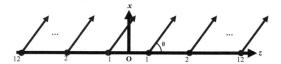

Figure 1. The geometry of the 24-element symmetrically, equally spaced linear array.

## I. INTRODUCTION

Pattern synthesis of antenna array is a vital issue in electromagnetics and antenna engineering. It has become popular for many years. Many techniques have been presented and used [1-6] for synthesizing array patterns. Some array patterns, such as sector beam patterns [7-8], cosecant beam patterns [9-11], and pencil beam patterns with very low sidelobe levels [12], are complicated and difficult, which are not easy to use traditional methods [13] such as Woodward-Lawson method, Taylor method, and Fourier transform method to synthesize them. Hence, using a global optimization algorithm to synthesize a complicated array factor pattern is an alternative way to obtain satisfactory results.

The particle swarm optimization (PSO) is a global optimized technique that can handle a problem with discontinuous and nondifferentiable, and multidimensional characteristics without depending on initial conditions. PSO was proposed in 1995 by Kenney and Eberhar [14], and it has been successfully applied to solve electromagnetic problems [15]. The basic idea of PSO is similar to that of animals to find the food cooperatively. The examples of this concept are the swarm of bees or birds. Bees or birds (particles) are allowed to fly in a finite area (solution space), looking for foods. Assuming there is only one location (the optimal solution) of the food. Once a bee (or bird) finds the location which is better than before, the bee informs other bees to change their location and velocity toward to the food. After certain time (iterations), all bees or birds will gather around the location which close the food (global optimum). This process continues until the location of most of the food is found. PSO algorithm is inspired from this model.

In this study, the beam shape synthesis of antenna array is designed and optimized by PSO. The proposed optimization procedure determines the excitation magnitude and phase of each element to synthesis the 24-element linear array to obtain its array factor with cosecant beam shape. Optimized results show that the satisfactory cosecant beam shape of the antenna array has been successfully achieved.

## II. COSECANT BEAM ARRAY FACTOR SYNTHESIS USING PSO TECHNIQUES

A 24-element symmetrically, equally spaced linear array aligns along the $z$-axis is shown in Figure 1. The elements are symmetric with respect to the $x$-axis. The excitation magnitudes of each symmetric element are the same. However, the phases of each symmetric element are reversed. The spacing between elements is a half-wavelength. For the linear array, there are 12 excitation magnitudes and 12 phases should be determined to synthesis the cosecant beam array factor. Hence, the array factor $AF(\theta)$ can be written as

$$
\begin{aligned}
AF(\theta) &= I_1 e^{j\beta_1} e^{jkd_1 \cos\theta} + I_2 e^{j\beta_2} e^{jkd_2 \cos\theta} + \ldots + I_{12} e^{j\beta_{12}} e^{jkd_{12}\cos\theta} \\
&+ I_1 e^{-j\beta_1} e^{-jkd_1 \cos\theta} + I_2 e^{-j\beta_2} e^{-jkd_2 \cos\theta} + \ldots + I_{12} e^{-j\beta_{12}} e^{-jkd_{12}\cos\theta} \\
&= 2\sum_{n=1}^{12} I_n \cos(kd_n \cos\theta + \beta_n) .
\end{aligned}
\tag{1}
$$

Where, $k$ is the wave number. $I_n$, $\beta_n$, and $d_n$ are the excitation magnitude, phase, and location of the $n$-th element, respectively. Since the spacing between elements is half-wavelength, the array factor can be simplified as

$$
AF(\theta) = 2\sum_{n=1}^{12} I_n \cos\left[\frac{(2n-1)}{2}\pi\cos\theta + \beta_n\right] .
\tag{2}
$$

978-7-5641-4279-7

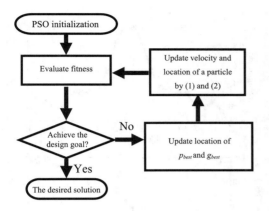

Figure 2. The PSO flow chart.

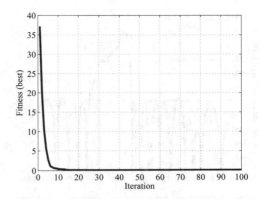

Figure 3. The fitness curve of the 24-element linear array for the cosecant pattern design by PSO.

The optimized ranges of all magnitudes and phases are set to 0 to 1 and 0 to $\pi$, respectively. The desired cosecant pattern is characterized by equations of (3) and (4).

$$f_1(\theta) = \begin{cases} 1 & , \text{for } 90° \le \theta < 97° \\ 1.122\,\csc(\cos\theta)\csc(\cos 99°) & , \text{for } 97° \le \theta < 120° \\ 1.122\,\csc(\cos 135°) - \csc(\cos 99°) & , \text{for } 120° \le \theta \le 127° \\ 10^{\frac{25}{20}} & , \text{elsewhere} \end{cases} \quad (3)$$

$$f_2(\theta) = \csc(\cos\theta) - \csc(\cos 99°)\,/\,1.122 \quad , \text{for } 95.8° \le \theta \le 120° . \quad (4)$$

As can be seen later in Figure 4, $f_1(\theta)$ is denoted by red dashed lines and $f_2(\theta)$ is denoted by blue dotted lines. The $f_1(\theta)$ allows 1.0 dB (1.122 in linear scale) tolerance between $97^0$ and $120^0$ and restricts sidelobe levels below -25 dB between $0^0$ and $90^0$ and between $127^0$ to $180^0$. The $f_2(\theta)$ allows 1.0 dB tolerance with $f_1(\theta)$ between $95.8^0$ and $120^0$.

The PSO optimization procedure flow chart is shown in Figure 2. In PSO, $p_{best}$ is the location of the best result of each particle; $g_{best}$ is the location of the best result of the entire particles in history. In each iteration, the velocity and location of each particle are updated by equations of (5) and (6), respectively.

$$v_n = w \cdot v_n + c_1\,rand_1()\,(p_{best} - x_n) + c_2\,rand_2()\,(g_{best} - x_n), \quad (5)$$

$$x_n = x_n + v_n . \quad (6)$$

Where, $v_n$ is a particle's velocity. $x_n$ is a particle's location. $c_1$ and $c_2$ are scaling factors, which are both set to 1.8. $w$ is an inertia weight, which is decreased linearly from 0.9 to 0.4 over the course of the iteration. $rand_1()$ and $rand_2()$ are uniform random values with the range between 0 and 1. In this study,

the number of particles is set to 81. The maximum number of iteration is set to 100. Once an $x_n$ of a particle is out of its optimization range, the $x_n$ is changed to the minimum or maximum boundary. Then, the sign of the particle's velocity $v_n$ is changed which forces the particle reflected back toward to the solution space.

A fitness function is applied to evaluate the performance of current optimization process by the obtained result. The following fitness function is used in this optimization

$$\text{Fitness} = w_1 \sum_{\theta=0°}^{\theta=180°} (AF(\theta) - f_1(\theta))[\frac{1 + \text{sgn}(AF(\theta) - f_1(\theta))}{2}]\Delta\theta \quad (7)$$
$$+ w_2 \sum_{\theta=95.8°}^{\theta=120°} (f_2(\theta) - AF(\theta))[\frac{1 + \text{sgn}(f_2(\theta) - AF(\theta))}{2}]\Delta\theta .$$

Where, the $\Delta\theta$ is the angular interval, which is set to $0.1°$. W1 and W2 are the weights of the fitness from the regions shown in (3) and (4), respectively. Both W1 and W2 are set to 1.0 in this study. The fitness shown in (7) accumulates the difference area between the desired pattern and the obtained pattern. The smaller the fitness reflects the better obtained pattern.

### III. OPTIMIZATION RESULTS

Figure 3 shows the fitness curve versus iteration. The fitness drops significantly during the first 10 iterations. After that, it is converged between 10-th to 100-th iterations. It finally reaches the value of 0.095 at the maximum number of iteration 100. The optimized cosecant array factor pattern is shown in Figure 4. The desired cosecant array factor pattern is successfully achieved. Sidelobe levels are below -25 dB. Compared with the results shown in [11], it can be found that there are 30 elements used in the same linear array to achieve the same desired goal by a the tabu search algorithm (TSA). The reduction in the number of array element is six, which shows that the optimization performance of the proposed PSO

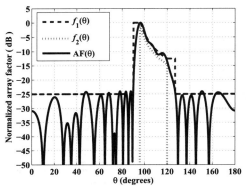

Figure 4. Optimized array factor (normalized) with cosecant array factor pattern obtained by PSO.

TABLE I
OPTIMIZED EXCITATION MAGNITUDES AND PHASES OF THE ELEMENTS

| Element numbers | Optimized magnitudes (Normalized) | Optimized phases (Degree) |
|---|---|---|
| 1 | 1.0000 | 20.6908 |
| 2 | 0.8433 | 57.5668 |
| 3 | 0.5438 | 86.2708 |
| 4 | 0.3609 | 91.4545 |
| 5 | 0.3115 | 87.1325 |
| 6 | 0.3029 | 124.6747 |
| 7 | 0.2977 | 138.9065 |
| 8 | 0.1539 | 143.8578 |
| 9 | 0.1454 | 148.8518 |
| 10 | 0.2045 | 179.9873 |
| 11 | 0.1050 | 151.3374 |
| 12 | 0.1584 | 179.9196 |

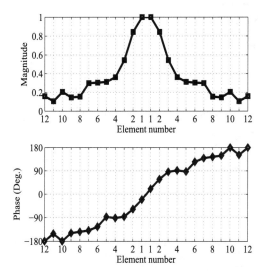

Figure 5. Optimized excitation magnitudes (normalized) and phases of the 24-element linear array for the cosecant array factor pattern.

show that six array elements reduction and less number of required iterations can achieve the same desired goal by the proposed PSO method.

ACKNOWLEDGMENT

This work was supported by the Taiwan Science Council under Grants 101-2221-E-260-020.

REFERENCES

[1] W. L. Stutzman, "Synthesis of shaped-beam radiation patterns using the iterative sampling method," *IEEE Trans. Antennas Propag.*, vol. 19, no. 1, pp. 36–41, Jan. 1971.
[2] R. S. Elliott and G. J. Stern, "A new technique for shaped beam synthesis of equispaced arrays," *IEEE Trans. Antennas Propag.*, vol. 32, no. 10, pp. 1129–1133, 1984.
[3] D. H. Werner and A. J. Ferraro, "Cosine pattern synthesis for single and multiple main beam uniformly spaced linear arrays," *IEEE Trans. Antennas Propag.*, vol. 37, pp. 1480–1484, Nov. 1989.
[4] B. P. Ng, M. H. Er, and C. Kot, "A flexible array synthesis method using quadratic programming," *IEEE Trans. Antennas Propagat.*, vol. 41, pp. 1541–1550, Nov. 1993.
[5] M. J. Buckley, "Synthesis of shaped beam antenna patterns using implicitly constrained current elements," *IEEE Trans. Antennas Propagat.*, vol. 44, pp. 192–197, Feb. 1996.
[6] B. P. Kumar and G. R. "Branner, Design of unequally spaced arrays for performance improvement," *IEEE Trans Antennas Propagat.*, AP-47, pp. 511–523, 1999.
[7] W. C. Weng, F. Yang and A. Z. Elsherbeni, *Electromagnetics and Antenna Optimization Using Taguchi's Method*, Morgan and Claypool Publishers, CA, USA, 2007.
[8] D. Gies and Y. Rahmat-Samii, "Particle swarm optimization for reconfigurable phase-differentiated array design," *Microwave and Opt. Tech. Lett.*, vol. 38, no. 3, pp. 168–175, Aug. 2003.
[9] O. M. Bucci, G. Franceschetti, G. Mazzarella, G. Panariello, "Intersection approach to array pattern synthesis," *IEE Proc.-Microw. Antennas Propag.*, vol. 137, no. 6, pp. 349–357, Dec. 1990.

is better in this problem. The optimized excitation magnitudes and phases of array elements are shown in Figure 5. Detailed values of magnitudes and phases of elements are listed in Table I.

## IV. CONCLUSION

The paper describes a cosecant beam pattern of the 24-element symmetrically, equally spaced linear array was successfully achieved by the proposed PSO method. The fitness value converges to the optimal result quickly with few numbers of iterations. The proposed optimization method is easy to implement. Results of the optimization comparison

[10] L. Josefsson and P. Persson, "Conformal array synthesis including mutual coupling," *Electronics Letters*, vol. 35, no. 8, pp. 625–627, Apr. 1999.

[11] A. Akdagli and K. Guney, "Shaped-beam pattern synthesis of equally and unequally spaced linear antenna arrays using a modified tabu search algorithm," *Microwave and Opt. Tech. Lett.*, vol. 36, no. 1, pp. 16–20, Jan. 2003.

[12] W. C Weng, F. Yang, and A. Z. Elsherbeni, "Linear Antenna Array Synthesis Using Taguchi's Method: A Novel ptimization Technique in Electromagnetics," *IEEE Trans. Antennas Propag.*, vol. 55, no. 3, pp. 723–730, Mar. 2007.

[13] C. A. Balanis, *Antenna Theory: Analysis and Design.* 3rd ed. New Jersey: John Wiley & Sons Inc., 2005.

[14] J. Kennedy and R. C. Eberhart, "Particle swarm optimization," in Proc. *IEEE Conf. Neural Networks IV*, Piscataway, NJ, 1995.

[15] J. Robinson and Y. Rahmat-Samii, "Particle swarm optimization in electromagnetics,"*IEEE Trans. Antennas Propag.*, vol. 52, no. 2, pp. 397–407, Feb. 2004.

# FA-2(C)

## October 32 (FRI) AM

## Room C

## Freq. Selective Surface

# Gain Enhancement for Multiband Fractal Antenna Using Hilbert Slot Frequency Selective Surface Reflector

Chamaiporn Ratnaratorn, Norakamon Wongsin, Chatree Mahatthanajatuphat, and Prayoot Akkaraekthalin

Department of Electrical and Computer Engineering, King Mongkut's University of Technology North Bangkok,
1518 Pracharat Sai 1 Rd., Bangsue, Bangkok, Thailand 10800,
email: r.chamaiporn@gmail.com, w.norakamon@gmail.com, cmp@kmutnb.ac.th, prayoot@kmutnb.ac.th

*Abstract*- In this paper, the gain enhancement for a multiband antenna with Hilbert fractal slot in a modified rectangular patch is presented. The composite structure including a wide slot, a Hilbert fractal slot with 1st iteration, and modified rectangular patch used as additional resonators producing multi-resonant frequencies are developed. The antenna gain can be enhanced by using frequency selective surface (FSS) reflector. The proposed FSS is optimized to control the phase and the magnitude of S-parameters of reflection and transmission characteristics over a bandwidth suitable for multiband antenna. FSS layer has been designed and employed with a unit cell of Hilbert fractal slot. Experimental results show good multiband operation at the frequency ranges of 870 – 960 MHz, 1710 – 1880 MHz, 1.85 – 1.99 GHz, 1.92 – 2.17 GHz, 2.3 – 2.36 GHZ and 2.4 – 2.485 GHz, respectively. Finally, the designed antenna can operate and cover the applications of GSM, DCS, PCS, UMTS, WiMAX and WLAN IEEE802.11 b/g. Moreover, the antenna gains at the operating frequencies of 0.9, 1.8, 1.92, 2.045 and 2.45 GHz are 7.607, 10.41, 9.54, 8.67 and 7.98 dBi, respectively.

## I.    INTRODUCTION

Over the years, we have seen unprecedented growth in wireless communication applications. Multiband and wideband antennas are compatible for wireless communication systems. Recent advances in the designing method of multiband antenna are recognized using several methods. Firstly, multi-resonators [1] generated multi-resonant frequencies. For instant in [2], the multiband antenna was created by adding bow-tie patches and a modified fractal loop to the sides and bottom of a strip line for the applications of DCS 1800, WLAN(IEEE802.11 b/g), WiMAX and IMT advanced system. Secondly, the antenna [3] was fabricated to improve the band-notch characteristic by cutting u-slots on the patch for dual and multiband applications. In [4], designing and analyzing the fractal antennas for multiband operation are researched by inserting Hilbert fractal slit in both sides of the rectangular stub in order to create the multiband antenna by using the technique of wideband antenna with multiple notch frequency. Another method, the development of the multiband antenna could be also achieved with the using of fractal concepts. The fractal geometries have been researched including Sierpinski, Koch and Hilbert shapes [5]. The Hilbert geometry is a space-filling curve, since with a larger iteration, and it was trying to fill the area. Mostly, self similarity property of fractal shapes is used to design of multiband operations.

There is a long history of development for technology of frequency selective surfaces (FSSs) [6-7]. FSSs are periodic structures of infinite identical cells which have different behaviors as low pass, high pass, band pass or band stop filter characteristics [8]. In [9], adequate designs of FSSs allow to control the phases of reflection and transmission characteristics over a wide bandwidth by designing with a slot antenna for enhancing the gain of ultra wideband antennas.

In this paper we present a gain enhancement for a multiband fractal antenna with FSS reflector. The multi-resonators are produced to response the multiband operation. A wide slot is employed on the ground plane of antenna designs for 1st operating frequency band. A modified rectangular patch and a modified Hilbert fractal slot are created for 2nd and 3rd operating frequency ranges, respectively. The operating frequencies can be controlled by varying electrical length of Hilbert fractal slot and the coupling value between both edges of wide slot and modified rectangular patch. However, impedance matching of the antenna is graceless for operating frequencies as without FSS reflector. Therefore, the designed FSS is applied to allow significant improvement in the gain over the all impedance bandwidth. The effective parameters of the proposed antenna will be investigated by using the CST software.

## II.    THE HILBERT FRACTAL SLOT FSS DESIGN

The two simply structures of FSS with the 0 iteration and the 1st iteration of Hilbert fractal slot are shown in Fig. 1(a) and Fig. 1(b), respectively. The magnitudes and phases of the s-parameters for the 0 iteration and the 1st iteration of Hilbert fractal slot are shown in Fig. 2(a) and Fig. 2(b), respectively. A unit cell has been designed to achieve the band stop responses. The highly reflective surface affects to the gain of the antenna. As seen in Fig. 2, the suitable magnitude and phase of the s-parameters are the key part for the gain enhancement. As the unit cell of Hilbert fractal FSS with 0 iteration designed, the magnitude of S-parameters reflect effectively at the frequency ranges of 1.06 – 2.05 GHz and 2.18 – 2.60 GHz. Furthermore,

978-7-5641-4279-7

the phases of Hilbert fractal FSS with 0 iteration at operating frequencies of 0.9, 1.8, 1.92, 2.045 and 2.45 GHz are -155.126, -178.552, 177.805, 165.381, and 179.131 degree, respectively as shown in Fig. 2(a). As the unit cell of Hilbert fractal FSS

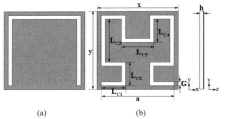

(a)                              (b)

Fig. 1 Unit cell of Hilbert fractal FSS layer with (a) the 0 iteration (b) the 1st iteration.

with 1 iteration designed, the magnitude of S-parameters conspicuously shifted to lower frequency due to the extending electrical length of Hilbert fractal slot. Especially, the magnitude can reflect effectively the frequency ranges of 670 – 1360 MHz, 1710 – 2.23 GHz and 2.4 – 2.51 GHz. Furthermore, the phases of Hilbert fractal FSS with 1st iteration at operating frequencies of 0.9, 1.8, 1.92, 2.045 and 2.45 GHz are -177.002, -165.940, -172.573, -177.350, and -172.072 degree, respectively, as shown in Fig. 2(b). Also, the 1st iteration of Hilbert fractal slot is chosen because the simulated results have the higher reflectivity and the suitable reflection phase (negative phase) that can satisfy for frequency ranges of 870 – 960 MHz, 1710 – 1880 MHz, 1.85 – 1.99 GHz, 1.92 – 2.17 GHz, and 2.4 – 2.485 GHz. Also, the physical parameters of Hilbert fractal slot on unit cell are achieved for $x = 60$ mm, $y = 60$ mm, $L_{c1} = 17$ mm, $L_{c2} = 18$ mm, $L_{c3} = 33$ mm, $L_{c4} = 16$ mm, $L_{c5} = 19$ mm and $g = 2$ mm. The FSS is printed on FR4 substrate with relative permittivity of $\varepsilon_r = 4.2$, thickness of $h = 1.6$ mm, and the unit cell with its size of $60 \times 60$ mm$^2$.

### III.    MULTIBAND ANTENNA WITH FSS REFLECTOR

The geometry of proposed antenna with FSS reflector is shown in Fig. 3(a). The antenna is printed on FR4 substrate with relative permittivity of $\varepsilon_r = 4.2$, thickness of $h = 1.6$ mm, the antenna size of $85 \times 72.5$ mm$^2$, and the Hilbert fractal slot FSS reflector with its size of $120 \times 120$ mm$^2$. The proposed antenna with Hilbert fractal slot FSS reflector consists of the wide slot, the modified rectangular patch with adding a modified Hilbert fractal slot, and Hilbert fractal slot FSS reflector. First, the wide slot is created by etching a slot on the ground plane to operate at the first operating frequency band (860 – 964 MHz). Next, the 2nd operating frequency band (1.71 – 2.17 GHz) is produced by compounding the harmonic of the first resonant frequency and the resonant frequency created from the modified rectangular patch with electrical length of $\lambda/4$. Additionally, the patch is still increasing the matching efficiency and bandwidth enhancement of 2nd operating frequency band. Finally, the modified fractal slot in rectangular patch with Hilbert geometry of 1st iteration is created to operate

at 3rd operating frequency band (2.25 – 2.49 GHz). The modified 1st iteration Hilbert shape is shown in Fig. 3(b). However, adding FSS reflector occurs a better matching and a

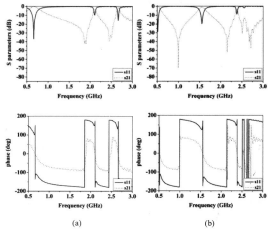

(a)                              (b)

Fig. 2   Simulated magnitudes and phase of S-parameters ($S_{11}$ and $S_{21}$) for unit cell of Hilbert fractal FSS layer with (a) the 0 iteration (b) the 1st iteration.

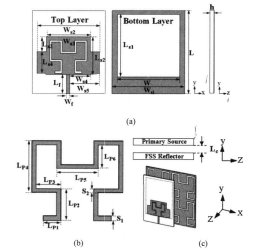

Fig. 3 Configuration of the proposed antenna with (a) the top layer and the bottom layer, (b) the modified Hilbert Fractal slot structure and (c) FSS reflector.

wide frequency bandwidth at whole resonant frequencies. The antenna with Hilbert fractal slot FSS reflector is shown in Fig. 3(c). The effect of the return loss for the proposed antenna is depicted in Fig. 4(a). The impedance bandwidths are comprehensive for the requirement bandwidth of 1st and 2nd operating frequency bands as the proposed antenna with FSS as incomprehensive bandwidth of 1st and 2nd operating frequency for the antenna with PEC. Furthermore, the gains of proposed antenna with Hilbert FSS reflector are higher than the PEC case as shown in Fig.4 (b). Moreover, the simulated far-field radiation patterns at the centre frequency of 2.45 GHz with the PEC reflector are graceless compared with the FSS reflector as illustrated in Fig.5. In order to study antenna parameters affecting to multiband operations, the optimum parameters of the proposed antenna are the following: $W = 72.5$ mm, $W_{s1} = 62$ mm, $W_{s2} = 54$ mm, $W_{s3} = 38$ mm, $W_{s4} = 22$ mm, $W_{s5} = 29.5$ mm, $L = 85$ mm, $L_{s1} = 64$ mm, $L_{s2} = 33.75$ mm, $L_{s3} = 13.5$ mm, $L_{s4} = 19.25$ mm, $L_f = 17$ mm, $W_t = 3$ mm, $L_{p1} = 6.2$ mm, $L_{p2} = 11.5$ mm, $L_{p3} = 9$ mm, $L_{p4} = 19$ mm, $L_{p5} = 10$ mm, $L_{p6} = 14$ mm, $S_1 = 2$ mm , $S_2 = 1.5$ mm, and the gap between the two layer of $L_c = 46.775$ mm ($L_c$ is cavity high).

In order to obtain optimized parameter values, the significant parameters mainly affecting to resonant frequencies including $L_{s3}$, and $L_c$ will be observed and varied as fixing the other parameters to investigate the effects on return loss, as depicted in Fig. 6. First, as the $L_{s3}$ varied ($L_c = 46.775$ mm) shown in Fig. 6(a), it can be seen that the parameter affects to the $3^{rd}$ operating frequency band resulting from altering coupling values and electrical length between both side edges of wide slot and modified rectangular patch. Additionally, the return losses of level $1^{st}$ and $2^{nd}$ frequency bands are altered by the coupling effect. Then, as the parameter $L_c$ varied ($L_{s3} = 13.5$ mm) depicted in Fig. 6(b), it can be clearly seen that the parameter affects to the all of operating frequency band due to the coupling value between the proposed antenna and the FSS reflector on impedance matching. Also, three resonant frequency ranges are exhibited with a good matching for $L_c = 46.775$ mm and covering the requirement bandwidth. As illustrated in Fig. 7, the gain alters as varying parameter $L_c$ due to the changing phase between the FSS reflector and the radiating antenna. Especially, the best result of gain is obtained by the sufficiently reflection phase between the FSS reflector and the radiating antenna.

## IV. RESULTS AND DISCUSSION

The antenna prototype is illustrated in Fig.8. The proposed antenna is placed at 46.775 mm above the Hilbert fractal slot FSS reflector, corresponding to $0.146\lambda_0$, $0.292\lambda_0$, $0.312\lambda_0$, $0.332\lambda_0$, and $0.398\lambda_0$ with the resonant frequency of 900 MHz, 1800 MHz, 1.92 GHz, 2.045 GHz, and 2.45 GHz, respectively. The simulated and measured return losses of the proposed antenna are illustrated in Fig. 9(a). It is explicitly seen that the difference between simulated and measured return losses of antenna occurred due to the etching process. The antenna gains at 900 MHz, 1800 MHz, 1.92 GHz, 2.045 GHz, and 2.45 GHz are 7.607, 10.41, 9.54, 8.67, and 7.98 dBi, respectively, as shown in Fig. 9(b). The measured far-field radiation patterns in X-Z and Y-Z planes at the centre frequency of each operating band are good directional radiation patterns, as illustrated in Fig.10. The maximum beams at all operating frequencies in X-Z plane are occurred at 0 degree. As the maximum beams in Y-Z plane at 900 MHz, 1800 MHz and 1.92 GHz are occurred at 0 degree, whereas at 2.045 MHz and 2.45 GHz are occurred at 30 degree.

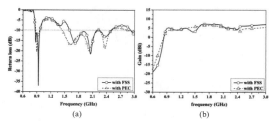

Fig. 4 (a) The return losses and (b) the gains of the proposed antenna with FSS and with PEC.

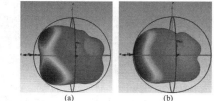

Fig. 5 The far-field radiation patterns at 2.45 GHz of (a) PEC and (b) FSS.

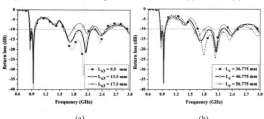

Fig. 6 Simulated return loss results of effective parameters (a) $L_{s3}$ and (b) $L_c$.

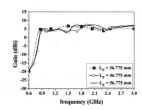

Fig. 7 The gains of effective parameter $L_c$.

Fig. 8 Prototype of the proposed antenna.

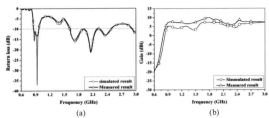

Fig.9 (a) The measured return losses, (b) The measured gains of the simulated and measured return losses of the proposed antenna.

Additionally, our antenna has high gain as improving the efficiency of radiation pattern with the operating frequency band of (859 - 962 MHz), (1605 - 2180 MHz), and (2.252 - 2.490 GHz) for the applications of GSM, DCS, PCS, UMTS, WiMAX, and WLAN IEEE802.11 b/g.

V.    CONCLUSION

In this paper, the gain and impedance bandwidth enhancement of the multiband antenna with Hilbert fractal slot in rectangular patch has been presented. The FSS is designed by using 1$^{st}$ iteration of Hilbert fractal geometry. The sufficiently reflection magnitude and phase affect to enhancement of the gains. Additionally, the gains are 7.607, 10.41, 9.54, 8.67, and 7.98 dBi at operating frequencies of 900 MHz, 1800 MHz, 1.92 GHz, 2.045 GHz and 2.45 GHz, respectively, covering applications of GSM(870–960 MHz), DCS(1710–1880 MHz), PCS(1.85–1.99 GHz), UMTS(1.92–2.17 GHz), WiMAX(2.3–2.36 GHz) and WLAN IEEE802.11 b/g(2.4–2.485 GHz). Moreover the radiation patterns are still directional patterns at all of operating frequency bands.

ACKNOWLEDGMENT

The researchers would like to gratefully thank the Wireless Communication Laboratory at Rajamangala University of Technology Thanyaburi for providing simulation software.

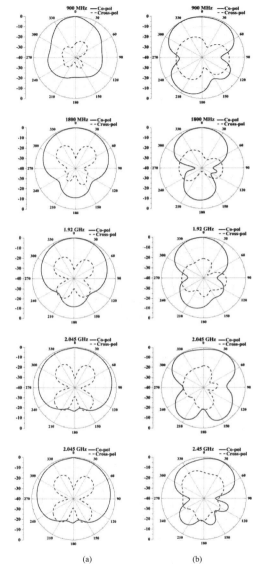

Fig.10 The measured radiation patterns in (a) X-Z plane and (b) Y-Z plane.

REFERENCES

[1]  Y. Xu, Y.-C. Jiao and Y.-C. Luan, "Compact CPW-fed printed monopole antenna with triple-band characteristics for WLAN/WiMAX applications," *Electronic Letters*, vol. 48, no. 24, 22 November 2012.
[2]  C. Mahatthanajatuphat, N. Wongsin and P. Akkaraekthalin, "A Multiband Monopole Antenna with modified fractal Loop Parasitic," *In International Symposium on Antenna and Propagation*, p. 57-74, November 2010.
[3]  K. Fong Lee, S. Yang and Kishk, A.A. "U-slot patch antennas for dual – band or multiband applications," *IEEE Antenna Technology*, 2-4 March 2009.
[4]  T. Hongnara, C. Mahatthanajatuphat, P. Akkaraekthalin, and M. Krairiksh, "A multiband CPW-Fed slot antenna with fractal stub and parasitic line," *Radioengineering*, vol. 21, no. 2, June 2012.
[5]  D. H. Werner and S. Gangul, "An overview of fractal antenna engineering research," *IEEE Antennas and Propagation Magazine*, vol. 45, no. 1, February 2003.
[6]  Y. Ranga, L. Matekovits, K. P. Esselle, and A. R. Weily, "Multioctave frequency selective surface reflector for ultrawideband antennas," *IEEE Antennas and Wireless Propagolion Letters*, vol. 10, 2011.

[7]  H. So, A. Ando, T. Seki, M. Kawashima and T. Sugiyama, "Directional multi-band antenna employing frequency selective surfaces," *Electronic Letters*, vol. 49, no. 4, 14 February 2013.
[8]  M. Kaerali, İ. Gungor, and B. Doken, "A new reflector antenna providing two different pattern," *General Assembly and Scientific Symposium*, 13-20 Aug. 2011.
[9]  Y. Ranga, L. Matekovits, K. P. Esselle, and Andrew R. Weily, "Enhanced gain UWB slot antenna with multilayer frequency-selective surface reflector," *International Workshop on Antenna Technology (iWAT)*, 7-9 March 2011.

# Unit Cell Structure of AMC with Multi-Layer Patch Type FSS for Miniaturization

Ming YING[#], Ryuji KUSE[#], Toshikazu HORI[#], Mitoshi FUJIMOTO[#]
Takuya SEKI[*], Keisuke SATO[*] and Ichiro OSHIMA[*]

[#]Graduate School of Engineering, University of Fukui, 3-9-1, Bunkyo, Fukui, 910-8507 Japan
[*]Denki Kogyo Co., Ltd., 13-4, Satsuki-cho, Kanuma, Tochigi, 322-0014 Japan

E-mail: omei@wireless.fuis.u-fukui.ac.jp

*Abstract* - This paper describes unit cell structure of AMC (Artificial Magnetic Conductor) with a multi-layer FSS for miniaturization. Two types of unit cell, "Patch type" and "Grid type" unit cell are treated.

Moreover, two types of FSS collocation method for multi-layer structure, "Stacked structure" and "Alternated structure" are considered. As the results of analysis, it is shown that the miniaturization effect is obtained only the case of "Patch type" FSS, and there is polarization dependence of AMC with the multi-layer FSS. It is also shown that the most miniaturization effect for unit cell size is obtained by "Alternated structure" when the layers are displaced just a half of the period of the unit cell, and the miniaturization effect is increased as the number of layers is increased.

*Index Terms* - AMC, patch type FSS, PMC characteristic, antenna reflecter.

## I. INTRODUCTION

An AMC has the PMC (Perfect Magnetic Conductor) characteristics in a specific frequency. The electromagnetic wave is reflected without phase rotation on the surface of the AMC with the PMC characteristics. The AMC is easily composed of the ground plane and FSS [1] [2]. The FSS is a surface which makes electromagnetic waves reflect or transmit in a specific frequency band [3]. As one of antenna applications, a low-profile antenna is realized by using the AMC reflector. Then, it is required that a unit cell size of the AMC reflector should be small [4].

This paper describes the optimal unit-cell structure of AMC equipped with the multi-layer patch type FSS for the miniaturization of unit-cell size.

## II. STRUCTURE OF AMC WITH DOUBLE LAYERED FSS

Figure 1 shows the nit cell of AMC. As shown in Fig. 1(a), the AMC is composed of using the metal patch FSS and the ground plane.

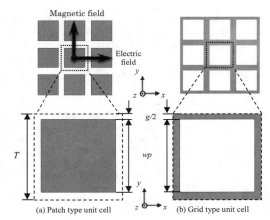

Figure 1. Unit cell of AMC

(a) Patch type unit cell    (b) Grid type unit cell

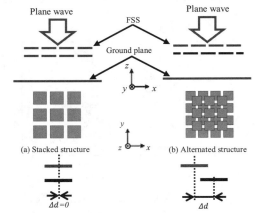

(a) Stacked structure    (b) Alternated structure

Figure 2. Structure of the AMC with double layer patch type and grid type FSS

The parameters of the patch type FSS structure are labeled as unit cell period $T$, patch width $wp$, gap of the patches $g$, patch type unit cell. The parameters of the grid type FSS structure is just opposite with patch type, as shown in Fig. 1(b). Figure 2 shows the structure of the AMC with double layer patch type FSS [5].

This section treats two type double layer structures. One is stacked structure as shown in Fig. 2(a), and the other is alternated structure as shown in Fig. 2(b). As shown in Fig. 2, $\Delta d$ is the displacement length between layers. When $\Delta d = 0$, the layers are not displaced (stacked structure). When $\Delta d$ is half of $T$, it is correspond to alternated structure. A plane wave is vertically incident from top to bottom as shown in Fig. 1. The electric field component and magnetic field component is directed in the x-axis direction and in the y-axis direction, respectively. Periodic boundary condition is used for the analysis of infinite structure, and the FDTD method is applied for the analysis. There is no dielectric between the double patch layers.

## III. REFLECTION CHARACTERISTICS OF PATCH AND GRID TYPE AMC

Reflection phase characteristics of the AMC with patch type FSS is shown in Fig. 3. The solid line, the dotted line and the dashed line denote the reflection phase of AMC with the single layer, the double layered (stacked) and the double layered (alternated) FSS, respectively. The horizontal axis indicates the frequency normalized by $f_p$. Here, $f_p$ is the frequency when the single layer patch type FSS shows the PMC characteristic. The vertical axis is reflection phase (we can see that the solid line of single layer patch type corresponds to the normalized frequency is 1 when the reflection phase is 0 (deg). It can be seen that the frequency with PMC characteristics of double-layered AMC is shifted to the low frequency side. It means that the unit cell size is miniaturized by using double layered patch type FSS.

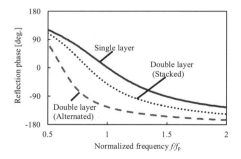

Figure 3. Reflection phase Characteristics of AMC with patch FSS

The reflection phase characteristics of the AMC with grid type FSS is shown in Fig. 4. The solid line, the dotted line and the dashed line denote the reflection phase of AMC with the single layer, the double layered (stacked) and the double layered (alternated) FSS, respectively. The horizontal axis indicates the frequency normalized by $f_g$. Here, $f_g$ is the frequency when the single layer grid type FSS shows the PMC characteristic. The vertical axis is reflection phase (we can see that the solid line of single layer grid type corresponds to the normalized frequency is 1 when the reflection phase is 0 (deg)). It can be seen that the frequency with PMC characteristics of double-layered is shifted to the high frequency side. So it is said that the unit cell size can't be miniaturized by using double layered grid type FSS. From Fig. 3 and Fig. 4, we can say that the unit cell size can be miniaturized by using only the case of the patch type FSS for the double layered AMC.

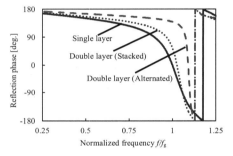

Figure 4. Reflection phase Characteristics of AMC with grid FSS

## IV. POLARIZATION DEPENDENCE OF THE UNIT CELL STRUCTURE

In this section, effect of displacement between the upper and lower patch is discussed, and further identify the polarization dependence of the AMC unit cell structure.

The direction of polarization is as shown in Fig. 1. There are two displacement directions concerned with the polarization. One side of double layer is displaced to the direction of the magnetic field or the electric field as shown in Fig. 5.

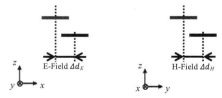

(a) Displaced to the electric field     (b) Displaced to the magnetic field

Figure 5. Displacement length $\Delta d$

E-Field $\Delta d_E$ and H-Field $\Delta d_H$ denote the displacement length to the electric field, and to the magnetic field, respectively.

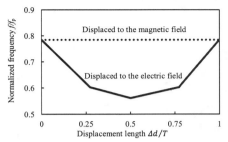

Figure 6. Relationship between $\Delta d$ and PMC characteristics with patch type FSS

Figure 6 shows the relationship between the displacement length and the PMC characteristics with the patch type FSS when the upper and lower patches are displaced. The solid line and dotted line show the case that the layers are displaced in the direction of electric field (x-axis), and magnetic surface (y-axis), respectively. The horizontal axis indicates the displacement length normalized by $T$. Here, $T$ is the period of the unit cell. The vertical axis indicates the frequency normalized by $f_p$. When the layers are displaced to the electric field direction of the incident wave (x-axis), it can be seen that the frequency with PMC characteristics is shifted to the low frequency side. It can be seen that the lowest frequency of PMC characteristics is obtained when it is displaced just a half of the period of the unit cell. On the other hand, if the layers are displaced to the magnetic field direction of the incident wave (y-axis), it can be seen that the frequency with PMC characteristics does not change.

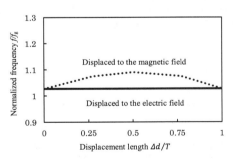

Figure 7. Relationship between $\Delta d$ and PMC characteristics with grid type FSS

Figure 7 shows the relationship between the displacement length and the PMC characteristics with the grid type FSS when the upper and lower patches are displaced. The solid line and dotted line show the case that the layers are displaced in the

direction of electric field (x-axis), and magnetic surface (y-axis), respectively. The horizontal axis indicates the displacement length normalized by $T$. The vertical axis indicates the frequency normalized by $f_g$. When the layers are displaced to the electric field direction of the incident wave (x-axis), it can be seen that the frequency with PMC characteristics does not change. On the other hand, if the layers are displaced to the magnetic field direction of the incident wave (y-axis), it can be seen that the frequency with PMC characteristics is shifted to the high frequency side. It can be seen that the highest frequency of PMC characteristics is obtained when it is displaced just a half of the period of the unit cell.

Therefore, when the patch type layers are displaced to the electric field direction of the incident wave (x-axis), the frequency with PMC characteristics is shifted to the low frequency side. But if the grid type layers are displaced to the magnetic field direction of the incident wave (y-axis), the frequency with PMC characteristics is shifted to the high frequency side.

We were confirmed that there is polarization dependence of AMC with the double-layer FSS, but the unit cell size can be miniaturized by using only the case of the double layered patch type, and the best miniaturization effect is obtained when it is displaced just a half of the period of the unit cell.

## V.  UNIT CELL STRUCTURE OF AMC WITH MULTI-LAYER FSS

### A.  RELATIONSHIP BETWEEN $\Delta d$ AND PMC CHARACTERISTICS BY THE COLLOCATION METHOD

According to the above discussion, the unit cell size can be miniaturized by using only the case of the double layered patch type, when the layers are displaced to the electric field direction. So we are considering the AMC with the multi-layer patch type FSS only the direction to the electric field. Figure 8 shows the model of the displacement length $\Delta d$ of the three-layer (lower and middle). Figure 8(a) shows the layers are displaced only the lowest layers, and Fig. 8(b) shows the layers are displaced only the middle layers.

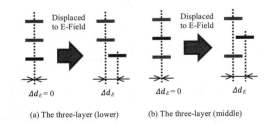

(a) The three-layer (lower)          (b) The three-layer (middle)

Figure 8. Displacement length $\Delta d$ of the three-layer (lower and middle)

The results are shown in Fig. 9. The solid line, the dotted line, the dashed lines denote the frequency with PMC characteristics of the AMC with the double layer FSS, the three-layer (lower) FSS and the three-layer (middle) FSS, respectively. The PMC characteristics of the AMC with three-layer patch type FSS is shifted to lower frequency than that of the double layer. And it is best effective when it has been displaced as the Fig. 8(b).

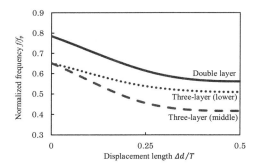

Figure 9. Relationship between $\Delta d$ and PMC characteristics by the collocation method

*B.* RELATIONSHIP BETWEEN $\Delta d$ AND PMC CHARACTERISTICS BY THE NUMBER OF LAYER

Figure 10 shows displacement length $\Delta d$ of the three-layer when it is displaced to electric field little by little. The double layer and four-layer are collocated as the same method as the three-layer. The collocation method of the three-layer AMC is the same as the Fig. 8(b), when $\Delta d = T/2$.

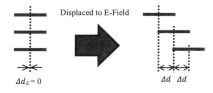

Figure 10. Displacement length $\Delta d$ of the three-layer

The results are shown in Fig. 11. The solid line, the dotted line, the dashed lines denote frequency with PMC characteristics of AMC with the double layer FSS, the three-layer FSS and the four-layer FSS, respectively. It can be seen that the frequency

with PMC characteristics is shifted to the low frequency side by increasing number of layers. It is also found that the most miniaturization effect for the unit cell size is obtained when the layers are displaced just a half of the period of the unit cell.

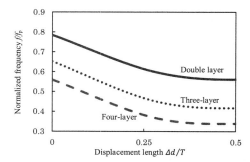

Figure 11. Relationship between $\Delta d$ and PMC characteristics by the number of layer

## VI. CONCLUSION

It was found the optimal structure of AMC with double layer patch type FSS for miniaturization of unit cell. As the result of studies, it was shown that the unit cell could be miniaturized, and further identify the polarization dependence of the AMC unit cell structure. The PMC characteristics of the AMC with multi-layer patch FSS could be obtained as the same as the double layer. We were confirmed that there was polarization dependence of AMC with the multi-layer patch FSS. It was also found that the most miniaturization effect for the unit cell size was obtained when the layers were displaced just a half of the period of the unit cell, and the effect of the miniaturization was increased as the numbers of layers was increased.

REFERENCES

[1] Ben A. Munk, Frequency Selective Surfaces: Theory and Design. New York, NJ John Wiley& Sons, Inc., 2000.
[2] E. A. Parker, R. J. Langley, R. Cahill, J. C. Vardaxoglou "Frequency Selective Surfaces," IEE Proc., ICAP'83, Norwich, UK, 1, pp. 459-463, Apr. 1983.
[3] D. Sievenpiper, "High-impedance electromagnetic surfaces," Ph.D. dissertation, Dept.Elect. Eng., Univ. California at Los Angeles, Los Angeles, CA, 1999.
[4] W. H. Cantrell, "Tuning analysis for the high-Q class-E power amplifier," IEEE Trans. Microw. Theory & Tech., vol. 48, no. 12, pp. 2397-2402, Fes. 2000.
[5] Y. Kawakami, T. Hori. M. Fjimoto, R. Yamaguchi and K. cho, "Low-profile design of meta-Surface by considering filtering characteristics of FSS," Proc. iWAT2010, Lisbon, Portugal, PS2.27, Mar. 2010.

# Scattering Analysis of Active FSS Structures Using Spectral-Element Time-Domain Method

H. Xu, J. Xi and R. S. Chen
Department of Communication Engineering
Nanjing University of Science and Technology , Nanjing , China, 210094

*Abstract-* Active frequency selective surface (FSS) structure is a kind of electromagnetic spatial filter, which has been widely used in the communication and radar systems. The spectral-element time domain method has been used to analyze the scattering characteristic of finite planar FSS structures. It has the advantages of spectral accuracy and block-diagonal mass matrix. This technique is based on Gauss-Lobatto-Legendre polynomials and Galerkin's method is used for spatial discretization. The absorbed boundary condition is employed to truncate the boundary. Numerical results demonstrate the accuracy of the method.

*Index Terms-* active frequency selective surface, spectral-element time domain method, scattering.

## I. INTRODUCTION

Frequency selective surface is usually made up of metal patches or periodic arrangement of aperture unit on the metal sheet, which has been widely used in communication and radar system. Usually the operating frequency band can not be changed once they are designed and manufactured. As a result, many people have been doing researches on active FSS structures, which has the ability of adjusting their transmission characteristics by using active devices such as PIN diodes[1]-[2].

Time-domain method has been more and more popular in recent years because of their abilities in the analysis of transient electromagnetic fields and the broadband properties of devices. Among these, one popular technique is the finite-element time-domain (FETD) method [3], which is famous for convenient modeling of complex geometries and materials. However, the inversion of the mass matrix makes it cost a lot in calculating. Finite difference time domain (FDTD), on the other hand, has the advantages of simple theory and good generality, but does not suit for complicated objects and media. The spectral-element time-domain (SETD) method, based on Gauss-Lobatto-Legendre polynomials, has the advantages of spectral accuracy and block-diagonal mass matrix and thus has received much attention recently. Moreover, the SETD can be regarded as a special kind of FETD method with different choices of nodal points and quadrature integration points [4].

In this paper, spectral-element time-domain method is employed to compute the scattering characteristic of finite planar active FSS structures with PIN diodes. RCS results of the FSS are presented respectively when it is in ON state and OFF state.

## II. FORMULATION

### A. Scattering formulation

For the proposed method, we can use the total-field/scattered-field formulation . In the total-field region

$$\nabla \times \left[ \frac{1}{\mu} \nabla \times \mathbf{E}^{t}(\mathbf{r},t) \right] + \varepsilon \frac{\partial^2}{\partial t^2} \mathbf{E}^{t}(\mathbf{r},t) = 0 \qquad (1)$$

while in the scattered-field region

$$\nabla \times \left[ \frac{1}{\mu_0} \nabla \times \mathbf{E}^{sc}(\mathbf{r},t) \right] + \varepsilon_0 \frac{\partial^2}{\partial t^2} \mathbf{E}^{sc}(\mathbf{r},t) = 0 \qquad (2)$$

where $\mathbf{E}^{t}(\mathbf{r},t) = \mathbf{E}^{sc}(\mathbf{r},t) + \mathbf{E}^{inc}(\mathbf{r},t)$

The absorbed boundary condition is employed to truncate the boundary

$$\hat{n} \times \left[ \frac{1}{\mu_0} \nabla \times \mathbf{E}^{sc}(\mathbf{r},t) \right] + \gamma \frac{\partial}{\partial t} \left[ \hat{n} \times \hat{n} \times \mathbf{E}^{sc}(\mathbf{r},t) \right] = 0 \qquad (3)$$

to discretize the equation , we introduce the Gauss-Lobatto-Legendre points. The Nth- order GLL basis functions in a one-dimensional standard reference element are defined as

$$\Phi_j^{(N)}(\xi) = \frac{1}{N(N+1)L_N(\xi_j)} \frac{(1-\xi^2)L_N'(\xi)}{\xi - \xi_j} \qquad (4)$$

for j = 0,...,N, where $L_N(\xi)$ is the Nth-order Legendre polynomial and $L_N'(\xi)$ is its derivative. The grid points $\xi_j$ are chosen as the GLL points, i.e.,the (N+1) roots of equation $(1-\xi_j^2)L_N'(\xi_j) = 0$

On a 3-D standard cubic reference element, we use vector basis functions

978-7-5641-4279-7

$$\Phi_{rst}^{\xi} = \hat{\xi}\phi_r^{(N\xi)}(\xi)\phi_s^{(N\eta)}(\eta)\phi_t^{(N\zeta)}(\zeta)$$

$$\Phi_{rst}^{\eta} = \hat{\eta}\phi_r^{(N\xi)}(\xi)\phi_s^{(N\eta)}(\eta)\phi_t^{(N\zeta)}(\zeta) \qquad (5)$$

$$\Phi_{rst}^{\zeta} = \hat{\zeta}\phi_r^{(N\xi)}(\xi)\phi_s^{(N\eta)}(\eta)\phi_t^{(N\zeta)}(\zeta)$$

as shown in Fig.1, where $N_\xi$, $N_\eta$ and $N_\zeta$ are the interpolation orders of the reference domain along $\xi$, $\eta$ and $\varsigma$ parametric coordinates, respectively [5].

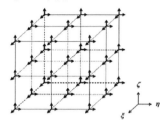

Fig. 1. Vector basis functions in reference element. The second-order basis functions are shown here

When constructing a curl-conforming vector basis functions , we use the appropriate mapping given by

$$\tilde{\Phi} = \mathbf{J}^{-1}\Phi$$

$$\nabla \times \tilde{\Phi} = \frac{1}{|\mathbf{J}|}\mathbf{J}^T\nabla\times\Phi \qquad (6)$$

where $\mathbf{J}$ is the Jacobian matrix defined as

$$\mathbf{J} = \begin{bmatrix} \frac{\partial x}{\partial \xi} & \frac{\partial y}{\partial \xi} & \frac{\partial z}{\partial \xi} \\ \frac{\partial x}{\partial \eta} & \frac{\partial y}{\partial \eta} & \frac{\partial z}{\partial \eta} \\ \frac{\partial x}{\partial \zeta} & \frac{\partial y}{\partial \zeta} & \frac{\partial z}{\partial \zeta} \end{bmatrix} \qquad (7)$$

In the reference element, the electric field can be represented by the tensor product of Lagrange–Legendre interpolation polynomials as

$$\mathbf{E}(\xi,\eta,\zeta) = \sum_{r=0}^{N_\xi}\sum_{s=0}^{N_\eta}\sum_{t=0}^{N_\zeta}\Phi_{rst}^{\xi}e^{\xi}(\xi_r,\eta_s,\zeta_t)$$

$$+ \sum_{r=0}^{N_\xi}\sum_{s=0}^{N_\eta}\sum_{t=0}^{N_\zeta}\Phi_{rst}^{\eta}e^{\eta}(\xi_r,\eta_s,\zeta_t) \qquad (8)$$

$$+ \sum_{r=0}^{N_\xi}\sum_{s=0}^{N_\eta}\sum_{t=0}^{N_\zeta}\Phi_{rst}^{\zeta}e^{\zeta}(\xi_r,\eta_s,\zeta_t) = \sum_{j=1}^{N}\Phi_j e_j$$

after employing the basis function and Galerkin's method,

$$\int\left\{\varepsilon\Phi_i\cdot\frac{\partial^2}{\partial t^2}\mathbf{E^t}+\frac{1}{\mu}(\nabla\times\Phi_i)\cdot(\nabla\times\mathbf{E^t})\right\}dV_t$$

$$+\int\left\{\varepsilon_0\Phi_i\cdot\frac{\partial^2}{\partial t^2}\mathbf{E^{sc}}+\frac{1}{\mu_0}(\nabla\times\Phi_i)\cdot(\nabla\times\mathbf{E^{sc}})\right\}dV_s$$

$$+\sqrt{\frac{\varepsilon_0}{\mu_0}}\int\left(\hat{n}\times\Phi_i\right)\cdot\left(\hat{n}\times\frac{\partial}{\partial t}\mathbf{E^{sc}}\right)dS - \frac{1}{\mu_0}\int\left(\hat{n}\times\Phi_i\right)\cdot\left(\nabla\times\mathbf{E^{inc}}\right)dS = 0$$

$$(9)$$

*B. Active frequency selective surface*

In this paper, the active frequency selective surface is mainly focused on FSS structure with PIN diodes.

We also start from the wave equation, for simplification, after the Galerkin's method, we can get

$$[A]E+[B]\frac{dE}{dt}+[C]\frac{d^2E}{dt^2}+[D]E^{'}+[E]E^{''}+[M]\frac{\partial J}{\partial t}=0 \qquad (10)$$

$$[M]_{ij} = \iint_{S\ l}(\int\bar{N}_i\cdot\hat{n}_y dy)dS \qquad (11)$$

Using the central difference,

$$\left(0.5\Delta tT_q+T\right)E_1^{n+1} = \left(2T-\Delta t^2(S+T_p)\right)E_1^{n}+\left(0.5\Delta tT_q-T\right)$$

$$E_1^{n-1}-\left(\Delta t^2S_1\right)E_1^{'n}-\left(\Delta t^2S_{th}\right)E_1^{''n}-\Delta t^2M\frac{\partial J_d}{\partial t} \qquad (12)$$

expanding the M matrix,

$$[TT]E_1^{n+1} = [SS]E_1^{n\cdots\cdots}-\Delta t^2 h\,\text{int}\,w_m\frac{\partial I_d}{\partial t} \qquad (13)$$

In the forward bias condition, the diode presents a resistance in series with the package inductance while in the reverse bias the circuit becomes a parallel combination of and in series with ,for simplification, we can also use a series RLC circuit model

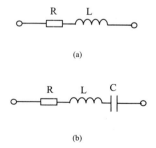

Fig. 2. The PIN diode (a) the forward bias equivalent circuit (b) the reverse bias equivalent circuit

In the forward bias condition,

$$V_d = L\frac{\partial I_d}{\partial t} + I_d R \qquad (14)$$

Using the forward difference,

$$\frac{V_d^{n+1}+V_d^n}{2} = L\frac{I_d^{n+1}-I_d^n}{\Delta t} + \frac{I_d^{n+1}+I_d^n}{2}R = \left(L+\frac{\Delta t}{2}R\right)\cdot\frac{\partial I_d}{\partial t} + I_d^n R \qquad (15)$$

Take (15) into (13), we ultimately obtain the time iteration scheme

$$[TT]e_1^{n+1} = [SS]e_1^{n\cdots} - \Delta t^2 h\,\mathrm{int}\,w_m\frac{\partial I_d}{\partial t}$$

$$= [SS]e_1^{n\cdots} - \Delta t^2 h\,\mathrm{int}\,w_m[\frac{V_d^{n+1}+V_d^n-2I_d^n R}{2L+\Delta tR}] \qquad (16)$$

The derivation of reverse bias condition is the same as that of the forward bias.

The above scheme calculates the near field. To obtain far field data, we introduce an artificial boundary $S_{far}$ inside the solution domain, which can be placed at or near the surface of the scatterer. The equivalent electric and magnetic currents can be determined from the fields calculated by SETD[6].

$$\mathbf{J}(\mathbf{r},t) = \hat{n}\times\mathbf{H}(\mathbf{r},t) = -\hat{n}\times\frac{1}{\mu}\int_0^t \nabla\times\mathbf{E}(\mathbf{r},t)dt$$

$$\mathbf{M}(\mathbf{r},t) = -\hat{n}\times\mathbf{E}(\mathbf{r},t) \qquad (17)$$

The following surface integrals are computed

$$\mathbf{L}(\mathbf{r},t) = \iint_{S_{far}} \mathbf{M}(\mathbf{r}',t+c^{-1}\hat{r}\cdot\mathbf{r}')ds'$$

$$\mathbf{N}(\mathbf{r},t) = \iint_{S_{far}} \mathbf{J}(\mathbf{r}',t+c^{-1}\hat{r}\cdot\mathbf{r}')ds' \qquad (18)$$

The scattered electric far-field is then readily obtained as

$$4\pi r E_\theta^{far}\left(t+c^{-1}r\right) = -c^{-1}\partial_t\left[L_\phi(\mathbf{r},t)+\eta N_\theta(\mathbf{r},t)\right]$$

$$4\pi r E_\phi^{far}\left(t+c^{-1}r\right) = -c^{-1}\partial_t\left[L_\theta(\mathbf{r},t)-\eta N_\phi(\mathbf{r},t)\right] \qquad (19)$$

Once the scattered field in the far region is known, the radar cross section(RCS) can be gotten from

$$\sigma = \lim_{r\to\infty} 4\pi r^2 \frac{\left|\mathcal{F}\{\mathbf{E}^{far}(\mathbf{r},t)\}\right|^2}{\left|\mathcal{F}\{\mathbf{E}^{inc}(\mathbf{r},t)\}\right|^2} \qquad (20)$$

### III. NUMERICAL RESULTS

One active frequency selective surface with PIN diodes is

demonstrated here to show the accuracy of the proposed method. This example is a quadrate metal patches printed on

a sheet of 0.04m with dielectric constant of 3.0. The typical values used for forward bias are R = 5Ω and L = 0.4 nH, while for reverse bias a series capacitance of 0.27 pF has been added to the circuit model.

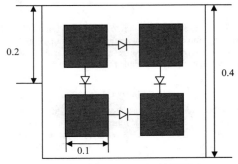

Fig.3. Structure of the active FSS with PIN diodes

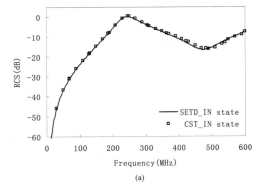

(a)

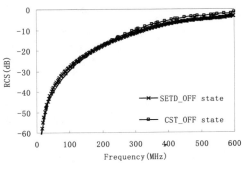

(b)

Fig.4. (a) Backscattered RCS of the active FSS when the PIN diodes are in ON state  (b) Backscattered RCS of the active FSS when the PIN diodes are in OFF state

From the result, we can clearly see that the results got from the spectral-element time domain method meet well with that of the commercial software.

## IV. CONCLUSION

The spectral-element time domain method has been employed to compute the scattering characteristic of the active FSS structures. Numerical results show that this method can analyze the PIN diodes-loaded FSS accurately.

## ACKNOWLEDGMENT

We would like to thank the support of Major State Basic Research Development Program of China (973 Program: 2009CB320201), Natural Science Foundation of 61171041, Jiangsu Natural Science Foundation of BK2012034.

## REFERENCES

[1] B.A.Munk.,Frequency Selective Surfaces,Theory and Design[M]. New York:Wiley,2000.

[2] C. Mias, "Waveguide and free-space demonstration of tunable frequency selective surface," Electronics Letters ,vol. 39, no. 11, pp. 850-852, May 2003.

[3] J.Jin, The Finite Element Method in Electromagnetics. New York:Wiley,1993.

[4] Joon-Ho Lee, Tian Xiao, and Qing Huo Liu, "A 3-D Spectral-Element Method Using Mixed-Order Curl Conforming Vector Basis Functions for Electromagnetic Fields," IEEE Transactions on Microwave Theory and Techniques.,vol.54,no.1,pp.437-444,Jan. 2006.

[5] Joon-Ho Lee and Qing Huo Liu, "A 3-D Spectral-Element Time-Domain Method for Electromagnetic Simulation," IEEE Transactions on Microwave Theory and Techniques.,vol.55,no.5,pp.983-991,May 2007.

[6] Dan Jiao, J.Jin,Eric, Michielssen, and Douglas J.Riley, "Time-domain finite-element simulation of three-dimensional scattering and radiation problems using perfectly matched layers," IEEE Trans. Antennas Propagat., vol. 51, no. 2 ,pp.296–305, Feb. 2003.

# A Novel Frequency Selective Surface for Ultra Wideband Antenna Performance Improvement

Hui-Fen Huang, Shao-Fang Zhang and Yuan-Hua Hu

South China University of Technology

Guangzhou, 510641 China

*Abstract-* **A novel frequency selective surface (FSS) is proposed in this paper. The FSS has four layers on both sides of two separate substrates with a distance 8 mm, which is much smaller than the wavelength. The FSS has wide bandwidth, ranging from 2.7 GHz to 13.2 GHz. The FSS cell for Electromagnetic Band Gap (EBG) is used as a reflector for ultra wideband antenna to improve its front-to-back ratio. Compared the performance for the reference antenna with and without the proposed reflector, the average improvement of the front-to-back ratio of the UWB antenna with designed FSS reflector is 15 dB from 3.1 GHz to 10.6 GHz and the peak even can get to 30 dB at 5 GHz.**

## I. INTRODUCTION

Rigorous research by academics and industries has focused on exploiting the EBG properties best suited for respective applications. Their shape, periods and number of layers as well as the substrate characteristics determine the performance of frequency selective surfaces (FSS) [1]-[3]. FSS are widely used as polarizer, space filter, reflector in dual frequency antennas and as a random for radar cross section (RCS) controlling as well as front-to-back ratio improvement [4]-[7].

In this paper, a frequency selective surface (FSS) cell is proposed. The FSS has four layers on two substrates with a distance 8 mm, which is smaller than the wavelength. The FSS has wide bandwidth, ranging from 2.7 GHz to 13.2 GHz. Then a reflector with the proposed EBG cells for ultra wideband antenna is developed. Compared the performance for the reference antenna with and without the proposed reflector, the average improvement of the front-to-back ratio of the UWB antenna with designed FSS reflector is 15 dB from 3.1 GHz to 10.6 GHz and the peak even can get to 30 dB at 5 GHz. The rest of the paper is organized as follows. Section II is the designed frequency FSS cell. Section III is the proposed reflector with the developed EBG cells for ultra band antenna. Section IV is conclusion.

## II. FREQUENCY SELECTIVE SURFACE DESIGN

### A. Equivalent Circuit for the Frequency Selective Surface

Fig. 1 is the designed FSS cell and its equivalent circuit. The parameters $d_1$, $d_2$, $g_1$, $g_2$, a, b, n, w, s, are marked in Table I. The FSS is constructed on the FR-4 substrate with dielectric constant of 4.4, loss tangent 0.02, thickness of 1.5 mm.

TABLE I

PARAMETERS OF THE PROPOSED FSS (UNITS: MM)

| $d_1$ | $d_2$ | $g_1$ | $g_2$ | a | b | n | w | s |
|---|---|---|---|---|---|---|---|---|
| 15 | 15 | 14 | 14 | 1.5 | 2 | 4 | 2 | 6 |

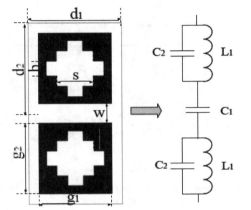

Figure 1. The corresponding equivalent circuits of the designed FSS element.

From the equivalent circuits of the FSS in Fig.1, C1, C2, L1 are the equivalent inductive and capacitive. The gap w between the squares forms the capacitance C1, and the capacitance C2 is formed by the gap b in the square as shown in Fig. 1. The capacitance value C1 is determined by the patch length g2, the gap width w between the two adjacent patches and the effective dielectric constant of the substrate. The inductance value L1 is determined by the length g1 and the width n of the metallic strip [8]-[9].

The bandwidth of monolayer FSS is very narrow and didn't meet the requirements of ultra bandwidth application. Wide stop band filters can be realized by cascading two or more layers of periodic arrays. A four layer FSS structures with the same FSS cell shape in each layer and the equivalent circuit for the four layer FSS structure is shown in Fig. 2. The dielectric substrates between the metallic layers are modeled as short transmission lines with characteristic impedances Z1, Z1 = Z0/ε, Z0=377 Ω, d is the thickness of FR4.

### B. Simulated Results for the Designed FSS Cell

The ultra wide stop band FSS structure is designed by Ansoft HFSS software. The simulated transmission coefficients

978-7-5641-4279-7

are in Fig. 3. It is observed that the proposed FSS has wider bandwidth, ranging from 2.7 GHz to 13.2 GHz.

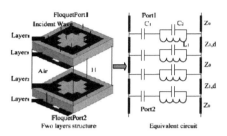

Figure. 2. Structures and Equivalent circuit of the designed layer FSS, T=1.5 mm, H=8mm.

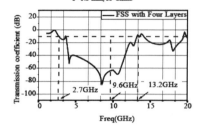

Fig.3. Transmission coefficient for FSS.

## III. REFLECTOR WITH THE DESIGNED FSS CELLS FOR ULTRA WIDEBAND ANTENNA

The performance for the ultra-wideband antenna with the four layers FSS reflector has been simulated. The structure for the antenna with the designed FSS is in Fig. 4, and Fig. 5 is the prototype of the antenna [11]. The size for the four layers FSS is 108 mm × 108 mm and the distance K between the antenna and FSS is 10 mm, which is approximately $\lambda$ / 4 at the central frequency 6 GHz.

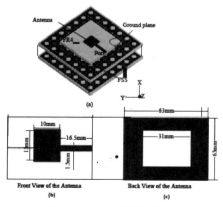

Fig.4. The antenna with FSS reflector

The four layers FSS acts as a reflector for the UWB antenna and is able to get low-backlobe level and high front-to-back ratio over ultra wideband from 3.1 GHz to 10.6 GHz. The prototype of the proposed antenna is fabricated and measured. S11 of the antenna with FSS is measured by using an Advantest R3770 network analyzer. Fig. 6 shows the simulated and measured S11 and Fig.7 shows the front-to-back ratio of the antenna. The FSS reflector has small effect on the impedance bandwidth, and the front-to-back ratio is significantly improved by the designed FSS reflector. The average improvement of the front-to-back ratio of the UWB antenna with FSS reflector is 15 dB from 3.1 GHz to 10.6 GHz and the peak even can get to 35 dB at 5 GHz. The discrepancy between the measured and simulated results is due to fabrication and measurement deviation.

(a)

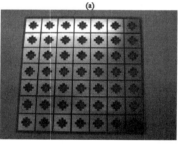

(b)

Fig.5. The fabricated antenna with FSS (a) Top view (b) bottom view.

Fig.6. The simulated and measured S11 of the UWB antenna with and without FSS.

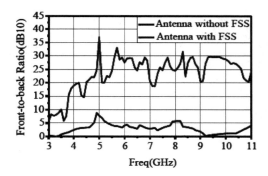

Fig.7. The front-to-back ratio of the UWB antenna with and without FSS

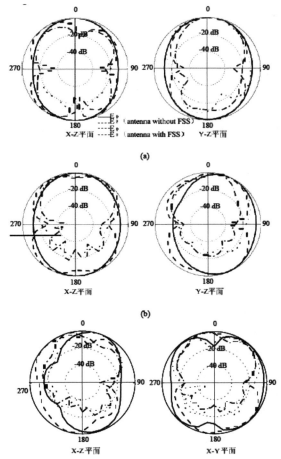

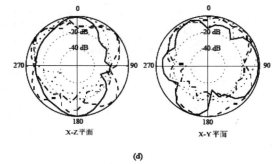

(d)

Fig.8. Radiation patterns on x-z plane and y-z plane (a) 3.5 GHz, (b) 5.3 GHz, (c) 7.9GHz, (d)10GHz

The far-field radiation characteristics for the antenna with FSS at the frequencies of 3.5, 5.3, 7.9 and 10 GHz are shown in Fig. 7 (a)–(d). Compared to the reference antenna [11], the antenna with the FSS reflector obviously reduces backward wave level from 3.1 to 10.6 GHz. It observed that the FSS reflector is an effective way to improve the front-to-back ratio of antenna.

## IV. CONCLUSION

In this paper, the four layer FSS structure with the proposed cells has wider bandwidth from 2.7 GHz to 13.2 GHz. The FSS is used as a reflector for an ultra band antenna. Compared the reference antenna with and without the proposed reflector, the average improvement of the front-to-back ratio of the UWB antenna with FSS reflector is 15 dB from 3.1 GHz to 10.6 GHz and the peak even can get to 30 dB at 5 GHz. The designed FSS can be used as a reflector to improve the antenna performance.

ACKNOWLEDGMENT

This work is supported by the National Natural Science Foundation of China under Grant 61071056.

REFERENCES

[1] B. A. Munk, Frequency Selective Surfaces: Theory and Design. New York: John Wiley & Sons Inc. 2000.
[2] M. A. Al-Joumayly and N. Behdad, "Low-profile, highly-Selective, dual-band Frequency Selective Surfaces with closely spaced bands of operation," IEEE Trans. Antennas Propag., vol. 58, no. 12, Dec. 2010.
[3] M. Salehi and N. Behdad, "A second-order dual X-/Ka-band Frequency Selective Surface," IEEE Microwave and Wireless Components Letters., vol. 18, no.12, pp. 248 -254, Feb. 2007.
[4] R. M. S. Cruz, A. G. D.Assunção and P. H. da F. Silva, "A new FSS design proposal for UWB applications," IWAT Antenna Technology., pp. 1-4, Mar. 2010.
[5] S. Genovesi, F. Costa and A. Monorchio, "Low-profile array with reduced radar cross section by using hybrid Frequency Selective Surfaces," IEEE Trans. Antennas Propag., vol. 60, no. 5, pp. 2327-2335, May. 2012.
[6] M.Pasian, S. Monni, A. Neto, M. Ettorre and G. Gerini, "Frequency Selective Surfaces for extended bandwidth backing reflector functions," IEEE Trans. Antennas Propag., vol. 58, no. 1, pp. 43-50, Jan. 2010.
[7] D. W. Woo, J. H. Kim ,J. K. Ji , G.H. Kim ,W. M. Seong and W. S. Park, "Design of a DSRR FSS for CDMA/RFID isolation," IEEE Antennas and Propagation Society International Symposium (APSURSI).. Jul. 2010.

[8]    M.Moallem and K. Sarabandi, "A single-layer metamaterial-based polarizer and bandpass Frequency Selective Surface with an adjacent
transmission zeros,"    IEEE Antennas and Propagation Society International Symposium (APSURSI), pp. 2649 – 2652, Jul. 2011.

[9]     K. Sarabandi and N. Behdad,"A Frequency Selective Surface with miniaturized elements," IEEE Trans. Antennas Propag., vol. 55, no. 5, pp.
1239-1245, May. 2007.

[10]   T. T. Thai, G. R. DeJean and M. M. Tentzeris, "Design and development
of a novel compact soft-surface structure for the Front-to-Back Ratio
improvement and size reduction of a microstrip yagi array antenna,"
IEEE Antennas Wireless Propag. Lett., vol. 7, pp. 369-373, Jun. 2008.

[11]   H. D. Chen, J. S. Chen and J.   N.   Li, "Ultra-wideband square-slot
antenna," Micro. Opt. Technol. Lett., vol. 48, no. 3, pp. 500-502, Jan.
2006.

# Terahertz Cassegrain Reflector Antenna

Xiaofei Xu*[1,2], Xudong Zhang[2], , Zhipeng Zhou[1,2], Tie Gao[1,2], Qiang Zhang[1,2], Youcai Lin[1,2], and Lei Sun[2]
[1]Science and Technology on Antenna and Microwave Laboratory (STAML), Nanjing, 210039
[2]Nanjing Research Institute of Electronics Technology (NRIET), Nanjing, 210039
*xuxiaofei@nriet.com

*Abstract*- **In this report, we present our recent work on terahertz Cassegrain reflector antenna with compact configuration and directive radiation pattern. The geometrical parameters and electrical performance of the Cassegrain reflector system are calculated and analyzed through folded optics approach, combined with feeding horn design through numerical package simulations. Analytic results of the antenna show high directivity over 50dB, side lobe less than -20dB and beam width of 0.5° from the far field radiation pattern, making the antenna an interesting candidate for high resolution imaging applications as surveillance equipment.**

## I. INTRODUCTION

Very recently, terahertz (THz) technology has attracted much attention due to the unique characteristics [1-3]. THz wave could transmit opaque material as clothing or heavy smoke, and thus provides a powerful tool for concealed object detection and imaging underneath [1-3]. Besides, for a diffraction-limited image application, very high image resolution could also be achieved with a compact and effective terahertz antenna aperture [4-5]. In this presentation a compact and directive reflector antenna for these application cases is discussed. The antenna is with high directivity to radiate and receive wave in 0.33THz frequency band.

In the following parts, we will first introduce the design approach of folded geometry optics method [5-6], and then investigate the antenna system, with results and analysis predicted through the method. Analysis shows that the antenna directivity is very sensitive to the surface smoothness of the reflector, which could be controlled and minimized by finish machining and techniques.

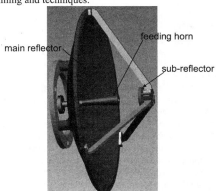

Figure 1. Configuration of the terahertz Cassegrain reflector antenna.

## II. DESIGN APPROACH

Maxwell's equations are generally acknowledged to describe and predict radiation and transmission properties of electromagnetic (EM) waves. However, for electrically large objects, it is time-consuming work to solve the EM fields over the objects with analytical expressions of the equations or their numerical resolutions with convergent accuracy. For these cases as electrically large antenna aperture, one special approximation method of classical folded geometry optics is introduced to facilitate design [5-6]. Fig.1 portrays the large aperture reflector antenna of Cassegrain type with compact configuration as a promising candidate. Two reflectors, one hyperbolic subreflector and parabolic main reflector combined with the feeding horn compose the antenna.

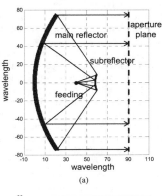

(a)

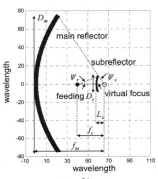

(b)

Figure 2. Scheme for the terahertz Cassegrain antenna

978-7-5641-4279-7

Terahertz wave from the feeding as the focus of the hyperbolical subreflector is folded to radiate onto the aperture plane, with ray path portrayed in Fig.2 (a). With equivalent geometry path theory, the multi reflector system could be regarded as a single parabolic reflector with feeding at the virtual focus. The aperture for the main reflector is about 150 wavelengths, while the dimension in the radiation direction for the antenna is about 60 wavelengths. Detailed description of the aperture sizes, the positions, and characteristic focal lengths are presented with parameters in Fig.2 (b). These parameters are related as following [5-6],

$$\begin{cases} \tan(\psi_v / 2) = D_m / (4 f_m) \\ 1/\tan\psi_r + 1/\tan\psi_v = 2 f_c / D_s \\ 1 - \sin((\psi_v - \psi_r)/2)/\sin((\psi_v + \psi_r)/2) = 2L_v / f_c \end{cases} \quad (1)$$

The relationship in (1) restricts terahertz ray trajectory of the reflector systems in Fig. 2(a). For large apertures, this geometry method provides rather accurate and effective approach to predict antenna performance [5-6].

### III. RESULTS AND ANALYSIS

With the folded geometry optics method described in the above literature, results and qualitative analysis of the reflector antenna will be shown in this section, together with the introduction of feeding antenna with satisfactory primary radiation patterns.

In reflector antenna system, conical hybrid mode horns as feedings are widely used for the desirable radiation properties as good pattern symmetry and low cross-polar sidelobe levels. In terahertz band, dual-mode Potter horns are preferable to corrugated horns for fabrication facility. The dual-mode Potter horn antenna with steps could converts the incident dominant TE11 mode to hybrid TE11 and TM11 modes. With proper length compensation, the hybrid two modes will be in phase on the radiation plane. Fig.3 gives the simulated radiation pattern with co-polar and cross polar levels of the dual-mode horn. We could see that good pattern symmetry is achieved in the subreflector coverage range. And the cross-polar radiation field is about 40dB level lower than the copolar field, which provides expected primary patterns for the reflector systems.

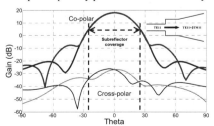

Figure 3. Farfield radiation pattern for dual-mode feeding horn.

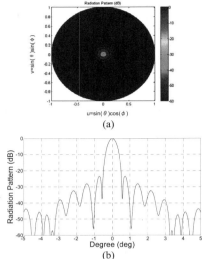

(a)

(b)

Figure 4. Farfield radiation pattern of the reflector antenna

The near field distribution for the reflector aperture is obtained from geometry optics method, which is then extrapolated to the farfield radiation pattern shown in Fig.4. The extrapolation procedure follows the equations as

$$\vec{E}(u,v) \propto \iint \vec{E}_a(x,y) \exp(jk_0(ux + vy))dxdy, \quad (2)$$

where $u=\sin(\theta)\cos(\varphi)$, $v=\sin(\theta)\sin(\varphi)$, and $\vec{E}(u,v)$ and $\vec{E}_a(x,y)$ denote vector farfield and aperture field separately. It should be noted that the aperture field $\vec{E}_a(x,y)$ in (2) includes the blockage effect of subreflector and mechanical support.

We could estimate the beamwidth and near-in sidelobe of the antenna as $0.5°$ and -23dB from the patterns in Fig.4. The directivity of the antenna could be further processed with the farfield data with the following equation

$$D = \frac{|E_{max}|^2}{\frac{1}{4\pi} \iint |E(\theta,\varphi)|^2 \sin\theta d\theta d\varphi} = \frac{|E_{max}|^2}{\frac{1}{4\pi} \iint \frac{|E(u,v)|^2}{\sqrt{1-u^2-v^2}} dudv}. \quad (3)$$

And the efficiency of the antenna excluding the loss and spillover energy is calculated by $\eta = D\lambda^2 /(4\pi A)$. Through these expressions, we could calculate the directivity as 51.1 dB and the efficiency as 57%.

Perturbations deriving from the manner the terahertz reflector antenna fabricated, assembled and installed, might occur in the radiation pattern. Generally, these errors departure from the theoretical design might reduce directivity; raise the sidelobe and cross-polar level. The RMS error associated with surface random fabrication has been analyzed in [6-7], while a quantitative effect on directivity performance due to small correlation interval deviation is given by

$$D/D_0 = \exp(-(4\pi\varepsilon/\lambda)^2)\,, \qquad (4)$$

where $D$ and $D_0$ are antenna directivity with and without errors, $\lambda$ is the wavelength and $\varepsilon$ is the RMS surface error. A detailed illustration could be found in Fig.5. Fabrication error should be controlled in 0.01mm as the loss might be less than 0.1dB. Assembly and installations errors should also be controlled with the same order, provided careful checkout procedure used.

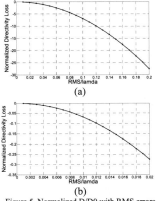

(a)

(b)

Figure 5. Normalized D/D0 with RMS errors

## IV. Conclusion

In conclusion, we present one compact Cassegrain type antenna system for terahertz band applications with high directivity. The reflector antenna system is designed with classical folded optics method, validated by desired geometry parameters and positions of the reflectors system, while the primary illumination for the reflectors from a dual-mode horn feeding is predicted with accurate numerical simulated radiation performance. Near aperture fields from the geometry optics are further extrapolated to obtain the far field distribution. Radiation performance is also investigated to obtain antenna beamwidth, sidelobe and directivity with surface error analysis.

## References

[1] J. C. Dickinson, et. al., "Terahertz imaging of subjects with concealed weapons," *Proc. SPIE*, vol. 6212, pp. 62120Q-1–62120Q-12, 2006.

[2] R. Appleby, and H. B. Wallace, "Standoff detection of weapons and contraband in the 100 GHz to 1 THz region," *IEEE Trans. Antennas Propag.*, vol. 55, no. 11, pp. 2944–2956, 2007.

[3] K. B. Cooper, et. al., "THz imaging radar for standoff personnel screening," *IEEE Trans. Terahertz Science and Technology*, 2011, vol. 1,no.1, pp.169-182. 2011.

[4] D. L. Mensa, *High Resolution Radar Cross Section Imaging*, Artech House, Boston, 1991.

[5] M. Skolnik, *Introduction to Radar Systems*. MA: McGraw-Hill, 2001.

[6] R. C. Johnson, and H. Jasik, *Antenna Engineering Handbook*, NY: McGraw-Hill, 1984.

[7] J. Ruze, "Antenna Tolerance Theory-A Review," *Proceedings of the IEEE*, vol 54, no. 4, pp: 633-640,1966.

# *FA-1(D)*

## October 28 (FRI) AM

## Room D

## EM in Circuits-1

# Simplified Modeling of Ring Resonator (RR) and Thin Wire Using Magnetization and Polarization with Loss Analysis

Dongho Jeon, Bomson Lee

Dept. Electronics and Radio Engineering, Kyung Hee University
Yongin-si, Gyeonggi-do, Republic of Korea
bomson@khu.ac.kr

**Abstract** — In this work, the ring resonator (RR) and thin wire are simply modeled using the concept of magnetization and polarization, respectively, with convenient expressions for its effective permeability and permittivity. Equivalent circuits for them are also provided with closed-form expressions for all circuit elements. The ring resonator and the thin wire were designed and their characteristics were examined in terms of S-parameters, effective permeability and permittivity, loss rate, bandwidth, etc. The theoretical, circuit-, and EM-simulated results are shown to be in an excellent agreement. Based on this modeling, loss analysis has been carried out.

## I.   INTRODUCTION

The realization of media having negative permittivity and permeability, so called left-handed media (or metamaterials), became feasible after 1991 when Pendry proposed to use an array of thin wires and split ring resonators (SRRs) [1-2]. A number of approaches followed in many aspects in an effort to realize similar left-handed characteristics [3-6]. However, most of them have been known to suffer from high losses and narrow bandwidth.   Even alleviating these problems could have a significant impact on many applications.

In this paper, we model the ring resonator (the original form of SRRs) excited by a magnetic field as an induced magnetic dipole and derive the effective permeability in a very convenient form. The mechanism of SRRs is explained with more familiar terms than in [1-2]. In addition, we model the thin wire excited by an electric field and the effective permittivity is derived using the similar procedure as used in the RR case. For both the ring resonator and the thin wire, we perform a loss analysis with some examples with respect to the size of unit cell and the radius of each structure. The medium having low loss can be achieved through the modification of geometry based on the loss analysis. The effects of employing these methods are discussed with some comparisons.

## II.   MODELING OF RING RESONATOR

We model the ring resonator using magnetization concept and provide an equivalent circuit. The ring resonator is shown with the orientations of incident fields in Fig. 1. The TEM wave travels in the z direction with the electric and magnetic fields oriented in the x and y direction, respectively. The radius of the loop is $r$ and the radius of the ring is $r_{ring}$. The side length of the unit cell is $a$. $C$ is the capacitance of a chip capacitor loaded

on the RR. It is inserted for resonance of the ring resonator at s design angular frequency $\omega_0$. The total resistance $R$ of the ring resonator is given by

$$R = R_r + R_l \tag{1}$$

where $R_r$ is the radiation resistance, $R_l$ is the ohmic resistance. As the magnetic field $H_0$ couples through the ring resonator, an induced voltage $V_{emf}$ is given by

$$v_{emf} = \frac{\partial\left(\mu_0 H_0 \pi r^2\right)}{\partial t} \xrightarrow{\frac{\partial}{\partial t}=j\omega} V_{emf} = j\omega\mu_0 H_0 \pi r^2 \tag{2}$$

based on Faraday's law.

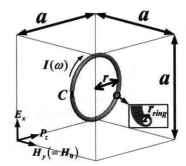

Fig.1. Structure of a ring resonator

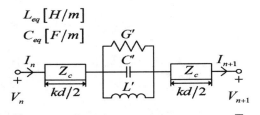

Fig.2. Equivalent circuit of ring resonator

978-7-5641-4279-7

Accordingly, the current $I$ can be determined simply by dividing $V_{emf}$ with the resonator impedance $Z(\omega)$ (series sum of $R$, $L$, and $C$). Using the magnetization defined by $\overline{M}(\omega) = \overline{m}(\omega)/a^3$, the relative effective permeability can be obtained as

$$\mu_{eff}(\omega) = 1 + \chi_m(\omega) = 1 - \frac{j\omega\mu_0\left(\pi r^2\right)^2}{a^3 R\left[1 + jQ\left(\dfrac{\omega}{\omega_0} - \dfrac{\omega_0}{\omega}\right)\right]} \quad (3)$$

where $\omega$ is the angular frequency, $\overline{m}(\omega)$ is the magnetic dipole moment ($\overline{m} = -I(\omega)\pi r^2 \overline{a_y}$), $Q$ ($Q = \omega_0 L/R$) is the quality factor, $\chi_m$ is the magnetic susceptibility, and $\mu_0$ is the permeability in free space. Now, we want to devise an equivalent circuit for the structure shown in Fig. 1. By multiplying the derived relative effective permeability with the geometrical factor ($g$) which is determined for the cross-sectional shape of a specific transmission line, we can obtain the effective inductance of the transmission line given by

$$L_{eff}(\omega) = \mu_0\mu_{eff}(\omega)g \; [H/m]. \quad (4)$$

For an instance, $g$ for the parallel plate waveguide T/L with width $W$ and height $h$ is given by $h/W$. Now, the effective series impedance can be expressed as

$$Z(\omega) = j\omega L_{eff}(\omega)d$$
$$= j\omega\mu_0 gd + \frac{1}{\dfrac{a^3}{\left(\omega\mu_0\pi r^2\right)^2 gd}\left[R + j\omega_0 L\left(\dfrac{\omega}{\omega_0} - \dfrac{\omega_0}{\omega}\right)\right]} \; [\Omega] \quad (5)$$

where $d$ is the physical length of the T/L unit cell. A close examination of (5) leads to the equivalent circuit as depicted in Fig. 2. The equivalent circuit consists of a right handed (RH) transmission line with $L_{eq}d$, $C_{eq}d$ and a parallel $G'$, $C'$, $L'$ resonator in the series branch. The RH T/L may be alternatively represented by $Z_c$ and $kd$ at $\omega_0$. The total series impedance of the equivalent circuit in Fig. 2 can be written as

$$Z_{eq}(\omega) = j\omega L_{eq}d + \frac{1}{G' + j\omega_0 C'\left(\dfrac{\omega}{\omega_0} - \dfrac{\omega_0}{\omega}\right)} \; [\Omega] \quad (6).$$

The value of parameters $G'$, $C'$, $L'$ in (6) can be obtained as (7) by comparing (5) and (6).

$$G' = \frac{a^3}{\left(\omega\mu_0\pi r^2\right)gd}R, \; C' = \frac{a^3}{\left(\omega\mu_0\pi r^2\right)gd}L, \; L' = \frac{1}{\omega_0^2 C} \quad (7)$$

## III. MODELING OF THIN WIRE

The same manner used in modelling of the ring resonator is employed for the case of thin wire and its equivalent circuit. The structure and dimensions of the thin wire are depicted in Fig. 3. The radius of the wire is $r'$ and the height is same as the one side length of the unit cell $a'$. Since the total electric field on the surface of the wire should be 0, the induced electric field is $-E_0 \overline{a_x}$ as shown in Fig. 3. With the current determined by $I = E_0 a/(R + j\omega L)$, the induced charge on the top of the thin wire is given by $q = I/(j\omega)$. Accordingly, the polarization $P$ can be written as

$$P = \frac{\overline{p}}{a'^3} = \frac{E_0}{j\omega(R + j\omega L)a'}\overline{a_x} \quad (8)$$

based on the electric dipole moment ($\overline{p} = qa'\overline{a_x}$). Through the electric flux density ($\overline{D} = \varepsilon_0 E_0 + P$), the effective permittivity can be obtained as

$$\varepsilon_{eff}(\omega) = 1 + \chi_e(\omega) = 1 - \frac{\omega_p^2\left(\omega + jR/L\right)}{\omega\left[\omega^2 + \left(R/L\right)^2\right]} \quad (9)$$

where $\varepsilon_0$ is permittivity in free space, $R$ is total resistance of wire, $L$ is the inductance of wire, $\chi_e$ is the electric susceptibility and $\omega_p$ is the plasma frequency and the square of it can be written as

$$\omega_p^2 = \frac{2\pi}{\mu_0\varepsilon_0 a'^2 \ln(a'/r')} \quad (10).$$

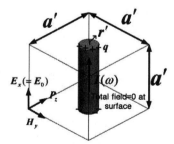

Fig.3. Structure of a thin wire

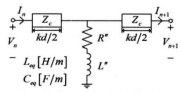

Fig.4. Equivalent circuit for thin wire

An equivalent circuit can be devised very conveniently. By dividing the effective permittivity with the geometrical factor, $g$, the effective capacitance of a transmission line is given by

$$C_{eff}(\omega) = \frac{\varepsilon_0 \varepsilon_{r,eff}(\omega)}{g} = \frac{\varepsilon_0}{g} - \frac{\omega_p^2}{\omega\left(\omega - j\dfrac{R}{L}\right)} \varepsilon_0 \cdot \frac{1}{g} \quad \left[\frac{F}{m}\right] \quad (11).$$

Now, the effective shunt admittance can be expressed as

$$Y(\omega) = j\omega C_{eff}(\omega)d = \frac{j\omega \varepsilon_0 d}{g} + \frac{1}{\dfrac{Rg}{L\omega_p^2 \varepsilon_0 d} + j\omega \dfrac{g}{\omega_p^2 \varepsilon_0 d}} \quad (\mho) \ (12)$$

where $d$ is the physical length of the T/L unit cell. The total shunt admittance of the equivalent circuit in Fig. 4 can be written as

$$Y_{eq}(\omega) = j\omega C_{eq}d + \frac{1}{R'' + j\omega L''}. \quad (13)$$

The values of the parameters $R''$ and $L''$ can be obtained as

$$R'' = Rg(\Omega), \quad L'' = Lg(H) \quad (14)$$

by comparing (12) and (13).

## IV. VARIFICATION OF MODELINGS AND DISCUSSION

The ring resonator used in the EM simulation is made of copper and is designed at 13.56 MHz. The unit cell size ($a$) is 12 cm (roughly $0.005\lambda_0$). The radii of the loop ($r$) and the ring ($r_{ring}$) are 5 cm and 1 mm, respectively. Let's name this as a standard structure. The ohmic resistance $R_l$ is 0.058 $\Omega$ and the radiation loss $R_r$ is negligible at the resonance frequency. The inductance of loop is 266nH. The value of capacitance for the resonance is 533 pF. Fig. 5 (a) and (b) show the extracted [4] real and imaginary part of permeability based on circuit and EM simulations when the TEM wave propagates in the z direction as illustrated in Fig. 1. The circuit parameters based on (7) and (14) are $G'$=0.0075 mho, $C'$=40.92 nF, $L'$=3.37 nH, $R''$=1 $\Omega$, $L''$=88.84nH. The loss is the most significant at the resonant frequency of 13.56 MHz as implied by the imaginary part. However, the loss tangent is about 0.05 when the real part of effective permeability crosses -1 at 14 MHz. Fig. 6 (a) and (b) show the real and imaginary part of the effective permittivity for the thin wire when $r'$ and $a'$ are 1mm and 12cm, respectively. The loss tangent is about 0.0011 when the real part of effective permittivity crosses -1 at 347 MHz. The high loss occurs at low frequencies as shown in Fig. 6 (b), but it decreases drastically as the frequency increases. The circuit and EM simulated results show an excellent agreement with the curve (theory) plotted based on (3) and (11), respectively. The excellent agreement validates the proposed modeling. Fig. 7 (a) and (b) show the extracted effective permeability of the ring resonator when the unit cell size, $a$, and the thickness, $r_{ring}$, are fixed with 12 cm and 1 mm while the radius, $r$, varies from 5 cm to 1 cm. The effect of magnetization is seen to decrease

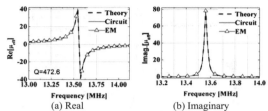

(a) Real  (b) Imaginary

Fig. 5. Extracted effective permeability when a=12 cm, r= 5 cm, $r_{ring}$= 1mm (based on Fig.1)

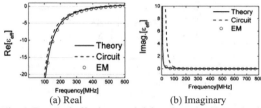

(a) Real  (b) Imaginary

Fig. 6. Extracted effective permittivity when $a' = 12$cm, $r' = 1$ mm (based on Fig. 3)

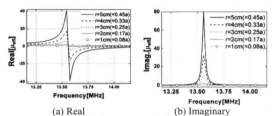

(a) Real  (b) Imaginary

Fig. 7. Extracted effective permittivity ($a$=12cm fixed)

as expected. Fig. 8 and 9 show the real and imaginary part of the effective permeability when the unit cell size ($a$) varies with respect to the wavelength with the ratio of $a$, $r$, and $r_{ring}$ maintained as the standard structure ($a$=12cm, $r$=5cm, and $r_{ring}$=1mm). For this case, $r/a$ and $r/r_{ring}$ ($R$) are maintained the same. As is expected from (3), the real part of the permeability becomes smoother as the size $a$ decreases. From the imaginary part of the permeability shown, we can see that the loss becomes more significant as $a$ increases near the resonance frequency. On the other hand, at some useful frequencies where $\mu_{eff}$= -1 or $\mu_{eff}$= -2, the loss becomes smaller as $a$ increases. Fig. 10 and 11 show the real and imaginary part of the effective permittivity with respect to the size of unit cell $a'$ and radius $r'$ of the thin wire. As $a'$ increases, the real part of the effective permittivity is seen to have stiff slopes when they are negative. In the low frequency region, the imaginary part becomes larger as $a$ decreases. However, it is shown to be very close to 0 at higher frequencies.

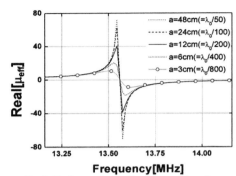

Fig. 8. Real part of $\mu_{\text{eff}}$ with respect to size of $a$

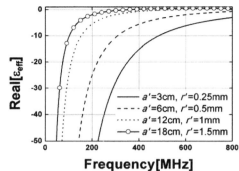

Fig. 10. Real part of $\varepsilon_{\text{eff}}$ with respect to size of $a'$

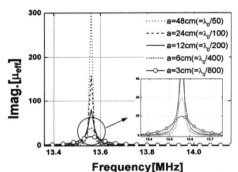

Fig. 9. Imaginary part of $\mu_{\text{eff}}$ with respect to size of $a$

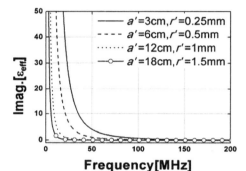

Fig. 11. Imaginary part of $\varepsilon_{\text{eff}}$ with respect to size of $a'$

## IV. CONCLUSION

The effective permeability of ring resonators (or SRRs) and the effective permittivity of thin wires have been formulated based on the concept of magnetization and polarization, respectively. Equivalent circuits of them have also been proposed and analyzed with necessary comparisons. The circuit- and EM-simulated results are in excellent agreement. With the provided modeling, the problem of synthesizing the effective medium can be engineered more systematically.

### ACKNOWLEDGEMENT

This work was supported by Mid-career Researcher Program through the National Research Foundation of Korea (NRF) grant funded by the Korea government (MEST) (No. 2012047938).

### REFERENCES

[1] J.B. Pendry, A.J. Holden, D.J. Robbins, and W. J. Stewart, "Magnetism from conductors and enhanced nonlinear phenomena," IEEE Transactions on Microwave Theory Technology, vol. 47, pp. 2075–2084, November 1999.

[2] J.B. Pendry, A.J. Holden, D.J. Robbins, and W. J. Stewart, "Low frequency plasmons in thin-wire structures," Journal of Physics, vol. 10, June 1998.

[3] K. Zhang, Q. Wu, J. Fu, F.Yimeng, and L. Li, "Metamaterials With Tunable Negative Permeability Based on Mie Resonance,", IEEE Transactions on Magnetics, vol. 48, pp. 4289–4292, November 2012.

[4] S. Mao, S. Chen, C. Huang, "Effective Electromagnetic Parameters of Novel Distributed Left-Handed Microstrip Lines,"IEEE Transactions on Microwave Theory Technology, vol. 53, no. 4, pp. 1515–1521, April, 2005

[5] K. Zhang, Q. Wu, J. Fu, F.Yimeng, and L. Li, "Metamaterials With Tunable Negative Permeability Based on Mie Resonance,", IEEE Trans. Magnetics, vol. 48, pp. 4289–4292, November 2012

[6] S.M. Rudolph, C. Pfeiffer, and A. Grbic, "Design and Free-Space Measurements of Broadband, Low-Loss Negative-Permeability and Negative-Index Media,", IEEE Transactions on Antennas and Propagation, vol. 59, pp. 2989–2997, August 2011.

# Transmission Characteristics of Via Holes in High-Speed PCB

He Xiangyang, Lei Zhenya, Wang Qing
School of Electronic Engineering, Xidian University
Xi'an Shaanxi 710071, China

*Abstract*-A method of analyzing transmission characteristics of via holes in the time domain is presented in this paper. A suitable analysis model is set up using CST (Computer Simulation Technology). With theoretical analysis, the effect of via structures on signal transmission characteristics could be acquired. The simulation results show that appropriate design for via holes can reduce transmission characteristics issues effectively.

## I. INTRODUCTION

As operating frequency and data rate increase, PCB(Printed Circuit Board) layout becomes increasingly dense. The via hole provides a good choice to high-speed multi-layered PCB. However, a discontinuity occurs as a result of the via hole, which is also the cause of some signal integrity problems[1-2], such as time delay, signal reflection and crosstalk. For high-speed PCB, signal integrity is as important as the circuit function and clock rate[3]. Hence, appropriate analysis of electrical characteristics and parasitic parameters of via holes are needed. Several types of via holes and their scattering parameters (S-Parameters) is discussed in [4]. Effects of the structure of via holes on scattering parameters and transmission loss is given in [5]-[6]. All the papers mentioned above are based on the frequency-domain which is less direct than the time-domain. Even though we could achieve transmission characteristics of via holes using TDR in practical problems, there are few papers which disscussed transmission characteristics of via holes using simulation software directly in the time-domain. The one-via hole model and three-via hole model are given in detail in this paper using the time-domain simulation software-CST. And, the transmission characteristics of via holes are studied.

## II. VIA HOLE

Via holes[7] are the common holes interconnecting transmission lines on different layers in two-layer or multi-layer PCB. A typical structure of a via hole is shown in Fig. 1.

As operating frequency increases, the parasitic parameters of each via hole should be considered, including parasitic capacitance and parasitic inductance. A typical equivalent circuit of a via hole is shown in Fig. 2.

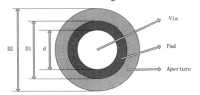

Fig. 1. Via structure. D2 is the aperture diameter, D1 is the pad diameter, d is the via diameter.

978-7-5641-4279-7

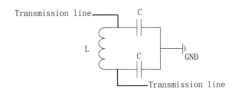

Fig. 2. Equivalent circuit. C is parasitic capacitance, L is the parasitic inductance.

### A. Parasitic Capacitance of Via Holes

Parasitic capacitance of a via hole[7] is defined as

$$C = \frac{1.41\varepsilon_r TD_1}{D_2 - D_1} \qquad (1)$$

where $\varepsilon_r$ is the dielectric constant, and T is the thickness of the PCB.

### B. Parasitic Inductance of Via Holes

While there is parasitic capacitance of via holes, there is parasitic inductance. Parasitic inductance of a via hole[7] is defined as

$$L = 5.08h\left[\ln\left(\frac{4h}{d}\right)+1\right] \qquad (2)$$

where $h$ is the height of via hole

## III. EFFECT OF VIA HOLES ON TIME DELAY

### A. Theoretical Analysis

A via hole inevitably puts influence on signal transimission characteristics in high-speed PCB. It could be considered as a transimission line. Its parasitic capacitance increases signal rising time and slows down circuit speed, leading to propagation delay. For instance, the parasitic capacitance of a PCB is 0.23pF from Equation (1), of which the thickness of the substrate is 24mil, the via diameter is 10mil, the pad diameter is 22mil, the aperture diameter is 36mil and the relative dielectric constant of the substrate is 4.3.

For a transmission line of characteristic impedance of $50\,\Omega$, the variation of the rising time caused by the parasitic capacitance is defined as

$$\Delta T_{10\%-90\%} = 2.2C(Z_0 / 2) = 13ps \qquad (3)$$

The time delay of one via hole is not notable. However, if numerous via holes are used in the design, the signal quality will not be guaranteed. Therefore, issues produced by considerable via holes should be taken into account.

As shown by Equation (1), in order to minimize the time delay, the size of via holes must be carefully studied. The parasitic capacitance is proportional to the via diameter and the thickness of the substrate, and inversely proportional to the aperture diameter. If dielectric materials are given, the via diameter should be minimized and the aperture diameter should be maximized if possible.

Materials and thickness of the substrate definitely have influence on the transmission characteristics of via holes. However, if materials or thickness of the substrate is changed merely, the characteristic impedance of the stripline will deviate from $50\,\Omega$. If so, this comparison will not based on the same conditions. Thus, this paper does its research on condition that materials and thickness of the substrate are seclected.

### B. Simulation Analysis

This paper develops a PCB in the CST MICROWAVE STUDIO, as shown in the Fig. 3(a). Input and output of the model are simulated in CST DESIGN STUDIO.

The model is a six-layered PCB where the via diameter is d=10mil, the pad diameter is D1=22mil, the aperture diameter is D2=36mil, the thickness of each layer is h=8mil, the width of the transmission line is 6.8mil and the thickness of copper layer is 0.7mil. The simulation result of this one-via hole is shown in Fig. 3(b).

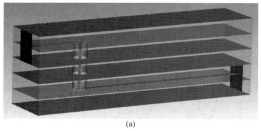

(a)

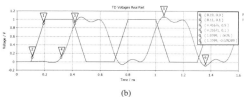

(b)

Fig. 3. One-via hole

In high frequency, the time delay is determined by materials of the substrate and the length of the transmission line. For the transmission line itself has time delay, the time delay of the single stripline should be studied so as to find out the results of the via holes caused. The whole length of the stripline contains the length of the via hole and the original stripline. The model and the result of the time delay are shown in Fig. 4(b).

(a)

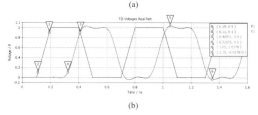

(b)

Fig. 4. Single transmission line

Fig. 3 and 4 showed the rising time $T_{10\%-90\%}$ of the excitation is 80ps and the rising time $T_{10\%-90\%}$ of the one-via hole model is 88ps. Then the change $\Delta T_{10\%-90\%}$ of the rising time caused by the single via hole is 12ps. From Equation (1) and (3), the change of the rising time caused by the single via hole is 13ps. The results of simulation are consistent with that of theoretical analysis.

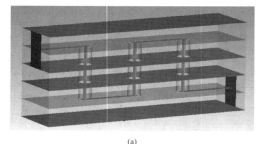

(a)

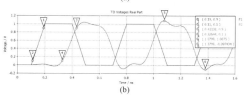

(b)

Fig. 5. Three-via hole

As shown in Figure 5, through modeling and simulating the three-via hole, the rising time of this three-via hole is achieved. The rising time is 107ps. Compared with the module with one-via hole, the time delay of the rising time increases.

Fig. 3, 4 and 5 give the time delay of the input and output at 0.1V. Due to transmission-line effect, the time delay of the single transmission line is 8ps. However, the time delay of the one-via hole is 20ps and the time delay of the three-via hole is 27ps. As the number of via holes increases, the time delay increases.

When the via diameter is 10mil, the aperture diameter is 36mil and the pad diameter is 18mil, 22mil and 26mil, results of the time delay are shown in Fig. 6.

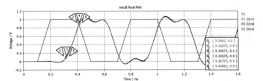

Fig. 6. Different pad diameters

When the via diameter is 10mil, the pad diameter is 22mil and the aperture diameter is 32mil, 36mil and 40mil, results of the time delay are shown in Fig. 7.

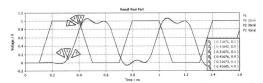

Figure 7 different aperture diameters

As shown in Fig. 6 and 7, the time delay decreases with the increase of via diameter and increases with the increase of aperture diameter. As a result of the combined effect of the complexity of the system and the parasitic parameters, these structures result in time delay indistinctively. But it did happen. From next session, effects caused by the parasitic capacitance is less significant than that caused by the parasitic inductance.

## IV. EFFECT OF VIA HOLES ON SIGNAL REFLECTION

### A. Theoretical Analysis

During signal transmitting process, signal feels transient impedance at all times. If the transient impedance is constant, signal propagation will proceed normally; if the transient impedance varies anywhere, signal reflection will happen. Via holes are discontinuities of the transmission line in high frequency and high-speed circuits. The reflection formula[8] is shown by

$$\Gamma = \frac{Z_L - Z_0}{Z_L + Z_0} = \frac{V_{\text{Reflected}}}{V_{\text{Incident}}} \qquad (4)$$

where $\Gamma$ is reflection coefficient, $Z_0$ is characteristic impedance, $Z_L$ is transient impedance, $V_{Reflected}$ is reflected voltage, $V_{Incident}$ is incident voltage.

In high-speed PCB, harm caused by the parasitic inductance of via holes is bigger than that of the parasitic

capacitance. The parasitic inductance spoils the effectiveness that shunt capacitors and decoupling capacititors have, thus impairing the filtering effect of the whole system. Still using the example above, the parasitic inductance of this via hole is

$$L=0.29\text{nH} \qquad (5)$$

If the signal rising time is 0.1ns, the equivalent impedance could be illustrated as follows

$$X_L = \pi L / (T_{10\%-90\%}) = 9.86\,\Omega \qquad (6)$$

The change in impedance is $9.86\,\Omega$. At this point the reflection coefficient is $\Gamma = 0.09$. The bigger change in impedance is, the more reflected signal is. In high frequency, this impedance could not be ignored.

Meanwhile, the signal is reflected when it arrive at the via hole as a result of the parasitic capacitance.

### B. Simulation Analysis

As shown in Fig. 3, 4 and 5, the overshoot of the single stripline changes by 3.78% and undershoot by 3.79%. The overshoot of one-via hole changes by 6.76% and undershoot of one-via hole changes by 7.83%. Overshoot of three-via hole changes by 8.75% and undershoot 9.74%. The reflection portion of signal increases with the via hole number. As for high-speed PCB, these changes are relatively big to the circuit performance.

The parasitic inductance and capacitance increase with the via hole number. And so is the signal reflection. The increase of the via hole number not only increases the time delay, but also the signal reflection. From the point of waveform, wave quality becomes bad when the via hole number increases.

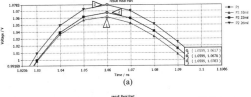

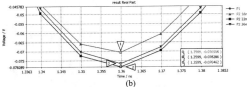

Fig. 8. Different pad diameters (a)overshoot (b)undershoot

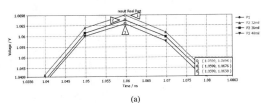

(a)

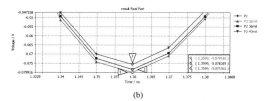

(b)

Fig. 9. Different aperture diameter (a)overshoot (b)undershoot

As shown in Fig. 8 and 9, the transmission line with a via hole changes the maximum of the 8 percent. When the via diameter changes 8mil, the overshoot changes 1.6% and the undershoot 0.8%. When the aperture diameter changes 8mil, the maximum changes 0.4%. Therefore, the reflection portion of the signal increases with the pad diameter and decreases with the aperture diameter. In addition, the effect of the via diameter on the maximum is greater than that of the aperture diameter.

V. CONCLUSION

In high-speed PCB, a proper design in via holes will save costs, decrease electromagnetic interference and reduce radiation. However, the effect of parasitic parameters of the via hole shouldn't be ignored. This paper discusses transmission characteristics of via holes in time-domain. Results reflect time delay and signal reflection directly. Hence, through appropriate design in via holes, the validity and reliability in high-speed PCB will be increased.

REFERENCES

[1]  Eric Bogatin. Signal Integrity: Simplified[M]. Beijing: Publishing House of Electronics Industry, 2007.
[2]  Li Chao, Chen Shaochang, Liu Renyang. Methods of Descending Effects of Via in High-speed PCB[J]. SAFETY & EMC, 2012(4): 57 – 60.
[3]  Lynne Green. Understanding the importance of signal integrity[J]. IEEE Circuits and Devices Magazine, 1999, 15(6): 7 - 10.
[4]  Giulio Antonini, Antonio Ciccomancini Scogna and Antonio Orlandi. S-Parameters Characterization of Through, Blind, and Buried Via Holes[J]. IEEE Transactions on mobile computing , 2003, 2(2):174 - 184.
[5]  Du Meizhu , Li Shufang , Qiu Xiaofeng. Via Design in Multi-layer PCB[C]. Hangzhou China : Asia-Pacific Conference on Environmental Elecltromagnetic , 2003.
[6]  Zhang Chenggang, Wang Liuchun and Zhang Debin. Through Hole Design of the High-speed Hybrid PCB[J]. Journal of Microwaves, 2010(8) :257 - 259.
[7]  Jiang Fupeng. Circuit Board Design Techniques for EMC Compliance[M]. Beijing: China Machine Press, 2011.
[8]  Liang Changhong, Xie Yongjun and Guan Boran. Concise Microwave[M]. Beijing: Higher Education Press, 2006

# Novel W-slot DGS for Band-stop Filter

Lin Chen, MinquanLi, Wei Wang, Jiaquan He, Wei Huang

*Abstract*-A novel W defected ground structure (WDGS) was presented. The structure of WDGS was simple, processing easy and performance excellent. After this unit structure was applied to the design of the band-stop filter, its band-stop characteristic was enhanced efficiently and the dimension was reduced. The effects of the WDGS geometric parameters on frequency and width were analyzed in detail. A band-stop filter was built by three cascaded WDGS. The bandwidth of the filter was increased effectively and the attenuation to signals was improved. The results show that the filter rejects signals from -25dB to -37dB in 2.6GHz-3.4GHz.

*Index Terms—W-slot DGS, band-stop filter, microstrip line*

## I. INTRODUCTION

In microwave and millimeter-wave band, the defected ground structure (DGS) opens up a new path for the design of microwave device [1],[2]. The DGS is to sculpture missing patterns to change the distributed capacitance and inductance of microstrip line to realize band-stop characteristic. Its structure is simple, the performance excellent, and it is easy to create the modeling of equivalent circuit and the analysis of electromagnetic theory. However, the conventional DGS band-stop filter does not apply to the circuit design of lower frequency and smaller size [3],[4].

The size of the WDGS microstrip band-stop filter proposed in this paper is small and it is more effective in rejecting signal. It can be better integrated in microwave communication system to reject signal in 2GHz-5GHz [5],[6]. The filter has such advantages as small size which is occupied by the circuit prone to the integration of circuits, high performance, low power dissipation and simple craftsmanship, thus it is of great value to be researched and applied.

## II. W-SLOT DGS AND CHARACTERISTIC ANALYSIS

Fig.1 shows the sketch of W-slot defected ground structure. For the dimensions *l*=7mm, *g*=0.2mm, and *w*=1.33mm, the transfer characteristics of the W-slot DGS are calculated. The characteristic impedance of the microstrip line is assumed to be 50 Ω and the simulation is performed by using HFSS12.

Manuscript received June 6, 2013. This work was supported in part by the NSF of Anhui Province of China under Grant KJ2011A007 and 1208085MF104, in part by the Key Program of NSFC under Grant 60931002.

Lin Chen, Minquan Li, and Wei Wang are with the Key Laboratory of Intelligent Computing & Signal Processing, Ministry of Education, Anhui University, Hefei 230039, China. (Corresponding author's mobile phone number:13865913448; e-mail:450084812@qq.com).

MinquanLin is a professor with the Department of Electronics from Anhui University(e-mail: "li_mq68"<li_mq68@yahoo.com.cn>;)
The substrate with the thickness of 1.2 mm and a dielectric constant of 10.2 was used for all simulation.

The single DGS unit has its attenuated nadir point and cut-off frequency. The frequency features resemble Butterworth low-pass filter frequency response. Besides LC parallel circuit possesses band rejection characteristic. They can be determined by resonant frequency and cut-off frequency.

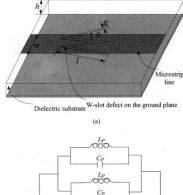

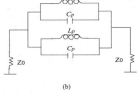

Fig.1 Sketch of WDGS (a) and equivalent circuit (b)

The impedance of LC equivalent circuit equals to the impedance of Butterworth low-pass filter, so equivalent capacitance and equivalent inductance can be attained. The circuit parameters of the equivalent circuit are extracted from the simulated scattering parameters as :

$$C = \frac{f_c}{2\pi Z_0 g_1 (f_0^2 - f_c^2)} \qquad (1)$$

$$L = \frac{1}{4\pi^2 f_0^2 C} \qquad (2)$$

Here, $f_0$ is the resonance frequency, $f_c$ is the 3-dB cutoff frequency, and $Z_0$ is the characteristic impedance of the microstrip line.

The simulated transfer characteristics for various angle ( $\theta$ ) are shown in Fig.2.The dimensions of the W-slot DGS are $w$=1.33mm, $h$=1.2mm, $l$=7mm and $g$=0.2mm. As $\theta$ increases, the equivalent parallel capacitor increases. When the angle equals to $\pi/2$ , $\pi/3$ , $\pi/6$ and $\pi/8$ , the resonant frequency changes over. In TABLE I, it is confirmed that the resonant frequency increases as the angle decreases.

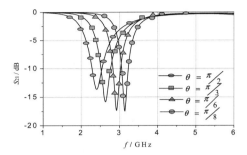

Fig.2 Simulated S21-parameters of the WDGS with different $\theta$ values
($l$=7mm, $g$=0.2mm, $w$=1.33mm, $\varepsilon_r$ of substrate = 10.2, $h$=1.2mm)

TABLE I Absolute attenuation with different $\theta$ values

| $\theta$ | Resonant Frequency (GHz) | Cutoff Frequency (GHz) | Absolute Attenuation (dB) |
|---|---|---|---|
| $\pi/2$ | 2.4 | 2 | -13.7 |
| $\pi/3$ | 2.63 | 2.28 | -16 |
| $\pi/6$ | 2.92 | 2.62 | -17.4 |
| $\pi/8$ | 3.15 | 2.85 | -17.5 |

In Fig.3, the simulated transfer characteristics for the W-slot DGS are plotted as functions of length ($l$). The dimensions of the W-slot DGS are $\theta=\pi/8$ and $g$=0.2mm. As the length increases, the resonant frequency decreases.

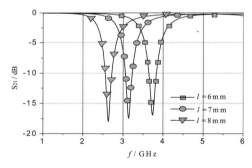

Fig.3 Simulated S21-parameters of the WDGS with different $l$ values
($g$=0.2mm, $\theta=\pi/8$, $w$=1.33mm, $\varepsilon_r$ of substrate = 10.2, $h$=1.2mm)

TABLE II Absolute attenuation with different $l$ values

| $l$/mm | Resonant Frequency (GHz) | Cutoff Frequency( GHz) | Absolute Attenuation (dB) |
|---|---|---|---|
| 6 | 3.78 | 3.42 | -17 |
| 7 | 3.15 | 2.82 | -17.5 |
| 8 | 2.62 | 2.4 | -18 |

Fig.4 shows the simulated transfer characteristic as functions of various widths ($g$). This parameter does not affect attenuation much, but it exerts pronounced effect on absolute bandwidth. The dimensions of the W-slot DGS are $l$=7mm and $\theta = \pi/6$. From TABLE III, one may clearly observe that the increased width causes rapid increase in absolute bandwidth.

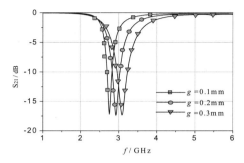

Fig.4 Simulated S21-parameters of the WDGS with different $g$ values
($l$=7mm, $\theta = \pi/6$, $w$=1.33mm, $\varepsilon_r$ of substrate = 10.2, $h$=1.2mm)

TABLE III Absolute bandwidth with different $g$ values

| $g$/mm | Resonant Frequency (GHz) | Cutoff Frequency( GHz) | Absolute bandwidth (GHz) |
|---|---|---|---|
| 0.1 | 2.75 | 2.55 | 0.45 |
| 0.2 | 2.91 | 2.62 | 0.67 |
| 0.3 | 3.08 | 2.65 | 0.95 |

### III. BAND-STOP FILTER DESIGN

For a design of a filter satisfying the required rejection bandwidth and attenuation, multiple DGS are cascaded along with the transmission line. Fig.5 illustrates the configuration of the band-stop filter with there cascaded W-slot DGSs. The dimensions for the W-slot are $l$=7mm, $g$=0.2mm, $\theta=\pi/8$ for each DGS and the characteristic impedance of the line is 50 Ω. The distance between the W slots are fixed as $s$=7mm.

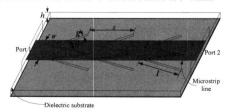

Fig.5 sketch of a filter consisted of three WDGS
($l$=7mm, $g$=0.2mm, $w$=1.33mm, $s$=7mm, $\theta=\pi/8$, $\varepsilon_r$ of substrate = 10.2, $h$=1.2mm)

The simulation results show that this band-stop filter has quite wide stop-band , its suppression over the signal within the range from 2.6GHz to 3.4GHz being less than -25dB. The transferring characteristics of W-slot DGS band-stop filter in pass-band are comparatively smooth and the attenuation being quite low. Thus this novel band-stop filter has such desirable advantages as excellent performance, small size, low price and easy to manufacture. So the filter can be independently used to replace traditional one [7],[8].

(a) Top view of the filter

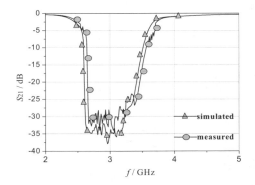

(b) Bottom view of the filter

Fig.6 The photos of top and bottom views of the fabricated filter PCB

Fig.7 Simulated and measured S21-parameters of filter

Fig. 6 shows the top and bottom views of the fabricated band-stop filter with three cascaded W-slot DGSs units. The dimensions for WDGS are: $l$=7mm, $g$=0.2mm, $\theta=\pi/8$ , $s$=7mm

and $w$=1.33mm. The substrate is a R03210 circuit board with a thickness of 1.2mm and a dielectric constant of 10.2. Fig.7 shows the comparative results between simulation and measurement on the fabricated three cascaded W-slot DGSs. It can be seen that they quite correspond to each other. It is also proved that the performance of WDGS is superior and filters of cascade connection can be widely used in small type microwave and radio frequency components.

## IV. CONCLUSION

A Novel W-slot DGS has been proposed in this paper and the influences that DGS parameters exert on transferring characteristics have been discussed. Three cascaded W-slot DGSs filter have been designed and fabricated, it rejects the signals at the frequencies from 2.6 GHz to 3.4GHz with less than -25dB suppression. The transfer characteristic in the pass-band shows low loss and flatness as a function of the frequency. The size of W-slot DGS filter is $0.16\lambda g \times 0.24\lambda g$ , which is even smaller than the Compact Band-stop Filter $(0.16\lambda g \times 0.35\lambda g)$ [9],[10]. The proposed structure may have wide applications in the design of microwave components and antenna arrays.

## REFERENCES

[1]  R. Li, D. I. Kim, C. M. Choi, "Compact structure with three attenuation poles for improving stopband characteristics," IEEE Microwave Wireless Compon..,vol. 16, pp. 663-665, Dec. 2006.

[2]  V. Radisic, Y. Qian, R. Coccioli, and T. Itoh, "Novel 2-D photonic bandgap structure for microstrip lines," IEEE Microw. Guided Wave Lett., vol. 8, pp. 69–71, Feb. 1998.

[3]  D. J. Woo, T. K. Lee, C. S. Pyo, and W. K. Choi, "High-Q band rejection filter by using U-slot DGS," in Proc. 35th Eur. Microw. Conf., Paris, France, pp. 1279–1282, Oct. 2005.

[4]  S. J. Wu, C. H, Tsai, T. L. Wu, T. Itoh. "A novel wideband common mode suppression filter for GHz differential signals using coupled patterned ground structure," IEEE Transactions on Microwave Theory and Techniques, Lett., vol. 57, pp. 848-855, Apr. 2009.

[5]  Duk-Jae Woo,Taek-Kyung Lee, Jae-Wook Lee, "Novel U-Slot and V-Slot DGSs for Bandstop Filter With Improved Q Factor," IEEE Transactions onMicrowave Theory and Techniques, vol. 54, pp. 2840-2847, Jun. 2006.

[6]  C. S. Kim, J. S. Park, D. Ahn, and J. B. Lim, "A novel 1-D periodic defected ground structure for planar circuits," IEEE Microw. Guide Wave Lett., vol.10, pp. 131–133, Apr. 2000.

[7]  S. Y. Huang, and Y. H. Lee, "A compact E-shaped patterned ground structure and its applications to tunable bandstop resonator," IEEE Trans. Microw. Theory Tech., vol. 57, pp. 657-666, Mar. 2009.

[8]  C.S.Kim, J.S.Lim, S.Nam, K.Y.Kang, D.Ahn, "Equivalent circuit modelling of spiral defected ground structure for microstrip line," Electron. Lett., vol.38, pp. 1109-1110, Sep.2002.

[9]  Maohui Yang, Jun Xu, Yuliang Dong, "A Novel Open-loop DGS for Compact Bandstop Filter With Improved Q Factor," IEEE Trans.on antennas propagation and EM theory, pp. 649-652, Nov. 2008.

[10]  N.C.Kamakar, S.M.Roy, "Quasi-static modeling of defected ground structure," IEEE Trans. Microw. Theory Tech.,vol.54, pp. 2160-2168, May 2006.

# Coupled-Mode Analysis of Two-Parallel Post-Wall Waveguides

Kiyotoshi Yasumoto[1], Hiroshi Maeda[1], and Vakhtang Jandieri[2]

[1]Faculty of Information Engineering, Fukuoka Institute of Technology, Fukuoka 811-0295, Japan
[2]School of Electrical Engineering and Computer Science, Kyungpook National University
Daegu 702-701, Republic of Korea

*Abstract-* **The guided modes supported in a coupled two-parallel post-wall waveguides are analyzed by using a coupled-mode theory for coupled two-dimensional photonic crystal wave guides. The coupled-mode equations, which govern the evolution of the modal amplitude of individual post-wall waveguides, are derived in self-contained way on the basis of eigenmode fields of each single waveguide in isolation. Numerical examples show that for various configurations of post-walls, the solutions of the coupled-mode equations are in very close agreement with those obtained by the rigorous numerical analysis.**

## I. INTRODUCTION

Post-wall waveguides [1], also called laminated waveguides [2] or substrate integrated waveguides [3], have received a growing attention because of their promising applications in planar circuit components operating in the microwave and millimeter wave frequency range. The modal properties of post-wall waveguides have been extensively investigated in the past years using various numerical or analytical techniques [1-5]. A number of components based on post-wall wave-guides have been also proposed and demonstrated, such as filters, couplers, and slot array antennas.

The post-wall waveguides are integrated waveguide-like structures composed of periodic rows of circular metallic posts in a grounded dielectric substrate. The structures are quite similar to those of two-dimensional photonic crystal wave-guides [6] consisting of layered periodic arrays of circular metallic cylinders with infinite length. Taking into account this similarity, we have recently proposed [7] a novel analytical model of post-wall waveguides based on the model of two-dimensional photonic crystal waveguides.

In this paper, we shall use the proposed model to analyze a coupled two-parallel post-wall waveguides which is a basic component to be used for a directional coupler. The coupled-mode theory, which has been developed [8] for dealing with coupled two-dimensional photonic crystal waveguides, is reformulated for the post-wall waveguide structures. The coupled-mode equations, which describe the evolution of the modal amplitude of individual post-wall waveguides, are derived in self-contained way on the basis of the eigenmode fields of each single waveguide in isolation. Numerical examples show that for various configurations of post-walls, the solutions of the coupled-mode equations are in very close agreement with those obtained by the rigorous numerical analysis.

## II. FORMULATION OF THE PROBLEM

The post-wall waveguide, as illustrated in Fig. 1, is composed of periodic arrays of conducting circular posts embedded in a dielectric substrate that connect two parallel conducting plates separated by a distance $d$. The radius of the posts is $r$, the pitch of the periodic arrangement of posts in the $z$-direction is $h$, and the material constants of the dielectric substrate are $\varepsilon_s$ and $\mu_0$. Although the post arrays in both sides may be $N$-layered, Fig. 1 shows the structure formed by a single layer array. The waveguide width in the $x$ direction is defined by the separation distance $w$ between the two innermost post arrays. Since the substrate is very thin ($d \ll \lambda$), the electric and magnetic fields are uniform ($\partial/\partial y = 0$) in the $y$ direction. Hence this periodic waveguide is quite similar to a two-dimensional photonic crystal waveguide [6] formed by parallel circular rods which are infinitely long in the $y$ direction.

The transversal view in the $x$-$z$ plane of a post-wall waveguide bounded by $N$-layered post arrays is illustrated in Fig. 2. If we assume an even TE mode whose $E_y$ field is symmetric with respect to $x = 0$, the guided field in the post-wall waveguide is expressed as follows:

$$E_y(x,z) = \sum_{m=-\infty}^{\infty} a_m \cos(\kappa_m x) e^{i\beta_m z} \qquad (1)$$

where $\beta_m = \beta + 2m\pi/h$, $\kappa_m = \sqrt{k_s^2 - \beta_m^2}$, $k_s = \omega\sqrt{\varepsilon_s\mu_0}$, $\omega$ is the angular frequency, $\beta$ is the mode propagation constant, and $\{a_m\}$ are unknowns. Let us define the column vector $a$ whose

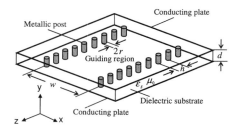

Fig. 1. Schematic of a post-wall waveguide.

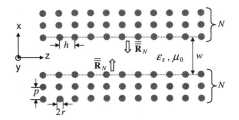

Fig. 2. Transversal view in the $x$-$z$ plane for the post-wall waveguide bounded by $N$-layered post arrays.

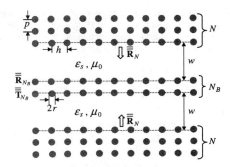

Fig. 3. Two-parallel post-wall waveguides "$a$" and "$b$" coupled through $N_B$-layered post-wall barrier.

elements are $\{a_m\}$. Using the model of two-dimensional photonic crystal waveguide [6] for Eq.(1), the linear equation to solve $\boldsymbol{a}$ is obtained as follows:

$$[\mathbf{I}-\mathbf{W}(\beta,\omega)\overline{\overline{\mathbf{R}}}_N(\beta,\omega)]\cdot\boldsymbol{a}=0 \tag{2}$$

with

$$\mathbf{W}(\beta,\omega)=[e^{i\kappa_m w}\,\delta_{mn}] \tag{3}$$

where $\overline{\overline{\mathbf{R}}}_N(\beta,\omega)$ is the generalized reflection matrix of $N$-layered post arrays viewed from the guiding region to the outward direction. $\overline{\overline{\mathbf{R}}}_N(\beta,\omega)$ can be calculated [9] using the T-matrix for a circular rod of perfect conductor and the lattice sums. From Eq.(2) the transcendental equation to determine the propagation constant $\beta$ is obtained as

$$\det[\mathbf{I}-\mathbf{W}(\beta,\omega)\overline{\overline{\mathbf{R}}}_N(\beta,\omega)]=0\,. \tag{4}$$

The solution to $\beta$ obtained from Eq.(4) is substituted into Eq.(2) to calculate the amplitude vector $\boldsymbol{a}$. Finally, the mode solution to the post-wall waveguide is expressed as follows:

$$\boldsymbol{a}=A e^{i\beta_0 z}\,\boldsymbol{f}(z) \tag{5}$$

where $\beta_0$ is a solution of Eq.(4), $A$ is the mode amplitude and $\boldsymbol{f}(z)$ is the normalized mode eigenvector, which is a periodic function of $z$ with period $2\pi/h$.

Figure 3 shows the transversal view of two-identical post-wall waveguides "$a$" and "$b$" which are situated in parallel and coupled through the $N_B$-layered post-wall barrier. Since the post-wall waveguide is a laterally open structure, a part of the electromagnetic energy may leak out through the gaps between adjacent posts. If the gap length $h-2r$ is chosen to be sufficiently smaller than the wavelength and the post-wall is multilayered in the $x$ direction, the leakage effect becomes negligible. However such a situation is not suitable for designing a post-wall waveguide coupler. We assume here that the two-parallel post-wall waveguides are bounded by the enough number of post-array layers in the upper and lower regions but are separated by a barrier consisting of a small number of post-array layers with $N_B < N$. In this case, the leakage of the guided field into the upper and lower half space is strongly suppressed, whereas the two guided modes supported by waveguides "$a$" and "$b$" can interact efficiently

to attain the expected power transfer between two waveguides.

Using the model [6] of two-dimensional photonic crystal waveguides, the rigorous dispersion equation for the coupled post-wall waveguides shown in Fig. 3 is derived as follows:

$$\det\Big\{\mathbf{I}-\mathbf{W}(\beta,\omega)\Big[\overline{\overline{\mathbf{R}}}_{N_B}(\beta,\omega)+\overline{\overline{\mathbf{T}}}_{N_B}(\beta,\omega)\Big]$$
$$\times\mathbf{W}(\beta,\omega)\overline{\overline{\mathbf{R}}}_N(\beta,\omega)\Big\}=0 \;\; \text{for } even \text{ mode} \tag{6}$$

$$\det\Big\{\mathbf{I}-\mathbf{W}(\beta,\omega)\Big[\overline{\overline{\mathbf{R}}}_{N_B}(\beta,\omega)-\overline{\overline{\mathbf{T}}}_{N_B}(\beta,\omega)\Big]$$
$$\times\mathbf{W}(\beta,\omega)\overline{\overline{\mathbf{R}}}_N(\beta,\omega)\Big\}=0 \;\; \text{for } odd \text{ mode} \tag{7}$$

where $\overline{\overline{\mathbf{R}}}_{N_B}(\beta,\omega)$ and $\overline{\overline{\mathbf{T}}}_{N_B}(\beta,\omega)$ denote the generalized reflection and transmission matrices of the $N_B$-layered post-wall barrier. We may directly solve Eqs.(6) and (7) with use of roots searching algorithm for transcendental equations. However such a direct method is very time consuming. If the transmission of guided fields through the barrier layer is weak enough, Eqs.(6) and (7) are approximated by the coupled-mode equations based on the eigenmode field of each single waveguide in isolation as discussed in what follows.

### III. COUPLED-MODE EQUATIONS

When two waveguides "$a$" and "$b$" are well separated, each waveguide behaves as a single waveguide and supports the guided mode independently in the form of Eq.(5). If the waveguides are placed in close proximity, the two guided modes interact through the barrier layer and their mode amplitudes slowly vary along the propagation in the $z$ direction. Under this situation, we express the amplitude vectors for guided field modes as follows:

$$\boldsymbol{a}=A(z)e^{i\beta_0 z}\boldsymbol{f}(z) \;\; \text{for waveguide "}a\text{"} \tag{8}$$

$$\boldsymbol{b}=B(z)e^{i\beta_0 z}\boldsymbol{f}(z) \;\; \text{for waveguide "}b\text{"} \tag{9}$$

where the slowly varying amplitudes $A(z)$ and $B(z)$ describe a small perturbation in the mode propagation constant and $|dA(z)/dz|, |dB(z)/dz| \ll \beta_0, 2\pi/h$. The first-order perturbation analysis under the weak coupling through $\overline{\overline{\mathbf{T}}}_{N_B}(\omega,\beta)$ is applied to Eqs.(6) and (7) with use of Eqs.(8) and (9). Following the same analytical procedure developed in [8], the coupled-mode equations for $A(z)$ and $B(z)$ are derived as follows:

$$\frac{d}{dz}A(z) = i\kappa B(z) \tag{10}$$

$$\frac{d}{dz}B(z) = i\kappa A(z) \tag{11}$$

with

$$\kappa = \frac{g^T \cdot \mathbf{D}_{ab}(\beta_0,\omega) \cdot f}{g^T \cdot \dfrac{\partial \mathbf{D}(\beta_0,\omega)}{\partial \beta_0} \cdot f} \tag{12}$$

$$\mathbf{D}(\beta,\omega) = \mathbf{I} - \mathbf{W}(\beta,\omega)\overline{\overline{\mathbf{R}}}_N(\beta,\omega) \tag{13}$$

$$\mathbf{D}_{ab}(\beta,\omega) = \mathbf{W}(\beta,\omega)\overline{\overline{\mathbf{R}}}_N(\beta,\omega)\mathbf{W}(\beta,\omega)\overline{\overline{\mathbf{T}}}_{N_B}(\beta,\omega) \tag{14}$$

where $\kappa$ denotes the coupling coefficient, $g$ is the right eigenvector of $\mathbf{D}(\beta_0,\omega)$ that satisfies $\mathbf{D}^T(\omega,\beta_0) \cdot g = 0$ and the superscript $T$ denotes the transpose of the indicated vector and matrix.

Equations (10) and (11) govern the evolution of the mode amplitudes in individual post-wall waveguides. Their solutions describe the power transfer characteristics between two-parallel post-wall waveguides "$a$" and "$b$" coupled through the $N_B$-layered barrier. From Eqs.(10) and (11) we have two solutions $\Delta\beta = \pm\kappa$ for the perturbed mode propagation constants due to the coupling. Then the propagation constants for the $even$ and $odd$ modes in the coupled waveguide system are obtained as follows:

$$\beta_{even} = \beta_0 + \kappa \quad \text{for } even \text{ mode} \tag{15}$$

$$\beta_{odd} = \beta_0 - \kappa \quad \text{for } odd \text{ mode}. \tag{16}$$

## IV. NUMERICAL RESULTS

To validate the coupled-mode analysis, we performed numerical computations on Eqs.(10)-(14). Although a substantial number of numerical examples could be generated, we consider here only a few examples of the coupled post-wall waveguides. We assume a square lattice with $h=p$ for the upper and lower post-walls in Fig. 3. The numbers of post-wall layers are fixed to be $N=3$ for the upper and lower layers and $N_B=1$ for the barrier layer. Then the structure of post-wall waveguides is characterized by four parameters $h$, $r/h$, $w/h$, and $\varepsilon_s/\varepsilon_0$.

TABLE I shows the comparison of the normalized propagation constants $\beta_{even}h/2\pi$ and $\beta_{odd}h/2\pi$ calculated by the coupled-mode equations (10) and (11) with those obtained from a rigorous mode analysis for Eqs.(6) and (7). The results are tabulated from ($a$) to ($e$) for five different structural parameters. The propagation constants of the single waveguide have been analyzed in [3-5] for the structures ($a$), ($b$), ($c$), and ($d$). The propagation constant and the coupling coefficient for the two-parallel waveguides has been reported

TABLE I
NORMALIZED PROPAGATION CONSTANTS $\beta h/2\pi$ OF $EVEN$ AND $ODD$ MODES CALCULATED FOR COUPLED TWO-PARALLEL POST-WAVEGUIDES WITH $N=3$ AND $N_B=1$.

($a$) $h = 2mm$, $r/h = 0.2$, $w/h = 3.6$, $\varepsilon_s/\varepsilon_0 = 2.33$

| $f$ [GHz] | Coupled-mode analysis (10), (11) | | Rigorous analysis (6), (7) | |
|---|---|---|---|---|
| | $\beta_{even}h/2\pi$ | $\beta_{odd}h/2\pi$ | $\beta_{even}h/2\pi$ | $\beta_{odd}h/2\pi$ |
| 15 | 0.04589 | 0.04051 | 0.04584 | 0.04045 |
| 20 | 0.14225 | 0.14054 | 0.14226 | 0.14055 |
| 25 | 0.20869 | 0.20747 | 0.20870 | 0.20748 |
| 30 | 0.26860 | 0.26760 | 0.26861 | 0.26761 |

($b$) $h = 5.165mm$, $r/h = 0.075$, $w/h = 2.2767$, $\varepsilon_s/\varepsilon_0 = 2.2$

| $f$ [GHz] | Coupled-mode analysis (10), (11) | | Rigorous analysis (6), (7) | |
|---|---|---|---|---|
| | $\beta_{even}h/2\pi$ | $\beta_{odd}h/2\pi$ | $\beta_{even}h/2\pi$ | $\beta_{odd}h/2\pi$ |
| 8 | 0.09068 | —— | 0.07510 | —— |
| 10 | 0.17252 | 0.14697 | 0.17158 | 0.14489 |
| 12 | 0.24278 | 0.22483 | 0.24226 | 0.22356 |
| 14 | 0.30601 | 0.29143 | 0.30564 | 0.29034 |

($c$) $h = 1.016mm$, $r/h = 0.3125$, $w/h = 3.9075$, $\varepsilon_s/\varepsilon_0 = 9.9$

| $f$ [GHz] | Coupled-mode analysis (10), (11) | | Rigorous analysis (6), (7) | |
|---|---|---|---|---|
| | $\beta_{even}h/2\pi$ | $\beta_{odd}h/2\pi$ | $\beta_{even}h/2\pi$ | $\beta_{odd}h/2\pi$ |
| 15 | 0.06742 | 0.06721 | 0.06741 | 0.06721 |
| 19 | 0.14152 | 0.14141 | 0.14152 | 0.14142 |
| 23 | 0.19782 | 0.19775 | 0.19783 | 0.19775 |
| 27 | 0.24878 | 0.24871 | 0.24878 | 0.24872 |

($d$) $h = 3.556mm$, $r/h = 0.25$, $w/h = 2.3088$, $\varepsilon_s/\varepsilon_0 = 1.0$

| $f$ [GHz] | Coupled-mode analysis (10), (11) | | Rigorous analysis (6), (7) | |
|---|---|---|---|---|
| | $\beta_{even}h/2\pi$ | $\beta_{odd}h/2\pi$ | $\beta_{even}h/2\pi$ | $\beta_{odd}h/2\pi$ |
| 25 | 0.16143 | 0.15850 | 0.16146 | 0.15846 |
| 30 | 0.25502 | 0.25301 | 0.25501 | 0.25301 |
| 35 | 0.33343 | 0.33170 | 0.33346 | 0.33176 |
| 40 | 0.40574 | 0.40405 | 0.40569 | 0.40399 |

($e$) $h = 12.5mm$, $r/h = 0.2$, $w/h = 2.0$, $\varepsilon_s/\varepsilon_0 = 1.0$

| $f$ [GHz] | Coupled-mode analysis (10), (11) | | Rigorous analysis (6), (7) | |
|---|---|---|---|---|
| | $\beta_{even}h/2\pi$ | $\beta_{odd}h/2\pi$ | $\beta_{even}h/2\pi$ | $\beta_{odd}h/2\pi$ |
| 9 | 0.26083 | 0.25334 | 0.26077 | 0.25325 |
| 10 | 0.31900 | 0.31237 | 0.31894 | 0.31229 |
| 11 | 0.37308 | 0.36675 | 0.37302 | 0.36668 |
| 12 | 0.42534 | 0.41877 | 0.42528 | 0.41868 |

in [10] for the structure (e).

From TABLE I we can see that for all structures the results obtained by the coupled-mode analysis are in close agreement with those of the rigorous mode analysis over a broad range of frequency. For the structure (e), since the *odd* mode of the rigorous analysis enters in a cutoff region at $f$ =8.0GHz, we have discarded the corresponding propagation constant of the coupled-mode analysis. Note that the coupled-mode analysis always yields two propagation constants whenever the single post-wall waveguide in isolation supports one guided mode. The results given for the structure (e) agree well with the theoretical and experimental results reported in [10].

The coupling length $L_c$, which is the characteristic length for the complete power transfer from one waveguide to another, is given by

$$L_c = \frac{\pi}{\beta_{even} - \beta_{odd}} = \frac{\pi}{2\kappa}. \tag{17}$$

TABLE I demonstrates that the coupling length tends to increase as $r/h$ increases and hence the gap width between the adjacent posts decreases. In order to design the directional coupler within a practical device length, we need to optimize the value $r/h$ for the assumed frequency band.

## V. CONCLUSION

We have presented a self-contained coupled-mode analysis for two-parallel post-wall waveguides based on the model of two-dimensional photonic crystal waveguides. The first-order coupled-mode equations have been systematically derived, which govern the evolution of the modal amplitudes in each individual post-wall waveguide. The coupling coefficients have been calculated by using the propagation constants and eigenmode solutions of the single post-wall waveguide in isolation. The proposed formulation provides a useful analytical and numerical technique for approximating the coupling between post-wall waveguides in close proximity with a good physical justification.

### ACKNOWLEDGMENT

This work was supported in part by Grant-in-Aid for Scientific Research (C) 24560430 from Japan Society for the Promotion of Science.

### REFERENCES

[1] J. Hirokawa, and M. Ando, "Single-layer feed wave-guide consisting of posts for plane TEM wave Excitation in parallel plates," *IEEE Trans. Antennas Propagat.*, vol. 46, no. 5, pp. 625-630, 1998.

[2] H. Uchiyama, T. Takenoshita and M. Fujii, "Development of a "laminated waveguide"," *IEEE Trans. Microwave Theory Tech.*, vol. 46, no. 12, pp. 2438-2443, 1998.

[3] Y. Cassivi, L. Perrengrini, P. Arcioni, M. Bressan, K. Wu, and G. Conciauro, "Dispersion characteristics of substrate integrated rectangular waveguide," *IEEE Microw. Wireless Compon. Lett.*, vol. 12, no. 9, pp. 333-335, Sep. 2002.

[4] D. Deslandes, and K. Wu, "Accurate modeling, wave mechanisms, and design considerations of a substrate integrated waveguide," *IEEE Trans. Microwave Theory Tech.*, vol. 54, no. 6, pp. 2516-2526, June 2006.

[5] M. Bozzil, F. Xu, L. Perregrinil, and K. Wu, "Circuit modeling and physical interpretation of substrate integrated waveguide structures for millimetre-wave applications," *Int. J. Microwave Opt. Technol.*, vol. 3, no. 3, pp. 329-338, July 2008.

[6] K. Yasumoto, H. Jia, and K. Sun, "Rigorous modal analysis of two-dimensional photonic crystal wave- guides," *Radio Sci.*, vol. 40, no. 6, RS6S02, 2005.

[7] K. Yasumoto, H. Maeda, and V. Jandieri, "Modal analysis of post-wall waveguides based on a model of two-dimensional photonic crystal waveguides," PIERS Abstracts, p. 466, Taipei, Mar. 2013.

[8] K. Yasumoto, V. Jandieri, and Y. Liu, "Coupled-mode formulation of two-parallel photonic crystal waveguides," *J. Opt. Soc. Am. A*, vol. 30, no. 1, pp. 96-101, Jan. 2013.

[9] K. Yasumoto, H. Toyama, and T. Kushta, "Accurate analysis of two-dimensional electromagnetic scattering from multilayered periodic arrays of circular cylinders using lattice sums technique," *IEEE Trans. Antennas Propagat.*, vol. 52, no. 10, pp. 2603-2611, Oct. 2004.

[10] K. S. E. Bankov and L. I. Pangonis, "Experimental investigation of waveguide components designed on the basis of EBG structures," *J. of Comm. Technol. Electron.*, vol. 53, no. 3, pp. 274-281, 2008.

# Systematic Microwave Network Analysis for Arbitrary Shape Printed Circuit Boards With a Large Number of Vias

Xinzhen Hu, Liguo Sun

Department of Electronic Engineering and Information Science, University of Science and Technology of China

Address: 443 Huangshan Rd, Hefei, Anhui, China, 230027

auto@mail.ustc.edu.cn

liguos@ustc.edu.cn

*Abstract*—**When calculating the admittance matrix of a singular via, a two-port microwave network is obtained instead of the three-port circuit network in the intrinsic via circuit model , without changing any accuracy. Combined with the novel impendence of a parallel plate pair; we can obtain a systematic method to analyze arbitrary shape printed circuit boards with a large number of vias.**

## I. INTRODUCTION

The increase of integrated circuit package density results in a high concentration of interconnecting lines, which makes the use of a multilayer printed circuit boards (PCBs) become necessary. In multilayer high-speed printed circuit boards or packages, the signal traces are usually on different layers. So vias are widely used to connect the signal traces on different layers and decrease the complexity of the high-speed digital electronic systems. But as discontinuities, vias may cause a lot of signal integrity problems such as crosstalk, ring, mismatch, and mode conversion [1], [2]. Some parasitic phenomenon ignored at low frequency will become more prominent and serious impact on the interconnect performance, when the systems working at Gb/s range. Therefore an accurate model of printed circuit boards with vias is critical for high-speed digital systems.

Certainly the numerical methods are generally used to address this problem. Moreover, in most of the methods the computational burden sharply grows as the size increases and the structure becomes more complex. This is the biggest limitation of the numerical methods to solve the particularly complex structure. To overcome this limitation, a physics-based via model was proposed based on physical institution in [3], [4]. In this model, each cavity of the multilayer structure is treated separately, ignoring the interaction between the different layers. In each cavity, the via-plate structure is equivalent to a π-type circuit which consists of a simple short circuit, representing the barrel of a via, and two shunt capacitors, representing the interaction between the via barrel and the top/bottom plate. Meanwhile, the plate-pair is modeled as the return current path impedance $Z_{pp}$.

Note that although this model offers a tremendous flexibility to model arbitrary shape PCBs, it does not satisfy the boundary conditions of the via structure rigorously.

Naturally, a more accurate and efficient method have been introduced in [5]. The intrinsic circuit model reckons in the contribution of the evanescent modes in the antipad region of vias, which is ignored in the physics-based via model. Three assumptions are proposed in the intrinsic via model, which will still be used in our work. With these assumptions, we can divide the whole plate-pair into two domain, via domain and plate domain, as shown in Fig. 1.

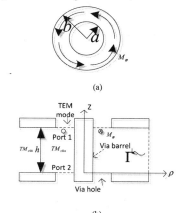

(a)

(b)

Fig. 1. (a) Top view of a via hole (b) side view of an irregular plate pair

The work of this paper is to first simply the three port network, the equivalent of a via domain, into two-port network. Then combined with the method to solve $Z_{pp}$ depicted in [6], we can obtain a systematic method to analyze arbitrary shape printed circuit boards with a large number of vias.

## II. SYSTEMATIC MICROWAVE NETWORK ANALYSIS

Under the first assumption in [5], we know that only $TM_{z0n} (n \geq 0)$ modes can be excited in the region between

978-7-5641-4279-7

the plate-pair when the displacement current flows through via barrel. Using the equivalence principle, the TEM mode in the via hole is represented by an angular magnetic current ring source $M_\varphi$ as

$$M_\varphi = -\frac{V_0}{\rho' \ln(b/a)} \delta(z - z') \qquad (1)$$

where $V_0$ is the voltage between the via barrel and antipad at $z' = 0$ or $z' = h$.

In this paper, we do not account the effect of the pad. In other words, only the magnetic field due to the magnetic frill current distributed at the region $a \le \rho \le b$ should be considered. In [7], the expression of the magnetic field as

$$H_\varphi(a,z) = -\frac{\omega \varepsilon \pi V_0}{h \ln(b/a)} \sum_{n=0}^{\infty} G_n^s(a) \cos\left(\frac{n\pi z}{h}\right) \cos\left(\frac{n\pi z'}{h}\right) \qquad (2)$$

where the auxiliary is defined as

$$G_n^s(a) = \frac{\left(1 - \Gamma_a^{(n)} \Gamma_b^{(n)}\right)^{-1}}{k_n(1 + \delta_{n0})}$$
$$\cdot \{[H_0^{(2)}(k_n b) - H_0^{(2)}(k_n a)]$$
$$+ \Gamma_b^{(n)}[J_0(k_n b) - J_0(k_n a)]\}$$
$$\cdot \left[J_1(k_n a) + \Gamma_a^{(n)} H_1^{(2)}(k_n a)\right] \qquad (3)$$

where $\Gamma_a^{(n)}$ and $\Gamma_b^{(n)}$ are the reflection coefficients for the nth cylindrical waves from the via barrel and the outer radial boundary, as

$$\Gamma_a^{(n)} = -\frac{J_0(k_n a)}{H_0^{(2)}(k_n a)} \qquad (4)$$

$$\Gamma_b^{(n)} = \begin{cases} -\dfrac{H_0^{(2)}(k_n b)}{J_0(k_n b)} & PEC \\ -\dfrac{H_1^{(2)}(k_n b)}{J_1(k_n b)} & PMC \\ 0 & PML \end{cases} \qquad (5)$$

and $k_n$ is the radial wavenumber of the $TM_{z0n}$ modes as

$$k_n = \sqrt{k_0^2 \varepsilon_r - \left(\frac{n\pi}{h}\right)^2} \qquad (6)$$

The current due to the magnetic field distributed on the via barrel as [8]

$$I(a,z) = 2\pi a H_\varphi(a,z)$$
$$= -\frac{2a\omega\varepsilon\pi^2 V_0}{h \ln(b/a)} \sum_{n=0}^{\infty} G_n^s(a) \cos\left(\frac{n\pi z}{h}\right) \cos\left(\frac{n\pi z'}{h}\right) \qquad (7)$$

The admittance matrix of the equivalent two-port network can be defined as [1]

$$\begin{bmatrix} I_t \\ I_b \end{bmatrix} = \begin{bmatrix} Y_{tt} & Y_{tb} \\ Y_{bt} & Y_{bb} \end{bmatrix} \begin{bmatrix} V_t \\ V_b \end{bmatrix} \qquad (8)$$

where $(V_t, I_t)$ and $(V_b, I_b)$ are the voltage and current pair of port1 and 2. The reciprocity of the material between the plate-pair results in the admittance matrix satisfies reciprocal, that is $Y_{bt} = Y_{tb}$. The admittance matrix of the two-port network can be obtained as follow:

1) Let $V_t = 0$ and $V_b = V_0$ ,then $I_b = I(a,0)$ and

$$I_t = -I(a,h)$$

$$Y_{tb} = \frac{I_t}{V_b}\bigg|_{V_t=0} = \frac{2\omega\varepsilon\pi^2 a}{h \ln(b/a)} \sum_{n=0}^{\infty} (-1)^n G_n^{(s)}(a) = Y_{bt} \qquad (9a)$$

$$Y_{bb} = \frac{I_b}{V_b}\bigg|_{V_t=0} = -\frac{2\omega\varepsilon\pi^2 a}{h \ln(b/a)} \sum_{n=0}^{\infty} G_n^{(s)}(a) \qquad (9b)$$

2) Let $V_t = V_0$ and $V_b = 0$ ,then $I_b = -I(a,0)$ and

$$I_t = I(a,h)$$

$$Y_{tt} = \frac{I_t}{V_t}\bigg|_{V_b=0} = -\frac{2\omega\varepsilon\pi^2 a}{h \ln(b/a)} \sum_{n=0}^{\infty} G_n^{(s)}(a) \qquad (10)$$

Now we have obtained the admittance of a single via. For a P-vias domain, top (bottom) voltage (current) vectors are represented as

$$\mathbf{V(I)}_{t(b)} = \left[V(I)_{t(b)}^{(1)}, V(I)_{t(b)}^{(2)}, \cdots, V(I)_{t(b)}^{(P)}\right]^T \qquad (11)$$

The admittance of the P-via network can be derived easily from the two-port admittance matrix of each via as

$$\begin{bmatrix} \mathbf{I_t} \\ \mathbf{I_b} \end{bmatrix} = \begin{bmatrix} \mathbf{Y_{tt}} & \mathbf{Y_{tb}} \\ \mathbf{Y_{bt}} & \mathbf{Y_{bb}} \end{bmatrix} \begin{bmatrix} \mathbf{V_t} \\ \mathbf{V_b} \end{bmatrix} \qquad (12)$$

where

$$\mathbf{Y_{\alpha\beta}} = diag\{Y_{\alpha\beta}\} \qquad \alpha, \beta = t, h \qquad (13)$$

Combine with the $\mathbf{Y_{pp}}$ obtained through the method depicted in [6], the admittance matrix of the whole plate-pair with P-vias can be derived as

$$\mathbf{Y}_{final} = \begin{bmatrix} \mathbf{Y}_{tt} & \mathbf{Y}_{tb} \\ \mathbf{Y}_{bt} & \mathbf{Y}_{bb} \end{bmatrix} + \begin{bmatrix} \mathbf{Y}_{pp} & -\mathbf{Y}_{pp} \\ -\mathbf{Y}_{pp} & \mathbf{Y}_{pp} \end{bmatrix} \quad (14)$$

Obviously, $\mathbf{Y}_{final}$ is a 2P-dimensional matrix. Then, the scattering matrix among multiple coaxial ports can be obtained as [6]

$$\mathbf{S} = \left(\mathbf{Y}_0 + \mathbf{Y}_{final}\right)^{-1}\left(\mathbf{Y}_0 - \mathbf{Y}_{final}\right) \quad (15)$$

Where $\mathbf{Y}_0$ is a diagonal matrix whose elements are characteristic admittances $0.02\Omega^{-1}$.

## III. VALIDATIONS AND DISCUSSIONS

To validate the systematic method discussed above, an example of an L-shape PCB with seven signal vias, whose dimensions are shown in Fig. 2 is studied here. The relative permittivity of the dielectric between the plate-pair is 4.2 and the loss tangent is 0.02. The radii of the via barrels and the antipads are 10mils and 15mils. Port 1-7 are defined on the top of the plate-pair, while Port 8-14 are on the opposite side. The separation of plates is 10mils. Obviously, the separation of the parallel is electrically small, so the edge boundary $\Gamma$ can be approximated as a PMC boundary. The results simulated by HFSS and the present method are shown in Fig. 3 and Fig .4. In the HFSS simulation, all S-parameters obtained are normalized to $50\Omega$.

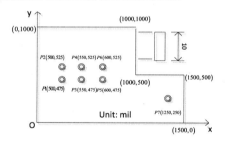

Fig. 2 Simulation model with arbitrary shape whose height is 10mils

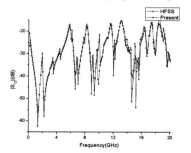

Fig. 3a Magnitude of the near-end crosstalk of the structure on the same side

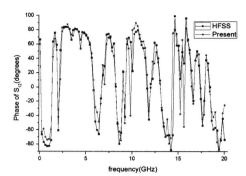

Fig. 3b Phase of the near-end crosstalk of the structure on the same side

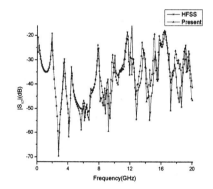

Fig. 4a Magnitude of the far-end crosstalk of the structure on the same side

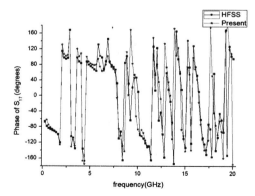

Fig. 4b Phase of the far-end crosstalk of the structure on the same side

Fig. 3a and Fig. 3b compare the magnitude and the phase of the near-end crosstalk on the same side, while Fig. 4a and Fig. 4b compare the magnitude and the phase of the far-end crosstalk on the same side. It can be seen that no matter the near-end crosstalk or the far-end crosstalk the systematic microwave network methods agrees very well with HFSS even up to 20GHz. While the presented method can achieve such anastomosis compared with the HFSS, it only took 8min for 200 frequency sampling points instead of 2h for the HFSS simulation at the same computer, which clearly demonstrates its efficiency in modeling the arbitrary shape plate-pair with dense vias.

## IV. CONCLUSION

When calculating the admittance matrix of the via domain, obviously we just need to obtain a 2P-dimensioal matrix, while 3P-dimensional matrix need to be calculated in the intrinsic via circuit model. When we calculate a via array, that is P is large, the method we presented can achieve similar accuracy, while the time spent is much less than the intrinsic via circuit model. So the method has a great potential in modeling arbitrary plate-pair structures with a large number of vias.

## ACKNOWLEDGMENT

The authors wish to acknowledge the assistance and support of the ISPA organizing committee.

## REFERENCES

[1] H.W. Johnson and M.Graham, *High-Speed Digital Design*: A Handbook of Black Magic. Englewood Cliffs, NJ: Prentice-Hall, 1993, ch. 7.

[2] S. H. Hall, G. W. Hall, and J. A. McCall, *High-Speed Digital System Design*—A Handbook Of Interconnect Theory and Design Practices.New York, Wiley, 2000, ch. 5.

[3] C. Schuster, Y. Kwark, G. Selli, and P.Muthana, *"Developing a 'physical'model for vias,"* presented at the DesignCon, Santa Clara, CA, Feb. 6–9,2006.

[4] G. Selli, C. Schuster, Y. H. Kwark, M. B. Ritter, and J. L. Drewniak, *"Developing a physical via model for vias—Part II: Coupled and Ground Return Vias,* " presented at the DesignCon, Santa Clara, CA, Jan. 29–Feb.1, 2007.

[5] Y.-J. Zhang and J. Fan, *"An intrinsic circuit model for multiple vias in an irregular plate pair through rigorous electromagnetic analysis,"* IEEE Trans. Microw. Theory Tech., vol. 58, no. 8, pp. 2251–2265, Aug. 2010.

[6] Y.-J. Zhang, G. Feng, and J. Fan, *"A novel impedance definition of a parallel plate pair for an intrinsic via circuit model,"* IEEE Trans. Microw. Theory Tech., vol. 58, no. 12, pp. 3780–3789, Dec. 2010.

[7] Y. Zhang, J. Fan, G. Selli,M. Cocchini, and F. D. Paulis, *"Analytical evaluation of via-plate capacitance for multilayer printed circuit boards and packages,"* IEEE Trans. Microw. Theory Tech., vol. 56, no. 9, pp. 2118–2128, Sep. 2008.

[8] H. Chen, Q. Lin, L. Tsang, C.-C.Huang, and V. Jandhyala, *"Analysis of a large number of vias and differential signaling in multilayered structures,"* IEEE Trans. Microw. Theory Tech., vol. 51, no. 3, pp. 818–829, Mar. 2003.

# *FA-2(D)*

## October 33 (FRI) AM

## Room D

## EM in Circuits-2

# A Novel Phase Shifter Based on Reconfigurable Defected Microstrip Structure (RDMS) for Beam-Steering Antennas

Can Ding[1,2], Y. Jay Guo[2], Pei-Yuan Qin[2], Trevor S. Bird[2], and Yintang Yang[1]

[1]School of Micro-electronics, Xidian University, Xi'an, China

[2]CSIRO ICT Center, Sydney, Australia

Email: Can.Ding@csiro.au

*Abstract*—A low cost phase shifter based on a cascaded reconfigurable defected microstrip structure (RDMS) is proposed. This RDMS unit is produced by etching a slot to introduce a defect on a microstrip line and then PIN diodes are inserted in the defected area. By switching the PIN diodes, the RDMS unit is able to operate in two different states with a phase shift of 17° at 5.2 GHz. The RDMS units can be cascaded for higher phase shift values that may be determined by array design requirements. Phase shifters cascading three and six RDMS units were designed, fabricated, and measured. The measured results show that the two phase shifters introduce 45° and 90° phase shifts, respectively, with low insertion loss. Finally, a four element patch array is proposed with a beamforming network employing the phase shifters and Wilkinson power dividers. The array is able to switch its main beam direction to 0° and ±20° in the H plane and the impedance bandwidth covers the overlapping wireless local area network (WLAN) bands in the vicinity of 5.2 GHz.

## I. Introduction

Phase shifters are critical components in phased array and beam steering antenna designs. In recent years, novel phase shifters have been proposed based on defected ground structures (DGS) [1], which was originally used in filter, coupler, and oscillator designs. The DGS was employed as a termination load in [2] which increased the phase shift range by 80° compared with the structure using a conventional load. A DGS based 45° phase shifter was proposed in [3], which increased the bandwidth and reduced the size comparing to the structure without a DGS. In [4], thin-film copper membranes were placed in a ground plane. By actuating these membranes, the space between the transmission line and the ground plane was changed, which resulted in a phase shift. In [5], the membranes were replaced by flexible micro-ribbons and significantly lowered the biasing voltage. In [6], the metallized polydimethylsiloxane (PDMS) elastomeric ground plane was selected to reconfigure the air-gap spacing which resulted in reduced cost. However, these phase shifter designs based on DGS or ground plane reconfiguration still suffer from high complexity and cost, which limits their applications in beamforming and phased array designs.

Here we propose phase shifters that based on the defected microstrip structure (DMS) [7] [8] which is the dual structure of DGS. Fig. 1 shows the layout of the proposed DMS unit.

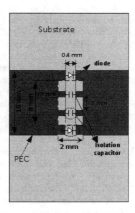

Fig. 1. Geometry of the RDMS unit.

The DMS unit is made by etching a slot in a microstrip line printed on a Rogers4003 substrate with the dielectric constant of 3.55 and substrate thickness of 1.524 mm. By inserting controllable PIN diodes into the defected area, the DMS unit is endowed with structural reconfiguration and can be turned into a reconfigurable DMS (RDMS). By turning the diodes "on" and "off", a phase shift is introduced. This is due to the current path flowing across the RDMS unit has been changed. Specifically, the surface current on a microstrip line is concentrated at the edges. As a consequence, the etched slot on the microstrip line has little effect on the current flow concentrating at the edges when the diodes are "on". When the diodes are "off", the diodes behave as an open circuit and the current mainly flows across the capacitors, which results in current path change and introduces phase shift. The dimensions of the RDMS unit are optimized by considering both the insertion loss and phase shift together. With the dimensions shown in Fig. 1, an RDMS unit is able to introduce a phase shift of 17° with 0.8 dB insertion loss at 5.2 GHz. When the RDMS units are cascaded, the insertion losses increase slightly, but the phase shift value

978-7-5641-4279-7

is significantly increased, almost linearly with the number of the RDMS units cascaded. Using this approach, phase shifters with variable and significant phase shifts can be obtained for different requirements. The phase shifter design using this method has advantages of low cost, easy to control, easy to integrate, and low insertion losses, which makes it suitable for low cost phased array feed network.

## II. PHASE SHIFTER BASED ON CASCADED RDMS UNITS

Figs. 2(a) and 2(b) show the phase shifters made by cascading three and six RDMS units, respectively. They are designated as 3-RDMS phase shifter and 6-RDMS phase shifter for simplicity. The phase shifters are composed of cascaded RDMS units for phase shift, tapered microstrip lines for impedance match, and DC bias networks to control the diodes. The 3-RDMS phase shifter has one DC bias network, which allows the phase shifter to work in two states when positive or negative DC voltages are applied on the pad. The two working states of the 3-RDMS phase shifter are defined as "All-on" and "All-off" states when all the diodes of the RDMS units are turned "on" and "off", respectively. For the 6-RDMS phase shifter, two DC bias networks are employed, with each bias network controlling 3 RDMS units. The 6 RDMS units are divided into two RDMS groups as depicted in Fig. 2(b). By applying different DC voltages on the bias pads, the 6-RDMS phase shifter is able to work in 4 possible states when the two RDMS groups are all "on", "on" and "off", "off" and "on", and all "off", respectively. The 4 states are defined as "All-on", "On-off", "Off-on", and "All-off" states. In addition, the "On-off" and "Off-on" states are expected to have the same performance since the configurations of the phase shifter in these two states are symmetric.

The phase shifters were not only simulated but they were also fabricated and measured. Fig. 3 shows the two fabricated phase shifter prototypes. To assess the performance, the insertion loss and phase shift were measured as shown in Figs. 4 and 5. The phase shifts shown in Figs. 4(b) and 5(b) are given by Phase (other states) – Phase (All-on state). The All-on state phase is used as a reference and not presented in the figures. As shown in Figs. 4 and 5, the simulated results agree well with the measured ones. For the 3-RDMS phase shifter, the measured insertion losses of the two states are below 1 dB for a phase shift of 45° at 5.2 GHz. For the 6-RDMS phase shifter, the measured insertion losses are <1.8 dB in all the four states. With reference to the phase of the "All-on" state, the measured phase shifts of the "On-off", "Off-on", and "All-off" states are 45°, 45°, and 90°, respectively, at 5.2 GHz.

The proposed 3-RDMS and 6-RDMS phase shifters are suitable for the antenna array design as described in the next section. In addition, there is greater flexibility to obtain different phase shift values and the step sizes for the proposed phase shifters. Various phase shifts are able to be achieved by tuning the dimensions of the RDMS units or by controlling the quantity of the RDMS units cascaded. As well the step-size of the phase shifters is mainly determined by the phase shift

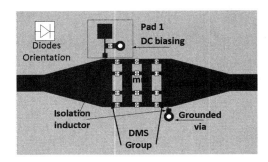

(a)

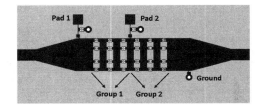

(b)

Fig. 2. Layout of the (a) 3-RDMS phase shifter and (b) 6-RDMS phase shifter.

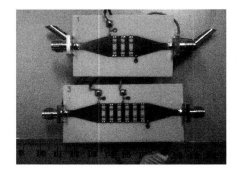

Fig. 3. The phase shifter prototypes.

of a single RDMS unit and the number of the RDMS units biased simultaneously.

## III. ANTENNA ARRAY DESIGN

Based on the aforementioned phase shifters, a 4-element patch antenna array for beam steering was designed for the 5.2 GHz WLAN band. The beamforming network of the

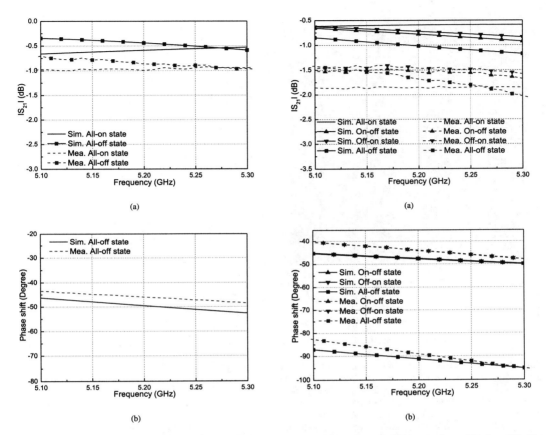

Fig. 4. Simulated and measured (a) insertion loss and (b) phase shift of the 3 RDMS phase shifter.

Fig. 5. Simulated and measured (a) insertion loss and (b) phase shift of the 6 RDMS phase shifter.

phased array consists of three Wilkinson power dividers, two 6-RDMS, and four 3-RDMS phase shifters. The prototype of the antenna array is shown in Fig. 6. With only two DC bias voltages $V_1$ and $V_2$ labeled in Fig. 6, the antenna array is able to operate in 4 states. When $V_1$ and $V_2$ are all positive or all negative, the two states are defined as "All-on" and "All-off" states, respectively. For these two states, all the phase shifters employed in the array work in the same state, which results in uniform phase excitations of the antenna elements. This leads to a boresight beam direction in the H plane (Z-Y plane). When $V_1$ is positive and $V_2$ is negative, the working state is defined as "Left-on-right-off" state with the phase shifters working at different states and results in 45° phase advances compared to that of the adjacent elements on the right. Therefore, the main lobe is shifted to $\theta = -20°$ in the H plane. When $V_1$ is negative and $V_2$ is positive, the state is defined as "Left-off-right-on" state with 45° phase delays

between the adjacent patches. As a consequence, the main lobe in the H plane is shifted to $\theta = 20°$.

Computed results for the reflection coefficient ($S_{11}$) and the far-field pattern in H plane for the four states are given in Figs. 7(a) and 7(b), respectively. As shown in Fig. 7(a), the overlapping impedance bandwidth for the four states covers the 5.2 GHz WLAN band (e.g. 5.15-5.35 GHz in the USA, 5.15-5.25 GHz in Japan, and 5.15-5.35 GHz in Europe). It is observed from Fig. 7(b) that the main beam of the antenna array is able to be switched in the H plane between 0° and ±20° by employing only two DC voltage levels. The maximum realized gain is 10 dB and the gain variation between the four states is below 1 dB. In addition, the maximum sidelobe level is less than -8.5 dB below the peak.

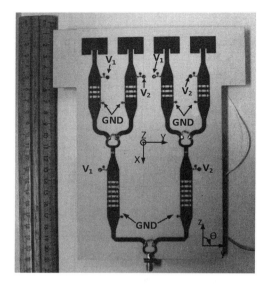

Fig. 6. The antenna array prototype.

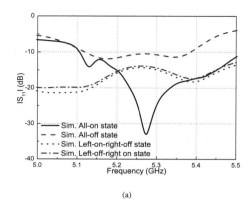

(a)

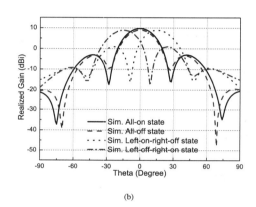

(b)

Fig. 7. Simulated (a) reflection coefficient and (b) far-field pattern of the antenna array in the four states.

## IV. CONCLUSIONS

An RDMS unit has been described for low cost phase shifting applications. The RDMS is able to realize a phase shift of 17° by controlling the working states of the PIN diodes. By cascading the RDMS units, the phase shift value increases quasi-linearly with the unit quantity. Phase shifters made by cascading three and six RDMS units were obtained with phase shifts of 45° and 90°, respectively, and these were used in an antenna array beamforming network. Employing the phase shifters and typical Wilkinson power dividers, a 4-element phased array antenna was built. A beam steering characteristic of the antenna array was observed from the simulated results, demonstrating the practicability of the phase shift method based on the RDMS. The RDMS offers a low cost and low complexity phase shifting solution that is suitable for low cost beamforming and phased array antennas.

## REFERENCES

[1] D. Ahn, J. S. Park, C. S. Kim, J. Kim, Y. Qian, and T. Itoh, "A design of the low-pass filter using the novel microstrip defected ground structure," *IEEE Trans. Microw. Theory Tech.*, vol. 49, no. 1, pp. 86–92, Jan. 2001.
[2] S. M. Han, C. -S. Kim, D. Ahn and T. Itoh, "Phase shifter with high phase shifts using defected ground structures," *Electron. Lett.*, vol. 41, no. 4, pp. 196–197, Feb. 2005.
[3] S. Y. Zheng, W. S. Chan, K. S. Tang, and K. F. Man,, "Broadband parallel stubs phase shifter using defected ground structure," in *Asia-Pacific Microw. Conf.*, Dec. 2008, pp. 1–4.
[4] C. Shafai, S. K. Sharma, L. Shafai, and D. D. Chrusch, "Microstrip phase shifter using ground-plane reconfiguration," *IEEE Trans. Microw. Theory Tech.*, vol. 52, no. 1, pp.144–153, Jan. 2004.
[5] C. Shafai, S. K. Sharma, J. Yip, and L. Shafai, "Microstrip delay line phase shifter by actuating integrated ground plane membranes," *IET Microw. Antennas Propag.*, vol. 2, no. 2, pp. 163–170, Mar. 2008.
[6] S. Hage-Ali, "A millimeter-wave elastomeric microstrip phase shifter," in *IEEE MTT-S international Microwave Symposium*, Jun. 2012, pp. 1–3.
[7] J. A. Tirado-Mendez, "A proposed defected microstrip structure (DMS) behavior for reducing rectangular patch antenna size," *Microw. Opt. Technol. Lett.*, vol. 43, no. 6, pp. 481–484, Dec. 2004.
[8] S. Ye, X. L. Liang, W. Z. Wang, R. H. Jin, J. P. Geng, T. S. Bird and Y. J. Guo, "High Gain Planar Antenna Arrays for Mobile Satellite Communications," *IEEE Antennas Propagat. Mag.*, vol. 54, no. 6, pp. 256–268, Dec. 2012.

# Transient Response Analysis of a MESFET Amplifier Illuminated by an Intentional EMI Source

Qi-Feng Liu[1], Jing-Wei Liu[2], Chong-Hua Fang[1]

[1]Science and Technology on EMC Laboratory, Wuhan, China, emclqf@126.com

[2]Wuhan Wuda Jucheng Strengthening Industrial Co. Ltd, Wuhan, China

*Abstract*—**Unintentional as well as intentional electromagnetic interference can cause improper functionality of microwave circuits or systems. In this paper, our attention is focused on transient response characterization of microwave MESFET amplifiers, which are used widely in the integration of communication circuits and systems, under the impact of an intentional electromagnetic interference(IEMI) source but with different waveforms, respectively. The mathematical treatment is based on two-port lumped networks FDTD method. Parametric studies are carried out to shows effects of the EMI waveforms, its magnitudes on the transient coupled voltages on the input-output of the microwave MESFET amplifiers, with sufficient information obtained for understanding the interaction between the IEMI source and the microwave MESFET amplifier.**

*Key Words*- Electromagnetic pulses (EMP), microwave amplifier, MESFET device, two-port lumped-network, FDTD.

## I. INTRODUCTION

Intentional electromagnetic interference (IEMI) refers to "intentional malicious generation of electromagnetic energy introducing noise or signals into electrical and electronic systems, thus disrupting, confusing, or damaging these systems for terrorist or criminal purposes"[1]. Such an IEMI can be generated by a high-power microwave source, such as a pulse antenna. On the other hand, it is known that the trend of microwave circuits is being moved towards highly hybrid integrated system, such as monolithic microwave integrated circuits (MMIC), *etc*. Physically, a microwave amplifier with active devices can be easily disturbed by a EMI source, such as an ultra-wideband EMP.

Generally speaking, there are two major approaches for incorporating passive and active lumped devices into finite-difference time-domain (FDTD) method to analyze the hybrid microwave circuit [2]. One approach is to use the S-parameters to represent the devices [3] and the other is to use an equivalent SPICE model to account for the devices [4]. In this paper, our attention is focused on using the two-port lumped-network FDTD (TP-LN FDTD) method to investigate transient responses of a microwave MESFET amplifier illuminated by an external EMI.

## II. FROMULATION

To incorporate the two-port lumped network (TP-LN) into the FDTD method, two electrical nodes are used to interface the FDTD mesh with two-port lumped networks. The two symmetric electrical $E_{x1}$ and $E_{x2}$ are used to interface with the lumped networks, as shown in Fig. 1(a), and the Ampere's equation, at nodes $E_{x1}$ and $E_{x2}$, is then complemented by adding current density term $J_{x1}$ and $J_{x2}$ to take the lumped networks into account [4], respectively. This term is discretized by using a time average, so we can obtain the following equations:

$$E_{x1}^{n+1} = E_{x1}^n + \frac{\Delta t}{\varepsilon}\left[\nabla \times \vec{H}\right]_{x1}^{n+\frac{1}{2}} - \frac{\Delta t}{2\varepsilon}(J_{x1}^{n+1} + J_{x1}^n) \qquad (1)$$

$$E_{x2}^{n+1} = E_{x2}^n + \frac{\Delta t}{\varepsilon}\left[\nabla \times \vec{H}\right]_{x2}^{n+\frac{1}{2}} - \frac{\Delta t}{2\varepsilon}(J_{x2}^{n+1} + J_{x2}^n) \qquad (2)$$

The two-port networks are defined in terms of its admittance matrix in the Laplace domain as follows:

$$\begin{bmatrix} I_1(s) \\ I_2(s) \end{bmatrix} = \begin{bmatrix} Y_{11}(s) & Y_{12}(s) \\ Y_{21}(s) & Y_{22}(s) \end{bmatrix} \begin{bmatrix} V_1(s) \\ V_2(s) \end{bmatrix} \qquad (3)$$

where $I_p$ and $V_p$, $p = 1, 2$ are the port terminal current and voltage, respectively.

In the Laplace domain, the V-I relation of a two-port lumped-network can be expressed as

$$I_p(s) = \sum_{q=1,2} Y_{pq}(s) \cdot V_q(s) \qquad (4)$$

where the admittance $Y_{pq}(s)$ is a rational function for one-port and a matrix consisting of rational functions for multi-port cases, respectively. Therefore, we have

$$Y_{pq}(s) = \sum_{m=0}^{M_{pq}} a_m^{(p,q)} s^m / \sum_{n=0}^{N_{pq}} b_n^{(p,q)} s^n \qquad (5)$$

978-7-5641-4279-7

where $a_m^{(p,q)}$ and $b_n^{(p,q)}$ are real-valued coefficients, $M_{pq}$ and $N_{pq}$ are the order numbers of the model, respectively. We can change the $Y(s)$ into the Z-domain using the bilinear transform [5]. In the Z-domain, (5) is turned into

$$Y_{pq}(Z^{-1}) = \sum_{m=0}^{M_{pq}} c_m^{(p,q)} Z^{-m} / \sum_{n=0}^{N_{pq}} d_n^{(p,q)} Z^{-n} \qquad (6)$$

where the coefficients $p_m^{(p,q)}$ and $q^{(p,q)}$ are obtained from $a_m^{(p,q)}$ and $b_n^{(p,q)}$, and this discretization procedure preserves the second-order accuracy of the conventional FDTD.

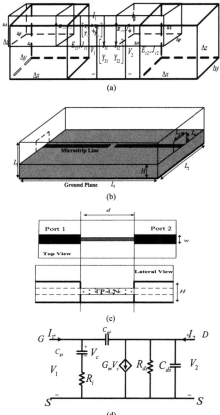

Fig. 1. The configuration of a MESFET microwave circuit. (a) The TP-LN connected to two FDTD cells, (b) three-dimensional structure, (c) Top and lateral view of the hybrid circuit, and (d) the equivalent circuit for MESFET.

To transfer (4) into the Z-domain, we have

$$I_{pq}^{n+1} + \sum_{m=1}^{M_{pq}} d_m^{(p,q)} I_{pq}^{n-m+1} = \sum_{m=0}^{M_{pq}} c_m^{(p,q)} V_q^{n-m+1} \qquad (7)$$

where $V_q^n \approx E_{xq}^n \Delta x$ and $I_p^{n+\frac{1}{2}} \approx J_{xp}^{n+1} \Delta y \Delta z$, then we can obtain from above equations that

$$\begin{bmatrix} \varepsilon E_{x1}^{n+1} \\ \varepsilon E_{x2}^{n+1} \end{bmatrix} = \begin{bmatrix} 1 + \frac{\Delta t}{2} \overline{c}_0^{(1,1)} & \frac{\Delta t}{2} \overline{c}_0^{(1,2)} \\ \frac{\Delta t}{2} \overline{c}_0^{(2,1)} & 1 + \frac{\Delta t}{2} \overline{c}_0^{(2,2)} \end{bmatrix}^{-1} \begin{bmatrix} T_{x1}^n \\ T_{x2}^n \end{bmatrix} \qquad (8a)$$

$$T_{xp}^n = \varepsilon E_{xp}^n + \Delta t (\nabla \times \vec{H})_{xp}^{n+1/2} - \frac{\Delta t}{2}(J_{xp1}^n + A_{xp1}^n + J_{xp2}^n + A_{xp2}^n) \qquad (8b)$$

$$A_{xpq}^n = -\sum_{m=1}^{M_{pq}} d_m^{(p,q)} J_{x,pq}^{n-m+1} + \sum_{k=N_{pq}}^{N_{pq}} \overline{c}_m^{(p,q)} \varepsilon E_{xp}^{n-k+1} \qquad (8c)$$

$$J_{x,pq}^n = -\sum_{m=1}^{M_{pq}} d_m^{(p,q)} J_{x,pq}^{n-m+1} + \sum_{k=0}^{N_{pq}} \overline{c}_m^{(p,q)} \varepsilon E_{xp}^{n-k+1} \qquad (8d)$$

Therefore, this equation allows a two-port lumped-network to be incorporated into the FDTD, but still preserving the explicit nature of the Yee's scheme.

### III. NUMERICAL RESULTS AND DISCUSSIONS

We consider the equivalent circuit of the linear model of a general MESFET microwave amplifier [4]. The MESFET device is placed at the microstrip gap, as shown in Figs. 1(b) and 1(c), respectively. The scattering parameters of the hybrid circuits are captured by the TPLN-FDTD algorithm. It spans several FDTD cells, which corresponds to the length of the microstrip gap. The four ideal wire of single cell are used to connect the MESFET device to the strips and to the ground. The microstrip substrate has dielectric constant $\varepsilon_r = 2.17$ and a thickness $H = 0.254$ $mm$. The line width is $w = 0.79$ $mm$, which corresponds to an impedance of approximately $50$ $\Omega$. The microwave amplifier is illuminated by an external EMP, and it is described by a triple-exponential function as follows:

$$E^{(inc)} = E_0 \times \sum_{p=1}^{3} a_i^p e^{-\left(\frac{t - t_i^p}{\tau_i^p}\right)^2} \qquad (9)$$

To check the accuracy of our two-port lumped-network FDTD code, we at first to calculate the S-parameters of the microwave MESFET amplifier, with no external EMI source implemented. The S-parameters are obtained using the Fourier transform at the end of FDTD iteration. The input and output voltages at points 1 and 2 are shown in Fig. 2, respectively. Fig. 3 shows the simulated S-parameters of the microwave amplifier, as compared with the ADS results given in [4]. It is obvious that excellent agreements are obtained between them.

Then, we pay our attention to the transient response of the microwave MESFET amplifier, illuminated by an external high-power EMP. The incident EMP direction is set to be $\varphi = 135^o$ and $\theta = 90^o$, with a polarization angle of $90^o$ assumed. In our simulation, the output and input ports are terminated by a $50\Omega$ matched load, respectively, when the amplifier is illuminated by an external EMI source. The induced voltages at the input/output ports of Fig.1 (c) and the two terminals of the MESFET device in Fig.1(d) are plotted in Fig. 4 for $E_0 = 58kV/m$. It is found that the external high-power wideband EMP can induce as high as 11.09 V and 2.1 V voltages at the output/input ports, and even as high as 16.98V and 1.48V at the two device terminals, respectively.

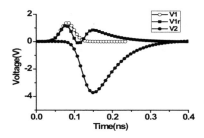

Fig. 2. The reflected and total voltage at input/output ports, respectively.

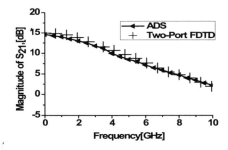

Fig. 3. Comparisons of S-parameters in the TPLN-FDTD and ADS.

We now change the magnitude $E_0$ of the incident intentional EMI source, and set it to be 2.9 $kV/m$, and 5.8 $kV/m$, respectively. For practical case, the amplifier will be shielded by a metallic enclosure. Under such circumstances, the inner input EMI signal magnitude will be much smaller than that of the external source.

In Fig. 5, it is shown that as the magnitude of the incident EMP decreases, low voltage signals are induced at two output ports, as compared with those in Fig. 4(a), respectively. Therefore, it is important to design the some packaging enclosure to further suppress the EMI signal impact on the amplifier.

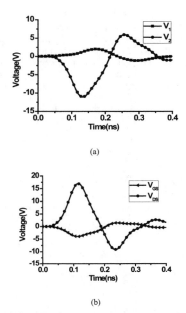

(a)

(b)

Fig. 4. (a)The induced voltages at the input/output ports of the amplifier, respectively; and (b) the induced voltages at the device terminals.

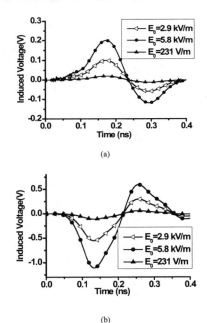

(a)

(b)

Fig. 5. The induced voltages of the amplifier for different magnitudes of the incident EMP recorded at (a) at input port (b) at output port.

## IV. CONCLUSION

The FDTD method is implemented to characterize EMI effects on the microwave MESFET amplifier caused by an external IEMI source. According to our simulation, the effects of different geometrical and physical parameters of the devices as well as the waveform parameters of the IEMI source on the interfere responses at both input and output ports can be predicted. Of course, certain protection method can be further implemented so as to suppress the IEMI impact on the wideband communication circuits and systems.

## REFERENCES

[1]  W. A. Radasky, C. E. Baum, and M. W. Wik, "Introduction to the special issue on high-power electromagnetics (HPEM) and intentional electromagnetic interference (IEMI)," *IEEE Trans. Electromagn. Compat.*, vol. 46, no. 3, pp. 314-321, 2004.

[2]  C. C. Wang, C. W. Kuo, "An efficient scheme for processing arbitrary lumped multiport devices in the finite-difference time-domain method," *IEEE Trans. Microwave Theory Tech.*, vol. 55, no.3, pp. 958-965, May 2007.

[3]  X.Ye and J. L. Drewniak, "Incorporating two-port networks with S-parameters into FDTD," *IEEE Microw. Wireless Compon. Lett.*, vol. 11,no. 2, pp. 77–79, Feb. 2001.

[4]  O. Gonzalez, J. A. Pereda, A. Herrera, and A. Vegas, "An extension of the lumped-network FDTD method to linear two-port lumped circuits," *IEEE Trans. Microw. Theory Tech.*, vol. 54, no. 7, pp. 3045–3051, Jul. 2006.

[5]  X. T. Dong, W. Y. Yin, and Y. B. Gan, "Perfectly matched layer implementation using bilinear transform for microwave device applications," *IEEE Trans. Microw. Theory Tech.*, vol. 53, no. 10, pp. 3098–3105, Oct. 2005.

# Crosstalk Analysis of Through Silicon Vias With Low Pitch-to-diameter ratio in 3D-IC

Sheng Liu[1], Jianping Zhu[1], Yongrong Shi[1], Xing Hu[1], Wanchun Tang[2]

[1]School of Electronic and Optical Engineering, Nanjing University of Science and Technology, Nanjing 210094, China
[2]Department of Communication Engineering, Nanjing Normal University, Nanjing 210023, China
eewctang@aliyun.com

*Abstract-* An equivalent circuit model for low pitch-to-diameter ratio (P/D) through silicon via (TSV) in three-dimensional integrated circuit (3-D IC) is proposed in this paper. The shunt admittance of this model is calculated based on the method of moments which can accurately capture the proximity effect for both a TSV pair and TSV array. The metal-oxide-semiconductor (MOS) capacitance of TSV is also considered. With this model, the crosstalk of TSV array can be fully analyzed regardless of the pitch. The results by this model agree well with those by the electromagnetic simulations up to 40GHz.

## I. INTRODUCTION

Through silicon via (TSV) technology has proven to be the key point of three-dimensional integrated circuit (3-D IC), especially for high performance, more functionality and compact 3-D ICs. In TSV-based 3D-IC, there are large amounts of signal/ground and power/ground TSV pair and TSV array, to connect chips vertically with shortened electrical delay and provide extremely dense I/O connections. With the fabrication technique being enhanced, TSVs with high aspect ratio and low pitch-to-diameter ratio (P/D) will become the main stream of the high performance system with wide I/O bus on 3-D IC [1].

In order to evaluate the electrical behavior for TSV-based 3D-IC, it is desirable to have an equivalent circuit model for TSVs. Usually the model of two parallel wire and coaxial line is employed (e.g., [2-3]) for high pitch-to-diameter ratio of a TSV pair and TSV array, under the assumption that the circular interface between the insulator layer and the silicon substrate of TSVs has equal potential or is like a metal interface. For low pitch-to-diameter ratio of TSVs, however, this assumption would not be valid because of the strong electric-field interaction. Hence, Cheng et al [1] used conformal mapping method to obtain the equivalent capacitance and conductance of a TSV pair in this case. However, it is difficult to find the conformal mapping formula for TSV array.

In this paper, an equivalent circuit model for TSV structure with low P/D is proposed. The shunt admittance of TSV without depletion region [4] is calculated by the method of moments previously used for the single layer dielectric-coated wires [5]. We extend this method to calculate the admittance of double layers dielectric-coated wire in this paper, resulting that both the oxide and metal-oxide-semiconductor (MOS) capacitance of TSVs can be considered in the equivalent

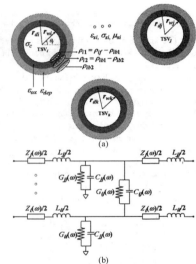

Figure 1. (a) The cross-section view of *n* TSVs with charge distributions at the three interfaces. (b) Equivalent circuit model for TSV structure (the mutual inductance is not shown for clarity).

circuit model. For both a TSV pair and TSV array with low P/D, the non-uniform charge distributions on conductor peripheries due to the proximity effect are completely considered. Based on this model, the crosstalk among TSVs can be fully analyzed regardless of the pitch.

## II. EQUIVALENT CIRCUIT MODEL FOR TSVS WITH LOW P/D

For an array of *n* TSVs in Fig. 1(a), the equivalent circuit model is proposed and shown in Fig. 1(b). Unlike those of [1-3], the admittance of this model can be directly calculated and it is the sum of the oxide capacitance, MOS capacitance, silicon substrate capacitance and conductance. Since the geometry size of TSV structure is small enough compared with the wavelength at the frequency of interest, these frequency dependent *RLCG* parameters are computed under the quasi-static assumption.

### A. Capacitance and Conductance

For each TSV in Fig. 1(a), there are three interfaces: conductor-oxide, oxide-depletion region and depletion region-

978-7-5641-4279-7

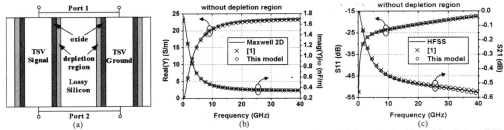

Figure 2. (a) Structure of a signal/ground TSV pair. Comparison of (b) the shunt admittance without depletion region by this model and Maxwell 2D, (c) the S parameters without depletion region by this model and HFSS. The results by conformal mapping method [1] are also added for comparison.

lossy silicon. The key point of handling this inhomogeneous structure is to replace the interface with the bound charge at these three interfaces.

As shown in Fig. 1(a), the charge distribution $\rho_{i1}(\theta_i)$ at the conductor-oxide interface contain the free charge $\rho_{if}(\theta_i)$ and oxide bound charge $-\rho_{ib1}(\theta_i)$, while the charge distribution $\rho_{i2}(\theta_i)$ at the oxide-depletion region interface contain the oxide and depletion region bound charge $\rho_{ib1}(\theta_i)$, $-\rho_{ib2}(\theta_i)$, respectively. These charge distribution can be expanded into a Fourier series [5] and must meet two boundary conditions: (a) potential at observation points of each conductor surface is equal, and (b) normal component of the displacement vector is continuous at the observation points on two dielectric interfaces. This will give the generalized capacitance matrix for $n$ TSVs with the same form [5]:

$$\begin{bmatrix} q_{1f} \\ q_{2f} \\ \vdots \\ q_{nf} \end{bmatrix} = \begin{bmatrix} C_{11} & C_{12} & \cdots & C_{1n} \\ C_{21} & C_{22} & & \vdots \\ \vdots & & \ddots & \vdots \\ C_{n1} & \cdots & \cdots & C_{nn} \end{bmatrix} \begin{bmatrix} \phi_1 \\ \phi_2 \\ \vdots \\ \phi_n \end{bmatrix} \tag{1}$$

where $q_{if}$, $\phi$ are the per-unit-length free charge and potential on the conductor of the $i$th TSV respectively. $C_{ij}$ is the capacitance coefficient established by the above two boundary conditions.

With the complex permittivity of the silicon

$$\varepsilon_{si}^* = \varepsilon_{si} - j \frac{\sigma_{si}}{\omega} \tag{2}$$

a complex capacitance matrix $C^*$ can be directly calculated from (1). And from this complex capacitance matrix, we have

$$C = \mathrm{Re}(C^*) \text{ and } G = -\omega \cdot \mathrm{Im}(C^*) \tag{3}$$

where $\omega = 2\pi f$ is the angular frequency of interest.

### B. Inductance and Resistance

The per-unit-length inductance matrix $L$ can be obtained from the duality of

$$L = \mu_0 \varepsilon_0 C_0^{-1} \tag{4}$$

where $C_0$ is the capacitance matrix with the oxide and silicon surrounding the TSVs removed.

The internal impedance of the TSV is calculated by the formula [6]:

$$Z = \frac{\sqrt{j\omega\mu\sigma_c}}{2\pi r_w} \frac{I_0(j\omega\mu\sigma_c)}{I_1(j\omega\mu\sigma_c)} \tag{5}$$

where $\sigma_c$ is the conductivity of the TSV metal. $I_0$ and $I_1$ are, respectively, the modified Bessel functions of order zero and one, and $r_w$ is the radius of the TSV.

### III. ANALYSIS OF TSVs WITH LOW P/D

#### A. A TSV Pair

A TSV pair with geometric parameters same as [1] are chosen to verify the accuracy of our model. As shown in Fig. 2(a), the TSV pair is 100 μm in height, while the diameter and pitch of each TSV are 30 μm and 40 μm, respectively. The conductivity of silicon is 10 S/m, where the relative permittivity of the silicon and oxide are 11.9 and 4, respectively. The maximum depletion width is 0.79 μm by the formula in [6].

To compare with the conformal mapping method [1] which does not consider the depletion region, the equivalent shunt admittance is obtained by this model and MAXWELL 2D [7], as shown in Fig. 2(b). Fig. 2(c) shows the corresponding S-parameters calculated and compared with HFSS [8]. As can be seen, all the results show good agreements.

To evaluate the effect of depletion region, Fig. 3(a) shows the comparison of the insertion loss (S21) of TSV pair with and without depletion region under different P/D at 10GHz. As can be seen, the difference between the S21 with and without depletion region will increase slightly from 0.01dB to 0.07dB as the P/D decreases from 3 to 1.2. One can also observe that the S21 is in the range of 0.3-0.46 dB, which means that the insertion loss is not very large when the P/D varies from 3 to 1.2.

If several chips are stacked and connected by the TSV pair, the total length of the TSV will affect the insertion loss S21, as illustrated in Fig. 3(b), which shows the tendency of S21 with the number of stacked chips ($N_c$) for P/D=1.33 at 10GHz. The microbumps are ignored in this paper for simplicity. For the case of without depletion region, the insertion loss increases from around 0.45 dB to 3.2 dB when the number of stacked chips increases. Meanwhile, the difference between the S21 with and without depletion region increases from 0.05 dB ($N_c$=1) to 0.3 dB ($N_c$=8).

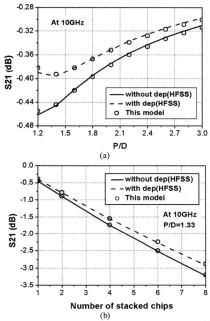

Figure 3. The S21 of the TSV pair with and without depletion region under (a) different P/D at 10GHz (b) different number of stacked chips with P/D=1.33 at 10GHz.

*B. TSV array*

As stated in section I, it is difficult to find a conformal mapping formula for TSV array with low P/D. Fig. 4(a) shows such a TSV array with 5 signal TSVs and 1 grounded TSV for reference. It is like that of [6], but the P/D of TSVs has been reduced from 10 in [6] to 1.33. The thickness of the oxide, depletion region and the diameter of each TSV are the same as that in the last example of section III A.

Though the significance of crosstalk noise depends on the products and applications, it becomes a concern when its level is in a range between -60 dB and -26 dB in the frequency domain or 0.1% and 5% of the supply voltage in time domain in general [9]. In frequency domain, the near-end crosstalk (NEXT) and far-end crosstalk (FEXT) between $TSV_1$ (aggressor) and $TSV_i$ ($i \neq 1$, victim) are evaluated using this model and given in Fig. 4(b). The results by this model are in good agreement with that by HFSS. It can be observed that NEXT is higher than FEXT for all cases. And both NEXT and FEXT affect the nearest victim most. Meanwhile, higher frequency strengthens both the capacitive and magnetic coupling between aggressor and victim, which leads to the crosstalk increasing with the frequency. As can be seen from Fig. 4(b), the NEXT and FEXT between $TSV_1$ and $TSV_2$ both exceed -60 dB at 0.2 GHz and reach to -16.6 dB and -22.3 dB at 40 GHz respectively.

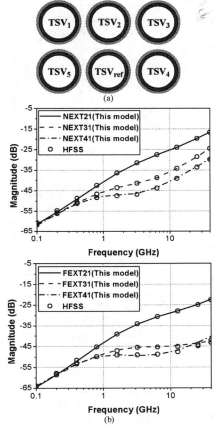

Figure 4. (a) TSV array with 5 signal TSVs and 1 grounded TSV for reference. (b) Magnitude of near-end crosstalk (NEXT) and far-end crosstalk (FEXT) between $TSV_1$ and $TSV_i$ ($i \neq 1$) with P/D=1.33.

In order to evaluate the crosstalk between $TSV_1$ and $TSV_2$ (P/D=1.33) in time domain, a step signal of 1V is injected into $TSV_1$. All TSVs are terminated at both ends by a 50 Ω resistance. With the rise time of the signal decreasing, the peak value of both the NEXT and FEXT on $TSV_2$ increase and reach beyond 1% of the supply voltage when the rise time is 30 ps, as shown in Fig. 5. This is due to that the higher frequency component contained in the signal of lower rise time will cause higher noise level on adjacent TSV, as can be seen in Fig. 4(b). It should be noticed that the noise margin of a gate depends on both the peak amplitude of noise and its duration. For example, digital circuits can often tolerate (and indeed filter out) spike-like crosstalk noise with a large peak amplitude and very small noise width, as indicated in [10]. Thus, comprehensive evaluation are needed for the crosstalk in TSV interconnect.

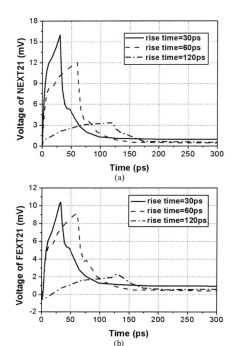

Fig. 5 Time domain voltage of (a) NEXT and (b) FEXT between TSV$_1$ and TSV$_2$ (P/D=1.33) for different rise time of a step signal.

## IV. CONCLUSION

The equivalent circuit model for low P/D TSV structure has been proposed. For both a TSV pair and TSV array with low P/D, the proximity effect on the shunt admittance of TSVs with depletion region is rigorously considered by the method of moments. The crosstalk behavior of TSV array with low

P/D has also been analyzed in both frequency domain and time domain. Results by this model show good agreements with those of electromagnetic simulations.

## ACKNOWLEDGMENT

This work is supported by the Specialized Research Fund for the Doctoral Program of Higher Education in China (no. 20103219110017) and the Major State Basic Research Development Program of China (973 Program: 2009CB320201).

## REFERENCES

[1] Tai-Yu Cheng, Chuen-De Wang, Yih-Peng Chiou, Member, IEEE, and Tzong-Lin Wu, "A New Model for Through-Silicon Vias on 3-D IC Using Conformal Mapping Method," *IEEE Microw. Wireless Compon. Lett.*, vol. 22, no. 6, pp. 303-305, Jun. 2012.

[2] Joohee Kim, et al., "High-Frequency Scalable Electrical Model and Analysis of a Through Silicon Via (TSV)," *IEEE Trans. Compon., Packag., Manuf. Technol.*, vol. 1, no. 2, pp. 181-195, Feb. 2011.

[3] Wei Yao, Siming Pan, Brice Achkir, Jun Fan, Lei He, "Modeling and Application of Multi-port TSV Networks in 3-D IC," *IEEE Trans. Computer-Aided Design Integr. Circuits Syst.*, vol. 32, no. 4, pp. 487-496, Apr. 2013.

[4] Dazhao Liu, Siming Pan, Brice Achkir, and Jun Fan, "Fast Admittance Computation for TSV Arrays," in *Proc. IEEE Electromagn. Compat.*, pp. 28-33, Aug. 2012.

[5] Clayton R. Paul, Arthur E. Feather, "Computation of the Transmission Line Inductance and Capacitance Matrices from the Generalized Capacitance Matrix," *IEEE Trans. Electromagn. Compat.*, vol EMC-18, no. 4, pp. 175-183, Nov. 1976.

[6] A. Ege Engin, and Srinidhi Raghavan Narasimhan, "Modeling of Crosstalk in Through Silicon Vias," *IEEE Trans. Electromagn. Compat.*, vol. 55, no. 1, pp. 149-158, Feb. 2013.

[7] Maxwell, Ansys Corporation, Canonsburg, PA, USA [Online]. Available: http:// www.ansys.com.

[8] High Frequency Structural Simulator, Ansys Corporation, Canonsburg, PA, USA [Online]. Available: http:// www.ansys.com.

[9] Zheng Xu, et al., "Crosstalk Evaluation, Suppression and Modeling in 3D Through-Strata-Via (TSV) Network," in *Proc. IEEE 3D Systems Integration Conference (3DIC)*, pp. 1-8, Nov. 2012.

[10] Payam Heydari and Massoud Pedram, "Analysis and Reduction of Capacitive Coupling Noise in High-Speed VLSI Circuits," in *IEEE Int. Conf. Comput. Design*, Austin, TX, pp. 104–109, Sep. 2001.

# Design of a Feed Network for Cosecant Squared Beam based on Suspended Stripline

Huiying Qi[1], Fei Zhao[2], Lei Qiu[1], Ke Xiao[1], Shunlian Chai[1]

[1]College of Electronic Science and Engineering, National University of Defense Technology,
Changsha, Hunan 410073, China

[2]Southwest Electronics and Telecommunication Technology Research Institute,
Chengdu, Sichuan 610041, China

*Abstract*-For the cosecant square beam pattern application of an 8-element antenna array, we designed a 8-way non-equal power-divider based on Wilkinson scheme. An improved suspended stripline(SSL) structure is proposed and analyzed here, by using which as planar transmission line, the designed power-divider has good performance in the insertion-loss, and the excitations for output ports agree well with the simulated results.

## I. INTRODUCTION

The cosecant square beam pattern are very significant in some applications (e.g., radars, wireless communications). A cosecant squared beam antenna is required elevation pattern in order for a target approaching at a constant height to be detected with constant power. In order to achieve the pattern, we always need a complex feed network which is made up of some power divided. The Wilkinson power divider is usually used. There are many planar transmission lines applicable for this usage such as microstrip line, air stripline, suspended stripline and so on. But the insertion loss for microstrip line is always large especially for high-frequency applications，and the air stripline has some problem for machining which is not easier assembly than printed transmission lines[1]. The suspended stripline provide a way to reduce the attenuation in a microstrip line at higher frequencies and yet retain some of the features of the microstrip line such as the quasi-TEM nature of the dominant-mode propagation[2], so it can decrease propagation losses and approve the input matching efficiently. Therefore, it is of paramount importance to use SSL on low-loss and highly efficient antenna systems[3-5].

SSL is widely used in filters to ensure quality, but it is rarely applied in power-divide circuits. In this study, we use the suspended stripline as the transmission lines to achieve a non-equal amplitude and non-equal phase feed network. Additionally, by using scheme of Wilkinson power divider and quasi-coaxial structure, high isolation between the outputs is obtained.

## II. ANALYSIS AND DESIGN PROCEDURE

Firstly, we give the objective pattern obtained by synthesized algorithm, then the structure of SSL is provided and analyzed, after which, the experiment results are shown and discussed.

### A. Target pattern and power divide

We propose to design the input of an 8-element array with cosecant square beam pattern, as shown in Fig. 1. By using the technique for power pattern synthesis proposed by Orchard et al[6], the pattern of cosecant square and the pattern is calculated and the excitation current $I_n$ of each element is shown in the table I.

In order to arrive the excitations of 8 ports, we use the structure as shown in Fig. 2, which is composed of 7 Wilkinson power dividers, and the magnitude of $K_i$ (i=1~7) are depicted in table II.

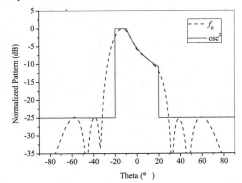

Figure 1. Target Pattern.

TABLE I
EXCITATION CURRENTS

| Number | Normalized Amplitude | Phase(°) | Normalized Phase(°) |
|---|---|---|---|
| 1 | 0.3213 | 129.9121 | -45.5094 |
| 2 | 0.4336 | 175.4215 | 0 |
| 3 | 0.7539 | -144.6394 | -320.0609 |
| 4 | 1 | -116.9071 | -292.3287 |
| 5 | 0.7817 | -93.7603 | -269.1818 |
| 6 | 0.3201 | -60.0165 | -235.4381 |
| 7 | 0.3200 | 55.2841 | -120.1375 |
| 8 | 0.3261 | 89.4477 | -85.9738 |

978-7-5641-4279-7

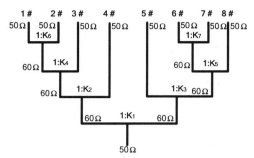

Figure 2. power divided.

TABLE II
POWER DIVIDED PLAN

| K₁ | K₂ | K₃ | K₄ | K₅ | K₆ | K₇ |
|---|---|---|---|---|---|---|
| $K_1$ | $K_2$ | $K_3$ | $K_4$ | $K_5$ | $K_6$ | $K_7$ |
| 0.7042 | 1.0786 | 0.7136 | 1.3969 | 0.7206 | 1.3493 | 0.9998 |

*B. The structure of suspended stripline*

Structure shown in Fig. 3 are applied for the SSL application, in upper and bottom layer, it is metal for ground of SSL, in medium layer, there is a the substrate with $\varepsilon_r$ equal to 2.2, and the thickness is $h_0 = 0.25mm$, other dimensions are as: $h_t = 3mm$, $h_d = 1.5mm$, $w_{air} = 10$mm. The width of the center conductor is $w$, which is considered due to different requirements of characteristic impedance.

It should be noted that, two sets of vias are designed in the two sides of the SSL structure, so the microwave energy is focused in the area including the middle substrate and upper and down air gap. Additionally, by observing the field distributions, we can find that the electric field mainly distributed between the center conductor and the outer ground, so the insertion loss for the transmission line can be lower than the traditional microstrip line and stripline.

### III. EXPERIMENTAL RESULTS AND DISCUSSION

According to the specifications given in the previous section, the feed network was fabricated as shown in Fig. 4. We compared the normalized amplitude and the phase of 8-element input in Fig. 5, the number on transverse axis means the output port number. From the results, we can conclude that, excellent agreement can be obtained between the measured and simulated results. Besides, the insertion loss for the power divide can reached 0.22dB at the center frequency 5.6 GHz by simulation which have not plus the ports loss.

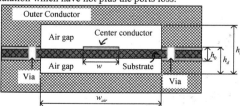

Figure 3. Geometry of the suspended stripline.

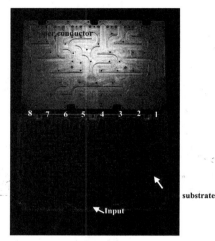

Figure 4. Photograph of the feed network.

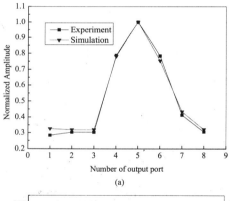

(a)

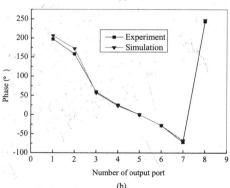

(b)

Figure 5. The compare of the simulation and the experiment.

## IV. CONCLUSION

An improved suspended stripline structure is applied in this paper for designing of low-loss non-equal power divider. The power-divider has one input and 8 non-equal output ports, Wilkinson power-divider is applied for each "T" junction, and by using such SSL, low insertion-loss results are obtained. Besides, not any tuning method is considered in this application, while excellent agreement between experiments and simulated results are obtained, which indicates that it is a good choice to design compact and low-loss power-divider based on this structure provided here. But we have not obtain the test insert loss, because the measure of the precision data is difficult and we can reference the simulation data.

## REFERENCES

[1] C. Chih-Chieh and G. M. Rebeiz, "A Three-Pole 1.2-2.6-GHz RF MEMS Tunable Notch Filter With 40-dB Rejection and Bandwidth Control," *Microwave Theory and Techniques, IEEE Transactions on*, vol.60, pp. 2431-2438, 2012.

[2] T. Itoh, "Overview of quasi-planar transmission lines," *Microwave Theory and Techniques, IEEE Transactions on*, vol.37, pp. 275-280, 1989.

[3] R. Glogowski, J. F. Zurcher, C. Peixeiro and J. R. Mosig, "*A Low-loss Planar Ka-band Antenna Subarray for Space Applications*," *Antennas and Propagation, IEEE Transactions on*, vol.PP, pp. 1, 2013.

[4] L. Song and M. Eron, "Development of an Ultra-Wideband Suspended Stripline to Shielded Microstrip Transition," *Microwave and Wireless Components Letters, IEEE*, vol.21, pp. 474-476, 2011.

[5] X. Jinxiong, "Suspended Stripline and Ka Band Integrated Mixer," in *Proc. 2000 Asia-Pacific Conference on Environmental Electromagnetics*, pp. 57-60.

[6] H. J. Orchard, R. S. Elliott and G. J. Stern, "Optimising the synthesis of shaped beam antenna patterns," *IEE Proceedings*, vol.132, pp. 63-68, 1985.

# The study on Crosstalk of Single Wire and Twisted-Wire Pair

Lijuan Tang, Zhihong Ye, Linglu Chen, Zheng Xiang , Cheng Liao

Institute of Electromagnetics, Southwest Jiaotong University

Chengdu, Sichuan 610031, P. R. China

*Abstract*- **A method based on time-domain Baum-Liu-Tesche (BLT) equations of transmission line model is presented in this paper for solving the crosstalk problem between a single wire and a twisted-wire pair. The single wire and the twisted-wire pair (TWP) together are modeled as a set of uniform multi-conductor transmission lines with abrupt interchanges of wire positions at the end of each loop of the twisted-wire pair, each group of multi-conductor transmission lines is solved by using the finite-difference time-domain (FDTD) method, the nodes of the twisted pair are corrected by using twisted-wire pair two-line hybrid iterative method. Finally, the time-domain results are obtained. It is found that the simulation results obtained by using BLT equations proposed in this paper agree very well with the results of commercial simulation software, the method is verified to be corrected.**

*Key words*- **Crosstalk, Twisted-wire pair, BLT Equations, FDTD**

## I. INTRODUCTION

A twisted-wire pair (TWP) consists of two identical wires which are smoothly twisted together. Aside from the obvious advantage of holding the two wires together, the twist also tends to reduce inductive coupling because of its special structure, which was shown in [1]. TWP is widely concerned in many fields of electromagnetic compatibility. Study of crosstalk problem involving TWP is practically valuable.

The analysis of crosstalk involving TWP is complex because of the special structure of TWP, it can not be solved via using transmission line equations directly. Traditional research about TWP is generally expanded in the frequency domain, the time-domain terminal response can not be obtained effectively. Investigation of crosstalk between single wire and TWP in specific frequency band is presented in [2]. Studies on TWP has generally concentrated on determining the magnetic field resulting from the excitation of an isolated twisted pair, which are shown in [3]-[5]. Study of the near end crosstalk in twisted pair cables is shown in [6].

A method based on time-domain Baum-Liu-Tesche (BLT) equations [7]-[9] of transmission line model is presented in this paper for solving crosstalk problem between a single wire and a

TWP. The TWP is modeled as a cascade of loops consisting of uniform two-wire sections with abrupt interchanges of wire positions at the end of each loop. The single wire and the TWP together are modeled as a set of uniform multi-conductor transmission lines with a specific way of cascade at the end of each loop of twisted pair, each group of multi-conductor transmission lines is solved by using the finite-difference time-domain (FDTD) method, the nodes of the twisted pair are corrected using twisted pair two-line hybrid iterative method. Finally, the time-domain results are obtained. It is found that the simulation results obtained by using BLT equation proposed in this paper agree very well with the results of commercial simulation software, the method is verified to be corrected.

## II. MODEL

The crosstalk configuration of a single wire and a twisted-wire pair is shown in Fig. 1. In Fig. 1(a), VS is the voltage source. The cross-sectional view is shown in Fig. 1(b). In Fig. 1, the single wire is designated as the generator wire, the TWP is designated as the receptor wire. The single wire and the twisted-wire pair (TWP) together are modeled as a set of uniform multi-conductor transmission lines with abrupt interchanges of wire positions at the end of each loop of twisted pair. Approximations about the twisted pair are shown as follow: (a) the twisting part is considered as uniform transmission line; (b) two twisted parts are approximately infinite small. One section of the TWP is shown in Fig. 2. The voltage control equation at the cascade can be written as:

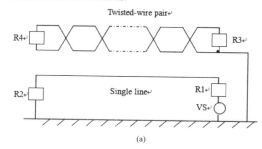

(a)

* Project supported by the Fundamental Research Funds for The Central Universities (Grant No.SWJTU12ZT08).

978-7-5641-4279-7

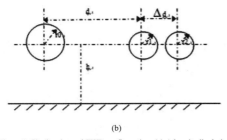

(b)

Figure 1. Single wire and TWP configuration. (a) A longitudinal view.
(b) A cross-sectional view.

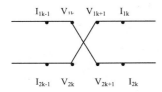

Figure 2. The twisted part of twisted-wire pair

$$\begin{bmatrix} V_{1k} \\ V_{2k} \end{bmatrix} = \begin{bmatrix} 0 & 1 \\ 1 & 0 \end{bmatrix} \begin{bmatrix} V_{1k+1} \\ V_{2k+1} \end{bmatrix} \quad (1)$$

The time-domain form of the First Telegraph Equation and Second Telegraph Equation obtained according to Maxwell's equations can be written as

$$\frac{\partial V(z,t)}{\partial z} + L(z)\frac{\partial I(z,t)}{\partial z} + R(z)I(z,t) = V_s(z,t) \quad (2a)$$

$$\frac{\partial I(z,t)}{\partial z} + C(z)\frac{\partial V(z,t)}{\partial z} + G(z)V(z,t) = I_s(z,t) \quad (2b)$$

In (2), $V_s(z,t)$、$I_s(z,t)$ are the column vector consisting of voltage source and current source of the single wire and TWP configuration separately. R、L、G、C are the cable per-unit-length distribution parameters matrix, which can be obtained through electromagnetic simulation. Equation (2) is discrete by using FDTD, the terminal responses of the multi-conductor transmission lines consisting of a single wire and a twisted-wire pair can be written as

$$V_k^{n+1} = (C\frac{\Delta z}{\Delta t} + \frac{G}{2}\Delta z)^{-1}((C\frac{\Delta z}{\Delta t} - \frac{G}{2}\Delta z)V_k^n - (I_k^{n+1/2} - I_{k-1}^{n+1/2})) \quad (3a)$$

$$I_k^{n+3/2} = (L\frac{\Delta z}{\Delta t} + \frac{R}{2}\Delta z)^{-1}((L\frac{\Delta z}{\Delta t} - \frac{R}{2}\Delta z)I_k^{n+1/2} - (V_{k+1}^{n+1} - V_k^{n+1})) \quad (3b)$$

Considered the terminal condition shown in Fig 1, the terminal response of the single wire and twisted-wire pair can be expressed as

$$V_1^{n+1} = (\frac{\Delta z}{\Delta t}R_S C + E)^{-1}[(\frac{\Delta z}{\Delta t}R_S C - E)V_1^n - 2R_S I_1^{n+1/2} + (V_S^{n+1} + V_S^n)] \quad (4a)$$

$$V_{NDZ+1}^{n+1} = (\frac{\Delta z}{\Delta t}R_L C + E)^{-1}[(\frac{\Delta z}{\Delta t}R_L C - E)V_{NDZ+1}^n + 2R_L I_{NDZ}^{n+1/2}] \quad (4b)$$

In (4), $R_S$ and $R_L$ are the beginning and end impedance matrix separately. Considered the iteration integrity of the nodes of the twisted part, the iteration equations of the nodes of the twisted part can be expressed as

$$V_{1k}^{n+1} = (C\frac{\Delta z}{\Delta t} + \frac{1}{2}G\Delta z)^{-1}((C\frac{\Delta z}{\Delta t} - \frac{G}{2}\Delta z)V_{1k}^n \quad (5a)$$
$$-(I_{2k}^{n+1/2} - I_{1k-1}^{n+1/2}))$$

$$V_{2k}^{n+1} = (C\frac{\Delta z}{\Delta t} + \frac{1}{2}G\Delta z)^{-1}((C\frac{\Delta z}{\Delta t} - \frac{G}{2}\Delta z)V_{2k}^n \quad (5b)$$
$$-(I_{1k}^{n+1/2} - I_{2k-1}^{n+1/2}))$$

### III. NUMERICAL SIMULATION

#### A. Example one

The crosstalk configuration of a single wire and a straight-wire pair (SWP) is shown in Fig. 3, the single wire is designated as the generator wire, the SWP is designated as the receptor wire. In Fig. 3, $R1=0, R2=R3=R4=50\ \Omega$, the length of the single line and SWP are 1 m, $r_0=r_1=r_2=0.69$ mm, h=5 cm, d=5 cm, $\triangle d$=1.75 mm. The waveform of the voltage source is shown in Fig. 4。The results are shown in Fig. 5-6.

Fig. 5-6 show two methods agree well with each other, which verifies the correctness of dealing with crosstalk problem of the single wire and SWP by using multi-conductor transmission line theory.

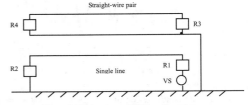

Figure 3. Single wire and SWP configuration.

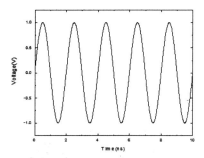

Figure 4. The waveform of voltage source

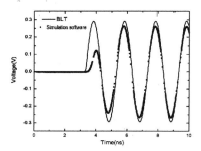

Figure 5. Results of the far end of the single wire.

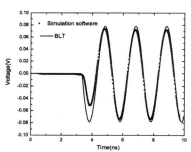

Figure 6. Results of the far end of the SWP.

### B. Example two

The crosstalk configuration of a single wire and a twisted-wire pair (TWP) is shown in Fig. 1, the single wire is designated as the generator wire, the TWP is designated as the receptor wire. In Fig. 1 $R1=0, R2 = R3 = R4 = 50 \ \Omega$, the length of the single line and TWP are 1 m, $r_0$=0.69 mm, $r_1$=$r_2$=0.564 mm, h=5 cm, d=5 cm, $\square$d=1.25 mm, the length of each twisted loop is 2 cm. The waveform of the voltage source is shown in Fig. 7. The results are shown in Fig. 8-9.

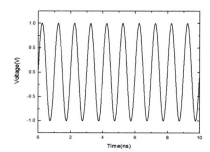

Figure 7. The waveform of voltage source

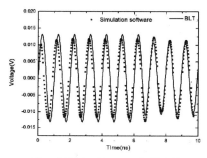

Figure 8. Results of the near end of the TWP

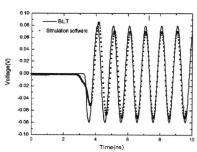

Figure 9. Results of the far end of the TWP

Fig. 8-9 show two methods agree well with each other, which verifies the correctness of dealing with crosstalk problem of the single wire and TWP by using BLT equations proposed in this paper. It is known that the crosstalk amplitude between a single wire and SWP has no matter with the frequency of voltage source, it is decided by the amplitude of voltage source. In Fig.6 and Fig.9, the response of SWP and TWP has the same level, although the distance of TWP is smaller, which shows the TWP can reduce inductive coupling.

## IV. CONCLUSION

The terminal response of the single wire and twisted-wire pair can be obtained accurately by using the method based on time-domain BLT equation of transmission line model presented in this paper for solving crosstalk problem between a single wire and a TWP which can not be solved via using transmission line equations directly. The results match well with the results of commercial simulation software.

## ACKNOWLEDGMENT

This work is supported by the Fundamental Research Funds for The Central Universities (Grant No. SWJTU12ZT08).

## REFERENCES

[1] C. R. Paul and J. W. McKnight , *IEEE Transactions on Electromagnetic Compatibility*, Vol. 21, May 1979, pp. 92-105.

[2] D. Welsh and J. M. Tealby, *1990 IEEE Int. Symp. Electromagn. Compat, Washington , DC*, Aug.21-23, pp. 478-482.

[3] C. A. Nucci and F. Rachidi, *IEEE Trans. Electromagn. Compat,* vol. 37, no. 4, Nov.1995, pp. 505–508.

[4] Ake Karsberg, Gustaf Swedenborg, and Kjell Wyke, *TELE (Stockholm, English Ed.)* no. 1, 1959, pp. 2841.

[5] Gustaf Swedenborg and Kjell Wyke, *TELE (Stockholm, English Ed.)* no. 1, 1959, pp. 4148.

[6] J. Poltz, J. Beckett, and M. Josefsson, *Proc.2005 IEEE Int. Symp. Electromagn. Compat, Chicago, IL*, Aug. 8–12, pp. 572–577.

[7] F. M. Tesche and C. M. Butler，*Kirtland AFB, Albuquerque，NM，Interaction Note588*，2003, pp.1-43.

[8] C. E. Baum，*Air Force Res. Lab，Wright-Patterson Air Force Base，OH，InteractionNote553*，1999.

[9] F. M. Tesche，M. V. Ianoz，and T. Karlsson，*NewYork:Wiley*，1997.

# *FA-P*

## October 29 (FRI) AM

## Room E

# Bow-tie Shaped Meander Slot on-body Antenna

ChenYang[1], Guang Hua[2], Ping Lu[3], Houxing Zhou[4]

2013 International Symposium on Antennas and Propagation

School of Information Science and Engineering, Southeast University

Nanjing, 210096, P. R. China

[1]chenyang@emfield.org  [2]guanghua@emfield.org

[3]pinglu@emfield.org  [4]houxingzhou@emfield.org

Abstract-In this paper a meander slot on-body antenna is proposed for the use of wireless body area network. In order to reduce the human body radiation, metallic foil has been used on the back of the antenna. To realize miniaturized antenna having large bandwidth, the bow-tie meander slot construction is combined in this paper. A hybrid-ring coupler is used to provide phase difference. Results show that the proposed antenna has 460MHz bandwidth on the central frequency of 17GHz.

## I. INTRODUCTION

Recently, the millimeter wave band has been identified as a highly attractive solution for future wireless body area networks (WBAN). Rigorous requirements for antennas such as small size, large bandwidth and stable performance are under consideration when the antenna is applied to the various parts of human body.

For the Engineering application of microwave and millimeter wave, on-body antenna is receiving more and more attention because of its low radiation to human body [1] ~ [4]. To protect human from radiation, metallic foil has been used to weaken the antenna on the human body injury [5]~[8]. In order to make antennas available to integrated circuits, high frequency and a suitable dielectric constant is proposed [9]~[12].One of the main challenges for the design of antennas for BANs involves the adaption of the antenna topology to the shape of the human body. For this purpose, bow-tie shaped meander slot antenna has been studied to realize miniaturized antennas having large bandwidth [13] ~ [19].

In this paper, metallic foil is reserved on the back of the antenna to reduce the radiation of the human body. Bow-tie shape has been used to realize wide bandwidth and small size. In order to apply the antenna to the integrated chip and minimize the size, high frequency of 17GHz and low dielectric constant of 2.6 have been used.

## II. ANTENNA DESIGN

The configuration of the proposed antenna is shown in Fig.1. The antenna is constructed by making meander slots in a perfectly conducting plane supported by a dielectric substrate of 1 mm thickness and relative dielectric constant of 2.6. On the back of the antenna, metallic foil has been used to reflect the radiation of antenna to protect the human body.

The values of design parameters are listed in Table 1.

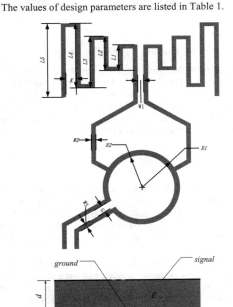

Fig. 1 The construction of the body antenna (top view and side view)

TABLE I

VALUES OF DESIGN PARAMETERS (ALL IN MILLIMETRES)

| L1 | L2 | L3 | L4 | L5 | W1 | W2 |
|------|------|------|------|------|------|------|
| 1.83 | 2.44 | 3.66 | 4.58 | 5.19 | 0.42 | 0.32 |
| W3 | K | R1 | R2 | G | d | $\varepsilon_r$ |
| 0.31 | 0.66 | 2.46 | 2.64 | 0.49 | 1 | 2.6 |

The use of the meander slot radiators is to reduce the size. As shown in Fig. 2, the radiation is mostly generated from the turnings. To generate different polarizations, the values of constant $H$, $L$ and $D$ can be modulated [20]~[22].

Correspondence should be addressed to Guang Hua, huaguang@seu.edu.cn

978-7-5641-4279-7

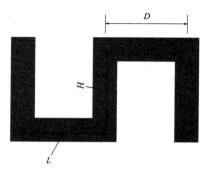

Fig. 2 The schematic diagram of meander slot

The meander slots and bilateral symmetry is divided into several sections. The slot width of the antenna is equal to the spacing between the parallel slots. To have the larger bandwidth, the construction of bow-tie slot and meander slot geometries can be used. The equivalent antenna length $L5$ are introduced and calculated according to:

$$L5 \approx \frac{0.5\lambda_0}{\sqrt{\varepsilon_r}} \quad (1)$$

where $\varepsilon_r$ is the dielectric constant of the substrate and $\lambda_0$ is the wavelength in free space at central frequency. The spacing between the parallel slots and the width of the slot are all 0.5 mm. The values of design parameters in Table 1 above are optimized by commercial electromagnetic simulation software finally.

The hybrid ring coupler, also called the rat-race coupler, is a four-port 3 dB directional coupler consisting of a $3\lambda/2$ ring of transmission line with four lines at the intervals shown in Fig.3. Power input at port 1 splits and travels both ways round the ring.

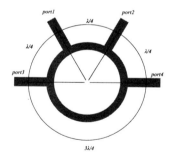

Fig. 3 The diagram of hybrid-ring coupler

The scattering matrix of hybrid-ring coupler is given by:

$$[S] = -\frac{j}{\sqrt{2}} \begin{pmatrix} 0 & 1 & 1 & 0 \\ 1 & 0 & 0 & -1 \\ 1 & 0 & 0 & 1 \\ 0 & -1 & 1 & 0 \end{pmatrix} \quad (2)$$

The hybrid ring is not symmetric on its ports; choosing a different port as the input does not necessarily produce the same results. If one of the ports is omitted to form a three-port network, as is shown in Fig.4, the scattering matrix of three-port hybrid-ring coupler is given by:

$$[S] = -\frac{j}{\sqrt{2}} \begin{pmatrix} 0 & 1 & -1 \\ 1 & 0 & 0 \\ -1 & 0 & 0 \end{pmatrix} \quad (3)$$

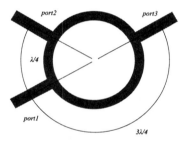

Fig. 4 The diagram of three-port hybrid-ring coupler

The impedance of the slot line ring is given by

$$Z_s = \sqrt{2}Z_{CPW} \quad (4)$$

where $Z_{CPW}$ is the impedance of the CPW feed, the radius of the slot line ring is designed by

$$2\pi r = \frac{3}{2}\lambda \quad (5)$$

where $\lambda$ is the guide wavelength of the slot line ring.

The hybrid-ring coupler shown in Fig.4 provides a hybrid ring coupler with a slot line ring by CPW feed to extend the bandwidth with structure and simple design procedure. The distances between two arms of the CPW feed are $3\lambda/4$ and $\lambda/4$, where $\lambda$ is the guide wavelength [23]~[25].

III. SIMULATED AND EXPERIMENTAL RESULTS

The front and back of the proposed antenna to be discussed is shown in Fig.5 and Fig.6, of which has the similar size with the coin.

Fig.5 the front of the proposed antenna

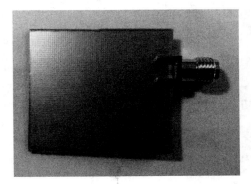

Fig.6 the back of the proposed antenna

The analysis of the body antenna is completed by using the Ansoft HFSS software. The simulated and measured return loss from 16GHz to 18GHz of the antenna is shown in Fig.7 and Fig.8. It can be seen that the antenna has an impedance bandwidth of 460MHz.

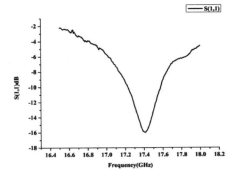

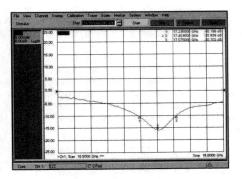

Fig. 8 the measured return loss of the proposed antenna

From the simulated and measured return loss shown in Fig.7 and Fig.8 respectively, the measured return loss have only 340MHz bandwidth. The central frequency has moved to 17.4GHz.The simulated and measured patterns of E-Plane and H-Plane at central frequency of 17 GHz are shown in Fig. 9 and Fig. 10 respectively.

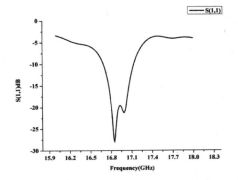

Fig.7 the simulated return loss of the proposed antenna

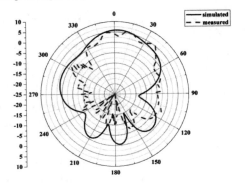

Fig. 9 E-plane radiation pattern at 17GHz

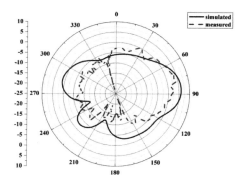

Fig.10 H-plane radiation pattern at 17GHz

The measured E-Plane and H-Plane patterns at 17 GHz are shown in Fig.9 and Fig.10 respectively. From the figure of radiation pattern above, the radiation of the front of the antenna have gain of 5dB, much higher than the back, which protect human body from radiation. The measured simulated results are not as good as simulated with several reasons of the influence of unideal test environment. From the E-plane and H-plane radiation pattern, lower radiation to human has been presented.

## IV. CONCLUSION

We proposed a bow-tie shaped meander slot on-body Antenna with 460 MHz bandwidth and central frequency of 17GHz. The proposed antenna takes advantage of low radiation to human by retaining metallic foil and small size applied on the integrated chip. As the limits of test measurement environment, the experimental results are not as good as simulated. All these deficiencies need to be considered seriously and improved in further research.

### ACKNOWLEDGEMENT

This work was supported in part by National 973 project 2010CB327400 and in part by the National Science and Technology Major Project of China under Grant 2010ZX03007-001-01.

## REFERENCES

[1] Patel, M.;Jianfeng Wang; Applications, challenges, and prospective in emerging body area networking technologies, IEEE wireless commun., vol.17, No.1, pp. 80-88, 2010.

[2] T.Salim; P.S.Hall Efficiency Measurement of Antennas for On-Body Communications, Microwave and Optical Technology Letters, 48, no.11, pp. 2256-2259, 2006.

[3] M.R.Kamarudin,Y.I. Nechayev, P. S. Hall. Performance of Antennas in the On-Body Environment, IEEE Antennas and Propagation Society International Symposium, 3A, pp. 475-478, 2005.

[4] Tuovinen Tommi,Kumpuniemi Timo,Yazdandoost Kamya Yekeh . et al, Effect of the antenna-human body distance on the antenna matching in UWB WBAN applications, Medical Information and Communication Technology (ISMICT), 2013 7th International Symposium,pp.193-197, 2013.

[5] Z.H.Hu;Y.I.Nechayev ;P.S.Hall .et al. Measure- ments and Statistical Analysis of On-Body Channel Fading at 2.45GHz, IEEE Antennas and Wireless Propagation Letters, vol.6, pp.612-615, 2009.

[6] Tommi Tuovinen; Kamya Yekeh Yazdandoost; Jari Iinatti, Ultra Wideband Loop Antenna for On-Body Communication in Wireless Body Area Network, Antennas and Propagation (EUCAP), 2012 6th European Conference, pp.1349-1352, 26-30 March 2012.

[7] Hu, Z.H. ; Gallo, M. ; Bai, Q. et al, Measurements and Simulations for On-Body Antenna Design and Propagation Studies , pp.1-7,2007.

[8] G. A. Conway and W.G. Scanlon, Antennas for Over-Body-Surface Communication at 2.45 GHz, in IEEE Transactions on Antennas and Propagation, vol. 57, no. 4, pp. 844-855, April 2009.

[9] Cheng, S. ; Hallbjorner, P. ; Rydberg, A. .et al, T-matched dipole antenna integrated in electrically small body-worn wireless sensor node. Microwaves, Antennas & Propagation IET, vol.3, no.5, pp.774-781, August 2009.

[10] Lilja, Juha ; Pynttari, Vesa ; Kaija, Tero, Body-Worn Antennas Making a Splash: Lifejacket-Integrated Antennas for Global Search and Rescue Satellite System , Antennas and Propagation Magazine, vol. 55, pp.324-341, April 2013.

[11] Rydberg, A.; van Engen, P. ; Shi Cheng .et al, Body surface backed flexible antennas and 3D Si-level integrated wireless sensor nodes for 17 GHz wireless body area networks, Antennas and Propagation for Body-Centric Wireless Communications, 2009 2nd IET Seminar , pp.1-4, April 2009

[12] Hasan, M.F. ; Islam, M.A. ; Muttalib, A.Z.M.S., A miniaturization of the quasi-self-complementary antenna with a wearable leather substrate and the use of specific ultra-wide-band frequency range for on-body communications , Informatics, Electronics & Vision (ICIEV), 2012 International Conference, pp.28-33, 2012

[13] Yazhou Dong, Guang Hua, Wei Hong, Uniplanar Bow-tie Shaped Meander Slot Antenna Fed by CPW Antennas, Propagation and EM Theory, 2008. ISAPE 2008. 8th International Symposium, pp 70-73,2-5 Nov. 2008.

[14] J.F Huang and C.W. Kuo, CPW-fed bow-tie slot antenna Microwave Opt. Tech. Lett. vol.19, no.5, pp.358-360, Dec. 1998.

[15] S. H. Wi, J. M. Kim, T. H. Yoo, and H. K. Park, Bow-tie-shaped meander slot antenna for 5 GHz application, Antennas and Propagation Society International Symposium, 2002. IEEE, vol.2, pp 456-459, 2002

[16] Hyungrak Kim ; Kwang Sun Hwang ; Ki Hun Chang .et al, Compact meander slot microstrip antenna with suppressed harmonics , Antennas and Propagation Society International Symposium, 2003. IEEE, vol.2, pp.577-580, 2003.

[17] X. Ding and A.F Jacob, CPW-fed slot antenna with wide radiating apertures, IEE Proc. Microwaves, Antennas and Propagat., vol.145,no. l, pp.104-108, Feb. 1998.

[18] H.D. Chen, Broadband CPW-fed square slot antennas with a widened tuning stub, IEEE Trans. Antennas Propagat., vol.51, no.8, pp.1982-1986, Aug. 2003.

[19] K. F. Tong, K. Li, T. Matsui and M. Izutsu, Wideband Coplanar Waveguide Fed Coplanar Patch Antenna, IEEE MU-S Int. Microwave Symp. Dig. pp. 406-409, 2001.

[20] H. Y. Wang, I. Simkin, C. Emson, and M. J. Lancester, Compact meander slot antennas, Microwave and Optrcal Tech Letters vol. 24 , pp.377-380, 2000

[21] Jui-Han Lu and Kin Lu Wong, Slot-loaded meandered rectangular microstrip antenna with compact dual frequency operation, Electronics Letters vol. 34, pp.1048-1050, 1998

[22] I. J. Bahl and P. Bhartia, Microstrip Antennas. MA, Artech Housse, 1998

[23] Kizilbey, O. ; Palamutcuogullari, A novel 180° hybrid ring coupler proposal and realization , Signal Processing and Communications Applications Conference (SIU), 2012 20th ,pp 1-4, 2012.

[24] Sung-Chan Kim ; Baek-Seok Ko ; Tae-Jon Baek .et al , Hybrid ring coupler for W-band MMIC applications using MEMS technology , Microwave and Wireless Components Letters, IEEE, Oct. 2005,vol.15, no.10,pp.652-654,2005.

[25] Ho, C.-H. ; Fan, L. ; Chang, K. ,A broad-band uniplanar slotline hybrid ring coupler with over one octave bandwidth , Microwave Symposium Digest, 1993., IEEE MTT-S International, vol.2,pp.585-588,1993.

# Evaluation Koch Fractal Textile Antenna using Different Iteration toward Human Body

M. E. Jalil, M. K. A. Rahim, N. A. Samsuri, N A Murad , B Bala

Radio Communication Engineering Department, Faculty of Electrical Engineering , University Technology Malaysia, Skudai 81310, Johor , Malaysia

ezwanjalil@gmail.com, mkamal@fke.utm.my, asmawati@fke.utm.my, asniza@fke.utm.my, bashirbala@yahoo.com

*Abstract-* **A Koch Fractal textile antenna are designed using denim material operating at 0.915 GHz. From the zero through second iteration, all the antennas performance in terms of return loss, bandwidth, realized gain and efficiency have been compared and analyzed. Then, on-body simulations was conducted using Voxel Human Body model in using CST Microwave software. At the same time, maximum Specific Absorption Rate (SAR) toward the backside body are discussed using two different orientations of the antenna; vertical and horizontal.**

## I. INTRODUCTION

Nowadays, the wearable system has been developed for monitoring, tracking and healthcare activity. The wearable system is capable of taking real-time information data including human body condition and user location .Therefore, textile antenna is introduced to give comfort to the user at the same time still perform well on the wearable activity [1]. Flannel, denim, felt and foam have been chosen as main element for producing textile antenna. Most fabrics are low profile with low permittivity, flexible and durable material. Although the fabric material has inconsistent surface layer and high loss tangent, the antennas have good performance similar to the rigid board such as FR4 and Duroid [2 and 3].

Small and compact size antennas are required for the wearable system. Implementation of the fractal geometry into the wearable antenna design can help to achieve a miniaturized design. However, producing small antenna with this technique, may limit the antenna performance. Then, incrementing Q factor will produce narrow bandwidth of antenna due to its addition iteration reduction in the size antenna [4]. Therefore, Koch curve geometry has property of approximately of filling a plane[5] which are introduced to overcome the limitations of the antenna performance unlike other Euclidean geometry . Besides, increasing iteration of fractal geometry likes Koch curve, triangular and rectangular meandered line will shift the resonant frequency to the lower frequency band. Hence the lower radiation resistance also decrease [6]. From previous research work, the patch-slotted, array log period antenna was presented . Then printed and wired monopole antenna have been designed using Koch curve geometry [7].

The effect of placing antenna on human body is another issue which is discussed. ICNRIP basic retraction stated that the maximum power of electric devices and component must have less than 2 W/kg in 10 g of tissue for Specific Absorption Rate (SAR) value toward the human body [8]. Furthermore, when the antennas are placed closed with the human body, the coupling affect antenna performance in terms of resonant frequency and efficiency.

In this paper, Koch fractal textile dipole antennas are designed for $0^{th}$, $1^{st}$ and $2^{nd}$ iteration at 0.915 GHz using CST Microwave Studio as shown in Figure 1. Planar and simple types of dipole antenna will provide omni-directional radiation patterns which are suitable for the wearable application. Mostly, the wearable computing system will communicates with the other system using Bluetooth, ZigBee and LAN network operating at 0.915 GHz band. Then, comprehensive analysis has been conducted to compare antenna performance and SAR value on human body between antenna with different iteration.

## II. ANTENNA DESIGN

Firstly, the textile antenna is produced using denim material as the substrate with thickness of 0.8 mm. For the conducting layer, the conventional copper tape has been used. The estimated conductivity of this copper tape is $5.88 \times 10^7$ S/m. The relative permittivity $\varepsilon_r$ and loss tangent $\tan \delta$ of denim material are discovered using the open ended coaxial probe that show the value of $\varepsilon_r = 1.71$ and $\tan \delta = 0.085$ at 0.9 GHz. The Koch fractal textile antennas are designed at 0.915 GHz using CST Microwave Studio using the estimated and measured properties of the materials.

Simple and planar dipole antenna structure are introduced that will be integrated with Koch fractal structure to miniaturize the length of antenna. All antennas are designed with flare angle of 45 degree. The theoretical formula of total length of the antenna for dipole structure are shown in equation (1)

$$l_0 = \frac{3 \times 10^8}{2f \sqrt{\varepsilon_{eff}}} \quad (1)$$

978-7-5641-4279-7

where; $l_0$ is theoretical length of antenna, $f$ is the determined resonant frequency and $\varepsilon_{eff}$ is the effective permittivity.

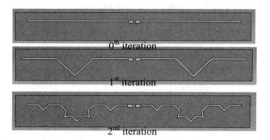

Figure 1 : Antenna design using the first, second and third iteration

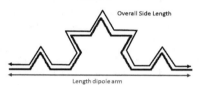

Figure 2: Overall Side Length and Length Dipole Arm

For the first case, three Koch fractal dipole antennas are designed with different iteration; zero, first and second with fixed length dipole arm = 132.5 mm. However, the overall side length increased related the value of iteration. Based on the formula, the overall side length of dipole antenna is calculated as shown in Equation 2.

$$l_{total} = l_0 \left( 4/3 \right)^n$$

(2)

where $l_{total}$ is total overall side length, $l_o$ is the length of antenna and n is the number of iteration . The optimization of the parameter is done using the simulation software. From the optimization, the width of 1 mm and the gap between antennas of 1.7 mm are obtained. The optimized overall side length and performance each antenna is summarized in Table 1. The results of the simulated return loss are shown in Figure 3. Figure 3 show the resonant frequency are shift to the low frequency due to the increasing iteration

TABLE 1: Comparison antenna performance based on different iteration antenna

| Antenna Iteration | Overall Side Length ($l_{total}$) | Resonant frequency ($f_c$) | Bandwidth (MHz) |
|---|---|---|---|
| 0 th | 132.5 mm | 1.029 GHz | 43.9 |
| 1 st | 176.1 mm | 0.959 GHz | 69.9 |
| 2 nd | 234.2 mm | 0.915 GHz | 72.5 |

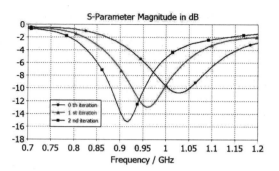

Figure 3: Simulated return loss with fixed length antenna parameter.

In the second case, all antennas are optimized at 0.915 GHz. The total length of dipole arm is 147.6mm for the zero iteration. Therefore, the length of the dipole arm will reduced to 139.08, and 132.48 mm for the first iteration and second iteration respectively as shown in table 2. The return loss for the second case is shown in Figure 4.

TABLE 2: Optimized Total length dipole arm at 0.915 GHz

| Antenna Iteration | Length dipole arm ($l_0$) | Reduction Compared with 0th iteration (%) |
|---|---|---|
| 0 | 149.28 | - |
| 1 | 139.08 | 6.8 |
| 2 | 132.48 | 11.3 |

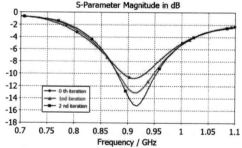

Figure 4: Simulated return loss of optimized at 0.915 GH

### III   RESULT AND DISCUSSION

*A.   Simulation of Radiation Pattern*

Radiation patterns of the antenna (E-plane and H-plane) for the optimized results are shown in Figure 5. An omni-directional and dipole-like patterns are obtained for both Hand E-plane respectively. The size of radiation pattern increased when the number of iteration of Koch fractal antenna increases. At the same time, addition iteration value on the Koch fractal antenna will increase the efficiency and the realized gain as shown in Table 3.

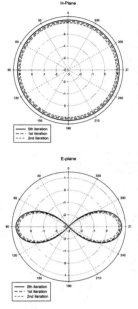

Figure 5: E-plane and H-Plane Radiation Pattern Antenna using Different Iteration

Table 3 : Realized Gain and Efficiency

| Antenna Iteration | Realized Gain (dB) | Efficiency (%) |
|---|---|---|
| 0 | 0.954 | 77.6 |
| 1 | 1.087 | 80.2 |
| 2 | 1.202 | 83.15 |

### B. On-body simulation

From this research, the backside, chest and arm part of the body is chosen for the placement of the antenna compared to other locations. The antenna are placed 5 mm from the centre backside of Gustav man model which has the height of 163 cm and weight of 63 kg as shown in Figure 6. The backside body location is far away from the high permittivity organs such as the heart ($\varepsilon_r = 54.8$), the muscle ($\varepsilon_r = 50.8$) and the kidney ($\varepsilon_r = 52.7$) that will affect the antenna performance.

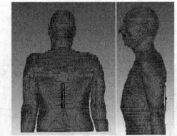

Figure 6: Placement Antenna at Backside Body

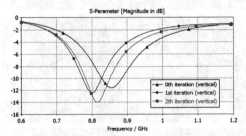

Figure 7: Return loss antenna performance using different iteration with vertical position.

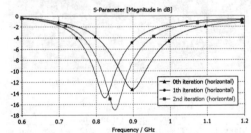

Figure 8: Return loss antenna performance using the different iteration with horizontal position.

Figure 7 and 8 show the simulated return loss for placement antenna with vertical and horizintal position at the backside body. The addition iteration for Koch fractal antenna will detune the resonant frequency to the low frequency. However, the Koch fractal antenna with first iteration more less affected on resonant frequency compared with the first iteration. Placement antenna at the backside body with vertical positon is more shifting resonant frequenc than the horizontal position

### C. Maximum SAR Value

In the section, the simulation result of the SAR value is discussed toward the backside body when surrounding the textile antenna. The value SAR is highest when the high conductivity tissue organ such as heart, muscle and kidney surrounding the antenna. So, the backside body is a suitable placement for the antenna. From Figure 9, the highest SAR value is achieved when the antenna are placed with horizontal orientation for the second iteration with only 2.05 W/kg.

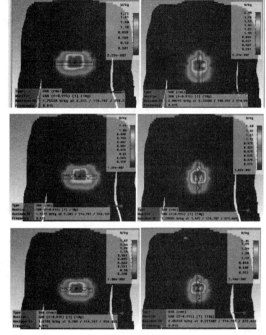

Figure 9: Maximum SAR Value on the human body

TABLE 4 : Comparison maximum SAR value based on different iteration antenna

| Maximum SAR | Horizontal | Horizontal |
|---|---|---|
| 0th iteration | 1.75 | 2.00 |
| 1st iteration | 1.13 | 1.60 |
| 2nd iteration | 1.62 | 2.05 |

## III. CONCLUSION

In this work, the Koch fractal textile antennas are designed and simulated using CST software. The antenna has been compared and analyzed with different iteration. Placement antenna at the backside body is under of basic restriction limit of SAR value.

### ACKNOWLEDGMENT

The authors thank the Ministry of Higher Education (MOHE) for supporting the research work, Research Management Centre (RMC), School of Postgraduate (SPS) and Communication Engineering Department Universiti Teknologi Malaysia (UTMJB) for the support of the research under grant no Q.J130000.2523.04H38.

### REFERENCES

[1]  M. A. R. Osman, M. K. A. Rahim, N. A. Samsuri, H. A. M. Salim, and M. F. Ali, "Embroidered fully textile wearable antenna for medical monitoring applications," Progress In Electromagnetics Research, Vol. 117, 321-337, 2011.

[2]  Jalil, M.E.; Rahim, M. K A; Abdullah, M.A.; Ayop, O., "Compact CPW-fed Ultra-wideband (UWB) antenna using denim textile material," Antennas and Propagation (ISAP), 2012 International Symposium on , vol., no., pp.30,33, Oct. 29 2012-Nov. 2 2012

[3]  Salvado, R.; Loss, C.; Gonçalves, R.; Pinho, P. Textile Materials for the Design of Wearable Antennas: A Survey. Sensors 2012, 12, 15841-15857.

[4]  Baliarda, C.P.; Romeu, J.; Cardama, A., "The Koch monopole: a small fractal antenna," Antennas and Propagation, IEEE Transactions on , vol.48, no.11, pp.1773,1781, Nov 2000

[5]  Mondal, A.; Chakraborty, S.; Singh, R.K.; Ghatak, R., "Miniaturized and dual band hybrid Koch fractal dipole antenna design," Computer, Communication and Electrical Technology (ICCCET), 2011 International Conference on , vol., no., pp.193,197, 18-19 March 2011

[6]  Best, S.R., "On the resonant properties of the Koch fractal and other wire monopole antennas," Antennas and Wireless Propagation Letters, IEEE , vol.1, no.1, pp.74,76, 2002

[7]  M. N. A Karim, M. K. A. Rahim, H. A. Majid, O. B. Ayop, M. Abu, and F. Zubir, "Log periodic fractal koch antenna for UHF band applications," Progress In Electromagnetics Research, Vol. 100, 201-218, 2010.

[7]  A. Ramadan, K. Y. Kabalan, A. El-Hajj, S. Khoury, and M. Al-Husseini, "A reconfigurable u-koch microstrip antenna for wireless applications," Progress In Electromagnetics Research, Vol. 93, 355-367, 2009.

[8]  Kurup, D.; Joseph, W.; Vermeeren, G.; Martens, L., "Specific absorption rate and path loss in specific body location in heterogeneous human model," Microwaves, Antennas & Propagation, IET , vol.7, no.1, pp.,, January 11 2013

# Compact UWB Antenna with Controllable Band Notches Based On Co-directional CSRR

Tong Li, Huiqing Zhai, Guihong Li and Changhong Liang
School of Electronic Engineering, Xidian University
Xi'an, 710071, China

*Abstract-* In this paper, compact ultra-wideband (UWB) antenna with controllable notched bands is proposed for UWB communication applications. The antenna utilizes co-directional complementary split ring resonator (CSRR) to achieve dual suitable notched bands in small enough size. The center frequencies of the notched bands can be electronically tuned by changing the effective electrical length of the CSRR, which is achieved by employing varactor diodes. The proposed antenna can also achieve reconfigurable notched bands by replacing the varactor diodes with two electronic switches. The total antenna dimensions are only $25 \times 33$ mm$^2$ and shows good environmental adaptation, which makes it a good candidate for UWB applications.

## I. INTRODUCTION

In recent years, ultra-wideband (UWB) technology has become a very promising wireless technology because of the attractive benefits it provides, such as low power consumption, high data transmission rates, resistant to severe multipath and jamming, etc.[1]. The high demands on such communication systems have stimulated research into UWB antenna designs. A UWB antenna should be capable of operating over a frequency band from 3.1 to 10.6 GHz, the commercial usage for UWB radio system approved by the Federal Communications Commission (FCC), and exhibit stable radiation patterns in the entire bandwidth. In addition, it needs to have a compact size and low manufacturing cost for consumer electronics applications. Most importantly, the antenna should overcome electromagnetic interference (EMI) problems. The EMI problems are quite serious since there exist several other wireless narrowband standards within UWB bandwidth, such as IEEE 802.16 world interoperability for microwave access (WiMAX) system from 3.3 to 3.6 GHz, and IEEE 802.11a wireless local area network (WLAN) in the frequency band of 5.15-5.825 GHz.

Lately, lots of antennas with band-notched characteristic have been discussed. The commonly used methods are embedding slots on the patch or on the ground plane and adding parasitic elements, as reported in [2]-[5]. However, these antennas still have some shortcomings in practical applications. The conventional method to implement multi-notched function is loading notch elements of different types, different numbers at different spaces, thus it requires too much space and results in complicated design. Moreover, once the antenna is fabricated, the notched bands are also fixed.

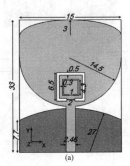

(a)

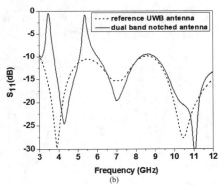

(b)

Figure 1. Dual band-notched UWB antenna. (a) Geometry with dimensions. (b) Simulated return loss.

However, the existing undesired narrowband radio signals vary from place to place and from time to time. Therefore, multiple band notches with frequency controllable capabilities in a compact antenna size are highly required.

In this paper, a compact UWB antenna with controllable notch band capability is presented. The antenna utilizes co-directional complementary split ring resonator (CSRR) to achieve dual suitable notched bands in small enough size. To achieve band notch controllability, varactor diodes and electronic switches are mounted across the slots, thus helping in the adaptation against environmental changes and increasing the antenna performance.

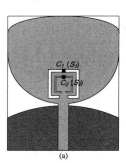

(a)

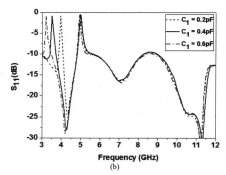

(b)

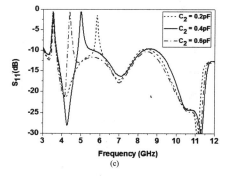

(c)

Figure 2. (a) Geometry of the UWB antenna with controllable band notches. (b) Effect of varactor diodes capacitance C1 and (c) C2 on return loss.

TABLE I
DIFFERENT SWITCHING CASES AND CORRESPONDING NOTCHED BANDS

| Case | Notched bands (GHz) | S1 | S2 |
|------|---------------------|----|----|
| 1 | None (UWB operation) | ON | ON |
| 2 | 5.26 | ON | OFF |
| 3 | 3.4 | OFF | ON |
| 4 | 3.5 , 5.26 | OFF | OFF |

miniaturization and ensures a good impedance match over the entire UWB frequency range.

To generate band notches, co-directional complementary split ring resonator (CSRR) are etched on the patch. Originally proposed by Pendry [6], SRR is an electrically small resonator with a very high quality factor. When applying a time varying external magnetic field along the axis of the ring, SRR exhibits a strong magnetic response and restrain signal propagation in a narrow band in the vicinity of the resonant frequency. Analogously, the corresponding CSRR structure has the similar characteristic. Compared to traditional CSRR, co-directional CSRR can exhibit dual distinct fundamental magnetic resonance frequencies for each ring due to the weaker mutual coupling between inner and outer rings, thus it is utilized here to achieve dual band-notches. The co-directional CSRR is arranged close to feedline with the gap opposite to y-axis. The sizes are optimized so that the outer split-ring generates a notch at 3.5 GHz and the inner one at 5.2 GHz. Fig. 1(b) exhibits the simulated return loss of the proposed dual band notched antenna, while the return loss of the original UWB antenna without any slots is also shown as a reference. As can be seen, the antenna provides a wide impendence bandwidth of 3 - 12 GHz with dual effective notched bands of 3.2-3.7 GHz and 5.1-5.93 GHz respectively, covering WiMax and WLAN successfully.

### B. Controllable Band Notches Design

Since the interference signals vary with environmental changes, it is more desirable to introduce tunable notched bands within operating band of the antenna. As shown in Fig. 2(a), two varactor diodes C1 and C2 are mounted across the co-directional CSRR to achieve this function. Each CSRR can be considered as a LC resonator and its effective capacitance varies by changing the capacitance of the varactor diode, thus producing tunable notched bands. A parametric analysis for the effect of tuning varactor diodes on the resonance frequencies is shown in Fig. 2 (b)(c), where several specific capacitance values are chosen to simulate the varactor diodes' changing process. As the capacitance of C1 increases from 0.2 pF to 0.6 pF, the central frequency of the lower notch varies from 3.9 to 3.2 GHz and almost has little effect on the central frequency of the higher notch. A same situation happens when we tune the capacitance of C2. By varying it from 0.2 pF to 0.6 pF, the higher notch changes from 5.8 GHz to 4.4 GHz. From Fig. 2, it is observed that the increase of capacitance results in the decrease of notch frequency and each notched band can be controlled independently.

## II. ANTENNA CONFIGURATION AND DESIGN

### A. Compact UWB Antenna with Dual Band Notches

Fig. 1(a) shows the geometry and dimensions of the proposed compact dual band-notched UWB antenna. The antenna is designed on a 1.5 mm-thick substrate with permittivity constant $\varepsilon_r = 2.65$, loss tangent of 0.0015, and the overall size is $25 \times 33$ mm$^2$. Both the radiating monopole and the ground plane have rounded edges to broaden the bandwidth and to produce smooth transitions from one resonant mode to another. This characteristic helps to achieve

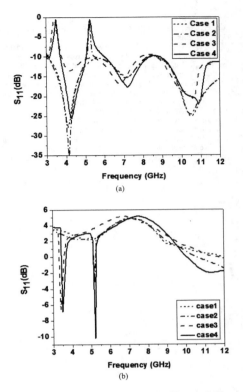

(a)

(b)

Figure 3. Simulated (a) return loss and (b) gains for different switching cases.

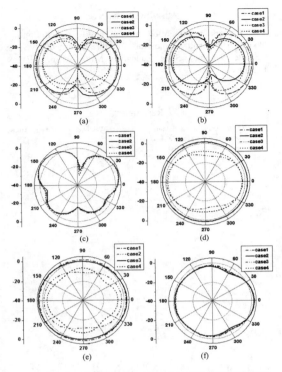

Figure 4. Simulated radiation patterns (a) E plane at 3.5 GHz (b) E plane at 5.2 GHz (c) E plane at 8 GHz, (d) H plane at 3.5 GHz (e) H plane at 5.2 GHz (f) H plane at 8 GHz.

For the uncertainty of interference, sometimes there might exist only one or no interference signal, thus it is necessary for a UWB antenna to have reconfigurable notched bands. To achieve reconfigurability, the varactor diodes in Fig. 2(a) were replaced by two electronic switches S1 and S2, the state of which determines whether or not a notched band is induced. When the switch is OFF, the corresponding slot behaves as a single-ring CSRR, causing a notch in its resonant band. While closing the switch (ON) shorts the slot and effectively eliminates the resonance, consequently the notched band disappears. When S1 is OFF, the outer CSRR resonates, thus inducing a notch in the 3.5 GHz band. This notch disappears when S1 is ON. For the inner CSRR, a notch in the 5.2 GHz band appears when S2 is OFF, and disappears when it is ON. The different switching cases lead to different band notch combinations which include one, two band notches, or no notch at all. Different switching cases listed in Table I are simulated, and the return loss plots of these cases are illustrated in Fig. 3(a). For case 1, both switches are ON, which is needed when there exists no interference. In this case, none of the slots resonates, resulting in a UWB response with no notches in the operating band of the antenna. In case 2 and 3, only one switch is ON, producing a single notched band at

5.26 GHz and 3.4 GHz respectively. With both switches OFF in case 4, the co-directional CSRR resonates normally and dual band notches are obtained.

Fig. 4 shows the simulated radiation patterns for the reconfigurable band-notched antenna in different cases at 3.5 GHz, 5.2 GHz and 9 GHz, respectively. It is seen that the state of the switches does not affect the radiation performance of the antennas in working frequencies [Fig. 4(c) and (f)]. However, within the notched band in each case, the radiated intensity degrades. This degradation in the band-notch range is also verified from the gain simulation shown in Fig. 3(b).

### III. CONCLUSION

In this paper, a compact UWB antenna with controllable band notches is proposed. Simulations show that the antenna can achieve dual independently band notches which are induced by co-directional CSRRs, and controlled by varactor diodes and electronic switches mounted across the CSRRs. The antenna shows good environmental adaptation and can be a good candidate for practical applications. Fabrication and further measurements are ongoing.

ACKNOWLEDGMENT

This work is supported by the NSFC under Contract No.61101066 and Foundation for the Returned Overseas Chinese Scholars, State Education Ministry and Shaanxi Province.

REFERENCES

[1] I. Oppermann, M. Hamalainen, J. Iinatti, UWB Theory and Applications, Wiley, New York, pp. 3–4, 2004.

[2] J. Kim, C. S. Cho, and J. W. Lee, "5.2 GHz notched ultra-wideband antenna using slot-type SRR", *Electron. Lett.*, vol. 42, No. 6, pp. 315–316, Mar. 2006.

[3] Y. Zhang, W. Hong, C. Yu, Z. Kuai, Y. Don, J. Zhou, "Planar ultra wideband antennas with multiple notched bands based on etched slots on the patch and/or split ring resonators on the feed line", *IEEE Trans. Antennas Propag.*, vol. 56, No. 9, pp. 3063–3068, Sep. 2008

[4] Q. X. Chu and Y. Y. Yang, "A compact ultra-wide band antenna with 3.4/5.5 GHz dual band-notched characteristics", *IEEE Trans. Antennas Propag.*, vol. 56, No. 12, pp. 3637–3644, Dec. 2008.

[5] K. S. Ryu, A. A. Kishk, "UWB antenna with single or dual band-notches for lower WLAN band and upper WLAN band", *IEEE Trans. Antennas Propag.*, vol. 57, No. 12, pp. 3942–3950, Dec. 2009.

[6] J. B. Pendry, A. J. Holden, D. J. obbins, W. J. Stewart, "Magnetism from conductors and enhanced nonlinear phenomena", *IEEE Trans. Microw. Theory Tech.*, vol. 47, No. 11, pp. 2075 –2084, Nov. 1999.

# ULTRA-WIDEBAND DUAL POLARIZED PROBE FOR MEASUREMENT APPLICATION

Yong Li, Meng Su, Yu-zhou Sheng, and Liang Dong

National Key Laboratory of Antenna and Microwave Technology，Xidian University

Xi'an, 710071, China

*Abstract*-A two-element ultra-wideband dual polarized antenna with resistance loaded is proposed in this paper. Vivaldi antenna is selected as the element and resistive loading in its slots leads to a compact structure. A bandwidth ranging from 0.9GHz to 10.5GHz(11.67:1) for VSWR less than 2.5 is achieved, and isolation between two ports is better than 20dB.A prototype is fabricated, and the measured results are in agreement with the simulated theory by Ansoft HFSS 13. It indicates that this antenna can be used as the probe in anechoic chamber.

Key Words: ultra-wideband; Vivaldi antenna; dual polarized; probe

## I. INTRODUCTION

Recent years have seen a rapid growth in the development of ultra-wideband(UWB) technology, primarily driven by the increasing requirements of modern radar, electronic warfare, wireless communication, and radio telescope systems that require very wide bandwidth. Ultra-wideband antennas and arrays specially act the key roles in ultra-wideband technology [1]. And Vivaldi antenna is one of the most familiar ultra- wideband antennas.

Vivaldi antenna, which was first introduced by P. J. Gibson, is one of the endfire antennas fed by microstrip line. With its continuous, non-periodic and gradual changing structure, Vivaldi antenna can theoretically radiate travelling wave in a quite wide band. The electric field vector of the Vivaldi antenna is parallel to the substrate, and it is linear polarized on its two major radiation planes [2]. As it has the advantages of good symmetry property, simple structure and high gain, and it is very easy to be integrated and manufactured, Vivaldi antenna is always widely used in many fields, such as ultra-wideband communication, radar systems, and antenna measuring system [3] [4].

In the antenna measuring process, sometimes, the performance of dual polarization or cross polarization of the antenna under test (AUT) is required. Then, using a dual polarized probe can be quite effective without turning the AUT. In order to further improve the performance of this kind of antenna, especially in impedance bandwidth and miniaturization for low frequency, previous researchers have done a lot of effort, and some special shapes for the radiation arms are designed, for example, corrugated ripples [5], bunny ear-shaped combline [6], composite function tapered slots [7]

and so on. In this paper, a cruciform dual polarized probe which is composed of two Vivaldi elements is presented, and the application of resistance loading leads to a compact structure conspicuously.

## II. CONFIGURATION AND DESIGN STRATEGY OF LOADED VIVALDI

As shown in Figure 1, a single-element Vivaldi antenna, which has the property of linear polarization, is primarily presented. The metallic radiator is shaped by two symmetrical exponential curves. Expressions of the curves are showed in Equation (1), as follows,

$$y = \pm(c_1 e^{Rx} + c_2), \qquad (1)$$

$$c_1 = \frac{y_2 - y_1}{e^{Rx_2} - e^{Rx_1}}, \qquad c_2 = \frac{y_1 e^{Rx_2} - y_2 e^{Rx_1}}{e^{Rx_2} - e^{Rx_1}} \qquad (2)$$

Where, both $c_1$ and $c_2$ are constants that can be calculated by Equation (2); ($x_1, y_1$) and ($x_2, y_2$) are respectively the top and bottom points on the exponential tapered curve; and $R$ is the exponential factor that regulates the breadth of the tapered slot.

Approximately, the cut-off wavelength at low frequency of the Vivaldi antenna is about twice the maximum width of the exponential tapered curve, the radiation performances at high frequency of the Vivaldi antenna is restricted by the minimum width of the exponential tapered slot, and the minimum width of the curve is about 2% the cut-off wavelength at high frequency [8]. Actually, the final accurate dimension can be obtained only by optimization. According to the expected

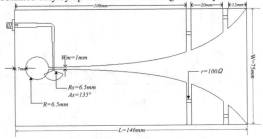

Figure 1. Geometry of the proposed antenna

978-7-5641-4279-7

operating frequency range, the radiation patch of the Vivaldi antenna is printed on a substrate of Teflon ($\varepsilon_r$=2.65) material, 75mm× 146mm, with the thickness of 1.0mm.

Then a classical microstrip-slot balun is made to feed to Vivaldi antenna. With the balun connected to the exponential tapered radiator, feeding is balanced and the Vivaldi antenna realizes the property of ultra wide-band. Optimization of the structure and geometric dimension of the balun is difficult but pivotal in the whole designing process.

The balun is made up of three parts. Firstly, a circular resonator which is used to achieve the impedance matching of the microstrip transmission line; secondly, the microstrip transmission which works as an impedance converter, is used to realize the matching from 50Ω-microstrip to the input impedance at the feeding point of the antenna. Generally, the theoretical impendence value of the middle segments can be calculated by looking up the Chebyshev impedance conversion table. And the third part is the fan-shaped structure which is used to realize the terminal load matching can be equal to grounding terminal.

Furthermore, two pairs of gaps are etched, and four 100Ω-chip resistors are loaded on the radiation arms symmetrically. It is clear that this structure can make the antenna achieve good impedance matching characteristics in the expected working frequency, especially at the low frequency. Figure 2(a) leads us to a conclusion that the proposed antenna has good VSWR less than 2.5 in 0.88 to 11.28GHz (12.82:1). On the other hand, part of the electromagnetic energy radiated by the antenna is absorbed by load resistance, and naturally the antenna gain decreases, especially in the low frequency range. The plot of the dependence of gain upon frequency is shown in Figure 2(b). Taking into account the fact that high gain is not strictly required when used as a probe in antenna measurement system, this design is still feasible. In this way, geometric dimension of the antenna can be reduced significantly.

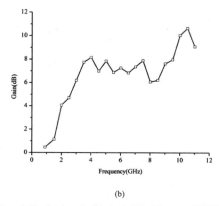

(b)

Figure 2. Simulated results of the loaded Vivaldi antenna. (a) VSWR, (b) Gain

## III. THE UWB DUAL POLARIZED PROBE

Based on the loading Vivaldi antenna expounded above, an ultra-wideband dual polarized probe is introduced in this paragraph. The probe is cross shaped by laying the two antenna elements orthogonally, and slots are designed particularly on two dielectric substrates, avoiding cutting off the feeder line. The fabricated prototype is shown in Figure 3.

Figure 4(a) reveals the VSWR graph of the simulation result by Ansoft HFSS 13 and the measurement result of the fabricated prototype. VSWR of port1 is a little higher than that of port2, and the measurement result conforms to the simulation result well. Deterioration of return loss is caused by the coupling effect between the two antenna ports. Though

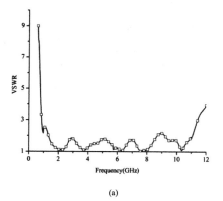

(a)

Figure 3. Photography of the fabricated proposed antenna.

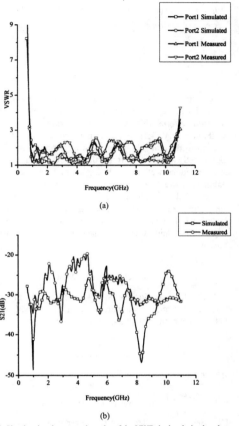

(a)

(b)

Figure 4. Simulated and measured results of the UWB dual-polarized probe.

(a) VSWRs, (b) $S_{21}$ parameters

VSWR becomes worse than the result of linear polarized antenna mentioned in Part I, especially at high frequency, it is acceptable in application in the bandwidth ranging from 0.9 to 10.5GHz for VSWR less than 2.5.

In Figure 4(b), it can be seen that the measured $S_{21}$ parameter is not as good as the simulation result, especially in low frequency band. Difference may be related to the precision in manufacture, and the error caused by uncertainty of instrument also should be taken into account. However, the measured $S_{21}$ parameter is better than 20dB in the whole band. Isolation described by the $S_{21}$ parameter, means the degree of coupling between the two ports for the dual polarized probe. High isolation reveals that weak coupling effect exists between the two independent antenna elements when they work respectively. It indicates that this design can be satisfied with the engineering requirement well.

The dual polarized probe has good performance in endfire. Far field radiation patterns are computed in the two principal planes, E-plane and H-plane. Figure 5 shows the gain patterns at some frequency points typically. It can be seen that gain in the principal radiation direction is 0.52dB at 1.0GHz, 4.19dB at 2.0 GHz, 6.78dB at 5.8GHz, 8.23dB at 7.0GHz, and 9.17dB at 9.8GHz. The E-Plane and H-plane radiation patterns of the two ports are symmetrical, they have wide beam performances and the beam width nearly decreases when the frequency increases.

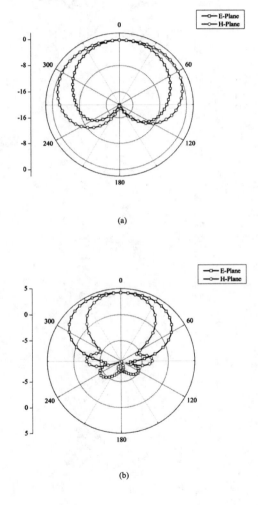

(a)

(b)

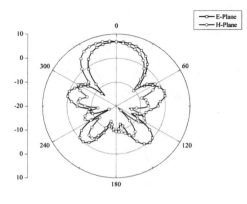

(c)

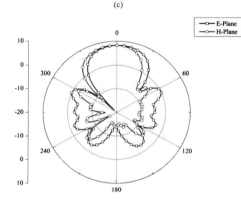

(d)

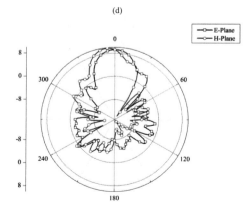

(e)

**Figure 5.** Radiation patterns in E-Plane and H-Plane of the dual-polarized probe at:(a) 1.0GHz, (b) 2.0GHz, (c) 5.8GHz, (d) 7.0GHz, and (e) 9.8GHz

(unit: dB)

## IV. CONCLUSION

An ultra-wideband dual polarized probe composed of two Vivaldi antenna elements is presented in this paper. Shown as the gain patterns, this antenna has a good property of endfire. Without a very rigorous requirement for gain, load resistance is used to decrease the dimension of the antenna effectively. Compared with the current designs in other papers, this structure does well in miniaturization. Moreover, a suitable balun is made to ensure that this antenna can work in the entire band of 0.9GHz to 10.5GHz. Conformity of the simulation results and measured results of VSWR and isolation indicates that this probe can meet the requirement of engineering, and it can be used in anechoic chamber for measurement.

## ACKNOWLEDGMENT

The author would like to acknowledge for his excellent teacher Li Yong; good partners and others who gave him helps in the course of study. Best wishes!

## REFERFENCE

[1]    T. Barrett, History of ultra wide band (UWB) Radar Communication: Pioneers and Innovators J. Progress in Electromagnetic Research Symposium Cambridge, 2000.

[2]    Gibson, P. J. "The Vivaldi aerial." Microwave Conference, 1979. 9th European. IEEE, 1979.

[3]    Nai-biao Wang, Design and Implementation of Ultra Wide Band Tapered Slot Antennas and Arrays, PHD Thesis, Xidian University, 2009

[4]    Jun-qian Niu, Shan-wei Lu, Juan Liu, and Ren-dong Nan, Vivaldi Antenna and its Use in Wideband Measurement System, Journal of Astronautic Metrology and Measurement, vol. 24, pp. 20-24, Jun., 2004

[5]    Han Xu, Juan Lei, Changjuan Cui and Lin Yang, UWB Dual-Polarized Vivaldi Antenna with High Gain, Microwave and Millimeter Wave Technology (ICMMT) International Conference, IEEE Conference Publications, vol.3, pp.1-4, 2012

[6]    Yong-wei Zhang, Antony K. Brown, Bunny Ear Combline Antenna for Compact Wide-Band Dual-Polarized Aperture Array, IEEE Transactions on Antenna and Propagation,vol.59, No.8,Aug., 2011

[7]    Yue Song, A Study of Multiband/Ultra-Wideband Printed Antenna and Tapered Slot Arrays, PHD Thesis, Xidian University, 2009

[8]    Li-zhong Song, Qing-yuan Fang, Design and Measurement of a Kind of Dual-Polarized Vivaldi Antenna, Cross Strait Quad-Regional Radio Science and Wireless Technology Conference (CSQRWC), vol.1, pp.494-497, 2011

# Conformal Monopulse Antenna Design Based on Microstrip Yagi Antenna

Chen Ding, Wenbin Dou
State Key Laboratory of Millimeter Waves, Southeast University,
Nanjing 210096, China
837932422@qq.com, njdouwb@163.com

*Abstract*-In this paper, a kind of microstrip Yagi antenna whose central frequency is 5.2GHz is designed, because of its planar and end-fire characteristics, it is sticked on a cylindrical supporter to form a conformal monopulse antenna; the performance of antenna is simulated in electromagnetic simulation software HFSS.

*Keywords*- cylinder; conformal antenna; monopulse

## I.    INTRODUCTION

Today, conformal antenna as a more and more popular direction of development in science and technology has a wide range of applications in the area of civil, military and aerospace. In the civilian area, it can be applied to communications [1] as well as the automotive sector [2]; it also has great development prospects in military area like Radar and Fighter. In literature [3], a kind of conformal microstrip antenna which can be affixed around the surface of cylinder is described, but its actual value is limited because it is an Omni-directional antenna. Monopulse antenna, also known as simultaneous multi-beam, is a kind of antenna which can produce several beams simultaneously; they are sum beam, azimuth difference beam and elevation difference beam. If we can combine monopulse antenna and conformal antenna together, may produce gorgeous results.

Until now, research on conformal monopulse antenna is still not enough, especially the end-fire and broadband conformal antennas. If the monopulse antenna can be made into conformal antenna and be conformal on the cylindrical body, it can be used for positioning and tracking. In this paper, we mainly focus on the researching of end-fire conformal antenna; our goal is the design of wideband conformal antenna, and the simulation of that.

## II.    DESIGN OF MICROSTRIP YAGI ANTENNA

Yagi antenna is invented by two Japanese Electrical Engineering Professor Shintaro Uda and Hidetsugu Yagi in the 1920 's, Yagi antenna is also known as To Antennas or Wave Channel Antenna. Yagi antenna consists of a feed and a few passive parasitic arrays which ranked side-by-side, it is an end-fire antenna widely used in meter-wave, decimeter wave band communication, radar, TV and other radio equipment. The development of traditional Yagi antenna is limited because of its large size. In the early of 1970 of the 20th century there was

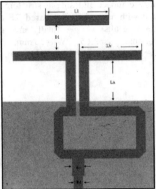

Figure 1.. Schematic of microstrip Yagi antenna

a microstrip Yagi antenna as Figure 1 shows. Compared to traditional Yagi antenna, it has lots of advantages such as small size, light weight, low profile, and flat structure. It has the advantages what traditional Yagi antenna has, and also can be made into the structure which can conformal on cylinder, satellites and other supporters.

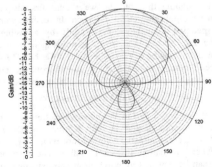

Figure 2..Radiation pattern of microstrip Yagi antenna when f=5.2GHz

Generally, the more directors antenna has, the higher gain antenna will have, but its bandwidth will decrease, in order to meet the requirements of wideband antenna, we use the structure of one director. Central frequency of antenna is 5.2GHz; $\varepsilon_r$ of the substrate material is 2.65. The antenna

978-7-5641-4279-7

consists of a printed dipole directors and a driver dipole fed by a broadband microstrip-to-coplanar strips (CPS) transition. The work band and matching performance of microstrip Yagi antenna is mainly depending on the driver dipole and reflector. The work frequency is depending on the length of the dipole director, and the S parameter s is depending on the gap of the coplanar stripline and La. The metallization on the bottom plane is a truncated microstrip ground, which serves as the reflector element for the antenna. The feature of this antenna design is the use of the truncated ground plane on the backside of the microstrip substrate as its reflecting element. This results in a very compact and simple structure that can be easily integrated with any microstrip based RF circuitry [4]. The optimization results we get finally are: thickness of substrate h=0.8mm, $\varepsilon_r$=2.65, Wd=2.2mm, Wc=2.9mm, La=8.9mm, Lb=14.1mm, L1=14.4mm, D1=6mm.

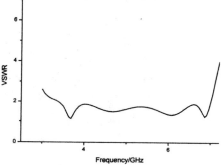

Figure 3.. VSWR of Yagi antenna

From Figure 3 we can get that: the central frequency of antenna is 5.2GHz, work band is 3.35GHz-7GHz, relative bandwidth is 70%, almost meet our design expectation.

When designing the antenna, we should first deign the balun, driver dipole and reflector to make the antenna working within specified frequency band, mainly by adjusting the parameter La and Lb. It is worth mentioning here that we should make the length of two balun arms differ by $\lambda/2$ so that it can feed the driver dipole. Then we can improve the gain of antenna as high as possible by optimizing other parameters.

III. CONFORMAL ANTENNA ARRAYS

The antenna we design in the precious section is a planar antenna; its section height is only 0.8mm, so it's easier attached on the cylindrical vectors. From paper [5] we can know, when simulating the conformal antenna, supporter cylinder can be replaced by polyhedron approximately. Here is schematic diagram of the antenna array. In Figure 4, there are totally 12 antenna units, every three antenna units form a group. In a real application, we can increase or decrease the number of units according to the size of cylinder. In this paper, we stimulate an monopulse antenna consists of eight units.

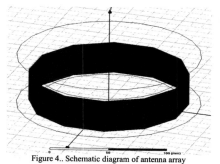

Figure 4.. Schematic diagram of antenna array

Simulated in HFSS, we can get VSWR of antenna.

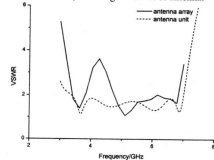

Figure 5..VSWR of antenna array

From Figure 5 we can get that when frequency range from 4.7GHz to 6.8GHz, the VSWR of antenna array is below 2, relative bandwidth of antenna is about 47.3%, decreased compared to single antenna when frequency is about 4GHz, this shows the antenna we designed is a broadband antenna.

Following is the sum beam, azimuth difference beam and elevation difference beam of antenna.

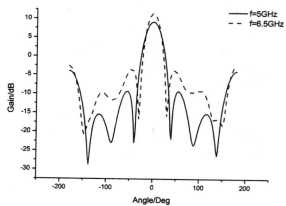

Figure 6.. The sum beam of antenna array

From Figure 6 we can know the gain of antenna towards the main radiation direction is about 10dB.

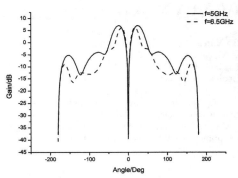

Figure 7.. The elevation difference beam of antenna array

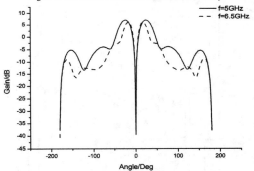

Figure 8.. The azimuth difference beam of antenna array

From Figure 7 and Figure 8 we can know that the elevation difference beam and azimuth difference beam of antenna is the same, it is not difficult to understand because our antenna is a symmetric antenna structure.

## IV. CONCLUSION

In this paper, our work is manly focus on the broadband conformal monopulse antenna. As an antenna, microstrip Yagi antenna has the characteristics of low profile and end-fire, so we selected it as the fundamental antenna unit. First, we design a microstrip Yagi antenna which only have one dipole director, its central frequency is 5.2GHz, relative bandwidth is about 70%. Next, we attach the antenna on a cylindrical supporter and simulate the performance of antenna in HFSS. Compared to antenna unit, bandwidth of the antenna arrays decreased to about 47.3%. We also have simulated the performance of antenna as a monopulse antenna; sum and difference beam of antenna is given. Next step we will take measures to further improve the bandwidth of the antenna.

## REFERENCES

[1] Joy An Lebaric. Ultra-Wideband Conformal Helmet Antenna. Us Ground Troop Helmet, 2000(3).
[2] Robert E. Munson Ball Aerospace. Conformal Antenna for Smart Cars. IEEE. Pro.1994, 29(3):325-331.
[3] Robert E. Munson. Conformal Microstrip Antennas and Microstrip Phased Antennas. IEEE Transaction on Antenna and Propagation. Communications 1974 (1):74-78.
[4] Pei Y Qin. A Reconfigurable Quasi-Yagi Folded Dipole Antenna. IEEE. Pro.2009.
[5] Josefsson L,Person P.Conformal Array Antenna Theory and Design [M] .Wiley-IEEE Press,2006.

# A Printed Monopole Antenna with Two Coupled Y-Shaped Strips for WLAN/WiMAX Applications

Zhihui Ma, Huiqing Zhai*, Zhenhua Li, Bo Yan and Changhong Liang

School of Electronic Engineering, Xidian University

No.2 South Taibai Road, Xi'an, Shaanxi 710071, China

*Corresponding email: hqzhai@mail.xidian.edu.cn

*Abstract*—In this paper, a new design of dual-band monopole antenna with two coupled Y-shaped strips for WLAN/ WiMAX applications is presented. By adjusting the geometries of the two Y-shaped strips, two separated impedance bandwidths of 17.5% (3.31~3.94 GHz) and 18.6% (4.98~5.94 GHz) ( $S_{11} \leq -10dB$ ) can be obtained. The proposed antenna has a low profile and can be easily fed by using a 50 $\Omega$ microstrip line. Good radiation characteristics of the antenna can also be obtained. The overall dimension of the proposed antenna can reach 17.7×26×1mm³.

## I. INTRODUCTION

Nowadays, the wireless local area net work (WLAN: 2.4-2.48GHz, 5.15-5.35GHz, and5.72-5.85GHz) and the worldwide interoperability for microwave access (WiMAX: 2.5-2.69GHz, 3.40-3.69GHz, and 5.25-5.85GHz) system are becoming increasingly popular. Thus, many researchers have beening paying much attention to design multi-bands, omidirectional pattern antennas for the mentioned above application, such as [1]-[5]. In [1], the antenna with dual band-notch characteristics can operate at two bands in WLAN system. A coupled dual-U-shaped antenna for WiMAX triple-band operation has been presented in [2]. In [3], by using defected ground structure, the design of triple-frequency antenna for WLAN/WiMAX applications is proposed. In [4], the double-T antenna for the dual-band WLAN operations has been reported, in which two T-shaped strips generate two different resonant frequencies. In [5], a monopole antenna with a shorted parasitic inverted-L wire for WLAN application is studied.

In this letter, a new design of dual-band monopole antenna with two coupled Y-shaped strips for WLAN/WiMAX applications is presented. The operation bands can be determined by the geometries of the two Y-shaped strips, the coupling of the two strips and the ground plane. The overall dimension of the proposed antenna can reach 17.7×26×1 mm³. The simulated and the measured results show that the antenna can effectively cover two separated impedance bandwidths of 630MHz (3.31~3.94 GHz) and 960MHz (4.98~5.94GHz), which satisfy both 5.2/5.8-GHz WLAN bands and 3.6/5.5-GHz WiMAX bands.

## II. ANTENNA DESIGN

Figure 1 illustrates the geometry of the presented dual-band antenna with two coupled Y-shaped strips for WLAN/ WiMAX applications. Both of the two coupled Y-shaped strips

and a 50 $\Omega$ microstrip line are printed on the same side of the substrate with dielectric constant of 2.65 and a thickness of 1mm. On the other side of the dielectric substrate, a ground plane is printed below the microstrip feed line.

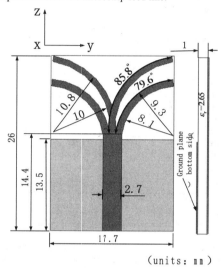

Figure 1.the geometry of the presented dual-band antenna

As shown in Figure 2, the top Y-shaped strip is designed to resonate at 3.56 GHz, whereas the bottom Y-shaped strip provides the high resonance frequency at 4.63 GHz. The length of each strip is about a quarter of the guided wavelength $\lambda_g$, which is calculated at the desired resonant frequency. The guided wavelength $\lambda_g$ is defined as:

$$\lambda_g = \frac{c}{\sqrt{\varepsilon_{eff}}f} \tag{1}$$

$$\varepsilon_{eff} \approx \frac{(\varepsilon_r + 1)}{2} \tag{2}$$

The Equation(1) and (2) are cited from [6]. Where $c$ is the speed of light, $f$ is the resonant frequency, $\varepsilon_{eff}$ is the effective relative permittivity and $\varepsilon_r$ is the relative permittivity. Given a resonant resonant frequency, the initial total length of the strip can be determined by (1). After the connection of the two Y-shaped strips, the proposed antenna can cover 3.31~3.94 GHz and 4.98~5.94GHz, which satisfies both bands 5.2/5.8-GHz

978-7-5641-4279-7

WLAN and 3.6/5.5-GHz WiMAX bands. It means that the operation bands can be determined by both the geometries of the two Y-shaped strips and the coupling of the two strips. Figure 3 shows the measured results of the proposed antenna, we can see that the simulated results of the antenna have good agreement with the measured results.

Figure 4 shows the results of the proposed antenna's surface current distribution for the two resonant frequencies at 3.60GHz and 5.17GHz. Obviously, from the Figure 4(a), much surface current distributes at the top Y-shaped strip, which means that its resonance mode has been excited at 3.60GHz. What's more, we also should note that there is a little current between the two strips. From the Figure 4(b), much current distributes at the top and bottom Y-shaped strips, which means that its resonance mode has been excited at 5.17GHz. In conclusion, both the two resonance modes has been excited at corresponding positions of the proposed antenna. Each Y-shaped strip and the couplings between the strips provide the contributions for the final two frequency bands.

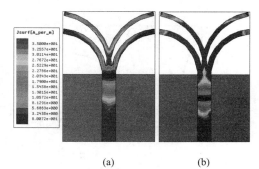

(a)                              (b)

Figure 4. Simulated results of the current distributions for the proposed monopole antenna at different frequencies (a) 3.60GHz (b) 5.17GHz, respectively

## III. SIMULATED RESULTS

Figures 5(a) and (b) show the simulated reflection coefficients changing with the central angle parameters of the proposed antenna for dual-band performance. Here, the central angle of the strip can be considered as one optimized parameter to obtain the two frequencies we need. From Figure 5(a), the low frequency band can be effectively adapted as the central angle of top Y-shaped strip is changed from 80.8 deg. to 90.8 deg. Clearly, from Fig.5 (b), we also know that as the central angle of bottom Y-shaped strip is changed from 74.6 deg. to 84.6 deg., the high frequency band is also tuned lower and lower. So we should also note that both the two operation bands can be determined by not only the geometries of the two Y-shaped strips but also the coupling of the two strips and the ground. Therefore, the final design needs the consideration of the two strips and the ground together.

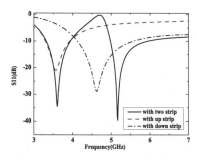

Figure 2. Simulated S-parameters of the proposed antenna without top or bottom strip and with two strip

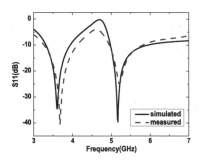

Figure 3. Measured S-parameter results of the presented antenna

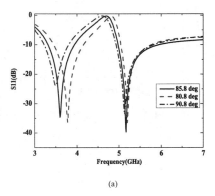

(a)

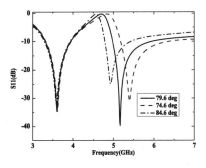

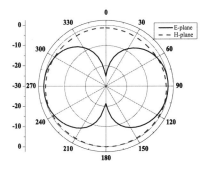

(b)

(a)

Figure 5. Simulated reflection coefficients of the proposed antenna with the central angle parameters: (a) central angle of the top Y-shaped strip (b) central angle of the bottom Y-shaped strip

According to Figure 6, we can see the simulated peak gains of the proposed antenna for WLAN/WiMAX bands. At the desired frequencies, their peak gain are about 3.51 dBi, 3.77 dBi, 3.58 dBi and 3.41 dBi at 3.6 GHz, 5.2 GHz, 5.5 GHz and 5.8 GHz, respectively.The far-field radiation patterns for the proposed antenna at 3.60 and 5.17GHz are given in Figure 7(a) and (b). It can be observed that the antenna gives nearly omnidirectional and stable radiation patterns in the H-plane at the two working frequency bands.

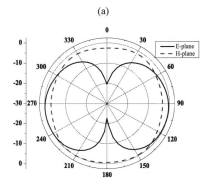

(b)

Figure 7. Simulated radiation patterns of the proposed antenna at (a) 3.60GHz E-plane and H-plane (b) 5.17GHz E-plane and H-plane

## IV. CONCLUSION

This paper presents a new design of dual-band monopole antenna with two coupled Y-shaped strips for WLAN/ WiMAX applications. By properly adjusting the geometries of the two Y-shaped strips, two separated impedance bandwidths can be obtained. The proposed antenna has good radiation characteristics for 5.2/5.8-GHz WLAN bands and 3.6/5.5-GHz WiMAX bands, as well. The overall dimension of the proposed antenna can reach $17.7 \times 26 \times 1$ mm$^3$. Therefore this novel simple antenna could be a better candidate for a WiMAX/WLAN dual-mode system.

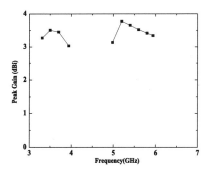

Figure 6. The peak gains of proposed antenna for the desired bands

## ACKNOWLEDGEMENTS

This work is supported by National Natural Science Foundation of China under Contract No. 61101066, and Foundation for the Returned Overseas Chinese Scholars, State Education Ministry and Shaanxi Province, and supported partly by the Program for New Century Excellent Talents in University of China, by National Natural Science Foundation of China under Contract No. 61072017, Fundamental Research

Funds for the Central Universities (K5051202051, K5051302025), National Key Laboratory Foundation.

REFERENCES

[1] Q. X. Chu and Y. Y. Yang, "3.5/5.5GHz dual band-notch ultra-wideband antenna," *Electron. Lett.*, vol. 44, no. 3, pp. 172–174, Jan. 2008.

[2] T. W. Koo, D. Kim, J. I. Ryu, J. C. Kim and J. G. Yook, "A COUPLED DUAL-U-SHAPED MONOPOLE ANTENNA FOR WiMAX TRIPLE-BAND OPERATION," *Microwave and Opt.Technol. Lett.*, vol. 53, no. 4, pp. 745–748, Apr. 2011.

[3] W. C. Liu and C. M. Wu, "Design of Triple-Frequency Microstrip-Fed Monopole Antenna Using Defected Ground Structure," *IEEE Trans. Antennas Propag.*, vol. 59, no. 7,pp. 2457–2463,Jul.2011.

[4] Y. L. Kuo and K. L. Wong, "Printed Double-T Monopole Antenna for 2.4/5.2GHz Dual-Band WLAN Operations," *IEEE Trans. Antennas Propag.*, vol. 51, no. 9, pp. 2187–2192, Sep. 2003.

[5] J. Y, Jan and L. C. Tseng, "Small Planar Monopole Antenna With a Shorted Parasitic Inverted-L Wire for Wireless Communications in the 2.4-,5.2-,and 5.8-GHz Bands," IEEE Trans. Antennas Propag., vol. 52, no. 7, pp.1903–1905, Jul. 2004.

[6] Y. Huang and K. Boyle, ANTENNAS FROM THEORY TO PRACTICE, 2008, pp. 190.

# Planar Circularly Polarized Antenna with Broadband Operation for UHF RFID System

Jui-Han Lu[1], Hai-Ming Chin[2] and Sang-Fei Wang[1]
[1]Department of Electronic Communication Engineering
[2]Department of Marine Engineering
Kaohsiung Marine University
Kaohsiung, Taiwan 811

*Abstract*- **Novel circular polarization (CP) design of planar broadband antenna with square slot for UHF RFID system is proposed and experimentally studied. By insetting the arc-shaped strip into the square slot, the proposed CP design can easily be achieved with the impedance bandwidth (RL □10 dB) of about 142 MHz (15.3% @ 931 MHz) and the 3 dB axial-ratio (AR) bandwidth of about 166 MHz (17.7 % @ 940 MHz) for UHF RFID applications. The measured peak gain and radiation efficiency are about 6.8 dBic and 98% across the operating band, respectively, with nearly bidirectional pattern in the XZ- and YZ-plane.**

## I. INTRODUCTION

UHF (860–960 MHz) band radio-frequency identification (RFID) system becomes more attractive for many industrial services such as supply chain, tracking, inventory management and bioengineering applications because it can provide longer reading distance, fast reading speed and large information storage capability. The RFID reader antenna is one of the important components in RFID system and has been designed with CP operation. The detection range and accuracy are directly dependent on the performance of reader / tag antennas. Since the RFID tags are always arbitrarily oriented in practical usage and the tag antennas are normally linearly polarized, circularly polarized (CP) antennas become the most popular candidates to receive the RF signal that emanates from arbitrarily oriented tag antennas for improving the reliability of communications between readers and tags. Moreover, circularly polarized antennas can reduce the loss caused by the multi-path effects between the reader and the tag antenna. However, the UHF frequencies authorized for RFID applications are varied in different countries and regions. Hence, a universal reader antenna with desired performance across the entire UHF RFID band operated at 860–960 MHz (a fractional bandwidth of 11.1%) would be beneficial for the RFID system configuration and implementation to overcome the operating frequency shift and impedance variations due to the manufacturing process errors. Circular polarization can be obtained by exciting the two orthogonal linearly polarized modes with a 90° phase offset. Numerous CP reader antennas for UHF RFID system have been presented such as the aperture-coupled annular ring patch antenna with thick high-dielectric substrate [1], a sequentially fed stacked corner-truncated CP patch antenna [2], the stacked patch antenna composed of two corner-truncated patches with a horizontally meandered strip [3], a circularly polarized patch antenna excited by an open circular ring microstrip line through multiple slots [4],

asymmetric-circular shaped slotted microstrip antenna [5]. Although, the above mentioned reader antennas are focused on the unidirectional radiation pattern, they have disadvantage of bulky volume [1, 3-4] or complex structure [2]. The reader antenna with bidirectional radiation can be introduced in the entry-way scanning system to minimize the number of needed reader antennas with unidirectional radiation, which definitely reduces the implemented cost. Two RFID reader antennas have also been presented such as the square-ring antenna fed by a Wilkinson power divider [6] and the annular-ring slot antenna with a slotline feed [7]. The former design is with less antenna gain while the latter design is with narrower 3-dB AR bandwidth which can't meet the overall bandwidth specification of global RFID system. Under the condition of the same antenna size, using a printed slot antenna is a certain method, which is due to the fact that printed slot antennas usually have a wider impedance bandwidth than the microstrip patch antennas [8-10], can be the effective way to improve the operating bandwidth. Several slot antennas with CP operation have been presented such as the slot antenna with the truncated corner and a grounded inverted-L strip at the two opposite corners [11], with a T-shaped stub and a microstrip T-junction [12], with multi parallel slots [13], a pair of orthogonally positioned radiating slots with a three-stub hybrid coupler as the feeding network [14], a stair-shaped slot antenna with a longitudinal slot etching at the middle part [15] and a number of CP slot antennas fed by an L-shaped strip [16-23]. However, wide slot antennas with CP design for UHF RFID system are very scant in the open literature. Therefore, in this article, we present a novel CP design of planar broadband UHF RFID reader antenna with bi-directional reading pattern. This RFID reader antenna is composed of the square slot antenna with the insetting grounded arc-shaped strip and fed by the F-shaped microstrip line to obtain the broadband CP operation. Due to the grounded arc-shaped strip to disturb the surface electric field distribution on the square slot, two near-degenerated resonant modes ($TE_{10}$ and $TE_{01}$ modes) with 90 degrees phase difference can be closely excited to form a wider CP operating bandwidth for UHF band. The obtained impedance bandwidth across the operating band can reach about 142 MHz (15.3 % centered at 931 MHz) and the 3 dB axial-ratio (AR) bandwidth of about 166 MHz (17.7 % centered at 940 MHz). With bi-directional reading pattern, the maximum antenna peak gain and radiation efficiency across the operating band are about 6.8 dBic and 98 %, respectively, which is more than that of the presented square-ring antenna [6]. Noted that a square metallic reflector can be arranged below the slot antenna to provide a unidirectional broadside patterns and reduce the backlobe radiation; its physical position is about one-quarter wavelength below the slots [15]. Details of the proposed UHF RFID reader antenna design is described and its experimental results from the obtained CP performance as operating at 900 MHz band are presented and discussed as well. A parametric study for the major parameters of the proposed CP antenna is also conducted.

Table 1 : The optimal dimensions of the proposed broadband circularly polarized antenna with square slot.

| Parameter | Value (mm) | Parameter | Value (mm) |
|---|---|---|---|
| L | 126 | W | 121 |
| L1 | 27.5 | W1 | 15 |
| L2 | 38 | W2 | 6.2 |
| L3 | 38 | W4 | 1.48 |
| L4 | 20 | R1 | 71 |
| G | 0 | R2 | 66 |

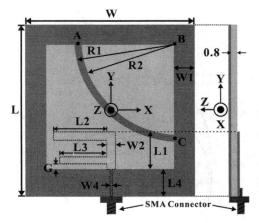

Figure 1. Geometry and photograph of the proposed planar broadband circularly polarized antenna with square slot for UHF RFID Reader.

## II. Antenna Design

The geometry and photograph of the proposed broadband circularly polarized antenna with square slot is shown in Fig. 1. The circularly polarized antenna with the total antenna size of $126 \times 121$ mm$^2$ is printed on an FR4 substrate ($\varepsilon_r = 4.4$, thickness = 0.8 mm, loss tangent = 0.0245). The optimal dimensions of this proposed broadband circularly polarized antenna with square slot are listed in Table 1. This radiating square slot is set to be the dimension of $91 \times 91$ mm$^2$ as to degenerate TE$_{10}$ and TE$_{01}$ components of CP. We obtain the operating frequency 915 MHz calculated by (1) and they are roughly the same as the frequency with the lowest axial ratio centered at 940 MHz in Fig. 2.

$$f_c = \frac{c}{2\pi\sqrt{\varepsilon_{eff}}} \times \sqrt{(\frac{m\pi}{a})^2 + (\frac{n\pi}{b})^2} \qquad (1)$$

In (1), $c$ is $3 \times 10^8$ m/s and $\varepsilon_{eff}$ is the effective permittivity of the FR4 substrate. Note that (1) is approximate formulas to predict the operating frequencies of two CP modes for this designed slot antenna and they would have obvious errors when the width (W1) of the ring ground plane is too narrow. An F-shaped strip is etched on the bottom-layer of the substrate and fed through a matching 50 Ω microstrip feed line which has a width (W4) of 1.48 mm and length (L4) of 20 mm. The vertical strip (L1) that succeeds the microstrip feed line is of 27.5 mm long. The first horizontal strip has a length (L2) of 38 mm and a 4.3 mm gap is made at the strip end and the ring ground plane as to provide capacitive effects. In addition, the second horizontal strip is placed above the ring ground plane with the gap (G) and has the length (L3) of 38 mm as the tuning stub to adjust the input impedance of this proposed CP antenna, which is different from the design using high characteristic impedance of the feeding microstrip line [17-18, 19-23]. The resonant length (L1 + L2) of the bent strip is 65.5 mm in corresponding to approximately 0.21 wavelength of the resonant mode at 940 MHz. The vertical and horizontal portions of the F-shaped strip have the same width (W2) of 6.2 mm. The vertical portion of the F-shaped strip is parallel to $y$-axis (side 1) contributing to TE$_{10}$ mode while the other side (parallel to $x$-axis, side 2) contributing to TE$_{01}$ mode. The current phase on side 2 lags behind that of side 1, which causes 90 degrees out of phases between $Ex$ and $Ey$ at the aperture.

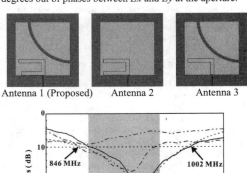

Antenna 1 (Proposed)    Antenna 2    Antenna 3

(a)

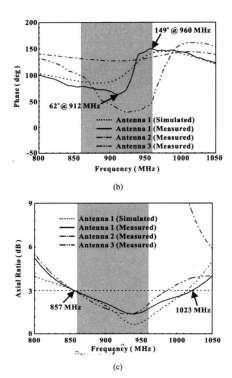

(b)

(c)

Figure 2. Simulated and measured results against frequency for the proposed planar broadband circularly polarized antenna with the grounded arc-shaped strip or not. (a) Return Loss. (b) Phase diagram. (c) Axial Ratio.

## III. RESULTS AND DISCUSSION

The proposed CP antenna is designed to operate at the centre frequency of about 940 MHz in the UHF band for RFID readers. The return loss is measured using an Agilent N5230A vector network analyzer. Fig. 2 shows the related simulated and experimental results of the return loss, phase diagram and axial ratio (in the boresight direction) for the proposed CP antenna of Fig. 1 fed by F-shaped or L-shaped microstrip line. The related results are listed in Table 2 as comparison. From the related results, the measured operating bandwidth (RL □ 10 dB) can reach about 142 MHz (860–1002 MHz) or 15.3 % centered at 931 MHz, which covers the entire UHF RFID band, and agrees well with the HFSS simulated results. Fig. 2(b) shows the simulated and measured phase diagram for the proposed CP antenna. It can be seen that these two orthogonal modes (912 MHz and 960 MHz) are excited, in 90° phase difference, resulting in good CP radiation. In Fig 2(c), this broadband CP antenna also provides a 3-dB AR over the UHF band of 857–1023 MHz or 17.7 % centered at 940 MHz. Meanwhile, it is found that the 3dB AR bandwidth of the square slot antenna fed by L-shaped microstrip line (Antenna 3) can cover the entire UHF RFID bandwidth of 860-960 MHz, however, with poor impedance matching. The CP radiation pattern measured at 940 MHz is plotted in Fig. 3, and good symmetry of bidirectional radiation has

been observed. Results show the coherent agreement between the measured and simulated results. Since a CP slot antenna radiates a bidirectional wave, the radiation patterns on both sides of the proposed CP antenna are almost the same, in which a contrary circular polarization is produced; the front-side radiates LHCP while the back-side radiates RHCP. By verification, this antenna structure has successfully achieved a cross polarization discrimination of 20 dB on a wide azimuth range, which is more than the related slot antennas fed by the L-shaped probe [16-23].

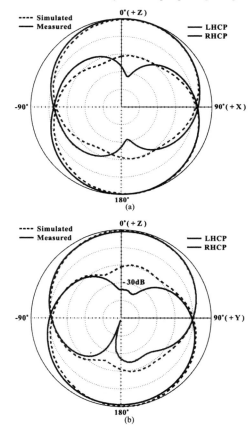

Figure 3. Simulated and measured normalized LHCP / RHCP radiation patterns for the proposed broadband CP antenna at 940 MHz. (a) X-Z plane. (b) Y-Z plane.

Plus, the radiation patterns also show that the 3-dB AR beamwidths are about 83 degrees, with symmetry in both the XZ and YZ planes. The related measured result against the operating frequency is shown in Fig. 4. It is found that the variation of the 3-dB AR beamwidth is less than 3 degrees across the overall UHF operating band. Then, a standard tag with Higgs™- 3 UHF RFID IC (ALN-9640) [27] for EPC Class1 Gen2 operating at 900 MHz is introduced for the measurement system. A commercial UHF RFID reader with the output power of EIRP 3.28 W connected to the LP antenna with peak gain of 8.5 dBi and the dimension of $200 \times 190 \times 40$ mm$^3$ as the reference antenna. In Fig. 4, the comparison of the detection ranges between the proposed CP antenna and the reference reader antenna can be observed to realize the performance sensitivity. It is easily found that the maximum reading distance for the proposed CP antenna is about 5.5 m and nearly the same as that of the reference reader antenna. Meanwhile, this proposed CP antenna has smaller volume than that of the reference antenna. Moreover, miniaturization is an important issue as we consider dexterity. Because of its low profile, the proposed CP antenna can provide compact operation with broadband operation.

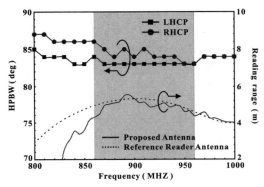

Figure 4. Measured 3-dB beamwidth and reading range across the operating frequencies for the proposed broadband CP antenna.

## IV. CONCLUSIONS

A novel broadband circularly polarized antenna with square slot is proposed for the application of UHF RFID system. The obtained impedance bandwidth across the operating band can reach about 142 MHz (15.3 % @ 931 MHz) and the 3 dB axial-ratio bandwidth of about 166 MHz (17.7 % @ 940 MHz), which can cover the entire UHF RFID band. The measured peak gain and radiation efficiency are about 6.8 dBic and 98% across the operating band, respectively, with nearly bidirectional pattern in the XZ- and YZ-plane.

## REFERENCES

[1] D. H. Lee, P. J. Park, J. P. Kim and J. H. Choi, "Aperture coupled UHF RFID reader antenna for a handheld application", *Micro. Opt. Technol. Lett.*, vol. 50, pp. 1261-1263, May 2008.

[2] Z. N. Chen, X. Qing and H. L. Chung, "A universal UHF RFID reader antenna", *IEEE Trans. Microwave Theory and Techniques*, vol. 57, pp. 1275-1282, May 2009.

[3] Z. Wang, S. Fang, S. Fu and S. Jia, "Single-fed broadband circularly polarized stacked patch antenna with horizontally meandered strip for universal UHF RFID applications", *IEEE Trans. Microwave Theory and Techniques*, vol. 59, pp. 1066-1073, April 2011.

[4] T. N. Chang and J. M. Lin, "A novel circularly polarized patch antenna with a serial multislot type of loading", *IEEE Trans. Antennas Propag.*, vol. 55, pp. 3345-3348, Nov. 2007.

[5] Z. N. Chen and X. Qing, "Asymmetric-circular shaped slotted microstrip antennas for circular polarization and RFID applications", *IEEE Trans. Antennas Propag.*, vol. 58, pp. 3821-3828, Dec. 2010.

[6] Y. F. Lin, H. M. Chen, F. H. Chu and S. C. Pan, "Bidirectional radiated circularly polarized square-ring antenna for portable RFID reader", *Electron. Lett.*, vol. 44, pp. 1383-1384, Nov. 2008.

[7] S. A. Yeh, Y. F. Lin, H. M. Chen and J. S. Chen, "Slotline-fed circularly polarized annular-ring slot antenna for RFID reader," *2009 Asia Pacific Microwave Conference*, pp. 618–621.

[8] H. D. Chen, "Broadband CPW-fed square slot antennas with a widened tuning stub", *IEEE Trans. Antennas Propagat.*, vol. 51, no. 8, pp. 1982–1986, Aug. 2003.

[9] A. U. Bhobe, C. L. Holloway, M. Piket-May, and R. Hall, "Coplanar waveguide fed wideband slot antenna", *Electron. Lett.*, vol. 36, pp. 1340–1342, Aug. 2000.

[10] J. Y. Sze, and K. L. Wong, "Bandwidth Enhancement of a Microstrip-Line-Fed Printed Wide-Slot Antenna", *IEEE Trans. Antennas Propagat.*, vol. 49, no. 7, pp. 1020–1024, July 2001.

[11] Y. Shen, C. L. Law and Z. Shen, "A CPW-fed circularly polarized antenna for lower ultra-wideband applications" *Micro. Opt. Technol. Lett.*, vol. 50, pp. 2365-2369, Oct. 2009.

[12] R. P. Xu, X. D. Huang and C. H. Cheng, "Broadband circularly polarized wide-slot antenna" *Micro. Opt. Technol. Lett.*, vol. 49, pp. 1005-1007, May 2007.

[13] H. A. Ghali and T. A. Moselhy "Broad-band and circularly polarized space-filling-based slot antennas," *IEEE Trans. microwave theory and techniques*, vol. 53, no. 6, pp. 1946–1950, 2005.

[14] X. Qing, Z. N. Chen and H. L. Chung, "Ultra-wideband circularly polarized wide-slot antenna fed by threestub hybrid coupler," *2007 IEEE International Conference on Ultra-Wideband*, pp. 487–490.

[15] C. J. Wang and C. H. Chen, "CPW-fed stair-shaped slot antennas with circular polarization," *IEEE Trans. Antennas Propagat.*, vol. 57, no. 8, pp. 2483–2486, 2009.

[16] T. Fukusako and L. Shafai, "Circularly polarized broadband antenna with L-shaped probe and wide slot", *2006 ANTEM/UJRSI*, pp. 445-448.

[17] S. S. Yang, A. A. Kishk and K. F. Lee, "Wideband circularly polarized antenna with L-shaped slot", *IEEE Trans. Antennas Propagat.*, vol. 56, no. 6, pp. 1780-1783, June 2008.

[18] J. S. Row and S. W. Wu, "Circularly-Polarized Wide Slot Antenna Loaded With a Parasitic Patch," *IEEE Trans. Antennas Propagat*, vol. 56, no. 9, pp. 2826-2832, 2008.

[19] J. Pourahmadazar, C. Ghobadi, J. Nourinia, N. Felegari and H. Shirzad, "Broadband CPW-fed circularly polarized square slot antenna with inverted-l strips for UWB applications," *IEEE Antenna and Wireless Propagation Letters*, vol. 10, pp. 369-372, 2011.

[20] Y. Muramoto and T. Fukusako, "Circularly polarized broadband antenna with L-shaped probe and L-shaped wide-slot", *2007 Asia-Pacific Microwave Conference*, pp. 1-4.

[21] T. Fukusako, R. Sakami and K. Iwata, "Broadband circularly polarized planar antenna using partially covered circular wide-slot and L-probe", *2008 Asia-Pacific Microwave Conference*, pp. 1-4.

[22] R. Joseph and T. Fukusako, "Broadband circularly polarized antenna with circular sot and separated L-probes", *2010 IEEE AP-S Int. Symp.*, pp. 1-4.

[23] T. Fukusako and R. Joseph, "Design of broadband circularly polarized aperture antennas using L-shaped bent probe", *2011 IEEE-APS Topical Conference on Antennas and Propagation in Wireless Communications*, pp.343-346

[24] Ansoft Corporation HFSS, http://www.ansoft.com/products/hf/hfss.

[25] IEEE Standard Test Procedures for Antennas: ANSI/IEEE-STD149-1979, Sec. 12-13, pp. 94-112.

[26] B. Y. Toh, R. Cahill and V. F. Fusco, "Understanding and measuring circular polarization", *IEEE Trans Edu.*, vol. 46, pp. 313–318, Aug. 2003.

[27] Alien Technology Corporation, http://www.alientechnology.com/

# A Frequency Selection Method Based on Fusion Algorithm in Bistatic HFSWR

Chen Weiwei, Yu Changjun, Chen Wentao
School of Information and Electrical Engineering
Harbin Institute of Technology at Weihai
Weihai, China
chenweiwei0320@126.com

*Abstract*-T/R-R High Frequency Surface Wave Radar can detect targets over the horizon. There is a large difference in working environment between stationary station and mobile station. The traditional frequency selection methods are hard to choose the best operating frequency for the T/R-R High Frequency Surface Wave Radar(HFSWR), and this flaw seriously affects detection performance and target positioning accuracy of the radar. This paper adopts the weighted fusion algorithm based on Kalman filter to fuse the spectrum information of the T/R-R bistatic radar in the electromagnetic environment, and selects the optimal working frequency band by average power and variance combination minimum, then finds the best operating frequency based on maximum signal-to-noise ratio(SNR) criterion of the target. Simulation results demonstrate the effectiveness of the adopted frequency method.

## I. INTRODUCTION

High Frequency Surface Wave Radar(HFSWR) transmits vertical polarization electromagnetic wave to detect and track maritime targets over the horizon [1][2][3]. HFSWR usually works in the short-wave band. As a result, there are a lot of interferences in this band. A lot of attentions have been attracted on the operating frequency selection of HFSWR [4][5]. For example, [6] established real-time frequency selection adaptive communication system for high frequency OTHR frequency selection. [7] analyzed the communication station frequency occupancy, and presented that ground wave OTH radar should adopt adaptive frequency conversion technology against radio interference. However, these methods only provides reference for the frequency selection of bistatic radar [8]. In addition, the traditional frequency selection methods of bistatic radar select suitable frequency by artificial methods. Although these methods are simple and easy to operate, the frequency selected is not the best operating frequency for the T/R-R bistatic HFSWR with two stations, As a result, the bistatic HFSWR could not work in the best performance, and even affects the quality of target detection and tracking.

In this paper, A novel method is proposed to select the best operating frequency of T/R-R HFSWR by data fusion [9][10][11]. The core idea of this method is that it combines and fuses the external electromagnetic spectrum information of T/R and R stations, then selects the best operating frequency. This method makes full use of the redundant information of each sensors in space and time according to some optimization criterion. Kalman filter weighted fusion algorithm incorporates and fuses the external electromagnetic

spectrum information [12][13] to get the optimal working frequency of bistatic radar. First, the silence working channel is confirmed based on average power and variance combination minimum. Secondly, the best operating frequency is determined based on maximum signal-to-noise ratio (MSNR) criterion of the target.

The rest of this paper is organized as follows. Section II describes the weighted fusion algorithm based on Kalman filter. The frequency selection principle and effect evaluation are presented in Section III. Conclusions are given in Section IV.

## II. WEIGHTED FUSION ALGORITHM BASED ON KALMAN FILTER

There are huge differences in the electromagnetic environment spectrum information between T/R and R stations. The spectrum information of two receivers should be fused with appropriate fusion model, so that a reliable operation frequency can be selected. The optimal spectrum band is selected with Kalman filter for each of the two receivers [14], respectively. The optimal operation frequency can be choosen with a weighted summation of the two optimal spectrum bands.

### A. Algorithm Overview

Kalman filter uses feedback control to achieve process estimation. The basic principle is to estimate the system state at certain time, and get feedback from the observed values (with noise).

So, the paper sees the external electromagnetic spectrum datas of T/R station and R station matriculated at Weihai as the measurement information that is $Z^k = \{Z(j) : j = 1, 2, ..., k\}$, then we get the state estimation informations of the two stations by Kalman filter that is $\hat{X}(k+1/k+1)$.

Fusion block diagram is shown in Figure.1. We can get the state estimations by Kalman filter, then the information after fusion is

$$\hat{x} = \omega_1 \hat{X}_1(k+1/k+1) + \omega_2 \hat{X}_2(k+1/k+1) \quad (1)$$

where $\Omega = (\omega_1, \omega_2)$ ——the weighted of each sensor.

The paper selects the weighted by the optimal weighted that is because the weighted can correct those divergent estimated value or the large deviation estimated value.

This work was supported by National Natural Science Foundation of China(61171188)

978-7-5641-4279-7

Weigthed fusion algorithm can make the mean-square error(MSE) minimum.

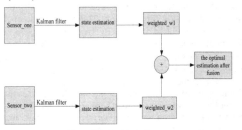

Figure.1 Fusion block diagram

So we can calculate the weighted by MSE, and the optimal weighted are

$$\omega_1^* = \frac{\sigma_2^2}{\sigma_2^2 + \sigma_1^2} \qquad (2)$$

$$\omega_2^* = \frac{\sigma_1^2}{\sigma_2^2 + \sigma_1^2} \qquad (3)$$

In the Kalman filter weighted fusion algorithm, we see the covariance estimates $\hat{P}(k/k)$ as the weighted factors .

### B. Simulataion Result

Spectrum data of T/R-R bistatic radar is matriculated in Weihai. The data is amplitude information of external electromagnetic spectrum and it can serve as the measurement information of Kalman filter. The measurement matrix is $H_1 = \begin{bmatrix} 1 & 0 \end{bmatrix}$ , $H_2 = \begin{bmatrix} 0 & 1 \end{bmatrix}$ , state-transition matrix is $\Phi = \begin{bmatrix} 1 & 0 \\ 0 & 1 \end{bmatrix}$, controlling signal is $\Gamma = \begin{bmatrix} 0 & 0 \end{bmatrix}^T$ . The initial value of state estimation is $x_0 = \begin{bmatrix} 0 & 0 \end{bmatrix}^T$ , and the initial value of error covariance matrix is $P_0 = \begin{bmatrix} 1 & 0 \\ 0 & 1 \end{bmatrix}$, Fusion results are shown in Figure.2.

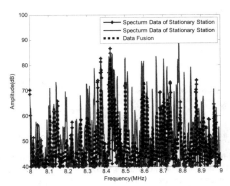

Figure.2 Data fusion results

### III. UNITS FREQUENCY SELECTION OF T/R-R HFSWR

#### A. The Frequency Selection Principle based on Average Power and Variance Combination Minimum

Silent spectrum band is essential for power performance of T/R-R HFSWR. Sliding window method is the prevailing method to confirm the location of the favor band. Specific frequency selection programs are divided into the following steps.

Step 1 Choose sliding window of smallest average power according to the fused spectrum information and the initial frequency of the window are confirmed. The method of average power is defined as

$$Average power = \frac{\sum_i power(i)}{length(Slidewinwidth)} \qquad (4)$$

The width of the sliding window is denoted as *Slidewinwidth* .

Step 2 Calculate the noise threshold of spectrum information:

$$Threshold = noise floor + cons \tan t \qquad (5)$$

It means that the higher the threshold, the longer the silence time of the band. Therefore, the constant is selected to be 4.7dB, and the noise floor is computed by the minimum average power.

Step 3 In the selected sliding window, select three channels with minimum average power to be candidate bands, and compare their average power values with the Threshold:

If the average power of the three bands *Averagepoewr* < *Threshold* , It means that three bands can be used theoretically. We can use variance to judge which is the best working band, and select the band with smallest variance to be the best band. The variance represents the degree of deviation average, and stands for stationarity of the band. It is defined as

$$Variance = \frac{\sum_i (a(i) - power(i))^2}{length(Slidewinwidth) - 1} \qquad (6)$$

Other cases, the preferred band is determined by the minimum average power criterion.

#### B. Simulation Results

(1) Silence sub-band duration comparison under different Threshold

Figure.3 shows that the silent time increases with the Threshold.

(2) The results of the frequency selection

The spectrum band we are interested is 8~9MHz, and Threshold=44.94.

TABLE I
SMILUTIONS

| parameter | value | | |
|---|---|---|---|
| Variance | 1.83 | 1.80 | 1.77 |
| Average power | 41.61 | 41.63 | 41.65 |

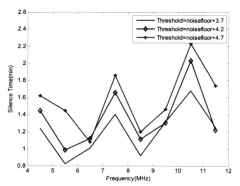

Figure.3 Silence subband duration under different threshold in 30kHz

From Table I, it shows that the averagepowers of the three candidate frequency bands are less than Threshold, So the optimal frequency band is determined by the variance. The results of the frequency selection are shown in Figure.4.

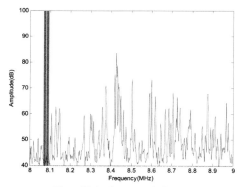

Figure.4 Fusion frequency selection results

From Figure.4, it demonstrates that the green belt is the optimal frequency band, but we still need to estimate this band by maximum signal-to-noise ratio (MSNR) criterion.

### C. Frequency Selection Effect Evaluation

The target detection performance is evaluated in T/R-R HFSWR, and atmospheric noise is defined as background noise. SNR is calculated as

$$SNR = \frac{P_r(f)}{P_0(f)} = \frac{P_r(f)}{KT_0F_a} \quad (7)$$

where, $P_r = \frac{6.22 \times 10^{-38} E^2(R_t)E^2(R_r)P_tG_tG_r\lambda^2\sigma}{l}$

$P_r$ is radar echo power equation [15][16]. $P_0(f)$ is atmospheric noise power. $E(R_t)$ and $E(R_r)$ are electric field strength with range $R_t$ and $R_r$ in the smooth conductive sphere, respectively. Field strength value are calculated by GRWAVE program [17].

The simulation parameters are as follows:

Averagepower is 1kW, system loss is 6dB, equivalent bandwidth is 30kHz, radar cross section(RCS)is 40dB/m², and coherent integration time is 100s. The operating frequencies are 5、7.5、9、11 MHz respectively. The duty cycle is 0.1. Minimum signal detection SNR is 14dB (Detection probability is 90%, false alarm probability is $10^{-6}$). Radar transmits wide beam in fixed direction, 3dB width is 60 degrees, and the normal direction is north west 50 degrees. The number of the receivers is 8, and the distance of the element is 14.5 meters. whereas normal direction of stationary receiver array is north west of 50 degrees, the normal direction of mobile receiver array is north west of 50 degrees.

Simulation results are shown in Figure.5 and Figure.6. In Figure.5 and Figure.6, coordinates of the stationary stations are (250, 0), coordinates of the mobile stations are (0,100).

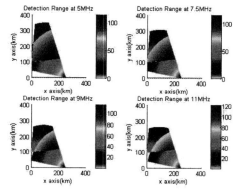

Figure.5 Maximum Detection Distance Criterion when the SNR threshold is given

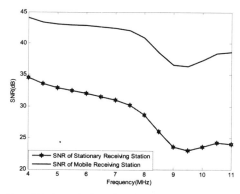

Figure.6 Maximum signal-to-noise ratio when target distance is given

From Figure.5, it demonstrates that the detection distance of stationary station and mobile station gradually reduces as the frequency increases, and the detected overlapping area of stationary and mobile stations are increasingly hardly.

Furthermore, detection distance of the T/R radar is increased. It demonstrates the superiority of the T/R-R radar. Under the view of maximum detection distance criterion with a given signal-to-noise ratio threshold, the lower operating frequency, the farther the detection range. If the study band can not reach the detection distance, study band shouled be reduced.

In Figure.6, coordinates of the target are (150, 190). SNR curves are presented at different frequencies of stationary and mobile stations in target location. Furthermore, the lower the frequency, the smaller the signal attenuation, and also the detection performance is improved.

## IV. CONCLUTION

(1) Simulation results show that the fusion algorithm can make full use of spectrum information of T/R and R statiions, and the band selected can make the bistatic HFSWR work worth preferable performance.

(2) The criterion of fusion frequency selection can be adjusted to meet the needs of practical application. We assess the effect of fusion frequency selection, and select the optimal operating frequency to match up the criterion. So, Maximum detection distance criterion when the SNR threshold is given is used to make the radar detect the farthest range and the maximum signal-to-noise ratio criterion when target distance is given is used to get the optimal effect of target detection and tracking in a given area.

## REFERENCES

[1] Wang W and Wyatt L R, "Radio frequency Interference Cancellation for Sea-state Remote Sensing by High-frequency Radar," *IET Radar, Sonar & Navigation*, vol. 5, no. 4, pp. 405-415, 2011.

[2] Maresca S, Greco M and Gini F et al, "The HF Surface Wave Radar WERA. Part I: Statistical Analysis of Recorded Data," *2010 IEEE Radar Conference*, 2010, pp. 826-831.

[3] Wyatt L R, Green J.J and Middleditch A et al, "Operational Wave, Current, and Wind Measurements with the Pisces HF Radar," IEEE Journal of Oceanic Engineering, vol. 31, no. 4, pp. 819-834, 2006.

[4] Su Hong-tao, Bao Zheng and Zhang Shou-hong, "OTHR Adaptive Selection of Operating Frequency Point," *Electronics & Information Technology*, vol. 27, no. 2, pp. 274-277, 2005.

[5] Lou Peng, Fan Jun-mei and Jiao Pei-nan et al, "The Operating Frequency Selection in HF Hybrid Sky-surface Wave System," *9th International Symposium on Antennas Propagation and EM Theory, 2010*, pp.513-516.

[6] Tian Jian-sheng and Wu Xiong-bin, "The Operating Frequency of the High-frequency Radar Online Real-time Optimal Choice," *Radar*, vol. 27, no. 7, pp. 8-10, 2005.

[7] Guan Rong-sheng, Xie Shu-Guo and Zhao Zheng-yu, "Ionosphere Return Oblique Detection Study 40 years in China," *Chinese Journal of Radio Science*, vol. 14, no. 4, pp. 479-484, 1999.

[8] Sun Xi-dong, "Ground Wave OTH Radar Anti-shortwave Radio Interference Study," *Aerospace Electronic Warfare*. no.5, pp. 56-58, 2004.

[9] Nichloas and J. Wills, "Bistatic Radar Technology Service Corporation. Silver Spring," *MI*. 1995, pp. 348-390.

[10] Chair Z and Varshney P. K, "Optimal Data Fusion in Multiple Sensor Detection System," *IEEE Trans. on AES*, vol. 22, pp. 98-101, 1986.

[11] Kang Yao-hong, "*Data Fusion Theory and Application*s," Xi 'an university of electronic science and technology press, 1997.

[12] J.Linas and E.Waltz, "Multisensor Data Fusion," *Artech House, Norwood, Massachusetts*, 1990.

[13] Deng Zi-li, "Information Fusion Filtering Theory and Application," Harbin: *Harbin Institute of Technology*, 2007.

[14] Quan Tai-fan, "*Information Fusion Theory and Application Based on NN-FR Technology*," Beijing, National Defense Industry Press, vol. 8, pp. 50-54, 2002.

[15] Liu Hai-bin and Gong Feng-xun, "An Improved Multi-sensor Weighted Fusion Algorithm," *The electronic products the world*, 2009, no.12.

[16] Xie Jun-hao, Li Bo and Chou Yong-bin et al, "The High-frequency Surface Wave Radar Ship Detection Range Analysis," *Radio Science*, vol. 25, no. 2, pp. 234-239, 2010.

[17] Zhou Wen-yu and Jiao Pei-nan, "Over-the-horizon Radar Technology," *Industry Press*, 2003, pp. 484-488.

# Effects of Antenna Polarization on Power and RMS Delay Spread in LOS/OOS Indoor Radio Channel

Z. Y. Liu, L. X. Guo, W. Tao, and C. L. Li
School of Science, Xidian University
Box 274, No.2, Taibai Road
Xi'an 710071, China

*Abstract*-The ray tracing (RT) method offers significant advantages in terms of the accurate and comprehensive prediction of radio channel characterization in indoor environments. As a consequence, this paper presents results of computer simulation based on a three-dimensional (3D) RT method for various directional polarized antennas, where the effects of antenna polarization on power and RMS delay spread are investigated in line-of-sight (LOS) and out-of-sight (OOS) indoor channel. These results indicate that the difference in predicted power between cross-polar case and copular case, under LOS environment, is more noticeable than that of OOS case. Whereas, the difference in predicted RMS delay spread between cross-polar and copular case, under OOS environment, is more noticeable than that of LOS case.

## I. Introduction

The forthcoming years, the market for indoor wireless networking facilities is expected to grow considerably in both commercial and domestic sectors [1]. Thus, a thorough investigation of indoor propagation channel characteristics represents a fundamental step toward the design and the implementation of such applications [2]. This makes it necessary to have an accurate propagation to predict propagation channel characteristics in indoor environment. Deterministic methods as ray tracing (RT) are a good alternative to be used [3]. RT is a commonly used computational method for site-specific prediction of the radio channel characteristics of wireless communication systems. The RT technique inherently provides time delay and angle of arrival information for multipath reception conditions [4].

Due to the advantage of the RT model, researchers have sought to use this model as a crucial reference to investigate the indoor and outdoor propagation channel characteristics. In [5], the influence of building shape near the corner and its electrical properties on the ray-tracing predictions are presented. The shape is shown to have an important role in accurately predicting both received power and delay spread. Rizk et al. [6] study the influence of database accuracy on two-dimensional (2D) RT-based predictions in urban microcells. The work reported in [7] presents results of computer simulation based on a three-dimensional (3D) RT method for various directional polarized antennas, where the effects of polarization, antenna directivity, and room size on delay spread are investigated in the line-of-sight (LOS) indoor channel.

The current study, based on the RT model that have been proven and proposed in [8], is focused on investigating effects of antenna polarization on power and RMS delay spread in LOS/ out-of-sight (OOS) indoor cases. This paper is organized as follows. The used indoor RT model is introduced in Section II. The main focus is described in Section III, where power and RMS delay spread for both transmitter (Tx) and receiver (Rx) with two polarizations (vertical and horizontal linear) is predicted. Finally, the conclusions are drawn in Section IV.

## II. Indoor Ray Tracing Model

The RT model used in the present study considers the following ray paths:
 1) direct rays as line-of-sight,
 2) ray paths with all possible combinations of reflected rays on vertical facets, transmitted rays on vertical indoor objects and diffracted rays on vertical edges,
 3) ray paths containing all ray paths belonging to type 1) with one additional reflection on the floor,
 4) ray paths involving all ray paths belonging to type 2) with one additional reflection on the ceiling,
 5) all possible combinations of reflected rays on both floor and ceiling and transmitted rays on indoor objects with one additional wall reflection or not,
 6) single-reflected ray paths on horizontal facets (roofs) belonging to indoor objects,
 7) single-diffracted ray paths on horizontal edges.

In order to improve the efficiency of finding propagation paths, all propagation paths mentioned above are divided into four major categories (i.e., ray category I, II, III, and IV). And, different methods are used to deal with different ray categories. Based on creating virtual source tree [9] in which the relationship between neighbor nodes is left-son-and-right-brother one, all the propagation paths belonging to ray category I can be found out. All the propagation paths included in ray category II can be easily and rapidly determined by using the geometry principle presented in [10] and doing necessary intersection tests. Due to only taking into account one order reflection or diffraction, the computational cost determining both ray category III and IV can be ignored. Other details about the the 3D RT model can be obtained in [8].

Once all ray paths from Tx to Rx are determined, the

contribution of each path can be expressed as

$$E = \frac{E_0 f_t f_r \cdot e^{-jkr}}{r} \cdot \prod_{i=1}^{n} R_i \cdot \prod_{s=1}^{u} T_s \cdot \prod_{l=1}^{m} \left( D_l A_l^d \right),$$  (1)

where

| | |
|---|---|
| $k$ | propagation constant; |
| $E_0$ | reference field; |
| $f_t$ and $f_r$ | transmitting and receiving antenna field radiation patterns in the direction of the ray, respectively; |
| $r$ | path length; |
| $A_l^d$ | spreading factor for $l$th diffraction; |
| $n, m,$ and $u$ | total number of reflections, diffractions, and transmissions, respectively; |
| $R_i, D_l,$ and $T_s$ | reflection coefficient for the $i$th reflector, diffraction coefficient for the $l$th diffracting wedge, and transmission coefficient for the $s$th transmission, respectively (the calculation of the three coefficients can be traced in [11], [12]). |

According to the results calculated in Eq. (1), the contribution of each propagation path to the received power can be calculated using the following expression [13]:

$$P_r = P_t G_t G_r \left( \frac{\lambda}{4\pi} \right)^2 \left| \frac{E_r}{E_i} \right|^2,$$  (2)

where

| | |
|---|---|
| $E_i$ | $\sqrt{(\eta_0 / 4\pi) P_t G_t}$ ; |
| $\eta_0$ | $\sqrt{\eta / \varepsilon} \approx 120\pi$ = intrinsic impedance; |
| $P_{t,r}$ | transmitted and received power, respectively; |
| $G_{t,r}$ | transmitter and received gain, respectively; |
| $E_r$ | total electric field at receiver. |

All rays contributing significantly to the channel characterization at the examined position must be traced, and the complex impulse response of the radio channel is then found as the sum of these contributions [14]

$$h(t) = \sum_{i=1}^{N} A_i \delta(t - \tau_i) \exp(-j\vartheta_i) .$$  (3)

Here, the received signal $h(t)$ is formed by $N$ time delayed impulses (rays), each represented by an attenuated and phase-shifted version of the original transmitted impulse. The amplitude $A_i$, arrival time $\tau_i$ and phase $\vartheta_i$ of each ray are calculated by Eq. (1) and path length $r$.

### III. SIMULATION AND DISCUSS

In order to analyze the effects of antenna polarization on RMS delay spread in LOS/OOS indoor radio channel based on the 3D RT model introduced in Section II, a practical complex indoor environment is investigated in this section. The geometrical model of this indoor environment is shown in Fig. 1, including the floor and ceiling in order to consider the reflections produced in both. The room dimensions are width 7.9 m, length 17.9 m, and height 3.85 m. For this study, not only are the windows and doors included in the environment

database, but also the tables and all major metallic objects. This indoor environment is also used to demonstrate the accuracy of the proposed 3D RT method, and details about the geometrical features can be obtained in [15].

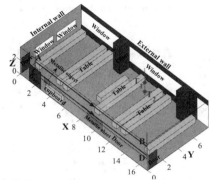

Fig. 1. 3D representation of the office.

In our calculation, two antenna positions are considered and are individually mounted at 1.8 m (located at $T_1$) and 0.5 m (located at $T_2$) above the floor. The carrier frequency of the two antennas are 2.4GHz and transmitted power are 20 dBm. Two sets of Rxs are distributed with constant spatial steps (chose as 0.2 m in the paper) along route AB (1.8 m high) and route CD (0.5 m high), respectively. For each route (15.0 m long) 76 prediction points are calculated. Both the transmitting and receiving antennas are typical half wavelength polarized dipoles. The two transmitting antennas are vertical polarization, while receiving antennas have both vertical polarization (parallel to the z-axe) and horizontal polarization (parallel to the x-axe). The material characteristics used in this study are shown in Table I. All the electrical properties and characteristics of the indoor environment elements, such as doors, windows, tables, cupboards and so on, are set by referring [1], [16].

TABLE I
THE MATERIAL CHARACTERISTICS USED IN THIS STUDY

| | Relative permittivity | Conductivity (S/m) |
|---|---|---|
| concrete (pillars, external wall, and the ceiling) | 9 | 0.1 |
| Plasterboard (internal walls) | 2.8 | 0.1 |
| Floor covered with marble | 7 | 0.00022 |
| Glass (windows) | 6.67 | $10^{-10}$ |
| Wood (doors) | 3.72 | $4.6 \times 10^{-4}$ |
| Metallic sheet | 3 | 100 |
| Metallic object | 1 | $3.23 \times 10^5$ |
| Chipboard (tables and cupboard) | 2 | 0.015 |

### A. Simulations for LOS Case

Fig. 2 shows the predicted power along the route AB in LOS indoor environment. It can be seen that the power curve for copular case have the same trend as that of cross-polar case. However, the average of predicted power for copular case is about 30 dBm more than that of cross-polar case.

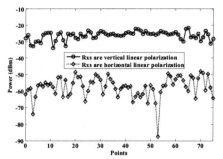

Fig. 2. Power for receivers along the route AB with two polarizations in LOS case, and Tx is located at $T_1$.

Fig. 3 illustrates the effects of antenna polarization on RMS delay spread in LOS indoor radio channel. Comparisons with results about predicted power shown in Fig. 2, the curves of delay spread for the two polarizations show better agreement, except for the middle of the route AB. This phenomenon occurs because the stronger signal with short delay is given up for cross-polar case.

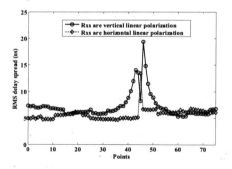

Fig. 3. RMS delay spread for receivers along the route AB with two polarizations in LOS case, and Tx is located at $T_1$.

### B. Simulations for OOS Case

As indicated in Fig. 4, the difference in predicted power between cross-polar and copular case is not noticeable in contrast to the simulation results shown in Fig. 2. So, we can presume that directions of the electromagnetic field for propagation paths are rotated through undergoing multiple reflections on the facet, diffractions on the edges, or the combinations of them.

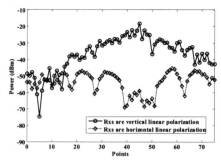

Fig. 4. Power for receivers along the route CD with two polarizations in OOS case, and Tx is located at $T_2$.

RMS delay spread for receivers along the route CD with two polarizations in OOS case is present. The trends of power curves for the two polarizations are opposition with each other. This proves that the directions of the electromagnetic field for propagation paths are rotated in OOS indoor environment. As for both LOS and OOS cases, the environment elements (tables, cupboard, and so on) more or less rotate the directions of the electromagnetic field, change the phase of each propagation path, and weaken the power taken by the path.

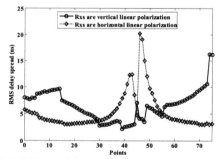

Fig. 5. RMS delay spread for receivers along the route CD with two polarizations in OOS case, and Tx is located at $T_2$.

### IV. CONCLUSIONS

In this paper, a new 3D indoor ray-tracing model, which can be applied to any complex indoor propagation environment, is presented. Based on a technique where multiple reflections, transmissions and diffractions are considered via the ray-path classification into four different categories, the used model implements different methods to deal with different ray categories. Utilizing this model, the effects of antenna polarization on power and RMS delay spread in LOS/OOS indoor radio channel are investigated. Simulation results indicate that for OOS cases, directions of the electromagnetic field for propagation paths are evidently rotated to varying degrees. Therefore, the signal arriving at Rx is not always the strongest for the copular case, especially in OOS indoor environment. Furthermore, the difference in predicted power between cross-polar and copular case, under LOS environment, is more noticeable than that of OOS case. Whereas, the difference in predicted RMS delay spread between cross-polar and copular case, under OOS environment, is more noticeable than that of LOS case.

### ACKNOWLEDGMENT

This work was supported in part by the Fundamental Research Funds for the Central Universities (Grant No. K50513100013), the National Science Foundation for Distinguished Young Scholars of China (Grant No. 61225002) and the Foundation of Huawei Technologies CO., Ltd. (Contract No. YBWL2010247).

### REFERENCES

[1] G. E. Athanasiadou and A. R. Nix, "A novel 3-D indoor ray-tracing propagation model: The path generator and evaluation of narrow-band and wide-band predictions," *IEEE Trans. Veh. Technol.*, vol. 49, no. 4, pp. 1152-1168, July 2000.

[2] V. Degli-Eposti, G. Lombardi, C. Passerini, and G. Riva, "Wide-band measurement and ray-tracing simulation of the 1900-MHz indoor propagation channel: Comparison criteria and results," *IEEE Trans. Antennas Propag.*, vol. 49, no. 7, pp. 1101-1110, July 2001.

[3] J. H. Jo, M. A. Ingram and N. Jayant, "Deterministic angle clustering in rectangular buildings based on ray-tracing," *IEEE Trans. Commun.*, vol. 53, no. 6, pp. 1047 - 1052, June 2005.

[4] K. A. Remley, H. R. Anderson and A. Weisshar, "Improving the accuracy of ray-tracing techniques for indoor propagation modeling," *IEEE Trans. Veh. Technol.*, vol. 49, no. 6, pp. 2350-2357, Nov. 2000.

[5] H. M. El-Sallabi, G. Liang, H. L. Bertoni, I. T. Rekanos, and P. Vainikainen, "Influence of diffraction coefficient and corner shape on ray prediction of power and delay spread in urban microcells," *IEEE Trans. Antennas Propag.*, vol. 50, no. 5, pp. 703-712, May 2002.

[6] K. Rizk, J. F. Wagen and F. Gardiol, "Influence of database accuracy on two-dimensional ray-tracing-based predictions in urban microcells," *IEEE Trans. Veh. Technol.*, vol. 49, no. 2, pp. 631-642, 2000.

[7] A. Kajiwara, "Effects of polarization, antenna directivity, and room size on delay spread in LOS indoor radio channel," *IEEE Trans. Veh. Technol.*, vol. 46, no. 1, pp. 169-175, Feb. 1997.

[8] Z. Y. Liu, L. X. Guo, X. Meng, and Z. M. Zhong, "A novel 3D ray-tracing model for propagation prediction in indoor environments," in *10th International Symposium on Antennas, Propagation & EM Theory*, Xi'an, China, 2012, pp. 428-431.

[9] Z. Y. Liu and L. X. Guo, "A quasi three-dimensional ray tracing method based on the virtual source tree in urban microcellular environments," *Prog. in Electromagn. Res.*, vol. 118, pp. 397-414, 2011.

[10] Z. Ji, B. Li, H. Wang, H. Chen, and T. K. Sarkar, "Efficient ray-tracing methmods for propagation prediction for indoor wireless communications," *IEEE Antennas Propag. Mag.*, vol. 43, no. 2, pp. 41 - 49, Apr. 2001.

[11] H. M. El-Sallabi and P. Vainikainen, "Improvements to diffraction coefficient for non-perfectly conducting wedges," *IEEE Trans. Antennas Propag.*, vol. 53, no. 9, pp. 3105-3109, Sept. 2005.

[12] S. Y. Seidel and T. S. Rappaport, "Site-specific propagation prediction for wireless in-building personal communication system design," *IEEE Trans. Veh. Technol.*, vol. 43, no. 4, pp. 879-891, Nov. 1994.

[13] S. Y. Tan and H. S. Tan, "A microcellular communications propagation model based on the uniform theory of diffraction and multiple image theory," *IEEE Trans. Antennas Propag.*, vol. 44, no. 10, pp. 1317-1326, Oct. 1996.

[14] G. L. Turin, F. D. Clapp, T. L. Johnston, S. B. Fine, and D. Lavry, "A statistical model of urban multipath propagation," *IEEE Trans. Veh. Technol.*, vol. VT-21, no. 1, pp. 1-9, Feb. 1972.

[15] Z. Y. Liu, L. X. Guo and X. Meng, "A novel 3D ray-tracing model for indoor propagation coverage prediction," *Journal of the Optical Society of America A*, (submitted).

[16] M. C. Lawton and J. P. McGeehan, "The application of a deterministic ray launching algorithm for the prediction of radio channel characteristics in small-cell environments," *IEEE Trans. Veh. Technol.*, vol. 43, no. 4, pp. 955-969, Nov. 1994.

# An RF Self-interference Cancellation Circuit for the Full-duplex Wireless Communications

Binqi Yang, Yunyang Dong, Zhiqiang Yu, Jianyi Zhou
State Key Laboratory of Millimeter Waves
School of Information and Engineering
Southeast University Nanjing, 211189, P. R. China

*Abstract-* **This paper presents the design of an RF self-interference cancellation circuit for the full duplex wireless communication systems. Excellent performance is achieved. The RF cancellation circuit provides more than 50dB self-interference cancellation across a 50MHz bandwidth. The transmitted power can be quite high for very low IMD products will be introduced in the circuit. The real time optimization algorithm is used to adaptive adjust the parameters of the circuit according to the change of the channel.**

*Keywords-* **Full-duplex wireless; Self-interference cancellation; Simplex Method**

## I. INTRODUCTION

The spectrum resource is very precious in wireless communication systems. Current wireless communication systems employ either a time-division duplexing or frequency-division duplexing approach to bidirectional communication, of which approaches are so-called Half-duplex communication techniques. The full-duplex is a more efficient way in terms of utilization of the spectrum resource. A full duplex radio can transmit and receive on the same frequency at the same time，so it can double the efficiency of spectrum.

The benefits of full duplex wireless communication have led researchers to explore how to build a practical full duplex radio. The main challenge in achieving full-duplex is to cancel the strong self-interference at the received side from the transmit antenna of a node, which can be 60-90dB stronger than received signal. Luckily, the self-interference is strongly correlated to transmit signals in the near field. Using knowledge of the transmission to cancel self-interference is feasible. Recent work has explored some self-interference cancellation techniques and practical implementations of full-duplex wireless nodes [1], [3], [5]. The antenna placement techniques, the RF self-interference cancellation and the digital cancellation are employed to achieve larger attenuation of the self-interfering signal.

This paper presents the design of an RF self-interference cancellation circuit which can realize adaptive self-interference cancellation in full duplex wireless. Both the antenna Separation and the RF cancellation technique are used and more than 52dB reduction is achieved across a 50MHz bandwidth. The operation frequency ranges from 2.3GHz to 2.4GHz. The maximum TX power can be up to 25dBm.

## II. DESIGN

The frequency response of the TX antenna and the RX antenna is very useful for the design and optimization of the cancellation circuit. Figure 1 shows the frequency response observed from separated transmit and receive antennas (commercial Omni-directional antennas, separated by 13cm). The loss is about 23dB.

Figure 2 shows the block diagram of the cancellation circuit. A vector modulator is used as the programmable phase shifter and attenuator to obtain the invert of self-interference signal. However, any radio that inverts a signal only through adjusting phase will always cause bandwidth constraint [1,2], like the *antenna cancellation* technique [2]. To obtain inverse of wideband signals, a fixed delay line is used for achieving same group delay in invert signal path and self-interference path. The vector modulator provides a fine-grained control to match amplitude and phase for invert signal path to optimal cancellation.

If the frequency response of self-interference channel is flat the channel frequency response of self-interference channel can be modeled as follows:

$$H_a(jw) = H_a e^{jw\tau_a} e^{j\varphi_a} \qquad (1)$$

where $H_a, \varphi_a, \tau_a$ are amplitude, phase offset and delay of self-interference channel respectively. The simplify frequency response of RF cancellation path is given by:

$$H_c(jw) = G e^{jw\tau_c} e^{j\varphi_c} \qquad (2)$$

where $G, \varphi_c, \tau_c$ are amplitude, phase offset and delay of cancellation path respectively. In a perfect cancellation:

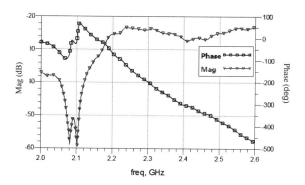

Fig 1. Frequency response observed from separated transmit and receive antennas in normal laboratory environment

978-7-5641-4279-7

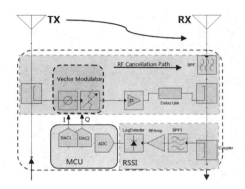

Fig 2. Block diagram of self-interference cancellation circuit. Passive delay line and vector modulator provide control to match amplitude and phase for the wideband interference

$$H_a(jw) + H_c(jw) = 0 \qquad (3)$$

For all $w$ in band, we have:

$$H_a e^{jw\tau_a} e^{j\varphi_a} = -G e^{jw\tau_c} e^{j\varphi}$$

$$-\frac{H_a}{G} e^{j(\varphi_a - \varphi)} = e^{jw(\tau_c - \tau_a)} \qquad (4)$$

Therefore, we have:

$$\tau_c = \tau_a$$

$$G = H_a \qquad (5)$$

$$\varphi = \varphi_a + \pi$$

It is obvious that to cancel in a width band, a radio needs to obtain the invert signal which is perfect negative of the transmit signal at all instants. The passive delay line is one of the key component for wide cancellation bandwidth.

A 15dB strip-line directional coupler is used to provide the transmit signal to the invert signal path as the self-interference reference. Another 15dB strip-line directional coupler is used as combiner to add invert signal and self-interference signal. The amplitude of the vector modulator can be controlled from a maximum of −7.5 dB to less than −37.5 dB at 2.35GHz. An additional 20dB RF amplifier is used in the RF cancellation path. Hence, the amplitude of RF cancellation path can be controlled from -17.5dB to -47.5dB, covering the range of self-interference channel.

The RF amplifier has very high P1dB and OIP3 specifications, so this cancellation design will not introduce too much non-linear distortion in a high transmit power. While this circuit can in theory provide a perfect cancellation to wideband signals, there is a practical limitation: the frequency response of self-interference channel is non-flat (Figure 1).It is hard to obtain a perfect invert of self-interference.

A logarithmic detector is used as the received signal strength indication (RSSI). A basic approach to estimate the state of cancellation is measuring the residual signal power after cancellation. This cancellation design can automatically adjust the phase and amplitude to minimize the residual signal

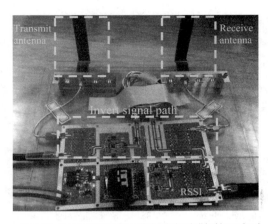

Fig 3. The implementation of cancellation design: build with two Omni-directional antennas and a cancellation circuit.

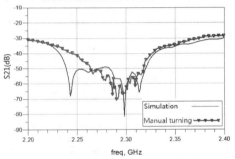

Fig 4. Cancellation performance with manual turning vs simulation.

power in response to channel changes by sampling RSSI.

Figure 3 shows the implementation of cancellation design. The invert signal path provides the inverse of self-interference. The RSSI value provides the residual signal energy after cancellation.

Figure 4 shows a cancellation result of manually turning the phase and amplitude of invert signal path to optimize cancellation at the received side. We use vector network analyzer N5230A to measure the S-parameter. The RF cancellation circuit provides around 50dB reduction over a 60MHz bandwidth. As a comparison, the figure also shows the simulation result using an ideal vector modulator model, measured S-parameter of amplifier and self-interference channel. The cancellation measured is imperfect across entire band. The key reason is the length deviation of delay line and that the frequency response of interference channel is non-flat (Figure 1).

III. AUTO-TURNING

The results in Figure 4 show that, if the amplitude and phase offset are set appropriate, the circuit can provide a good cancellation across a wide bandwidth. In this section we describe an algorithm that can quickly and accurately self-turn the cancellation circuit.

The I and Q inputs to the vector modulator set the gain and phase be-tween input and output. We have built a RSSI model with changing I and Q inputs to the vector modulator, vi and vq. For brevity we give the equation directly:

$$H(jw) = H_a \cdot e^{j\varphi_a} \cdot e^{jw\tau_a}$$

$$+ H_{coupler} \cdot H_{VM} \cdot H_{Amp} \cdot e^{jw\tau_a} \cdot H_{coupler}$$

$$= e^{jw\tau_a} \cdot (H_a \cdot e^{j\varphi_a} +$$

$$0.317 \times 10^{\left. 11.59 \left[ \sqrt{(\frac{V_i - 0.5}{0.5})^2 + (\frac{V_q - 0.5}{0.5})^2} \right]^{0.03895} -11.81 \right.} e^{j \arctan(\frac{V_i - 0.5}{V_q - 0.5})})$$

$$(6)$$

where $H_a, \varphi_a, \tau_a$ are amplitude, phase offset and delay of self-interference channel. The RSSI value is the magnitude of Equation (6). We add random amplitude changes and phase jitter presenting the dynamic environment when simulation. We plot the RSSI output with changing I and Q inputs to vector modulator in Figure 5. There is a deep null exists at the optimal point. The simplex method is used for finding the optimal point. For two variables, a simplex is a triangle and the method is a pattern search that compares values at the three vertices of triangle. The worst vertex will be replaced with a new vertex. The greatest advantage is that simplex method does not use derivatives. The algorithm can converge to the optimal point in 8-15 iterations and each iteration requires 1-4 measurements, less than 2 measurements required in most cases. Each measurement involves an amplitude and phase adjustment followed by RSSI sampling. It is effective and computationally compact. We omit the mathematics and plot the result of 8 iterations after the initialization of algorithm in Figure 6. Obviously, the best vertex of triangle is close to the optimal point. An efficient algorithm should perform less measurement in each iteration.

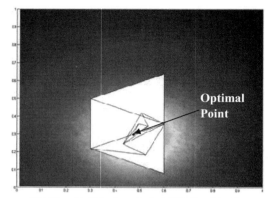

Fig6. The result of 8 iterations after the initialization of algorithm

In implementation of cancellation design, auto-turning algorithm is executed on a mixed signal microcontroller with 12-bit ADC and 12-bit DAC.

## IV. EXPERIMENTAL RESULTS

In this section, we provide experimental results for performance of the adaptive self-interference cancellation. We program a vector signal generator N5182A to generate a wideband 15MHz digital modulation signal with a center frequency of 2.35GHz which include MSK, QPSK, 16-QAM, and 64-QAM. Using spectrum analyzer FSL6 to observe signals at the received side. The results are summarized in Table 1. Figure7 shows the spectrum of residual signal after auto-tuning cancellation at the received side. The circuit provides isolation of 52dB for the self-interference.

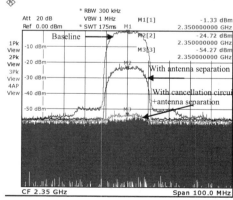

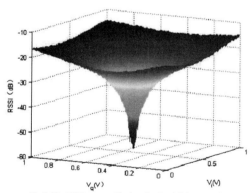

Fig 5. The RSSI output with changing I and Q inputs to the vector modulator, vi and vq.

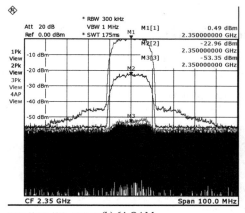

Date: 23.MAY.2013 14:39:44 (b) 64-QAM

Fig 7.Spectrum snapshots showing the effect of adaptive cancellation:
(a) 16-QAM (b) 64-QAM

*TABLE I: The results of adaptive cancellation*

|  | MSK | QPSK | 16-QAM | 64-QAM |
|---|---|---|---|---|
| Baseline | -0.27 dBm | -1.01 dBm | -1.33 dBm | 0.49dBm |
| Antenna Separation | -22.24dBm | -24.18dBm | -24.72dBm | -22.96dBm |
| Total Reduction | -49.24dBm | -53.75dBm | -54.27dBm | -53.35dBm |
| RF Cancellation | 27dB | 29.57dB | 29.55dB | 30.39dB |

## V. CONCLUSION

This paper presents an RF interference cancellation circuit for the full duplex wireless communication with strip-line directional couplers, a vector modulator and a passive delay line. This circuit provides around 52dB self-interference cancellation in a very wide frequency band. It can accurately and quickly optimize the cancellation for dynamic environment. With this RF cancellation mechanism, if further reduction is obtained by digital interference cancellation techniques, a practical full duplex communication will be established soon.

## ACKNOWLEDGEMENT

This work is supported in part by the National Natural Science Foundation of China under Grant 60702163, in part by the National Science and Technology Major Project of China under Grant 2010ZX03007-002-01, and 2011ZX03004-003.

## REFERENCES

[1] Jain, M., Choi, J. I., Kim, T., Bharadia, D., Seth, S., Srinivasan, K., ... & Sinha, P. (2011, September). Practical, real-time, full duplex wireless. In Proceedings of the 17th annual international conference on Mobile computing and networking (pp. 301-312). ACM.

[2] J. I. Choi, M. Jain, K. Srinivasan, P. Levis, and S. Katti. Achieving single channel, full duplex wireless communication. In Proceedings of the sixteenth annual international conference on Mobile computing and networking,MobiCom '10, pages 1–12, New York, NY, USA, 2010. ACM.

[3] M. Duarte and A. Sabharwal. Full-duplex wireless communications using off-the-shelf radios: Feasibility and first results. In Forty-Fourth Asilomar Conference on Signals, Systems, and Components, 2010.

[4] Sahai, A., Patel, G., & Sabharwal, A. (2011). Pushing the limits of full-duplex: Design and real-time implementation. arXiv preprint arXiv:1107.0607.

[5] Hong, S. S., Mehlman, J., & Katti, S. (2012). Picasso: flexible RF and spectrum slicing. ACM SIGCOMM Computer Communication Review, 42(4), 37-48.

[6] John H.Mathews ,Kurtis D.Fink. Numerical Methods Using Matlab.Forth Edition,Chapter8,Numerical Integration,Page399-425

# Ad Hoc Quantum Network Routing Protocol based on Quantum Teleportation

Cai Xiaofei[1) †] ,Yu Xutao[2)], Shi Xiaoxiang[3)], Qian Jin[4)], Shi Lihui[5)], Cai Youxun[5)]

1) 2) 5) 6) State Key Laboratory of Millimeter Waves
Southeast University
*Nanjing, P. R. China*
3) Nanjing Guobo electronics Co., Ltd   4) Jiangsu Province Water Conservancy Internet Data Center

*Abstract*-**In this paper, a quantum communication routing protocol is designed for quantum ad hoc network. This protocol is on-demand routing based on EPR numbers shared by adjacent nodes, concerning that it is a limited source. When quantum channel is established, quantum states from one quantum device can be teleport to another even when they do not share EPR pairs wirelessly. Part of information transferred by classic channel can be dealt with using simple logics. In this way, the goal of safety communication between source and destination is realized, improving the weakness of ad hoc network such as Eavesdropping and Active attacks. In terms of time complexity, the mechanism transports a quantum bit in time almost the same as the quantum teleportation does regardless of the number of hops between the source and destination.**

*Index Terms*-**EPR pair, quantum route, quantum entanglement, ad hoc network, quantum teleportation.**

## I. INTRODUCTION

In quantum mechanics, there is some kind of entangled relationship between two microscopic particles from same source, no matter how far away they are separated from each other, the state of a particle would immediately change according to another particle's change, which is called quantum entanglement. There is a certain degree of confidentiality in quantum teleportation. Based on the above characteristics, there is a article proposes a quantum routing mechanism in hierarchical network[1]. Another article designs a two-way secure communication protocol, which can set up the secure routing path from sender to receiver when they already share EPR pairs[3]. Communication Protocol Based on Classical Network Coding is also proposed[4].

This article introduces quantum communication technologies into an ad hoc network protocol to increase communication security[2]. In order to realize the quantum routing mechanism in a peer-to-peer network, first part of the article introduces how to build up a quantum network. The rest of this paper describes a routing process. Section IV discussed the some of the simulation results. And Section V discussed some related issues. Finally, conclusion are drawn in Section VI.

## II. THE MODEL OF QUANTUM AD HOC NETWORK

Base on ordinary ad hoc network, this network assumes each mobile device has a quantum communication functions, including the capacity of holding and manipulate EPR pairs, and they support classical communication at the same time. The scenario assumed as follows:

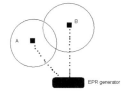

Figure 1. Skeleton diagram for EPR pairs generator

Therefore, a model of the ad hoc network can be simplified as follows:

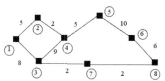

Figure 2. Model for quantum ad hoc network

Connection between points represents that entangled quantum channel has been established between them in the figure. Numbers marked in the solid line are the remaining number of EPR pairs between the two nodes. As shown, the path from node 1 to node 5, we tend to choose the path1-3-4-5, since there is a smaller number of the EPR pairs between node 2 and 4. Otherwise, a path is lost after EPR pairs are consumed out.

## III. QUANTUM PROTOCOL

### A. Route establishment process

When a node can not find an available route to a node, it broadcasts a RREQ message. This can happen if the destination is previously unknown to the node, or if a previously valid route to the destination expires or is broken .

The Originator Sequence Number in the RREQ message is the node's own sequence number. The RREQ ID field is

*This work has been supported by Science Project of Ministry of Transport of China (No. 2012-364-222-203).

978-7-5641-4279-7

incremented by one from the last RREQ ID used by the current node. Each node maintains only one RREQ ID.

When a node receives a RREQ, it first checks to determine whether it has received a RREQ with the same Originator IP Address and RREQ ID within at least the last PATH_TRAVERSAL_TIME milliseconds. If such a RREQ has been received, the node discards the received RREQ[1].

In quantum communication, we need to adopt the path of more EPR pairs, so weights will be modified: First, we take the reciprocal of the remaining number of EPR pairs to replace the path length as a measure of algorithms. Consider this, a transfer between two nodes will the consume the least EPR pairs. In consideration of the fairness of each node, if adjacent nodes has a high number of EPR pairs, an nodes with few remaining EPR pairs may be picked , this design is incomplete.

Therefore, redo some processing on the weight case: the weight can be reset as $n^{1/\alpha}$ ( $n$ represent the number of remaining EPR, $\alpha > 1$) In addition, the number 1 is special in inverse proportion, so if a node keeps only one EPR pair, its adjacent nodes automatically set its own weight to, but when the node itself need to communicate, it assumes the channel exists. Then normalized EPR numbers are used, and each received valid message will be recorded and compared of the EPR numbers.

After EPR number is processed, the node will record the source node IP address of the reverse routing. If necessary, this route will be created, or update according to the serial number of the source node in the RREQ message. The initiating node serial number in RREQ message will be used to compare with the corresponding destination node serial number in the reverse routing table, if bigger than that in the table, it will be added into the routing table. Entry "Normalized EPR number " in reverse route table is added directly from the RREQ message. Then the node further forwarded this RREQ message to other neighbors.

After a certain period of time, if an intermediate node received a second RREQ message from a different path, it compares the source node serial number first, found serial number is same to that in the table, further examine if the new path correspond to more EPR pairs number, then update the reverse routing in routing table, and change the "Normalized EPR number", then once again forwarded the message to other neighbors.

When destination node finally generates a RREP message, it propagates along the reverse route. The prior established quantum channel is usually a classical less hops or shorter distance path. The rest of the RREP message reach the destination node one after another will be checked about the serial number and EPR number. If the RREP is enough fresh, compare remaining EPR pairs number, if more than the existing line, namely it corresponds to a smaller "Normalized EPR number ", destination note will broadcasts new RREP of the same serial number to its neighbor node, informing that there is a better path. Since time delay is limited, before the optimal path is found, messages can transmitted through earlier

paths. When better path is found, source node received the RREQ message again, then a new quantum channel is established. Source node will take the new quantum channel.

### B.  Routing example

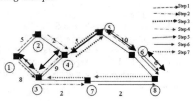

Figure 3. Route discovery process

The routing process is as follows:

*1)* Node 1 wants to communicate with node 8, first generate a RREQ message. And the serial number is set to 1.

*2)* Neighbor nodes 2, 3 received the RREQ, generate the reverse routing tables to record a path to the source node, then add remaining EPR numbers on the "Normalized EPR number ", record the sequence number as 1. Node 2 forwards the RREQ message to node 4 (EPR normalized number = 1/5). Node 3 forwards the RREQ message to node 4 (EPR normalized number = 1/8).

*3)* Node 4 receives two RREQ message, assuming that first received the message from node 2, it repeats the previous process: generate reverse routing table, record the path to node 2, then add the reciprocal of remaining EPR number to "Normalized EPR number ", record the serial number 1. Then the RREQ is forwarded to node 5. At this time the node received a message from node 3. It compares the serial number, found the serial number is same, so further compare the "Normalized EPR number ", found that there remains more EPR pairs in the path from nodes 3 (EPR normalized quantity = 1/8 +1 / 9 <1/5 +1 / 2), then forwards a new RREQ message to the downstream node, and change upstream path node in reverse routing table to node 3. Node 3 will also sent the RREQ message from node 1 to node 7. (EPR normalized quantity = 1/8 +1 / 2)

*4)* Node 5 twice received the RREQ message from node 4, the remaining amount in the second EPR records is more (EPR normalized quantity = 1/8 +1 / 8 +1 / 5), node 5 also update its routing table, sent a new RREQ message to the neighbor node 6. Node 8 also received the RREQ from node 7, then find out that the destination IP address in message matches its own IP address, prepare to send a RREP message to node 7.

*5)* Node 8 reply a RREP message along the prior arrival path.. In each hop, the node verifies the serial number first, and then record the positive route. The source node find the sequence number in RREP message sequence number meet that pre-issued serial number of the destination node set in RREQ message, so the path is set up.

Node 8 will receive two other RREQ messages from path 5-6-8, the destination node find that new message have the same serial number, but there are less  remaining EPR pairs. Node 8 will send a new RREP message to the source node along the new reverse route. If there is other RREQ message through a different path to node 8, but the remaining EPR pairs did not outnumber the previous one, the message will not be processed.

*6)* Node 8 sends a new RREP message.

The first RREQ choose the path 8-6-5-4-2-1.

The second RREQ choose the path 8-6-5-4-3-1.

Source node already communicate with node 8 in other path , but after it received RREP message from node 2, found the serial number is same while the path is better, source node choose the new paths.

## C. Communication example

There are two communication cases:

*1) Case of Quantum Relay*

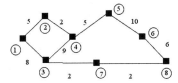

Figure 4. Model for Quantum Relay

The optimal path is 1-3-4-5-6-8, in the quantum relay scheme, communication process between node 1 and 8 should be as follow:

A qubit 0 can be symbolized in the form of

$$|y>=a|0>_0 +b|1>_0 \qquad (1)$$

Node 1 hope to sent the quantum state $|y>$ to node 8. And the shared entanglement between node 1 and node 3 is :

$$|\phi>_{13}=1/\sqrt{2}(|0>_1|1>_3 -|1>_1|0>_3) \qquad (2)$$

The statement of 3 qubits is shown as follows:

$$|\varphi>_{130}=|y>_0 \otimes|\phi>_{13}$$
$$=(a|0>_0 +b|1>_0)\otimes1/\sqrt{2}(|0>_1|1>_3 -|1>_1|0>_3)$$
$$=1/\sqrt{2}(a|0>_0|0>_1|1>_3 +b|1>_0|0>_1|1>_3$$
$$-a|0>_0|1>_1|0>_3 -b|1>_0|1>_1|0>_3)$$
$$=a/\sqrt{2}(|0>_0|0>_1|1>_3 -|0>_0|1>_1|0>_3)$$
$$+b/\sqrt{2}(|1>_0|0>_1|1>_3 -|1>_0|1>_1|0>_3) \qquad (3)$$

In order to complete quantum teleportation, node 1 must measure qubit 1 and 3. Thus, the wave function of the three particle system can be expressed as:

$$|\varphi>_{130}=1/2[|\theta^->_{10} (-a|0>_3 -b|1>_3)$$
$$+|\theta^+>_{10} (-a|0>_3 +b|1>_3)$$
$$+|\phi^->_{10} (a|1>_3 +b|0>_3)$$
$$+|\phi^+>_{10} (a|1>_3 -b|0>_3)] \qquad (4)$$

Node 1 uses the analyzer that can identify Bell base to measure qubit 1 and EPR qubit 0 together, and then transport measurement result to node 3.

Node 3 accordingly implement unitary transformation on qubit 3 correspond to the results, so as to achieve the quantum teleportation.

So that the information is transmitted from node 1 to node 3, and similarly in turn transmitted to the destination node 8.

As can be seen, the shortcoming of quantum relay plan is: the target qubit is transmitted through intermediate nodes, which results in the lack of security and confidentiality. In addition, the time it requires of data transfer is proportional to the number of routing hops. In order to overcome the above drawbacks, EPR-pair bridging program is proposed.

*2) Case of EPR-Pair Bridging*

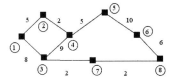

Figure 5. Model for RPR-pair Bridging

The optimal path is 1-3-4-5-6-8.

Quantum circuit makes quantum teleportation and measurement transmission of classical information be parallel, which is equivalent to transmit results for once.

The Quantum circuit is as below[1]:

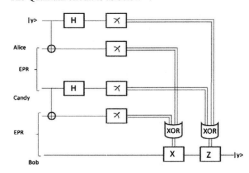

Figure 6. Quantum circuit for RPR-pair Bridging

Node 1 make $|y>$ serve as the control bit of the NAND gate control terminal to act on the EPR pair shared by node 1 and node 3. Then H gate is applied on $|y>$ and this measurement result is transmitted to XOR logic before Z gate of destination node. The other measurement result of EPR pairs is sent to XOR logic before X gate.

Node 3 make the EPR-pair shared with node 1 serve as the control bit of the NAND gate control terminal to act on the

EPR pair shared by node 3 and node destination node. Then make H-transform on the first EPR pair shared with node 1, send the measurement results to the Z gate XOR logic of destination node 8, and the other EPR pair measurement result to the X gate XOR logic of the destination node.

Then destination node exclusive XOR processing on measurement results transferred through classical channel, and send then into the quantum gates.

Finally, the EPR pair that destination node 8 shared with node 6 transformed into $|y>$ through X gate and Z gate ,which realizes the secure communications between two nodes that do not share EPR-pairs directly.

## IV. SIMULATION

Assume there are 8 nodes in a communication net. Source node and destination node are generated randomly. Different length of destination and different numbers of EPR pairs are distributed to each nodes.

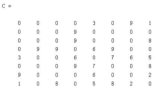

Figure 7. Model for network

Numbers in Matrix C shows length of destination between nodes. Then source node is node 4 and destination node is node 7. All the routes from node 4 to node 7 should be found out. EPR pairs [2 4 5 2 1 6 3 7].

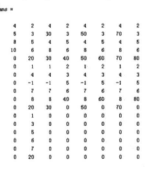

Figure 8. Transpose for result matrix

When element is -1,it means destination is found out. And when element is greater than 10, it means following elements are belong to another upstream node. There are 4 paths together. Path discovery order is according to distance between nodes: 4-5-7, 4-6-8-7,4-3-8-7,4-5-6-8-7.

Then first path to be found out is 4-5-7. Time is 2*(6km+6km)/c. At first communication is using this path.

Traditional routing protocol will always use this path.While quantum protocol will take EPR numbers into concern. So the final path is 4-6-8-7.

## V. DISCUSSION

When quantum channel is established, messages can be transmitted through the new path. There are usually two ways to teleport a quantum state from one quantum device to another that do not share EPR pairs wirelessly.

Due to space constraints, the route maintenance mechanism is not completely explained. And we can further consider to add the quantum communication technology to Routing Protocol contents. In addition, random routing on multistage interconnection networks can also be taken into concern[5]. The security and credibility of the intermediate nodes shall also be concerned, since teleportation through a long path rely much on the path[6].The concept of control qubits is a new idea[7].

## VI. CONCLUSION

This paper introduced quantum communication technology into Ad hoc network to improve its security of communication, Then proposed a quantum routing mechanism in peer-to-peer network, which enables a quantum mobile device to teleport a quantum state to a remote site even if they do not share EPR pairs mutually. In terms of the time complexity, the time that quantum network takes to teleport a quantum state is independent of the number of routing hops. Among the route protocol, remaining EPR pair number is also taken into concern since EPR pairs is a limited source.

REFERENCES

[1] Sheng-Tzong Cheng, Chun-Yen Wang, and Ming-Hon Tao, "Quantum Communication for Wireless Wide-Area Networks," IEEE JOURNAL ON SELECTED AREAS IN COMMUNICATIONS, vol. 23, NO.7, JULY 2005, pp. 1426–1430.

[2] Ad hoc On-Demand Distance Vector (AODV) Routing. IETF RFC 3561, 19 July 2002, pp 10-11.

[3] Tien-Sheng Lin, Tien-ShengLin, Sy-Yen Kuo, QUANTUM WIRELESS SECURE COMMUNICATION PROTOCOL, ICCST 2007, pp 146-154.

[4] Hirotada Kobayashi, Franc¸ois Le Gall, "Perfect Quantum Network Communication Protocol Based on Classical Network Coding," ISIT 2010, Austin, Texas, U.S.A., June 13 - 18, 2010

[5] Rahul Ratan, Manish K. Shukla, A. Yavuz Oruc, "On Random Routing and its Application to Quantum Interconnection Networks," 40th Annual Conference on Information Sciences and Systems, pp. 1744 - 1749 .

[6] Stefan Rass and Peter Schartner, "A Unified Framework for the Analysis of Availability, Reliability and Security, With Applications to Quantum Networks," IEEE TRANSACTIONS ON SYSTEMS, MAN, AND CYBERNETICS—PART C: APPLICATIONS AND REVIEWS, VOL. 41, NO. 1, JANUARY 2011.

[7] Tien-Sheng Lin, Tien-ShengLin, Sy-Yen Kuo. SECURE QUANTUM PACKET TRANSMISSION MECHANISM FOR WIRELESS NETWORKS. IEEE2008, ICCST2008, pp29-35.

# The Service Modeling and Scheduling for Wireless Access Network Oriented Intelligent Transportation System (ITS)*

Wang Xiaojun, Dai Haikuo, Chen Xiaoshu

NCRL, Southeast University

Sipailou, Nanjing 210096 China

*Abstract*-This paper first describes the architecture and the main services of the wireless network oriented ITS (Intelligent Transportation System). The services of CVIS (Cooperative Vehicle-Infrastructure System) and PTT (Push To Talk) voice are analyzed and modelled. Pointed scheduling schemes are proposed for different services according to their QoS requirements. For CVIS service, vehicle speed adapted dynamic scheduling is adopted. For service of PTT voice, scheduling based on state transition is adopted. For video and other services which are not sensitive to delay, scheduling request is generated based on buffer status. Finally, the scheduling schemes are simulated on OPNET, from which the availability is verified with performance curve of delay and packet loss rate. And also, this provides reference for ITS access network design.

*Key words* - Intelligent Transportation System (ITS), wireless access networks, QoS, Scheduling schemes, OPNET

## I. INTRODUCTION

In ITS[1] advanced information technology such as electronic control technology, data communication technology and computer technology in vehicle and road management is adopted to reduce traffic congestion and accident, and to save resource for improving environment. As one of ITS foundation, communication technology plays an important role in ITS, which is used to exchange information among vehicle and infrastructure.

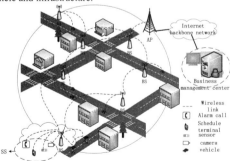

Figure 1. ITS access network structure

The ITS communication network structure[2] which is adopted in this paper is shown in figure 1. There are two types of SS (Subscriber Station) which are fixed type and mobile type. The fixed terminals include emergency telephones,

*This work has been partially supported by science and technology projects of China Ministry of Transport ( No. 2012-364-222-203).

978-7-5641-4279-7

sensors and cameras. And the mobile terminals include OBUs (On Board Units) in vehicle for CVIS and PTT terminals. The BSs (Base Stations) are in charge of data forwarding and form a wireless MESH network. Multiple SSs access the nearest BS through wireless link, which is called PMP wireless access network. The AP (Access Point) is the gateway between wireless MESH network and ITS backbone network.

## II. SERVICES ANALYSIS FOR ITS WIRELESS ACCESS NETWORK

In ITS, QoS guarantee of the access network helps to provide timely and reliable information exchange. Table 1 describes the classification and characteristics[3] of services in ITS.

TABLE I
TYPES AND CHARACTERISTICS OF BUSINESS IN ITS

| Types | content | characteristics |
|---|---|---|
| The service of CVIS | Vehicle identifier, vehicle GPS position, speed | Highest priority, related to vehicle speed |
| PTT voice | PTT voice | requires low latency |
| Video | Surveillance camera | Volatility, latency requirements |
| Other data | Sensor data | Low priority, low speed, stable |

The data of CVIS service, which is most important in ITS, need to be timely exchanged among vehicles and infrastructure to improve traffic safety such as crash avoidance, and to reduce traffic congestion. PTT voice is used in trunking communication, the end-to-end delay of which less than 0.1s will not be aware of. The video service needs more bandwidth, but is not very sensitive with time delay because it's not interactive. The other data service has the lowest priority and low generating speed.

### A. Service analysis and Modeling for CVIS

Service of CVIS has the highest priority. Enough channel resource must be allocated to it in scheduling design. Usually, higher upload frequency is needed for vehicle with higher speed.

Suppose the number of the vehicles within the range of a BS is $n$, which maximum is $N_{max}$ when the road is in completely congested status. When $n$ is very small the speed of vehicle $v$ reaches its maximum: $V_{max}$.

When $n$ is small, the average speed $\mu$ of vehicles is large. With the increase of $n$, $\mu$ begins to decrease, which will fall to zero on the congestion status. The relationship between $n$ and $\mu$ can be represented by traffic flow theory [4], which is:

$$\mu = V_{max}(1 - \frac{n}{N_{max}})$$
(1)

According to traffic flow theory, in a certain range, all vehicle speed follows normal distribution, that is:

$$f(v) = \frac{1}{\sigma\sqrt{2\pi}} e^{-\frac{(v-\mu)^2}{2\sigma^2}}$$
(2)

Where $\mu$ is average speed and $\sigma$ is standard variances.

According to the above discussion, the upload frequency $f$ should be determined by the speed $v$, high upload frequency with high speed. For study convenience, the relationship between $f$ and $v$ is defined as:

$$f = \frac{F_{max} - F_{min}}{V_{max}} v + F_{min}$$
(3)

Now the network traffic of CVIS can be concluded as:

$$s = n \int_0^{v_{max}} (\frac{F_{max} - F_{min}}{V_{max}} v + F_{min}) \frac{1}{\sigma\sqrt{2\pi}} e^{-\frac{(v-\mu)^2}{2\sigma^2}} dv$$
(4)

According to traffic flow theory, $\sigma \ll \mu$. Then:

$$s = n(F_{max} - \frac{F_{max} - F_{min}}{N_{max}} n)$$
(5)

The result above is that the offered traffic of CVIS and the number of vehicles have a quadratic function relationship. The offered traffic first increased and then decreased. When

$$n = \frac{F_{max} N_{max}}{2(F_{max} - F_{min})}$$, the offered traffic reaches its maximum:

$$S_{max} = \frac{F_{max}^2 N_{max}}{4(F_{max} - F_{min})}$$
(6)

### B. Service analysis for PTT voice

PTT is introduced in ITS because it can play a important role in regular and emergency scheduling. To achieve push to talk, the access time must be less 1s. And the delay of speech frame on the link of wireless access network must be far below 0.1s.

PTT voice basically has two states: activation state and silent state [5]. These two states appear alternately. Voice packets are created periodically on the activation state, while packets of silent frame is created in far lower period on the silent state, as shown in figure 2.

Figure 2. Voice of quantitative group

### III. ITS MULTI-REQUEST(ITS-MR) SCHEDULING SCHEME

For downlink service scheduling, BS knows all arrived service data, and can complete scheduling alone. For uplink service scheduling, each SS sends the request of resource to BS, according to which BS allocate correspond resource to each SS. The uplink scheduling needs more control overhead and more coordination, which make it more complex than downlink scheduling.

ITS scheduling model with QoS guarantee is designed as in figure 3, reference to that of CHU [6].

In the uplink scheduling, the services is classified by SS. Different services flow into respective buffer queue. Resource request message is generated by SS for each buffered service and is sent to the BS in uplink sub-frame. For service of CVIS, the request message is implicit in each service data. PTT SS only need to report the status transition to BS with piggyback. For other two types of services, the queen length is reported to BS periodically using piggyback, for which EDF (Earliest Deadline First) [7] is adopted that means timeout packets need to be discarded.

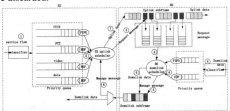

Figure 3. ITS scheduling model with QoS guarantee

BS receives the request messages from SS and allocate the resource accordingly. Then BS send these results in UL-MAP in downlink sub-frame. Each SS sends these packets which BS has allocate resource for.

The uplink scheduling of CVIS is shown in figure 4. Each OBU transfers one CVIS packet or not in the next uplink sub-frame according to received UL-MAP.

The packet of CVIS includes the identifier, position and speed of the vehicle. Once BS gets the speed, it calculates the serial number of frames away from this frame by the formula (3).

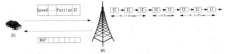

Figure 4. Uplink scheduling of CVIS

BS maintains a chained list, the node of which contains SS ID and the serial number of the frame to be allocated. Each calculation result is to be attached at the end of the list. BS will fill and broadcast the UL-MAP one frame in advanced according to the head of the list.

For PTT voice, the traffic in silent state is less than in activation state. So resource utilization will be low if fixed resource is allocated for both of the states. A scheduling based on state transitions is designed as shown in figure 5, reference to that of Dou [9].

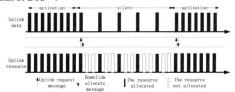

Figure 5. State transitions scheduling of PTT voice

This scheme takes advantage of the voice service features: the rate is unchanged during the activated state or silence state. SS only need to send the change of state to BS. And BS allocates different TDMA resource to SS accordingly. Only one downlink allocation message is needed in one state duration, which means very little overhead and no extra delay. The disadvantage of state transition scheduling is that it may introduce a small amount of packet loss on transition from silent state to active state. This small packet loss could be more tolerable because of the continuity of PPT voice.

ITS video mainly refers to monitoring video, which are not interactive. So it may need more bandwidth, and has a lot of volatility, but no restrict time delay is required. According to these characteristics, a buffer is set for generated video data. SS send the queen length of video data in the buffer, as shown in figure 6. The scheduling of other data adopt the same way to video.

BS first allocates the resources to services of CVIS and PTT voice, then allocates the rest to video and data services.

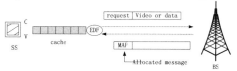

Figure 6. ITS video and data business scheduling scheme

## IV. SIMULATION OF ITS-MR SCHEDULING

ITS-MR scheduling scheme is simulated on OPNET, to verify whether ITS-MR can meet the QoS requirements of ITS services.

In the scenarios, there are one BS, 336 vehicles with OBU, 3 PTTs, 1 video and 1 data. The simulation duration is 30min, in which the number of available OBUs increases from 0 to the maximum linearly.

Figure 7 shows the receiving rate of CVIS data. As the core Service of ITS, CVIS packet has highest priority to get the transfer resource. The throughput first increased and then decreased as which has been analyzed previously. The small

fluctuations of the curve is the result of normal distribution of vehicle speed.

Figure 8 shows the simulation result of PTT voice delay, it can be seen from the figure that most of the PTT packet delay is less than 0.01s, which meets the requirements of PTT service.

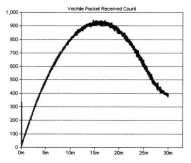

Figure 7. simulate results of CVIS service

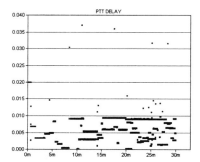

Figure 8. simulate results of PTT voice business

The comparison of generating and receiving rate of video service is shown in figure 9. The receiving rate curve is smoother than generating rate curve because of buffer adopted. It also can be seen that the delay is larger when CVIS service rate is larger in middle part of simulation duration.

The left part of figure 10 shows the additive comparison of generation amount and the reception amount of other data. Packet loss occurs in middle interval. The right part of the figure shows the packages delay has reached the maximum (5s) also in the middle interval. These are because that larger amount of CVIS data occupy more bandwidth.

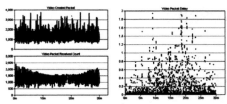

Figure 9. simulate results of video service

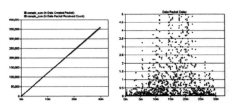

Figure 10. simulate results of data service

From the above simulation results, it can be concluded that ITS-MR scheduling scheme effectively satisfy the QoS requirements of each service. ITS-MR scheduling scheme provides a reference for the actual network design.

### REFERENCES

[1] http://www.intellidriveusa.org/.
[2] Chen Wenhui. Research and implement on physical layer bearer technology of ITS-Oriented wireless access network [D]. Nanjing: Southeast University, 2011.
[3] Jin Shengbo. The MAC Layer Research and Design of ITS-Oriented Wireless Mesh Access Network [D]. Nanjing: Southeast University, 2010.1.
[4] Daniel, Matthew. Traffic flow theory [M].Beijing: China Machine Press, 2007.
[5] Yang T, Tsang H.K.A novel approach to estimating the cell loss probability in a multiplexer with homogeneous ON-OFF sources. [J]. IEEE Trans Commun, 43(1), pp. 117- 126, 1995.
[6] Chu GuoSong, Wang Deng, Mei Shunliang. A QoS Architecture for the MAC Protocol of IEEE 802.16 BWA System[C]. Communications, Circuits and Systems and West Sino Expositions, IEEE2002 International Conference on, Volume 1, pp. 435-439, 29 June-1 July 2002.
[7] Chipalkatti R., Kurose J.F., Towsley D.. Scheduling policies for real-time and non-real-time traffic in a statistical multiplexer. In IEEE INFOCOM'89, Volume 3, pp. 774-783, April 1989.
[8] Golestani S.J.. A Self-clocked Fair Queuing Scheme for High Speed AGplications. Proc.INFOCOM, 1994, 4.
[9] Dou Huiru,Bi Haizhou,Xie Yongbing. Scheduling research of LTE system voice business [J]. Digital communications, 2007.1.

# Radio Channel Modeling and Measurement of a Localization Rescue System

Lun-Shang Chai, Jiao He, Xing-Chang Wei

Dept of Information Science and Electronic Engineering
Zhejiang University
Hangzhou, China
E-mail: Chailunshang@126.com

*Abstract*—Nowadays with the frequent occurrence of natural disasters, there is a great demand for accurate and fast novel localization technique for rescue. Thanks for the advancements of modern wireless communication technology, lots of people own a mobile phone today, and the most important thing is that the phone is often the only device people carried when such disasters happen, which can send a radio signal to prove their exist. Thus, we can take use of this fact to search and rescue the victims in disaster areas. This paper presents the characterizing radio channel modeling in collapse and measurement experiment, which is a key component for developing a novel passive radio localization system. The behavior of wireless channel can be regarded as a stochastic process, the channel is composed of many complex random obstacles, their influence on the mobile signal propagation is simulated and measured to set up a database, the statistical characteristics of channel can be determined through plenty of simulation and measurement data.

## I. INTRODUCTION

Last two decades have witnessed a dramatic boom in the wireless communications industry, hence increasing the number of users of mobile communication devices. Recent survey shows that the majority of buried victims carry their mobile phones in natural disasters like avalanches, earthquakes or landslides, this fact can be taken advantages for smart search and rescue applications when disasters happen. Of course, there are already several localization methods, but many are active, what's more, the victims usually can't do anything when buried and in faint, so they are not always suitable for search and rescue applications [1]. It is necessary to develop a new passive approach for mobile localization. Such development requires that the information of mobile signal attenuation and time delay is measured and modeled, through signal process technique, the position of victim within phone will be determined.

In order to capture the accurate information about the located phones, the characteristics of radio channel between the victim and location system must be determined, which will form the foundation for realization of location algorithm and hardware detection cell in the next rescue system development. Conventional studies have been mostly performed in indoor [2], outdoor-to-indoor environment, except an earthquake rescue project made in Chengdu University of Technology and the famous "I-LOV" project made in Germany , almost no much study is performed in complex ruin environment after big disasters [3]. In disasters, the radio channel becomes pretty complex due to the complex ruin structures. The reflection, scattering and diffraction of electromagnetic waves introduce the attenuation and frequency dispersion character of the radio channel, even at the same position, the RF signals can become different at the next time interval. This actually yet becomes a challenge problem to deal with.

The method for modeling radio channel propagation can be divided in three main categories. The first one is using the ideal statistical model, typically assumed a complex Gaussian Channel. The second one is using ray-tracing method based on the establishment of geometric distribution of scatters in wireless channel [4], and the third method is based on measurement in actual physical environment. Due to the goodness of reality, simulation based on ray-tracing method and measurement method are employed in our work. Based on high precision measurement of channel response in time and frequency domain, a parametric model of radio channel is established. The channel impulse response power delay spectrum and other physical properties are employed to set channel model with parametric mathematical expressions, then using parameter estimation method to extract the model parameters from large amounts of measured data, thus the statistical character of channel in collapse can be analyzed and determined. As the two important parameters of the channel, the pass loss and the Root-Mean-Spread delay will suffer from different changes in different collapsed structures or at different frequency range, their statistic distribute characteristics can be analyzed in the channel model [5]. Taking account to small scale fading caused by multipath effects, the multipath delay and amplitude characteristics are also analyzed, respectively, spectral estimation techniques can be employed to model the channel [6]. Due to the limitation of our measurement condition, only the signal strength and channel received power are measured in this paper, the delay spread information will be measured and characterized in our future work.

978-7-5641-4279-7

## II. CHANNEL MECHANISM

### A. Small Scale Fading Mechanism

The channel in collapse first suffers from small scale fading caused by multipath propagation [7]. Our study is mainly performed on the mobile communication channel which works in the frequency range of 890MHz to 960MHz (GSM) and 1805 to 1880MHz in DCS network.

The radio channel in collapse can be regarded as fast time invariant in a extremely short time interval, so the multipath fading channel can be described as follows:

$$h(t) = \sum_{k=0}^{M-1} a_k \cdot \delta(t - \tau_k) \qquad (1)$$

where M is the number of significant multipath components, $a_k$ is the complex amplitude of the k'th component arriving at receiving end. Channel Frequency Response (CFR) can be obtained utilizing the Fourier transform:

$$H(f) = \sum_{k=0}^{M-1} a_k \cdot e^{-j2\pi f \cdot \tau_k} \qquad (2)$$

Radio channel is fully determined by $h(t)$ in time domain or by $H(f)$ in frequency domain.

The power delay profile (PDP) is the average of $h(t)$ [8]:

$$p(t) = E\left(\left|h(t)\right|^2\right) = \left|\sum_{k=0}^{M-1} a_k \cdot \delta(t - \tau_k)\right|^2 \qquad (3)$$

where E(...) denotes the mathematical expectation. The mean excess delay $D$ and $\tau_{rms}$ can be calculated through $p(t)$.

### B. Large Scale Fading Mechanism

Considering over a longer time frame, the received signal strength is variable due to different radio channel environment, this fact reflects the large scale fading of the radio channel [9]. Pass loss in collapse can be divided into two loss mechanism.

$$P_L = L_{free} + L_{obstacle} \qquad (4)$$

In line-of-sight condition, $L_{free}$ is approximately yielded with the dB form as below:

$$L_{free} = 32.4 + 20 \lg r + 20 \lg f \qquad (5)$$

where the units of $r$ and $f$ is km and MHz, respectively.

$L_{obstacle}$ is related to the ruin structure and the material property of obstacles. Research shows that the attenuation constant ($\alpha$) through obstacle is determined by the complex permittivity and permeability of the material, as well as the frequency of electromagnetic waves:

$$\alpha = f(\mu, \tilde{\varepsilon}, \omega) \qquad (6)$$

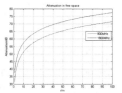

Fig.1. Attenuation in free space

For plane wave incident into obstacles with infinite transverse size, the attenuation factor can be derived from plane wave theory. Simulation can be done based on this result. Research shows that this approximate method is significant if the phone is fully covered with obstacles.

$$\alpha = \omega \sqrt{\frac{\mu\varepsilon}{2}\left(\sqrt{1 + (\frac{\sigma}{\omega\varepsilon})^2} - 1\right)} \qquad (7)$$

where $\sigma$ is the conductivity of obstacle material.

If the ruin is stratified structure, it can be described as following:

$$L_{obstacle} = \sum_k \alpha_k \cdot h_k \qquad (8)$$

$h_k$ is the thickness of $k$ layer barrier. Attenuation constant is studied in many literatures in fact. Thus if the thickness of every obstacle is measured, the total attenuation is calculated.

The received signal strength can be studied from statistical perspective. Let $S_n$ denote the signal strength through $n$ layer obstacle, then

$$S_n = S_0 \cdot e^{-\sum_{k=1}^{n} \alpha_k h_k} \qquad (9)$$

$\alpha_k$ and $h_k$ can be regarded as mutually independent random variables since they are different for different obstacles, and not relevant with each other. In the case of heavy disasters, large numbers of complex obstacles appear in ruins, so according to the Central-Limit-Theorem for independent identical distribution, when the number of obstacles $n$ is large enough, the sum $\sum_{k=1}^{n} \alpha_k \cdot h_k$ obeys normal distribution approximately. Thus the logarithm of received signal strength $S = 10 \lg s$ is a normal Gauss process, obeying normal distribution $N(\mu_s, \sigma_s^2)$:

$$p(S) = \frac{1}{\sqrt{2\pi}\sigma_s} e^{-(S-\mu_s)^2/2\sigma_s^2} \qquad (10)$$

The statistical characteristic of signal strength suffered from ruin attenuation obeys lognormal distribution in some case, this fact can be validated through future measurement.

## III. SIMULATION

In the disaster debris area, mobile phones may be buried or blocked by all sorts of obstacles, such as collapsed wall, ferroconcrete, soil, clay, cement brick, sand, glass, pipes, furniture, etc. Their electromagnetic parameters and distribution structure situation seriously affect the spread of mobile signal, duel to penetration, reflection, absorption, diffraction and scattering phenomenon. Through large numbers of simulation experiments and actual measurement, a database which contains the geometry and electromagnetic information of obstacles will be established to analysis their barrier effect.

The electromagnetic parameters of common obstacle materials have been researched by many literatures. Most barrier materials are non-magnetic materials, their complex permittivity, conductivity and loss angle are the main parameters which affect signal propagation.

Ferroconcrete is one of the main material of modern architecture, as well as the most common obstruction material in collapse. Different material of concrete and different number of steel bars make different effect on wireless communication. As a reference, we choose ferroconcrete with relative dielectric constant 6.5 and loss tangent 0.2 (at 900MHz) for simulation. The penetration loss is about 0.46 dB/cm for GSM signal as shown in fig 2. This value is very close to the measurement of ferroconcrete wall in IMTEK buildings in "I-LOV" project [9]. Cement brick with relative dielectric constant 5 and conductivity 0.05 is also simulated.

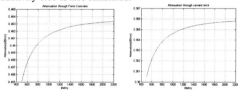

Fig.2. Penetration loss in ferroconcrete and bricks

As a contrast, Table I shows the difference between our simulation and measurement results in "I-LOV" project. Simulation results are higher than measurement value duel to the infinite transverse size of obstacle in simulation and the finite obstacle size in actual field measurement, and their electric parameters are not same yet.

TABLE I.
THE ATTENUATION FACTORS FOR GSM900 SIGNAL BETWEEN SIMULATION AND MEASUREMENT

|  | Our Simulation | Measurement in "I-LOV" project |
|---|---|---|
| $\alpha\_ferroconcrete$ [dB/cm] | 0.46 | 0.42 [9] |
| $\alpha\_brick$ [dB/cm] | 0.36 | 0.27 [10] |

Simulation shows that the conductivity of obstacle material is the most important parameter which affects the penetration loss of wireless signal. Metal has large conductivity, for metal furniture materials, due to the influence of skin effect of electromagnetic wave, signal penetration depth is limited, which causes electromagnetic shielding effect, leading to serious signal attenuation. Other obstacle materials will be investigated in the future.

## IV. MEASUREMENT

Field measurement is the most popular mean to estimate and analyze the characteristics of radio channel. Analysis of the radio channel in collapse strongly depends on the original database acquired from measurement [8][9]. Channel measurement can be conducted in both time domain and frequency domain. This paper presents the measurement results conducted in frequency domain as shown in Fig.3. which is set up based on a vector network analyzer or spectrum analyzer.

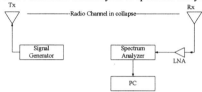

Fig .3. Measurement platform

The transmitting antenna is buried in ruins, which simulates the buried phone with weak signal launched, while the receiving antenna is fixed in a certain height. The LNA is utilized to increase the dynamic range of received signal if possible. Taking account of the behavior of antennas, cables, and measurement equipment itself, calibration aimed at reducing the effect of other unexpected channels must be on the way. The method is conducting the same measurement without ruins, after removing this response component, it is the behavior of ruin channel itself.

For the limitation of experimental conditions, experiment is conducted in a $7 \times 4 \times 3m$ electromagnetic shielding absorbing chamber at first stage. Agilent analog signal generator N5183A is used to generate the signal with 1W power. An omni-directional double-cone antenna with 500MHz to 3GHz bandwidth is employed as transmitting antenna, and a directional log-periodic antenna with 200MHz to 2GHz bandwidth is employed as receiving antenna. Agilent signal analyzer N9020A is utilized to measure the signal strength outside the chamber. Some bricks are used as obstacles. Two frequency band and two polarization mode are conducted. The distance between two antennas is 3m.

Figures 4, 5 and 6 show the measurement environment and results. The measured channel power spectrum density is shown in Tab. II.

The measurement shows two results. Firstly, the signal vertically received is larger than that of horizontally received both at 900MHz and 1800MHz, it is caused by the vertical polarization mode of two antennas. Secondly, obstacles show different effect on 900MHz and 1800MHz. There is an interesting phenomenon that the maximum signal strength with bricks is a little bigger than without bricks at 900MHz, while it doesn't happen at 1800MHz. At 900MHz, the signal wave length is close to the obstacle size, so the diffraction phenomenon is obvious, while for 1800MHz signal, wave length is smaller, it is more difficult to diffract. As a result, the channel in 900MHz is more stable than in 1800MHz. Channel with obstacles introduces multipath effect, received signal is the superposition of every path signal, coupled with the influence of antenna pattern, the total maximum signal is weaken someplace while it may be enhanced at another place.

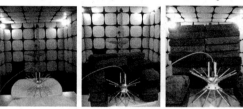

(a) No obstacles    (b) Bricks surrounding    (c) Bricks stop
Fig.4. Measurement environment

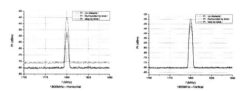

Fig.5. Measured received power at 900MHz

Fig.6. Measured received power at 1800MHz

TABLE II.
CHANNEL POWER SPECTRUM DENSITY (PSD) OF MEASUREMENT

| Freq. | Polarization | PSD ( dBm/Hz ) | | |
|---|---|---|---|---|
| | | *(a)* | *(b)* | *(c)* |
| 900MHz | *Horizontal* | -116 | -109.7 | -110.1 |
| | *Vertical* | -96.9 | -96.1 | -96.5 |
| 1800MHz | *Horizontal* | -103.1 | −115.2 | -117.5 |
| | *Vertical* | -93 | -96.8 | -100 |

## V. CONCLUSION

This paper presents a modeling and simulation method of radio channel in collapse when disasters happen for rescue, which makes up the most important component for our radio localization system. The channel suffers small scale fading and large scale fading, its behavior can be processed as a stochastic process. As a statistical result in theory, the received signal strength obeys lognormal distribution in some case, future measurement is on the scheme to validate this result. Simulation based on numerical calculation is done to analyze the loss of some typical obstacles in collapse area. Ferroconcrete and bricks are investigated first, simulation results are similar to measurement results in "I-LOV" project. Measurement in dark chamber shows that the radio channel performs different characteristics at different frequency range. Coupled with the multipath effect and diffraction phenomenon, obstacles in channel show quite different effect on 900MHz and 1800MHz frequency. As a result, signal at 900MHz frequency shows diffraction more easier than that of 1800MHz. More field measurement will be conducted in future.

ACKNOWLEDGMENT

The research is partially supported by the Fundamental Research Funds for the Central Universities (2013xzzx008-2), Zhejiang Nature Science Foundation No. Z1110330 and China National Science Fund Grant No.61274110. The authors would like to thank the colleagues of RF & Nano-Electronic Research Centre for their scientific discussions and guidance.

REFERENCES

[1] He Suqin,Wang Jian-Lin. Research on Positioning Technique Based on Measurement of TDOA[C].//ISTM/2005 :.2005:8468-8471.

[2] G. Morrison, "Measurement, characterization and modelling of the indoor radio propagation channel", Ph.D. Thesis, The University of Calgary, November 2000.

[3] S. Zorn, R. Rose, A. Goetz, and R. Weigel, "A novel technique for mobile phone localization for search and rescue applications," in Proc. Int Indoor Positioning and Indoor Navigation (IPIN) Conf, 2010, pp. 1–4.

[4] J. W. McKown and R. L. Hamilton, Jr., "Ray Tracing as a Design Tool for Radio Networks," IEEE Network Magazine, pp. 27-30, Nov. 1991.

[5] L. Chen,M. Loschonsky, and L.M. Reindl, "Characterization of delay spread for mobile radio communications under collapsed buildings," in IEEE PIMRC Symp., Istanbul, Turkey, Sep. 2010, vol. PHY 12.3, pp.329–334.

[6] W. Gersch and D. R. Sharpe, "Estimation of power spectra with finite-order autoregressive models," IEEE Trans. Autom. Control, vol.AC-18, no. 8, pp. 367–369, Aug. 1973.

[7] A. Saleh and R. A. Valenzuela, "A statistical model for indoor multipath propagation," IEEE Journal on Selected Areas in Communications, vol. 5, pp. 128–137, February 1987.

[8] L.Chen,T.Ostertag,M.Loschonsky,andL.M.Reindl,"Measurement of mobile radio propagation channel in ruins," in IEEE Int. Wireless Commun., Network., Inform. Security Conf., Peking, China, Jun. 2010,vol. 1, pp. 252–256.

[9] L. Chen,M. Loschonsky, and L.M. Reindl, "Large-scale fading model for mobile communications in disaster and salvage scenarios," Wireless Communications and Signal Processing (WCSP), 2010 International Conference on , vol., no., pp.1,5, 21-23 Oct. 2010.

[10] S. Zorn, G. Bozsik, R. Rose, A. Goetz, R. Weigel, A. Koelpin, "A power sensor unit for the localization of GSM mobile phones for search and rescue applications," Sensors, 2011 IEEE , vol., no., pp.1301,1304, 28-31 Oct. 2011.

# A Weighted OMP Algorithm for Doppler Super-resolution

WU Xiao-chuan, DENG Wei-bo, DONG Ying-ning

Institute of Electronic Engineering Technology, Harbin Inst. Of Technology, Harbin 150001, China

*Abstract-* Radar target speed resolution is related to the accumulation time. In order to obtain higher frequency resolution, we need longer coherent accumulation time. However, long accumulation will bring many problems. Since the targets are sparse in the Doppler domain, Doppler super-resolution can be implemented in shorter accumulation time. Doppler frequency of different targets may be correlative, as well as data obtained by single sample, which can be solved effectively by using compressed sensing theory. For the time-frequency transform dictionary generated from frequency resolution, a weighted OMP (Orthogonal matching pursuit) algorithm is proposed to reconstruct the sensing dictionary. This algorithm is superior to OMP etc. greedy algorithm. Especially in the case of the target Doppler frequency close, it shows the high resolution accuracy.

*Index Terms-* Doppler super-resolution; compressed sensing; orthogonal matching pursuit; sensing dictionary

## I. INTRODUCTION

Doppler resolution of radar targets generally is implemented by virtue of the echo signals through matched filtering, distinguished by Doppler filter bank for different range unit. The length of coherent accumulation time directly affects the performance of Doppler resolution. In practice, to enhance the radar detection performance, more coherent accumulation cycle is needed. Because radar antenna beam has a certain width, $M$ target echo pulses will be received when there exists targets in the beam and $M$ detection pulses have been transmitted. If the targets move so fast that cross-range unit occurs, the coherent accumulation time will be greatly limited. How to use less accumulation time to achieve Doppler super-resolution is the research background in this paper.

For super-resolution problem, there have been many methods in the field of DOA (direction of arrival) estimation, but the target Doppler super-resolution method is relatively less studied.

The relationship between Doppler and DOA super-resolution has their similarities, but also differences.

Similarities: 1) both of them achieve high-precision estimation with fewer amount of data. Doppler super-resolution separates multiple targets radial velocity by fewer pulses, and DOA super-resolution uses fewer array elements to locate multiple targets spatial angle. 2) both of them can be distinguished by super-resolution method.

Differences: 1) Doppler estimation is for frequency domain, but DOA for spatial domain. 2) Doppler estimation usually uses single sample in the same range unit. Furthermore, the same target response is coherent and there exists partial correlation for similar speed targets. DOA generally is for multiple snapshots, and the correlation may exist between different targets.

Following the similarities and differences, several novel methods appeared for super-resolution by CS theory in recent years such as $\ell_1 - SVD$ [1], $ISL0$ [2], greedy algorithm etc. Particularly, greedy algorithm is widely applied due to its computational efficiency and similar performance with convex optimum in high SNR. When greedy algorithm is used in time-frequency domain transform dictionary (i.e. redundant Fourier matrix), OMP[3] algorithm is superior with respect to the other greedy algorithms, such as StOMP (stagewise orthogonal matching pursuit)[4], ROMP (regularized orthogonal marching pursuit)[5], CoSaMP (compressive sampling matching pursuit)[6]. This is mainly induced by strong correlation between dictionary atoms. OMP only selects an optimal atom each time, which is more adaptive than other greedy algorithm. However, OMP also calculates the inner product between dictionary atom and observed data, which is severely affected by the correlation of atoms and noise. To address the impact of these negative factors, one sort of dictionary precondition method by alternating projection is proposed by Tropp and Schnass[7]. They construct the sensing dictionary to get the mutual coherence achieve Welch bound, however, this method does not guarantee the existences of equiangular tight frame for any size. Bo L. et al. proposed a similar method[8] which constructs the sensing dictionary and measurement dictionary to improve performance. These two methods above are both aimed at reducing mutual coherence or accumulative mutual coherence of dictionary, and the defection is that the ratio between number of columns and rows of dictionary (redundancy ratio) need relatively to be smaller. Therefore, coherence accumulation time also need longer for these methods.

In practice, optimum sensing dictionary not only depends on redundant dictionary, but also on observed signal. If observed signal can be used effectively, resolution performance will greatly be improved. Based on the assumption, we propose a weighted OMP algorithm, which constructs the weight vector through the cycle iteration to the observed signal, as well as need less accumulation time.

## II. DOPPLER RESOLUTION SPARSE MODEL

For coherent radar system, if the adjacent $M$ detection has the same cycle $T_r$ and the carrier frequency of the transmitted signal is a constant $f_0$, $M$ echo signals from the same distance unit of the same target constitute $M$ points frequency coherent signal. Assumed the entire process is divided into $P$

978-7-5641-4279-7

range units, synchronization pulse is transmitted in each radar probe cycle. Then the $p^{th}$ range unit, $m^{th}$ probe cycle signal can be expressed as $x_p(m)$ $(1 \le p \le P, 0 \le m \le M-1)$, as shown in Figure 1. Traditional approach solving the target Doppler frequency is making $L \ge M$ points FFT to the signal $x_p(m)$, which also can be implemented by FIR digital filter.

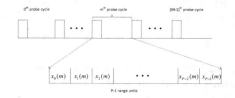

Figure 1 the adjacent $M$ detection cycle

Unambiguous Doppler range is $0 \sim f_s$, where $f_s = 1/T_r$. For the same range, spatial angle units, Doppler resolution of multiple targets can be transformed into the situation of Doppler frequency range is divided into $N$ discrete frequency points $f_i$, $0 \le i \le N$. Different Doppler frequency vector is expressed as:

$$\varphi(f_i) = \exp\left\{ j\frac{2\pi}{f_s}[0,1,\cdots,M-1]^T f_i \right\}, \quad i = 1, \cdots, N \quad (1)$$

Which constitute the measurement dictionary

$$\Phi = [\varphi(f_1), \varphi(f_2), \cdots, \varphi(f_N)] \in \mathbb{C}^{M \times N} \quad (2)$$

Assumed there exists $K$ targets corresponding to different frequency points among $0 \sim f_s$. Based on the discrete grid division, if one target exists at the frequency point, the target amplitude that corresponds to the frequency is nonzero, otherwise the corresponding amplitude is zero. This allows the target amplitudes in the entire unambiguous frequency range constitute a sparse signal by sparsity is $K$, i.e. $x$ has $K$ non-zero elements. After matched filtering, residual noise of each pulse corresponds to the given range unit is denoted $n$, then the signal from the unit after matched filter can be expressed as

$$y = \Phi x + n \quad (3)$$

This issue meets the basic model of the CS theory, greedy algorithm can be used to reconstruct the Doppler frequency.

### III. ALGORITHM IMPLEMENTATION

For a given overcomplete dictionary $\Phi$, the possibility of each atom being the optimum atom depends on observed signals. Once the possibility is obtained, OMP can be improved by weighted matrix which is related to optimum atoms.

**Theorem 1** Let observed signal $y$ be $K$-sparse in $\Phi$, i.e. $y = \Phi x = \sum_{i \in \Lambda_0} \varphi_i x_i$, the optimum atoms index $|\Lambda_0| = K$, OMP using the sensing matrix $\Psi$ will always select components of the true index $\Lambda_0$ if

$$\mu_1(K, \Phi, \Psi) + \mu_1(K-1, \Phi, \Psi) < \beta \quad (4)$$

Where $\mu_1(K, \Phi, \Psi) = \max_{|\Lambda_0|=k} \max_{i \in \Lambda_0} \sum_{j \in \Lambda_0} |\langle \varphi_i, \psi_j \rangle|$, i.e. accumulative mutual coherence, and $\beta = \min_i |\langle \varphi_i, \psi_i \rangle|$, i.e. diagonal coherence parameter. The proof refers to [7].

By the theorem above, optimum sensing dictionary should make $\mu_1$ as possible as small and $\beta$ as possible as big[9]. Considered the dictionary is normalized, the maximum value of $\beta$ is one. Noted that the correlation between sensing dictionary atoms and other non-optimal atoms do not affect reconstruction performance. Therefore, construct the optimum sensing dictionary as follows.

$$\min \left\| \Psi^H \Phi_{\Lambda_0} \right\|_F^2 \quad s.t. \quad \psi_i^H \varphi_i = 1, i = 1, \cdots, N \quad (5)$$

The optimization problem in (5) requires the optimum atoms index $\Lambda_0$, however, the index is unknown in practice. Instead of $\Phi_{\Lambda_0}$ by $\Phi W$ in the optimization process, then

$$\min \left\| \Psi^H \Phi W \right\|_F^2 \quad s.t. \quad \psi_i^H \varphi_i = 1, i = 1, \cdots, N \quad (6)$$

Where $W = diag\{w_1, \cdots, w_N\}, w_i \in [0,1]$ denotes weighted matrix. Since the absolute inner product between observed signal and redundant dictionary reflects the optimum atoms index information in a certain extent. Therefore, we can initialize $W = diag\{|\Phi^H y|\}$. In extreme case, the weighted value is one corresponding to optimum atoms, otherwise zero.

For (6), by Lagrangian function

$$L(\lambda_i, \psi_i) = \frac{1}{2} \left\| W \Phi^H \psi_i \right\|_2^2 + \lambda_i \left( \psi_i^H \varphi_i - 1 \right) \quad (7)$$

based on the definition of matrix norm, the first term on the right of (7) is written as

$$\min \left\| \Psi^H \Phi W \right\|_F^2 = \min \sum_{i=1}^{N} \left\| W \Phi^H \psi_i \right\|_2^2 \quad (8)$$

Differentiating to $\psi_i$ and $\lambda_i$ respectively.

$$\frac{\partial L(\lambda_i, \psi_i)}{\partial \psi_i} = \Phi W^2 \Phi^H \psi_i + \lambda_i \psi_i = 0 \quad (9)$$

$$\frac{\partial L(\lambda_i, \psi_i)}{\partial \lambda_i} = \psi_i^H \varphi_i - 1 = 0 \quad (10)$$

By (9) and (10),

$$\psi_i = \frac{R^{-1} \varphi_i}{\varphi_i^H R^{-1} \varphi_i} \quad (11)$$

where $R = \Phi W^2 \Phi^H$ .

The main steps of the weighted OMP algorithm are summarized below.

---

**Input:** $K$ , $\Phi$ , $y$ , *iterations: $J = 20$ (select the iterations to meet the accuracy requirements)*

---

**Process 1**: calculate the sensing dictionary

Initialization:

$$W = diag\left\{\left|\Phi^H y\right|\right\}$$

Iteration: at the $j^{th}$ iteration ( $1 < j \leq J$ ), go through the following steps

(1) calculate $R = \Phi W^2 \Phi^H$

(2) update $\psi_i = \dfrac{R^{-1}\varphi_i}{\varphi_i^H R^{-1}\varphi_i}$

(3) update $W = diag\left\{\left|\Psi^H y\right|\right\}$

End of the iteration, the sensing dictionary is $\Psi$ .

**Process 2**: Doppler resolution

Using the constructed sensing dictionary $\Psi$ renew OMP algorithm.

Initialization: $a = 0$ , $r = y$ , $I = \phi$

Sensing : $i = \arg\max_j \left|\langle \psi_j , r \rangle\right|$

Reconstruction: $I = I \bigcup i$ , $a = \Phi_I \Phi_I^\dagger y$ , $r = y - a$

Where $\Phi_I^\dagger := \left(\Phi_I^H \Phi_I\right)^{-1} \Phi_I^H$ denotes the pseudo-invserse of the matrix $\Phi_I$ , and " $H$ " stands for matrix conjugate transpose.

---

### IV. SIMULATION RESULTS

In this section, we present several experimental results for our Doppler resolution method. First, we compare the resolution performance of our approach to FFT, OMP. Next, we show the bias of OMP and our approach and analyze the reason.

We consider three targets exist in the same range unit and angle unit. Furthermore, coherence accumulation pulses $M = 32$, pulse repetition frequency $f_s = 500$Hz, frequency domain grids $N = 256$, SNR= 10dB, the real Doppler frequencies are $f_1 = 309.8039$Hz, $f_2 = 329.4118$Hz, $f_3 = 466.6667$Hz. The grid numbers in the frequency domain shows that the frequency point number between $f_1$ and $f_2$ is five.

The result in Figure 2 shows that FFT method do not distinguish $f_1$ and $f_2$ due to their close proximity. Although OMP gives sharp peaks, but $f_1$ estimated is biased. Our method locates the three frequencies accurately, only the amplitude fluctuates compared to real signals since the effect of Gaussian white noise.

In the second experiment, we investigate bias by considering localization of two signals more closely and varying the frequency separation between them. We plot the bias of each of the two signal location estimates as a function of the frequency separation when one frequency is held fixed at the 55th frequency point, and the other change from 57th to 68th frequency point. After a Monte Carlo experiment, the estimation bias is calculated. The definition of bias is

$$bias = \frac{1}{Mon}\sum_{mon=1}^{Mon}\left|\hat{f}_{num} - f_{num}\right| , \quad num = 1,2 . \text{ Where } Mon = 50 ,$$

$\hat{f}_{num}$ , $f_{num}$ represent the estimated and real frequency points respectively. The SNR is 10dB, $M = 32$, $N = 256$, $f_s = 500$Hz. The simulation result in Figure 3 illustrates when our approach is applied in the close proximity of the frequency resolution, the performance is much better than OMP. To facilitate the viewing, the sign of the longitudinal axis in Figure 3 does not represent the frequency deviation direction, only denotes the size of the deviation.

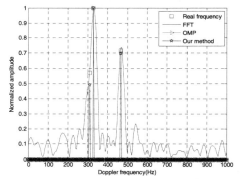

Figure 2 Doppler estimation of three targets

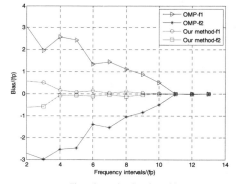

Figure 3 two signals estimate bias
(coordinate units: fp represents frequency point intervals)

In addition, within seven frequency point interval, OMP has obvious fluctuation. This can be explained by the following two reasons. Firstly, the result is partially caused by high cycle correlation character of the measurement dictionary atoms, which is corresponding to cycle fluctuation of Gram matrix[10] constructed by redundant Fourier matrix. Secondly, under the condition of the simulation above, the resolution have not located such close target frequencies accurately. Beyond seven frequency interval, OMP is gradually stable.

## V. DISCUSSION AND CONCLUSIONS

In this paper, we construct the optimum sensing dictionary by using of the observed signal information, and analyze the rationality of weighted matrix theoretically. Next, a weighted OMP algorithm is established. The simulation shows our approach is applicable to the case of the target Doppler frequency close. Compared with the existing algorithms, especially for greedy algorithm, the estimate accuracy has been further improved. Since the observed signal is introduced to generate the weight matrix in the process, the complexity of algorithm will increase. Therefore, future work will focus on the computational efficiency.

## REFERENCES

[1] Malioutov D, "A sparse signal reconstruction perspective for source location with sensor arrays", *IEEE Trans. Signal Processing*. vol. 53, no. 8, pp. 3010-3022, 2005.

[2] Hyder M M, Mahata K, "An improved smoothed $\ell_0$ approximation algorithm for sparse representation", *IEEE Trans. Signal Processing*. vol. 58, no. 4, pp. 2194-2205, 2010.

[3] Tropp J A, Gilbert A C, "Signal recovery from random measurements via orthogonal matching pursuit", *IEEE Trans. Inf. Theory*, vol.53, no. 12, pp. 4655-4666, 2007.

[4] Donoho D L, Tsaig Y, Drori I, "Sparse solution of underdetermined systems of linear equations by stagewise orthogonal matching pursuit", *IEEE Trans. Inf. Theory*, vol.58, no. 2, pp. 1094-1121, 2012.

[5] Needell D, Vershynin R, "Signal recovery from incomplete and inaccurate measurements via regularized orthogonal matching pursuit", *IEEE Journal of Selected Topics in Signal Processing*, vol. 4, no. 2, pp. 310-316, 2010.

[6] Needell D, Tropp J A, "CoSaMP: iterative signal recovery from incomplete and inaccurate sample", *Applied and Computational Harmonic Analysis*, vol. 26, no. 3, pp. 301-321, 2008.

[7] Karin S, Pierre V, "Dictionary preconditioning for greedy algorithm", *IEEE Trans. Signal Processing*, vol. 56, no. 5, pp. 1994-2002, 2008.

[8] Bo L, Yi S, Jia L, "Dictionaries construction using alternating projection method in compressed sensing", *IEEE Signal Processing Letters*, vol. 18, no. 11, pp. 663-666, 2011.

[9] Zhang S L, Wan Q, "DOA estimation in mechanical scanning radar system using sparse signal reconstruction methods", *Wireless Communications, Networking and Mobile Computing (WiCOM)*, pp. 1-4, 2011.

[10] Tropp J A, Dhillon I S, "Designing structured tight frame via an alternating projection method", *IEEE Trans. Inf. Theory*, vol.51, no. 1, pp. 188-209, 2005.

# Developing RSR for Chinese Astronomical Antenna and Deep Space Exploration

J. Ping, M. Wang

Key Laboratory of Lunar and Deep Space, National Astronomical Observatories, CAS, 20A Datun Rd., Beijing, 10012, China

Q. Meng, C. Chen, S. Qiu

Faculty of Space Science, South-East University, Sipailou, Nanjing, 210018, China

N. Jan, L. Fung, S. Zhang, K. Shang

Shanghai Astronomical Observatory, CAS, 80 Nandan Rd., Shanghai, 200030, China

Z. Wang

Xinjiang Astronomical Observatory, CAS, Urumuqi, China

*Abstract*-Following the fast development of Chinese lunar and deep space exploration science 2004, VLBI OD technique and planetary radio science experiments have been involved in the Chinese lunar exploration missions and other deep space missions. In order catch up with above progress and to satisfy the requirements from both of S/C VLBI OD tracking and multi-bands lunar and planetary radio science experiments, we have been developing a software radio defined Radio Science Receiver (RSR) system. The first version of RSR can be used as 4-channel astronomical VLBI receiver with maximum bandwidth of 16 MHz for each channel. It can also work at differential one way range and Doppler model with minimum bandwidth of 50 KHz for each Channel of 8 bits digital resolution. To enhance the application of RSR, real-time Doppler counting mode was developed for signal site observation. Under the Doppler model, it can track 4 separated carrier wave sent from spacecraft(s) simultaneously, with frequency or phase resolution as high as 1/100 circle, when tracking the ESA MEX Martian orbiter and VEX Venus orbiter. In laboratory test, the resolution is about 1 order better at X-band. Additionally, the RSR can work as planetary radio science experiment mode, which will record and retrieve the spacecraft carrier wave signal variation of frequency, phase, amplitude and polarization. The radio science experiment makes a virtue out of a necessity by using the radio propagation techniques that convey data and instructions between the spacecraft and Earth to investigate the planet's atmosphere, ionosphere, rings, and magnetic fields, surface, gravity field, GM and interior, as well as testing the theories of relativity. The experiments have been conducted by most planetary missions and are planned for many future ones. This is almost like getting science for free some way, and has been used also in Chinese lunar orbiting missions to carry out POD and estimate the lunar gravity field successfully.

4 RSRs had been developed for a joint Chinese-Russian Martian mission in order to carry out the orbiter determination of Yinghuo-1 Martian mission. They were installed and connected to the IF and H-Maser system of Chinese VLBI stations. A series of radio occultation experiments have been carried out. In the techniques, radio frequency transmission from MEX and VEX spacecraft, occulted by center planets, and received on Earth probe the extended atmosphere of the planets. The radio link is perturbed in phase and amplitude, the perturbation is converted into an appropriate refractivity profile, from which, information is derived about the electron distribution in the ionosphere, temperature-pressure profile in the neutral atmosphere, or particle size distribution of the ring material surrounding the center planet, in the case of a ring occultation. In Chang'e-3/4 and Luna-Glob landing missions, the two radio downlinks, S/X-bands or X/Ka-bands, will be used for different types of investigation about the physical mechanisms that brought the changes about. We will use the signal to measure the lunar physical liberations, so as to improve the lunar interior studies together with LLR data. Above method may be used in Chinese Martian mission.

Recently, research group are updating a new version of RSR, which can recode a Maximum VLBI bandwidth of 4 GHz of totally 4 Channels . The new systems will be installed at antennas of Chinese VLBI Network and at other Chinese astronomical large antennas in the future as standard instrument of lunar and planetary radio science experiments, and to support the VLBI experiments.

978-7-5641-4279-7

# Microwave Attenuation and Phase Shift in Sand and Dust Storms

Qunfeng Dong [#1], Yingle Li [#2], Jiadong Xu [*3], Mingjun Wang [#4]

[#]Xianyang Normal University, Xianyang, Shaanxi, 712000, China
[1]qunfengdong1028@163.com
[2]Liyinglexidian@yahoo.com.cn
[4]wmjxd@yahoo.com.cn

[*]The Electronic Information Institute, Northwestern Polytechnical University, Xi'an, Shaanxi 710072, China
[3]jdxu@nwpu.edu.cn

*Abstract*—Based on the forward scattering amplitude function for sand particle under the Rayleigh approximation, the calculation models of microwave attenuation and phase shift due to sand and dust particles is given in terms of lognormal size distribution. The model shows that attenuation and phase shift are found to be dependent on visibility, frequency and complex dielectric constant. Obtained results show that moisture content of particle has significant impact on the attenuation and phase shift, and the higher frequency is, the greater are the influence on attenuation and phase shift. The attenuation and phase shift increase with the increase of moisture content.

## I. INTRODUCTION

The theory of microwave propagation in sand storms has received much attention in the literature owing to the importance of radio relay, communication and remote sensing. Usually, when sandstorms occur, sand particles can rise to such heights above ground level that these particles interfere with microwave or millimeter-wave radio. Consequently, the absorption and scattering effect happens which causes loss in signal energy and additional phase shift. This would greatly affect the coverage, quality of Communication and can even interrupt the local area communications [1].

A number of papers have addressed the problem of prediction of the amount of attenuation and phase shift in sand storms[2]-[11]. These investigators concluded that attenuation by dust storms is not serious except for very dense storms. For different size distribution of sand storms, millimeter propagation has been analyzed by Ahmed [12] and Ali [13]. Based on the Maxwell-Garnett equation [14], millimeter attenuation through sand storms has been investigated by Ruike [15]. Goldhirsh and Bashir obtained the formulas of wave attenuation induced by sand and dust storms in terms of mass of dust per unit volume of air. Dazhang [16] And Zhou [17] also calculated microwave attenuation in sand and dust storms with lognormal size distribution, the models require data for the number of dust particles N or dust concentration, which statistically is difficult to measure accurately.

In this paper, using the relationship of the number of dust particles and optical visibility, the number of dust particles for lognormal size distribution is obtained. The formulas of attenuation and phase shift are presented for microwave propagation in storms, suitable for different frequency. The effects of different frequency and moisture content of particle on microwave attenuation are discussed.

## II. PARTICLE DISTRIBUTION IN SAND AND DUST STORMS

Ahmed et al. [12] indicated that log-normal functions fit more adequately than power law or exponential functions for modeling particle size distributions. Analysis of particle sizes carried out in Tengger Desert and Yellow River Beach in China [16] indicates that the particle radii fit log-normal distribution. The log-normal distribution may expressed by

$$N(a) = N_0 p(a) \qquad (1)$$

$$p(a) = \frac{1}{\sqrt{2\pi}\sigma a} \exp\left(-\frac{(\ln a - m)^2}{2\sigma^2}\right) \qquad (2)$$

where $a$ is particle radius, $N_0$ is the mean number density of particles of all sizes, $p(a)$ is the function of size distribution. $m$ is the mean of $\ln a$, $\sigma$ is the variance of $\ln a$.

## III. MICROWAVE ATTENUATION AND PHASE SHIFT IN SAND AND DUST PARTICLES

The complex refractive index of a scattering medium given by Van De Hulst [19] can

$$n_e = 1 - i2\pi k_0^{-3} \int_0^\infty S(0) N(a) da \qquad (3)$$

$$k_0 n_e = \alpha + j\beta' \qquad (4)$$

where $k_0$ is the propagation constant in free-space, $S(0)$ is the the is the forward scattering amplitude and N(a)=N₀P(a) is the particle size distribution per unit volume (cm³) having radii in the region $a \to a + da$. The attenuation and phase shift can then be determined directly as

$$\alpha = 8.686 \times 10^3 \frac{2\pi}{k_0^2} \int_0^\infty \mathrm{Re}\left[S(0)\right] N(a) da \qquad (5)$$

$$\beta = 57.296 \times 10^3 \frac{2\pi}{k_0^2} \int_0^\infty \mathrm{Im}\left[S(0)\right] N(a) da \qquad (6)$$

As the sand particle size is relatively small and the frequency is not too high, i.e. $ka \ll 1$, using the Rayleigh approximation, the forward scattering amplitude of the sand [3] is

$$S(0) = jk_0^3 \left( \frac{\varepsilon_m^* - 1}{\varepsilon_m^* + 2} \right) a^3 \qquad (7)$$

where $a$ is the radius of sand particle, $\varepsilon_m^*$ is the dielectric constant of sand particle, $\varepsilon_m^* = \varepsilon' - j\varepsilon''$.

Most meteorological observations of sand and dust storms are made in terms of optical visibility rather than concentration. It is further useful to adopt the optical attenuation coefficient $\alpha_0$, which is inversely proportional to visibility [13]

$$V_b = \frac{15}{\alpha_0} \qquad (8)$$

where $\alpha_0$ is measured in decibels per kilometre and is related to the concentration via [18]

$$N_0 = 5.5 \times 10^{-4} \frac{1}{V_b a_e^2} \qquad (9)$$

$$a_e = \left[ \int_0^\infty a^2 p(a) da \right]^{\frac{1}{3}} \qquad (10)$$

where $a_e$ is the equivalent particle radius.

Substituted (2) into (9), the value of $N_0$ (number of particles of dust) is given as

$$N_0 = 5.5 \times 10^{-4} \frac{1}{V_b \exp\left[ 2m + 2\sigma^2 \right]} \qquad (11)$$

Substituted (6), (7), (11) into (3), and the attenuation coefficient can then be expressed as

$$\alpha = 1.88 \cdot \frac{F}{V_b} \cdot \frac{\varepsilon''}{(\varepsilon' + 2)^2 + \varepsilon''^2} \cdot \exp\left( m + 1.5\sigma^2 \right) \qquad (12)$$

Similarly, the phase shift coefficient can be shown that

$$\beta = 4.154 \cdot \frac{F}{V_b} \cdot \frac{(\varepsilon' - 1)(\varepsilon' + 2) + \varepsilon''^2}{(\varepsilon' + 2)^2 + \varepsilon''^2} \cdot \exp\left( m + 1.5\sigma^2 \right) \qquad (13)$$

where F is the frequency (GHz), $V_b$ is the visibility (km).

## IV. CALCULATIONS AND RESULTS

Microwave attenuation and phase shift with different moisture content at 14, 24 GHz are calculated using (12), (13). The following parameters [3] are $\varepsilon_m^* =2.8-j0.035$, $\varepsilon_m^* =3.9-j0.62$, $\varepsilon_m^* =5.5-j1.3$ at 0.3% , 5.0%, 10.0% moisture content for 14 GHz, and $\varepsilon_m^* =2.5-j0.028$, $\varepsilon_m^* =3.6-j0.65$, $\varepsilon_m^* =5.1-j1.4$ at 0.3% , 5.0%, 10.0% moisture content for 24 GHz, respectively.

Results are obtained for 0.3%, 5.0%, 10.0% moisture content as shown in Fig. 1 and Fig. 2. From Figs.1-2, moisture content has a significant impact on the wave attenuation. Increase in attenuation is observed at higher moisture content. It is because that the sand particle will absorb moisture in the atmosphere and acts as a water vapor the more humidity in the atmospheric the more condensation nucleus. This induces the significant changes of the dielectric constant of sand particles and causes higher attenuation. So sandstorms with a high humidity have a greater attenuation and signal fading. And the

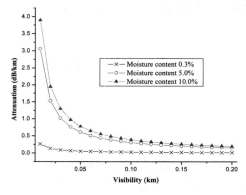

Fig. 1 Variation of attenuation with different moisture content at 14GHz

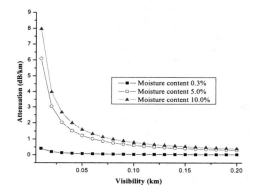

Fig. 2 Variation of attenuation with different moisture content at 24GHz

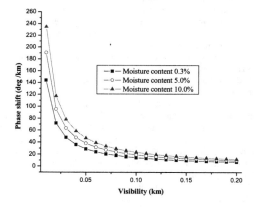

Fig. 3 Variation of phase shift with different moisture content at 14GHz

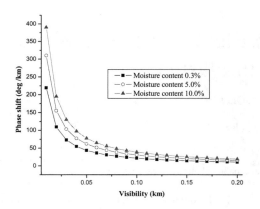

Fig. 4  Variation of phase shift with different moisture content at 24GHz

wave attenuation in sand and dust storms increases as the visibility decreases because wave attenuation is directly proportional to the particle number. Furthermore they have a great influence on the microwave signal for a higher frequency. Figs.3-4 show the relationship between microwave phase shift and visibility under the moisture content 0.3%, 5.0%, 10.0% at 14, 24 GHz.

From Figs.3-4, moisture content has also a significant impact on phase shift. Phase shift increases with the increase of the moisture content; for the same moisture content, the phase shift decreases with the increase of visibility. Under the same moisture content and visibility, the phase shift increases with the increase of frequency.

## V.  CONCLUSION

This work investigates microwave propagation in dust and dust storms. The mathematical models for evaluating attenuation and the phase shift in dust storms are developed. The proposed model suitable for frequency and different visibility range is based on the method of Rayleigh approximation. The model shows that attenuation and phase shift are found to be dependent on visibility, frequency and complex dielectric constant for the lognormal size distribution. The results of the calculation show that moisture content has a significant impact on the wave attenuation, and microwave attenuation and Phase shift increase with the increase of the moisture content. And the wave attenuation and phase shift in sand and dust storms increase as the visibility decreases because wave attenuation is directly proportional to the particle number. Furthermore they have a great influence on the microwave signal for a higher frequency.

ACKNOWLEDGMENT

This works is supported by the National Natural Science Foundation of China (Grant No. 61102018, 61271110)

REFERENCES

[1]  T.S. chu, "Effect of sand storms on microwave propagation," *Bell Syst. Tech.J.*, Vol.58, pp. 549-555, 1979.
[2]  A. O. Bashir and N. J. Mcewan, "Microwave propagation in dust storms: A review." *IEEE Proceedings H*, vol. 133(3), pp. 241-247, 1986.
[3]  A. J. Ansari and B. G. Evans, "Microwave propagation in sand and dust storms," *Inst. Elec. Eng. Proc.*, *VI*, vol. 129, pp. 315-322, no. 5,1982.
[4]  S. I. Ghobrial, "The effect of sand storms on microwave propagation," in *Proc. Nat. Telecommunication Conf.*, vol. 2, Houston, Texas, 1980, pp. 43.5–43.5.4.
[5]  X. Y. Dong, H. Y. chen, "Microwave and Millimeter wave attenuation in sand and dust storms," *IEEE Antennas and Wireless Propagation Letters*, vol. 10, pp. 469-471, 2011.
[6]  H. Y. Chen, and C. C. Ku, "Calculation of wave attenuation in sand and dust storms by the FDTD and turning bands methods at 10–100 GHz," *IEEE Trans. Antennas Propagat.*, vol. 36, pp. 2951-2960, 2012.
[7]  S. I. Ghobrial and S.M.Sharief, "Microwave attenuation and cross polarization in dust storms," *IEEE Trans. Antennas Propagat.*, vol. 35, pp. 418-427, 1987.
[8]  Dong, Q. F., Ying-Le Li, Jia-dong Xu,Hui Zhang and Ming-jun Wang, "Effect of Sand and Dust Storms on Microwave Propagation," *IEEE Transactions on Antennas and Propagation*, Vol.61, 910-916, 2013.
[9]  B. R.Vishvakarma and C. S. Rai, "Limitations of Rayleigh scattering in the prediction of millimeter wave attenuation in sand and dust storms," Geoscience and Remote Sensing Symposium, IEEE Inter., 1993.
[10]  Wu. C. M, "Attenuation and phase shift of sand and dust storms on microwave and millimeter wave," *Advances in Electric Mathematics*, vol.8, pp.284-290, 2001.
[11]  Z. Elabdin, M. R. Islam. et al, "mathematical model for the prediction of microwave signal attenuation due to dust storms," *PIER M*, vol. 6, pp.139–153, 2009.
[12]  A. S. Ahmed, "Role of particle-size distributions on millimeter-wave propagation in sand/duststorms," *IEE Proceedings*, vol. 134, 55-59, 1987.
[13]  A. A. Ali, "Effect of particle size distribution on millimeter wave propagation into sandstorms," *Int. J. Infrared and millimeter waves*, vol.7, no.6, pp. 857–868, 1986.
[14]  A. H. Sihvola and J. A. Kong, "Effective permittivity of dielectric mixtures," *IEEE Trans on Geoscience and Remote Sensing*, vol.26, pp. 420-429, 1988.
[15]  H. Dazhang, "Measurements of diameter distribution and shape of sand at the Beach of the Yellow River," *Chinese Journal of Radio Science*, vol.8, pp.63-69, 1993.
[16]  Y. Ruike, W. Zhensen, and Y. Jinguang, "The study of MMW and MW attenuation considering multiple scattering effect in sand and dust storms at slant paths," *Int. J. Infrared and millimeter waves*, vol. 24, pp.1383-1392, 2003.
[17]  Z. Wang, "Calculation and simulation of sand and dust attenuation in microwave propagation," *High Power Laser and Particle Beams*, vol. 17, pp. 1259-1262, 2005.
[18]  A. A. Shakir, "Particle-Size Distribution of Iraqi Sand and Dust Storms and Their Influence on Microwave Communication Systems," *IEEE Trans. Antennas Propagat.*, vol. 36, pp. 114-126, 1988.
[19]  H.C.Van DeHulst, " Light Scattering by Small Particles," New York: John Wiley, 1957.

# Experimental Research on Electromagnetic Wave Attenuation in Plasma

Li Wei[1,2], Suo Ying[1,2], Qiu Jinghui[1]

(1.School of Electronics and Information Engineering, Harbin Institute of Technology, Harbin, 150001; 2.Electronic Science and Technology Postdoctoral Station, Harbin Institute of Technology, Harbin, 150001)

*Abstract-* **The power loss of electromagnetic wave in plasma is analyzed and calculated in this paper. An experiment scheme is proposed to measure the electromagnetic wave attenuation in plasma. The experiment scheme includes a plasma flat slab consist of several plasma tubes. The attenuation characteristics of electromagnetic waves in plasma are measured by horn antenna.**

## I. INTRODUCTION

When a high speed spacecraft flies in reentry aerosphere stage, the plasma is generated by the interaction between spacecraft and aerosphere, and a phenomenon of electromagnetic wave attenuation increase is able to occur. The plasma is a main cause of electromagnetic wave attenuation. Plasma is a kind of electric liquid with high electron density, collision frequency and dispersive characteristic. The electromagnetic wave attenuation caused by plasma is consisted with the electromagnetic wave absorption and reflection of plasma, and gain reduced by mismatch of antenna impedance.

The attenuation of GPS signal by reentry plasma is analyzed in [1]. In [2,3] Kim calculates the attenuation values of remote sensing signal in reentry, and he gets the refractive index of electromagnetic wave with different electron density and collision frequency, and attenuation values on 1.575GHz. The attenuation effect in plasma layer with incident frequency, electron density and collision frequency is studied in [4]. In [5] the electromagnetic wave attenuation and reflection is measured by a experimental system, and the results are analyzed.

To verify the results of theory and simulation analysis, an experiment scheme is proposed to measure the electromagnetic wave attenuation of plasma. The experimental scheme includes a plasma flat slab consist of several plasma tubes by low voltage gas discharge. The plasma flat slab can simulate the reentry plasma characteristics. After determine the average electron density of the plasma flat slab, the plasma slab is put on the transmitting antenna, and the electromagnetic attenuation values are measured by transmitting and receiving horn antenna.

## II. ELECTROMAGNETIC WAVE ATTENUATION CALCULATION

The electromagnetic wave attenuation caused by plasma is consisted with the electromagnetic wave absorption and reflection of plasma, and gain reduced by mismatch of antenna impedance. The electromagnetic wave attenuation can be calculated.

If the electromagnetic wave incidence is vertical, the attenuation of electromagnetic wave is

$$A = 8.68\alpha d - 10\lg(1 - \left|\frac{1-\sqrt{\varepsilon_r}}{1+\sqrt{\varepsilon_r}}\right|^2) \tag{1}$$

$\alpha$ is attenuation coefficient,

$$\alpha = \frac{\omega}{c\sqrt{2}}\sqrt{\sqrt{(1-\frac{\omega_p^2}{\omega^2+v_e^2})^2 + (\frac{\omega_p^2}{\omega^2+v_e^2}\cdot\frac{v_e}{\omega})^2} - (1-\frac{\omega_p^2}{\omega^2+v_e^2})}$$

where $\omega$ is angular frequency of incident electromagnetic wave, $c$ is velocity of light, $V_e$ is collision frequency, $\omega_p$ is characteristic frequency of plasma, and

$$\omega_p = \sqrt{\frac{N_e q_e^2}{\varepsilon_0 m_e}} \tag{2}$$

$N_e$——electron density,
$q_e$——electronic charge,
$m_e$——electron mass.

To describe the macro graphic characteristic of plasma, the Drude dispersion model is able to employ to analyze the interaction of electromagnetic wave and plasma. In Drude dispersion model, the relative dielectric constant is

$$\varepsilon_r = \varepsilon_\infty - \frac{\omega_p^2}{\omega(\omega - jv_e)} \tag{3}$$

where $\varepsilon_\infty$ is relative dielectric constant when operating frequency is infinite, generally $\varepsilon_\infty = 1$.

The electromagnetic wave attenuation with different electron density in plasma is shown in figure 1, where the thickness of homogeneous plasma layer is $d$=0.02m, and the collision frequency is $v_e$=1GHz. And the electron density $N_e$ is $1\times10^{18}$/m$^3$、$3\times10^{18}$/m$^3$ and $1\times10^{19}$/m$^3$. As shown in figure 1, the higher the electron density of plasma is, the larger the value of electromagnetic wave attenuation is. The values of attenuation increase first and decreased afterwards, and there is a maximum value of attenuation. Figure 2 show the values of electromagnetic wave attenuation when $N_e$=3×10$^{18}$/m$^3$ and $d$=0.02m with different collision frequency $v_e$, and $v_e$ is 1GHz、5GHz and 10GHz. As shown in figure 2, the higher is the collision frequency of plasma, the smaller the values of electromagnetic wave attenuation is when the operating frequency is less than 14GHz. The higher the collision

978-7-5641-4279-7

frequency of plasma is, the larger the value of electromagnetic wave attenuation is when the operating frequency is more than 14GHz. The values of attenuation increase first and decreased afterwards, and there is a maximum value of attenuation.

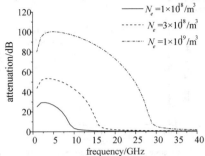

Figure 1.Electromagnetic wave attenuation with different electron density in plasma

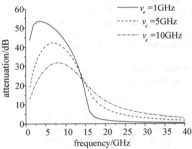

Figure 2.Electromagnetic wave attenuation with different collision frequency in plasma

The electromagnetic wave attenuation with different thickness of plasma layer is shown in figure 3, where the electron density is $N_e=3 \times 10^{18}/m^3$, and the collision frequency $v_e$=1GHz. And the thickness is $d$ is 0.02m, 0.03m and 0.05m. As shown in figure 3, the larger the thickness of plasma layer is, the larger the value of electromagnetic wave attenuation is. The values of attenuation increase first and decreased afterwards, and there is a maximum value of attenuation.

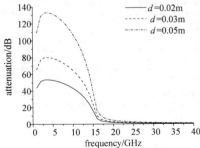

Figure 3.Electromagnetic wave attenuation with different thickness in plasma

III. ELECTROMAGNETIC WAVE ATTENUATION MEASUREMENT

In experimental research of electromagnetic wave attenuation measurement, the plasma is generated by direct current discharge under low atmospheric pressure. The glass tube is used in the experiment and the radius is 15mm. The air in the tube is extracted by a vacuum pump. The metal electrode is installed on two ends of the tube after the gas pressure is less than $10^{-2}$Pa. The mixed air with Ar gas and Hg vapor is filled in the glass tube until the gas pressure reaches 10mmHg. A high voltage DC power (the output voltage is about 7000V and the power is about 60W) is connected with both ends of the electrode. The mixed gas is broken down when the electrodes discharge, and the plasma is generated by Penning Effect. The length of each plasma tube is about 1m. And a plasma flat slab is required in this experiment, so several plasma tubes can be arranged with series feeding. Figure 4 shows the plasma flat slab in the experiment. The plasma flat slab is made by 10 plasma flat slabs where the interval of two tubes is about 10mm.

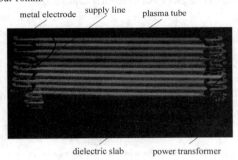

Figure 4.Slab made by plasma tubes

The measurement system of electromagnetic wave transmission is shown in figure 5. The system is consisted with a signal generator, a vector network analyzer, a directional coupler, a transmitting antenna and a receiving antenna. The electromagnetic wave attenuation by plasma is

$$10\lg\frac{P_{R2}(f)}{P_{R1}(f)} = 10\log\frac{P_{R2}(f)}{P_T(f)} - 10\log\frac{P_{R1}(f)}{P_T(f)}$$
$$= |S_{21b}| - |S_{21a}| (dB) \qquad (4)$$

where $S_{21a}$ is the $S$ parameter measured by the vector network analyzer when plasma is generated, and $S_{21b}$ is the $S$ parameter measured by the vector network analyzer when plasma is not generated. $P_T(f)$ is the power at the plasma interface transmitted by the transmitting antenna, $P_{R1}(f)$ is the power received by receiving antenna when plasma is generated, and $P_{R2}(f)$ is the power received by receiving antenna when plasma is not generated,

$$|S_{21a}|(dB) = 20\log[S_{21a}(f)] = 10\lg\frac{P_{R1}(f)}{P_T(f)},$$

$$|S_{21b}|(dB) = 20\log[S_{21b}(f)] = 10\lg\frac{P_{R2}(f)}{P_T(f)}.$$

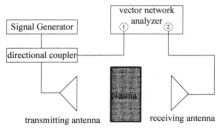

Figure 5.Measurement system of electromagnetic wave transmission

The electromagnetic wave attenuation is calculated where $N_e=5 \times 10^{16}/m^3$, $v_e=1 \times 10^8$Hz, $d$=50mm, and the operating frequency in from 0.9GHz to 1.3GHz. The value of attenuation is shown in figure 6. And the gain of the horn antenna with plasma layer can be simulated by the model in figure 7 by CST MWS studio. The value of the gain with plasma and the gain without plasma is shown in table 1.

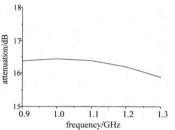

Figure 6.Calculation of electromagnetic wave attenuation in plasma

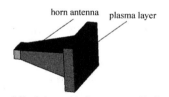

Figure 7.Simulation model of horn antenna with plasma

Table 1 Gain of horn antenna with or without plasma

| Frequency/GHz | 0.9 | 1.0 | 1.1 | 1.2 | 1.3 |
|---|---|---|---|---|---|
| Gain without plasma/dB | 13.0 | 14.6 | 16.2 | 17.2 | 18.9 |
| Gain with plasma /dB | 4.5 | 7.6 | 9.4 | 10.6 | 11.8 |

The total calculated value of electromagnetic wave attenuation is shown in figure 8. The value includes electromagnetic wave attenuation and the decrease of antenna

gain. The measured value shown in figure 8 is got by formula (4). There is difference between calculated value and measured value for the electromagnetic wave reflection on the boundary of plasma, and the plasma is unstable and inhomogeneous.

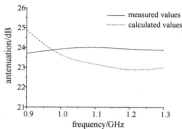

Figure 8.Measurement value of total electromagnetic wave attenuation

## IV. CONCLUSION

An experiment system is presented to measure the electromagnetic wave attenuation of plasma. The system is consisted with a signal generator, a vector network analyzer, a directional coupler, a transmitting antenna and a receiving antenna. The experimental scheme includes a plasma flat slab consist of several plasma tubes by low voltage gas discharge. The calculated value and measured value of electromagnetic wave attenuation is given.

## ACKNOWLEDGMENT

The authors would like to express their sincere gratitude to CST Ltd. Germany., for providing the CST Training Center (Northeast China Region) at our university with a free package of CST MWS software. The authors would like to express their sincere gratitude to funds supported by "the National Natural Science Funds" (Grant No. 61201014), and "the Fundamental Research Funds for the Central Universities" (Grant No. HIT.NSRIF.2012025).

## REFERENCES

[1] D. S. Frankel, P. E. Nebolsine, M. G. Miller and J. M. Glynn. "Re-entry plasma induced pseudorange and attenuation effects in a GPS simulator". SPIE Defense and Security Symposium, 2004:12~16
[2] M. Kim, M. Keidar and I. D. Boyd. "Effectiveness of an electromagnetic mitigation scheme for reentry telemetry through plasma". 46th AIAA Aerospace Sciences Meeting and Exhibit, 2008:1~11
[3] M. Kim, M. Keidar and I. D. Boyd. "Analysis of an electromagnetic mitigation for reentry telemetry through plasma". Journal of Spacecraft and Rockets. 2008,45(6): 1223~1229
[4] Liu Minghai, Hu Xiwei, Jiang Zhonghe, Liu Kefu, Gu Chenglin, Pan Yuan. "Property of electromagnetic wave attenuation in the artificial plasma of atmosphere". Acta Physica Sinica. 2002,51(6):1317~1320
[5] Yuan Zhongcai, Shi Jiaming, Wang Jiachun. "Experimental studies of the interaction of microwaves with mixture burning plasmas in the atmosphere". High Power Laser and Particle Beams. 2005,17(5):225~228
[6] D. M. Pozar Microwave Engineering 2005:555~556

# Modulation Recognition Based on Constellation Diagram for M-QAM Signals

Zhendong Chou[1], Weining Jiang[1], Min Li[2]

[1]The 41st Institute of China Electronics Technology Group Corporation, Qingdao, Shandong, 266555, China

[2]University of Electronic Science and Technology of China, Chengdu, Sichuan, 610054, China

*Abstract*-In this paper, the author proposed a modulation recognition algorithm for M-QAM signals by the constellation diagram which does not require the prior information. Firstly, this scheme estimates the modulation parameters. Secondly, it reconstructs the received signals' constellation and use k-means cluster algorithm to compute the number of the signal constellation points which is as a recognition feature used for classification. The simulation results show that this method is experimentally approved effective for M-QAM signals, and the correct recognition ratio reaches 100% under the condition that the SNR is higher than 15dB. So it has relatively preferable practical value.

*Index Terms*-modulation recognition, clustering, constellation, K-means clustering algorithm, M-QAM.

## I. INTRODUCTION

Signal modulation recognition is an important part in the Communication Signal Analysis filed. It has broad application prospects and occupies an important position in the field of reconnaissance, communications confrontation, and electronic warfare. It is one of the key technologies for software radio. With the increase of communicational service, M-QAM has a high bandwidth ratio, and QAM is a better choice. Therefore, the identification of signals is very necessary [3].

Practically, in this paper, the author starts from the modulation identification methods of statistical pattern recognition, by signal preprocessing and feature extraction by classifying, and adjudging modulation schemes of signal. Recognition method of M-QAM signals including 16QAM, 32QAM and 64QAM is researched with feature extraction of the signal constellation. This scheme does not require the prior information about baud rate and carry frequency of the received signals, but it directly estimates these parameters, computing the number of the signal constellation points by blind clustering algorithm. Consequently, the modulation type is identified [1].

## II. M-QAM MODULATION SIGNAL CONSTELLATION CLUSTERING FEATURES

Any kind of digital amplitude phase modulated signals can be represented with a unique constellation diagram. By this correspondent relationship, the constellation diagram can be used for the identification of modulation method. The identification process includes symbol sequence matches with constellation graph, constellation point counting, reconstruction of the constellation diagram match, clustering, shape normalization, the size and location normalization, shape modeling, se-

lection of the mesh size, and shape classification of maximum likelihood estimation, etc.

The ISI (Inter-symbol interference) caused by multi-path effects has great impact on the shape of the constellation diagram in modulation method of QAM (Quadrature Amplitude Modulation).Therefore, before the signal identification, firstly, using blind equalization technique to overcome channel multipath effects and the system synchronization error. Secondly, k-means clustering on the signal, normalized the constellation diagram, and thereby completed the reconstruction of the constellation diagram. Finally, it matches with the ideal constellation diagram model, to achieve the M-QAM modulation method identification.

QAM is a frequent use of digital communication techniques in digital modulation. It is dual-baseband digital signals on the two mutually orthogonal to the same frequency carrier suppressed carrier double sideband modulation, and using the modulated signals within the same bandwidth spectrum orthogonal nature to achieve a dual-way parallel transmission of digital information. The QAM signal recognition can be mainly divided into three categories. First, modulation classification methods use maximum likelihood ratio. The second one are commonly used identification methods based on statistical moment feature. The third recognition methods are based on transform domain feature. According to searching signal amplitude's probability distribution function (Estimated approximate by Histogram) Fourier transform spectrum's first zero position for category QAM signals, we look upon QAM signal constellation as a graphical with grid-like distribution, constellation point uniformly distributing in the plane with the same intervals. Therefore, the dual-dimensional histogram can be transformed by rotating. The QAM constellation is shown in Fig. 1.

Clustering is a data set which is divided into several groups that cause the similarity within the group is greater than different groups. To achieve such a division needs a similarity measure, which input two vectors, and returned the value which reflects the similarity between these two vectors [4]. Since most of the similarity measure is very sensitive to the elements' value range of the input vectors, each input variable must be normalized, and its value belongs to interval [0, 1].

From the M-QAM signal constellation, we know that constellation point is very obvious clustering, and this feature can be used for identification of QAM signals. In the present QAM signal recognition, identification M-ary is generally up to 64, which can also be seen from the constellation. In the

978-7-5641-4279-7

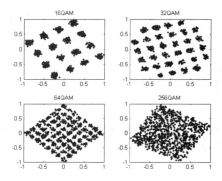

Fig. 1. QAM signal constellation.

256QAM constellation map, constellation points are too much. Even signal has high SNR and large number of symbols in the case, it is still difficult to achieve the desired effect of cluster identification. Therefore, we mainly achieve signal identification for 16QAM, 32QAM and 64QAM.

## III. BASED ON M-QAM CONSTELLATION DIAGRAM CLUSTERING RECOGNITION

After using blind equalization techniques to overcome channel's multipath effects and system synchronization error, intuitive performance of the signal constellation of data points gathered in the respective modulated state. Continue processing of output signals from equalizer can identify the type of modulation of the signal.

M-QAM signal identification methods are currently mainly based on wavelet transform modulus sequence recognition, those based on the recognition of the magnitude's statistical moments, those based on the recognition of the log-likelihood function, etc. These methods either need to know the baud rate of the sender or be sensitive to frequency offset or have to get the modulation interval from the sender, in short, all these must get more of prior parameter info. However, in non-collaborative communication, frequency offset, baud rate, and prior knowledge are unknown, so these identification methods have not too high practical value, but the proposed method does not require prior knowledge to complete the M-QAM signal identification.

Designed in this paper based on the k-means clustering, QAM constellation diagram identification is achieved through the identification of the signal constellation cluster characteristics, so it need to restore the signal constellation diagram, but the quality of restored constellation diagram, directly affect the level of signal recognition rate. Since the constellation map recovery is achieved in the baseband signal, in non-cooperative communication, under the premise of advance unknown with the offset, the baud rate, and the timing error. Firstly, it need estimate these parameters from the received signal then process them and identify. This method is implemented in Fig.2, it can be seen from the figure, estimating the

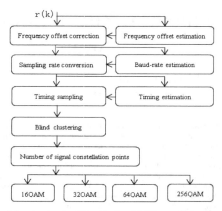

Fig. 2. Based on the signal identification of QAM constellation diagram.

received signal's frequency offset for one thing; carrying out the baud rate estimation and timing error estimate, thereby estimating the frequency offset and symbol rate, and timing information, to achieve carrier synchronization and symbol synchronization, and then recover the signal constellation diagram.

After acquire signal constellation diagram, the rest of the work need an identifier analysis through cluster and automatically determine the specific number of points in the constellation diagram, where use k-means clustering algorithm, the algorithm is a typical representative of partition clustering algorithms, in real terms, the algorithm based on the average of the objects in the cluster. Partition-based clustering requires to enumerate all possible division in order to achieve global optimal. The algorithm can use effective range of the data according to pre-specified, in accordance with the minimum mean square error criteria to get the final clustering points and the position of the cluster center. Algorithm input objects of data and the number of clusters $k$, to obtain the minimum of the square error criterion k clusters [2]. Algorithm is shown as follows:

*1) Set the initial category center and the number of categories.* Arbitrarily select $k$ objects from the entire sample $n$ as initial cluster center $m_i (i = 1,2,...,k)$.

*2) Partition the data according to the category center.* Use (1) to calculate the distance from each of the data set $p$ to the center of $k$ clusters $d(p,m_i)$.

*3) Recalculate the center the category in the current category partition.* Find the minimum $d(p,m_i)$ of each object $p$, put this $p$ into the cluster the same with $m_i$.

*4) Partition category in the gained category center.* After traversing all the objects, use (2) to recalculate the value of $m_i$, as the new cluster center.

*5) Re-assign the entire data object set to the most similar cluster.* This process is repeated until gain a minimum value of the squared error criterion. If two consecutive category partition results are same, stop the algorithm. Otherwise, loop 2 to 5.

$$d(i,j) = \sqrt{(x_{i1} - x_{j1})^2 + (x_{i2} - x_{j2})^2 + ... + (x_{in} - x_{jn})^2} \quad (1)$$

In (1), $i = (x_{i1}, x_{i2}, ..., x_{in})$ and $j = (j_{i1}, j_{i2}, ..., j_{in})$ are two n-dimensional data objects.

$$m_k = \sum_{i=1}^{N} \frac{x_i}{N} \quad (2)$$

In (2), $k$ represents the cluster center in $k$ clusters, $N$ represents the number of data objects of the $k$ clusters.

Squared error criterion attempt to make clustering results as independent and compact as possible, in the other word, the similarity of the objects within the cluster is as high as possible. In (3), $E$ represents the sum of all the square errors of the object, $p$ represents the space object, and $m_i$ represents the average value of the cluster $C_i$.

$$E = \sum_{i=1}^{k} \sum_{p \in C_i} |p - m_i|^2 \quad (3)$$

## IV. FIGURES SIMULATION AND EXPERIMENT RESULTS OF M-QAM SIGNAL IDENTIFICATION

### A. Recognition Rate under the Gaussian Channel Environment

Fig. 3 shows the 16QAM, 32QAM and 64QAM signal recognition of the computer simulation results, the SNR range used in simulation is 0dB ~ 30dB, normalized baud rate set 1.0 or 0.5, shaping filter roll-off factor α set 0.35,0.6,0.8 respectively. Simulate each signal with every modulate parameter (including the carrier frequency, roll-off factor, signal to noise ratio) 100 times. Testing the effect on SNR for signal identification. Simulation result shows that the signal can get a higher recognition ratio in a condition of SNR is10dB.

### B. Recognition rate under the Multipath channel environment

Fig. 4 shows the simulation with the same signal set, to test the effect on SNR for signal identification. The result indicates

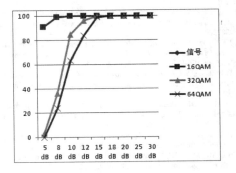

Fig. 3. Signal recognition of the computer simulation results under the Gaussian channel environment.

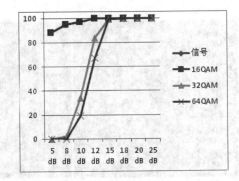

Fig. 4. Signal recognition of the computer simulation results under the Multipath channel environment.

that it can also achieve a high recognition rate in the case of the SNR is higher than 15dB.

In simulation, select a dual-diameter radio channel, channel impulse response is delayed of the raised cosine pulse function.

$$h(t) = (0.2c(t,0.11) + 0.4c(t - 2.5,0.11))w_{6T} \quad (4)$$

In (4), $c(t,a)$ represents a single pulse, $a$ is the Roll off factor, $W_{6T}$ is rectangular window with 6T wide. Oversampling the received signal with four times baud rate.

In the experiment, we found that 16QAM signals can achieve 100% recognition rate, when SNR equals 10dB. However, 32QAM and 64QAM signals only have 63% and 81% recognition rate respectively even the SNR equals 25dB. From the recognition results (see Table I), it can be seen, identifying QAM signal use constellation, which is more sensitive to noise. When SNR equals 5dB, the recognition rate of 16QAM, 32QAM and 64QAM signals respectively drop to 89%, 5% and 3%. Moreover, the constellation points of 32QAM and 64QAM signal were more than 16QAM obviously, using k-means clustering method to cluster, increase the points lead to an elevated recognition rate. Might as well, comparing the recognition rate of 32QAM and 64QAM signals, we can see, the recognition rate of 64QAM signal is higher than 32QAM. The constellation diagram of 32QAM signal is a cross-shaped, however, that of the 64QAM is a square, when fewer symbols, the symbols estimate error rate of 32QAM signal is greater than 64QAM, which leading to the reconstructed constellation fuzzy, sequentially the recognition rate of 32QAM is lower than 64QAM.

TABLE I
QAM SIGNAL RECOGNITION RATE (%)

| SNR(dB) Signals | 5 | 8 | 10 | 12 | 15 | 18 | 20 |
|---|---|---|---|---|---|---|---|
| 16QAM | 91 | 98 | 100 | 100 | 100 | 100 | 100 |
| 32QAM | 3 | 24 | 57 | 63 | 97 | 100 | 100 |
| 64QAM | 1 | 36 | 66 | 81 | 99 | 100 | 100 |

Fig. 5.    16QAM, 32QAM, 64QAM Signal correct recognition ratio reaches 100% with SNR 15dB.

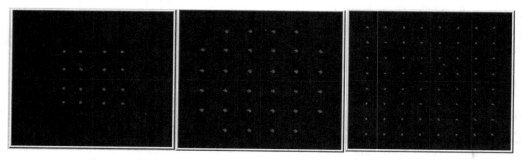

Fig. 6.    16QAM, 32QAM, 64QAM Signal correct recognition ratio reaches 100% with SNR 30dB.

Experimental results show that this method is experimentally approved effective for M-QAM signals, and the correct recognition ratio reaches 100% under the condition that the SNR is higher than 15dB (see in Fig. 5).

However, as Fig. 6 shown, only when the SNR must be higher than 30dB, the recognition ratio can reaches 100% which uses Genetic Algorithm and Hierarchical Clustering identification method [6], and which also need 20-25dB to reaches 100% [7][8].

## V.  CONCLUSION

In this paper, the author presents an identification method based on constellation for QAM signals, the K-means clustering algorithm used is simple and easy to implement, and it can determine the specific constellation points in the constellation map in the better way. Experimental results show that the algorithm when SNR is greater 15dB, the QAM signal recognition rate can reach 100%, which has a high application value. Since the algorithm is more sensitive to noise, identification of the constellation normally require a higher signal to noise ratio, and high-level QAM signal transmission also requires a higher signal to noise ratio, Therefore, identification of QAM signals mainly in case of signal recognition with high SNR, and pre-process signal with low-pass filter, reducing the timing phase dithering caused by noise.

### REFERENCES

[1] Y. F. Zhan, Z. G. Cao, "Modulation classification of M-QAM signals," *Journal of China Institute of Communications*, vol. 25, no. 2, pp. 68-74, 2004.

[2] T. Huang, S. H. Liu, "Research of Clustering Algorithm Based on K—means," *Computer Technology and Development*, vol. 21, no.7, pp.54-57, 2011.

[3] J Hou, H K Wang, "MQAM Recognition Based on Research of Constellation Clustering," *Radio Communications Technology*, vol.35, no.3, pp.35-38, 2009.

[4] J X Wang, H Song, "Digital Modulation Recognition Based On Constellation Diagram," *Journal of China Institute of Communications*, vol.25, no.6, pp.166-173, 2004, vol.48, no.2, pp.189-193, 2000.

[5] D P Zhang, Y F Chen, "QAM Signal Based on Constellation Diagram," *Communication Countermeasures*, vol. 1, no.1, pp.8-11, 2006.

[6] N AHMADI, "Modulation Classification of QAM and PSK from Their Constellation Using Genetic Algorithm and Hierarchical Clustering," *IEEE ICITA Cairns*, 2008, pp.98-103.

[7] P Bradley, U M Fayyad, "*Refining Initial Points for KM Clustering,*" MS Technical Report MSR-TR-98-36, May.1998.

[8] B G MOBASSERI, "Digital modulation classification using constellation shape," *Signal Processing*, vol.80, no.2, pp.251-277, 2000.

# Broadband Four-Way Power Divider for Active Antenna Array Application

Lei Zhang, Xiaowei Zhu, *Member, IEEE,* Peng Chen, Ling Tian and Jianfeng Zhai, *Member, IEEE*

State Key Lab of Millimeter Wave, Southeast University, Nanjing, 210096, P. R. China

*Abstract-*A novel four-way power divider based on substrate integrated waveguide is presented. Broadband performance is obtained by using a stepped coaxial line transformer together with tapered substrate integrated waveguides. Simulated and measured results show that good performance of insertion loss and impedance matching is achieved over a broad bandwidth from 4 GHz to 11 GHz. When the proposed power divider is applied for active antenna array, it can be shown that by using appropriate matching networks, gain fluctuation caused by non-ideal return loss and isolation can be alleviated.

## I. INTRODUCTION

Active antenna arrays [1] are widely used in modern communication system while the loss of the coaxial cable will have a severe impact on performance. In active antenna system, power dividers/combiners are key components for microwave signal distribution, especially in application of wideband power amplification where power level of a single device is not high enough. Air-filled metal waveguide power dividers show their advantages such as low insertion loss and high-power handing capacities, but the relatively expensive manufacturing and complex transitions to planar circuits limits their application. On the other hand, planar transmission-line (microstrip for example) based power divider suffered from high insertion loss and low power handing capacities. Substrate Integrated Waveguide (SIW) and Half Mode Substrate Integrated Waveguide (HMSIW) have been proposed as attractive techniques for their inherently low loss, low cost, compactness and easy integrated with planar components. Some SIW power dividers have been proposed with good performance [2-9]. In [10], resonant structure-based SIW power dividers are described, but their bandwidth are very low. Broadband travelling-wave four-way power divider is presented [2], but the input matching is rather complicated for manufacture.

In this paper, a broadband SIW four-way power divider is presented. The topology is similar to travelling-wave power divider, as described in [2], but the input matching structure is modified for easy manufacture while providing broadband impedance matching. Simulation results show that proposed power divider exhibits low insertion loss and high return loss over wide bandwidth. When this power divider is connected with active array system, the transfer function from input to individual antenna element is given so wideband gain flatness can be optimized.

## II. DESIGN PRINCIPLE

### A. Design of SIW four-way power divider

The top view of proposed SIW four-way power divider is shown in Fig. 1. This structure is axially symmetric. It is centrally fed by a current probe through a stepped coaxial line [2]. Four SIWs are used as arms for signal distribution. In each SIW, side walls are realized by arrays of metallic via in relatively thin dielectric substrate. Via spacing of three times of the radius is chosen to minimize leakage losses while staying away from overloading the substrate.

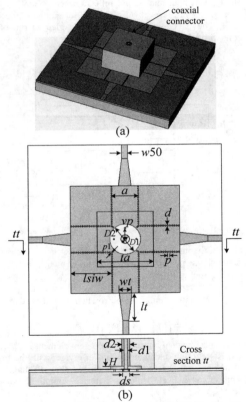

Figure 1. Layout of proposed SIW four-way power divider. (a) 3-D View (b) Top view and cross section of symmetrical plane tt.

In previous literature, dual-disk probe is proposed to provide a broadband impedance matching from the input coaxial line to

978-7-5641-4279-7

radial line [2]. But the structure is rather complicated as more mechanical process is needed to form a closed space on the back side of the substrate. In this paper, stepped coaxial line transformer together with tapered SIWs is used for input matching, as shown in Fig. 1(b). Tapered SIW is formed by guiding posts, as illustrated in [8]. Four additional vias (named as middle via) located just in front of each SIW arm are incorporated to improve return loss. The diameter of inner conductor is also stepped for broadband matching. Input return loss can be optimized by changing $vp$ and spacing between guiding posts. Compared with those described in [2], feeding structure in this paper is simpler while broadband input match is keeping.

### B. Power divider connected with antenna array

In active antenna system, power divider is connected with active antenna unit, as shown in Fig. 2. Each antenna unit comprises a power amplifier (PA) and a passive antenna. For simplicity, the input and output reflection coefficients of all PAs are $\gamma_{in}$ and $\gamma_{out}$, respectively. Their linear gain is supposed to be $A$. The passive antenna can be regarded as one-port network with input reflection coefficient $\gamma_L$.

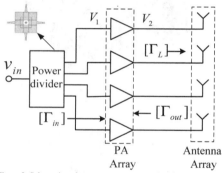

Figure 2. Schematics of antenna array connected with power divider.

In Fig. 2, the input port of the power divider is driven by an incident voltage $v_{in}$. Let $V_1$ and $V_2$ denote voltage vectors at input and output of PA array, i.e. $V_1 = [v_{in}\ v_{in}\ v_{in}\ v_{in}]^T$, $V_2 = [v_{out}\ v_{out}\ v_{out}\ v_{out}]^T$, $t$ represents the transfer coefficient of the power divider from input to either output branch. It can be derived that:

$$V_1 = \left([I]+[S][\Gamma_{in}]\right)v_{in}T \qquad (1)$$

Where $[\Gamma_{in}]$ is diagonal matrix with non-zero element $\gamma_{in}$, $T=[t\ t\ t\ t]$, $[S]$ is scattering matrix of the power divider with the first row and column excluded. Obviously, the off-diagonal elements of $[S]$ indicate isolation between outputs of power divider. $[I]$ is identity matrix.

Similarly, the voltage vector at antenna input is:

$$V_2 = A\left([I]+[\Gamma_{out}][\Gamma_L]\right)V_1 \qquad (2)$$

Where $[\Gamma_{out}]$ and $[\Gamma_L]$ are diagonal matrixes with non-zero element $\gamma_{in}$, and $\gamma_{out}$, respectively.

By substituting (1) into (2), the above equation can be rewritten as:

$$\begin{aligned} V_2 &= A\left([I]+[\Gamma_{out}][\Gamma_L]\right)V_1 \\ &= A\left([I]+[\Gamma_{out}][\Gamma_L]\right)\left([I]+[S][\Gamma_{in}]\right)v_{in}T_1 \end{aligned} \qquad (3)$$

In practice, the amplitude of all reflection coefficients is better than -10dB (lower than 0.32), so the high-order term in (3) can be neglect:

$$V_2 \approx Av_{in}T+\left([\Gamma_{out}][\Gamma_L]+[S][\Gamma_{in}]\right)v_{in}T \qquad (4)$$

The first term of (4) is ideal output, the second term is frequency-depending interfering signal caused by multiple-reflection and non-ideal isolation. This term should be minimized to improve wideband response of active antenna system.

### III. SIMULATE RESULTS AND DISCUSSIONS

A SIW four-way power divider is designed on a Taconic TLX-8 substrate with relative dielectric constant of 2.55 and thickness of 60mil. CST MICROWAVE STUDIO 2011 software is used for full-wave simulation and optimization. The layout parameters of power divider are listed in Table I.

TABLE I
PARAMETERS OF STRUCTURE IN FIG. 1

| w50 | 4.1 mm | a | 25 mm |
|------|--------|------|--------|
| lsiw | 30 mm | d | 1 mm |
| wt | 12 mm | p | 1.5 mm |
| lt | 26 mm | d1 | 1.3 mm |
| d2 | 4.4 mm | D1 | 5.3 mm |
| H | 0.5 mm | ds | 3.4 mm |
| p1 | 3.3 mm | ta | 60 mm |
| vp | 10 mm | D2 | 26 mm |

The simulated frequency responses of power divider with and without matching vias (guiding posts and middle via) are shown in Fig. 3. Compared with power divider without matching vias, the proposed power divider exhibits lower insertion loss and better return loss over wideband, especially in low frequency range below 6 GHz and high frequency range over 10.3 GHz.

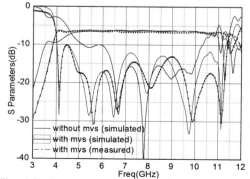

Figure 3. Simulated and measured S parameters of power divider with and without matching vias (mvs)

The photograph of designed four-way power divider is shown in Fig. 4. Four SMA connectors are added for testing. The measured results are also shown in Fig. 3 which includes the loss of SMA connectors. Good agreement between simulated and measured results can be observed in wideband. The discrepancy between the two results is mainly attributed to unexpected tolerance of fabrication and SMA connectors assembling. The measured minimum insertion loss is 0.35dB. The measured and simulated bandwidth over which return loss is better than -10 dB and insertion loss is less than 1 dB is about 7 GHz (form 4 GHz to 11 GHz).

(a)                              (b)

Figure 4. Photograph of broadband four-way power divider (a) before and (b) after adding stepped coaxial line transformer and SMA connectors.

In Fig. 5, the simulate characteristics of two ports network, from power divider input to individual antenna unit, is plotted. For simplicity, the amplitude of all reflection coefficients is set to be 10 dB, their phase is assumed to zero degree. As explained in (4), the non-zero return loss and non-ideal isolation between output ports of power divider cause gain fluctuation. This frequency response can be improved by incorporating appropriate matching networks with power amplifier, as shown in Fig. 6. In order to preserve output power of PA, only input matching network is added. By optimizing this matching network, gain fluctuation in frequency response can be alleviated, as depicted in Fig. 5. The optimized parameters of matching network are summarized in Table II.

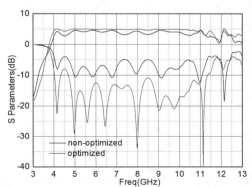

Figure 5. Frequency responses of signal path from power divider input to antenna unit.

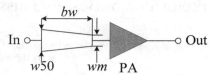

Figure 6. Schematics of power amplifier with input matching taper.

TABLE II
PARAMETERS OF STRUCTURE IN FIG. 6

| bw | 26 mm | wm | 0.8 mm |
|---|---|---|---|

## IV. CONCLUSIONS

A novel four-way SIW power divider has been proposed. Low insertion loss and good return loss were achieved over wideband from 4 GHz to 11 GHz. This structure has been cooperated with active antenna array and the transfer function was deduced. Gain fluctuation could be improved by optimizing the matching network. It is expected that these power dividers can be widely used in wideband millimeter-wave communication circuit especially active antenna system.

### ACKNOWLEDGMENT

This work was supported by National Science and Technology Major Project (2013ZX03001017-003), and Research Fund for the Doctoral Program of Higher Education of China under grant (20100092120013).

### REFERENCES

[1] D. M. Pozar, "The active element pattern," *IEEE Trans. Antennas Propag.*, vol. 42, no. 8, pp. 1176–1178, August 1994.
[2] K. Song, and Y. Fan, "Broadband travelling-wave power divider based on substrate integrated rectangular waveguide," *Electronics Letter.*, vol. 45, no. 12, pp. 631–632, December 2009.
[3] X. Zou, C. M. Tong, and D. W. Yu, "Y-junction power divider based on substrate integrated waveguide," *Electronics Letter.*, vol. 47, no. 12, pp. 1375–1376, December 2011.
[4] D. S. Eom, J. Byun, and H. Y. Lee, "Multilayer Substrate Integrated Waveguide Four-Way Out-of-Phase Power Divider," *IEEE Trans. Microwave Theory Tech.*, vol. 57, no. 12, pp. 3469–13476, December 2009.
[5] L. Zhang, X. W. Zhu, and J. Zhai, "Design of wideband planar power dividers/combiners," *Microwave Workshop Series on Millimeter Wave Wireless Technology and Applications (IMWS), 2012 IEEE MTT-S International*, pp. 173–175, 2012.
[6] Q. X. Chu, and J. M. Yan, "A two-layer planar spatial power divider/combiner," *Microwave Symposium Digest, 2009. IEEE MTT-S International*, pp. 989–992, June 2009.
[7] H. Jin and G. Wen, "A novel four-way ka-band spatial power combiner based on HMSIW," *IEEE Microwave Wireless Component Letter.*, vol. 18, no. 8, pp. 515–517, August 2008.
[8] T. Y. Seo, J. W. Lee, S. C. Choon, and T. K. Lee, "Radial guided 4-way unequal power divider using substrate integrated waveguide with center-fed structure," *Proc. Asia-Pacific Microw. Conf.*, pp. 2758 - 2761, December 2009.
[9] B. Liu, B. tian, Z. S. Xie, et.al, "A novel image transition in half mode substrate integrated waveguide power divider design," *Cross Strait Quad-Regional Radio Science and Wireless Technology Conference (CSQRWC)*, pp. 621–624, 2011.
[10] K. Song, Y. Fan and Y. Zhang, "Eight-way substrate integrated waveguide power divider with low insertion loss," *IEEE Trans. Microwave Theory Tech.*, vol. 56, no. 6, pp. 1473–1477, June 2008.

# Production of Bessel-Gauss Beams at THz by Use of UPA

Yanzhong Yu, Yanfei Li and Yunyan Wang

College of Physics & Information Engineering, Quanzhou Normal University, Quanzhou, 362000 China

yuyanzhong059368@gmail.com

*Abstract*- Applying the principle of antenna pattern synthesis, a uniform planar array (UPA) is designed to generate an arbitrary-order Bessel-Gauss beam (including zero-order and high-order) at Terahertz (THz) range. Numerical results show that the designed Bessel-Gauss beams are in excellent agreement with the desired ones and the project of creating Bessel-Gauss beam is practicable. The generated beams can be applied in THz systems.

*Index Terms*–Bessel-Gauss beam; uniform planar array (UPA); Terahertz (THz)

## I. INTRODUCTION

Physical generation of an ideal diffraction-free beam or Bessel beam can not be made, as it is unbounded in a transverse plane and would require an infinite amount of energy. In order to conquer this difficulty, a Bessel-Gauss beam was thus proposed [1]. It may be considered as a Bessel beam modulated by a Gaussian function so that its boundary does not extend infinitely and its field carries a limit of energy [2]. Therefore, the production of this beam becomes much easier practically. Lots of approaches [3, 4] have been suggested to create a Bessel-Gauss beam of zero order or high order. In fact, these approaches can be divided into two kinds. One is known as a passive way, which implies a Bessel-Gauss beam generated by transforming an incident Gaussian beam using an optical element, such as computer-generated holograms (CGHs) [5], axicons [6], and diffractive phase elements (DPEs) [7]; the other is an active project in which a resonator should be constructed to shape a Bessel-Gauss beam and output directly from it. These schemes have been reported in articles [8, 9]. In the present paper, a novel way is proposed to generate a Bessel-Gauss beam of arbitrary order. It is known that the pattern of array antenna can be synthesized by adjusting the weight of each element. We can get inspired from the technique of pattern synthesis of array antenna. Accordingly a uniform planar array is employed to form a Bessel-Gauss beam by optimally selecting its weights. The numerical results demonstrate that Bessel-Gauss beams with arbitrary orders can be produced effectively.

## II. OPTIMAL DESIGN OF UPA

### A. Bessel-Gauss Beam

In the cylindrical coordinates system, the complex amplitude distribution of the *nth* -order Bessel-Gauss beam is given by

$$E_n(\rho,\varphi,z=0) = E_0 J_n(k_\perp \rho)\exp(in\varphi)\exp(-\rho^2/w_0^2) \quad (1)$$

where $E_0$ is a constant, $J_n$ represents the *nth* -order Bessel function of the first kind, $\rho^2 = x^2 + y^2$, $k_\perp^2 + k_z^2 = k^2 = (2\pi/\lambda)^2$, $k_\perp$ and $k_z$ denote respectively the transverse and longitudinal wave numbers, $\lambda$ is a wavelength in free space, the waist of Gaussian beam is denoted by $w_0$. Therefore, the amplitude distribution of the *nth* -order Bessel-Gauss beam can be written as

$$U_n(\rho,\varphi,z=0) = U_0 J_n(k_\perp \rho)\exp(-\rho/w_0^2) \quad (2)$$

It can be seen from Eq. (2) that the Bessel-Gauss beam of order *n* may be considered as a Gaussian modulation had been added on the Bessel beam. Unlike the ideal Bessel beam, the extension of its lateral oscillation is localized and it thus carries a limited energy. Thereupon, the creation of this beam becomes much easier in practice when compared with the Bessel beam. Fig. 1 displays the radial profiles of the zero-order Bessel-Gauss beam and ideal Bessel beam, respectively. And the high-order distributions of them are illustrated in Fig. 2.

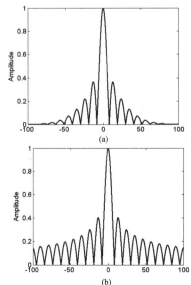

(a)

(b)

Fig. 1. The amplitude distributions of the zero-order Bessel-Gauss beam (a) and Bessel beam (b).

This work is supported by State Key Lab of Millimeter Waves (No. K201307), and the Science and Technology Project of Fengze District of Quanzhou City (2012FZ46).

978-7-5641-4279-7

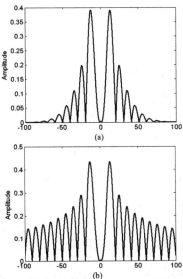

Fig. 2. The amplitude profiles of the high-order Bessel-Gauss beam (a) and Bessel beam (b).

## B. Uniform Planar Array (UPA)

A UPA is formed by placing elements along a rectangular grid, as illustrated in Fig. 3. It can be employed to control and shape the radiation pattern of array antenna. It have been demonstrated that it is more versatile and can provide more symmetrical pattern with lower side lobes [10]. Provided that $M \times N$ elements are positioned along the x-axis and y-axis, and the element in the top left corner of array is set as the reference point, and the azimuth and pitch angels of the incident signal are represented by $\varphi$ and $\theta$ (see Fig. 3), its radiation pattern can thus be expressed as [10, 11]

$$G_0 = \sum_{m=1}^{M} I_{mx} \exp(j\beta_{mx}) \exp[-j(m-1)(kd \sin\theta \cos\varphi)]$$
$$\times \sum_{n=1}^{N} I_{my} \exp(j\beta_{my}^{\backslash}) \exp[-j(n-1)(kd \sin\theta \sin\varphi)] \quad (3)$$
$$= G_x \times G_y$$

Eq. (3) indicates that the total eradiation pattern $G_0$ of a UPA is the product of the pattern of the arrays in the $x$ and $y$ directions. The amplitude and phase of initial excitation current along the x-axis and y-axis are denoted by $I_{mx}$, $I_{my}$, $\beta_{mx}$, $\beta_{my}$, respectively. Therefore the desired pattern of the UPA can be shaped by choosing optimally the weight factor of $I_{mx}$, $I_{my}$, $\beta_{mx}$, $\beta_{my}$. To radiate the Bessel-Gauss beams, a genetic algorithm is used in this paper to select the optimal coefficients of each element. The flow of executing genetic algorithm program is depicted detailedly in ref. [12], and

consequently will not be covered herein for avoiding repetition.

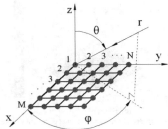

Fig. 3. The geometry of a UPA

## III. SIMULATION RESULTS

Using the optimal algorithm introduced above, we have successfully synthesized several UPAs to radiate Bessel-Gauss beams. In the first instance, the 9×9 -element UPA, with $d = 0.9\lambda$ and $\lambda = 1.0mm$, is designed to form the zero-order Bessel-Gauss beam by adjusting the amplitude and phase of exciting current of each element, as illustrated in Fig. 4. At the visible scope, the generated beam, with $k_\perp = 3318mm^{-1}$ and $w_0 = 0.009mm$, has 10 null points and 9 maximums. We can readily see from Fig. 4(a) that the desired beam exhibits an excellent agreement with the designed one. For the sake of understanding the properties of the constructed Bessel-Gauss beam clearly, the transverse amplitude distributions in two- and three-dimensional figures are depicted in Figs. 4(b) and 4(c), respectively. The normalized amplitude and phase of the exciting current of each element are listed in Table 1.

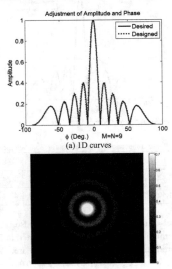

(a) 1D curves

(b) 2D pattern

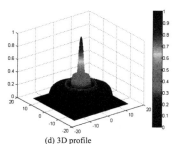

(d) 3D profile

Fig. 4. Zero-order Bessel-Gauss pattern shaped by 9×9 -element UPA

Table 1. Amplitudes and phases of the exciting current for 9×9 -element UPA

| No. | 1 | 2 | 3 | 4 | 5 | 6 | 7 | 8 | 9 |
|---|---|---|---|---|---|---|---|---|---|
| $I_{mx}$ | 0.89 | 0.26 | 0.26 | 0.72 | 0.58 | 0.07 | 0.41 | 0.79 | 0.29 |
| $I_{my}$ | 0.56 | 0.31 | 0.43 | 0.45 | 0.62 | 0.55 | 0.46 | 0.37 | 0.81 |
| $\beta_{mx}$ | 0.23 | 3.1 | 2.07 | 3.14 | 0.31 | 2.80 | 0.22 | 3.01 | 0 |
| $\beta_{my}$ | 3.13 | 3.0 | 3.04 | 0.13 | 3.1 | 3.05 | 0.12 | 3.06 | 3.13 |

Fig. 5 shows another zero-order Bessel-Gauss pattern having $k_\perp = 10529mm^{-1}$ and $w_0 = 0.0022mm$, which is generated by the 8×8 -element UPA through selecting the phase-only components of exciting current. The related parameters utilized in Fig. 5 are : $\lambda = 0.32mm$, $d = 0.5\lambda$. Likewise, as can be observed distinctly from Fig. 5, the designed beam almost overlaps with the desired one. The optimal results of the phase of the current are given in Table 2.

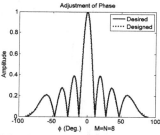

Fig. 5. Bessel-Gauss beam of zero order radiated by 8×8 -element UPA

Table 2. Phase components of the current for 8×8 -element UPA

| No. | 1 | 2 | 3 | 4 | 5 | 6 | 7 | 8 |
|---|---|---|---|---|---|---|---|---|
| $\beta_{mx}$ | 2.82 | 2.60 | 1.49 | 3.14 | 0.01 | 2.20 | 1.24 | 3.01 |
| $\beta_{my}$ | 1.50 | 2.22 | 0.48 | 2.06 | 2.06 | 0.47 | 2.22 | 1.51 |

The Bessel-Gauss beam of zero order can also be shaped by altering the amplitude components, as illustrated in Fig. 6. The relevant parameters used in this example are: $M = N = 5$, $\lambda = 0.30mm$, $d = 0.5\lambda$, $k_\perp = 11539mm^{-1}$, $w_0 = 0.0015mm$. From

Fig. 6 one can see that two curves display a good accord. Table 3 lists the optimal amplitude components.

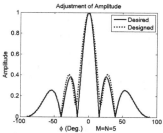

Fig. 6. Zero-order Bessel-Gauss beam eradiated by 5×5 -element UPA

Table 3. Amplitude components of the current for 5×5 -element UPA

| No. | 1 | 2 | 3 | 4 | 5 |
|---|---|---|---|---|---|
| $I_{mx}$ | 0.7557 | 0 | 0 | 0.1746 | 0.8951 |
| $I_{my}$ | 0.9966 | 0.2561 | 0.0207 | 0.9999 | 0.718 |

In the last instance, the 4×4 -element UPA is synthesized, by the way of tuning simultaneously the amplitude and phase, to eradiate the first-order Bessel-Gauss beam, whose amplitude at the origin is zero. Figs. 7 (a), (b) and (c) give the 1D, 2D and 3D distributions of the first-order Bessel-Gauss beam, respectively. The other parameters used in Fig. 7 are : $\lambda = 0.30mm$, $d = 0.6\lambda$, $k_\perp = 11303mm^{-1}$, $w_0 = 0.0018mm$. We can observe obviously that the created beam is almost identical with the expected one.

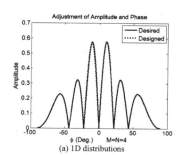

(a) 1D distributions

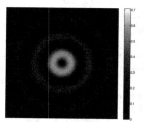

(b) 2D distribution

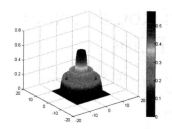

(d) 3D distribution

Fig. 7. First-order Bessel-Gauss beam produced by $4\times4$ -element UPA

Table 4. Amplitudes and phases for $4\times4$ -element UPA

| No. | 1 | 2 | 3 | 4 |
|---|---|---|---|---|
| $I_{mx}$ | 0.8926 | 0.6979 | 0.5171 | 0.483 |
| $I_{my}$ | 0.4626 | 0.2847 | 0.2551 | 0.4754 |
| $\beta_{mx}$ | 0 | 0.188 | 3.1416 | 3.1416 |
| $\beta_{my}$ | 0 | 3.1345 | 0.0173 | 3.1416 |

IV. CONCLUSION

In this paper, a new approach to generate efficiently Bessel-Gauss beam of arbitrary order at THz range is suggested. It employs the UPA to shape a desired beam by adjusting the amplitude-only, phase-only, amplitude and phase components of the exciting current of each element. Simulation results demonstrate that the proposed method is a feasible scheme.

The produced beam can find its applications in communication, measuring, power transmission, and imaging system at THz spectrum range.

REFERENCES

[1] F. Gori, G. Guattari, C. Padovani, "Bessel-Gauss beams," *Opt. Comm.* vol. 64, no. 6, pp. 491-495, Dec. 1987.
[2]. V. Bagini, F. Frezza, M. Santarsiero, G. Schettini and G. Schirripa Spagnolo, "Generalized Bessel-Gauss beams," *J. of Modern Optics*, vol. 43, no. 6, pp. 1155-1166, 1996.
[3]. F. T. Wu, Y. B. Chen, and D. D. Guo, "Nanosecond pulsed Bessel-Gauss beam generated directly from a Nd:YAG axicon-based resonator," *Appl. Opt.*, vol. 46, no. 22, pp. 4943-4947, Aug. 2007.
[4]. M. Dallaire, C. Fortin, M. Piché, N. McCarthy, "Generation of spatiotemporal Bessel-Gauss beams," *Lasers and Electro-Optics/Quantum Electronics and Laser Science Conference*: 2010 Laser Science to Photonic Applications, CLEO/QELS 2010.
[5]. A. Vasara, J. Turunen, and A. T. Friberg, "Realization of general nondiffracting beams with computer-generated holograms," *J. Opt. Soc. Am. A*, vol. 6, no. 11, pp. 1748-1754, 1989.
[6]. R. M. Herman and T. A. Wiggins, "Production and uses of diffractionless beams," *J. Opt. Soc. Am. A*, vol. 8, no. 6, pp. 932- 942, 1991.
[7]. W. X. Cong, N. X. Chen, and B. Y. Gu, "Generation of nondiffracting beams by diffractive phase elements," *J. Opt. Soc. Am. A*, vol. 15, no. 9, pp. 2362-2364, 1998.
[8]. P. Pääkkönen, and J. Turunen, "Resonators with Bessel-Gauss modes," *Opt. Commun.*, vol. 156, pp. 359-366, 1998.
[9]. J. Rogel-Salazar, G. H. C. New, and S. Chávez-Cerda, "Bessel-Gauss beam optical resonator," *Opt. Commun.*, vol. 190, pp. 117–122, 2001.
[10]. D. G. Fang, "*Antenna theory and microstrip antenna*," Beijing: Science Press, 2006.
[11]. M. Liu, C. Yuan, N. Jia, and T. Huang, "*Technology and application of smart antennas*," Beijing: China Machine Press, 2007.
[12]. Y. Z. Yu and W. B. Dou, "Optimal design of ULA for generating quasi-Bessel pattern using genetic algorithm," *J. Comput. Inf. Syst.*, vol. 7, no. 13, pp. 4716-4723, Dec. 2011.

# Electromagnetic Scattering from Rough Sea Surface Covered with Oil Films

Xincheng REN, Wenli LEI, Xiaomin ZHU, Wei TIAN

IEEE Conference Publishing

Shengdi Road, No. 580

Yanan, SHAAXI 716000 CHINA

*Abstract-* A composite random rough surface model is presented for describing rough sea surface covered with oil films, the electromagnetic scattering from this sea surface is studied based on the Stratton-Chu integral equations. A general expression for the radar cross section is derived taking into account a modulation of the rough surface by long surface waves, and the formulae of bistatic scattering coefficient is obtained further. The curves of the bistatic scattering coefficient of HH polarization with varying of the scattering angle are obtained by numerical implementation, the influence of the root mean square and correlation function of small scale roughness, the ratio of the root mean square and correlation function of large scale roughness, wave number of space, the amplitude of spatial fluctuation, the electromagnetic wave irradiation area, the root mean square and correlation function of large scale roughness and the frequency of the incident wave on the bistatic scattering coefficient is discussed. The numerical results show that the influence of these on the bistatic scattering coefficient is very complex.

## I. INTRODUCTION

To detect and monitor oil films at sea is becoming increasingly important, because of the threats posed by such pollution to marine and wildlife [1]. In recent years, remote-sensing techniques and corresponding processing techniques have been developed for this purpose [2]. Gabriel Soriano et al. proposed a cutoff invariant Two-Scale Model in electromagnetic scattering from sea surfaces [3], Joel T. Johnson et al. proposed a numerical study of the retrieval of sea surface height profiles from low grazing angle radar data[4]. But few studies have been reported on electromagnetic wave scattering from sea surface covered with oil films.

In this letter, an approach for a description of the composite random rough surface is developed in order to have some progress in the solution of electromagnetic scattering from rough sea surface covered with oil films. The basis for the analysis is an approximate solution of the integral equation.

## II. FORMULATION

Consider electromagnetic scattering from a perfectly conducting random rough surface. Assume that a large surface wave with some random parameters modulates this surface. Here we consider only the case when both incident and scattered waves show horizontal polarization. The electric and magnetic fields inside a closed surface may be determined by the Stratton-Chu integral equations [5].

According to the reference [6], the first term of the solution of the integral equation for the electric field shows the form

$$E = \frac{ik_z e^{i\vec{k}\vec{R}}}{4\pi|\vec{R}|} \int dx e^{-i\gamma_x x - i\gamma_z f(x)} \tag{1}$$

Here $|\vec{R}|$ means distance between a point at the surface and the observation point, $\gamma_x = |\vec{k}|(\sin\theta_i - \sin\theta_s)$,

$\gamma_z = |\vec{k}|(\cos\theta_i + \cos\theta_s)$, $k_z = |\vec{k}|\cos\theta_i$, where $\theta_i$ and $\theta_s$ are the incidence and scattering angles, $\vec{k}$ is the wave vector of the incident wave, and $f(x)$ means surface height at the horizontal position $x$.

The intensity of the scattered field can readily write

$$I \sim \iint dx dx' e^{-i\gamma_x(x-x') - i\gamma_z[\overline{M}(x)-\overline{M}(x')] - i\gamma_z[\widetilde{M}(x)-\widetilde{M}(x')]} \tag{2}$$

Here $\overline{M}(x)$ is the mean part of the large-scale roughness, and $\widetilde{M}(x) = f_1(x) + f_2(x)$ where $f_1(x)$ and $f_2(x)$ mean height of the small- and large-scale roughness.

The amplitudes of the statistically homogeneous small-scale components of the sea surface may be modulated by the large-scale components of the surface waves. As a result, a statistically inhomogeneous surface appears.

Let $f_1(x)$ be a statistically homogeneous random field with correlation length $l_1$, and $f_2(x)$ a modulating function with a scale $l_2 \gg l_1$. Hence we can choose its mean part as

$$\overline{M}(x) = A\cos Kx.$$

For averaging we note that, one can write for normally distributed random variables $f_k$

$$\left\langle \exp\left\{i\sum_{k=1}^{n} q_k f_k\right\}\right\rangle = \exp\left\{-\frac{1}{2}\sum_{r,s=1}^{n} W(f_r, f_s) q_r q_s\right\} \tag{3}$$

Where $W(f_r, f_s) = \langle f_r f_s \rangle$

We introduce new variables of integration $u = x - x'$ and $v = x + x'$ substitute them into (2), and average analytically by using (3). As result we obtain

$$\overline{I} \sim \int_{-L}^{L} dv \int_{-\infty}^{\infty} du e^{-i\gamma_x u + 2i\gamma_z A\sin(kv/2)\sin(ku/2)} \cdot$$
$$e^{-\gamma_z^2[\langle h_1^2\rangle(1-R_1)+\langle h_2^2\rangle(1-R_2)]} \tag{4}$$

978-7-5641-4279-7

Here $2L$ is the linear size of the illuminated area, $R_1$, $R_2$ are the correlation functions, and $h_1^2 = \langle f_1^2 \rangle$, $h_2^2 = \langle f_2^2 \rangle$ mean rms height of the small- and large-scale roughness.

Consider the case of a Gaussian distribution of both the small- and large-scale components of the surface waves. It is worth noting that one can use another probability density of the spectrum, that is, a Pierson-Moscowitz spectrum.

In order to simplify Eqn. (4), we expand the term taking into account the large-scale roughness into a Taylor series up to the second term. We obtain instead of (4)

$$\bar{I} \sim e^{-\gamma_z^2 \langle h_1^2 \rangle} \int_{-L}^{L} dv \int_{-\infty}^{\infty} du e^{-i\gamma_x u + 2i\gamma_z A \sin(kv/2)\sin(ku/2)} \cdot$$
$$e^{\gamma_z^2[\langle h_1^2 \rangle e^{-u^2/L_1^2} + \langle h_2^2 \rangle (u/L_2)^2]} \quad (5)$$

Expanding the integrand of (5) into an infinite Taylor series, the radar scattering cross section is obtained

$$\sigma^0 = \sqrt{\pi} L e^{-\gamma_z^2 h_1^2} \sum_{m=0}^{\infty} \frac{1}{m!} (\gamma_z h_1)^{2m} S_0^{-1/2}$$
$$\sum_{r=0}^{\infty} (\gamma_z A / 2)^{2r} \varphi(r) T(r) \quad (6)$$

Where

$$\phi(r) = \frac{1}{(r!)^2} + 2(-1)^r \sum_{n=0}^{r-1} \frac{(-1)^k}{n!(2r-n)!} \sin c[KL(r-n)]$$

$$S_0 = \gamma_z^2 \frac{h_2^2}{l_2^2} + \frac{m}{2l_1^2}$$

$$T(r) = \sum_{s=0}^{2r} \frac{(-1)^s (2r)!}{s!(2r-s)!} e^{-[\gamma_x - K(r-s)]^2/(4S_0)}$$

When the short-wave range of the electromagnetic waves is used for remote sensing of the sea surface, it is possible to simplify (6) by taking into account that the inequality $|r-s|K \ll \gamma_x$ holds for a wide range of variations of $r$ and $s$, we suppose that $\gamma_x K/(\gamma_z \gamma)^2 \ll 1$. Then one can write Eqn. (6) as

$$\sigma^0 = \sqrt{\pi} L e^{-\gamma_z^2 h_1^2} \sum_{m=0}^{\infty} \frac{1}{m!} (\gamma_z h_1)^{2m} S_0^{-1/2} e^{-\gamma_x^2/(4s_0)} M(m) \quad (7)$$

Where

$$M(m) = \sum_{r=0}^{\infty} [\gamma_x \gamma_z AK/(4S_0)]^{2r} e^{-\gamma_x Kr/(4S_0)} \phi(r)$$

If $KL = m_1 \pi, m_1 = 1, 2, \cdots$, the following identity can be used to simplify (7)

$$M(m) = J_0(y) \quad (8)$$

Where $J_0(y)$ is the Besssel function of zeroth order, whose argument is given by

$$y(m) = i\gamma_x \gamma_z AK \exp[-\gamma_x K/(4S_0)]/(2S_0)$$

Note that for the cases of most practical importance, the value of $KL$ is usually so large that the second term in the above expression for $\phi(r)$ (sum over $n$) is negligibly small as compared to the first one. Then one can use (7) with $M(m)$ given by (8) even if $KL \neq m_1 \pi$.

In this way, we can obtain the scattering coefficient of rough sea surface covered with oil films as

$$\sigma = 10 \log_{10} \sigma^0 \quad (9)$$

### III. NUMERICAL RESULTS AND DISCUSSION

The sampling of the incident frequency is $f = 900$ GHz, the sampling of the incident angle is $\theta_i = 20^0$. For length, HH polarization of bistatic scattering is numerically calculated only.

1. The influence of the root mean square of small-scale $h_1$ on the scattering coefficient

Fig.1 depicts the distribution of $\sigma$ with $\theta_s$ for different $h_1$ with $l_1 = 0.42\lambda$, $L = 100\lambda$, $A = 0.1/K$, $K = 2\pi$, and $h_2/l_2 = 0.1$.

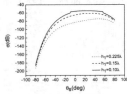

Fig.1 Distribution of $\sigma$ with $\theta_s$ for different $h_1$

It is obvious that the influence of $h_1$ on the scattering coefficient is big and obvious, the bigger of $h_1$ is, the smaller of $\sigma$ is, and scattering angle is closer the incident angle, the scattering coefficient with the more obvious changes in $h_1$, the more deviation incident angle, the scattering coefficient with the more obscure changes in $h_1$.

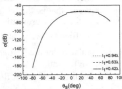

Fig.2 Distribution of $\sigma$ with $\theta_s$ for different $l_1$

2. The influence of the correlation length of small-scale $l_1$ on the scattering coefficient

The distribution of $\sigma$ with $\theta_s$ for different $l_1$ with $h_1 = 0.10\lambda$, $L = 100\lambda$, $A = 0.1/K$, $K = 2\pi$, and $h_2/l_2 = 0.1$ is depicted in Fig.2.

All the curves of Fig.2 are almost coincide, and it is obvious that the influence of $l_1$ on the scattering coefficient is small and obscure, and in the scattering angle is equal on both sides of the vicinity of the incident angle, the scattering coefficient with $l_1$ changes slightly apparent than some.

3. The influence of $h_2/l_2$ on the scattering coefficient

Fig.3 depicts the distribution of $\sigma$ with $\theta_s$ for different $h_2/l_2$ with $h_1 = 0.10\lambda$, $l_1 = 0.42\lambda$, $L = 100\lambda$, $A = 0.1/K$, $K = 2\pi$.

There is stated regularity in distribution of $\sigma$ with $\theta_s$ for different $h_2/l_2$ from Fig.3, that is, when $\theta_s < \theta_i$, the bigger of $h_2/l_2$ is, the bigger of $\sigma$ is, the scattering coefficient with $h_2/l_2$ is very obvious changes in, in the vicinity of both sides of the scattering angle is equal to incident angle, the scattering coefficient almost no change with the change in $h_2/l_2$, when $\theta_s > \theta_i$, the bigger of $h_2/l_2$ is, the bigger of $\sigma$ is, the scattering coefficient with $h_2/l_2$ is very obvious changes in, but it is not clear that such the case of $\theta_s < \theta_i$.

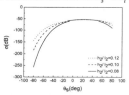

Fig.3 Distribution of $\sigma$ with $\theta_s$ for different $h_2/l_2$

4. The influence of the wave number of space $K$ on the scattering coefficient

Distribution of $\sigma$ with $\theta_s$ for different $K$ with $h_1 = 0.10\lambda$, $l_1 = 0.42\lambda$, $h_2/l_2 = 0.1$, $L = 100\lambda$, $A = 0.1/2\pi$ is depicted in Fig.4.

It is obvious that the influence of $K$ on the scattering coefficient is big and obvious, that is, the bigger of $K$ is, the bigger of the frequency of the curve oscillating is, and when $\theta_s = \theta_i$, the scattering coefficient has a sharp increase, but there is not stationary regularity for the influence of $K$ on the magnitude of scattering coefficient.

5. The influence of the amplitude of spatial fluctuation $A$ on the scattering coefficient

Fig.5 depicts the distribution of $\sigma$ with $\theta_s$ for different $A$ with $h_1 = 0.10\lambda$, $l_1 = 0.42\lambda$, $h_2/l_2 = 0.1$, $L = 100\lambda$, $K = 2.0\pi$.

It is obvious that the influence of $A$ on the scattering coefficient is big and obvious, that is, the bigger of $A$ is, the bigger of the frequency of the curve oscillating is, it is obvious that the influence of $A$ on the scattering coefficient is same

with the influence of $K$. Similarly, when $\theta_s = \theta_i$, the scattering coefficient has a sharp increase, but there is not stationary regularity for the influence of $A$ on the magnitude of scattering coefficient.

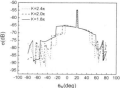

Fig.4 Distribution of $\sigma$ with $\theta_s$ for different $K$

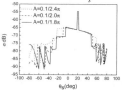

Fig.5 Distribution of $\sigma$ with $\theta_s$ for different $A$

6. The influence of the electromagnetic wave irradiation area $L$ on the scattering coefficient

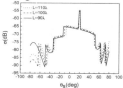

Fig.6 Distribution of $\sigma$ with $\theta_s$ for different $L$

The distribution of $\sigma$ with $L$ in the condition that $h_1 = 0.10\lambda$, $l_1 = 0.42\lambda$, $h_2/l_2 = 0.1$, $K = 2.0\pi$, $A = 0.1/K$ is depicted in Fig.6.

The curve of the distribution of $\sigma$ with $L$ is oscillatory, and the bigger of $L$ is, the bigger of the frequency of the curve oscillating is, but it is unconspicuous with $L$ than with $K$ and. Similarly, when $\theta_s = \theta_i$, the scattering coefficient has a sharp increase, but there is not stationary regularity for the influence of $L$ on the magnitude of scattering coefficient.

7. The influence of rms of large scale roughness $h_2$ on the scattering coefficient

Fig.7 depicts the distribution of $\sigma$ with $\theta_s$ for different $h_2$ with $h_1 = 0.10\lambda$, $l_1 = 0.42\lambda$, $l_2 = 127\lambda$, $L = 100\lambda$, $K = 2.0\pi$, $A = 0.1/K$.

It is obvious that the influence of $h_2$ on the scattering coefficient is big and obvious, the bigger of $h_2$ is, the bigger of $\sigma$ is, and scattering angle the closer the incident angle, the scattering coefficient with the more obscure changes in the $h_2$, the more deviation incident angle, the scattering coefficient with the more obvious changes in $h_2$, this is differ from the influence of rms of small scale roughness $h_1$ on the scattering coefficient.

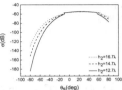

Fig.7 Distribution of $\sigma$ with $\theta_s$ for different $h_2$

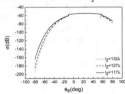

Fig.8 Distribution of $\sigma$ with $\theta_s$ for different $l_2$

8. The influence of correlation function of large scale roughness $l_2$ on the scattering coefficient

The distribution of $\sigma$ with $l_2$ in the condition that $h_1 = 0.10\lambda, l_1 = 0.42\lambda, h_2 = 12.7\lambda, L = 100\lambda$, $K = 2.0\pi, A = 0.1/K$ is depicted in Fig.8.

It is obvious that the influence of $l_2$ on the scattering coefficient is big and obvious, the bigger of $l_2$ is, the smaller of $\sigma$ is, and scattering angle the closer the incident angle, the scattering coefficient with the more obscure changes in the $l_2$, the more deviation incident angle, the scattering coefficient with the more obvious changes in $l_2$, this is differ from the influence of correlation length of small scale roughness $l_1$ on scattering coefficient.

9. The influence of the frequency of the incident wave on the scattering coefficient

The variation of $\sigma$ with $f$ under the condition that $h_1 = 0.10\lambda, l_1 = 0.42\lambda, h_2/l_2 = 0.1, L = 100\lambda$, $K = 2.0\pi, A = 0.1/K$ (i.e. a certain rough surface), $\theta_i = 20^0, \theta_s = 10^0, 40^0$ is depicted in Fig.9.

The curve of the variation of $\sigma$ with $f$ is oscillatory, generally speaking, when the incident frequency increases, the magnitude of the scattering coefficient is decreases, but this change is relatively slow, the oscillation frequency of curve decreases, the oscillation frequency of curve at $\theta_s = 40^0$ is bigger than at $\theta_s = 10^0$, the curve is continuous, but is non-differentiable in certain frequency point. In the point of the same frequency, the corresponding scattering coefficient when

the scattering angle is equal to $40^0$ is bigger than that when the scattering angle is equal to $10^0$.

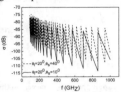

Fig.9 Distribution of $\sigma$ with $f$

It is obvious that the influence of $h_1$ on the scattering coefficient is big and obvious, the bigger of $h_1$ is, the smaller of $\sigma$ is, and scattering angle the closer the incident angle, the scattering coefficient with the more obvious changes in $h_1$, the more deviation incident angle, the scattering coefficient with the more obscure changes in $h_1$.

IV. CONCLUSIONS

In this paper, the electromagnetic scattering from the rough sea surface covered with oil films is studied using a composite random rough surface model. A general expression for the radar cross section is obtained taking into account a modulation of the rough surface by long surface waves. The curves of the bistatic scattering coefficient of HH polarization with varying of the scattering angle are obtained by numerical implementation, the influence of the root mean square and correlation function of small scale roughness, the ratio of the root mean square and correlation function of large scale roughness, wave number of space, the amplitude of spatial fluctuation, the electromagnetic wave irradiation area, the root mean square and correlation function of large scale roughness and the frequency of the incident wave on the bistatic scattering coefficient is discussed. These results will be applicable for solving many engineering, technical, and scientific problems.

REFERENCES

[1] M. Gade, W. Alpers, H. Hühnerfuss, V. R. Wismann, and A. Lange. On the reduction of the radar backscatter by oceanic surface films: Scatterometer measurements and their theoretical interpretation. Remote Sens. Environ., 66(1998), 52–70.
[2] A. H. S. Solberg, C. Brekke, and P. O. Husøy. Oil spill detection in envisat SAR images. IEEE Trans. Geosci. Remote Sens., 45(2007), 746–755.
[3] Gabriel Soriano and Charles-Antoine Guérin. A Cutoff Invariant Two-Scale Model in Electromagnetic Scattering From Sea Surfaces. IEEE Trans. Geosci. Remote Sens., 5(2008), 199-203.
[4] Joel T. Johnson, Robert J. Burkholder, Jakov V. Toporkov, David R. Lyzenga, and William J. Plant. A Numerical Study of the Retrieval of Sea Surface Height Profiles From Low Grazing Angle Radar Data. IEEE Trans. Geosci. Remote Sens., 47(2009), 1641-1650.
[5] Stratton, J. A., Electromagnetic Theory. New York, McGraw-Hill, 1941: 135-156.
[6] Holiday, D., L. L. DeRaad Jr., and G. J. St-Cyr. New equations for electromagnetic scattering by small perturbations of a perfectly conducting surface. IEEE Trans. Antennas Propag., 46 (1998), 1427 – 1432.

# A FBLP Based Method for Suppressing Sea Clutter in HFSWR

Yongpeng Zhu, Chao Shang, Yajun Li
Institute of Electronic Engineering Technology
Harbin Institute of Technology
Harbin, 150001, China

*Abstract-* **This paper presents a novel method for suppressing sea clutter in the high frequency surface wave radar (HFSWR). The proposed algorithm is based on the combination of the linear prediction technique and the multidimensional feature of the sea clutter. In order to ensure the accurate suppression of the first order sea clutter, the feature detection matrix (FDM) has been defined and constructed. Eventually, the performance of the derived algorithm is testified by the experimental data.**

## I. INTRODUCTION

High frequency surface wave radar, which is based on the surface wave diffraction, provides a unique capacity to detect the target far beyond the conventional microwave radar coverage. Therefore, it has been widely used in remote surveillance [1]. However, the sea clutter, constituting the major target detection background, has deteriorated the target detection performance significantly, since the amplitude of the first order sea clutter (Bragg Peak), which constructs the dominate component of the sea clutter, often masks the target. Besides, the Doppler frequency of the Bragg peak is similar with the targets with low velocity such as vessels [2]. Thus, how to suppress the strong sea clutter appears to be a fairly critical issue in order to enhance the capacity of HFSWR.

Recently, some methods have been put forward to address this problem. Specifically, after studying the time varying behavior of the two dominant narrowband frequency components in sea clutter echo, which corresponds to the Bragg peak Doppler frequency, Khan proposed a Hankel rank reduction method based on SVD to suppress the first order clutter [3-5]. However, the selection of the rank lacks in the theoretical support and the computational complexity deteriorates as the rank increases, which limits the real time processing efficiency of the radar system. In [6-7], Root solved this problem through clutter cancellation, which is similar with the CLEAN algorithm, while the performance depends on the estimation accuracy of the sinusoid parameters. Inspired by the inverse synthetic aperture radar imaging technique, an idea combining the adaptive chirplet transform with one-class SVM has been developed to separate the target and the first order sea clutter because of the difference in the chirp rate [8-10]. While this algorithm fails as a matter of fact that the target could not be modeled as a chirp signal and the difference existed in the chirp rate appears too tinny to be identified. Besides other interference could also express the feature of a chirp signal; therefore making this method invalid in the real system.

Different from the present methods, the one described in this paper takes full account of the multidimensional feature of sea clutter expressing in range and Doppler frequency, so as to maintain an accurate detection performance. Specifically, the signal model has been studied firstly. Afterwards, the sea clutter echo signal is decomposed into many sinusoidal signals by means of the forward backward linear prediction (FBLP) algorithm. Then, with the help of feature detection matrix, we could identify the signal parameters corresponding to the Bragg peaks, eventually we could make the corresponding sea clutter amplitude into zero in the temporal domain, realizing the purpose of suppressing the sea clutter.

## II. THE PARAMETER ESTIMATION OF THE SEA CLUTTER SIGNAL

### A. Sea Clutter Signal Model

In [12], it is argued that the dominant first order sea clutter exhibits a time varying characteristic which can be modeled with two narrowband signals, interpreted as two independent angular modulated components. And it could be demonstrated that the spectrum of such an angular modulated signal corresponds with that of the experimental sea clutter echo signal perfectly.

The time varying model could be expressed as the signals, composed of $M$ superimposed complex sinusoids.

$$y(n) = \sum_{i=1}^{M} a_i e^{j(2\pi f_i n + \varphi_i)} \qquad (1)$$

Where $a_i$, $f_i$ and $\varphi_i$ are the corresponding signal parameters indicating amplitude, frequency, and phase respectively.

### B. Forward and Backward Linear Prediction Method

The linear prediction based method could be used to estimate the parameters of sinusoidal signals, when the signal to noise is high enough, the parameters of which can be predicted as the weighted sum of $L$ previous values.

$$y(n) = -\sum_{i=1}^{L} y(n-i) \cdot \alpha_i \qquad (2)$$

Where $\alpha_i$ stands for the weight coefficients.

Besides, so as to track the time varying behavior of the sea clutter, the coefficients of the prediction error filter must be estimated over short data segments so that the filter coefficients could be updated adaptively. And the prediction equation matrix, which is defined in [15], could be expressed as follows:

978-7-5641-4279-7

$$\begin{bmatrix} y(L) & y(L-1) & \cdots & y(1) \\ y(L+1) & y(L) & \cdots & y(2) \\ \vdots & \vdots & \ddots & \vdots \\ y(N-1) & y(N-2) & \cdots & y(N-L) \\ y^*(2) & y^*(3) & \cdots & y^*(L+1) \\ y^*(3) & y^*(4) & \cdots & y^*(L+2) \\ \vdots & \vdots & \ddots & \vdots \\ y^*(N-L) & y^*(N-L+1) & \cdots & y^*(L) \end{bmatrix} \begin{bmatrix} \alpha_1 \\ \alpha_2 \\ \vdots \\ \alpha_L \end{bmatrix} = - \begin{bmatrix} y(L+1) \\ y(L+2) \\ \vdots \\ y(N) \\ y^*(1) \\ y^*(2) \\ \vdots \\ y^*(N-L) \end{bmatrix} \quad (3)$$

Simply, we denote as:

$$A \cdot \alpha = -y \quad (4)$$

And the weight coefficient matrix $\alpha$ could be estimated by the following equation:

$$\alpha = -A^{-1} y \quad (5)$$

As the linear equations above are over-determined, the total least square method is used to solve this problem [13], namely:

$$\alpha = -A^{-1} y = -(A^H A)^{-1} A^H y \quad (6)$$

Where $H$ indicates the conjugate transpose. Simultaneously, the order of prediction error filter polynomial $L$ could be determined by [14], which satisfies the inequality as follows:

$$M \le L \le (N - M / 2) \quad$$

That is to say, the order of prediction error filter should exceed the estimated signal number. Afterwards, we define frequency estimation matrix as:

$$f = [f_1, f_2, \cdots, f_M] \quad (7)$$

Where

$$f_k = (1, e^{-s_k}, e^{-2s_k}, \cdots, e^{-Ls_k}) \quad k = 1, 2, \cdots, M$$

It is easy to observe that each row in $A$ is a linear combination of $L$ linearly independent vectors in $f_k$. That is to say, the rank of $A$ is $M$ as long as $A$ has at least $M$ rows. Thus, the dimension of null space in $A$ is $L+1-M$ dimension. In addition, as $\alpha_k$ lies in the null space of $A$, we have:

$$\alpha_0 + \alpha_1 e^{-s_k} + \alpha_2 e^{-2s_k} + \cdots + \alpha_L e^{-Ls_k} = 0 \quad (8)$$

The signal frequency could be estimated from the roots of (8). Besides, in order to obtain the amplitude and initial phase of each signal, we define the following matrix equation as:

$$\begin{bmatrix} \beta_1^0 & \beta_2^0 & \cdots & \beta_M^0 \\ \beta_1^1 & \beta_2^1 & \cdots & \beta_M^1 \\ \vdots & \vdots & \ddots & \vdots \\ \beta_1^{N-1} & \beta_2^{N-1} & \cdots & \beta_M^{N-1} \end{bmatrix} \begin{bmatrix} h_1 \\ h_2 \\ \vdots \\ h_M \end{bmatrix} = \begin{bmatrix} y(1) \\ y(2) \\ \vdots \\ y(N) \end{bmatrix} \quad (9)$$

Simply, we denote as:

$$R \cdot h = Y \quad (10)$$

Where $\beta_i^k = e^{j2\pi f_i k / f_s}$, $h = (R^H R)^{-1} R^H Y$, $f_s$ represents the sampling frequency and $N$ is the sampling number. In addition, the amplitude and initial phase could be obtained after taking the manipulation of the absolute and angular value of $h$ respectively.

$$a_k = |h_k| \quad k = 1, \cdots, M \quad (11)$$

$$\varphi_k = acrtan(Im(h_k) / Re(h_k)) \quad k = 1, \cdots, M \quad (12)$$

Where $Im(\cdot)$ and $Re(\cdot)$ represents taking the image and real part of the signal respectively.

## C. The Multidimensional Feature of the First Order Sea Clutter

The detection and extraction of the first order sea clutter appears a much tougher problem especially in the background with the ocean current, the ionosphere interference and atmosphere noise, resulting in the Doppler shift and split to the Bragg peak. Therefore, it is hardly to approach a desiring clutter suppressing performance without taking the characteristics of sea clutter into consideration.

Actually, according to Fig. 1, it is straightforward to observe that the first order of sea clutter expresses a multi-dimension feature, not only in the range bins but also in the Doppler bins, which could be used to provide more effective and accurate identification of Bragg peak and therefore make it possible to suppress the sea clutter without canceling the target by mistake.

After studying many batches of experimental data, we could conclude the spectral multi-dimensional feature of the Bragg peak as follows:

1) Given a single range bin, the Doppler frequency of the sea clutter is nearly symmetry about the zero frequency. And the ocean current would give rise to the Doppler shift to the first order sea clutter, while the offset towards the positive and negative Bragg peaks is consistent.

2) Given the continuous range bins, the position of the Bragg peaks maintain fairly good continuity along the continuous range bins which could be evidently observed in Fig.1.

3) As for the spectral magnitude, the first order sea clutter exhibits a local maximum feature, which makes it possible to estimate the clutter parameter exactly.

4) In some range bin, sea clutter may appear to be dominant in either the positive or the negative Doppler frequency. That is to say, the sea clutter is asymmetric in some range bin.

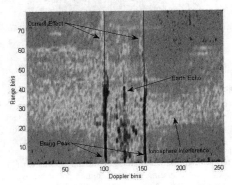

Figure 1. The range and Doppler map of the sea clutter in HF surface wave radar including different kinds of clutter and interference.

## III. SUPPRESSION OF THE FIRST ORDER SEA CLUTTER

### A. The Derivation of the Algorithm

The complexity of first order sea clutter distribution prompts us to take the multi-dimension feature into account to ensure a robust identification performance. Based on which, we would like to obtain each signal domain parameters so that we could suppress the sea clutter according to the feature expressed in Fig.1.

However, in order to use the FBLP method, the number of interested signals should be determined firstly. Instead of conventional signal number estimation method [11], we put forward a SNR criterion to limit the number of estimated signals in a simple way. Specially, we suppose $S$ as the number of signals, and the total power of sea clutter echo could be obtained by

$$P = \sum_{n=1}^{N} |y_n|^2 \tag{13}$$

Where $N$ is the length of signal sampling points. Then the power of each signal and noise is:

$$P_k = \sum_{n=1}^{N} \left| a_k \beta_k^{n-1} \right|^2 = |a_k|^2 \frac{1-|\beta_k|^{2N}}{1-|\beta_k|^2} \quad (k=1,2,\cdots,S) \tag{14}$$

$$P_{nosie} = \sum_{n=1}^{N} \left| y_n - \sum_{k=1}^{S} a_k \beta_k^{n-1} \right|^2 \tag{15}$$

The SNR of each signal is given by

$$SNR_k = P_k / P_{noise} \quad (k=1,2,\cdots,S) \tag{16}$$

The number of signal $M$ could be determined by letting the $SNR_k$ exceed a fixed threshold, thus reducing the number of the estimated signals to a large extent.

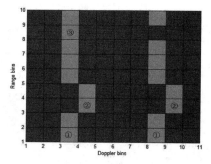

Figure 2. The schematic illustration of sea clutter spatial distribution

As a matter of fact, the feature of sea clutter expressed in Fig.1 could be simplified as Fig.2. In order to detect the location of the first order sea clutter distributed as in Fig. 2, a clever way is proposed to get around this based on the derivation of Feature Detection Matrix (FDM).

In terms of the sea clutter signal model defined in (1), the frequency corresponding to each signal could be estimated from the roots of polynomial. Now that the position of the Bragg peak exhibits a symmetric feature, even when the influence of ocean current exits; Therefore, the detection of the Bragg peak could be summarized as the maximum symmetry identification.

So as to construct the FDM, we calculate the relative offset towards the theoretical Bragg frequency $f_{MB}$ by manipulating as follows:

$$f_i' = \begin{cases} f_i + f_{MB} & f_i < 0 \\ f_i - f_{MB} & f_i > 0 \end{cases} \tag{17}$$

Where $f_{MB} = \sqrt{g\pi/\lambda}$ , $\lambda$ indicates the wave length , $g$ is the acceleration of gravity. Afterwards, we define the Feature Detection Matrix as:

$$F = \begin{bmatrix} f(1,2) & f(1,3) & f(1,4) & \cdots & f(1,M) \\ 0 & f(2,3) & f(2,4) & \cdots & f(2,M) \\ 0 & 0 & f(3,4) & \cdots & f(3,M) \\ 0 & 0 & 0 & \cdots & f(M-1,M) \end{bmatrix} \tag{18}$$

Where $f(i,j) = \left| f_i' - f_j' \right|$ $(i=1,\cdots,M-1; j=i+1,\cdots,M)$

For the upper triangular matrix $F$ , we search for the minimum value, the coordinates of which correspond with the Doppler frequency in (7) satisfying the best symmetry. We denotes the coordinates as $(m,n)$ .Based on the feature proposed previously, we suppose that the Doppler frequency detected is in conformity with that of the first order sea clutter. Afterwards, we let the amplitude of $a_m$ and $a_n$ zero, which correspond to the detected Doppler frequency $f_m$ and $f_n$ . Then the residual $M-2$ signals would be used to reconstruct the time series data with the help of (7), (11) and (12). The new time series data are the one with the clutter signal suppressed.

Additionally, given the third condition illustrated in Fig.2, the method derived would be revised by taking the adjacent range bin as a reference now that the sea clutter maintains continuous along the range dimension.

### B. The Procedure of the Algorithm

Based on the analysis mentioned before, the procedures of the algorithm could be interpreted as follows:

1) Construct the prediction equation matrix in (3) based on the sea clutter echo from each range bin and estimate the sinusoidal parameters including the amplitude, frequency and initial phase under the assumption of the signal model in (1).

2) According to the SNR criterion, we determine the number of the interest signals according to the SNR threshold.

3) Construct the feature detection matrix $F$ in (18) based on the frequency estimation matrix in (7) and we search $F$ for the values below a symmetric threshold. Then we choose the coordinates of the minimum value as the index of the first order sea clutter.

4) If there is not a value satisfying the symmetric threshold, we would compare it with the reference range bin to cancel the sea clutter failing to be dominant both in the negative and positive Doppler bins (the third condition in Fig.2).

5) We make the amplitude of the signals corresponding to the index obtained in procedure 3) and 4) to be zero and then reconstruct the signals after suppression with the residual parameters.

### IV. EXPERIMENT AND RESULT ANALYSIS

The performance of the algorithm could be testified with the experimental data in the HFSWR. Fig.3 illustrates the contrast result before and after suppression.

It is obvious to observe that the Bragg peak is fairly evident in the range and Doppler map of the experimental sea clutter echo data. While the dominant first order sea clutter is nearly canceled by virtue of the algorithm derived, which can be easily observed in Fig.4. It is noticeable to spot that the suppressed signal still maintains the characteristics of other clutter and interference so as to ensure the validity of other clutter and interference algorithms. Besides, as the method is mainly based on the symmetry detection, the signal of the target would be reserved, even when the Doppler frequency of target and the first order sea clutter is closed in the Doppler.

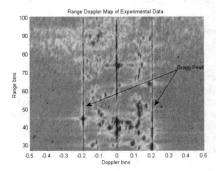

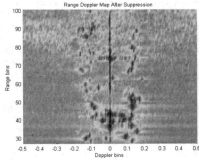

Figure 3. The contrast result before and after the first order sea clutter suppression based on the experimental data.

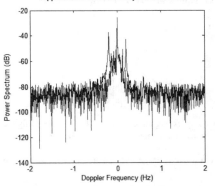

Figure 4. The suppression of the first order sea clutter in a certain range bin.

## V. CONCLUSIONS

It has been demonstrated that the method, based on the combination of forward and backward linear prediction and the multidimensional feature of sea clutter, can identify the first order sea clutter accurately even in the complicated detection background.

While the feature of sea clutter depends on the sea state and the ocean current field significantly, the method present would work only when the sea clutter is in conformity with the feature proposed. And the algorithm also exits some limitations especially when the peak of the first order sea clutter is splitting, which should be improved in the later work.

## REFERENCES

[1] B. J. Dawe, R. H. Khan, J. Walsh and J. R. Benoit, "A long range ground wave radar system for offshore surveillance," in Proc. Canad. Symp. Remote Sensing (IGARSS '89), Vancouver,B.C., Canada, July 1989, pp. 2950-2952.

[2] Leong, H. Ponsford, A., "The effects of sea clutter on the performance of HF Surface Wave Radar in ship detection," Radar Conference, 2008. RADAR '08. IEEE, vol., no., pp.1-6, 26-30 May 2008.

[3] C. L. DiMonte and K. S. Arun, "Tracking the frequencies of superimposed time-varying harmonics" in Proc.ICASSP90, Albuquerque, NM, Apr 1990, pp. 2536-2542.

[4] Martin W. Y. Poon, Rafaat H. Khan and Son Le-Ngoc, "A singular Value Decomposition(SVD) Based Method for Suppressing Ocean Clutter in High Frequency Radar," IEEE Trans. On Signal Processing, vol. 41, pp.1421-1425.

[5] Khan, R. Power, D. Walsh, J., "Ocean clutter suppression for an HF ground wave radar," Electrical and Computer Engineering, 1997. IEEE 1997 Canadian Conference on, vol.2, no., pp.512-515 vol.2, 25-28 May.

[6] Root, B., "HF radar ship detection through clutter cancellation," Radar Conference, 1998. RADARCON 98. Proceedings of the 1998 IEEE , vol., no., pp.281-286, 11-14 May 1998.

[7] Root, B., "High-frequency over-the-horizon radar ship detection through clutter cancellation: an alternative to high-resolution spectral estimation," Part of the SPIE Conference on Advanced Signal Processing Algorithms, July 1998 vol.3461, pp.488-500.

[8] Z. Bao, G. Wang and L. Luo, "Inverse synthetic aperture radar imaging of maneuvering target, Optical Engineering vol.37, no.5, pp. 1582-1588, 1998.

[9] Yajuan Tang, Xiapu Luo and Zijie Yang, "OCEAN CLUTTER SUPPRESSION USING ONE-CLASS SVM,"2004 IEEE Workshop on Machine Learning for Signal Processing.pp.559-568

[10] Ren-Zhou Gui;, "Utilization of Support Vector Machine based on Neural Network to Suppress Ocean Clutter and Zero Frequency Disturbances," Vehicular Electronics and Safety, 2006. ICVES 2006. IEEE International Conference on , vol., no., pp.496-501, 13-15 Dec. 2006

[11] Wax, M. Kailath, T.; , "Detection of signals by information theoretic criteria," Acoustics, Speech and Signal Processing, IEEE Transactions on , vol.33, no.2, pp. 387- 392, Apr 1985

[12] Khan, R.H., "Ocean-clutter model for high-frequency radar," Oceanic Engineering, IEEE Journal of , vol.16, no.2, pp.181-188, Apr 1991

[13] G. H. Golub and C. F. Van Loan, Matrix Computations. Baltimore, MD: John Hopkins University Press, 1983

[14] Kumaresan, R.; , "On the zeros of the linear prediction-error filter for deterministic signals," Acoustics, Speech and Signal Processing, IEEE Transactions on , vol.31, no.1, pp. 217- 220, Feb 1983

[15] Tufts, D.W.Kumaresan, R. "Estimation of frequencies of multiple sinusoids: Making linear prediction perform like maximum likelihood," Proceedings of the IEEE , vol.70, no.9, pp. 975- 989, Sept. 1982

# Radar HRRP Adaptive Denoising via Sparse and Redundant Representations

Min Li，Gongjian Zhou，Bin Zhao, Taifan Quan

School of Electronics and Information Engineering

Harbin Institute of Technology

Harbin, 150001, China

*Abstract*-We address the radar high resolution range profile (HRRP) denoising problem for improving the recognition rate of HRRP at low signal-to-noise ratio (SNR). Gaussian white noise in HRRP return is suppressed by an approach based on sparse representation. A Fourier redundant dictionary is established for sparsely representing HRRP returns. An adaptive signal recovering algorithm, Orthogonal Matching Pursuit-Modified Cross Validation (OMP-MCV), is proposed for obtaining denoised HRRP without requiring any knowledge about the noise statistics. As a modification to OMP-CV, OMP-MCV modifies the cross validation iteration condition, which can prevent the iteration procedure from terminating at local minimum impacted by noise. Simulation results show that OMP-MCV achieves better performance than OMP-CV and some other traditional denoising method, like discrete wavelet transform, for HRRP returns denoising.

## I. INTRODUCTION

High resolution range profile (HRRP) automatic target recognition (ATR) has received much attention in recent years [1-3]. In practice, the HRRP signatures are usually distorted due to the presence of noises, such as system noise, environmental noise and complicated battle interference etc. And this results in the recognition performance degradation. Several methods have been proposed to settle this issue [4-6]. The noise-robust bispectrum signatures have been extracted for target recognition [4]. Noise-robust factor analysis models based on multitask learning (noise-robust MTL-FA) has been developed to improve the recognition performance at low signal-to-noise ratio (SNR) [5]. HRRP denoising is another feasible choice, e.g., combine bispectrum-filtering has been used to suppress the noise of HRRP [6]. Nevertheless, there are several limitations of these methods. First, the noise-robust features are limited and maybe not optimal for classification. Secondly, the noise-robust MTL-FA classifier suffers from a large computational burden, impeding its practical application. Lastly, the combine bispectrum-filtering requires a certain number observations to reduce noise, increasing the time cost of HRRP recognition. It is therefore desired to develop new methods to handle the issue.

This research was supported by the National Science Foundation of China (no. 61201311).

Recently, sparse representation (SR) has attracted much attention in signal processing [7, 8]. In sparse and redundant dictionary, the signal energy is concentrated on minority atoms, whereas the noise energy is evenly dispersed over the atoms. This makes the signal part achieve a strong anti-noise ability. Thus in this paper, for extracting HRRP denoised features at low SNR, we seek to suppress the Gaussian white noise in HRRP return by sparse representation with only one observation. A sparse and redundant dictionary is established for sparsely representing the HRRP returns. The OMP-MCV method, a modification to Orthogonal Matching Pursuit-Cross Validation (OMP-CV) [9], is proposed to adaptively denoise the HRRP return without requiring any knowledge about the noise statistics. Simulation results show that the OMP-MCV algorithm outperforms over the original OMP-CV and discrete wavelet transform [10] for HRRP returns denoising.

## II. RADAR HIGH RESOLUTION RANGE PROFILING

Usually, radar HRRP is achieved through wideband signal, such as Liner Frequency Modulation (LFM) signal and Stepped-frequency (SF) signal etc. In this paper, LFM signal is taken as an example to discuss HRRP denoising. The principle described here can be extended to other waveforms.

The transmitted wideband LFM signal can be represented as

$$s(t) = \text{rect}\left(\frac{t}{T_p}\right) e^{j2\pi(f_c t + \frac{1}{2}\gamma t^2)}, \quad (1)$$

where

$$\text{rect}(u) = \begin{cases} 1 & |u| \le \frac{1}{2} \\ 0 & |u| > \frac{1}{2} \end{cases}, \quad (2)$$

denotes complex signal envelope, $T_p$ denotes pulse width, $f_c$ is carrier frequency, and $\gamma$ is frequency modulation slope.

The return from a target located at range $R_t$ is given by

$$s_r(t) = A \cdot \text{rect}\left(\frac{t - 2R_t/c}{T_p}\right) e^{j2\pi\left[f_c\left(t - \frac{2R_t}{c}\right) + \frac{1}{2}\gamma\left(t - \frac{2R_t}{c}\right)^2\right]}, \quad (3)$$

where $A$ is the amplitude of the return. With some

straightforward manipulations, the de-chirping output can be written as

$$s_{if}(t) = A \cdot \text{rect}\left(\frac{t - 2R_t/c}{T_p}\right) e^{-j\frac{4\pi}{c}\gamma R_\Delta t} e^{j\frac{8\pi R_{ref}}{c^2}\gamma R_\Delta} e^{-j\frac{4\pi}{c}f_c R_\Delta} e^{j\frac{4\pi\gamma}{c^2}R_\Delta^2}, \quad (4)$$

where $R_\Delta = R_t - R_{ref}$, and $T_{ref}$ denotes the pulse width of reference signal, usually lager than $T_p$. The later three phase terms in (4) are constants, with no contribution to HRRP. If the sum of the later three phase terms is denoted as $\varphi$, it follows

$$s_{if}(t) = A \cdot \text{rect}\left(\frac{t - 2R_t/c}{T_p}\right) e^{-j2\pi f_d t} e^{j\varphi}, \quad (5)$$

where $f_d = 2\gamma R_\Delta/c$. Equation (5) shows that after de-chirping, the return of every scatterer is a complex sinusoidal signal with a frequency proportional to the relative range of the scatterer.

III. HRRP Adaptive Denoising by Sparse Representation

In this section, we start the discussion of HRRP denoising via sparse and redundant representations by first discussing how redundant dictionary is established. Then, the OMP-MCV method is proposed to adaptively denoise the noisy returns without requiring any knowledge about noise statistics.

A. HRRP Return Sparse Representation

In noisy circumstance, assuming that a target contains $K$ scatterers situating at different ranges and a single pulse contains $M$ sampling points. From (5), the time domain sampling sequence of the de-chirping output pulse can be represented as

$$\begin{aligned} y(m) &= s(m) + n(m) \\ &= \sum_{k=1}^{K} A_k \cdot e^{-j2\pi f_k m} e^{j\varphi_k} + n(m), \quad (6) \\ m &= 0, 1, \cdots, M-1 \end{aligned}$$

where $A_k$, $f_k$, $\varphi_k$ are the amplitude, relative frequency normalized by sampling rate and constant phase of the return from the $k$th scatterer respectively. $s(m)$ and $n(m)$ denote signal sequence and Gaussian white noise sampling sequence respectively. Let $y = [y(0), y(1), \cdots, y(M-1)]^T$, $s = [s(0), s(1), \cdots, s(M-1)]^T$, $n = [n(0), n(1), \cdots, n(M-1)]^T$, then (6) can be rewritten as

$$y = s + n = \sum_{k=1}^{K} u_k \cdot v_k + n, \quad (7)$$

where $u_k = A_k e^{j\varphi_k}$ and $v_k = [1, e^{-j2\pi f_k}, \cdots, e^{-j2\pi f_k}(M-1)]^T$. Equation (7) indicates that the de-chirping output sequence is superposed of multiple complex sinusoidal signals and noise component. Usually, the number of the main scatterers from a target is much less than that of the range cells in HRRP. Thus $s$ is sparse in frequency domain and can be sparsely represented by complex Fourier redundant dictionary, which is constructed as

$$A = \{\phi_1, \phi_2, \cdots, \phi_N\} \in \mathbb{C}^{M \times N} \quad (8)$$

where

$$\phi_i = \exp\{-j2\pi \cdot f_N(i) \cdot m\}, \quad i = 1, \cdots, N, \quad (9)$$

$m = [0, 1, \cdots, M-1]^T$ and $N > M$. In (9), $f_N = [0, 1/N, \cdots, (N-1)/N]$ is the normalized frequency. Then (7) can be represented as

$$y = Ax + n \quad (10)$$

where $x$ is a sparse vector, composed of the decomposition coefficients of signal $s$ in $A$. Sparse representation theory shows that if $x$ satisfies $\|x\|_0 < (1/2)\text{spark}(A)$, where $\text{spark}(A)$ denotes the minimum number of columns of $A$ that are linearly dependent and $\|\ \|_0$ denotes the $l_0$ norm of a vector (i.e., the number of its non-zero components), $x$ can be stably recovered by the following $l_0$ optimization problem [11]

$$\hat{x} = \arg\min \|x\|_0 \ \text{s.t.} \ \|y - Ax\|_2 \leq \delta, \quad (11)$$

where $\delta$ is the representation error boundary dictated by noise level $\|n\|_2$. The solution of (11) is NP hard. Approximated solution can be acquired by greedy algorithms, e.g., OMP [12]. In this paper, OMP is utilized to solve (11) for its simplicity and efficiency. When obtaining $\hat{x}$ from (11), the denoised HRRP return can be acquired by $\hat{s} = A\hat{x}$.

B. Selection of Representation Error Boundary

For denoising by sparse representation, $\delta$ is an important parameter. If $\delta$ is not selected properly, the denoised HRRP either misses some scatterers information (underfitting) or still contains some noise part (overfitting). $\delta$ is determined by noise statistics, which are unknown and need to be estimated in most cases. Cross-validation (CV) is an effective method to recover sparse signal from noisy measurements in compressed sensing (CS) without estimating noise statistics. In this work, we modify OMP-CV to solve (11) for HRRP denoising. The sampling points of HRRP return are randomly separated into two sets: the estimation set and the CV set. The estimation set is used to solve (11) with an selection of $\delta$, and the CV set is employed to verify the solution. The HRRP denoising procedure consists of the following steps:

1) **Initialize:**
   Randomly separate the sampling points of HRRP return into two sets, $y_E \in \mathbb{C}^{M_1 \times 1}$ and $y_{CV} \in \mathbb{C}^{M_2 \times 1}$, where $M_1 + M_2 = M$. And separate the redundant dictionary $A$ into two sub-dictionary $A_E$ and $A_{CV}$ corresponding to $y_E$ and $y_{CV}$ respectively.
   Set $\alpha = \sqrt{M_1/M_2}$, $\delta = \|y_E\|_2$, $\varepsilon = \|y_E\|_2$, $\tilde{x} = 0$, $i = 1$.

2) **Estimate:**
   Estimate the sparse solution $\hat{x}$ of (11) by OMP using $A_E$ and $y_E$.

3) **Cross validate:**

If $\alpha\|y_{CV} - A_{CV}\hat{x}\|_2 < \varepsilon$, set $\varepsilon = \alpha\|y_{CV} - A_{CV}\hat{x}\|_2$, $\delta = \alpha\|y_{CV} - A_{CV}\hat{x}\|_2$ and $\tilde{x} = \hat{x}$, else continue to judge: if $\alpha\|y_{CV} - A_{CV}\hat{x}\|_2 > \lambda \cdot \delta$, terminate the algorithm, else set $\delta = \beta \cdot \delta$, where $\lambda$ and $\beta$ can be selected as $\lambda = 1.3$ and $\beta = 0.97$ respectively.

**4) Iterate:**

Increase $i$ by 1 and iterate from Step 2).

In the step 3), we increase the judgment of $\alpha\|y_{CV} - A_{CV}\hat{x}\|_2 > \lambda \cdot \delta$ and continuously reduce $\delta$ by $\delta = \beta \cdot \delta$. This is because that, as $\delta$ gradually decreasing, the CV error is not guaranteed to continuously decrease to the global minimum, but it may present a little fluctuation because of the influence of noise. Step 3) can effectively prevent the algorithm from terminating at local minimum. We refer to this modification as OMP-MCV modification to the OMP-CV algorithm. Finally, the denoised HRRP return is obtained by $\tilde{s} = A\tilde{x}$.

## IV.  EXPERIMENTAL RESULTS

In this section, the simulated experiments are conducted to investigate the denoising performance of OMP-MCV for HRRP.

### A.  Experiment Setup

The HRRP returns of a target with 7 scatterers located at different range, 12000m, 12001m, 12002.1m, 12002.3m, 12005.3m, 12006.9m, 12008.7m respectively, are simulated. And the amplitudes of returns from the scatterers are set as 0.2, 0.3, 0.3, 1, 0.3, 0.5, and 0.8 separately. The radar system parameters are set as follows: carrier frequency $f_0 = 5.52GHz$, bandwidth $B = 400MHz$, pulse width $T_p = 25us$, reference pulse width $T_{ref} = 25.6us$, sampling rate for de-chirped output $f_s = 10M$. Under these parameters, every sampling sequence of a single pulse is a 256-length vector. The redundant dictionary is established with $M = 256$, $N = 1024$ for sparsely representing the HRRP returns. After de-chirping processing, the noise-free HRRP obtained by FFT is illustrated in Fig. 1.

### B.  CV Error Variation in the Iteration Procedure

To investigate the variation of CV error in the iteration procedure, the representation error $\delta$ is gradually decreased with a fixed step, and the CV error is tested in this process. And the SNR is set to 5dB. The sampling sequence is randomly separated two sets: 176 sampling points for estimating the original noise-free signal and the rest 80 sampling points for the CV test. The variation of CV error $\|y_{CV} - A_{CV}\hat{x}\|_2$ and signal reconstruction error $\|s - A\hat{x}\|_2$ in the iteration procedure are shown in Fig. 2.

The results of Fig. 2 indicate that as $\delta$ gradually decreasing, the CV error is stepped down to the global minimum. Besides, the CV error presents a little influence nearby the minimum. In other words, CV error may encounter

local minimum in the procedure of converging to the global minimum. Compared with OMP-CV, the step 3) in OMP-MCV can effectively prevent the algorithm from terminating at local minimum. Fig. 2 also shows that the signal reconstruction error converging to the rock bottom as the CV error reaches the global minimum.

### C.  Selection of Number of CV

Usually, the signal pulse width and sampling rate are fixed in radar system, thus the length of the sampling sequence is fastened, i.e., $M_1 + M_2$ is a constant. There is a tradeoff between the size of the estimation set $M_1$ and the size of the CV set $M_2$. Increasing $M_1$ will improve estimation accuracy of the original noise-free signal, whereas, increasing $M_2$ will enhance CV estimation accuracy, also important for the OMP-MCV algorithm. Thus the influence of the CV size is investigated in this subsection. The SNR is set to 5dB. The root-mean-square error (RMSE) of the denoised signal is surveyed, which is defined as

$$\text{RMSE} = \sqrt{\frac{1}{I}\sum_{i=1}^{I}\|\hat{s}_i - s_0\|_2^2}, \qquad (12)$$

where $s_0$ is the original noise-free HRRP return, $\hat{s}_i$ is the denoised HRRP return, and $I$ is the times of Monte Carlo experiments which is selected as 100. The RMSE of the denoised returns varying with the number of cross validation is shown in Fig. 3.

The result of Fig. 3 demonstrates that as $M_2$ increase, the denoising performance improves at first, and then worsens. Thus in practice, a tradeoff should be made for selecting $M_1$ and $M_2$. In this experiment, the best balancing point is selected by $M_1 = 196$ and $M_2 = 60$.

### D.  Denoising Performance

In the subsection, the denoising performance of OMP-MCV is verified by compared with other methods. In this experiment, the sampling sequence is randomly separated into two sets with $M_1 = 196$, $M_2 = 60$. The RMSE of the denoised returns is investigated as SNR varying from 0dB to 30dB. For reference, we also assume that the exactly noise level $\|n\|_2$ is known and setting $\delta = \|n\|_2$ for solving (11) by OMP, referring to it as OMP-δ for short. Meanwhile, the original OMP-CV [9] and the discrete wavelet transform denoising (DWTDN) [10] are utilized for comparing. In DWTDN, the 'db8' wavelet basis is chosen due to its better denoising performance than other 'db' wavelet basis in our experiments. The returns are decomposed into 10 layers, and then the noise is rejected by Heursure threshold. RMSE of the denoised HRRP returns by various methods are surveyed with 100 times Monte Carlo experiments. The results are exhibited in Fig. 4.

The results of Fig. 4 demonstrate that OMP-MCV achieves much better denoising performance than DWTDN for HRRP. This is because the radar return is composed of complex sinusoidal components after de-chirping processing, and the return energy is more concentrated in Fourier redundant than wavelet basis, leading to a stronger anti-noise performance. The denoising performance of OMP-MCV is also better than that of original OMP-CV, close to that of OMP-$\delta$, because of its capability of preventing the iteration procedure from terminating at local minimum.

## V. CONCLUSION

For enhancing the performance of HRRP recognition at noise circumstance, this paper has presented an adaptive denoising method, OMP-MCV, for suppressing Gaussian white noise in HRRP return via sparse representation. This method is a modification to OMP-CV and can effectively prevent iterating estimation from terminating at local minimum. The HRRP returns are denoised by OMP-MCV without needing any knowledge about the noise statistics. Simulation results show that OMP-MCV achieves better denoising performance than original OMP-CV and DWTDN.

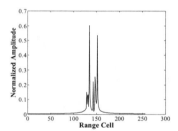

Fig. 1 HRRP of simulated target

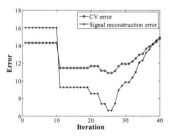

Fig. 2 Evolution of the error with $\delta$ gradually decreasing

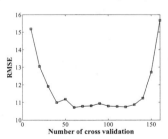

Fig. 3 RMSE of denoised HRRP returns varying with tradeoff of the number for CV set and estimation set.

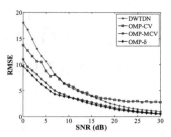

Fig. 4 RMSE of the denoised HRRP by various methods varying with SNR

## REFERENCES

[1] K. Copsey and A. Webb, "Bayesian Gamma Mixture Model Approach to Radar Target Recognition," *Ieee Transactions on Aerospace and Electronic Systems*, vol. 39, pp. 1201-1217, Oct 2003.

[2] J. Gudnason, J. J. Cui, and M. Brookes, "HRR Automatic Target Recognition from Superresolution Scattering Center Features," *Ieee Transactions on Aerospace and Electronic Systems*, vol. 45, pp. 1512-1524, Oct 2009.

[3] L. Du, P. H. Wang, H. W. Liu, M. Pan, F. Chen, and Z. Bao, "Bayesian Spatiotemporal Multitask Learning for Radar HRRP Target Recognition," *Ieee Transactions on Signal Processing*, vol. 59, pp. 3182-3196, Jul 2011.

[4] I. Jouny, F. D. Garber, and R. L. Moses, "Radar Target Identification Using the Bispectrum - a Comparative-Study," *Ieee Transactions on Aerospace and Electronic Systems*, vol. 31, pp. 69-77, Jan 1995.

[5] L. Du, H. W. Liu, P. H. Wang, B. Feng, M. Pan, and Z. Bao, "Noise Robust Radar HRRP Target Recognition Based on Multitask Factor Analysis With Small Training Data Size," *Ieee Transactions on Signal Processing*, vol. 60, pp. 3546-3559, Jul 2012.

[6] V. Lukin, A. Totsky, D. Fevralev, A. Roenko, J. Astola, and K. Egiazarian, "Adaptive Combined Bispectrum-filtering Signal Processing in Radar Systems with Low SNR," in *Circuits and Systems, 2006. ISCAS 2006. Proceedings. 2006 IEEE International Symposium on*, 2006, p. 4 pp.

[7] S. S. B. Chen, D. L. Donoho, and M. A. Saunders, "Atomic Decomposition by Basis Pursuit," *Siam Review*, vol. 43, pp. 129-159, Mar 2001.

[8] M. Elad and M. Aharon, "Image Denoising via Sparse and Redundant Representations over Learned Dictionaries," *Ieee Transactions on Image Processing*, vol. 15, pp. 3736-3745, Dec 2006.

[9] P. Boufounos, M. F. Duarte, and R. G. Baraniuk, "Sparse Signal Reconstruction from Noisy Compressive Measurements using Cross Validation," in *Statistical Signal Processing, 2007. SSP'07. IEEE/SP 14th Workshop on*, 2007, pp. 299-303.

[10] M. Jansen, *Noise Reduction by Wavelet Thresholding* vol. 161: Springer USA, 2001.

[11] M. Babaie-Zadeh and C. Jutten, "On the Stable Recovery of the Sparsest Overcomplete Representations in Presence of Noise," *Ieee Transactions on Signal Processing*, vol. 58, pp. 5396-5400, Oct 2010.

[12] J. A. Tropp and A. C. Gilbert, "Signal Recovery from Random Measurements via Orthogonal Matching Pursuit," *Ieee Transactions on Information Theory*, vol. 53, pp. 4655-4666, Dec 2007.

# GPU based FDTD method for investigation on the electromagnetic scattering from 1-D rough soil surface

C.-G. Jia, L.-X. Guo, J. Li
School of Science, Xidian University
No. 2, Taibai Road
Xi'an, Shaanxi 710071 China

**Abstract- In this paper, the graphic processor unit (GPU) implementation of the finite-difference time domain (FDTD) algorithm is presented to investigate the electromagnetic (EM) scattering from one dimensional (1-D) Gaussian rough soil surface. The FDTD lattices are truncated by uniaxial perfectly matched layer (UPML), in which the finite-difference equations are carried out for the total computation domain. Using Compute Unified Device Architecture (CUDA) technology, significant speedup ratios are achieved for different incident frequencies, which demonstrates the efficiency of GPU accelerated the FDTD method. The validation of our method is verified by comparing the numerical results with these obtained by CPU, which shows favorable agreements.**

## I. INTRODUCTION

Nowadays, the investigation on the statistical characteristic of electromagnetic scattering from random rough surface has attached considerable interest owing to its significant applications in the fields of remote sensing, target identification and radar detection [1]. Many kinds of methods, including analytical and numerical approaches, have been carried out to deal with the electromagnetic scattering model. Taking the approximate analytical methods for example, the Kirchhoff approximation [2], the small-slope approximation (SSA) [3], and the small-perturbation method (SPM) [4] have been studied, but they are usually limited by roughness, incident angle, and low precision. Numerical methods are widely employed to calculate the model, such as the parallel method of moment (MoM) as well as its accelerated method[5], the Generalized Forward-Backward Method (GFBM) , and finite-difference time domain (FDTD) method [6].

Compared with other numerical methods, the FDTD method has its own advantages in analyzing the scattering from rough surface [6]. The traditional FDTD method is too week to deal with the electrically large problem due to limitation of computation time. The MPI-based parallel FDTD was presented by J. Li *et al.* [6]. Using the method, the computation time is extremely reduced compared to sequential implementation. However, the speedup ratio of MPI-based method is limited by numbers of Central Processing Unit (CPU) processes. Fortunately, Compute Unified Device Architecture (CUDA) technology based on GPU has been extensively implemented for the large scale FDTD simulations

successfully [7]. Compared to the MPI technology, graphic processor unit (GPU) can achieve huge speedup ratios at low cost for its powerful computing capability, which motives us into adopt the GPU-based FDTD technology for analyzing the scattering from rough surface. Up to now, to our knowledge, few works have been reported to solve this problem using the GPU-based FDTD implementation. Additionally, a uniaxial perfectly matched layer (UPML) medium is employed to truncate the FDTD lattices, and the finite-difference equations in the UPML medium are used for the total computation domain making the parallel algorithm convenient to implement. In this paper, the precision of calculations performed on both GPU and CPU is single precision arithmetic.

## II. THEORETICAL ANALYSIS

### A. Gaussian rough surface model

In order to investigate the characteristic of EM scattering from rough surface, the first step is to generate the profile of 1-D rough surface, which is simulated by the Monte Carlo method. Taking the *TM* incidence for example, the scattering model for one dimensional (1-D) random rough surface with height profile function $y = f(x)$ is shown in Fig. 1, where $f(x)$ is a Gaussian distributed rough surface with the Gaussian power spectrum density function $W(K)$ expressed as follows [8]

$$W(K) = \frac{\sigma^2 l}{2\sqrt{\pi}} \exp\left(-\frac{K^2 l^2}{4}\right) \tag{1}$$

Where $\sigma$ and $l$ are root mean square (rms) height and correlation length, respectively, by which the profile of rough surface is determined. $L$ is the length of rough surface we simulate. In order to model the scattering from an infinite surface, Gaussian window function is introduced and expressed as [9]

$$G(x, y) = \exp\left\{-\left[(x - x_{cen})^2 + (y - y_{cen})^2\right]\left(\frac{\cos\theta_i}{T}\right)^2\right\} \tag{2}$$

Where $x_{cen}$ and $y_{cen}$ are the center coordinates of the connective boundary. $T$ is a constant which determines the tapering width of the window function so chosen that the

978-7-5641-4279-7

tapering drops from unity to $10^{-3}$ at the edge, as well as $cos\theta_i/T=2.6/\rho_m$, where $\rho_m$ is the minimum distance from the center coordinate to the edge of the connective boundary .

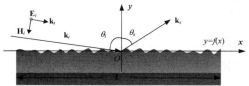

Fig. 1. Geometry for EM scattering from 1-D rough surface (TM wave).

*B.   FDTD method for rough surface*

Fig. 2 depicts the division model of computation region for the FDTD algorithm in calculating electromagnetic scattering from rough surface. In order to simulate the infinite free space in the finite computing field, a virtual absorbing boundary is employed outside the FDTD region. There have been many absorbing boundary conditions adopted in the FDTD method. In this paper, the UPML absorbing medium  are employed to truncate the FDTD lattices. The connective boundary is needed to divide the computation region into the total field region and the scattered field region [10], where the incident wave is generated. After the near fields are obtained, far fields can be achieved by doing a near-to-far field transformation at the output boundary [10]. Finally, the bistatic scattering coefficient in the far zone is calculated by [6].

$$\sigma = \lim_{r\to\infty} \frac{2\pi r}{L} \frac{|E_s|^2}{|E_i|^2} \qquad (3)$$

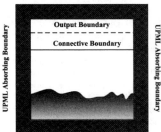

Fig. 2. FDTD model of 1-D rough surface

III.   GPU IMPLEMENTATION OF FDTD ALGORITHM

In this section, the parallel computing platform and programming model CUDA is introduced. The parallelization strategy of the GPU based FDTD method for EM scattering from rough surface is clarified in detail, which includes the management of device memory such as the global, shared and texture memory, as well as the kernels implementation.

*A.   CUDA programming model*

The introduction of NVIDIA' GPU based on CUDA architecture gave rise to a new era of graphics computing, without esoteric knowledge of graphics computation models. CUDA is a highly parallel and efficient computing architecture, with which GPUs can solve many complex problems by built-in streaming multiprocessors executing a number of threads in parallel [11]. The CUDA programing model assumes that the sequential code executes on the host (CPU) while the instruction with high data parallelism executes on the device (CUDA-enabled GPU).

As illustrated by Fig. 3, a CUDA program begins with serial execution on the host, including CPU and GPU memory allocation, initialization as well as de-allocation. Kernels defined as functions are executed on the device by a large amount of threads in parallel. The memories on the two platforms (host and device) are separated physically in the CUDA programming model. For further instructions on the CUDA technology, one can refer to [11].

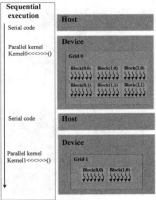

Fig. 3. The CUDA programing model [11]

*B.   Memory management*

From the point of view of performance optimization for the FDTD approach, it is of great significance to optimize the usage of device memory to obtain best performance on GPU. CUDA device memory consists of various memory spaces, including global, local, constant, shared, texture, and registers memory, which have their own characteristic. In this paper, global, shared, and texture memory is utilized to achieve high GPU performance. Global memory can be written and read by the host by the application programming interface (API) functions. The transactions of global memory access should be coalesced and bandwidth is low due to residing in uncached off-chip memory. In order to boost the performance of the kernels, the on-ship shared memory is utilized to eliminate the uncoalesced access. Shared memory is available to the thread block, in which threads share their results and the execution of threads in the threadblock can be synchronized in the block level. Taking TM$^z$ case for example, as demonstrated in Fig.

4, data is first loaded from global memory to shared memory when the $E$ field and the $H$ field update is executed. When the $H$ components ($H_x$, $H_y$) calculated, the $E_z$ values of the current block of threads are not only copied to shared memory, values of the left column threads of the right adjacent block and the up row thread of the down adjacent block are also loaded. When the $E$ field iteration function is invoked, the $H_x$ and $H_y$ values of the current block are not only transferred from global memory to shared memory, but $H_x$ values of the down row threads of the up adjacent block and $H_y$ values of right column of the left adjacent block are also delivered. Finally, further improvement of the speedup ratio is achieved by utilizing texture memory, which is read-only cache

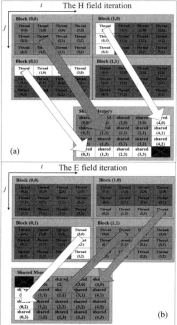

Fig. 4. Data transfers from global memory to shared memory

## C. Device kernels

First, as illustrated by Fig. 5, Gaussian rough model is built by the Monte Carlo method presented in Section 2.1. And then, CPU assigns the host and device memory, as well as grid and block size based on the model. Parallel implementation is carried out when referring to the near field iteration, which is extremely time-consuming in the whole FDTD computation. It is necessary to synchronize for some threads to share data with each other. The threads in the same block synchronize by using __syncthreads () though shared memory, while a new kernel function is invoked to synchronize though global memory for the threads belonging to different blocks. In order to force synchronization on the grid level, five *kernels* are

utilized to achieve the functions, including *IncidentHKernel* (the incident magnetic field update), *IncidentEKernel* (the incident electric field update), *ConnectionKernel* (introducing the incident wave at the connective boundary), *eKernel* (the electric field component(s) update), and *hKernel* (the magnetic field component(s) update). When the near field iteration is finished by GPU, the far field can be obtained with great ease on the CPU platform

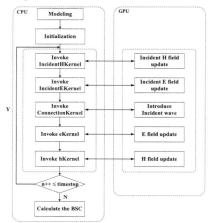

Fig. 5. The flowchart of GPU based FDTD algorithm for rough surface

## IV. NUMERICAL RESULTS AND DISCUSSION

. In this section, the numerical results of EM scattering from 1-D soil surface by GPU based FDTD method are discussed in detail. To ensure the accuracy and stability of the FDTD method, the spatial increment and time increment are taken as $\Delta x = \Delta y = \Delta = \lambda / 20$ and $\Delta t = 0.5 \times \Delta / c$, respectively. $\lambda$ is the incident wavelength and $c$ is the light speed in vacuum. The UPML thickness is $8\Delta$.

The accuracy of the CUDA implementation of FDTD algorithm is illustrated by comparing the numerical results with those obtained by sequential execution on CPU. Fig. 6 demonstrates the bistatic scattering from a Gaussian soil surface with characteristic parameters $\delta = 0.1\lambda$ and $l = 1.0\lambda$ under the incident angle $\theta_i = 35°$ at the incident frequency of $f = 8\text{GHz}$. The generated length of rough surface is $L = 1638.4\lambda$ ($32768\Delta$). The relative permittivity of surface corresponding to the soil with 3.8% moisture is taken as $\varepsilon_r = (2.5, 0.18)$. The results averaged by 20 surface realizations are in good agreement by the two implementations for both *TM* and *TE* incidence, demonstrating the accuracy of our FDTD-CUDA implementation. The time consuming of traditional FDTD scheme are approximately 112.91 and 119.67 hours for *TM* and *TE* case respectively. By contrast, the computation time of GPU based FDTD are 1.51 and 1.54 hours for the two incident cases. Additional, we can easily find that the time cost reduces dramatically by the GPU implementation.

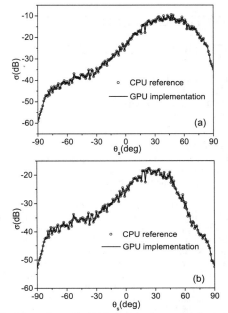

Fig. 6. Comparisons of the bistatic scattering from a soil surface by two implementations: (a) *TM* case; (b) *TE* case.

Furthermore, the efficiency of our implementation is demonstrated by comparison of the CPU and GPU time for calculating the EM scattering from rough surface as incident frequency increases from 1GHz to 16GHz under the incident angle of $\theta_i=55°$. And the mash along $x$ direction increases from 4096 to 65536 by keeping the length of rough surface $L = 61.44$m . The C program is executed on Intel Core2 Quad Q8300, 2.5GHz and 3.5GB of RAM. The kernels are performed on a NVIDIA GeForce GTX570 GPU card, which has 480 stream processors and 1280 MB video memory. Table I compares computation time of serial FDTD method for rough surface with one surface realization with that of GPU implementation. As illustrated by the table, it is obviously observed that speedup factor increases with the increase of unknowns, which demonstrates that huge computations can make full use of thousands of threads on the GPU. In addition, the GPU accelerated FDTD implementation has also striking advantage versus the message-passing-interface (MPI)-based FDTD scheme [6] that the speedup factor is in direct proportion to the number of CPU process used for computation, especially when solving the problem of electrically large rough surface.

V. CONCLUSIONS

In this paper, GPU accelerated FDTD method is extended to investigate the electromagnetic scattering from 1-D rough soil surface. Global, shared, and texture memory is utilized to optimize the performance of GPU. The accuracy of the method is demonstrated by comparing the result obtained by sequential execution on the CPU platform. Compared to the previous work, MPI-based FDTD method, our implementation can achieve favorable speedup factor. In our future work, the investigation will be focused on the electromagnetic scattering from 2-D randomly rough surface, which has more practical applications.

Table I
Comparison of CPU and GPU time (TM case).

| Incident frequency (GHz) | Mesh along $x$ | Mesh along $y$ | CPU time(s) | GPU time(s) | speedup |
|---|---|---|---|---|---|
| 1 | 4096 | 100 | 265.94 | 7.82 | 34.00 |
| 4 | 16384 | 100 | 3960.13 | 63.45 | 62.41 |
| 8 | 32768 | 100 | 15726.96 | 214.97 | 73.15 |
| 16 | 65536 | 100 | 62961.59 | 790.25 | 79.67 |

ACKNOWLEDGEMENT

This work was supported by the National Science Foundation for Distinguished Young Scholars of China (Grant No.61225002), the Specialized Research Fund for the Doctoral Program of Higher Education (Grant No.20100203110016), and the Fundamental Research Funds for the Central Universities (Grant No.K50510070001).

REFERENCES

[1] Ku, H.C., Awadalah, R.S., Mcdonald, R.L., and Woods, N.E., "Fast and accurate algorithm for electromagnetic scattering from 1-D dielectric ocean surface," *IEEE Trans. Antennas Propag.*, vol. 54, pp. 2381-2391, 2006.

[2] A. K. Sultan-Salem and G. L. Tyler, "Validity of the Kirchhoff approximation for electromagnetic wave scattering from fractal surfaces," *IEEE Trans. Geosic. Remote Sens.*, vol. 42, pp. 1860-1870, 2004.

[3] Y. Q. Wang and S. L. Broschat, "A systematic study of the lowest order small slope approximation for a Pierson-Moskowitz spectrum," *IEEE Geosci. Remote Sens. Lett.*, vol. 8, pp. 158-162, 2011.

[4] L. X. Guo, Y. Liang, J. Li, and Z. S. Wu, "A high order integral SPM for the conducting rough surface scattering with the tapered wave incidence TE case," *Progress In Electromagnetics Research, PIER*, vol. 114, pp. 333-352, 2011.

[5] L. X. Guo, A. Q. Wang, and J. Ma, "Study on EM scattering from 2-D target above 1-D large scale rough surface with low grazing incidence by parallel MOM based on PC clusters," *Progress In Electromagnetics Research, PIER*, vol. 89, pp. 149-166, 2009.

[6] J. Li, L. X. Guo, H. Zeng, and X. B. Han, "Message-passing-interface-based parallel FDTD investigation on the EM scattering from a 1-D rough sea surface using uniaxial perfectly matched layer absorbing boundary," *J. Opt. Soc. Am. A*, vol. 23, pp. 359-369, 2009.

[7] Piotr Sypek, Adam Dziekonski, and Michal Mrozowski, "How to render FDTD computations more effective using a graphics accelerator," *IEEE Trans. Magn.*, vol. 45, pp. 1324-1327, 2009.

[8] Thorsos, E.I., "The validity of the Kirchhoff approximation for rough surface scattering using a Gaussian roughness spectrum," *J. Acoust. Soc. Am.*, vol. 83, pp. 78-92, 1988.

[9] Adrian K. Fung, Milind R. Shah, Saibun Tjuatia, "Numerical simulation of scattering from three-dimensional random rough surface," *IEEE Trans. Geosic. Remote Sens.*, vol. 32, pp. 986-994, 1994.

[10] Taflove, A. and S. C. Hagness, *Computational Electrodynamics: The Finite-difference Time-domain Method*, Artech House, Boston, 2005.

[11] "*NVIDIA CUDA C Programming Guide*," ver. 4.2, NVIDIA Corporation, May. 2012.

# Research on the GPS Signal Scattering and Propagation in the Tropospheric Ducts

G.C. Lee[#1], L. X. Guo[#1,2], J.J.Sun[#1], and J. H.Ge[#2]

[1] School of Science, Xidian University

[2] State Key Lab. of Integrated Service Networks, Xidian University

Box 274, No.2, Taibai Road, Xi'an 710071, China

*Abstract*— In this paper, the method of moment (MoM) is used to investigate the characteristics of the GPS (Global Positioning System) signal scattering from the rough sea surface under low grazing incidence, and then on the basis of the results, the GPS signal could be received as the initial field by the improved Discrete Mixed Fourier Transform algorithm (DMFT) method to study the propagation characteristics in the tropo-spheric ducts. The advantages of MoM are shown the validity in computing the EM scattering at low grazing incident angle and also shown by the comparison of the Bistatic Scattering Coefficient (BSC) with MoM and Kirchhoff Approximation (KA). Finally, the propagation properties of GPS scattering signal in the evaporation ducts with different evaporation duct heights and elevation angles of GPS are discussed by the improved discrete mixed Fourier transform with taking into account the sea surface roughness.

## I. INTRODUCTION

This paper deals with works on the GPS signal propagation in the tropospheric ducts using the improved Discrete Mixed Fourier Transform algorithm (DMFT) for the parabolic equation method[1] [2]. In recent years, radio wave propagation in the troposphere has been attended extensively by many researchers, especially for the radio wave propagation in the ducting environment and its effect on radar system and electronic system. When the refractivity gradient that is less than $-157\,\mathrm{N}$-units per-kilometer will result in radio wave that refracts towards the surface of the earth with a curvature that exceeds the curvature of the earth, thus radio wave will be trapped into a thin atmospheric layer, this phenomenon is similar to that of the metal waveguides. Radio system and radar of maritime operation are greatly influenced by the existence of atmospheric ducts, it can give a rise to over-the-horizon detection and form the radar detection shadow zone.

The GPS is a space-based global navigation satellite system that provides reliable location and time information in all weather and at all times and anywhere on or near the Earth when and where there is an unobstructed line of sight to four or more GPS satellites, it is freely accessible by anyone with a GPS receiver[3~4]. The GPS satellite transmits, worked at 1575.42 MHz (L1 frequency), an unique binary pseudo-random code (C/A code) and the minimum received power specified (in L1) is -160 dBW, about -20 dB under the local thermal noise. The pseudorange is estimated through correlation between the received signal and its replica generated by the receiver. The GPS receiver demodulates L1 signals with power levels between -130 and -160 dBW (in an optimal case) corresponding to a dynamic range of 30 dB and its antenna gain ranges from about 0 (horizon) to 10 (zenith) dB in order to eliminate some multipath effects.

Currently, methods used to study the radio wave propagation in the troposphere mainly included the Parabolic Wave Equation (PWE) [2] method, Ray Tracing approach (RTA) [1] and Mode theory[5]. The PWE method have been developed and used extensively over the past twenty years to calculate the radio wave propagation problems due to its accuracy and stability, what is more important, it can consider the influence of refractive index on radio propagation in ducting environment easily. The PWE method has been widely used to model the propagation of electromagnetic and acoustic waves through inhomogeneous media. Most applications of the PWE consider low-grazing angle or near-horizontal propagation of radar or acoustic waves. In many cases, the horizontal boundary, be it terrain, ocean surface, or ocean bottom, plays a significant role. In this paper, the method of moment (MoM) [6] is used to investigate the characteristics of the GPS (Global Positioning System) signal scattering from the rough sea surface under low grazing incidence, and then on the basis of the results, the GPS signal could be received as the initial field by the SSPE method to study the propagation characteristics in the tropo-spheric ducts.

## II. NUMERIC SOLUTIONS OF THE TROPOSPHERIC DUCTS PROPAGATION

The DMFT method used to solve parabolic-type equations is popular for modeling EM wave propagation in the troposphere. Leontovich and Fock [7]were pioneers who introduced the use of the PE method for solving the EM wave propagation in the atmospheric ducts. However, this approach become famous after introduced the SSPE algorithm by Tappert[8].

### A. Split-step parabolic Equation Method Solution

The standard parabolic wave equation is obtained from the two-dimensional Helmholtz equation by separation the rapidly varying phase term to obtain an amplitude factor, which varies in range which is the direction of the propagation axis. The standard parabolic wave equation is given by

$$\frac{\partial^2 u}{\partial z^2} + 2ik\cdot\frac{\partial u}{\partial x} + \kappa^2(m^2-1)\cdot u = 0 \qquad (1)$$

Where $k$ is the wave number in vacuum and $m$ is the modified refractive index, $x$ axis is the direction of the wave propagation, $z$ axis is the height direction.

Equation (1) can be solved by a step technique where the initial field distribution $u(0,z)$ is specified at an open boundary, the solution at $(\Delta x, z)$ is obtained as a function of the initial (incident) field and of the boundary conditions at the top and bottom of the domain. After some tedious derivation, the solution of the parabolic equation is described as follow[2].

$$u\left(x+\Delta x, z\right)$$
$$= e^{ikm(x,z)\Delta x/2} F^{-1}\left\{ e^{i\sqrt{k^2-p^2}} F\left\{ e^{ikm(x,z)\Delta x/2} u\left(x,z\right)\right\}\right\} \quad (2)$$

where $p$ is the p-space, $p = k\sin\theta$, $\theta$ is the incident angle, $\Delta x$ is the step of $x$, follow the Nyguist theorem, $z_{\max}$ and $p_{\max}$ should satisfy $z_{\max} p_{\max} = N\pi$.

For the rough dielectric ocean surface studied in the current work, the improved Discrete Mixed Fourier Transform algorithm (DMFT) method introduced by Kuttler and Dockery (1991) is employed for the simulation of GPS propagation because it incorporates an impedance boundary into the split-step solution.

From the theory of the microwave propagation, the propagation loss includes two parts: free space propagation loss and the propagation loss caused by the absorption, scattering and diffraction. Once propagation factor $F$ is obtained, the propagation loss [8] in dB is determined by the basic radio wave theory in the form of

$$L = L_f + 20\lg F \quad (3)$$

where $L_f$ and $F$ stand for free space propagation loss and propagation factor, respectively. Here, propagation loss in the free space is given by the following formula [8]

$$L_f = 32.45 + 20\lg f(\text{MHz}) + 20\lg r(\text{km}) \quad (4)$$

And the propagation factor $F$ [2] is defined as

$$F = \left| E/E_0 \right| \quad (5)$$

where $E$ is the field strength at a point, and $E_0$ is the field strength that would occur at that point under free space conditions. In rectangular coordinate system, propagation factor [2] can be calculated by

$$F = \sqrt{x}\left| u(x,z) \right| \quad (6)$$

$x$ is the distance between the transmitting antenna and receiving antenna, and $z$ is the altitude measured with respect to the mean sea surface.

### B. Initial field and the boundary condition

It is well known that solving the parabolic equation is related to an initial value problem, and the field in range $x + dx$ is dependent on the field at the previous range step, so one must begin with an initial field at range zero for propagation the filed forward. In general, the far-field antenna pattern is used to calculate the initial field distribution by FFT. In this paper the GPS signal is introduced regarded as the

initial field, and the initial field is calculated by the MoM[9] with different direct incident and reflected angles and observation distances for height mesh space interval. The geometry of the initial field[10] is depicted in Fig.1.

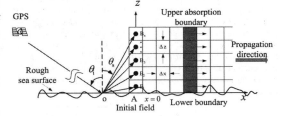

Fig.1. Initial GPS signal field with the propagation region.

The initial GPS signal field with the upper absorption region is attenuated by a Hanning window in the upper region of the implementation-domain. For the lower boundary the effective reflection coefficient is implemented through the use of FFT's in the angular spectrum domain. The incident field is used to form the source image where the 180 degree phase shift is then implemented.

For the purpose of comparing the numerical results of the bistatic scattering by KA and MoM[6], we introduce the tapered wave into the classical KA, and redefine the definition of bistatic scattering coefficient calculated by KA. In Fig. 2, the wind speed is $U = 5$ m/s, the incidences are small incident angle $\theta_i = 20°$ and large incident angle $\theta_i = 80°$, respectively. From Fig. 2, it is observed that for the small incidence, $\sigma$ of the results by the two methods agree with each other very well over the whole scattering region, but for the large incidence, an obvious discrepancy appearing in the non-specular direction. Therefore, the results shows that the MoM are shown the validity in computing the EM scatte-ring at low grazing incident angle.

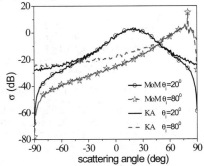

Fig.2. Comparison of $\sigma$ by MoM and KA

To further examine the scattering characteristics of different wind speeds at low grazing incident angle, the bistatic scattering at incident angle $\theta_i = 85°$ with different wind speeds $U = 3$ m/s, $U = 5$ m/s and $U = 8$ m/s are presented and compared in Fig. 3(a). And $\sigma$ near the specular direction $\theta_s = 80° \sim 90°$ is specially shown in Fig. 3(b). It is obviously seen that there is a peak in the specular direction for the smaller wind speed, which proves the point that the specular peak of the coherent wave depends on the surface roughness, we attribute this behavior to the fact that the roughness of the sea surface depends on the wind speed, the smaller wind speed, the flatter is the sea surface, leading to the obviously peak in the specular direction. Oppositely, with larger wind speed, the sea surface becomes rougher and will cause $\sigma$ increasing in the direction far from the specular direction.

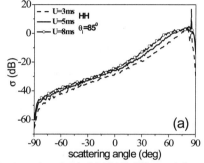

(a)   Comparison of $\sigma$ by MoM with different wind speeds

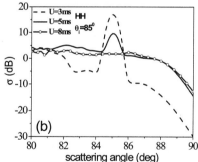

(b)   Comparison of $\sigma$ by MoM with different wind speeds at large scattering angles.

Fig.3.   The distribution of $\sigma$ for different wind speeds at low-grazing incident angle.

Fig.4 shows that the GPS scattering initial field at low grazing incident angle with different winds. The initial filed obtained by calculation the Bistatic Scattering Coefficient (BSC) with MoM.

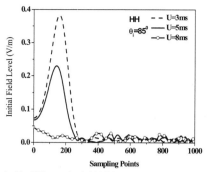

Fig.4.   The GPS scattering initial field at low grazing incident angle with different winds.

### C. Tropospheric duct structure

The tropospheric duct is formed by small variations in the index of refraction of the troposphere. These variations are due to the anomalous vertical gradients of temperature and humidity.The evaporation duct created by a rapid decrease in moisture immediately adjacent to the ocean surface is a nearly permanent propagation mechanism for the radar waves propagating over the ocean surface. The modified refractivity profile (M-profile) model commonly used for evaporation ducts is the log–linear Paulus–Jeske model.This model is parameterized using only the evaporation duct height (EDH) as follows:

$$M(z) = M(0) + c_0 (z - d \ln \frac{z + z_0}{z_0}) \qquad (7)$$

where h is the height (m) above the mean sea level; $d$ denotes the EDH (m); M0 represents the modified refractivity at the sea surface, taken as 330 M-units in this paper.

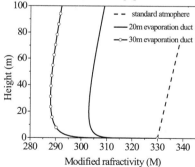

Fig.5.   Modified refractivity profiles for three different atmosphere environment.

Fig.5 shows that the Modified refractivity profiles for three different atmosphere environment, $d = 0$ shows the standard atmosphere, $d = 20$ shows 20m height evaporation duct and $d = 30$ shows 30m height evaporation duct.

## III. NUMERICAL RESULTS AND DISCUSSIONS

In this section, the propagation of GPS signals at different conditions are presented in Fig.6 standard atmosphere, 20m evaporation duct and 30m evaporation duct. Fig.7 gives comparison of the GPS signal propagation Loss various with range increasing and Fig.8 gives comparison of the GPS signal propagation Loss various with height increasing.

*A. Analysis of the GPS signal propagation in the evaporation ducts*

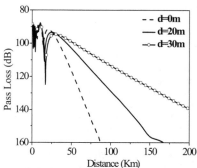

Fig.6. Comparison of the GPS signal propagation Loss various with different ducts

Fig.6 shows that the GPS signal propagation characteristics in different conditions, standard atmosphere, 20m evaporation duct and 30m evaporation duct. It is found that the great effect on the propagation characteristic in the duct condition, microwave could transform longer than the standard condition, and the higher the evaporation duct height is, the smaller the propagation loss for a fixed propagation distance will be.

*B. Analysis of the GPS signal propagation loss in the tropospheric ducts with different winds*

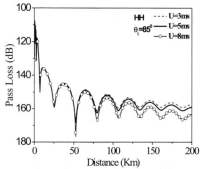

Fig.7. Comparison of the GPS signal propagation Loss various with range increasing

Fig.7 depicts the GPS signal propagation loss in the 20m evaporation duct with different winds. It is shows that all the curves of the propagation loss increase with the propagation distance increasing, and it is obviously seen that the wind do great influence on the microwave propagation in the duct, with the wind increasing, the propagation loss increased then.

Fig.8 presents the GPS signal propagation characteristics various with height increasing at range 80 Km, it is clearly seen that the wind is higher, the propagation loss is larger at 20m evaporation duct.

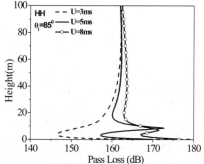

Fig.8. Comparison of the GPS signal propagation Loss various with height increasing

## IV. CONCLUSION

This paper presents the GPS signal propagation characteristics in different conditions. Future investigation is the propagation characteristics in more areas, which is more important to practical application.

## ACKNOWLEDGEMENT

This work was supported by the National Science Foundation for Distinguished Young Scholars of China (Grant No. 61225002), the Specialized Research Fund for the Doctoral Program of Higher Education (Grant No. 20100203110016), and the Fundamental Research Funds for the Central Universities (Grant No. K50510070001).

REFERENCES

[1]  D. E. Kerr, ed. Propagation of Short Radio Waves., London: IEE Electromagnetic Waves Series, Peter Peregrinus, 1987.
[2]  G. D. Dockery, Modeling electromagnetic wave propagation in the troposphere using the parabolic equation. IEEE Trans Ant & Prop., vol. 36, pp.1464-1470, 1988.
[3]  http://en.wikipedia.org/wiki/Global_Positioning_System
[4]  Kaplan E D. Understanding GPS Principles and Applications [M]. Artech House, Norwood,1996.
[5]  J. R. Wait, Coupled mode analysis for a nonuniform tropospheric waveguide, Radio Sci., vol.15, pp.667-673, 1980.
[6]  Ku H C, Awadallah R S, McDonald R L, and Woods N E 2006 IEEE Trans. Antennas Propag. 54 238
[7]  Leontovich M. A. and Fock V. A. Solution of the problem of propagation of electromagnetic waves along the earth's surface by method of parabolic equations. J. Phys. USSR, 1946, 10(1): 13-23.
[8]  Hardin R. H. and Tappert F. D. Applications of split-step Fourier method to the numerical solution of nonlinear and variable coefficient wave equations. SIAM Rew, 1973, 15:423.
[9]  D. E. Kerr, ed. Propagation of Short Radio Waves., London: IEE Electromagnetic Waves Series, Peter Peregrinus, 1987.
[10] G.C.Balvedi and F.Walter, Analysis of GPS signal propagation in tropospheric ducts using numerical methods. 11th URSI Commission F Open Sysposium on radio Wave Propagation and Remote Sensing Proceedings, Rio de Janerio, RJ, Brazil, 2007.

# Bandwidth Enhancement of PIFA with Novel EBG Ground

Y.Cao, Q.N.Qiu, Y.Liu, S.X.Gong

National Laboratory of Science and Technology on Antennas and Microwaves

Xidian University，xi'an 710071,Peoples R China

**Abstract-A Planar Inverted-F Antenna (PIFA) with novel compact double-layer Electromagnetic Band-Gap (EBG) structure is presented. The upper layer of the EBG structure is formed by an interdigital structure and a cross-slot is embedded in a metal patch as the lower layer. Compared with a conventional PIFA, the simulated -10dB impedance bandwidth of the proposed PIFA is broadened by 48% with similar directivity pattern characteristics. In order to verify the simulated analysis, antenna prototypes are fabricated. From the measured results, the bandwidth has been improved from 550MHz to 800MHz, which means an expansion of 45%. The measured results basically show agreement with the simulated results, which illustrates the superiority of this PIFA with the EBG ground plane.**

## I. INTRODUCTION

In recent years, more and more attention has been paid to the development of antennas with small size and wide bandwidth for mobile communication. Conventional planar inverted-F antennas (PIFA) have attracted so much attention as they have many advantages namely low cost, small size and low backward radiation, compared with conventional microstrip antennas [1]. However, they have limitations of low efficiency, narrow bandwidth and surface wave loss. A large amount of effort has been made to broaden its bandwidth, such as a T-shaped ground plane [2] and meandered shorting strip [3]. But these methods need to be further optimized. Another way to improve antenna performance is to use Electromagnetic Band-Gap (EBG) as the antenna's ground plane. EBG structures are periodic structures with the characteristic of high-impedance electromagnetic surfaces within a certain frequency band-gap [4]. A conventional PIFA do not function effectively when it is applied on perfect electric-conductor (PEC) ground plane. However, it works efficiently above a high-impedance surface. Because the high surface impedance is capable of prohibiting the propagation of electromagnetic waves [5] and has the characteristic of in-phase reflection [6]. This property can improve the overall performance of the antenna without changing the antenna's dimension.

However, practical applications of EBG structures have difficulties in accommodating their physical sizes, because the dimension of an EBG lattice has to be half of the free space wavelength. The study of a mushroom-like EBG structure changes this situation. The length of the periodic EBG unit approaches 10% of the wavelength. Compared with other EBG structures such as dielectric rods and holes, this structure has a superior feature of compactness [7]. In [8], a mushroom-like EBG structure is used as the ground plane of the PIFA, so that the original relative bandwidth is increased from 7.1% to 10%. However, if the layout of components is required to be very

compact, the mushroom-like EBG structure is still too large. In [9], a new JCSS (Jerusalem Cross) geometry is applied to the mushroom-like EBG structure. It greatly improves the bandwidth of the PIFA. However, because the frequency of band-gap generated by the EBG structure is high, the antenna's bandwidth only increases at 5GHz, while at 2.5GHz its bandwidth is even narrower than the conventional PIFA.

In this paper, in order to further extend the band-gap width and reduce the band-gap frequency，the concept of interdigital capacitor is introduced into the design of the mushroom-like EBG structure. The connection of two adjacent elements constructs an interdigital structure, which results a larger fringe capacitor [10], [11]. The design of a double-layer EBG structure [12] makes another step towards compact EBG configuration. In this structure, a metal patch is inserted between the upper patch of the EBG structure and the ground plane. The double-layer EBG structure increases the equivalent capacitance of EBG units by increasing the effective area of the EBG unit. A novel type of EBG structure is designed for much lower frequencies by combining the concept of interdigital capacitor with the design of a double-layer EBG structure. The PIFA by using this EBG structure as the antenna's ground plane is proposed and analyzed using Ansoft HFSS. The simulation and measurement results reveal that the novel EBG structure can significantly improve the antenna's performance. The proposed PIFA has a similar radiation pattern shape but the bandwidth is extended by 48% at 3.2GHz, compared with a conventional PIFA of the same size.

## II. ELECTROMAGNETIC BAND-GAP STRUCTURE DESIGN

Within the EBG structure, the resonance effect of the periodical unit plays a leading role when the band-gap is formed. The inductance L results from the current flowing through the connecting via. The gap between the conductor edges of two adjacent cells introduces equivalent capacitance C. Thus a two dimensional periodic LC network is realized which results in the frequency band-gap and the center frequency of the band-gap. The central frequency of the band-gap ( $f_0$ ) is approximately determined by the unit's equivalent inductance ($L$) and equivalent capacitance ($C$):

$$f_0 = \frac{1}{2\pi\sqrt{LC}} \tag{1}$$

978-7-5641-4279-7

As (1) shows, in order to achieve an even more compact EBG structure, the equivalent capacitance and inductance should be increased. But in the EBG design procedure, if the dielectric material and its thickness have been chosen, the inductance cannot be altered. Therefore, only the capacitance can be enlarged. According to the analysis above, a new planar EBG structure is proposed. As is shown in Figure 1(a) and (b), it is a double-layer EBG structure. The upper layer is formed by an interdigital structure, whose branches crosses over with each other and constructs an interdigital capacitor. The length of the interdigital structure is 2.1mm, the width is 0.2mm and the distance between two interdigital units is 0.6mm. An interdigital EBG structure is connected with the ground plane by a metal via hole with the radius of 0.5mm. This interdigital structure greatly increases the coupling path of two units so it can realize a larger capacitance. The design of a double-layer EBG structure is another improvement in this paper. As is shown in Figure 1(c), a metal patch is inserted between the upper patch of the EBG structure and the ground patch. To avoid its contact with the metal via hole, there is a via hole on the metal patch, with the radius of 1.4mm. At the same time, to increase the coupling capacitance, another cross slot whose width is 0.7mm is added to the metal patch. The spaces between the upper EBG structure and the metal patch, the metal patch and the ground plane are all filled with FR4 substrate whose relative permittivity is 4.4. Both of the two substrate boards have a thickness of 1mm.

In the array of the EBG structure, as is shown in Figure 1 (d), the metal patches under adjacent interdigital EBG units connect with each other. In this way, the coupling capacitance between two adjacent EBG units can be increased and the resonant frequency is further lowered. Thus, the EBG structure can be miniaturized and used in low-frequency PIFA.

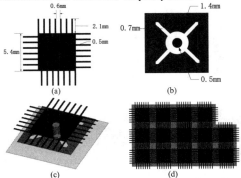

(a)  (b)

(c)  (d)

Figure 1.The view of the proposed EBG: (a)The upper layer of interdigital EBG structure (b)The metal patch inserted in the EBG structure (c)The EBG unit cell (d)The array of the EBG structure.

## III. PIFA WITH EBG GROUND PLANE

The configurations of the PIFA with the EBG ground plane are shown in Figure 2and 3. The ground of the proposed antenna is composed of two parts. The first part is a new EBG structure (45mm×25mm) of 14 unit cells. These periodic cells are printed on a dielectric slab with permittivity of 4.4 and a thickness of 2mm. The second part is a dielectric slab (45mm×60mm) with permittivity of 4.4 and a thickness of 2mm.The radiation patch (18mm×10mm) is placed above the ground with a height of 8.5 mm. The proposed antenna is fed by 50Ω coaxial probe feeding structure underneath the ground plane. The frequency of operation ($f_r$)is calculated using (2), where $c$ is the velocity of light (a constant approximately equal to $3 \times 10^8$m/s), $A$ and $B$ are, respectively, the width and the length of the radiation patch. The dimension of the radiating patch:

$$f_r = \frac{c}{4(A+B)} \qquad (2)$$

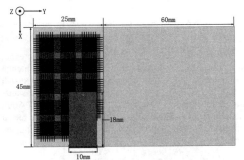

Figure 2.The top view of the PIFA with the EBG ground plane

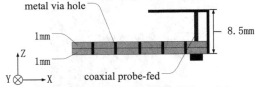

Figure 3.The side view of the PIFA with the EBG ground plane

## IV. ANALYSIS OF SIMULATION AND MEASUREMENT

For comparison, a conventional PIFA is also constructed with a radiating patch of 18mm×10mm. A short-circuit probe and a coaxial probe is placed on the ground plane (45mm×85mm×2mm) with a height of 8.5mm. The simulated return losses of both the proposed antenna and the conventional PIFA with the same dimension are shown in Figure 4. The simulated frequency range of the conventional PIFA for the return loss ≤ -10 dB (VSWR ≤ 2) is from 3GHz to 3.43GHz with the bandwidth of about 430MHz at the centre frequency of 3.2GHz. The simulated frequency range of the PIFA with the EBG ground plane is from 2.94GHz to 3.58GHz with the bandwidth of about 640MHz at the same centre frequency. From the results, the bandwidth has been improved from 430MHZ to 640MHz, which means an expansion of 48%. That is because the EBG structure makes the ground plane of PIFA as a high-impedance surface which leads to a change in the input impedance of the proposed antenna.

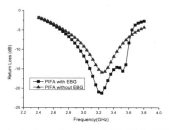

Figure 4.The simulated return losses of the PIFA with the EBG ground plane and the conventional PIFA

In order to verify the theoretical analysis of the antenna, according to the specific dimensions referred before, an antenna prototype is fabricated. As shown in Figure 5, a dielectric slab with permittivity of 4.4 is used, the radiating patch and a short-circuiting piece is made by copper with the thickness of 0.5mm. The antenna is fed by SMA connector coaxial. Here the return loss is measured by Agilent vector network analyzer and the results are shown in Figure 6.

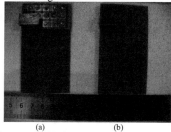

Figure 5.The prototypes (a) The PIFA with the EBG ground plane (b) the conventional PIFA

In Figure 6, the measured frequency range of the conventional PIFA for the return loss ≤ -10 dB (VSWR ≤ 2) is from 3.1GHz to 3.65GHz with the bandwidth of about 550MHz at the centre frequency of 3.35GHz. The proposed antenna which is fabricated in the same size has a bandwidth as large as

800MHz form 3.05 to 3.85GHz at the same center frequency. Compared with the conventional PIFA, the bandwidth of the proposed antenna is expanded by 45%. The measured results basically show agreement with the simulated results, which illustrates the superiority of this PIFA with the EBG ground plane. But the measured frequency is lower than the simulated one. The discrepancy is mainly brought by the fabrication tolerance, because the resonance of the EBG is very sensitive to the dimension. Another reason is the measured errors in the experiment.

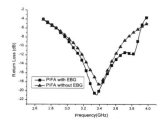

Figure 6.The measured return losses of the PIFA with the EBG ground plane and the conventional PIFA

The simulated far-field radiation pattern of the conventional PIFA and the proposed PIFA with the EBG ground plane at 3.35GHz and 3.6GHz are given in Figure 7. The shape of the pattern for the proposed PIFA is similar to that of the conventional PIFA. The simulated maximum radiation gain of the proposed PIFA and a conventional PIFA is 4.56dB, 4.41dB at 3.35GHz, and 4.81dB, 4.48dB at 3.6GHz, respectively. So it can be applied to the ground plane to expand bandwidth while maintaining a similar radiation pattern with the conventional PIFA.

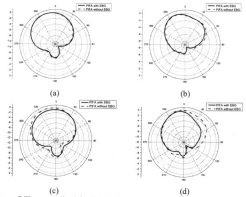

Figure 7.The normalized far-field radiation patterns of the PIFA with the EBG ground plane and the conventional PIFA. (a)the xz-plane of 3.35GHz, (b)the yz-plane of 3.35GHz, (c)the xz-plane of 3.6GHz, (d)the yz-plane of 3.6GHz

## V. CONCLUSION

For application, a novel compact double-layer EBG structure in interdigital shape is investigated. Using the idea of the high-impedance surface to construct the ground plane, a novel type of compact PIFA is proposed. Compared with a conventional PIFA of the same size, the measured and the simulated results show that the EBG structure can significantly improve the patch antenna's bandwidth and the two antennas have similar directivity pattern characteristics. So the proposed antenna demonstrated with the new EBG structure is suitable for mobile-communication application.

ACKNOWLEDGMENT

The work is supported by Program for New Century Excellent Talents in University (NCET-11-0690) and the Fundamental Research Funds for the Central Universities (K5051202049)

REFERENCES

[1] D.B. Lin, I.T. Tang, and M.Z. Hong, "a compact quad-band PIFA by tuning the defected ground structure for mobile phones," *Progress In Electromagnetics Research B*, Vol. 24, 173-189, 2007.

[2] P. S. Hall, C. T. P. Song, H. H. Lin, H. M. Chen, Y. F. Lin, and P. S.Cheng, "Parametric study on the characteristics of planar inverted-F antenna," *Microwaves, Antennas and Propagation*, Vol. 152, No. 6, 534-538, Dec 2005.

[3] P. W. Chan, H. Wong, and E. K. N. Yung, "Wideband planar inverted-F antenna with meandering shorting strip," *Electronics Letters*, Vol. 44, No. 6, 395–396, Mar 2008.

[4] Y. Qian, V. Radisic, and T. Itoh, "Simulation and experiment of photonic band-gap structures for microstrip circuits," *Proc. 1997 Asia-Pacific Microwave Conf*, 585–588, Dec 1997.

[5] S.H. Kim, T. T. Nguyen, and J.H. Jang, " Reflection characteristics of 1-D EBG ground plane and its application to a planar dipole antenna," *Progress In Electromagnetics Research*, Vol. 120, 51-66, 2011.

[6] D. Sievenpiper, L. Zhang, R. F. J. Broas, N. G. Alexópolous, and E.Yablonovitch, "High-impedance electromagnetic surfaces with a forbidden frequency band," *IEEE Trans. Microwave Theory Tech*, Vol. 47, 2059–2074, Nov 1999.

[7] F. Yang, and Y. Rahmat-Samii, "Microstrip antennas integrated with electromagnetic band-gap structures: a low mutual coupling design for array applications," *IEEE Trans. Antennas and Propagation*, Vol. 51, 2936-2946, Oct 2003.

[8] L. C. kretly, Alexandre M. P. Alves S, "The effect of an electromagnetic band-gap structure on a PIFA antenna array," *PIMRC 2004. 15th IEEE International Symposium on*, Vol. 2, 1268-1271, 2004.

[9] C.R.Simovski, P.D.Maagt and I.V.Melchakova, "High-Impedance Surfaces Having Stable Resonance With Respect To Polarization and Incident Angle", *IEEE Transactions on Antennas and Propagation*, Vol. 53, No. 3, March 2005

[10] Y. Fu, N. Yuan and G. Zhang, "Compact high-impedance surfaces incorporated with interdigital structure," *Electronic Letters*, Vol. 40, 310-311, 2004.

[11] C.M. Lin, C.C. Su, S.H. Hung, and Y.H. Wang, "A compact balun based on microstrip EBG cell and interdigital capacitor," *Progress In Electromagnetics Research Letters*, Vol. 12, 111-118, 2009.

[12] Mu-Shui Zhang, Yu-Shan Li, Chen Jia, "A Double-Surface Electromagnetic Bandgap Structure With One Surface Embedded in Power Plane for Ultra-Wideband SSN Suppression," *Microwave and Wireless Components Letters*, Vol. 17, 706-708, 2007.

# Artificial Magnetic Conductor and Its Application

Hongyuan Zhou and Feng Xu

School of Electronic Science and Engineering

Nanjing University of Posts and Telecommunications

Nanjing 210003 China

*Abstract*-In this paper, the reflection phase diagram along with frequency changes of artificial magnetic wall (AMC) is studied and we can determine the frequency band at which the AMC can be achieved. The concept of virtual electric/magnetic walls which are formed by combining perfect electric conducting parallel plates and perfect magnetic conducting parallel plates is cited. In practice, the perfect magnetic wall can be replaced with an artificial magnetic conductor surface. The implement of the Quasi-TEM waveguide with uniplanar compact electromagnetic band gap (UC-EBG) structure is simulated. The field distribution in the waveguide is presented with two simulation methods in HFSS which verifies the existence of the VMW. The phenomena in connection with AMC and resonance are also discussed.

Index Terms-Artificial magnetic wall (AMC), virtual magnetic/electric Walls (VMWs/VEWs) , quasi-TEM waveguide, resonance phenomena.

## I. INTRODUCTION

The perfect electric conductor (PEC) surface is well known that fully reflects incident waves with reflection phase of 180 while the perfect magnetic conductor (PMC) introduces a zero degree phase shift. But perfect magnetic conductor doesn't exist in nature. We use artificial magnetic conductors (AMC) instead of the perfect magnetic conductor [1]. The AMC is an equivalent magnetic conductor with a limited bandwidth and it works as an ideal equivalent magnetic conductor just at a fixed frequency. Recently, there has been a growing in realization of artificial magnetic conductors through the use of periodic surfaces [2]. In practice, the reflection phase of AMC surface cross zero at just only one frequency point (for one resonant mode) because of the resonant nature of such an AMC structure. The useful bandwidth of an AMC is in general within the range of +90 degree to -90 degree on either side of the central frequency [3].

One application of AMC surface can be seen in development of quasi-TEM waveguide in which phase velocity is equal to the speed of light in free space and field distribution becomes uniform in the center of the waveguide at a specific frequency point [4]. Another application of AMC surface is to construct a TEM waveguide by replacing the side PEC surfaces in conventional rectangular waveguide with the virtual magnetic Walls (VMWs) [5]. Two important advantages are observed, namely, a loss reduction since there are no metallic side walls, and also, an easy way to integrate or create rectangular waveguide in planar circuit. The field distribution in a cross section of the waveguide was simulated.

978-7-5641-4279-7

## II.TEM WAVEGUIDES BASED ON VE/MWS

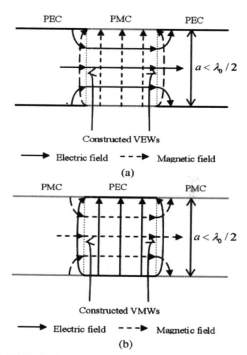

Fig.1. Field distributions of TEM modes and VE/MWs (a) TEM mode waveguide based on VEWs. (b) TEM mode waveguide based on VMWs.

The virtual magnetic/electric Walls (VMWs/VEWs) can be used to construct TEM waveguides as shown in Fig. 1. In Fig. 1(a), a PMC parallel plate waveguide is connected to two PEC parallel plate waveguides at both sides. In this case, two vertical VEWs are formed at the PEC/PMC interfaces. Only the TEM mode with E-fields parallel to the plates can be supported in the PMC parallel plate waveguide when the distance of the two parallel plates is smaller than a half of wave-length. The electromagnetic fields are confined in the center region bounded by the two side VEWs. Different from the real electrical conductor, the VEW allows the fields to penetrate into the PEC parallel plate waveguide region. However, the fields will decay very quickly when it passes through the VEW. Fig. 1(b) shows a complementary TEM waveguide based on VMWs. It shares the same properties as that in Fig. 1(a) except that the electric and magnetic fields are interchanged.

### III. REFLECTION PHSE OF AMC

As all we know, the frequency selective surface (FSS) is a periodic structure which shows a band-pass or band-stop characteristics at a fixed frequency [6]. This frequency is a point where the structure resonance occurs. The AMC is also a frequency selective surface but different from it which has a ground plane on the back of the periodic structure. The principle of AMC operation is based on the resonance of cavities formed between the periodic array and the ground plane. The AMC has a limited bandwidth because of the nature of resonance and its characteristics vary in the bandwidth.

A uniplanar compact electromagnetic band gap (UC-EBG) st structures, first presented by Itoh's group [7], is used as AMC su rfaces. The UC-EBG surface is fabricated on a conductor-base d Duroid substrate (Duroid 6010) with dielectric constant of 10 .2 and thickness of 25 mil. As in Fig. 2, the unit cell design of t he EBG surface with d=180mil,b=162mil,h=45mil,g=s=18mil i s showed.

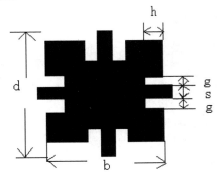

Fig. 2  A unit cell of the PBG lattice

The full-wave finite-element-based simulation package Ansoft-HFSS [8] is applied to obtain the reflection phase characteristics of the AMC structure in figure 3 and 4. Two methods were used to achieve the results we want. In figure 3(a), two pairs of master-slave boundaries of assumed the scan angles difference was placed at the front and back boundaries and the left and the right boundaries of the unit cell along the x-direction and the y-direction. The Floquet excitations were added to both the upper and the lower surfaces. The result is showed in figure 3 (b). As in figure3 (b), the reflection phase changes along with the change of frequency and crosses the 0 degree (for one resonant mode) at the frequency of 9.66GHz. The useful bandwidth is 9.3~9.9GHz in which the reflection

phase is within the range of +90 degree to -90 degree. In general, we have achieved the artificial magnetic conductor in this bandwidth.

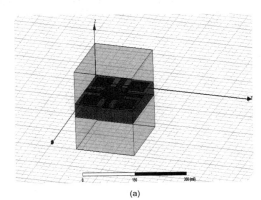

(a)

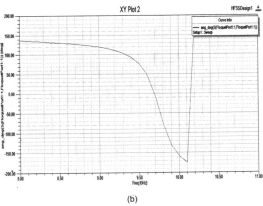

(b)

Fig.3  Floquet method of reflection phase (a) The model of UC-PBG in HFSS
(b) The result of the reflection phase with the frequency

### IV. FIELD DISTRIBUTIONS OF THE TEM WAVEGUIDE BASED ON AMC TECHNIQUE

In order to prove the existence of the virtual magnetic wall, the full-wave finite-element-based simulation package Ansoft-HFSS is applied while the existence of the virtual electric wall has been proved in another paper. The waveguide consists of two parallel PCB with four pairs of periodic on both sides. The length of the PCB is about 10 times the length of UC-PBG used above. A standard rectangular waveguide whose sidewalls are removed with the length of 900mil and width of 400 mil is in the middle of the model .Two standard rectangular waveguides with the same parameters are placed at the front and back boundaries in order to eliminate the impact of discontinuities. The whole model is covered by air box whose four walls are set to radiation boundary condition. The part we are interested in is the field distribution between the two parallel PCB. One must be pointed out that the distance between the two parallel must be less than half of a wavelength in the frequency band we are interested in. The whole model is presented in figure 4 (a).

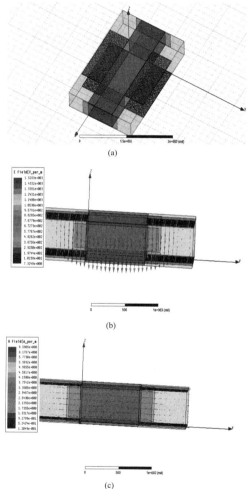

(a)

(b)

(c)

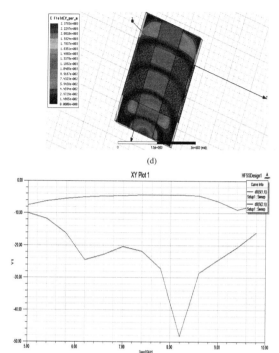

(d)

(e)

Figure 4 Calculation of the VMWs with the Driven mode (a) The model of the Quasi-TEM waveguide in AMC technology (b) Electric field distribution on the middle cross section (c) Magnetic field distribution on the middle cross section (d) The magnitude of the electric field distribution throughout the model (e) Magnitude of S(1,1) and S(2,1) in dB

Figure 4 (b) and (c) show the electric and magnetic distributions in the middle of cross section of the model. The solution frequency is 8GHz. The field distributions are close to that of figure 1 (b).The field almost keeps uniform in the entire center region of the waveguide and decrease very quickly outside the VMWs. The characteristics of the field distribution indicate that two VMWs exist on both sides of the waveguide. The magnitude of the electric is also showed in figure 5(d). We can easily find that the magnitude of electric decrease very quickly outside the VMWs. All of these can prove the existence of the VMWs. Figure 4 (e) shows the magnitude of S(1,1) and S(2,1) in dB. Most of the energy is transmitted to the port 2 in a large band which is probably between 6.8GHz and 8.6GHz. This frequency band doesn't fall within the resonance frequency band which can prove that the Quasi-TEM with UC-PBG structure doesn't depend on the resonance phenomenon and has a wider bandwidth.

The Eigen mode method in HFSS is also applied to achieve the virtual magnetic wall. The structure of the waveguide is similar to the above but with only one periodic. The parameters of the waveguide is the same to the Driven mode. Figure 6 (a) shows the model of the Quasi-TEM waveguide based on AMC technology. A pair of master-slave boundaries of assumed phase difference was placed at the front and back boundaries of the unit cell along the longitudinal axis (x-direction) to compute a corresponding frequency of the eigen value. The electric and magnetic distributions in the cross section of one cell of 30 degree phase difference in the master-slave boundaries are showed in figure 6 (b) and (c) which are similar to the distributions of the Driven mode. The corresponding frequency is 8.2GHz which is very close to that of the Driven mode.

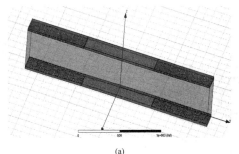

(a)

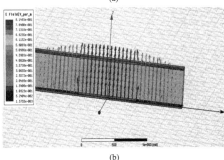

(b)

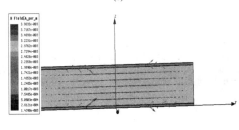

(c)

Figure 6 Calculation of the VMWs with the Eigen mode (a) The model of the Quasi-TEM waveguide in AMC technology with only one periodic (b) Electric field distribution on the middle cross section (c) Magnetic field distribution on the middle cross section

## V. CONCLUSION

In this paper, two methods of implementing the phase reflection diagram were presented. We can easily determine the frequency band at which the AMC can be achieved. The implementation of the AMC is concerning to the resonance phenomenon. The existence of the VMW was also proved with two kinds of simulation methods in HFSS. In both methods, the Quasi-TEM waveguide was also achieved because of the VMWs. The field distributions in the cross section were simulated. The different frequency points between the AMC which is the resonance point in general and the Quasi-TEM waveguide with the VMWs show that the realization of the Quasi-TEM is not based on the traditional resonance theory and have a wider bandwidth.

## REFERENCES

[1] Sievenpiper D, Zhang L, Broas R F J, et al. High-impedance electromagnetic surfaces with a forbidden frequency band[J]. Microwave Theory and Techniques, IEEE Transactions on, 1999, 47(11): 2059-2074.

[2] Zhang Y, von Hagen J, Younis M, et al. Planar artificial magnetic conductors and patch antennas[J]. Antennas and Propagation, IEEE Transactions on, 2003, 51(10): 2704-2712.

[3] Li D, Wu K. Geometric characteristics, physical mechanism, and electrical analysis of quasi-TEM waveguides with dipole-FSS walls[C]//Microwave Symposium Digest (MTT), 2010 IEEE MTT-S International. IEEE, 2010: 17-20.

[4] Yang F R, Ma K P, Qian Y, et al. A novel TEM waveguide usinguniplanar compact photonic-bandgap (UC-PBG) structure[J]. Microwave Theory and Techniques, IEEE Transactions on, 1999, 47(11): 2092-2098.

[5] Li D C, Boone F, Bozzi M, et al. Concept of virtual electric/magnetic walls and its realization with artificial magnetic conductor technique[J]. Microwave and Wireless Components Letters, IEEE, 2008, 18(11): 743-745.

[6] Cucini A, Caiazzo M, Bennati P, et al. Quasi-TEM waveguides realized by FSS-walls[C]//Antennas and Propagation Society International Symposium, 2004. IEEE. IEEE, 2004, 1: 807-810.

[7] Ma K P, Hirose K, Yang F R, et al. Realisation of magnetic conducting surface using novel photonic bandgap structure[J]. Electronics Letters, 1998, 34(21): 2041-2042.

[8] Li M Y. The detail design and application of electromagnetic simulation in HFSS [M].People's Posts and Telecommunications Press,2011.

[9] Goussetis G, Feresidis A P, Vardaxoglou J C. Tailoring the AMC and EBG characteristics of periodic metallic arrays printed on grounded dielectric substrate[J]. Antennas and Propagation, IEEE Transactions on, 2006, 54(1): 82-89.

# Working Principles of A Broadband/Dual-band Frequency Selective Structure with Quasi-Elliptic Response

Tan Jianfeng, Liu Peiguo, Huang Xianjun, Li Guohua, Yang Cheng
College of Electronic Science and Engineering
National University of Defense Technology
Changsha 410073,China

*Abstract*-The working principles of a novel quasi-elliptic frequency selective structure (FSS) based on apertured cavity with six ring slots are investigated completely. Similar with the elliptic filter circuits design, cross coupling is essential in elliptic FSS. The cross-coupling and the mechanism of producing transmission zeros are discussed. Besides, from the aspects of both wave construction and equivalent circuit, the operating principles are studied in details, and further verified with simulations of power flow at resonance frequencies. With the groundwork laid, two dual-band quasi-elliptic FSS are reported utilizing the same basic unit structures. The work in this paper would be beneficial to the research of frequency selective surfaces, and is especially meaningful to the development of elliptic FSS.

## I. INTRODUCTION

Frequency Selective Surface is a constantly hot research topic in the last several decades for its remarkable application value in areas such as radar cross-section (RCS) reduction, electromagnetic compatibility (EMC), communication systems, etc [1][2]. Ben.A.Munk proposed the Periodic Moment Method (PMM) which endows structural parameters with clear physical meanings. With this method, the theory, design and application of traditional FSSs have been comprehensively studied and were systematically summarized in a classical treatise [2].

The filtering responses of FSSs are dominated by unit structures. The traditional unit structures could be arranged into four groups: center connected type such as the famous Jerusalem cross and simple straight element; loop type like circular or square loops; plate types and the combination of abovementioned structures [2]. Except for these basic types, the development of computational technology boosts the designs of complicated unit structures. Rahmat-Samii et al introduced the concept of fractal to designs of multi-band FSSs[3]. Wener,D.H and Shigesawa combined the Genetic Algorithm (GA) with FSS design and optimization, and presented some heteroclite structures[4][5].

However, the traditional FSS designs are mainly with two dimensional planar unit structures. The application of single-layered traditional FSSs is limited for the drawbacks of poor filtering responses and narrow bandwidth. With the method of cascading 2-D FSSs and thin dielectric layers, the filtering responses and bandwidth could be substantially enhanced. The combined multi-layer FSSs nevertheless follow Butterworth or Chebyshev responses whose attenuation of stopband increases slowly, leading to a wide transition band[2][6]. To cut down the interference of "useless signal", a speedy transition from passband to stopband, namely rapid rolloff of transmission responses is preferred in applications. In contrast, Elliptic response is superior in high selectivity, rapid rolloff and high attenuation in stopband, due to the transmission zeros placed near passband [7].

To design FSSs with elliptic responses, people break through the restriction of traditional planar FSSs, and proposed the innovative concept of "three dimensional frequency selective structures" [8]. A.Abbaspour developed an array of antenna-filter-antenna modules to design elliptic FSS with high-order poles [9]. Luo et al presented a class of quasi-elliptic FSS designs by introducing cross coupling in a dual-mode FSS based on substrate integrated waveguide (SIW) [10]. Shen et al contributed a series of elliptic FSS designs employing dielectric-filled waveguides, and shielded microstrip lines [11][12].

In our previous work, a broadband quasi-elliptic FSS adopting a novel apertured cavity with six ring slots was reported [13]. In this paper, its working principles would be discussed in details, and a dual-band quasi-elliptic design would be developed with the similar unit.

## II. TRANSMISSION ZERO AND CROSS-COUPLING

In filter circuit, elliptic responses are commonly designed with mature theory. Transmission zeros at finite frequencies are normally obtained by introducing cross coupling between resonators [14]. Fig.1 gives the basic concepts of cross coupling and illuminates a way to design the elliptic FSSs.

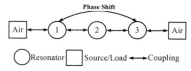

Figure 1. Cross coupling between resonators.

978-7-5641-4279-7

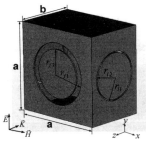

Figure 2. Unit structure and simulation configuration.
a=12mm,b=8mm rf1=4.3mm, rf2=3.4mm, rl1=3.4mm,
rl2=3.2mm, and thickness of the metal t=0.2mm.

With the introduction of cross-coupling, at least two parallel paths are provided for signals, and every path is with different phase shift to signal at certain frequency. Signals from different paths combine out of phase at the load or receiver, resulting to transmission zeros, while combination in phase contributes transmission poles. With the above principle, it is essential to built up parallel paths for electromagnetic waves in elliptic FSS. The unit structure in Fig.2 is a typical one. This FSS is proved to be with high selectivity, broad bandwidth, and rapid rolloff at upper band edge.

### III.  WORKING PRINCIPLES

The working principles of the novel structure in Fig.2 could be discussed from the view of wave propagation. The following Fig.3 is the up view inside the unit and the propagation situation is sketched. As can be seen, there are three main paths for electromagnetic waves from the front to the back of FSS: i. the incident wave passes through the front slots and directly to the free space; ii. Before transmitting to the back slots, waves are reflected by lateral slots. iii. Waves transmitting to the back slots are from the nearby units.

From the described paths, it can be found that, compared with traditional planar FSSs, the lateral rings bring another two paths and work as secondary frequency selectors. Incident waves go through the front slots, and then come to the back slots through plenty of parallel paths, thus waves to the free space are with different phase shift. Constructive superposition of these waves contributes to the passband response of the FSS. On the other hand, destructive superposition of these waves corresponds to the stopband response of the FSS.

FSS is essentially a filter in space, thus the working principles of this elliptic FSS can also be studied from the view of filter circuit. Fig.4 is the equivalent circuit of the novel FSS. As is revealed, this equivalent circuit consists of five parts, three resonators and their mutual couplings. Three resonators correspond with the three main paths in Fig.3, and their mutual influences are represented with mutual coupling circuits. For a signal at certain frequency, the resonator and coupling circuits determine its amplitude and phase together, in accordance with superposition principle in FSS.

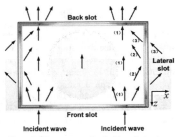

Figure 3. Wave propagation inside unit structure.

From the described paths, it can be found that, compared with traditional planar FSSs, the lateral rings bring another two paths and work as secondary frequency selectors. Incident waves go through the front slots, and then come to the back slots through plenty of parallel paths, thus waves to the free space are with different phase shift. Constructive superposition of these waves contributes to the passband response of the FSS. On the other hand, destructive superposition of these waves corresponds to the stopband response of the FSS.

FSS is essentially a filter in space, thus the working principles of this elliptic FSS can also be studied from the view of filter circuit. Fig.4 is the equivalent circuit of the novel FSS. As is revealed, this equivalent circuit consists of five parts, three resonators and their mutual couplings. Three resonators correspond with the three main paths in Fig.3, and their mutual influences are represented with mutual coupling circuits. For a signal at certain frequency, the resonator and coupling circuits determine its amplitude and phase together, in accordance with superposition principle in FSS.

To examine the validity of the equivalent circuit, its response is compared with the proposed FSS in Fig.5. As depicted, the responses of the equivalent circuit are almost the same with FSS, which reveals the correctness of this equivalence.

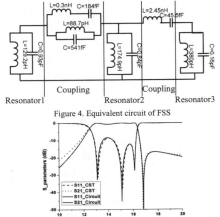

Figure 4. Equivalent circuit of FSS

Figure 5. S-Parameters of equivalent circuit and FSS.

The filtering response of the proposed FSS is also displayed in Fig.5. In the reflection response, three resonances at 13.16GHz, 15.14 GHz and 16.12 GHz, are exhibited, giving rise to a broad 3dB passband from 12.45GHz to 16.38GHz. Besides, this FSS has a rapid rolloff near upper band edge, and its bandwidth ratio of -3dB to -0.5dB is as low as 1.13, which is much lower than the typical ratio value 3.5 of traditional FSSs. Moreover, the stopband attenuation at higher frequency is generally higher than 20dB. In conclusion, this novel FSS has a high selectivity, rapid rolloff and high attenuation in stopband.

## IV. VERIFICATION

To verify the abovementioned principles, the power flow at three resonance frequencies 13.16GHz, 15.14 GHz and 16.12 GHz, is given. Fig.6 is the top view situation of power flow at 13.16GHz. As shown in the right color bar, the depth of the red color represents the intensity of power flow, namely the darker the color, the bigger the power flower. From the figure, most of the power flows from the front rings to the back slot at 13.16GHz, which corresponds to the path (1) in Fig.3. As explained in part.2, response at certain frequency is the consequence of construction of different paths, and hence some minor power is from the lateral rings resembling path (2)(3), as observed.

The power flow at 15.14GHz in Fig.7 is obviously different from Fig.6. The intensity of power flow at 15.14GHz from the path (1) and path (2) are almost equal. That is to say, waves pass through these two paths with different phase shift, and have a constructive superposition, forming the resonance at reflection response.

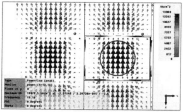

Figure 6. Power flow at resonance of 13.16GHz.

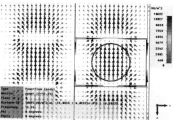

Figure 7. Power flow at resonance of 15.14GHz.

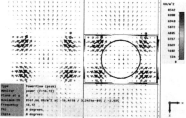

Figure 8. Power flow at resonance of 16.12GHz.

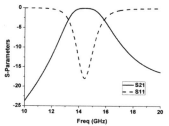

Figure 9. Responses without coupling between nearby units.

As is depicted in Fig.8, power flow at 16.12GHz is nearly all from lateral rings. However, it is hard to tell whether the wave is from the path (2) or path (3), as these two paths are very much different. Waves in path (2) have a phase shift not only from the propagation, but also have another phase shift when reflected by lateral ring slots. This additional phase shift is determined by the frequency/phase response of lateral selector.

As known, under the stimulation in Fig.2, responses of FSS could be calculated with a unit and proper boundary setting, i.e., perfect electric wall at XZ plane and perfect magnetic wall at YZ plane. According to equivalent theory, waves in path (3) could be replaced by magnetic conducting wall placed at lateral wall of the unit plus the equivalent magnetic current on it. The magnetic conducting wall and equivalent source produce another parallel path for cross-coupling.

To elucidate the function of coupling between nearby units in bringing cross-coupling and transmission zeros, the situation in which the above magnetic boundary is replaced by electric boundary is simulated. As revealed in the simulated S-parameters in Fig.9, without the coupling between nearby units, the reflection response has only one resonance. Moreover, the response is characterized with Butterworth response which is typical in multi-layered FSS.

## V. DUAL-BAND QUASI-ELLIPTIC RESPONSE

Based on the working principles analyzed above, a dual-band quasi-elliptic FSS is presented by tuning the structural parameters of the unit in Fig.2. As seen in the transmission responses in Fig.10, the response in solid curve has a quasi-elliptic passband with rolloff at right edge, and an elliptic response at higher band. Two transmission zeros appear at

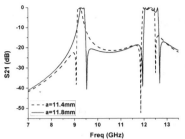

Figure 10. Dual-band quasi-elliptic responses.b=10.3mm rf1=4.78mm, rf2=2.63mm, rl1=4.47mm, rl2=0.95mm.

stopband edges between two passbands, leading to a high stopband attenuation. This unit could also be designed to achieve a quasi-elliptic passband with left-edge rapid rolloff and an elliptic passband at higher frequency band, by simply change of parameter a. These two superior responses are depicted and compared in Fig.10, and it illuminates a possible way to reach the dual-band elliptic filtering response by combining these two quasi-elliptic FSSs.

## VI. CONCLUSION

This paper firstly introduced the development and limitations of traditional frequency selective surfaces, and showed examples of most recent progress in quasi-elliptic and elliptic FSS. The theory of cross-coupling and transmission zeros has been discussed and guided the analysis of working principles of the novel FSS. Connected with the unit structure, its operating principles have been fully investigated from aspect of construction of waves, as well as aspect of equivalent filter circuit. Both methods reached an agreement that the quasi-elliptic filtering responses are constructed with three resonance modes and their mutual coupling, and the parallel paths is essential for transmission zeros. The conclusions were then validated with simulations of power flow at resonance frequencies. When the coupling between nearby units is eliminated, the FSS exhibits a typical Butterworth response, which proves the necessity of parallel paths to yield transmission zeros. Moreover, two dual-band quasi-elliptic FSSs are designed with the foundation of working principles.

Further related works would be emphasized on the research of elliptic FSS and multi-band elliptic FSSs based on same basic unit structure.

### ACKNOWLEDGMENT

The authors would like to give thanks to the help of CST China in simulation, and valuable suggestions and discussions from Dr. Dongming Zhou.

### REFERENCES

[1] T. K. Wu, *Frequency Selective Surface and Grid Array*, New York: Wiley, 1995.

[2] B.A.Munk, *Frequency Selective Surface: Theory and Design*, New York: Wiley-Interscience, 2011.

[3] J.P.Gianvittorio, J.Romeu, S.Blanch, Y.Rahmat-Samii, "Self-similar prefractal frequency selective surfaces for multiband and dual-polarized applications," IEEE *Trans. on Antennas and Propag.,* vol. 51, pp. 3088-3096, 2003.

[4] J.A.Bossard, D.H.Werner, T.S.mayer, R.P.Drupp, "A novel Design methodology for reconfigurable frequency selective surfaces using genetic algorithms," IEEE *Trans. on Antennas and Propag.,* vol. 53, pp. 1390-1400, 2005.

[5] M.Ohira, H.Deguchi, M.Tsuji, h.Shigesawa, "Multiband single layer frequency selective surface designed by combination of genetic algorithm and geometry refinement rechnique," IEEE *Trans. on Antennas and Propag.,* vol. 52, pp. 2925-2931, 2004.

[6] G.Bianchi, R.Sorrentino, *Electronic Filter Simulation & Design,* New York: Mc Graw-Hill, 2007.

[7] W. Alan Davis, *Radio Frequency Circuit Design,* New York: John Wiley & Sons, 2011.

[8] A. K. Rashid and Z. Shen, "Three-dimensional frequency selective surfaces," presented at the *Int. Conf. Communications, Circuits, and Systems (ICCCAS),* July 2010.

[9] A. Abbaspour, K. Sarabandi, and G. M. Rebeiz, "Antenna- filter-antenna array as a class of bandpass frequency selective surfaces," IEEE *Trans. on Microwave Theory and Tech.,* vol. 52, pp. 1781-1789, 2004.

[10] G.Q.Luo,W.Hong,Q.H.Lai,K.Wu,and L.L.Sun, "Design and experimental verification of compace frequency-selective surface with quasi-elliptic bandpass response," IEEE *Trans. on Microwave Theory and Tech.,* vol. 55, pp. 2481-2487, 2007.

[11] Y. Zuo, A. K. Rashid, Z. Shen, Y. Feng, "Design of dual-polarized frequency selective structure with quasi-elliptic bandpass response," IEEE *Antennas and Wireless Propagation Letters,* vol. 8, pp. 624-626, 2012.

[12] A. K. Rashid and Z. Shen, "Bandpass frequency selective surface based on a two-dimensional periodic array of shielded of microstrip lines," in *Proc. IEEE Antennas Propag. Symp.,* pp. 1-4, July 2010.

[13] X.-J. Huang, C. Yang, Z.-H. LU and P.-G. LIU, "A Novel Frequency-Selective Structure with Quasi-Elliptic Bandpass Response," IEEE *Antennas and Wireless Propagation Letters,* vol. 11, pp. 1497-1500, 2012.

[14] A. K. Rashid, Z. Shen, B.Li, "An Elliptical Bandpass Frequency Selective Structure Based on Microstrip Lines," IEEE *Trans. on Antennas and Propag.,* vol. 60, pp. 4661-4669, 2012.

# Anisotropic Photonic Band Gaps in Three-dimensional Magnetized Plasma Photonic crystals

Hai-Feng Zhang[1,2], Shao-Bin Liu[1], Huan Yang[1]

[1] College of Electronic and Information Engineering, Nanjing University of Aeronautics and Astronautics, Nanjing China
[2] Nanjing Artillery Academy, Nanjing 211132, China
Email: hanlor@163.com

*Abstract-* **The band Faraday effects in photonic band gaps (PBG) for the three-dimensional (3D) magnetized plasma photonic crystals (PCs) consisting of the uniaxial material with face-centered-cubic (fcc) lattices are theoretically investigated by a modified plane wave expansion (PWE) method, which the homogeneous anisotropic dielectric spheres (the uniaxial material) immersed in the magnetized plasma background, as the Faraday effects of magnetized plasma are considered. The anisotropic PBGs and a flatbands region can be obtained. The effects of the anisotropic dielectric filling factor on the properties of first two anisotropic PBGs are investigated in detail. The numerical results show that the anisotropy can open partial band gaps in fcc lattices at $U$ and $W$ points, and the complete PBGs can be achieved compared to the conventional 3D dispersive PCs composed of the magnetized plasma and isotropic material. It also is shown that the first two anisotropic PBGs can be tuned by the filling factor.**

## I. INTRODUCTION

Photonic crystals (PCs) have been attracting a lot of interest since the concept was firstly proposed by Yablonovitch [1] and John [2]. The PCs is a kind of artificial material with a periodically modulated dielectric constant in space, and can produce the magic regions named photonic band gaps (PBGs) [3], where the propagation of electromagnetic wave (EM wave) is forbidden. Thus, PCs can be used to design the applications in the optics. Recently, various dispersive materials have been introduced into PCs to obtain the tunable PBGs, such as ferrofluids [4], plasma [5], superconductor [6], and metal [7]. The tunable PBGs means the PBGs can be manipulated by the different stimuli, such as the magnetic field, temperature or electric field, and not need change the topology of PCs. Investigating of the tunable PCs becomes a new research focus. The plasma photonic crystals (PPCs) is a typical of tunable PCs, which is recently a hot research area. The idea of PPCs was firstly proposed by Hojo and Mase [8], and can be looked as metamaterial [9]. The PPCs also can be used to design some novel tunable devices [10-12] which can realize in the microwave region. Therefore, the PPCs have been extremely investigated in detail. Compared to the conventional PCs, the PPCs can display strong spatial dispersion [13, 14] and also can be tuned by the external magnetic field [15]. If the external magnetic field is introduced into the PPCs, the magnetized plasma photonic crystals (MPPCs) will be realized. As we know, if the external magnetic field is introduced into the plasma, two kinds of magneto-optical effects can be achieved. One configuration, in which the external magnetic field is perpendicular to the EM wave vector, gives rise to so-called Voigt effects. The other is that the EM wave vector is parallel to the external magnetic field. In this case, the Faraday effects can be obtained [16]. If different magneto-optical effects of magnetized plasma are considered, the different dispersive properties of MPPCs will be obtained. Recently, the properties of MPPCs have intensively been studied in theory and experiment by many research groups [17-20].

The most works on MPPCs are 1D or 2D cases. A few published reports about 3D MPPCs until Zhang *et al.* [21, 22] investigated the dispersive properties of 3D MPPCs with simple-cubic and diamond lattices as the Faraday effects of magnetized plasma are considered. As mentioned in their works, these 3D MPPCs suffer from high symmetry and dielectric constant of dielectric must be sufficiently large so that the resonant scattering of EM waves is prominent enough to open a band gap [23]. Unfortunately, technological difficulties in fabricating the 3D MPPCs with the large dielectric constant of dielectric can be found to achieve the complete PBGs. Therefore, if we want to obtain the complete PBGs in 3D MPPCs, the symmetry reduction [24] and anisotropy in dielectric may be good choices [25]. We note that many previous studies on the MPPCs considered that the filling dielectric only is isotropic, and the Faraday effects of magnetized plasma are not considered for 3D case at same time. Therefore, the aim of the present paper is to perform a study of the band Faraday effects in the 3D dispersive PCs consisting of the magnetized plasma and uniaxial material with fcc lattices based on a modified PWE method. The proposed 3D MPPCs are that the anisotropic dielectric spheres are immersed in the magnetized plasma background periodically with fcc lattices.

## II. THEORETICAL MODEL AND FORMULATION

As we know, the expression of the plasma function $\varepsilon_p$ is determined by the angle $\theta$ between the wave vector and the external magnetic field [16, 17]. If the Faraday effects of magnetized plasma are considered ($\theta=0°$), the dielectric function of magnetized plasma is anisotropic and has this form:

$$\varepsilon_p(\omega) = 1 - \frac{\omega_p{}^2}{\omega^2 - jv_c\omega \pm \omega\omega_c} \quad (1)$$

where $\omega_p$, $v_c$ and $\omega_c$ are the plasma frequency, the plasma collision frequency and plasma cyclotron frequency, respectively. Plasma frequency $\omega_p = (e^2 n_e / \varepsilon_0 m)^{1/2}$ in which $e$, $m$, $n_e$ and $\varepsilon_0$ are electric quantity, electric mass, plasma density and dielectric constant in vacuum, respectively.

978-7-5641-4279-7

$\omega_c = (eB/m)$ in which $B$ is the external magnetic field. Here, the "+" sign in the third term of denominator involving plasma cyclotron frequency $\omega_c$ is effective dielectric function for the left circular polarization whereas it is the case for right circular polarization if the "−" sign is taken [21, 22]. In this paper, we shall limit our consideration to right circular polarization case since the resonant behavior can appear only in right circular polarization [21, 22].

In order to calculate the dispersive curves of such 3D MPPCs, the modified PWE method has been used [17, 26]. As we know, the uniaxial material is one kind of anisotropic material, which can be found in the nature. The uniaxial material has two different principal-refractive indices named as ordinary-refractive and extraordinary-refractive indices, and we consider the ordinary-refractive and extraordinary-refractive indices are $n_o$ and $n_e$, respectively. For the uniaxial material, the dielectric constant $\varepsilon_a$ is a dyadic and can be written as

$$\varepsilon_a = \begin{pmatrix} \varepsilon_x & 0 & 0 \\ 0 & \varepsilon_y & 0 \\ 0 & 0 & \varepsilon_z \end{pmatrix} \quad (2)$$

where $\varepsilon_x = n_x^2$ , $\varepsilon_y = n_y^2$ , $\varepsilon_z = n_z^2$ .

Therefore, the dielectric dyadic has only three cases for diagonal element permutation as [23] (a) $n_x=n_e$, $n_y=n_z=n_o$; (b) $n_y=n_e$, $n_x=n_z=n_o$; (c) $n_z=n_e$, $n_x=n_y=n_o$. We call them type-1, type-2 and type-3 uniaxial materials, respectively. In order to simplify, we just deduce the equations for calculating the dispersive curves of such 3D MPPCs containing the type-1 uniaxial material. According to the technique as mentioned in Ref.[21], the band structure of 3D MPPCs can be obtained from following equation and the definitions of parameter also can be found in Refs.[21, 22]:

$$\mu^4 \vec{I} - \mu^3 \vec{T} - \mu^2 \vec{U} - \mu\, \vec{V} - \vec{W} = 0 \quad (3)$$

$$\vec{T}(G\,|\,G') = (j\frac{v_c}{c} + \frac{\omega_c}{c})\delta_{G\cdot G'},$$

$$\vec{U}(G\,|\,G') = \{\sum_{i=x,y} (\frac{\omega_p^2}{c^2} + (f + \frac{1}{\varepsilon_i}(1-f)) \cdot |k+G|^2 \cdot \vec{F}_i)\}\delta_{G\cdot G'}$$
$$+ \sum_{i=x,y}(1-\frac{1}{\varepsilon_i})\overline{M}_i$$

$$\vec{V}(G\,|\,G') = \{\sum_{i=x,y} (-(j\frac{v_c}{c} + \frac{\omega_c}{c})(f + \frac{1}{\varepsilon_i}(1-f)) \cdot |k+G|^2 \cdot \vec{F}_i)\}\delta_{G\cdot G'}$$
$$+ \sum_{i=x,y}(-(j\frac{v_c}{c} + \frac{\omega_c}{c})(1-\frac{1}{\varepsilon_i})\overline{M}_i)$$

$$\vec{W}(G\,|\,G') = \{\sum_{i=x,y} (-\frac{\omega_p^2}{c^2}\frac{1}{\varepsilon_i}(1-f) \cdot |k+G|^2 \cdot \vec{F}_i)\}\delta_{G\cdot G'}$$
$$+ \sum_{i=x,y}(\frac{\omega_p^2}{c^2}(\frac{1}{\varepsilon_i})\overline{M}_i)$$

where

$$\overline{M}_i = |k+G|^2 \cdot \vec{F}_i \cdot 3f(\frac{\sin(|G|R) - (|G|R)\cos(|G|R)}{(|G|R)^3})(i=x,y),$$

the element of the $N \times N$ matrices are $\vec{T}$, $\vec{U}$, $\vec{V}$ and $\vec{W}$. The

This polynomial form can be transformed into a linear problem in $4N$ dimension by $\vec{Q}$ that fulfills

$$\vec{Q}z = \mu z, \qquad \vec{Q} = \begin{bmatrix} 0 & \vec{I} & 0 & 0 \\ 0 & 0 & \vec{I} & 0 \\ 0 & 0 & 0 & \vec{I} \\ \vec{W} & \vec{V} & \vec{U} & \vec{T} \end{bmatrix} \quad (4)$$

The complete solution of Eq.(3) is obtained by solving for the eigenvalues of Eq.(4). Of course the dispersion relation can be determined by the real part of such eigenvalues. The analogue equation to Eq.(3) for another two types of cases also can be easily derived.

### III. NUMERICAL RESULTS AND ANALYSIS

A symmetric set of primitive vectors for the fcc lattice is $a_1=(0.5a, 0.5a, 0)$, $a_2=(0, 0.5a, 0.5a)$, $a_3=(0.5a, 0, 0.5a)$. The reciprocal lattice vector basis can be defined as $b_1=(2\pi/a, 2\pi/a, -2\pi/a)$, $b_2=(-2\pi/a, 2\pi/a, 2\pi/a)$, $b_3=(2\pi/a, -2\pi/a, 2\pi/a)$. The high symmetry points have the coordinate as $\Gamma=(0, 0)$, $X=(2\pi/a, 0, 0)$, $W=(2\pi/a, \pi/a, 0)$, $K=(1.5\pi/a, 1.5\pi/a, 0)$, $L=(\pi/a, \pi/a, \pi/a)$, and $U=(2\pi/a, 0.5\pi/a, 0.5\pi/a)$. As a total number of 729 plane waves are used to calculate, the convergence accuracy is better than 1% for the lower energy bands [23]. Without loss of generality, we plot $\omega a/2\pi c$ with the normalization convention $\omega_{p0}a/2\pi c=1$. Thus, we can define the plasma frequency as $\omega_p=\omega_{pl}=0.3\pi c/a=0.15\omega_{p0}$ to make the problem scale-invariant. We also choose the plasma collision frequency and plasma cyclotron frequency as $v_c=0.02\omega_{pl}$ and $\omega_c=0.8\omega_{pl}$, respectively. Here, we only focus on the first two PBGs in the frequency domain 0-$2\pi c/a$.

#### A. The anisotropic PBG for 3D MPPCs

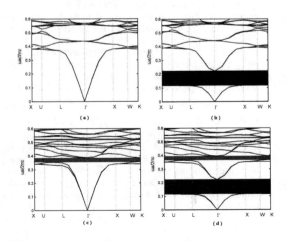

Fig.1. Band structures for 3D fcc MPPCs with $f$=0.3 but with different $n_x$, $n_y$, $n_z$, $\omega_p$, $\omega_c$ and $v_c$, respectively. (a) $n_z=n_x=n_y=4.8$, $\omega_p=0$, $v_c=0$, $\omega_c=0$; (b) $n_z=n_x=n_y=4.8$, $\omega_p=0.15\omega_{p0}$, $v_c=0.02\omega_{pl}$, $\omega_c=0.8\omega_{pl}$; (c) $n_x=n_e=6.2$, $n_y=n_z=n_o=4.8$, $\omega_p=0$, $v_c=0$, $\omega_c=0$, and (c) $n_x=n_e=6.2$, $n_y=n_z=n_o=4.8$, $\omega_p=0.15\omega_{p0}$, $v_c=0.02\omega_{pl}$, $\omega_c=0.8\omega_{pl}$, respectively. The red shaded regions indicate PBGs.

In Figs.1(a) and (b), we plot the dispersive curves for 3D MPPCs with fcc lattices containing a refractive index contrast of $n$=4.8 ($n_z$=$n_x$=$n_y$=4.8), $f$=0.3, but with different $\omega_p$, $\omega_c$ and $v_c$, respectively. As shown in Fig.1(a), if $\omega_p$=0, $\omega_c$=0 and $v_c$=0, the magnetized plasma can be looked as the air. The complete PBGs can not be found. The cases of 3D MPPCs ($\omega_p$=0.15$\omega_{p0}$, $v_c$=0.02$\omega_{pl}$, $\omega_c$=0.8$\omega_{pl}$) also can be seen in Figs.1(b). The similar conclusion also can be draw, there are not the complete PBGs. The band structures close at $U$ and $W$ points in fcc lattices. Of course, the some stop band gaps can be found in the some directions of symmetry such as $\Gamma$-$X$ and $\Gamma$-$L$ directions. This can be explained by the relative dielectric constant of isotropic dielectric is not enough to open a band gap [23]. In order to obtain the complete PBGs, we can use the uniaxial material (anisotropic material) to replace the isotropic dielectric to form 3D MPPCs with fcc lattices. In Figs.1(c) and (d), we display the dispersive curves of 3D MPPCs doped by the type-1 uniaxial material as $n_x$=$n_e$=6.2, $n_y$=$n_z$=$n_o$=4.8 and $f$=0.3 but with different plasma frequency, plasma cyclotron frequency and plasma collision frequency, respectively. In Fig.1(c), it is clearly seen that the two complete PBGs can be obtained as the type-1 uniaxial material is introduced. The first two PBGs cover 0.3577-0.3849 ($2\pi c/a$) and 0.5179- 0.5242 ($2\pi c/a$), respectively. The bandwidths are 0.0272 and 0.0063 ($2\pi c/a$), respectively. . Fig.1(d) shows that the edges of PBGs are upward to higher frequencies and a flatbands region can be obtained as the Faraday effects of magnetized plasma are considered (the external magnetic field is introduced). The main reason for formed the flatbands is because the existence of surface plasmon modes which stem from the coupling effects between the plasma [21, 22]. Compare to Fig.1(a), the edges of first two PBGs shift upward to higher frequencies, and the bandwidths of first two PBGs are tuned notably The first two PBGs span from 0.3677 to 0.3945 ($2\pi c/a$), from 0.5264 to 0.5327 ($2\pi c/a$), and bandwidths are 0.0268 and 0.0063 ($2\pi c/a$), respectively. The flatbands region is located 0.12-0.2068 ($2\pi c/a$). This can be explained by the cutoff frequency of right circular polarization and left cutoff frequency ( $f_R$ and $f_L$ ) since the $f_L$ and $f_R$ are nearly corresponding to the lower and upper edge frequencies of flatbands region, respectively [21, 22]. As the Faraday effects are considered, the cutoff frequency of right circular polarization and left cutoff frequency are $f_R$=0.2068 ($2\pi c/a$) ( $f_R = \omega_c / 2 + \sqrt{\omega_c^2 / 4 + \omega_p^2}$ ) and $f_L$=0.12 ($2\pi c/a$) ( $f_L = \omega_c$ ) [21, 22], respectively. Therefore, the flatbands are caused by the magnetized plasma itself. As mentioned above, it is found that the inclusion of type-1 uniaxial material in 3D MPPCs with fcc lattices can open band gaps near high-symmetry points and the anisotropic PBGs (complete PBGs) can be achieved compared to the 3D MPPCs containing the conventional isotropic dielectric. As the magnetized plasma is introduced in 3D dielectric PCs (the Faraday effects of magnetized plasma are considered), the PBGs can be tuned. Consequently, introducing the uniaxial material into the 3D MPPCs with high-symmetry lattices can obtained the complete PBGs.

### B. Effects of the filling factor on anisotropic PBGs

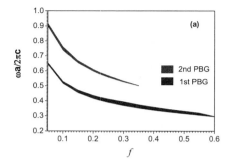

Fig.2. The effects of filling factor on the first two PBGs for such 3D MPPCs with $\omega_p$=0.15$\omega_{p0}$, $v_c$=0.02$\omega_{pl}$, $\omega_c$=0.8$\omega_{pl}$, $n_y$=$n_e$=6.2, $n_x$=$n_z$=$n_o$=4.8, respectively. The shaded region indicates the PBGs.

In Fig.2, we plot the dependences of the properties of anisotropic PBGs for 3D MPPCs with fcc lattices containing type-1 uniaxial material on the anisotropic dielectric filling factor $f$ as $\omega_p$=0.15$\omega_{p0}$, $v_c$=0.02$\omega_{pl}$, $\omega_c$=0.8$\omega_{pl}$, $n_o$=4.8 and $n_e$=6.2, respectively. The shaded regions indicate the PBGs. Fig.2 reveals that the edges of first two PBGs are downward to lower frequencies with increasing $f$. The both bandwidths of first two PBGs increase first then decrease with increasing $f$ If anisotropic dielectric filling factor is less than 0.05, there does not exist the 1st PBG, and will disappear as $f$ is larger than 0.6. The 2nd PBG will never appear until $f$ is larger than 0.35. As $f$ is increased from 0.05 to 0.6, the maximum bandwidths of first two PBGs are 0.0268 and 0.0242 ($2\pi c/a$), which can be found at the cases of $f$=0.3 and 0.1, respectively. Compared to the case of $f$=0.35, the frequency ranges of both PBGs are increased by 0.0028 and 0.0219 ($2\pi c/a$), respectively. Thus, the first two PBGs can be tuned by the filling factor of anisotropic dielectric spheres. This can be explained in physics that increasing anisotropic dielectric filling factor means the space averaged dielectric constant of such 3D MPPCs becoming larger [21, 22]. Fig.2 also reveals that the relative bandwidth of both PBGs increase first then decrease with increasing $f$. The maximum relative bandwidths of first two PBGs are 0.0703 and 0.0324, which can be found at the cases of $f$=0.3 and 0.1, respectively. As mentioned above, the filling factor of anisotropic dielectric spheres is an important parameter which need be chosen. It also is noticed that if anisotropic dielectric filling factor is small enough and close to null, the 3D MPPCs can be looked as a magnetized plasma block. The flatbands regions will disappear.

### IV. CONCLUSION

In summary, the band Faraday effects in the 3D dispersive PCs composed of the homogeneous anisotropic dielectric spheres (the uniaxial material) immersed in the magnetized plasma background with fcc lattices are theoretically investigated by the plane wave expansion (PWE) method, as the Faraday effects of magnetized plasma are considered. The equations for calculating the anisotropic PBGs in the first irreducible Brillouin zone are theoretically deduced. Based on the calculated results, some conclusion can be drawn. Compared with the same structure composed by the isotropic dielectric spheres in the air, the 3D MPPCs doped by the uniaxial material can obtain complete PBGs and one flatbands region also can be achieved. The flatbands caused by the existence of surface plasmon modes which stem from the coupling effects between the magnetized plasma. The relative bandwidths of first two PBGs will increase first then decrease as the anisotropic dielectric filling factor is increased in a certain range. It also is noticed that if plasma filling factor is large enough and close to one, the 3D MPPCs can be seen as a magnetized plasma block. The flatbands region will disappear. The 1st PBG has a larger relative bandwidth compared to 2nd PBG. As mentioned above, we can acquire the PBGs in expected frequency and take advantage of the uniaxial material to obtain the complete PBGs for 3D MPPCs with fcc lattices by selecting the appropriate parameters as the Faraday effects are considered.

## ACKNOWLEDGMENT

This work was supported by the Fundamental Research Funds for the Central Universities and the Funding of Jiangsu Innovation Program for Graduate Education (Grant No.CXZZ11_0211).

## REFERENCES

[1] E.Yablonovitch, "Inhibited spontaneous emission of photons in solidstate physics and electronies," *Phys.Rev.Lett.*, Vol.58, pp.2059-2061,1987.

[2] S.John, "Strong localization of photons in certain disorder ed dielectric superlattices," *Phys.Rev.Lett.*, Vol.58, pp.2486-2489,1987.

[3] A. Chutinan , and S. Noda, "Highly confined waveguides and waveguide bend in three-dimensional photonic crystals," *Appl. Phys. Lett.* Vol.75, pp.3739-3741, 1999.

[4] C. Fan, J. Wang, S. Zhu, J. He, P. Ding and E. Liang, "Optical properties in one-dimensional graded soft photonic crystals with ferrofluid", *J. Opt.* Vol.15, 2013, Art.ID 055103.

[5] H. F. Zhang, S. B. Liu and X. K. Kong, "Defect mode properties of two-dimensional unmagnetized plasma photonic crystals with line-defect under transverse magnetic mode," *Acta Phys. Sin,* Vol. 60, pp 025215-1-025215-8, 2011.

[6] H. F. Zhang, S. B. Liu, X. K. Kong, B. R. Bian and Y. Dai, "Omnidirectional photonic band gaps enlarged by Fibonacci quasi-periodic one-dimensional ternary superconductor photonic crystals," *Solid State Commun.* Vol.152, pp2113-2119, 2012.

[7] V. Kuzmiak, and A. A. Maradudin, "Distribution of electromagnetic field and group velocities in two-dimensional periodic systems with dissipative metallic components," *Phy. Rev. B*, Vol.58, pp.7230-7251, 1998.

[8] H. Hojo and A. Mase, "Dispersion relation of electromagnetic waves in one dimensional plasma photonic crystals," *Plasma Fusion Res.*, Vol. 80, pp.89-90, 2004.

[9] O. Sakai and K. Tachibana, "Plasma as metamaterials: a review," *Plasma Sources Sci. Technol.*, Vol.21, pp. 013001-1-013001-18, 2012.

[10] H. F. Zhang, S. B. Liu, X. K. Kong, L. Zou, C. Z. Li, and W. S. Qing, "Enhancement of omnidirectional photonic band gaps in one-dimensional dielectric plasma photonic crystals with a matching layer," *Physics Plasmas*, Vol. 19, pp. 022103-1-022103-7, 2012.

[11] L. Shiverhwari, "Zero permittivity band characteristics in one-dimensional plasma dielectric photonic crystals," *Optic-Int.J. Light and Electron Optic*, Vol.122, pp.1523-1526, 2011.

[12] H. F. Zhang, S. B. Liu, X. K. Kong, L. Zhou, C. Z. Li and B. R. Bian, "Enlarged omnidirectional photonic band gap in heterostructure of plasma and dielectric photonic crystals," *Optic-Int.J. Light and Electron Optic*, Vol.124, pp. 751-756, 2013.

[13] H. F. Zhang, S. B. Liu, X. K. Kong, L. Zhou, B. R. Bian, and H. C. Zhao, "Properties of omnidirectional photonic band gap in one-dimensional staggered plasma photonic crystals," *Optic Commun.* vol.285, pp.5235-5241, 2012.

[14] H. F. Zhang, X. K. Kong, and S. B. Liu, "Analsys of the properties of tunable prohibited band gaps for two-dimensional unmagnetized plasma photonic crystals under TM mode," *Acta Phys. Sin,* Vol. 60, 2011, 055209.

[15] H. F. Zhang, S. B. Liu and X. K. Kong, "Photonic band gaps in one-dimensional magnetized plasma photonic crystals with arbitrary magnetic declination," *Phys.Plasma*, Vol. 19, 2012, Art.ID 122103.

[16] V. L. Ginzburg, *The Propagation of Electromagnetic Wave in Plasma*. Oxford, U.K: Pergamon, 1970.

[17] H. F. Zhang, L. Ma and S. B. Liu, "Study of periodic band gap structure of the magnetized plasma photonic crystals," *Optoelectr. Lett.*, vol.5, pp112-116, 2009.

[18] H. F. Zhang, L. Ma and S. B. Liu., "Defect mode properties of magnetized plasma photonic crysrals", *Acta Physica Sinica*, vol. 58, pp01071-01075, 2009.

[19] L. Qi, Z. Yang and T. Fu, "Defect modes in one-dimensional magnetized plasma photonic crystals with a dielectric defect layer," *Physics Plasma*, Vol. 19, 2012, Art.ID 012509.

[20] S. M. Hamidi, "Optcal and magneto-optical properties of one-dimensional magnetized coupled resoner plasma photonic crystals," *Phys. plasma*, vol.19, 2012, Art.ID 012503.

[21] H. F. Zhang, S. B. Liu, and B. X. Li, "The properties of photonic band gaps for three-dimensional tunable photonic crystals with simple-cubic lattices doped by magnetized plasma," *Optics & Laser Technol.*, Vol. 50, pp.93-102, 2013.

[22] Zhang H. F., S. B. Liu, H. Yang and X. K. Kong, "Analysis of band gap in dispersive properties of tunable three-dimensional photonic crystals doped by magnetized plasma," *Phys.Plasma*, Vol. 20, 2013, Art.ID 032118.

[23] Z. Y. Li, J. Wang, and B. Y. Gu, "Creation of partial gaps in anisotropic photonic-band-gap structures," *Phy. Rev. B*, vol.58, pp.3721-3729, 1998.

[24] N. Malkova, S. Kim, T. Dilazaro and V. Gopalan, "Symmetrical analysis of complex two-dimensional hexagonal photonic crystals, " *Phys.Rev.B*, Vol.67, 2003, Art.ID 125203.

[25] Z. Y. Li, B. Y. Gu and G. Y. Yang, "Large absolute band gap in 2D anisotropic photonic crystals," *Phys.Rev.Lett.* Vol.81, pp.2574-2577, 1998.

[26] H. F. Zhang, S. B. Liu, X. K. Kong, L. Zhou, C. Z. Li, and B. R. Bo, "Comment on "Photonic bands in two-dimensional microplasma array. I. Theoretical derivation of band structures of electromagnetic wave"[J.Appl.Phys. 101,073304 (2007)]," *J.Appl. Phys.*, Vol.110, 2011, 026104.

# Miniaturized Frequency Selective Surface with Bionic Structure

W. Jiang, T. Hong, and S. X. Gong
Xidian University
Taibai south road 2#
Xi'an, Shaanxi 710071 China

*Abstract*- **A novel miniaturized frequency selective surface (FSS) is proposed. The size reduction is achieved by using the model of leaf arrangement. Compared with conventional cross FSS, the operation frequency of this novel bionic FSS is changed from 10.4 GHz to 4.74 GHz at the same size. Both theoretical and experimental investigations are carried out. It is observed that the miniaturized FSS presents excellent frequency stability with the increase of the incident angle. The proposed FSS can be a candidate for FSS whose miniaturization is required.**

## I. INTRODUCTION

Frequency selective surface (FSS) has attracted much attention in the past few decades for their spatial filtering characteristic. FSS is infinite periodically arranged metallic-patch elements or aperture elements within a metallic screen. It can be used in various applications, such as sub-reflectors of the frequency reuse system, band-pass radomes for radar cross section controlling and so on[1-2]. FSS should be an infinite array theoretically but in practical applications FSS is finite. It is very necessary to use sufficient elements to keep the characteristics of infinite FSS. However, when the operation frequency is low, the size of elements will be so large that it is difficult to fill enough elements in a reasonable size. Miniaturization is of great significance at low frequency [3-6]. The proposed novel miniaturized FSS in this letter is designed based on the model of leaf arrangement. Both theoretical and experimental investigations are carried out. The results show that compared to the reference cross FSS, the novel bionic FSS has lower resonance frequency and favorable related performances. Hence, applying bionics principle to FSS is proved feasible, which will serve as a good candidate for the future FSS design.

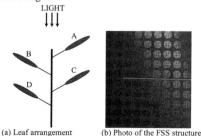

(a) Leaf arrangement    (b) Photo of the FSS structure

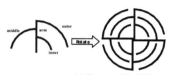

(c) Element of the FSS
Figure 1. Gemetry of the proposed bionic FSS

## II. FSS DESIGN

A novel miniaturized frequency selective surface (FSS) is proposed. The size reduction is achieved by using the model of leaf arrangement. The proposed element of FSS is shown in Fig. 1. It is the transformation of traditional cross cell. Its arm has three branches: outer, middle, and inner ones. The three branches are arranged as the way that leaves generate along the arm of a tree. The arc shaped branches lengthen the electrical length of FSS. Then the single arm is rotated $90°$, $180°$, and $270°$, another three arms will appear. The four arms are joined together, and the FSS element is obtained. As shown in the Fig. 1, the middle one located in the opposite side of the other two branches, which expands the distance not only between the outer and inner branches, but also between itself and the other two branches. Hence, the coupling between the three branches is avoided within the small area. A set of optimal dimensions is shown as: $L = 11mm$ ; $w = 0.25mm$ ; $w_1 = 4.5mm$ ; $w_2 = 3.39mm$ ; $w_3 = 2.28mm$ ; $\theta_1 = \theta_2 = \theta_3 = 10°$ . The proposed FSS array has a rectangular lattice and a substrate with relative permittivity of 2.65, a loss tangent of 0.001 and a thickness of 1mm. The picture of the fabricated FSS is also shown in Fig. 1.

## III. RESULTS AND DISCUSSION

The compared frequency response of the proposed FSS and the reference cross FSS with a same size are presented in Fig. 2. The figure shows that the center frequency of the two FSS is 3.9 GHz and 10.4GHz, separately. Compared with the reference cross FSS, the proposed bionic FSS has lower resonance frequency with the same size. It is obtained that the introduction of the branches expands the electrical size of the FSS greatly, which makes the resonance frequency of bionic FSS lower than that of cross FSS magnificently. The application of interactive branches increases the distance and avoids the coupling between the adjacent branches. The transmission characteristics at different oblique incident

978-7-5641-4279-7

angles of 0°, 30°, 45°, and 60° are also obtained and shown in Fig. 3. The results show that the proposed bionic miniaturized FSS has excellent center frequency stability. Fig. 4 shows the schematic picture of the FSS measurement setup. Measurement is implemented in the anechoic chamber. Measured and simulated frequency response of the proposed FSS is shown in Fig. 5. The theoretical and experimental results are in good agreement except for the noise. The measured results validate the correctness of the theoretical analysis.

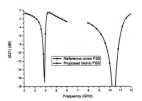

Figure 2. Frequency response of the two FSS

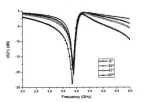

Figure 3. The transmission characteristics at different angles

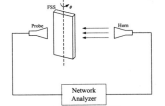

Figure 4. The schematic picture of the FSS measurement setup

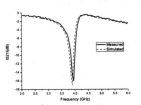

Figure 5. Measured and simulated results of the proposed FSS

## IV. CONCLUSION

In this paper a novel miniaturized FSS with bionic structure is proposed. Compared to the conventional cross FSS, the centre frequency of the proposed FSS is reduced 62.5%. The size reduction is realized by the introduction of the bionic structure. Experimental and theoretical results agree well with each other. The miniaturized FSS has excellent centre frequency stability. It will be a good candidate for FSS radomes. This paper supplies a novel direction to the future design of FSS with or without a requirement of miniaturization.

### ACKNOWLEDGMENT

The authors are especially grateful to Anechoic Chamber of National Laboratory of Antennas and Microwave Technology of China for providing measuring facilities. This work is supported by "the Fundamental Research Funds for the Central Universities" (No. K5051302024，K5051202010) and the financial support from national natural science fund of P. R. China (No. 61201018).

### REFERENCES

[1] Martynyuk, E., and Lopez, J. I. M., "Frequency-selective surfaces based on shorted ring slots," *Electron. Lett.*, 2001, 37, (5), pp. 268-269.

[2] Zheng, S. F., Yin, Y. Z., Zheng, H. L., Liu, Z. Y., Sun, A. F., "Convoluted and interdigitated hexagon loop unit cells for frequency selective surfaces," *Electron. Lett.*, 2011, 47, (4), pp. 233-235.

[3] Yang, H. Y., Gong, S. X., Zhang, P. F., Zha, F. T., and Ling, J., "A novel miniaturized frequency selective surface with excellent center frequency stability," *Microwave Opt. Technol. Lett.*, 2009, 51, (10), pp. 2513-2516.

[4] Yang, H. Y., Gong, S. X., Zhang, P. F., and Guan Y., "A compound frequency selective surface with quasi-elliptic band-pass response," *Electron. Lett.*, 2010, 46, (1), pp. 7-8

[5] Munk, B. A., "Frequency selective surfaces: theory and design," (Wiley, New York, 2000).

[6] Sarabandi, K., and Behdad, N., "A frequency selective surface with miniaturized elements," *IEEE Trans. Antennas and Propag.*, 2007, 55, (5), pp. 1239-1245.

# Artificial Magnetic Conductor based on InP Technology

Yan Jun Dai, Guo Qing Luo*, and Ya Ping Liang

The Key Laboratory of RF Circuits and System of Ministry of Education,

Hangzhou Dianzi University, 310018, China

*Abstract-* **Artificial magnetic conductor (AMC) is a kind of periodic structures and it introduces a zero degrees reflection phase shift to incident wave. In order to reduce size of artificial magnetic conductor and increase their working bandwidth, InP process is used to design AMC in this paper. InP material has advantages of large forbidden bandwidth, high electron mobility, negative resistance effect etc. AMC based on InP process has been analyzed and results show that the proposed AMC has a good bandwidth compared with that of CMOS process for its smooth reflection phase variation and its whole size is greatly reduced.**

*Keywords-* **AMC, InP, reflection phase.**

## I.    INTRODUCTION

Photonic band-gap (PBG), as a type of artificial periodic dielectric structure, has been involved in research fields of optical, electromagnetic, and acoustic. It is also called as electromagnetic band-gap (EBG) structures in microwave and millimeter wave fields. Research of EBG technology was beginning in the late 1990s. D. Sievenpiper presented high impedance surface (HIS) by using periodic arrangement of metalized via to connect with the ground plane. It has unique performance of surface wave propagation prohibition within the specific frequency band. For this characteristic of high impedance surface similar to magnetic conductor, it also be called artificial magnetic conductor (AMC) [1]. Artificial magnetic conductor can introduce effective isolation between antenna and its lossy dielectric substrate. Antenna radiation efficiency is improved and back lobe level is reduced for its characteristics of surface wave suppression. Artificial magnetic conductor shows a zero degrees reflection phase shift to incident wave within a certain frequency band. This characteristic make AMC can be used as reflector with nearly zero thickness replacing a conventional backed ground with a thickness of one-quarter wavelength [3]-[6]. AMC has been investigated by many researchers. Some novel AMCs have been presented, such as uniplanar compact photonic band gap (UC-PBG), which is easy to be fabricated because metallized vias have been removed.

In this paper, an AMC structure based on InP technology is presented. InP is one of the most important compound semiconductor materials. It is a new generation of functional materials after Si, GaAs. The InP material preparation is difficult and expensive. The researches of components and devices based on InP technology are far less than those of Si, GaAs, and GaN, etc. However, due to the unique advantages of InP materials, such as large forbidden bandwidth, high electron mobility, significantly negative resistance effect, more its investigations become more and more popular. InP materials is widely used in optical fiber communication. It is also an ideal substrate material of microwave and millimeter wave devices and devices for high speed and high frequency applications.

## II.    DESIGN

InP process is simpler than the common CMOS process, whose cross section is shown in Fig.1. There are only two metal layers in an InP structure and the proposed AMC is constructed at the bottom metal layer M1. From Fig.1 it can be found that the bottom layer is an InP substrate, its dielectric constant $\varepsilon_r$=12.4. BCB is a high performance dielectric material, whose curing temperature is relatively low and its dielectric constant is low and stable. Its dielectric constant is $\varepsilon_r$=2.65, loss tangent $\delta$=0.0008. Thicknesses of each layer are listed in Table I.

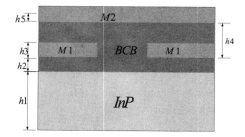

Fig.1 Cross section of an InP structure.

978-7-5641-4279-7

Table 1. Thicknesses of different materials.

| material | Thickness (um) |
|---|---|
| BCB | h2+h4=4.5 |
| M1 | h3=1 |
| M2 | h5=2 |
| INP | h1=620 |

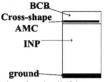

Fig. 2(a).Side view of a unit cell of the proposed AMC.

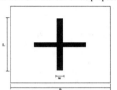

Fig.2(b).Top view of a unit cell of the proposed AMC.

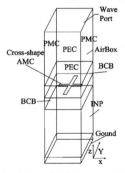

Fig.3. Simulation model of the proposed AMC.

In this paper, a periodic cross-shaped array presented in [7] is adopted in AMC design. Its side view and top view are shown in Fig.2 (a) and (b). In practice an AMC is constructed by finite periodic element array. But in simulations it should be treated as infinite periodic element array. It can be accomplished by introducing image boundary conditions around a single unit cell. Simulation model of the proposed AMC is shown Fig.3, in which the incident plane wave is propagating in the negative z direction with its electric field in the positive y direction. The image boundary conditions are defined by setting the two parallel x-z surfaces as perfect electric conductors (PEC) and setting the two parallel y-z surfaces as perfect magnetic conductors (PMC). Wave port at the negative z direction end is used as excitation source. A

reference plane calculating the phase of reflection coefficient is set at the surface of the cross-shape AMC.

III. RESULTS AND ANALYSIS

AMC constructed by periodic metallic cross-shape elements at the bottom metal layer M1 has been investigated and a sample operating at 35GHz has been presented. Its geometrical parameters are listed in Table 2.

Table 2. Parameters of the proposed InP AMC (Unit: um).

| L | W | P |
|---|---|---|
| 40 | 20 | 100 |

Phase responses of reflection coefficient of the proposed AMC are shown in Figs.4. From those figures it can be found that reflection coefficient phase of the proposed AMC is approaching to zero at the frequency of 34.5GHz. AMC operating bandwidth is defined as the frequency range in which its reflection coefficient variation from $-90^0$ to $90^0$ [8]. From these figures we can find that bandwidth of the proposed AMC is more than 12GHz (from 28GHz to 40GHz). Variations of the reflection coefficient phase with different geometrical parameters of the proposed AMC have been studied. From Figs.4 it can be found that length of the cross shape L has the largest influence on the reflection coefficient phase and width of the cross shape W has the least influence on the reflection coefficient phase. Operating frequency and bandwidth of the proposed AMC increase with length L or width W of the cross shape reduction, or periodic space P between two elements increment.

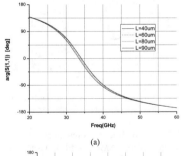

(a)

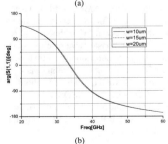

(b)

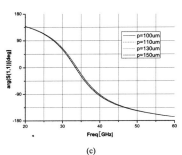

(c)

Fig.4. Phases of reflection coefficients vs. different parameters.

(a) L: length of cross shape.

(b) W: width of cross shape.

(c) P: periodic space between two cross shape element.

## IV. CONCLUSION

AMC constructed by a periodic cross shape array based on InP technology has been studied in this paper, whose phase variation curve is smoother than that of a corresponding AMC based on CMOS technology. The whole size of an InP AMC is far less than that of the corresponding CMOS AMC. These advantages make the AMC based on InP technology be more suitable for application with a lower cost.

## V. ACKNOWLEDGEMENT

This work was supported in part by ZJNSF-R1110003, NCET-09-0910, FANEDD-201045 and Zhejiang Province Oversea Returnee Research Project Fund.

As a reference, an AMC comprised of periodic cross shape element based on standard CMOS process is presented, whose geometrical parameters are listed in Table 3. Comparison between phases of reflection coefficients of the proposed two AMC based on InP and CMOS processes is shown in Fig.5.

Table 3. Parameters of the compared CMOS AMC (Unit: um).

| L | W | P |
|---|---|---|
| 300 | 200 | 302 |

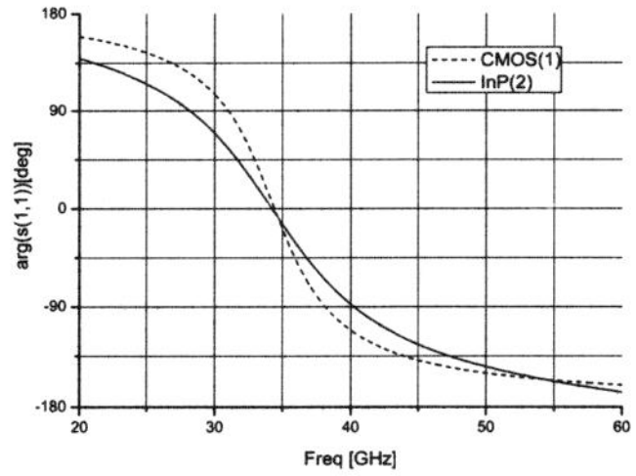

Fig.5. Phases of reflection coefficients of the proposed cross shape AMC based on InP and CMOS technology

In Fig.5, the dot line represents reflection coefficient phase of the cross shape AMC based on CMOS technology, and the solid line represents reflection coefficient phase of the cross shape AMC based on InP technology. From the figure it can be found that the phase curve of cross shape AMC based on InP technology is more smoother than that of the AMC based on CMOS technology. Operating bandwidth of the InP AMC is 12GHz (28~40GHz), which is about 50% more than 8GHz (30~38GHz) of the CMOS AMC. Furthermore, by comparing Table 2 and 3 it also can be found that the whole size of the proposed InP AMC is far less than that of the proposed CMOS AMC.

## REFERENCES

[1] D. Sievenpiper, "High-impedance electromagnetic surfaces with a forbidden frequency band," *IEEE Trans. Microw. Theory Tech.*, vol. 47, no.11, pp.2059-2074, Nov. 1999.

[2] Li Z, Rahmat-Samii Y, "PBG, PMC and PEC surface for antenna application: a comparative." *IEEE AP-S Dig.* pp. 674-677

[3] F. Yang, Y. Rahmat-Samii, "A low-profile circularly polarized curl antenna over an electromagnetic band-gap ( EBG ) surf ace ." *Microw. Opt. Technol. Lett.*, vol.31, no.4, pp.264-267, Apr. 2001.

[4] Zhang Y, Hagen J, Younis M, Fischer C, Wiesbeck W. "Planar artificial magnetic conductors and patch antennas," *IEEE Trans. Antenna Propag.*, vol 51, pp.2704-2712, 2003.

[5] Du Z, Fu J S, Gao B, Feng Z, "A compact planar inverted-F antenna with a PBG-type ground plane form mobile communications." *IEEE Trans. Vehicular Tech.*, vol.52, pp.483-489, 2003.

[6] Mckinzie III WE, Fahr R. R., "A low profile polarization diversity antenna built on an artificial magnetic conductor," *IEEE Antennas Propagat. Soc. Int. Symp.*, vol.1, pp.762-765, 2002.

[7] M. A. Hiranandani, A. B Yakovlev, A. A. Kishk, "artificial magnetic conductor realised by frequency-selective surfaces on a grounded dielectrical slab for antenna applications," *IEE Proc. Microw., Antennas Propag.*, pp.487-493, 2006.

[8] D. Sievenpiper, L. Zhang, et al. "High-impedance electromagnetic surfaces with a forbidden frequency band," *IEEE Trans. Microw. Theory Tech.*, vol.47, no.11, pp.2059-2074, Nov. 1999.

# A Compact Triple mode Metamaterial Inspired-Monopole Antenna for Wideband Applications

Bashir D. Bala, Mohamad Kamal A. Rahim, and Noor Asniza Murad

Communication Engineering Department,
Faculty of Electrical Engineering,
Universiti Teknologi Malaysia (UTM),81310, Johor Bahru, Malaysia.
bashirdbala@yahoo.com, mkamal@fke.utm.my, asniza@fke.utm.my,

*Abstract*— A compact triple mode metamaterial (MTM) inspired antenna for wideband is presented. The antenna is based on composite right/left handed transmission line (CRLH-TL) which employs MTM loading on a conventional monopole to attain a certain degree of miniaturization. The antenna has triple mode of operations- negative order resonance (NOR), zero order resonance (ZOR) and positive order resonance (POR) modes. Two modes are merged into a single pass band (3.1 - 6.0 GHz) to obtain a dual band. The over size of the antenna is 33 x 26 x 1.6 mm³. A high level of miniaturization is obtained when the return loss of loaded and unloaded metamaterial antenna is compared. Simulated return loss shows that the proposed antenna is suitable for Wi-Fi (2.4 GHz) and WiMAX (3.5, 5.2-5.8 GHz).

*Index terms* — metamaterial, dual band, Monopole, triple mode.

## I. INTRODUCTION

In recent years, the concept of composite right/left handed (CRLH) metamaterial design have been widely applied to RF devices [1]. Basically, metamaterials (MTM's) represents novel electromagnetic materials with negative refractive index (NRI) over a specify range of frequency [2–5]. Microwave researchers use these unusual electromagnetic phenomena to design antennas and other RF components for improved performance. [6]. MTM antennas can be designed based on resonant and non-resonant approach [7]; the resonant approach requires resonant elements such as split ring resonators (SRR) and complementary split ring resonators (CSRR). This approach requires a very large number of elements which makes the design bulky [8] and undesirable where miniaturization is a priority. However the non-resonant or transmission line approach offers a greater advantage of miniaturization due to its zeroth-order resonance frequency [9]. Several CRLH transmission line (TL) MTM antennas have been presented in literature which employed MTM loadings. In [10] a triple-band monopole antenna loaded with CRLH unit cell was presented. The first two narrow bands 0.925 and 1.227 GHz which represent the negative and zero order modes occurred due to the loaded unit cell and the third band 2.5 GHz was due to the monopole itself. Because of decrease in the resonance frequency when loaded with MTM unit cell, the design is considered to be compact. In [11] a wideband zero-order MTM antenna was presented. The antenna utilizes mushroom unit cell where the negative and zero order modes are merged to obtain a wider bandwidth although the gain is low. A printed monopole antenna that employed NRI-TL loading was reported in [12]. In the design a single MTM unit cell was integrated directly onto the antenna which transforms it to a folded monopole with resonance frequency around 5.5 GHz and the unit cell contribute the lower frequency bands.

In this paper, a compact, dual band, triple mode MTM antenna is proposed. The design is inspired by [12] with modification on the top monopole patch. An inter-digital capacitor (IDC) is etched on the top patch to create a negative order resonance (NOR) at 2.4 GHz. Hence with proper tuning of the constitutive parameters, the zero order resonance (ZOR) can be merged to the positive order resonance (POR). The antenna resonates at NOR (2.4 GHz) and a single pass band (3.1 to 6.0 GHz) when ZOR and POR are merged. Hence a triple mode and dual band is thus obtained.

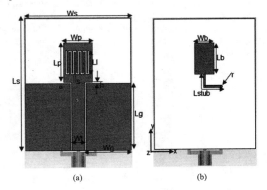

(a)         (b)

Fig.1. Geometry of the proposed metamaterial antenna. (a) front view (b) back view

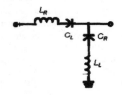

Fig. 2. Equivalent circuit diagram

978-7-5641-4279-7

Table 1. Parameters of the proposed antenna in (mm)

| Ls | Ws | Lp | Wp | Lg | Wg | h |
|----|----|----|----|----|----|----|
| 33 | 26 | 10 | 7 | 16.7 | 11.1 | 0.3 |
| Wt | LI | S | Lb | Wb | Lstub | r |
| 3.2 | 6 | 0.3 | 8 | 5 | 8.1 | 0.5 |

## II. ANTENNA DESIGN

### A. ANALYSIS OF THE LOADED UNIT CELL

The geometrical configuration of the proposed metamaterial antenna is shown in Fig. 1. The antenna is simulated on FR4 substrate of permittivity 4.3, tangential loss of 0.025 and thickness of 1.6 mm. The antenna is a printed monopole type with loaded MTM unit cell. The idea of loading MTM unit cell onto a conventional monopole is to impose a left handed property onto the antenna so that the overall design will have a reduced form factor. The antenna is coplanar waveguide (CPW) fed by a 50 Ω transmission line. Fig. 2 show the equivalent circuit model of the proposed design, where an inter-digital capacitor is loaded on top of the patch which is responsible for the series capacitance $C_L$. The shunt capacitance $C_R$ was formed between the top monopole and the bottom patch (Lb x Wb). The series inductance $L_R$ is the inductance along the monopole length and the shunt inductance $L_L$ is the inductance of the stub (Lstub) and the via. Table 1 shows the parameters of the antenna. Four resonance frequencies can be determined from the equivalent circuit model [7]:

$$f_{LH} = \frac{1}{4\pi\sqrt{L_L C_L}} \qquad (1)$$

$$f_{sh} = \frac{1}{2\pi\sqrt{L_L C_R}} \qquad (2)$$

$$f_{se} = \frac{1}{2\pi\sqrt{L_R C_L}} \qquad (3)$$

$$f_{RH} = \frac{1}{\pi\sqrt{L_R C_R}} \qquad (4)$$

Where $f_{LH}, f_{sh}, f_{se}$ and $f_{RH}$ are left handed, shunt, series and right handed resonance frequencies respectively. $f_{sh}$ and $f_{se}$ are the frequencies where ZOR occurs.

### B. ANTENNA PARAMETERS TUNING

In order to show the effect of the loaded MTM unit cell on the antennas small form factor or miniaturization, a conventional or unloaded monopole antenna is design and simulated. As a control experiment, the unloaded antenna has same dimension and material properties as the loaded antenna. The return loss of loaded and unloaded antennas is compared as shown in Fig 3. Three resonant modes are obtained: first NOR, ZOR and first POR modes. The first NOR resonates at 2.4 GHz for the Wi-Fi and the other two modes are merged at

3.1 GHz and 6.0 GHz which is suitable for WiMAX (3.5, 5.2-5.8 GHz).

Figure 4 shows the effect of increasing the capacitance $C_L$ on the resonance frequency of the first NOR mode. From equation (1), an increase in the length LI of the inter-digital capacitor from 5.2-6.5 mm shifts the resonance frequency from 2.6- 2.2 GHz. While the ZOR is unaffected by $C_L$ but only suppressed. Length of the stub Lsub together with the via accounts for the shunt inductance $L_L$. Equation (2) and (3) gives the resonance frequency of the ZOR mode. Decrease in the length of the stub from 11.1 – 8.1 mm shifts the ZOR mode from 2.8- 3.25 GHz. However, ZOR mode is largely affected by $L_L$. Hence ZOR mode is merged to the POR mode and this result into a wider bandwidth as shown in Fig. 5.

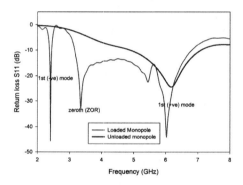

Fig. 3. Simulated return loss of the loaded and unloaded MTM antenna

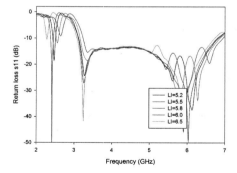

Fig. 4. Effect of $C_L$ of first NOR mode

GHz respectively. Low gain at 2.4 GHz is as a result of out of phase current along the inter-digital capacitor which cancels out the radiation in the far field.

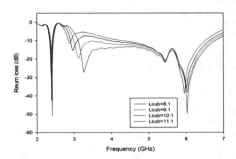

Fig. 5. Effect of the $L_L$ on the ZOR mode

### III. DISCUSSION OF THE RESULTS

Fig. 6 shows the simulated radiation patterns of the proposed antenna. Patch and monopole-like radiation patterns are obtained as shown in Fig. 6 (a) and (b) for both E and H-plane respectively. Figure 7 shows the surface current distributions of the proposed antenna for different frequency modes. At the first NOR mode (2.4 GHz), high concentration of surface current can be noticed across the inter-digital capacitor as shown in Fig. 7 (a). This shows the dependency of the resonance frequency largely on the length of the IDC. Also at ZOR mode (3.1 GHz), the shunt stub Lsub is responsible for the resonance at that frequency as shown in Fig. 7(b). However, Fig 7(c) shows the first POR mode where the concentration of surface current is along the entire monopole length.

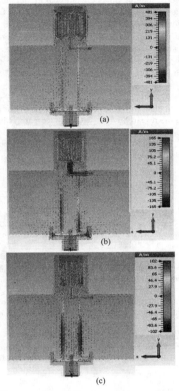

Fig.7. surface current distributions at (a) 2.4 GHz (b) 3.1 GHz (c) 6.0 GHz

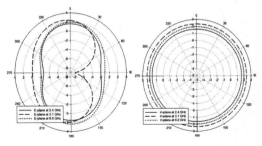

Fig. 6. Simulated radiation patterns

The simulated realized gain according to the operation band is presented in Fig. 8. The antenna gains vary from (-11.42 - 2.53) dBi within the frequency band of 2 - 7 GHz. The peak realized gains of -2.53 and 1.43 dBi are obtained at 2.4 and 3.1

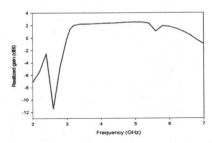

Fig.8. Simulated peak realized gains

## IV. CONCLUSION

A compact triple mode MTM inspired antenna for wideband is presented. The antenna is based on CRLH-TL which employs MTM loading on a conventional monopole to attain a certain degree of miniaturization. The antenna has triple mode of operations: first NOR, ZOR and first POR modes. Two modes are merged into a single pass band to obtain a dual band. The over size of the antenna is 33 x 26 x 1.6 mm3. Simulated return loss shows that the proposed antenna is suitable for Wi-Fi (2.4 GHz) and WiMAX (3.5, 5.2-5.8 GHz).

## ACKNOWLEDGMENT

The authors thank the Ministry of Higher Education for supporting the research work (MOHE), Research Management Centre (RMC) and Communications Engineering Department, Universiti Teknologi Malaysia (UTM) for support of the research paper under the grant no 4S007.

## REFERENCES

[1] H. Yu and Q. Chu, "An Omnidirectional Small Loop Antenna Based on First-Negative- Order Resonance," *Journal of Electromagnetic Waves and Applications*, vol. 26, no. February 2012, pp. 111–119, 2012.

[2] H.-X. Xu, G.-M. Wang, M.-Q. Qi, and Z.-M. Xu, "Theoretical and experimental study of the backward-wave radiation using resonant-type metamaterial transmission lines," *Journal of Applied Physics*, vol. 112, no. 10, p. 104513, 2012.

[3] D. R. Smith and S. Schultz, "Determination of effective permittivity and permeability of metamaterials from reflection and transmission coefficients," *Physical Review B*, vol. 65, pp. 1–5, 2002.

[4] G. V Eleftheriades, S. Member, A. K. Iyer, S. Member, and P. C. Kremer, "Planar Negative Refractive Index Media Using Periodically L – C Loaded Transmission Lines," *IEEE Transactions on Microwave Theory and Techniques*, vol. 50, no. 12, pp. 2702–2712, 2002.

[5] D. Smith, W. Padilla, D. Vier, S. Nemat-Nasser, and S. Schultz, "Composite medium with simultaneously negative permeability and permittivity," *Physical review letters*, vol. 84, no. 18, pp. 4184–7, May 2000.

[6] C. Caloz and T. Itoh, "Novel microwave devices and structures based on the transmission line approach of meta-materials," in *Microwave Symposium Digest, 2003 IEEE MTT-S International*, 2003, vol. 1, pp. 195–198.

[7] C. Caloz and T. Itoh, *Electromagnetic metamaterials: transmission line theory and microwave applications*. Wiley-IEEE Press, 2005.

[8] D. K. Upadhyay, C. Engineering, and S. Pal, "Design Of Novel Improved Unit- Cell for Composite Right Left Handed Trasmission Line Based," *International Journal of Engineering Science and Technology*, vol. 3, no. 6, pp. 4962–4967, 2011.

[9] C. Caloz, A. Sanada, and T. Itoh, "A novel composite right-/left-handed coupled-line directional coupler with arbitrary coupling level and broad bandwidth," *Microwave Theory and Techniques, IEEE Transactions on*, vol. 52, no. 3, pp. 980–992, 2004.

[10] A. A. Ibrahim, A. M. E. Safwat, and H. El-hennawy, "Triple-Band Microstrip-Fed Monopole Antenna Loaded With CRLH Unit Cell," *IEEE Antennas and wireless progation letters*, vol. 10, pp. 1547–1550, 2011.

[11] S. Mok, S. Kahng, and Y. Kim, "A wide band metamaterial ZOR antenna of a patch coupled to a ring mushroom," *Journal of Electromagnetic Waves and Applications*, vol. 26, no. January 2013, pp. 1667–1674, 2012.

[12] M. A. Antoniades and G. V. Eleftheriades, "A Broadband Dual-Mode Monopole Antenna Using NRI-TL Metamaterial Loading," *IEEE Antennas and Wireless Propagation Letters*, vol. 8, pp. 258–261, 2009.

# Field-circuit co-simulation for microwave meta-materials with nonlinear components

Delong Li,[1] Jinfeng Zhu,[1*] Shanshan Wu,[1] Xiaoping Xiong,[1] Yanhui Liu,[1] Qing H. Liu[1,2*]

[1]Department of Electronic Science, Xiamen University, Xiamen 361005, China
[2]Department of Electrical and Computer Engineering, Duke University, Durham, NC 27708, USA
*Email: nanoantenna@hotmail.com (J. Zhu), qhl@xmu.edu.cn (Q. H. Liu)

*Abstract*-We demonstrate an analysis of the metamaterials medium consisting of SRRs (split ring resonators) integrated with nonlinear microwave components based on a circuit and 3D electromagnetic wave co-simulation method. The simulations are performed by using Wavenology EM, which is an efficient multiphysics and multiscale wave simulator. Our investigations show that the resonant frequency of the SRR loaded by a varactor reduces slightly when incident wave power increases, which is consistent with previous experimental research. We propose to connect an adjustable capacitor in parallel with the varactor to tune the resonant frequency. The simulation results indicate the tuning range is from 12.30 to 14.92 GHz, and the resonant frequency and quality factor decrease as the parallel capacitance increases.

## I. Introduction

Microwave metamaterials are usually periodic arrays of artificial structures which have negative refractive index, and they have diverse potential applications, such as cloaking [1], superlens [2]. A split ring resonator (SRR) is a commonly used array unit that delivers strong magnetic coupling for metamaterials [3, 4]. Recently, the nonlinear metamaterials based on SRRs has attracted surge of interest [5-7], due to the strong local electromagnetic field enhancement in a sub-wavelength unit cell [5, 8]. Da Huang et al. have done the fundamental research about the nonlinear metamaterial consisting of varactor-loaded SRRs, and they have used an analytical method to investigate the power dependent tuning at microwave frequency. However, this method is inconvenient and inefficient, especially for the engineering design of circuit-loaded metamaterials with complex structures. In this paper, the field-circuit problem of SRR-based metamaterials with nonlinear components is co-simulated by using Wavenology EM, which is an efficient multiphysics and multiscale wave simulator software. We propose to make the nonlinear metamaterials intelligent by paralleling an adjustable capacitor without changing the structure geometry of SRRs, and investigate its resonant properties by using the co-simulation method.

## II. Modeling and simulation

### A. Physical model and co-simulation using Wavenology EM

One of the highlighted features of Wavenology EM is the ability to co-simulate both complex circuits and microwave systems together. It provides a flexible and fast route to solving field-circuit coupling problem. A full wave transient 3D EM solver based on finite difference time domain method and a SPICE module for circuit design are hybridized to simulate microwave response of SRRs loaded by nonlinear components.

The SRR is a single copper square with an outer edge length of 2.2 mm, an inner edge length of 1.8 mm, and a slit width of 0.3 mm, placed on a 0.25 mm thick FR4 ($\varepsilon = 4.4$) substrate. In the modeling work, the SRR is integrated with a circuit module (including an SMV1231-079 varactor) into the gap of the SRR [5], as shown in Fig. 1. The circuit-loaded SRR unit cell is located at the center of the computation domain, and perfect magnetic conductor and perfect electric conductor boundary conditions are used at the top and bottom walls of the simulation region. Two wave ports perpendicular to x axis are used to investigate the plane wave excitation and response. Such a configuration in the co-simulation model can reflect the resonant properties of a microwave nonlinear metacrystal consisting of SRRs with nonlinear microwave components.

### B. Circuit model

The RLC equivalent circuit model of a varactor-loaded SRR is presented in Fig. 2(a). The varactor is integrated into the capacitive gap of the SRR. We also propose to connect an adjustable capacitor Cp in parallel with the varactor to manipulate resonant properties of the circuit-loaded SRR, as shown in Fig. 2(b).

Fig.1 Simulation model of an SRR with a circuit module.

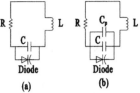

Fig.2. Equivalent circuit: (a) a varactor-loaded SRR, and (b) a varactor-loaded SRR with a parallel capacitor.

## III. RESULTS AND DISCUSSION

### A. Nonlinear response of changing incident wave power

Here, we use the shift of resonant frequency by increasing incident power to characterize the nonlinear response of varactor-loaded SRRs. The varactor has a nonlinear capacitance that varies with the applied voltage as $C(V_D)= C_0(1-V_D/V_P)^{-M}$, where $V_D$ is the bias voltage, $C_0$ is the zero bias capacitance, $V_P$ is the intrinsic potential, M is the gradient coefficient, and the reverse breakdown voltage is 15V. The nonlinear response of the microwave metacrystal is originated from the performance of the nonlinear capacitance.

We first discuss the co-simulation result of the model shown in Fig. 2(a). It is observed that the amplitude of incident excitation voltage is 0.05V, but voltage amplitudes of more than 0.15V at P and N nodes of the varactor are generated, as shown in Fig. 3. This simulation result demonstrates that the strong electric filed enhancement at the gap of the SRR, which excites the nonlinear response in the varactor-loaded SRR. Fig. 4 shows that the resonant frequency of S21 decreases slightly as the incident power $V_i$ increases, which demonstrates the nonlinear properties of the investigated microwave metacrystal, and the simulation result is consistent with previous research work in Ref. [5] and [7].

### B. Effects of tunable circuit

Next, we discuss the proposed model shown in Fig. 2(b). An adjustable capacitor in parallel with the varactor can be used to tune the resonant response in the RLC circuit [9], and therefore make the unit cell intelligent and flexible for research of nonlinear metamaterials. As shown in Fig. 5, the incident voltage amplitude is set to 0.01V, and the transmission properties are tuned obviously as the value of $C_p$

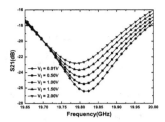

Fig.4. S21 of incident voltage amplitudes from 0.01V to 2.00V, for the model shown in Fig. 2 (a).

Changes from 0.05pF to 0.10pF [10]. The result in Fig. 5 indicates that the quality factor and resonant frequency decrease as the parallel capacitance increase. As $C_p$ increase from 0.05pF to 0.10pF, the quality factor is gradually reduced, but the resonant frequency is shifted from 14.92GHz to 12.30GHz, which is a broad tunable range for microwave metamaterials.

## IV. CONCLUSION

We have demonstrated the tunability, strong local field enhancement, and adjustability of circuit-loaded SRRs by using field-circuit co-simulation. The co-simulation result for microwave metamaterials with nonlinear circuit components is in agreement with previous research analysis. Without changing the physical geometry of the SRR unit cell, the permeability of the nonlinear metacrystal is adjustable through manipulating the nonlinear circuit. In future work, the co-simulation approach for tunable circuit-loaded SRRs will be experimentally verified by S-parameter measurement using a vector network analyzer and applied in the design of nonlinear metamaterials.

## ACKNOWLEDGMENT

Jinfeng Zhu acknowledges the support from the Institute of Electromagnetics and Acoustics at Xiamen University. The authors thank Dr. Bo Zhao from Wave Computation Technologies, Inc. for technical support and helpful discussions.

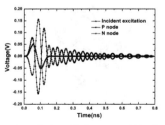

Fig.3. Incident wave excitation and voltages at P and N nodes of the varactor, for the model shown in Fig. 2(a).

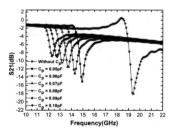

Fig.5 S21 of controlling $C_p$, for the model shown in Fig. 2(b).

## REFERENCES

[1] W. Cai, U. K. Chettiar, A. V. Kildishev, and V. M. Shalaev, "Optical cloaking with metamaterials," *Nature photonics,* vol. 1, no. 4, pp. 224-227, 2007.

[2] X. Rao, and C. Ong, "Subwavelength imaging by a left-handed material superlens," *Physical Review E,* vol. 68, no. 6, pp. 067601, 2003.

[3] D. Smith, D. Vier, T. Koschny, and C. Soukoulis, "Electromagnetic parameter retrieval from inhomogeneous metamaterials," *Physical Review E,* vol. 71, no. 3, pp. 036617, 2005.

[4] D. Huang, E. Poutrina, H. Zheng, and D. R. Smith, "Design and experimental characterization of nonlinear metamaterials," *JOSA B,* vol. 28, no. 12, pp. 2925-2930, 2011.

[5] D. Huang, E. Poutrina, and D. R. Smith, "Analysis of the power dependent tuning of a varactor-loaded metamaterial at microwave frequencies," *Applied Physics Letters,* vol. 96, no. 10, pp. 104104-104104-3, 2010.

[6] E. Poutrina, D. Huang, and D. R. Smith, "Analysis of nonlinear electromagnetic metamaterials," *New Journal of Physics,* vol. 12, no. 9, pp. 093010, 2010.

[7] B. Wang, J. Zhou, T. Koschny, and C. M. Soukoulis, "Nonlinear properties of split-ring resonators," *arXiv preprint arXiv:0809.4045,* 2008.

[8] J. Pendry, A. Holden, D. Robbins, and W. Stewart, "Magnetism from conductors and enhanced nonlinear phenomena," *Microwave Theory and Techniques, IEEE Transactions on,* vol. 47, no. 11, pp. 2075-2084, 1999.

[9] I. V. Shadrivov, S. K. Morrison, and Y. S. Kivshar, "Tunable split-ring resonators for nonlinear negative-index metamaterials," *arXiv preprint physics/0608044,* 2006.

[10] M. Gorkunov, and M. Lapine, "Tuning of a nonlinear metamaterial band gap by an external magnetic field," *Physical Review B,* vol. 70, no. 23, pp. 235109, 2004.

# Metamaterial Absorber and Polarization Transformer Based on V-shape Resonator

Song Han, Helin Yang[1]

*College of physical Science and Technology, Central China Normal University Wuhan 430079, Huibei*
*People's Republic of China*
*1emyang@mail.ccnu.edu.cn*

*Abstract*-A metamaterial (MM) resonator composed of double V-shape metal wires shows the different manipulations of the transmission and reflection. Parametric studies on the MM reveal that it can work as dual-direction absorber or polarization transformer based on the electric and magnetic resonance between the metal layers. The investigation shows that the absorption of MM resonator is 95.3% at 12.1GHz; and the MM resonator can turn vertical polarized EM waves into horizontal polarized EM waves with polarization conversion ratio (PCR) of 78% at 10.2 GHz. The potential application for microwave dual-directional ideal absorbing and polarization transformation are discussed respectively.

## I. INTRODUCTION

Since their inception in 2001 [1], metamaterials (MMs) have ushered in a new era of electromagnetism. These composite MMs have electromagnetic properties transcending those naturally-occurring media, such as negative refractive index [1-3], invisibility cloaks [4-6], super lens [7-8], perfect electromagnetic (EM) wave absorber [9], and so on. The fundamental purpose of MMs is to realize the manipulation of EM wave, such as the MM absorbers, polarization transformers, Electromagnetically Induced Transparency (ETI) MMs and Negative Refractive Index (NRI) MMs. MM absorber, as one of the most important applications, has been investigated since 2008 [9]. Since then, MM absorbers have progressed significantly with designs shown across the electromagnetic spectrum, from microwave to optical spectrum [9-17]. Recently, dual-directional MM absorbers have been investigated as an available MM, but the disadvantages of the MM absorbers are that the polarization status is sensitive and the structure is bulky [12, 18]. Polarization transformer is another important application in many areas, such as antennas, astronavigation, and communication [19-21]. Thus, it is highly desirable to efficiently control the polarization of EM waves [22]. One of the most efficient methods is the asymmetric transmission (AT) [23-24].

In this paper, a double V-shape MM resonator is designed, which consists of metal layers separated by the substrate (FR-4). The dual-directional MM absorber and the linear polarization transformer can be obtained by taking the advantage of the arrangement of the V-shape gap direction.

Numerical simulation shows that the absorption of the MM absorber is perfect at dual-direction. Further investigation shows that the absorber performs insensitive under different polarized EM waves. In addition, the resonant character is verified that only dielectric losses affect the optimized absorbers' overall performance. Then a polarization transformer is investigated by rotating the V-shape's direction. The MM transformer can convert vertical polarization wave into horizontal polarization wave based on AT, and the polarization conversion ratio (PCR) reach 78% at 10.2 GHz. The MM resonator is tunable so that realize the manipulation of EM wave [24].

## II. STRUCTURE DESIGN

A schematic layout of the implemented unit cell of the MM, including all geometrical parameters, is shown in Fig. 1. The model consists of double V-shape copper wires on the FR-4 dielectric substrate. The conductivity of the copper is $\sigma=5.8\times10^7$S/m and the permittivity of the FR-4 is $\varepsilon=4.9+i0.025$.The inner gap of the inner V-shape $a$=3.5mm; the width of the metal wire $w_1=w_2$=0.5mm, and the thickness is $t$=0.04mm, the gap between the two V-shapes $g$=0.3mm, the arm of the V-shape $L_1$=4.6mm, $L_2$=6.35mm.

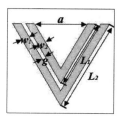

Figure 1. The schematic of the matamaterial resonator consists of double V-shape copper wires.

## III. DEMONSTRATION OF DUAL-DIREACTION METAMATERIAL ABSOBER

978-7-5641-4279-7

To obtain dual-directional absorbing, the double V-shape is arranged on both of the dielectric layer sides. In consideration of the polarization factor, the gap orientation is open to four perpendicular directions to each other, as shown in Fig. 2. The dielectrics $b_1$=13mm, the distance of V-shape peak $d_1$=0.5mm, $d_2$=0.75mm, and the thickness of dielectric substrate is 1.2mm, Computational simulation install periodic boundary as unit cell on $x$-$y$ plane. Along $z$ axis is the EM wave vector **k**.

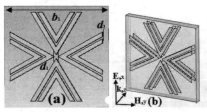

Figure 2. (a) The planar sketch of the MM absorber. (b) The 3-D model.

Based on finite difference time domain method (FDTD), we use CST simulating the S-parameter of the reflection and transmission. The results are shown in Fig. 3. Comparing (a) with (b), it is seen clearly that the absorbing performances are totally coincident when the incident waves propagate from $+z$ and $-z$ direction. The resonant band occurs at 12.1GHz, and the absorption peak reach 95.3%. Then the different polarized EM wave is simulated, which is shown in Fig. 4. The simulation result shows that the MM absorber is insensitive to different polarization status.

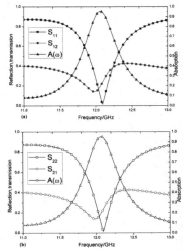

Figure 3. Simulated reflection, transmission and absorption curve of the V-shape resonator. (a) $+z$ direction propagation EM waves. (b) $-z$ direction propagation EM waves.

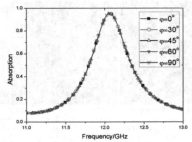

Figure 4. Absorption curves of the MM absorber at different linear polarization angles ($\varphi$).

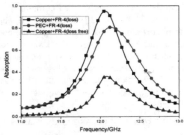

Figure 5. Investigate the physical origin by using different combination of metal and substrate.

To reveal the physical original mechanism of the absorption in the MM absorber, the connection between the metal and substrate are investigated. We use loss free dielectric substrate and loss substrate (FR-4) to simulation, the result is shown in Fig. 5. When Copper loads loss free FR-4, the maximum absorption is 35.9%; PEC and loss FR-4, the absorption peak is 80.7%; and Copper match loss FR-4, the absorption is 95.3%. That means the energy loss origin from the loss dielectric. Then the electronic and magnetic energy distribution on the substrate is displayed, as shown in Fig. 6. Note that when the EM wave illuminates the absorber along $+z$ axis (see Figure 2), the electronic and magnetic energy concentrate on the back of the substrate, mainly on the wedge and the arms of the V-shape. That is completely different from the normal resonant absorber. The transmitted power concentration increases considerably on the back, mimicking the angle-based sub-wavelength scale focusing of EM energy. Such a resultant ability to focus EM waves will find a number of new applications.

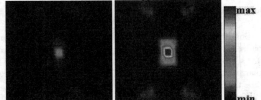

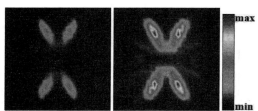

Figure 6. The energy distribution on the substrate, EM waves propagate along +z axis. (a) Electric energy on the front substrate.(b) Electric energy on the back substrate.(c) Magnetic energy on the front substrate. (d) Magnetic energy on the back substrate.

We demonstrate and simulate a double V-shape dual-directional absorber which was utilized combination method, the absorptivity is 95.3% at 12.1 GHz, and the absorber is insensitive to polarization status. According to the energy loss distribution, we find that the energy is backward concentration on dielectric substrate when the EM wave propagates along +z axis. That is to say, we may use the MM absorber to realize energy control. Furthermore, the model can also be applied on microwave absorbing materials, EM wave clacking and some other related fields.

## IV. DEMONSTRATION OF POLARIZATION TRANSFORMER

We realign the double V-shape resonator to make the V-shape gap twisting 90° which is shown in Fig. 7. Every unit holds a quarter of the spatial position, shown in Fig. 7 (a), the dielectric width is $b_2$=14mm and thickness is 1.0mm, $d_3$=0.75mm, $d_4$=0.65mm, $d_5$=0.325mm. Fig. 7 (b) shows the 3-D polarization transformer model. The computational simulation has been performed based on the standard time domain (FDTD) method by using CST Microwave Studio's Frequency Domain Solver, installing periodic boundary as unit cell on x-y plane. Along +z axis is the linear polarized EM wave vector $k$.

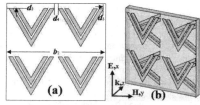

Figure 7. The model of linear polarization transformer. (a) The planar sketch. (b) The 3-D model.

Based on the simulation, the reflection and transmission coefficients are shown in Fig. 8. When the linear polarized EM waves propagate along the +z direction (see Fig. 7), the co-polarization transmission $t_{xx}$ and $t_{yy}$ are always remain same, while the cross-polarization transmission components $t_{xy}$ and $t_{yx}$ differ significantly across the whole frequency range. That means there existing a strong asymmetric transmission of the linear polarized wave. The amplitudes for the co-polarization transmission reduce to a minimum of about 0.14 at 9.9 GHz.

One can clearly observe that the amplitude of the cross-polarization transmission $t_{xy}$ achieves a maximum of 0.6 in the simulation around the resonance frequency of 10.2 GHz, while $t_{yx}$ is stay below 0.1 in the entire frequency range.

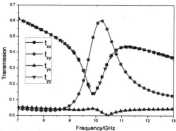

Figure 8. Simulated transmission spectra of the polarization transformer.

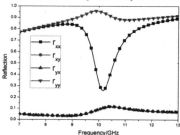

Figure 9. Simulated reflection spectra of the polarization transformer.

Fig.9 shows the simulation reflection for propagation along +z direction (see Fig. 7); the cross-polarization reflection $r_{xy}$ and $r_{yx}$ stay small and below 0.12 in the entire frequency range. While the co-polarization $r_{xx}$ reduces to a minimum of about 0.26 at the frequency 10.2 GHz, $r_{yy}$ is near unity across the whole frequency range. These results of the reflection further verify that the incident x-polarization wave will be forbidden to transmit for propagation along the +z direction. That is, the coupling of the front and back double V-shape each other is very low for this polarization, and nearly without magnetic coupling, resulting in a high reflectivity, and no polarization transformation can be observed.

We demonstrate the double V-shape resonator whose direction's permutation is not uniform may obtain a linear polarization transformer. The transmission and reflection spectra show that the resonator emerges the phenomenon of asymmetric transmission at about 10.2 GHz. The x-polarization (vertical polarization) wave is transformed to y-polarization (horizontal polarization) wave. The transmission spectra reach a maximum 0.6 and the PCR is 78%. Such a design may find potential application in optical isolators, microwave wave plates, or other EM control devices.

## V. CONCLUSION

In conclusion, we demonstrate a double V-shape MM resonator, which can work as microwave absorber when the front and back V-shape direction is coincident. While the back

V-shape gap relative to the front is transformed 90°, the resonator is a 90° linear polarization transformer. As a MM absorber, the absorption peak is 95.3% at 12.1 GHz, and the absorber is polarization insensitive to the incident EM microwave. As a polarization transformer, the vertical polarization wave can be transformed to horizontal polarization. This design may apply on dual-directional invisibility technology, radar absorbing materials, optical isolator and other EM control devices because of the tunable property.

## REFERENCES

[1] R. A. Shelby, D. R. Smith, S. Schultz. "Experimental Verification of a Negative Index of Refraction," *Science*, vol. 292, no. 5514, pp. 77-79, 2001.

[2] J. Lezec , J. A. Dionne , H. A. Atwater , "Negative Refraction at Visible Frequencies," *Science* , vol. 316, no. 5823, pp. 430-432, 2007.

[3] N. Engheta, R. W. Ziolkowski, "A positive future for double-negative metamaterials," *IEEE Trans. Microwave Theory Techniques*, vol.53, no. 4, pp. 1535-1556, 2005.

[4] J. B. Pendry, D. Schurig, D. R. Smith, "Controlling Electromagnetic Field," *Science*, vol. 312. no. 5781, pp. 1780-1782, 2006.

[5] U. Leonhardt, "Optical Conformal Mapping," *Science*, vol. 312. no. 5781, pp. 1777-1780, 2006.

[6] D. Schurig, J. J. Mock, B. J. Justice, S. A. Cummer, J. B. Pendry, A. F. Starr, D. R. Smith, "Metamaterial Electromagnetic Cloak at Microwave Frequencies," *Science*, vol. 314, no. 5801, pp. 977-980.

[7] J. B. Pendry, "Negative Refraction Makes a Perfect Lens," *Phys. Rev. Lett*, vol. 85, no. 18, pp. 3966-3969, 2000.

[8] N. Fang, H. Lee, C. Sun, and X. Zhang, "Sub- Diffraction-Limited Optical Imagine with a Silver Superlens," *Science*, vol. 308, no. 5721, pp. 534-537, 2005.

[9] N. I. Landy, S. Sajuyibe, J. J. Mock, D. R. Smith, and W. J. Padilla, " Perfect Metamaterial Absorber," *Phys. Rev. Lett*, vol. 100, no 20, 2008.

[10] Qiwei Ye, Ying Liu, Hai Lin, Minhua Li, Helin Yang, "Multi-band metamaterial absorber made of multi-gap SRRs," *Appl. Phys. A*, vol. 107, no. 1, pp. 155-160.

[11] Xiaojun Huang, Helin Yang, Shengqing Yu, Jixin Wang, Minhua Li, and Qiwei Ye, "Triple-band polarization-insensitive wide-angle ultra-thin planar spiral metamaterial absorber," *J. Appl. Phys*, vol. 113, no 21, 2013.

[12] ChengGang Hu, Xiong Li, Qin Feng, Xu'Nan Chen, and XianGang Luo, " Introducing dipole-like resonance into magnetic resonance to realize simultaneous drop in transmission an reflection at terahertz frequency," *J. Appl. Phys*, vol. 108, no. 5, 2010.

[13] Li Jiu-Sheng, "TERAHERTZ-WAVE ABSORBER BASED ON METAMATERIAL," *Microwave and Optical Technology Letters*, vol. 55, no. 4, pp. 793-796.2013.

[14] Ashish Dubey, A. Jain, C. G. Jayalakshmi, T. C. Shami, N. Awari, and S. S. Prabhu, "MULTILAYER BROAD BAND ABSORBER STRUACTURES FOR TERAHERTZ REGIN," *Microwave and Optical Technology Letters*, vol. 55, no. 2, pp. 393-395.2013.

[15] Na Liu, Martin Mesch, Thomas Weiss, Mario Hentschel, and Harald Giessen, "Infrared Perfect Absorber and Its Application As Plasmonic Sensor," *NANO. LETTERS*, no. 10, pp. 2342-2348, 2010.

[16] Zhi Hao Jiang, Seokho Yun, Fatima Toor, Douglas H. Werner, and Theresa S. Mayer, "Conformal Dual-Band Near-Perfectly Absorbing Mid-Infrared Metamaterial Coating," *ACS. NANO*, vol. 5, no. 6, pp. 4641-4647, 2011.

[17] Nan Zhang, Peiheng Zhou, Dengmu Cheng, Xiaolong Weng, Jianliang Xie, and Longjiang Deng, "Dual-band absorption of mid-infrared metamaterial absorber based on distinct spacing layers," *OPTICS LETTERS*, vol. 38, no. 7, pp. 1125-1127, 2013.

[18] Lu Lei, Qu Shao-Bo, Xia Song, Xu Zhuo, Ma Hua, Wang Jia-Fu, Yu Fei, "Simulation and experiment demonstration of a polarization-Independent dual-directional absorption metamaterial absorber," *Acta Phys.Sin*, vol. 62, no. 1, 2013.

[19] V. A. Fedotov, P. L. Mladyonov, S. L. Prosirnin, A. V. Rogacheva, Y. Chen, N. I. Zheludev, "Asymmetric Propagation of Electromagnetic Waves Through a Planar Chiral Structure," *Phys. Rev. Lett*, vol. 97, no. 16, 2006.

[20] C. Menzel, C. Rockstuhl, F. Lederer, "Advanced Jones calculus for the classification of periodic metamaterial," *Phys. Rev. A*, vol. 82, no. 5, 2010.

[21] C. Menzel, C. Helgert, C. Rockstuhl, E.-B. Kley, A. Tunnermann, T. Pertsch, F. Lederer, "Asymmetric Transmission of Linearly Polarized Light at Optical Metamaterials," *Phys. Rev. Lett*, vol. 104, 2010.

[22] M. Kang, J. Chen, H. Cui, Y. Li, H. Wang, "Asymmetric transmission of linearly polarized electromagnetic radiation," *Opt. Express*, vol. 19, no. 9, pp. 8347-8356, 2011.

[23] C. Huang, Y. Feng, J. Zhao, Z. Wang, T. Jiang, "Asymmetric electromagnetic wave transmission of linear polarization via polarization conversion through chiral metamaterial stuctures," *Phys. Rev. B*, vol. 85, no. 19, 2012.

[24] Yongzhi Chen, Yan Nie, Xian Wang, Rongzhou Gong, "An ultathin transparent metamaterial polarization transformer based on a twist-ring resonator," *Appl. Phys. A*,vol. 111, no.1, pp. 209-215, 2013.

# Design of Broadband Metamaterial Absorber Based on Lumped Elements

Xiaojun Huang[*#], Helin Yang[*l], Song Han[*]

*College of Physical Science and Technology, Central China Normal University, Wuhan 430079, Hubei
People's Republic of China
# Department of Physics, Kashgar Teachers College, Kashgar 844000, Xinjiang
People's Republic of China
l emyang@mail.ccnu.edu.cn

*Abstract-* An enhanced broadband metamaterial based on lumped elements is presented. The structure is composed of lumped elements and two conductive layers with a single substrate (FR-4) between them. The simulation results show that the bandwidth of absorption of 80% is about 10.58 GHz and the full width at half maximum (FWHM) can be up to 92%. The further simulations of the proposed MA can operate at a wide range of incident angles under both transverse electric and transverse magnetic polarizations.

## I. INTRODUCTION

Electromagnetic metamaterials (MMs) have produced many exotic effects and devices, which are usually defined as a class of artificial media with unusual properties not found in nature, such as negative refraction [1], sub-wavelength imaging super-lens [2], and cloaking [3]. Recently, Landy et al. have proposed a thin metamaterial absorber (MA), in which electric and magnetic resonance makes the absorber possess matched impedance to eliminate the reflection and strongly absorb the incident wave [4]. Since then, many MAs have been proposed and demonstrated from microwave to optical frequencies [5-7]. However, the bandwidth of these MAs is usually narrow, and is not applicable in some areas. Generally, MAs are composed of periodic arrays of sub-wavelength metallic elements. Various approaches have been proposed to extend the absorption band, the purpose of which is to make the MA units resonate at several neighbouring frequencies [8-12]. These MMs obtain high absorption properties primarily through dielectric loss and impedance matching at resonance for incident EM waves. MA could be founded have various potential applications in thermal imaging, thermal bolometer, wavelength-selective radiators, stealth technology, and nondestructive detection.

In this paper, in contrast to the broadband absorbers in the previous studies, we proposed a simple broadband MA based on lumped elements. The proposed MA is composed of lumped elements and two conductive layers with a single substrate (FR-4) between them. The top layer consists of a ring and a deformed cross (RDC) embedded in the ring which loads the lumped elements, whereas the bottom layer has a metallic ground plate without patterning. The absorption bandwidth can be broadened by changing the lumped elements. The design

and simulations for the proposed MA have been presented in the paper.

## II. STRUCTURES AND DESIGN GUIDELINES

According to the effective medium theory, MMs can be characterized by a complex permittivity $\tilde{\varepsilon}(\omega)$ and permeability $\tilde{\mu}(\omega)$. In practice, loss is measured by the amount of electromagnetic power absorbed, where absorbed power is $A(\omega) = 1 - R(\omega) - T(\omega)$, where $R(\omega)$ and $T(\omega)$ is the reflection and the transmission as functions of frequency $\omega$, respectively. To achieve unity absorption of the MA, minimizing $R(\omega)$ and $T(\omega)$ to near zero simultaneously is necessary. The frequency dependent $R(\omega)$ and $T(\omega)$ are dependent on the complex refraction and relative wave impedance on the complex refraction $\tilde{n}(\omega) = \sqrt{\tilde{\mu}(\omega) \cdot \tilde{\varepsilon}(\omega)}$ and relative wave impedance $\tilde{z}(\omega) = \sqrt{\tilde{\mu}(\omega)/\tilde{\varepsilon}(\omega)}$. Therefore, it is possible to absorb both the incident electric and magnetic field tremendously by properly tuning $\tilde{\varepsilon}(\omega)$ and $\tilde{\mu}(\omega)$, and achieving perfectly impedance-matched to the free space.

In order to design the broadband MA, we firstly proposed a two-band MA; the two-band MA is composed of two conductive layers with a single substrate (FR-4) between them. The top layer consists of a ring and a deformed cross (RDC) embedded in the ring, whereas the bottom layer has a metallic ground plate without patterning. A substrate (FR-4) with a relative permittivity $\varepsilon_r$=4 (loss tangent, 0.025) and thickness $h$= 0.8 mm is chosen. The chosen metal is copper with the thickness of 0.035 mm and the electric conductivity of $5.8 \times 10^7$ s/m. The optimal parameters of the two-band MA are as follows: the outer diameter of the ring is $r$=3.9mm, and the distances of the inner cross shown in the structure are $d$=0.3mm and $t$=0.2mm. The width of the structure $W$=0.2mm, and the length of the inner cross is $l$=2.9mm, which is shown in Fig.1 (a). And then the lumped elements are loaded on the two-band MA to expand the bandwidth, eventually, the broadband MA can be achieved, which is shown in Fig.1 (b). Because of loading the lumped elements, the thickness of the substrate (FR-4) is increased to $h$=3.2mm. As to the two-band MA, the two resonant peaks can be shifted closely through optimizing the geometric parameters carefully.

978-7-5641-4279-7

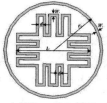

Figure 1. (a) Perspective of the two-band MA; (b) broadband MA based on lumped elements

### III. NUMERICAL RESULTS

The proposed MA is designed and optimized based on the standard finite-difference time domain (FDTD) method. Periodic boundary condition is set along the lateral directions of the structure and open boundary condition is set along the $z$ direction. Frequency domain solver and hexahedral mesh are applied in our design. The optimized simulate results of the two-band MA are shown in Fig.2, the absorptions of two resonant frequencies are 98.68% and 98.57% at 7.145 GHz and 7.43GHz for structure I; and the absorptions at 6.84GHz and 7.024GHz are 97.24%, 99.47% for structure II, respectively. The two resonant peaks can be shifted closely through optimizing the geometric parameters carefully. The optimized geometric parameters are shown in Tab.1.

Table.1. optimized geometric parameters of structures (unit:mm)

| Structure | $r$ | $l$ | $W$ | $d$ | $t$ |
|---|---|---|---|---|---|
| Structure I | 3.7 | 2.7 | 0.2 | 0.2 | 0.1 |
| Structure II | 3.9 | 2.9 | 0.2 | 0.3 | 0.2 |

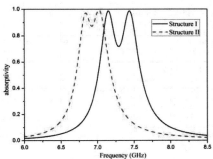

Figure 2. absorptions of the two-band MA.

To achieve broadband MA, the device is composed of the dielectric substrate sandwiched with metal RDC loaded with lumped elements (lumped resistance R )and continuous metal film. The unit cell of composite MA is shown in Fig.1 (b). We load the resisters on the two-band MA. Fig.3 shows the absorptivities of the broadband MA with different lumped resistances. From Fig.3, we can clearly see that when $R$=200Ω, the absorptivity is highest and the absorptivity exceeds 80% from 6.16GHz to 16.74 GHz and the FWHM is up to 92%. We

can also see that the lumped resistance mainly influences the magnitude of the absorptivity. These results further confirm the role of the lumped elements to improve performance of the MA and also indicate that lumped parameters have a great influence on the absorbing properties.

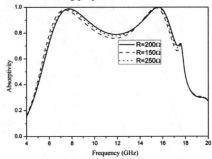

Figure 3. absorptions of the broadband MA loaded on the lumped elements.

We also examine the sensitivity of the broadband MA to the polarization states. The reflection characteristics under a normal incident planar EM wave with different polarizations are plotted in Fig.4. From the figures, the intensity of these reflection dips is nearly unchanged when φ changes from 0° to 45°.

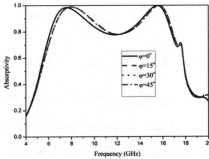

Figure 4. Polarization independence of the reflectivity of the proposed MA

In addition, it is necessary to discuss the situations of large angles of oblique incidence with instable polarization states. Fig.5 shows the simulated results of the proposed MA with frequency for different oblique incident elevation angles (θ, defined as the angle between the wave vector and the normal) for both TE and TM polarizations, respectively. Following the figure, the proposed MA still retains a good performance with broad bandwidths and high absorptions in TE and TM mode, when θ changes from 0°to 45°. When the incident electric field propagates across the MA, the FWHM at normal incident exceeds 80% is up to 92% in TE case; the FWHM at normal incident exceeds 80% is up to 93% in TM case, respectively. Furthermore, when the incident angle reached to 45°, the FWHM at normal incident exceeds 70% is up to 90% in TE case; the FWHM at normal incident exceeds 65% is up to 88% in TM case, respectively. This can be explained by the fact that the position of magnetic-resonance peak is not changed much by increasing the angle of incidence. It is found that the function of this absorber is rarely limited by the direction of incident electric field in a wide angle range.

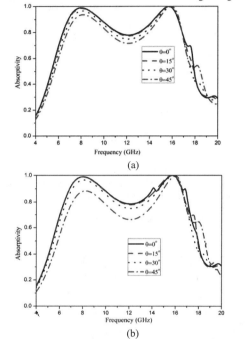

(a)

(b)

Figure 5. Angular independence of the absorptivities of the proposed MA (a) TE and (b) TM modes

To better understand the physical mechanism of the proposed broadband MA, the electric energy density was plotted in Fig.6. Referring to the figure, it can be seen that the electric energy density distributions on the outer ring at the low frequency for the two-band MA, whereas the electric energy density distributions on the lumped elements loaded on outer ring at the low frequency for the broadband MA.

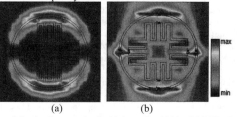

(a)          (b)

Figure 6. the electric energy density (a) the two-band MA at 7.145GHz (b) the broadband MA at 7.58GHz.

## IV. CONCLUSIONS

In conclusion, we have designed and simulated broadband MA with lumped elements. Simulation results demonstrate that the absorptivity of nearly 80% is about 10.58 GHz and the FWHM is up to 92%, and the absorptivity is nearly unchanged for different polarization angles. Compared the RDC structure MA, our design has a wider bandwidth of absorption. Furthermore, the proposed MA is not limited by the quarter-wavelength thickness and relatively thin. The further simulated results indicate that lumped parameters have a great influence on the absorbing properties and there exist optimal values, where the performance of the composite MA is best.

## REFERENCES

[1] J. B. Pendry, A. T. Holden, W. J. Stewart, I. Youngs, "Extremely Low Frequency Plasmons in Metallic Mesostructures," Phys. Rev. Lett. vol.76, no. 25, pp. 4773, (1996).

[2] J. B. Pendry, "Negative Refraction Makes a Perfect Lens," Phys. Rev. Lett. vol. 85, no. 18, pp. 3966-3969, 2000.

[3] J. B. Pendry, D. Schurig, D. R. Smith, "Controlling Electromagnetic Fields," Science vol. 312, no. 5781, pp. 1780-1782 ,2006

[4] N. I. Landy, S. Sajuyigbe, J. J. Mock, D. R. Smith, W. J.Padilla, "Perfect Metamaterial Absorber," Phys.Rev. Lett. vol. 100, no. 20, pp. 207402, 2008.

[5] N. I. Landy, C. M. Bingham, T. Tyler, N. Jokerst, D. R.Smith, W. J.Paddila, "Design, theory, and measurement of a polarization-insensitive absorber for terahertz imaging," Phys. Rev. B. vol. 79, no. 12, pp. 125104, 2009.

[6] X. Liu, T. Starr, A. F. Starr, W. J. Padilla, "Infrared Spatial and Frequency Selective Metamaterial with Near-Unity Absorbance," Phys. Rev. Lett. vol. 104,no. 20, pp. 207403, 2010.

[7] C. H. Lin, R. L. Chern, and H. Y. Lin, "Polarization-independent broadband nearly perfect absorbers in the visible regime," Opt. Express vol. 19, no. 2, pp. 415-424, 2011.

[8] J. B. Sun, L. Y. Liu, G. Y. Dong, J. Zhou, "An extremely broad band metamaterial absorber based on destructive interference," Opt. Express, vol. 19, no. 22, pp. 21155-21162, 2011.

[9] Y. H. Liu, S. Gu, C. R. Luo, X. P. Zhao, "Ultra-thin broadband metamaterial absorber," Appl. Phys. A. vol. 108, no. 1, pp. 19-24, 2012.

[10] F. Ding, Y. X. Cui, X. C. Ge, Y. Jin, S. L. He, "Ultra-broadband microwave metamaterial absorber," Appl. Phys. Lett. vol. 100, no. 10, pp. 103506, 2012.

[11] Y. Z. Cheng, Y. Wang, Y. Nie, R. Z. Gong, X. Xiong, X. Wang, "Design, fabrication and measurement of a broadband polarization-insensitive metamaterial absorber based on lumped elements," J. Appl. Phys. vol. 111, no. 4, pp. 044902, 2012

[12] A. Dimitriadis, N. Kantartzis, T. Tsiboukis, "A polarization-/angle-insensitive, bandwidth-optimized metamaterial absorber in the microwave regime," Appl. Phys. A. vol. 109, no. 4, pp. 1065-1070, 2012.

# A novel RF resonator using microstrip transmission line for human body MRI at 3T

Hyeok-Woo Son[1], Young-Ki Cho[1], Hyungsuk Yoo[2]

[1]School of Electronics Engineering, Kyungpook National University
80 Daehakro, Buk-gu, Daegu, 702-701, Korea
[2]Department of Biomedical Engineering, University of Ulsan
Department of Biomedical Engineering, University of Ulsan, Nam-gu, Ulsan, 680-749, Korea

*Abstract*- A square-slots loaded (SSL) radio frequency (RF) resonator using microstrip transmission line (MTL) is designed for a human body 3T MRI. The SSL RF resonator shows greater penetrated RF magnetic fields near the center of the phantom than traditional RF resonators using MTL. A multichannel coil using SSL RF resonators was also simulated and provides good parallel excitation performance. In addition, RF shimming for homogenization can be effectively controlled by adjusting the inputs of eight resonators. Numerical results were obtained using a spherical phantom and a realistic human body model at 3T to calculate $B_1^+$ fields.

## I. INTRODUCTION

High magnetic resonance imaging (MRI) systems (> 3 T) have good intrinsic SNR (signal-to-noise ratio), high resolution, and are used important instrument for clinical diagnosis of the human body [1]-[3]. One of the main challenges for these systems is $B_1^+$ inhomogeneity. To alleviate this problem, multi-channel radio frequency (RF) coils with parallel excitations are used [4,5]. They can provide higher $B_1^+$ fields in the region of interest (ROI) and can optimize the $B_1^+$ fields by driving the currents of individual coil elements. In this paper, these multichannel RF coils are applied to 3 T MRI systems, and a new RF resonator is proposed to enhance the multichannel coil performance. A general microstrip transmission line (MTL) resonator and a proposed square-slot loaded (SSL) resonator are simulated with a head phantom, to obtain the $B_1^+$ distribution. Then, an eight channel body coil with MTL resonators is replaced by the proposed SSL resonators for human body simulations. To determine the excitation parameters of the RF multichannel coils, convex optimization is used, and the SSL multichannel coil provides better $B_1^+$ fields. To the best of our knowledge, there are no previous analyses of multichannel RF coils in 3 T MRI systems. SSL resonators based on multichannel RF body coils can be used at hospitals for higher quality images.

## II. METHOD

### A. The MTL RF resonator

Figure 1 shows a RF resonator based on the use of microstrip transmission line(MTL). Low loss dielectric material Teflon ($\varepsilon_r$ =2.08, loss tangent=0.004) was used as a substrate of microstrip with height (h) and length (l) of 2.0 cm and 15 cm, respectively, and microstrip line width (w) is 18 mm. The microstrip line is used as $\lambda/2$ resonator with its

ends terminated capacitors. The terminated shunt capacitors are used to reduce the physical length and to match the desired Larmor frequency(123.5 MHz, 3T).

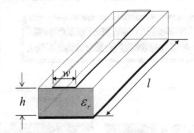

Figure 1. Illustration of microstrip transmission line (MTL)

### B. The SLL RF resonator

The SSL RF resonator has a same dielectric material and size such as a general RF resonator in Fig. 2. The width and length of a square slot is 6 mm and 18 mm, respectively. The distance between adjacent square-slots and between square-slots along a longitudinal direction is 2 mm and 6 mm, respectively.

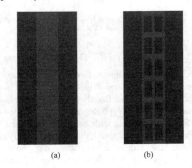

(a)                    (b)

Figure 2. RF resonators based on the use of microstrip transmission line. (a) a general RF resonator (b) squares-slot loaded(SSL) RF resonator

In this paper, the purpose of RF resonator design is to increase the penetrated magnetic field into an object. To increase penetrated magnetic field, the surface current density on strip conductor is also increased. The impact of loading squares-slot on strip conductor can be seen surface current density profiles in Fig. 3. In the general RF resonator, the surface current density is concentrated at the edges of strip conductor. The surface current density of the SSL RF resonator is not only concentrated at the edges of strip conductor, but also in the center of strip conductor. As a result, penetrated magnetic field of a SSL RF resonator is anticipated more greater than a general RF resonator. Fig. 4 shows the penetrated RF magnetic field of a general RF resonator and a SSL RF resonator in a spherical phantom ($\varepsilon_r$=58.1, $\sigma$=0.539). The penetrated magnetic field of the SSL RF resonator is stronger than the general RF resonator at the center of the phantom.

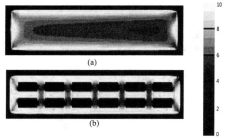

(a)

(b)

Figure 3. Surface current density ($10^4$ A/m$^2$) distribution of a general RF resonator (a) and a SSL RF resonator (b).

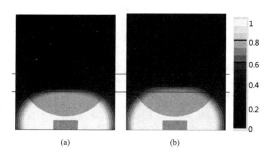

(a)                    (b)

Figure 4. The penetrated magnetic field in the phantom of a general RF resonator (a) and a SSL RF resonator (b).

## III. SIMULATION AND RESULTS

SEMCAD based on finite difference time-domain method was used to obtain $B_1^+$ and $E$ field with human model [6]. Duke model from Virtual Family Models is used to simulate a realistic body model. A photograph of 8-channle body coil using SSL RF resonators with human model and a realistic body model including many different tissue types (e.g., heart, liver, lung, kidney, intervertebral disc, bone, fat, and so on) is shown Fig. 5. Before obtaining $B_1^+$ and $E$ of 8-channel volume coil, the RF resonators with a spherical phantom ($\varepsilon_r$ =58.1, $\sigma$ =0.539 Siemens/m) were tuned to desired Larmor frequency by terminated shunt capacitors. When RF resonators are tuned, these capacitors are different capacitance between a general RF resonator and the SSL RF resonator. The frequency response (S11) of RF resonators is shown in Fig. 6. The SSL RF resonator is better than the general RF resonator. It means that the SSL

RF resonator has better matching condition compared with the general RF resonator.

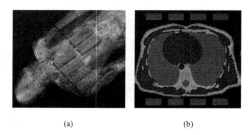

(a)                    (b)

Figure 5. 8-channel body coil with human model (a) and a realistic body model including many different tissue types on axial view.

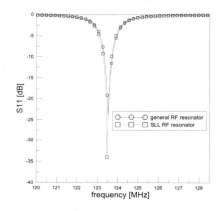

Figure 6. Frequency response of RF resonators.

Fig. 7 shows the $B_1^+$ field distribution of 8-channel volume coil using the general RF resonator with no optimization and with optimization on axial view. The $B_1^+$ field is calculated to control effectively the magnitude and phase of each RF resonator for optimization. To determine the input excitation values of each resonator element, convex optimization is used for the $B_1^+$ shimming [7]. When The $B_1^+$ field was calculated with parallel excitations, the input power of RF resonators is also normalized to the same input power (1 Watt). The total peak $B_1^+$ value with no optimization and optimization is 0.036 (T) and 0.041 (T), respectively. The $B_1^+$ field distribution of the 8-channel volume coil using the SSL RF resonator with no optimization and optimization is shown in Fig. 8. The total peak $B_1^+$ with no optimization and optimization is 0.036 (T) and 0.042 (T), respectively. Compared to the total $B_1^+$ value after optimization, the general RF resonator and SSL RF resonator are 12.6 % and 17.4 % larger than those before optimization. Table I shows the $B_1^+$ total, mean, and standard value of each 8-channel body coil. In the optimization results, the $B_1^+$ total value of the 8-channel body coil using SSL RF resonators is 3.29 % larger than it is using general RF resonator. The 8-channel body coil with

SSL RF resonators at the average value is also 3.27 % increased.

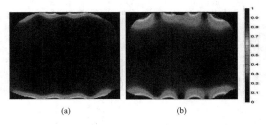

(a)          (b)

Fig. 7. $B_1^+$ field distribution of 8-channel volume coil using the general RF resonator with no optimization (a) and optimization (b)

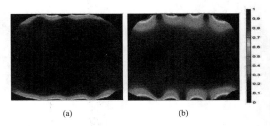

(a)          (b)

Fig. 8. $B_1^+$ field distribution of 8-channel volume coil using SSL RF resonator with no optimization (a) and optimization (b) on axial view.

TABLE I

COMPARISON OF THE $B_1^+$ TOTAL, AVERAGE, AND STANDARD VALUE BETWEEN BODY COILS

| Body coil | No optimization | | | Optimization | | |
|---|---|---|---|---|---|---|
| | Sum (T) | Mean (μT) | Std. (μT) | Sum (T) | Mean (μT) | Std. (μT) |
| General RF resonator | 0.036 | 0.5589 | 0.7144 | 0.0410 | 0.629 | 0.5267 |
| SSL RF resonator | 0.036 | 0.5538 | 0.6971 | 0.0423 | 0.650 | 0.5368 |

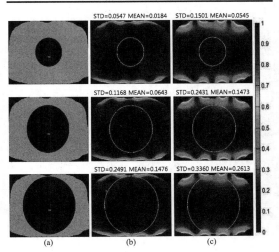

(a)      (b)      (c)

Fig. 9. The $B_1^+$ standard and mean value of the 8-channel body coil using general RF resonator are obtained for three different region of interest(ROI) (a) with no optimization (b) and optimization (c).

Fig. 9 and Fig. 10 illustrate that the $B_1^+$ standard and mean value of the 8-channel body coil using general RF resonator and SSL RF resonator are obtained for three different region of interest (ROI) under no optimization and optimization. The ROIs size are circles with radius 5 cm, 8 cm, and 10 cm at the center of transaxial slice, respectively. In case of three different ROI (5 cm, 8 cm, and 10 cm), the $B_1^+$ mean value of that is 4 % larger approximately than the 8-channel body coil using general RF resonator. As the results, the $B_1^+$ field of the 8-channel body coil using SSL RF resonators penetrated deep into human body and homogenized compared with the 8-channel body coil using general RF resonators.

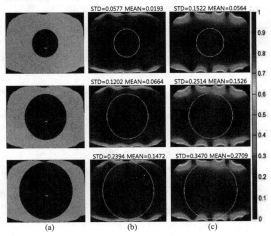

(a)      (b)      (c)

Fig. 10. The $B_1^+$ standard and mean value of the 8-channel body coil using SSL RF resonator are obtained for three different region of interest (ROI) (a) with no optimization (b) and optimization (c).

## IV. CONCLUSION

3 T MRI systems have become a standard technique for diagnosis imaging of the human body. However, compared to ultra-high field MRIs (> 7 T), 3 T MRI have a worse intrinsic SNR. To overcome this problem, we propose a multichannel coil with eight SSL RF resonators and compared to traditional RF resonator based coils. From numerical simulations based on FDTD method, it was estimated that the RF magnetic field of the SSL RF resonator is stronger than the general RF resonator. After optimization, the $B_1^+$ total and average value are 4 % larger. As the result, the volume coil using SSL RF resonator is controlled effectively to parallel excitations for increasing the $B_1^+$ field and mitigating the inhomogeneous $B_1^+$ field at 3 T.

## REFERENCES

[1] J. Vaughan, M. Garwood, C.M. Collins, W. Liu, L. DelaBarre, G. Adriany, P. Andersen, H. Merkle, R. Goebel, M.B. Smith, and K. Uğurbil, "7T vs. 4T: RF power, homogeneity, and signal-to-noise comparison in head images," *Magn Reson Med*, vol. 46, no. 1, pp. 24-30, 2001.

[2] J. Vaughan, L. DelaBarre, C. Snyder, J. Tian, C. Akgun, D. Shrivastava, W. Liu, C. Olson, G. Adriany, J. Strupp, P. Anderson, A. Gopinath and P. Moortele, Magn. Reson. Med.56,1274(2006)

[3] J. Vaughan, G. Adriany, C. J. Snyder, J. Tian, T. Thiel, L. Bolinger, H. Liu, L. DelaBarre and K. Ugurbil, Magn. Reson. Med. 52, 851(2004)

[4] X. Zhang, K. Ugurbil and W. Chen, Magn. Reson. Med. 46, 443(2001)

[5] G. Adriany, P-F. V. Moortele, F. Wiesinger, S. Moeller, J. P. Strupp, P. Anderson, C. Snyder, X. Zhang, W. Chen, K. P. Pruessmann, P. Boesiger, T. Vaughan and K. Ugurbil, Magn. Reson. Med. 53, 434( 2005)

[6] SEMCAD X by SPEAG, www.speag.com

[7] H. Yoo, A. Gopinath and J. T. Vaughan, IEEE Trans. Biomed. Eng. 59(12), 3365(2012)

# FP-1(A)

## October 34 (FRI) PM

## Room A

## Antennas for RFID

# A Low-Profile Dual-Band RFID Antenna Combined With Silence Element

Yongqiang Chen, Huiping Guo, Xinmi Yang, Xueguan Liu

School of Electronics and Information Engineering, Soochow University, SuZhou, China 215006

*Abstract*—**A low-profile dual-band three-dimensional antenna covering bands of 915 MHz and 2.45 GHz is proposed for radio frequency identification (RFID) applications. The eighth-wavelength antenna utilizes the planar triangular patches and a disk to reduce the profile. Planar tuning patches are adopted to replace the conventional tuning poles for improving the impedance matching. Silence element, defined as a radiator with little influence on others, is composed of quarter-wavelength patch working at higher band in this study. The antenna with a merit of good production consistency exhibits an impedance bandwidth of 22% from 842 to 1049 MHz in the lower band, while the upper band covers 100 MHz (from 2.4 to 2.5 GHz). The measured peak gain is 3dBi and 5dBi at 915 MHz and 2.45 GHz, respectively.**

*Index Terms*—**Dual-band, low-profile, gap-coupled feed, RFID, three-dimensional, silence element.**

## I. INTRODUCTION

The Radio-frequency identification (RFID) technology has had a profound impact on human society in recent years. More and more enterprises and government utilize the RFID technology to collect and manage information. There is a fact that a lot of the frequency bands, such as 915 MHz (ISO 18000-6) and 2.45 GHz (ISO 18000-4) [1], are used in RFID applications. One antenna that meets different frequency bands not only saves space, but also improves the versatility of the system. Combined antenna, which usually consists of different radiators to realize different frequency bands, is one of the effective ways to achieve multi-band. Furthermore, silence element can reduce the interaction between these bands remarkably [2].

A number of research efforts have been devoted to the patch dual-band antenna [3]–[5], but a few studies pay attention to three-dimensional (3D) dual-band antenna. Three-dimensional omni-directional antenna is a common type of base station antennas, which have many advantages on the large-scale deployment of RFID applications. In [6], an eighth-wavelength 3D antenna composed of a conical monopole with metallic parasitic elements and a capacitive disk was proposed to achieve the characteristics of broadband and low-profile. Its drawback is too expensive to fabricate. A modified 3D RFID antenna [7] using printing technology is much cheaper. However, there are four tuning poles in that antenna. These tuning poles make it complex to manufacture and can be damaged easily.

In this study, a low-profile dual-band 3D antenna covering bands of 915 MHz and 2.45 GHz is presented. Two wedge patch tuning elements, contributing to the production consistency and reliability, are used to replace the four tuning poles in [7]. Independent quarter-wavelength patch radiators at 2.45 GHz, namely silence elements, are added to realize dual-band.

The paper is organized as follows. Section II gives the principle of the combined antenna with silence elements and describes the configuration of the proposed antenna in details. Section III offers the design process and simulation discussion. Results are presented in Section IV. The conclusions are outlined in Section V.

## II. ANTENNA PRINCIPLE AND CONFIGURATION

Radiating structures in combined antenna are usually distinct, but the coupling caused by the adjacent radiators will change the single radiator's performance such as input impedance and radiation patterns. Antenna with silence elements can reduce these problems. Fig. 1 shows an ideal combined antenna with silence elements. Suppose $f1$ and $f2$ are the center frequencies in two bands. $A_{f1}$ and $A_{f2}$ are two separate antennas designed to work at $f1$ and $f2$, respectively. When the antenna operates at $f1$, all of the input energy should be fed into $A_{f1}$. On this occasion, $A_{f1}$ is the active radiator in this band, while $A_{f2}$ is the silence element, and vice versa. In fact, due to the inevitable presence of mutual coupling between radiators, the silence element also has a few negative impacts on the effective radiated body [2].

According to the concept of silence element, a new dual-band 3D antenna is proposed. The configuration of the antenna is shown in Fig. 2 (a). The antenna is fed by coplanar wave guide with ground, and a metal reflector ground plane is adopted to enhance the gain. There are three parts used for 915 MHz. Two identical printed triangular patchs perpendicularly connect to each other which can increase the current path and enhance the bandwidth. A disc loading on the top further reduces the height of the antenna. Wedge-shaped tuning patches are placed on the both sides of the triangular patch in order to improve production consistency and ameliorating the impedance matching. Two quarter-wavelength patches fed by gaps work at 2.45 GHz, shown in Fig. 2 (b). All the parts of the antenna are fabricated on FR4 substrate with a relative dielectric constant of 4.4 and thickness of 1.6mm.

## III. DESIGN AND SIMULATION

The proposed antenna is designed to improve the production consistency and cover two bands required for RFID applications. In addition, mutual coupling between radiators is the key point in combined antenna design too. These issues are discussed below.

978-7-5641-4279-7

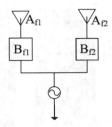

Fig. 1. Combined Antenna [2].

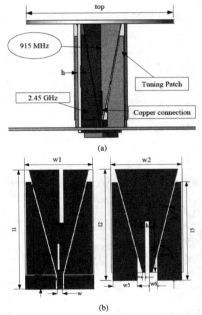

(a)

(b)

Fig. 2. Model of the antenna (a) whole (b) 2.45 GHz side on left and tuning side on right.

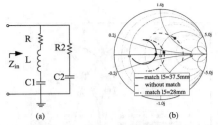

(a)                                    (b)

Fig. 3. Equivalent circuit and smith chart (a) equivalent circuit (b) input impedance with square marks at 915 MHz.

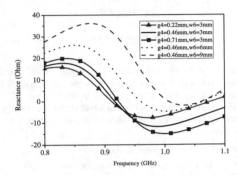

Fig. 4. Effect of $g4$ and $w6$ on reactance of input impedance.

### A. Patch Tuning Elements

Though the triangular monopole and the top loading disk can cut down the height of antenna, the input impedance of antenna in [7] still exhibits inductive characteristic without tuning poles. By introducing some capacitive elements, inductive reactance can be compensated. In this paper, two wedge tuning patches are adopted to improve the matching and promote the production consistency.

Two adjacent electrical conductors can constitute a capacitor. As a result of the small gap between triangular monopole and wedge patches, the capacitive reactance is significantly increasing. The equivalent circuit of matching network is given in Fig. 3 (a). The antenna, consisting of components $R$ and $L$,

will be modeled as a series resonator. By selecting the values of $C1$, $R2$ and $C2$, a complete cancellation of the reactive part of the input impedance at 915 MHz can be accomplished, as indicated in Fig. 3 (b). What's more, the impedance bandwidth is expanding too.

According to the principle of equivalence [8], width ($g4$) and length ($w6$) of the gap have an observable effect on the value of capacitor $C1$. The reactance is decreasing by reducing $g4$ and $w6$ that can be proved in Fig. 4. Owing to the gradual change structure, the matching patches have become a part of antenna and also affect the real part of input impedance. Here we introduce shunt-connected parasitic resistance $R2$ and parasitic capacitance $C2$. From simulation data in Fig. 3 (b), we can see the altitude of wedge patches is one of the factors for $R2$. The higher the altitude is, the smaller the real part of the input impedance is. The reasonable height is optimized for 37.5 mm.

### B. Radiator at 2.45GHz

Two symmetric wedge-shaped quarter-wavelength patches are intended to add a frequency band of 2.45 GHz. The current on triangular patches is mainly distributed along edges [7], so energy will feed into wedge-shaped patches through gaps. These two wedge-shaped patches are basically planar monopoles. So it's crucial that the length determines

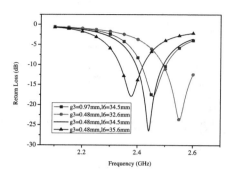

Fig. 5.  Effect of $g3$ and $l6$ on return loss of antenna.

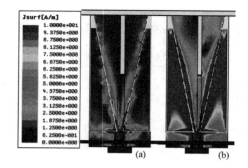

Fig. 6.  Current distribution of antenna (a) 915 MHz (b) 2.45 GHz.

the position of the resonance frequency. In Fig. 5, the centre frequency moves to a higher frequency when the length $l6$ is decreasing. The optimal dimensions will be $l6$=34.5mm at 2.45 GHz. The centre frequency moves to lower frequency when g3 is decreasing because the coupling gaps also have an influence on this band.

*C. Mutual Coupling Between Radiators*

The mutual coupling between radiators is an interesting problem in combined antenna. As in Fig. 1, if we take the 915 MHz radiator of the antenna as $A_{f1}$, then $B_{f1}$ is considered in the straight-through state. On the other hand, $A_{f2}$ is the wedge-shaped quarter-wavelength patches and $B_{f2}$ is a complex coupling network. In most cases, the coupled $A_{f2}$ will contribute to $A_{f1}$ in the same frequency. But in this study, the appropriate gaps can hinder energy transferring to $A_{f2}$ at 915 MHz approximately like a high-pass filter. Fig. 6 shows the current distribution of the antenna at 915 MHz and 2.45 GHz using HFSS. It is obvious that when the antenna is operating at 915 MHz, the current density of 2.45 GHz patches is extremely tiny. While in band of 2.45 GHz, the current density is quite large. So the 2.45 GHz patches mainly work in the higher frequency. After removing the 2.45 GHz patches, the return loss is investigated in Fig. 7. The result shows that the lower frequency band has little change and the higher frequency band totally disappears. It demonstrates that the 2.45 GHz patches are the silence elements and are transparent to the frequency band centered at 915 MHz.

## IV. MEASURED RESULTS

After the analysis and optimization, the design parameters are selected for the proposed antenna and they are reported in Table I. The photograph of the fabricated antenna is shown in Fig. 8. The overall height is $59mm \times 59mm \times 45mm$. The comparison between simulated and measured return loss is given in Fig. 9. The solid and the dashed lines denote the measured and simulated return loss, respectively. The antenna exhibits a measured -10 dB return loss bandwidth of 915 MHz,

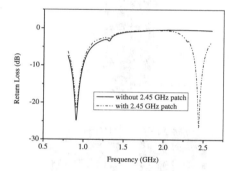

Fig. 7.  Comparison of return loss.

TABLE I
DESIGN PARAMETERS OF THE PROPOSED ANTENNA (UNIT: MILLIMETERS)

| $w$ | $w1$ | $w2$ | $w3$ | $w4$ | $w5$ | $w6$ | $top$ |
|------|------|------|------|------|------|------|-------|
| 2 | 24.9 | 26.5 | 11 | 1 | 9 | 3 | 59 |
| $l1$ | $l2$ | $l3$ | $l4$ | $l5$ | $l6$ | $l7$ | $l8$ |
| 44.5 | 42.5 | 20 | 19.5 | 37.5 | 34.5 | 3 | 10 |
| $l9$ | $g$ | $g1$ | $g2$ | $g3$ | $g4$ | $h$ | $theta$ |
| 5 | 0.6 | 0.4 | 0.8 | 0.48 | 0.46 | 1.6 | 76° |

from 842 MHz to 1.049 GHz, and a high-frequency bandwidth of 2.45 GHz, from 2.4 GHz to 2.5 GHz. Fig. 10 and Fig. 11 show the measured and simulated H-plane and E-plane radiation patterns at 915 MHz and 2.45 GHz, respectively. Both of them are omni-directional in H-plane and tilt by around 30 degrees in E-plane due to the reflection of disk ground. The measured peak gain is 3 dBi at 915 MHz and 5 dBi at 2.45 GHz, as shown in Fig. 12.

## V. CONCLUSION

In this work, a low-profile combined RFID antenna operating at 915 MHz and 2.45 GHz has been designed and fabricated.

Fig. 8.   Photograph of the fabricated antenna.

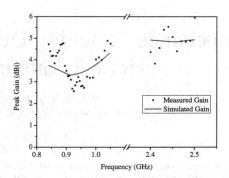

Fig. 12.   Simulated and measured gain of the proposed antenna.

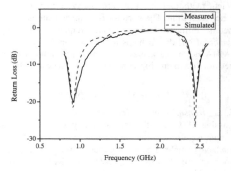

Fig. 9.   Measured and simulated return loss of the proposed antenna.

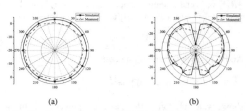

(a)                              (b)

Fig. 10.   Radiation pattern at 915 MHz (a) H-plane (b) E-plane.

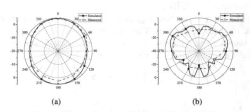

(a)                              (b)

Fig. 11.   Radiation pattern at 2.45 GHz (a) H-plane (b) E-plane.

The overall antenna height is about eighth-wavelength at 915

MHz. The silence element is introduced to realize dual-band and it has little effect on the band of 915 MHz. Instead of conventional tuning poles, the wedge-shaped tuning patches make the antenna easy to design and fabricate. Radiation pattern in both bands is omni-directional and the gain is acceptable, which is very competitive for applications of wireless RFID systems.

### ACKNOWLEDGMENT

This work was supported in part by Suzhou Key Laboratory of RF and Microwave Millimeter Wave Technology, and in part by the Natural Science Foundation of the Higher Education Institutions of Jiangsu Province under Grant No. 12KJB510030. The authors would like to thank the staffs of the Jointed Radiation Test Center of Soochow University.

### REFERENCES

[1] M. Grilo, F. Arnold, M. Gonçalves, L. Bravo-Roger, A. Moretti, and I. Lima, "Novel dual-band rfid antenna configuration with independent tuning adjustment," *Microwave and Optical Technology Letters*, vol. 54, no. 9, pp. 2214–2217, 2012.

[2] H. Guo, W. Cai, C. Zhou, and X. Liu, "Design and realization of a new combined multi-band antenna," *Chinese Journal of Radio Science*, vol. 2, p. 039, 2008.

[3] Z. L. Ma, L. J. Jiang, J. Xi, and T. Ye, "A single-layer compact hf-uhf dual-band rfid tag antenna," *Antennas and Wireless Propagation Letters, IEEE*, vol. 11, pp. 1257–1260, 2012.

[4] A. Mobashsher and R. Aldhaheri, "An improved uniplanar front-directional antenna for dual-band rfid readers," *Antennas and Wireless Propagation Letters, IEEE*, vol. 11, pp. 1438–1441, 2012.

[5] F. Paredes, G. Zamora, F. Herraiz-Martinez, F. Martin, and J. Bonache, "Dual-band uhf-rfid tags based on meander-line antennas loaded with spiral resonators," *Antennas and Wireless Propagation Letters, IEEE*, vol. 10, pp. 768–771, 2011.

[6] S. Palud, F. Colombel, M. Himdi, and C. Le Meins, "A novel broadband eighth-wave conical antenna," *Antennas and Propagation, IEEE Transactions on*, vol. 56, no. 7, pp. 2112–2116, 2008.

[7] L. Tian, H. Guo, X. Liu, Y. Wang, and X. Yang, "Novel 3d rfid antenna with low profile and low cost," in *Antennas, Propagation EM Theory (ISAPE), 2012 10th International Symposium on*, 2012, pp. 69–72.

[8] X. Liu and H. Guo, *Microwave technology and antennas (Chinese)*. Xidian University Press, 2007.

# Impedance Matching Design of Small Normal Mode Helical Antennas for RFID Tags

Yi Liao, Yuan Zhang, Kun Cai, and Zichang Liang

Shanghai Key Laboratory of Electromagnetic Environmental Effects for Aerospace Vehicle

No. 846 Minjing RD

Yangpu District, Shanghai  200438 China

*Abstract*-**The normal mode helical antenna (NMHA) is widely used on RFID applications because it fits well with the requirements of smaller physical size and high inductive impedance. In this paper, a power transmission coefficient method of mapping a modified impedance function onto the conventional Smith chart is applied to conjugate matching design between NMHA and chips. Moreover, the effect of various geometrical parameters of NMHA on resonant frequency and input impedance is investigated. Some significant characteristics of NMHA in the Smith chart are obtained by varying each parameter. It could provide an effective guidance to tune the antenna to desirable complex impedance in the Smith chart. Finally, impedance matching design of a specific NMHA is conducted according to the previously mentioned method.**

## I. INTRODUCTION

The normal mode helical antenna (NMHA) is the kind of helical antenna with the circumference of helix much smaller than a wavelength [1]. The radiation pattern of NMHA is generally similar to that of the dipole antenna and the physical size, with its self resonant structure, is much shorter compared to the dipole antenna [2]. It is even electrically smaller when placed in a dielectric medium. Therefore, the NMHA is widely used for many RF identification (RFID) applications such as portable equipments, TPMS in vehicles, and access control [3]-[5].

A typical RFID tag consists of an antenna and a chip whose impedance is strongly capacitive. Most of the available RFID chips in the ultra-high-frequency (UHF) band exhibit a reactance roughly ranging from -100 Ω to -600 Ω, while the real part is much smaller [6]. Generally, conjugate matching is achieved between antenna and chip for maximum power transfer [7]. A very straightforward way for impedance matching is to use the Smith chart which is usually normalized to 50 Ω. Nevertheless, a method where modified impedance function is mapped onto the conventional Smith chart to determine a power transmission coefficient is much more convenient for the case when both generator and load impedances are complex [8].

Although helical antennas have been known for more than half a century, it seems that reliable formulae for helical antennas do not exist in the open literature [9]. Therefore, numerical tools are essential to helical antenna design and analysis [10]. Resonant frequency and impedance characteristics of NMHA are the functions of its various physical parameters. This paper uses a full-wave numerical modeling tool to calculate the resonant frequency and the input impedance of NMHA [11]. The impedances are mapped onto the Smith Chart based on the power transmission coefficient method. The results are analyzed by varying one parameter each time. The purpose is to investigate how these parameters influence the resonant frequency and input impedance of NHMA. It helps to give a significant guidance for NMHA design to achieve the inductive input reactance required for the microchip conjugate impedance matching in practical cases.

## II. IMPEDANCE MATCHING FOR NMHA

### A. Power transmission coefficient method

In passive RFID tags, the chip is usually directly connected to the input terminals of the tag antenna and both the chip and antenna have complex input impedances. An equivalent circuit is shown in Fig. 1, which represents the RFID tag in the receiving mode.

$V_S$ is the equivalent voltage source in the antenna port, $Z_A = R_A + jX_A$ is the antenna impedance, $Z_L = R_L + jX_L$ is the chip impedance. The reflection coefficient can be expressed by

$$\Gamma = \frac{Z_L - Z_A^*}{Z_L + Z_A}. \tag{1}$$

And the power transmission coefficient

$$\sigma = 1 - |\Gamma|^2 = \left| \frac{Z_L - Z_A^*}{Z_L + Z_A} \right|^2. \tag{2}$$

It shows what fraction of the maximum power available from the generator is delivered to the load and determines the read distance of the RFID device.

Kurokawa described a graphically intuitive way to obtain the reflection coefficient $\Gamma$ plotted in the Smith chart [12]. It makes a transformation to move the $Z_L$ to the center of the Smith chart by using the normalized impedance

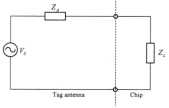

Figure 1. Equivalent circuit for RFID tag.

978-7-5641-4279-7

$$r + jx = \frac{R_A}{R_L} + j\frac{X_A + X_L}{R_L}. \tag{3}$$

The contours of constant reflection coefficient are concentric circles centered around the origin of the Smith chart, which corresponds to a perfect complex conjugate match. Therefore, the power transmission coefficient $\sigma$ can be easily calculated from the Smith chart.

### B. NMHA impedance in the normalized Smith chart

Fig.1 shows the geometry of a typical helical dipole antenna. The helix is uniformly wound with a constant pitch, $S$. The diameter of a helix is $D$. The helix's conductor is a wire of radius, $a$. The antenna is fed at the midpoint of the coil winding with a constant diameter, and uniform circular cross section of the conductor.

In this section, three parameters, $N$, $S$, and $D$ are considered. The geometrical parameters of the antennas grouped in three sets are given in Table 1. Resonant frequencies, as well as the input impedances over the frequency range from resonance to antiresonance in the Smith chart are presented when one of physical parameters is varied. All the impedances shown in Smith Chart are normalized with the reference impedance of 50-j50 $\Omega$. The analysis frequency range is from 700 MHz to 1400 MHz, which covers the UHF RFID bands.

Figure 2. A typical helical dipole antenna.

TABLE I
GEOMETRICAL PARAMETERS OF NMHA

| No | Geometry (a=0.1 mm) | Resonant Frequency |
|----|---------------------|--------------------|
| Different number of turns | | |
| 1 | D=1.2 mm, S=0.7 mm, N=110 turns | 846 MHz |
| 2 | D=1.2 mm, S=0.7 mm, N=115 turns | 813 MHz |
| 3 | D=1.2 mm, S=0.7 mm, N=120 turns | 783 MHz |
| 4 | D=1.2 mm, S=0.7 mm, N=125 turns | 756 MHz |
| 5 | D=1.2 mm, S=0.7 mm, N=130 turns | 730 MHz |
| Different pitch | | |
| 6 | D=1.2 mm, N=120 turns, S=0.5 mm | 827 MHz |
| 7 | D=1.2 mm, N=120 turns, S=0.6 mm | 809 MHz |
| 8 | D=1.2 mm, N=120 turns, S=0.7 mm | 783 MHz |
| 9 | D=1.2 mm, N=120 turns, S=0.8 mm | 755 MHz |
| 10 | D=1.2 mm, N=120 turns, S=0.9 mm | 725 MHz |
| Different diameter | | |
| 11 | S=0.7 mm, N=120 turns, D=1.10 mm | 851 MHz |
| 12 | S=0.7 mm, N=120 turns, D=1.15 mm | 816 MHz |
| 13 | S=0.7 mm, N=120 turns, D=1.20 mm | 783 MHz |
| 14 | S=0.7 mm, N=120 turns, D=1.25 mm | 752 MHz |
| 15 | S=0.7 mm, N=120 turns, D=1.30 mm | 724 MHz |

The number of turns varies from 110 to 130 with an interval of 5 as shown in Table 1. As the number of turns increases the resonant frequency decreases. Fig. 3 indicates that two input impedance locus curves are almost the same in the Smith Chart. Basically, changing the number of turns will not influence the input impedance locus in a significant way as long as $N$ exceeds a critical number [13].

When pitch is varied from 0.5 mm to 0.9 mm the resonant frequency decreases. Fig. 4 depicts two impedance curves for case 6 and case 10. The dashed curve ($S$=0.5 mm) moves outwards to the unity cycle in the Smith Chart. It also indicates that smaller pitch of the antenna has the smaller resistance at resonant frequency.

The increasing diameter lowers the resonant frequency as shown in Table 1. Fig. 5 gives the results when varying $D$. The input impedance is very sensitive to the helix diameter, the dashed curve moves inwards as $D$ increasing.

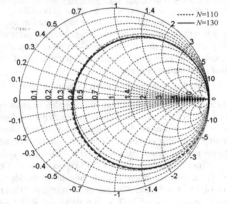

Figure 3. NMHA input impedance mapped onto the Smith chart normalized to 50-j50 $\Omega$ for case 1(dashed curve) and case 5 (solid curve).

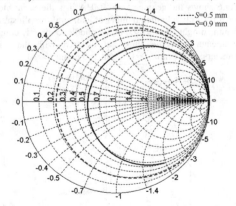

Figure 4. NMHA input impedance mapped onto the Smith chart normalized to 50-j50 $\Omega$ for case 6(dashed curve) and case 10 (solid curve).

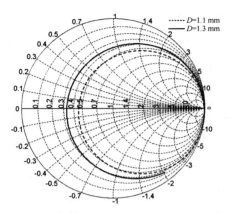

Figure 5. NMHA input impedance mapped onto the Smith chart normalized to 50-j50 Ω for case 11(dashed curve) and case 15 (solid curve).

It can be seen that the behaviors of impedance in the Smith chart are regular within a certain range for each parameter. It is very convenient to get the desired resonant frequency and impedance by changing geometrical parameters for small normal mode helical antennas.

### III. IMPEDANCE MATCHING DESIGN EXAMPLE

Since every chip for RFID tag has complex impedance and there is no consistency for different manufactures, designers usually have to adjust the impedance to realize conjugate matching with chip complex impedance to maximize antenna performance. As an example, consider a UHF tag designed for tire pressure monitoring system (TPMS) in vehicle. The tag employs a normal mode helical antenna embedded in a dielectric block with the dimensions of $100 \times 20 \times 20$ mm$^3$ and relative permittivity of 5. The impedance of RFID chip is $Z_L=27\text{-}j201$ Ω, approximately constant in the frequency range of interest (902 MHz ~ 928 MHz).

Fig.6 shows the geometry of NMHA in a dielectric block. For clarity, only a limited number of helix coils are illustrated. A 3-mm straight wire is used in the middle to directly connect the antenna with the chip. Let us start with the initial parameters: $a=0.1$ mm, $D=1.1$ mm, $S=0.6$ mm, $N=80$.

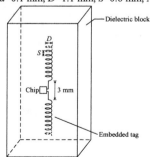

Figure 6. NMHA embedded in the dielectric block.

Fig. 7 shows the frequency-dependent NMHA impedance mapped onto the Smith chart normalized to 27-j201 Ω. Locus (a) is the impedance of initial geometry of NMHA over the frequency range from 660 MHz to 1060 MHz. The reflection coefficient within the considered bands (902 MHz ~ 928 MHz) is far away from the origin of the Smith chart ($|\Gamma| \approx 1$).

For the sake of achieving the conjugate match, we have to adjust the geometrical parameters to make the impedance locus curve move closer to the center of Smith chart. Since the important information we obtain previously is that the impedance locus is not sensitive to the number of turns, the parameters $D$ and $S$ are firstly adjusted to make the NMHA impedance locus move inward and go through the origin of the Smith chart. Then the helices are cut shorter uniformly in two sides of NHMA to obtain the tag resonant frequency and impedance matching. Locus (b) in Fig.7 describes the results of final NMHA with the geometrical parameters: $a=0.1$ mm, $D=1$ mm, $S=0.785$ mm, $N=69$.

The power transmission coefficient over the frequency range of interest (902 MHz~928 MHz) is also calculated from the Smith chart by observing the distance between the origin and the mapped impedance point, as shown in Fig.8. It can be seen that conjugate matching is achieved around 915 MHz where the maximum σ is obtained.

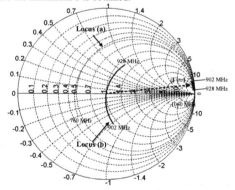

Figure 7. Impedance mapped onto the Smith chart normalized to 27-j201Ω for matching design.

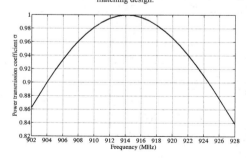

Figure 8. Power transmission coefficient σ versus frequency.

## IV. CONCLUSION

Small normal mode helical antennas are usually directly connected to the chip in RFID tags due to cost and fabrication issues. In this paper, a power transmission coefficient method is employed to get conjugate impedance matching in the Smith chart for NMHA. The resonant frequency and impedance characteristics of NMHA depending on the configuration of the physical parameters are also investigated. Three parameters are considered. It helps to provide effective guidance for designers to tune the impedance of NMHA to desirable value in the Smith chart. Finally, a specific NMHA is given and demonstrated for impedance matching design.

## REFERENCES

[1]   J. D. Kraus, *Antennas*, 2nd ed., New York: McGraw-Hill, 1988, pp. 274-276.

[2]   Y. Hiroi and K. Fujimoto, "Practical usefulness of normal helical antenna," *IEEE AP-S Int. Symp.*, pp. 238-241, 1976.

[3]   H. Morishita, Y. Kim, H. Furuuchi, K. Sugita, Z. Tanaka, et al., "Small balance-fed helical dipole antenna system for handset," *51st IEEE Vehicular Technology Conference Proc.*, Toyko, vol. 2, pp. 1377-1380, May 2000.

[4]   N. Q. Dinh, N. Michishita, Y. Yamada, and K. Nakatani, "Electrical characteristics of a very small normal mode helical antenna mounted on a wheel in the TPMS application," *IEEE AP-S Int. Symp.*, pp. 1-4, 2009.

[5]   Y. Yamada, W. G. Hong, W. H. Jung, and N. Michishita, "High gain design of a very small normal mode helical antenna for RFID tags," *IEEE Region 10 Conference*, pp.1-4, 2007.

[6]   G. Marrocco, "The art of UHF RFID antenna design: impedance-matching and size-reduction techniques," *IEEE Antennas and Propagation Magazine*, vol. 50, no. 1, pp. 66-79, February 2008.

[7]   K. V. S. Rao, P. V. Nikitin, and S. F. Lam, "Impedance matching concepts in RFID transponder design," *4th IEEE Automat. Identification Adv. Technol. Workshop*, pp. 39–42, October 2005.

[8]   P. V. Nikitin, K. V. S. Rao, S. F. Lam, V. Pillai, R. Martinez, et al., "Power reflection coefficient analysis for complex impedances in RFID tag design," *IEEE Trans. on Microw. Theory and Techn.*, vol. 53, no. 9, September 2005.

[9]   A. R. Djordjevic, A. G. Zajic, M. M. Ilic, and G. L. Stuber, "Optimization of helical antennas," *IEEE Trans. on Antennas and Propagation*, vol. 46, pp. 525-530, 1998.

[10]  C. Su, H. Ke, and T. H. Hubing, "A simplified model for normal mode helical antennas," *ACES Journal*, vol. 25, no. 1, pp.32-40, January 2010.

[11]  *"FEKO User's Manual*," EM Software & Systems-S.A. (Pty) Ltd, 32 Techno Avenue, Technopark, Stellenbosch, 7600, South Africa, 2011.

[12]  K. Kurokawa, "Power waves and the scattering matrix," *IEEE Trans. Microw. Theory Tech.*, vol. MTT-13, no. 3, pp. 194–202, March 1965.

[13]  K. H. Awadalla, S. H. Zainud-Deen, and H. A. Sharsher, "Analysis of normal mode helical antenna and scatterer", *IEEE AP-S Int. Symp.*, pp. 1731-1734, 1992.

# Circularly Polarized Antenna with Circular Shaped Patch and Strip for Worldwide UHF RFID Applications

Yi Liu, Xiong-Ying Liu

School of Electronic and Information Engineering, South China University of Technology

Guangzhou 510640, China

liuxy@scut.edu.cn

*Abstract*– A circularly polarized Radio Frequency Identification (RFID) reader antenna is proposed for worldwide Ultra-High Frequency (UHF) (840–960 MHz) applications. The antenna is composed of two circular shaped patches, a ring shaped strip, and a suspended conducting strip with open-circuited termination. The simulations show that the antenna has an impedance bandwidth (S11 < –25 dB) of about 17.4% (807–961 MHz), 3-dB axial ratio (AR) bandwidth of about 14.4% (833-962 MHz), 2-dB AR bandwidth of 12.9% or 842-958 MHz, and gain level of about 9.1 dBic or larger within the 3-dB AR bandwidth. The proposed antenna can be a good candidate for universal UHF RFID readers at the UHF band of 840–960 MHz, which has excellent impedance matching (S11 < –25 dB), high gain (Gain > 9.1 dBic), and good AR (AR< 2 dB) performance.

*Index Terms*— Broadband antenna, circularly polarized (CP), axial ratio (AR), impedance matching, ultra-high frequency (UHF).

## I. INTRODUCTION

Radio Frequency identification (RFID) is a technology that can be used to identify any object that carries an electronic tag by electromagnetic waves [1]. The design of antennas for RFID application has been playing a crucial role in the continuous development of this technology. Owing to the advantages of long detection range, fast reading speed and high data transfer rate, passive RFID systems at ultra-high frequency (UHF) band are preferred in many applications, such as animal tagging, supply chain management, and electronic payment. However, the UHF frequencies authorized for RFID applications are particular in different countries and regions [2-3], for example, 840.5–844.5 and 920.5–924.5 MHz allocated for China, 866–869 MHz for Europe, 920–926 MHz for Australia, 866–869 MHz, 923–925 MHz for Singapore, 902–928 MHz for North and South of America, and 952–955 MHz for Japan. Therefore, if the entire band of 840-955 MHz is covered, the implementation and cost of RFID systems can be simplified and reduced. As is known, the RFID system consists of reader and tag, and the antennas in tag are normally linearly polarized. Considering the arbitrary orientation of the tags, circularly polarized (CP) antennas are required in UHF RFID readers to avoid severe polarization mismatch between readers and tags [4-5]. CP antennas can be obtained by exciting two spatial orthogonal modes of equal amplitude with a 90° phase difference [6]. Various CP reader antennas for UHF RFID system have been researched. A planar reader antenna covering

the ultra-high frequency band of 860–960 MHz is presented in [7] by insetting grounded arc-shaped strip to a square slot and feeding it with F-shaped microstrip line. A crossed dipole reader antenna with phase delay lines is designed in [8], which has a 3-dB axial ratio (AR) band of 905–935 MHz. An aperture-coupled patch antenna for RFID reader with a Wilkinson power divider feeding network is realized in [9], whose 3-dB bandwidth is 909–921 MHz. The mentioned antennas either have a bidirectional radiation pattern or suffer narrow bandwidth, low gain, and complex feeding structure. Hence, a number of stacked patch antenna have been studied recently [1], [6] to achieve broadband CP antenna with high gain and the compromise between performance and complexity.

Comparing with the common square shaped patch CP antenna [6], in this paper, a stacked circular shaped patch CP antenna for UHF RFID applications is designed. The antenna comprises two suspended circular shaped patches, a ring shaped strip and a conducting strip. The main patch is sequentially fed by four probes. A parasitic patch is positioned right above the main patch for improving the bandwidth. In particular, to enhance the performance of AR and impedance matching, the central parts of the two patches are cut off, two slots and a ring shaped strip are added to the main patch on the diagonal and the edge of the parasitic patch, respectively. Finally, the proposed antenna achieves better performance in terms of AR (AR < 2 dB), impedance matching (S11 < – 25 dB), and gain (Gain > 9.1dBic) in the band of 840–960 MHz.

## II. ANTENNA CONFIGURATION

Fig. 1 shows the configuration of the proposed antenna in detail. The antenna is composed of two radiating patches, a suspended conducting feed strip, a ring shaped strip, and a finite- size ground plane. The conducting feed strip with a width of 26 mm is suspended above the ground plane (250 mm × 250 mm) at a height of h1 = 5 mm. Two cuts (15 mm × 2 mm) near the feed point are introduced to enhance the impedance matching. According to previous studies [10], thick air substrate can be used to enhance bandwidth and gain. And the large reactance caused by the thick air spacing and the long probes can be effectively cancelled out with the effect of the strong electromagnetic coupling between the circular shaped conducting strip and the radiating patch [11]. Also, this is an

978-7-5641-4279-7

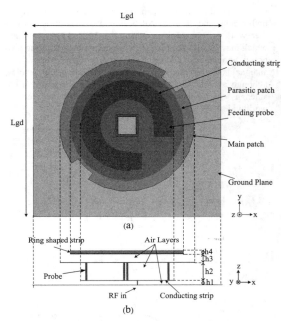

(a)

(b)

Figure 1. Configuration of the proposed antenna. (a) Top view. (b) Side view.

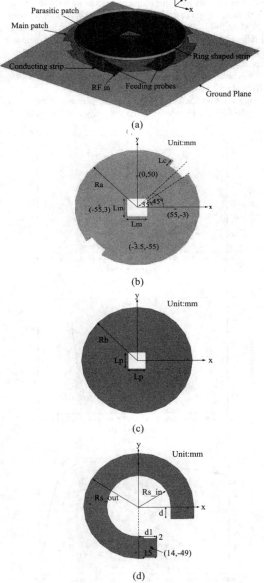

(a)

(b)

(c)

(d)

Figure 2. Antenna detailed dimensions. (a) Antenna structure. (b) Main patch.
(c) Parasitic patch. (d) Conducting feed strip.

important approach to achieve a good impedance matching and broad AR bandwidth. Along this train of thought, large spacing of h2 = 20 mm between the feeding line and the main patch of Ra = 90 mm is adopted. Two slots with radian of 20° and Lc = 10.2 mm is made on the diagonal of the main patch to further widen the AR bandwidth. The main patch and conducting strip are connected by four probes with diameter of 2.2 mm. The position of the probes are optimized along the conducting strip to create 90° phase difference between the probes and two spatial orthogonal modes of equal amplitude that are two necessary conditions to achieve circular polarization. To broaden the bandwidth, a parasitic patch with a dimension of Rb = 76 mm is placed above the main patch with the spacing of h3=10 mm. A squared patch is removed with a dimension of Lm = 30 mm and Lp = 24 mm in the center of the main patch and parasitic patch, respectively, strengthening interaction between the main patch and parasitic patch to further enhance the performance of AR. In particular, a ring shaped strip of h4 = 4 mm are added along the edge of parasitic patch to improve the ration pattern and the AR performance. Fig.2 shows the detailed dimensions with separated parts.

### III.    SIMULATION RESULTS AND ANALYSIS

Simulations of the designed structure were carried out by using Ansoft HFSS software based on finite element method (FEM). The simulation results show that the antenna has exc-ellent impedance matching, good circular polarization, high gain, and hemispherical radiation pattern.

Fig.3 (a) gives the S11 of the antenna. The simulated S11 is less than –25 dB over the frequency range of 807–961 MHz. As shown in Fig.3 (b), the 3-dB axial ratio bandwidth is about 129 MHz from 833–962 MHz or 14.4%, with a minimum axial ratio of 0.36 dB at 860 MHz, and most of the AR is below 2 dB in the operating bandwidth (840–960 MHz). Fig.3(c) exhibits the antenna gain. It is observed that the antenna gain is 9.1 dBic or even larger within the operating bandwidth (840–960 MHz), with a peak gain of 9.8 dBic.

Fig. 4 shows the simulated radiation patterns at 840, 910, and 950 MHz, respectively, in the x-z and y-z planes. It is ob-

served that the radiation patterns are almost symmetrical in both planes, and the 3-dB AR beamwidth is about 69°. A further work can be done to widen the 3-dB AR beamwidth, which may be disturbed by some asymmetric parts of the proposed antenna. In addition, the front-to-back ratio of the antenna is better than 15 dB at all operating frequencies, although a finite size ground plane is used. Therefore, a CP antenna with good performance has been designed for the worldwide UHF RFID reader.

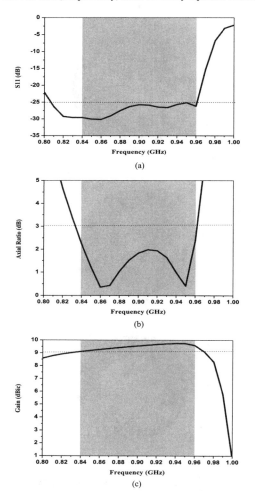

(a)

(b)

(c)

Figure 3. Simulated results of the proposed antenna. (a) S11. (b) AR. (c) Gain.

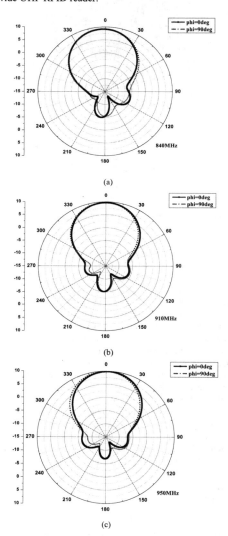

(a)

(b)

(c)

Figure 4. Simulated radiation patterns of the proposed antenna. (a) 840 MHz. (b) 910 MHz. (c) 950 MHz.

## IV. CONCLUSIONS

Comparing with many of the previous works focusing on the design of feeding network, this paper attempts to do some research on the structure of the radiating patches to achieve better performance of the antenna in RFID reader. A broadband circular shaped patch antenna for worldwide UHF (840–960 MHz) RFID applications has been presented. By using sequential feed, stacked structure and added ring shaped strip, the simulation results show that the proposed antenna has an impedance bandwidth (S11 <–25 dB) of 17.4% or 807–961 MHz, 3-dB AR bandwidth of 14.4% or 833–962 MHz, 2-dB AR bandwidth of 12.9% or 842-958 MHz and gain level of 9.1 dBic or even larger within the operating bandwidth. Therefore, the proposed CP antenna can be universal UHF RFID reader antenna.

The next work will be done to make physical antenna model and measure reading rang.

### ACKNOWLEDGMENT

This work is supported by National natural science foundation of China (No.60802004) and Guangdong natural science foundation (No.9151064101000090).

### REFERENCES

[1] Z. Wang, S. Fang, S. Fu and S. Jia, "Single-fed broadband circularly polarized stacked patch antenna with horizontally meandered strip for universal UHF RFID applications," *IEEE Trans. Microw. Theory Tech.*, vol. 59, pp. 1066-1073, Apr. 2011.

[2] Nasimuddin, Z. N. Chen, and X. M. Qing, "Asymmetric-circular shaped slotted microstrip antennas for circular polarization and RFID applications," *IEEE Trans. Antennas Propag.*, vol. 58, no. 12, pp.3821–3828, Dec. 2010.

[3] Z. B. Wang, S. J. Fang, S. Q. Fu, and M. J. Fan, "Single-fed single-patch broadband circularly polarized antenna for UHF RFID reader applications," in *Int. Ind. Inform. Syst. Conf.*, Dalian, China, 2010, pp. 87–89.

[4] K. V. S. Rao, P. V. Nikitin, and S. F. Lam, "Antenna design for UHF RFID tags: A review and a practical application," *IEEE Trans. Antennas Propag.*, vol. 53, no. 12, pp. 3870–3876, Dec. 2005.

[5] E. Mireles, and S. K. Sharma, "A broadband microstrip patch antenna fed through vias connected to a 3dB quadrature branch line coupler for worldwide UHF RFID reader applications," in *Antennas Propag. IEEE Int. Symp.*, Spokane, WA, America, 2011, pp.529-532.

[6] Z. N. Chen, X. Qing and H. L. Chung, "A universal UHF RFID reader antenna," *IEEE Trans. Microw. Theory Tech.*, vol. 57, pp.1275-1282, May 2009.

[7] J. H. Lu, and S. F. Wang, "Planar Broadband Circularly Polarized Antenna with Square Slot for UHF RFID Reader," *IEEE Trans. Antennas Propag.*, vol. 61, no. 1, pp. 45-53,Jan. 2013

[8] Y. F. Lin, Y. K. Wang, H. M. Chen, and Z. Z. Yang, "Circularly Polarized Crossed Dipole Antenna With Phase Delay Lines for RFID Handheld Reader," *IEEE Trans. Antennas Propag.*, vol. 60, no. 3, pp. 1221–1227, Mar. 2012.

[9] D. Yu, Y. Ma, Z. Zhang, and R. Sun, "A Circularly Polarized Aperture Coupled Patch Antenna for RFID Reader," in *Wireless Commun. Networking and Mobile Computing, WiCOM' 4th Int. Conf.*, Dalian, China, 2008, pp. 1-3.

[10] F. S. Chang, K. L. Wong, and T. Z. Chiou, "Low-cost broadband circularly polarized patch antenna," *IEEE Trans. Antennas Propag.*, vol.51, no. 10, pp. 3006–3009, Oct. 2003.

[11] Z. B. Wang, S. J. Fang, S. Q. Fu, and X. J. Li, "Circularly polarized antenna with U-shaped strip for RFID reader operating at 902–928 MHz," in *Signals Syst. Electron. Int. Symp.*, Nanjing, China, 2010, pp.1-3.

# Material Property of On-metal Magnetic Sheet Attached on NFC/HF-RFID Antenna and Research of Its Proper Pattern and Size On

Naoki OHMURA[†], Eriko TAKASE[†] and Satoshi OGINO[†], Yoshinobu OKANO[‡] and Syhota ARAI[‡]

† Microwave Absorbers Inc.
5-4-5, Asakusabashi, Taito-Ku, Tokyo, Japan
‡ Tokyo City University
1-28-1, Tamazutsumi, Setagaya-Ku, Tokyo, Japan

*Abstract*-When a payment system with NFC/HF-RFID is installed in smart-phone, their communication with 13.56MHz is blocked by such metal as battery, coil and PCB. To solve this problem, magnetic sheet with high permeability is set between NFC/HF-RFID antenna and near-by metal. So far thin compound magnetic sheet or thin sintered ferrite has been used.

In order to make thickness of these materials thinner, we propose amorphous magnetic sheet. Besides this amorphous magnetic sheet causes to increase the transmission effect.

## I. INTRODUCTION

NFC and HF-RFID system are used in such traffic toll collection system as SUICA and PASMO in Japan, or in such electric money as EDY [1]. This NFC and HF-RFID system have started being installed in mobile phone and smart phone. Its newly developed applications increase recently. This system uses 13.56MHz band and is blocked by metals nearby NFC/HF-RFID antenna of reader/writer or RF tag [2], [3]. The reason is that flowing electric current on NFC/HF-RFID antenna is lowered by anti - flowing electric current generated by these near-by metal. When there is 10mm gap between NFC/HF-RFID antennas and near-by metal, its communication works.

However smartphone must be thin and at the same time NFC/HF-RFID antenna must be thin, too. Thus it is difficult to make 10mm gap in it. Then as usual, thin sheet with high permeability is installed instead of making gap between them. The reasons are described as below; First- to suppress the anti - flowing electric current, Second – to make magnetic flux flow through this magnetic sheet. By the way, as on-metal magnetic sheet, thin compound magnetic sheet or thin sintered ferrite has been used. But thin compound magnetic sheet cannot be thinner. And thin sintered ferrite cost much because it must be covered not to scatter ferrite powder of cut edge. Besides when both sheet are used as square shape, central square part should be cut off. It causes disposal problems. We has found out higher permeability amorphous sheet and developed how to use them to solve these existing problems. We die cut this amorphous sheet into 4 smaller pieces and set them on each sides. In addition there should be some gap between each piece to reduce anti-flowing electric current. As its permeability exceeds to both compound magnetic sheet and

sintered ferrite, this amorphous sheet can be thinner and lighter than them. In this paper, we will report the experimental results of NFC/HF-RFID communication with amorphous sheet when near or on metal.

## II. TEST MODEL

The system configuration of communication distance measurement system is indicated in Fig.1. We use a smartphones with NFC function and measure the communication distance when magnetic sheet is set between NFC antenna and battery. In this report, the measurement is executed by following two situations; (1) NFC antenna is used as Tx-antenna, (2) NFC antenna is used as Rx-antenna. Frequency of NFC is set to 13.56 MHz. Measurement conditions are indicated in TABLE I. To conjugate match with NFC IC's capacitance, this NFC antenna has large inductance at 13.56 MHz [4]. The amorphous sheet consists of PET cover film, amorphous magnetic sheet and adhesive. Total amount thickness of NFC antenna and amorphous sheet is 0.1 mm.

In the situation of Fig. 1 (1) (hereafter, it is called "read model 1"), the card type passive tag (width X length =50mm X 80mm) according to ISO 14443 [5]-[8] is used. In Fig. 1(2) (hereafter, it is called "read model 2"), NFC antenna communicates with the Reader/Writer antenna in accordance with ISO 14443 as Rx-antenna The communication distance in "read model 1" and "read model 2" is described as $d1$ and $d2$, respectively.

TABLE I CONDITIONS OF BOTH TESTS

| Frequency | 13.56 MHz |
|---|---|
| Size of NFC antenna | 40 × 40 mm |
| Width of NFC antenna pattern | 6 mm |
| Structure of amorphous sheet | PET cover film, amorphous magnetic sheet and adhesive. |
| Total thickness of NFC antenna and amorphous sheet | 0.1 mm |

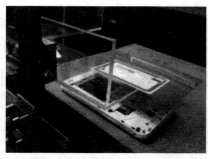

(1)Read model 1

(2) Read model 2

Fig.1 Communication property evaluation procedure for NFC-Tag. (a) In case of NFC-Tag in free space. (b) In case of NFC-Tag equipped RF device.

The characteristic and dimension of magnetic sheet used for this measurement are listed in Tabel II. There are three kinds of magnetism sheet used for the measurement. That is compound magnetic sheet, sintered ferrite and amorphous sheet. 3 types of sheet size and shape are indicated in Fig.2
(1) Patch type : square shape covers whole antenna.
(2) Loop type : square without central part covers only antenna lining pattern.
(3) Strip type : Loop type to be got rid of each corner sheet. When strip type magnetism sheet is used, the same length strips are put on each sides of the antenna. The length of these pieces is all the same. Its length is indicated as below;$l_g$[mm]

TABLE II TYPES OF MAGNETIC SHEET

| Composition | Thickness [μm] | Permeability $\mu$ (Real part) | Permeability $\mu'$ (Imaginary part) |
|---|---|---|---|
| Compound magnetic sheet | 200 | 50 | 1.5 |
| Sintered ferrite | 100 | 100 | 1 |
| Amorphous sheet | 20 | 250 | 400 |

A correlation between the input impedance and the communication distance when the NFC antenna is attached to a metallic object is described in the following chapter. The variation of input impedance and the communication distance when magnetic sheet is inserted between the NFC antenna and a metallic object is also described in next chapter.

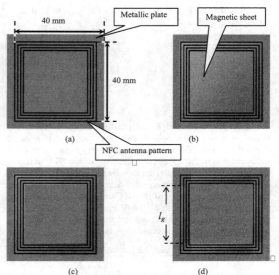

Fig.2 Location procedure of magnetic substance sheet. (a) Basic situation (not magnetic substance sheet loadings). (b) Magnetic substance patch loadings case. (c) Loop-shape magnetic substance sheet loadings case. (d) Magnetic substance strip loadings case.

III. IMPEDANCE MATCHING AND COMMUNICATION DISTANCE

*A.Impedance Matcing*

The permeability (=μ) of magnetic sheet led from the impedance measurement result with a coaxial waveguide is shown by the expression (1).

$$\mu = \frac{Z_S - Z_0}{j\omega\mu_0} \frac{2\pi}{h \ln \frac{c}{b}} + 1 \qquad (1)$$

where,
Zo: Intrinsic impedance of coaxial waveguide
Zs: Input impedance of coaxial waveguide when magnetic substance sheet is inserted
h: Thickness of magnetic substance sheet
c: Internal conductor's radius of coaxial waveguide.
b: External conductor's radius of coaxial waveguide.
On the other hand, 'Zs' is shown by using Rs (=equivalent resistance of magnetic sheet) and Ls (equivalent inductance of magnetic sheet) like expression (2).

$$Z_S = R_S + j\omega L_S \qquad (2)$$

Moreover, μ is separated to real part (=μ') and imaginary part (=μ'') as shown in the expression (3).

$$\mu = \mu' + j\mu'' \qquad (3)$$

Therefore, Ls and Rs are given as a function of μ' or μ" like the expression (4) and (5).

$$L_S = \frac{(\mu'-1)}{\gamma} \qquad (4)$$

$$R_S = Z_0 - \frac{\omega}{\gamma}\mu'' \qquad (5)$$

Fig. 3 shows Ls and Rs as the function of permeability based on actual measurement data. As results, it is guessed that the permeability control of magnetic sheet is indispensable to the impedance matching of the NFC antenna. However, the permeability control by the composition of the magnetic substance is not so easy. Accordingly, it is attempted to control the impedance of the NFC antenna by intermittently attaching magnetic sheet. TABLE III shows the input impedance measurement result of the NFC antenna when procedure for magnetic sheet loading is changed.

TABLE III TEST RESULT OF IMPEDUNCE

| Loading material | Type | Resistance [Ω] | Inductance L [nH] |
|---|---|---|---|
| Cf. Antenna in free space | - | 6.0 | 1220 |
| Without magnetic sheet | - | 5.8 | 200 |
| Compound magnetic sheet | Patch | 12.6 | 912 |
| Compound magnetic sheet | Loop | 11.2 | 872 |
| Compound magnetic sheet | Strip | 9.4 | 768 |
| Sintered ferrite | Patch | 10.4 | 1634 |
| Sintered ferrite | Loop | 7.0 | 988 |
| Sintered ferrite | Strip | 6.0 | 840 |
| Amorphous sheet | Patch | 60 | 1800 |
| Amorphous sheet | Loop | 42 | 1480 |
| Amorphous sheet | Strip | 16 | 1100 |

### B. Communication distance

Fig. 4 shows the comparison measurement results of the NFC communication distance about three magnetic sheet insertion procedures between the antenna and metallic object. The comparison measurement results by two types read model are also shown in Fig. 4.

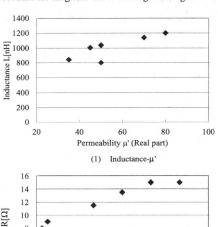

(1) Inductance-μ'

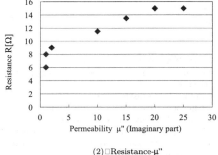

(2) Resistance-μ"

Fig 3. Relation with impedance and permeability.

The result of verifying the influence that the impedance matching for the NFC antenna gives to the communication distance is shown in the next paragraph.

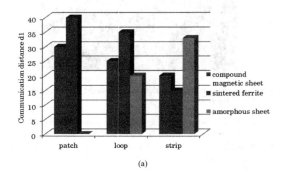

(a)

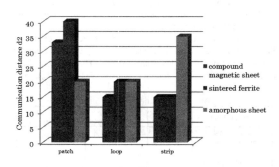

(b)

Fig.4 Communication distance transition for location procedure of magnetic substance sheet. (a) In case of read model 1. (b) In case of read model 2.

In these results, length of the strip type magnetic sheet is fixed to $l_g$=25mm. About compound magnetic sheet and sintered ferrite, as the size of sheet is bigger, so its communication distance becomes longer. On the other hand, as the size of amorphous sheet is bigger, so its communication distance becomes shorter. The Table II indicates that amorphous sheet has higher permeability compound magnetic sheet and sintered ferrite and at the same time has higher loss property. Therefore when amorphous sheet used, the size must be as small as possible to minimize the resistance. However this causes decrease of inductance. The shape and size of amorphous sheet to obtain the desire resistance and inductance is important task. TABLE IV shows test result of communication distance test and impedance of strip type when $l_g$ is changed. Furthermore, the relation with sheet size and impedance of amorphous sheet is summarized in the Fig.5. From the viewpoint of the antenna input reactance reduction, it predicted that the communication distance expand when the amorphous sheet inserted antenna's inductance approaches the inductance in free space (L=1220 [nH]). In this figure, it is shown that the resistance mismatch quantity is an important parameter for the communication distance besides the remaining reactance. When magnetic sheet size is set to $l_g$=25, the communication distance becomes 30 - 50% longer at the loop type pattern.

TABLE V    TEST RESULT OF IMPEDUNCE AND COMMUNICATION
DISTANCE ABOUT STRIP TYPE

| $l_g$[mm] | R [Ω] | L [nH] | $d_1$ [mm] | $d_2$ [mm] |
|---|---|---|---|---|
| 20 | 15 | 840 | 18 | 18 |
| 23 | 16 | 910 | 24 | 21 |
| 25 | 22 | 1160 | 33 | 35 |
| 27 | 26 | 1200 | 27 | 31 |
| 30 | 30 | 1500 | 24 | 30 |

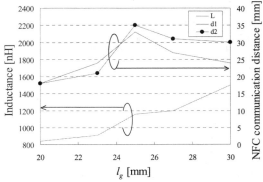

Fig.5. Relation with inductance and communication distance
about strip type .

## IV. CONCLUSION

In this report, impedance matching and communication distance of NFC system was mainly measured with respecting for the permeability and shape of on-metal magnetic sheet. It was confirmed that the amorphous sheet, even if it had not

been used because of its higher permeability at 13.56 MHz, was applicable for NFC antenna with near-by metal by shaping different size of this sheet. The distance of it is as long as theirs of existing compound sheet and sintered ferrite. The longer the communication distance is the wider the size of these 2 sheets are. As their imaginary part of permeability is very low, the resistance is hardly be changed. On the other hand amorphous sheet with higher permeability indicates higher loss property. Therefore it is necessary that with downsizing sheet and suppressing its resistance intrinsic impedance of sheet match to the preferable impedance of this sheet. Before this amorphous sheet is used, it is inevitable to evaluate the impedance of this sheet. Finally, this amorphous sheet is useful for downsizing and cost down for smartphone because even smaller size and thinner thickness is applicable for NFC/HF-RFID on-metal communication. From now on, we will get more data of test and simulation at the same time to optimize the amorphous sheet size and thickness according to NFC/HF-RFID antenna of smartphone.

### REFERENCES

[1]  I.Lacmanovic, B.Radulovic, D.Lacmanovic, "Contactless Paymemt system on RFID Technology"in *proc.33rd Int. Convention. MIPRO* 2010,pp. 1114 - 1119

[2]  P.Marc, Reynaud. J, Rosenberger.C, " Secure payment with NFC mobile phone in the SmartTouch project" in*proc,int. Collaborative Technologies and Systems(CTS) symp.* 2008. pp.121-126

[3]  Ceipidor.U.B, Medaglia.C.M, Marino. A, Sposato. S, Moroni.A," A protocol for mutual authentication between NFC phones and POS terminals for secure payment transactions"in*proc.9th Int. Information Security and Cryptology (ISC) conference.* 2012, pp.115-120

[4]  L.Li,Z.Gao,Y.Wang,"NFC Antenna Research and A Sample Impedance Matching Method" *Electronic and Mechanical Engineering and Information Technology (EMEIT),* vol.8,2011,pp. 3968 – 3972

[5]  ISO/IEC 14443-1:2008 Identification cards – Contactless integrated circuit cards – Proximity cards – Part 1: Physical

[6]  ISO/IEC 14443-2:2001 Identification cards – Contactless integrated circuit (s) cards – Proximity cards – Part 2: Radio frequency power and signal interface

[7]  ISO/IEC 14443-3:2001 Identification cards – Contactless integrated circuit(s) cards – Proximity cards – Part 3: Initialization and anticollision

[8]  ISO/IEC 14443-4:2008 Identification cards – Contactless integrated circuit cards – Proximity cards – Part 4: Transmission protocol

# A Low-Profile Planar Broadband
# UHF RFID Tag Antenna for Metallic Objects

Zhen-Kun Zhang and #Xiong-Ying Liu

School of Electronic and Information Engineering, South China University of Technology

Guangzhou 510640, China

liuxy@scut.edu.cn

*Abstract*—A low-profile planar broadband tag antenna mountable on metallic objects for UHF RFID systems is presented. The antenna has a planar structure without the need of shorting pins or plates, which lowers the cost in mass production. By embedding the odd symmetric slots to the proposed antenna, two adjacent resonant modes are excited to enhance the operating bandwidth. The simulated half-power impedance bandwidth of the proposed antenna is 132 MHz (838-970 MHz) that covers the worldwide UHF RFID frequency bands (840-960 MHz).

## I. Introduction

Radio frequency identification (RFID) in the ultra-high frequency (UHF) has been widely used in access control, electronic ticketing, goods flow systems, and many other emerging applications [1]. The frequency allocation for RFID systems in the UHF band varies in different regions. In China, for example, the frequency range is 840-845 MHz and 920-925 MHz. In Europe, the frequency range 865-868 MHz is allocated to the UHF RFID system whereas it is 902-928 MHz in the US. So it is necessary to design a broadband RFID tag antenna that covers the entire UHF RFID frequency bands (840-960 MHz).

In many practical applications, tags need to be placed on metallic objects. However, some problems must be solved due to the electromagnetic wave scattering from the metallic objects. Some metal tag antennas mountable on metallic objects have been studied and presented [2-4]. These antenna structures, however, have some shortcomings, such as high cost or difficulty of fabrication, because they require shorting pins or plates. For mass production of the tag, the cost of the tag should be as low as possible. Therefore a planar antenna design with no shorting pins or plates is a good candidate for a low cost RFID tag. In [5-6], fully planar antennas have been proposed, but the bandwidth is not wide enough to cover the worldwide UHF RFID frequency bands. Therefore, a UHF RFID tag mountable on metallic objects with completely planar structure and broadband characteristic is desirable and worth developing.

In this paper, a completely planar broadband UHF RFID tag patch antenna with odd symmetric slots is proposed. Two resonant modes are generated by the odd symmetric slots to broaden the half-power bandwidth of the proposed antenna. Details of the antenna design and obtained simulated results are presented and discussed.

## II. Antenna Design

Fig. 1 shows the configuration of the proposed antenna. The tag antenna consists of a radiation U-shape patch with a horizontal slot and odd symmetric slots perpendicular to the horizontal one, which is coupled inductively to the feeding loop. The rectangular feeding loop with a gap is reserved for placing the chip at the center. The feeding loop provides the required input reactance for conjugate matching to highly capacitive chip input impedance. The gap between the loop and the U-shape patch is crucial for the tuning of the input resistance. By appropriately adjusting the dimension of loop size $l6$ and the gap width $g$, the conjugate matching between the antenna impedance and the chip impedance is achieved. By embedding the odd symmetric slots and horizontal slot on the U-shape patch, two adjacent resonant frequencies are excited in the operating bands. The resonant frequencies can be easily tuned by the length of the odd symmetric slots $l4$ and $l5$.

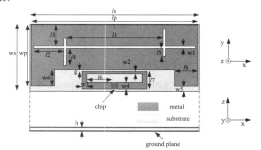

Figure 1. Configuration of the proposed antenna.

The antenna is etched on a thin 1.6 mm FR4 substrate ($\varepsilon_r = 4.4, \tan \delta = 0.02$), which is placed on a 73×30 mm$^2$ ground plane. The simulated results are obtained using Ansoft HFSS based on the finite element method. The antenna is designed for microchip Impinj Monza 3, which has an input impedance of 32-j216 Ω at 915 MHz and a minimum power sensitivity of -15 dBm. The optimized value of each parameter is given in Table 1. In order to study the effect of the parameters, only the chosen parameters are changed at one time, while the others are kept as in Table 1.

978-7-5641-4279-7

TABLE 1. Optimized dimension for the proposed antenna. (unit: mm)

| Parameter | Value | Parameter | Value | Parameter | Value |
|-----------|-------|-----------|-------|-----------|-------|
| ls | 73 | ws | 30 | lp | 72 |
| wp | 29 | l1 | 40 | l2 | 14 |
| l3 | 9.5 | l4 | 7.7 | l5 | 14 |
| l6 | 11 | l7 | 8 | l8 | 10 |
| w1 | 1 | w2 | 1 | w3 | 2 |
| w4 | 3 | w5 | 2 | w6 | 7.4 |
| g | 1.3 | h | 1.6 | | |

### III. RESULTS AND DISCUSSIONS

Fig. 2 shows the simulated input impedance of the proposed antenna with odd symmetric slots and even symmetric slots. As can be seen in the figure, one resonant frequency is generated by the even symmetric slots while two resonant frequencies are generated by the odd symmetric slots due to the perturbation of current flows, making the bandwidth enhancement possible. Fig. 3 shows the simulated current distribution of the radiation patch at 882 MHz and 928 MHz. It is observed that the current paths are controlled by the slots on the patch and we can deduce that the resonant frequencies can be tuned by the length of the odd symmetric slots $l4$, $l5$. The resonant frequency of the antenna with different lengths of $l4$, $l5$ is given in Fig. 4. It can be illustrated by the figure that as the length of $l4$ and $l5$ are increased, the resonant frequencies are moved to the lower frequencies due to the longer current path.

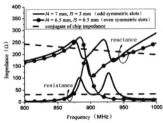

Figure 2. The simulated impedance of the proposed antenna with odd symmetric slots and even symmetric slots.

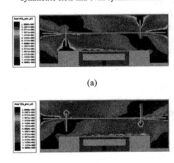

(a)

(b)

Figure 3. Simulated current distribution of radiation patch: (a) at 882 MHz and (b) at 928 MHz.

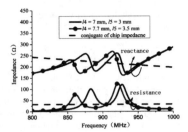

Figure 4. The simulated impedance of the antenna with different lengths of $l4$ and $l5$.

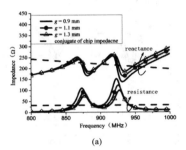

(a)

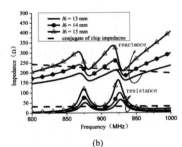

(b)

Figure 5. Simulated impedance characteristics (l4 = 7 mm, l5 = 3 mm): (a) impedance of the antenna with different g and (b) impedance of the antenna with different l6.

The conjugate matching between the antenna and the chip can be easily achieved by adjusting the loop dimension ($l6$) and the gap width ($g$) between the loop and the radiation patch. Fig. 5 depicts the simulated impedance characteristics of the antenna with different $l6$ and $g$. Fig. 5(a) shows the simulated impedance versus frequencies with different $g$. As $g$ increases, the resistance of the antenna is decreased. Fig. 5(b) shows the simulated impedance with different $l6$. As $l6$ increases, the impedance of the proposed antenna is increased. Also, we can see that coarse tuning of the impedance is achieved via $l3$ and fine tuning via $g$. Then, adjusting $l3$ and $g$, the proposed antenna can be easily designed for different microchips with different chip impedance.

The power reflection coefficient, $|s|^2$, which is used to determine the power transfers from the antenna to the chip, can be calculated as:

$$|s|^2 = \left|\frac{Z_c - Z_a^*}{Z_c + Z_a}\right|^2 \qquad (1)$$

where $Z_c$ is the impedance of the chip, $Z_a$ the impedance of the antenna. The simulated power reflection coefficients of the proposed antenna when placed on different sizes of ground plane are presented in Fig. 6. The simulated half-power impedance bandwidth of the designed antenna (ground size: $200 \times 200$ mm$^2$) is 132 MHz from 838 MHz to 970 MHz (fractional bandwidth is 14.6%), which covers the whole bandwidth requirements of UHF RFID.

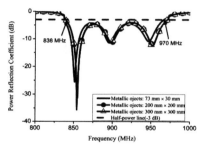

Figure 6. Simulated power reflection coefficient against frequency of the proposed antenna with different sizes of ground plane.

According to the free space Friis formula, the theoretical prediction read range, r, can be obtained by [7]:

$$r = \frac{\lambda}{4\pi} \times \sqrt{\frac{EIRP \times G_{tag} \times (1 - |s|^2)}{P_{th}}} \qquad (2)$$

where $\lambda$ is the free-space wavelength, $EIRP$ the effective isotropic radiated power of the reader, $G_{tag}$ the tag antenna gain, $|s|^2$ the power reflection coefficient, $P_{th}$ the threshold power necessary to provide enough power to the tag chip.

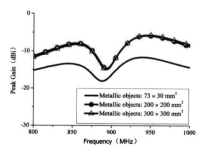

Figure 7. Simulated gains against frequency of the proposed antenna with different sizes of ground plane.

The peak gains of the proposed antenna when it is placed on different sizes of ground plane are shown in Fig. 7. The normalized radiation pattern of the tag when it is set on a $200 \times 200$ mm$^2$ metallic object is shown in Fig. 8. The maximum read distance (*EIRP* is set to 3.2 W) when the tag is set on different sizes of metallic objects at some chosen frequencies are given in Table 2. As shown in Table 2, the maximum read range of the proposed antenna, which is placed on a $200 \times 200$ mm$^2$ metallic object, is up to 4.3 m.

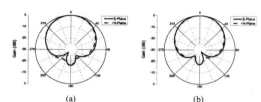

(a)          (b)

Figure 8. Simulated normalized radiation patterns of the proposed antenna: (a) 866 MHz and (b) 915 MHz

TABLE 2. Calculated maximum read range when antenna is placed on different sizes of metallic objects at some chosen frequencies.

| Country | Band centre (MHz) | Metallic object: 73×30 mm$^2$ | Metallic object: 200×200 mm$^2$ | Metallic object: 300×300 mm$^2$ |
|---------|------|------|------|------|
| | | Read range (m) | | |
| China | 843 | 1.8 | 2.8 | 2.6 |
| Europe | 866 | 1.4 | 3.0 | 2.8 |
| USA | 915 | 1.6 | 2.6 | 2.6 |
| China | 923 | 1.8 | 3.0 | 3.0 |
| Japan | 953 | 2.1 | 4.3 | 4.3 |

## IV. CONLUSIONS

A compact broadband UHF RFID tag antenna using odd symmetric slots is designed for metallic objects application. The proposed tag is easy for mass production as it does not need any shorting pins or plates. By embedding the odd symmetric slots in the patch antenna, two resonant modes are excited to achieve a broader bandwidth. The half-power impedance bandwidth of the proposed antenna is 132 MHz (838-970 MHz), i.e. fractional bandwidth 14.6%, covering the worldwide frequency band of UHF RFID (840-960 MHz). The maximum read range of the proposed antenna is about 4.3 m when it is placed on a $200 \times 200$ mm$^2$ metallic object.

ACKNOWLEDGEMENT

This work was supported by the National Natural Science Foundation of China (60802004), in part by the Guangdong Province Natural Science Foundation (9151064101000090).

REFERENCES

[1] K. Finkenzeller, *RFID handbook*, 3nd ed.. Wiley, New York, 2010.
[2] P.H. Yang, Y. Li, W.C. Chew, and T.T. Ye, "Compact metallic RFID tag antennas with a loop-fed method," *IEEE Trans. Antennas and Propag.*, vol. 59 no. 12, pp. 4454-4462, 2011.
[3] M.Y. Lai, and R.L. Li, "Broadband UHF RFID tag antenna with parasitic patches for metallic objects," *Microwave Opt. Technol. Lett.*, vol. 53 no. 7, pp. 1467-1470, 2011.

[4]  L. Xu, B.J. Hu, and J. Wang, "UHF RFID tag antenna with broadband characteristic," *Electron Lett.*, vol. 44 no. 2, pp. 79-80, 2008.

[5]  L.F. Mo, and C.F. Qin, "Planar UHF RFID tag antenna with open stub feed for metallic objects," *IEEE Trans. Antennas and Propag.*,vol. 58 no. 9, pp. 3073-3043, 2010.

[6]  A.A. Eldek, "Miniaturized patch antenna for RFID tags on metallic surfaces," *Microwave Opt. Technol. Lett.*, vol. 53 no. 9, pp. 2170-2174, 2011.

[7]  K.V.S. Rao, P.V. Nikitin, and S.F. Lam, "Antenna design for UHF RFID tags: A review and a practical application," *IEEE Trans. Antennas and Propag.*,vol. 53 no. 12, pp. 3870-3876, 2005.

# *FP-1(B)*

## October 35 (FRI) PM

## Room B

## Inversed Scattering

# Inversion of the dielectric constant from the co-polarized ratio and the co-polarized discrimination ratio of the scattering coefficient

Yuanyuan Zhang[1], Yaqing Li[1], Zhensen Wu[1], and Xiaobing Wang[1,2]

[1]School of Science, Xidian University, Xi'an, Shaanxi, China
[2]National Key Laboratory of Electromagnetic Environment Research, Shanghai, China

*Abstract* –Based on the co-polarized ratio and the co-polarized discrimination ratio of the back scattering coefficient, the inversion of the dielectric constant is discussed in this paper. The value of the co-polarized ratio and the co-polarized discrimination ratio are calculated for minimum effect of surface roughness for the retrieval of the dielectric constant. The ratio of the co-polarized and co-polarized discrimination of the SSA method extends the scope of the application of the ratio method. Combined with the minimum squares techniques and the GA method, two different ratios of the backscattering coefficients were proposed to retrieve the dielectric constant. The retrieved results of the real parts and the imaginary parts are in good agreement with the original results, which proves the validity of the theory primely. What is most significative is that it shows an alternative method of great help for us to retrieve the surface dielectric constant in a wide range by the ratio of the SSA method.

## I. INTRODUCTION

The research for the electromagnetic scattering and inverse scattering plays an extremely important role in the inversion of surface physical parameters and geometric parameters in the microwave remote sensing. The surface parameters inversion has been developed for multi-band multi-polarization, multi-angle surface inversion. Soil moisture retrieval is one of the important applications of microwave remote sensing in recent decades. Soil moisture is a key parameter to the environment, climate change and the grown status of the agriculture. Retrieval of the soil moisture is possible if the dielectric constant of the rough surface is known.

A lot of work has been done in the retrieval of the dielectric constant, and different retrieval algorithms have been designed. Some empirical model and semi-empirical models for the inversion of the bare soil roughness parameters and the dielectric constant are derived, such as the Oh mode [1], Dubois model [2] and Shi model [3], but these models are limited to know a lot of the priori information, and some ignores the retrieval of the correlation length. In addition, some optimization algorithms combined with the electromagnetic scattering theoretical model are used to retrieve the rough surface parameters. Yuequan Wang and Ya-Qiu Jin used the measured data of the backscattering coefficient combined with the Genetic Algorithm to

simultaneously retrieve the soil roughness and dielectric constant, and the retrieved results are in good agreements with the measured data[4]. Ceraldi using the polarization ratio of the Bistatic scattering coefficient combined with the genetic algorithm to retrieve the dielectric rough surface information, but the inversion requires a lot of times to get the true value of the dielectric constant[5]. Based on the measured scattering coefficients of different types of soil, R. Prakash established an empirical formula of C-band polarization ratio, and the inversion results are better than the Kirchhoff approximation, on which the soil composition have a less effect[6]. 2011, Mingquan Jia uses an AIEM method combined with the neural network to retrieve the parameters of the soil surface , with the measured data of the L / S / C / X-band of the soil surface, single frequency dual-polarization and dual-frequency single co-polarization backscattering coefficient, the retrieved results and experimental data are in good agreement, but the process of the retrieval is complex and time costing [7].

The present paper shows the co-polarized ratio and the co-polarized discrimination ratio are alternative methods to retrieve the surface dielectric constant fast and precisely. The co-polarized ratio and the co-polarized discrimination ratio are calculated to minimize the surface roughness effect, so the number of the unknowns are reduced. Additional, the comparison between the two ratio equations are analyzed. The retrieval and the original data are in great agreements, confirming the Correctness of the methods.

## II THEORY FORMULAS

*A. The equations of the co-polarized ratio and the co-polarized discrimination ratio*

In the theory of the rough surface scattering, the SPM, KA, and SSA are the most common and accuracy methods. Based on these methods, the co-polarized ratio and the co-polarized discrimination ratio will be derived in this section.

1) The small perturbation method (SPM)

The SPM method can predicte the scattering coefficient of the Small fluctuation rough surface well, in the condition of which the height standard of the rough surface is much smaller than the incident wavelength and the slope of the rough surface is not high. The scattering coefficient of the SPM can be written as follows:

978-7-5641-4279-7

$$\sigma_{pq}^0 = 8\left(k^2 \cos\theta_s \cos\theta_i \left|\alpha_{pq}\right|\right)^2 W\left(q_x, q_y\right) \quad (1)$$

The co-polarized ratio in the direction of the backscattering is:

$$f_{SPM}^C(\theta,\varepsilon) = \frac{\sigma_{hh}^0}{\sigma_{vv}^0} = \left|\frac{R_{hh}}{R'_{vv}}\right|^2 \quad (2)$$

The co-polarized discrimination ratio in the backscattering direction can be expressed as:

$$f_{SPM}^D(\theta,\varepsilon) = \frac{\sigma_{vv}^0 - \sigma_{hh}^0}{\sigma_{vv}^0 + \sigma_{hh}^0} = \frac{\left(R_{vv}'^2 - R_{hh}^2\right)}{\left(R_{vv}'^2 + R_{hh}^2\right)} \quad (3)$$

where

$$R'_{vv} = \frac{(\varepsilon-1)\left(\sin^2\theta - \varepsilon(1+\sin^2\theta)\right)}{\left(\varepsilon\cos\theta + \sqrt{\varepsilon - \sin^2\theta}\right)^2} \quad (4)$$

$$R_{hh} = \frac{\left(\cos\theta - \sqrt{\varepsilon - \sin^2\theta}\right)}{\left(\cos\theta + \sqrt{\varepsilon - \sin^2\theta}\right)} \quad (5)$$

2）The Scalar approximation Kirchhoff equations

When the radius of the curvature is much great than the incident length, the rough surface scattering plane can be treated as an infinite plane tangent. In this case, the method of the Scalar Kirchhoff approximation can be used to calculate the scattering coefficient of the rough surface.

Supposing the rough surface satisfy the Gaussian random process and Gaussian correlation, the scattering coefficient can be written as three components:

$$\sigma_{pqc}^r = \pi k_1^2 \left|a_0\right|^2 \delta(q_x) \delta(q_y) e^{-q_z^2 \sigma^2} \quad (6)$$

$$\sigma_{pqn}^r = \left(\left|a_0\right| k_1 l/2\right)^2 \exp(-q_z^2 \sigma^2)$$
$$\cdot \sum_{n=1}^{\infty} \frac{(q_z^2 \sigma^2)^n}{(n!n)} \exp(-\frac{(q_x^2 + q_y^2)l^2}{4n}) \quad (7)$$

$$\sigma_{pqs}^r = -(k_1 \sigma l)^2 \left(q_z/2\right) \exp(-q_z^2 \sigma^2)$$
$$\cdot \mathrm{Re}\left\{a_0\left(q_x a_1^* + q_y a_2^*\right)\right\} \times \sum_{n=1}^{\infty} \frac{(q_z^2 \sigma^2)^{n-1}}{n!n} \quad (8)$$
$$\cdot \exp(-\frac{(q_x^2 + q_y^2)l^2}{4n})$$

The co-polarization ratio in the backscattering direction is:

$$f_{KA}^C(\theta,\varepsilon) = \frac{\sigma_{hh}^0}{\sigma_{vv}^0} = \left|\frac{R_{hh}}{R_{vv}}\right|^2 \quad (9)$$

Which ignores the incoherent scattering component $\sigma_{pqs}^r$.
The co-polarization discrimination ratio in the backscattering direction can be expressed as:

$$f_{KA}^D(\theta,\varepsilon) = \frac{\sigma_{vv}^0 - \sigma_{hh}^0}{\sigma_{vv}^0 + \sigma_{hh}^0} = \frac{\left(R_{vv}^2 - R_{hh}^2\right)}{\left(R_{vv}^2 + R_{hh}^2\right)} \quad (10)$$

where the value of $R_{hh}$ can refer to the equations (5) and the value of $R_{vv}$ can be written as:

$$R_{vv} = \frac{\varepsilon\cos\theta - \sqrt{\varepsilon - \sin^2\theta}}{\varepsilon\cos\theta + \sqrt{\varepsilon - \sin^2\theta}} \quad (11)$$

3）The method of the Small slope approximation

The height standard of the rough surface is not limited in the SSA method. But it requires that the rms slope of the rough surface is low. The SSA method can respectively degenerate to the Kirchhoff approximation and the small perturbation method. With SSA-1, the bistatic scattering coefficient can be expressed as:

$$\Sigma(\bar{k}, \bar{k}_0) = \frac{1}{\pi}\left(\left|\frac{2qq_0}{q+q_0} B(\bar{k}, \bar{k}_0)\right|\right)^2 \quad (12)$$
$$\cdot \int\langle \exp\left[j(q+q_0)\cdot(z_2 - z_1)\right]\rangle \exp\left[-j(\bar{k} - \bar{k}_0)\cdot \bar{r}\right] d\bar{r}$$

where $W(\bar{r})$ is the Autocorrelation function and $W(0) = \sigma^2$.
$(k_0, k)$ are the horizontal projections of the incident wave vector and the scattering wave vector. $(q_0, q)$ are the vertical components of the incident wave vector and the scattering wave vector. $B(\bar{k}, \bar{k}_0)$ is associated with the dielectric constant and the incident vector wave.
Where

$$B_{vv}(\bar{k}, \bar{k}_0) = \frac{\varepsilon-1}{(\varepsilon q^{(1)} + q^{(2)})(\varepsilon q_0^{(1)} + q_0^{(2)})}(q^{(2)} q_0^{(2)} \frac{\bar{k}\cdot\bar{k}_0}{kk_0} - \varepsilon kk_0) \quad (13)$$

$$B_{hh}(\bar{k}, \bar{k}_0) = -\frac{\varepsilon-1}{(q^{(1)} + q^{(2)})(q_0^{(1)} + q_0^{(2)})}\frac{\omega^2}{c^2}\frac{\bar{k}\cdot\bar{k}_0}{kk_0} \quad (14)$$

So the co-polarization ratio in the backscattering direction is:

$$f_{SSA}^C(\theta,\varepsilon) = \frac{\sigma_{hh}^0}{\sigma_{vv}^0} = \left|\frac{B_{hh}}{B_{vv}}\right|^2 \quad (15)$$

The co-polarization discrimination ratio in the backscattering direction can be expressed as:

$$f_{SSA}^D(\theta,\varepsilon) = \frac{\sigma_{vv}^0 - \sigma_{hh}^0}{\sigma_{vv}^0 + \sigma_{hh}^0} = \frac{B_{vv}^2 - B_{hh}^2}{B_{vv}^2 + B_{hh}^2} \quad (16)$$

## III RETRIEVAL ALGORITHM

According to the above co-polarization ratio and the co-polarization discrimination ratio in the backscattering direction, the retrieval of the dielectric constant can be conducted based on the minimum squares theory. The evaluation function of the inversion can be written as

$$g(\varepsilon',\varepsilon'') = \sum_i \left|f(\varepsilon',\varepsilon'',\theta_i) - \hat{f}(\theta_i)\right|^2 \quad (17)$$

Where $\hat{f}(\theta_i)$ is the experimental data, and $f(\varepsilon',\varepsilon'',\theta_i)$ is the theoretical results of the co-polarization ratio or the co-polarization discrimination ratio in the backscattering direction, and $\varepsilon'$, $\varepsilon''$ refer to the real and imaginary parts of the dielectric constant. The Genetic Algorithm is used to find the minimization of the function(17). The co-polarization ratio and the co-polarization discrimination ratio in the backscattering direction can be used to retrieve the dielectric constant in a wider range of the rough surface roughness.

## IV  NUMERICAL RESULTS AND DISCUSSIONS

### A  For the small fluctuant rough surface

For the soil surface, as shown in the Fig. 1, the SPM method can give a good description of the small fluctuant rough surface, with the dielectric constant of the soil surface (15.3,-3.7). Now we can use the co-polarized ratio and the co-polarized discrimination ratio of the backscattering coefficient to retrieve the dielectric constant of the soil minimizing the surface roughness.

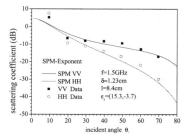

Figure 1. The backscattering coefficient of soil surface via the incident angle with SPM method

The co-polarized ratio formula and the measured data were put together in a least-squares equation with GA method yields the retrieved value of the dielectric constant. In the inversion process, the population size of each generation is 6000.The maximum number of iterations is 10000. The chromosome mutation probability is 2%, and the crossover probability is 90%. Finally, the dielectric constant of the retrieved results is (15.29,-3.70), and the evaluation value is 5.03e-14. As shown in the Fig. 2, the predicted and actual values of the ratio are in good agreements.

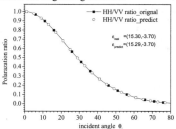

Figure 2. The co-polarized ratio via the incident angle

For the co-polarized discrimination ratio, the same method is applied for the retrieval of the dielectric of the soil surface. It can also minimize the effect of the roughness of the soil surface. The parameters of the soil surface are shown in the Fig. 1. The retrieved results of the dielectric is (15.29, -3.71), the value of evaluation is 1.747e-13. The co-polarized discrimination ratio of the original and the retrieved were compared in Fig. 3. The good agreements show the correctness of the method.

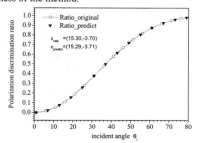

Figure 3. The co-polarized discrimination ratio via the incident angle

### B.  For the moderate fluctuant rough surface

The Small slope approximation method had a wide range of the surface roughness. It can predict the scattering coefficient effectively and precisely. As shown in the Fig. 4, the SSA method is used.

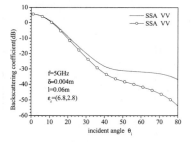

Figure 4. The backscattering coefficient of surface via the incident angle with SSA method

Referring to the equation (15), (16) and (17), the co-polarized ratio and the co-polarized discrimination ratio for the SSA method are used to retrieve the dielectric of the rough surface in the same way, but the equations of the ratio are different from that of the SPM. The retrieved results of the dielectric constant, using the co-polarized ratio, are shown in the Fig. 5. The retrieved values of the dielectric constant are (6.79,-2.80), and the evaluation value is 4.17e-12. In the same way, the retrieved results of the dielectric constant, using the co-polarized discrimination ratio, is shown in the Fig. 6. The retrieved value of the dielectric constant is (6.80,-2.80), and the evaluation value is 2.56e-15.The good agreement proves the validity of the theory primely.

To demonstrate this method is very effective in retrieving the dielectric constant, more dielectric constant values will be analyzed and retrieved. The real parts and imaginary parts of the retrieval results are shown in Fig. 7.

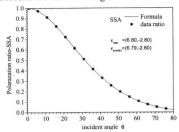

Figure 5. The co-polarized ratio of the SSA method via the incident angle

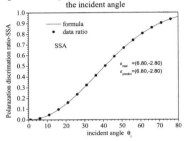

Figure 6. The co-polarized discrimination ratio of the SSA method via the incident angle

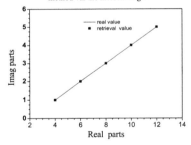

Figure 7. The original and the retrieved dielectric constant

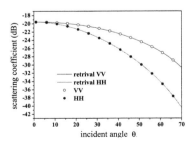

Figure 8. The comparison between the retrieval results and the original results with SPM method

In the Fig. 7, the inverse results are very accurate using the co-polarization ratio equations and the GA method. Put one of the inversion results, (3.99998, -1.00005), into the SPM method and calculate the backscattering coefficients, as shown in the Fig. 8, the calculation results are in good agreement with the original data of the backscattering coefficients, which the true dielectric constant is (4.0, -1.0).

## V. CONCLUSION

In this paper, the ratio of the co-polarized and co-polarized discrimination of the SSA method extends the scope of the application of the ratio method. Combined with the minimum squares techniques and the GA method, two different ratios of the backscattering coefficients were proposed to retrieve the dielectric constant in this paper, with the method of SPM, KA and SSA. The ratio of the co-polarized and the co-polarized discrimination of the backscattering coefficients minimized the roughness of the surface and it was only related to the dielectric constant. In a wide range of the roughness, the ratio exists, for the method of the SPM, KA, and SSA. The great agreements on the retrieved results and the original results of the dielectric constant, confirm the present retrieved method. It is a alternative method of great help for us retrieve the surface dielectric constant precisely.

### ACKNOWLEDGMENT

This work was supported by the National Natural Science Foundation of China under Grant 61172031.

### REFERENCES

[1] Y. Oh, K. Sarabandi, and F. T.Ulaby, "An empirical model and an inversion technique for radar scattering from bare soil surfaces " *IEEE Transactions on Geoscience and Remote Sensing,* vol. 30, pp. 370-381, 1992.

[2] P. C. Dubois, J. Van Zyl, and T. Engman, "Measuring soil moisture with imaging radars," *Geoscience and Remote Sensing, IEEE Transactions on,* vol. 33, pp. 915-926, 1995.

[3] J. Shi, J. Wang, A. Y. Hsu, *et al.*, "Estimation of bare surface soil moisture and surface roughness parameter using L-band SAR image data," *Geoscience and Remote Sensing, IEEE Transactions on,* vol. 35, pp. 1254-1266, 1997.

[4] Yuequan Wang, Yaqiu Jin," The roughness and the soil moisture of the land surface simultaneously are retrieved with the Genetic Aglorithm," Journal of Remote Sensing, vol. 4, 2000.

[5] D. Singh, "Polarization discrimination ratio approach to retrieve bare soil moisture at X-band ", 2005.

[6] R.Prakash, S. D, and N.P.Pathak, "The Effect of Soil Texture in Soil Moisture Retrieval for Specular Scattering at C-Band," *PIER,* vol. 108, pp. 177-204, 2010.

[7] J. Mingquan, C. Yan, T. Ling, *et al.*, "Land-Based Scatterometer Measurements and Retrieval of Surface Parameters Using Neural Networks," in *Education Technology and Training, 2008. and 2008 International Workshop on Geoscience and Remote Sensing. ETT and GRS 2008. International Workshop on,* 2008, pp. 448-452.

# Microwave Radiation Image Reconstruction Based on Combined TV and Haar Basis

Lu Zhu   Jiangfeng Liu   Yuanyuan Liu

School of Information Engineering East China Jiaotong University, NanChang, 330013 China

*Abstract*—Due to the complicated structure of microwave radiometric imaging system and the massive amount of data collection in one snapshot, it is difficult to achieve good performance by interferometry based on the Nyquist sampling and conventional microwave radiation imaging method. In this paper, we use the random observation matrix to sparsely sample microwave radiation image on the basis of digging compressible information of microwave radiation image, reducing the amount of data collection. Considering that sparseness of microwave radiation image on traditional sparse basis (such as: TV and Haar wavelet sparse basis) can't meet the requirements, we combine TV and Haar wavelet sparse basis to sparsely represent microwave radiation image, adopt the method of OMP algorithm finding the optimal atom to reconstruct the original microwave radiation image. We use synthesis data to simulate the three methods of combined TV and Haar wavelet, TV, Haar wavelet. The simulation results show that the proposed method to reconstruct microwave radiation image is better than the reconstruction method of single orthogonal basis.

*Index Terms*—Aperture Synthesis Radiometers; Combined TV and Haar wavelet; OMP; Compressed Sensing (CS).

## I. INTRODUCTION

PASSIVE microwave satellite-borne imaging has become of increasing interest for scientific, military and commercial applications over the last years. If we can get precise soil moisture data by inverting the obtained microwave radiation image, it would enhance the accuracy of weather forecasts by the analysis of soil moisture data, and would effectively monitor drought, flood and other geological disasters. Interferometric synthetic aperture microwave radiometry[1] integrates the small-bore array into large observation bore, doesn't need a mechanical scanning for imaging, and solves the disadvantages of real aperture microwave radiometers. But spaceborne ISAMR of L-Band still need diameter up to 9 m antenna array for 50 km spatial resolution[2]. And with the development of refinement and structuralization of the image, we must increase the diameter of the antenna array to meet the needs of high resolution, ISAMR has evolved into an enormous and complex system, data collected easily reaches tens of millions in one snapshot. So it is difficult to achieve good performance by interferometry based on the Nyquist theory and conventional microwave radiation

imaging method. Because of complex structure and low imaging resolution, present microwave radiometric imaging system seriously limits its practical application in regional soil moisture remote sensing.

Compressed Sensing (CS) theory is a great breakthrough in the field of information processing in recent years[3]-[5], it has changed people's traditional concept of information acquisition. The core idea of CS theory is that it applies the sparse prior knowledge of signal representation to the process of signal reconstruction, uses far less than the Nyquist sampling rate to reconstruct original signal, and thus effectively reduces the complexity of the sensor and the sampling system. That is to say, assuming that interferometry on Nyquist sampling collects 100 datum to accomplish the microwave radiation imaging, for example, sparseness is 0.1, using the CS method, it only need about 40 datum to achieve the same imaging resolution. If the signal is sparse enough or compressible, the actual random collection data can be less. With the increasing amount of data acquisition, the advantages of CS method are highlighted increasingly.

In CS method, the sparse representation of the image is a precondition for image reconstruction. In general, natural signal in time domain or the image in space domain isn't sparse, but it can be sparse in Wavelet, Ridgelet, Curvelet and Contourlet transform domain, etc. However, the usage of single orthogonal basis is difficult to sparsely represent the soil microwave radiation image of complex scene. In this paper, we utilize the combined TV and Haar wavelet sparse basis to sparsely represent microwave radiation image, use the observation matrix $[I \ R]$ [6] to sparsely sample microwave radiation image, adopt the method of OMP algorithm finding the optimal atom to reconstruct the data of sparse sampling, acquire actual microwave radiation image.

The rest of this paper is organized as follows. Section II discusses the basic theory. Signal reconstruction method based on combined TV and Haar wavelet is detailed in Section III. Section IV demonstrates the performance of our method, and verifies the performance of three kinds of sparse basis. In Section V, we conclude our work.

## II. THE BASIC THEORY

### A. Microwave radiation imaging method

In the Aperture Synthesis Microwave Radiation Image (ASMRI), measurement visibility function V(u) and the brightness temperature distribution of the observed scene satisfy the following relation:

The work was supported by the Natural Science Foundation of Jiangxi Province (No. 2010GQS0033) and National Nature Science Foundation of China (No. 61162015, No.31101081).

Corresponding author: Lu Zhu (e-mail: luyuanwanwan@163.com)

978-7-5641-4279-7

$$V(u) \propto \iint_{\|\xi\|<1} \left[ T(\xi) \Big/ \sqrt{1-\|\xi\|^2} \right]$$
$$\times F_k(\xi) F_l^*(\xi) r(-u\xi / f_0) e^{-j2\pi u\xi} d\xi \qquad (1)$$

where $F_{k,l}$ is the normalized antenna voltage pattern, $f_0$ is the center frequency, r is stripe function, u is the baseline coordinate of normalized wavelength, $\xi$ is direction cosine. Assuming that the number of zero-base line of ASMRI measurement is M, the M equations set as shown in (1) can be expressed as by matrix

$$V = G \cdot T \qquad (2)$$

where V is the column vector of $(M+1) \times 1$ visibility function sampling, referred to as visibility sampling, T is the $J \times 1 (J > M)$ discrete brightness temperature distribution, model operator G is the shock response of ASMRI. From the view of visibility sampling reconstruction of ASMRI measurement, the brightness temperature distribution of the observed scene is an ill-posed inverse problem [7].

### B. The basic theory of CS

The essence of CS theory is a kind of non-adaptive and nonlinear compressible signal reconstruction method, its main content is that on certain basis (called sparse basis) $\Psi = [\psi_1, \psi_2, \cdots, \psi_N]$, the m - sparse description and N sampling signal $x \in R^N$, namely $x = \sum_{n=1}^{N} \alpha(n) \psi_n$, can be precisely reconstructed by the M( $m \leq M << N$ ) linear projection $y(i) = \langle x, \phi_i^T \rangle$, $i \in \{1, 2, \cdots, M\}$, on the another incoherent basis $\Phi = [\phi_1^T, \phi_2^T, \cdots, \phi_M^T]$ (referred to as the measurement basis). Using the matrix form, measurement process can be referred to as $y = \Phi x = \Phi \Psi \alpha$. Where $\alpha$ is the m - sparse transformation coefficient vector, M dimension column vector y is measurement vector, $M \times N$ dimension matrix $\Phi$ is measurement matrix. Its goal is accurately reconstructing or approximating signal x by the measurement vector y obtained by M (M<<N) measurements.

The condition $m \leq M << N$ shows that CS theory is mainly to solve signal reconstruction problem under the undersampling condition. Condition M<<N makes the reconstruction of the signal x essentially become a ill inverse problem, but the first reason it is possible to recover signals from a small amount of measurement is the sparsity of original signal x itself, while the second reason is incoherence between the measurement basis with sparse basis, therefore, in order to ensure the reconstruction, measurement matrix must satisfy certain constraints. Candes, etc[4] gives the following conclusion: in order to reconstruct sparse or compressible signal, measurement matrix must meet Restricted Isometry Property (RIP) of certain parameters. Here we first present the definition of RIP.

Definition 1(RIP): For the matrix $\Phi \in R^{M \times N}$, if all meet the index set of $|I| \leq m < M$, $I \subset \{1, 2, \cdots, N\}$ and arbitrary vector

$v \in R^{|I|}$, there is constant $0 < \delta < 1$ making that:

$$(1-\delta) \|v\|_{l_2} \leq \|\Phi_I v\|_{l_2} \leq (1+\delta) \|v\|_{l_2} \qquad (3)$$

true, so calling the matrix meets the RIP of parameters $(m, \delta)$, where $\Phi_I$ is the sub-matrix composed by the column vectors of index set $I \subset \{1, 2, \cdots, N\}$ referring to $\Phi$, and the infimum of all parameters $\delta$ making the type (3) set up refers to as Restricted Isometry Constant (RIC), refers to as $\delta_m$ [8].

Let us denote measurement matrix $\Phi = [I \ R]$, the $\Phi$ matrix can be written as:

$$\Phi = \begin{pmatrix} 1 & 0 & \cdots & 0 & \phi_{1\,M+1} & \cdots & \phi_{1\,N} \\ 0 & 1 & \cdots & 0 & \phi_{2\,M+1} & \cdots & \phi_{2\,N} \\ & & \ddots & & \vdots & \vdots & \vdots \\ 0 & 0 & \cdots & 1 & \phi_{M\,M+1} & \cdots & \phi_{M\,N} \end{pmatrix}$$

By using $[I \ R]$ as the measurement matrix, the first M sensor nodes do not have any computation load, and the rest of nodes have the same computation and communication load as in the basic random measurement matrix.

Theorem 1: Let R be a $M \times (N-M)$ matrix with elements drawn according to $N(0, 1/M)$ and let I be an $M \times M$ identity matrix. If $M \geq C_1 K \log \left( \dfrac{N}{K} \right)$, then $[I \ R]$ satisfies the RIP of order K with probability exceeding $1 - 3e^{-C_2 M}$, where $C_1$ and $C_2$ are constants[6].

### III. SIGNAL RECONSTRUCTION METHOD BASED ON COMBINED TV AND HAAR WAVELET

### A. Sparse representation

The precondition of CS theory is that the signal is sparse or compressible, in order to make the model simplification, only considering length N and discrete real signal x, named x(n), $n \in [1, 2, \cdots, N]$. Known by the signal theory, x can be expressed with linear combination of a set of basis $\Psi = [\psi_1, \psi_2, \cdots, \psi_N]$,

$$x = \sum_{n=1}^{N} \alpha(n) \psi_n, \text{ where } \alpha(n) = \langle x, \psi_n \rangle, \alpha \text{ and } x \text{ are } N \times 1$$

matrix, $\Psi$ is a $N \times N$ matrix. When the signal x on certain basis $\Psi$ has only K<<N nonzero coefficients $\alpha(n)$, named $\Psi$ as the sparse matrix of signal x[9].

The image is not sparse in time domain, so we use TV, Haar wavelet basis and combined TV and Haar wavelet to transform the image to sparse domain, observe three sparse transformation results and compare the sparsity between them.

Using these three sparse basis to analyze the sparse prior knowledge of the image, using threshold processing method, the sparse ratio of TV transformation is 0.206, the sparse ratio of the third-order Haar wavelet transformation is 0.282, the sparse ratio of TV + third-order Haar wavelet transformation is

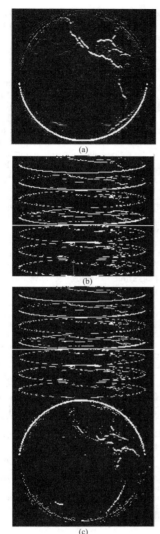
(a)

(b)

(c)

Fig. 1. The transformation of image using sparse basis: (a) The transformation of the image using TV, (b) The transformation of the image using third-order Haar wavelet, (c) The transformation of image using TV + third-order Haar

0.178, we can conclude that sparse property of Combined TV and Haar wavelet sparse basis is the best.

### B. Image reconstruction method

For a signal $x \in R^{N \times 1}$, via the M linear measurement, $f = \Phi x$, $\Phi \in R^{M \times N}$, here each row of $\Phi$ can be regarded as a sensor, it acquires a part of the signal information by multiplying the signal. Analyzing the sparse prior knowledge of

the signal, namely $\theta = \Psi x$, $\Psi \in R^{N \times N}$, we obtain a strict mathematical optimization problem[10]:

$$\min_x \|\theta\|_0 \quad Subject \ to \quad f = \Phi x \qquad (4)$$

Extracting the image blocks, regarding each column of the image as a vector, namely an image block $x_i$, we can get an optimization problem:

$$\min_x \sum_i \|\theta - \Psi x_i\|_2^2 + \lambda \|f - \Phi x_i\|_2^2 \quad s.t. \quad \|\theta\|_0 \le s \quad \forall i. \quad (5)$$

This problem is a NP hard problem, but it can be solved by greedy algorithm. The current epidemic greedy algorithm is OMP algorithm. We analyze sparsity of the image using TV, Haar wavelet basis and combined TV and Haar wavelet basis, finding the sparse coefficient and its position of the sparse representation of the observation vector, so as to reconstruct the image, and then combine all the blocks together, gaining the reconstructed image.

Algorithm steps are organized as follows:

**Input:** *measurement matrix* $\Phi$ *, sparse matrix* $\Psi$ *, measurement vector* $f$ *, the number of iterations* $K$ *;*

**Output:** *reconstructed original signal* $x_i$ *;*

**Initialization:** *residual* $r_0 = f$ *, Index set* $\Lambda_0 = \varnothing$ *, t = 1, restoration matrix* $T = \Phi \Psi^{-1}$ *,* $\hat{x} = 0$ *;*

**Iterative steps 1-6:**

- *Step 1: finding the foot mark of the maximum product between residual r with column $T_q$ of restoration matrix, namely* $\lambda_i = \arg \max_{j=1...M} |\langle r, T_j \rangle|$ *;*

- *Step 2: updating the index set* $\Lambda_t = \Lambda_{t-1} \cup \{\lambda_i\}$ *, recording the atom set* $W_t = [W_{t-1}, T_{\lambda_i}]$ *found in the restoration matrix;*

- *Step 3: getting* $\hat{x}_t = \arg \min \|f - W_t \hat{x}\|_2^2$ *through the least-square method;*

- *Step 4: updating the residual* $r_t = f - W_t \hat{x}_t$ *, t = t+1;*

- *Step 5: judging whether t>K, if met, it will stop, if not satisfy, then perform step 1;*

- *Step 6: the reconstruction of the original signal* $x_i = \Psi^{-1} \hat{x}_t$ *.*

Finally we gain the original image $x$ via the integration of the reconstructed image block $x_i$.

### IV. EXPERIMENT

In this section, we conduct simulating experiments of three kinds of sparse basis, comparing the performance of three kinds of sparse basis, finally concluding that what kind of sparse basis is better to reconstruct the signal. We use the microwave radiation image, its size is $180 \times 180$, firstly observe signal by observation matrix $[I \ R]$, and the size of the observation matrix is $140 \times 180$, then analyze the sparse prior knowledge of the image, based on the optimization problem. We use the OMP

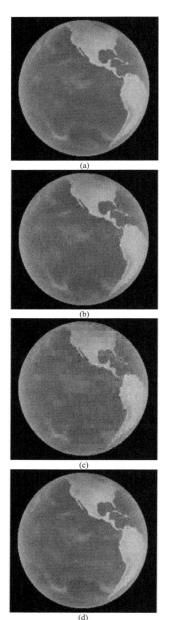

(a)

(b)

(c)

(d)

Fig. 2. Performance of method: (a) The original image, (b) The reconstructed image using TV, (c) The reconstructed image using third-order Haar wavelet, (d) The reconstructed image using TV + third-order Haar wavelet

TABLE 1 The comparison of three sparse basis performance

| Sparse basis | MSE | PSNR | M | Running time |
|---|---|---|---|---|
| TV | 5.3473 | 40.8494 | 140 | 9.8594 |
| HAAR | 22.4322 | 34.6221 | 140 | 12.4219 |
| TV+HAAR | 1.7334 | 45.7419 | 140 | 118.9688 |

reconstruct the sparse image by combined TV and Haar wavelet. In simulation, the directed OMP algorithm can't find the atoms of TV basis in the process of finding the atoms, so we use the alternate iterative method to find atoms. Finally, we can find the atoms to minimize residual, accelerating the convergence speed and reducing the number of iterations.

By comparing the performance of the three kinds of sparse basis, under the same sampling points, the PSNR of sparse basis for combined TV and Haar wavelet is highest, but also its running time is longest; sparse basis for TV is the center, the PSNR is high, it also has a shorter running time; the PSNR of sparse basis for Haar is minimum, and running time is slightly longer than TV. So we pay attention to both quality and speed, and improve reconstruction algorithm to achieve better results.

V. CONCLUSION

In this paper, we propose and analyze the microwave radiation image reconstruction based on combined TV and Haar Basis, when the sampling is below Nyquist theory. Via simulating experiments, we can know that the the proposed method better than signal TV, Haar wavelet sparse basis. We will pay attention to both quality and speed, and improve reconstruction algorithm to achieve better results.

REFERENCES

[1]  C T Swift, D M Le Vine, C S Ruf, "Aperture synthesis concepts in microwave remote sensing of the Earth," IEEE Trans on Microwave Theory and Tech., vol. 39, no. 12, pp. 1931-1935, Dec. 1991.

[2]  Y H Kerr, P Waldteufel, J-P Wigneron, et al, "The SMOS mission: New tool for monitoring key elements ofthe global water cycle," IEEE Proceeding Magazine, vol. 98, no. 5, pp. 666-687, May. 2010.

[3]  D. Donoho, "Compressed sensing," IEEE Trans. Inform. Theory, vol. 52, no. 4, pp. 1289–1306, April 2006.

[4]  E. Cand` es, J. Romberg, and T. Tao, "Robust uncertainty principles: Exact signal reconstruction from highly incomplete frequency information," IEEE Trans. Inform. Theory, vol. 52, no. 2, pp. 489–509, Feb. 2006.

[5]  R G Baraniuk. "Compressive sensing,". IEEE Signal Processing Magazine, vol. 24, no. 4, pp. 118-121, July 2007.

[6]  C Luo, F Wu, J Sun, C W Chen, "Efficient measurement generation and pervasive sparsity for compressive data gathering," IEEE Trans Wireless Communications, vol. 9, no. 12,pp. 3728-3738, Dec. 2010.

[7]  F M He, Q X Li, Z H Zhao, K Chen, "A sparse prior based statistical inversion approach for aperture synthesis radiometric imaging of extended source," Acta Electronic Signal Journal, vol. 41, no. 3, pp. 417-423, Mar. 2013.

[8]  H Fang, H R Yang, "Greedy algorithm and compressed sensing theory," Acta Automation Signal Journal, vol. 37, no. 12, pp.1413-1421, Dec. 2011.

[9]  L J Yu, X C Xie, "An introduction to compressed sensing theory," TV Technology Journal, vol. 32, no. 12, pp. 16-18, Nov. 2008.

[10] http://www.eee.hku.hk/~wsha/Freecode/freecode.htm

# Imaging of object in the presence of rough surface using scattered electromagnetic field data

Pengju Yang, Lixin Guo, Chungang Jia

School of science, Xidian University, Xi'an 710071, China

*Abstract*-Electromagnetic (EM) scattering from a perfect electrically conducting object above lossy dielectric half-space with rough surface is investigated, for both TE and TM polarizations, employing a parallel fast multiple method. Then, based on the scattered electromagnetic field data at multiple-incidence angles and frequencies, a back-projection tomography technique is applied to generate two-dimensional (2-D) synthetic aperture radar images. The randomly rough surfaces are modeled as realizations of a Gaussian random process with the Gaussian spectrum, while the tapered incident wave is chosen to reduce the truncation error.

## I. INTRODUCTION

The problem of detecting and imaging objects with electromagnetic sensors relies on accurate computational modeling, which is a challenging task duo to the existence of strong ground clutter and low electromagnetic wave penetration into lossy media. Over the past few decades, a significant amount of research effort has been spent toward developing accurate and efficient forward solvers for the scattering problem, which is valid over a wide range of incident and scattering angles. Rigorous numerical methods, such as method of moments (MoM) [1] and its fast algorithms [2] [3] have been implemented to analyze electromagnetic scattering from target embedded in lossy dielectric half-space with a flat interface or rough surface.

A variety of techniques have been provided in recent years to solve this category of problems of the detection of buried objects below rough surface or flat interface. The angular correlation function technique has been applied to the detection of radar object in the presence of rough surface clutters, indicating that ACF technique is more effective than the RCS method in suppressing the clutter due to rough surface scattering effect [4]. Radar imaging techniques have been widely used in the detection of buried targets with the development of wide-band high-resolution synthetic aperture radar (SAR) technology, which has shown its ability to imaging the buried objects. For example, ultra-wideband (UWB) ground penetrating radar (GPR) has been applied extensively to detect the subsurface objects. The radar imaging of target embedded in dielectric half-space has been conducted both theoretically and experimentally [5] [6] [7], which demonstrate the performance of UWB radar in the detection of buried objects.

In this paper, the surface integral equations of scattering problem involving a perfect electric conductor (PEC) object

embedded in dielectric media with rough interface formulated by employing Helmholtz equation and Green's theorem. Due to the considerable computation time in tomographic processing which involves the need for data at multiple frequencies and aspect angles, a parallel fast multiple method is applied as a forward solver to solve the surface integral equation. Based on the scattered fields, a back-projection tomography technique is employed to generating SAR images of 2-D cylinder above rough surface.

## II. SCATTERING THEORY IMAGEING TECHNIQUE

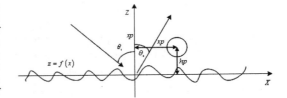

Fig. 1. Geometry of a PEC object above rough surface.

Fig. 1 illustrates the basic geometry considered in this paper: a PEC object is above a rough interface described by $z = f(x)$. The upper medium is vacuum and the lower medium is considered to be lossy with relative complex permittivity $\varepsilon_1$. $\bar{k}_i$ and $\bar{k}_s$ are the incident and scattering wave vector, respectively. $\theta_i$ and $\theta_s$ are the incident and scattering angle, respectively. $xp$ denotes the horizontal distance of between center of the cylinder and the origin, and $hp$ represents the height of the object. Let $\psi_0$ and $\psi_1$ denote the fields in region 0 and region 1, respectively. Let $\bar{r}' = x'\hat{x} + z'\hat{z}$ and $\bar{r} = x\hat{x} + z\hat{z}$ represent source and field points, respectively.

The randomly rough surfaces are modeled as realizations of a Gaussian random process with the Gaussian spectrum [8]

$$S(k) = \frac{\delta^2 l}{2\sqrt{\pi}} \exp(-k^2 l^2 / 4) \tag{1}$$

where $\delta$ and $l$ are the rms height and correlation length, respectively.

978-7-5641-4279-7

According to Helmholtz equation and Green's theorem, one can obtain the following coupled integrals for the TM case.

$$\frac{1}{2}\psi_0\left(\overline{r}\right)-PV\int_{S_r}ds'\psi_0\left(\overline{r}'\right)\hat{n}'\cdot\nabla'G_0\left(\overline{r},\overline{r}'\right)$$
$$+\int_{S_r}ds'G_0\left(\overline{r},\overline{r}'\right)\hat{n}'\cdot\nabla'\psi_0\left(\overline{r}'\right)$$
$$-\int_{S_o}ds'\psi_0\left(\overline{r}'\right)\hat{n}'\cdot\nabla'G_0\left(\overline{r},\overline{r}'\right)$$
$$=\psi_{inc}\left(\overline{r}\right)\qquad\overline{r}\in S_r \qquad (2)$$

$$\frac{1}{2}\psi_0\left(\overline{r}\right)-\int_{S_r}ds'\psi_0\left(\overline{r}'\right)\hat{n}'\cdot\nabla'G_0\left(\overline{r},\overline{r}'\right)$$
$$+\int_{S_r}ds'G_0\left(\overline{r},\overline{r}'\right)\hat{n}'\cdot\nabla'\psi_0\left(\overline{r}'\right)$$
$$-PV\int_{S_o}ds'\psi_0\left(\overline{r}'\right)\hat{n}'\cdot\nabla'G_0\left(\overline{r},\overline{r}'\right)$$
$$=\psi_{inc}\left(\overline{r}\right)\qquad\overline{r}\in S_o \qquad (3)$$

$$\frac{1}{2}\psi_1\left(\overline{r}\right)+PV\int_{S_r}ds'\psi_1\left(\overline{r}'\right)\hat{n}'\cdot\nabla'G_1\left(\overline{r},\overline{r}'\right)$$
$$-\int_{S_r}ds'G_1\left(\overline{r},\overline{r}'\right)\hat{n}'\cdot\nabla'\psi_1\left(\overline{r}'\right)=0\quad\overline{r}\in S_r \qquad (4)$$

Note that $\psi_{inc}$ is the incident field. $G_0\left(\overline{r},\overline{r}'\right)$ is the Green's function in region 0 with $G_0\left(\overline{r},\overline{r}'\right)=(i/4)H_0^{(1)}(k_0\,|\,\overline{r}-\overline{r}'\,|)$ and $G_1\left(\overline{r},\overline{r}'\right)$ is the Green's function in region 1 with $G_1\left(\overline{r},\overline{r}'\right)=(i/4)H_0^{(1)}(k_1\,|\,\overline{r}-\overline{r}'\,|)$, where $H_0^1\left(\cdot\right)$ is the zeroth-order Hankel function of the first kind. The aforementioned integral equations can be used for TE polarization with a slight modification.

When $\overline{r}$ is on the rough surface, the total fields satisfy the following boundary conditions

$$\psi_0\left(\overline{r}\right)=\psi_1\left(\overline{r}\right) \qquad (5)$$

$$\hat{n}\cdot\nabla\psi_1\left(\overline{r}\right)=\rho\hat{n}\cdot\nabla\psi_0(\overline{r}) \qquad (6)$$

where $\rho=\varepsilon_1/\varepsilon_0$ for TM incidence, and $\rho=1$ for TE incidence.

To avoid artificial edge diffraction resulting from the finite length of the simulated rough surface, the following Thorso's tapered plane wave rather than the generally used plane wave is chosen as the incident field.

$$\psi_{inc}\left(\overline{r}\right)=\exp[ik_0(x\sin\theta_i-z\cos\theta_i)(1+w(\overline{r}))]\cdot$$
$$\exp\left(-(x+z\tan\theta_i)^2/g^2\right) \qquad (7)$$

where $g$ is the tapering parameter controlling the tapering length of the incident wave. The additional factor in the phase

is $w(\overline{r})=\left[2(x+z\tan\theta_i)^2/g^2-1\right]/(k_0 g\cos\theta_i)^2$.

Upon solving the matrix equation by using a parallel fast multiple method which is based on the message passing interface (MPI), the surface fields and their normal derivatives can be obtained. When the point $\overline{r}$ is located in the far field, the scattered field $\psi^s\left(\overline{r}\right)$ in space $\Omega_0$ is [8]

$$\psi_s\left(\overline{r}\right)=\frac{i}{4}\sqrt{\frac{2}{\pi k_0 r}}\exp\left(-i\frac{\pi}{4}\right)\exp\left(ik_0 r\right)\psi_s^{(N)}\left(\theta_s\right) \qquad (8)$$

Where

$$\psi_s^N\left(\theta_s\right)=\int_{S_1}ds[-i(\hat{n}\cdot\overline{k}_s)I_1(x)-I_2(x)]\exp(-i\overline{k}_s\cdot\overline{r}) \qquad (9)$$

with $\overline{k}_s=k_0\left(\sin\theta_s\hat{x}+\cos\theta_s\hat{z}\right)$.

A 2-D SAR image of a deterministic surface can be constructed from a set of frequency and angular swept complex backscatter field data. This corresponds to a "spotlight" SAR image in which the incident beam is oriented to illuminate a fixed surface area. Tomographic processing using an inverse Fourier transform with back projection [9] is employed to generate the images of this paper.

Range and cross-range resolutions of the image can be determined by the frequency and angular bandwidths, respectively. The range and cross-range resolutions, $r_d$ and $r_c$, are given by

$$r_d=\frac{c}{2B}\qquad r_c=\frac{c}{2f_0\sin\theta} \qquad (10)$$

where $c$ is the velocity of light, and $B$ and $\theta$ represent the frequency bandwidth centered on $f_0$ and the angular rotation, respectively. To resolve surface variations on the order of a wavelength, backscatter data were collected over a 4GHz frequency bandwidth (3-7GHz) and a 40° angular bandwidth corresponding 3.75cm range and 4.67cm cross-range resolution in the image domain, respectively.

The unambiguous range and cross-ranges, $D_d$ and $D_c$, can be obtained by the following equations

$$D_d=\frac{c}{2\delta f}\qquad D_c=\frac{c}{2f_0\delta\theta} \qquad (11)$$

where $\delta f$ and $\delta\theta$ denote the steps in frequency and angle, respectively.

To reduce the side-lobe level, a proper choice of windows is necessary. Since image formation is closely related to the Fourier transform, there is a tradeoff between side-lobe level and spatial resolution. Windows functions in both frequency and angle are chosen to set the relationship between these quantities. A rectangular window has optimum resolution, but the first side-lobe level is relatively high (-13 dB) so that minor scattering events other than strong single scattering can be completely obscured by the side lobes. The Hamming window has a low side-lobe level (-43 dB) with a wider main lobe. The disadvantage of the Hamming window is that the side-lobe level does not decrease significantly at wide ranges. Throughout this paper a Blackman window is selected as an appropriate choice for the windowing function, resulting in a fast decaying side-lobe level at the expense of degrading image resolution.

### III. NUMERICAL RESULTS AND DISCUSSIONS

First, one investigate the spot-mode SAR images of the deterministic Gaussian rough surface of length $L = 2.2\,\text{m}$ with correlation length $l = 12.8\,\text{cm}$ and rms $\delta = 1.3\,\text{cm}$. Fig. 2 shows the radar images constructed from Kirchhoff approximation backscattered prediction for both TE and TM polarization. Each image is expressed within the dynamic range of 60 dB ($-60\,\text{dB} \sim 0\,\text{dB}$) and composed of $200 \times 200$ pixels in range $2.2\,\text{m} \times 2\,\text{m}$ so that each pixel size is much smaller than the range resolution. The pixel resolution depends on the choice of windowing function as well as the angular and frequency bandwidths. It is observed that the scattering centers are on the entire domain of rough surface due to the plane wave illumination on the rough surface rather than concentrating on the center of the rough surface resulting from the tapered wave illumination. The radar images for both TE and TM polarization exhibit similar distribution of scattering centers show little polarization dependence.

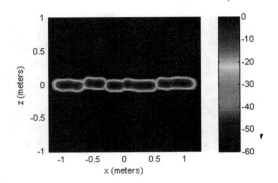

(b)

Fig. 2. Radar images of Gaussian rough surface. (a) TE (b) TM

Fig. 3 exhibit the radar image of object above rough surface for both TE and TM polarization, in which the Blackman window is utilized. The length of the rough surface is $L = 1.98\,\text{m}$ with correlation length $l = 7.5\,\text{cm}$ and rms $\delta = 1.25\,\text{cm}$. The position parameters of the object is $xp = 0.0\,\text{m}$, $hp = 0.3\,\text{m}$ and the radius of the cylinder is $R = 0.1\,\text{m}$. The images are expressed within the dynamic range of 40 dB ($-40\,\text{dB} \sim 0\,\text{dB}$) and composed of $200 \times 200$ pixels in range $2\,\text{m} \times 2\,\text{m}$ so that each pixel size is much smaller than the range resolution. The object can be clearly observed in the presence of rough surface which will result in clutter and obscure the object. It is readily found that scattering centers are at the central domain which is attributed to the taperd wave illumination, while the scattering centers are distributed on the entire domain of rough surface resulting from the plane wave illumination as depicted in Fig. 2.

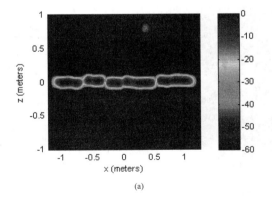

(a)

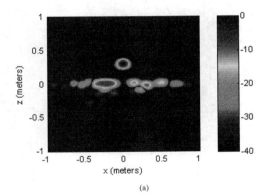

(a)

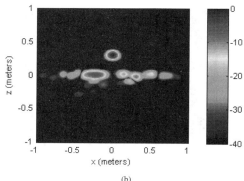

(b)

Fig. 3. Radar images for different polarization. (a) TE (b) TM

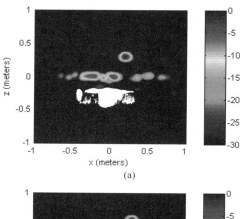

(a)

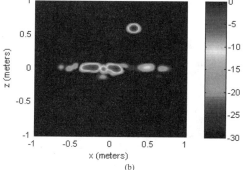

(b)

Fig. 4. Radar images with different object positions.

Fig.4 present the radar image of object with different positions in the presence of rough surface, and the incident wave is TM polarization. The Blackman window is utilized in all the following results. In Fig. 4(a), the position parameters of the object is $xp = 0.2\,\mathrm{m}$, $hp = 0.2\,\mathrm{m}$ and the radius of the

cylinder is $R = 0.1\mathrm{m}$, and in Fig. 4(b) the position parameters of the object is $xp = 0.2\,\mathrm{m}$, $hp = 0.3\,\mathrm{m}$. The radius of the cylinder is $R = 0.1\mathrm{m}$ in Fig. 4. It is readily observed that the position of the object is different from the one in Fig. 3.

## IV. CONCLUSION

In this paper, mono-static radar images of 2-D composite model involving an object above rough surface have been investigated. A parallel fast multiple method is applied as a forward solver to solve the composite electromagnetic scattering problem. Based on the scattered fields, a back-projection tomography technique is employed to generating SAR images of 2-D cylinder above rough surface. The results show that radar images for both TE and TM polarization can be obtained. The studies of this paper demonstrate that radar images provide a means for better understanding of composite scattering problems.

## ACKNOWLEDGMENT

This work was supported by the National Science Foundation for Distinguished Young Scholars of China (Grant No. 61225002), the Specialized Research Fund for the Doctoral Program of Higher Education (Grant No. 20100203110016), and the Fundamental Research Funds for the Central Universities (Grant No. K50510070001). The authors would like to thank the reviewers for their helpful and constructive suggestions.

## REFERENCES

[1] J. T. Johnson and R. J. Burkholder, "A study of scattering from an object below a rough surface," *IEEE Transactions on Geosciences and Remote Sensing*, vol. 42, pp. 59-66, 2004.

[2] N. Geng, A. Sullivan, and L. Carin, "Fast multipole method for scattering from an arbitrary PEC target above or buried in a lossy half space," *IEEE Transactions on Antennas and Propagation*, vol. 49, pp. 740-748, 2001.

[3] L. Yu, G. Li-Xin, and W. Zhen-Sen, "The fast EPILE combined with FBM for electromagnetic scattering from dielectric targets above and below the dielectric rough surface," *IEEE Transactions on Geosciences and Remote Sensing*, vol. 49, pp. 3892-3905, 2011.

[4] G. Zhang and L. Tsang, "Application of angular correlation function of clutter scattering and correlation imaging in target detection," *IEEE Transactions on Geosciences and Remote Sensing*, vol. 36, pp. 1485-1493, 1998.

[5] A. Sullivan, R. Damarla, N. Geng, Y. Dong, and L. Carin, "Ultrawide-band synthetic aperture radar for detection of unexploded ordnance: modeling and measurements," *IEEE Transactions on Antennas and Propagation*, vol. 48, pp. 1306-1315, 2000.

[6] L. Lin, A. E. C. Tan, K. Jhamb, and K. Rambabu, "Buried Object Characterization Using Ultra-Wideband Ground Penetrating Radar," *IEEE Transactions on Microwave Theory and Techniques*, vol. 60, pp. 2654-2664, 2012.

[7] I. Catapano, L. Crocco, Y. Krellmann, G. Triltzsch, and F. Soldovieri, "A Tomographic Approach for Helicopter-Borne Ground Penetrating Radar Imaging," *IEEE Geoscience and Remote Sensing Letters*, vol. 9, pp. 378-382, 2012.

[8] L. Tsang, *Scattering of Electromagnetic Waves: Numerical Simulations* vol. 2. New York: Wiley-Interscience, 2001.

[9] D. C. Munson, Jr., J. D. O"Brien, and W. Jenkins, "A tomographic formulation of spotlight-mode synthetic aperture radar," *Proceedings of the IEEE*, vol. 71, pp. 917-925, 1983.

# A Simple and Accurate Model for Radar Backscattering from Vegetation-covered Surfaces

Yisok Oh and Soon-Gu Kweon

Department of Electronic, Information, and Communications Engineering
Hongik University
72-1 Sangsu-Dong, Mapo-Gu, Seoul, Korea

*Abstract-* A simple and accurate microwave backscattering model for vegetation-covered soil surfaces is developed in this study. A vegetated surface is modeled as a two-layer structure comprising a vegetation layer and an underlying ground layer. This scattering model includes five main scattering mechanisms including the first-order multiple scattering. The vegetation layer is modeled by random distribution of mixed scattering particles, such as leaves, branches and trunks, and its radar backscatter is computed using the first-order vector Radiative Transfer model. The number of input parameters has been minimized to simplify the scattering model, and its accuracy has been improved by comparing with the experimental measurements.

## I. INTRODUCTION

Radar backscattering from vegetated surfaces involves complicated electromagnetic wave interactions, because of the randomly oriented complex geometries of the various scattering particles. Therefore, it is very difficult to develop an accurate polarimetric radar scattering model for a vegetation layer over an underlying soil surface. Moreover, a scattering model will get many input parameters to increase the computation accuracy, which leads to a very complicate model for a practical usage [1-2].

In this study, we developed a simple and accurate scattering model, which has only ten input parameters and shows a reasonably good accuracy. This model employs the iterative vector Radiative transfer theory to compute the backscattering coefficients of a vegetation canopy over an underlying soil surface [3]. We modeled the vegetation canopy with randomly oriented and positioned leaves, branches and trunks having random size distributions. Existing scattering models for dielectric disks, spheroids and cylinders are examined, and their validity regions are determined. A polarimetric semi-empirical model (PSEM) for bare soil surfaces in [4] was used for this model development.

## II. SCATTERING MATRICES FOR PARTICLES

### A. Lossy Dielectric Cylinders Disks

Deciduous leaves can be modeled by thin disks. The generalized Rayleigh-Gans (GRG), and the physical optics (PO) models are commonly used for calculation of scattering matrices of leaves [5]. The scattered field vector is related to the incident field vector in terms of a dyadic scattering amplitude $\overline{\overline{S}}$ as follows:

$$\overline{E}^s(\overline{r}) = \frac{e^{ikr}}{r} \overline{\overline{S}}(\hat{k}_s, \hat{k}_i) \cdot \hat{q}_i E_0 \qquad (1)$$

where the scattering amplitude $S_{pq}$ for a $q$-polarized incident wave and a $p$-polarized scattered wave can be written by

$$S_{pq} \equiv \hat{p}_s \cdot \overline{\overline{S}}(\hat{k}_s, \hat{k}_i) \cdot \hat{q}_i . \qquad (2)$$

It was found that both of the GRG and the PO approximation can be applied for scattering from thin leaves at microwave frequencies [5]. In this paper, the PO model is used. For the PO model, the leaf is assumed as a resistive sheet. Then, the equivalent current $\overline{J}(\overline{r}')$ can be approximated to a surface current distribution $\overline{J}_s^R(\overline{r}')$ on the resistive sheet lying on x-y plane as

$$\overline{J}_s^R(\overline{r}') = \overline{J}_s^{pc}(\overline{r}') \Gamma_q . \qquad (3)$$

The PO surface current on a perfect conductor $\overline{J}_s^{pc}(\overline{r}')$ can be obtained using the equivalence principle. The horizontal and vertical reflection coefficients ( $\Gamma_h$ and $\Gamma_v$ ) for a resistive sheet can be derived using the impedance boundary conditions.

$$\Gamma_h = \left[1 + \frac{2R\cos\theta_0}{\eta_0}\right]^{-1} \text{ and} \qquad (4a)$$

$$\Gamma_v = \left[1 + \frac{2R}{\eta_0 \cos\theta_0}\right]^{-1}, \qquad (4b)$$

with $R = \dfrac{i\eta_0}{k_0 t(\varepsilon_r - 1)}$, where $R$ is the resistivity of the leaf, $\theta_0 = \pi - \theta_i$, and $t$ is the leaf thickness.

The relative permittivity $\varepsilon_r$ for vegetation particles ( $\varepsilon_v$ ) has been computed by the following empirical formula [6].

$$\varepsilon_v = \varepsilon_r + v_{fw}[4.9 + \frac{75.0}{1 + jf/18} - j\frac{18\sigma}{f}]$$
$$+ v_b[2.9 + \frac{55.0}{1 + (jf/0.18)^{0.5}}] \qquad (5)$$

where $\varepsilon_r = 1.7 - 0.74M_g + 6.16M_g^2$ (residual permittivity), $v_{fw} = M_g(0.55M_g - 0.076)$ (volume fraction of free water),

978-7-5641-4279-7

$v_b = 4.64 M_g^2 / (1 + 7.63 M_g^2)$ (volume fraction of the bulk vegetation bound water mixture), and $\sigma = 1.27$.

### B. Lossy Dielectric Cylinders

Branches and trunks usually have cylindrical shapes. The physical optics model is used for scattered field of a finite cylinder with arbitrary cross section and orientation [3]. For a single branch, the scattered matrix can be computed for a arbitrarily oriented cylinder.

$$\bar{\bar{S}} = Q \begin{bmatrix} (\hat{v}_s^{c} \cdot v_s) & (\hat{h}_s^{c} \cdot v_s) \\ (\hat{v}_s^{c} \cdot \hat{h}_s) & (h_s^{c} \cdot h_s) \end{bmatrix} \cdot \begin{bmatrix} T'_{vv} & T'_{vh} \\ T'_{hv} & T'_{hh} \end{bmatrix} \\ \cdot \begin{bmatrix} (\hat{v}_i^{c} \cdot v_i^{c}) & (\hat{h}_i \cdot v_i^{c}) \\ (\hat{v}_i \cdot \hat{h}_i^{c}) & (h_i \cdot h_i^{c}) \end{bmatrix} \quad (6)$$

where

$$Q = \frac{-il_c \cos\phi_s}{\pi \cos\phi_i} \left\{ \frac{\sin[k_0(\sin\phi_i + \sin\phi_s)l_c/2]}{[k_0(\sin\phi_i + \sin\phi_s)l_c/2]} \right\} \quad (7)$$

with

$$T'_{vv} = \sum_{n=-\infty}^{\infty} (-1)^n C_n^{TM} e^{in\phi'} \quad (8a)$$

$$T'_{vh} = \sum_{n=-\infty}^{\infty} (-1)^n \bar{C}_n e^{in\phi'} \quad (8b)$$

$$T'_{hv} = -\sum_{n=-\infty}^{\infty} (-1)^n \bar{C}_n e^{in\phi'} \quad (8c)$$

$$T'_{hh} = \sum_{n=-\infty}^{\infty} (-1)^n C_n^{TE} e^{in\phi'} . \quad (8d)$$

The coefficients in above equations are given in [7].

## III. First-Order Radiative Transfer Model

In the 1st-order RTM, the backscattering coefficients can be computed from five scattering mechanisms in the ground and vegetation layers of a vegetation field as shown in Fig. 1.

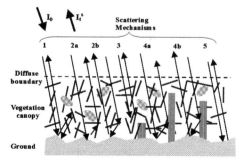

Fig 1. Scattering mechanisms.

Five scattering mechanisms consist of (1) ground forward, vegetation backward, and ground forward scattering, (2) vegetation and ground forward scattering, (3) direct vegetation backscattering, (d) ground and trunk forward scattering, and (5) direct ground backscattering, as shown in Fig. 1. The ground backscattering is calculated using Oh et al.'s model which is one of the most well-known empirical scattering model for bare soil surfaces [8].

The total backscattered intensity $\bar{I}_t^s(\mu_0, \phi_0)$ is related to the intensity $I_0$ incident upon the canopy through the transformation matrix $\bar{T}_t(\mu_0, \phi_0)$ by the equation,

$$\bar{I}_t^s(\mu_0, \phi_0) = \bar{T}_t(\mu_0, \phi_0) \bar{I}_0(-\mu_0, \phi_0) . \quad (9)$$

where $\mu_0 = \cos\theta_0$, $\theta_0$ and $\phi_0$ are the vertical and horizontal incidence angles. The transformation matrix $\bar{T}_t(\mu_0, \phi_0)$ is a 4×4 matrix which the elements have the following relationships with the scattering coefficients.

$$\sigma_{vv}^0 = 4\pi \cos\theta_0 [T_t(\mu_0, \phi_0)]_{11} \quad (10a)$$

$$\sigma_{hh}^0 = 4\pi \cos\theta_0 [T_t(\mu_0, \phi_0)]_{22} \quad (10b)$$

$$\sigma_{hv}^0 = 4\pi \cos\theta_0 [T_t(\mu_0, \phi_0)]_{21} \quad (10c)$$

$$\sigma_{vh}^0 = 4\pi \cos\theta_0 [T_t(\mu_0, \phi_0)]_{12} \quad (10d)$$

The transformation matrix comprises two components:

$$\bar{T}_t(\mu_0, \phi_0) = \bar{T}_c(\mu_0, \phi_0) + \bar{T}_g(\mu_0, \phi_0) \quad (11)$$

where $\bar{T}_c(\mu_0, \phi_0)$ and $\bar{T}_g(\mu_0, \phi_0)$ are the vegetation canopy and ground backscattering transformation matrices, respectively. The expression for $\bar{T}_g$ is given by

$$\bar{T}_g(\mu_0, \phi_0) = \exp(-\bar{k}_e^+ d/\mu_0) \bar{G}(\mu_0) \cdot \exp(-\bar{k}_e^- d/\mu_0), \quad (12)$$

where $\bar{k}_e^+$ and $\bar{k}_e^-$ are the 4×4 extinction matrices of the vegetation layer for upward and downward propagation, respectively. $\bar{G}(\mu_0)$ is the ground backscattering matrix given by

$$\bar{G}(\mu_0) = \frac{1}{\cos\theta_0} \bar{M}_m, \quad (13)$$

where $\bar{M}_m$ is the modified Stokes scattering operator, which describes ground backscatter [3].

The expression for $\bar{T}_c$ is given by [3, 7].

$$\bar{T}_c = 1/\mu_0 \exp(-\bar{k}_e^+ d/\mu_0) \bar{R}(\mu_0) \bar{\varepsilon}_c(-\mu_0, \phi_0 + \pi) \\ \cdot \bar{A}_1 \bar{\varepsilon}_c^{-1}(\mu_0, \phi_0) \bar{R}(\mu_0) \exp(-\bar{k}_e^- d/\mu_0) \\ + 1/\mu_0 \exp(-\bar{k}_e^+ d/\mu_0) \bar{R}(\mu_0) \\ \cdot \bar{\varepsilon}_c(-\mu_0, \phi_0 + \pi) \bar{A}_2 \bar{\varepsilon}_c^{-1}(-\mu_0, \phi_0) \\ + 1/\mu_0 \bar{\varepsilon}_c(\mu_0, \phi_0 + \pi) \bar{A}_3 \bar{\varepsilon}_c^{-1}(\mu_0, \phi_0) \bar{R}(\mu_0) \\ \cdot \exp(-\bar{k}_e^- d/\mu_0) \quad + 1/\mu_0 \bar{\varepsilon}_c(\mu_0, \phi_0 + \pi) \bar{A}_4 \bar{\varepsilon}_c^{-1}(-\mu_0, \phi_0) \\ + 1/\mu_0 \exp(-\bar{k}_e^+ d/\mu_0) \bar{R}(\mu_0) \\ \cdot \bar{\varepsilon}_c(-\mu_0, \phi_0 + \pi) \bar{A}_5 \bar{\varepsilon}_c^{-1}(-\mu_0, \phi_0) \exp(-\bar{k}_e^- d/\mu_0) \\ + 1/\mu_0 \exp(-\bar{k}_e^+ d/\mu_0) \bar{\varepsilon}_c(\mu_0, \phi_0 + \pi) \bar{A}_6 \bar{\varepsilon}_c^{-1}(\mu_0, \phi_0) \\ \cdot \bar{R}(\mu_0) \exp(-\bar{k}_e^- d/\mu_0) \quad (14)$$

where the matrices $A_1$, $A_2$, $A_3$, $A_4$, $A_5$, $A_6$ corresponding the scattering mechanisms 1, 2a, 2b, 3, 4a and 4b, respectively, can be obtained by integration of a function of the phase matrices and the extinction matrices. $\bar{R}(u)$ is the reflectivity matrix of the ground surface. The matrix $\bar{\varepsilon}_c$ is the 4×4 eigen-matrix of the vegetation layers [3]. The phase matrix can be computed by integration of the Mueller matrix multiplied with the distribution function of the particles. The

Mueller matrix functions are calculated using the scattering matrices of scatterers.

## IV. NUMERICAL RESULTS

At first, we tried to examine the sensitivities of the scattering coefficients on each input parameters. Then, we selected only ten most important input parameters, which are (1) the volumetric moisture content $m_v$ ($cm^3/cm^3$) and (2) the rms surface height $s$ ($cm$) of the ground surface, (3) the height $h$ ($m$) of the vegetation layer, (4) leaf density $n_l$ ($m^{-3}$), (5) leaf length $l_l$ ($cm$), (6) leaf width $W_l$ ($cm$), (7) branch density $n_b$ ($m^{-3}$), (8) branch length $l_b$ ($cm$), (9) trunk density $n_t$ ($m^{-3}$), and (10) trunk length $l_t$ ($m$).

Other input parameters are appropriately induced from the ten major input parameters. For example, all of the probability distribution functions for particle (leaf, branch and trunk) are assumed to be uniform for simplicity. The minimum and maximum boundaries of the particles are computed from the given mean values. The trunk diameter is assumed to be 0.015 times of the trunk length. These assumptions are based on the experimental observations.

We tested the variation of the model on various each parameter values. As an example, the backscattering coefficients for a vegetation canopy with $m_v = 0.15$ $cm^3/cm^3$, $s = 0.5$ $cm$, $h = 5$ $m$, $n_l = 100$ $m^{-3}$, $l_l = 6$ $cm$, $W_l = 3$ $cm$, $n_b = 10$ $m^{-3}$, $l_b = 0.5$ $m$, $n_t = 0.1$ $m^{-3}$, and $l_t = 2.5$ $m$, are computed at 5.3 GHz at vv- and hh-polarizations.

Fig. 2 shows the computation results. The letters G, C, and T in Fig. 2 denote ground, crown and trunk, respectively. It was found that the backscattering coefficient is dominated by the ground scattering at low incidence angles, and by the crown-direct scattering at higher incidence angles for both polarizations at 5.3 GHz as shown in Figs. 2 (a) and (b).

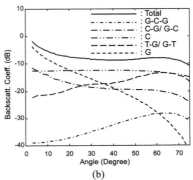

(b)

Fig. 2. Computation results for a typical forest at 5.3 GHz: (a) *vv*- and (b) *hh*-polarizations.

## V. COMPARISON WITH MEASUREMENTS

The backscattering coefficients are computed by the new scattering model and compared with the JPL/AirSAR measurements obtained at Non-San area for the PACRIM-2 campaign in Korea in 2000. The computation results are based on the ground truth data measured *in situ* at the same time.

Fig. 3 (a) shows the comparison between the computations and the measurements of the backscattering coefficients of ten rice fields at 5.3 GHz. The estimated backscattering fields were computed using the ground truth data measured *in situ*. A good agreement is shown in Figs. 3 (a) and (b) with only ten input parameters. The discrepancy of about 2~3dB may be from the errors of the radar calibration, the ground data measurements, and/or the scattering model. Fig. 3 (b) shows the comparison between the computations and the measurements for three forest areas.

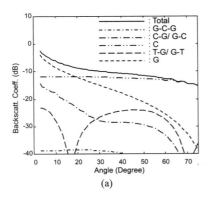

(a)

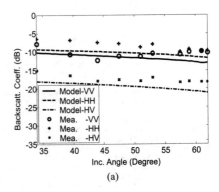

(a)

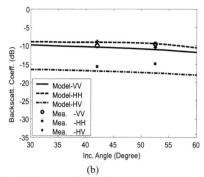

(b)

Fig. 3. The comparison with SAR measurement data and computation results for (a) rice fields and (b) forests at 5.3GHz.

## VI. CONCLUDING REMARKS

A simple and accurate model was developed for microwave backscattering from a vegetation canopy over an underlying soil surface. The vegetation canopy was modeled with randomly distributed and oriented particles having random size distributions. The model was tested for many kinds of vegetation canopies, and verified with the JPL/AirSAR measurement data sets for rice fields and forest areas. Good agreements are shown between the computed backscattering coefficients for the fields and the measurements.

### ACKNOWLEDGMENT

This work was supported by the Basic Research Program through the National Research Foundation of Korea(NRF) funded by the Ministry of Education, Science and Technology (2013R1A1A2005336).

### REFERENCES

[1] F. T. Ulaby, K. Sarabandi, K. McDonald, M. Whitt and M. C. Dobson, "Michigan microwave canopy scattering model," Int. J. Remote Sensing, vol. 11, no. 7, pp. 1223-1253, 1990.

[2] M. A. Karam, A. K. Fung, R. H. Lang, and N. S. Chauhan, "A microwave scattering model for layered vegetation," IEEE Trans. Geosci. Remote Sensing, vol. 30, no. 4, pp. 767-784, July 1992.

[3] L Tsang, J. A. Kong, R. T. Shin, Theory of Micirowave Remote Sensing, John Wiley & Sons, 1985

[4] Y. Oh, K. Sarabandi, F.T. Ulaby, "Semi-Empirical Model of the Ensemble-Averaged Differential Mueller Matrix for Microwave Backscattering from Bare Soil Surfaces", IEEE Trans. Geosci. Remote Sensing, Vol. 40, No. 6, pp.1348-1355. June 2002.

[5] Y. Oh, "Comparative evaluation of two analytical models for microwave scattering from deciduous leaves," Korea Journal of Remote Sensing, Vol.20, No.1, pp.39-46, Feb. 2004.

[6] F. T. Ulaby and M. A. El-rayes, "Microwave dielectric spectrum of vegetation -Part II :Dual-dispersion model", IEEE Trans. Geosci. Remote Sensing, vol. GE-25, pp. 550-557, 1987.

[7] F. T. Ulaby, K. Sarabandi, K. McDonald, M. Whitt, M. C. Dobson, Michigan Microwave Canopy Scattering Model (MIMICS), Tech. Report, Radiation Lab., University of Michigan, 1988.

[8] Y. Oh, K. Sarabandi, and F. T. Ulaby, "An empirical model and an inversion technique for radar scattering from bare soil surfaces," IEEE Trans. Gesci. Remote Sensing, vol. 30, no. 2, pp. 370-381, Mar 1992.

# The Analysis of Sea Clutter Statistics Characteristics Based On the Observed Sea Clutter of Ku-Band Radar

Zhuo Chen[1], Xianzu Liu, Zhensen Wu[1], and Xiaobing Wang[1,2]

[1]School of Science, Xidian University, Xi'an, Shaanxi, China 710071
[2]National Key Laboratory of Electromagnetic Environment Research, Shanghai, China 200082

*Abstract-* The model of GTI，TSC，NRL distribution the most fundament characteristic of sea clutter, as used in radar performance evaluation. The model of Rayleigh，LogNormal，Weibull and K distribution radar clutter are analyzed and modeled and some kinds of radar clutter，such as ground，weather，chaff and sea clutter are modeled and simulated. The analysis focuses on amplitude characteristic of sea clutter. The analysis would contribute to designing and implementation of radar filter and increasing the ability of suppressing sea clutter and ensuring the detection ability of radar itself.

## I. INTRODUCTION

Ideally, the accurate modeling of sea clutter should include both temporal and spatial characteristics and may ultimately require probabilistic descriptions. The most important characteristic of sea clutter, however, is its average reflectivity defined in the dimensionless unit of square-meters of radar cross section per square-meter of surface area illuminated by the radar, often denoted by $\sigma_0$. Some amount of smoothing of experimental errors leads to a need for empirical models, which, while supported by experimental data, allows computations to be performed over a continuum of parameter values. Radar environmental clutter plays an important role in the simulation of radar environment，statistical characterization of which can be described by the scattering coefficient model．

## II. THEORY FORMULAS

### A. Average scattering coefficient model of sea clutter

The radar designer requires estimates of the range of different clutter characteristics likely to be encountered. Empirical models for these are also available. The study of average scattering coefficient model of sea clutter is mainly about the average scattering coefficient of sea clutter with the change of grazing angle and the variation of parameters of the marine environment. It plays an important role for radar range performance prediction.

Internationally, typical semi-empirical model of sea clutter average scattering coefficient are GIT model, TSC model and the NRL model.

GIT model [1] is a function which describes the sea clutter backscatter coefficient of three factors: (1) interference factor $G_i$; (2) wind factor $G_u$; (3) wind factor $G_w$.

GIT model sea clutter scattering coefficient is:

$$\sigma_{HH}^0 = 10\log_{10}\left[3.9\times10^{-6}\lambda\phi_{gr}^{0.4}G_iG_wG_u\right] \quad (1)$$

$$\sigma_{VV}^0 = \begin{cases} \sigma_{HH}^0 - 1.05\ln\left(h_{av}+0.015\right)+1.09\ln\left(\lambda\right)+ \\ \quad 1.27\ln\left(\phi_{gr}+0.0001\right)+9.7, \\ \quad\quad 3GHz < f < 10GHz \\ \sigma_{HH}^0 - 1.73\ln\left(h_{av}+0.015\right)+3.76\ln\left(\lambda\right)+ \\ \quad 2.46\ln\left(\phi_{gr}+0.0001\right)+22.2, \\ \quad\quad f < 3GHz \end{cases} \quad (2)$$

where $\lambda$ is the radar wavelength (m), $\phi_{gr}$ is the grazing angle (radians), $h_{av}$ is average wave height (m).

TSC model [2] is a function which describes sea clutter backscatter coefficient of three factors: (1) small grazing angle factor $G_A$; (2) wind factor $G_u$; (3) wind factor $G_w$.

TSC model sea clutter scattering coefficient is:

$$\sigma_{HH}^0 = 10\log_{10}\left[1.7\times10^{-5}\phi_{gr}^{0.5}G_AG_wG_u/\left(\lambda+0.5\right)^{1.8}\right] \quad (3)$$

$$\sigma_{VV}^0 = \begin{cases} \sigma_{HH}^0 - 1.73\ln\left(2.507\sigma_z+0.05\right)+ \\ 3.76\ln\left(\lambda\right)+2.46\ln\left(\sin\left(\phi_{gr}\right)+0.0001\right) \\ +19.8, \; f < 2GHz \\ \sigma_{HH}^0 - 1.05\ln\left(2.507\sigma_z+0.05\right)+ \\ 1.09\ln\left(\lambda\right)+1.27\ln\left(\sin\left(\phi_{gr}\right)+0.0001\right) \\ +9.65, \; f > 2GHz \end{cases} \quad (4)$$

where $\lambda$ is the radar wavelength (m), $\phi_{gr}$ is the grazing angle (radians), $\sigma_z$ is the sea surface height standard deviation.

NRL model [3] of the sea clutter scattering coefficient is:

978-7-5641-4279-7

$$\sigma_{HH,VV}^{0} = c_1 + c_2 \cdot \log_{10}\left(\sin\left(\phi_{gr}\right)\right)$$
$$+ \frac{\left(c_3 + c_4 \cdot \phi_{gr}\right) \cdot \log_{10}\left(f\right)}{\left(1 + c_5 \cdot \phi_{gr} + c_6 \cdot s\right)} \tag{5}$$
$$+ c_7 \cdot \left(1 + s\right)^{1/\left(2 + c_8 \cdot \phi_{gr} + c_9 \cdot s\right)}$$

where $\phi_{gr}$ is the grazing angle (degrees), $s$ is the sea state, $f$ is the radar operating frequency (GHz).The parameter of NRL model is in table I.

TABLE I
PARAMETER OF NRL MODEL

| Parameter | Estate of Polarization | |
|:---:|:---:|:---:|
| | HH-pol | VV-pol |
| c1 | -72.76 | -48.56 |
| c2 | 21.11 | 26.30 |
| c3 | 24.78 | 29.05 |
| c4 | 4.917 | -0.5183 |
| c5 | 0.6216 | 1.057 |
| c6 | -0.02949 | 0.04839 |
| c7 | 26.19 | 21.37 |
| c8 | 0.09345 | 0.07466 |
| c9 | 0.05031 | 0.04623 |

B.  *Average distribution model of sea clutter amplitude*

Sea clutter amplitude distribution characteristics have a great significance to the radar target detection, simulation and system design and performance evaluation. The main sea clutter amplitude distribution models are Rayleigh-distribution model[4], lognormal-distribution model[5], Weibull-distribution model [6] and K-distribution[7-10] model.

The probability density function (PDF) of Rayleigh-distribution is:

$$p(E) = \frac{2E}{x}\exp\left(-E^2/x\right) \tag{6}$$

Cumulative probability density function (CDF) of Rayleigh-distribution is:

$$P(E) = 1 - \exp\left(-E^2/x\right) \tag{7}$$

N-order moments of Rayleigh-distribution is:

$$M_n = x^{n/2}\Gamma\left(1 + n/2\right) \tag{8}$$

where x is the average square value of the amplitude of the clutter, $E$ is the instantaneous value of the clutter amplitude, $\Gamma(\cdot)$ is gamma function.

The probability density function (PDF) of Lognormal-distribution is:

$$p(E) = \frac{1}{E\sqrt{2\pi\sigma^2}}\exp\left(-\frac{\left[\ln(E/m)\right]^2}{2\sigma^2}\right) \tag{9}$$

Cumulative probability density function (CDF) of Lognormal-distribution is:

$$P(E) = 1 - \frac{1}{2}erfc\left(\frac{\ln\left(E - m\right)}{\sqrt{2\sigma^2}}\right) \tag{10}$$

N-order moments of Lognormal-distribution is:

$$M_n = \exp\left(n\ln(m) + \left(n\sigma\right)^2\right) \tag{11}$$

where $\mathrm{erfc}(\cdot)$ is the residual error function, $m$ is the average noise amplitude, $\sigma^2$ is the variance of $\ln E^2$.

The clutter probability density function (PDF) of Weibull-distribution is:

$$p(E) = \beta\frac{E^{\beta-1}}{\alpha^{\beta}}\exp\left(-\left(E/\alpha\right)^{\beta}\right) \tag{12}$$

Cumulative probability density function (CDF) of Weibull-distribution is:

$$P(E) = 1 - \exp\left(-\left(E/\alpha\right)^{\beta}\right) \tag{13}$$

N-order moments of Weibull-distribution is:

$$M_n = \alpha^n\Gamma\left(1 + n/\beta\right) \tag{14}$$

where $\beta$ is the shape parameter, $\alpha$ is the scale parameter.

The clutter probability density function of K-distribution is:

$$p(E) = \frac{4b^{(v+1)/2}E^v}{\Gamma(v)}K_{v-1}\left(2E\sqrt{b}\right) \tag{15}$$

Cumulative probability density function (CDF) of K-distribution is:

$$P(E) = 1 - \frac{2}{\Gamma(v)}\left(E\sqrt{b}\right)^v K_v\left(2E\sqrt{b}\right) \tag{16}$$

N-order moments of K-distribution is:

$$M_n = b^{-n/2}\frac{\Gamma\left(1 + n/2\right)\Gamma\left(v + n/2\right)}{\Gamma(v)} \tag{17}$$

where $b$ is the scale parameter, $v$ is the shape parameter, $K_v(\cdot)$ is $v^{th}$-order Bessel modified function of second kind.

III.    ANALYSIS OF EXPERIMENTAL DATA AND RESULTS OF EXPERIMENTSS

The following experimental data is collected in Qingdao on July 23, 2011 and July 24, 2011.

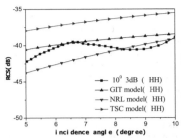

Figure 1 comparison with data of scattering coefficient model and the experimental data

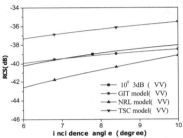

Figure 2 comparison with data of scattering coefficient model and the experimental data

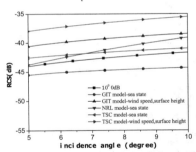

Figure 3 comparison with data of scattering coefficient model and the experimental data

Figure 1 and Figure 2 shows the comparison with the sea clutter data and the experimental data of GIT, TSC and NRL model. In Figure 1, the incidence angle of the data is 10 degree 3dB, HH polarization. In Figure 2, the incidence angle of the data is 10 degree 3dB, HH polarization. As comparison is shown in Figure 1 between the GTI, NRL and TSC sea clutter model and data for horizontal. While the agreement with NRL model at 8-10 degree is reasonable good, the agreement with GTI model at 6.5 degree is good. As comparison is shown in Figure 2 between the GTI, NRL and TSC sea clutter model and data for vertical. The agreement with GTI model at 6-10 degree is reasonable good.

Figure 3 shows the comparison with the sea clutter data and the experimental data of GIT, TSC and NRL model. The

incidence angle of the data is 10 degree 0dB. The parameters of GIT model are wind speed, wave height and sea state level. TSC model also have two parameters. The comparison is shown in Figure 3 between the five sea clutter models. The agreement with TSC model with parameter of sea state level at 5-10 degree is reasonable good.

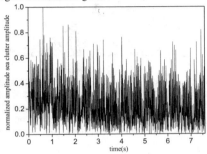

Figure 4 normalized sea clutter amplitude of time series

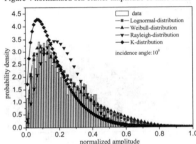

Figure 5 normalized amplitude of the probability density distribution

Figure 4 is the normalized sea clutter amplitude of time series. Figure 5 is the normalized amplitude of the probability density distribution of Figure 4. The incidence angle is 10 °, the wind speed is 3.2 m / s. As comparison is shown in Figure 5, the agreement with Weibull-distribution model is reasonable good; the agreement with Lognormal-distribution model is less good.

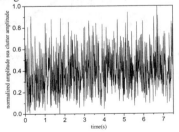

Figure 6 normalized sea clutter amplitude of time series

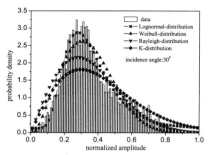

Figure 7 normalized amplitude of the probability density distribution

Figure 6 is the normalized sea clutter amplitude of time series. Figure 7 is the normalized amplitude of the probability density distribution of Figure 6. The incidence angle is 30°, the wind speed is 8.1 m/s. As comparison is shown in Figure 7, the agreement with Lognormal-distribution model is reasonable good; the agreement with Weibull-distribution model is less good.

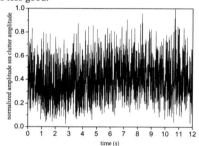

Figure 8 normalized sea clutter amplitude of time series

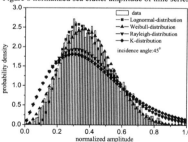

Figure 9 normalized amplitude of the probability density distribution

Figure 8 is the normalized sea clutter amplitude of time series. Figure 9 is the normalized amplitude of the probability density distribution of Figure 8. The incidence angle is 45°, the wind speed is 6.3 m / s. As comparison is shown in Figure 9, the agreement with Lognormal-distribution model is reasonable good; the agreement with Weibull-distribution model is less good; the agreement with Rayleigh-distribution is not that good.

IV.  CONCLUSIONS

Three kinds of scattering coefficient sea clutter models and four kinds of distribution model of sea clutter amplitude are introduced in this paper comparing with the experimental data. This paper analysis the scattering coefficient model and amplitude distribution model of sea clutter, based on the observed sea clutter of Ku-band Radar. This analysis would help to design the radar filter and increase the ability of suppressing sea clutter.

ACKNOWLEDGMENT

This work was supported by the National Natural Science Foundation of China under Grant 61172031.

REFERENCES

[1]  M. Horst, et al., "Radar sea clutter model," in Antennas and Propagation, 1978, pp. 6-10.
[2]  Z. Yu-shi, et al., "Applicability of sea clutter models in nonequilibrium sea conditions," 2009.
[3]  V. Gregers-Hansen and R. Mital, "An empirical sea clutter model for low grazing angles," in Radar Conference, 2009 IEEE, 2009, pp. 1-5.
[4]  J. A. Cadzow, "Spectral estimation: An overdetermined rational model equation approach," Proceedings of the IEEE, vol. 70, pp. 907-939, 1982.
[5]  G. Trunk and S. George, "Detection of targets in non-Gaussian sea clutter," Aerospace and Electronic Systems, IEEE Transactions on, pp. 620-628, 1970.
[6]  F. Fay, et al., "Weibull distribution applied to sea clutter," Radar-77, pp. 101-104, 1977.
[7]  E. Jakeman and P. Pusey, "A model for non-Rayleigh sea echo," Antennas and Propagation, IEEE Transactions on, vol. 24, pp. 806-814, 1976.
[8]  K. Ward, "Compound representation of high resolution sea clutter," Electronics letters, vol. 17, pp. 561-563, 1981.
[9]  K. Ward, "A radar sea clutter model and its application to performance assessment," in Inter. Conf., Radar, 1982.
[10] J. Jao, "Amplitude distribution of composite terrain radar clutter and the κ-Distribution," Antennas and Propagation, IEEE Transactions on, vol. 32, pp. 1049-1062, 1984.

# *FP-1(C)*

## October 36 (FRI) PM

## Room C

## Slot Antennas

# Design of slot antenna loaded with lumped circuit components

Chichang Hung[#], Tayeh Lin[#], Hungchen Chen[#], and Tsenchieh Chiu[#], and Dachiang Chang[*]

[#]Electrical Engineering, Central University, No.300, Jhongda Rd., Jhongli City, Taoyuan County 32001, Taiwan
[*]Chip Implementation Center, 7F, No. 26, Prosperity Rd. 1, Science Park, Hsinchu City 300, Taiwan
985401021@cc.ncu.edu.tw

*Abstract* - A design of slot antenna loaded with circuit components is presented in this paper. The antenna is accommodated at the edge of the printed circuit board and loaded with circuit components. Using the folded slot and loaded with circuit components can miniaturize the antenna area. The slot antenna is deposed into several sub-networks, and the corresponding equivalent circuit is presented. The design equation can be acquired from the equivalent circuit by using the resonant condition. The antenna is operated for both Bluetooth and WLAN applications at 2.45 GHz. The design procedure of the antenna will be described in details. The experimental results of the antenna are shown for validation.

## I. INTRODUCTION

With the development of modern wireless communications, the feature of compactness is important and strongly demanded in handheld devices. In many conventional designs, in order to achieve compact configuration, these antennas could occupy large circuit areas. Therefore, many compact designs are proposed. In [1], the metamaterial ring antenna is proposed, which features a low profile. In [2], LTCC process is used for reducing the antenna size. The antenna can be miniaturized by using capacitive or inductive loading. In [3], a PIFA (planar inverted F antenna) is a compact design using capacitive load. In [4], the antenna is fed by using fork like-tuning stub for reducing the antenna area.

In the paper, a design of slot antenna loaded with circuit components is proposed for Bluetooth and WLAN at 2.45 GHz. For reducing the size of the slot antenna, circuit components can be used across the slot. Antenna measurements regarding with return loss, and radiation patterns are conducted for design validation. It is found that the simulation and measurement are in good agreement. In what follows, antenna design will be mentioned in Section II, and the experimental results are shown in Section III.

## II. ANTENNA DESIGN

The proposed folded slot antennas are designed and fabricated on the FR4 printed circuit board (PCB) with dielectric constant of 4.4 and loss tangent of 0.0125. Similar to many mobile phones, the dimension of the PCB is $70 \times 30 \times 0.8$ mm$^3$, as shown in Fig.1. The antenna is deployed within the

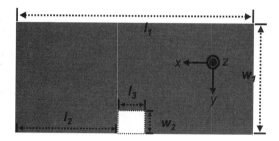

Fig. 1. The configuration of the proposed antenna. $l_1$ =70, $l_2$ =30, $l_3$ =8, $w_1$ =30 and $w_2$ =6.5. All dimensions are in mm.

antenna region in the middle of the upper edge, and is fed with 50-$\Omega$ coaxial cable. The full-wave and circuit simulation of antennas are carried out by using High Frequency Structure Simulator (HFSS) and Advanced Design System (ADS), respectively.

Figure 2(a) shows the configuration of the proposed single-band folded slot antenna loaded with circuit elements. By using the folded configuration, it is flexible to adjust the length of the antenna. The circuit elements are placed along the edge of the slot. It is also found that the component can be placed across the slot in other applications [5]. Due to the high-pass (capacitor) or low-pass (inductor) characteristics of the components, placing the components across the slot could significantly shorted the length of the slot, thus affecting the antenna gain. For single-band operation of antenna, two components are used in the design. Although one component is sufficient for the design, using two components with provide more flexibility and it will be explained later.

To conduct the circuit analysis and determine the circuit components of the antenna, the antenna structure is decomposed into three circuit sub-networks, as shown in Fig. 2(b). From circuit point of view, the antenna is considered as a three-port circuit, as shown in Fig. 3(a). Conducting full-wave simulation by using HFSS, scattering matrices ([$S_1$], [$S_2$] and

978-7-5641-4279-7

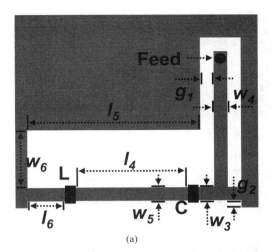

(a)

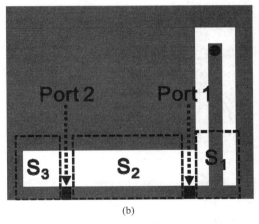

(b)

Fig. 2. The configuration of the single band antenna (a). The corresponding network decomposition (b).

TABLE I
ALL DIMENSIONS ARE IN MM.

| Parameter | $l_4$ | $l_5$ | $l_6$ | $w_3$ | $w_4$ |
|---|---|---|---|---|---|
| Value (mm) | 4.2 | 6.5 | 1.4 | 0.5 | 0.5 |
| Parameter | $w_4$ | $w_5$ | $w_6$ | $g_1$ | $g_2$ |
| Value (mm) | 0.5 | 0.5 | 4.2 | 0.5 | 0.2 |

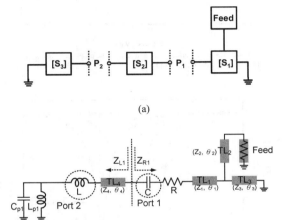

(a)

(b)

Fig. 3. Block diagram (a) and the equivalent circuit (b) of the antenna in Fig. 2.

circuit mentioned above. In Fig. 3 (b), a reference line is set between Port #1 and sub-network $S_2$. When the antenna is at the center frequency, it is operating at resonance. One of the conditions of resonant circuit is that, at any point in the circuit, the sum of the input impedances seen looking into either side must be zero. Therefore, as shown in Fig. 3 (b) at the reference line, it is found that

$$\text{Im}\{Z_{L1}\} + \text{Im}\{Z_{R1}\} = 0 , \qquad (1)$$

where

$$Z_{L1} = Z_1 \cdot \frac{\left[ \dfrac{j2\pi f L_{p1}}{1 - (2\pi f)^2 L_{p1} C_{p1}} + j2\pi f L \right] + jZ_1 \tan \theta_1}{Z_1 + j\left[ \dfrac{j2\pi f L_{p1}}{1 - (2\pi f)^2 L_{p1} C_{p1}} + j2\pi f L \right] \cdot \tan \theta_1} , \qquad (2)$$

[ $S_3$ ]) for those three sub-network can be obtained. Three scattering matrices are can be further transformed into impedance (or admittance) matrices [6]. With the impedance (or admittance) matrices, equivalent circuits for three sub-network can be built by using T (or $\pi$) equivalent circuits [6]. The proposed equivalent circuit for the antenna in Fig. 2 is shown in Fig. 3 (b). A capacitor ( C ) is placed in Port #1, while an inductor ( L ) is placed in Port #2. The sub-network $S_1$ is basically a T-connection of three transmission lines and a resistor. The sub-network $S_2$ is simply a uniform transmission line. The sub-network $S_3$ obtained by using the T-equivalent

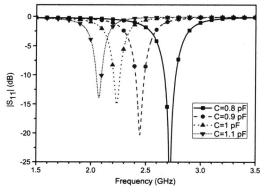

Fig. 4. The simulated reflection coefficients of the proposed antenna by adjusting the C from 0.9 to 1.1 pF.

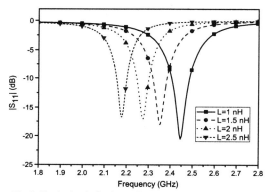

Fig. 5. The simulated reflection coefficients of the proposed antenna by adjusting the L from 1 to 2.5 nH.

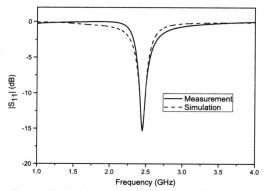

Fig. 6. The simulated and measured reflection coefficients of the proposed antenna.

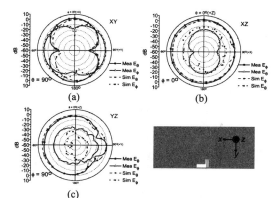

Fig. 7. The simulated and measured radiation patterns of the single band antenna. (a) XY-plane, (b) XZ-plane and (c) YZ-plane.

$$Z_{R1} = \frac{1}{j2\pi fC} + R + Z_4 \cdot \frac{\dfrac{jZ_2 \cdot Z_3 \tan\theta_2 \cdot \tan\theta_3}{Z_2 \tan\theta_2 + Z_3 \tan\theta_3} + jZ_4 \tan\theta_4}{Z_4 - \dfrac{Z_2 \cdot Z_3 \cdot \tan\theta_2 \cdot \tan\theta_3}{Z_2 \tan\theta_2 + Z_3 \tan\theta_3} \cdot \tan\theta_4}. \quad (3)$$

where R=76.25, $Z_1$=105, $\theta_1$=15.8, $Z_2$=46.5, $\theta_2$=12.5, $Z_3$=46.5, $\theta_3$=17, $Z_4$=30.48, $\theta_4$=78.14, $L_1$=1.683, $C_1$=0.982.

In (1), $f$ is set at the 2.45 GHz and two unknown parameters, C and L, are to be determined. Since there is only one equation in (1), either L or C should be determined in advance. In this design, L is first chosen with 1.7 nH. By using (1), C is calculated as 0.982 pF. It should be noted that the values of the inductor and capacitor obtained from (1) could be impractical, that is, those components my not be

commercially available. With two parameters, C and L, the designer can choose practical values to satisfy (1). In order to validate the calculated values of C and L, the Full-wave simulation is applied. In Fig. 4 and Fig. 5, the value of capacitance and inductance are varied from 0.8 to 1.1 pF and 1 to 2.7 nH, respectively. In the figure, the values of C and L are closely meet with the calculated values. Even though, the final value of C is set in 1 pF, which is slightly different from the calculated value. To make the antenna operate at exactly 2.45 GHz, the dimension of the antenna could be slightly adjusted.

III. RESULTS

Figure 6 shows the simulated and measured reflection coefficients of the antenna in Fig. 2 (a). In Fig. 6, the center frequency in both simulation and measurement are at 2.47 GHz, which is due to the discrepancy between the real (1 pF)

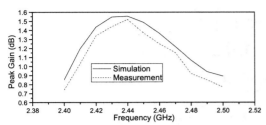

Fig. 8.   The simulated and measured peak gains of the proposed antenna.

and calculated (0.982 pF) values of C. Figure 7 shows the simulated and measured radiation patterns. In the XZ-plane pattern, the electric field is mainly polarized in $\phi$ direction. In the YZ-plane pattern, the electric field is mainly polarized in $\theta$ direction. Considering that the electric field over the slot aperture of the unfolded section is in $y$ direction, it is concluded that the unfolded section of the slot is responsible for the most of the radiation. Figure 6 shows the simulated and measured peak gain of the antenna in Fig. 2 (a). Based on the results shown in Fig. 6, 7, and 8, the proposed design is validated.

## IV. CONCLUSIONS

In this paper, a design of slot antenna loaded with circuit components has been proposed. The proposed configuration is based on a typical folded slot with an edge branch of strip loaded with inductors and capacitors. The equivalent circuit and design equations are conducted. The values of the circuit components can be determined by using the design equations. The proposed designs also have been realized and validated by measurement. Both experimental and simulation results has good agreement with each other.

## REFERENCES

[1]  F. Qureshi, M. A. Antoniades, and G. V. Eleftheriades, "A compact and low-profile metamaterial ring antenna with vertical polarization," *IEEE. Antennas Wireless Propag. Lett.,* vol. 4, pp. 333-336, 2005.

[2]  E. Tentzeris, R.L. Li, K. Lim, M. Maeng, E. Tsai, G. DeJean, and J. Laskar, "Design of compact stacked-patch antennas on LTCC technology for wireless communication applications," *Antennas and Propagation Society International Symposium, 2002. IEEE* , vol.2, no., pp.500,503 vol.2, 2002.

[3]  C. R. Rowell, and R. D. Murch, "A capacitively loaded PIFA for compact mobile telephone handsets," *IEEE Trans. Antennas and Propag.*, vol. 45, no. 5, pp. 837-842, May. 1997.

[4]  S. Sadat, M. Fardis, G. Dadashzadeh, N. Hojjat, and M. Roshandel, "A compact microstrip square-ring slot antenna for UWB applications," *Antennas and Propagation Society International Symposium 2006, IEEE* , vol., no., pp.4629,4632, 9-14 July 2006.

[5]  N. Behdad, and K. Sarabandi, "A varactor-tuned dual-band slot antenna," *IEEE Trans. Antennas and Propag.*, vol. 54, no. 2, pp. 401-408, Feb. 2006.

[6]  D. M. Pozar, *Microwave Engineering*, 3rd ed. New Work: Wiley, 2003.

# Narrow-wall confined slotted waveguide structural antennas for small multi-rotor UAV

D. Gray*, K. Sakakibara† and Y. Zhu‡

* The University of Nottingham, Ningbo, Zhejiang, China
† Nagoya Institute of Technology, Nagoya, Aichi, Japan
‡ Ningbo Institute of Materials Technology and Engineering, Chinese Academy of Sciences, Ningbo, Zhejiang, China

*Abstract-* **In preparation for a cross-discipline study of slotted waveguide antennas to be used as the load-bearing rods connecting the motor units to the central hub of a multi-rotor unmanned aerial vehicle, past work on carbon fiber reinforced plastic slotted waveguide is reviewed, and a novel narrow wall slot design is presented. The new slot is a distorted Z-shape which fits entirely within the narrow wall, and consequently does not require cutting of the corners of the rectangular tube which would significantly weaken the structure. The Z-slot was used in a 10-slot 15dBi antenna for 10GHz, which was shown to have comparable gain and return loss performance to a conventional narrow wall slotted waveguide antenna.**

## I. INTRODUCTION

Small multi-rotor "helicopter" unmanned aerial vehicles (UAVs) are self-stabilizing flying platforms capable of hovering or low speed flight for periods of up to 40 minutes. These flight characteristics make these small UAVs perfect for high quality photography, such as for real estate. To date, these small UAVs have not been used for radio frequency applications, possibly due to the limited endurance. However, in an educational context where restricted flight time is not an operational impediment, these small UAVs are attractive for student projects on radio frequency sensing and radar; students are engaged by been able to interact with an actual UAV and likewise find the opportunity to work on a system to be integrated with the UAV exciting. Commercially available small mass Ultra-WideBand (UWB) and X-band FMCW radar units make radio frequency systems and machine vision projects possible for undergraduate students.

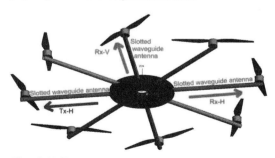

Figure 1. Multi-rotor unmanned aerial vehicle, with possible transmit (Tx) and receive (Rx) polarizations marked; diameter approximately 1 metre.

Applications such as radiometry, synthetic aperture radar (SAR) and ground moving target indicator (GMTI) require high gain antennas. While a means of using the full circular aperture of a 1 meter diameter multi-rotor UAV for a downward facing antenna is far from clear, a first practical step is to use the booms connecting the motor-propeller units to the central hub as antennas, Figure 1. Structural antennas are one approach to implementing high gain antennas without significantly adding to the weight or aerodynamic drag of an aircraft [1].

The main load-bearing structure of multi-rotor UAVs are hollow Carbon Fiber Reinforced Plastic (CFRP) tubes which connect the central box (housing sensors, battery and controller) to the outlaying electric motors and propellers, Figure 1. The hollow composite material tubes are roughly the same dimensions as X-band rectangular and circular waveguide, suggesting that slotted waveguide antennas can be used as arms without changing the design of an existing multi-rotor UAV nor adding to the mass. For the benefit of adding radio application without reducing flight endurance, there is the cost of weakening the booms by cutting the radiating slots.

The aim of this initial work was to produce a series of equivalent 10GHz 15dBi slotted waveguide antenna designs, which will be tested for deformation and instability in the future in a similar fashion to [1]. Attention was paid to the minimum slot-end to slot-end distance which is assumed to relate to maximum mechanical strength. In the following section, prior work on broad-wall slotted waveguide antennas made from CFRP is reviewed, with consideration given to the distance between the slot-ends. Narrow-wall slotted waveguide antennas are then considered, and a novel slot design discussed.

## II. PRIOR WORK ON BROAD-WALL SLOTTED WAVEGUIDE

As part of a project to integrate large aperture antennas into the load bearing structures on a fixed wing aircraft, a series of dry-layup CFRP rectangular slotted waveguide antennas were designed, built and tested for 9.375GHz Ground Moving Target Indicator or airborne weather radar [2]. For the layup of the CFRP WR-90 sized tubes, prepreg tape was wrapped around Teflon coated mandrels which were removed after curing [3]. This was a successful partial repeat of some historical work [4, 5].

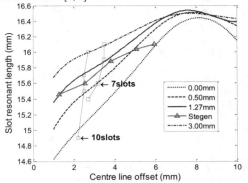

Figure 2. Single longitudinal broad-wall slot resonant length dependence on offset distance from broad-wall centre line at 9.375GHz for different waveguide wall thicknesses; derived from Figure 3 of [2], "Stegen" trace is experimental data taken from [6], with 7 and 10 slot antenna designs, from FEKO™.

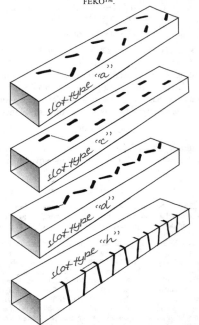

Figure 3. Sketches of WR-90 slotted waveguide antennas, based on Figure 9-2 of [6]; red line indicating minimum slot-end to slot-end distance.

As the aim of this work was to produce a series of slotted waveguide designs which will later be tested as load-bearing beams, the 9.375GHz 7-slot and 10-slot slotted waveguide design work of [2] was extended to include infinitely thin walled WR-90 waveguide, Figure 3. The "0.00mm" wall thickness results from FEKO™ were in agreement with the trend from the other wall thicknesses. For 10-slot antennas, the broad-wall center line offset remained at 2.1mm with the only difference been a shortening of the resonant slot length, Figure 2.

With FEKO™ simulations of infinitely thin walled 10-slot slotted waveguide antennas having proven to be quick and convenient of getting indicative antenna designs, the guided wavelength was changed to 39.7mm (10GHz) and a series of designs produced to gauge the maximum slot-end to slot-end distance, which was assumed to give maximum mechanical strength.

It is noted in [7] that longitudinal slots in the broad-wall of rectangular waveguide can be offset and parallel to the center line (type "c"), centered and rotated (type "d") or offset and rotated (type "a"), Figure 3. Examples of all 3 were designed in FEKO™ for 10GHz using 1.6mm width slots, Table 1. Both rotated slot designs had maximum cross-polarized radiation at 15.5dB below peak. The centered and rotated design gave the smallest slot-end to slot-end distance, and is considered a mechanical worst case; the array of slots look like a zipper-tear line. In contrast, the offset and rotated design gave an increased slot-end to slot-end distance. The 3 designs will be mechanically tested in the future and the effect of decreased slot-end to slot-end on weakening the rectangular waveguide tubes noted.

TABLE I
10GHz 10-SLOT BROAD-WALL SLOTTED WAVEGUIDE ANTENNAS

| Slot type | Slot length (mm) | Center line offset (mm) | Slot angle to center line (°) | Minimum distance (mm) |
|---|---|---|---|---|
| a | 13.1 | 3.5 | 12.5 | 10.0 |
| c | 13.7 | 2.5 | 0 | 8.0 |
| d | 13.7 | 0 | 15 | 6.6 |

III. NARROW-WALL SLOTTED WAVEGUIDE

Having found that a less common broad-wall slotted waveguide antenna design gave increased slot-end to slot-end distance, and presumably increased mechanical strength, narrow wall type "h" slotted waveguide designs were investigated. As structural beams for a multi-rotor UAV, narrow-wall slotted waveguide antennas are preferred as a lesser cross-section is presented to the primary flight air flow from the propellers giving minimal drag as well as least bending moment. Four different waveguide wall thicknesses were used, Table 2. It is assumed that wall thicknesses approaching 3.00mm will be needed for mechanical strength.

TABLE II
WALL THICKNESSES FOR WR-90 SLOTTED WAVEGUIDE ANTENNAS

| Wall thickness (mm) | Description |
|---|---|
| 0.00 | aluminum or brass foil inner layer of non-conductive fiber reinforced plastic tube |
| 0.50 | 4-ply Carbon Fiber Reinforced Plastic (CFRP) dry layup as per [2] |
| 1.27 | standard WR-90 wall thickness for aluminum or copper waveguide |
| 3.00 | approximate face sheet or aircraft CFRP panel thickness |

TABLE III
10GHZ 10-SLOT NARROW-WALL SLOTTED WAVEGUIDE ANTENNAS

| Wall thickness (mm) | Slot angle (°) | Slot extension (mm) | Minimum distance (mm) |
|---|---|---|---|
| 0.00 | 17.8 | 2.4 | 14.5 |
| 0.50 | 19.0 | 2.2 | 14.2 |
| 1.27 | 17.5 | 1.8 | 14.6 |
| 3.00 | 21.7 | 0.6 | 13.7 |

TABLE IV
10GHZ 10-SLOT Z-SLOT SLOTTED WAVEGUIDE ANTENNAS

| Wall thickness (mm) | Transverse dimension (mm) | Longitudinal dimension (mm) | Offset (mm) | Minimum distance (mm) |
|---|---|---|---|---|
| 0.00 | 9.6 | 6.9 | 1.7 | 11.1 |
| 0.50 | 9.6 | 6.6 | 1.6 | 11.4 |
| 1.27 | 9.6 | 6.4 | 1.5 | 11.7 |
| 3.00 | 9.6 | 5.9 | 1.1 | 11.7 |

Past work on narrow-wall slots appears to have been restricted to maximum slot angles of 15° [7, 8], Figure 4A. Here, slot angles of 17.5° to 22° were found to be best, designs for all 4 wall thicknesses giving satisfactory return loss, Table 3 and Figure 5A. The radiation pattern characteristics of all 3 designs were likewise satisfactory, with around 15.5dBi directivity across the useful bandwidth of 9.8 to 10.1GHz, although the peak cross-polarized component increased with wall thickness, Figures 6A to 8A. The 8.5dB peak cross-polarized component of the 3.00mm wall thickness design here is unacceptably high, Figure 7A.

From a mechanical stand-point, the minimum slot-end to slot-end distances around 14.5mm were significantly better than the broad-wall designs, been a 45% increase in that dimension, Tables 1 and 3. Despite this improvement over the best broad-wall design, albeit at the cost of increased cross-polarization, the narrow wall slots do cut 2 corners of the rectangular waveguide. With the results of [1] in mind, cutting the corners of the waveguide is expected to significantly increase the susceptibility of the rectangular waveguide to both deformation and instability. Consequently, a slot design that is entirely confined to the narrow-wall was sort.

Considering the fundamental mechanism behind radiation from narrow wall slots, the slots sit at an angle to the narrow-wall currents, the angled current flow caused by a slot could be thought of as a "vector sum" of both longitudinal and transverse currents. Offsetting the ends of a square Z-slot was found to produce the required disturbance to the narrow wall currents, Figure 4B. A set of 4 designs was prepared for different wall thicknesses at 10GHz with the maximum transverse dimension been fixed at 9.6mm, Table 4. The return loss and radiation characteristics of the Z-slot antennas were satisfactory across the 9.8 to 10.1GHz band, Figures 5B to 8B. Comparing to the conventional narrow wall slot design, the cross-polarized component was noticeably lower, Figures 7 and 8. The only potential disadvantage been the lesser minimum slot-end to slot-end distance around 11.5mm, which,

however, is an improvement over the original broad-wall designs, Tables 1 and 4.

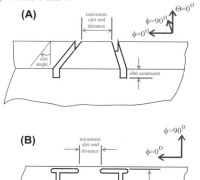

Figure 4. Sketches of slot pairs of narrow-wall slotted waveguide antenna; (A) conventional "h type" slots, (B) Z-slots.

IV. CONCLUSIONS

A series of 10GHz 15dBi slotted waveguide antennas were designed in the commercially available simulator FEKO™, which will be tested as load bearing beams in the future. In order to avoid corner-cutting in narrow-wall slotted waveguide antennas, a Z-slot that was confined within the narrow wall was designed, without loss of performance.

ACKNOWLEDGEMENT

The authors are indebted to EMSS SA for providing academic licenses of FEKO™ and a high level of technical support.

REFERENCES

[1] J.W. Sabat, "Structural response of the slotted waveguide antenna stiffened structure components under compression," M.Sc. thesis of Air Force Institute of Technology, Wright-Patterson Air Force Base, Ohio, March, 2010.

[2] D. Gray, K. Nicholson, K. Ghorbani & P. Callus, "Carbon Fibre Reinforced Plastic Slotted Waveguide Antenna", Asia Pacific Microwave Conference (APMC 2010), Yokohama, December 2010.

[3] P.J. Callus & K.J. Nicholson, "Standard operating procedure – manufacture of carbon fibre reinforced plastic waveguides and slotted waveguide antennas, version 1.0," Defence Science and Technology Organisation, Australia, document DSTO-TN-0937, June 2011.

[4] R. Wagner & H.M. Braun, "A slotted waveguide array antenna from carbon fibre reinforced plastics for the European space SAR", Acta Astronautica, vol. 8, no. 3, March 1981, pp. 273-282.

[5] L. Knutsson, S. Brunzell & H. Magnusson, "Mechanical design and evaluation of a slotted CFRP waveguide antenna", Proceedings of the Fifth International Conference on Composite Materials, San Diego, CA, July 29-August 1, 1985, pp. 475-481.

[6] R.J. Stegen, "Longitudinal shunt slot characteristics", Hughes Aircraft Company, Technical Memorandum no. 261, Nov. 1951.

[7] M.J. Ehrlich, "Slot-antenna arrays," Chapter 9 in Antenna Engineering Handbook, H. Jasik, ed., 1961.

[8] W.H. Watson, "Resonant slots," Journal of the IEE, pt. IIIA: Radiolocation, vol. 93, iss. 4, 1946, pp. 747 – 777.

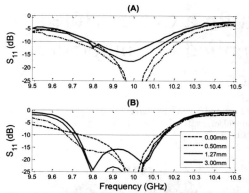

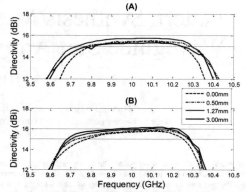

Figure 5. Return loss of 10 slot narrow-wall slotted WR-90 waveguide antennas designs, from FEKO™; (A) conventional slots, (B) Z-slots.

Figure 6. Directivity of 10 slot narrow-wall slotted WR-90 waveguide antennas, from FEKO™; (A) conventional slots, (B) Z-slots.

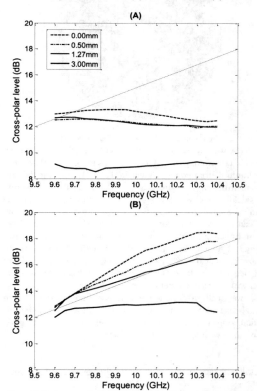

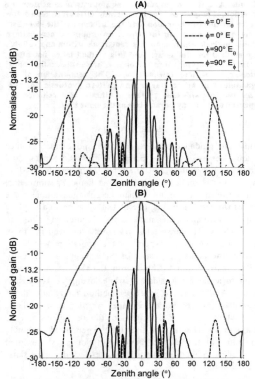

Figure 7. Peak cross-polarized component of radiation patterns of 10 slot narrow-wall slotted WR-90 waveguide antennas, from FEKO™; (A) conventional slots, (B) Z-slots..

Figure 8. Radiation patterns of 1.27mm wall thickness 10 slot narrow-wall slotted WR-90 waveguide antennas at 10GHz, from FEKO™; (A) conventional slots, (B) Z-slots.

# Pattern Synthesis Method for a Center Holed Waveguide Slot Array Applied to Composite Guidance

Huang Jingjian, Xie Shaoyi, Wu Weiwei and Yuan Naichang

College of Electronic Science and Engineering, National University of Defense Technology, Changsha 410073, China

*Abstract*-The circular waveguide slot array with a hole in the center is one of the key techniques to realize the shared aperture composite guiding. Due to the influence of the hole, a sound realization of low side-lobe pattern is the difficulty and key for this antenna. The waveguide broadside shunt slot array only can realize low side-lobe by controlling the distribution of the amplitude. And it needs to control the slot offsets to achieve the distribution of amplitude. But it is already difficult to realize low side-lobe for this configuration, and the influence of the slot offset will not be ignored. So in the paper, we propose a new pattern synthesis method, which combines genetic algorithm and Elliott's design method for slot array. And this method takes slot offsets and mutual coupling into account, then adjust the slots' parameter to achieve low side-lobe. The combination of the designation of array parameter and pattern synthesis not only realizes the low side-lobe pattern synthesis, also achieves the designation of the array parameters at the same time.

## I. INTRODUCTION

Composite guidance is the research hot spot of the guidance technique, especially, the composite of the optics and microwave. Waveguide slot array has been widely used in bomb due to its excellent electric and mechanical performance. To fulfill the shared aperture composite guidance, a hole in the center of the circular waveguide slot array is needed to place the optical guidance equipment. But the center key slots are loss, which would increase the sidelobe level. And the more elements lost, the more exasperate of the far-field pattern.

It is hard to adjust the phases of the waveguide slots. Only amplitude can be adjusted by altering the slot offsets. So it is difficult to achieve low sidelobe while losing the key center elements. And traditional synthesis method rarely considered the slot offset effects. So in this paper, we propose a synthesis method which considers the effects of the slot offsets and the mutual coupler. This method combine the genetic algorithm[1] (GA) and Elliott's classic waveguide slot array designed method[2][3]. While the pattern is synthesized, the array slots' parameter will get.

## II. FAIRFIELD WITHOUT CONSIDER SLOT OFFSETS

Taking an X band (center frequency is 10GHz) waveguide slot array as example. The array model is shown in Fig1. The diameter of this circular antenna is 260mm, and the diameter of center hole is 40mm. So the four center slots are lost.

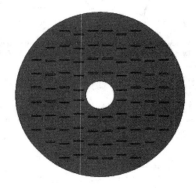

Fig.1. Center holed circular waveguide slot array.

Firstly, we synthesize a pattern with -20dB sidelobe level without considering the slot offsets using GA method. The amplitude distribution is shown in Table 1. And the farfield pattern with zero slot offsets is shown in Fig2. Fig2. (b) shows the sidelobe level is below -20dB. But in reality, the slot offset cannot be neglected. According to the amplitude distribution, we can calculate the slot parameters of slot offsets and slot length by using Elliott's method. And the real pattern is shown in Fig3. The sidelobe level is up to -17.6dB. It is not easy to realize low sidelobe while center key elements are losing. If we still neglect the slot offsets in pattern synthesis, the sidelobe level will not meet the demand in engineering.

So we need to design a pattern synthesis method which should take slot offsets into account, and also consider the mutual coupler. Then the designed farfield pattern can be good accordance with the real farfield pattern.

Table I

Amplitude distribution of the quarter array without considering slots offsets.

| | | | | |
|---|---|---|---|---|
| 0.399 | 0.422 | 0.279 | - | - |
| 0.401 | 0.182 | 0.231 | 0.370 | - |
| 0.700 | 0.681 | 0.470 | 0.214 | 0.299 |
| 1.000 | 0.535 | 0.675 | 0.200 | 0.409 |
| - | 0.995 | 0.685 | 0.396 | 0.374 |

978-7-5641-4279-7

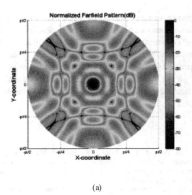

(a)

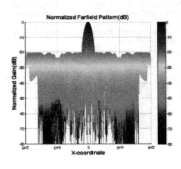

(b)

Fig.2. Antenna farfield pattern with zero offsets.

(a) top view (b) side view

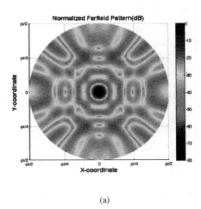

(a)

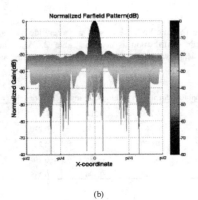

(b)

Fig.3. Antenna farfield pattern with slot offset.

(a) top view (b) side view

## III.  FARFIELD SYNTHESIS CONSIDERING SLOT OFFSETS

To take the slot offset into account, we design a pattern synthesis method. The flow chart is shown in Fig4. And the main process contains three steps.

Step 1： Calculate the amplitude distribution **Vs**, slots offsets **Ds** and slots length **Ls** by Using GA method. And we use formula (1) to calculate the initial slot offsets. This formula is deduced by Stevenson[4]. And gr is the normalized conductance, ds is the slot offset.

$$ds = \frac{a}{\pi}\operatorname{asin}((\frac{gr}{2.09}\cdot\frac{b}{a}\cdot\frac{\lambda}{\lambda_g}\Big/\cos^2(\frac{\pi\lambda}{2\lambda_g}))^{0.5}) \quad (1)$$

Step 2: calculating the array mutual coupling matrix and using Elliott's three formulas[3] to iterative compute the slots parameters and compensating the mutual coupling. Then new slot length matrix **Lsm** and slot offsets matrix **Dsm** are got.

Step 3: Judging if the results are convergent? If yes, the final results are got. If no, we turn to step 1 and substitute **Dsm** for **Ds** to recalculate the Vs. And till convergent.

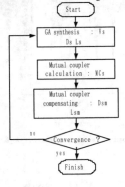

Fig.4. Pattern synthesis flow chart.

Using this method, we design a waveguide slot array. The amplitude distribution is list in Table 2. And the farfield pattern is shown in Fig.5. The sidelobe levels are all below -20dB.

Table II
Amplitude distribution of the quarter array considering slots offsets.

| 0.280 | 0.230 | 0.200 | - | - |
|-------|-------|-------|------|-------|
| 0.298 | 0.309 | 0.200 | 0.220 | - |
| 0.597 | 0.470 | 0.427 | 0.237 | 0.200 |
| 1 | 0.523 | 0.586 | 0.209 | 0.333 |
| - | 0.897 | 0.596 | 0.335 | 0.351 |

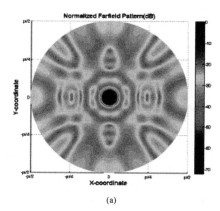

(a)

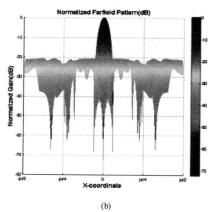

(b)

Fig.5. Farfield pattern consider slot offsets. (a) top view, (b) side view.

IV. CONCLUSION

This paper designs a pattern synthesis method for center holed waveguide slot array which is applied to composite guidance. This method combines GA and Elliott's method, and fully considers the possible factors which could affect the pattern property. Taking slot offsets and mutual coupling into account, so the results can be most close to the real condition. It is a effective method for center holed waveguide slot array designation.

REFERENCE

[1] F. J. Ares-Pena, J. A. Rodriguez-Gonzalez, E. Villanueva-Lopez, and S. R. Rengarajan, Genetic algorithms in the design and optimization of antenna array patterns, [J]. IEEE Trans. Antennas Propagat. , 1999, vol.47, no.3, pp.506–510.

[2] R. S. Elliott, Improved design procedure for small arrays of shunt slots, [J]. IEEE Trans. Antennas Propagat. , Jan. 1983, vol. AP-31, pp. 48-53.

[3] R. S. Elliott and W. R. O'Loughlin, The Design of Slot Arrays Including Internal Mutual Coupling, [J]. IEEE Trans. Antennas Propagat. Sep. 1986, Vol AP-34(9), pp. 1149-1154.

[4] J. L. Volakis. Antenna Engineering Handbook [M]. Fourth Edition, McGraw-Hill, 2007: Ch9.

# Design and Measurement of a Parallel Plate Slot Array Antenna Fed by a Rectangular Coaxial Line

[#]Hajime Nakamichi[1], Makoto Sano[1], Jiro Hirokawa[1], Makoto Ando[1], Katsumori Sasaki[2], Ichiro Oshima[2]

[1]Dept. of Electrical and Electronic Eng., Tokyo Institute of Technology
S3-19, 2-12-1 O-okayama, Meguro-ku, Tokyo 152-8552, JAPAN
{nakamichi, msano, jiro, mando}@antenna.ee.titech.ac.jp
Denki Kogyo Co., Ltd.
13-4 Satsuki-cho, Kanuma-shi, Tochigi 322-0014, JAPAN
{ka-sasaki, i-oshima}@denkikogyo.co.jp

*Abstract-* **We propose a parallel plate slot array antenna fed by a rectangular coaxial line, which can simplify the electrical contact between the bottom plate of the parallel plate and the rectangular coaxial line. We adopt hard walls at the both side walls of the parallel plate in order to excite a quasi-transverse electromagnetic (TEM) wave. We analyze the overall structure of the antenna, and measure its prototype. The measured bandwidth for VSWR less than 1.5 is 2.1%, and the antenna efficiency is 50.5% at 4.859 GHz where the reflection is minimized.**

## I. Introduction

Wireless access system in the 5 GHz band has expected to have a potential for high speed and long distant wireless transmission because it can use high power in outside. This system is needed to register and the number of the stations is small, so that wireless communication system which has less interference and enables to communicate stably can be created. Wireless LAN can be suitable for disasters, places difficult to construct wired networks and temporary usages. We propose a parallel plate slot array antenna which has a very simple structure and low loss as one of the antennas for this system.

We proposed a parallel plate slot array antenna fed by a rectangular waveguide [1]. The feed slot pairs on the rectangular waveguide for the reflection suppression needed to be spaced by not a half of the guide wavelength but one guide wavelength. The relative permittivity of the dielectric in the feed waveguide should have a range in comparison with that in the parallel plate waveguide. Since a current of the dominant TE10 mode in the feed waveguide flows between the bottom plate of the parallel plate and the side walls of the feed waveguide, the electrical contact between them needed firmly. In this paper, we propose a rectangular coaxial line as the feed waveguide. The transverse electromagnetic (TEM) wave is dominant in the parallel plate waveguide and a current flows in parallel to the axis of the rectangular coaxial line, which can simplify the electrical contact between the between the bottom plate of the parallel plate and the rectangular coaxial line [2].

We demonstrated the feasibility of this antenna which used the feed slot pairs on the rectangular coaxial line for the reflection suppression. However, when the coupling got larger, the distance between the slots in a pair got smaller due to the mutual coupling, which could not be realized. We propose one

feed slot and the step structure in the inner conductor of the rectangular coaxial line for the reflection suppression. In order to excite a quasi-TEM wave, we adopt hard walls, which is filled with the dielectric which has a large relative permittivity at the both side walls of the parallel plates [3][4]. In this paper, we design the feed slots and the radiating slot pairs, and present the analysis result of the overall antenna structure and measurement of its prototype.

## II. Antenna Structure and Operation

Figure 1 shows the antenna structure of a parallel plate slot array antenna fed by a rectangular coaxial line. A radiating waveguide consists of two square conductor plates. An $x$-directed rectangular coaxial line is mounted on the back of the parallel plates in the center. Both the parallel plate waveguide and the rectangular coaxial line are filled with the dielectric of the permittivity $\varepsilon_r$ and $\varepsilon_f$, respectively. Radiating slot pairs are arrayed on the top plate of the parallel plates. The two slots in each pair are spaced by about a quarter of a guide wavelength. Reflections from the two slots in a pair are canceled. The two slots in a pair are shifted from each other in the $x$ direction in order to decrease the mutual coupling.

Feed elements are arrayed between the parallel plate and the rectangular waveguides. The feed element consists of one slot and a step structure in the inner conductor of the rectangular coaxial line. The reflection from the slot is canceled by that from the step structure.

A TEM mode, excited by a probe at the center of the feed waveguide, propagates in the $\pm x$ direction in the feed waveguide. It couples to the parallel plate waveguide through the feed slots which are spaced by the guide wavelength $\lambda_f$ to excite in equal amplitude and phase. A quasi-TEM wave excited in the parallel plate waveguide propagates in the $\pm y$ direction. The quasi-TEM wave couples to the radiating slot pairs, spaced by $a_r$ and $\lambda_r$ in the $x$ and the $y$ directions, respectively. The slot pairs radiate a linearly polarized wave in equal amplitude and phase [1].

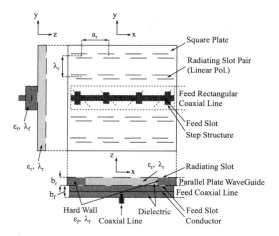

Figure 1: Antenna structure

## III. DESIGN OF THE FEED ELEMENT

Figure 2 shows the design model of the feed element. The parallel plate waveguide is a rectangular waveguide whose narrow walls satisfy the periodic boundary condition. A feed slot is inclined 45° to the axis of the rectangular coaxial line and provides coupling both the rectangular coaxial line and the parallel plate waveguide. Because of the symmetry structure, the power at the ports 3 and 4 are of equal amplitude but of opposite phase.

This antenna is designed at 4.85 GHz. The width of the feed slot is 3.0 mm and the plate thickness is 1.5 mm. The width and the thickness of the rectangular coaxial line are 24.0 mm and 10.0 mm. The thickness of the parallel plate waveguide is 7.0 mm. The relative permittivity $\varepsilon_f$, $\varepsilon_r$ of the dielectric in the rectangular coaxial line and the parallel plate waveguide are 1.16 and 1.09, respectively. Hard walls at the both side walls of the parallel plate are filled with the dielectric of the relative permittivity of 1.16.

The design parameters are the slot length $l_f$, the length $d$ and the position $p$ of the step structure. Firstly, the slot length $l_f$ is given and the length $d$ and the position $p$ of the step structure are determined to minimize the reflection to port 1.

Figure 3 shows the design parameter of the feed element for the desired coupling when the width of the parallel plate waveguide is 0.86 $\lambda_f$. The slot length $l_f$ gets longer as the required coupling gets larger and the length $d$ of the step structure gets longer because the reflection gets larger as the slot length $l_f$ gets longer.

## IV. EXPERIMENTAL RESULTS

11×10 radiating slot pairs placed in the top plate are designed according to [1]. The number of the feed element is 10. These slots are designed for uniform excitation. We analyze the overall structure of the antenna by HFSS, and measured its prototype.

Figure 4 shows the frequency characteristic of the reflection of the overall structure of the antenna. The calculated and the measured bandwidth for VSWR less than 1.5 is 1.5% and 2.1%, respectively. A frequency shift of 45 MHz is observed, since the calculated and the measured reflections are minimized at 4.905 GHz and 4.859 GHz, respectively.

Figure 5 and 6 shows the H plane and the E plane radiating patterns where the calculated and the measured reflections are minimized respectively. The good agreements are obtained not only in the main lobe but also in the side lobes.

Figure 7 shows the frequency behavior of the gain and efficiency. The calculated directivity is 26.3 dBi at 4.905 GHz where the reflection is minimized and the aperture efficiency is 53.8%. The calculated gain is 26.2 dBi and the antenna efficiency is 53.6%, while the measured gain is 25.9 dBi at 4.859 GHz where the reflection is minimized, and the antenna efficiency is 50.5%. The frequency shift of 45 MHz is also observed about the gain, but the bandwidth is almost same.

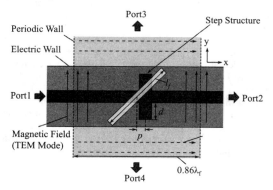

Figure 2: Design model of the feed slot

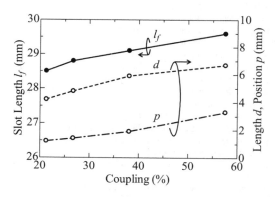

Figure 3: Design parameter of the feed slot for the desired coupling

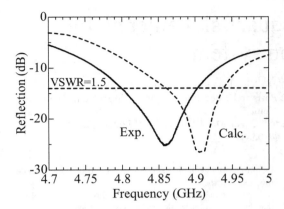

Figure 4: Frequency characteristic of the reflection

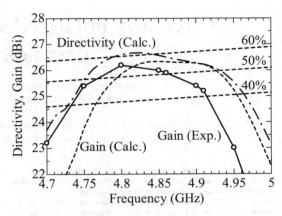

Figure 7: Frequency behavior of the gain and the efficiency

## V. CONCLUSION

We have designed the feeding part of a parallel plate slot array antenna fed by a rectangular coaxial line. We have analyzed the overall structure of the antenna, and measured its prototype. The measured bandwidth for VSWR less than 1.5 is 2.1% and the antenna efficiency is 50.5% at 4.859 GHz where the reflection is minimized. The frequency shift of 45 MHz is observed between the measurement and the calculation.

We will consider the bandwidth enhancement of the reflection and the antenna efficiency by changing the relative permittivity of the dielectric in the parallel plate waveguide and the rectangular coaxial line and the angle of the feed slot inclined to the axis of the rectangular coaxial line as the future studies.

### REFERENCES

[1] J.Hirokawa, M.Ando and N.Goto, "Waveguide-Fed Parallel Plate Slot Array Antenna," IEEE Trans. Antennas Propagat., vol.40, no.2, pp.218-223, Feb.1992.

[2] S.Yamaguchi, Y.Tahara, T.Takahashi and T.Nishino, "Inclined Slot Array Antennas on a Rectangular Coaxial Line," EuCAP, POS2-36, Apr.2011.

[3] P.-S. Kildal, "Artificially Soft and Hard Surfaces in Electromagnetics," IEEE Trans. Antennas Propagat., vol.38, no.10, pp.1537-1544, Oct.1990.

[4] M.Samardzija, T.Kai, J.Hirokawa and M.Ando, "Single-Layer Waveguide Feed for Uniform Plane TEM-Wave in Oversized-Rectangular Waveguide With Hard-Surface Sidewalls," IEEE Trans. Antennas & Propagat., vol.54, no.10, pp.2813-2819, Oct.2006.

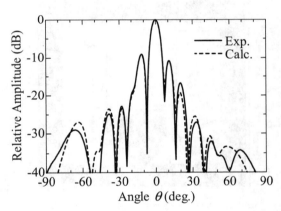

Figure 5: Radiation pattern in the H-plane
(Calc.: 4.905 GHz, Exp.: 4.859 GHz)

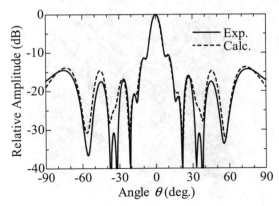

Figure 6: Radiation pattern in the E-plane
(Calc.: 4.905 GHz, Exp.: 4.859 GHz)

# Circularly polarized square slot antenna for navigation system

Yixing Zeng#, Yuan Yao#, Junsheng Yu#,  Xiaodong Chen*, Youbo Zhang#

\# School of Electronic Engineering, Beijing University of Posts and Telecommunications, No.10 Xitucheng Road, Beijing, China

\* School of Electronic Engineering and Computer Science, Queen Mary, University of London, London, UK

yixing_zeng@163.com

**Abstract-This letter presents a circularly polarized square slot antenna used for navigation system. A tuning stub protruded into the slot from the narrowed stripe at the end of the signal strip of the CPW is used to obtain good impedance matching. A measurement showed that the bandwidth below -10dB of the antenna varied from 1.15GHz to 1.71GHz , and the Axial Ratio below 3dB is from 1.14 GHz to 1.75GHz. The radiation pattern is bi-directional.**

## I.    INTRODUCTION

For generating circular polarization(CP) radiation, many solutions have been reported[1]. The CP antennas have lost of different types and structures including dielectric resonator antennas, reconfigurable patch antennas, monopole antennas and the most prevalent microstrip antennas. Now days CPW-fed square slot antenna has got much attentions due to their simple structure, wide bandwidth and easy to fabricated.

In [2], Jia-Yi Sze has designed the antenna by protruding a T-shaped metallic stripe from the ground plane toward the slot center and feeding the square slot antenna using a 50 Ω CPW with a protruded signal stripe at 90 to the T shaped stripe. In [3], he designed the antenna with two inverted-L grounded stripes around two opposite corners of the slot ad a widened tuning stub protruded into the slot from a signal stripe of CPW. In [4], Yue-Ying Chen designed the antenna with two opposite corners of the square slot. This paper presents a wideband circularly antenna which is printed on a square board. The antenna is fed by a coaxial cable through SMA. The bandwidth of return loss of the antenna varied from 1.15GHz to 1.71GHz . The bandwidth AR of the antenna varied from 1.15GHz to 1.75GHz.

## II.    ANTENNA DESIGN

The structure of the proposed circularly polarized antenna design for automobile is illustrated in fig 1. The proposed CPW-fed antenna is printed on a square substrate with a side size of G=90mm, a height of h=0.8mm, a dielectric constant of 2.3, Etched at the center of the top-side ground plane is a 75mm*75mm square slot, which is fed by a CPW line. the resistance of the CPW line is 50 Ω . The signal tripe of the CPW is protruded into the slot by a line. The CP operation of

the proposed antenna is mainly attributed to the two grounded inverted-L stripe.

The CPW-fed square slot antenna is shown in fig 2.. Notice that the x-axis is horizontal and z-axis is vertical.

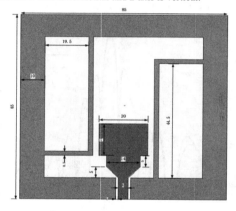

Figure 1.  A structure of antenna configuration

Figure 2.  Fabricated CPW-fed Square slot  antenna

### III. SIMULATED AND MEASURED RESULT

The performance of this CPW-fed slot antenna is very sensitive to many parameters, which would affect the performance of the antenna, such as the size of the L-stripe. The simulation results of the antenna was carried out via HFSS, which is a full wavelength numeric electromagnetic simulation tool. A comparison between simulated and measured return loss is illustrated in fig 3. The measurement was carried out in chamber. It is evident that the antenna has a good match and good accuracy between measurement and simulated results is observed. The -10dB bandwidth is from 1.1GHz to 1.7GHz, which is broad enough to cover several frequency bands for the navigation system.

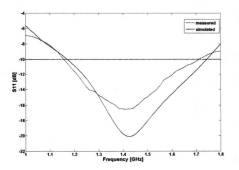

Figure 3. Measured and simulated return loss(S11) of the CPW-fed square slot antenna

Fig.4 compare the measured and simulated axial ratio(AR). We can see the AR is below -3dB from 1.14GHz to 1.75GHz, which means that the antenna has a good CP performance.

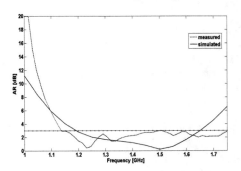

Figure 4. Measured and simulated Axial Ratio of the CPW-fed square slot antenna

Fig 5 and fig 6 shows the 2-D radiation pattern at 1.227GHz and 1.575GHz. The maximum gain is about 4dBi, which is found in the y and –y direction. It can be seen that the antenna radiates RHCP from the top of the antenna and LHCP from the bottom of the antenna. The measured result has good match with simulated result. Fig 7 is the simulated gain of the antenna at 1.227GHz and 1.575GHz.

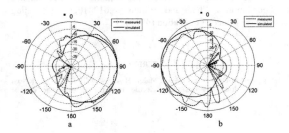

Figure 5. Measured and simulated 2-D radiation pattern of y-z plane at 1.227GHz
(a) RHCP (b) LHCP

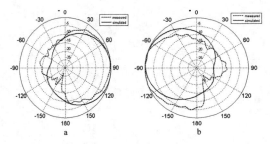

Figure 6. Measured and simulated 2-D radiation pattern of y-z plane at 1.575GHz
(a) RHCP (b) LHCP

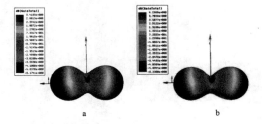

Figure 7. Simulated 3-D radiation pattern of the proposed antenna
(a) at 1.227GHz (b) at 1.575GHz

### CONCLUSION

A CPW-fed slot antenna for GPS, GNSS, Compass and Galileo was designed using CPW-fed square slot structure. The simulation result obtained by HFSS show good agreement with the measured results. The measured radiation patterns of the proposed antenna are bi-directional.

## ACKNOWLEDGMENT

This work is supported by the Fundamental Research Funds for the Central Universities, the National Natural Science Foundation of China under Grant No.61201026, and Beijing Natural Science Foundation (4133091).

## REFERENCES

[1]  Zainud-Deen S H, El-Azem Malhat H A, Awadalla K H. A single-feed cylindrical superquadric dielectric resonator antenna for circular polarization[J]. Progress In Electromagnetics Research, 2008, 85: 409-424.

[2]  Sze, Jia-Yi, Kin-Lu Wong, and Chieh-Chin Huang. "Coplanar waveguide-fed square slot antenna for broadband circularly polarized radiation." Antennas and Propagation, IEEE Transactions on 51.8 (2003): 2141-2144..

[3]  Sze, Jia-Yi, and Chi-Chaan Chang. "Circularly polarized square slot antenna with a pair of inverted-L grounded strips." Antennas and Wireless Propagation Letters, IEEE 7 (2008): 149-151.

[4]  Chen Y Y, Jiao Y C, Zhao G, et al. A novel compact slot antenna for broadband circularly polarized radiation[C]//Antennas Propagation and EM Theory (ISAPE), 2010 9th International Symposium on. IEEE, 2010: 201-204

# *FP-1(D)*

## October 37 (FRI) PM

## Room D

## A&P in Mata-structures

# Radiation from a Metahelical Antenna

H. Nakano, M. Tanaka, and J. Yamauchi
Hosei University
Koganei
Tokyo, Japan

*Abstract-* A novel helical antenna, designated as the metahelical antenna, is proposed, where the helical arm has electromagnetic right-handed (RH) and left-handed (LH) properties. The antenna design process is described. It is found that the metahelical antenna radiates a circularly polarized (CP) wave within two frequency bands. The rotational sense of the CP wave within one of the frequency bands is opposite to that of the CP wave within the other frequency band. The VSWR is less than 2, as desired, when the gain is at the maximum value.

## I. Introduction

A conventional helical antenna [see Fig. 1(a)] radiates a circularly polarized (CP) wave in the forward direction under the conditions that (i) the arm peripheral length S is approximately one guided wavelength ($1\lambda_g$) [1][2], and (ii) the current on the helical arm flows toward the arm end with no current reflected toward the feed point. This radiation is called end-fire radiation.

The current distribution for the conventional end-fire helical antenna was revealed experimentally by Kraus [1]; it was later revealed theoretically by Nakano, with the solution of an integral equation [3] using the method of moments [4]. It was found that the current distribution has two distinct regions: the C-current region and the S-current region [5]. The C-current region generates a CP wave, and the S-current region acts as a director for the electromagnetic wave from the C-current region. An array composed of low-profile helical antennas having only the C-current region has been used in Japan for direct broadcasting satellite antennas and as the primary feed for a Cassegrain reflector (VERA project by NAOJ). Such an array has also been adopted for use on the Mercury magnetospheric orbiter (BepiColombo project by JAXA and ESA).

The aforementioned conventional end-fire helical antenna, shown in Fig. 1(a), radiates a CP wave of either left-handed (LH) or right-handed (RH) polarization, determined by the arm winding direction (either clockwise or counter clockwise). This paper presents a novel end-fire helical antenna, shown in Fig. 1(b), which differs from the conventional end-fire helical antenna in that it radiates an LH CP wave within a specific frequency band and an RH CP wave within a different frequency band. This antenna is designated as the metahelical antenna and is distinguished from the conventional helical antenna. The radiation characteristics of the metahelical antenna are presented and discussed in this paper.

## II. Structure

Figs. 1(a) and (b) show the conventional helical antenna and the novel helical antenna, respectively. The conventional antenna consists of a continuous arm wound on a dielectric layer, while the novel antenna has a noncontinuous arm wound on a grounded dielectric layer. Both antennas are upright above a conducting plate [ground plate (GP)].

The noncontinuous arm of the novel antenna is composed of segments, each having length $L_{seg}$, and each being loaded by an inductor (inductance $L_L$). Neighboring segments, separated by a gap g, are connected through a capacitor (capacitance $C_L$). The length $L_{seg} + g$ is designated as the periodicity P. Note that the width and pitch of the helical arm are denoted as w and a, respectively, and the grounded substrate has a thickness of B and a permittivity of $\varepsilon_r$. The notation for the other structural parameters is included in Fig. 1.

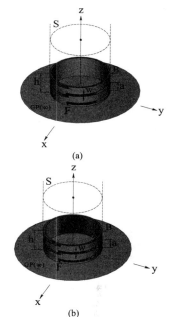

Figure 1: Helical antennas. (a) Conventional helical antenna. (b) Novel helical antenna.

978-7-5641-4279-7

## III. DISCUSSION

The parameters for the metahelical antenna are determined using the following three steps. First, the relative permittivity $\varepsilon_r$ and thickness B of the grounded dielectric layer are specified. Second, the width w of the helical arm printed on this dielectric layer is determined, so that the helical arm without gaps (g = 0) and without the $C_L$ and $L_L$ loading has a specified intrinsic impedance of $Z_R$. Third, values of $C_L$ and $L_L$ are determined such that a specified balanced frequency of $f_{bal}$ is obtained for a given segment length $L_{seg}$ and gap g.

Thus, parameters w, $C_L$, and $L_L$ are determined for a given $\varepsilon_r$, B, $L_{seg}$, g, and $f_{bal}$. Using these parameters, the dispersive relation between $\beta/k_0$ and the operating frequency f is revealed, where $\beta$ is the phase constant for the current flowing along the helical arm and $k_0$ is the phase constant for electromagnetic wave propagation in free space. The radiation occurs between frequencies $f_L$ and $f_U$, where $f_L$ and $f_U$ are the lower and upper frequency bounds for a fast wave.

A negative phase constant ($\beta/k_0 < 0$) appears below the balanced frequency $f_{bal}$ [6] in the dispersion diagram. This predicts that the metahelical antenna will radiate an LH CP wave when the peripheral length of the arm, normalized to the guided wavelength, is approximately one. Conversely, the metahelical antenna will radiate an RH CP wave above $f_{bal}$, due to a positive phase constant ($\beta/k_0 > 0$) when the normalized peripheral length is approximately one, as in the LH CP radiation. Note that the guided wavelength $\lambda_g$ normalized to the free-space wavelength $\lambda$ for LH CP radiation differs from that for RH CP radiation.

Analysis is performed by increasing the number of helical turns N up to N = 6. As N is increased, the gain in the bore sight direction (z-direction) increases. When the number of helical turns is N > 2, the maximum gain for principal LH CP radiation, $G_{max-LH}$, for our antenna model, having a balanced frequency of $f_{bal}$ = 3 GHz, appears at a frequency of approximately 2.5 GHz $\equiv f_{N-LH}$, and the maximum gain for principal RH CP radiation, $G_{max-RH}$, appears at a frequency of approximately 3.6 GHz $\equiv f_{H-RH}$. Note that $f_{N-LH}$ is a frequency below the balanced frequency $f_{bal}$, and $f_{H-RH}$ is a frequency above $f_{bal}$. This is consistent with the aforementioned prediction.

The radiation field is decomposed into two components: an LH CP wave component $E_L$ and an RH CP wave component $E_R$. It is found that, as the number of helical turns N is increased, the cross-polarization component becomes smaller. This is illustrated in Fig. 2, where the radiation patterns for N = 2 and N = 6 are compared. Note that these radiation patterns are obtained at a frequency above the balanced frequency $f_{bal}$.

The helical arm is not symmetric with respect to the z-axis. Asymmetry is seen in the cross-polarization component, indicated in Fig. 2 by the dotted line. On the other hand, as N is increased, the principal component becomes relatively symmetric with respect to the z-axis (see the solid line). Although the principal LH CP radiation pattern is not shown

in Fig. 2, it is found that the principal RH CP and LH CP radiation patterns for the same number of helical turns N have different half-power beam widths, due to the dispersive characteristic of the antenna arm.

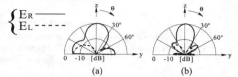

Figure 2: Comparison of the radiation patterns. (a) Number of helical turns N = 2. (b) N = 6.

The input impedance $Z_{in}$ for the number of helical turns N > 2 is investigated at the frequency where the LH CP radiation shows maximum gain. It is found that the VSWR relative to 50 ohms, calculated based on $Z_{in}$, is less than 2, as desired. A similar frequency response for the VSWR is obtained at the frequency where the RH CP radiation shows maximum gain.

## IV. CONCLUSIONS

A novel helical antenna, designated as the metahelical antenna, has been proposed. The metahelical antenna has two electromagnetic properties: the LH and RH properties. With these properties, the metahelical antenna radiates both LH CP and RH CP waves in the bore-sight direction (i.e., axial mode radiation), when the peripheral length of the helical arm corresponds to one guided wavelength. These axial-mode counter-CP waves, which cannot be obtained by conventional axial-mode helical antennas, occur within a low frequency band below the balanced frequency and a high frequency band above the balanced frequency. In other words, the metahelical antenna acts as a dual-band counter CP radiation element. The radiation pattern within the low frequency band is wider than that within the high frequency band. The VSWR is low within these frequency bands, as desired.

## ACKNOWLEDGMENT

We thank V. Shkawrytko for his assistance in the preparation of this manuscript.

## REFERENCES

[1] J. D. Kraus, *Antennas* (second edition), Chapter 7, McGraw-Hill, 1988.
[2] H. Nakano, *Helical and spiral antennas*, Research Studies Press, Hertfordshire, England, 1987
[3] E. Yamashita edit, *Analysis methods for electromagnetic wave problems*, Chapter 3, Artech House, Boston, 1996.
[4] R. F. Harrington, *Field Computation by Moment Methods*, Macmillan, New York, 1968.
[5] H. Nakano, J. Yamauchi, and H. Mimaki, "Backfire radiation from a monofilar helix with a small ground plane," IEEE Trans., Antennas and Propagat., vol. 36, no. 10, pp. 1359-1364, October 1988.
[6] N. Engahta and R. Ziolkowski, *Metamaterials*, Wiley, 2006.

# Transformation Optical Design for 2D Flattened Maxwell Fish-Eye Lens

Guohong Du[1, 2], Chengyang Yu[1], and Changjun Liu[1]

[1] School of Electronics and Information Engineering, Sichuan University, Chengdu 610064, China
[2] School of Electronic Engineering, Chengdu University of Information Technology Chengdu, 610225, China
Corresponding author: Changjun Liu email:cjliu.cn@gmail.com

*Abstract*-In this paper, a 2D Flattened Maxwell fish-eye (MFE) lens is proposed using transformation optics, which transforms a traditional MFE lens into a flattened MFE lens. The coordinate transformation is presented from a circular to a rectangular domain. We also discussed the rotated ray trajectories in MFE lens. The required permittivity and permeability tensors are derived for the flattened MFE lens and full-wave simulations are obtained. The simulated results show that the flatted lens can also provide the image focus opposite to the object focus on the boundary of rectangle.

## I. INTRODUCTION

Transformation optics (TO) as a new and powerful technique for microwave and optical device design has drawn great attention since it was reported by Leonhardt and Pendry et al. in 2006 [1,2]. Because Maxwell's equations possess the form-invariance regardless of the coordinate systems used, TO can provide the permittivity and permeability tensors which are functions of spatial coordinates. Then metamaterials with the rebuilt constructive parameters can guide and manipulate light. There have been various transformation devices which are usually fantastic, such as invisibility cloak [3], illusion generators and so on.

In 1968, the electromagnetic and optical behavior of two classes of dielectric lenses including Luneburg lens and Maxwell fish-eye lens was studied through optical ray path by Uslenghi et al. [4]. The refractive index distribution of these dielectric lenses are all functions of radius and the constructive parameters are all isotropic. These lenses have found success in commercial applications, but heavily restricted from broad use in imaging tools for two reasons. One of those is the spherical locus of focal points represents an inherent mismatch to conventional detector/receiver arrays, which are generally planar. To deal with the problem, Kundtz et al. suggested to flatten the some part of spherical surface into a planar one using the quasi-conformal mapping technique [5] for Luneburg lens.

In this paper, TO is employed to transform a 2D conventional MFE lens into a flatten MFE which has a planar surface. A circular domain is transformed into a rectangular domain and the ray trajectories are also presented when the

object point is moved on the original circumference of the circle. The required permittivity and permeability are derived and evaluated when the original space is free space. Combining the refractive index distribution of the original MFE and the constructive parameters derived above, the rebuilt parameters are obtained for flattened MFE. The full-wave simulated results show the proposed MFE lens can give the image focus opposite to the object point on the boundary square or rectangle.

## II. EQUATIONS UNDER TRANSFORMATION OPTICS

In TO, geometry and coordinate transformations play the dominant role in the design process. The coordinate transformation is interpreted as anisotropic expansion and compression of the original space. In Figure 1, the circular domain with the radius $a$ in the original coordinate system $(x, y, z)$ is transformed into a rectangular domain with the width $w$ and the height $h$ in the transformed system $(x', y', z')$. To flatten the lens, the following transformation is chosen

$$x' = \frac{wx}{a}, \tag{1}$$

$$y' = \frac{ly}{\sqrt{a^2 - x^2}}, \tag{2}$$

$$z' = z, \tag{3}$$

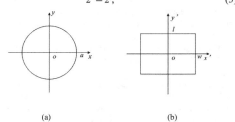

Figure 1 Spatial coordinate transformation
(a) the original coordinate system (b) the transformation coordinate system

This work was supported by Program for New Century Excellent Talents in University under Grant No.NCET-12-0383, and Sichuan Provincial Educational Department Project Research Fund under Grant No. 11ZA268.

978-7-5641-4279-7

When the geometry is specified by $a = 1$, $w = 1$, $l = 0.5$, the ray trajectories in spherical MFE is transformed into those in flattened MFE, shown in Figure 2. If the object points on the circumference of the circle is rotated $\theta$ around the origin, the original coordinate system for 2D should be multiplied by a coordinate rotation matrix firstly as follows

$$\begin{pmatrix} x_{rt} \\ y_{rt} \end{pmatrix} = \begin{pmatrix} \cos\theta & -\sin\theta \\ \sin\theta & \cos\theta \end{pmatrix} \begin{pmatrix} x \\ y \end{pmatrix} \tag{4}$$

where $x_{rt}$ and $y_{rt}$ are in the rotated coordinate system. In Figure 3, the flattened region and the ray trajectories are illustrated with $\theta = 30°$.

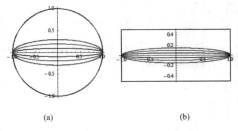

(a)  (b)

Figure 2 Ray trajectories in Flattened MFE
(a) the original coordinate system (b) the transformation coordinate system

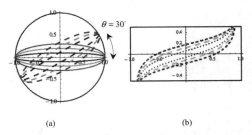

(a)  (b)

Figure 3  Rotated ray trajectories in Flattened MFE
(a) the original coordinate system (b) the transformation coordinate system

Spherical MFE, as a gradient index (GRIN) lens, has a radial distribution of index that varies as

$$n(r) = \frac{n_0}{1 + (r/r_0)^2} \tag{5}$$

where $n_0$ is the index of refraction at the center of the lens and $r_0$ is the radius of the lens [6]. Then the permittivity and permeability tensors can be derived from the usual TO [7] as

$$\varepsilon' = \frac{\Lambda\varepsilon\Lambda^T}{|\Lambda|} n^2(x, y) \tag{6}$$

$$\mu' = \frac{\Lambda\varepsilon\Lambda^T}{|\Lambda|}$$

Where $n(x, y)$ is the initial index distribution, and the tensor $\Lambda$ is the Jacobian matrix relating differential distance between the two coordinate systems which has the explicit form as

$$\Lambda = \begin{pmatrix} \dfrac{\partial x'}{\partial x} & \dfrac{\partial x'}{\partial y} & \dfrac{\partial x'}{\partial z} \\ \dfrac{\partial y'}{\partial x} & \dfrac{\partial y'}{\partial y} & \dfrac{\partial y'}{\partial z} \\ \dfrac{\partial z'}{\partial x} & \dfrac{\partial z'}{\partial y} & \dfrac{\partial z'}{\partial z} \end{pmatrix} \tag{7}$$

Using (1)-(3) in (6), the constructive parameters can be obtained in terms of the transformed system coordinates as

$$\varepsilon'_{xx} = \frac{\sqrt{w^2 - x'^2}}{l} \tag{8a}$$

$$\varepsilon'_{xy} = \varepsilon'_{yx} = \frac{x'y'}{l\sqrt{w^2 - x'^2}} \tag{8b}$$

$$\varepsilon'_{yy} = \frac{x'^2 y'^2}{l(w^2 - x'^2)^{3/2}} + \frac{l}{\sqrt{w^2 - x'^2}} \tag{8c}$$

$$\varepsilon'_{zz} = \frac{a^2\sqrt{w^2 - x'^2}}{w^2 l} \tag{8d}$$

and $\varepsilon'_{xz} = \varepsilon'_{zx} = \varepsilon'_{yz} = \varepsilon'_{zy} = 0$. For TE wave, the permittivity tensors in (8a)-(8d) should be replaced by permeability tensors.

III.   SMILATION AND ANALYSIS

Considering the original circular domain is free space, the values of tensors are compared in Figure 4 with a=1, w=1, l=0.125.

When the line source is embedded in the center of the rectangular region with $a = w = 60$mm and $l = 7.5$mm, the total electric field distribution is simulated for TE wave at 10GHz by full-wave software Comsol Multiphysics for both free space and transformation material in Figure 5(a) and 5(b), which indicates that when the rectangular region is full of transformed permeability tensors, it can be used as a wave collimator.

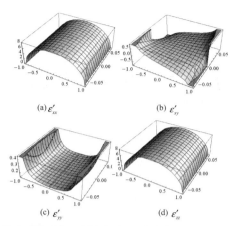

(a) $\varepsilon'_{xx}$      (b) $\varepsilon'_{xy}$

(c) $\varepsilon'_{yy}$      (d) $\varepsilon'_{zz}$

Figure 4 Material permittivity in the transformation coordinate system

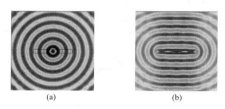

(a)        (b)

Figure 5 Electric field distribution with a line source located at the coordinate origin: (a) for free space and (b) for transformation material

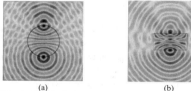

(a)        (b)

Figure 6 Electric field distribution (a) for sphere MFE (b) for flattened MFE

Combining (5) and (6), the required permittivity and permeability for flattened MFE can be derived. The electric field distributions for sphere MFE and flattened MEF are simulated respectively shown in Figure 6(a) and 6(b) at 10GHz, where $a = w = 60$mm and $l = 30$mm. It is obvious that the flattened MFE can also provide the image point on the planar surface opposite to the object point.

## IV. CONCLUSIONS

A flattened Maxwell fish-eye lens is proposed in this paper, which has a full and compressed flattened structure. The required constructive parameters are derived for the coordinate transformation from sphere to flattened structure. The components of tensors are compared quantitatively. Furthermore, the ray trajectories in flattened MFE lens is also presented using coordinate rotation transformation. The flattened MFE can be believed to have a potential application in imaging system with detector/receiver arrays.

REFERENCES

[1] U. Leonhardt, "Optical conformal mapping," *Science* 312, pp.1777–1780 , 2006.

[2] J. B. Pendry, D. Schurig, and D. R. Smith, "Controlling electromagnetic fields," *Science* 312, 1780–1782, 2006.

[3] McCall M. W., Favaro A., Kinsler P. and Boardman A., "A spacetime cloak, or history editor," *J. Opt.*, vol.13, No.2, 024003, 2011.

[4] Uslenghi and P.L.E., "Extreme-angle broadband metamaterial lens," *IEEE Trans. Antennas Propag.*, vol. AP-17, No.2, pp.235-236, Mar. 1969.

[5] Nathan Kundtz and David R. Smith, "Experimental verification of a negative index of refraction," *Science, Nature Materials*, vol.9, Feb. 2010.

[6] E. W. Marchand, *Gradient Index Optics*, Academic Press, New York, 1978.

[7] D. Schurig, J. B. Pendry, and D. R. Smith, "Calculation of material properties and ray tracing in transformation media," Opt. Express vol.14, No.21, pp.9794–9804, 2006.

# Tunable Electromagnetic Gradient Surface For Beam Steering by Using Varactor Diodes

Jungmi Hong[#], Youngsub Kim[#], and YoungJoong Yoon[#]

[#]Department of Electrical & Electronic Engineering, Yonsei University 134 Shinchon-dong, Seodaemun-Gu, Seoul 120-749, Korea

yjyoon@yonsei.ac.kr

*Abstract* - In this paper, tunable beam steering has been realized by arranging Electromagnetic Band- gap (EBG) that has constant phase difference, Δφ. The proposed structure is the active periodic Electromagnetic Gradient Surface (EGS) that consist of the 4 by 4 matrix form of unit cells to easily control the several varactor diodes. The phase gradient ±120° of the proposed EGS is selected. Moreover this structure uses only one diode per four unit cells. The conventional loading of a single varactor for each of the unit cells of the EBG is overcome by connectdkgkding the unit cells to a feeding network directly without via. Thus, the EGS can steer reflected beam to -9° at Δφ = 120° and to +9° at Δφ = -120° by actively changing the biasing voltage of varactor diode.

## I. INTRODUCTION

Nowadays, the electromagnetic band gap (EBG) structures have attracted increasing interests because of their desirable electromagnetic properties that cannot be observed in natural materials. Conventional EBG, also called as High impedance surface (HIS), is typically constructed as an array of small (≪λ) metal patches on the top layer and a metal ground plane on the bottom layer. The EBG have been applied to improve performances of many devices such as low profile antennas[1], absorbers[2], TEM waveguides[3], and microwave filters[4].

Another interesting approach is phased reflector which is physically flat surface but electromagnetically gradient surface. A mushroom-type EBG structure with a linearly graded reflection phase shows electromagnetic gradient properties across its surface. Its ability to create an EGS of a physically flat structure allows it to reflect incident plane wave to other direction [5]. By decreasing or increasing the length of each lattice, the phase curve of the reflection coefficient moves right or left respectively, on frequency domain. In [5], difference size of metallic patches with via can make the reflection phase for steering beam pattern of incident wave. However this structure has phase discontinuity error and steered beam pattern is fixed.

On the other hand, the resonance frequency and reflection phase of EBG unit cell is controlled actively by using phase shifters, which increase the complexity and cost of the system. The lumped device such as varactor diodes between each of the elements instead of by changing structure physically also can change the resonance frequency and reflection phase of a tunable high impedance surface[6]. It has also been demonstrated in [7] that tunable EBGs are feasible for beam steering applications. The disadvantages of these structures are

that the active element is sensitive to environment, thus more errors by using many active devices become serious problems. Moreover, in order to make the linear phase gradient, system for controlling the various voltages for biasing varactor diodes is too complex.

The aim of this paper is to illustrate the advantage of using reduced varactor diodes in a simple EBG structure to tune the reflection phase and to use this property in tunable tilting the reflected beam pattern. Moreover, if we consider the unit price of microwave varactors, reduced lumped devices can decrease the cost and complexity for manufacture.

## II. THEORY AND DESIGN

### A. Tunable reflection phase of the 4x4 group cell

The resonance frequency and the reflection phase of the EBG are tuned by changing the capacitance, C, the inductance, L, or both. From this principle, we can shift the reflection phase without varying the patch size. Generally, the L is difficult to tune because it is a product of magnetic permeability μ and thickness of the substrate t. The C is easier to control by changing the geometry and arrangement of the metal plates, or by adding tunable lumped capacitors.

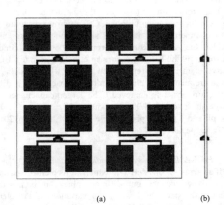

(a)          (b)

Figure 1. Configuration of the tunable EBG Structure
(a) Top view. (b) Side view

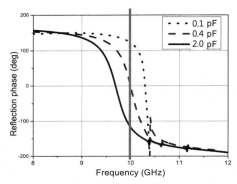

Figure 2. Reflection phase of the EBG for various capacitances

The active EBG structure is shown in Fig.1. It consists of a ground, matrix of grouped unit cells, a dielectric substrate of Tarconic TLY-5 with a thickness of 0.79 mm and a dielectric constant of 2.2 and tanδ=0.0009, a varactor diode, and thin line with width of 0.5mm. The distance between neighboring cells is 14 mm, satisfied the condition of < λ/2 at 10 GHz. Therefore, the EBG's physical dimension is 56 mm x 56 mm.

Each unit cell between neighboring patches of the conventional tunable EBG should be connected by varactor diodes[6]. For the purpose of reducing the lumped elements, four unit cells are grouped by one using only one diode. If 4x4 array structures in [6] use 12 diodes, the proposed EBG structure uses only four diodes in the same array. Therefore when the active EBG is implemented, the design complexity of its feeding network can be reduced, and also can be controlled the four cells by one bias at one time. Then, 4 unit cells operate one group cell that varies all together. The width of the thin line is not so serious for phase curve

The reflection phase of the structure for various capacitances is shown in Fig. 2. CST microwave studio, which is a commercial 3D full field simulation tool, was used to model the EBG using appropriate boundary conditions surrounding the EBG to ensure an effective infinite EBG. The active elements are represented by the RLC models. For a fixed frequency, by changing diode capacitance, a different reflection phase is achieved. The larger capacitance, phase curve shifts to left. When the value of capacitance is 0.1 pF, 0.4 pF and 2 pF, the phase curve is +120°, 0° and -120° at 10 GHz, respectively.

*B. Periodic Electromagnetic Gradient Surface (EGS)*

If an array is considered, in which all its elements are placed at a fixed distance but different phase, it is possible to electrically steer the beam by a variation of the elements' phase. The normal EGS requires a full phase reflection range of -180° ~ +180°. The passive EGS can change reflected direction of incident plane wave[5]. However, when the EGS

is arranged to cover large surface area, it cannot have the linearly reflection phase gradient, because the limited reflection phase range with one via in unit cell and it cannot be change the reflected beam direction.

The proposed active periodic EGS consists of the 4 by 4 matrix form of unit cells to easily control the several varactor diodes. When the active EBGs arrange by introducing a phase difference along the E-field direction, a beam steering is obtained. By adjusting the phase gradient, Δφ, and the distance, Δs, between neighboring EBGs, the proposed periodic EGS can determine the reflected angle of the incident plane wave[8]. Assuming normal incident of electromagnetic wave to the EGS, the reflected angle θ_r is given by

$$\theta_r = \sin^{-1}\left(-\frac{\Delta\varphi}{\Delta s} \cdot \frac{\lambda}{2\pi}\right) \qquad (1)$$

To satisfy the reflection phase continuity, the periodic arrangement along the x-axis, the phase gradients ±180°, ±120° and ±90° can be selected. In this case, full range of reflection phase is not necessary.

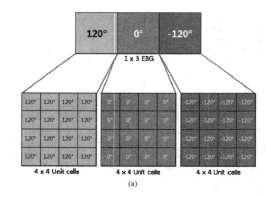

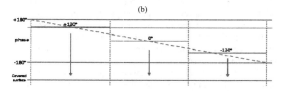

Figure 3. Concept of the periodic EGS
(a) Configuration of periodic EGS (ex: Δφ=-120°) (b) Reflection phase of periodic EGS

The larger the phase difference is and the smaller size of the EGS is, the larger the tilted reflection angle is achieved. In particular, by applying a linear variation with respect to the position, a voltage gradient is applied along the E-plane, obtaining the desired beam deflection.

For example, the phase gradient of Fig. 3 is 120°, thus, the proposed active EGS can cover the large surface area using the only three biasing, +120°, 0°, and -120°, with the matrix form to make linearly reflection phase gradient. And this EGS can be expanded periodically.

In case of the active EBG, the diode junction capacitance can be independently changed for each EBG. In particular, by applying a linear variation with respect to the row position, a voltage gradient is applied along the E-plane, obtaining the desired beam deflection

### C.  Reconfigurable Reflected beam pattern

By using the easy tuning characteristic of varactor diode it simply change the reflection phase gradient direction of the EGS with difference bias voltage, and then the reflection phase gradient turns reversely at the fixed frequency. As a result, reconfigurable beam steering gets accomplished.

The proposed active EGS is illustrated in Fig. 4. The active EGS structure can have phase gradient ± 120 by controlling the bias voltage. The distance between the neighboring EBG is 56 mm and overall dimension is 168 mm x 56 mm. Thus, in the equation (1), expected reflection angle is -9° at $\Delta\varphi$ = 120° and to +9° at $\Delta\varphi$ = -120°.

### III.  Simulation Result

Fig. 5 shows the normalized beam reflection patterns obtained in correspondence of 10 GHz. Since the radar cross section (RCS) of the EGS is used to characterize the reflective behavior of the phased reflector, we use RCS pattern in this study. As shown in the 2D pattern graph, the simulated reflection angle is -9° at $\Delta\varphi$ = 120° and to +9° at $\Delta\varphi$ = -120° and 0° when the structure is PEC. Since the phase gradient changes opposite, the tilting angle also is symmetry.

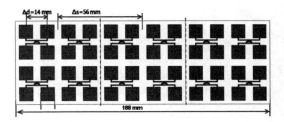

Figure 4.  The proposed active EGS structure with $\Delta\varphi$ = ± 120°

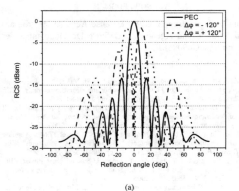

(a)

(b)

Figure 5. The steered reflection beam patterns
(a) The 2D pattern $\Delta\varphi$ = ± 120° and PEC (b) The 3D pattern $\Delta\varphi$ = + 120°

### IV.  Conclusion

The active periodic EGS for reconfigurable tiling the reflected beam is proposed and studied. The active EGS is composed of three EBGs with 4 by 4 matrix form and constant phase gradient. This matrix form makes easy to control the bias voltage for varactor diodes. In this paper, only three values of capacitance are needed. Moreover, since the each four unit cells are connected by one varactor diode, we can reduce the number of the active devices. Because the active element is sensitive, reducing the diode can decrease the errors and the cost of system. The size of the proposed EGS reflector is 168 mm x 56 mm and the number of the varactor diodes is 12. The reflection angles -9° at $\Delta\varphi$ = 120° and to +9° at $\Delta\varphi$ = -120° are obtained in correspondence of 10 GHz.

Although the structure in this study can use the two changeable angles, it is expanded to 2D structure then more various angles can be accomplished with further studies on the EGS.

ACKNOWLEDGMENT

This work was supported by Defense Acquisition Program Administration and Agency for Defense Development under the contract UD090088JD.

REFERENCES

[1] F. Costa, A. Monorchio, S. Talarico, and F. M. Valeri, "An active high impedance surface for low profile tunable and steerable antennas,"*IEEE Antennas Wireless Propag. Lett.*, vol. 7, pp. 676–680, 2008.

[2] S. A. *Tretyakov and S. I. Maslovski*, "Thin absorbing structure for all incidence angles based on the use of a high-impedance surface," *Microw. Opt. Technol. Lett.*, vol. 38, no. 3, pp. 175–178, Aug. 2003.

[3] F. R. Yang, K. P. Ma, Y. Qian, and T. Itoh, "A novel TEM waveguide using uniplanar compact photonic-bandgap (UC-PBG) structure," *IEEE Trans. Microw. Theory Tech.*, vol. 47, pp. 2092–2098, Nov. 1999.

[4] S. Y. Huang and Y. H. Lee, "Compact U-Shaped Dual Planar EBG Microstrip Low-Pass Filter," *IEEE Trans. Microw. Theory Tech.*, vol.53, no.12, pp.3799–3805, Dec. 2005.

[5] K. Chang, "Electromagnetic Gradient Surface and Its Application to Flat Reflector Antennas," *Ph. D. dissertation*, Department of Electrical Engineering, Univ. Yonsei, Seoul Korea, 2009

[6] Lee, H.-J.,Langley, R.J., Ford, K.L. *"Tunable Active EBG", Antennas and Propagation*, 2007. EuCAP 2007. 11-16 Nov. 2007

[7] Sievenpiper, D. F., J. H. Scha®ner, H. J. Song, R. Y. Loo, and G. Tangonan, "Two-dimensional beam steering using an electrically tunable impedance surface," *IEEE Transactions on Antennas and Propagation*, Vol. 51, 2713-2722, 2003.

[8] M. Ko, "Preiodic Electromagnetic Gradient Surface for Monostatic Radar Cross Section Reduction," *M. D. dissertation*, Department of Electrical Engineering, Univ. Yonsei, Seoul Korea, 2011

# Asymmetric Electromagnetic Wave Polarization Conversion through Double Spiral Chiral Metamaterial Structure

Linxiao Wu, Bo Zhu, Junming Zhao,Yijun Feng*

Department of Electronic Engineering, School of Electronic Science and Engineering, Nanjing University, 210093

*yjfeng@nju.edu.cn

*Abstract-*This paper presents the design, simulation and measurement of a chiral metamaterial with double spiral unit cell structure that can be used to convert the polarization states of linearly polarized electromagnetic (EM) wave. The metamaterial is composed of a dielectric slab sandwiched by a double spiral metallic structure. By full-wave EM simulations and free space measurements, a strong polarization conversion of linearly polarized EM wave has been demonstrated in the microwave band. Moreover, the polarization conversion is asymmetric due to the chirality of the structure, therefore the structure could allow propagation of linearly polarized EM wave along one direction, while almost no propagation along the opposite direction. The proposed chiral metamaterial structure could find applications in designing polarization control devices.

## I. INTRODUCTION

Chirality is a geometric concept, indicating lack of geometric symmetry between the substance and its mirror image and lack of chance of overlapping via both translation and rotation operations. Chiral structures can be found in natural materials; however their chirality is rather weak. Recently, chiral metamaterial structures have been proposed and exotic electromagnetic (EM) properties have been experimentally validated including negative refraction characteristics [1]. With the rapid progress in the study of the electromagnetic properties of chiral metamaterial, some of its other inherent exotic features are constantly revealed, and experimentally verified, such as the optical activity [2], nonlinearity, circular dichroism [3]. Polarization is an important characteristic of electromagnetic wave. It is always desirable to have full control of the polarization states of EM waves in many applications. Since the mirror symmetry of chiral metamaterial is broken either in the propagation direction or in the perpendicular plane, it leads to the interaction of electromagnetic wave radiation with the structural chirality in the metamaterial. These chiral structures could be utilized to control the polarization states of EM waves. In particular, asymmetric EM propagations have been realized with particularly designed chiral metamaterial associated with the conversion between two independent linear polarizations [4-8].

In our previous work, we established a theoretical analysis on a kind of bilayered metamaterial structure with specific structure asymmetry that enables the asymmetric EM wave propagation only for linear polarization. We also proposed and

experimentally verified a kind of chiral metamaterial structure that has substantial asymmetric propagation for a certain linearly polarized EM wave, but none for circular polarizations [9].

In this paper, we propose a chiral metamaterial structure composed of a dielectric slab sandwiched by a double spiral metallic structure which can achieve strong asymmetric polarization conversion for linearly polarized EM wave. We will demonstrate that the chiral metamaterial can achieve cross-polarization conversion with an efficiency of over 90% for a certain linearly polarized EM wave. Such phenomenon is more interesting in the application for designing polarization control devices. We will present the design, simulation and measurement of the chiral metamaterial structure, and validate its strong polarization conversion ability with transmission measurement in the microwave band (X band).

## II. THEORETICAL ANALYSIS

Asymmetric transmission of EM waves in the chiral metamaterial is usually caused by the partial polarization conversion of the incident EM radiation into one of the opposite polarization, which is asymmetric for the opposite directions of propagation [10]. To analyze such phenomenon we consider an incoming plane wave propagating along the negative $z$ direction, with electric field as

$$E^{in}(r,t) = \begin{pmatrix} I_x \\ I_y \end{pmatrix} e^{i(-kz-\omega t)}, \tag{1}$$

where $\omega$, $k$, $I_x$, $I_y$ represent the frequency, wave vector, and complex amplitudes, respectively. The transmitted electric field through the slab can be then described as

$$E^{trans}(r,t) = \begin{pmatrix} T_x \\ T_y \end{pmatrix} e^{i(-kz-\omega t)}. \tag{2}$$

To understand better the cross-polarization conversion due to the chirality of the metamaterial structure, we invoke the transmission matrix expression for the EM fields, which relates the incident and the transmitted electric fields in terms of linearly polarized components. The subscripts $i$ and $j$ correspond to the polarization base states of the transmitted and incident waves, which could be either $x$ or $y$ linear polarization (assuming EM wave propagates along $-z$ direction). The transmission matrix element includes information of both the amplitude $|t_{ij}| = |E_i^{trans}|/|E_j^{inc}|$, and phase

$\arg(t_{ij}) = \arg(E_i^{trans} / E_j^{inc})$ , then the EM wave transmission perpendicularly propagating through a slab of metamaterial structure can be described by the so-called complex Jones matrix $T$ as

$$\begin{pmatrix} E_x^{trans} \\ E_y^{trans} \end{pmatrix} = \begin{pmatrix} t_{xx} & t_{xy} \\ t_{yx} & t_{yy} \end{pmatrix} \begin{pmatrix} E_x^{in} \\ E_y^{in} \end{pmatrix} = T_{lin}^f \begin{pmatrix} E_x^{in} \\ E_y^{in} \end{pmatrix}. \qquad (3)$$

The superscript $f$ and subscript $lin$ indicate the forward propagation (along -z direction) and a special linear base, respectively. If the medium does not contain magneto-optical material, the reciprocity theorem can be applied and the transmission matrix for $T_{lin}^b$ propagation in the backward direction (+z direction) can be derived as [11]

$$T_{lin}^b = \begin{pmatrix} t_{xx} & -t_{yx} \\ -t_{xy} & t_{yy} \end{pmatrix}. \qquad (4)$$

When the propagation direction is reversed, the off-diagonal elements $t_{xy}$ and $t_{yx}$ not only interchange their values, but also get an additional $180^0$ phase shift.

We can also define the polarization conversion ratio (PCR) for x or y polarization as

$$PCR_y = t_{xy}^2 / (t_{yy}^2 + t_{xy}^2), \qquad (5)$$

$$PCR_x = t_{yx}^2 / (t_{xx}^2 + t_{yx}^2), \qquad (6)$$

If there is no absorption and reflection, we could have $t_{xy}^2 + t_{yy}^2 = t_{yx}^2 + t_{xx}^2 = 1$ and thus the PCR can be further reduced to $PCR_y = t_{xy}^2$ and $PCR_x = t_{yx}^2$.

### III. DESIGN AND CHARACTERISTICS

The proposed chiral structure is designed to work at 8 - 12 GHz, composed of a dielectric substrate sandwiched by two thin copper layers. The dielectric substrate is chosen as FR4 (a glass-reinforced epoxy printed circuit board) with a permittivity of 4.6, loss tangent of 0.01, and thickness of 1 mm. The top and bottom copper layers have the same thickness of 17 μm. Fig. 1(a) shows the schematic diagram of the metamaterial unit cell with side length $a = 12$ mm, while the whole slab sample is composed of 16 × 17 unit cells. The right part of Fig. 1(a) indicates metallic patterns on the top and bottom copper layers. The patterns on top and bottom layers are two copper rings with gaps respectively, which have inner and outer radii of $r_1 = 3.5$ mm and $r_2 = 4$ mm. The gap angle corresponding to the ring center is $\theta = 135^0$. The point $p_1$ indicates center of the unit cell and the $p_2$ and $p_3$ points indicates centers of two copper rings on the top layer. If the $p_1$ point is set as the origin of x - y plane, $p_3$ and $p_2$ will have coordinates of (1 mm, 1 mm) and (-1 mm, -1 mm), respectively. The cooper rings on top and bottom layers with same position in x direction and y direction are connected through one metallic via hole of radius of 0.5 mm and form a double spiral structure. The twisted metallic patterns form a chiral structure and have strong strength to convert polarization of EM wave due to the electric and magnetic mutual couplings between them. Fig. 1(b) shows the

photograph of the fabricated sample. The orientation of the incident wave is indicated in fig. 1(c) for forward (along −z direction) or backward (along +z direction) propagation.

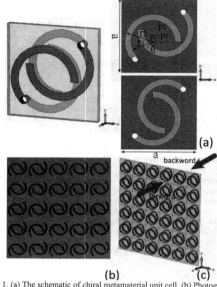

Figure 1. (a) The schematic of chiral metamaterial unit cell. (b) Photograph of fabricated sample slab. (c) Its orientation with the incident wave.

The proposed metamaterial slab has been analyzed through full-wave electromagnetic simulations with commercial software (CST Microwave Studio™) based on finite-difference time-domain method. The simulation is carried out upon the unit cell as shown in Fig. 1(a). Open boundary conditions are employed along the propagation direction, and unit cell boundary conditions are used along the directions perpendicular to the propagation direction. Therefore, the structure is assumed to be periodic and infinite along the directions that are perpendicular to the propagation direction. After the parameters study and optimization through the simulations, a sample slab (with outer dimension of 200 × 220 mm$^2$) is fabricated by printed circuit board technique and characterized through free space electromagnetic transmission measurement in a microwave anechoic chamber. Two linearly polarized horn antennas with Teflon lenses are used to emit and receive microwave signal, and a vector network analyzer (Agilent N5244A) is employed to measure the transmission. In the experiment, transmission measurement is calibrated to the case where the sample is left as hollow (as the unit transmission). By rotating the transmitting or receiving horn antenna 90° around the main radiation direction to generate and receive EM waves with different linear polarizations (either x-polarization or y-polarization), all the four components of the electromagnetic wave transmission (the complex Jones matrix) for different polarizations have been measured.

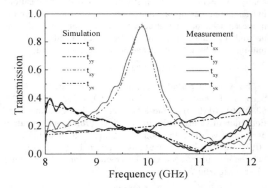

Figure 2. The measurements (solid lines) and simulations (dashed lines) of the four transmission matrix elements of the slab for forward propagations.

Fig. 2 shows the simulated and measured results of the four transmission matrix elements of the slab. The measurements and simulations agree with each other quite well across the whole frequency range. As the frequency increases, the cross-polarization transmission $t_{xy}$ experiences a resonant peak and reaches a maximum of around 0.9 at the frequency of 9.88 GHz, while the cross-polarization transmission $t_{xy}$ maintains around 0.2 and the co-polarization transmission $t_{xx}$ and $t_{yy}$ are very close to each other across whole frequency range.

According to Eq. (5) and Eq. (6), PCR is calculated for both $x$- and $y$-polarized incident EM wave for forward and backward propagations. As illustrated in Fig. 3, the PCR parameter for $y$-polarized incident wave reaches around 0.95 from 9.8 GHz to 11 GHz, indicating that the $y$-polarization converts mostly to $x$-polarization after transmission through the slab, while PCR parameter for $x$-polarized incident wave reaches the maximum of 1 at approximate 11GHz, corresponding precisely to the resonance at which $t_{xx}$ and $t_{yy}$ tend to zero. Either polarization experiences a pure polarization conversion when propagates through this metamaterial slab. The PCR parameters for opposite propagations show strong asymmetric characteristics due to the chirality of the structure.

The PCR parameter for linear polarization in the proposed structure leads to a strong asymmetric transmission. As a result the total transmission for a certain linear polarized wave is quite different along opposite directions. We measured both the forward and backward total transmissions (including both the co- and cross-polarizations) for $y$-polarized wave, and the result is shown in Fig. 4. The forward transmission reaches above 0.85, while the backward transmission is below 0.1 at around 9.88 GHz.

To look into the mechanism of the polarization conversion that is associated with the chiral metamaterial, we have analyzed the distribution of electric energy density of top and

bottom copper layers and mid-plane of dielectric substrate and the surface current distribution of top and bottom cooper rings of $y$-polarized incident wave at 9.88 GHz. Through careful analysis of Fig. 5 and the dynamic graphics of surface current across whole phase, following phenomenon can be observed. The incident $y$-polarized wave induces a strong electric field in the gaps of the top cooper layer (Fig. 5(a)). Surface currents are mainly distributed in parts of top and bottom rings that having overlap with each other. Surface currents on top and bottom parts of the cooper rings with overlap to each other are in opposite directions across whole phase (Fig. 5(d) and (e)).

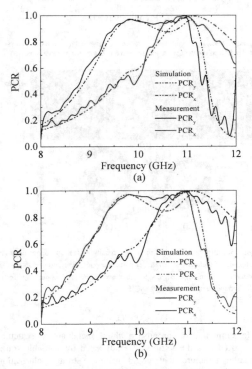

Figure 3. PCR for linear polarization determined by measured (solid lines), and simulated (dashed lines) data for (a) forward and (b) backward propagations.

The result demonstrates clearly that when a $y$-polarized wave at 9.88 GHz normally incidents into the structure along the $-z$ direction, the wave is strongly coupled to the slab by inducing a strong electric field in the gaps of the top SRR layer; meanwhile due to significant mutual coupling between the top and bottom metallic rings through the overlap parts, $y$-polarized wave is converted mostly to $x$ polarization when it pass through the structure. The backward propagation can be understood in a similar manner.

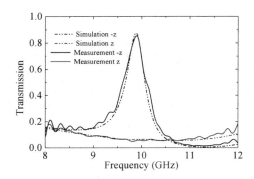

Figure 4. Measured (solid lines), calculated (dashed lines) total transmission for $y$-polarized incident wave for forward and backward propagations.

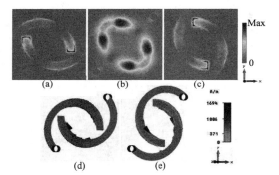

Figure 5. Distribution of electric energy density of (a) top copper layer (b) mid-plane of dielectric substrate (c) bottom copper layer and distribution of surface current of (d) upper cooper rings and (e) bottom cooper rings for $y$-polarized incident EM wave at 9.88GHz.

## IV. CONCLUSIONS

In summary, we propose a chiral metamaterial structure composed of a dielectric slab sandwiched by a double spiral metallic structure that could achieve strong polarization conversion for linearly polarized EM wave. After analyzing and optimizing the metamaterial structure through full-wave EM simulations, we fabricate and test the structure at the microwave band. The experimental results have confirmed the strong polarization conversion and asymmetric transmission ability. We believe that such structure could find applications in designing polarization control devices. By scaling down the proposed metamaterial structure, the concept could also be utilized to function at other frequency bands, such as millimeter, sub-millimeter wave or even terahertz band.

ACKNOWLEDGMENT

This work is partially supported by the National Nature Science Foundation of China (60990322, 60990320, 60801001, 61101011), the Key Grant Project of Ministry of Education of China (313029), the Ph.D. Programs Foundation of Ministry of Education of China (20100091110036, 20120091110032), and partially supported by Jiangsu Key Laboratory of Advanced Techniques for Manipulating Electromagnetic Waves.

REFERENCES

[1] B. Wang, J. Zhou, T. Koschny, and C. M. Soukoulis, "Nonplanar chiral metamaterials with negative index," *Appl. Phys. Lett.*, vol. 94, 151112, 2009.
[2] M. Decker, et al. "Strong optical activity from twisted-cross photonic metamaterials," *Opt. Lett.*, vol. 34, pp. 2501-2503, 2009.
[3] M. Decker, M. W. Klein, M. Wegener, and S. Linden, "Circular dichroism of planar chiral magnetic metamaterials," *Opt. Lett.*, vol. 32, pp. 856-858, 2007.
[4] A. S. Schwanecke, et al. "Nanostructured metal film with asymmetric optical transmission," *Nano Lett.*, vol. 8, pp. 2940-2943, 2008.
[5] R. Singh, et al. "Terahertz metamaterial with asymmetric transmission," *Phys. Rev. B*, vol. 80, 153104, 2009.
[6] C. Menzel, et al. "Asymmetric transmission of linearly polarized light at optical metamaterials," *Phys. Rev. Lett.*, vol. 104, 253902, 2010.
[7] E. Plum, V. A. Fedotov, and N. I. Zheludev, "Planar metamaterial with transmission and reflection that depend on the direction of incidence," *Appl. Phys. Lett.*, vol. 94, 131901, 2009.
[8] V. A. Fedotov, A. S. Schwanecke, N. I. Zheludev, V. V. Khardikov, and S. L. Prosvirnin, "Asymmetric transmission of light and enantiomerically sensitive plasmon resonance in planar chiral nanostructures," *Nano Lett.*, vol. 7, pp. 1996-1999, 2007.
[9] C. Huang, Y. Feng, J. Zhao, Z. Wang, and T. Jiang, "Asymmetric electromagnetic wave transmission of linear polarization via polarization conversionthrough chiral metamaterial structure," *Phys. Rev. B*, vol. 85, 195131, 2012.
[10] V. A. Fedotov, P. L. Mladyonov, S. L. Prosvirnin, A. V. Rogacheva, Y. Chen, and N. I. Zheludev, "Asymmetric propagation of electromagnetic waves through a planar chiral structure," *Phys. Rev. Lett.*, vol. 97, 167401, 2006.
[11] C. Menzel, C. Helgert, C. Rockstuhl, E. B. Kley, A. Tünnermann, T. Pertsch, and F. Lederer, "Asymmetric transmission of linearly polarized light at optical metamaterials," *Phys. Rev. Lett.*, vol. 104, 253902, 2010.

# Metamaterial Absorber with Active Frequency Tuning in X-band

Hao Yuan, Bo Zhu*, Junming Zhao, Yijun Feng

Department of Electronic Engineering, School of Electronic Science and Engineering
Nanjing University, Nanjing, 210093 China
*bzhu@nju.edu.cn

*Abstract*-**This paper presents the design, fabrication, and measurement of an active frequency tunable metamaterial absorber. The unit cell of the metamaterial absorber consists of a metallic strip on the top layer of the fully grounded dielectric substrate, with a varactor loaded at the slit in the middle of the central strip. Simulation and measurement results show that by tuning the bias voltage applied on the varactors, the peak absorption frequency can be tuned about 0.44 GHz with the peak absorption greater than 95%. Field analysis reveals that the unit cell of the absorber works as the microstrip resonator which exhibits matched input resistance to the incident waves.**

## I. INTRODUCTION

Electromagnetic (EM) wave absorber is a functional material that absorbs incident EM energy. Conventional EM absorbers are mainly realized by using material with high ohmic loss and magnetic polarization loss [1-3].

The metamaterial absorber (MA) is another new type of EM absorber proposed recently [4]. It is normally composed of electrical small unit cells arranged periodically in a 2D plane. Each unit cell can be regarded as the resonant circuit. At resonance, MA exhibits purely effective surface resistance matched to incident EM wave impedance to achieve efficient wave absorption. The MA enjoys the features of light weight and low profile, but suffers narrow absorbing frequency bandwidth [5-6].

Since MA is realized through artificial design of the unit cell structure, one unique advantage of MA is that more functions than absorbing waves can be combined into MA, such as absorbing frequency tuning, polarization dependent absorbing, etc [7-8].

In this paper, an active frequency tunable MA in X-band is presented employing the microstrip resonator mode. It is realized by integrating varactor into the unit cells of the MA. Simulation and measurement show that by tuning the bias voltage applied on the varactors, the peak absorption frequency can be tuned about 0.44GHz with the peak absorption greater than 95%. Field distribution is analyzed to explore the working mode and physical origin of the property.

## II. STRUCTURE AND WORKING MECHANISM

The unit cell of the proposed tunable MA is shown in Fig. 1(a). It consists of a metallic strip on the top layer of the fully grounded dielectric substrate. In order to dynamically control the absorbing frequency, a varactor is loaded at the slit in the middle of the central strip. The bias circuit is designed and positioned at the upper and lower edge of the unit cell to supply bias voltages onto the varactors.

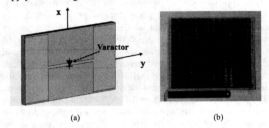

Fig. 1. (a) A schematic of the unit cell of the actively tunable MA. (b) The fabricated sample board.

Since the whole MA is a periodic structure, we only need to analyze one unit cell using the periodic boundary condition according to Floquet theorem. Because the unit cell structure is symmetric with respect to both $x$ and $y$ axes, the PEC and PMC boundaries can be used as the periodic boundary condition for normal incidence case. This boundary condition actually constitutes a PEC-PMC waveguide which supports the plane wave as the fundamental mode with zero cut off frequency. The unit cell structure can be regarded as the microstrip resonator, which is loaded at the termination of the waveguide. The waveguide works as the feeding line for the resonator. The two opposite PEC boundaries are insulated by the two lateral PMC boundaries since they don't conduct electric current, hence the PEC-PMC boundaries can also be regarded as the dual conductor transmission line with a certain characterization impedance. When the unit cell resonates, it exhibits pure resistance to the feeding line, and its total input reactive impedance is zero. By adjusting the size of the microstrip line, and the thickness of the substrate, the resonance frequency can be designed. In order to achieve high absorption, the unit cell outer dimension should be optimized to adjust the effective input resistance exhibited by the microstrip resonator to the waveguide. When the effective input resistance is equal to the characterization impedance of the feeding line, perfect wave absorption is achieved. A varactor is integrated into the microstrip resonator, as shown in Fig. 1(a). Through tuning the bias voltage applied on the varactor, the junction capacitance of the varactor is changed so that the resonance frequency, i.e. the absorbing frequency, can be shifted.

978-7-5641-4279-7

Fig. 2(a) shows the electric field distribution at the *xz* plane. The unit cell can be regarded as two segments of microstrip lines with short circuit termination. They are coupled through the varactor. Under the excitation of the fundamental mode of the PEC-PMC waveguide, i.e. plane wave excitation, the two segments of microstrip line resonate out of phase, as shown in the Fig. 2(a). Hence, the *z* component of the electric field is zero at *yz* plane so that there is a virtual PEC plane at *yz* plane. As a result, the left or right half of the unit cell can be modeled as the circuit depicted in Fig. 2(b). In the X-band, the varactor's parasitic inductance dominates so that the varactor shall be regarded as a tunable inductor. The short-terminated microstrip line exhibits capacitance to the varactor when it is larger than its quarter guided operation wavelength. By optimizing the dimensions of the microstrip line, a proper effective input capacitance of the microstrip line can be obtained, which will resonate with the effective inductance of the varactor at a certain frequency. When the unit cell resonates, it exhibits purely effective input resistance to the feed waveguide, while the total reactive impedance is zero. The resistance originates from the energy ohmic loss in the unit cell. Its value is related to the resonance strength, and can be adjusted by optimizing the unit cell dimensions, such as the unit cell periodicity. In order to achieve efficient absorption, the effective input resistance should match the characteristic impedance of the PEC-PMC waveguide if we regard the PEC-PMC waveguide as a dual conductor transmission line.

subjected to an incident plane wave. The PEC-PMC boundary conditions are employed. The electric field polarization is along the *x* axis, and magnetic field polarization along the *y* axis. The dielectric substrate is chosen as the FR4 with the permittivity of 4.1, the loss tangent of 0.025 and the thickness of 0.8 mm. The metal on the top and bottom layer are modeled as the copper film with the conductivity of $5.8 \times 10^7$ S/m. The size of the unit cell is 16 mm × 10 mm. The width of the central conductor is 8 mm, and the gap where varactor is loaded is 0.4 mm. In simulation, the varactor is modeled as the resistor, inductor and capacitor in series. The capacitance range is from 2.350 pF at 0 V to 0.466 pF at 15 V. The resistor is 2.5 ohm, and the inductor is 1.1 nH.

Experiment is also carried out to verify the performance of the proposed tunable absorber. A 210 mm × 244 mm sample board consisting of 20 × 14 unit cells is fabricated using printing circuit board technology, as shown in Fig. 1(b). The SMV1231-011 varactor is used in experiment.

The experiment is performed in the microwave anechoic chambers. Fig. 3(a) shows the schematic of the measurement arrangement. A vector network analyzer (Agilent E8363C) and two horn antennas are used to transmit EM waves onto the sample board and receive the reflected signals. Since the sample absorber has a metallic ground on the bottom layer so that EM transmission is zero, we only measure the reflection coefficient $S_{11}$ of the sample to obtain its absorption, which is calculated as $1 - |S_{11}|^2$. The measurement is calibrated by replacing the sample with the aluminum board of the same size as the perfect reflector (unit reflection).

## IV. RESULTS AND DISCUSSION

The measured and simulated absorption for normal incident EM wave at different varactor bias voltage is exhibited in Fig. 3(b).

Tunability of the diode capacitance is achieved by varying the width of its specifically doped P-N junction through the application of the DC bias voltage. Changes in diode capacitance alter the resonant and absorbing frequency of the MA.

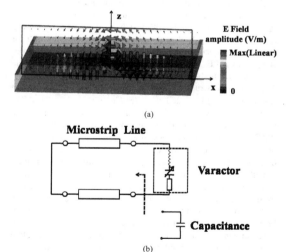

(a)

(b)

Fig. 2. (a) Simulated electric field at *xz* plane at the peak absorption frequency. (b) The equivalent circuit of the left or right half of the unit cell.

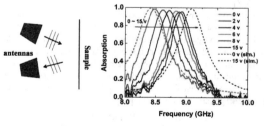

(a)                              (b)

Fig. 3. (a) Schematic of the measurement setup. (b) Simulated (dash) and measured (solid) absorption of the tunable MA at different bias voltage for normal incident EM waves.

## III. SIMULATIONS AND MEASUREMENTS

Simulations are performed to design and optimize the unit cell using the full wave EM solver based on the finite integration technique. In the simulation setup, the structures are

When we increases the DC bias voltage from 0.0 V to 15.0 V, the width of the doped P-N junction of the varactor becomes larger and the capacitance becomes smaller, leading to the rise of the absorbing frequency, as shown in Fig. 3(b). When the DC bias voltage is 0.0 V, the measured absorption is 98.3% at 8.48 GHz with a full width at half magnitude (FWHM) of 5%. As increasing the bias voltage, the absorption peak shifts to higher frequencies. At 15.0 V, the frequency of the peak absorption is 8.92 GHz with the absorption rate of 95.2% and the FWHM of 5%. The tuning range of the absorbing frequency is 0.44 GHz.

The peak absorption rate changes slightly as the peak absorbing frequency varies because we optimize the unit cell for the best absorption rate only at one capacitance value. The dash lines in Fig. 3(b) show the simulated absorption at 0.0 V and 15.0 V for comparison. They differ from the measurement results a little bit. This is mainly due to two reasons. First, the varactor parameters used in simulation is measured at the frequency far lower than the X-band, so that the parasitic impedance effect of the diode at high frequency such as X-band is not taken into account. Second, the varactor soldering position may also affect the resonance frequency. However, the agreements are good overall between the experimental results and the numerical simulations.

The resonance mode used in the proposed absorber is the second higher order mode of the unit cell, i.e. microstrip resonator mode. Its fundamental mode is the LC parallel resonance mode which is commonly used in other reported absorber designs [6]. By using the second higher order mode, the unit cell is not much smaller than the working wavelength such that less varactors will be used in the proposed absorber board. If the fundamental LC mode is used to design absorbers in X-band, the physical size of the unit cell have to be very small. This is because the resonance and peak absorption frequency is related to the product of the unit cell inductance and the integrated varactor capacitance. The smallest typical varactor capacitance in market is a few tenths picofarad. Therefore, in order to design high frequency absorber, e.g. in X-band, unit cell shall be small to achieve small structure inductance, which leads to a large number of varactors to be used. However, the frequency tuning range based on the microstrip resonator mode is generally smaller than that based on the fundamental LC mode because the effective input capacitance provided by the microstrip line is dispersive.

Furthermore, the period of the proposed absorber is smaller than the half working wavelength so that the reflected energy concentrates on the specular direction for oblique incidence.

## V. CONCLUSION

An active frequency tunable MA is proposed by employing microstrip resonator loaded with varactors. Simulation and experiment have shown that by tuning the bias voltage on the varactors, the absorption peak frequency can be adjusted actively in the X-band. The frequency tuning range is around 0.44 GHz with the peak absorption greater than 95%. Compared with the conventional absorbers based on the LC resonance mode, using the second higher order mode, the unit cell size can be large such that less varactors will be consumed. However, the tuning range gets smaller with this resonance mode. Currently, the proposed design is polarization sensitive. It can be further developed to achieve polarization insensitive EM wave absorbing by using symmetric unit cell structure in the future.

## ACKNOWLEDGMENT

This work is partially supported by the National Nature Science Foundation of China (60990322, 60990320, 60801001, 61101011), the Key Grant Project of Ministry of Education of China (313029), the Ph.D. Programs Foundation of Ministry of Education of China (20100091110036, 20120091110032), and Partially supported by Jiangsu Key Laboratory of Advanced Techniques for Manipulating Electromagnetic Waves.

## REFERENCES

[1] E.F. Knott, J.F. Shaeffer, and M.T. Tuley, *Radar Cross Section*, Artech House, Norwood, 1985.

[2] K. Hatakeyama, and T. Inui, "Electromagnetic wave absorber using ferrite absorbing material dispersed with short metal fibers," *IEEE Trans. Magn.*, vol. 20, no. 5, pp. 1261-1263, September 1984.

[3] M. Matsumoto, and Y. Miyata, "Thin electromagnetic wave absorber for quasi-microwave band containing aligned thin magnetic metal paticles," *IEEE Trans.Magn.*, vol. 33, no. 6, pp. 4459-4464, November 1997.

[4] N.I. Landy, S. Sajuyigbe, J.J. Mock, D.R. Smith, and W.J. Padilla, "Perfect metamaterial absorber," *Phys. Rev. Lett.*, vol. 100, 207402, 2008.

[5] B. Zhu, Z. Wang, C. Huang, Y. Feng, J. Zhao, and T. Jiang, "Polarization insensitive metamaterial absorber with wide incident angle," *PIER,* vol. 101, pp. 231-239, 2010.

[6] B. Zhu, Z. Wang, Z. Yu, Q. Zhang, J. Zhao, Y. Feng, and et al., "Planar metamaterial microwave absorber for all wave polarizations," *Chin. Phys. Lett.*, vol. 26, no.11, 114102, 2009.

[7] B. Zhu, C. Huang, Y. Feng, J. Zhao, and T. Jiang, "Dual band switchable metamaterial electromagnetic absorber," *PIER B*, vol. 24, pp. 121-129, 2010.

[8] B. Zhu, Y. Feng, J. Zhao, C. Huang, and T. Jiang, "Switchable metamaterial reflector/absorber for different polarized electromagnetic waves," *Appl. Phys. Lett.*, vol. 97, 051906, 2010.

# *FP-P*

## October 38 (FRI) PM

## Room E

# A Novel Circular Polarized SIW Square Ring-slot Antenna

Fang-fang Fan, Wei Wang, Ze-hong Yan

National Laboratory of Science and Technology on Antennas and Microwaves

Xidian University, Xi'an, Shaanxi, 710071, China

*Abstract-* A single-layer circularly polarized (CP) substrate integrated waveguide (SIW) square ring-slot antenna at X-band is proposed in this letter. By using two shorted square ring-slot in the top wall of SIW and two shorting vias, the CP wave is obtained. To verify the design method, the SIW antenna is fabricated and measured. The measure impedance bandwidth and AR bandwidth are 11% (9.62-10.74GHz) and 1.2% (10.08-10.2GHz), respectively. The antenna also has good radiation pattern and high gain than other similar SIW antennas.

## I. INTRODUCTION

Circularly polarized (CP) antennas are extensively used in space applications such as satellite communication, radar system and so on, because they can resolve problems in wireless channels such as polarization mismatch generated by Faraday effect and interference generated by multi-path effect.

Substrate integrated waveguide (SIW) technology, which is firstly proposed by Wu [1], has the advantage of easy integration with planar circuits by replacing the conventional microstrip and strip line. Also it has the merits of low loss, high power capacity, high Q-factor, low cost than conventional waveguide. SIW-based CP antennas have been reported by many researchers [2-7]. A 16-element top wall SIW slot antenna in Ref. 2 proposed two-compounded slot pairs to obtain CP wave centered at 16GHz, but the usable bandwidth is just 2.3%. In Ref. 3, the authors proposed an X-band cavity backed crossed-slot antenna fed by a single grounded coplanar waveguide. The antenna also met the same problem: a narrow impedance bandwidth of less than 3% and a narrow axial ratio (AR) bandwidth about 1% for AR less then 3 dB. In Ref. 4, a circular ring-slot antenna embedded in a single-layered SIW with a microstrip-to-SIW transition was presented. The antenna had a wider AR bandwidth 2.3% and a wide impedance bandwidth 18.74% than the cavity backed crossed-slot antenna [2-3], but its gain was less than 6 dBi, and the antenna had a high level of cross-polarization. In Ref. 5, the authors studied SIW cavity backed antennas using microstrip-to-SIW and coax-to-SIW transitions. A good level of cross-polarization was obtained with the coax-to-SIW transition; the antenna reached a gain about 7 dBi. The AR bandwidth and the impedance bandwidth were almost the same as for the circular ring-slot antenna [4]. However, the antenna occupies a relative large area, and its design is complex. In Ref.4-6, the authors employed shorting via to connect the metal area bounded by the ring-slot at the top wall and the bottom wall of SIW to obtain CP, so the location of the shorting via and fabrication errors affects the CP characteristic largely.

In this letter, two square ring-slots SIW CP antenna for X-band application is proposed. The shorted strips and shorting vias are used at the same time to obtain CP wave with good performance, the design is very simple.

## II. ANTENNA CONFIGURATION

The geometrical configuration for the antenna (side view and top view) is shown in Fig.1. The overall size of the SIW antenna is 34mm×34mm, it is composed of three parts: SIW-based rectangular waveguide, two square ring-slot, two shorting vias and coax feeding probe. P and d are the distance between the two vias and their diameter of SIW structure. W and L are the width and length of the SIW structure. These values of SIW should certify that it operates in the fundamental mode $TE_{10}$. g1 and g2 is the length of the two shorted strips and its width is as long as the width of the ring-slot denoted as ws and ws1. Then the distance between the feeding probe and the bottom of square ring-slot is denoted as s2, and it is in the center of the SIW in the y-direction. s1 is the distance between the top of square ring-slot and the shorted end , the inner dimension of the square ring-slot is ls. And the other square ring-slot has the side length ls1 and width ws1. The distance between the two square ring-slot is d1. The two shorting vias are located in the middle of two square ring-slots. dy is the distance between the shorting via a-

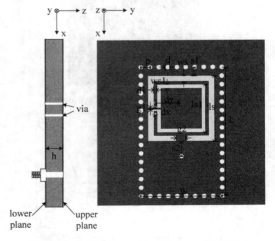

Figure 1 The SIW antenna geometry (side view and top view)

978-7-5641-4279-7

nd the y-direction, dx is the distance between the shorting via and the center of square ring-slot in the x-direction. The two shorting vias is at both side of the center of square ring-slot in the x-direction.

The side length of square ring-slot ls and ls1 mainly influence the resonating frequency of the SIW antenna, the location and the diameter of the two shorting vias and the parameters g1, g2 determine the CP wave performance. The antenna is simulated and analyzed in simulation software HFSS (High Frequency Simulation software). The final optimized parameter values are shown in Table1.

TABLE 1
The value of parameters for SIW antenna

| Parameters | W | L | ls | ws | p | d | dx | dy |
|---|---|---|---|---|---|---|---|---|
| Value(mm) | 16.2 | 26.4 | 13.4 | 1.2 | 1.8 | 1 | 1.2 | 5.2 |
| Parameters | g1 | g2 | s1 | s2 | ls1 | ws1 | d | |
| Value(mm) | 0.3 | 2 | 1.7 | 3.1 | 9.8 | 0.8 | 0.6 | |

### III. FABRICATION AND THE MEASURED RESULTS

Taconic TLX-8 with relative permittivity of 2.55, height of 1.52mm is used for the antenna. The fabricated SIW antenna is shown in Fig.2. The simulated and measured VSWR is shown in Fig.3 with the help of the vector network analyzer E8363B from Agilent Technologies, it can be seen that the bandwidth with VSWR less than 2 is 11% (9.62GHz-10.74GHz), the simulated and measured impedance bandwidth are in good agreement.

Figure 2 The photo of fabricated SIW antenna

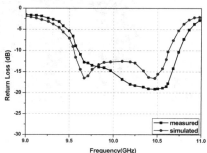

Figure 3 Simulated and measured VSWR versus frequency

The AR versus frequency is depicted in Fig. 4; the measured AR bandwidth below 3dB is 1.2% (10.08-10.2GHz). There is a discrepancy, which appears as a frequency shift to higher frequency for the measured AR, due to the tolerance error in the manufacturing process with the location and diameter of shorting via.

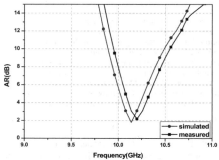

Figure 4 The simulated and measured AR versus frequency

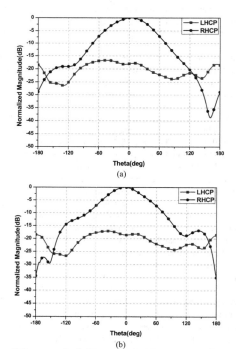

(a)

(b)

Figure 5 The measured radiation pattern at 10.2GHz (xz plane, yz plane)

Fig.5 delineates simultaneously the RHCP and LHCP radiation pattern in xz- and yz-planes at 10.2GHz of the SIW antenna. It can be seen that the proposed antenna satisfies the RHCP generation with a lower cross-polarisation at the boresight direction. The maximum gain at the boresight direction is 8.0dBi at 10.2GHz.

## IV. CONCLUSION

In short, a CP antenna based on SIW structure with high gain is presented here. Through the shorted strips and shorting vias, a RHCP wave is obtained; the design is simple and easy for fabrication. The SIW antenna achieves 11% impedance bandwidth and 1.2% AR bandwidth, respectively. The maximum gain at the boresight direction is 8.0dBi at 10.2GHz. The antenna can be used in satellite communication system.

## ACKNOWLEDGMENT

This work was supported by "the Fundmental Research Funds for the Central Universities". (K50511020018).

## REFERENCES

[1] Xu, F., and Wu, K.: 'Guided-wave and leakage characteristic of substrate integrated waveguide', IEEE Transactions on Microwave and Theory Techniques, 2005, 53, (1), pp. 66–73.

[2] Chen Peng, Hong Wei, Kuai Zhenqi and et al., A substrate integrated waveguide circular polarized slot radiator and its linear array, IEEE Transactions on Antennas and Propagation, 2009, 59, (8), pp. 1398–1403.

[3] G.Q. Luo, Z.F. Hu, Y. Liang, L.Y. Yu, and L.L. Sun, Development of low profile cavity backed crossed slot antennas for planar integration, IEEE Trans Antennas Propag 57 (2009), 2972–2979.

[4] D. Kim, J.W. Lee, C.S. Cho and et al., X-band circular ring-slot antenna embedded in single-layered SIW for circular polarization, Electronics Letters, 2009,45,13, pp.668-669 .

[5] D. Kim, J.W. Lee, T.K. Lee and et al., Design of SIW cavity-backed circular-polarized antennas using two different feeding transitions, IEEE Transactions on Antennas and Propagation, 2011, 59, (4), pp. 1398–1403.

[6] Jaroslav Lacik, Circularly polarized SIW square ring-slot antenna for X-band applications, Microwave and Optical Technology Letters, 2012, 54, (11), pp. 2590–2593.

[7] Yue Li, Zhi NingChen, Xianming Qing, and et al., Axial ratio bandwidth enhancement of 60-GHz substrate integrated waveguide-fed circularly polarized LTCC antenna array, IEEE Transactions on Antennas and Propagation, 2012, 60, (10), pp. 4619–4626.

# Design and Optimization of Broadband Single-Layer Reflectarray

Kai Zhang, Jianzhou Li, Gao Wei, Yangyu Fan,
Jiadong Xu

School of Electronics and Information
Northwestern Polytechnical University
Xi'an, P. R. China
zhangkai@nwpu.edu.cn

Steven Gao
School of Engineering and Digital Arts
University of Kent
Canterbury, UK
s.gao@kent.ac.uk

*Abstract*—**A novel broadband single-layer reflectarray for satellite communications has been presented. The design and simulation results are obtained and discussed. The element in the reflectarray provides a nearly 360° linear phase range. The reflectarray is fed by a printed microstrip log-periodic dipole antenna. The broadband characteristic of the reflectarray is obtained due to the sub-wavelength of the element space. In order to increase the gain bandwidth of this reflectarray further, an optimization technique was utilized to minimize the frequency dispersion. Finally, a prime-focus 256-element reflectarray at C band is designed and simulated. The obtained 3-dB gain bandwidth reaches 30.8% (from 4.4 to 6.0GHz).**

*Keywords*—*broadband reflectarray; log-periodic dipole antenna; single-layer; satellite communication*

## I. INTRODUCTION

Microstrip reflectarray antennas have been widely investigated because of their potential advantages over microstrip arrays and parabolic reflectors [1]. The radiation element in the reflectarray can scatter the incident field with a designated phase to achieve a specific shaped beam. It is well known that printed reflectarray have some advantages such as low profile, low cost, easy fabrication and possibilities integrated with electronic beam control circuits. These characteristics make the reflectarray technology a suitable choice for satellite and wireless communication systems. However, the most severe drawback of the microstrip reflectarray is its limited bandwidth performance. This shortcoming is apparent for single-layer microstrip reflectarrays especially. The bandwidth performance of a microstrip reflectarray is limited primarily by two factors. One is inherent narrow bandwidth behavior of microstrip elements themselves. Another one is differential spatial phase delay [1]. In order to extend the range of linear phase-frequency response as wide as possible, several novel element structures, including multi-resonant loop element [2], square cross and modified Malta cross [3], multi-dipole element [4], windmill-shaped element [5], disk element with attached phase-delay line [6], circular rings with open-circuited stubs [7], have been proposed in the literature. The objectives of these methods are to obtain linear phase-frequency response in a phase shift range larger than

360°, which can be used as an additional degree of freedom for a further improvement of bandwidth. However, these methods can only address the first factor of the bandwidth limitation. The second factor is more significant for large reflectarrays. One effective technique to compensate for the spatial phase delay is using sub-wavelength coupled-resonant elements [8] instead of conventional $\lambda/2$ ones. It has been shown that this method can realize a similar S-shape reflection phase response by reducing the phase error at different frequencies. Another method was introduced in [9] to overcome the bandwidth limitation of large reflectarrays by using an optimization routine to adjust the parameters of the radiation element to minimize the frequency dispersion.

In this paper, the combination of square patch and ring with thin loop boundary is adopted as the essential reflectarray element structure. The geometry of this element exhibits a cycle evolutionary characteristic [10]. The inter-element space of unit-cell is investigated using commercial software HFSS [11]. In order to increase the gain bandwidth further, the optimization method is also adopted in this paper. The advantages of this method are demonstrated in the design of a prime-focus 256-element reflectarray operating at C-band. Compared to other single layer reflectarray designs in Table 1, the gain bandwidth of reflectarray in this paper show a significant improvement across the band. Its 3-dB gain bandwidth (BW) can reach 30.8%.

TABLE I.    PERFORMANCE OF DIFFERENT BROADBAND SINGLE LAYER REFLECTARRAY

| Ref. | [2] | [3] | [4] | [5] | [6] | [7] | [8] |
|---|---|---|---|---|---|---|---|
| 1-dB Gain BW (%) | 9 | 19 | NAN | 20 | NAN | NAN | 17 |
| 3-dB Gain BW (%) | NAN | NAN | 14.1 | NAN | 18 | 17.8 | NAN |

## II. ANALYSIS OF REFLECTARRAY ELEMENT

The structure of individual radiation element is presented in Fig. 1. It is built from a square ring slot of length $L_s$ and width $W_s$. A metallic ring is inserted into the slot, and its length and width are $L_R$, $W_R$ respectively. The required phase shift reflected by the element is obtained by changing the position parameter $L_R$ of the metallic ring. $L_R$ changes from $L_{R\min} = L_p$

978-7-5641-4279-7

to $L_{R\,\mathrm{max}} = L_s + 2W_R$. In the following simulations, the default inter-element space $L_s$ is 20mm, which is $\lambda_0/3$ ($\lambda_0$ is the wavelength in vacuum at the center frequency $f_0 = 5\mathrm{GHz}$). The radiation element is fabricated on a 1.6 mm thickness dielectric substrate (FR4) of $\varepsilon_{r1} = 4.4$ and $\tan\delta_1 = 0.02$. The substrate is suspended above ground plane with a distance of $h = 18\mathrm{mm}$. The substrate layer and the ground plane are separated by Rohacell foam ($\varepsilon_{r2} = 1.06$). The width of metallic ring is $W_R = 0.5\mathrm{mm}$. The sidelength of the center square patch is $L_p = 6\mathrm{mm}$.

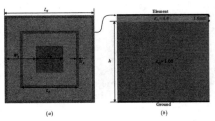

Fig. 1. (a) Geometry of the radiation element. (b)Side view of the element.

In general, the reflection phase characteristics of radiation elements are angle-dependent, which can be observed in Fig. 2. It shows phase of the reflected wave with respect to $L_R$ for different incident angles in the range of 0°~40°. The maximum phase discrepancy with respect to normal incident is 70° at θ=40°. Hence, oblique incidence for each element is needed to take into account in the design procedure. To obtain the phase response, periodic boundary conditions are introduced to consider the mutual coupling between the identical neighbor elements in HFSS. The element is excited by the Floquet port with linear polarized electric field.

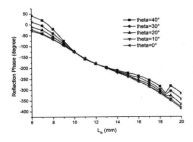

Fig. 2. Phase responses of the radiation element with respect to $L_s$ for different incident angle.

According to the reflectarray design method, the position of the metallic ring in the radiation element is calculated to produce the required reflection phase at the center frequency. However since the reflection phase of the elements is a function of frequency, as the frequency changes, phase errors are introduced in the array and results in bandwidth limitation of the reflectarray. The phase variation corresponding to the inter-

element space is shown in Fig. 3. On the one hand, it can be seen that the sub-wavelength unit cells can almost achieve the same phase range as half wavelength unit cell which is typically around 360° for single layer designs. This factor will not deteriorate the bandwidth of reflectarray. On the other hand, a set of phase curves with better parallelism over a wide frequency range can be achieved by adjusting sub-wavelength element spacing. It is well known that this can result in an improved antenna bandwidth. Although there is no theoretical limit on using smaller unit cells in reflectarray designs, the fabrication tolerance of the elements becomes a critical factor. In this paper, the inter-element space is chosen as $\lambda_0/3$.

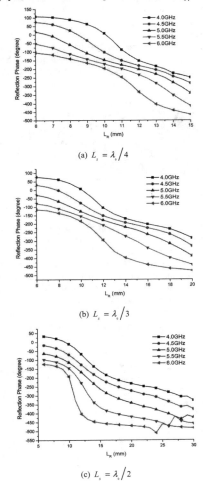

(a) $L_s = \lambda_0/4$

(b) $L_s = \lambda_0/3$

(c) $L_s = \lambda_0/2$

Fig. 3. Phase of the reflected wave with respect to $L_R$ for different values of $L_s$.

## III. FEED DESIGN

Conventionally, most of the reflectarrays are excited by a horn antenna. In order to minimize the aperture blockage effect in the center-feed reflectarray, a broadband printed log-periodic dipole arrays (PLPDA) is designed as the feed. In this paper, a printed LPDA is designed and shows better performances. Actually, in the range of 3GHz ~ 9 GHz, its gain is remarkably constant versus frequency. The array structure is based on a pair of parallel balanced printed transmission lines on the two sides of a thick 0.8mm dielectric slab (FR4). The array dipoles are connected to the two printed lines in an alternate way. A coaxial infinite balun is used to feed the parallel lines and enlarge the impedance bandwidth. The outer conductor of the coaxial cable is soldered to one of the parallel line, and the inner pin is bent and welded to the opposite parallel line. The characteristic impedance of the balanced feeding line is selected as $50\Omega$. In order to improve the impedance matching in the whole operating bandwidth, the feed line is terminated with a $187\Omega$ resistor load. All the details are shown in Fig. 4.

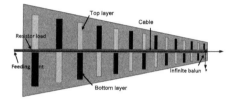

Fig. 4. Printed LPDA layout.

The simulated results of return loss and gain of the PLPDA are revealed in Fig. 5. It is demonstrated that the overlap between impedance bandwidth ( $S_{11} \le -10dB$ ) and 1-dB gain bandwidth is 3.1GHz~8.6GHz. The relative bandwidth of the proposed antenna is more than 94%. The simulated radiation patterns also show that the 10-dB beam width in E- and H-planes are almost the same around 60°. This kind of antenna is suitable for use as the feed of reflectarray. The beam-width of 10-dB gain-drop can used to determine the ratio of focus to diameter.

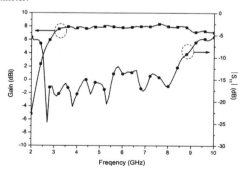

Fig. 5. Simulated S11 and gain.

## IV. FULL ARRAY DESIGN AND OPTIMIZATION

In order to produce a beam in the given direction $(\theta_b, \varphi_b)$, the phase shift required at each element is expressed as:

$$\phi_{Ref}^{required}(x_i, y_i, f) = \phi_{Inc}(x_i, y_i, f) - k_0(f)\sin\theta_b(\cos\varphi_b x_i + \sin\phi_b y_i)$$

(1)

where $k_0(f)$ is the propagation constant in vacuum at frequency $f$, and $(x_i, y_i)$ is the coordinate of element $i$, $\phi_{Inc}(x_i, y_i, f)$ is the phase of incident field on element $i$ at frequency $f$, $\phi_{Ref}^{required}(x_i, y_i, f)$ is the phase of the coefficient for element $i$ at frequency $f$. Generally, the phase of incident field is calculated as:

$$\phi_{Inc}(x_i, y_i, f) = -k_0(f)d_i$$

(2)

where $d_i$ is the distance between the phase center of the feed and the cell $i$. For the horn feed, the phase center is in the vicinity of radiation aperture. But for the PLPDA feed, it is difficult to determine the phase center accurately. In order to avoid this problem, the phase of incident field on the element's surface can be calculated directly using HFSS.

In order to minimize the frequency dispersion, the following error function can be defined firstly [9]:

$$e(x_i, y_i) = \sum_{m=l,c,u} \left| \phi_{Ref}^{required}(x_i, y_i, f_m) - \phi_{Ref}^{achieved}(x_i, y_i, f_m) \right|$$

(3)

where $\phi_{Ref}^{required}(x_i, y_i, f)$ and $\phi_{Ref}^{achieved}(x_i, y_i, f_m)$ are the required and achievable phase delay, respectively, for the $i$th element at the frequencies $f_c$, $f_l$ and $f_u$. $f_c$, $f_l$ and $f_u$ are center, lower and upper band edge frequencies, respectively. In practice, the error function introduced in (3) is calculated for all elements within the cell parameters search space. The element with the lowest error function is selected as the optimum element from the point of view of bandwidth performance.

Based on the above considerations, a C-band center-fed pencil beam microstrip reflectarray is designed for center operating at 5GHz. The lower and upper band edge frequencies are 4GHz and 6.5GHz respectively. As shown in Fig. 6, the prime-focus reflectarray antennas have a rectangle aperture. The dimension of this array is 32cm×32cm and the total number of radiating elements is 256 (16×16). The distance between the feed (LPDA) and the reflectarray aperture surface is 160mm (the ratio of focus $F$ to diameter $D$ is $F/D = 0.5$ ). The reflectarray is analyzed by using HFSS. Fig. 7 shows the phase of incident field on the designed reflectarray with PLPDA feed by changing the material of the whole reflectarray to vacuum.

The simulated radiation patterns at 5GHz are shown in Fig. 8, where the gain is 21.5dBi and the side lobe level is below -18.1dB and -20.2dB in E- and H-plane, respectively.

The simulated gain from 4GHz to 6.5GHz for reflectarray is shown in Fig. 9. It is indicated that the maximum gain is 22dBi at 5.2GHz. The 3-dB gain bandwidth of the reflectarray antenna is about 30.8%.

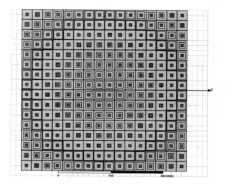

Fig. 6.   Geometry of PLPDA feed reflectarray.

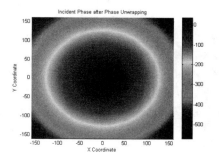

Fig. 7.   Phase of incident field in degree.

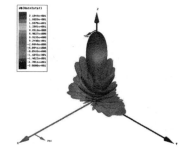

Fig. 8.   Simulated radiation pattern at 5GHz.

## V.   CONCLUSIONS

A single-layer microstrip reflectarray has been designed and analyzed in this paper. This reflectarray was designed for broadband satellite communications at C-band. In order to reduce the blockage effects and increase aperture efficiency, a printed broad-band log-periodic dipole arrays is adopted as the feed. By using sub-wavelength radiation element and optimized method, the frequency dispersion on the level of gain is minimized. The performance of a prime-focus 256-element reflectarray was verified using HFSS. The simulated results have a significant improvement in bandwidth of the reflectarray. The 3-dB gain bandwidth of 30.8% was achieved.

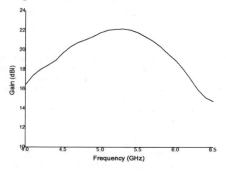

Fig. 9.   Simulated gain against frequency.

## REFERENCES

[1]   J. Huang and J. A. Encinar, "Reflectarray Antenna", New York: Wiley-Interscience, 2007.

[2]   M. R. Chaharmir, J. Shaker, and M. Cuhaci, et al., "A Broadband Reflectarray Antenna with Double Square Rings," Microwave and Optical Technology Letter, vol. 48, no. 7, pp. 1317-1320, July 2006.

[3]   P. D. Vita, A. Freni, and G. L. Dassano, et al., "Broadband Element for High-gain Single-layer Printed Reflectarray Antenna," Electronics Letters, vol. 43, no. 23, pp. 1247-1248, Nov. 2007.

[4]   L. Li, Q. Chen, and Q. W. Yuan, et al., "Novel Broadband Planar Reflectarray with Parasitic Dipoles for Wireless Communication Applications," IEEE Antenna and Wireless Propag. Letters, Vol. 8, pp. 881-885, 2009.

[5]   H. Li, B. Z. Wang, and P. Du, "Novel Broadband Reflectarray Antenna with Windmill-shaped Elements for Millimeter-wave Application," International Journal of Infrared and Millimeter Waves, Vol. 28, Issue 5, pp 339-344, May 2007.

[6]   H. Hasani, M. Kamyab, and A. Mirkamali, "Broadband Reflectarray Antenna Incorporating Disk Elements with Attached Phase-delay Lines," IEEE Antenna and Wireless Propag. Letters, Vol. 9, pp. 19-29, 2011.

[7]   Y. Z. Li, M. E. Bialkowski, and A. M. Abbosh, "Single Layer Reflectarray with Circular Rings and Open-Circuited Stubs for Wideband Operation," IEEE Trans. Antennas Propagat., Vol. 60, No. 9, pp. 4183-4189, Sep. 2012.

[8]   G. Zhao, Y. C. Jiao, and F. Zhang, et al., "A Subwavelength Element for Broadband Circularly Polarized Reflectarrays," IEEE Antenna and Wireless Propag. Letters, Vol. 9, pp. 330-333, 2010.

[9]   M. R. Chaharmir, J. Shaker, and H. Legay, "Broadband Design of a Single Layer Large Reflectarray Using Multi Cross Loop Elements," IEEE Trans. Antennas Propagat., Vol. 57, No. 10, pp. 3363-3366, Oct. 2009.

[10]   L. Moustafa, R. Gillard, and F. peris, et al., "The Phoenix Eell: A New Reflectarray Cell with Large Bandwidth and Rebirth Capabilities," IEEE Antenna and Wireless Propag. Letters, Vol. 10, pp. 71-74, 2011.

[11]   HFSS 15, Ansys [Online], Available: http://www.ansys.com.

# A Simple Broadband Patch Antenna

Guoming Gao[1], Hong Yuan[1], Ling Jian[1], Min Ma[2]

1.Nanjing Marine radar Institute, Nanjing, 210003 Jiangsu Province, P.R.China

2. China Mobile Group Jiangsu Co., Ltd. Nanjing Branch, 210029 Jiangsu Province, P.R.China

*Abstract*-The design and development of a simple broadband patch antenna for X-band application is introduced. The antenna consist of 2 layers substrate, which can be manufactured easily. The measured bandwidth is more than 20% for voltage standing wave ratios<2. The measured radiation patterns of the $1 \times 8$ element linear subarray are presented and discussed in detail. It can be used as a conformal antenna.

## I. INTRODUCTION

A great deal of research has been devoted in the last years to improve microstrip antenna bandwidth to meet the radar antenna demands[1-4]. However, most antenna use aperture-coupled microstrip patch arrays, which is a multilayer structure, need the support of the foam[5-6]. For these antenna, the manufacturing is complex and the cost is expensive.

In the present article, the broadband patch antenna element consists of dual-stacked patches, this results in a low profile. The antenna is easy to manufacture, and can be used as a conformal antenna. The simulations were performed using Ansys HFSS electromagnetic software. Experimental results are presented for a $4 \times 8$ elements array in X-band.

## II. ANTENNA DESIGN

The radiation element geometry is shown in Fig.1. Arlon Diclad880 with a relative dielectric constant of 2.2 has been selected as the material for the microstrip antenna substrate. The excited patch, which fed by T-shaped microstrip lines, is placed on the top side of the lower substrate, and the ground is placed on the bottom side. The T-shaped microstrip line is on the same layer of the excited patch. Coaxial connector to the microstrip line adopts vertical transition form in this antenna. The parasitic patch is placed on the top side of the upper substrate, so that increase the antenna bandwidth. Thickness of both the substrates is 1.524mm, which leads to a low profile as well as low weight and low cost.

The designed antenna works from 9 to 11GHz, and the simulated bandwidth of the antenna is more than 20% for voltage standing wave ratios(VSWR)<2, as shown in Fig.2. Its simulated radiation patterns are shown in Fig.3, which indicates good pattern performance. As the radiation patterns of horizontal polarization (HP) and vertical polarization (VP) are similar, only horizontal polarization radiation patterns are given in Fig.3.

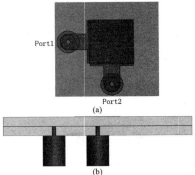

(a)

(b)

Figure 1. Geometry of the propose antenna. (a) Top view and (b) Cross-sectional view.

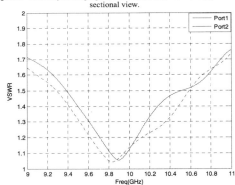

Figure 2. Simulated VSWR of the propose antenna.

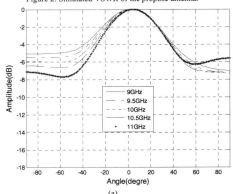

(a)

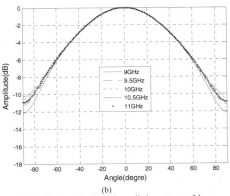

(b)

Figure 3. Simulated horizontal polarization radiation patterns of the propose antenna. (a) Elevation cuts and (b) Azimth cuts at 9, 9.5, 10.0, 10.5 and 11 GHz.

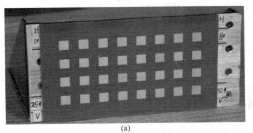

(a)

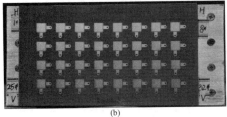

(b)

Figure. 4. Photograph of the 4×8 element array. (a) parasitic patch layer and (b) excited patch layer.

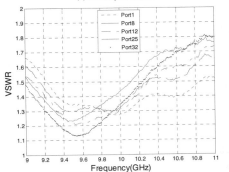

Figure. 5. Measured VSWR of the propose antenna

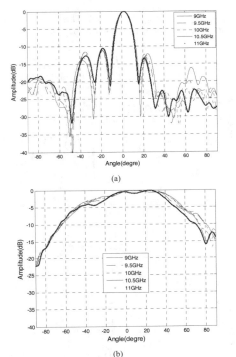

(a)

(b)

Figure. 6. Measured radiation patterns for H-pol. (a) Elevation cuts and (b) Azimth cuts at 9, 9.5, 10.0, 10.5 and 11 GHz.

## III. MEASURMENTS RESULTS

A 4×8 element array arranged as foresaid configuration is designed, manufactured and measured. The photograph of the 4×8 element array is presented in Fig.4. Two layers of the substrate have to be fixed together using RO4400 series prepreg flake, under high temperature. The measured results suggest that the prepreg flake layer does not degrade the RF performance of the antenna.

Fig.5 shows the measured VSWR curves of the array elements, VSWR is less than 2 from 9 GHz to 11 GHz for both polarizations. In order to see clearly, only typical curves are given in Fig.5. The radiation patterns of the array were measured in an anechoic chamber. For the 4×8 element array, 8 elements in azimuth plane and 4 elements in elevation. As a result, attentions are placed on the azimuth plane. Two additional 1:8 microstrip divider networks are used for constructing the 1×8 element linear subarray. For HP of the 1 ×8 element linear subarray, the uniformly radiation pattern measured at 9, 9.5, 10, 10.5 and11 GHz for the 0°-tilt case in azimuth plane (all linear arrays are excited equally) is given in Fig.6, where it shows a near-in sidelobe peak at -11.5dB level. For VP of the 1×8 element linear subarray, the radiation pattern in azimuth plane is given in Fig.7, where it shows a near-in sidelobe peak at -13dB level. This is because in the azimuth plane, vertical polarization feed network is symmetric.

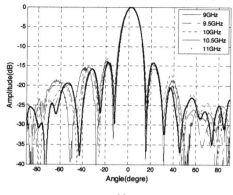

(a)

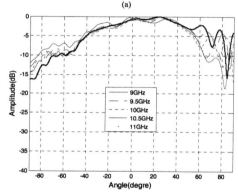

(b)

Figure. 7. Measured radiation patterns for V-pol. (a) Elevation cuts and (b) Azimth cuts at 9, 9.5, 10.0, 10.5 and 11 GHz.

## IV. CONCLUSION

A simple broadband patch antenna, consisting of 2 layers substrate, has been developed to improve microstrip antenna bandwidth. A T-shaped feeding structure is used to excite the patch antenna, and a parasitic patch is employed on the top side of the upper substrate. The measured bandwidth is more than 20% for voltage standing wave ratios<2, the radiation patterns indicate good performance. The thickness of total antenna is about $\lambda/10$ , which means low profile as well as low weight and low cost. Consequently, It can be used as a conformal antenna.

REFERENCES

[1] S.-C. Gao and S.-S. Zhong, Dual-polarized microstrip antenna array with high isolation fed by coplanar network, Microwave Opt Technol Lett 19 (1998), 214 – 216.

[2] F. Rostan and W. Wiesbeck, Design consideration for dual polarized aperture-coupled microstrip patch antennas, In Proceedings of IEEE International Antennas and Propagation Symposium, 1995, Newport Beach, CA pp. 2086 – 2089.

[3] L.-X. Ling, S.-S. Zhong, and W. Wang, Design of a high isolation dual-polarized slot-coupled microstrip antenna, Microwave Opt Technol Lett 47 (2005), 212 – 215.

[4] S.B. Chakrabarty, M. Khanna, and S.B. Sharma, Wideband planar array antenna in C band for synthetic aperture radar application, Microwave Opt Technol Lett 33 (2002), 52 – 54.

[5] Wei Wang, Xian-Ling Liang, Yu-Mei Zhang,Shun-Shi Zhong, and Yan Guo. Experimental characterization of a broadband dual- polarized microstrip antenna for X-band SAR applications. Microwave Opt Technol Lett 479(2007), 649 – 652.

[6] D.M. Pozar, S.D.Targonski, A Shared-Aperture Dual-Band Dual-Polarized Microstrip Array [J]. IEEE Trans Antennas and Propagat, 2001, 49(2) : 150-157.

# A Compact Dual-Band Assembled Printed Quadrifilar Helix Antenna for CNSS Application

K. Chen[1], H. Wang[1], Y. Huang[2] and J. Wang[2]

[1]Millimeter Wave Technique Laboratory, Nanjing University of Science&Technology,Nanjing,210094, China

[2]Suzhou Bohai Microsystem CO., LTD., Suzhou, 215000, China

haowangmwtl@gmail.com

*Abstract-* This paper presents a novel compact dual-band printed quadrifilar helix antenna (PQHA) for application in Compass Navigation Satellite System (CNSS) of China. This antenna is consisted of two quadrifilar helix elements that are arranged at inner and outer cylinder. It works at the B1 (1.561GHz) and S (2.492GHz) bands. Each four spiral arms is excited by a series feed network in equal magnitude and successive 90° phase difference. It achieves the axial radiation pattern at B1 band, furthermore, it produces the shaped-conical radiation pattern at S band that is benefit for receiving the navigation signal from low elevation angle. Simulations show that 10dB impedance bandwidth of the proposed novel compact CNSS dual-band PQHAs are 15.4% at B1 band and 14.2% at S band. In addition, the axial ratio bandwidth is less than 3 are 12.8% and 2.8% at B1 and S band, respectively. The isolation between two quadrifilar helix elements is higher than 27dB.

## I. INTRODUCTION

Compass Navigation Satellite System (CNSS) or "BeiDou" in its Chinese name is the first regional satellite system in the world, which has been widely used in navigation, mapping, time service, communications, etc. Antenna is quite important in CNSS system as the media to transfer electromagnetic signals. It is required that CNSS antennas should be circularly polarization, and have broad impedance and axial ratio bandwidths.

A very attractive candidate for these applications is the resonant quadrifilar helical antenna (QHA) and more recently the conventional printed quadrifilar helical antenna (PQHA) due to their performances in terms of circular polarization, good axial ratio (AR), light weight, high dimensional stability, ease of fabrication and low cost [1]. Although some PQHA structures are already small, further size reduction is necessary to satisfy the space limitations of CNSS terminal. As a result of the service function of the CNSS is more and more, so, we hope that antenna can work in dual-band or multi-band. In addition, it is desirable in many satellite and ground station applications to produce low elevation radiation pattern. Compact printed quadrifilar helical antenna with Iso-Flux-Shaped pattern was introduced in [2], but it only working at one frequency. Assembled QHAs for CNSS are proposed to achieve duplex communication. The two QHAs are stacked together from top to bottom as an entity, which

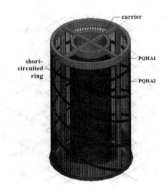

Fig 1. Structure of the proposed compact CNSS PQHAs.

will lead to antenna's size more larger and is not satisfied for the space limitations of CNSS terminals [3]. A compact PQHA with integrated feed network was realized for satellite mobile communication systems at L1 band [4].

In this paper, we present a novel compact dual-band printed quadrifilar helix antenna suitable for CNSS. The two PQHAs are arranged from inside to outside. The inside PQHA is working at S band (2.492GHz) and the outside one is working at B1 band (1.561GHz). For the S satellite navigation signal always from the low elevation direction, the S PQHA achieves the shaped-conical radiation pattern, which is benefit for receiving the low elevation signal. And the B1 PQHA maintains axial radiation pattern. The software Ansoft HFSS has been used to optimize the design. In section Ⅱ, the design of radiation element and power divider network are introduced. The simulated results are discussed in section Ⅲ.

## II. ANTENNA DESIGN

This CNSS dual-band antenna comprises of two PQHAs with the employed feeding networks working at center frequency 1.561GHz (PQHA2) and 2.492GHz (PQHA1) respectively, as shown in Fig. 1. The inside PQHA1 works at high frequency band and the outsider PQHA2 works at low frequency band. Its polarization type are right-handed circularly polarization (RHCP).

### A. Radiation Element Design and Pattern Control

It is seen from Fig. 1 that each PQHA has four radiation elements connecting to a short-circuited ring in their top and a series feeding network on their bottom, forming an antenna unit. The thin substrate that supports PQHAs is rolled up with cylindrical structure, its relative permittivity of 2.2 and thickness of 0.125mm. Based on design theories in [5], the parameters of the PQHA can be set by the following equation:

$$H = N\sqrt{(1/N^2)(L - 2r_o)^2 - 4\pi^2 r_o^2}$$

Where $H$ means the axial length of the helix; $L$ means the length of the helix element; $r_o$ means the radius of the helix; $N$ means the number of turns for one element.

The detail parameters of PQHAs are showed following in Fig. 2.

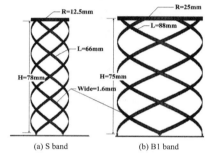

(a) S band          (b) B1 band

Fig 2. Parameters of PQHAs.

To achieve the shaped-conical pattern at S band and axial pattern at B1 band, the radius of PQHA is a key parameter that should be optimized. The effects of radius for QHA was analyzed in [5] . the radiuses of PQHAs are 12.5mm at S band and 25mm at B1 band. The corresponding radiations are shown in Fig.3 (a) and (b), respectively. The maximum gain direction of shaped-conical pattern is at 74deg. The investigations of radiation patterns that affected by the mutual coupling are also given in Fig.3. It is seen that the B1 radiation pattern remains stable in the existence of the S PQHA. However, the S radiation pattern is deteriorated by the mutual coupling. The maximum gain direction of shaped-conical pattern is from 74deg to 45deg, for the existence of the B1 PQHA. While the S PQHA radius is defined, the pitch angle of helix can be adjusted to optimize the peak gain direction slightly. The optimization results are listed in Table I, which can be served as a guideline to the design. Therefore, in order to achieve the radiation requirement of CNSS at 2.492GHz, The pitch angle 38deg is selected.

### B. The Design of Power Divider Network

Two series-type feed networks are printed on the substrate, its relative permittivity of 2.2 and thickness of 0.5mm. Fig. 4

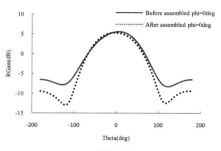

(a) B1 band

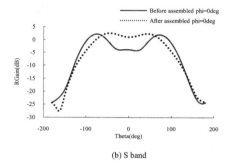

(b) S band

Fig 3. Gain patterns before and after assembled.

TABLE I
DETAIL DATA AT 2.492GHz FOR DIFFERENT PITCH ANGLES

| Pitch angle (deg) | Maximum gain direction(deg) | Maximum gain (dB) | Gain in phi=0deg(dB) |
|---|---|---|---|
| 35 | θ=48 | 2.36 | 0.58 |
| 38 | θ=45 | 2.58 | 1.04 |
| 41 | θ=42 | 2.62 | 1.40 |

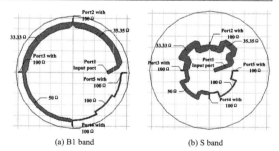

(a) B1 band          (b) S band

Fig 4. Topologies of the antenna feed network [6].

shows the topologies of the feed networks. Starting from the 50-Ω input microstrip line, the power is divided into four outputs in an equal amount using the branch-line power divider. The feeding length between the two ports is a quarter wavelength so as to provide 90° phase differential. The feed

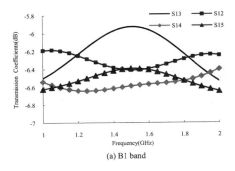

(a) B1 band

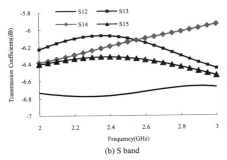

(b) S band

Fig 5. Simulated transmission coefficients of the network.

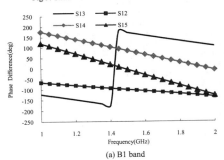

(a) B1 band

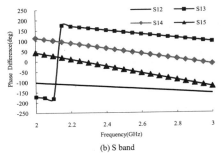

(b) S band

Fig 6. Simulated phase differences between four output ports of the network.

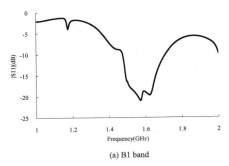

(a) B1 band

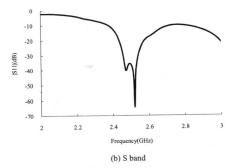

(b) S band

Fig 7. Simulated |S11| of the proposed CNSS compact PQHAs.

network provides an equal distribution of RF signal power to the four helix windings in quadrature phase rotation with 0°, 90°, 180° and 270° phase differentials.

Figs. 5-6 show simulated transmission coefficients and phase differences of the designed feed network. Transmission coefficients from the input port to four output ports, are ranging from -6 to -6.7dB. The phase differences are about 90°±2° at working frequency. As a result, the power divider delivers quadrature phase-shifting.

III.  SIMULATION RESULTS

CNSS dual-band PQHAs with the employed feed network are modeled in HFSS, as showed in Fig. 1. All the simulation results are achieved by HFSS software.

Fig. 7 shows the results of reflection coefficient for the proposed PQHAs on working frequency bands. The PQHA1 exhibits simulated 10dB impedance (|S11|<10dB) bandwidth of about 14.2%, from 2.37 to 2.73GHz and the PQHA2 exhibits simulated 10dB impedance (|S11|<10dB) bandwidth of about 15.4%, from 1.47 to 1.71GHz.

Fig. 8 shows the isolation between two PQHA elements. It is showed that the PQHAs have good isolation on two frequency bands, -43dB at 1.561GHz and -27dB at 2.492GHz. Simulated RHCP radiation patterns at $\varphi=0°$ and $\varphi=90°$ of the proposed CNSS compact PQHAs are shown in Fig. 9. The gain of B1 PQHA can achieve 5.3dBi in axial direction. The maximum

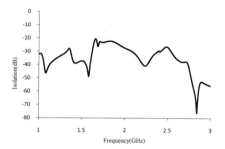

Fig 8. Isolation between the two PQHAs.

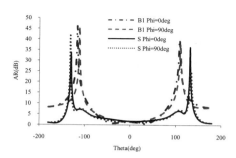

Fig 10. Simulated AR of the proposed CNSS compact PQHAs.

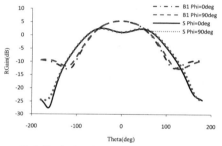

Fig 9. Simulated gains of the proposed CNSS compact PQHA.

gain of S PQHA can obtain at $\theta$=45°, which is 2.6dBi.

Fig. 10 shows the results of AR for the PQHA. The proposed CNSS compact PQHA exhibits excellent Circular Polarized (CP) radiation. The simulated AR values of PQHA are less than 3dB at bore sight through the whole working frequency bands. The PQHA1 exhibits simulated AR less than 3dB bandwidth of about 2.8%, from 2.44 to 2.51 GHz. The PQHA2 exhibits simulated AR less than 3dB bandwidth of about 12.8%, from 1.4 to 1.6 GHz. Good AR values also indicate that the feed network proposed could generate stable quadrature feed phases for PQHA.

IV. CONCLUSION

In this paper, a novel compact dual-band quadrifilar helix antenna is proposed for CNSS application. The size of antenna is reduced by arranging two elements in outer and inner carriers. It obtains axial radiation pattern at B1 band and shaped-conical radiation pattern at S band, through optimizing the radiuses of PQHAs.

The simulation shows the bandwidth of return loss for the proposed PQHAs are 15.4% in B1 band and 14.2% in S band. The gain of PQHA1 is 2.6dBi at $\theta$=45° and PQHA2 is 5.3dBi at axial direction. The AR bandwidths of 3dB are 12.8% in low frequency and 2.8% in high frequency. This design is very suitable for application on vehicles like trains, cars and so on. It can be widely used in CNSS. More measured results will be presented at the conference.

REFERENCES

[1] J Rabemanantsoa and A Sharaiha, "Size reduced multi-band printed quadrifilar helical antenna," *IEEE Transaction on Antennas and Propagation*, vol. 59, pp. 3138-3143, September 2011.

[2] S. Hebib, N.J. G. Fonseca, and H. Aubert, "Compact printed quadrifilar helical antenna with Iso-Flux-Shaped pattern and high cross-polarization discrimination," *IEEE Antennas and Wireless Propagation Letters*, vol. 10, pp. 635-638, July 2011.

[3] Q.X. Chu, W. Lin, W.X. Lin, and Z.K. Pan, "Assembled dual-band broadband quadrifilar helix antennas with compact power divider networks for CNSS application," *IEEE Transaction on Antennas and Propagation*, vol. 61, pp. 516-523, October 2013.

[4] S.Q. Fu, S.J. Fang, K. Lu, and Z.B. Wang, "Printed quadrifilar helix antenna with integrated feed network," *Microwave Antenna Propagation and EMC International Symposium*, Beijing, pp. 67-69, October 2009.

[5] C.C. Kilgus, "Shaped-Conical radiation pattern performance of the backfire quadrifilar helix," *IEEE Transaction on Antennas and Propagation*, vol. 23, pp. 392-397, May 1975.

[6] B. Jae-Hoon, B. Enkhbayar, and M. Dong-Hyun, "A compact GPS antenna for artillery projectile applications," *IEEE Antennas and Wireless Propagation Letters*, vol. 10, pp. 266-269, April 2011.

# A Novel Broadband High-Isolation Cross Dipole Utilizing Strong Mutual Coupling

Zengdi Bao, Xianzheng Zong and Zaiping Nie

University of Electronic Science and Technology of China

NO.4, Section 2, North Jianshe Road

Chengdu, Sichuan 610054 China

*Abstract*-A novel broadband cross dipole with high isolation and low cross polarization is presented in this paper. By skillfully utilizing the strong mutual coupling between the two dipoles, very good impedance match, sufficiently high isolation and low cross polarization can be achieved simultaneously. To get a directional radiation pattern, a metal reflector is placed under the antenna. The simulated measured results show that a more than 56% impedance bandwidth for SWR<1.5 from 1.66GHz to 2.96GHz is achieved, within which the port isolation is larger than 30dB. As an important advantage, the input impedance of the antenna can be changed within certain limits by properly adjusting the configuration, thus the antenna can be fed directly by 50-ohm coaxial cables without using impedance transformers. Particularly, the mechanisms for wide impedance bandwidth and high port isolation are discussed in this paper.

## I. INTRODUCTION

With the space resources for base stations decreasing, it is very important to demand that base station antennas can cover a wider frequency range and accommodate to more communication standards so as to avoid the repetitive constructions of telecom equipment. In recent years, the ±45° dual-polarized antennas were widely used to increase the communication capability and quality. Though significant development in the design of ±45° dual-polarized antennas has been observed [1]-[6], it is still difficult to achieve wide impedance bandwidth, high isolation, and low cross polarization simultaneously.

In this paper, we present a new design for dual-polarized antenna and discuss the mechanisms for wide bandwidth and high isolation from new points of view. The proposed antenna works in the frequency range from 1.66GHz to 2.96GHz, which can cover the DCS, PCS, UMTS, and part of LTE bands. And the input impedance of the antenna can be changed by properly adjusting its configurations, thus it can be fed directly by coaxial cables without using impedance transformers. From the structure point of view, the antenna is simple and compact and can be used to form an antenna array for base stations in wireless communications.

## II. ANTENNA CONFIGURATION

As shown in Fig. 1 (a), the proposed antenna consists of two orthogonally situated dipoles, a dielectric post, and a metal reflector. If this antenna is used as an element to form an array antenna for base stations, a certain number of elements would be linearly arranged in a reflector which should be some

longer than the array itself. In this paper, only one antenna with a truncated reflector is studied. The configuration details of the dipoles and the dielectric post are shown in Fig. 1 (b). The dielectric post is made of Teflon with relative permittivity $\varepsilon_r$=2.1 and is just used to support and fix the two dipoles above the reflector. Two through holes are drilled in the dielectric post, through which coaxial cables pass and then feed the dipoles respectively. On the metal reflector plane, there are also two holes, corresponding with the two above mentioned through holes, which are used to guide the cables behind the reflector. For either dipole, the outer conductor of its feeding coaxial cable is connected to one arm, and the inner connector is connected to the so-called copper connector, the far end of which is jointed with the other arm. The E-planes for dipole1 and dipole2 are φ=45°and φ=135° respectively according to the coordinate system shown in Fig. 1 (a), and correspondingly the H-planes are φ=135°and φ=45° respectively.

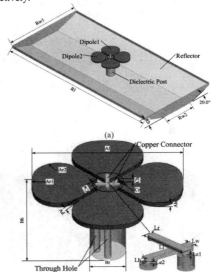

Figure 1. Configuration of the antenna: (a) overall diagram and (b) detail view; *Rl*=300, *Rw1*=145, *Rw2*=60, *Rh*=20, *Ah*=2.2, *Ad*=1.8, *Al*=62.8, *Ar1*=31.4, *Ar2*=10.5, *Br*=7.5, *Bh*=33.4, *Ch*=1.8, *Cw*=4.2, *Cr*=1.8, *Ll*=10.4, *Lw*=1.6, *Lh*=0.7, *Le1*=0.3, *Le2*=0.5, *Lr*=0.8, *Ld*=0.9 (Units: mm).

978-7-5641-4279-7

Since the two dipoles are orthogonally situated, the feeding structures for them have to be properly arranged in order to avoid intersecting. The feeding points of the dipole2 are lowered with a height of *Ch* comparatively with that of the dipole1. This is the only difference between the two dipoles. Since the configurations of the two dipoles are highly symmetrical, the port and radiation characteristics of them are highly symmetrical too.

### III. DISCUSSION OF WORKING MECHANISM

#### A. Impedance Bandwidth Enhancement by Coupling

In order to explain the reasons why the antenna we proposed has such a wide impedance bandwidth clearly, a comparison between the dual-polarized antenna (DPA) and the linear-polarized antenna (LPA) was made. The DPA is the antenna we proposed, the LPA is obtained simply by deleting the dipole2 (shown in Fig. 1 (b)) from the DPA while remaining all other configurations unchanged.

Fig. 2 shows the simulated S11 curves of the LPA and the DPA: in the frequency range from 1.65GHz to more than 3.4GHz, the DPA has a much better impedance match performance. Fig. 3 shows the current distribution on the DPA at center frequency 2.31GHz when only one dipole is fed and the other is terminated with a 50-ohm load. It can be found that current distribution with large magnitude appears on the unfed dipole. So there is strong mutual coupling between the two dipoles. As mentioned above, the only difference between the LPA and the DPA is that the LPA has only one dipole, while the DPA has two orthogonal dipoles. Therefore, the reason why the DPA has a much wider impedance bandwidth becomes clear. That the orthogonally situated dipole2 works

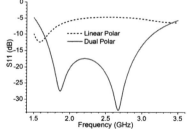

Figure 2. Comparison of the simulated S11 curves of the LPA and DPA.

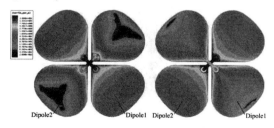

(a)                          (b)

Figure 3. Current distribution on the two dipoles at 2.31GHz when (a) only dipole1 is fed and (b) only the dipole2 is fed and the other dipole is terminated with a 50-ohm load.

as a parasitic element for dipole1 greatly extends the impedance bandwidth of the latter: due to the existence of parasitic element, another resonant frequency is introduced. And vice versa, the dipole1 will also extend the impedance bandwidth of the dipole2.

In broadband antenna design, various bandwidth enhancement techniques based on passive parasitic structures are widely used. However, in this paper, both of the two dipoles have active feed-port, and they are parasitic to each other simultaneously. The effective utilizing of the mutual coupling between the dipoles can improve the bandwidth without the aid of other passive parasitic structures, which is helpful to make the design more compact.

#### B. Isolation Analysis

As analyzed above, with the aid of the strong mutual coupling between the two dipoles, the impedance bandwidths of both of them are greatly extended. In traditional designs, strong mutual coupling between the two dipoles will lead to poor isolation between them. However, high isolation still can be achieved in this design. See Fig. 4, apparently, the current distribution on the dipole2 is coupled from the dipole1. For convenience, we call the currents on dipole1 *Excited Current Vectors* (ECV), and call the currents on dipole2 *Coupled Current Vectors* (CCV). The distribution is extracted at t=T/8, where T is the length of one oscillation period. As illustrated in Fig. 4, the CCV distribution on the two arms of the dipole2 are strictly symmetrical with respect to the center of the dipole2, which means that there is no potential difference between the two feeding points of the dipole2. Therefore, the induced currents on the dipole2 cannot enter the feeding port. Vice versa, when the dipole2 is fed, the induced currents will not enter the feeding port of the dipole1 for the same reasons.

Though only the current distributions at t=T/4 is presented, distributions at other time are similar: the CCV distribution on the dipole2 is strictly symmetrical with respect to the center of the dipole2 all the time. So, high isolation between the dipole1 and the dipole2 can be achieved in the entire period. Though only the case at 2.31GHz is discussed, current distributions at other frequencies are similar. Hence, sufficiently high isolation can be achieved in the entire working frequency range.

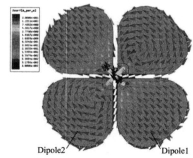

Dipole2                          Dipole1

Figure 4. Vector-form current distribution on the two dipoles at 2.31GHz when only the dipole1 is fed and the dipole2 is terminated with a 50-ohm load.

## IV. PARAMETER STUDY

Since dipole1 and dipole2 have highly symmetric port and radiation characteristics, only dipole1 is studied. When one parameter is studied, all the other uncorrelated parameters keep constant. The results of these studies offer a useful guideline for practical design.

### A. Fillet Radius Ar2

Shown in Fig. 1 (b), *Ar2* is the Fillet radius on the arms of the two dipoles. *Ar2* is more sensitive in effects on impedance match than the others parameters. As shown in Fig. 5, it influences the port isolation slightly, but can make the impedance match change remarkably. The upper resonant frequency moves from 2.5GHz to 3.05GHz as *Ar2* increasing from 9.0mm to 12.0mm and a wider impedance bandwidth can be achieved. However, the impedance match in the middle part of the frequency band becomes worse at the same time, which must be taken into consideration when adjusting *Ar2* to get a wider impedance bandwidth.

### B. Arm Spacing Distance Ad

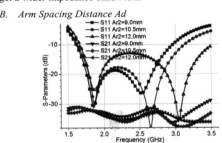

Figure 5. Effects of *Ar2* on S11 and S21 of the antenna.

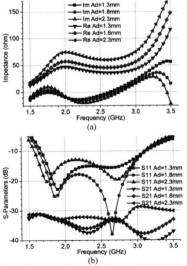

(a)

(b)

Figure 6. Effects of *Ad* on (a) input impedance and (b) S11 and S21 of the antenna.

By changing *Ad*, *Ah*, or the angel of the two parallel adjacent sides of two arms, the input impedance of the antenna can be changed. For instance, Fig. 6 (a) shows the input impedance of the dipole1 with varied *Ad*. Shown in Fig. 1 (b), *Ad* is the spacing distance between the two parallel adjacent sides. See Fig. 6, the impedance real-part curve moves upward remarkably when *Ad* increases. For *Ad*=1.8mm case, in the working frequency band from 1.66GHz to 2.96GHz, the impedance real-part is closer to 50 Ohms and the imaginary-part is closer to zero, which is helpful to get a better match with general coaxial cables. By properly adjusting the configuration details, the input impedance can be changed conveniently, which indicates the antenna can be fed directly by coaxial cables without using impedance transformers. For this reason, the antenna has the advantage of compactness. Fig. 6 (b) shows S11 and S21 of the antenna corresponding to Fig. 6 (a). The S11 curves vary remarkably when *Ad* is changing. The upper resonant frequency around 2.7GHz disappears when *Ad* is too small, which should be avoided.

## V. SIMULATED AND MEASURED RESULTS

To verify the design, a prototype of the proposed antenna was fabricated and measured. The photograph of the fabricated antenna is shown in Fig. 7. Limited by the manufacture conditions, the manufacture precision of the antenna is not very good. The port characteristics and the radiation characteristics of the antenna are measured by Agilent E5071C network analyzer and SATIMO StarLab antenna near field measurement system respectively.

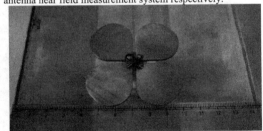

Figure 7. Photograph of the fabricated antenna

### A. Port Characteristics

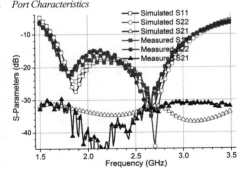

Figure 8. Simulated and measured S-Parameters of the antenna.

Fig. 8 shows the simulated and measured S-Parameters of the antenna. It can be observed that the simulated and measured impedance matches agree very well. The simulated and measured S21 do not agree very well, but both of them are less than -30dB across the working frequency band. The measured results show that more than 56% impedance bandwidth from 1.66GHz to 2.96GHz with SWR<1.5 and isolation>30dB (i.e., S11, S22<-14dB and S21<-30dB) is achieved for both dipole1 and dipole2 of the antenna. Very good impedance match and sufficiently high isolation are achieved.

### B. Port Characteristics

The dipole 2 is terminated with a 50-ohm load when the radiation patterns of the dipole 1 are being tested, and vice versa. Since the radiation patterns of the two dipoles are highly symmetrical, only the radiation patterns of the dipole1 are shown here. As illustrated in Fig. 9 (a)-(f), the measured and simulated radiation patterns agreed well except for the cross polarization patterns and backward co-polarization patterns at 1.66GHz. The peak gain of the antenna at 2.31GHz is about 8.3dBi and varies within ±0.5dBi across the working frequency range. The 3dB beamwidth varies from 58.2° to 72.6° in E-plane, and varies from 75.2° to 84.3°in H-plane across the working frequency range. Relatively stable radiation patterns are achieved. Larger than 25dB front-to-back ratio is achieved at 2.31GHz and 2.96GHz. And at 0° direction, the cross polarization at 2.31GHz is lower than -28dB and lower than -24dB at 2.95GHz. Low cross polarization is achieved.

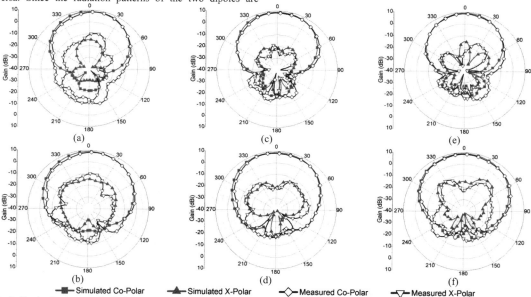

Fig. 9. Simulated and measured radiation patterns of dipole1 at (a) 1.66GHz E-plane, (b) 1.66GHz H-plane, (c) 2.31GHz E-plane, (d) 2.31GHz H-plane, (e) 2.96 E-plane, (f) 2.96GHz H-plane.

## VI. Conclusion

In this paper, a broadband cross dipole with high isolation and low cross polarization is proposed. It has a compact and simple structure and can be fed directly by coaxial cables without using impedance transformers. The presented design can be used to form an antenna array for base stations in mobile wireless communications. The wide impedance bandwidth and high port isolation mechanisms are also discussed in this paper, which are also appropriate for other antennas with similar patterns.

### References

[1] H. Wong, K. L. Lau, and K. M. Luk, "Design of Dual-Polarized L-Probe Patch Antenna Arrays With High Isolation," *IEEE Trans Antennas Propag.*, vol. 52, no. 1, pp. 45-52, Jan. 2004.

[2] C. Sim, C. Chang and J. Row, "Dual-Feed Dual-Polarized Patch Antenna With Low Cross Polarization and High Isolation," *IEEE Trans. Antennas Propag.*, vol. 57, no. 10, pp. 3321-3324, Oct. 2009.

[3] B. Li, Y. Z. Yin, W. Hu, Y. Ding, and Y. Zhao, "Wideband Dual-Polarized Patch Antenna With Low Cross Polarization and High Isolation," *IEEE Antennas Wireless Propag. Lett.*, vol. 11, pp. 427-430, 2012.

[4] H. Huang, Z. Niu, B. Bai, and J. Zhang, "Novel broadband dual-polarized dipole antenna," *Micorwave Opt. Technol. Lett.*, vol. 53, no. 1, pp. 148-150, Jan 2011.

[5] B. Q. Wu and K. M. Luk, "A Broadband Dual-Polarized Magneto-Electric Dipole Antenna With Simple Feeds," *IEEE Antennas Wireless Porpag. lett.* vol. 8, pp. 60-63, 2009.

[6] G. Adamiuk, T. Zwick, and W. Wiesbeck, "Compact, Dual-Polarized UWB-Antenna, Embedded in a Dielectric," *IEEE Trans. Antennas Propag.*, vol. 58, no. 2, pp. 279-286, Fab. 2010.

# Design of Feed Horn Integrated 380 GHz Sub-Harmonically Pumped Mixer Cavity

Xiaofan Yang[1],* Liandong Wang[1], Bo Zhang[2] and Xiong Xu[1]

[1]State Key Laboratory of Complex Electromagnetic Environment Effects on Electronic and Information System
Luoyang Electronic Equipment Test Center, Luoyang, 471003, China

[2]EHF Key Laboratory of Fundamental Science
University of Electronic Science and Technology of China, Chengdu, 611731, China

* E-mail: xiaofan_uestc@sina.com

*Abstract-* **This paper presents the design and fabrication of feed horn antenna integrated 380 GHz sub-harmonically pumped mixer's split cavity block. The mixer circuit is fully integrated with the microstrip circuit and the flip-chipped diode (planar GaAs air-bridged Schottky anti-parallel diode, which was designed and fabricated by millimeter technology group, Rutherford Appleton Laboratory, UK) on suspended 50μm thick quartz substrate. The mixer's split cavity block is designed using ANSOFT's three-dimensional full-wave electromagnetic simulation software HFSS, as well as 3D CAD design software SOLIDWORKS. The mixer cavity and its circuit are fabricated and assembled at MMT's Precision Development Facility, UK Rutherford Appleton Laboratory. Mixer circuit configuration and cavity block are detailed in the paper. The simulation investigation shows state-of-the-art results obtained. A best conversion loss of 6.5dB is achieved with 3mW LO power at 383 GHz.**

## I. INTRODUCTION

Terahertz (THz) technology is a new research area which developed rapidly during last two decades. This topic involves electromagnetism radiation [1], spectroscopy [2], imaging [3], semiconductor physics [4], materials science [5], and biology [6] etc. THz covers wideband electromagnetic radiation area from 100 GHz to 10 THz, which ends connect to the microwave/millimeter wave and infrared/visible light, respectively. For a long time, there formed the "THz gap" in the application electromagnetic spectrum due to the lack of effective THz sources and detection methods.

When THz wave propagates in the atmosphere, there would be certain attenuation due to resonance with gas molecules, thus formed THz atmospheric attenuation characteristics. Between the atmosphere THz wave absorption lines as shown in Fig. 1, there are some relatively minor atmospheric absorption bands, these minimum value attenuation peaks are called atmospheric window. In the millimeter/THz band, there are several atmospheric windows, distribute at the center frequencies of 94 GHz, 118 GHz, 140 GHz, 183 GHz, 225 GHz, 380 GHz etc., and the relative bandwidth reaches about 20%, even 70% [7]. These bands will make great potential

application in receivers. In this paper, 380 GHz was selected as our sub-harmonic mixer's working frequency.

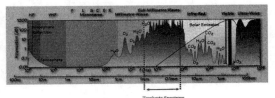

Fig. 1 THz Gap and atmosphere window frequencies

As shown in Fig. 2, among various THz application systems, such as communication, radar, astronomical observation System etc., the most important and also the first problem should be solved is how to realize THz signal down-transformation, which requires THz receiver front-end. Sub-harmonically pumped (SHP) mixers employing an anti-parallel Schottky diode pair are key components for millimeter/ THz wave heterodyne receivers, which using non-linear effect to realize THz signal down-transformation.

**THz Communication、Radar、Astronomical observation System**

Firstly question

**THz signal down-conversion**

Realized

**Heterodyne Receiver**

Key component

**Planar Schottky diode based THz mixer**

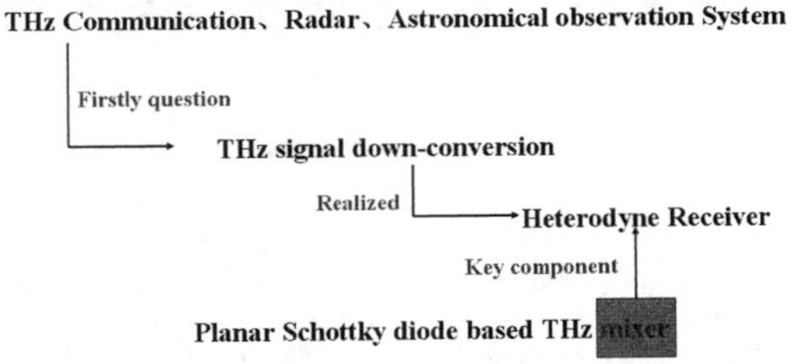

Fig. 2 THz system signal down-transformation using mixer

This paper presents the design and fabrication of feed horn antenna integrated suspended low-loss fixed-tuned 380GHz sub-harmonically pumped mixer's split cavity block. A best double-sideband mixer simulation loss of 6.5dB was achieved at 383GHz with 3mW LO power.

978-7-5641-4279-7

## II.  INTEGRATED FEED HORN DESIGN

The 380 GHz feed horn antenna integrated mixer is fixed-tuned, using least parts to minimize the cost, as well as maximize its potential convenience for circuit and block manufacture. The mixer's RF input port connects to the feed horn. The feed horn antenna modeling is shown in Fig. 3.

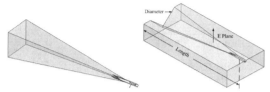

Fig. 3 Feed horn antenna modeling

**Tab. 1 VDI feed horn parameters under certain standard waveguide**

| Waveguide Type | Frequency Range (GHz) | Horn Type | Horn Length (mm) | Aperture Diameter |
|---|---|---|---|---|
| WR-4.3 | 170-260 | Conical | 16.5mm | 7.1mm |
| WR-2.8 | 260-400 | Diagonal | 21.4mm | 4.6mm |

| Waveguide Type | TaperHalf-Angle | Full 3dB Bandwidth | Gain | Beam Waist |
|---|---|---|---|---|
| WR-4.3 | 12.1deg | 13deg | 21dB | 2.7mm |
| WR-2.8 | 6.1deg | 10deg | 26dB | 1.9mm |

Reference to U.S. Virginia Diodes, Inc. feed horn style and parameters under certain standard waveguide shown in Tab. 1, this feed horn is design and fabricated, using additional special feed horn attached, as shown in Fig. 4.

Fig. 4 Feed horn integrated 380GHz mixer

## III.  MIXER CAVITY DESIGN

Thanks to the improvement of 3D electromagnetic solvers and nonlinear circuit simulators, fix tuned SHP mixers using discrete or integrated planar Schottky diodes have already demonstrated lower conversion loss than traditional mixers using mechanically tunable back-shorts [8]. The discrete planar diode could be used for low conversion loss fixed-tuned mixers while providing significant cost reduction, up to 600GHz [9].

The SHP mixer cavity would be realized by a two way split block, split among the interface of the 50μm thick quartz substrate (Relative dielectric constant 3.78) is located. The designed SHP Mixer's split cavity block is shown in Fig. 5.

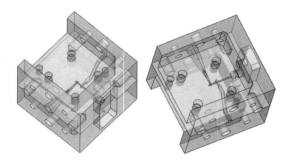

Fig. 5 The designed SHP mixer's split cavity block

The anti-parallel diode pair is flip-chipped onto the quartz substrate, connected with RF/LO relevant circuits. Two step-impedance line low-pass filters are used to block the RF and LO frequency, respectively. The output IF signal would be via a sparkplug-style K connector with glass bead and sliding contact through the low-pass LO filter. This feed horn antenna integrated suspended low-loss fixed-tuned 380GHz sub-harmonically pumped mixer's two ways split block shown in Fig. 6, its circuit configuration shown in Fig. 7.

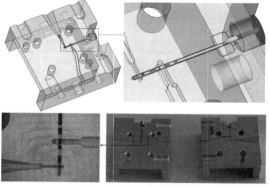

Fig. 6 The suspended quartz circuit in mixer split cavity

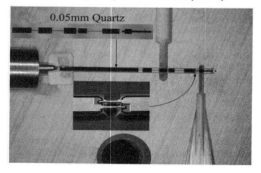

Fig. 7 The designed 380 GHz SHP mixer circuit configuration

## IV. SIMULATION RESULTS

The 380 GHz SHP mixer after assemble shown in Fig. 8. The whole mixer cavity block dimension is 20mm×20mm× 20mm. Cavity up-side is integrated rectangular RF feed horn antenna, down-side is LO input waveguide and left-side is sparkplug-style K connector to the output IF signal.

Fig. 8 The 380 GHz SHP mixer after assemble

The performance of the mixer has been unite-simulated by Agilent's ADS and Ansoft's HFSS. 380 GHz SHP mixer's system level modeling is studied and the state-of-the-art simulation results are obtained. The best simulation conversion loss achieved is about 6.5dB, at 383 GHz with 3mW LO power, as shown in Fig. 9. This state-of-the-art simulation results [10, 11] are attributed to low parasitic UK RAL planar GaAs air-bridged Schottky anti-parallel pair diode AP1/G2, low-loss quartz suspended microstrip circuit, and system level mixer modeling.

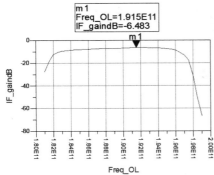

Fig. 9 Conversion loss simulation results of the designed SHP mixer

## IV. CONCLUSION

This paper presents the design method of feed horn antenna integrated suspended low-loss fixed-tuned 380GHz sub-harmonically pumped mixer's split cavity block. The assembled mixer cavity and inner circuit are detailed. Simulation results show best double-sideband mixer loss of 6.5dB at 383GHz with 3mW LO power.

## ACKNOWLEDGMENT

The authors thank Dr. Byron Alderman from Teratech Componets Ltd. U. K. for the diodes and technical support. This work is sponsored in part by the National High-Technology Research and Development Program of China (863 Program) and in part by the National Natural Science Foundation of China under Grant Nos. 60671034 and 60901022.

## REFERENCES

[1] Michele Ortolani, Alessandra Di Gaspare, Ennio Giovine, Florestano Evangelisti and Vittorio Foglietti, et al., "Study of the Coupling of Terahertz Radiation to Heterostructure Transistors with a Free Electron Laser Source", *Journal of Infrared, Millimeter and Terahertz Waves*, Volume 30, Number 12, 2009, pp. 1362-1373.

[2] Svilen Petrov Sabchevski and Toshitaka Idehara, "Design of a Compact Sub-Terahertz Gyrotron for Spectroscopic Applications", *Journal of Infrared, Millimeter and Terahertz Waves*, Volume 31, Number 8, 2010, pp. 934-948.

[3] Christian am Weg, Wolff von Spiegel, Ralf Henneberger, Ralf Zimmermann and Torsten Loeffler, et al., "Fast Active THz Cameras with Ranging Capabilities", *Journal of Infrared, Journal of Infrared, Millimeter and Terahertz Waves*, Volume 30, Number 12, 2009, pp. 1281-1296.

[4] Dietze. D, Darmo. J, Unterrainer. K, "Guided Modes in Layered Semiconductor Terahertz Structures", *IEEE Journal of Quantum Electronics*, Volume: 46 , Issue: 5, 2010, pp. 618-625.

[5] Atakaramians. S, Vahid. S.A, Ebendorff-Heidepriem. H, Fischer. B.M, Monro. T, Abbott. D, "Terahertz Waveguides and Materials", *14th International Conference on Teraherz Electronics*, IRMMW-THz, 2006, pp. 2262-2266.

[6] Alfonsina Ramundo Orlando and Gian Piero Gallerano, "Terahertz Radiation Effects and Biological Applications", *Journal of Infrared, Millimeter and Terahertz Waves*, Volume 30, Number 12, 2009, pp. 1308-1318.

[7] T.W Crowe, W.L Bishop, and D.W Porterfield, "Opening the Terahertz window With Integrated Diode Circuits, "*IEEE Journal of Solid-State Circuits*, Volume 40, Issue 10, 2005, pp.2104-2110.

[8] J. Hesler, W.R. Hall, T.W. Crowe, R.M. Weikle, B.S. Deaver, Jr., R.F. Bradley, and S.K. Pan, "Fixed-tuned submillimeter wavelength waveguide mixers using plannar schottky-barrier diodes," *IEEE Trans. on Microwave Theory Tech.*, Volume 45, Issue 5, 1997, pp. 653-658.

[9] D. Poterfield, J. Hesler, T.W. Crowe, W. Bishop, and D. Woolard, "Intergrated terahertz transmit/receive modules", *Proc. 33rd Euro. Microwave Conf.*, Munich, Germany, 2003, pp. 1319-1322.

[10] B. Thomas, A. Maestrini, and G. Beaudin, "A low-noise fixed-tuned 300-360-GHz sub-harmonic mixer using planar Schottky diodes," *IEEE Microwave and Wireless Components Letters*, Volume 15, Issue 12, 2005, pp. 865-867.

[11] S. Marsh, B. Alderman, D. Matheson, and P. de Maagt, "Design of low-cost 183 GHz subharmonic mixers for commercial applications," *IET Circuits Devices Syst.*, Volume 1, Issue 1, 2007, pp. 1-6.

**Xiaofan Yang** Research Assistant of State Key Laboratory of Complex Electromagnetic Environment Effects on Electronic and Information System. He received PhD degree from UESTC (University of Electronic Science and Technology of China) at 2012. He has been as visiting scientist to Space Science and Technology Department, Rutherford Appleton Laboratory, UK. His research interests including millimeter-wave/terherz-wave componets and communication systems, as well as information system's electromagnetic environment effects & mechanism.

# Millimeter-Wave Cavity-Backed Patch-Slot Dipole for Circularly Polarized Radiation

Xue Bai and Shi-Wei Qu

School of Electronic Engineering, University of Electronic Science & Technology

Chengdu, China

dyon.qu@gmail.com

*Abstract*—A novel millimeter-wave antenna for circularly polarized (CP) radiation is presented. The proposed antenna is composed of a patch dipole and a slot dipole, and it is realized by a plated through hole printed technique on a microwave substrate. Simulations show that the antenna can present a bandwidth from 55.8 to 65.3 GHz for a reflection coefficient ($S_{11}$) ≤ -10 dB as well as for an axial ratio (AR) ≤ 3dB, and the 9.6 dBi average gain is achieved. This design yields good directional radiation patterns and low fabrication cost. Studies on several critical parameters are performed for practical designs.

*Index Terms*—Cavity-backed antennas, circular polarization, millimeter wave.

## I. INTRODUCTION

Due to large capacity and high speed [1], the electromagnetic spectrum of 60-GHz band has drawn more and more attention. Meanwhile, circularly polarized (CP) radiation is very desirable for 60-GHz antennas because of its capabilities to reduce polarization mismatch and to suppress multipath interferences [2], for example they allow more flexible orientation of the transmitting and receiving antennas.

In microwave frequency bands, there are many ways to construct CP antennas, e.g., patches [3]-[6], crossed dipoles [7]-[10], and loop antennas [11]-[13]. However, it is difficult to construct these antennas in millimeter-wave (mm-wave) frequency bands, firstly because the very fine structures, especially the feeding network, cannot be realized. Moreover, a 90° power divider [8] or a balun [10] is necessary for the crossed dipoles, which obviously complicates the design.

On the other hand, on-chip CP antennas are popular due to the ease of system integration [14]. However, most of the on-chip antennas would suffer from low radiation gain. Other mm-wave off-chip CP antennas are typically expensive, owing to their tiny element size or preciseness in feed structures [15], [16] or high cost in micromachining process [17]. Although some low-cost designs have been proposed [18], the radiation patterns are unstable across the operating frequency band.

In this letter, a new implementation of cavity-backed patch-slot antenna on a substrate using plated through hole printed technology [19] is introduced for mm-wave CP applications. The CP radiation of the proposed antenna is realized by overlapping a patch dipole and a slot dipole to achieve two orthogonal electric field components and then by adjusting the connection between the patch and the slot to introduce 90° phase difference. The design yields good directional radiation patterns, low fabrication cost, and a 3-dB axial ratio (AR)

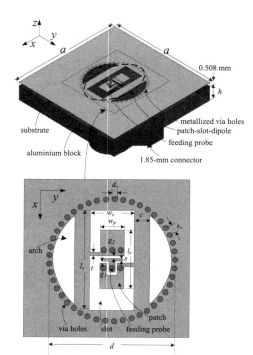

Fig. 1. Geometry of the proposed cavity backed patch-slot antenna for CP radiation. $a$ = 16 (3.2$\lambda_o$), $d$ = 8 (1.6$\lambda_o$), $h$ = 3.23, $w_s$ = 2.8 (0.56$\lambda_o$), $l_s$ = 6.5 (1.3$\lambda_o$), $w_p$ = 1.5 (0.3$\lambda_o$), $l_p$ = 3.9 (0.78$\lambda_o$), $c$ = 1, $t$ = 0.1, $s$ = 0.15, $g_1$ = 0.05, $g_2$ = 0.25, $d_v$ = 0.3, $s_v$ = 0.56 (in mm, $\lambda_o$ is the free-space wavelength at 60 GHz).

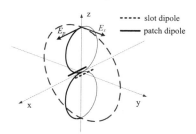

Fig. 2. Superposition of the electric fields of the patch and slot dipoles.

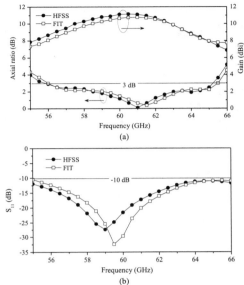

(a)

(b)

Fig. 3. Simulated results by FIT and HFSS, (a) simulated axial ratio and broadside gain, (b) simulated $S_{11}$.

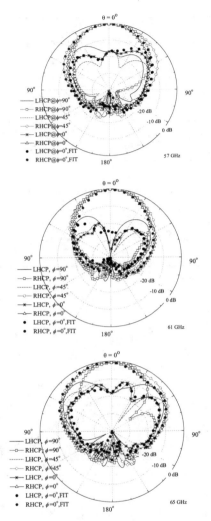

Fig. 4. Simulated radiation patterns by FIT and HFSS of the proposed antenna at 57, 61, and 65 GHz, respectively.

bandwidth covering from 55.8 to 65.3 GHz, in which the reflection coefficient ($S_{11}$) is kept below -10 dB.

## II. GEOMETRY

Fig. 1 shows geometry of the proposed cavity-backed patch-slot dipole antenna, and it consists of a patch-slot dipole, a cavity formed by metalized via holes, an aluminum block and a 1.85-mm connector for measurements. The patch dipole in [19] is introduced. Five metallic vias located at the center of the dipole are shorted to the ground plane. A special via hole acting as the feeding pin, is connected with a T-shaped coupled strip. The width and the length of the patch are $w_p$ and $l_p$ while those of the slot are $w_s$ and $l_s$, respectively. To excite the slot dipole, two narrow strips with a width $t$ connect the two broad edges of the slot to each arm of the patch dipole. Two arches are cut away beside the patch-slot dipole to make full advantage of the cavity. They are etched on a Rogers RT/duroid 5880(tm) substrate with an area of $a \times a$ (relative permittivity 2.2, thickness 0.508 mm, and 9μm cladding copper layer). The diameters of the cavity is $d$, and the height of the cavity shares the same value with the substrate, i.e., 0.508 mm.

Additionally, an aluminum block for supporting the antenna is mounted on the bottom of the substrate using conductive glue. A 50-Ω glass insulator with a probe (0.3 mm in diameter) is used to make sure accuracy of the feeding structure. Therefore, the height of the block should be designed according to the probe length of the glass insulator, i.e., $h =$ 3.23 mm.

Fig. 2 depicts the design principle, in which a patch dipole and a slot dipole with complementary radiation characteristics are superposed. The radiation pattern of the patch dipole is in a shape of character "8" in the E plane (solid line) and of character "O" in the H plane, whereas the pattern of the slot dipole looks like "O" in the E plane (dash line) and "8" in the H plane. Once the two dipoles are excited with proper amplitudes and phases, the CP radiation can be achieved in z-axis direction.

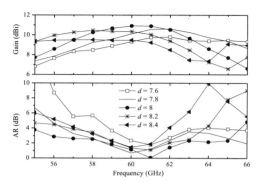

Fig. 5.  Influence of $d$ on broadside gain and axial ratio.

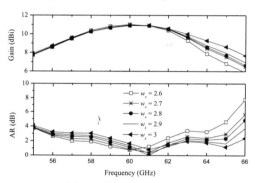

Fig. 6.  Influence of $w_s$ on broadside gain and axial ratio.

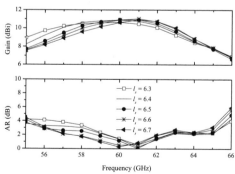

Fig. 7.  Influence of $l_s$ on broadside gain and axial ratio.

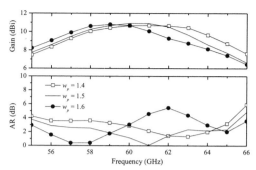

Fig. 8.  Influence of $w_p$ on broadside gain and axial ratio.

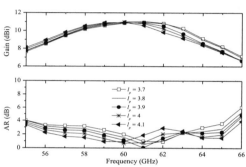

Fig. 9.  Influence of $l_p$ on broadside gain and axial ratio.

## III.  SIMULATED RESULTS

All simulations and optimizations are performed by the commercial software HFSS$^{TM}$ using the finite element method (FEM). The HFSS results are verified by the finite integration technique (FIT) additionally. The optimized parameters are shown in the caption of Fig. 1. The simulated broadside gains and ARs are shown in Fig. 3(a). The simulated frequency bandwidth for AR $\leq$ 3 dB by FEM is 15.7%, covering from 55.8 to 65.3 GHz, in which the broadside gain is between 8 and 11.2 dBi with a maximum at 60.5 GHz. Comparatively, the FIT result is from 56.2 to 65.5 GHz, meanwhile the broadside gain is lower by around 0.5 dBi at frequencies below 61.5 GHz and matches the FEM results well at higher frequencies. Fig. 3(b) shows the simulated $S_{11}$ of the proposed antenna, and the simulated $S_{11}$ by FIT matches the FEM results reasonably. They are kept below -10 dB across the corresponding 3-dB AR bandwidth.

The simulated radiation patterns at 57, 61, and 65 GHz are shown in Fig. 4. The simulated patterns by FIT in $\varphi = 0°$ plane agree well with FEM results at these frequencies. It can be seen from the figure that the antenna presents symmetrical radiation patterns, small backward radiation (over -20 dB), without side lobes at lower frequencies. However, at higher frequencies, e.g., at 64 GHz, due to the increase of antenna electric size, the pattern in $\varphi = 90°$ plane is slightly distorted, but it still features a broadside radiation.

## IV.  PARAMETRIC STUDIES

To clearly show how each parameter controls the performance of the antenna, five critical parameters are studied. When one parameter'is studied, the others are kept to their optimized values in the caption of Fig. 1. Generally, the radiation performances, e.g., radiation patterns, broadside gain, axial ratio, and aperture efficiency are directly determined by the electric field distribution in the antenna aperture. Therefore, variations of the aperture dimensions will dramatically change its performances, as shown in Fig. 5. Too large or too small value of $d$ not only significantly degrades the 3-dB AR bandwidth, but reduces the antenna gain. Note that according to our studies, a small variation of the parameters does not cause significant effects on the antenna impedance matching. Therefore, during the parametric studies, $S_{11}$ are not given.

Fig. 6 shows the broadside gain and AR with different values of the width of the slot dipole $w_s$. At frequencies below 61 GHz, variation of $w_s$ does not cause many effects on the broadside gain, but a larger $w_s$ will worsen the CP performance. At frequencies above 61 GHz, a larger $w_s$ is of benefit to both higher gain and lower AR.

The influence of the length of the slot dipole $l_s$ on broadside gain and AR is shown in Fig. 7. As $l_s$ is decreased, the maximum broadside gain is shifted downwards. Meanwhile, for a smaller $l_s$, it features a good AR at higher frequencies but a poor AR at lower frequencies.

The influence of the width of the patch dipole $w_p$ on AR and gain is shown in Fig. 8. It can be seen that the CP performance is sensitive to the value of $w_p$, e.g., at 62 GHz as $w_p$ is varied from 1.5 to 1.6 mm, the AR is significantly enlarged by 4 dB.

Fig. 9 shows the broadside gain and AR with different $l_p$, i.e., the length of the patch dipole. A larger $l_p$ can obviously increase the 3-dB AR bandwidth, but at the middle frequencies it will worsen the CP performance. Meanwhile, at frequencies below 60 GHz the variation of $l_p$ does not cause much effects on the broadside gain at lower frequencies, but at higher frequencies a smaller $l_p$ is slightly of benefit to higher gain.

## V. CONCLUSION

A novel CP cavity-backed antenna composed of a patch dipole and a slot dipole for millimeter wave applications is investigated. The vertical walls of the cavity are realized by a low-cost fabrication process of plated through hole printed technology on a microwave substrate. Its available 3-dB AR bandwidth is from 55.8 to 65.3 GHz, in which the $S_{11}$ is kept below -10 dB, and the 9.6 dBi average gain has been achieved.

## ACKNOWLEDGMENT

This work was partly supported by the Natural Science Foundation of China (NSFC) Projects under No. 61101036, and partly by the General Financial Grant from the China Postdoctoral Science Foundation under No. 2012M521682.

## REFERENCES

[1] T. Manabe, K. Sato, H. Masuzawa, K. Taira, K.-C. Huang, and D.J. Edwards, *Millimeter Wave Antennas for Gigabit Wireless Communications*. Chichester, U.K.: Wiley, 2008.

[2] T. Ihara, Y. Kasashima, K. Yamaki, "Polarization dependence of multipath propagation and high-speed transmission characteristics of indoor millimeter-wave channel at 60 GHz," *IEEE Trans. Vehicular Technol.*, vol. 44, pp. 268–274, May 1995.

[3] Nasimuddin, Z.-N. Chen, and X. Qing, "A compact circularly polarized cross-shaped slotted microstrip antenna," *IEEE Trans. Antennas Propag.*, vol. 60, pp. 1584–1588, Mar. 2012.

[4] K. L. Chung, "A wideband circularly polarized H-shaped patch antenna," *IEEE Trans. Antennas Propag.* vol. 58, pp.3379–3383, Sep. 2010.

[5] T.-N. Chang and J.-M. Lin, "Circular polarized ring-patch antenna," *IEEE Antennas Wirel. Propag. Lett.*, vol. 11, pp. 26-29, 2012.

[6] K.-Y. Lam, K.-M. Luk, K.-F. Lee, H. Wong and K. B. Ng, "Small circular polarized u-slot wideband patch antenna," *IEEE Antennas Wirel. Propag. Lett.*, vol. 10, pp. 87–90, 2011.

[7] J. Zhang, H.-C. Yang and D. Yang, "Design of a new high-gain circularly polarized antenna for inmarsat communication," *IEEE Antennas Wirel. Propag. Lett.*, vol. 11, pp.350–353, 2012.

[8] K. M. Mak and M.-M. Luk, "A Circularly polarized antenna with wide axial ratio beamwidth," *IEEE Trans. Antennas Propag.* vol. 57, pp.3309–3312, Oct. 2009.

[9] R. Li, D. C. Thompson, J. Papapolmerou, J. Lasker and M. M. Tentzeris "A circularly polarized short backfire antenna excited by an unbalance-fed cross aperture," *IEEE Trans. Antennas Propag.*, vol. 54, pp. 852–859, Mar. 2006.

[10] S.-W. Qu, C. H. Chan, and Q. Xue, "Wideband and high-gain composite cavity-backed crossed triangular bowtie dipoles for circularly polarized radiation," *IEEE Trans. Antennas Propag.*, vol. 58, pp. 3157–3164, Oct. 2010.

[11] Q. Yang, X. Zhang, N. Wang, X. Bai, J. Li, and X. Zhao, "Cavity-backed circularly polarized self-phased four-loop antenna for gain enhancement," *IEEE Trans. Antennas Propag.*, vol. 59, pp. 685–688, Feb. 2011.

[12] S.-W. Qu, J.-L. Li, C. H. Chan, and Q. Xue, "Cavity-backed circularly polarized dual-loop antenna with wide tunable range," *IEEE Antennas Wirel. Propag. Lett.*, vol. 7, pp.761–763, 2008.

[13] R. Li, B. Pan, J. Papapolmerou, J. Lasker and M. M. Tentzeris, "Development of a cavity-backed broadband circularly polarized slot/strip loop antenna with a simple feeding structure," *IEEE Trans. Antennas Propag.*, vol. 56, pp. 312–318, Feb. 2008.

[14] X. Y. Bao, *et al.*, "60-GHz AMC-based circularly polarized on-chip antenna using standard 0.18- μm CMOS technology," *IEEE Tran. Antennas Propagat.*, vol. 60, no. 5, pp. 2234–2242, May 2012.

[15] C. Liu, Y.-X. Guo, X. Bao, and S.-Q. Xiao, "60-GHz LTCC integrated circularly polarized helical antenna array," *IEEE Tran. Antennas Propagat.*, vol. 60, no. 3, pp. 1329–1336, Mar. 2012.

[16] Y. Li, Z. N. Chen, X. Qing, Z. Zhang, J Xu, and Z Feng, "Axial ratio bandwidth enhancement of 60-GHz substrate integrated waveguide-fed circularly polarized LTCC antenna array," *IEEE Tran. Antennas Propagat.*, vol. 60, no. 10, pp. 4169–4177, Oct. 2012.

[17] M. Sun, Y.-O. Zhang, Y.-X. Guo, M. F. Karim, O. L. Chuen, and M. S. Leong, "Axial ratio bandwidth enhancement of 60-GHz substrate integrated waveguide-fed circularly polarized LTCC antenna array," *IEEE Tran. Antennas Propagat.*, vol. 59, no. 8, pp. 3083–3089, Aug. 2011.

[18] A. D. Nesic and D. A. nesic, "Printed planar 8 × 8 array antenna with circular polarization for millimeter-wave application," *IEEE Antennas Wirel. Propag. Lett.*, vol. 11, pp. 744–747, 2012.

[19] K. B. Ng, H. Wong, K. K. So, C. H. Chan, and K. M. Luk, "60 GHz plated through hole printed magneto-electric dipole antenna," *IEEE Trans. Antennas Propag.*, vol. 60, pp. 3129–3137, Jul. 2012.

# Compact Wideband Millimeter-Wave Substrate Integrated Waveguide Fed Interdigital Cavity Antenna Array

Yang Cai, Zuping Qian, Yingsong Zhang, and Dongfang Guan

Institute of Communications Engineering, PLA University of Science and Technology Nanjing, Jiangsu 210007, China

*Abstract*-**In this paper, a compact wideband millimeter-wave substrate integrated waveguide (SIW) interdigital cavity antenna based on a novel coplanar waveguide (CPW) fed power divider is presented. First, through combining the CPW-slotline transition and the slotline-SIW transition, a two-way power divider with $S_{11}$ <-15dB from 34.6 GHz to 37.1 GHz and good balance performance is designed. Then, the substrate integrated cavities (SIC) are placed interdigitally based on the power divider, which makes the antenna array structure much compact. In the end, this antenna array is fabricated and measured. The results show that the proposed antenna yields an impedance bandwidth of 33.3-38 GHz (13.1%). The compact wideband millimeter-wave SIW interdigital cavity antenna indicates a wide application foreground as its distinguish compact-size feature especially when the cavity number is relatively big.**

## I. INTRODUCTION

Millimeter-wave communication system operating at 30-300GHz have attracted increasing attention for many years. In order to meet the communication requirement, the millimeter-wave antennas should obtain the capabilities of low-cost, high gain, easy fabrication and so on.

As planar embedded waveguide, substrate integrated waveguide (SIW) has attracted much attention for its advantages of low loss, high power handling capability, wideband operation, coplanar integration, and so on. Based on the novel waveguide structure, many antenna arrays fed by the broad-wall slots of SIW have been proposed. The impedance bandwidth in [1]-[3] were typically smaller than 7%. A 4×4 SIW slot antenna array [4] achieved 10.7% impedance bandwidth for 10-dB return loss. However, the gain performance suffered a large variation of up to 11dB in the impedance bandwidth.

To overcome the drawbacks mentioned above, [5] proposed an SIW slotted narrow-wall fed cavity antenna. As the dual resonance generated from both the slot and the SIW slotted-cavity, the working bandwidth of the proposed antenna could be broadened effectively. Furthermore, a 2×2 antenna array working at 35GHz was designed. The measured results showed that the frequency range for S11<-10dB was 32.7GHz-37.4GHz and the maximum gain reached 10.8dB.

Considering the power divider for the SIW cavity antenna array, there are basically two types of feeding networks: T-type and Y-type [6]. Based on the basic types, multi-way broadband SIW power divider was designed. It is obvious that the width

of the SIW slot antenna doubles at least as the feeding way number doubles. Besides, the length has to increase as well so as to realize the transition for more feeding ways. Therefore, because of the multiplication of the antenna array size, the traditional feeding ways have much limit to the increase of the antenna cells.

This paper introduces a novel method to design a compact wideband millimeter-wave SIW-fed interdigital cavity antenna array, which can make the structure more compact and maintain the capacities of broadband, high gain and easy integration as well. Firstly, a novel in-phase power divider fed by CPW is designed. Secondly, the SICs are arranged interdigitally in the middle of the two adjacent feeding SIW. In the end, this novel SIW interdigital cavity antenna array is fabricated and measured. All the structures in this letter are simulated with the full-wave CAD software Ansoft-HFSS and designed on Duroid 6010 substrate with a dielectric constant of 10.2 and a thickness of 0.635mm.

## II. ANTENNA ARRAY ELEMENT DESIGN

Fig. 1 depicts the physical configuration of the proposed coplanar waveguide (CPW)-fed in-phase power divider, where D and S are the diameter and period of metallic vias, and $W_{siw}$ stands for the SIW width that determines its cut-off frequency and the working frequency range of $TE_{10}$ mode. $W_{cpw}$ and $W_c$ are the CPW width and the CPW slotline width designed for 50 Ω operation, respectively. The rectangular with light color in Fig. 1 represents the no-ground area in the bottom layer. Besides, the vias painted with light color are used to suppress the undesired energy leakage. This power divider consists of a CPW-slotline transition and a slotline-SIW transition.

By taking the discontinuity effect and mode conversion effect between even and odd CPW modes into account, [7] created a new input-impedance-based circuit model for CPW-to-slotline transition. The extended CPW signal strip into the slotline is viewed as a probe for exciting the slotline. By this reasonable equivalent, a specific impedance formula was given to characterize the T-junction. Based on the impedance formula, the slotline width $W_s$ is calculated to realize the impedance matchment between CPW and the slotline.

978-7-5641-4279-7

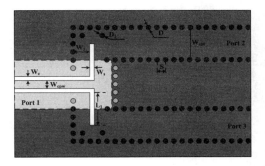

Fig 1. Geometry of the CPW fed power divider

In the second transition part, the electric field of the slotline is horizontally-polarized, which is perpendicular to that in the SIW. Because of the overlapped metallic covers on the top and bottom of the SIW, the horizontally-polarized electric field can be easily converted to the vertically-polarized field of the HMSIW [8]. Four via posts with the diameter of $D_1$ are used to optimize the resonance of the transition. By combining the two transition parts mentioned above, a novel CPW-fed power divider is produced. TABLE. I lists the optimized parameters of the structure. Fig. 2 shows the simulated $S_{11}$ result of the power divider. Within the frequency range of 34.6-37.1 GHz, the return loss is less than -15 dB. Fig. 3 and Fig. 4 show that the maximum phase and amplitude imbalances within the working frequency range are less than 1.1 and 0.17 dB, respectively. These results indicate that the proposed power divider can operate in a wide band with good balance performances for both the amplitude and the phase.

Besides, the SIC size [9] comes given by a (length) and b (width) to realize $S_{11}$ <-10dB from 33.5 GHz to 36.9 GHz.

TABLE I

THE PARAMETERS OF THE POWER DIVIDER (UNIT: mm)

| $D_1$ | D | $W_s$ | $W_c$ | $W_{cpw}$ | $W_{siw}$ | S | $W_1$ | $L_1$ |
|------|------|------|------|------|------|------|------|------|
| 0.3 | 0.3 | 0.1 | 0.1 | 0.3 | 2.4 | 0.4 | 1.35 | 2.6 |

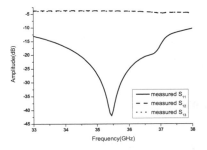

Fig 2. The simulated return loss of the power divider

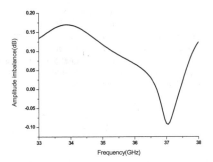

Fig 3. The amplitude imbalance of the power divider

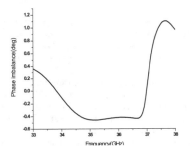

Fig 4. The phase imbalance of the power divider

III. ANTENNA ARRAY DESIGN

As shown in Fig. 5, the proposed antenna array is composed of $2 \times 2$ SICs and a compact four-way power divider mentioned above. The attracting design consideration here is that the SIC elements are arranged interdigitally in the middle of the two feeding SIW based on the feeding networks, which can make the structure much more compact.

This unique design could reduce the width by approximate 25% comparing with the traditional structure. Besides, the length can also be reduced in some degree by adopting the mentioned CPW-fed power divider. Moreover, this design method will exhibit great advantage, especially when the number of the SIC increases. In traditional design method for the SIW cavity antenna array, the feeding structure will become more complex when more SICs are added, therefore the whole

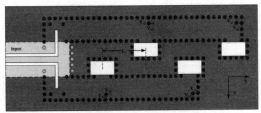

Fig 5. Geometry of the proposed antenna array

size of the antenna array will become very large. Through applying the method introduced in this design, it is easy to increase the SIC number by just lengthen the feeding SIW without changing the feeding structure at all.

To make the power entering into four SICs in-phase, the distance $L_x$ between two adjacent SICs should be nearly waveguide wavelength $\lambda_g$. According to [10]:

$$\lambda_g = \frac{\lambda_s}{\sqrt{1-(\frac{\lambda_s}{2W_e})^2}} \quad (1)$$

$$\lambda_s = \frac{\lambda_0}{\sqrt{\varepsilon_r}}, W_e = W_{siw} - \frac{D^2}{0.95S} \quad (2)$$

Since the length of the SIC, a, is longer than that of $\lambda_g$, $L_x$ has to set approximately $2\lambda_g$. As a result, the final optimized distance between the adjacent elements is 8.2 mm （$0.95\lambda_0$）, where $\lambda_0$ is the wavelength at 35GHz in the free space.

Besides, four vias are adjusted to ensure the power entering into four SICs in-phase. After optimizing with the HFSS, the proposed antenna array is fabricated. Fig. 6 shows the photograph of the proposed antenna array operating at 35 GHz. TABLE. II lists the parameters of the antenna array structure.

Fig. 7 gives the result of the simulated and measured $S_{11}$ result of the antenna array. It can be observed that the frequency range for $S_{11}<$-10dB is 33.3-38.0 GHz(13.1%). The measured result shows well agreement with the simulated one.

The simulated radiation patterns at 35GHz are depicted in Fig. 8. The radiation pattern follow the trend depicted in [5] but low level of the side lobe in the E-plane is obtained. This may attribute to the decreasing radiation loss from the slotline.

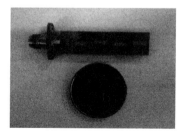

Fig 6. Photograph of the antenna array.

TABLE II

THE PARAMETERS OF THE ANTENNA ARRAY (UNIT: mm)

| Vx1 | Vy1 | Vx2 | Vx2 | Vx3 | Vy3 | Vx4 | Vy4 | $L_s$ |
|-----|-----|-----|-----|-----|-----|-----|-----|-----|
| 0.4 | 0.3 | 0.3 | 0.4 | 0.4 | 0.3 | 0.3 | 0.4 | 8.2 |

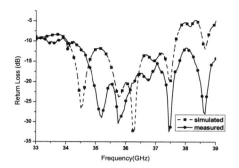

Fig 7. The simulated and measured return loss of the antenna array

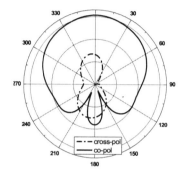

(a) E-plane

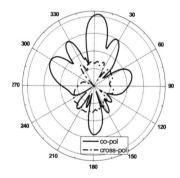

(b)H-plane

Fig 8. The simulated radiation patterns of the antenna array at 35GHz.

## IV. CONCLUSION

In this paper, a compact wideband SIW interdigital cavity antenna array has been proposed. By using the CPW-fed power divider, the SICs can be arranged in a more compact way comparing with the traditional design method. Therefore this antenna array can be fabricated in a smaller PCB board while keeping the good capacities in impedance width and the gain. This technology exhibits much more advantages when the number of the elements increases, thereby it is promising for high gain antenna in millimeter-wave applications.

## REFERENCES

[1]  J. Hirokawa and M. Ando, "Efficiency of 76-GHz post-wall wave-guide-fed parallel-plate slot arrays," *IEEE Trans. Antenna Propag.*, vol. 48, no. 11, pp. 1742-1745, Nov. 2000.

[2]  Y. J. Cheng, W. Hong, and K. Wu, "Millimeter-wave substrate integrated waveguide multibeam antenna based on the parabolic reflector principle," *IEEE Trans. Antenna Propag.*, vol. 56, no. 9, pp. 3055-3058, Sep. 2008.

[3]  P. Chen, W. Hong, Z. Q. Quai, J. F. Xu, H. M. Wang, J. X. Chen, H. J.Tang, J. Y. Zhou, and K. Wu, "A multibeam antenna based on substrate integrated waveguide technology for MIMO wireless communications," *IEEE Trans. Antenna Propag.*, vol. 57, no. 6, pp. 1813-1821, Jun. 2009.

[4]  S. Cheng, H. Yousef, and H. Kratz, "79 GHz slot antenna based on sub-strate integrated waveguides (SIW) in a flexible printed circuit board," *IEEE Trans. Antenna Propag.*, vol. 57, no. 1, pp. 64-71, Jan. 2009.

[5]  Y. Zhang, Z. N. Chen. X. M. Qing, and W. Hong, "Wideband millime-ter-wave substrate integrated waveguide slotted narrow-wall fed cavity antenna for 60-GHz," *IEEE Trans. Antenna Propag.*, vol. 59, no. 5, pp. 1488-1496, May. 2011.

[6]  Germain. S, D. Deslandes, K. Wu, "Development of substrate integrated waveguide power dividers," *IEEE Conf on Electrical and Computer Engineering. Canadian*, vol. 3, 4-7, May 2003.

[7]  C. H. Wang, Y. S. Lin. M. C. Tsai, C. Huai, and C. H. Chen, "An in-put-impedance-based circuit model for coplanar waveguide-to-slotline T-junction," *IEEE Trans. Microwave Theory Tech.*, vol. 52, no. 6, pp. 1585-1590, Jun. 2004.

[8]  F. F. He, K. Wu. W. Hong, H. J. Tang, H. B. Zhu and J. X. Chen, "A planar magic-T using substrate integrated circuits," *IEEE Micro. Wireless Compon. Lett.*, vol. 18, no. 6, pp. 386-388, June. 2008.

[9]  K. Gong, Z. N. Chen. X. M. Qing, P. Chen, and W. Hong, "Substrate integrated waveguide cavity-backed wide slot antenna for 60-GHz Bands," *IEEE Trans. Antenna Propag.*, vol. 60, no. 12, pp. 6023-6026, Dec. 2012.

[10]  Pozar D M. *Microwave engineering,3rd*. New York: Wiley, 2004.

# Polarization Reconfigurable Cross-Slots Circular Patch Antenna

M. N. Osman, M. K. A. Rahim, M. F. M. Yussof, M. R. Hamid, H. A. Majid

Communication Engineering Department, Faculty of Electrical Engineering,
Universiti Teknologi Malaysia, 81310 Johor Bahru, Johor, Malaysia

nasrun_osman@yahoo.com, mkamal@fke.utm.my, fairus@fke.utm.my, rijal@fke.utm.my, huda_set@yahoo.com

*Abstract*—In this paper, an antenna with polarization reconfigurability is presented. The antenna consists of a two cross-slots diagonally positioned in the centre of a circular patch antenna. Two pair of switches are placed in the slots and by controlling the length of the slots, three different polarizations can be reconfigured. The proposed antenna is capable to reconfigure between linear polarization (LP), right-hand circular polarization (RHCP) and left-hand circular polarization (LHCP). The linear polarization is achieved when the lengths of both slots are equal and the circular polarization type is obtained due to the inequality of the slot length. Details of the design and simulation results are discussed. This reconfigurable antenna with polarization diversity can be used for wireless local area network (WLAN) application.

*Index terms*—polarization reconfigurable, patch antenna, cross-slot

## I. INTRODUCTION

Reconfigurable antennas plays an important role towards smart, cognitive and modern wireless communication systems. Recently, polarization diversity antenna has received a significant attention due to its ability to reduce the multipath fading, realizing frequency reuse and immunity to interference signals. Apart from that, this type of reconfigurability are capable to improve the gains offered by multiple-input multiple-output (MIMO) systems hence provide better channel reliability and capacity [1,2]. To design an antenna with polarization reconfigurability, the antenna structure, material properties, or feed configurations have to be modified to alter the current flows of the antenna.

Several antennas with polarization reconfigurations have been proposed in the literature. The polarization diversity was successfully achieved through the modification to the feeding network [3,4]. In [5], a compact polarization reconfigurable antenna is proposed by switching transition between coplanar waveguide mode and slot line mode to achieve both orthogonal linear polarizations.

Single-fed patch antenna with polarization reconfigurable is designed by using perturbation method such as introduction of loop slots in the ground plane [6,7]. For reconfigurability, the slots are connected through p-i-n diodes, which act as a switch. Another solution proposed is based on the establishment of the slots on the patch itself [8-10]. By using this method, the slot

dimension is manipulated to obtain the desired polarization mode.

This paper proposed an antenna with polarization reconfigurability based on the use of circular patch with two crossing diagonally slots. Two pair of switches are used to control the slot length. Consequently, depending on the switch configuration the antenna can excited with RHCP, LHCP and LP polarization mode. The simulated results are presented and key parameters for the proposed antenna are investigated. In this study, copper strips are used as switches for proof of concept.

## II. ANTENNA STRUCTURE AND APPROACH

In this section, the structure of the proposed antenna is described. Figure 1 shows the geometry of the proposed antenna. A circular patch, having a radius $r$, is designed on the top layer of the substrate and finite ground plane on the bottom layer. A dielectric used in the design is Taconic RF35 with thickness, $h$ of 1.52 mm, permittivity of 3.54 and tangent loss of 0.0018. The antenna is excited with a coaxial feed with a distance of $d$ from the center of circular patch and incorporating at $45°$ to the two arms of the cross slots. Two slots, having a width of $w1$, is located diagonally cross on the same layer of circular patch. Four copper strips, which act as diode, are located at specific position to control the length of the slot in order to achieve polarization reconfigurability. The presence of the copper strips means the state of diode is ON while represent OFF with the absence of copper strips.

Due to the existence of slots, the equivalent excited patch surface current path along the direction perpendicular to the narrow slot is lengthened while on the parallel to the slot orientation is only have slightly affected. Hence, by further selecting the proper slot length difference and suitable position for feeding, the two near-degenerate orthogonal resonant modes can have equal amplitudes and 90 degree out of phase, and CP operation can thus be obtained. Alternatively, when the length of the both slot arms are equal, there is no phase difference between modes and therefore the antenna will excite LP operation.

978-7-5641-4279-7

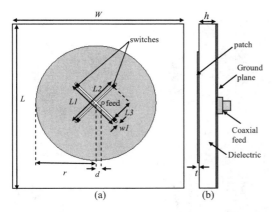

(a)                              (b)

Figure 1: Geometry of the proposed antenna (a) front view and (b) side view

### III. STUDIES OF KEY PARAMETERS

The effect of slot length towards the result of return loss and axial ratio are investigated. In this analysis, one of the slot length, $L1$ is varied from 15 mm to 17.4 mm while the other slot length, $L2$ is remained at 15 mm. The effects are shown in Figure 2. It shows that by lengthening the slot, the return loss is improved and produces wider bandwidth as shown in Figure 2(a). At certain length, two resonant frequencies are obtained due to the two unequal slot lengths.

Meanwhile, in Figure 2(b), it is found that the slot length greatly influence the CP characteristics. When the lengths of both slots are equal, which means the length difference is zero, the axial ratio is at constant value and the antenna is excited with horizontal LP. However, as the length difference is varied from zero to 2.4 mm, the polarization is changed from LP to elliptic polarization. The axial ratio decreases as $L1$ increases. At the specific slot length, starting from 16.2 mm, the axial ratio value is dropped to below than 3 dB, which proved that the CP is excited. The range of slot length between 16.2 mm to 17mm is where the antenna is excited with two-orthogonal equal magnitude modes with 90 degree out of phase. The polarization is switched again to the elliptic polarization    (AR > 3 dB) when the slot length $L2$ is greater than 17 mm.

The parameter $r$, influencing the operating frequency and the distance $d$ is optimized to obtain a good impedance matching. The optimized values for the parameters are given in Table I. The dimension of the copper switches used in the design is  0.5 mm x 0.5 mm.

TABLE I
DIMENSIONS OF THE PROPOSED ANTENNA

| Parameters | $L$ | $W$ | $L1$ | $L2$ | $L3$ |
|---|---|---|---|---|---|
| Value (mm) | 55 | 55 | 16.4 | 16.4 | 15 |
| Parameters | $r$ | $D$ | $w1$ | $h$ | $t$ |
| Value (mm) | 17 | 2 | 0.5 | 1.52 | 0.035 |

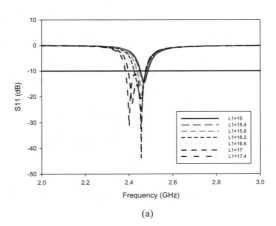

(a)

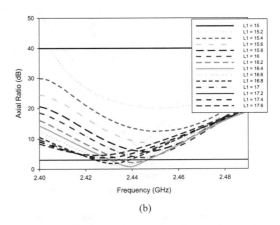

(b)

Figure 2: Effect of the slot length, L1 (mm) to the ;(a) S11 (dB) and ;(b) AR (dB)

### IV. SIMULATION RESULTS AND DISCUSSION

Two pair of switches is used to control the length of the slots. Pair 1, consist of switch 1 (S1) and switch 2 (S2) are for $L1$ meanwhile pair 2 which is a combination of switch 3 (S3) and switch 4 (S4) are for $L2$. By changing the state of the switch, either ON or OFF, consequently the polarization of the antenna can be changed. When slot length $L1$ is less than $L2$, RHCP mode is achieved while in contrast it, LHCP mode is excited when $L1$ is greater than $L2$. When both slot lengths are equal, horizontal LP is excited. There are four configurations involved; S1&S2 is ON and S3&S4 is OFF; S1&S2 is OFF and S3&S4 is ON; all switches are ON, and; all switches are OFF.

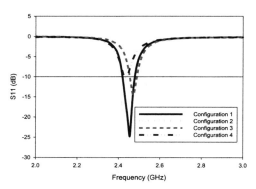

Figure 3: Simulated return loss, S11 (dB) for all switch configurations

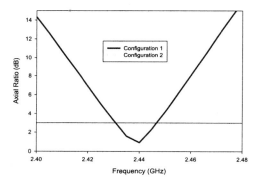

Figure 4: Simulated axial ratio (dB) for CP

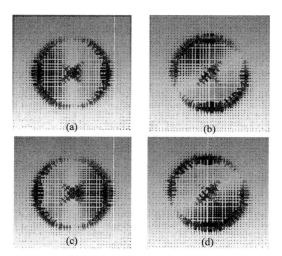

Figure 5: Electric field distribution on the patch for configuration 1 at 2.44 GHz (a) 0°; (b) 90°; (c) 180° ; (d) 270°

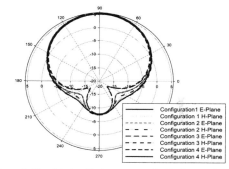

Figure 6: Simulated radiation patterns for E-plane and H-plane for all switch configurations at resonant frequency

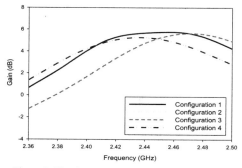

Figure 7: Simulated gain over frequency for all switch configurations

The simulated S-parameters for all configurations are shown in Figure 3. The simulated 10-dB return loss impedance bandwidth, are from 2.418 to 2.478 GHz, 2.454 to 2.488 GHz, and 2.427 to 2.440 GHz for both type of CP (configuration 1 and configuration 2), configuration 3 and configuration 4, respectively. The percentage of bandwidth for CP is 2.4% (60 MHz) with respect to the center frequency of 2.454 GHz. The bandwidth of the proposed antenna is narrow due to the characteristics of the patch antenna itself.

For antenna configuration 3 and configuration 4, it excites with LP. However, the resonant frequency is different as the length of the slots give an effect to the resonating frequency. Lengthened the slot length will decreased the center frequency. Hence, the presence of the copper strips will radiate at higher frequency compare to when all switches are at OFF state.

Figure 4 presents the axial ratio against frequency graph for the CP modes. The axial ratio value for LP is constant for any frequency. The antenna excites with CP mode between frequency range of 2.431 GHz to 2.447 GHz, which is about

TABLE II
SIMULATION RESULTS OF THE PROPOSED ANTENNA

| Configuration | Switch state | | | | Polar Type | $F_r$ (GHz) | S11 (dB) | Bandwidth (GHz) | | AR Bandwidth (GHz) | | Gain (dB) |
|---|---|---|---|---|---|---|---|---|---|---|---|---|
| | S1 | S2 | S3 | S4 | | | | $f_L$ | $f_H$ | $f_L$ | $f_H$ | |
| Configuration 1 | ON | ON | OFF | OFF | RHCP | 2.454 | -24.812 | 2.418 | 2.478 | 2.431 | 2.447 | 5.7 |
| Configuration 2 | OFF | OFF | ON | ON | LHCP | 2.454 | -24.812 | 2.418 | 2.478 | 2.431 | 2.447 | 5.7 |
| Configuration 3 | ON | ON | ON | ON | LP | 2.471 | -14.554 | 2.454 | 2.488 | NA | | 5.6 |
| Configuration 4 | OFF | OFF | OFF | OFF | LP | 2.434 | -10.360 | 2.427 | 2.440 | NA | | 5.2 |

16 MHz of 3-dB bandwidth. The percentage of bandwidth for CP is 0.7% with respect to the center frequency 2.471 GHz when $L1 = 16.4$ mm.

The return loss and axial ratio for both RHCP and LHCP give similar results. The different is on how the electric field flow on the patch antenna, either on clockwise or anticlockwise rotation. Figure 5 shows the electric field distribution on the radiating patch at different phase for RHCP which illustrates the rotation of an electric field with right-hand circular path.

The simulated radiation pattern for E-plane and H-plane are plotted in Figure 6. The gain obtained at the resonant frequency for CP is 5.7 dB whilst 5.6 dB and 5.2 dB for antenna configuration 3 and configuration 4 respectively. The variation value of gain over frequency is shown in Figure 7. Good broadside radiation patterns with 3-dB beamwidths with more than 90° are observed for all configurations. The summary for the simulated results are tabulated in Table II.

## V. CONCLUSIONS AND FUTURE WORK

A cross-slots microstrip patch antenna with polarization reconfigurability has been presented in this paper. From the simulation results, it shows that the polarization reconfigurability can be obtained by controlling the length of the slots. To achieve this, four switches are used to enable the antenna to radiate between LP, RHCP and LHCP. The simulated bandwidth extends from 2.418 to 2.478 GHz for CP. Bandwidth for LP is 34 MHz and 13 MHz for antenna configuration 3 and configuration 4 respectively. An axial ratio bandwidth of 16 MHz is achieved for CP mode. Throughout the simulation process, copper strip is used to represent switch. In future, the proposed antenna will be fabricated and PIN diodes will be used as switches to replace the copper strips. This proposed antenna can support various applications in WLAN systems.

ACKNOWLEDGMENTS

The authors thank the Ministry of Higher Education (MOHE) for supporting the research work, Research Management Centre (RMC), School of Postgraduate (SPS) and Radio Communication Engineering Department (RACeD) Universiti Teknologi Malaysia (UTM) for the support of the research under grant no (Vote No: 04H38/4L811/05J64).

REFERENCES

[1] D. Piazza et al., "Design and evaluation of a reconfigurable antenna array for MIMO systems", IEEE Transactions on Antenna and Propagation, vol. 56, pp. 869-879, March 2008.
[2] D. Piazza, P. Mookiah, M. D'Amico, and K.R. Dandekar, "Pattern and polarization reconfigurable circular patch for MIMO systems", European Conference on Antenna and Propagation, pp. 1047-1051, March 2009.
[3] W. Cao et al., "A reconfigurable microstrip antenna with radiation pattern selectivity and polarization diversity", IEEE Antennas and Wireless Propagation Letters, vol. 11, pp. 453-456, 2012.
[4] S.-Y. Lin, Y.-C. Lin, C.-Y. Li, and Y.-M. Lee, "Patch antenna with reconfigurable polarization", Asia Pacific Microwave Conference Proceedings, pp. 634-637, 2011.
[5] Y. Li, Z. Zhang, W. Chen, and Z. Feng, "Polarization reconfigurable slot antenna with a novel compact CPW-to-slotline transition for WLAN application", IEEE Antennas and Wireless Propagation Letters, vol. 9, pp. 252-255, 2010.
[6] R.-H. Chen and J.-S. Row,"Single-fed microstrip patch antenna with switchable polarization", IEEE Transactions on Antennas and Propagation, vol. 56, no. 4, pp. 922-926, April 2008.
[7] Yang, B. Shao, F. Yang, A. Elsherbeni and B. Gong, "A Polarization Reconfigurable Patch Antenna with Loop Slots on the Ground Plane," IEEE Antennas and Wireless Propagation Letters, vol. 11, pp. 69-72, Jan 2012.
[8] C.-C. Wang, L.-T. Chen, and J.-S. Row, "Reconfigurable slot antennas with circular polarization", Progress In Electromagnetics Research Letters, vol. 34, pp. 101-110, 2012.
[9] E. A. Soliman, W. D. Raedt, and G. A. E. Vandenbosch, "Reconfigurable slot antenna for polarization diversity", Journal of Electromagnetic Waves and Application, vol. 23, pp. 905-916, 2009.
[10] B. Kim, B. Pan, S. Nikolaou, Y.-S. Kim, J. Papapolymerou and M.M. Tentzeris, "A novel single-feed circular microstrip antenna with reconfigurable polarization capability", IEEE Trasactions on Antennas and Propagation, vol. 56, No. 3, pp. 630-638, March 2008.

# A Configurable Dual-H Type Planar Slot Antenna Applicable for Communication Well

Shengjie Wang, Xueguan Liu, Xinmi Yang, Huiping Guo

School of Electronics and Information Engineering, Soochow University, SuZhou, China 215006

*Abstract*—A communication well is a well digged underground for data communication with base station using a specific antenna. To use communication wells, there is a need to design the specific antenna working at 836 MHz and 881 MHz frequency with each bandwidth of 50 MHz. The antenna should be embedded on the side of the well so that it does not take much room and the antenna should have a certain radiation direction. To meet requirements, a configurable dual-H type planar slot antenna with adjustable antenna directivity has been designed, fabricated and measured as part of out work. In this paper we studied the influence of edge structure on the antenna directivity and find out that antenna's radiation directivity is adjustable. By carefully designing the edge structure, we obtained the desired radiation pattern. All antenna parameters are optimized. Our work shows that the return loss of proposed antenna is -19 dB at 836 MHz and -15 dB at 881 MHz. The bandwidth spans from 800 MHz to 1 GHz. The measured maximum gain at each frequency is around 3 dB at a fixed direction. The measured results show that the propoesed antenna is applicable for communication wells.

*Index Terms*—communication well; planar slot antenna; adjustable antenna radiation patterns; antenna edge structure.

## I. INTRODUCTION

Antenna has been widely used in modern society. It plays an important role in wireless communication system. Antenna's working frequency varies from one to one so that it meets the requirements of different wireless system for communication. The proposed antenna in this paper is based on the engineering application for communication wells. Generally a communication well is a well digged underground for data communication with base station using a specific antenna. In order to use a communication well, there is a need to design the specific antenna working at 836 MHz and 881 MHz frequency with each bandwidth of 50 MHz. The antenna should be embedded on the side of the well and able to communicate with base station faraway. So the antenna radiation pattern should have a fixed direction. Fig. 1 shows the working environment of antenna briefly. By comprehensively considering the working environment and the size of the antenna, planar antenna [1] using slotted technique is most suitable for the case and the proposed antenna ultimately meet the need of practical application.

In recent years slotted technique is widely used in planar antenna design. These antennas can be called planar slot antennas [2], [3]. As is known that planar slot antenna has the advantages of planarization, small volume, light weight, low profile, easy manufacture, low cost and easy shaped with carrier. These advantages match well with the antenna that is to be applied in wells. So in the beginning planar slot antenna design is adopted.

Reference [4] proposes a H shaped slot antenna, it has no other spurious band, but the bandwidth is relatively narrow. Reference [5] presents a H-shaped slot antenna fed by microstrip coupling in the ground plane, the antenna can operate at triple frequency while the bandwidth and gain is unsatisfactory. In Reference [6] by adopting U shaped slot in H shaped radiation patch, broadband operating is achieved. This paper adopts the double-layer dielectric substrate. On the radiation patch of antenna, two H-shaped slot are applied to extend the current path without increasing its physical size thus miniaturizing the size of the antenna. On the ground plane there is a rectangular slot which can greatly broaden the bandwidth at target frequency. At the same time connecting the ground plane on negative side and positive side of the antenna by loading shorting pin at the edge of antenna, the antenna radiation pattern comes to one main directon.

## II. CONFIGURATION OF THE ANTENNA

The detailed configuration of the proposed antenna is shown in Fig. 2. Two H-shaped slot is adopted in radiation patch to reduce the antenna size as the slot can extend the current path in limited space. On the back side of antenna, a rectangular slot in ground plane can extend antenna operating bandwidth at target frequency. The substrate is FR4 with thickness of 3mm and relative dielectric constant of 4.4. In this design the origin of coordinates coincides with the center of the dielectric substrate, as shown in Fig. 2 (a), (b), (c), the dot in red symbolizes the origin point. For fabrication convenience all the shorting pins are chosen to have a diameter of 0.8 mm and the center-to-center spacing are 4mm for short distance and 7mm for longer distance. The shorting pins at the edge of antenna connect the ground on the back side of antenna and ground on the front side

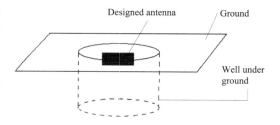

Fig. 1.   The working environment of proposed antenna.

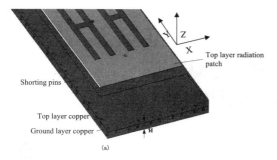

Shorting pins

Top layer copper

Ground layer copper

Top layer radiation patch

(a)

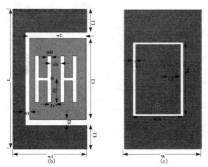

(b)          (c)

Fig. 2.   Configurations of the proposed slot structures (a) overall 3D view of proposed antenna, (b) the front side of antenna with two H shaped slot, (c) the back side of antenna with rectangular slot.

TABLE I
OPTIMIZED PARAMETERS OF THE PROPOSED ANTENNA (UNIT: MILLIMETERS)

| W | L | H | w1 | w2 | w3 | w4 | L1 |
|---|---|---|----|----|----|----|----|
| 50 | 120 | 3 | 47 | 37.5 | 6.5 | 2 | 20 |
| L2 | L3 | S | RW | RL | dx | dy | dh |
| 70 | 11.5 | 5 | 33 | 64 | 3.5 | 22 | 6 |

together. The proposed antenna is fed by coaxial probe and the feeding position has a great effect on the input impedence of the antenna. The optimized position is (-dx,dy) shown in Fig. 2. Extensive studies have been carried out in HFSS ver11 to optimize parameters to come up with the best values for the desired antenna. The optimized antenna parameters are shown in Table.

III.  DESIGN SCHEME AND SIMULATION

In order to realize dual frequency operation, broaden bandwidth of antenna and make antenna radiation pattern adjustable, three techniques are used, namely dual H shaped slot on the radiation patch, rectangular slot in the ground plane and antenna edge structure. These techniques are discussed in the following.

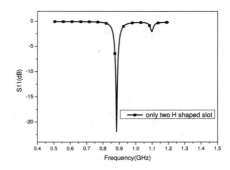

Fig. 3.   Return loss where there are only dual H-Shaped slot in the radiation patch.

### A.  Dual H-shaped slot in the radiation patch

By adopting H slot in the radiation patch, the current path is extented without increasing its physical size and the resonant frequency decreases [7]. This helps to reduce the antenna size. In this paper dual H-shaped slot is applied in the radiation patch. Fig. 3 shows simulated S11 when there are only two H-shaped slot in the radiation patch. It can be seen that the resonant frequency is at 883 MHz with good resonant depth. However the bandwidth is narrow. The bandwidth at -10 dB is about 13 MHz. So it is necessary to improve the antenna structure.

### B.  Rectangular slot in ground plane

The antenna bandwidth can be broadened by slotting on the ground plane because the slot enables the current direction on the ground plane is in the opposite of that in radiation patch [8]. Based on this theory, a rectangular slot is applied in the ground plane. As the antenna structure changed, some antenna parameters have been optimized to get the result we want. Fig. 4 gives the comparison of S11 with and without rectangular slot in the ground plane. From the figure, it is noted that the bandwidth increases by a large margin.

### C.  Antenna edge structure

As is shown in Fig. 2, at the edge of antenna, shorting pins connect the ground plane on the back side and ground plane on the front side together. This edge structure enables the proposed antenna radiate energy concentratelly at a certain direction. During our work, we find out that the antenna radiation pattern is adjustable by adjusting the size of antenna edge structure. The comparision about the effect of different antenna edge structure size on antenna radiation pattern is carried out in this paper. Fig. 5 and Fig. 6 shows the simulated radiation pattern of antenna1, 2, 3 in 836 MHz and 881 MHz in XOZ plane respectively. Antenna1, 2, 3 depicted in Fig. 5 and Fig. 6 represents three types of antenna. Antenna1 has no edge structure, the width of it is 0mm. Antenna2 has middle size edge structure, the width of it

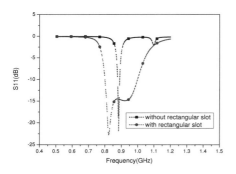

Fig. 4. Comparison of S11 with and without H shaped slot in the ground plane on the back side of antenna.

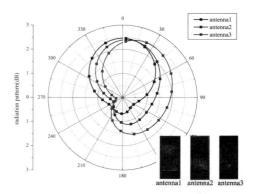

Fig. 6. Simulated radiation patterns of antenna1, 2, 3 at 881 MHz in XOZ plane.

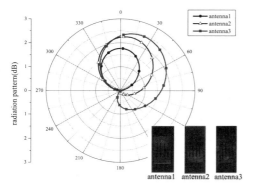

Fig. 5. Simulated radiation patterns of antenna1, 2, 3 at 836 MHz in XOZ plane.

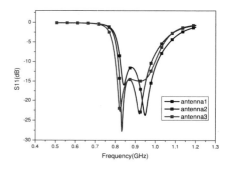

Fig. 7. Simulated S11 of antenna1, antenna2 and antenna3.

is 4mm. Antenna3 has the biggest size edge structure, the detail size is shown in Table I. From the figure it can be seen that antenna1's main radiation direction has no offset. Antenna2's main radiation direction is tilted about 15° with Z axis while antenna3's main radiation direction is tilted about 30° with Z axis, besides the max gain is bigger when the edge structure is larger. This illustrates that the antenna's directivity is adjustable. Fig. 7 gives the simulated S11 of antenna1, 2 and 3. The operating band of three antennas satisfy the requirement of communication well and as the size of edge structure increases, the whole band move to relatively lower frequency.

## IV. MEASURED RESULTS AND ANALYSIS

The proposed antenna with optimized structure is fabricated on FR4 substrate with r=4.4 and the thickness is H = 3 mm. The photo of the fabricated antenna is shown in Fig. 8.

The measured return loss is compared with simulation results in Fig. 9. Good agreement is achieved. The measured band-width is about 20% which is from 0.8 GHz to 1 GHz. Fig. 10 (a) and (b) shows the simulated and measured radiation patterns respectively in XOZ plane of the antenna at 836 MHz and 881 MHz. Obviously this antenna has an directional radiation pattern. The main beam in XOZ plane has been tilted by around 30 degrees due to the reflection of edge structure. The difference between simulated and measured radiation pattern both in 836MHz and 881MHz mainly due to the unwanted radiation by the feeding cable and the response of the anechoic chamber and the instability of the anechoic chamber also leads to a deviation. In general the edge structure has an impact on the radiation pattern, the bigger size of edge structure, the bigger angle tilted with the Z axis.

## V. CONCLUSION

This paper presented a planar slot antenna with adjustable antenna directivity. The antenna is proposed to achieve good bandwidth and directivity by using slot technique and edge structure. The return loss of proposed antenna is -19 dB at

Fig. 8.  Photo of the fabricated antenna.

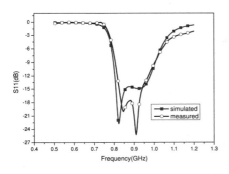

Fig. 9.  Measured and simulated return loss of the proposed antenna.

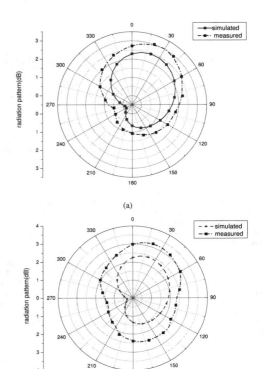

(a)

(b)

Fig. 10.  Measured and simulated radiation pattern in XOZ plane (a) 836MHz, (b) 881MHz.

836 MHz and -15 dB at 881 MHz. -10 dB bandwidth spans from 800 MHz to 1GHz. This work also focus on the impact of edge structure on the antenna radiation pattern. By comparing three types of antenna with different size of edge structure, our work shows that the directivity is proportion to the size of edge structure and by carefully designing the edge structure, we obtained the desired radiation pattern. The simulated maximum gain at each frequency is around 2.4 dB at the fixed direction and the measured gain of the proposed antenna is around 3 dB at each frequency. In general the proposed antenna provides directional radiation with reasonable gain and is applicable for communication wells.

## ACKNOWLEDGMENT

The author would like to thank all the members of Suzhou Key Laboratory of RF and Microwave Millimeter Wave Technology for their help during designing, fabricating and testing process.

## REFERENCES

[1] C. Kalialakis, "Planar antennas for wireless communications [book review]," *Antennas and Propagation Magazine, IEEE*, vol. 46, no. 1, pp. 110–110, 2004.

[2] X. Liu and H. Guo, *Microwave technology and antennas (Chinese)*. Xidian University Press, 2007.

[3] H. Nakano and J. Yamauchi, "Printed slot and wire antennas: A review," *Proceedings of the IEEE*, vol. 100, no. 7, pp. 2158–2168, 2012.

[4] M. Jais, M. Jamlos, M. Malek, and M. Jusoh, "Conductive e-textile analysis of 1.575 ghz rectangular antenna with h-slot for gps application," in *Antennas and Propagation Conference (LAPC), 2012 Loughborough*, 2012, pp. 1–4.

[5] T. Chang and J. Kiang, "Compact multi-band h-shaped slot antenna," *Antennas and Propagation, IEEE Transactions on*, vol. PP, no. 99, pp. 1–1, 2013.

[6] V. Tarange, T. Gite, P. Musale, and S. Khobragade, "A u slotted h-shaped micro strip antenna with capacitive feed for broadband application," in *Emerging Trends in Networks and Computer Communications (ETNCC), 2011 International Conference on*, 2011, pp. 182–184.

[7] H. Li, Y. Li, and W. Wang, "Wireless ad-hoc network mac protocol based on directional antenna," *Journal of Chongqing University of Technology:Natural Science Edition*, vol. 11, pp. 75–79, 2010.

[8] R. Waterhouse, "Small microstrip patch antenna," *Electronics Letters*, vol. 31, no. 8, pp. 604–605, 1995.

# Quantum State Propagation in Quantum Wireless Multi-hop Network based on EPR pairs

Kan Wang†　　　Xutao Yu *　　　Shengli Lu†

† National ASIC System Engineering Research Center, Southeast University, Nanjing 210096, China
* State Key Lab. of Millimeter Waves, Southeast University, Nanjing 210096, China
E-mail: wangkan@seu.edu.cn

*Abstract*--Quantum wireless multi-hop network (QWMN) is one emerging area which is relative to quantum communication and information. Long distance delivery of quantum information in network is an important problem in future. This results in our study on quantum state propagation in QWMN by using EPR pairs. In this paper, we present and prove method to transfer quantum state in QWMN. With this method, quantum state can be teleported from source to destination and intermediate nodes can complete teleportation in parallelism and independently in QWMN. The time complexity of this method is independent of number of hops. In addition, we verify that EPR-pair bridging is still applicable for transmitting quantum state in n-hop routing and extend the logical relation of result to n-hop case. Based on these studies, we can propagate quantum states instantaneously in QWMN and destination node can recover the state only with the measurement outcomes of intermediate nodes.

*Keywords:* Quantum state propagation, teleportation, quantum wireless multi-hop network, EPR pairs.

## I. INTRODUCTION

Quantum communication network is the combination of quantum and information theory. The advanced communications and quantum networking technologies offered by quantum information processing will revolutionize the traditional communication and networking methods [1]. Due to its potential impact and significance, quantum communication has drawn more and more attentions in the decade.

The fascinating way of propagation in quantum communication is teleportation, which is also the key to quantum network. Quantum teleportation, proposed by Bennett [2], employing a particular property of quantum called quantum entanglement, allows quantum devices to transmit a quantum state to another node at a distance. Shared entanglement is essential to achieve quantum teleportation, which can be accomplished by generating an EPR pair and distributing the pair to the source and destination in advance [3].

An important problem of quantum communications in future is the long distance delivery of quantum information. The fidelity of the entanglement decreases during the transmission through the noisy quantum channel. Considering the quantum entanglement state can be generated and distributed between smaller segments, quantum repeaters are used as a router to sharing and distributing quantum entanglement [4]. However, the repeaters are connected by optical fibers while they are static. In the field of wireless network, where nodes usually are mobile, the EPR pairs cannot be distributed to wireless

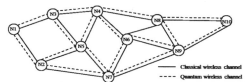

Fig.1 Quantum wireless multi-hop network

quantum device through the air. As a result, wireless quantum devices cannot set up EPR pairs instantaneously. In the meantime, it is unreasonable for a limited storage mobile device to keep many EPR pairs entangled with all possible communication parties for the future use in wireless environment. So, new quantum mechanisms are established to realize the teleportation even when the two sites do not share EPR pairs mutually. Cheng et al. [5] introduced the first quantum routing mechanism in wireless communication network using quantum relay and EPR-pair bridging, and analyzed the process of the propagation in two hops after setting up the routing path. Yu et al. [6] presented one on-demand quantum routing protocol in wireless ad-hoc network, employing the both approximation method to set up the quantum channel in three hops. Quantum communication in distributed wireless sensor network is discussed in [7], but it mainly solved the quantum key distribution (QKD) and quantum cryptography problem in WSN.

The aim of this paper is to solve the information propagation problem in quantum wireless multi-hop network. By analyzing and deducing, this paper extends the EPR-pair bridging to n-hop situation and proves that teleportation mechanism is also applicable in wireless multi-hop quantum networks. In the meanwhile, the paper analyzes the process of teleportation and demonstrates the XOR relation between measured outcomes of intermediate nodes.

The rest of this paper is organized as follows. In section II, the model of quantum wireless multi-hop network is given. Teleportation in quantum wireless multi-hop network is discussed in section III. More general cases of teleportation are deduced in section VI. Finally, in section V, conclusions are drawn.

## II. QUANTUM WIRELESS MULTI-HOP NETWORK MODEL

Quantum wireless multi-hop network, as is shown in Fig.1, is composed of *wireless quantum nodes* (WQN) with both quantum and wireless communication capabilities. There are

978-7-5641-4279-7

two kinds of communication channels between WQNs: classical channel for transmitting classical radio data with traditional radio transceivers and quantum channel for teleporting quantum state from a source node to a faraway destination node. A quantum bit (*qubit*) is a unit of quantum information, two possible states for a qubit are $|0\rangle$ and $|1\rangle$ so that one qubit can be symbolized in the linear combinations form of $|\phi\rangle = \alpha|0\rangle + \beta|1\rangle$, called superposition, where $\alpha$ and $\beta$ are complex numbers. We cannot examine a qubit to determine its quantum states, only can get the result 0 with probability $|\alpha|^2$, or the result 1 with probability $|\beta|^2$. Naturally, $|\alpha|^2 + |\beta|^2 = 1$, since the probabilities must sum one. Every state in a superposition would be transformed simultaneously.

In quantum networks, the quantum state is propagated by the means of quantum teleportation, applying the property of EPR pairs. An EPR pair is a two-qubit system in maximally entangled state and can be denoted by $1/\sqrt{2}(|00\rangle + |11\rangle)$. The quantum communication channel exits between arbitrary two nodes, which share at least one couple of entangled EPR pairs. Each node owns one qubit of the EPR pair. With the aid of the EPR pair, by only performing local operations and classical communication, these two nodes can propagation quantum states with each other.

In quantum wireless multi-hop network, routing mechanism is needed to choose one routing path from source to destination before quantum teleportation starts. This routing includes one classical communication path and one quantum communication path. Here, we discuss quantum state propagation rather than the routing, so we assume that the routing path has been selected appropriately to form one line of multi-hop nodes.

### III. TELEPORTATION THROUGH BRIDGING

In quantum wireless multi-hop network, it is unrealistic and impossible to share EPR pairs with all the other potential destination nodes in network. To achieve quantum teleportation even if the source and destination do not share any EPR pair mutually, quantum relay and EPR-pair bridging are proposed.

The principle of quantum relay is to perform the quantum teleportation hop by hop from source to destination when there are intermediate nodes equipped with EPR-pair shared with upstream and downstream nodes in between. In other words, every intermediate node on the routing path would recover the state of the quantum teleported by the upstream nodes, then teleport it again to the downstream nodes until it reaches the destination. The quantum relay indeed teleport the quantum state in multi-hop network, but the target qubit is transmitted through each intermediate node, this is lack of security and privacy. In the meantime, the relay must be performed hop by hop so that the complexity depends linearly on the number of the intermediate nodes.

The EPR-pair bridging speeds up the whole quantum relay process by performing the quantum teleportation at each intermediate nod in parallelism. The intermediate nodes are able to establish the entanglement-assisted quantum channel between its upstream and downstream nodes.

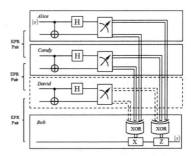

Fig.2 Quantum circuit for EPR-pair bridging

The entire quantum circuit for EPR-pair bridging as mention in [5] is given in Fig.2. The circuit is composed with quantum utility gates and measuring tool. The quantum gate is the similar to the logic gate in classical computer circuit to transform quantum state. *X*, *Z*, *Hadamard* and *CNOT* gates are four important quantum gates used in quantum circuit. The quantum *X* gate can be used to flip its state $|0\rangle$ and $|1\rangle$. The *Z* gate is used to flip the phase of the state $|1\rangle$. In other words, it can exchange $\alpha|0\rangle + \beta|1\rangle$ and $\alpha|0\rangle - \beta|1\rangle$. The *Hadamard* gate is one of the most useful quantum gates, which can turn $|0\rangle$ into $1/\sqrt{2}(|0\rangle + |1\rangle)$ and turn $|1\rangle$ into $1/\sqrt{2}(|0\rangle - |1\rangle)$. The *Controlled-NOT* (*CNOT*) gate is used to process two quits. When the first qubit is in state $|0\rangle$, it does nothing. Conversely, when the first qubit is in state $|1\rangle$, then it would flip the second qubit state. In other words, the *CNOT* gate is used to exchange $|10\rangle$ and $|11\rangle$.

In Fig. 2, single lines represent quantum data and the double lines stand for classical measurement results. Without the dotted portion, *Alice* is the source and *Bob* is the destination, while *Candy* is intermediate node between *Alice* and *Bob*. It can be seen from Fig.2 that *Alice*'s second qubit and *Candy*'s first qubit shared the EPR-pair as $1/\sqrt{2}(|00\rangle_{A2C1} + |11\rangle_{A2C1})$ and EPR-pair $1/\sqrt{2}(|00\rangle_{C2B1} + |11\rangle_{C2B1})$ is shared between *Candy*'s second qubit and *Bob*. The suffixes (e.g. A2,C1, etc.) stand for the ordinal number at the owner's side. For example, the state with the A2 represents *Alice*'s second qubit. When *Alice* intends to deliver a qubit $|y\rangle = \alpha|0\rangle + \beta|1\rangle$ to *Bob*, *Alice* and *Candy* send their second qubits through a *CNOT* gate. After that they perform *H* gate on their first qubits, then they measure the qubits and transmit the measurement outcomes to *Bob* through classical channel. Once *Bob* receives the two classical bits, he knows that *Alice* has teleported a qubit state to him. According to the two classical bits received, the quantum state $|y\rangle$ can be reconstructed by appropriate operation.

In the beginning, the five-qubit system can be represented as:

$$\frac{1}{2\sqrt{2}}\left(|00\rangle_{A1C1} + |11\rangle_{A1C1}\right) \otimes \left(|00\rangle_{A1C1} + |11\rangle_{A1C1}\right) \tag{1}$$

After *Alice* and *Candy* both send their second qubits through a *CNOT* gate and their first qubits through a *H* gate, the entire system would turn into:

$$\frac{1}{4}\begin{bmatrix}\left(|00\rangle_{A1C1}+|11\rangle_{A1C1}\right)\otimes\left(|00\rangle_{A2C2}+|11\rangle_{A2C2}\right)\otimes\left(\alpha|0\rangle+\beta|1\rangle\right)_{B1}\\ +\left(|00\rangle_{A1C1}+|11\rangle_{A1C1}\right)\otimes\left(|01\rangle_{A2C2}+|10\rangle_{A2C2}\right)\otimes\left(\alpha|1\rangle+\beta|0\rangle\right)_{B1}\\ +\left(|01\rangle_{A1C1}+|10\rangle_{A1C1}\right)\otimes\left(|00\rangle_{A2C2}+|11\rangle_{A2C2}\right)\otimes\left(\alpha|0\rangle-\beta|1\rangle\right)_{B1}\\ +\left(|01\rangle_{A1C1}+|10\rangle_{A1C1}\right)\otimes\left(|01\rangle_{A2C2}+|10\rangle_{A2C2}\right)\otimes\left(\alpha|1\rangle-\beta|1\rangle\right)_{B1}\end{bmatrix} \quad (2)$$

Considering the logical relation and utilizing the digital logic method, the expression can also be rewritten as:

$$\frac{1}{4}\begin{bmatrix}\overline{A1\oplus C1}\otimes\overline{A2\oplus C2}\otimes(\alpha|0\rangle+\beta|1\rangle)_{B1}\\ +\overline{A1\oplus C1}\otimes(A2\oplus C2)\otimes(\alpha|1\rangle+\beta|0\rangle)_{B1}\\ +(A1\oplus C1)\otimes\overline{A2\oplus C2}\otimes(\alpha|0\rangle-\beta|1\rangle)_{B1}\\ +(A1\oplus C1)\otimes(A2\oplus C2)\otimes(\alpha|1\rangle-\beta|0\rangle)_{B1}\end{bmatrix} \quad (3)$$

From expression (3), we can see that *Bob* would have the information of his qubit state once he is informed of *Alice's* and *Candy's* two qubit states. After receiving the measurement outcomes, *Bob* can fix up his own qubit to recover by either applying nothing, $X$ gate, $Z$ gate, or both $X$ and $Z$ gate. We can depict the relation as following: if 1) the result of *Alice's* first qubit XOR *Candy's* first qubit is 0 and 2) the result of *Alice's* second qubit XOR *Candy's* qubit is 1, then *Bob* has to fix up his qubit by applying quantum $X$ gate. Likewise, if 1) the result of *Alice's* first qubit XOR *Candy's* qubit is 1 and 2) the result of *Alice's* second qubit XOR *Candy's* qubit is 0, then *Bob* can fix up his state by applying quantum $Z$ gate. The basic quantum circuit just takes 2-hop case into consideration; we will verify its validity in more general cases.

## IV. TELEPORTATION IN QUANTUM WIRELESS MULTI-HOP NETWORK

In the long distance wireless transmission, the fidelity of the entanglement decreases as the distance increases and the propagation radius of classical wireless communication is limited, it is more practical to teleport quantum state in multi-hop way from source to destination with more than one intermediate. Therefore, it is necessary to extend the EPR-pair bridging to n-hop network and verify the validity and feasibility.

### A. 3-hop case

We start the work from two intermediate nodes, adding intermediate node *David* (i.e dotted portion) on the routing path. *David* also shares EPR-pair with *Candy* and *Bob*. So the system can be represented as follow:

$$\frac{1}{2\sqrt{2}}|y\rangle\otimes\left(|00\rangle_{A1C1}+|11\rangle_{A1C1}\right)\otimes\left(|00\rangle_{A2C2}+|11\rangle_{A2C2}\right)\otimes\left(|00\rangle_{C2D2}+|11\rangle_{C2D2}\right) \quad (4)$$

Because the system can perform asynchronously, source and intermediate nodes do not need to coordinate before they perform each gate. So After *Alice, Candy and David* all send their second qubits through *CNOT* gates and their first qubits through $H$ gates, the entire system state can be seen as the result of system state of 2-hops direct product result of *Da-*

*vid's* qubits going through *CNOT* and $H$ gate respectively. The expression would change into:

$$\begin{aligned}&1/4\sqrt{2}\cdot a\otimes B\left[\left(\alpha|0\rangle+\beta|1\rangle\right)_{D1}\otimes\left(|00\rangle_{D2B1}+|11\rangle_{D2B1}\right)\right]\\ &+1/4\sqrt{2}\cdot b\otimes B\left[\left(\alpha|0\rangle-\beta|1\rangle\right)_{D1}\otimes\left(|00\rangle_{D2B1}+|11\rangle_{D2B1}\right)\right]\\ &+1/4\sqrt{2}\cdot c\otimes B\left[\left(\alpha|1\rangle+\beta|0\rangle\right)_{D1}\otimes\left(|00\rangle_{D2B1}+|11\rangle_{D2B1}\right)\right]\\ &+1/4\sqrt{2}\cdot d\otimes B\left[\left(\alpha|1\rangle-\beta|0\rangle\right)_{D1}\otimes\left(|00\rangle_{D2B1}+|11\rangle_{D2B1}\right)\right]\end{aligned} \quad (5)$$

In expression (5), factors $a$, $b$, $c$, $d$ represent the four possible terms entangled qubits. For example, $a$ replaces the possible term $\left(|00\rangle_{A1C1}+|11\rangle_{A1C1}\right)\otimes\left(|00\rangle_{A2C2}+|11\rangle_{A2C2}\right)$. Also, operator $B$ is introduced here to denote $(H\otimes I)CNOT$ operation that the first qubit goes through *CNOT* gate and second qubit $H$ gate. Expression (5) can be expanded into:

$$\begin{aligned}&1/8\cdot\begin{pmatrix}a\otimes|00\rangle_{D1D2}+b\otimes|01\rangle_{D1D2}\\ +c\otimes|10\rangle_{D1D2}+d\otimes|11\rangle_{D1D2}\end{pmatrix}\otimes\left(\alpha|0\rangle+\beta|1\rangle\right)_{B1}\\ &+1/8\cdot\begin{pmatrix}a\otimes|01\rangle_{D1D2}+b\otimes|00\rangle_{D1D2}\\ +c\otimes|11\rangle_{D1D2}+d\otimes|10\rangle_{D1D2}\end{pmatrix}\otimes\left(\alpha|1\rangle+\beta|0\rangle\right)_{B1}\\ &+1/8\cdot\begin{pmatrix}a\otimes|10\rangle_{D1D2}+b\otimes|11\rangle_{D1D2}\\ +c\otimes|00\rangle_{D1D2}+d\otimes|01\rangle_{D1D2}\end{pmatrix}\otimes\left(\alpha|0\rangle-\beta|1\rangle\right)_{B1}\\ &+1/8\cdot\begin{pmatrix}a\otimes|11\rangle_{D1D2}+b\otimes|10\rangle_{D1D2}\\ +c\otimes|01\rangle_{D1D2}+d\otimes|00\rangle_{D1D2}\end{pmatrix}\otimes\left(\alpha|1\rangle-\beta|0\rangle\right)_{B1}\end{aligned} \quad (6)$$

The expression (6) can be rewritten and then we can get the final formula expression:

$$\frac{1}{8}\begin{bmatrix}\overline{A1\oplus C1\oplus D1}\otimes\overline{A2\oplus C2\oplus D2}\otimes(\alpha|0\rangle+\beta|1\rangle)_{B1}\\ +\overline{A1\oplus C1\oplus D1}\otimes(A2\oplus C2\oplus D2)\otimes(\alpha|1\rangle+\beta|0\rangle)_{B1}\\ +(A1\oplus C1\oplus D1)\otimes\overline{A2\oplus C2\oplus D2}\otimes(\alpha|0\rangle-\beta|1\rangle)_{B1}\\ +(A1\oplus C1\oplus D1)\otimes(A2\oplus C2\oplus D2)\otimes(\alpha|1\rangle-\beta|0\rangle)_{B1}\end{bmatrix} \quad (7)$$

Similarly, when *Bob* got the measurement outcomes of intermediate nodes including *David's*, he can apply specific quantum gate to recover$|y\rangle$ and we can get the similar relation to 2-hop situation.

### B. More general cases

According to the result of the 2-hop and 3-hop cases, we consider system expression at n-1-hop situation as following:

$$\frac{1}{2^{n-1}}\begin{bmatrix}\overline{N_{1,1}\oplus N_{2,1}\cdots\oplus N_{n-1,1}}\otimes\overline{N_{1,2}\oplus N_{2,2}\cdots\oplus N_{n-1,2}}\otimes(\alpha|0\rangle+\beta|1\rangle)_n\\ +\overline{N_{1,1}\oplus N_{2,1}\cdots\oplus N_{n-1,1}}\otimes(N_{1,2}\oplus N_{2,2}\cdots\oplus N_{n-1,2})\otimes(\alpha|1\rangle+\beta|0\rangle)_n\\ +(N_{1,1}\oplus N_{2,1}\cdots\oplus N_{n-1,1})\otimes\overline{N_{1,2}\oplus N_{2,2}\cdots\oplus N_{n-1,2}}\otimes(\alpha|0\rangle-\beta|1\rangle)_n\\ +(N_{1,1}\oplus N_{2,1}\cdots\oplus N_{n-1,1})\otimes(N_{1,2}\oplus N_{2,2}\cdots\oplus N_{n-1,2})\otimes(\alpha|1\rangle-\beta|0\rangle)_n\end{bmatrix} \quad (8)$$

There are $n$ nodes on the n-1 hops route. In expression (8), $N$ represents the qubit measurement outcome of the nodes on the route while the subscript denotes the number of the nodes and the number of the qubit. So, $N_{2,1}$ denotes the measurement outcome of the second node's first qubit. Similarly, $N_{n-1,2}$ denotes the outcome of the (n-1)th node's second qubit.

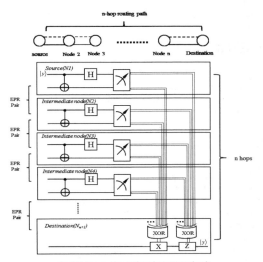

Fig.3 N-hop routing path and quantum circuit

TABLE I
LOGIC RELATION BETWEEN THE OUTCOMES AND OPERATIONS

| The outcomes of all intermediate nodes' first qubit XOR($R1$) | The outcomes of all intermediate nodes' second qubit XOR($R2$) | Applied operations |
|---|---|---|
| 0 | 0 | Nothing |
| 0 | 1 | $X$ gate |
| 1 | 0 | $Z$ gate |
| 1 | 1 | Both $X$ and $Z$ gate |

The relation between outcomes and applied operations can be found in Table I. We can see the applied operations only depend on the result of outcomes' XOR calculations. 1) If the XOR result of all nodes' first qubits ($R1$) is 0 and the result of all second qubits ($R2$) is 0, the destination should do nothing to his qubit. 2) If $R1$ is 0 while the $R2$ is 1, the destination can fix up his own qubit by applying $X$ gate. 3) Conversely, if the $R1$ is 1 and $R2$ is 0, $Z$ gate can be used. 4) Both $R1$ and $R2$ are 1, the destination node should send his own qubit through both $X$ gate and $Z$ gate. Just with the information of qubits' measurement outcomes in intermediate nods, destination node can do the analysis and choose the right operation to recover the quantum state which is transmitted at source node.

## V. CONCLUSION

We present study on the quantum state propagation in quantum wireless multi-hop network. Our study shows that quantum states can be transmitted from source node to destination node based on EPR pairs. We verify that EPR-pair bridging is still applicable in n-hop case even if they do not share EPR pairs mutually. Through quantum circuit of n-hop network, intermediate nodes can make measurement independently and the quantum state can be transferred successfully. Therefore, the time complexity of teleporting a quantum state is independent of the number of routing hops. In addition, we also demonstrate that to recover the transmitted quantum state, the destination node just need to know measurement outcomes of all intermediate nodes on the routing path and fix up its own qubit to corresponding operation and XOR relation of measurement outcomes and operations is illustrated as well.

When we add one intermediate node, the route is n-hop long, as shown in Fig.3.We just need to use the result of n-1-hop case to direct product the measurement overcome of the new node after going through $CNOT$ and $H$ gate, so that the entire system state turn into:

$$\frac{1}{2^{n-1}\sqrt{2}}\begin{bmatrix}\overline{N_{1,1}\oplus\cdots N_{n-1,1}}\otimes\overline{N_{1,2}\oplus\cdots N_{n-1,2}}\otimes B(\alpha|0\rangle+\beta|1\rangle)_{n1}\\ \otimes(|00\rangle_{n2,B}+|11\rangle_{n2,B})+\overline{N_{1,1}\oplus\cdots N_{n-1,1}}\otimes(N_{1,2}\oplus\cdots N_{n-1,2})\\ \otimes B(\alpha|1\rangle+\beta|0\rangle)_{n1}\otimes(|00\rangle_{n2,B}+|11\rangle_{n2,B})+(N_{1,1}\oplus\cdots N_{n-1,1})\\ \otimes\overline{N_{1,2}\oplus N\cdots N_{n-1,2}}\otimes B(\alpha|0\rangle-\beta|1\rangle)_{n1}\otimes(|00\rangle_{n2,B}+|11\rangle_{n2,B})\\ +(N_{1,1}\oplus\cdots N_{n-1,1})\otimes(N_{1,2}\oplus\cdots N_{n-1,2})\otimes B(\alpha|1\rangle-\beta|0\rangle)_{n1}\\ \otimes(|00\rangle_{n2,B}+|11\rangle_{n2,B})\end{bmatrix} \quad (9)$$

We can rewrite the expression (9) according to the way of expression (3) and (8):

$$\frac{1}{2^n}\begin{bmatrix}\overline{N_{1,1}\oplus\cdots N_{n-1,1}\oplus N_{n,1}}\otimes\overline{N_{1,2}\oplus\cdots N_{n-1,2}\oplus N_{n,2}}\otimes(\alpha|0\rangle+\beta|1\rangle)_{n+1}+\\ \overline{N_{1,1}\oplus\cdots N_{n-1,1}\oplus N_{n,1}}\otimes(N_{1,2}\oplus\cdots N_{n-1,2}\oplus N_{n,2})\otimes(\alpha|1\rangle+\beta|0\rangle)_{n+1}+\\ (N_{1,1}\oplus\cdots N_{n-1,1}\oplus N_{n,1})\otimes\overline{N_{1,2}\oplus\cdots N_{n-1,2}\oplus N_{n,2}}\otimes(\alpha|0\rangle-\beta|1\rangle)_{n+1}+\\ (N_{1,1}\oplus\cdots N_{n-1,1}\oplus N_{n,1})\otimes(N_{1,2}\oplus\cdots N_{n-1,2}\oplus N_{n,2})\otimes(\alpha|1\rangle-\beta|0\rangle)_{n+1}\end{bmatrix} \quad (10)$$

Note that the expression (10) has the same form with expression (8), so we can conclude that adding one intermediate node on the route has no effect on the logical relation. For fixing up its own qubit to recover the state $|y\rangle$, the destination node only needs to know the measurement outcomes of intermediate nodes on the routing path. These outcomes were transmitted in classical channel by each intermediate node respectively and independently, not relying on each other.

## REFERENCES

[1] Imre, Sandor, and Laszlo Gyongyosi. *Advanced quantum communications: An engineering approach.* Wiley-IEEE Press, 2012.
[2] Bennett, Charles H., et al. "Teleporting an unknown quantum state via dual classical and EPR channels." *Physical Review Letters* 70.13,1993.
[3] Nielsen, Michael A., and Isaac L. Chuang. *Quantum computation and quantum information.* Cambridge university press, 2010.
[4] Van Meter, R.; Ladd, T.D.; Munro, W.J.; Nemoto, K., "System Design for a Long-Line Quantum Repeater," *Networking, IEEE/ACM Transactions on*, vol.17, no.3, pp.1002,1013, 2009.
[5] Sheng-Tzong Cheng; Chun-Yen Wang; Ming-Hon Tao, "Quantum communication for wireless wide-area networks," *Selected Areas in Communications, IEEE Journal on*, vol.23, no.7, pp.1424,1432, 2005.
[6] Yu Xu-tao, Xu Jin, Zhang Zai-Chen, "routing protocol for wireless ad hoc quantum communication network based on quantum teleportation " *Acta.Phys.Sin*, vol.61, no.22, pp220303-220308, 2012.
[7] Jung-Shian Li; Ching-Fang Yang, "Quantum communication in distributed wireless sensor networks," *Mobile Adhoc and Sensor Systems, IEEE 6th International Conference on*, pp.1024,1029, 2009.

# The Restoring Casimir Force between Doped Silicon Slab and Metamaterials

Xue-Wei Li[a], Zhi-Xiang Huang[a*], Xian-Liang Wu[a,b*]

[a]Key Laboratory of Intelligent Computing and Signal Processing, Anhui University, Hefei 230039,China
[b]School of Electronic and Information Engineering, Hefei Normal University, Hefei 230061,China
*Email: zxhuang@ahu.edu.cn,xlwu@ahu.edu.cn

*Abstract*-Casimir effect is an observable macroscopic quantum effect. It has significant influence on micro-machine and nano-machine. Based on a generalization of the Lifshitz theory, we calculate Casimir force between the doped silicon slab and metamaterials. From simulating, we can see that the magnitude and the direction of the Casimir force can be changed by varying doping level, the thickness of doped silicon slab, and filling factor of metamaterials. Thus the force can be controlled by tuning these parameters. It can be restoring force in a certain range of distance, which can provide a method to deal with the stability problem, and bring much meaningful results from the practical views.

*Key words*: Repulsive and Restoring Casimir Force; Doped Silicon; Metamaterials

## I. Introduction

Casimir effect results from the change of zero-point energy due to the existence of the boundaries and may be observed as a vacuum force. In 1948, this effect, the attractive interaction of a pair of neutral perfectly conducting parallel plates placed in the vacuum, was first proposed and theoretically derived by Casimir [1]. In recent years, the Casimir effect has attracted considerable attention. Especially, with the development of nanotechnology, Casimir force has a significant impact on the micro-electromechanical systems (MEMS) and nano-electromechanical (NEMS) [2-4]. It can be applied to dynamic devices, but also can be lead to stiction problem in MEMS and NEMS. Therefore, Casimir force has become a current research hot spot.

Boyer pointed out a repulsive Casimir force may rise between a perfect conductor plate and a material plate with infinite permeability [5]. But the material in the nature usually has not strong magnetic response in near-infrared and optical spectrum. However, with the development of artificial metamaterials, strong magnetic response material can be obtained, which provides the conditions to get a repulsive Casimir force. Various calculation techniques have been developed, and Lifshitz theory coincides well with the experimental results [6, 7]. Therefore, based on Lifshitz theory, we calculate the Casimir forces between doped silicon slab and metamaterials in vacuum, and study the changes of Casimir forces due to the changes of the doping level, the thickness of slab, and the filling factor of metamaterials.

## II. THE INTERACTION BETWEEN DOPED SILICON SLAB AND METAMATERIALS

Let us consider the configuration with an infinite doped silicon slab and thick metamaterials plate separated by a distance $d$, in the vacuum. The thickness of doped silicon slab is $D$. Based on Lifshitz theory, the resulting Casimir force per unit area $A$ is eventually expressed as [8, 9]

$$\frac{F(d)}{A} = \frac{\hbar}{2\pi^2} \int_0^\infty d\xi \int_0^\infty k_0 k_\parallel dk_\parallel \times \sum_\alpha \frac{r_\alpha^{(1)} r_\alpha^{(2)}}{e^{2k_0 d} - r_\alpha^{(1)} r_\alpha^{(2)}} \ , (1)$$

where $k_\parallel$ is the component of the wave vector parallel to the slab surface, and $r_\alpha$ ($\alpha$ = TE,TM ) is the reflection coefficient for TE- or TM-polarized waves. In $r_\alpha^{(j)}$, superscripts $j$=1, 2 stands for the doped silicon slab and metamaterials, respectively. $k_0 = \sqrt{k_\parallel^2 + \xi^2/c^2}$ , $c$ is the speed of light in the vacuum.

The reflection coefficients for $p$-doped silicon are gave as

$$r_{TM}^{(1)} = \left(\varepsilon_1 k_0 - \sqrt{k_\parallel^2 + \varepsilon_1 \xi^2/c^2}\right) \Big/ \left(\varepsilon_1 k_0 + \sqrt{k_\parallel^2 + \varepsilon_1 \xi^2/c^2}\right), (2a)$$

$$r_{TE}^{(1)} = \left(k_0 - \sqrt{k_\parallel^2 + \varepsilon_1 \xi^2/c^2}\right) \Big/ \left(k_0 + \sqrt{k_\parallel^2 + \varepsilon_1 \xi^2/c^2}\right). (2b)$$

Here we characterize the electric response of doped silicon as [10]

$$\varepsilon_1(i\xi) = \varepsilon_{si}(i\xi) + \frac{\omega_p^2}{\xi^2 + \gamma\xi} \ , \tag{3}$$

$$\varepsilon_{si}(i\xi) = \varepsilon_\infty + \frac{(\varepsilon_0 - \varepsilon_\infty)\omega_{si}^2}{\xi^2 + \omega_{si}^2} \ , \tag{4}$$

$$\omega_p = \sqrt{\frac{Ne^2}{\varepsilon_0 m^*}} = \frac{2\pi c}{\lambda_p}, \gamma = \frac{Ne^2 \rho}{m^*} \ , \tag{5}$$

where $\varepsilon_{si}$ is the dielectric function of intrinsic silicon, and $\varepsilon_\infty = 1.035$ , $\varepsilon_0 = 11.87$ , $\omega_{si} = 6.6 \times 10^{15} rad/s$ . $m^* = 0.34 m_e$ is the effective mass of the charge carrier in the silicon crystal, and $m_e$ is the electron mass. $\rho$ is the resistivity of silicon, and $N$ corresponds to the doping level. The different values of the plasma frequency $\omega_p$ and the relaxation ratio $\gamma$ for various densities are given in Table 1 [10].

Considering the role of the thickness $D$ of the doped silicon slab on the Casimir interaction, the reflection coefficients $r_{TM}^{(1)}$ and $r_{TE}^{(1)}$ in Eq. (2a) and Eq. (2b) become [9, 11]

978-7-5641-4279-7

Table 1 the plasma frequency $\omega_p$ and relaxation ratio $\gamma$ for various doping levels

| $N/cm^{-3}$ | $\omega_e / \times 10^{12} \, rad / s$ | $\gamma / \times 10^{13} \, s^{-1}$ | $\rho / \Omega cm$ |
|---|---|---|---|
| $1.1 \times 10^{15}$ | 3.1899 | 1.18482 | 13 |
| $1.3 \times 10^{18}$ | 110.1275 | 3.75193 | $3.5 \times 10^{-2}$ |
| $1.4 \times 10^{19}$ | 361.522 | 7.86842 | $6.8 \times 10^{-3}$ |
| $1.0 \times 10^{20}$ | 966.084 | 9.917551 | $1.2 \times 10^{-3}$ |

$$r_{TM}^{(1)} \to r_{TM}^{(1)} \frac{1 - e^{-2\sqrt{k_{\parallel}^2 + \varepsilon_1 \xi^2 / c^2} D}}{1 - r_{TM}^{(1)2} e^{-2\sqrt{k_{\parallel}^2 + \varepsilon_1 \xi^2 / c^2} D}}, \quad (6a)$$

$$r_{TE}^{(1)} \to r_{TE}^{(1)} \frac{1 - e^{-2\sqrt{k_{\parallel}^2 + \varepsilon_1 \xi^2 / c^2} D}}{1 - r_{TE}^{(1)2} e^{-2\sqrt{k_{\parallel}^2 + \varepsilon_1 \xi^2 / c^2} D}}. \quad (6b)$$

The reflection coefficient for metamaterials are gave as [8]

$$r_{TM}^{(2)} = \left( \varepsilon_2 k_0 - \sqrt{k_{\parallel}^2 + \varepsilon_2 \xi^2 / c^2} \right) / \left( \varepsilon_2 k_0 + \sqrt{k_{\parallel}^2 + \varepsilon_2 \xi^2 / c^2} \right), (7a)$$

$$r_{TE}^{(2)} = \left( \mu_2 k_0 - \sqrt{k_{\parallel}^2 + \mu_2 \varepsilon_2 \xi^2 / c^2} \right) / \left( \mu_2 k_0 + \sqrt{k_{\parallel}^2 + \mu_2 \varepsilon_2 \xi^2 / c^2} \right). (7b)$$

For metamaterials with partly metallic, we assume that the dielectric function has a Drude background response in addition to the resonant part. Here the optical parameters of metallic-based metamaterials are characterized by

$$\varepsilon_2 (i\xi) = 1 + f \frac{\Omega_D}{\xi^2 + \gamma_D \xi} + (1-f) \frac{\Omega_e}{\xi^2 + \omega_e^2 + \gamma_e \xi}, \quad (8a)$$

$$\mu (i\xi) = 1 + \frac{\Omega_m^2}{\omega_m^2 + \xi^2 + \gamma_m \xi}, \quad (8b)$$

where $\Omega_e$ and $\Omega_m$ are plasma frequency, $\omega_e$ ( $\omega_m$ ) is the electric (magnetic) resonance frequency, and $\gamma_e$ ( $\gamma_m$ ) is the metamaterials electric (magnetic) dissipation parameter. $\Omega_D$ and $\gamma_D$ are the Drude parameters of this metallic structure. $f$ is the filling factor that accounts for the fraction of structure contained in the metamaterials. We choose $\omega_0 = 10^{14} \, rad / s$. For metamaterials with silver material, the corresponding parameters are: $\Omega_e = 0.04\Omega_D$, $\Omega_m = 0.1\Omega_D$, $\omega_e = \omega_m = 0.1\Omega_D$, $\gamma_D = 0.002\Omega_D$, $\gamma_e = \gamma_m = 0.005\Omega_D$, $\Omega_D = 137.036\omega_0$.

III. THE SIMULATION AND THE RESULT ANALYSIS

We first consider the relative Casimir force $10^3 Fd^3 / AhcK_0$ between metamaterials and doped silicon slab with different thickness. We choose the doping level $N = 1.3 \times 10^{18} \, cm^{-3}$. As is shown in the Fig.1, we can see that there is not much change when the silicon thickness $D > 0.005\lambda_0$. The reason is that if $D > 0.005\lambda_0$, so $\exp(-2\sqrt{k_{\parallel}^2 + \varepsilon_1 \xi^2 / c^2} D) \ll 1$. The influence of thickness on this interaction disappears. It can be also

explained with the skin effect of electromagnetic. The skin depth of the material is $\delta = \sqrt{2/\omega\mu\sigma}$. When $D \gg \delta$, the thickness of material almost doesn't affect the transmission and reflection of electromagnetic field. Therefore, the influence of thickness on this interaction disappears.

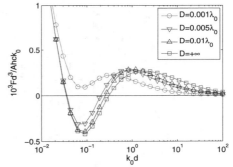

Figure 1 the Casimir interaction between doped silicon slab ( $N = 1.3 \times 10^{18} \, cm^{-3}$ ) and metamaterials ( $f = 10^{-4}$ )

The dependence of the relative force on $N$ is illustrated in Fig.2 by varying $D$. We can see that the doped level has a significant impact on the Casimir effect when $D < 0.005\lambda_0$. The higher doping level, the greater the attractive force and the repulsive force is smaller; and the position of the equilibrium point that the Casimir force is zero will be different if the doping level is different. Thereby we can control the strength of the force and the equilibrium point by changing the doping level. When $D \geq 0.01\lambda_0$, the doping level almost does not affect the interaction, but the restoring force is significant in the range of $150 \, nm - 30 \, \mu m$.

The dependence of relative Casimir forces on the filling factor of metamaterials is shown in Fig.3. The filling factor affects the direction of the force. The smaller the filling factor, the greater the repulsive Casimir force. When the filling factor increases, that is, the proportion of metal structure in metamaterials increases, the Casimir interaction is only attractive.

IV. THE CONCLUSIONS

Based on the Lifshitz theory, we have studied the Casimir forces between doped silicon slab and metamaterials. It is found that the direction and the magnitude of the force are related to the doping level, the thickness and the filling factor. When the thickness of silicon slab is very small, we can control the position of equilibrium point by changing the

doping level. The smaller the filling factor and the greater the repulsive force; and there exists the restoring force, which may provide a method to deal with the stiction problems.

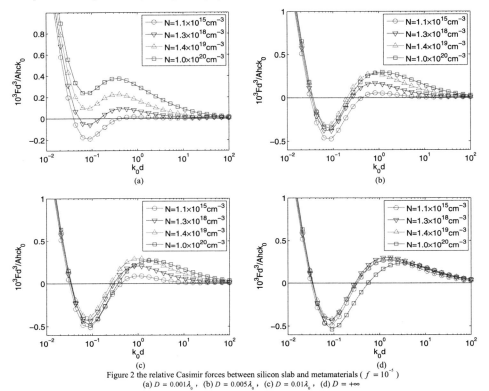

Figure 2 the relative Casimir forces between silicon slab and metamaterials ($f = 10^{-5}$)
(a) $D = 0.001\lambda_0$, (b) $D = 0.005\lambda_0$, (c) $D = 0.01\lambda_0$, (d) $D = +\infty$

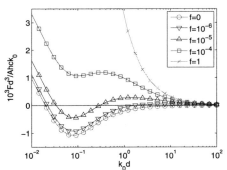

Figure 3 the relative Casimir forces between the doped silicon slab ($N = 1.3 \times 10^{18} cm^{-3}$, $D = 0.1\lambda_0$) and metamaterials

ACKNOWLEDGMENT

This work was supported by the National Natural Science Foundation of China under Grant (Nos. 60931002, 61101064, 51277001, 61201122), DFMEC (No.20123401110009) and NCET (NCET-12-0596) of China, Distinguished Natural Science Foundation (No.1108085J01), and Universities Natural Science Foundation of Anhui Province (No. KJ2011A002), and the 211 Project of Anhui University.

REFERENCES

[1] H. B. G. Casimir, "On the attraction between two perfectly conducting plates," *Proc Kon Ned Akad Wetenschap*, vol. 51, pp. 793-795, 1948.
[2] G. L. Klimchitskaya, U. Mohideen, V. M. Mostepanenko, "The Casimir force between real materials: Experiment and theory," *Rev. Mod. Phys.*, vol. 81, pp. 1827-1880, 2009.
[3] Alejandro W. Rodriguez, Federico Capasso, Steven G. Johnson. "The Casimir effect in microstructured geometries," *Nature Photonics*, vol. 5, pp. 211-221, 2011.
[4] Guo jian-gang, Zhao ya-pu. "Dynamic stability of torsional NEMS actuators with Casimir effect," *Chinese Journal of Sensors and Actuators*, vol. 19, pp.1645-1648, 2006.

[5] Timothy H. Boyer, "Van der Waals forces and zero-point energy for dielectric and permeable materials," *Phys. Rev. A,* vol. 9, pp. 2078-2084. 1974.

[6] E M. Lifshitz, "The theory of molecular attractive forces between solids," *Sov. Phys. JETP,* vol. 2, pp. 73-83, 1956.

[7] P.J. van Zwol, G. Palasantzas, "Repulsive Casimir force between solid materials with high-refractive-index intervening liquids," *Phys. Rev. A,* vol. 81, 062502, 2010.

[8] F.S.S. Rosa, D.A.R Dalvit, P.W. Milonni "Casimir-Lifshitz theory and metamaterials," *Phys. Rev. Lett.,* vol. 100, 183602, 2008.

[9] R. Zhao, Th. Koschny, E.N. Economou, C.M. Soukoulis, "Repulsive Casimir forces with finite-thickness slabs," *Phys. Rev. B,* vol. 83, 075108, 2011.

[10] I. Pirozhenko, A. Lambrecht , "Influence of slab thickness on the Casimir force ," *Phys. Rev. A,* vol. 77, 013811, 2008.

[11] Pochi. Yeh, Optical Waves in Layered Media, 2$^{nd}$ ed., *John Wiley & Sons, Inc. Hoboken, New Jersey,* 2005, pp.83-90.

# Fast Electromagnetic Simulation by Parallel MoM Implemented on CUDA

Ming Fang[a], Kai-Hong Song[a], Zhi-Xiang Huang[a], Xian-Liang Wu[a,b]

[a]Key Laboratory of Intelligent Computing and Signal Processing, Anhui University, Hefei 230039, China
[b]School of Electronic and Information Engineering, Hefei Normal University, Hefei 230061, China
Email: sk_hong@sina.com, zxhuang@ahu.edu.cn

*Abstract*- The keys to electromagnetic simulation of arbitrary configuration of conducting surfaces by method of moments (MoM) are filling of impedance matrix elements and solving of linear equations. A CUDA (Compute Unified Device Architecture) enabled graphics processing unit (GPU) launched by NVIDIA company to accelerate implementation of the fast filling of impedance matrix in MoM based on Rao-Wilton-Glisson (RWG) basis functions was presented. Parallel LU decomposition method is applied for the solving of linear equations, and a parallel method looping over squares is proposed in CUDA parallel platform, furthermore. The GPU numerical results for a user-created benchmark structures are checked with comparison to CPU results. A noticeable speedup of the filling of impedance matrix (hundreds times) and solving of linear equations (about 10 times) is achieved due to the employing of GPU.

*Keywords*- Method of Moments; GPU; LU Decomposition; CUDA

## I. INTRODUCTION

The MoM method was introduced to electromagnetic simulation by R.F.Harrington[1]. As a numerical method for strictly computing electromagnetic problems, it is widely applied and its precision of computed results is high. However, as the dimension of object to be computed increases, the dimensionality of the impedance matrix to be filled is higher, and gives rise to computation and memory capacity, which is the bottleneck of computation by MoM[2]. It has been difficult for the traditional computing element CPU competent for massive data processing.

Fortunately, the MoM method is by its nature a massively parallelizable algorithm and thus can benefit greatly from most advances in multi-core and parallel computing. In this paper, the acceleration of MoM method is implemented with parallel programming of GPUs using Nvidia's CUDA programming platform. We extend MoM's electromagnetic scattering problem from two aspects. Firstly, the filling computation process of each element in impedance matrix is the same, meeting the feature of GPU parallel computation. Each thread of the multi threads of GPU can fill an element, and then tens of thousands of elements are computed parallelly, so it has obvious computation advantages compared with traditional computing element in CPU. Secondly, as the computation method of linear equations, LU decomposition, conjugate gradient (CG) method and Gaussian elimination can be used. These methods may change matrix element during iteration

process, which make these methods unsuited for parallel processing. We will use parallel LU decomposition method. It is proposed for parallel solution of linear equations based on CUDA. In this method, traditional LU decomposition method is partitioned in blocks. Following from the two points, filling time of the impedance matrix and optimizing computation of linear equations become the key concerns and are the subjects of this paper.

## II. THEORIES

### A. Acceleration Techniques in CUDA

In general, applying more effective algorithms and using powerful computers are carried out for speeding up an application. It's a naturally rising possibility to increase the operational speed of the traditional application processor-CPU. However, the computing capacity is under the impact of the manufacturing restrictions. The other possibility is to increase the number of CPUs or the number of cores. Due to the complexity of CPU cores junction and the architectural design problems, it's an expensive attempt.

Taking into account the above factors, GPU launched by NVIDIA supports CUDA techniques which initially designed for graphic processing, and the new hardware design, which is well suited for general parallel computation, and consisting of hundreds of processor cores is capable to execute thousands of threads parallelly at the same time. As we can see in Fig. 1, GPU has more ALUs(Arithmetic Logical Units) than CPU[3].

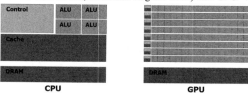

Figure 1. GPU has more transistors for arithmetic logical processing

### B. MoM with RWG Basis Functions

Solving an electromagnetic scattering problem involving perfect electric conduction (PEC), the well-known electric field integral equation (EFIE) is to be solved, which can be expressed as

978-7-5641-4279-7

$$\mathbf{E}^{inc}(\mathbf{r})|_{\tan} = jk_0\eta \left[ \mathbf{J}(\mathbf{r'}) + \frac{1}{k^2}\nabla(\nabla'\cdot\mathbf{J}(\mathbf{r'})) \right] G(\mathbf{r},\mathbf{r'})ds' \Big|_{\tan} \quad (1)$$

and $k_0$ is wave number in the air medium, $\eta$ is intrinsic impedance of the free space, $G(\mathbf{r},\mathbf{r'})$ is the free space Green's function.

With respect to computation of scattering characteristics of three-dimensional PEC body, RWG basis functions adopt triangle pairs to conduct modeling mesh generation on the conductions surface, and the application of triangle pairs can well simulate three-dimensional conductions with arbitrary shapes [4]. RWG basis functions adopt two adjacent triangles as surface element to define the current, the current vector to common edge by the triangle flowing through is as shown in Fig. 2.

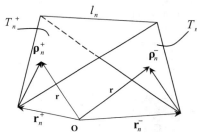

Figure 2. RWG basis function

Assuming $A_n^+$ and $A_n^-$ are areas of triangles $T_n^+$ and $T_n^-$, $l_n^+$ is the common edge, the vector draw from the vertex triangle $T_n^+$ to the observation point is $\boldsymbol{\rho}_n^+$, and $\boldsymbol{\rho}_n^-$ is the vector draw from observation point to the vertex of triangle $T_n^-$. So the expression of basis function can be defined as

$$\mathbf{f}_n(\mathbf{r}) = \begin{cases} \dfrac{l_n}{2A_n^+}\boldsymbol{\rho}_n^+ & \mathbf{r} \in T_n^+ \\ -\dfrac{l_n}{2A_n^-}\boldsymbol{\rho}_n^- & \mathbf{r} \in T_n^- \end{cases} \quad (2)$$

Expand the scattered surface by RWG basis functions. The following linear equation system can be obtained by testing the EFIE by the RWG basis functions

$$\sum_{n=1}^{N} Z_{mn} I_n = V_m \quad m = 1,2,\cdots,N \quad (3)$$

where the computational formulas of EFIE-MoM impedance matrix can be gotten

$$Z_{mn} = jk_0\eta \left( \iint_{S} \int_{S'-r} f_m(\mathbf{r}) \cdot f_n(\mathbf{r'})G(\mathbf{r},\mathbf{r'})dS'\,dS \right.$$
$$\left. - \iint_{S} \int_{S'-r} \frac{1}{k^2}\nabla_S \cdot f_m(\mathbf{r})\nabla'_S \cdot f_n(\mathbf{r'})G(\mathbf{r},\mathbf{r'})dS'\,dS \right) \quad (4)$$

*C. Doolittle LU Decomposition*

The acquisition of impedance matrix and excitation-vectors is followed by a key step which is the consideration of solving linear equations. For *LU* decomposition method, as a solution

of linear equation $Ax = b$, if $A$ is invertible matrix, then matrix $A$ can be resolved to a product of lower triangular matrix $L$ and upper triangular matrix $U$[5].

$$\begin{bmatrix} a_{11} & a_{12} & \cdots & a_{1n} \\ a_{21} & a_{22} & \cdots & a_{2n} \\ \vdots & \vdots & \ddots & \vdots \\ a_{n1} & a_{n2} & \cdots & a_{nn} \end{bmatrix} = \begin{bmatrix} 1 & & & \\ l_{21} & 1 & & \\ \vdots & \vdots & \ddots & \\ l_{n1} & l_{n2} & \cdots & 1 \end{bmatrix} \begin{bmatrix} u_{11} & u_{12} & \cdots & u_{1n} \\ & u_{22} & \cdots & u_{2n} \\ & & \ddots & \vdots \\ & & & u_{nn} \end{bmatrix} \quad (5)$$

Elements of matrix $L$ and matrix $U$ are easily solved

$$u_{1j} = a_{1j}, j = 1,2,\cdots,n \quad (6)$$

$$l_{i1} = a_{i1}/a_{11}, i = 1,2,\cdots,n \quad (7)$$

$$u_{ij} = a_{ij} - \sum_{k=1}^{i-1} l_{ik}u_{kj}, j = i,i+1,\cdots,n \quad (8)$$

$$l_{ij} = \left( a_{ij} - \sum_{k=1}^{i-1} l_{ik}u_{kj} \right)\Big/u_{ii}, i = j,j+1,\cdots,n \quad (9)$$

Substituting $L$, $U$ for $A$, let $UX = Y$, and then we two back substitution problems $Ax=b$ are the things to be resolved.

$$\left.\begin{array}{l} Ly=b \\ Uy=y \end{array}\right\} \quad (10)$$

Solving $Ly = b$ can get $y$,

$$\left.\begin{array}{l} y_1 = b_1 \\ y_1 = b_i - \sum_{j=1}^{i-1} l_{ij}y_j \quad i = 2,\cdots,n \end{array}\right\} \quad (11)$$

Taking the values for $y$ and solving the equation $Ux=y$. And the solution to the system $Ax=b$ can be acquired.

$$\left.\begin{array}{l} x_n = y_n/u_{nn} \\ x_i = \left[ y_i - \sum_{j=i+1}^{n} u_{ij}x_j \right]\Big/u_{ii}, \quad i = n-1,\cdots,1 \end{array}\right\} \quad (12)$$

### III. PARALLEL COMPUTATION IMPLEMENTATION CUDA

3D Studio Max software is used to conduct modeling and mesh generation on three-dimensional PECs. In the CUDA program, the vertexes and common edges data got by 3D Studio Max are required to conduct numbering and matching treatment. GPU reads post-processing information of the mesh from the internal memory, the needed threads are allocated and computed on GPU according to the data volume, and the mesh data read from host-CPU need to be placed on shared memory in device-GPU. Each thread realizes the filling of an impedance matrix element [6].

After filling impedance matrix and excitation vertex, when adopting LU decomposition subprogram to compute induced current coefficient, it requires to share the data above and the left side of diagonal elements to other threads, then we can read the data to complete a circulation according to right-angle circulation. These computations are conducted successively in line with number sequence, and multi-threads may not read shared data at the same time. Therefore, the shared data can be stored in shared memory. To prevent multi-threads reading shared memory at the same time, the "__syncthreads()"

CUDA order can be used to avoid race conditions when loading shared memory[7]. This is quite dangerous in CUDA platform. According to the induced current coefficient, the scattering features of scattered field, induced current at conductor surface and radar cross section (*RCS*) etc. can be achieved.

It can be seen from Eq. (8) and Eq. (9) that computation of $u_{ij}$ needs $a_{ij}$, $l_{ik}$ and $u_{kj}$, while computation of $l_{ij}$ needs $a_{ij}$, $l_{ik}$ and, i.e. computation of $u_{ij}$ needs all data above the $j^{th}$ line, and computation of $l_{ij}$ needs all data at the left of the $i^{th}$ row, as shown in the Fig. 3. Therefore, if the traditional *LU* decomposition method is used, the computations of $u_{ij}$ and $l_{ij}$ can be done only after data relied on by computation are computed, so it is unable to realize parallel computation. Doolittle *LU* decomposition adopts right-angle circulation pattern to distribute the computation process, thus making *LU* decomposition process able to parallelly compute.

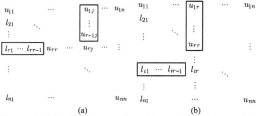

Figure 3. Traditional LU decomposition: (a)Elements required for calculation of $u_{rj}$ ,(b) Elements required for calculation of $l_{ir}$

What shown in Fig. 4 are three processes of distribution according to right-angle circulation. Allocate the elements at the left side of and above the diagonal elements into a process, through loop computation, till the whole matrix is worked out. The method features that when computing $u_{mm}$, ..., $u_{mn}l_{m+1}$, ..., $l_{nm}$, other processes can be computed through sharing all data at the left side of and above umm to them after finishing computing $u_{mm}$. The whole computation process can be completed through data sharing for n times. Data volume needed to be shared at $m^{th}$ step is $2 \times (m - 1)$ data, while the data needed to be transmitted at $m^{th}$ step of traditional *LU* decomposition method is $(m - 1) \times 2$, while greatly reduces the data volume of data transmission.

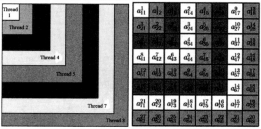

Figure 4. Distribution and numbering according to right-angle circulation of simple eight process in parallel LU decomposition

## IV. RESULTS

With the same hardware platform and solution method, the solutions to scattering features of three-dimensional conductor are realized respectively on CPU and GPU. The configuration of the computer used in the text is: Intel Core i5-2410M dual core CPU main frequency 2.3GHz, host memory 4Gbyte, GPU is NVIDIA GeForce GT 550M, there are totally 1,480,000 thread blocks, and each thread block has 1024 threads, and the video memory is 2Gbyte.

Taking normal plane wave incidence with frequency $3.0 \times 10^8$Hz to PEC sphere of the radius as 0.45 times of wavelength, and compute the bistatic RCS of the PEC sphere. Using C++ serial program and GPU parallel programs for computation, the result through comparison validation between GPU computed result, C++ serial program and analytic solution [8] is shown in Fig. 5.

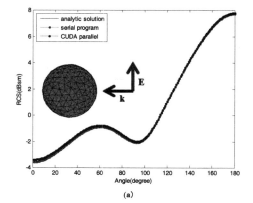

(a)

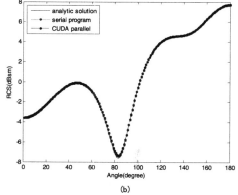

(b)

Figure 5. Comparisons among the bistatic RCS of PEC sphere illuminated by a normal incident vertically polarized plane wave obtained by analytic solution, serical program and parallel program, respectively. (a) Bistatic RCS of E-plane, (b) bistatic RCS of H-plane.

From Fig.5, it can be perceived that, the results solved by parallel program based on CUDA and serial MoM program almost match the analytic solution perfectly. So that the accuracy of CUDA parallel program is verified. To compare the serial program based on CPU and GPU parallel program, the speedup ratio is introduced, i.e. the ratio of times needed for computing same issues by means of GPU parallel program and CPU serial program.

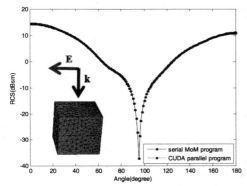

Figure 6. H-plane RCS for PEC cube illuminated by a plane wave (depicted in the figure).Excellent agreements between analytic solution and CUDA parallel program are observed.

The speedup ratios are shown in the Table. I and Table. II for the impedance matrix with different dimensions for the cube structure shown in the insert of Fig. 6. It can be seen that as impedance matrix dimensionality increases, the speed-up ratio may also increase, and the speedup effect is more obvious [9].

Table I:
The speedup ratios of parallel matrix filling

| Matrix dimensions | CUDA parallel program (s) | CPU serial program (s) | Speed-up ratio |
|---|---|---|---|
| 1440 | 0.0373 | 23.98 | 328.49 |
| 2280 | 0.1192 | 63.55 | 533.14 |
| 5220 | 0.5852 | 358.23 | 612.15 |
| 9360 | 1.7430 | 1257.48 | 748.71 |

Table II:
The speedup ratios of parallel matrix filling

| Matrix dimensions | Parallel LU decomposition(s) | Traditional LU decomposition(s) | Speed-up ratio |
|---|---|---|---|
| 1440 | 0.8912 | 8.98 | 10.08 |
| 2280 | 1.2351 | 13.28 | 10.75 |
| 5220 | 3.0729 | 38.34 | 12.48 |
| 9360 | 8.1984 | 127.27 | 15.52 |

## V. CONCLUSION

A computational method was presented for facilitating the fast solution of scatting problems due to PEC surfaces. The proposed method was aiming at solving of the two difficulties i.e. the filling of impedance matrix and solving of linear equations in MoM method. The RWG basis functions were used to expand currents on the conductor surfaces, and GPU was used as parallel programming computing platform to realize parallel filling of impedance matrix elements and parallel computation of LU decomposition, respectively. The proposed method for parallel filling of impedance matrix is found hundreds times faster than CPU serial program, and parallel LU decomposition method can decrease communication quantity on CUDA platform and accelerate computing speed efficiently.

## ACKNOWLEDGMENT

This work was supported by the National Natural Science Foundation of China under Grant (Nos. 60931002, 61101064, 51277001, 61201122), DFMEC (No. 20123401110009) and NCET (NCET-12-0596) of China, Distinguished Natural Science Foundation (No. 1108085J01), and Universities Natural Science Foundation of Anhui Province (No. KJ2011A002), Graduate Academic Innovate Research Foundation (No.60931002, 61101064), and the 211 Project of Anhui University.

## REFERENCES

[1] R. F. Harrington, "Field Computation by Moment Methods". *Macmillan New York*, pp. 30-36, 1968.
[2] Walton C. Gibson, "The Method of Moments in Electromagnetics", Chapman & Hall/CRC, pp.270-281, March 2007.
[3] [Online], Link: http://www.nvidia.cn/object/cuda_home_new_cn.html
[4] D. R. Wilton, S. M. Rao, A. W. Glisson, "Electromagnetic scattering by surface of arbitrary shape", *IEEE Transactions on Antennas and propagation*, vol. 30, pp. 409-418, 1982.
[5] Gilbert Strang, "Introduction to Linear Algebra, 3th edition", Massachutts Institute of Technology, vol. 2, pp. 21-96, 2003.
[6] D. David and E.Lezar, GPU acceleration of method of moments matrix assembly using RWG basis functions, *International Conference On ICEIE*, Amsterdam, pp. 56-60, 2010.
[7] Z. Badics, I. Kiss, Parallel realization of the element-by-element fem technique by CUDA, *IEEE Transactions on Magnetics*, vol. 48, pp. 507-510, 2012.
[8] S.L. Grand, M. Garland, and J. Hardwick, Parallel Computing Experiences With CUDA, *IEEE Transactions on Antennas and propagation*, vol.28, pp. 13-27, 2008.
[9] C. Leat, N. Shuley and G. Stickley, Triangular-patch model of bowtie antennas, *IEE Proceeding Microwaves Antennas and Prop.*, vol. 145, pp. 465-470, 1998.

# An Efficient DGTD Implementation of the Uniaxial Perfectly Matched Layer

Da Peng[1], Lin Chen[2], Wenlu Yin[2], Hu Yang[1]

[1]College of Electronic Science and Engineering, National Univ. of Defense Technology
Changsha, Hunan 410073, China
[2]Southwest Electronics and Telecommunication Technology Research Institute
Chengdu, Sichuan 610041, China

*Abstract*- we proposed a straightforward and highly efficient DGTD implementation of the Uniaxial Perfectly Matched Layer (UPML). Discontinuous Galerkin method (DGM) is imposed on both the UPML Maxwell's curl equations and the auxiliary differential equations (ADE). By recording the intermediate results in the updated processes of E-fields and H-fields, the computation time for auxiliary variables in ADE can be reduced greatly within each time step. Moreover, the auxiliary variables and Riemann flux on each elemental interface are treated dependently here, using which, an excellent stable PML scheme can be obtained. Some examples demonstrate the effectiveness of this UPML, and the proposed method behaves much more stable than the well-posed PML, which is commonly used as PML absorber in DGTD algorithm currently.

## I. INTRODUCTION

Discontinuous Galerkin Time-Domain (DGTD) method is a very useful full-wave numerical algorithm to solve large-scale time-dependent electromagnetic problems with complex geometries in which high accuracy and efficiency are required. It employs discontinuous piecewise polynomials as basis and test functions, and then applies a Galerkin test procedure for each element to obtain the spatial discretization. The solution is not enforced continuous across interface of any two adjacent elements. Instead, a unique numerical flux is constructed to provide the coupling mechanism between elements, which gives rise to a highly parallel formulation.

One of the greatest challenges for DGTD has been obtaining the efficient and accurate solution of electromagnetic wave interaction problems in an open region. In general, there are two classes techniques for such problems: analytical absorbing boundary conditions (ABC) and perfectly matched layer (PML). For the former, a first-order Silver-Müller ABC [1] is employed in the early researches, which is equivalent to Shankar et al.'s [2] straightforward approach of setting the incoming flux to zero in Riemann flux, and a high-order ABC [3] is developed subsequently by William F. Hall and Adour V. Kabakian. Such approaches require putting the outer boundary at least a few wavelengths away from the scatterer in order to get reasonable reflections. This condition leads to an increase of the computational domain scale and CPU time. In the latter case, the well-posed PML is adopted for computational domain truncation by Tian Xiao and Qing H. Liu [4]. In [5,6,15], the auxiliary fileds are redesigned to make the curl operator vanish in the auxiliary differential equations

(ADE). Unfortunately, these PML suffers from some problems of stability [7] when incorporated with DGTD, especially in the case of existence of some very distorted cells in the PML mesh.

In this paper, we proposed a straightforward and highly efficient DGTD implementation of the Uniaxial Perfectly Matched Layer (UPML). Discontinuous Galerkin method (DGM) is imposed on both the UPML Maxwell's curl equations and the auxiliary differential equations (ADE). In ADE, the auxiliary variables and Riemann flux [8] on each elemental interface are treated dependently, using which, an excellent stable PML scheme can be obtained. Finally, the UPML is validated through some examples. Numerical results show this UPML is effective, and much more stable than the well-posed PML, which is commonly used as PML absorber in DGTD algorithm currently.

## II. THEORY

To model electromagnetic waves in an unbounded physical domain, an anisotropic form of the PML [9] is applied here to the DGTD formulation. Maxwell's curl equations are expressed as follows in the PML:

$$\frac{\partial(\mu H)}{\partial t} + K(\nabla \times E) = CB - \sigma_m H \tag{1}$$

$$\frac{\partial(\varepsilon E)}{\partial t} - K(\nabla \times H) = CD - \sigma E \tag{2}$$

where the diagonal matrices $K$, $C$, $\sigma$ and $\sigma_m$ are defined as

$$K = Diag\left[\frac{\kappa_x}{\kappa_y \kappa_z}, \frac{\kappa_y}{\kappa_x \kappa_z}, \frac{\kappa_z}{\kappa_x \kappa_y}\right],$$

$$C = \frac{1}{\varepsilon} Diag\left[\frac{\sigma_x}{\kappa_z} - \frac{\kappa_x}{\kappa_y \kappa_z}\sigma_y, \frac{\sigma_y}{\kappa_x} - \frac{\kappa_y}{\kappa_x \kappa_z}\sigma_z, \frac{\sigma_z}{\kappa_y} - \frac{\kappa_z}{\kappa_x \kappa_y}\sigma_x\right],$$

$$\sigma = Diag\left[\frac{\sigma_z}{\kappa_z}, \frac{\sigma_x}{\kappa_x}, \frac{\sigma_y}{\kappa_y}\right], \quad \sigma_m = \frac{\mu}{\varepsilon}\sigma,$$

and the flux densities $B$ and $D$ satisfy the following ADE:

$$\frac{\partial(\kappa B)}{\partial t} + \nabla \times E = -\gamma B \tag{3}$$

$$\frac{\partial(\kappa D)}{\partial t} - \nabla \times H = -\gamma D \tag{4}$$

where

978-7-5641-4279-7

$$\kappa = Diag\left[\kappa_y, \kappa_z, \kappa_x\right], \quad \gamma = \frac{1}{\varepsilon} Diag\left[\sigma_y, \sigma_z, \sigma_x\right].$$

Note that the UPML Maxwell's equations can reduce to the original Maxwell's equations by setting the PML coefficients $\sigma_x = \sigma_y = \sigma_z = 0$ and $\kappa_x = \kappa_y = \kappa_z = 1$. Therefore, in terms of spatial discretization, it permits a unified treatment of both the lossless interior working volume and the UPML slabs.

For the convenience of using DGM, the above equations are written compactly in a conservation form:

$$Q\frac{\partial q}{\partial t} + \tilde{K}\nabla \cdot F(q) = f \tag{5}$$

$$\tilde{\kappa}\frac{\partial p}{\partial t} + \nabla \cdot F(q) = g \tag{6}$$

where the material matrix $Q$, the state vectors $q$ and $p$, the flux $F(q) = \left[F_x, F_y, F_z\right]^T$, and the body forces $f$, $g$ are defined as follows:

$$Q = \begin{bmatrix} \mu & \\ & \varepsilon \end{bmatrix}, \quad \tilde{K} = \begin{bmatrix} K & \\ & K \end{bmatrix}, \quad \tilde{\kappa} = \begin{bmatrix} \kappa & \\ & \kappa \end{bmatrix} \tag{7}$$

$$q = \begin{bmatrix} H \\ E \end{bmatrix}, \quad p = \begin{bmatrix} B \\ D \end{bmatrix} \tag{8}$$

$$F_i(q) = \begin{bmatrix} e_i \times E \\ -e_i \times H \end{bmatrix} \tag{9}$$

$$f = \begin{bmatrix} CB - \sigma_m H \\ CD - \sigma E \end{bmatrix}, \quad g = \begin{bmatrix} -\gamma B \\ -\gamma D \end{bmatrix}. \tag{10}$$

Here, $e_i$ denotes the three Cartesian unit vectors.

The spatial discretization of (5) and (6) is based on a Galerkin discretization procedure for the spatial derivatives. First, the space domain needs to be partitioned into nonoverlapping elements. To capture the complex geometries, an unstructured grid is used to divide the computational domain into a number of tetrahedrons. Then, within each tetrahedral element, the Lagrange interpolation polynomials $L_j(x)$, associated with a set of nodal points $x_j$, are chosen for the local basis function sets of order $n$ (we refer to this scheme as DGTD-$p_n$), and the state vectors $q$ and $p$ can be represented as

$$q(x,t) \approx \sum_{j=1}^{N} q(x_j, t) L_j(x) = \sum_{j=1}^{N} q_j(t) L_j(x) \tag{11}$$

$$p(x,t) \approx \sum_{j=1}^{N} p(x_j, t) L_j(x) = \sum_{j=1}^{N} p_j(t) L_j(x) \tag{12}$$

where $N$ denotes the total number of nodes in a tetrahedron and is equal to $(n+1)(n+2)(n+3)/6$. More details can be found in [10].

The test functions are chosen to be the same as the basis functions. Multiplied by the test functions and integrated over each element, two weak forms of Eqs. (5) and (6) are obtained over each element $D$

$$\int_D Q\frac{\partial q}{\partial t} L_i(x)dx = -\int_D \tilde{K}\nabla \cdot FL_i(x)dx + \int_D fL_i(x)dx$$

$$\int_D \tilde{\kappa}\frac{\partial p}{\partial t} L_i(x)dx = -\int_D \nabla \cdot FL_i(x)dx + \int_D gL_i(x)dx.$$

Applying the divergence theorem to the first term on the right-hand side, we obtain

$$\int_D Q\frac{\partial q}{\partial t} L_i(x)dx = -\int_{\partial D} \tilde{K}(\hat{n} \cdot F) L_i(x)dx \tag{13}$$
$$+ \int_D \tilde{K}F \cdot \nabla L_i(x)dx + \int_D fL_i(x)dx$$

$$\int_D \tilde{\kappa}\frac{\partial p}{\partial t} L_i(x)dx = -\int_{\partial D} \hat{n} \cdot FL_i(x)dx \tag{14}$$
$$+ \int_D F \cdot \nabla L_i(x)dx + \int_D gL_i(x)dx$$

where $\partial D$ is the boundary of element $D$ and $\hat{n}$ is the outward unit normal vector. The Riemann flux $\hat{n} \cdot F$ is obtained by solving exactly the one-dimensional Riemann problem on each elemental interface [8]:

$$\hat{n} \cdot F|_{\partial D} = \begin{bmatrix} \hat{n} \times \dfrac{(YE - \hat{n} \times H)^- + (YE + \hat{n} \times H)^+}{Y^- + Y^+} \\ -\hat{n} \times \dfrac{(ZH + \hat{n} \times E)^- + (ZH - \hat{n} \times E)^+}{Z^- + Z^+} \end{bmatrix} \tag{15}$$

where $Z$ and $Y$ are the impedance and admittance of the medium respectively. The superscript "-" refers to the outgoing flux and the superscript "+" to the incoming flux across the interface. For the special case of a PEC boundary, the boundary flux can be obtained by setting $Y^+ = \infty$ because the PEC boundary behaves as a material with an infinite admittance. Similarly, a PMC boundary corresponds to the case of setting $Z^+ = \infty$. As for the absorbing boundary, a straightforward approach of setting the incoming flux to zero is employed [2,11].

Substituting Eqs. (7)~(12) into Eqs. (13) and (14), and on the assumption that $\varepsilon$, $\mu$ are elementwise constant, we obtain the semi-discrete scheme in a matrix form

$$\frac{\partial H}{\partial t} = \frac{1}{\mu}\left(-M^{-1}\sum_{m=1}^{4} F^m\left[K\hat{n} \times \frac{(YE - \hat{n} \times H)^- + (YE + \hat{n} \times H)^+}{Y^- + Y^+}\right]_{\partial D_m} + M^{-1}KS \times E + [CB - \sigma_m H]\right) \tag{16}$$

$$\frac{\partial E}{\partial t} = \frac{1}{\varepsilon}\left(M^{-1}\sum_{m=1}^{4} F^m\left[K\hat{n} \times \frac{(ZH + \hat{n} \times E)^- + (ZH - \hat{n} \times E)^+}{Z^- + Z^+}\right]_{\partial D_m} - M^{-1}KS \times H + [CD - \sigma E]\right) \tag{17}$$

$$\kappa\frac{\partial B}{\partial t} = -M^{-1}\sum_{m=1}^{4} F^m\left[\hat{n} \times \frac{(YE - \hat{n} \times H)^- + (YE + \hat{n} \times H)^+}{Y^- + Y^+}\right]_{\partial D_m} + M^{-1}S \times E - \gamma B \tag{18}$$

$$\kappa\frac{\partial D}{\partial t} = M^{-1}\sum_{m=1}^{4} F^m\left[\hat{n} \times \frac{(ZH + \hat{n} \times E)^- + (ZH - \hat{n} \times E)^+}{Z^- + Z^+}\right]_{\partial D_m} - M^{-1}S \times H - \gamma D. \tag{19}$$

Here we have

$$M_{ij} = \int_D L_i(x)L_j(x)dx, \quad S_{ij} = \int_D \nabla L_i(x)L_j(x)dx,$$

$$F_{ij}^m = \int_{\partial D_m} L_i(x)L_j(x)dx$$

and $H$, $E$, $B$, $D$ are vectors whose entries associate with corresponding nodes. Note that there is a similarity between (16) and (18), (17) and (19). By recording the intermediate results in the updated processes of **H** and **E**, the computation time for **B** and **D** can be reduced greatly, which brings an efficient implementation. Moreover, the auxiliary variables of **B** and **D** are dependent with Riemann flux on each elemental interface. It leads to an excellent stable PML scheme.

Following the spatial discretization, we use an explicit time integration scheme, namely a $2^{th}$-order two-stage Runge-Kutta low-storage method. The scheme to solve equation, $d\mathbf{U}/dt = R(t,\mathbf{U})$, is represented as [12]

$$
\begin{aligned}
\mathbf{Q}^0 &= \mathbf{U}^n \\
k_j &= a_j k_{j-1} + \Delta t \cdot R\left((n+c_j)\Delta t, \mathbf{Q}^{j-1}\right) \\
\mathbf{Q}^j &= \mathbf{Q}^{j-1} + b_j k_{j-1}
\end{aligned}
\quad \bigg\} \, \forall j \in [1,2] \qquad (20)
$$
$$
\mathbf{U}^{n+1} = \mathbf{Q}^2
$$

where $a_1 = c_1 = 0$, $b_1 = c_2 = 0.5$, $a_2 = -0.5$, and $b_2 = 1$.

### III. NUMERICAL RESULTS

Because the convergence of DGTD with a set of fixed grids by increasing the order of the approximation within each tetrahedron is beyond our discussion in this paper, unless stated otherwise, all examples are done using a second-order scheme (DGTD-$p_2$) and an average element size of $\lambda/4$ is adopted for tetrahedron mesh generation.

To demonstrate the effectiveness of UPML as an ABC, the radiation of current source in an unbounded 3D region is considered firstly. As shown in Figure 1, an electric current source **J** is surrounded by the air with PML as the outer boundary condition and has the time signature of a differentiated Gaussian pulse

$$
J_z = -2\left[(t-t_0)/\tau\right]\exp\left\{-\left[(t-t_0)/\tau\right]^2\right\} \qquad (21)
$$

where $\tau = 1.25ns$ and $t_0 = 4.5\tau$.

The finite element grids consist of 22867 tetrahedrons with an average element size of 0.25m. The component $E_z$ is probed at two points, $A$ and $B$, as shown in the Figure 1. A relative error is defined as

$$
Rel.error\big|_{(x,y,z)}^{n\Delta t} = 20\log_{10}\left(\left|E^{n\Delta t} - E_{ref}^{n\Delta t}\right|/\left|E_{ref\ max}\right|\right)\Big|_{(x,y,z)} \ (dB) \quad (22)
$$

where $E_{ref\ max}$ is the maximum amplitude of the reference field obtained at spatial location $(x,y,z)$ by running a sufficiently large grids. Figure 2 graphs the relative error calculated using (22) at points $A$ and $B$. Here, the key UPML parameters [14] are $\sigma_{max} = \sigma_{opt}$, $\kappa_{max} = 1$ and $m = 4$. The perpendicular reflection error is set as $R(0) = e^{-8}$.

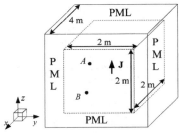

Figure 1. $z$-directed electric current source located at (0.023, -0.134, 0.016 m) in a 3-D DGTD grid. E-fields are probed at points A (0.122, -0.16, -0.66 m) and B (-0.147, 0.575, -0.664 m)

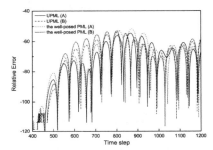

Figure 2. Relative error at points A and B of Fig. 1 over 1200 time-steps for two PML approaches

From this example, it was observed that UPML has a reflection error comparable to or even better than the well-posed PML for the outgoing waves absorption.

Subsequently, let us consider plane wave scattering by perfectly conducting sphere, the analytic solution of which is given by a Mie series. We use both the well-posed PML and the UPML to truncate the computational domain for the same finite element discretization, and choose the PML parameters for the former as recommended in [13]. The thickness of perfectly matched layer is set to half a wavelength.

The perfectly conducting sphere with a diameter of $15\lambda$ is illuminated by a Gaussian derivative pulse whose waveform is depicted by (21), and the component $E_z$ is probed at the same position for both PML approaches. The received signal becomes unstable after 4300 time steps when using the well-posed PML (Figure 4(a)), whereas no sign of instability is observed after 8000 time steps for UPML. The E-plane bistatic cross sections are shown in Figure 4(b). It is in good agreement with the exact solution. The finite element grids consist of 807950 tetrahedrons (Figure 3), and a memory size of 1.53G Bytes is required.

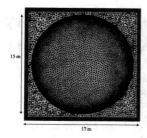

Figure 3. The finite element grid, consisting of 807950 tetrahedrons, used for computing scattering by a PEC sphere with a diameter of $15\lambda$

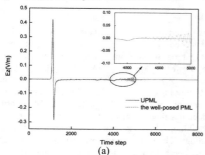

(a)

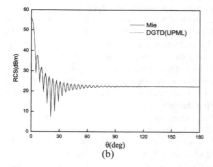

(b)

Figure 4. (a) Comparison of $E_z$ probed at point (-7.5, -1.63, 0.11 m); (b) Bistatic RCS for a PEC sphere with a diameter of $15\lambda$

## IV. CONCLUSIONS

An effective and stable perfectly matched layer approach has been introduced for DGTD to truncate the computational domain. It applies Discontinuous Galerkin method to both the UPML Maxwell's curl equations and the auxiliary differential equations (ADE). By recording the intermediate results in the updated processes of E-fields and H-fields, the computation time for auxiliary variables in ADE can be reduced greatly within each time step. Moreover, because the auxiliary variables and Riemann flux on each elemental interface are linked together, the UPML scheme is more stable than the well-posed PML, whose auxiliary variables are only dependent on E-fields and H-fields at the same discrete nodes. Some examples demonstrate the effectiveness of UPML, and the proposed method behaves much more stable than the well-posed PML, which is commonly used as PML absorber in DGTD algorithm currently.

### REFERENCES

[1] P. Bonnet, X. Perrieres, F. Issac, F. paladian, J. Grando, J. C. Alliot, and J. Fontaine, Finite-Volume Time Domain Method, in Time Domain Electromagnetics. San Diego,CA: Acade Press, 1999, pp. 307-368.

[2] V. Shankar, C. Rowell and W. F. Hall, Unstructured grid-based parallel solver for time-domain maxwell's equations, *Proceedings of the 1997 IEEE Antennas and Propagation Society*, 1997, p. 98-100.

[3] W. F. Hall and A. V. Kabakian, sequence of absorbing boundary conditions for Maxwell's equations, *Journal of Computational Physics*, pp. 140-155, 2004.

[4] T. Xiao and Q. H. Liu, Three-dimensional unstructured-grid discontinuous Galerkin method for Maxwell's equations with well-posed perfectly matched layer, *MICROWAVE AND OPTICAL TECHNOLOGY LETTERS*, vol.46, pp. 459-463, 2005.

[5] T. Lu, W. Cai and P. Zhang, Discontinuous galerkin time-domain method for gpr simulation in dispersive media, *IEEE T Geosci Remote* 43 (2005), no. 1, 72-80.

[6] X. Ji, T. Lu, W. Cai and P. Zhang, Discontinuous galerkin time domain (dgtd) methods for the study of 2-d waveguide-coupled microring resonators, *J Lightwave Technol.* vol. 23, no. 11, pp. 3864-3874, 2005.

[7] G. Cohen and X. Ferrieres, A spatial high-order hexahedral discontinuous Galerkin method to solve Maxwell's equations in time domain, *Journal of Computational Physics*, pp. 340-363, 2006.

[8] A. H. Mohammadian, V. Shankar and W. F. Hall, Computation of electromagnetic scattering and radiation using a time-domain finite-volume discretization procedure, *Computer Physics Communications*, vol.68, pp. 175-196, 1991.

[9] S. D. Gedney, An anisotropic perfectly matched layer absorbing media for the truncation of FDTD lattices, *IEEE Trans. Antennas Propagat.*, vol.44, pp. 1630-1639, 1996.

[10] J. S. Hesthaven and Warburton, Nodal High-Order Methods on Unstructured Grids I. Time-Domain Solution of Maxwell's Equations, *Journal of Computational Physics*, pp. 186-221, 2002.

[11] A. V. Kabakian, Unstructured Grid-Based Discontinuous Galerkin Method for Broadband Electromagnetic Simulations, *Journal of Scientific Computing*, vol.20, 2004.

[12] M. H. Carpenter and C. A. Kennedy, Fourth-Order 2N-Storage Runge-Kulla Scheme, Tech. Rpt. NASA-TM-109112. 1994.

[13] G. Zhao, THE 3-D MULTI-DOMAIN PSEUDOSPECTRAL TIME-DOMAIN METHOD FOR ELECTROMAGNETIC MODELING, Doctor dissertation, Department of Electrical and Computer Engineering, Duke University, 2005.

[14] A. Taflove and S. C. Hagness, *Computational Electrodynamics The Finite-Difference Time-Domain Method* (Third Edition), BOSTON: ARTECH HOUSE, 2005.

[15] S. D. Gedney et al., A Discontinuous Galerkin Finite Element Time Domain Method with PML, *Antennas and Propagation Society International Symposium*, 2008.

# A Graph-Theoretic Approach to Building Layout Reconstruction from Radar Measurements

Bo Chen[1], Tian Jin[1], Biying Lu[1], Zhimin Zhou[1], Pu Zheng[2]

[1] College of Electronic Science and Engineering, National University of Defense Technology,
Changsha, Hunan, P. R. China, 410073
chenbo_nudt@163.com
[2] Southwest Electronic and Telecommunication Technology Research Institute,
Chengdu, Sichuan, P. R. China, 610041
pufkk_zp@sina.com

*Abstract*-Motivated by the desire to obtain the interior layout of a building from through-the-wall radar measurements, we have developed a building layout reconstruction algorithm based on the minimum spanning tree (MST), which is a graph-theoretical method. Based on the extraction of all the wall-wall-floor trihedrals from radar measurements, we have defined the vertex and edge set of the building layout graph. Then the edge weight is determined according to actual conditions. Finally, the MST method is applied to reconstruct the building layout. Simulation results have shown effectiveness of the method. The techniques in this paper are intended to serve as an exploration into the graph theoretical solution on the building interior layout reconstruction problem.

## I. INTRODUCTION

In recent years, the technology of sensing through walls has received considerable attention. Through-the-wall radar imaging (TWRI) is considered very effective to achieve the objectives of "seeing" through walls. TWRI is highly desirable for a range of organizations, including police, fire and rescue personnel missions, surveillance, first responders, and defense forces [1, 2].

At present TWRI mainly focus on behind-the-wall targets. These through-the-wall radars often require a close position to the wall or they have to be pressed against the wall. However, they are only available for single-wall penetration, providing range, direction, and motion information of moving objects [3-5].

Nowadays there has been growing interest in techniques for determining the layout of a building's interior from radar measurements made from that building's exterior. There are many applications for such an ability, including firefighting and lawenforcement operations.

The newly works on this topic are the Synthetic Aperture Polarimetric Phased Array Interferometer Radar Equipment (SAPPHIRE) and the visibuilding program. The SAPPHIRE is developed by the Netherlands Organization for Applied Scientific (TNO), whose operational principle relies on detecting and identifying dominant scatterers inside a building by exploiting specific phase relations in the measured 3D data cube[3,4]. SAPPHIRE focuses on 2D reconstruction of an empty room on the ground floor. However, the reconstruction method is sensitive to model inaccuracies. Additional filtering step is needed to cluster the detected scatterers. The visibuilding program was funded by the Defense Advanced Research Projects Agency (DARPA) [6]. The program addresses the desire to see inside structure to determine the layout of buildings, using a model-based signal processing method. In addition, the U.S. Army Research Laboratory (ARL) has conducted a field experiment to map an abandoned Army barrack building. The images were obtained from two sides of the building. They also have performed some computer simulations with Xpatch to explain some phenomenon observed in the measured SAR images [7, 8].

In literatures, there is consensus on the fact that a map of a building can be best obtained by detecting and identifying principal scatterers especially trihedrals inside a building [9, 10]. As long as we have got the position and orientation of each trihedral inside a building, the layout of the building can be reconstructed subsequently. Motivated by the desire to estimate the interior layout more accurately, we have proposed an iterative deducing method. Due to compensating the effect of wall mentioned in the literature [11], the estimation results produced by the present iterative loop will be closer to the true layouts of the building than the previous one. Thus, the estimation layout will approach the true structure loop by loop. In order to improve the efficiency of the iterative method and to keep away from mannual intervention, auto-reconstruction method needs to be developed.

The treatment in this paper considers, without the loss of too much generality, that the floorplan of the building is rectangular in shape. Namely, the interior walls are assumed to be either parallel or perpendicular to the front wall. Moreover, any closed structure is not allowed to exist inside the building. The basis for this graph-theoretical approach is drawn from a paper on the subject of layout estimation by Lavely et al. They employ graph theory in its problem formulation [12]. The way that graphs are employed in that paper is that the edges of the graph are used to indicate the walls that are present or not present, and the nodes are used to represent the location and length of walls. In the end of this paper, we have carried some computer simulations to validate the graph method.

978-7-5641-4279-7

## II. GRAPH-BASED MODELS FOR BUILDING INTERIOR LAYOUT

The building interior layout reconstruction approach considered by us is based on a minimum spanning tree (MST) method. This technique has its roots in graph theory. In this section, we will define the graph elements in advance and make relationship between the layout reconstruction problem and the MST problem.

### A. Graph Elements

This paper intends to deduce the interior layout of a building. We treat the interior layouts of a building as a weighted undirected graph. It is called building layout graph in the following text. Usually, a graph is represented as $G(V, E)$, where $V$ is the set of vertices (or nodes) and $E$ is the set of weighted edges connecting them. The vertex set and the edge set of the building layout graph are defined as follows:

*1) Vertices (or nodes):* The wall-wall-floor trihedrals inside the building are treated as nodes of the building layout graph.

*2) Edges:* Walls existing between two adjacent nodes, the edges are undirected.

*3) Degree:* It is used to describe a node. It means the number of walls which relates to current wall-wall-floor trihedral node. For undirected graph, in-degree and out-degree are both treated as uniform degree with no difference.

*4) Order:* It is used to describe a graph. It means the number of trihedral nodes in the building graph and is denoted as $|V|$.

Positions and orientations of the wall-wall-floor trihedrals are attributes of the vertex set. A weighted undirected graph can be established if we have got all the trihedrals including their attributes through radar measurements. How to obtain these attributes has been mentioned in the literature [13]. From the literature we have known that the pose angles of all the trihedrals in a radar image can be estimated by using a virtual aperture imaging model.

### B. MST in Building Layout Graph

In this subsection, we will present the concept of MST and the relationship between MST and the interior layout of a building.

Given a complete weighted undirected graph $G(V, E)$ with $|V| = N$, the number of trees (a subgraph of $G$ without closed loops) that connects all the nodes of the graph is $N^{N-2}$. The MST is the tree with the minimum total weight, defined as the sum of the weight of each tree's edge. As is mentioned afore, in a radar measurement data set consisting of principal scatterers, we consider the wall-wall-floor trihedrals as the nodes of a graph, the edges being the wall lines joining the nodes. If we have defined reasonable edge weight to translate the building interior layout reconstruction process into searching the MST of the complete weighted undirected graph.

Hereto, the last job we have not finished to construct the weighted undirected graph is the definition of edge weight. In the following sections, we will present the definition process of edge weight so as to establish the equivalence between the MST and the correct building interior layout. The keypoint of solving the MST problem and the reconstruction problem lies in defining the edge weight.

### III. BUILDING LAYOUT RECONSTRUCTION USING MST METHOD

#### A. Edge Weight

Before we define the edge weight, we will give the definition of angle range of each trihedral node in advance. The following figure gives the angle range definition when the degree of a trihedral node equals 2, 3 and 4. $\theta_s$ means the starting angle and $\theta_e$ means the ending angle. The angle range is denoted as $[\theta_s, \theta_e]$. Due to the rectangular shape of the building. The orientation of trihedrals are divided into four quadrants.

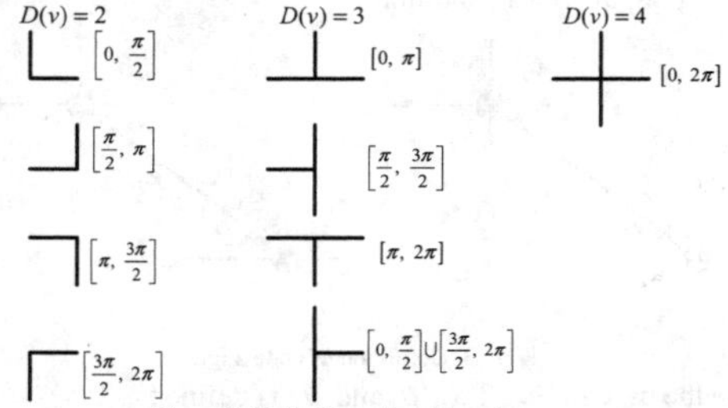

Figure 1. Definition of node angle range.

Next, we will define the edge weight between a couple of trihedral nodes. According to the actual situation, the definition of the edge weight $\omega_{ij}$ between a couple of nodes which are denoted as $v_i$ and $v_j$ should subject to the following three conditions:

1) The edge weight that drops within the angle range $[\theta_s, \theta_e]$ should be smaller than that drops out of the angle range. Moreover, the edge weight that drops within the angle range should better be much smaller as it is not reasonable that an edge is outside the angle range.

2) To keep the monotonicity of the Euclid distance in the angle range, the edge whose Euclid distance is smaller should also be smaller in weight.

3) If the weight function is in exponential form, the base should not be equal to 1, and the exponent should not be equal to 0. Otherwise, the weight can not be distinguished from angle range and Euclid distance.

If we have constructed an edge weight that satisfying the above three conditions, the building interior layout reconstruction process can be translated into searching the MST of the complete weighted undirected graph.

From the above three conditions, we could see that the edge weight between a couple of nodes should consider not only their Euclid distance but also the angle constraint. The edge weight we considered is expressed in exponential form as follows:

$$\omega_{ij} = \left(\frac{d_{ij}}{d_{max}}\right)^{\gamma(\theta_i,\theta_j)} \qquad (1)$$

where $d_{ij}$ is the Euclid distance between node $v_i$ and node $v_j$ .

$$d_{ij} = \sqrt{(x_i - x_j)^2 + (y_i - y_j)^2} \qquad (2)$$

$(x_i, y_i)$ represents the position attribute of node $v_i$ and $(x_j, y_j)$ has the similar meaning. $\gamma(\theta_i, \theta_j)$ is the exponential term of the edge weight, the definition of $\theta_i$ and $\theta_j$ is shown in Fig. 2. $d_{max} = \alpha L_{max}$ $(\alpha > 1)$ , $L_{max}$ represents the biggest Euclid distance among all possible node pair which is prior known. $L_{max}$ is defined as the diameter of a graph in this paper. The existence of factor $\alpha$ is to keep the base from equaling to 1, satisfying the third condition.

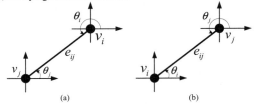

(a)        (b)

Figure 2. Definition of node angle.

For the case in Fig. 2(a), $\theta_i$ and $\theta_j$ is defined as:

$$\begin{cases} \theta_i = \arccos \dfrac{x_i - x_j}{d_{ij}} + \pi & (\theta_i \in [\pi, 2\pi]) \\ \theta_j = \arccos \dfrac{x_i - x_j}{d_{ij}} & (\theta_j \in [0, \pi]) \end{cases} \qquad (3)$$

For the other case in Fig. 2(b), $\theta_i$ and $\theta_j$ is defined as:

$$\begin{cases} \theta_i = \arccos \dfrac{x_j - x_i}{d_{ij}} & (\theta_i \in [0, \pi]) \\ \theta_j = \arccos \dfrac{x_j - x_i}{d_{ij}} + \pi & (\theta_j \in [\pi, 2\pi]) \end{cases} \qquad (4)$$

According to the above description and conditions, the exponential term of the edge weight can be expressed as follows when $\theta_i \in [\theta_{is}, \theta_{ie}]$ and $\theta_j \in [\theta_{js}, \theta_{je}]$ :

$$\gamma(\theta_i, \theta_j) = -|\sin(2\theta_i)| - |\sin(2\theta_j)| + \beta \qquad (5)$$

otherwise, when the edge $\omega_{ij}$ drops out of the angle range, the exponential term can be written as:

$$\gamma(\theta_i, \theta_j) = \eta \qquad (6)$$

The parameter $\beta$ is introduced in order to keep the exponential term above zero. Thus, $\omega_{ij}$ will increase with the Euclid distance increase when $\theta_i$ and $\theta_j$ are within the angle range. This is a reply to the second condition. The exponential term will be a negative constant $\eta$ when either $\theta_i$ or $\theta_j$ is out of the angle range. The negative property of $\eta$ will increase the weight dramatically and be greater than the edge weight

when both $\theta_i$ and $\theta_j$ are within the angle range, satisfying the first condition. The greater the absolute value of $\eta$ , the bigger the edge weight. The parameter $\eta$ and $\beta$ adjust the difference between the edge weight that drops in the angle range and the one out of the angle range.

*B. MST Solving Algorithm*

We have give the definition of edge weight of the building layout graph. Now we should focus on how to find the MST of the building graph so as to obtain the interior layout of the building.

The frequently used MST searching algorithms are the Kruskal algorithm and the Prim algorithm. Both of these two algorithms come from greedy ideas. However, proof has been shown that these algorithms would obtain the global optimization results instead of being driven to the local optimization results. The Kruskal method needs to sort the edge weight only one time, while the Prim method needs to sort the edge weight more than one times. As a result, we have used the algorithm of Kruskal to construct our MSTs for computation efficiency. The process of the Kruskal algorithm in building layout reconstruction is described with four steps:

*Step 1:* Using definition in (3), sort the edge weight of the weighted undirected building graph in ascending order;

*Step 2:* Set $i = 1$ and let the initial edge be $E_0 = \varnothing$ ;

*Step 3:* Select an edge $e_i$ of minimum weight value not in $E_{i-1}$ such that $T_i = <E_{i-1} \cup \{e_i\}>$ is acyclic and define $E_i = E_{i-1} \cup e_i$ . If no such edge exists, let $T = <E_i>$ and stop;

*Step 4:* Replace $i$ by $i+1$ . Return to Step 3.

After $N-1$ iterations, where $N$ is the order of the building layout graph, the complete MST is found. Thus, the building interior layout is obtained subsequently.

IV. SIMULATION RESULTS

In the above sections, we have translate the building layout reconstruction problem into the MST searching problem. With the edge weight defined as (1)~(6) and the Kruskal algorithm, some simulations have been implemented and we have got some useful results. We have simulated two buildings which have different layouts as is shown in the following figure.

We have simulated a simple building consisting of four rooms at first to validate the reconstruction algorithm based on MST. Without loss of generality, the building size is assumed to be $2m \times 2m$ and the size of each room is $1m \times 1m$ . The layout is shown in Fig. 3.

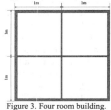

Figure 3. Four room building.

Fig. 4(a) is the simulation results when we consider all the wall-wall-floor trihedrals appearing in the radar image, including those trihedrals located along the building outline corners. The red line represents the generated MST based on the wall-wall-floor trihedral nodes. The other four kind colors denote the quadrant at which the wall-wall-floor trihedral was oriented. In this case, we are unable to obtain the correct layout of the building. Fig. 4(b) is the simulation results after we eliminate the outmost trihedrals of the building. The dashed line represents the outline of the building. It can be seen that the MST based on the remaining trihedrals represent the correct interior layout. It is reasonable to assume that the building outline could be obtained in advance. Thus, combinning the prior outline information, we are able to obtain the whole layout of the building.

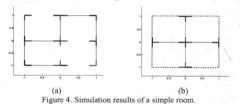

(a)          (b)

Figure 4. Simulation results of a simple room.

A little more complex residential building layout shown in Fig. 5 is also simulated in this paper. The size of the building is $8m \times 8m$. The outmost trihedrals of the building are eliminated and the final results are presented in Fig. 6. The dashed line represents the outline of the building.

Figure 5. A residential building layout.

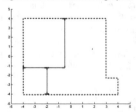

Figure 6. Simulation results of a residential building.

The above simulation results has shown that the MST based reconstruction algorithm is capable of deducing the interior layouts of a rectangular shape building. The whole structure of a building can be obtained if the outline information is known prior. Comparing with other exhaustive attack method, the MST based method will greatly reduce the computation burden.

It is more suitable to be applied in the iterative process and enhance the accuracy of the deducing results.

## V. CONCLUSION

We have proposed a building layout reconstruction algorithm assuming that complete position and orientation information has been obtained. It is hoped that based on these preliminary results, we are able to predict the building layouts using radar measurements. However, the following three situations should be further explored in practical radar measurements. The first one is some trihedrals may not be detected in the radar image. The second one is some behind-the-wall targets may be treated as trihedrals by mistake. The last one is the position of the trihedrals will deviate from their original location. These cases will be discussed in our future research work.

## ACKNOWLEDGMENT

This work was supported by the National Natural Science Foundation of China under Grants 61271441 and the Foundation for the Author of National Excellent Doctoral Dissertation of China under Grant 201046.

## REFERENCES

[1] S. E. Borek, "An Overview of Through the Wall Surveillance for Homeland Security," in *Proceedings of the 34th Applied Imagery and Pattern Recognition Workshop (AIPR05)*, 2005, pp.

[2] J. Clerk Maxwell, *A Treatise on Electricity and Magnetism*, 3rd ed., vol. 2. Oxford: Clarendon, 1892, pp.68-73. L. Frazier, "Surveillance through walls and other opaque materials," *IEEE Aerosp. Electron. Syst. Mag.*, vol. 11, no. 10, pp. 6-9, 1996.

[3] J. J. M. de Wit, W. L. van Rossum and F. M. A. Smits, "SAPPHIRE-a Novel Building Mapping Radar," in *Proceedings of the 39th European Microwave Conference*, 2009, pp. 1896-1899.

[4] J. J. M. de Wit, L. Anitori, W. L. van Rossum and R. G. Tan, "Radar Mapping of Buildings using Sparse Reconstruction with an Overcomplete Dictionary," in *Proceedings of the 8th European Radar Conference*, 2011, pp. 9-12.

[5] F. Soldovieri and R. Solimene, "Through-Wall Imaging via a Linear Inverse Scattering Algorithm," *IEEE Geosci. Remote Sens. Lett.*, vol. 4, no. 4, pp. 513-517, 2007.

[6] E. J. Dr. Baranoski, "Through Wall Imaging: Historical Perspective and Future Directions," in *Proc. IEEE ICASSP*, 2008, pp. 5173-5176.

[7] C. Le, T. Dogaru, L. Nguyen and M. A. Ressler, "Ultrawideband (UWB) Radar Imaging of Building Interior: Measurements and Predictions," *IEEE Trans. Geosci. Remote Sens.*, vol. 47, no. 5, pp. 1409-1420, 2009.

[8] T. Dogaru, A. Sullivan, C. Kenyon and C. Le, "Radar Signature Prediction for Sensing-Through-the-Wall by Xpatch and AFDTD," in *DoD High Performance Computing Modernization Program Users Group Conference*, 2009, pp. 339-343.

[9] P. C. Chang, R. J. Burkholder and J. L. Volakis, "Adaptive CLEAN With Target Refocusing for Through-Wall Image Improvement," *IEEE Trans. Antennas Propag.*, vol. 58, no. 1, pp. 155-162, 2010.

[10] E. Ertin and R. L. Moses, "Through-the-Wall SAR Attributed Scattering Center Feature Estimation," *IEEE Trans. Geosci. Remote Sens.*, vol. 47, no. 5, pp. 1338-1348, 2009.

[11] T. Jin, B. Chen and Z. Zhou, "Image-Domain Estimation of Wall Parameters for Autofocusing of Through-the-Wall SAR Imagery," *IEEE Trans. Geosci. Remote Sens.*, vol. 51, no. 3, pp. 1836-1843, 2013.

[12] E. M. Lavely, Y. Zhang, E. H. Hill III, Y. Lai, P. Weichman and A. Chapman, "Theoretical and Experimental Study of Through-wall Microwave Tomography Inverse Problems," *Journal of the Franklin Institute*, vol. 345, no. 6, pp. 592-617, 2008.

[13] B. Chen, T. Jin, Z. Zhou and B. Lu, "Estimation of Pose Angle for Trihedral in Ultrawideband Virtual Aperture Radar," *Progress In Electromagnetics Research*, vol. 138, pp. 307-325, 2013.

# The Imaging Approach of Sparse Interferometry to Microwave Radiation

Yuanyuan Liu  Suhua Chen  Lu Zhu

School of Information Engineering East China Jiaotong University, NanChang, 330013 China

*Abstract*—**High efficiency image reconstruction and inversion algorithm is one of the key technologies for interference synthetic aperture microwave radiometer. Due to the fact that the brightness temperature of the Earth has a local smoothness characteristic, it could be random sparse interferometry. Based on compressive sensing, this paper proposes a novel imaging approach of sparse interferometry to microwave radiation. According to the sparity of the image and the characteristic of the interferometry, we set up the microwave radiation sparse interferometric imaging model using Total Variation constraints on the basis of the traditional microwave radiation imaging. In the model, we use a new sparse interferometry to sample frequency information on the basis of the sparse antenna array. During the process of microwave radiation inversion imaging, we use the steepest descent method and the alternate iteration method reconstruct. Experimental results show that the proposed approach is able to rapid, accurate and efficient inverse microwave radiation image.**

*Index Terms*—*interference synthetic aperture microwave radiometer, compressive sensing, random sparse interference method, TV reconstruction algorithm, steepest descent method, Alternating Direction Algorithm, Total Variation, image inversion*

## I. INTRODUCTION

Interference synthetic aperture microwave radiometer (ISAMR) was suggested in the 1980s as an alternative to real aperture radiometry for Earth observation at low microwave frequencies with high spatial resolution [6]. ISAMR integrated small antenna array into a large observation aperture, and imaging without mechanical scan, so it can solves the disadvantages of real aperture microwave radiometer [12]. How to efficient and accurate inverse the synthetic aperture microwave radiation image is the key. According to the Fourier transform relationship between inversion bright temperature of ISAMR and visibility function, He Yuntao [2] and other scholars put forward various inversion algorithm based on Fourier

transform, but this algorithm is very high requirement for the hardware system and the imaging error is larger. The present synthetic aperture image inversion methods are mostly deterministic inversion method [7], so it is rarely make full use of the local smoothness characteristics of brightness temperature image. For then now L-Band satellite-based ISAMR, it still needs a diameter of 9m antenna array to achieve 50 km spatial resolution [3], with the microwave radiation imaging to refinement and structured development, we must increase the diameter of the antenna array to satisfy the need of high resolution. ISAMR collect tens of millions data during one observation [8]. So the Nyquist spatial interferometry and conventional microwave radiation imaging method are difficult to achieve.

Compressed Sensing (CS) has become the research hot spots focus in various fields [13], which use the adaptive linear projection to preserve the original structure of the signal, and accurately reconstruct the original signal through numerical optimization [10]. In this method, it is also an increasing dimension problem from measured values to the original signal, which is similar to super-resolution image reconstruction.

Therefore, to against the problem of huge data and less resolution of the traditional microwave radiation interference measuring method, by fully exploiting the spatial structure information of the microwave radiation imaging, from the aspects of CS, using the sparse random interference measuring method of microwave radiation imaging to study [9], this paper put forward a imaging approach of sparse interferometry to microwave radiation, reduce the complexity of the imaging system structure and hardware cost, and break through the inherent limits of the imaging system's spatial resolution to get the high spatial resolution image, which is close to the expensive big aperture imaging system. After several simulation experiments, the results show that the approach can get good image inversion result.

## II. PRINCIPLE THEORY

### A. Principles of CS

The area of CS was initiated in 2006 by two ground breaking papers by Candès, Romberg, and Tao [5], and by Donoho [4]. CS theory mainly includes signal sparse representation, encoding measurement and reconstruction algorithm. If the signal has only a few nonzero elements, then the signal is sparse, so it can project onto an orthogonal transformation matrix, use far lower sampling rate than the traditional Nyquist sampling to sample and compress, and apply the prior information of the original

The work was supported by the Natural Science Foundation of Jiangxi Province (No. 2010GQS0033) and National Nature Science Foundation of china (No. 61162015, No.31101081).

Corresponding author: Lu Zhu (e-mail: luyuanwanwan@163.com)

978-7-5641-4279-7

signal to accurately reconstruct it. The mathematical model of compressed sensing is shown as follows:

The N dimensional real signal $X \in R^{N \times 1}$ is being unfold under a set of orthogonal basis $\{\Psi_i\}_{i=1}^{N}$ $\Psi_i$ (is N dimensional Column vector):

$$x = \sum_{i=1}^{N} \theta_i \Psi_i \qquad (1)$$

The matrix form is:

$$x = \Psi \theta \qquad (2)$$

$\psi = [\Psi_1, \Psi_2, ...., \Psi_N] \in R^{N \times N}$ is the orthogonal basis, and satisfies $\Psi \Psi^T = \Psi^T \Psi = I$ . $\theta = [\theta_1, \theta_2, ...., \theta_N]^T$ is expansion coefficient vector. Assuming that the coefficient is K-sparse vector, and the number of nonzero coefficient is K, K<<N, so we can use another observation matrix $\Phi \in R^{M \times N}$ (M<<N) which is not relevant with the orthogonal basis dictionary to observe

$$y = \Phi x \qquad (3)$$

M linear observations (or projections) $y \in R^{M}$ can be get. These small amount of linear projection contains enough information to reconstruct signal x. The observations y can be reconstructed from signal x When $\Theta = \Phi \Psi$ satisfies the constraint of restricted isometry property (RIP) [1]. The reconstruction can be expressed by solving the following optimization problem

$$\min_{x} \left\| \Psi^T x \right\|_0 \quad s.t \quad y = \Phi x \qquad (4)$$

But the optimization problem (4) is $l_0$ norm, and it's an NP - Hard problem. In order to solve the problem, we usually use the $l_1$ norm instead of $l_0$ norm, namely:

$$\min_{x} \left\| \Psi^T x \right\|_1 \quad s.t \quad y = \Phi x \qquad (5)$$

### B. ISAMR sparse interferometry

ISAMR has the high perceptive ability of the earth observation at low microwave frequencies with high resolution [11]. It measures the complex correlation between the signals collected by pairs of spatially separated antennas which have overlapping fields of view, yielding samples of the visibility function V (also termed complex visibilities) of the brightness temperature distribution T of the observed scene. The relationship between $V(u)$ and $T(\xi)$ is given by [8]:

$$V(u_{kl}) \propto \frac{1}{\sqrt{\Omega_k \Omega_l}} \iint F_k(\xi) \overline{F_l}(\xi) T(\xi)$$

$$\times \tilde{r}_{kl} \left( \frac{-u_{kl} \xi}{f_0} \right) e^{-2j\pi u_{kl} \xi} \frac{d\xi}{\sqrt{1 - \|\xi\|^2}} \qquad (6)$$

where $u_{kl}$ is the spatial frequency associated with the two

antennas $A_k$ and $A_l$ (namely, the spacing between the antennas normalized to the central wavelength of observation), the angular position variable $\xi$ is direction cosine($\theta$ and $\phi$ are the traditional spherical coordinates, $F_k(\xi)$ and $F_l(\xi)$ are the normalized voltage patterns of the two antennas and with equivalent solid angles $\Omega_k$ and $\Omega_l$, $\tilde{r}_{kl}$ is the so-called fringe-wash function, which accounts for spatial decorrelation effects, $u_{kl} \xi / f_0$ is the spatial delay, and $f_0$ is the central frequency of observation. When the nonzero base line number of ISAMR is M, M equations as shown in (6) can be expressed by matrix:

$$V = GT \qquad (7)$$

Where V is the column vector of the visibility function sampling, T is the discrete brightness distribution, and G is model operator matrix. We know that the relationship between V and T is actually Fourier transform, so we begin to disperse for the space frequency on the basis of (7), and then randomly selected spatial Fourier frequency component, the microwave radiation sparse random interference measurement model is as follows:

$$V = F_\wedge T \qquad (8)$$

$F_\wedge$ is the sparse observation matrix. When T is sparse, we use $F_\wedge$ to interferometry for T on the basis of sparse antenna. By doing this, we can get the useful information which is far less than the amount of microwave radiation image data. While the traditional interferometry imaging method based on Nyquist sampling theory, so the amount of sampling information is relatively big. Fig.1 is the principle of traditional interferometry imaging method, in this method, the antenna array are arranged to get the visibility function V, which must be consistent in Nyquist theory. We, then get the brightness temperature V through the linear reconstruction to the visibility function. Fig. 2 is the principle of new random sparse interferometry. Firstly, we get the microwave radiation V from very sparse antenna arrays just as Fig. 1, similarly using the random sparse observation matrix $F_\wedge$ to sparsely sample the spatial frequency. Finally we can use the nonlinear reconstruction algorithm to inversion the brightness temperature.

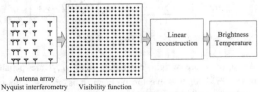

Antenna array
Nyquist interferometry      Visibility function

Fig.1.Traditional interferometry imaging method

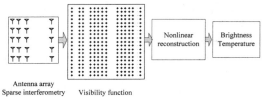

Antenna array
Sparse interferometry    Visibility function

Fig.2. Random sparse interferometry image method

### C. The reconstruction algorithm

According to the area smoothness of microwave radiation image, with the thought of CS, we use Total variation (TV) to sparse representation the image, so as to save all the information. The definition of TV is shown as follows:

$$TV(T) = \sum_{i,j} \sqrt{(T_{i,j} - T_{i-1,j})^2 + (T_{i,j} - T_{i,j-1})^2} \qquad (9)$$

So in the foundation of (2), the microwave radiation image's sparse representation model can be presented as follows:

$$V = F_\wedge \Psi \alpha \qquad (10)$$

While $\alpha$ is the useful information which is far less T, $A = F_\wedge \Psi$, $A \in R^{M \times N}$ is the random observation matrix, in fact, it is the part of random Fourier observation matrix. The part of random Fourier observation matrix is that to generate a random $N \times N$ Fourier matrix firstly, then to sample of the M line randomly to form a new matrix, and unitized for each column lastly. Equation (11) is a morbid linear equation from the mathematical perspective. In order to solve the problem, we build the following TV model based on the principle of CS and the characteristic of temperature image

$$\begin{cases} \min_{w_i \in R^2, T_b \in R^n} \sum_i \|w_i\| \\ \text{s.t. } V=AT \text{ and } D_i T = w_i \text{ for all i} \end{cases} \qquad (11)$$

While $W_i$ is a secondary variable, $\|\cdot\|$ can be either 1-norm or 2-norm. $D_i$ is the differential operator. How to accurately resolve the formula (11) is the key. The (11) is a convex optimization problem, so we use the Lagrange principle to transfer the constrained problem into unconstrained problem as follows:

$$\min_T \ell_k(T) = \sum_i \left( -v_i^T (D_i T - w_i) + \frac{\beta}{2} \|D_i T - w_i\|^2 \right) \\ - \lambda^T (GT - V) + \frac{\mu}{2} \|GT - V\|^2. \qquad (12)$$

While, $v_i$, $\lambda$ are weight coefficient, and $\mu$, $\beta$ are penalty factor. In order to solve the problem, it is necessary to set appropriate value for weight coefficient $v_i$, $\lambda$. However the above problem is still difficult to solve, because it is non-differentiable and non-linear, therefore we utilize the

separable structure of the variables and alternating direction algorithm (ADM) [14] to optimize. So the sub-problem of T and W are as follows:

$$\min_T \ell_k(T) = \sum_i \left( -v_i^T (D_i T - w_i) + \frac{\beta}{2} \|D_i T - w_i\|^2 \right) \\ - \lambda^T (GT - V) + \frac{\mu}{2} \|GT - V\|^2. \qquad (13)$$

$$w^{t+1} = \arg\min_w \left( \sum_i w_i - v_i^T (D_i T - w_i) + \frac{\beta}{2} \|D_i T - w_i\|^2 \right) \qquad (14)$$

In order to resolve W, we firstly fix T, and we use the following 2D shrinkage-like formula [15]

$$w_i = \max \left\{ \left\| D_i T - \frac{v_i}{\beta} \right\| - \frac{1}{\beta}, 0 \right\} \frac{D_i T - v_i/\beta}{\|D_i T - v_i/\beta\|} \qquad (15)$$

For fixed $w_i$, the minimization $\ell_k$ of with respect $T_b$ to becomes a least squares problem, its gradient is

$$d_k(T) = \left( \sum_i \beta_i D_i^T (D_i T - w_{i,k+1}) - D_i^T v_i \right) \\ + \mu A^T (AT - V) + \frac{\mu}{2} \|AT - V\| \qquad (16)$$

Forcing $d_k(T) = 0$. Theoretically, it is ideal to accept the exact minimizer as the solution of the u-subproblem by directly resolving the (14). However, it is too costly to implement numerically. Therefore, the steepest descent method is highly desirable. The method is able to solve iteratively by applying the recurrence formula

$$T^{k+1} = T^k - \alpha_k d_k \qquad (17)$$

The initial point $T_0$ is $A^T y$, $d_k$ is the gradient direction of the objective function, $\alpha_k$ is determined by the inexact search.

### III. SIMULATION AND EXPERIMENT RESULTS

In this section, we present several simulations to proposed method. All simulations were performed under Windows XP and MATLAB R2009a running on a Acer laptop with an Intel Core i3 CPU at 2.4GHz and 2GB of memory. We generated our test from the image moon. The dimension of the moon is $64 \times 64$. We introduce the concept of compressed radio, which is the radio between the number of nonzero elements and the total number of elements. In our experiments, we simply set $\mu = 2^8$ and $\beta = 2^5$, the radio are set 0.4, 0.5, 0.6, 0.7, 0.8. The reconstruction result of different radio is given in TABLE 1, and the following quantities are list the number of iteration (Iter), the CPU time (T) in seconds, the peak signal to noise ratio (Psnr), root mean square error (Rmse) and Bias. The quality parameters of the image have been shown in the following TABLE 1.

TABLE 1
PARAMETER FOR TV RECONSTRUCTION ALGORITHM

| Radio | 0.4 | 0.5 | 0.6 | 0.7 | 0.8 |
|-------|-----|-----|-----|-----|-----|
| Iter | 123 | 99 | 109 | 107 | 94 |
| T | 8.753259 | 9.759920 | 11.864406 | 13.549631 | 14.832024 |
| Bias | 1.0071 | 0.8687 | 0.7744 | 1.0471 | 0.8821 |
| Rmse | 2.1626 | 1.9728 | 1.8009 | 2.2986 | 1.9267 |
| Psnr | 41.4311 | 42.2290 | 43.0211 | 40.9016 | 42.4345 |

The inversion images of different radio have been shown as follows:

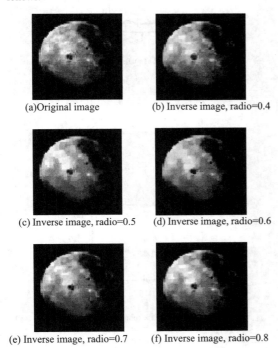

(a)Original image  (b) Inverse image, radio=0.4

(c) Inverse image, radio=0.5  (d) Inverse image, radio=0.6

(e) Inverse image, radio=0.7  (f) Inverse image, radio=0.8

From the above form and the inversion images, we can be seen that when the ratio is greater than or equal to 0.4, the effect of the inversion image is better, and the Psnr value is bigger. With the increase of the ratio, the time that used to reconstruction becomes longer, and the value of Psnr becomes larger, so the quality of the inversion image becomes better. When the radio is 0.6, we can get the best inversion image. From the table1, we can see that the CPU time is between 8s and 15s, it is very small. So the algorithm is very fast for the image reconstruction.

## IV. CONCLUSION

In this paper, we propose the imaging approach of sparse interferometry to microwave radiation, and apply the TV reconstruction algorithm to reconstruct the image. The simulation results show that the proposed method is effective.

This approach is better than the traditional algorithm of using G matrix for image reconstruction and can break through the problem by traditional microwave radiation imaging method.

REFERENCES

[1] Shutao Li, Dan Wei. "A Survey on Compressive Sensing," Acta Automatic Sinca., vol.35, no. 11, pp. 1369-1377, 2009.

[2] Qiusheng Lian, Shuzhen Chen. "Image Reconstrcution for Compressed Sensing Based on the Combined Sparse Image Representation," Acta Automatic Sinca., vol.36, no. 3, pp. 385-391, 2010.

[3] Ulaby, Moore and Fung. "Microwave Remote Sensing," Active and Passive., vo.1, Chapter 6. Addison-Wesley: Reading MA, 1981.

[4] D.L.Donoho. "Compressed Sensing," IEEE Trans.Inf.Theory., vol.52, no. 4, pp. 1289-1306, 2006.

[5] E. Candès, J. Romberg, T. Tao. "Robust uncertainty principles: exact signal reconstruction from highly incomplete frequency information," IEEE Trans. Inf. Theory., vol.52, no. 2, pp. 489-509, 2006.

[6] Yuntao He, Yuesong Jiang, Haiting Chen. "Studies of Optimation and Imaging Properties of Two-dimensional Circle Array for mm-wave Synthetic Aperture System," JOURNAL OF REMOTE SENSING., vol.11, no. 1, pp. 33-38, 2007.

[7] Kun Chao, Houcai Chen, Zhenwei Zhao. "Image Reconstruction and Inversion Algorithm of Synthetic Aperture Radiometer," CHINESE JOURNAL OF RADIO SCIENCE., vol.26, no. 5, pp. 881-886, 2011.

[8] Camps A, V Mercè, C Ignasi, et al. "Improved Image Reconstruction Algorithms for Aperture Synthesis Radiometers," IEEE Trans. Geoscience and Rem. Sen., vol.46, no.1, pp. 146-158, 2008.

[9] Fangmin He, Qingxia Li, Zhihua Zhao. "A Sparse Prior Based Statistical Inversion Approach for Aperture Synthesis Radiomtric Imaging of Extened Source," ACTA ELECTRONICA SINCA., vol.41, no.3, pp. 417-422, 2013.

[10] Jérôme Bobin, Jean-Luc Starck. "Compressed Sensing in Astronomy," IEEE Signal Process. Mag., vol.32, no. 5, pp. 718-726, 2008.

[11] Shiyong Li, Xi Zhou. "A Compressive Sensing Approach For Synthetic Aperture Imaging Radiometers," Progress In Electromagnetics Research., vol.135, pp. 583-599, 2013.

[12] Xi Zhou, Houjun Sun. "NUFFT-Based Iterative Reconstruction Algorithm for Synthetic Aperture Imaging Radiometers," IEEE Remote Sensing Leters., vol.6, no.2, pp. 273-276, 2009.

[13] http://dsp.rice.edu/cs

[14] J.F. Yang, Y Zhang, and W.T. Yin. "A Fast Alternating Direction Method for TVL1-TVL2 Signal Reconstruction From Partial Fourier Data," IEEE J. of Selected Topics in Signal Process., vol.4, no.2, pp. 288-297, 2010.

[15] http://www.caam.rice.edu/~optimization/L1/TVAL3/.

# A Hybrid Method for the study of the Mono-static Scattering from the Rough Surface and the Target Above It

R. Wang, S. R. Chai, Y. W. Wei, L. X. Guo

School of Science, Xidian University, Xi'an, China, 710071

*Abstract*-The purpose of this study is to describe a new hybrid method based on the reciprocity theorem, the physical optics (PO) and the Kirchhoff approximation (KA) for calculating the composite electromagnetic scattering from a target above a one-dimensional rough interface. The KA method is used to investigate characteristics of electromagnetic scattering from the rough interface (including the equivalent electric current densities and the scattered field from the rough interface). The scattered field from the isolated target was simulated by the PO method. Based on the reciprocity theorem, the multiple scattering up to 3rd order by the target and the underlying randomly rough interface was considered. The validity of our methods is shown by comparing our results with that of Method of Moments (MoM). It is found that our methods are in good agreement with MoM and has a higher computational efficiency.

## I. INTRODUCTION

EM scattering form a target situated above a rough surface is of great interest in recent years, with application in remote sensing, oceanic surveillance, target detecting, etc. The numerical methods such as the method of moments (MoM) [1], the finite difference time domain (FDTD) method [2], the finite element method (FEM), etc. have been developed. These techniques have a good accuracy, but their computational efficiency could not satisfy the requirement, especially when the composite scattering model is electronically large. The analytical methods and the high frequency methods, such as KA, the small perturbation method (SPM), the small slope approximation (SSA), the PO method, etc. have a high efficiency, but unable to calculate the coupling scattering field between target and rough surface. The numerical-analytical or high frequency-numerical combined methods, which includes the hybrid MoM/KA, the hybrid MoM/PO, the hybrid MoM/GTD, the hybrid FDTD/GTD, the FEM/KA [3], etc. are accurate and has high efficiency. But if the target has a complex shape or the mono-static scattering coefficient is expected, these hybrid methods are time consuming.

In this paper we discussed a new method based on reciprocity theorem, which combines KA and PO, to calculate the composite scattering from a target above a randomly rough surface.

## II. SCATTERING MODELS AND FORMULATIONS

### A. Scattering models

The geometric model is shown in Fig.1, a perfectly electronic conducting (PEC) target is located above 1D PEC rough interface.

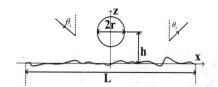

Figure 1. Geometric model of a target above 1D rough interface

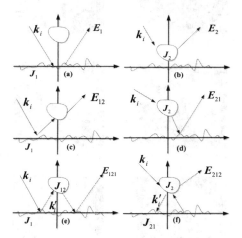

Figure 2. scattering mechanic of the composite scattering model (TE case)

Assuming that a plane wave impinging upon the model shown in Fig.1, the scattered field can be decomposed into three scattering terms: 1) direct scattered field from the rough interface $E_1$ and from the target $E_2$, as shown in Figure 2(a) and (b); 2) $2^{nd}$ order coupling scattered field $E_{12}$ and $E_{21}$, as shown in Fig.2(c) and (d); 3) $3^{rd}$ order coupling scattered field $E_{121}$ and $E_{212}$, as shown in Fig.2(e) and (f). So the total scattered field can be written as:

$$E_{total} = E_1 + E_2 + E_{12} + E_{21} + E_{121} + E_{212} \quad (1)$$

In this paper, the time dependence is set to be $e^{-i\omega t}$, so, the two dimension scalar Green's function becomes

$$g(\boldsymbol{\rho}, \boldsymbol{\rho}') = \frac{i}{4} H_0^{(1)}(k|\boldsymbol{\rho} - \boldsymbol{\rho}'|) \quad (2)$$

978-7-5641-4279-7

The formula for the incident wave in our paper can be written as $\varphi_i = \exp(ik_i \cdot \rho)$, where for TM incident wave $E_i = \hat{y}\varphi_i$, and it is $H_i = \hat{y}\varphi_i$ for TE polarization. $\rho = x\hat{x} + z\hat{z}$ is the position vector where we are interested. $k_i$ is wave-number vector of the incident wave.

*B. Computation of direct scattered filed*

Let incident wave illuminates on the rough interface, if the condition of validity of the standard KA is satisfied in modeling the EM scattering from the rough interface, the surface equivalent electric current density induced by the incident wave in the absence of the target can be calculated by PO, which yields:

$$J_1(\rho) = 2\hat{n}_s(\rho) \times H_i(\rho)$$

$$= \begin{cases} -\dfrac{2k}{\omega\mu} \hat{y}(\hat{n}_s \cdot \hat{k}_i)\exp(ik\rho_s \cdot \hat{k}_i) & \text{for TE case} \\ 2\exp\left[ik\rho_s \cdot \hat{k}_i\right]\hat{n}_s(\rho_s) \times \hat{y} & \text{for TM case} \end{cases} \quad (3)$$

When $\rho$ is in light region , where $\hat{n}_s$ is the unit normal vector on the rough interface , then the direct scattered field from the rough surface $E_1$ can be given by the Huygens' Principle, the scattered electromagnetic field has forms[4] as below

$$E_1 = i\omega\mu \int_s ds' \overline{\overline{G}}(\rho, \rho') \cdot J_1(\rho') \quad (4)$$

$$H_1 = \nabla \times \int_s ds' \overline{\overline{G}}(\rho, \rho') \cdot J_1(\rho') \quad (5)$$

Where $\overline{\overline{G}}(\rho, \rho')$ indicates dyadic Green's function, which can be written as

$$\overline{\overline{G}}(\rho, \rho') = [\overline{\overline{I}} + \dfrac{1}{k^2}\nabla\nabla]g(\rho, \rho') \quad (6)$$

Similarly, when the isolated target is illuminated by the incident wave, the direct scattered electric field has the same form with that from the rough interface:

$$E_2 = i\omega\mu \int_s ds' \overline{\overline{G}}(\rho, \rho') \cdot J_2(\rho') \quad (7)$$

$$H_2 = \nabla \times \int_s ds' \overline{\overline{G}}(\rho, \rho') \cdot J_2(\rho') \quad (8)$$

where

$$J_2(\rho) = 2\hat{n}_c(\rho) \times H_i(\rho)$$

$$= \begin{cases} -\dfrac{2k}{\omega\mu} \hat{y}(\hat{n}_c \cdot \hat{k}_i)\exp(ik\rho_c \cdot \hat{k}_i) & \text{for TE case} \\ 2\exp\left[ik\rho_c \cdot \hat{k}_i\right]\hat{n}_c(\rho_c) \times \hat{y} & \text{for TM case} \end{cases} \quad (9)$$

*C. Computation of $2^{nd}$ order coupling scattered filed*

The directed scattered field $E_1$ and $E_2$ can be evaluated by Huygens' Principle in (4). The challenge here is the calculation of the coupling scattered fields: $E_{12}$ , $E_{21}$ , $E_{121}$ and $E_{212}$ . Here the reciprocity theorem was applied to solve this problem.

For TE case, according to the reciprocity theorem, consider an elementary electric current source $J_e = \hat{y}\delta(\rho - \rho_0)$ placed at the observation point illuminating the target and the rough interface while the current source $J_1$ and $J_2$ is removed. The far-field generated by $J_e$ is approximated as

$$E_{ed}(\rho) = -\hat{y}\dfrac{\omega\mu}{4}\sqrt{\dfrac{2}{\pi k\rho_0}}e^{ik\rho_0}e^{-i\pi/4}\exp(-ik\rho \cdot \hat{k}_s) \quad (10)$$

The equivalent electric current density on the rough interface and target induced by $J_e$ are derived as

$$J_{e1}(\rho_s) = 2\hat{n}_s \times H_{ed}(\rho_s) = 2\hat{n}_s \times \dfrac{\nabla \times E_{ed}(\rho_s)}{i\omega\mu}$$

$$= -\hat{y}\dfrac{ik}{2}\sqrt{\dfrac{2}{\pi k\rho_0}}e^{-i\pi/4}e^{ik\rho_0}e^{-i\pi/2}J_{e1}^{(N)} \quad (11a)$$

$$J_{e2}(\rho_s) = 2\hat{n}_c \times H_{ed}(\rho_c) = 2\hat{n}_c \times \dfrac{\nabla \times E_{ed}(\rho_c)}{i\omega\mu}$$

$$= -\hat{y}\dfrac{ik}{2}\sqrt{\dfrac{2}{\pi k\rho_0}}e^{-i\pi/4}e^{ik\rho_0}e^{-i\pi/2}J_{e2}^{(N)} \quad (11b)$$

where

$$J_{e1}^{(N)} = (\hat{n}_s \cdot \hat{k}_s)\exp(-ik\rho_s \cdot \hat{k}_s) \quad (12a)$$

$$J_{e1}^{(N)} = (\hat{n}_s \cdot \hat{k}_s)\exp(-ik\rho_s \cdot \hat{k}_s) \quad (12b)$$

The induced currents in the target produce an electric field on the rough interface, which can be expressed in following terms:

$$E_{e21}(\rho_s) = i\omega\mu\left(\overline{\overline{I}} + \dfrac{\nabla\nabla}{k^2}\right)\int_c J_{e2}(\rho_c)g(\rho_s, \rho_c)dc$$

$$= \hat{y}\dfrac{k\omega\mu}{8}\sqrt{\dfrac{2}{\pi k\rho_0}}e^{-i\pi/4}e^{ik\rho_0}E_{e21}^{(N)} \quad (13)$$

where

$$E_{e21}^{(N)} = \int_c J_{e2}^{(N)}(\rho_c)H_0^{(1)}(k|\rho_s - \rho_c|)dc \quad (14)$$

Applying the reciprocity theorem [5][6], the secondary scattered fields $E_{12}$ (Fig.2 (c)) can be expressed as

$$\hat{y} \cdot E_{12} = \int_s J_1 \cdot E_{e21}ds \quad (15)$$

Where $J_1$ is the equivalent surface current density on the rough interface induced by the incident wave.

For the scattered field $E_{21}$ (Fig.2 (d)), the solving process is similarly to the scattered field $E_{12}$ , which can be expressed as

$$\hat{y} \cdot E_{21} = \int_c J_2 \cdot E_{e12}dc \quad (16)$$

where

$$E_{e12}(\rho_c) = i\omega\mu\left(\overline{\overline{I}} + \dfrac{\nabla\nabla}{k^2}\right)\int_s J_{e1}(\rho_s)g(\rho_c, \rho_s)ds$$

$$= \hat{y}\dfrac{k\omega\mu}{8}\sqrt{\dfrac{2}{\pi k\rho_0}}e^{-i\pi/4}e^{ik\rho_0}E_{e12}^{(N)} \quad (17)$$

$$E_{e12}^{(N)} = \int_s J_{e1}^{(N)}(\rho_s)H_0^{(1)}(k|\rho_c - \rho_s|)ds \quad (18)$$

For TM case, the elementary electric current source is replaced by an elementary magnetic current source $M_e = \hat{y}\delta(\rho - \rho_0)$, The far-field generated by $M_e$ is:

$$H_{md}(\rho) = -\hat{y}\dfrac{\omega\varepsilon}{4}\sqrt{\dfrac{2}{\pi k\rho_0}}e^{ik\rho_0}e^{-i\pi/4}\exp(-ik\rho \cdot \hat{k}_s) \quad (19)$$

The equivalent electric current density on the rough surface induced by the elementary magnetic current source $M_e$ is:

$$J_{m1}(\rho_s) = 2\hat{n}_s(\rho_s) \times H_{md}(\rho_s)$$

$$= -\sqrt{\frac{2}{\pi k \rho_0}} \frac{\omega \varepsilon}{2} e^{ik\rho_0} e^{-i\pi/4} \hat{n}_s(\rho_s) \times \hat{y} J_{m1}^{(N)}(\rho_s) \quad (20a)$$

$$J_{m2}(\rho_c) = 2\hat{n}_c(\rho_c) \times H_{md}(\rho_c)$$

$$= -\sqrt{\frac{2}{\pi k \rho_0}} \frac{\omega \varepsilon}{2} e^{ik\rho_0} e^{-i\pi/4} \hat{n}_c(\rho_c) \times \hat{y} J_{m2}^{(N)}(\rho_c) \quad (20b)$$

where

$$J_{m1}^{(N)}(\rho_s) = \exp(-ik\rho_s \cdot \hat{k}_s) \quad (21a)$$

$$J_{m2}^{(N)}(\rho_s) = \exp(-ik\rho_c \cdot \hat{k}_s) \quad (21b)$$

The electric field $E_{m21}$ on the rough interface produced by $J_{m2}$ can also be calculated by Huygens' Principle, then the secondary scattered fields in Fig.2 (c) and Fig.2 (d) becomes:

$$\hat{y} \cdot H_{12} = -\int_s J_1 \cdot E_{m21} ds \quad (22)$$

$$\hat{y} \cdot H_{21} = -\int_s J_2 \cdot E_{m12} dc \quad (23)$$

### D. Computation of 3rd order coupling scattered filed

In order to apply reciprocity theorem for the situation shown in Fig.2 (e), $J_{12}$ and $J_e$ ($J_{12}$ and $M_e$ for TM) are treated as these two reciprocity sources, then the scattered field in this condition can be obtained as

$$\hat{y} \cdot E_{121} = \int_c J_{12} \cdot E_{e12} dc \quad \text{for TE case} \quad (24)$$

$$\hat{y} \cdot H_{121} = -\int_c J_{12} \cdot E_{m12} dc \quad \text{for TM case} \quad (25)$$

Using the same method we can get the scattered field of Fig.2 (f), which is [7][8]:

$$\hat{y} \cdot E_{212} = \int_s J_{21} \cdot E_{e21} ds \quad \text{for TE case} \quad (26)$$

$$\hat{y} \cdot H_{212} = -\int_s J_{21} \cdot E_{m21} ds \quad \text{for TM case} \quad (27)$$

where the induced equivalent electric density

$$J_{12} = \begin{cases} \dfrac{\hat{y} k^2}{i\omega\mu} \int_s (\hat{n}_s \cdot \hat{k}_i)(\hat{n}_c \cdot \hat{R}_{21}) \cdot \exp(ik\rho_s \cdot \hat{k}_i) H_1^{(1)}(kR_{21}) ds & \text{for TE} \\ ik\hat{n}_c \times \hat{y} \int_s (\hat{n}_s \cdot \hat{R}_{21}) \cdot \exp(ik\rho_s \cdot \hat{k}_i) H_1^{(1)}(kR_{21}) ds & \text{for TM} \end{cases}$$

Since the length of this paper limited, some specific expressions have to be omitted.

So far, up to the directed scattered field from the target and that from the rough interface, as well as the coupling field up to 2rd and 3rd between them, the analytical expressions of normalized radar scattering cross section (NRCS) and the difference NRCS (DNRCS) can be defined as:

$$\sigma(\theta_s) = \begin{cases} 10\lg_{10}\left(\dfrac{2\pi r}{L} \left|E_{total(diff)}\right|^2\right) & \text{for TE case} \\ 10\lg_{10}\left(\dfrac{2\pi r}{L} \left|H_{total(diff)}\right|^2\right) & \text{for TM case} \end{cases} \quad (28)$$

where $E_{total}$, $H_{total}$, $E_{diff}$ and $H_{diff}$ are:

$$E_{total} = \hat{y} \cdot (E_1 + E_2 + E_{12} + E_{21} + E_{121} + E_{212}) \quad (29)$$

$$E_{diff} = \hat{y} \cdot (E_2 + E_{12} + E_{21} + E_{121} + E_{212}) \quad (30)$$

$$H_{total} = \hat{y} \cdot (H_1 + H_2 + H_{12} + H_{21} + H_{121} + H_{212}) \quad (31)$$

$$H_{diff} = \hat{y} \cdot (H_2 + H_{12} + H_{21} + H_{121} + H_{212}) \quad (32)$$

### III. SIMULATION RESULTS

Fig.3-5 shows the results of the mono-static scattering from the composite model consists of a Gaussian rough surface with an infinitely cylinder above it. The incident frequency $f = 0.3\text{GHz}$, the length of rough interface is $L = 102.4\lambda$.

Fig.3 give the comparisons of our method and MoM for two different polarizations. The root mean square and correlation length of the underlying rough surface are $l = 1.5\lambda$ and $\sigma = 0.2\lambda$, respectively. The radius of infinitely cylinder is $r = 2\lambda$, the distance between the axial line of the target and the rough surface is $h = 5\lambda$.

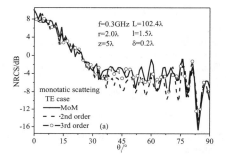

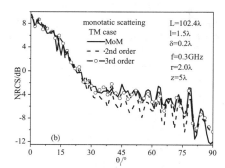

Figure 3. Comparison of the hybrid method and MoM for the mono-static scattering (a) TE case (b) TM case

From Fig.3, it is found that the mono-static NRCS of 3rd order for TM and TE by our method are in good agreement with that by MoM. Moreover, the results of 3rd order have higher accuracy than that of 2nd order. It is because the results of 3rd order take much higher coupling field between rough surface and the target into account.

Table.1 gives the simulating time of different unknowns for one same surface realization. It is obvious that our method has a higher computational efficiency than the MoM method.

TABLE.1. THE SIMULATING TIME OF DIFFERENT UNKNOWNS FOR ONE SAME SURFACE REALIZATION

| Polarization of incident wave | Number of Unknowns surface+target | Time elapsed(s) | |
|---|---|---|---|
| | | MoM | Our methods |
| TE | 512+100 | 163 | 35 |
| | 1024+100 | 577 | 70 |
| | 2048+100 | 2138 | 136 |
| TM | 512+100 | 166 | 34 |
| | 1024+100 | 585 | 70 |
| | 2048+100 | 2151 | 138 |

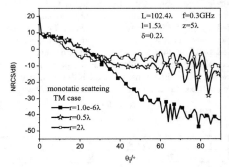

Figure 4. The mono-static scattering for different radius cylinder

Fig.4 gives the mono-static NRCS of $3^{rd}$ order with different cylinder radius for TM polarization. The root mean square and correlation length of the underlying rough surface are $l = 1.5\lambda$ and $\sigma = 0.2\lambda$, respectively. The distance between the axial line of the target and the rough surface is $h = 5\lambda$. It is observed that the mono-static NRCS increases with increasing the cylinder radius. This is because that the coupling scattering becomes stronger with the large cylinder radius.

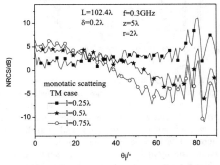

Figure 5. The mono-static scattering for different correlation length of rough surface

In Fig.5, the effect of the correlation length on the mono-static NRCS of $3^{rd}$ order for TM polarization is examined. The root mean square of the rough surface and the cylinder radius

are $\sigma = 0.2\lambda$ and $r = 2\lambda$, respectively. The distance between the axial line of the target and the rough surface is still $h = 5\lambda$. It is observed that the mono-static NRCS decreases over the big angular range, but increases over the small angular range with increasing $l$. It should be pointed out that the specular scattering becomes strong with the increasing $l$. For large angular, the mono-static NRCS will be weak when the results of specular scattering becomes strong. But for small angular, the results of mono-static scattering is close to that of specular scattering, so the mono-static NRCS increases with the increasing $l$.

## IV. CONCLUSION

In this paper, an EM scattering solution for evaluating the scattering interaction between rough interface and the target is presented. The reciprocity theorem was utilized to reduce the difficulty in formulating the $2^{nd}$ and $3^{rd}$ order scattered fields from the composite model. The validity of this work was demonstrated by comparing the our results with that of MoM. Finally, the $3^{rd}$ order coupling fields $E_{121}$ and $E_{212}$ ($H_{121}$ and $H_{212}$ for TM case) are discussed, which found that there are not reversible.

ACKNOWLEDGMENT

This work was supported by the Specialized Research Fund for the Doctoral Program of Higher Education (Grant No.20120203120023), the National Science Foundation for Distinguished Young Scholars of China (Grant No.61225002), the Postdoctoral Science Foundation of China (Grant No. 2011M501447), and the Fundamental Research Funds for the Central Universities.

REFERENCES

[1] X. Wang, C. F. Wang, and Y.B. Gan "Electromagnetic scattering from a circular target above or below rough surface," *Progress In Electromagnetics Research*, vol.40, pp. 207-227, 2003.

[2] J. Li, L. X. Guo, and H. Zeng, "FDTD investigation on the electromagnetic scattering from a target above a randomly rough sea surface," *Waves in Random and Complex Media*, vol.18, pp.641–650, 2008.

[3] J. Li, L. X. Guo, and Q. He, "Hybrid FE-BI-KA method in analyzing scattering from dielectric object above sea surface," *Electronics Letters*, vol.47 , pp. 1147–1148, 2011.

[4] Kong, J.A., *Electromagnetic Wave Theory*, John wiley & sons, 2002.

[5] R. Wang, and L. X. Guo, "Study on electromagnetic scattering from the time-varying lossy dielectric ocean and a moving conducting plate above it," *JOSA A*, vol.26, pp.517-529, 2009.

[6] T. Chiu, and K. Sarabandi, "Electromagnetic scattering interaction between a dielectric cylinder and a slightly rough surface," Antennas and Propagation, *IEEE Transactions on*, vol.47, pp.902-913, 1999.

[7] L. X. Guo, Y. H. Wang, and R. Wang, "Investigation on the electromagnetic scattering of plane wave/Gaussian beam by adjacent multi-particles," *Progress In Electromagnetics Research B*, vol.14, pp.219-245, 2009.

[8] Y. H. Wang, Y. M. Zhang, and M. H. Xia "Solution of scattering from rough surface with a 2D target above it by a hybrid method based on the reciprocity theorem and the forward–backward method," *Chinese Physics B*, vol.17, pp. 3696, 2008.

# A reconfigurable frequency selective surface for tuning multi-band frequency response separately

Jialin Yuan[1], Shaobin Liu[1*], Xiangkun Kong[1,2], Huan Yang[1].

[1]College of Electronic and Information Engineering, Nanjing University of Aeronautics and Astronautics, Nanjing 210016, China

[2]Jiangsu Key Laboratory of Meteorological Observation and Information Processing, Nanjing University of Information Science and Technology, Nanjing 210044, China

*Abstract*- **A reconfigurable frequency selective surface (FSS) is presented in this paper. The tunable FSS is based on a common FSS comprising annular ring slot and cruciform structure cells. The tuning of the reconfigurable surface is shown numerically to be possible by incorporating tuning varactors into the structure. Using varactors on both layers, a reconfigurable frequency response is achieved, which has three modes of center frequency. Numerical simulations show that lower and upper frequency can be tuned separately by altering the capacitance of the loaded varactors. So that multiple-band frequency response is electronically tunable separately. Besides that, the FSS is stable with different polarizations and incident angles.**

## I. INTRODUCTION

Frequency selective surface (FSS) are periodic surfaces, which provide spatial filtering have been widely used for electromagnetics, microwave, antennas, radar and satellite communications[1-8]. In the past few decades, tunable frequency filters have attracted a lot of interest over the years as research commenced. A frequency selective surface with miniaturized elements were proposed in[4,5]. The center frequency tuning with a fixed bandwidth and bandwidth tuning with a fixed center frequency can be achieved by mounting varactors[6]. A dual-band frequency selective surface with large separation can be achieved when varactors are employed[7] . Plasma frequency selective surface has been studied in [8], they replace metal with plasma in the FSS so that can be tunable by varying the density in the plasma elements, but it's a complicated and numerically expensive process. A novel symmetric anchor-shaped active frequency selective surface (AFSS) with a wide tuning frequency range is proposed in [9], a wide tunable frequency range from 1.76 to 2.45 GHz with a relative bandwidth of 33% is achieved.

A reconfigurable frequency selective surface (FSS) is presented in this paper whose multiple-band frequency response is electronically tunable separately. This FSS has two arrays of structures printed on either side of a thin flexible substrate by connecting the structures on each layer. The frequency tuning capability of the proposed FSS is demonstrated through numerical simulations using Ansys HFSS software.

## II. DESIGN PROCESS AND RESULTS

### A. Design process

The proposed tunable FSS structure is shown in Fig.1, including the design parameters. A three-dimensional (3D) view of the unit cell is shown in Fig. 1(a). As we can see,

the structure is composed of two layers separated by a thin substrate. The substrate relative permittivity is $\epsilon_r$=2.2. On the top surface of the unit cell, four chip capacitors are loaded symmetrically around the circle, on the bottom surface, there is a wire grid and four chip capacitors are loaded between the gaps of the grid. For this particular structure,when the unit cell is expanded periodically in the x- and the y-axis directions, a tunable FSS is obtained.

Geometrical dimensions of the structure as follow:

TABLE I

GEOMETRICAL DIMENSIONS OF THE STRUCTURE

| $L_1$ | h | $R_0$ | $R_i$ | $L_2$ | w |
|---|---|---|---|---|---|
| 18.2mm | 0.7mm | 8mm | 6.5mm | 8.6mm | 0.5mm |

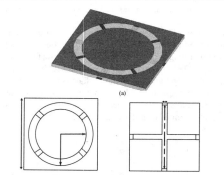

Fig.1. One-unit cell is shown.(a) 3D view of a unit cell. (b) top view (b) back view.

We use Ansys HFSS to calculate the results, we should choose appropriate capacitance values that are incorporated into the design. The simulation model is one unit cell periodic boundary and excited by an incident plane wave, which is set at different incident angles and with different polarizations. C1 is inserted in the circular slot of the top surface symmetrically and C2 is loaded between the gaps of the grid. The transmission coefficients at normal incidence are shown as the simulated results in Fig.2. There are three band-pass responses. When we keep C2 unchanged, and change C1 from 0.3 pf to 0.5 pf, the upper center frequency moves toward lower frequency. The second case is that we keep the C1 unchanged but set C2 as 1.6 pf, 2.5 pf, 4.0pf, the lower center frequency moves toward to the lower frequency. The result shows that the lower and upper center frequency can tuned separately by changing

978-7-5641-4279-7

the capacitances C1 and C12, but the performance of band-pass is well kept.

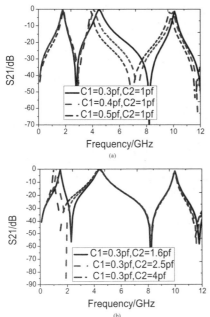

(a)

(b)

Fig.2.the S parameter of the FSS with the different values of the capacitance C1 and C2

### B. Analysis

Fig.3 shows the electric field distributions at $1.8GHz$ and $4.75GHz$ which correspond to the lower and upper center frequency. It can be found that the electric resonance occurs in the crossed patch on the bottom and the circular slot on the top. Based on this current distribution, the corresponding equivalent circuit model is given in Fig3(c)[10].

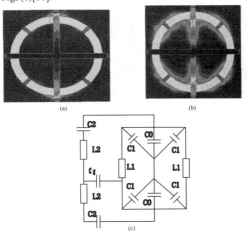

Fig.3. electric field distribution of the proposed FSS (a) at 1.80GHZ (b) at the 4.75GHz (c)equivalent circuit model

L1 can be understood as the equivalent inductance of each vertical slits on both sides of the metal patch on the top. C0 can be understood as the equivalent capacitance of each ring slot gap on both sides of the metal on the top. Cg is the equivalent capacitance of the gap between the upper and lower layers. Apparently, L2 models the traces of the bottom grid and C1, C2 are the capacitances which are loaded on the both surfaces.

### C. Stable performance

The stable performance of the FSS is very important in some special application. Fig.4 gives the simulation results. We can find that incident angle changes have greater impact with TM polarization. But there is no more difference when the polarization is changed. To sum up, the ring slot surface is stable for incident angles within 30 deg and different polarizations.

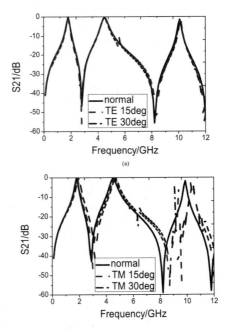

(a)

(b)

Fig.4 . the S parament of the FSS with different polarizations and incident angles. (a) TE (b) TM

### III. CONCLUSION

In this paper, a multiple-band tunable frequency selective surface is proposed. The FSS unit is consisted of a ring slot and cruciform structure which are loaded with varactors in parallel. The frequency tuning capability is achieved by changing the value of capacitance through numerical simulations. The lower and upper center frequency can be tuned separately. By the way, a equivalent circuit is given through the field analysis. Such a FSS has broad application prospects in radar and satellite communications .

ACKNOWLEDGMENT

This work was supported by the supports from Chinese Specialized Research Fund for the Doctoral Program of Higher Education(grant No. 20123218110017), the Jiangsu Province Science Foundation (Grant No.BK2011727), and the Foundation of Aeronautical Science (No. 20121852030), Outstanding Doctoral Dissertation in NUAA (Grant No.BCXJ11-05) and the Open Research Program in Jiangsu Key Laboratory of Meteorological Observation and Information Processing(Grant No.KDXS1207).

REFERENCE

[1]  T. K. Wu , Frequency-Selective Surface and Grid Array . New York:Wiley, 1995.

[2]  B. A. Munk , Frequency-Selective Surfaces: Theory and Design.NewYork: Wiley, 2000.

[3]  Farhad Bayatpur ,Kamal Sarabandi "Design and Analysis of a Tunable Miniaturized-Element Frequency-Selective Surface Without Bias Network" IEEE Trans. Antennas Propag.vol .58.no.4,pp:1214-1219, Apirl 2010

[4]  K. Sarabandi and N. Behdad, "A frequency selective surface withminiaturized elements,"IEEE Trans. Antennas Propag.vol. 55, no. 5,pp. 1239–1245, May 2007

[5]  Farhad Bayatpur and Kamal Sarabandi, "A Tunable, Band-Pass, Miniaturized-Element Frequency Selective surface:  Design and Measurement" Antennas and Propagation Society International Symposium, 2007 IEEE,pp.3964-3967,June 2007

[6]  Farhad Bayatpur and Kamal Sarabandi, "A Tunable Metamaterial Frequency-Selective Surface With Variable Modes of Operation" IEEE Trans .Microwave theory and technology vol 57, no. 6, pp.1433-1438.June 2009

[7]  Zhou hang ,Qu shao-bo, "Dual-band frequency selective surface with large band sepration and separation and stable performance". Chin. Phys.B.vol .21,no.5(2012) ,pp.054101,1-053101,4

[8]  Ted Anderson, Igor Alexeff, James Raynolds, " Plasma Frequency Selective Surfaces" IEEE Trans , Plasma  Science .vol 35,No .2 APRIL 2007

[9]  Liang Zhang, Guohui Yang , "A Novel Active Frequency Selective Surface With Wideband Tuning Range for EMC Purpose" IEEE Trans ,Magnetics ,vol .48,No .11,Novermber ,2012

[10] Marcello D'Amore,Valerio De Santis, Mauro Feliziani , "Equivalent Circuit Modeling of Frequency-Selective Surfaces Based on anostructured  Transparent  Thin  Films"  IEEE  Trans  on Magnetics.vol.48.no.2, pp.703-706 .Febuary 2012

# A Wideband Bandstop FSS with Tripole Loop

Ping Lu[1], Guang Hua[2], Chen Yang[3], Wei Hong[4]

State Key Laboratory of Millimeter Waves

School of Information Science and Engineering, Southeast University

[1]pinglu@emfield.org,[2]guanghua@emfield.org, [3]chenyang@emfield.rog, [4]weihong@emfield.rog

*Abstract-* In this paper, a microwave spatial band stop radome operating in x-band based on frequency selective surfaces (FSS) is presented. To meet the demand of wideband and large incident wave angle, a tripole loop FSS is designed as a multilayer sandwiched structure. The two FSS layers are developed to increase bandwidth of the stop band for the TE and TM mode wave. The simulation results show that the FSS has a wideband which up to 50% and 20% for TE and TM mode, respectively. A practical the FSS structure has also been fabricated and tested. Measurements indicate that the FSS has a band stop characteristics at large incident angle which range from 0° to 60° for TE and TM at X-band in the microwave anechoic chamber.

## I. INTRODUCTION

Frequency selective surface（FSS）structures are periodic structures that can be used to control and manipulate the propagation of electromagnetic waves for more than four decades[1-4]. Frequency selective surfaces are usually constructed from periodically arranged metallic patches of arbitrary geometries or their complimentary geometry having aperture elements similar to patches within a metallic screen, and these surfaces exhibit total reflection or transmission, for the patches and apertures respectively[5]. A number of FSS structures have been investigated, which can function as pass band or as stop band over the desired frequency range. The reflection and transmission performance of each FSS is determined by the shape of the metal patches or slots, the dimension of the structure, the periodicity of the array, the lattice of the array and the supporting dielectric characteristics of the substrate[6]. In the classical book on the subject of frequency selective surface written by Ben A. Munk, various applications of FSS have been mentioned. They have been adoped in many applications such as radomes, dichroic subreflectors, RCS augmentation, photonic band gap structures, and in low-probability of intercept systems[7]. The antenna mounted on a mask etched with FSS structure which works as spatial band pass filter can deflect out of band signals while the desired signals go through. For building wireless security, the conventional reflection / transmission type FSS may be placed in the building wall so as to provide isolation and reduce interference between adjacent vicinities[8]. The frequency selective property of FSS makes them potential candidates for military and civilian applications.

One of the most important challenges behind the design of the wideband bandstop FSS is its bandwidth for the large incident angle. In order to meet the specifications of frequency band steady at the situation of large incident angle, a hexagon is chosen for the cell shape which can make the configuration of tripole loop more compact than the square cell. The periodicity of the printed patterns is about half of the wavelength of operating frequency. The dielectric profile of the FSS panel is a multilayer sandwiched construction to meet the bandwidth specifications. All simulations have been done by ANSOFT HFSS software based on FEM.

## II. TRIPOLE LOOP FSS LAYER DESIGN AND SIMULATION

Figure 1 shows the dielectric profile of the proposed FSS screen. FSS structure is printed on a FR4 dielectric layer, the distance between two FSS layers is H1 which is about one-fourth wavelength. The stop band bandwidth has been effected by the distance between the FSS structure layers. FSS configuration designed and fabricated with standard PCB process, and the photograph is shown in Figure 2. Table I lists the corresponding values length of each section figure out in Figure 2. In order to minimize the effects of dielectric on the properties of the periodic surface, the thickness of dielectric substrate chosen for development of two identical FSS PCB layers is very thin (0.127mm). Hexagonal lattice structure is selected as the variation in the resonant frequency of the grid are smaller, resulting in a stable resonance frequency, particularly in the case of large incidence angle. It also helps to maintain the radome surface periodicity required by the application.

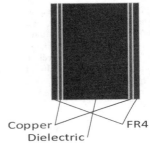

Figure 1 FSS layer structure

Tripole loop FSS is one of the best choice as spatial bandstop filter blocking transmission of the arbitrary polarization signal in large incident angle which is typical of the nose cone cover. It is considerably more broadbanded than the four-legged case[7]. Compared to the Jerusalem Cross and the cross structure, tripole loop show better performance in a large incident angle .The basic design of a Tripole element FSS is taken from [4].Ref [9] designs a broadband band pass FSS

Correspondence should be addressed to Guang Hua, huaguang@seu.edu.cn

structure with hexagonal cell and tripole slot which reached a 1Ghz transparent bandwidth at X-band.

Cell size A, the length of L and L1 play a significant role for the resonance frequency, while the width of the loop determines the width of stop band. The distance between two FSS layers is also one of the important factors affecting the bandwidth. The circumference of the loop is approximately the length of the wavelength of the operating frequency. A symmetric structure can be obtained for the angle of each pole

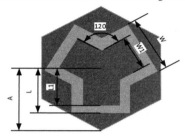

Figure 2  The parameters of FSS dimensions（top view）

TABLE I
Values of design parameters（unit：mm）

| A | H1 | L | W | L1 | W1 |
|---|---|---|---|---|---|
| 7.2 | 7.6 | 5.05 | 6 | 3.8 | 4.5 |

is set to 120 degrees. The symmetry of the loop makes the stable performance at different azimuth angles available. The analysis of the proposed FSS structure is completed by using the Ansoft HFSS software. As shown in Figure 3 and Figure 4, a wide stop band can be obtained which is 5 GHz under -10 dB for TE mode wave and more than 2 GHz for TM mode. It can be seen that a relative bandwidth up to 50% and 20% for TE and TM mode waves have been achieved for incidence angle is 60°, respectively. For different pitch angles, TE wave exists a small frequency offset, and the bandwidth of TM wave becomes narrow as the pitch angle increases. In a large incident angle, the electrical length of cell is different for different polarization resulting bandwidth difference. Compared to the large angle of incidence, periodic structures characteristics are more clearly demonstrated for a small angle of incidence.

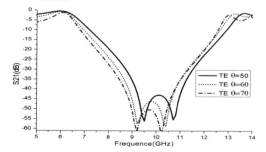

Figure 3 Simulated FSS performance curves for TE mode

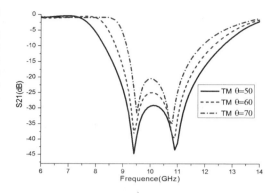

Figure 4 Simulated FSS performance curves for TM mode

Figure 5. Photograph of the tripole loop FSS PCB

III.  EXPERIMENTAL VERIFICATION

A photograph of the tripole loop FSS PCB layer of the radome is shown in Figure 5. Tripole FSS structure is etched on a 300mm × 300mm dielectric layer. Thickness of intermediate dielectric layer sandwiched between FSS about 1cm.The measurements of tripole FSS screen are carried out in the microwave anechoic chamber where the spacing between the two horns antennas is adjusted so that the FSS radome panel sample is irradiated a plane wave in far field of the transmitting antenna. As shown in Figure 6, FSS screen covers in the receiving horn antenna which is placed on the turntable to test the bandstop characteristics of the periodic structure.

The measurement of S-parameters in the case of incident angle Ɵ from 0° to 90° for TE mode is presented in Figure 7 and TM mode in Figure 8. There is a significant stopband from 0 degrees to 60 degree at 10 GHz both for TE and TM mode wave and also a blocking band at 8 GHz and 12 GHz. Because of the limits of manufacturing technology and measurement environment, the experimental results are not as good as simulated. The distance between the two FSS structure layers influences the bandwidth and the resonance frequency.As shown in Figure 7 and Figure 8，dielectric thickness of the intermediate layer is slightly larger than the simulation, results in the experimental are not as good as simulation when incident angle is 60 degree for TE mode. The offset of receiving antenna result in a shift for stopband in testing. The transmission coefficient is greater than 0 in the presence is caused by the existence of the pole.The antenna pattern exist some poles at certain angle while the transmission coefficient of FSS is relatively flat, which result in the differences are greater than 0dB at poles，as shown in Figure 9.Improvements can be done by adding absorbing material around FSS structure to make more accurate measurements.

Figure 6 Test environment

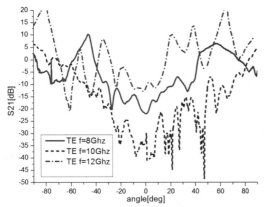

Figure 7 Measured transmission coefficient of TE mode wave

## IV. ACKNOWLEDGEMENT

This work was supported in part by National 973 project 2010CB327400 and in part by the National Science and Technology Major Project of China under Grant 2010ZX03007-001-01

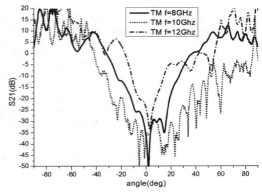

Figure 8 Measured transmission coefficient of TM mode wave

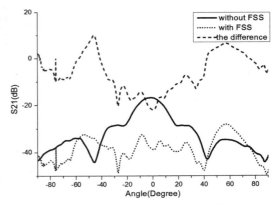

Figure 9 Measured transmission coefficient with/without FSS and the difference at 8 GHz

## REFERENCES

1. Chen, C.-C., *Scattering by a two-dimensional periodic array of conducting plates*. Antennas and Propagation, IEEE Transactions on, 1970. **18**(5): p. 660-665.
2. Munk, B., R. Kouyoumjian, and L. Peters Jr, *Reflection properties of periodic surfaces of loaded dipoles*. Antennas and Propagation, IEEE Transactions on, 1971. **19**(5): p. 612-617.
3. Ott, R., R. Kouyoumijian, and L. Peters Jr, *Scattering by a two-dimensional periodic array of narrow plates*. Radio Science, 1967. **2**: p. 1347.

4. Pelton, E. and B.A. Munk, *A streamlined metallic radome.* Antennas and Propagation, IEEE Transactions on, 1974. **22**(6): p. 799-803.

5. Sarabandi, K. and N. Behdad, *A frequency selective surface with miniaturized elements.* Antennas and Propagation, IEEE Transactions on, 2007. **55**(5): p. 1239-1245.

6. Meng, X. and A. Chen. *Influence of cross-loop slots FSS structure parameters on frequency response.* in *Microwave, Antenna, Propagation and EMC Technologies for Wireless Communications, 2009 3rd IEEE International Symposium on.* 2009. IEEE.

7. Munk, B.A., *Frequency selective surfaces: theory and design.* 2005: Wiley-Interscience.

8. Philippakis, M., et al., *Application of FSS structures to selectively control the propagation of signals into and out of buildings.* Tech. Rep.AY4464A,ERA Technology Ltd, 2004.

9. Sudhendra, C., et al., *A Tripole Slot FSS Based Band Pass Radome for Out-Of-Band RCS Reduction of Low RCS Antennas in Stealth Applications.*

# Wide Angle and Polarization Insensitive Circular Ring Metamaterial Absorber at 10 GHz

O. Ayop, M. K. A. Rahim, and N. A. Murad

Department of Communication Engineering,
Faculty of electrical Engineering, Universiti Teknologi Malaysia,
81300 Johor Bahru, Malaysia
osman@fke.utm.my, mkamal@fke.utm.my, asniza@fke.utm.my

*Abstract-*This report presents a design of wide angle and polarization insensitive circular ring metamaterial absorber working at 10 GHz. The structure is designed on lossy FR4 substrate sandwiched by thin copper layers. From the simulation, the circular ring metamaterial absorber can operates at wide incident angle, where they maintain high absorbance, which is more than 87% for incident angles as large as 70°. Due to the geometrical properties, this kind of absorbance is insensitive to any polarization state.

## I. INTRODUCTION

Metamaterials are composite structured materials, structured at sub-wavelength scales, and depend on the structure to give rise to electromagnetic resonances. [1]. The unique properties of metamaterials have attracted much attention among researchers. Then, many sub-areas of metamaterials have been developed such as artificial magnetic conductor [2], frequency selective surface [3], electromagnetic band gap [4] and left-handed metamaterials [5]. Metamaterials generally appear as periodic elements that are a combination of many unit cell and they work in sub-wavelength of their operating frequency. Due to this characteristic, they can be fabricated using very thin substrate and small surface area compared to their operating wavelength. Metamaterials, until this moment have shown that they can improve the performance of other electromagnetic devices, especially antenna [6] and filter [7] if these structures are combined together. For example, using left-handed metamaterial, the gain of microstrip antenna can be increased [6].

AMC (artificial magnetic conductor) type of metamaterial has been widely used in antenna design to improve the radiation pattern of antennas, especially for horizontal dipole antenna. This kind of antenna obtains poor radiation pattern due to the appearance of out-of-phase image current on the ground plane that may cancel the real current of the antenna. Using AMC structure as artificial ground plane, they provide in-phase image current that may reinforces the real current and improves the radiation pattern of the antenna [8]. AMC ground plane in this case acting as a reflector of the antenna structure where it reflects most of the backward radiation.

In reverse, there are increasing much interest on metamaterials as electromagnetic absorber [9] other than electromagnetic reflector. To obtain this, the material properties such as permittivity and permeability are manipulated so that they can obtain high absorption. Perfect absorbance achieved when both reflectance and transmittance value are simultaneously 0. To obtain this, proper structure should be used to ensure the electromagnetic waves do not reflected or pass through the structure. This involves the impedance matching between the absorbing structures with the free space impedance. Theoretically, the maximum absorber occurs when the surface impedance of the metamaterial absorber is the same as free space impedance (377Ω).

## II. PROPOSED DESIGN OF CIRCULAR RING METAMATERIAL ABSORBER

The conventional electromagnetic absorbers use thick substrate for their design. For example, the usage of Salisbury screen as an electromagnetic absorber need to be designed with the thickness of λ/4 compared to the operating frequency [10]. The thickness of electromagnetic absorber can be reduced using planar substrate like FR4 board. In this report, the circular ring metamaterial absorber as shown in Figure 1 is designed using lossy FR4 substrate, which has substrate thickness, h of 0.8 mm (0.027λ₀), relative permittivity of 4.6 and tangent loss of 0.019. The structure consists of single copper ring, which has inner radius, $r_i$ of 2.56 mm and outer radius, $r_o$ of 2.80 mm. The width of the ring is 0.24 mm with average circumference of circle of 16.85 mm. At the back of the structure, the full copper layer is used to minimize the transmission. The size of unit cell of the absorber is 9 mm x 9 mm. The ring shape is selected due to their symmetrical property for all rotational angles. That means the absorbance properties will remain the same for all rotational angles and make this structure insensitive to any polarization state.

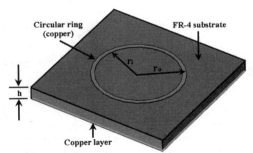

Figure 1. A unit cell of circular ring metamaterial absorber

### III. SIMULATION RESULT AND DISCUSSION

The structure as shown in Figure 1 is simulated using CST software with floquet solver. The structure, which is in a unit cell, is simulated by setting the boundary condition to be in periodic boundary condition while the top and bottom boundary to be open add space. The simulated result is given in term of $S_{11}$ (reflection) and $S_{21}$ (transmission).

#### A. Reflectance, Transmittance and Absorbance

Reflectance, $R(\omega)$ is a measure of the reflection ability of the structure with the incident electromagnetic waves. This value can be obtained by $R(\omega)=|S_{11}|^2$ where the $S_{11}$ magnitudes is given from simulation. The small value of reflectance indicated that the structure is not a reflector. For example, $R(\omega)=1$ means that the structure reflect all the electromagnetic waves while $R(\omega)=0$ means otherwise. But to ensure the structure is a good absorber, the value of transmittance, $T(\omega)$ should be determined. Transmittance is a measure of transmission ability of electromagnetic waves through the structure. This value can be determined by $T(\omega)=|S_{21}|^2$ where the $S_{21}$ magnitudes are obtained from simulation result. The small value of transmittance shows that the structure may not let the electromagnetic waves to pass through the structure. For many absorbance designs, the small value of transmittance can be easily obtained using full metal plane at the bottom of the absorbance structure. To determine the absorbance value, $A(\omega)$ this formula can be used; $A(\omega)=1-R(\omega)-T(\omega)$ where all value is linear. If both $R(\omega)$ and $T(\omega)$ are simultaneously 0, $A(\omega)=1$ indicates that the metamaterial structure can absorb 100% of incident electromagnetic waves.

Figure 2 shows the simulated result of circular ring metamaterial absorber in term of absorbance, reflectance and transmittance for normal incident EM waves. From the figure,

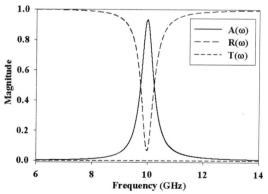

Figure 2. Simulated reflectance, transmittance and absorbance

$T(\omega)$ is 0 for all frequency range since the full copper layer that is used in the design will prevent the incident electromagnetic waves to pass through the structure. So the absorption is only depend on the value of $R(\omega)$. Through optimization and other design consideration, $R(\omega)=0.069$ (6.9%) is achieved and then gives $A(\omega)=0.931$ (93.1%). The full width half maximum, FWHM is given by 0.51 GHz (9.74 GHz-10.25GHz).

#### B. Current Distribution and Surface Current

To better understand the physical mechanism of the circular ring metamaterial absorber, the power loss distribution on the metals and surface current distribution is simulated. Figure 3 shows the power loss distribution for E-field and H-field of circular ring metamaterial absorber. For E-field, the concentration of power loss distribution is high at +y and −y-axis of the ring structure since it is parallel to the electric component of the incident EM waves. For H-field, the concentration of power loss distribution is high at +x and −x-axis of the ring structure since it is parallel to the magnetic component of EM waves.

The surface current distribution is then investigated for the structure. On the circular ring shape, the dipolar respond is noticed where the current is going up and down on the ring at the left side and the right side of the ring as shown in Figure 4. There is also a magnetic response associated by circulating displacement currents between the two metallic elements. Most currents are concentrated at the two sides of the copper layer indicate that the major loss for this absorber comes from copper losses compared to dielectric loss which is occur in the dielectric substrate between the metal layers.

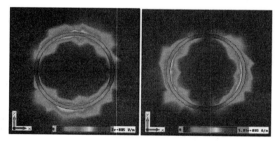

Figure 3. Power loss distribution for E-field (left) and H-field (right)

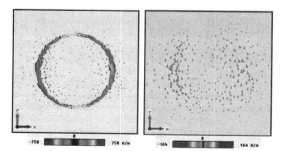

Figure 4. Surface current on ring structure at the top substrate (left) and copper plane at the bottom substrate (right)

*C. Effect of Absorbance for different Angle of Incident EM Waves*

From the previous result, the circular ring metamaterial absorber shows that it has good absorbance for normal incident of EM waves and the geometrical property make this structure insensitive to all polarization angles. To determine the absorbance characteristic of this kind of absorber for different angles of incident EM waves, simulation is done for TE and TM polarization incident waves. For TE polarization incident waves, the angles of incident of EM waves are varied where the electric component of EM waves always tangential to the surface of the metamaterial absorber. For TM polarization incident waves, the magnetic component of EM waves should be tangential to the surface of absorber for all incident angles of EM waves.

Figure 5 shows the magnitude of absorbance of circular ring metamaterial absorber for TE polarization incident waves. The angle of incident is varied from normal incident ($0^0$) to an angle where the absorbance drops to 0.5 (50%). The magnitudes of absorbance for $0^0$, $20^0$, $40^0$, $60^0$, $70^0$ and $82^0$ are 0.9310, 0.9494, 0.9889, 0.9781, 0.8775, and 0.5471 respectively. It shows that the circular ring structure manages to maintain high absorbance (more than 87%) for large incident angles ($70^0$) of EM waves for TE polarization incident waves.

Figure 6 shows the magnitude of absorbance of circular ring metamaterial absorber for TM polarization incident waves. The angle of incident is varied from $0^0$ to $83^0$. The magnitudes of absorbance for $0^0$, $20^0$, $40^0$, $60^0$, $70^0$ and $83^0$ are 0.9310, 0.9440, 0.9882, 0.9785, 0.8840, and 0.5056 respectively. The result shows that this structure can maintain high absorbance (more than 88%) for large incident angles ($70^0$) of EM waves for TM polarization incident waves. The results for TE and TM polarization incident waves are almost the same due to the characteristic of the structure that is highly symmetrical.

## IV. CONCUSION

In this report, a circular ring matamaterial absorber, which is independent from any polarization state and working at wide incident angles of electromagnetic waves, has been presented. Peak absorbance of 93.1% with 0.51GHz FWHM has been achieved for normal incident waves. The investigation of power loss distribution and surface currents in the unit cell of the structure reveals the behavior of power loss for both E-field and H-field.

### ACKNOWLEDGMENT

The authors thank the Ministry of Higher Education (MOHE) for supporting the research work, Research Management Centre (RMC), School of Postgraduate (SPS) and Communication Engineering Department (COMM) Universiti Teknologi Malaysia (UTM) for the support of the research under grant no R.J130000.7923.4S007.

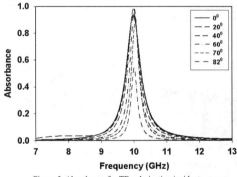

Figure 5. Absorbance for TE polarization incident waves

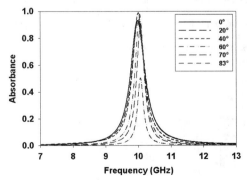

Figure 6. Absorbance for TM polarization incident waves

### REFERENCES

[1] S. A. Ramakrishna and T. M. Grzegorczyk, *Physics and Applications of Negative Refractive Index Materials* (CRC Press, Boca Raton, 2008).

[2] Maisarah Abu, Mohamad Kamal A Rahim, M.K. Suidi, I.M. Ibrahim, N.M.Nor, 'Dual band artifial magnetic conductor', *2009 IEEE International Conference on Antenna, propagation and Systems (INAS 2009)*, Johor Bahru Malaysia, 3- 5 Dec 2009.

[3] Yunus E. Erdemli, Kubilay Sertel, Roland A. Gilbert Daniel E. Wright, and John L. Volakis, "Frequency-selective surfaces to enhance performance of broad-band reconfigurable arrays", *IEEE transaction on antennas and propagation*, vol 50, no 12, 2002.

[4] O. Ayop, M. K. A. Rahim, M. R. Kamarudin, M. Z. A. Abdul Aziz and M. Abu "Dual band electromagnetic bamd gap structure incorporated with ultra-wideband antenna," *4th European Conference on Antenna and Propagation (EUCAP2010)*, Barcelona, 12-16 April 2010.

[5] H. A. Majid, M. K. A Rahim, T. Masri, "Left Handed metamaterial incorporated with circular polarized microstrip antenna", *International Symposium on Antennas and Propagation (ISAP 2009)*, Bangkok, Thailand, 20-23 October 2009.

[6] H. A. Majid, M. K. A. Rahim, T. Masri. 2009. " Micrsotrip Antenna's Gain Enhancement using Left Handed Metamaterial structure", *Progress in Electromagnetic Research M*, vol 8, pp 235 – 247, 2009.

[7] G. Jang and S. Kahn, "Design of a dual-band metamaterial band-pass filter using zeroth order resonance", *Progress In Electromagnetics Research C*, Vol. 12, 149{162, 2010.

[8] Ying Huang, Arijit De, Yu Zhang, Tapan K. Sarkar and Jeffrey Carlo, "Enhancement of radiation along the ground plane from a horizontal dipole located close to it", *IEEE Antennas and Wireless Propagation Letters,* Vol. 7, 2008.

[9] Landy, N. I., S. Sajuyigbe, J. J. Mock, D. R. Smith, and W. J. Padilla, "Perfect metamaterial absorber," *Physical Review Letters*, Vol. 100, 207402, 2008.

[10] R. L. Fante and M. T. McCormack, "Reflection properties of the Salisbury screen", *IEEE Trans Antennas Propaga*tion. vol 36, pp 1443-1454, 1988.

# A Plasmonic Multi-directional Frequency Splitter

Y. J. Zhou[1], X. X. Yang[1], and T. J. Cui[2]

[1]School of Communication and Information Engineering,
Shanghai University, Shanghai 200072, China

[2]State Key Laboratory of Millimeter Waves, Department of Radio Engineering, Southeast University,
Nanjing 210096, China

*Abstract* -Based on plasmonic metamaterials, we propose a plasmonic multi-directional frequency splitter with band-stop filters. Such a plasmonic splitter consists of three specially-designed metal gratings with finite thickness. Electromagnetic waves at the designed frequencies are confined and guided along the three grating structures and better isolation between different gratings is achieved with the band-stop filters. The experimental verification of the frequency splitter has been implemented in microwave frequencies with excellent agreements to full-wave simulations.

## I. INTRODUCTION

At visible frequencies, the electromagnetic (EM) surface waves supported at the metal surfaces are referred to surface plasmon polaritons (SPPs) [1]. SPPs can be tightly confined within distances of the order of wavelength in the dielectric and is much less than wavelength in the metal. Recently, researches have shown that metal surfaces drilled with sub-wavelength holes or grooves will produce field confinement in the microwave frequency [2-3]. The structured metal surface in microwave behaves like a planar metallic one at optical frequencies [3]. Thus the structural metallic surface with holes or grooves is capable of supporting surface EM modes known as "spoof" or "designer" SPPs which provides an effective method to guide and manipulate EM waves. The structural metallic surface belongs to plasmonic metamaterials.

Efficient unidirectional nanoslit couplers were proposed to ensure a unique propagation direction for SPPs [4]. The directional control of SPPs waves propagating through an asymmetric plasmonic Bragg resonator was demonstrated in [5]. A bidirectional subwavelength slit splitter for THz waves was proposed to guide the electromagnetic waves at different frequencies in the predetermined opposite directions [6]. The experimental investigation to split SPPs waves has been conducted in both microwave frequencies [7] and visible frequencies [8]. A grating with triangular grooves has been proposed due to the ability of higher confinement and lower bending loss [9]. A multi-directional surface wave splitter excited by a cylindrical wire was then proposed [10]. More recently, an ultrathin dual-band plasmonic frequency splitter based on a composite grating structure with nearly zero thickness was presented [11].

The isolation between the three grating branches of the mulit-directional surface wave splitter [10] is too low at low frequency, such as 5GHz. In this paper, we combine one specially designed band-stop plasmonic filter with such

frequency splitter to improve the isolation. The simulation and measurement results agree well.

## II. DESIGNS AND VERIFICATIONS

In [11], periodic grooves with two different depths are etched on a metallic strip with nearly zero thickness. According to the authors, the dispersion relation of the spoof SPPs in the low frequency band is mainly dominated by the deeper groove, while the spoof SPPs in the high-frequency band is primarily determined by the shallow groove. In this work, we construct a similar structure as shown in Fig. 1. The red rectangular faces on both ends are defined as port 1 and port 2, which are used to add excitation source (TM mode) in the CST microwave studio. The groove depths are denoted as $d_1$ and $d_2$, respectively. The groove period, groove width and grating depth are $p$, $w$ and $D$, where $p = 5$mm, $w = 2$mm and $D = 5$mm.

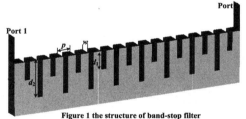

**Figure 1 the structure of band-stop filter**

The transmission spectra ($S_{21}$) is calculated and the results are shown in Fig. 2. Here $d_2$ is fixed to 14mm, and $d_1$ varies from 4mm to 7mm. It can be seen that the structure actually works as a band-stop filter. For the dual-band frequency splitter in [11], both working frequencies fall out of the stop-band. More works will be done to explore the deep physical mechanism for the stop-band filter.

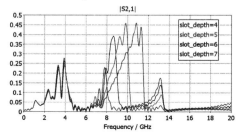

**Figure 2 The transmission spectra ($S_{21}$) of the band-stop filter**

978-7-5641-4279-7

We use such stop-band filter to improve the isolation between the three branches for multi-directional plasmonic frequency splitter at low frequencies. The detailed design principles for multi-directional SPPs splitter are explained in details [10]. The novel structure is given in Fig.3 (a), where the previous band-stop filter is put before each waveguide branch. The groove depths for these three branches are 4mm, 7mm and 11mm, respectively. For the stop-band filter, the deeper groove depth is 14mm. The fabricated sample and experimental setup are also shown in Fig. 3(a). For the multi-directional frequency splitter proposed in [10], the simulation result at 5GHz is shown in Fig. 3(b). It can be seen that the isolation between the three branches is bad at 5GHz. The simulation result and measurement result for the splitter in this paper are shown in Fig. 3(c) and (d). We can see that the isolation is much better and the simulation result agrees well with the experimental result. It shows that the plasmonic three-way frequency splitter with stop-band filters has better performance.

## III. CONCLUSIONS

We have proposed and fabricated a three-way plasmonic frequency splitter with band-stop filters, which has improved the isolation between the three branches at low frequencies and increase the performance of previous multi-directional surface wave splitter[10].The frequency splitter in the microwave frequency has been modeled by using the full-wave simulation method and experiments have been conducted for verification. The experiment and simulation results are in good agreements. Next we will try to explain why the band-stop filter works and to fabricate more realistic plasmonic multiplexer for application.

### ACKNOWLEDGMENT

This work was supported in part by the National Science Foundation of China under Grant Nos. 60990320, 60990324, 61271062.

### REFERENCES

[1] R. H. Ritchie, "Plasma Losses by Fast Electrons in Thin Films", *Phys. Rev*, vol. 106, pp. 874-881, 1957.

[2] J. B. Pendry, L. Matrin-Morenzo, and F. J. Garcia-Vidal, "Mimicking Surface Plasmons with Structured Surfaces", *Science*, vol. 305, pp. 847-848, 2004.

[3] A. P. Hibbins, B. R. Evans, and J. R. Sambles, "Experimental Verification of Designer Surface Plasmons", *Science*, vol. 308, pp. 670-672, 2005.

[4] F. López-Tejeira, S. G. Rodrigo, L. Martín-Moreno, F. J. García-Vidal, E.Devaux, T. W. Ebbesen, J. R. Krenn, I. P. Radko, S. I. Bozhevolnyi, M. U.González, J. C. Weeber, and A. Dereux, "Efficient unidirectional nanoslit couplers for surface plasmons", *Nat. Phys.*, vol. 3, pp. 324-328, 2007.

[5] S. B. Choi, D. J. Park, Y. K. Jeong, Y. C. Yun, M. S. Jeong, C. C. Byeon,J. H. Kang, Q.-H. Park, and D. S. Kim, "Directional control of surface plasmon polariton waves propagating through an asymmetric Bragg resonator", *Appl. Phys. Lett.*, vol. 94, pp. 063115-063117, 2009.

[6] Q. Gan, Z. Fu, Y. J. Ding, and F. J. Bartoli, "Bidirectional subwavelength slit splitter for THz surface plasmons", *Opt. Express*, vol. 15, pp. 18050-18055, 2007.

[7] H. Caglayan and E. Ozbay, "Surface wave splitter based on metallic gratings with sub-wavelength aperture", *Opt. Express*, vol. 16, pp. 19091-19096, 2008.

[8] Q. Gan and F. J. Bartoli, "Bidirectional surface wave splitter at visible frequencies", *Opt. Lett.*, vol. 35, pp. 4181-4183, 2010.

[9] A. I. Fernandez-Dominguez, E. Moreno, L. Martin-Moreno, and F. J.Garcia-Vidal, "Guiding terahertz waves along subwavelength channels", *Phys. Rev. B*, vol. 79, pp. 233104-233107, 2009.

[10] Y. J. Zhou and T. J. Cui, "Multidirectional surface-wave splitters", *Appl. Phys. Lett.*, vol. 98, pp. 221901-221903, 2011.

[11] Xi Gao, Jin Hui Shi, Xiaopeng Shen, Hui Feng Ma, Wei Xiang Jiang, "Ultrathin dual-band surface plasmonic polariton waveguide and frequency splitter in microwave frequencies", *Appl. Phys. Lett.*, vol. 102, pp. 151912-151915, 2013.

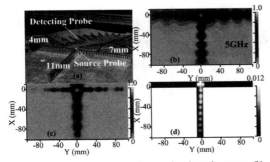

Figure 3 (a) the three-way frequency splitter sample and experiment setup; (b) simulation result for the splitter [10] at 5GHz; (c) and (d) simulation and experimental result for the novel splitter at 5GHz

# Channel Models for Indoor UWB Short Range Communications

Xiongwen Zhao, *Senior Member IEEE*, and Suiyan Geng
School of Electrical and Electronic Engineering, North China Electric Power University
102206, Beijing, China

*Abstract*—The cluster-based tapped-delay-line (TDL) models, and the novel path loss models vs. the transceiver distance and frequency band are developed in the paper. The TDL and path loss models are very useful for the UWB radio link level simulations and for the radio coverage predictions at any given distance with any frequency band.

## I. INTRODUCTION

Ultrawideband (UWB) radio is important in future short range communications, especially in indoor and rural environments for the fast traffic connections between a UWB transmitter and a computer, and among personal computers. For UWB radio system design, two models are critical. One is the path loss models for radio coverage, another is the cluster based tapped-delay-line (TDL) channel models for the link level simulations. In most of the open literature [1]-[7], path loss models are either related to the transceiver distance or the frequency band. In [2], UWB path loss models with the transceiver distances are summarized in detail from the open literature for indoor channel. However for UWB radios, path loss models related to both the distance and frequency band are do required in a specific system design and its coverage predictions.

## II. MEASUREMENT SETUP AND CAMPAIGN

### A. Measurement Setups

Frequency domain measurement was built by using a vector network analyzer (VNA). In the measurements, the VNA used 3200 tones to sweep a frequency band of 3-10 GHz which was segmented into two pieces of 3-6.5 GHz and 6.5-10 GHz for enhancing the delay range. The channel impulse responses were
obtained by the inverse Fourier transformation of the frequency domain measurement data. To remove the small-scale fading effect, spatial averaging was adopted, i.e., at each measurement position the measurement was performed at 25 grid points with 10 cm grid space in a square [7].

### B. Measurement Campaign

Fig. 1 shows the measurement layout in a corridor. Both line-of-sight (LOS) and non-LOS (NLOS) measurements were performed. In the LOS measurements, the transmitter TX was
located in the corridor with two meters away from point 1, the receiver was placed in the fixed positions from point 1 to point 12. The measurement started from point 1 and took one meter step till point 8 with the door open. Point 9 is at the same position as point 8, however, from point 9 till point 12, the door is closed and the distance step changed as two meters away from each other. The measurements were done after working hours to avoid the effect of moving people. In NLOS measurements, the transmitter was located in the room shown in Fig. 1, and the receiver is placed at different positions from point 1 to 10 by way of point 9' with one meter distance away from each other and the door closed. The distance from the transmitter to point 1 is about 5.1 meters.

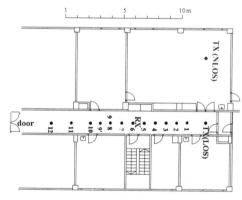

Fig. 1. Corridor Measurement layout

## III. CLUSTER BASED DELAY LINE CHANNEL MODELS

The power delay profiles (PDPs) for the LOS measurements are plotted in Fig. 2(a) where the receiver is located at different positions as described in II and shown in Fig. 1. The 30 dB dynamic range is used for noise cut. From RX1 to RX8, the receiver is away from the transmitter TX(LOS) with the distances from two to nine meters, basically two clusters are observed. At RX8, one more cluster has just started to appear. RX9 is the same position as RX8, the only difference is that starting from RX9 onwards the door is closed. At RX9, three clusters can be obviously seen. The 3rd cluster is much strong compared to the one at RX8. The 2nd cluster at RX8 is now

978-7-5641-4279-7

divided into two subclusters at RX9, the two subclusters can be regarded as in one cluster due to small delay range (same as at RX10 and RX12). It's seen that UWB radio is greatly affected by the change of the environment, e.g. the cluster structures and tap powers are changed when the door is open and closed.

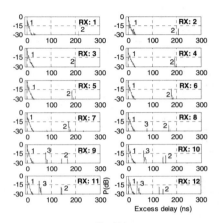

Fig. 2(a)

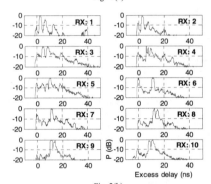

Fig. 2(b).

Fig. 2 The PDPs at different locations in the corridor. (a) LOS. (b) NLOS.

From Fig, 2(a), another interesting phenomenon is that the clusters (e.g. clusters 2 and 3) are moving when the distances between the transceiver are increasing. The clusters are obviously located at the side of the corridor with the door, and they follow the Saleh-Valenzuela model [8].

Table I (A) shows the cluster based TDL models for the LOS cases corresponding to the PDPs shown in Fig. 2(a). Cluster 1 is the main cluster which can contain more taps. Clusters 2 and 3 have less taps, but taps with large excess delays can be observed, which are the critical taps to affect the data transmission rate of the system. In Table I (A), the taps with * before mean that they are from the cluster 2, while with ** before, the taps are from cluster 3. With no sign before, the taps are from the main cluster. The TDL models are extracted by counting the local peaks in a cluster.

The PDPs for the NLOS measurements are plotted in Fig.

2(b) where the transmitter was located at the room and the receiver was placed from point 1 to point 10 with one meter away from each other by way of point 9'. In the NLOS environment, 20 dB dynamic range is applied for noise cut. From the NLOS PDPs, it's seen that only one cluster is found, with about 50 ns delay range. The first tap with zero excess delay is contributed by the transmission through the wall. The taps with the maximum powers are due to the diffraction from the two edges of the room door or contributed by other multipath in most of the RX positions. The NLOS TDL models are listed in Table I (B). Only the main cluster is found in the NLOS case.

## IV. NOVEL PATH LOSS MODELS

Path loss models related to both the distance between the transceiver and the frequency are derived by the steps described in A and B.

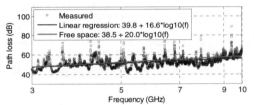

Fig. 3 Path loss and frequency band at RX1 in LOS

### A. Path Loss Frequency Dependence

The path loss frequency dependences at the locations from RX1 to RX12 in the LOS, and RX1 to RX10 in the NLOS are derived first, where the linear relationship (1) is applied. As an example, Fig. 3 shows the path loss vs. frequency band at position RX1 in the LOS measurements. In (1), $b$ is called the intercept and $n$ the exponent, and $f$ is in GHz. By performing the similar linear regressions at the other locations, the intercepts and 10*exponents are listed in Tables II (A) and (B) for the LOS and NLOS environments, respectively. The fitting standard deviations (stds) for the LOS and NLOS locations are within 4.9 ~6.3 dB and 5.3 ~ 6.2 dB, respectively.

$$PL = b + 10 * n * \log 10(f) \tag{1}$$

### B. Distance Dependences of the Intercept and Exponent

In Table II, the corresponding measurement distances between the transceiver are listed in the first column, the next step is to find the relationships of the intercepts and exponents with the distances. An examples, Figs. 4 (a) and (b) show the linear fittings for the intercepts and exponents for the LOS measurements. Similar ways can be applied in the NLOS case as well. The final path loss model are, therefore, related to both the distance and frequency. For LOS case, the path loss model is expressed as

$$PL = 31.4 + 18.1 * \log 10(d) + (22.1 - 0.27d) * \log 10(f) \tag{2}$$

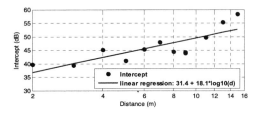

Fig. 4 (a)

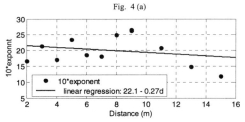

Fig. 4 (b)

Fig. 4 Relationships of the intercepts and exponents with the distances for the LOS measurements. (a) Intercept. (b) Exponent.

The distance can be in 15 meters. In the NLOS case, the path loss model can be derived as

$$PL = 28.3 + 41.9 * \log 10(d) + (28.5 - 0.76d) * \log 10(f) \quad (3)$$

The distance can be within 12 meters. (2) and (3) are the path loss models which are expected to be derived. However, the complicated issue is to estimate the shadow fading (SF), e.g. what are the SF standard deviations for (2) and (3) due to different fitting steps for the path loss and their related parameters. SF depends strongly on the measured environ-ments, but one measurement campaign cannot cover all the possible environments. The SF std can therefore be regarded as an empirical parameter. By open literature, the std for LOS can be within $1 - 4$ dB, and $2 - 8$ dB for NLOS environments [1][2][9]. At last, shadow fading in dB can be modeled as random variable with normal distribution with a specific standard deviation.

## V. TESTS OF THE PATH LOSS MODELS

The new models (2) and (3) are tested by using them to compare with the path loss model (1) at different receiver locations. Based on (2) and (3), the corresponding intercept and 10*exponent are tabulated in Table II as well to compare with the values obtained by (1). The fitting stds for different receiver positions are listed in the last columns in Table II (A) and (B) by using (1) and (2) for the LOS, and (1) and (3) for the NLOS, respectively. Very good agreement can be found in the most measurement locations, and the final mean stds are 1.0 and 2.0 dB for the LOS and NLOS, respectively. In [1], the path loss frequency dependence is described as $f^K$, where $K$ lies in $1 - 3$ ($K = 2$ is for free space). In this work, we can find that $K$ (exponent) lies in $1.2 - 2.7$ for the LOS and $1.1 - 3.3$ for the NLOS environments, respectively by (1)(2)(3).

### TABLE II
PATH LOSS FREQUENCY FITTING. (A) LOS. (B) NOLOS

#### TABLE II (A)

| | Path loss vs. frequency (LOS) | | | | |
|---|---|---|---|---|---|
| | Eqn. (1) | | Eqn. (2) | | std (dB) |
| d (m) | intercept (dB) | 10*exponent | intercept (dB) | 10*exponent | using (1) and (2) |
| 2 | 39.8 | 16.6 | 36.8 | 21.6 | 1.2 |
| 3 | 39.4 | 21.3 | 39.9 | 21.3 | 0.6 |
| 4 | 45.1 | 17 | 42.2 | 21 | 0.7 |
| 5 | 41.2 | 23.4 | 44 | 20.8 | 0.9 |
| 6 | 45.3 | 18.5 | 45.4 | 20.5 | 1.8 |
| 7 | 48 | 18.1 | 46.6 | 20.2 | 0.5 |
| 8 | 44.4 | 24.9 | 47.7 | 19.9 | 0.9 |
| 9 | 44.3 | 26.3 | 48.6 | 19.7 | 1.3 |
| 9 | 44 | 26.5 | 48.6 | 19.7 | 1.2 |
| 11 | 49.8 | 20.8 | 50.2 | 19.1 | 0.9 |
| 13 | 55.4 | 14.8 | 51.5 | 18.6 | 1 |
| 15 | 58.4 | 11.9 | 52.6 | 18.1 | 1.3 |

#### TABLE II (B)

| | Path loss vs. frequency (NLOS) | | | | |
|---|---|---|---|---|---|
| | Eqn. (1) | | Eqn. (3) | | std (dB) |
| d (m) | intercept (dB) | 10*exponent | intercept (dB) | 10*exponent | using (1) and (3) |
| 5.08 | 60.6 | 14.9 | 57.9 | 24.6 | 5.2 |
| 5.41 | 52.1 | 32.8 | 59 | 24.4 | 1.3 |
| 5.89 | 63.5 | 23 | 60.6 | 24 | 2.1 |
| 6.48 | 63.8 | 23.6 | 62.3 | 23.6 | 1.5 |
| 7.17 | 63.9 | 21.9 | 64.1 | 23.1 | 1.2 |
| 7.93 | 69.8 | 12.5 | 66 | 22.5 | 4.3 |
| 8.73 | 60.7 | 32.3 | 67.7 | 21.9 | 2 |
| 9.57 | 77.4 | 10.7 | 69.4 | 21.2 | 1.6 |
| 10.44 | 65.8 | 25.6 | 71 | 20.6 | 1.4 |
| 11.33 | 73.3 | 20.6 | 72.5 | 19.9 | 1.4 |

So far, the path loss related to both the distance and frequency are derived. It's good to show how the received power changes with both of the parameters. Let's define the received power is the minus value of the path loss and no antenna gains are included. Figs. 5(a) – (c) show the received powers vs. the transceiver's distances and the frequency bands in free space, the LOS and the NLOS environments, It's seen that the received power in the LOS corridor is close the power in free space, but a bit higher power can be received due to guided wave effect. The received power in the NLOS is much lower than that in the free space and in the LOS, especially in larger distances and with higher frequencies.

## VI. CONCLUSION

The TDL and path loss models are derived in this paper based on the corridor measurements. The TDL models are useful for the link level simulations, which is seldom to see in open literature. In most of the locations, the channel are Ricean and Rayleigh for the LOS and NLOS corridor, respectively. The traditional UWB path loss models are either distance or frequency dependence. In this paper, the new path loss models as the function of both the distance and frequency are derived, with which the UWB radio coverage can be predicted at any given distance and any interested frequency band.

[2] B. M. Donlan, D. R. McKinstry, and R. M. Buehrer, , "The WUB Indoor Channel Large and Small Scale Modelling," *IEEE Trans. Wireless Commun.*, vol. 5, no. 10, pp. 2863-2873, Oct. 2006.

[3] S. S. Ghassemzadeh, R. Jana, C. W. Rice, W. Turin, and V. Tarokh, "Measurement and Modeling of an Ultra-Wideband Indoor Channel," *IEEE Trans. Communi.*, vol. 52, no. 10, pp. 1786- 1796, Oct. 2004.

[4] J. A. Dabin, A. M. Haimovich, and H. Grebel, "A Statistical Ultr-Wideband Indoor Channel Model and the Effects of Antenna Directivity on Path Loss and Multipath Propagation," *IEEE J. Select. Areas Communi.*, vol. 24, no. 4, pp. 752 - 758, Apr. 2006.

[5] S. Thiagarajah, B. M. Ali, and M. Ismail, "Empirical UWB Path Loss Models for Typical Office Environments," in *Proc. 13th IEEE Int. Conf. Networks*, Malaysia, vol. 2, Nov. 2005, pp. 16-18.

[6] R. Saadane, M. Wanbi, A. Hayar, and D. Aboutajdine, "Path Loss Analysis Based on UWB Channel Measurements," in *Proc. IEEE Int. Conf. Computer Systems Applications*, Rabat, Morocoo, May 10-13, 2009, pp. 970-974

[7] S. Geng, S. Ranvier, X. Zhao, J. Kivinen, and P. Vainikainen, "Multipath Propagation Characterization of Ultra-wide Band Indoor Radio Channels," in *Proc. IEEE Int. Conf. Ultra-Wideband*, Zurich, Switzerland, Sept. 5-8, 2005, pp. 11-15.

[8] A. A. M. Saleh and R. A. Valenzuela, "A statistical model for indoor multipath propagation," *IEEE J. Select. Areas Commun.*, vol. 5, pp. 128-137, Feb. 1987.

[9] *WINNER II: Channel Models*, IST-4-027756 WINNER II, D1.1.2 V1.1, 2007.

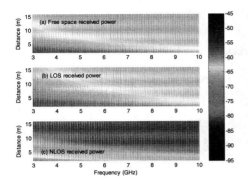

Fig.5 Path loss vs. both of the transceiver distance and frequency. (a) Free space. (b) the LOS corridor. (C ) the NLOS corridor.

## REFERENCES

[1] A.F. Molisch, "Ultra-Wide-Band Propagation Channels," *Proc. IEEE*, vol. 97, no. 2, pp. 353-371, Feb. 2009.

TABLE I  CLUSTER BASED TAPPED DELAY LINE MODELS. (A) LOS. (B) NOLOS

| | RX1 | | RX2 | | RX3 | | RX4 | | RX5 | | RX6 | | RX7 | | RX8 | | RX9 | | RX10 | | RX11 | | RX12 | |
|---|---|---|---|---|---|---|---|---|---|---|---|---|---|---|---|---|---|---|---|---|---|---|---|---|
| Delay (ns) | P (dB) | | | | | | | | | | | | | | | | | | | | | | | |
| 0 | 0 | 0 | 0 | 0 | 0 | 0 | 0 | 0 | 0 | 0 | 0 | 0 | 0 | 0 | 0 | 0 | 0 | 0 | 0 | 0 | 0 | 0 | 0 |
| 0.7 | -0.8 | 2 | -5.5 | 1.3 | -4.9 | 1 | -4.1 | 3.9 | -12 | 3.6 | -11 | 3.1 | -8.6 | 3 | -7.9 | 3 | -8 | 1.6 | -7.3 | 4 | -14.1 | 3.9 | -11.8 |
| 3.2 | -7.1 | 7.7 | -12.9 | 6.7 | -11.7 | 7.4 | -11.1 | 7.4 | -9.4 | 6.6 | -11.9 | 6.7 | -9.6 | 6.7 | -10.4 | 6.7 | -10.3 | 6.7 | -10.5 | 7.2 | -11.6 | 7.4 | -11.5 |
| 5.4 | -11.4 | 11.7 | -19.5 | 14.3 | -17.5 | 4.4 | -12.6 | 10.4 | -17.4 | 9.7 | -15.2 | 9.9 | -16.9 | 11.6 | -18.5 | 13.6 | -22.3 | 8.4 | -12.3 | 9.9 | -18.9 | **36,4 | **-9,1 |
| 9 | -15.5 | 21.9 | -21.1 | 19.2 | -23.8 | 12.1 | -18.2 | 16 | -20.3 | 12.9 | -20.5 | 12.5 | -19.3 | 16.6 | -22.6 | **76,1 | **-13,5 | 11.3 | -19.4 | 11.3 | -21.3 | **43,3 | **-20,1 |
| 12 | -21 | 25.9 | -25.4 | **199 | *-17,9 | 23.4 | -25.6 | 23.3 | -26 | 17.4 | -24.2 | 16.3 | -22.3 | *165,1 | *-13,9 | **83 | **-25,2 | 15.4 | -23.1 | **49,2 | **-9 | *126,2 | *-18 |
| 27.4 | -26.3 | *206,7 | *-20,5 | | | *192,4 | *-16,3 | *185,4 | *-15,7 | 21.7 | -25.9 | *172 | *-14,2 | | | *142 | *-23 | **62,9 | **-10 | **56,3 | **-18,6 | *142,9 | *-22,4 |
| *213,4 | *-24 | | | | | | | | | *178,7 | *-14,5 | | | | | *166,1 | *-21,2 | **70,1 | **-20,8 | *138,1 | *-18,2 | | |
| | | | | | | | | | | | | | | | | | | *142,3 | *-20,6 | *142,4 | *-21,4 | | |
| | | | | | | | | | | | | | | | | | | *152,7 | *-18,6 | | | | |

\* The taps in the 2nd cluster. \*\* The taps in the 3rd cluster.

TABLE I (A)

| | RX1 | | RX2 | | RX3 | | RX4 | | RX5 | | RX6 | | RX7 | | RX8 | | RX9 | | RX10 | |
|---|---|---|---|---|---|---|---|---|---|---|---|---|---|---|---|---|---|---|---|---|
| Delay (ns) | P (dB) | | | | | | | | | | | | | | | | | | | |
| 0 | -10.6 | 0 | -8.1 | 0 | -7.6 | 0 | -6.7 | 0 | -5.8 | 0 | -1 | 0 | -5.4 | 0 | -11.1 | 0 | -15.1 | 0 | -9.7 |
| 3 | 0 | 2.7 | 0 | 5 | -0.8 | 6.3 | 0 | 4.4 | -5.8 | 11.1 | 0 | 5.7 | -11.4 | 4.4 | -13.2 | 2.6 | -14.4 | 4.9 | -14.1 |
| 6.6 | -4.2 | 7.3 | -6.7 | 8.3 | 0 | 7.7 | -1.7 | 7.9 | -6.2 | 17.7 | -11 | 11.7 | 0 | 8.9 | -12.9 | 11 | 0 | 8.6 | -1.7 |
| 13.4 | -9.2 | 11.3 | -11.5 | 12.6 | -5.9 | 10.2 | -2.7 | 10 | 0 | 22.9 | -12.9 | 16 | -7.8 | 12.7 | 0 | 14.2 | -1.7 | 10.9 | 0 |
| 20.9 | -17.1 | 14.3 | -15.4 | 19 | -6.4 | 13.4 | -7.4 | 13.7 | -6.4 | 28.9 | -16.4 | 24.1 | -10.4 | 15 | -1.4 | 17.3 | -11.5 | 19.2 | -12.2 |
| 38.7 | -14.5 | 19.7 | -13.6 | 31.2 | -12.3 | 20.7 | -8.6 | 18.3 | -5.7 | 33.3 | -17.7 | 29.7 | -13.7 | 17.9 | -10.3 | 25.4 | -14.3 | 32 | -18.1 |
| | | 23.9 | -19.4 | 48.3 | -19.2 | 26.4 | -10.5 | 22.2 | -8.5 | | | 35.4 | -16.9 | 24.1 | 0 | 29.3 | -19.6 | | |
| | | | | | | 31.9 | -16.1 | 26.7 | -13.3 | | | | | 29.4 | -17 | | | | |
| | | | | | | 42.9 | -19.6 | 32.9 | -14 | | | | | 36.7 | -19.1 | | | | |
| | | | | | | | | 43.9 | -19.3 | | | | | | | | | | |

TABLE I (B)

# Design of a Compact UWB Diversity Antenna for WBAN Wrist-Watch Applications

*Seungmin Woo, Jisoo Baek, Hyungsang Park, Dongtak Kim and #Jaehoon Choi

Department of Electronics and Computer Engineering, Hanyang University 17 Haengdang-Dong, Seongdong-Gu, Seoul, 133-791, Korea

Email : *wsmint7@hanyang.ac.kr, #choijh@hanyang.ac.kr (corresponding author)

*Abstract-* This paper presents a compact ultra-wideband diversity antenna for wireless body area network wrist-watch applications. The antenna is designed based on a folded monopole antenna with a stub in order to achieve wideband characteristic. The proposed antenna has a size of 40 mm × 40 mm × 5 mm and operates at a frequency range of 2.9-5.1 GHz. The analysis of the antenna performance was carried out when it was placed on the human body equivalent flat phantom. Simulated results show that $S_{11}$ is lower than -10 dB and the isolation ($|S_{21}|$) is better than 15 dB across the UWB low band (3.1-4.8 GHz).

*Index Terms — Ultra-wideband, diversity antenna, wireless body area network, on-body communication*

## I. INTRODUCTION

There has been a lot of interest in wireless body area network (WBAN) applications such as biomedical, military and commercial services. An antenna used for WBAN system is desired to have small size, wideband characteristic and low specific absorption rate (SAR). These applications need high data rate to perform processing much data including high-definition video. At the same time, wearable devices are required to observe small power consumption (~ 1 mW/Mbps) [1]. When an antenna operates near the human body which has high dielectric constant, a high loss tangent and low conductivity, the performance of the antenna can be deteriorated. The multipath fading also is occurred by the movement of body parts, shadowing and scattering over the body. In order to overcome the human body effect and improve the antenna performance, UWB diversity technique is a good candidate for applying to WBAN systems [2].

UWB antennas have been studied for WBAN system especially for on-body applications recently [3-5]. In [3], the loop antenna with high gain characteristic was proposed. However, the antenna is sensitive so that the radiation property may not be maintained if the antenna was installed on a human wrist. The on-body directional antenna was presented in [4]. The antenna size was not adaptable to use for watches. In order to miniaturize the antenna size, a dual-port antenna with pattern diversity was reported in [5]. It had a good isolation property, but the feeding structure of the antenna was complex since the antenna has a central strip that extends vertically from the ground plane.

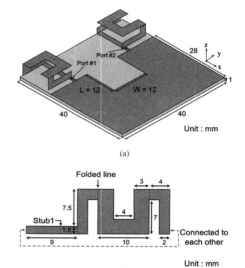

Figure 1. The proposed UWB diversity antenna:
(a) perspective view, (b) radiator structure

In this paper, we proposed a compact UWB diversity antenna for WBAN wrist-watch applications. To achieve a compact size of antenna and ultra-wide bandwidth, a stub1 is connected between side ends˜ of each radiator. For the simulated result, the human equivalent flat phantom was set using dispersive physical properties which varied in terms of frequency [6].

## II. ANTENNA DESIGN

The configuration of the proposed antenna for WBAN applications is shown in Fig. 1. The proposed antenna has a dimension of 40 mm × 40 mm × 5 mm. An FR-4 dielectric with a relative permittivity of 4.4 is used as a substrate. The proposed antenna consists of two radiators which are placed symmetrically on the top corner of the ground plane. Each antenna element has a size of 10 mm × 9 mm × 4 mm and is fed by a 50 Ω coaxial cable.

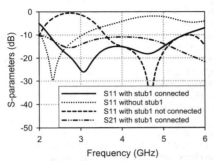

Figure 2. Simulated s-parameters for the variable stub1
in free space (without the slot)

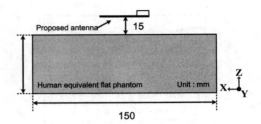

Figure 4. Human equivalent flat phantom for simulation

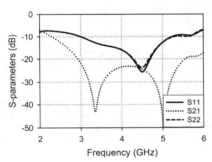

Figure 3. Simulated S-parameters in free space ($W$=$L$=12mm)

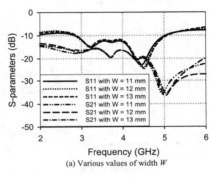

(a) Various values of width $W$

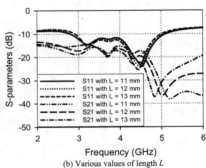

(b) Various values of length $L$

Figure 5. Simulated S-parameter characteristics of the proposed antenna
on the flat phantom.

To analyze the effect of stub1, the s-parameters of the proposed antenna with the variable stub1 is illustrated in Fig. 2. The stub1 which has a dimension of 9 mm × 1.5 mm is connected both each end of the folded line. When the stub1 is absent on the radiator, the antenna is operated as a folded monopole at 2.3 GHz. When the stub1 exist on the radiator, but not connected to port, the resonance frequency is changed to lower than 2.3 GHz. However, when the stub1 is connected to the port, the resonance frequency of lower band becomes higher. Consequently, the wideband characteristic of the antenna can be achieved.

Fig. 3 shows the s-parameter of the proposed antenna in free space. By adding a slot placed on a center between port #1 and #2, the isolation property of the antenna is improved from 10 dB to 23 dB at 4 GHz. The 10dB impedance bandwidth of the proposed antenna is 53% and the isolation is less than 20 dB at the entire UWB low band (3.1 – 4.8 GHz).

III. ANALYSIS OF ON-BODY PERFORMANCE

Fig. 4 shows the human equivalent flat phantom having the dispersive property from [6]. Considering a wrist-watch application, a human wrist phantom (70 mm × 70 mm × 150

mm) is used and placed away from the antenna by a distance of 15 mm.

Fig. 5 shows the s-parameters characteristic for various lengths $L$ and width $W$ of the slot. Choosing the proper dimension of the slot is important because the isolation is dependent on the width and the length of the slot. When the length $L$ and the width $W$ were increased, the isolation is improved over the frequency of interest. To obtain optimized isolation characteristic, the proper width W and length L were chosen that of 12 mm and 12 mm, respectively.

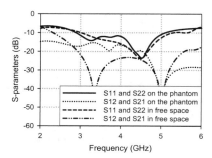

Figure 6. Simulated S-parameters characteristics.

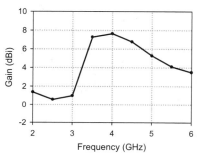

Figure 7. Simulated peak gain and radiation efficiency.

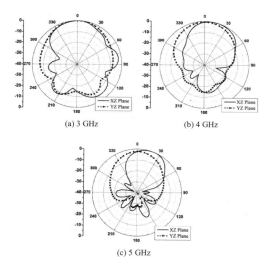

(a) 3 GHz          (b) 4 GHz

(c) 5 GHz

Figure 8. Normalized the radiation patterns.

In Fig. 6, the simulated s-parameters of the proposed antenna when the antenna is placed on the phantom and in free space are shown. 10 dB impedance bandwidth of the antenna is 55 % (2.9-5.1 GHz) and the isolation is greater than 15dB in the required frequency band.

Fig. 7 shows the simulated peak gain of the proposed antenna placed on the flat phantom. At 3 GHz, the peak gain is deteriorated in comparison with at 4 and 5 GHz because the electrical distance between the antenna and the phantom becomes larger as operating frequency decreases.

Fig. 8 shows the normalized radiation pattern of the antenna on flat phantom at 3, 4, and 5 GHz. The maximum power is delivered outward from the body and the front-to-back ratio of the antenna becomes larger as operating frequency increases. The reason of the phenomenon is according to the dispersive property of the phantom since conductivity increases when the frequency becomes higher.

## IV. CONCLUSION

A compact UWB diversity antenna for WBAN wrist-watch applications is proposed in this paper. The proposed antenna was simulated on the wrist-shaped human equivalent flat phantom. $S_{11}$ property satisfies lower than -10 dB and the isolation characteristic is higher than 15 dB over the whole UWB low band of 3.1-4.8 GHz. Radiation patterns of the antenna are toward off-body. Consequently, the proposed antenna can be a good candidate for UWB WBAN wrist-watch applications.

ACKNOWLEDGMENT

This research was funded by the MSIP(Ministry of Science, ICT & Future Planning), Korea in the ICT R&D Program 2013

REFERENCES

[1]  "IEEE 802.15 WPAN Task Group 6 (TG6) Body Area Networks," *IEEE Standard 802.15.6*, 2011.

[2]  Qiong, Wang, Hahnel, R., Hui, Zhang, Plettemeier, D., "On-body directional antenna design for in-body UWB wireless communication," *6th European Conference on Antenna and Propagation (EUCAP)*, 26-23 March 2012, pp.1011-1015.

[3]  See, T. S. P., Chiam, T. M., Ho, M. C. K., Yuce, M. R., "Experimental Study on the Dependence of Antenna Type and Polarization on the Link Reliability in On-Body UWB Systems," *IEEE Transactions on Antennas and Propagationin*, vol. 60, Nov. 2012, pp. 5373-5380.

[4]  W. C. Jakes, *Microwave Mobile Communications*. New York:Wiley, 1974.

[5]  Tuovinen, T., Yazdandoost, K. Y., Iinatti, J., "Ultra Wideband loop antenna for on-body communication in Wireless Body Area Network," *6th European Conference on Antenna and Propagation (EUCAP)*, 26-23 March 2012, pp.1349-1352.

[6]  Akimasa Hirata, Toshihiro Nagai, Teruyoshi Koyama, Junya Hattori, Kwok Hung Chan and Robert Kavet, "Dispersive FDTD analysis of induced electric field in human models due to electrostatic discharge" *Physics In Medicine and Biology*, vol. 57, June. 2012, pp. 4447-4458

# A Tablet MIMO Antenna with a Wave-Trap Slot for LTE/WiMAX Applications

Wen-Hsiu Hsu, Chung-Hsuan Wen, Shan-Cheng Pan and Huan-Yu Jheng
Department of Computer and Communication, Shu-Te University NO. 59, Hengshand Rd., Yanchao Dist., Kaohsiung City, 82445 Taiwan

*Abstract-* **A tablet MIMO antenna with a wave-trap for LTE/WiMAX applications is presented. To improve the isolation between the two radiating elements of MIMO antenna, added a wave-trap slot on the ground plane. The isolation is improved by approximately 20 dB at the LTE700 frequency. The design MIMO antenna satisfies a 6 dB return loss requirement and the obtained envelope correlation coefficient (ECC) is lower than 0.1 from the LTE (700) and LTE/WiMax (2300-2700). The method can be accurate operation at LTE/WiMax communication system.**

*Index Terms* — MIMO, LTE, WiMax, coupled-fed, antennas, ECC

## I. INTRODUCTION

Current wireless communication systems require higher bit rate transmission to support various multimedia services. A multi-input multi-output (MIMO) system has been regarded as a promising solution, since it can increase the channel capacity without sacrificing spectrum efficiency or consuming additional transmitted power [1]. In MIMO system, two or more antennas are used on both the transmitter and receiver sides. A critical point is to arrange compact antenna elements without impairing antenna performance and system requirement within a compact space. To do that, low mutual coupling or high isolation between adjacent antennas is a key factor. However, the antenna elements are strongly coupled with each other as well as with the ground plane, because they share the common surface current. Several techniques have been introduced to improve the isolation characteristic, such as employing protruded T-shaped ground plane [2], protruded ring strip ground plane [3], protruded ground plane and a spiral open slot [4], protruded a wave-trap ground plane[5], quarter wavelength slot on the ground plane [6], and notches on the ground plane as resonator [7]. These structures provide conspicuous decoupling effect, but suffer from complicated structures and large structural area. They cannot be applied for LTE mobile application with restricted space available for the antennas.

In this paper, we propose a MIMO antenna with application. To improve the isolation characteristic at the LTE700, a wave-trap slot is added on the ground plane.

## II. ANTENNA DESIGN

The configuration of the proposed MIMO antenna is shown in Figure.1. The proposed MIMO antenna include two radiating elements, the wave-trap slot on the ground plane, and FR4 substrate (= 4.4) with thickness of 0.8 mm.

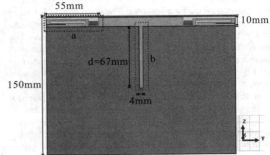

Figure.1 Tablet MIMO system for LTE/WiMax applications.

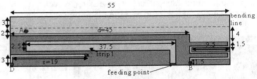

Figure.2 Antenna body.

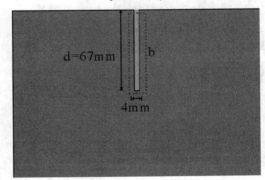

Figure.3 The wave-trap slot slot.

The two radiating elements of the MIMO antenna were symmetrically placed with respect to the center and were located near the corners of the top edge of the ground plane as shown in Figure1.consists of a wave-trap slot on the ground plane.Figure.2 shows the geometry of the proposed LTE/WiMax coupled-fed loop antenna. The loop antenna comprises an inverted L-shape radiating feed and dual coupled shorting-strip connects to display ground. One is for lower

978-7-5641-4279-7

band LTE (700) and the other is for higher bands (LTE2300~2700,WiMAX), The antenna is mainly printed on a 0.8mm thick FR4 substrate of dimension(10 x 55 x 3 mm³), relative permittivity 4.4, and loss tangent 0.024.

On the top edge of the FR4 substrate (cleance area), the printed metal pattern mainly includes the inverted L-shape radiating element, the main section and banding section. The banding section of the radiation element has a size of 3 x 55 mm², which is connected orthogonally to the printed pattern main section on the FR4 substrate. The main section has three parts , the first part is the coupled-fed (length 37.5 mm and width 2.5 mm) connect to the feeding point and second is lower band strip which is connect banding section and meandering line, the coupling gap to the inverted L radiating feed is 1.5 mm. From the major part of the coupled shorted strip, this has a length of about 79.5 mm (AB) which can operate at its quarter-wavelength mode as the lower resonant mode, which occurs at about 735 MHz in the proposed design. The last one is another coupled shorted strip, which has a length of about 22 mm (CD) and is coupled fed through the coupling section (length s ¼ 15.5 mm) and the coupling gap (g ¼ 1 mm) by the inverted L radiating feed. With the coupling feed, the antenna can operate at its quarter-wavelength mode which occurs at about 2300 MHz in the proposed design

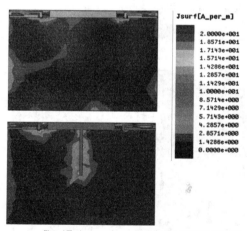

Figure.4 The time-averaged current distribution and ground plane at 750 MHz.

(a) without the wave-trap slot (b) with the wave-trap slot.

The configuration of the proposed a wave-trap slot on the ground plane is shown in Figure.3. To improve the isolation between two symmetrical radiators, a slot on the ground plane is used. The wave-trap slot can hinder the current of the antenna, so that the antenna current interference becomes small and increasing the isolation. In designing the wave-trap slot, a wave-trap slot on the ground plane of the FR4 substrate is symmetrically printed with respect to the center of the ground which size is 67 x 4 mm².

To investigate the effect on the isolation characteristic, the current distributions at 750 MHz with and without the wave-trap slot on the ground plane are shown in Figure.4. When the wave-trap slot on the ground plane was exited, a substantial current was induced at the other element. After the wave-trap slot on the ground plane was added, the induced current on the non-excited element become very weak.

## III. EXPERIMENTAL RESULTS

Figure 5 shows the measured and simulated S-parameter of the proposed antenna. The simulated are obtained using the three-dimensional full-wave electromagnetic field simulation HFSS [8].The measured results are very similar to simulated ones. The fabricated MIMO antenna has an impedance bandwidth of the return loss < 6 dB over the whole LTE 700、 LTE/WiMax (2300-2700) and the isolation is higher than 20 dB at the center frequency.

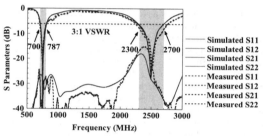

Figure.5 Measured and simulated S-parameter of the proposed antenna.

Antenna correlation calculation procedure is provided by appropriate methods of analysis [9-15]. The method of calculating envelope correlation of elements in each antenna array configuration is based on a fundamental Equation (1) that requires 3-dimensional radiation pattern considerations.

$$\rho_e = \frac{\left| \iint_{4\pi} \left[ \overrightarrow{F_1}(\theta,\phi) \bullet \overrightarrow{F_2}^*(\theta,\phi) d\Omega \right] \right|^2}{\iint_{4\pi} \left| \overrightarrow{F_1}(\theta,\phi) \right|^2 d\Omega \iint_{4\pi} \left| \overrightarrow{F_2}(\theta,\phi) \right|^2 d\Omega}$$

Equation (1)

The parameter is the field radiation pattern of the antenna system when only the port 1 is excited and the port 2 is terminated to 50 Ω load. The symbol " ● "also denotes the Hermitian product [9] [15].

Recent research activities have shown that the envelope correlation can be well defined by a simple closed-form equation that relates the scattering parameters of the elements in an antenna array configuration. Especially, in case of a multipath indoor environment with a uniform distribution of

Equation (2) is proved to be a good approximation [9]. For two antenna elements this equation using the scattering parameters becomes:

$$\rho_e = \frac{\left|S_{11}^{*}S_{12} + S_{21}^{*}S_{22}\right|^2}{\left(1-\left|S_{11}\right|^2-\left|S_{21}\right|^2\right)\left(1-\left|S_{22}\right|^2-\left|S_{12}\right|^2\right)}$$

Equation (2)

It is obvious that radiation pattern in Equation (1) makes the calculation more complicated than the envelope correlation calculations based on in Equation (2). The practical advantage of the third method that is based on second equation is that not only is quite simple to use it experimentally, but also provides sufficiently accurate results in many experimental environments such as in-door environments with rich multipath propagation performance.

Figure.6 show the ECC of two antennas, the obtained ECC is lowers than 0.1 from measured S-parameters and is sufficient for MIMO applications.

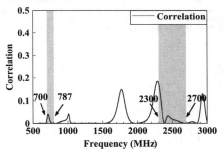

Figure.6 ECC of two antennas given by the obtained from measure S-parameters.

Figure.7 shows the measured antenna efficiency of the proposed antenna. The measurement is conducted in a far-field anechoic chamber, and the measured antenna efficiency includes the mismatching loss. The antenna efficiency is about 80–60% and 75–40% in the antenna's lower and upper bands, respectively, which are acceptable for practical applications.

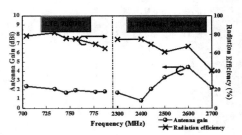

Figure.7 Measured antenna efficiency (mismatching loss included) of the proposed antenna

The measured 3D-radiation pattern of fabricated MIMO antenna was measured in the ETS chamber[16] and the 746、2500 results are presented in Figure.8.

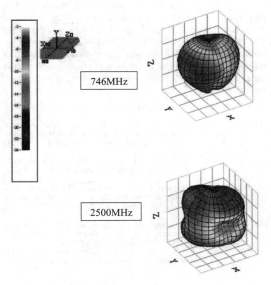

Figure.8 Measured 3D radiation patterns of the proposed antenna.

Figure.9 shows the S-parameter characteristics for various lengths (d) of wave-trap slot. Choosing the proper dimension of the wave-trap slot is very important because the isolation is strongly dependent on the length of the wave-trap slot. When the length d is increased, the improved isolation frequency band shifts to a lower frequency band. The optimized value of d is 67 mm.

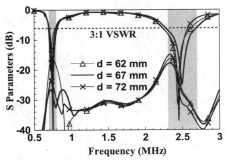

Figure.9 Simulated S-parameters characteristics for the proposed MIMO antenna for various values of "d"

## CONCLUSION

In this paper, MIMO antennas with improved isolation using a wave trap slot on the ground plane for LTE mobile application are proposed. The wave-trap slot is added on the ground plane to improve the isolation characteristic. The proposed MIMO antenna has a 6 dB return loss bandwidth and the isolation characteristic is higher than 11 dB over the whole LTE 700、LTE/WiMax (2300-2700). The obtained envelope correlation coefficient(ECC) was less than 0.4. Therefore, the proposed a wave-trap slot on the ground plane, can be a good candidate for MIMO antenna system requiring high isolation characteristic.

### REFERENCES

[1] Gesbert, D., Shafi, M., Shiu, D.S., Smith, P., and Naguib, A.,"From theory to practice : an overview of MIMO space-time coded wireless systems" IEEE Journal on Selected Areas in Communication., vol. 21, no.3, pp. 281-302, Mar. 2003.

[2] T.-Y. Wu, S.-T. Fang, and K.-L. Wong. "Printed diversitymonopole antenna for WLAN operation", Electronics Letters.,vol. 38, no. 25, pp. 1625-1626, Dec 2002.

[3] K.L. Wong, P.W. Lin and H.J. Hsu, "DecoupledWWAN/LTE antennas with an isolation ring stripembedded therebetween for smartphone application,"Microwave Opt. Technol. Lett., Vol. 55, pp. 1470-1476, Jul. 2013.

[4] K.L. Wong, H.J. Jiang and Y.C. Kao, "High-isolation2.4/5.2/5.8 GHz WLAN MIMO antenna array for laptopcomputer application," Microwave Opt. Technol. Lett.,Vol. 55, pp. 382-387, Feb. 2013.

[5] T.W. Kang and K.L. Wong, "Isolation improvement of2.4/5.2/5.8 GHzWLAN internal laptop computerantennas using dual-band strip resonatoras awave-trap,"Microwave Opt. Technol. Lett., Vol. 51, pp. 58-64, Jan.2010.

[6] Y.Ge, K.P. Esselle, and T.S. Bird, "Compact diversity antenna for wireless devices", Electronics Letters., vol. 41,no. 2, pp.52-53, Jan 2005.

[7] K.-J. Kim and K.-H. Ahn, "The high isolation dual-band inverted F antenna diversity system with the small N-section resonators on the ground plane", Microwave and Optical Technology Letters., vol. 49, no. 3, pp. 731-734, March 2007.

[8] Available at: http://www.ansys.com/products/hf/hfss/,ANSYS HFSS, Ansoft Corp., Pittsburg, PA 15219

[9] R.G.Vaughan,"Signals in Mobile Communications,"IEEE Transactions on Vehicular Technology, Vol. 35,1986, pp. 133-145.

[10] G. Lebrun, S. Spiteri and M. Faulkner, "MIMOComplex- ity Reductionthrough Antenna Selection,"Proceedings on Australian Telecommun CooperativeResearch Center, ANNAC'03, Vol. 5, 2003.

[11] S. Jacobs and C. P. Bean, "Fine Particles, Thin Films and ExchangeAnisotropy," In Magnetism, G. T. Rado and H.Suhl, Eds., Academic, New York, Vol. 3, 1963, pp. 271-350.

[12] R. H. Clarke, "A Statistical Theory of Mibile Recep-tion," Bell System Technology Journal, 1968, pp. 957- 1000. R. Nicole, "Title of PaperwithOnly First Word Capitalized," Journal Name StandardAbbreviations, in Press.

[13] K. Boyle, "Radiation Pattern and Correlation of CloselySpaces Linear Antennas," IEEE Transactions on Anten-nas Propagation, Vol. 50, 2002, pp.1162-1165.

[14] H. T. Hui, W. T. OwYong and K. B. Toh, "Signals Cor-relation between Two Normal-Mode Helical Antennas for Diversity Reception in a Multipath Environment," IEEE Transactions on Antennas Propagation, 2004, pp. 572-577.

[15] J. Blanch, J. Romeu and I. Cordella, "Exact Representa-tion of Antenna System Diversity Performance from In-put Parameter Description," Electronics Letters, Vol. 39, 2003, pp. 705-707.

[16] Available at: http://www.ets-lindgren.com/chambers,ETS Chamber.

# The MIMO Antenna Design for a TD-LTE Mobile Phone

Wang Wei [1, a], Wei Chongyu [2], Wei Weichen[3]

[1,2]Qingdao University of Science & Technology, Qingdao, 266061, China

[3]Melbourne University, VIC, Australia

E-mail: [a]wengain147@163.com

*Abstract*: **TD-LTE mobile communication systems use MIMO technology. In this paper, a mobile phone MIMO antenna design with two elements having high isolation is presented. The antenna can work in 2300/2500MHz frequency bands. The method of distance isolation is used to reduce the coupling between the two antenna elements. HFSS is used in the simulation analysis and optimization of the antenna. Test results show that antenna return loss, gain and isolation between the two elements meet the TD-LTE phone needs.**

*Key words: TD-LTE; MIMO; mobile phone antenna; HFSS*

## I .INTRODUCTION

TD-LTE uses MIMO technology to further improve the efficiency of using radio spectrum. MIMO technology can realize high-data-rate and large user capacity. In developing a TD-LTE mobile terminal, designing a MIMO antenna of high isolation is a key factor. This paper presents a TD-LTE dual band PIFA (planar inverted-F antenna) antenna covering 2300/2500MHz LTE bands. Generally, small separation distance between antenna elements increases the mutual coupling and reduces isolations between them. In this paper, the simple method of increasing the separation distance between antenna elements is used to provide high isolation. The whole size of the PIFA antenna designed is 18mm*9mm*4mm.

## II . ANTENNA CONFIGURATION

In this paper, the antenna is designed as dual-branch structure and each branch operates in different frequency bands. Figure 1 shows a single antenna element structure and measured dimensions. The single antenna element is made up of a radiating unit, a ground board, a short patch and a feed patch. The radiating unit consists of two G-branches, their lengths are about λ/4 respectively corresponding their required operating frequency bands. By changing the length of antenna branches and using the bending side of the structure, the operating frequency of the antenna can cover the two TD-LTE bands of 2300-2400MHz and· 2570-2620MHz. The whole ground floor size of the mobile terminal is 60 * 110mm. Antenna height from the floor is 4mm. The feed port is marked with different color from the short patch. Upper part of the antenna in Figure 1 is designed for frequency band 2500MHz, and the lower is for 2300MHz. In the figure, the gray area is a folding part perpendicular to the antenna floor, which is designed to increase the effective area of the antenna.

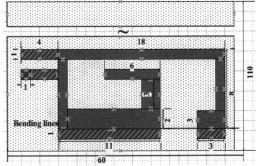

Fig . 1 . Configuration of a signal element.

## III. ANTENNA SIMULATION

Simulations are performed on Ansoft's HFSS. The MIMO antenna single model is shown in Figure 2. The working frequency is set in 2.3GHz and the sweeping frequency range is from 2GHz to 3GHz. Figure 3 shows the output S11 result, the reflection loss at an input port of a single radiator. As can be seen from the figure, S11 can be lower than -19 dB in the frequency range 2268-2407MHz and 2533-2472MHz. 3D pattern of the MIMO antenna single model is shown in Figure 4.

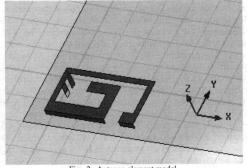

Fig . 2 . Antenna element model.

978-7-5641-4279-7

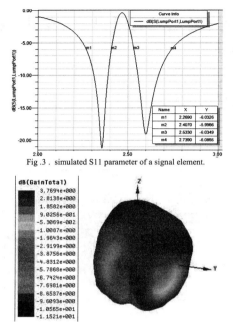

Fig .3 . simulated S11 parameter of a signal element.

Fig .4 . Simulated 3D pattern of the designed antenna.

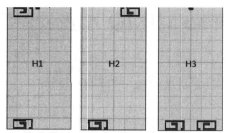

Fig .5 . Three models with different relative positions.

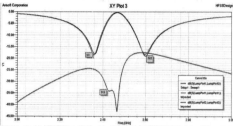

Fig .6 . Simulated S-parameters H1.

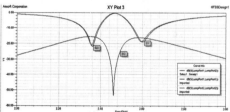

Fig .7 . Simulated S-parameters of H2.

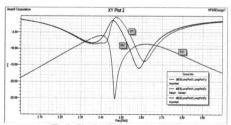

Fig .8 . Simulated S-parameters of H3.

A MIMO antenna in a mobile terminal is generally using the form of dual antenna elements. Measures must be taken to improve isolation between two elements because of the mutual coupling. Measures used now are that a ground slot, an extending ground, adjusting the relative position of the two antenna elements, using large-impedance materials, adding decoupling unit and adding parasitic elements, etc. Typically, the first two methods will change the original size and structure of the ground floor. The last three methods increase the number of components on the ground floor, so that increase the complexity of mobile phone. In this paper, we use the method that is always called distance isolation combining with a fold fide to adjust the place of antenna elements. By putting antenna elements as far away as possible from another, the isolation between the two elements can be improved. In this way, the design requirement can be satisfied without increasing the ground floor size and adding other material components. According to the relative position between antenna elements and ground floor, three models shown in Figure 5 called H1, H2 and H3 respectively were simulated. In the mobile terminal, if the separation distance between the two single antenna elements is larger than $\lambda/2$, the lower cross-correlation can be obtained. In the three models, H1 and H2 have a larger antenna separation than $\lambda/2$, the separation in H3 is approaching to $\lambda/2$. The simulated S-parameter results are respectively shown in Figure 6, Figure 7 and Figure 8.

Through analyzing and comparing the simulation results of H1, H2 and H3 models, the following results can be concluded. In H3, the difference between the S11 parameters of the two antenna elements is under 3dB. S11 parameter can reach -8dB in 2300MHz band and -15dB in 2500MHz. S12 is only -6dB and can't meet the needs of TD-LTE MIMO antenna isolation. The situation is similar in H2 that S11 can reach -22dB in 2300MHz band and -18dB in 2500MHz. But its S12 can reach -16dB and this meets the requirement of TD-LTE MIMO. In H1, S11 parameters of the two single MIMO antennas are also approximately equal. S11 in 2300MHz can reach -22dB and -

18dB in 2500MHz. S12 can reach -19dB and meets the required isolation. By comparing, H1 has both the lower S11 and gives high isolation, so H1 is chosen as the MIMO antenna design of mobile terminal.

## IV. TEST OF THE MIMO ANTENNA

A prototype of the antenna is fabricated based on model H1. Fig.9 gives the hand form block in comparing a Hisense T96 mobile phone. The prototype is tested with an Agilent's E5071C network analyzer and the results are shown in Figure 10. By comparing testing results with simulation, they are in good agreement. This designed antenna can cover 2300-2400MHZ and 2570-2620MHz and the S12 can reach -20dB.

Fig .9 . Prototype of the proposed antenna in comparing with HisenseT96.

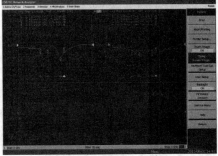

Fig .10 . S-parameters tested

## V. CONCLUSION

According to the simulations performed in Ansoft's HFSS and the S-parameter results tested in network analyzer, we know that H1 model with appropriate relative position between antenna and radiation ground is the most suitable MIMO antenna layout in these three models. This small MIMO antenna can cover 2300/2500MHz bandwidth with high isolation and works as a mobile phone MIMO antenna in TD-LTE communication system.

## REFERENCES

[1] K. L. Wong. Design of Nonplanar Microstrip Antennas and Transmission Lines, Wiley,New York, 1999.
[2] John D.Kraus and Ronald J.Marhefka.Antennas:For All Appllications Third Edition,2006.8.
[3] K. L. Wong. Compact and Broadband Microstrip Antennas, John Wiley & Sons, Inc,2002.
[4] Z. D. Liu, P. S. Hall, and D. Wake, "Dual-frequency planar inverted F antenna," IEEE Trans. Antennas Propagat. 45, 1451–1458, Oct. 1997
[5] M. Karaboikis, C. Soras, G. Tsachtsiris, and V. Makios, "Compact dual-printed inverted-F antenna diversity systems for portable wireless devices," IEEE Antennas Propag. Lett., vol. 3, pp. 9–14, 2004.
[6] K.Wong, T.Kang, andM. Tu, "Internalmobile phone antenna array for LTE/WWAN and LTE MIMO operations," Microw. Opt. Tech. Lett.,vol. 53, no. 7, pp. 1569–1573, Jul. 2011.
[7] H. T. Hui, "Practical dual-helical antenna array for diversity/MIMO receiving antennas onmobile handsets," IEE Proc.—Microw. Antennas Propag., vol. 152, no. 5, pp. 367–372, Oct. 2005.
[8] Meshram, M.K.Dept. of Electron. Eng., Banaras Hindu Univ., Varanasi, India Animeh, R.K. ; Pimpale, A.T. ; Nikolova, N.K. A novel quad-band diversity antenna for LTE and Wi-Fi applications with high isolation, Antennas and Propagation, IEEE Transactions on (Volume:60 , Issue: 9 ).

# High Port Isolation Co-Located Patch Antenna

W. W. Li[*1, 2], B. Zhang[1], Y. H. Liu[2], and B. Q. You[1]
[1] Department of Electronic Engineering
[2] Institute of Electromagnetics and Acoustics
Xiamen University
Xiamen, Fujian 361005, P. R. China
E-mail: wwl@xmu.edu.cn

*Abstract*-To fulfill co-located dual-polarized directional radiation, a dual-port patch antenna is presented. Constructed on a microwave PCB, the patch element is a foursquare of two open slots. The designed antenna can radiate the patch antenna mode while fed by the coaxial line (port 1), and the folded dipole mode while fed by the coplanar stripe (port 2). The experimental results show that the operating bandwidth defined by $S_{11}$ or $S_{22}$ less than −10 dB is 2.38–2.48 GHz for port 1, and 2.41–2.47 GHz for port 2. Over the operating band the $S_{21}$ or $S_{12}$ parameters are less than −23 dB.

## I. INTRODUCTION

Using multiple antennas at both the transmitter side and the receiver side can increase the channel capacity without additional frequency spectrum and transmitted power. However, due to the limited space at the size-limited terminal devices, the most critical problem in designing multiple antennas is the severe mutual coupling among them [1]–[3]. So in a small terminal device, polarization diversity antennas are applied to realize the MIMO function [4], [5].

In [6]–[8], the low-profile co-location antennas with coplanar waveguide slots and monopole of ground frame are adopted for the dual-polarized radiation. However, they all are omnidirectional or bidirectional antenna. Fed by downside slot and upside L-shaped feed line the patch antenna has the performance of directional dual-polarized pattern [9]. But with this antenna the back radiation level is large owing to the slots in the ground plane and the L-shaped feed line also enlarges the volume of the antenna.

Here we present the slot-loaded square antenna, which can simultaneously act as the patch antenna and half-wave folded dipole while excited by different ports. This antenna is a co-located dual-port dual-polarized antena of compact structure. The results show that this antenna has high port isolation.

## II. ANTENNA STRUCTURE

The proposed antenna, as shown in Fig. 1, is manufactured on the FR-4 microwave substrate of relative dielectric constant 4.3 and loss tangent 0.01. The thickness of the substrate, which is coated with copper layer of 0.035 mm on the both surfaces, is 3 mm and its size is 80 mm × 80 mm. As present in Fig. 1, the radiation element is cut from the square of side length W. A coaxial feed line port, here defined as port 1, is on the vertical diagonal line and the distance apart from the up apex is Lf. A shallow cut of width Wt is opened on the down apex of the square. Meanwhile two rectangular slots of width

Wc and length Lc are extended to the both sides from the top part of the shallow cut. Two conductors of width Wp, length Lp and spacing Wd between them are added as the feed sides of a coplanar stripe (CPS) port, here defined as port 2. The electromagnetic analysis is applied to optimize the antenna structure parameters with the commercial software XFDTD. All the optimized structure parameters are listed in Table I.

From Fig. 1 it is can be conceived that the surface electric current on the radiation patch will mainly flow along the vertical direction when port 1 is excited. However, due to the rectangular slots and the varied longitudinal dimension of the radiator, there will be some horizontal current components near both sides of the radiation patch. But the horizontal components of the radiation electric field over the central line can cancel out of phase. So the vertical radiation polarization can be expected when port 1 is fed.

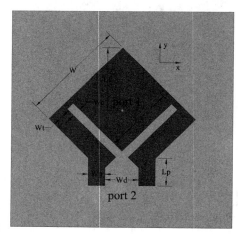

Figure 1. Geometry of the proposed patch antenna (top view).

TABLE I
OPTIMIZED ANTENNA STRUCTURE PARAMETERS

| Parameter | W | Lf | Wt | Wc | Lc | Wp | Lp | Wd |
|---|---|---|---|---|---|---|---|---|
| Value(mm) | 37.5 | 23.3 | 7.2 | 3.1 | 25.1 | 6.0 | 10.2 | 12.2 |

978-7-5641-4279-7

On the other hand, when port 2 is excited the patch radiates the patterns as the half-wave folded dipole works. However, this dipole bends up 90 degree angle at symmetric position and the upper rectangular arms are replaced by a small square patch. The reason for bending slots is that the cut position of port 2 is located at the radiation edge of port-1 excited microstrip antenna. If the slots are composed of single horizontal rectangular ring, the coupling between port 1 and port 2 will get strong. On this dipole the current will distribute along the inner edges of the two rectangular slots with the standing wave mode and the square patch arm, which will change the characteristic impedance of port 2, has no effect on the resonant frequency. The main polarization of folded dipole antenna is parallel to the arms. So the bent arms will introduce some vertical component. However, owing to the symmetric structure the vertical component polarization will cancel in spite of the bent arms. Therefore, the dipole antenna will mainly radiate horizontal polarization mode when port 2 works.

According to the proposed structure as shown in Fig. 1 and the structure parameters listed in Table I, a prototype antenna is manufactured with the etching method and the port performances are measured by vector network analyzer AV3629.

### III. IMPEDANCE CHARACTERISTICS

Fig. 2 shows the S-parameter curves. It is seen that the simulation operating frequency range is from 2.42 GHz to 2.47 GHz with the $S_{11}$ less than $-10$ dB while the measurement range is from 2.38 GHz to 2.48 GHz for port 1. Meanwhile with port 2 the simulation band is from 2.42 GHz to 2.47 GHz and the measurement band is from 2.41 GHz to 2.47 GHz.

With port 1 the simulation result shows that there is a notched step in the upper frequency. Fig. 3 shows the input impedance of port 1. It is seen that within the operating frequency band the resistance value in lower frequency is larger than in upper frequency. Although at the upper frequency the reactance approach to zero, the resistance value is also small due to the port-2 effect. So the high frequency resonant effect of port 1 is very weak and the measurement results don't show the dual-resonant frequency. But the measurement bandwidth is enlarged and shifts to the lower frequency. The reason is that the dielectric loss is not considered in simulation process.

For port 2, the simulation and measurement results of the port S parameters fit very well. However, it is noted that the measurement operating frequency shifts down a bit relative to the simulation. Comparatively speaking, the impedance matching performance of port 1 is better than of port 2. This is because of ground plane effect on the folded dipole, which low the input resistance of port 2. For this reason, the spacing of the coplanar stripes has to be widened to enhance the characteristic impedance to match port impedance.

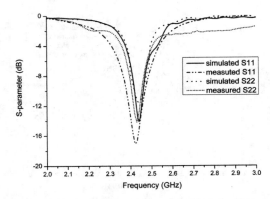

Figure 2. Simulated and measured S parameters of the designed antenna.

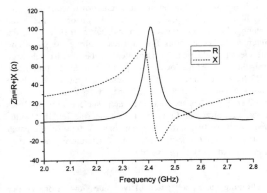

Figure 3. Input impedance of port 1.

With the help of the simulation results (not shown here), it is also seen that when port 2 is excited over the operating frequency the weak inductive reactance is present. At this time the antenna works in the half-wave dipole mode. In other words, this antenna is the deformation of a ring antenna.

### IV. PORT ISOLATION

As shown in Fig. 4, it is observed that within the frequency band range from 2.4 GHz to 2.5 GHz, the simulation $S_{12}$ / $S_{21}$ parameters all are less than $-32$ dB while the measurement results are only less than $-23$ dB. The reason for the large difference between the simulation and measurement results is that the simulation is processed in ideal conditions. However, it is also can be noticed that the variation tendency of the simulations and measurements is alike but for frequency points of the maximum isolation.

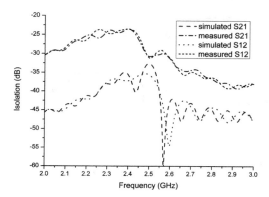

Figure 4. Simulated and measured isolation between ports.

It is the different radiation mechanism of the different port that helps to realize the high port isolation of the presented antenna. When port 1 is excited, as shown in Fig. 5 (a), the antenna surface currents flow vertically along the patch and the electric field mainly distributes on the small patch edge near the slots. At the location of port 2, tow conductors of the CPS have the same direction currents and the same voltage values. So the influence of port 1 on port 2 is little.

When port 2 is excited, the patch surface current flows along the horizontal direction and the electric field distributes mainly on the two side edges. The electric field on the patch central diagonal line is very weak. Therefore, the port 1 impact on the port 2 also is little.

## V. CONCLUSIONS

Utilizing the square patch of two side slots in one corner and fed by coaxial line and CPS, the proposed dual-port low-profile antenna can simultaneously work as the microstrip antenna mode and half-wave folded dipole mode. The measurement results show that its operating band defined by $S_{11}$ less than $-10$dB is from 2.4 GHz to 2.47 GHz and over the working band the port isolation value is larger than 23 dB. This proposed antenna can be the candidate for MIMO WLAN systems.

## ACKNOWLEDGMENT

This work was supported by Specialized Research Fund for the Doctoral Program of Higher Education (grant 20120121120027) and Xiamen Sci-Tech Development Foundation (grant 3502Z20123012).

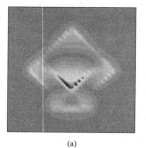

(a)

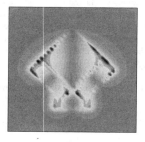

(b)
Figure 5. Simulation electric field distribution when (a) port 1 excited and (b) port 2 excited.

## REFERENCES

[1] K. Yu, and B. Ottersten, "Models for MIMO propagation channels: a review," *Wirel. Commun. Mob. Comput.*, vol. 2, no. 7, pp. 653-666, Nov. 2002.

[2] W. Weichselberger, M. Herdin, H. Ozcelik, and E. Bonek, "A stochastic MIMO channel model with joint correlation of both link ends," *IEEE Trans. Wireless Commun.*, vol. 5, no. 1, pp. 90-100, Jan. 2006.

[3] J. Thaysen, and K. B. Jakobsen, "Envelope correlation in (N, N) MIMO antenna array from scattering parameters," *Microw. Opt. Technol. Lett.*, vol. 48, no. 5, pp. 832-834, May 2006.

[4] L. Dong, H. Choo, R. W. Heath Jr, and H. Ling, "Simulation of MIMO channel capacity with antenna polarization diversity," *IEEE Trans. Wireless Commun.*, vol. 4, no. 4, pp. 1869-1873, Jul. 2005.

[5] H. Li, J. Xiong, Z. Ying, and S. L. He, "Compact and low profile co-located MIMO antenna structure with polarisation diversity and high port isolation," *Electron. Lett.*, vol. 46, no. 2, pp. 108-110, Jan. 2010.

[6] C. Lee, S. Chen, and P. Hsu, "Isosceles triangular slot antenna for broadband dual polarization applications," *IEEE Trans. Antennas Propag.*, vol. 57, no. 10, pp. 3347-3351, Oct. 2009.

[7] Y. Li, Z. Zhang, W. Chen, et al., "A dual-polarization slot antenna using a compact CPW feeding structure," *IEEE Antennas Wireless Propag. Lett.*, vol. 9, pp. 191-194, Mar. 2010.

[8] E. A. Soliman, M. S. Ibrahim, and A. K. Abdelmageed, "Dual-polarized omnidirectional planar slot antenna for WLAN applications," *IEEE Trans. Antennas Propag.*, vol. 53, no. 9, pp. 3093-3097, Sep. 2005.

[9] T. Chiou, and K. Wong, "A compact dual-band dual-polarized patch antenna for 900/1800-MHz cellular systems," *IEEE Trans. Antennas Propag.*, vol. 51, no. 8, pp. 1936-1940, Aug. 2003.

# Sensitivity Study for Improved Magnetic Induction Tomography (MIT) Coil System

Ziyi Zhang[1,2], Hengdong Lei[1,2], Peiguo Liu[2], Dongming Zhou[2]

[1] College of Science, National University of Defense Technology, Changsha, Hunan 410073, China
[2] College of Electronic Science and Engineering, National University of Defense Technology, Changsha, Hunan 410073, China
E-mail: ziyizhang@nudt.edu.cn

*Abstract*-**The improved magnetic induction tomography (MIT) coil system which consists of the two-arm Archimedean spiral coil (TAASC) as excitation coil and the solenoid as receiver coil has much better performance in coil system sensitivity than the conventional MIT coil system which uses the solenoids as excitation coil and receiver coil. In this paper the theoretical sensitivity property for improved MIT coil system are studied fully. The magnetic fields produced by TAASC and solenoid are derived approximately based on the Biot-Savart law. The relations between the coil system sensitivity and the parameters (amplitude of excitation current, maximum outer radius and number of turns of TAASC, and number of turns and length of solenoid) of coil system are calculated. The results show that the sensitivity for improved MIT coil system is proportional to the number of turns of solenoid and electric current of TAASC, and can be improved with the increase of maximum outer radius of TAASC and radius of solenoid. The sensitivity is decreased as the length of solenoid increasing, and not significantly associated with the number of turns of TAASC.**

*Keywords*-**Magnetic induction tomography (MIT), biological tissues, coli system sensitivity, two-arm Archimedean spiral coil (TAASC).**

## I. INTRODUCTION

Magnetic induction tomography (MIT) is an emerging imaging method to reconstruct the distribution of electrical conductivity within biological tissues. Compared with the earlier electrical impedance tomography (EIT) technique, MIT has a definite advantage of non-contact, and hence is more suitable for medical applications such as the haemorrhagic cerebral stroke or brain oedema diagnosis [1].

The design of coil system is much important for MIT, which can strongly affect the sensitivity of measurement signal [2]-[4]. A typical MIT coil system includes an array of excitation coils and an array of receiver coils. As illustrated in Fig. 1, the excitation coil carrying a harmonic current $I$ emits a primary excitation magnetic field $B_0$, and then a primary voltage $V_0$ is sensed in the receiver coil. With the action of $B_0$, the eddy current can be induced due to the electrically conductive object and a secondary magnetic field perturbation $\Delta B$ is produced, causing a secondary voltage perturbation $\Delta V$ sensed in the receiver coil. $\Delta V$ is of our interest, which contains the information about the electrical conductivity $\sigma$ of object. Unfortunately, because the values of $\sigma$ for human tissues are usually small, $\Delta V$ can be very weak in comparison with $V_0$, meaning low coil system sensitivity [1], [3].

The solenoid is the commonly used type of excitation coil and receiver coil in MIT coil system [5]-[8]. However, our recent study shows that this conventional MIT coil system has a low sensitivity for measuring the change of $\sigma$ inside biological tissues compared to the improved MIT coil system with the two-arm Archimedean spiral coil (TAASC) as excitation coil and the solenoid as receiver coil [9]. This paper aims to investigate the sensitivity property for our improved MIT coil system comprehensively. The expressions for $x$, $y$ and $z$ component of the magnetic fields produced by TAASC and solenoid are derived approximately based on the Biot-Savart law, and then the coil system sensitivity is obtained. The maximum coil system sensitivity is tested at different amplitudes of the excitation current, different radiuses and numbers of turns of the TAASC and the solenoid, and different length of the solenoid via the MATLAB tool.

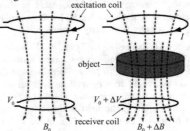

Figure 1. Schematic diagram for principle of MIT

## II. METHODS

### A. Model for Improved MIT Coil System

Fig. 2 shows the model for improved MIT coil system. In this model the excitation coil is a two-arm Archimedean spiral coil (TAASC) and the receiver coil is a solenoid. The TAASC is constructed by two opposite Archimedean planar spirals connected at its center. For a TAASC placed in the x-y plane with its center at the origin, the equation in polar coordinate system for it is

$$\begin{cases} L_1: r_1 = \dfrac{r_0}{2n\pi}\varphi_1 & 0 \le \varphi_1 \le 2n\pi \\[2mm] L_2: r_2 = \dfrac{r_0}{2n\pi}(\varphi_2 - \pi) & \pi \le \varphi_2 \le (2n+1)\pi \end{cases} \quad (1)$$

978-7-5641-4279-7

where $r_0$ and $n$ are the maximum outer radius and the number of turns of TAASC, respectively [9], [10].

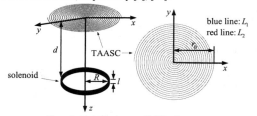

Figure 2. Model for improved MIT coil system.

### B. Sensitivity Calculation in MIT Coil System

In MIT coil system, the sensitivity $S$ is usually defined as

$$S = \frac{\Delta V}{I \Delta \kappa} \quad (2)$$

where $\kappa = \sigma + j\omega\varepsilon_0\varepsilon_r$ is the complex electrical conductivity of perturbing object and $\Delta\kappa$ represents the change of $\kappa$. According to the extensive Geselowitz relationship the coil system sensitivity $S$ can be calculated strictly, while it is a complicated and time-consuming work [11]. However, for an isolated perturbation in the empty space $S$ can be expressed briefly based on the reciprocity theorem as

$$S = kB_1 \cdot B_2 \quad (3)$$

where $B_1$ and $B_2$ are the magnetic fields produced by excitation coil and receiver coil respectively, and $k$ is a constant factor only related to the perturbation characteristics [12]. Due to the unknown $k$, only a scaled sensitivity map could be obtained for the improved MIT coil system, but it can still reflect the sensitivity property accurately.

### C. Magnetic Field Expressions for TAASC and Solenoid

For MIT the frequency of electric current in the excitation coil is generally less than 10 MHz. Therefore, the dimension of excitation coil could be very small in comparison with the wavelength of electromagnetic wave emitted by the excitation coil, and the electromagnetic field produced by the excitation coil could be treated as the static field in the region not far away from the excitation coil. According to the Biot-Savart law, for the TAASC in Fig. 2 when the excitation current $I_1$ in it is low-frequency, its three components of magnetic field $B_1$ at arbitrary point $(x, y, z)$ in the empty space are

$$B_{1x} = p\left( \int_0^{2n\pi} \frac{(\sin\varphi_1 + \varphi_1\cos\varphi_1)z}{((x-q\varphi_1\cos\varphi_1)^2 + (y-q\varphi_1\sin\varphi_1)^2 + z^2)^{3/2}} d\varphi_1 + \int_0^{2n\pi} \frac{(\sin\varphi_2 + \varphi_2\cos\varphi_2)z}{((x+q\varphi_2\cos\varphi_2)^2 + (y+q\varphi_2\sin\varphi_2)^2 + z^2)^{3/2}} d\varphi_2 \right) \quad (4)$$

$$B_{1y} = p\left( \int_0^{2n\pi} \frac{-(\cos\varphi_1 - \varphi_1\sin\varphi_1)z}{((x-q\varphi_1\cos\varphi_1)^2 + (y-q\varphi_1\sin\varphi_1)^2 + z^2)^{3/2}} d\varphi_1 + \int_0^{2n\pi} \frac{-(\cos\varphi_2 - \varphi_2\sin\varphi_2)z}{((x+q\varphi_2\cos\varphi_2)^2 + (y+q\varphi_2\sin\varphi_2)^2 + z^2)^{3/2}} d\varphi_2 \right) \quad (5)$$

$$B_{1z} = p\left[ \int_0^{2n\pi} \left( \frac{(\cos\varphi_1 - \varphi_1\sin\varphi_1)(y - q\varphi_1\sin\varphi_1)}{((x-q\varphi_1\cos\varphi_1)^2 + (y-q\varphi_1\sin\varphi_1)^2 + z^2)^{3/2}} - \frac{(\sin\varphi_1 + \varphi_1\cos\varphi_1)(x - q\varphi_1\cos\varphi_1)}{((x-q\varphi_1\cos\varphi_1)^2 + (y-q\varphi_1\sin\varphi_1)^2 + z^2)^{3/2}} \right) d\varphi_1 + \int_0^{2n\pi} \left( \frac{(\cos\varphi_2 - \varphi_2\sin\varphi_2)(y + q\varphi_2\sin\varphi_2)}{((x+q\varphi_2\cos\varphi_2)^2 + (y+q\varphi_2\sin\varphi_2)^2 + z^2)^{3/2}} - \frac{(\sin\varphi_2 + \varphi_2\cos\varphi_2)(x + q\varphi_2\cos\varphi_2)}{((x+q\varphi_2\cos\varphi_2)^2 + (y+q\varphi_2\sin\varphi_2)^2 + z^2)^{3/2}} \right) d\varphi_2 \right] \quad (6)$$

where $p = \mu_0 r_0 I_1/8\pi^2 n$ and $q = r_0/2n\pi$.

If a solenoid with radius $R$, length $l$, number of turns $N$ and electric current $I_2$ is placed at the position in Fig. 2, the magnetic fields produced by it can be derived similarly based on the Biot-Savart law as

$$B_{2x} = m\int_0^{2\pi} d\theta \int_0^l \frac{\cos\theta(Z-t)}{((x-R\cos\theta)^2 + (y-R\sin\theta)^2 + (Z-t)^2)^{3/2}} dt \quad (7)$$

$$B_{2y} = m\int_0^{2\pi} d\theta \int_0^l \frac{\sin\theta(Z-t)}{((x-R\cos\theta)^2 + (y-R\sin\theta)^2 + (Z-t)^2)^{3/2}} dt \quad (8)$$

$$B_{2z} = m\int_0^{2\pi} d\theta \int_0^l \frac{R - x\cos\theta - y\sin\theta}{((x-R\cos\theta)^2 + (y-R\sin\theta)^2 + (Z-t)^2)^{3/2}} dt \quad (9)$$

where $m = \mu_0 N I_2 R/4\pi l$ and $Z = z - d$.

### III. RESULTS

#### A. Normalized Sensitivity Map for Reference Coil System

The improved MIT coil system which is used as a reference is based on Fig. 2. The values of related parameters for the reference coil system are listed in Table I. The calculation region is in the plane $x = 0$ with $-100 \text{ mm} \leq y \leq 100 \text{ mm}$ and $10 \text{ mm} \leq z \leq 210 \text{ mm}$. Fig. 3 shows the normalized sensitivity map for the reference coil system. The maximum sensitivity appears at the point $x = 0$, $y = 20 \text{ mm}$, $z = 210 \text{ mm}$.

TABLE I
VALUES OF PARAMETERS FOR REFERENCE COIL SYSTEM

| Parameters | $I_1$ | $n$ | $r_0$ | $N$ | $R$ | $l$ | $d$ |
|---|---|---|---|---|---|---|---|
| Values | 1 A | 1 | 50 mm | 1 | 20 mm | 10 mm | 220 mm |

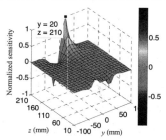

Figure 3. Normalized sensitivity map for reference coil system.

*B. Maximum Sensitivity versus Number of Turns of TAASC and Solenoid*

Fig. 4 shows the scaled maximum sensitivity versus the number of turns $n$ of TAASC and $N$ of solenoid. The maximum sensitivity is scaled to the maximum sensitivity of reference coil system. The variation range for $n$ is 1 to 10 and for $N$ is 5 to 50. In the process of changing $n$, the maximum sensitivity point is $x = 0$, $y = 20$ mm, $z = 210$ mm when $n$ is 2 to 6 while at the position $x = 0$, $y = -20$ mm, $z = 210$ mm when $n$ is 7 to 10. However, the maximum sensitivity is always at the position $x = 0$, $y = 20$ mm, $z = 210$ mm when changing $N$ in the given range.

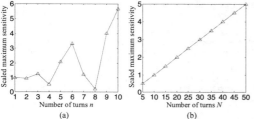

Figure 4. Scaled maximum sensitivity versus number of turns $n$ (a) of TAASC and $N$ (b) of Solenoid.

*C. Maximum Sensitivity versus Maximum Outer Radius of TAASC and Radius of Solenoid*

Fig. 5 shows the scaled maximum sensitivity versus the maximum outer radius $r_0$ of TAASC and radius $R$ of solenoid. The maximum sensitivity is scaled to the maximum sensitivity of reference coil system. The change range for $r_0$ and $R$ is both 10 mm to 100 mm. The maximum sensitivity is at the position $x = 0$, $y = 20$ mm, $z = 210$ mm when changing $r_0$ in the given range. For the parameter $R$, the maximum sensitivity is obtained at the point with $x$-coordinate 0, $z$-coordinate 210 mm, and $y$-coordinate listed in Table II.

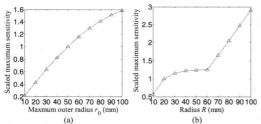

Figure 5. Scaled maximum sensitivity versus maximum outer radius $r_0$ (a) of TAASC and radius $R$ (b) of solenoid.

TABLE II
Y-COORDINATE OF MAXIMUM SENSITIVITY POINT VERSUS RADIUS OF SOLENOID IN THE CALCULATION REGION

| $R$ | 10 mm | 20 mm | 30 mm | 40 mm | 50 mm |
|---|---|---|---|---|---|
| $y$-coordinate | 10 mm | 20 mm | 30 mm | 40 mm | 50 mm |
| $R$ | 60 mm | 70 mm | 80 mm | 90 mm | 100 mm |
| $y$-coordinate | −20 mm | −20 mm | −20 mm | −20 mm | −20 mm |

*D. Maximum Sensitivity versus Electric Current of TAASC and Length of Solenoid*

Fig. 6 shows the scaled maximum sensitivity versus the electric current $I_1$ of TAASC and the length $l$ of solenoid. The maximum sensitivity is scaled to the maximum sensitivity of reference coil system. The change range for $I_1$ is 0.2 A to 2 A and for $l$ is 5 mm to 50 mm. The maximum sensitivity is at the position $x = 0$, $y = 20$ mm, $z = 210$ mm when changing $I_1$ and $l$ in the given range.

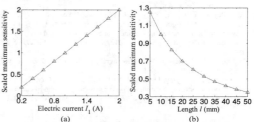

Figure 6. Scaled maximum sensitivity versus electric current $I_1$ (a) of TAASC and length $l$ (b) of solenoid.

IV. DISCUSSIONS AND CONCLUSIONS

In the improved MIT coil system, the sensitivity is a quantity related to the position of perturbation. For the reference coil system, its sensitivity map is anti-symmetric about the plane $y = 0$ and the maximum value appears at the point $x = 0$, $y = 20$ mm, $z = 210$ mm in the calculation region, as shown in Fig. 3.

It can be seen from Fig. 4 that the sensitivity for improved MIT coil system does not specifically relate to the number of turns of TAASC, but is proportional to the number of turns of solenoid. The position of maximum sensitivity is unfixed with the number of turns of TAASC increasing, and yet invariable no matter how the number of turns of solenoid changes.

In Fig. 5, the coil system sensitivity can be improved by increasing the maximum outer radius of TAASC or the radius of solenoid. The position of maximum sensitivity is unaltered with the increase of maximum outer radius of TAASC. However, the $y$-coordinate of maximum sensitivity point is variable when the radius of solenoid increases, as show in Table II.

The coil system sensitivity is proportional to the electric current of solenoid, and decreases as the length of solenoid increases, as shown in Fig. 6. The position of maximum sensitivity is unfixed regardless of the change for electric current of TAASC and length of solenoid.

In conclusion, the sensitivity for improved MIT coil system has its maximum value in the region of interest, and the perturbing object ought to be placed near the maximum sensitivity point. The coil system sensitivity could be improved by increasing the maximum outer radius of TAASC, electrical current of TAASC, number of turns of solenoid and radius of solenoid. For the actual design of improved MIT coil system, all the parameters should be considered synthetically in order to make the coil system sensitivity optimum.

ACKNOWLEDGMENT

This work was supported by the Specialized Research Fund for the Doctoral Program of Higher Education of China (No. 20114307110022) from Ministry of Education of the People's Republic of China.

REFERENCES

[1] H.-Y. Wei and M. Soleimani, "Electromagnetic tomography for medical and industrial applications: challenges and opportunities," *Proc. IEEE*, vol. 101, pp. 27–46, Mar. 2013.

[2] A. J. Peyton, Z. Z. Yu, G. Lyon, S. Al-Zeibak, J. Ferreira, J. Velez, F. Linhares, A. R. Borges, H. L. Xiong, N. H. Saunders and M. S. Beck, "An overview of electromagnetic inductance tomography: description of three different systems," *Meas. Sci. Technol.*, vol. 7, pp. 261–271, Mar. 1996.

[3] H. Griffiths, "Magnetic induction tomography," *Meas. Sci. Technol.*, vol. 12, pp. 1126–1131, Aug. 2001.

[4] S. Watson, C. H. Igney, O. Dössel, R. J. Williams and H. Griffiths, "A comparison of sensors for minimizing the primary signal in planar-array magnetic induction tomography," *Physiol. Meas.*, vol. 26, pp. S319–S331, Apr. 2005.

[5] A. Korjenevsky, V. Cherepenin and S. Sapetsky, "Magnetic induction tomography: experimental realization," *Physiol. Meas.*, vol. 21, pp. 89–94, Feb. 2000.

[6] S. Watson, R. J. Williams, W. Gough and H. Griffiths, "A magnetic induction tomography system for samples with conductivities below 10 S m$^{-1}$," *Meas. Sci. Technol.*, vol. 19, pp. 045501, Apr. 2008.

[7] M. Vauhkonen, M. Hamsch and C. H. Igney, "A measurement system and image reconstruction in magnetic induction tomography," *Physiol. Meas*, vol. 29, pp. S445–S454, Jun. 2008.

[8] H.-Y. Wei and M. Soleimani, "Hardware and software design for a National Instrument-based magnetic induction tomography system for prospective biomedical applications," *physiol. Meas.*, vol. 33, pp. 863–879, May 2012.

[9] Z. Zhang, P. Liu, D. Zhou and H. Lei, "Biomedical magnetic induction using two-arm Archimedean spiral coil: a feasibility study," unpublished.

[10] Z. Zhang, P. Liu, L. Ding and L. Zhang, "A new type of excitation coil for measurement of liver iron overload by magnetic induction method," in *Proc. 5th Int. Conf. BioMedical Engineering and Informatics*, Chongqing, China, 2012, pp. 679–683.

[11] R. J. Mortarelli, "A generalization of the Geselowitz relationship useful in impedance plethysmographic field calculations," *IEEE Trans. Biomedical Engineering*, vol. 27, pp. 665–667, Nov. 1980.

[12] J. Rosell, R. Casañas and H. Scharfetter, "Sensitivity maps and system requirements for magnetic induction tomography using a planar gradiometer," *Physiol. Meas.*, vol. 22, pp. 121–130, Feb. 2001.

# An Improved TR Method for the Measurement of Permittivity of Powder and Liquid Samples with Slabline

Licun Han*[1], Zhijun Xiang[2], Bingjie Tao[1], Minhui Zeng[1], Yu Zhang[3] and En Li[1].

1.    University of Electronic Science and Technology of China, Chengdu, 611731, P. R. China
2.    Southwest Institute of Electronic Equipment of China ( SWIEE), Chengdu, 610054, P. R. China
3.    Shandong Non-Metallic Material Institute, Jinan, 250031, P. R. China
E-mail:han88606@163.com

*Abstract*-An improved TR (transmission/reflection) method for measuring the complex permittivity of powder and liquid samples with slabline is provided in this paper. The method can afford quite high accurate measurement results without singularities, and broaden the operating frequency band to 2-18GHz. Some experiments are carried out to verify the feasibility of the improved TR method for permittivity measurement. The proposed method is validated by the measurement of the permittivity of Teflon from 2GHz to 18 GHz. Some useful conclusions are obtained for the development of measurement methods.

## I.    INTRODUCTION

It is of high practical value in modern medical research field that applying the biological electromagnetic effects of microwave radiation in clinical application, such as, physical therapy and cancer hyperthermia. The study [1]-[3] found, the best curative is closely related with microwave frequency and radiation, and different biological tissue has different electromagnetic properties. In order to select the best frequency and the amount of radiation, it is necessary to comprehensively understand the electromagnetic properties of biological tissue or the microwave equivalent phantom model of simulated biological tissues. Therefore, measuring the complex permittivity of the powder and liquid materials is of great significance for the development of bio-pharmaceutical.

Both at home and abroad, researchers adopt the scattering parameter method for measuring the complex permittivity of solid matter, which can't measure the paste-like substance. In order to broaden the measurement bandwidth, TR method with one sample in coaxial line [4] is proposed which covers 4-8GHz. But, there are several other issues to resolve, such as difficulty of manufacturing the sample under test, and the air gap which has great effect on the measuring accuracy. Actually, the TR method with one sample in rectangular waveguide can not only broaden the measurement bandwidth, but also reduce degree of difficulty of processing of the sample. Nonetheless, the measurement system is very difficult to clean up. To sum up, we proposed an improved TR method with one sample in slabline.

In this paper, we introduced an improved TR method by using famous SOLT calibration method in section II. The improved TR method using equivalent techniques can eliminate the errors coming from mismatches between connectors, the affection of Teflon blocks which is used to fasten the powder and liquid samples being test on the measure system, and the instabilities of the calibration kit. To sum up, the proposed method can accurately measure the relative permittivity of microwave materials.

## II.    THE IMPROVED TR METHOD

### A.    Slabline measurement system

The slabline is a transmission line where a thin wire replaces a printed trace for an inductor. In order to obtain small return loss in the measurement system, the slabline fixture is optimized for getting the characteristic impedance 50 Ohm in the whole operating frequency band (2-18GHz). The real fixture of slabline is shown in Fig.1. From the picture, it can be seen that the slabline fixture connected with VNA (vector network analyzer) by APC-7 converter. Furthermore, it is necessary to design a transition between slabline and cable. In this paper, we adopted impedance linear transformation in this transition.

Figure 1. The real measurement system.

978-7-5641-4279-7

To sum up, the permittivity of powder and liquid sample can be calculated from a new fixture which is quite similar to the proposed system. The system is added two Teflon blocks which is fixed the powder and liquid samples. Furthermore, we can calibrate the affection of Teflon blocks by using the equivalent theory.

### B. The TR method

In this paper, it is assumed that the dielectric sample under test is isotropic, symmetric, and homogeneous. It is also assumed that only the dominant mode (TEM) is presented in the slabline structure and there is no air gap between the sample and the measurement cell. For an isotropic and symmetric sample region, the reflection coefficient at the port and the transmission coefficient of the network [6] can be expressed as the following equations:

$$\Gamma = \frac{\eta_r - 1}{\eta_r + 1} \tag{3}$$

$$T = \exp(-jkd) \tag{4}$$

$$k = k_0 \sqrt{\mu_r \varepsilon_r} \tag{5}$$

$$\eta_r = \sqrt{\frac{\mu_r}{\varepsilon_r}} \tag{6}$$

where $\varepsilon_r$ and $\mu_r$ are, respectively, the relative complex permittivity and the relative complex permeability of the dielectric sample under test, and d is the thickness of sample, $\eta_r$ is the wave impedance in the dielectric region, k is the propagation constant of electromagnetic wave in the sample region, $k_0$ is the propagation constant of electromagnetic wave in vacuum.

According to the relationship between scattering parameters and transmission and reflection coefficients, the scattering parameters of the measurement cell can be written as:

$$S_{21s} = \frac{(1 - \Gamma^2)T}{1 - T^2 \cdot \Gamma^2} \tag{7}$$

$$S_{11s} = \frac{(1 - T^2)\Gamma}{1 - T^2 \cdot \Gamma^2} \tag{8}$$

where $S_{11s}$ and $S_{21s}$ are the measured value of the sample end. It is important that the $S_{11s}$ and $S_{21s}$ must be the data which has calibrated conductor loss of the entire slabline fixture [7] as shown in Fig.2. It can be calculated from the real measured data, $S_{11}$ and $S_{21}$, which are the value of the whole slabline structure. The calculating formula can be written as the following equations, where l and d are, respectively, the equivalent length of the slabline fixture and the sample thickness.

$$S_{11S} = S_{11} \cdot \exp[jk_0(l - d)] \tag{9}$$

$$S_{21S} = S_{21} \cdot \exp[-jk_0(l - d)] \tag{10}$$

From (7) and (8), we can get T and $\Gamma$. And then, from (3) and (4), k and $\eta_r$ can be obtained. The formula of k can be written as the following Equation (11), and where $\theta$ is the phase angle of T.

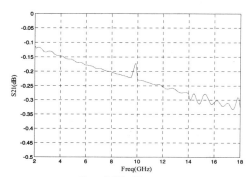

Figure 2. S21 of the structure.

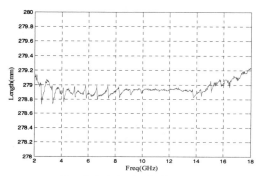

Figure 3. Equivalent length of the structure.

It is also assumed that only the dominant mode (TEM) is presented in the slabline, the metallic material of slabline is ideal conductor, and the measurement frequency (2-18GHz) is limited so that the higher-order modes cannot propagate. According to the definition of S-parameters, the $S_{21}$ of a part of slabline can be written as [5]:

$$S_{21} = \exp(-j\theta) \tag{1}$$

$$\theta = 2\pi l / \lambda \tag{2}$$

where l and $\lambda$ are, respectively, the length of a slabline and operating wavelength. If there is a lossless network which is operating on its dominant mode (TEM), it is feasible to consider this network as a transmission line.

In this paper, the slabline structure is assumed lossless, and it is easy to find that the dominant mode operating in this slabline structure is TEM. Hence, we apply this equivalent length theory to the calculation of the equivalent length of the slabline fixture. The measurement results of S21 and the equivalent length of slabline fixture are, respectively, shown in Fig.2 and Fig.3. The value of S21 is not equal to zero in the operating frequency band. And the return loss in the fixture is from the unideal conductor. The fluctuation in the line of Fig.3 is due to the unperfected match between slabline and coaxial line. In the following calculation, the average value 279mm of measurement data is chosen as the equivalent length.

$$k = \theta / d + j \ln|T| / d \qquad (11)$$

In conclusion, $\varepsilon_r$ and $\mu_r$ of the dielectric sample under test can be given.

$$\varepsilon_r = \frac{k(1-\Gamma)}{k_0(1+\Gamma)} \qquad (12)$$

$$\mu_r = \frac{k(1+\Gamma)}{k_0(1-\Gamma)} \qquad (13)$$

## III. MEASUREMENT RESULTS

In order to verify the feasibility of the improved TR method for permittivity measurement, a number of experiments were carried out. In the experiments, the VNA E8363B (Agilent) and a slabline fixture shown in Fig.1 are used for measurements. The dielectric sample under test in this system is Teflon placed in the center of the slabline fixture. Based on the principle [6] that the performance of Teflon is quite stable, Teflon is chosen as the standard sample to test the accuracy of the measurement system by considering the cost of the test, as shown in Fig.5. Fig.6 shows the calculated permittivity of air with certain thickness (10mm).

Figure 5. The Teflon samples.

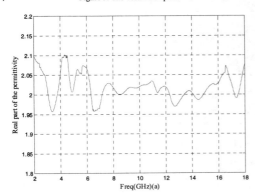

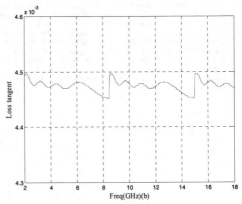

Figure 6. The permittivity of Teflon sample.

The real part of permittivity of Teflon (Fig.6) ranges from 1.96 to 2.11, and the loss tangent of Teflon is range from 0.0043 to 0.0046. The relative complex permittivity of the Teflon is 2.02-j0.0001 [7] in the operating frequency band 2-18GHz. As we know, the error of TR method is larger than any other methods. Furthermore, the error of the real part of complex permittivity is 0.01 [8] in TR method, the proposed TR method for measuring the permittivity is quite effective and acceptable, as well as high accurate.

## IV. CONCLUSION

We have shown the improved TR method suited to measure the dielectric material with low loss. The experiments of air and Teflon are proved that the proposed method can be used in the test system effectively and quite simply. From the measured data, there are errors in TR method, which come from mismatch at the connections between slabline fixture and the joint of VNA, imperfection of the metallic material of slabline, imperfection of Teflon sample processing, inaccuracy of the thickness measurement of the sample, etc. With the manufacture process of composite sample [4] and the application of logarithm principle of Lichtenecker [9], the proposed TR method can be used to measure the permittivity of powder and solid samples. The application of TR method for permittivity measurement in biological pharmacy is of practical significance in environmental protection and universal health.

## ACKNOWLEDGMENT

This work is supported by The Fundamental Research Funds for the Central Universities No. 2672011ZYGXJ017.

## REFERENCES

[1] Yang S H, Xing D, Lao Y Q, etal, Noninvasive monitoring of traumatic brain injury and post traumatic rehabilitation with laser induced photoacoustic imaging, Appl Phys Lett, 2007, 90(24): 243902.

[2] Leuschner C, Kumar CS, Hansel W, et al. "LHRH-conjugated magnetic iron oxide nanoparticles for detection of breast cancer metastases", Breast Cancer Res Treat, 2006, 99:163-176.

[3] Wunderbaldinger P, Josephson L, Weissleder R, "Tat peptide directs enhanced clearance and hepatic permeability of magnetic nanoparticles", Bioconjug Chem. 2002, 13(2):264—268.

[4] L. C. Han, E. Li, G. F. Guo and H. Zheng, "APPLICATION OF TRANSMISSION/REFLECTION METHOD FOR PERMITTIVITY MEASUREMENT IN COAL DESULFURIZATION", *Progress in Electromagnetics Research Letters*, Vol. 37, pp. 177-187, February 2013.

[5] David M. Pozar, *Microwave Engineering*, 4^ed ed., vol. 2. Hamilton: Printing, 2012, pp.150-156.

[6] L. F. Chen, C. K. Ong, C. P. Neo, V. V. Varadan and V. K. Varadan, *Microwave electronics-Measurement and material characterization*, John Wiley & Sons Ltd, 2004.

[7] Michael D. Janezic, Jeffrey A. Jargon, "Complex Permittivity Determination from Propagation Constant Measurements," Proc. IEEE, vol. 9, no. 2, Feb, 1999.

[8] Jing Shenhui, Ding Ding, Jiang Quanxing, "Measurement of electromagnetic properties of materials using transmission /reflection method in coaxial line," Proc. IEEE, pp. 590-595, Nov, 2003.

[9] Ray Simpkin, "Derivation of Lichtenecker's Logarithmic Mixture Formula From Maxwell's Equations," Proc. IEEE, vol. 58, no. 3, pp. 545 - 550, Mar. 2010.

# Design of Novel Slot UHF Near-Field Antenna for RFID Applications

Qichao Yang[1], Jingping Liu[1], Safieddin Safavi-Naeini[2]

[1]School of Electronic and Optical Engineering,Nanjing University of Science and Technology
Nanjing 210094,Jiangsu,China
[2]Electrical and Computer Engineering, University of Waterloo
N2L 3G1, ON, Canada

*Abstract*-A novel slot near-filed antenna is proposed for UHF(ultrahigh frequency) RFID(radio frequency identification) applications in this paper. The antenna overcomes the limitations in large dimension and weakness of environmental impact of slot antenna.The antenna prototype is printed onto a piece of FR4 substrate, with an overall size of 108mm × 72mm × 0.8mm while the slot area dimension is 54mm × 40mm.This antenna can achieve strong and uniform magnetic field in column interrogation zone with the radius of 30cm and height of 60cm, and it has good read efficiency in a wide frequency band of 900MHz-950MHz.

**Index Terms:near-field , slot antenna , UHF(ultrahigh frequency) , RFID(radio frequency identification)**

## I. INTRODUCTION

RFID（radio frequency identification）is a kind of automatic identification technology that uses RF (radio frequency)signal with the space coupling mode to achieve non-contact transmission of information and information processing, so as to recognize the objects. Due to its various superiorities such as high reading speed, large storage space, the capability of wireless working, adaptability in multiple environment, the RFID technology receives a lot of attention. It is currently in a stage of rapid evolvement. More and more industries put emphasis on its applications research[1].Technologies such as Internet of Things, radio frequency identification, near-field communication technology will be several key technologies in the future which can change people's daily life and make our life more convenient.

The paper proposes the design of the antennas those are smaller, better performance in metallic environment than other near-field antennas. Perhaps in the majority opinions, the stronger and bigger coverage of the field of the antenna should be the better, while a small range of reading is not functional. In fact, some applications situation just like engineering manufacture requires a wider range of reading the tag data. Imagine the scene in family or the shelf in library, there are many household appliances can be used on RFID technology to make our lives more comfortable. The affection each other in a small range area will be especially prominent. The near-field technology can be a good solution, to this problem. Interrogator zone of near-field antenna can be limited in a particular region with a very uniform and strong electroma-gnetic field without dead area, avoid the mutual interfere nce between the respective antennas, and ensure the effect-ive reliability of the region.

Generally, there are three methods to achieve near-field antenna:
1、Use antenna cavity.
2、Reduce the transmit power.
3、Segmented structure.[2]

In this paper we construct a kind of antenna with uniform magnetic field and with the second method to achieve near-field antenna. Compared to other near-field antennas, this antenna has the advantages such as novel structure, small size, uniform magnetic field and fabrication easily, the most important thing is that it can work well in metallic environment while other near-field antennas can hardly work closing to metal.

## II. ANTENNA STRUCTURE ANALYSIS

Slot antenna has an uniform magnetic field, but its dimension is large. What we must do is to optimize its structure.

According to Babinet Theorem: the dipole and slot antenna is the complementary structure so we can predict the gap length and the input impedance of the slot antenna corresponding to the dipole with the same resonant frequency.[3] As shown in Fig. 1.

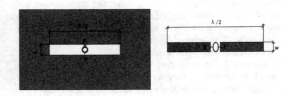

Figure 1.Slot antenna and complementary dipole

978-7-5641-4279-7

On the other hand, when port 2 is excited the patch radiates the patterns as the half-wave folded dipole works. However, this dipole bends up 90 degree angle at symmetric position and the upper rectangular arms are replaced by a small square patch. The reason for bending slots is that the cut position of port 2 is located at the radiation edge of port-1 excited microstrip antenna. If the slots are composed of single horizontal rectangular ring, the coupling between port 1 and port 2 will get strong. On this dipole the current will distribute along the inner edges of the two rectangular slots with the standing wave mode and the square patch arm, which will change the characteristic impedance of port 2, has no effect on the resonant frequency. The main polarization of folded dipole antenna is parallel to the arms. So the bent arms will introduce some vertical component. However, owing to the symmetric structure the vertical component polarization will cancel in spite of the bent arms. Therefore, the dipole antenna will mainly radiate horizontal polarization mode when port 2 works.

According to the proposed structure as shown in Fig. 1 and the structure parameters listed in Table I, a prototype antenna is manufactured with the etching method and the port performances are measured by vector network analyzer AV3629.

### III. IMPEDANCE CHARACTERISTICS

Fig. 2 shows the S-parameter curves. It is seen that the simulation operating frequency range is from 2.42 GHz to 2.47 GHz with the $S_{11}$ less than −10 dB while the measurement range is from 2.38 GHz to 2.48 GHz for port 1. Meanwhile with port 2 the simulation band is from 2.42 GHz to 2.47 GHz and the measurement band is from 2.41 GHz to 2.47 GHz.

With port 1 the simulation result shows that there is a notched step in the upper frequency. Fig. 3 shows the input impedance of port 1. It is seen that within the operating frequency band the resistance value in lower frequency is larger than in upper frequency. Although at the upper frequency the reactance approach to zero, the resistance value is also small due to the port-2 effect. So the high frequency resonant effect of port 1 is very weak and the measurement results don't show the dual-resonant frequency. But the measurement bandwidth is enlarged and shifts to the lower frequency. The reason is that the dielectric loss is not considered in simulation process.

For port 2, the simulation and measurement results of the port S parameters fit very well. However, it is noted that the measurement operating frequency shifts down a bit relative to the simulation. Comparatively speaking, the impedance matching performance of port 1 is better than of port 2. This is because of ground plane effect on the folded dipole, which low the input resistance of port 2. For this reason, the spacing of the coplanar stripes has to be widened to enhance the characteristic impedance to match port impedance.

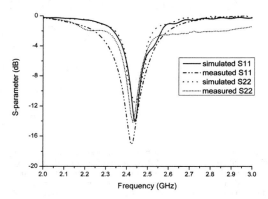

Figure 2. Simulated and measured S parameters of the designed antenna.

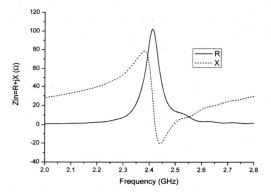

Figure 3. Input impedance of port 1.

With the help of the simulation results (not shown here), it is also seen that when port 2 is excited over the operating frequency the weak inductive reactance is present. At this time the antenna works in the half-wave dipole mode. In other words, this antenna is the deformation of a ring antenna.

### IV. PORT ISOLATION

As shown in Fig. 4, it is observed that within the frequency band range from 2.4 GHz to 2.5 GHz, the simulation $S_{12} / S_{21}$ parameters all are less than −32 dB while the measurement results are only less than −23 dB. The reason for the large difference between the simulation and measurement results is that the simulation is processed in ideal conditions. However, it is also can be noticed that the variation tendency of the simulations and measurements is alike but for frequency points of the maximum isolation.

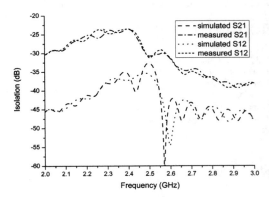

Figure 4. Simulated and measured isolation between ports.

It is the different radiation mechanism of the different port that helps to realize the high port isolation of the presented antenna. When port 1 is excited, as shown in Fig. 5 (a), the antenna surface currents flow vertically along the patch and the electric field mainly distributes on the small patch edge near the slots. At the location of port 2, tow conductors of the CPS have the same direction currents and the same voltage values. So the influence of port 1 on port 2 is little.

When port 2 is excited, the patch surface current flows along the horizontal direction and the electric field distributes mainly on the two side edges. The electric field on the patch central diagonal line is very weak. Therefore, the port 1 impact on the port 2 also is little.

## V. CONCLUSIONS

Utilizing the square patch of two side slots in one corner and fed by coaxial line and CPS, the proposed dual-port low-profile antenna can simultaneously work as the microstrip antenna mode and half-wave folded dipole mode. The measurement results show that its operating band defined by $S_{11}$ less than −10dB is from 2.4 GHz to 2.47 GHz and over the working band the port isolation value is larger than 23 dB. This proposed antenna can be the candidate for MIMO WLAN systems.

## ACKNOWLEDGMENT

This work was supported by Specialized Research Fund for the Doctoral Program of Higher Education (grant 20120121120027) and Xiamen Sci-Tech Development Foundation (grant 3502Z20123012).

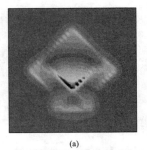

(a)

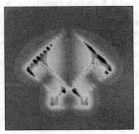

(b)

Figure 5. Simulation electric field distribution when (a) port 1 excited and (b) port 2 excited.

## REFERENCES

[1] K. Yu, and B. Ottersten, "Models for MIMO propagation channels: a review," *Wirel. Commun. Mob. Comput.*, vol. 2, no. 7, pp. 653-666, Nov. 2002.

[2] W. Weichselberger, M. Herdin, H. Ozcelik, and E. Bonek, "A stochastic MIMO channel model with joint correlation of both link ends," *IEEE Trans. Wireless Commun.*, vol. 5, no. 1, pp. 90-100, Jan. 2006.

[3] J. Thaysen, and K. B. Jakobsen, "Envelope correlation in (N, N) MIMO antenna array from scattering parameters," *Microw. Opt. Technol. Lett.*, vol. 48, no. 5, pp. 832-834, May 2006.

[4] L. Dong, H. Choo, R. W. Heath Jr, and H. Ling, "Simulation of MIMO channel capacity with antenna polarization diversity," *IEEE Trans. Wireless Commun.*, vol. 4, no. 4, pp. 1869-1873, Jul. 2005.

[5] H. Li, J. Xiong, Z. Ying, and S. L. He, "Compact and low profile co-located MIMO antenna structure with polarisation diversity and high port isolation," *Electron. Lett.*, vol. 46, no. 2, pp. 108-110, Jan. 2010.

[6] C. Lee, S. Chen, and P. Hsu, "Isosceles triangular slot antenna for broadband dual polarization applications," *IEEE Trans. Antennas Propag.*, vol. 57, no. 10, pp. 3347-3351, Oct. 2009.

[7] Y. Li, Z. Zhang, W. Chen, et al., "A dual-polarization slot antenna using a compact CPW feeding structure," *IEEE Antennas Wireless Propag. Lett.*, vol. 9, pp. 191-194, Mar. 2010.

[8] E. A. Soliman, M. S. Ibrahim, and A. K. Abdelmageed, "Dual-polarized omnidirectional planar slot antenna for WLAN applications," *IEEE Trans. Antennas Propag.*, vol. 53, no. 9, pp. 3093-3097, Sep. 2005.

[9] T. Chiou, and K. Wong, "A compact dual-band dual-polarized patch antenna for 900/1800-MHz cellular systems," *IEEE Trans. Antennas Propag.*, vol. 51, no. 8, pp. 1936-1940, Aug. 2003.

# UHF Electrically Large Near-Field RFID Reader Antenna Using Segmented Loop Unit

Longlong Lin[1], Jin Shi [1], Xianming Qing [2], Zhi Ning Chen [3]

[1]*School of Electronics and Information, Nantong University, China*
*linlonglong111@hotmail.com, jinshi0601@hotmail.com*
[2]*Institute for Infocomm Research, A\*STAR, Singapore*
*qingxm@ i2r.a-star.edu.sg*
[3]*Institute for Infocomm Research, National University of Singapore, Singapore*
*chenzn@ i2r.a-star.edu.sg*

*Abstract-* **An ultra high frequency (UHF) electrically large near-field RFID reader antenna using segmented loop unit is proposed to enlarge the interrogation zone. The proposed antenna is composed of two segmented loops with one common side and a feeding network. Each loop has a uniform and single direction flowing current itself. By providing reverse-direction flowing current to the two segmented loops, the strong and uniform magnetic field distribution over a large interrogation zone can be achieved.**

## I. INTRODUCTION

Ultra-high frequency (UHF) near-field radio frequency identification (RFID) receives a lot of attention due to its promising opportunities in item-level RFID applications, such as sensitive products tracking, pharmaceutical logistics, transport, medical products, and bio-sensing applications [1-2]. The top challenge of the UHF near-field RFID reader antenna is to generate a strong and uniform magnetic field distribution over an electrically large interrogation zone area.

In most near-field RFID applications, the inductive coupling between the reader antennas and tags is preferred because it is capable of operating in close proximity to metals and liquids [3]. Some designs have been reported to address the design of electrically large single-loop antennas for UHF near-field RFID readers. The key design principle of such works is to ensure that the current is of equal magnitude and in-phase along the loop. Dobkin et al. presented a segmented loop antenna loaded with lumped capacitors [4]. Oliver conceptually proposed three broken-loop antenna patents using triple lines, double lines and single lines [5]. Qing et al. proposed the segmented loops using distributed capacitors [6] or dash lines [7]. The reported single segmented loop antennas are with a limited interrogation zone, the perimeter of the interrogation zone is less than $2\lambda$ ($\lambda$ is the operating wavelength at 915 MHz in free space).
In this paper, the segmented loop with a side length of about $0.5\lambda$ is selected as a unit to configure an UHF near-field antenna with an interrogation zone of double area of that of the segmented loop unit and the perimeter of the interrogation zone achieves $2.8\lambda$. Comparison between the proposed antenna and the single-loop antenna with the same area is provided. The procedure to implement this antenna prototype is addressed with a practical guideline.

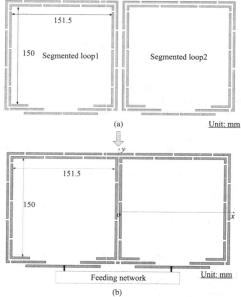

Fig. 1. Configuration of (a) two separated segmented loop units with the size; (b) the proposed electrically large antenna.

## II. THE PROPOSED NEAR-FIELD ANTENNA

### A. Antenna configuration

Fig. 1(a) shows the two separated segmented loops. The segmented loop, which is similar to that in Ref. 7, can generate the strong and uniform magnetic field distribution because the current along the segmented loop is in-phase. To get the strong and uniform magnetic field distribution over a larger interrogation zone, the segmented loop is set as a unit to combine a new UHF near-field antenna as shown in Fig. 1(b), where the two segmented loop units sharing a common segmented side. A Cartesian coordinate system is built, and the origin of the coordinate system is the center between the two loops at the upper surface of the substrate. The internal area of the proposed antenna is indicated as the interrogation zone with

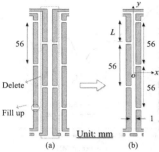

(a)                    (b)

Fig. 2. Change from two separated segmented sides into one common segmented side of the proposed antenna.

a perimeter of 910 mm or about 2.8 $\lambda$ at 915 MHz.

### B. Design procedure

The design procedure of the proposed antenna is summarized in the following two steps.

### Step 1 Design a segmented loop as the unit

The first is to configure a segmented loop with in-phase current along the loop, as shown in Fig. 1(a). The methodology to configure the segmented loop is similar to that in Ref. 7.

### Step 2 Configure the proposed antenna

The second is to configure the electrically large antenna by combining the two separated segmented loop units. The key is to configure the common side. The other three sides of each segmented loop will not be changed. Fig. 2 exhibits how to change the two separated segmented lines into the common segmented side of the proposed antenna.

As shown in Fig. 2(a), to configure the common segmented side, the separated segmented lines of the two segmented loops are moved close to each other. Then the two segmented lines in the middle are deleted, and the lower gap on the left side segmented line is filled up to get the common side.

To compensate the central area of the proposed antenna, there should be a strong current along the common side. Therefore, the two ports of the feeding network should be 180° out-of-phase so that the currents along the two loops are superposed each other when flowing into the common side.

However, the length $L$ in Fig.2 (b) should be optimized to keep the magnetic field distribution of the proposed antenna is the best. Fig. 3 shows magnetic field distribution of the electrically large antenna at 915 MHz along $y = 45$ mm when $L$ is changed. As shown in Fig. 3, the electrically large antenna has the strongest magnetic field distribution ($|H_z|$) when $L$ is 28 mm, which is exactly half of the distance (56 mm) of the segmented line sections between the two gaps. The length of 56 mm is also the main length of the single segmented loop unit to keep the current along the unit in-phase at 915 MHz.

### C. Results

The antenna is optimized and prototyped with an overall size of 318 mm × 182 mm × 0.5 mm and offers an interrogation

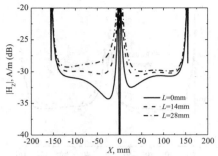

Fig. 3. The simulated magnetic field distributions of the proposed antenna with two segmented loop units with varying $L$ at 915 MHz ($z = 0.5$ mm) along $y = 45$ mm.

Fig. 4. Photo of the antenna prototype using the FR4 substrate.

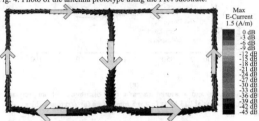

Fig. 5. Simulated current distribution of the proposed antenna at 915MHz at 915MHz

zone of 308 mm × 150 mm. The antenna prototype fabricated onto an FR4 PCB is shown in Fig. 4, where the feeding network is realized by the double-sided parallel-strip line printed onto the opposite sides of the substrate.

### Current and magnetic field distribution

Fig. 5 exhibits the simulated current distribution of the proposed antenna operating at 915 MHz. From Fig. 5, it can be seen that the currents along each loop are still in-phase, but the two currents are reverse-direction, one is clockwise, and the other is counter clockwise. Since the currents along the two loops are superposed each other when the currents from the two loops flow into the common side, the current along the common side is stronger than those currents along other sides except the sides with excited line sections

Fig. 6 compares the simulated and measured magnetic field distribution at 915 MHz. The measurement method is the same as that in Ref. 8. The near-field magnetic field probe was placed on the surface of the antenna prototype and the interval of detection points is 5mm.

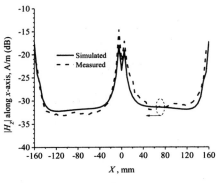

(a)

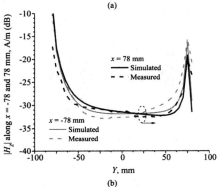

(b)

Fig.6. Simulated and measured magnetic field distribution of the proposed antenna prototype at 915 MHz ($z = 0.5$ mm) along (a) x-axis and (b) $x = -78$ and 78 mm.

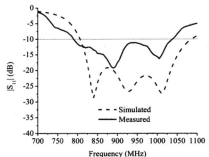

Fig. 7. Simulated and measured $|S_{11}|$ of the proposed antenna prototype.

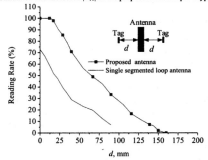

Fig. 8. Measured reading rate against reading range of the proposed antenna prototype.

## Impedance matching

The measured return loss of the proposed antenna was carried out using Agilent E5230A vector network analyzer. Fig. 7 shows the simulated and measured $|S_{11}|$ of the proposed antenna. From 790 to 1040 MHz, the $|S_{11}|$ is smaller than -10 dB. The slight shift of $|S_{11}|$ is produced by the fabrication error of the gaps and the thickness of the substrate.

## Reading range

To further verify the performance of the proposed antenna, the antenna prototype was used as the reader antenna in the UHF near-field RFID system to detect UHF near-field tags (J12, $15 \times 8$mm$^2$). The measured reading rate against the reading range is exhibited in Fig. 8. The proposed antenna offers the bi-directional detection along the $\pm z$ axis. A 100% reading rate is achieved within a maximum distance of 13.5 mm, while the reading rate of the single segmented loop antenna with the identical size is greatly reduced at the same distance.

## III. CONCLUSION

Designing electrically-large UHF near-field RFID reader antennas is a big challenge, especially when the perimeter of interrogation zone reaches three wavelengths or more. The proposed antenna using segmented loop unit can produce strong and uniform magnetic field distributions in the near-field region of the antenna with the perimeter of the interrogation zone up to 2.8 times of operating wavelength. The proposed antenna is promising for UHF near-field RFID reader applications.

## ACKNOWLEDGMENT

The work was partially supported by Agency for Science, Technology and Research (A*STAR), Singapore, A*STAR SERC Metamaterial Program: Meta-Antennas (092 154 0097), by the National Natural Science Foundation of China under Grants 60901041, 61101002, 61271136, by Program for New Century Excellent Talents in University (NCET-11-0993), by the Natural Science Foundation of Jiangsu Province, China (Grant No. BK2012657), and by Graduate Research and Innovation Plan Project of the University of Jiangsu Province under Grant YKC13004.

## REFERENCES

[1] P. Harrop, "New field UHF vs. HF for item level tagging" [Online]. Available: http://www.eurotag.org/?Articles_and_Publications.

[2] D. Desmons, "UHF Gen2 for item-level tagging," presented at the RFID World 2006, [Online]. Available: www.impinj.com/files/Impinj_ILT_RFID_World.pdf.

[3] P. V. Nikitin, K. V. S. Rao, and S. Lazar, "An overview of near field UHF RFID," in *Proc. IEEE Int. Conf. on RFID*, pp. 167–174, Mar. 2007.

[4] D. M. Dobkin, S. M. Weigand, and N. Iyec, "Segmented magnetic antennas for near-field UHF RFID," *Microw. J.*, vol. 50, no. 6, Jun. 2007.

[5] R. A. Oliver, "Broken-loop RFID reader antenna for near field and far field UHF RFID tags," U.S. design patent D570, 337 S, Jun. 3, 2008.

[6] Y. S. Ong, X. Qing, C. K. Goh, Z. N. Chen, "A segmented loop antenna for UHF near-field RFID," *Antennas and Propagation Society International Symposium*, pp. 1–4, 2010 IEEE.

[7] X. Qing, C. K. Goh, and Z. N. Chen, "A broadband near-field UHF RFID antenna," *IEEE Trans Antennas and Propagation*, vol. 58, no. 12, pp. 3829–3838, Dec. 2010.

[8] J. Shi, X. Qing, Z. N. Chen and C. K. Goh, "Electrically large dual-loop antenna for UHF near-field RFID reader," accepted by *IEEE Trans Antennas and Propagation,* November 2012.

# Resonator Bandpass Filter Using the Parallel Coupled Wiggly Line for Supurious Suppression

P. Arunvipas

Telecommunications Engineering Department, Mahanakorn University of Technology
Bangkok, 10530, Thailand
parunvipas@yahoo.com

*ABSTRACT* — **This paper describes the resonator bandpass filter design using the non-uniform parallel coupled "wiggly-line" technique. The proposed resonator is consisted of three sections quarter wavelength coupler and each of sections has perturbation and $Z_0$ connected in the end of each ports. This resonator can reduce the spurious response at $2f_0$ and $3f_0$ by its structure. The directivity enhanced parallel-coupler should be employed to solve this problem. The experimental results and designs of all type are 0.9 GHz to implement bandpass filters. There were evidently shown the second and third spurious response suppression improvement. Both the theoretical and experimental performance is presented.**

*Index Terms* — **Parallel coupled line, non-uniform coupled line, bandpass filter, spurious suppression, wiggly-line.**

## I. INTRODUCTION

A parallel-coupled (edge-coupled) microstrip bandpass and bandstop filters which are widely used a half wave length line resonator. The microstrip filter parameters can be derived by using both Chebyshev and Butterworth prototypes [1]. This type of filter suffers from spurious bandpass at the harmonic frequency. The even-mode and the odd-mode phase velocities for a coupled pair of microstrip lines are unequal [2]. Because the microstrip is a non-homogeneous medium (it is consisted of air above and dielectric below medium) [3]. In the past, the capacitive and inductive compensating techniques were proposed to improve the symmetry of bandpass, spurious free response of the microstrip coupled lines based bandpass filters [4], and other methods have been also proposed in the literature to eliminate this first spurious passband [5], [6]. To suppress or reject this spurious response, in this paper, we introduce the resonator bandpass filter design using parallel coupled "wiggly-line". This technique is proposed to more suppress the spurious response. This is achieved by designing the parallel coupled (edge coupled) line by Chebyshev prototypes and step of design of the parallel coupled "wiggly-line". Although this structure is very good to suppress the spurious response but it is not miniature circuit for using the limited area in microwave equipment. In section II, we present a comprehensive circuit theory, the propose technique of the resonator bandpass filter is based on parallel coupled "wiggly–line". This structure of resonator can suppress the second and third spurious response by itself. The step design and all results of the proposed parallel coupled "wiggly-line" bandpass filters are described in section III. The finally of this paper is a conclusion in section IV.

## II. THEORY AND DESIGN

The resonator bandpass filter based on parallel coupled "wiggly-line" is composed of three sections of parallel coupled line with asymmetry shape of perturbation. Input and output ports are tapped in the end of the coupled line. This configuration of input and output port feeding causes the magnitude of input/output impedances pulled raise up. Therefore, step impedance transmission lines are employed both in input and output port to step down the port's impedance close to the magnitude of characteristic impedance $Z_0$. Parallel-coupled (or edge-coupled) bandpass filters, for this method, maximum coupling are obtained between physically parallel microstrip when the length of the coupled region is $\lambda_g/4$ therefore the microstrip circuit must have the general layout shown in Fig. 1(a), where $l_1$, $l_2$, $l_3 \approx \lambda_g/4$.

The parallel coupled "wiggly line" microstrip bandpass filters with spurious suppression are designed from Chebyshev bandpass filters with centered at $f_0$ =0.9 GHz. The substrate employed has relative dielectric constant $\varepsilon_r$ = 6.0 and strictness $h$ = 1.52 mm. The first stage in the design process of the "wiggly line" filter is calculated from conventional parallel coupled microstrip geometry. The schematic of the parallel coupled line is depicted in Fig. 1(a), while the schematic of the proposed filters are depicted in Fig. 1(b) and 1(c). The outlining a design steps are following below

For the first coupling structure
Ripple = 0.1 dB, $g_0$ =1, $g_1$ =0.843, $g_2$ = 0.662, $g_3$= 1.3554, $\delta$ = the fractional bandwidth

$$J_{0,}Z_0 = \sqrt{\frac{\pi\delta}{2g_0 g_1}} \qquad (1)$$

$$\delta = \frac{f_2 - f_1}{f_0} \qquad (2)$$

For the intermediate coupling structures, $\omega_c' = 1$

$$J_{j,j+1}Z_0\big|_{j=1to(n-1)} = \frac{\pi\delta}{2\omega_c' \sqrt{g_j g_{j+1}}} \qquad (3)$$

For the final coupling structure

978-7-5641-4279-7

$$J_{n,n+1}Z_0 = \sqrt{\frac{\pi\delta}{2g_n g_{n+1}}} \qquad (4)$$

When n=2 (order), $g_0$ =1, $g_1$ =0.843, $g_2$=0.622, $g_3$ =1.3554

$$(Z_{0e})_{j,j+1} = Z_0(1 + aZ_0 + a^2 Z_0^2) \qquad (5)$$

$$(Z_{0o})_{j,j+1} = Z_0(1 - aZ_0 + a^2 Z_0^2) \qquad (6)$$

$a = J_{j,j+1}$

$Z_{0e}$ = odd mode coupled line impedance
$Z_{0o}$ = even mode coupled line impedance

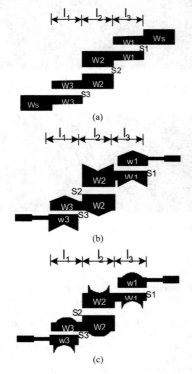

(a)

(b)

(c)

Figure. 1 (a) Parallel coupled line and (b) Resonator bandpass filter using the parallel coupled "wiggly-line" line type I and (c) type II

TABLE I
THE ELECTRICAL PARAMETER FOR THE PARALLEL COUPLED LINE SECOND-ORDER BANDPASS FILTER.

| j | $J_{j,j+1}Z_0$ | $(Z_{0e})_{j,j+1}$ | $(Z_{0o})_{j,j+1}$ |
|---|---|---|---|
| 0 | 0.43175 | 80.9079 | 37.7329 |
| 1 | 0.21701 | 63.2051 | 41.5041 |
| 2 | 0.43173 | 80.9060 | 37.7330 |

TABLE II

THE PHYSICAL PARAMETERS FOR THE SECOND-ORDER PARAMETERS CHEBYSHEV BANDPASS FILTER CENTERED AT 0.9 GHz WITH 10% FRACTIONAL BANDWIDTH

| j | W (mm) | S (mm) | L (mm) |
|---|---|---|---|
| 0 | 1.8630 | 0.425 | 46.894 |
| 1 | 2.6011 | 0.8497 | 45.875 |
| 2 | 1.9236 | 0.2918 | 47.045 |

Table I shows the electrical parameter for the parallel coupled line second-order bandpass filter. While the table II shows the physical parameters chebyshev bandpass filter centered at 0.9 GHz with 10% fractional bandwidth. The coupler which have $Z_0$ = 50Ω, C = - 8.2 dB, $Z_{0e}$, $Z_{0o}$, $\theta = \pi/2$, $l = \lambda_g/4$ We get three of microstrip line dimensions are finally obtained as follows (substrate permittivity $\varepsilon_r$ = 0.6) The second stage in the design process is calculated the period of the sinusoidal perturbation to adjust it to suppress the spurious response at $2f_0$, for $f_0$ is the design frequency. This employed rejection correspond to backward coupling of the kind of Bragg reflection in the same, but counter propagation mode is the coherence relationship reduce to

$$\Delta\beta = 2\beta_1 = 2.\frac{4\pi}{\lambda_g} = \frac{2\pi}{\lambda_B} \qquad (7)$$

where $\lambda_g$ is the guide wavelength at the design frequency, $\lambda_B = \frac{\lambda_g}{4}$ is the beat wavelength, $\beta_1$ is the mode phase constant at the frequency to be rejected and $\Delta\beta$ is the difference between the unperturbed phase constants of two interaction modes. Each coupled-line section of the filter has its own mean phase constant value and its own guided wavelength at the design frequency. In every coupled line section have the same electrical length of 90 degree at the design frequency. The last stage in the design process consists on the introduction of the perturbation in the conventional filter previously designed. This perturbation is introduced in an asymmetrical way. The constructor strip-width variation $w_i(z)$ in the $i$ th coupled-line section can be expressed as follows:

$$w_i(z) = w_i\left(1 + \frac{1}{2}\frac{M_i\%}{100}\cos\left(\frac{2\pi z}{\lambda_{B,i}} + \phi_i\right)\right) \qquad (8)$$

$\phi_i$ are theirs initial phases (0 and 180 degree), $w_i$ is the constant width calculated for the conventional filter, $\lambda_{B,i}$ is the beat wavelength and $M_i$ is the strip-width modulation parameter expressed in percentage.

III. SIMULATED AND MEASURED RESULTS

The simulated and the experimental results were presented for the conventional parallel coupled-line microstrip filter and two types of parallel coupled "wiggly-line" prototypes with perturbations $M$ = 37.5 %. The prototype circuits were designed and fabricated on RF60 microwave substrate. The three parallel coupled have the same coupling factor synthesized from -8.2 dB. The prototype with first coupled lines have factor electrical parameters $Z_{0e} = 80.9079$ Ω and $Z_{0o} = 37.7329$ Ω. The second coupled lines have a factor electrical parameters $Z_{0e} = 63.2051$ Ω and $Z_{0o} = 41.5041$ Ω.The last coupled was the same as a factor electrical parameter approximately the first coupled. The EM simulated results of the two types of bandpass filter based on the periodically non-uniform coupled line uncompensated were shown in Fig. 2. The proposed bandpass filter's magnitudes type I of $S_{21}$ and $S_{11}$ around $f_0$ are about -0.43 dB and less than -25.256 dB, while the suppression performances at $2f_0$, and $3f_0$ compare with the bandpass filter are approximately 44.1 dB, and 37.613 dB, respectively. The proposed bandpass filter's magnitudes type II of $S_{21}$ and $S_{11}$ around $f_0$ were about -0.947 dB and less than -20.705 dB, while the suppression performances at $2f_0$, and $3f_0$ compare with the bandpass filter were approximately 36.10 dB, and 43.254 dB, respectively.

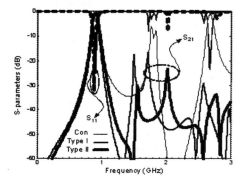

Figure.2 The EM simulated results of conventional parallel coupled line, parallel coupled "wiggly-line" type I and type II.

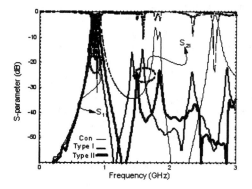

Figure.3 The measured simulated results of conventional parallel coupled line, parallel coupled "wiggly-line" type I and type II.

The measured results of the proposed microstrip coupled-feed filter were shown in Fig. 3. The magnitudes type I of $S_{21}$

and $S_{11}$ around $f_0$ are about -0.65 dB and -20.16 dB, while the suppression performances at $2f_0$, and $3f_0$ are approximately 42.5 dB, and 54.2 dB, respectively. The magnitudes type II of $S_{21}$ and $S_{11}$ around $f_0$ were about -0.97 dB and -18.6 dB, while the suppression performances at $2f_0$, and $3f_0$ are approximately 36.7 dB, and 48.6 dB, respectively. The photograph of the print circuit board of the resonator bandpass filter type I and type II based on parallel coupled "wiggly-line" were shown in Fig. 4 and Fig. 5, respectively.

Figure.4 Photographs of parallel coupled "wiggly-line" type I

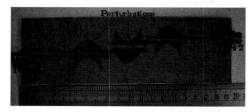

Figure.5_Photographs of parallel coupled "wiggly-line" type II

## IV. CONCLUSION

In this paper, the simple design procedures for the resonator bandpass filters based on the parallel coupled "wiggly-line" type I and type II were presented. These resonators give more efficiency of performance to reduce spurious response $2f_0$ and shift $3f_0$ to higher frequency by their structures. The closed form design equations of parallel coupled lines were suitable to use in many wireless and microwave applications. This proposed bandpass filter can be used in many wireless and microwave systems.

## ACKNOWLEDGEMENT

The author is grateful to KOREA TACONIC Co.,Ltd. for supplying Taconic RF60-0600 microwave substrates

REFERENCES

[1]  R.K. Mongia, I.J. Bahl, P.Bhartia  and J.Hong,  RF and Microwave Coupled-Line Circuits,  2 nd  edition. Artech  house, 2007.

[2]  Wenzel, R.J. and W.G. Erlinger,    "Problems in Microstrip Filter Design", IEEE Int. Microwave Symp. Digest, 1981, pp. 203-205

[3]  S. L. March, "Phase velocity compensation in parallel-coupled microstrip," in IEEE MTT-S Int. Microw. Symp. Dig., Jun. 1982, pp. 581-584

[4]  I. J. Bahl, "Capacitively compensated performance parallel coupled microstrip filter", IEEE MTT-S Digest, pp. 679-682, 1989.

[5]  J.-T. Kuo, S.-P. Chen, and M. Jiang, "Parallel-coupled-microstrip filers with over-coupled end stages for suppression of spurious responses", IEEE Microwave Wireless Compon, Lett., vol. 13, pp. 440-442. Oct. 2003.

# Robust Optimization of PCB Differential-Via for Signal Integrity

Shi-lei Zhou
Institute of Information Engineering,
Communication University of China, Beijing 100024,
China
thinking_lei@cuc.edu.cn

Gui-zhen Lu
Institute of Information Engineering,
Communication University of China, Beijing 100024,
China
luguizhen@cuc.edu.cn

*Abstract*—This paper describe a electromagnetic optimization technique using Taguchi's method. Taguchi's method was developed on the basis of the orthogonal array (OA) concept, which offers systematic and efficient characteristics. In manufacturing, the dielectric constant of High-speed PCB's dielectric material is uncontrollable factor. The paper carries out a comprehensive study of the impacts of the different dielectric constant in various differential-via design parameters on signal integrity (SI) using DOE of Taguchi method. To optimize the differential-via's parameters so that there is small effect on SI in different dielectric material.

*Keywords—Signal Integrity; Robust Optimization; Design of Experiments(DOE); HFSS*

## I. INTRODUCTION

In this paper, We mainly analysis under different structure parameters and dielectric constant of the tiny changes impact on differential-via's signal integrity. And we compare the S-parameters in different cases and a lot of simulations have been done to demonstrate how those parameters affect the signal integrity with the help of a full-wave 3-D electromagnetic solver (HFSS). In 1980, Taguchi's introduction of robust design to several major American industries resulted in significant quality improvements in product and manufacturing process design. His parameter design, in a narrow sense, where the levels (values) of design variables (control factors) minimize the effect of noise on the product's quality, must be determined to find the optimum levels. This paper also use the Taguchi method to optimize the differential-via's parameters so that there is small effect on Signal Integrity in different dielectric constant.

### A. Signal Integrity

"Signal Integrity (SI) ensures that a signal is moved from point A to point B with sufficient quality or integrity to allow effective communication" [1]. Now, as technology innovation marches forward, new kinds of devices, media formats, and large inexpensive storage are converging. They require significantly more bus bandwidth and transfer rate to maintain the interactive experience users have come to expect. Signal Integrity analysis becomes important to ensure reliable high-speed data transmission. In this paper, we use the differential to common mode conversion ($S_{dc}$dB: Driver is common mode and receiver is differential) as the main performance of differential-via's signal integrity.

### B. Design of Experiments(DOE)

Design of Experiments is a systematic method for determining the effect of factors and their possible interactions in a design or a process towards achieving a particular output of the quality characteristic(s)[2]. DOE is used to ensure the value of the selected output parameter (which is called Quality Parameters) within defined range, when the system has unwanted and uncontrollable design variations. Thus, DOE is a method to design a system in a robust way as well as meeting the system output requirements. The 'treatments' are the well-defined procedures or experiments to examine a system for its output characteristics. An 'interaction' is the variation among the differences between means for different levels of one factor over different levels of the other factor.

### C. Taguchi method

Taguchi method is a methodology based on orthogonal arrays (OAs) concept, which effectively reduces the number of test iterations required in an optimization process [3]. Taguchi optimization technique is predominantly used in industrial engineering, in which the experiments are planned for designing of a product efficiently and reliably. Based on these experiments, the effects of various factors on design are calculated. Therefore, by controlling dominant factor, product is optimized to achieve better quality. This study shows that the method can quickly converge to the desired design since it takes less computational resource. Most important of all, Taguchi method is easy and straightforward to implement.

The orthogonal array, which has a profound background from statistics [4], plays an essential role in the method. The selection of an orthogonal array depends on the input parameters of an optimization problem.

### D. Orthogonal array

A matrix experiment consists of a set of experiments. Each experiment is performed by changing the setting of the various product or process parameters which has to study. After conducting matrix experiments, the data from all experiments in the set taken together is analyzed to determine the effects of various parameters. Matrix experiments are conducted using special matrices, called "orthogonal arrays" [5]. Number of arrays was designed. Every array is directed to a special type of experimental situation. The title orthogonal expresses that the array is balanced. The word balance expresses that every column is balanced within itself and that any two columns in

978-7-5641-4279-7

the arrays are also balanced. That means, at first, that within a column, there are an equal number of levels. At second, that means that any two columns in the arrays are also balanced. This balance between any two columns assures that all possible factor combinations exist in equal numbers [6].

*E. Robust Optimization*

Robust optimization means to optimize a design in such a way that it will certainly work according to the specifications for which it is desired to work. Taguchi Methods are used for general Robust Optimization. There are various other methods also that perform Risk Analysis for robust optimization [7].The approach used in the paper is to make the robust optimization of differential-via's parameters so that there is small effect on signal integrity in different dielectric constant.

## II. MOLDING HFSS

The differential-vias has been designed in 3D full wave modeling (HFSS). The boards are 5mil thick, the layers are 1.2mil thick and the high speed trace is 5mil wide as shown in the Figure 1 and Figure 2.

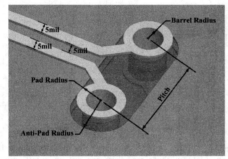

Fig. 1. HFSS Differential-via Modeling

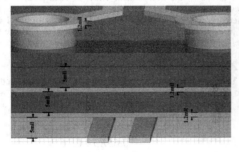

Fig. 2. Differential-via Parameters

In this DOE, we used three dielectric materials. Their dielectric constants respectively are $\varepsilon_1=4$, $\varepsilon_2=4.4$ and $\varepsilon_3=4.6$. Their loss tangent are 0.02. The physical dimensions considered as variables into the experiments were the next:

R1=Barrel Radius;

R2=Pad Radius;

R3=Anti-Pad Radius;

P=Pitch(as shown in the Figure 1);

In our case the random variables or factors are 4 and the levels are 3, so the Taguchi Array of $L_9(3^4)$ is used. The experiments or the orthogonal simulations performed are planned according to this array. The $L_9(3^4)$ Taguchi Array is used from [6]. The factor selected and level values are illustrated in Table I. Table II shows the values of orthogonal array selected $L_9(3^4)$

TABLE I. FACTOR AND LEVEL SELECTION

| Factors | Level 1 | Level 2 | Level 3 |
|---|---|---|---|
| Barrel Radius(R1) | 5mil | 6.5mil | 8mil |
| Pad Radius(R2) | 9mil | 10.5mil | 12mil |
| Anti-pad Radius(R3) | 13mil | 14.5mil | 16mil |
| Pitch(P) | 34mil | 38mil | 42mil |

TABLE II. VALUES OF ORTHOGNAL ARRAY

| Experiment | R1 | R2 | R3 | P |
|---|---|---|---|---|
| Case1 | 1 | 1 | 1 | 1 |
| Case2 | 1 | 2 | 2 | 2 |
| Case3 | 1 | 3 | 3 | 3 |
| Case4 | 2 | 1 | 2 | 3 |
| Case5 | 2 | 2 | 3 | 1 |
| Case6 | 2 | 3 | 1 | 2 |
| Case7 | 3 | 1 | 3 | 2 |
| Case8 | 3 | 2 | 1 | 3 |
| Case9 | 3 | 3 | 2 | 1 |

## III. SIMULATION

In this subsection, the modeling of multimode S-Parameters has been obtained after simulating each case in HFSS.

The differential-to-common mode S-parameters($S_{dc}21$), where the energy is being transmitted in the odd mode and received in the even mode. Used the dielectric materials which dielectric constants is $\varepsilon_2=4.4$, the $S_{dc}21$ results are shown in Figure 3.

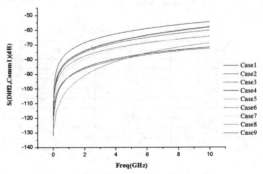

Fig. 3. The Differential-to-common mode S-parameters ($S_{dc}21dB$)

As lower curves as better signal quality can be expected. In Figure 3, it can be observed that cases Case1, Case6 and Case8 have the lowest values as the frequency increases. The cases showing the highest was Case9.

When using the same method, we also can get the differential-to-common mode S-parameters($S_{dc}21$) in different dielectric constant($\varepsilon_1$=4, $\varepsilon_3$=4.6).

Using the simulation results, we calculated average S-parameters by Eq. (1). The outputs are $Y_1$, $Y_2$ and $Y_3$, as shown in Table III.

$$Y_n = \overline{S_{dc}21}_{CaseN}(dB) \tag{1}$$

TABLE III.  SIMULATION RESULTS OF EXPERIMENT

| Simulation | Y1($\varepsilon$1)dB | Y2($\varepsilon$2)dB | Y3($\varepsilon$3)dB |
|---|---|---|---|
| Case1 | -72.39 | -80.17 | -75.52 |
| Case2 | -70.16 | -69.29 | -71.33 |
| Case3 | -69.53 | -72.54 | -66.53 |
| Case4 | -66.74 | -67.88 | -64.3 |
| Case5 | -74.63 | -68.13 | -72.41 |
| Case6 | -80.75 | -80.75 | -68.07 |
| Case7 | -69.86 | -73.58 | -75.7 |
| Case8 | -76.8 | -81.88 | -86.52 |
| Case9 | -68.77 | -63.18 | -64.16 |

$$\mu = \frac{1}{3} \sum_{i=1}^{n} Y_n (dB) \tag{2}$$

$$\sigma^2 = \frac{1}{3} \sum_{i=1}^{3} \left(Y_n - \mu\right)^2 \tag{3}$$

Using the Eq. 2 and Eq. 3, we calculate the mean($\mu$) and variance($\sigma^2$) as shown in Table IV.

TABLE IV.  CALCULATED RESULTS

| Simulation | $\mu$(dB) | $\sigma^2$ |
|---|---|---|
| Case1 | -76.02 | 15.3 |
| Case2 | -70.26 | 1.06 |
| Case3 | -69.53 | 9.03 |
| Case4 | -66.3 | 3.36 |
| Case5 | -71.72 | 10.9 |
| Case6 | -76.52 | 53.63 |
| Case7 | -73.05 | 8.75 |
| Case8 | -81.73 | 23.63 |
| Case9 | -65.37 | 8.88 |

## IV.  STATISTICAL ANALYSIS

The purpose of this robust optimization is select the differential-via's parameters design, in a narrow sense, where the levels of design variables (control factors) minimize the effect of different dielectric constant(un-control factor) on the differential-via's signal integrity, must be determined to find the optimum levels. The parameters are R1(Barrel Radius), R2(Pad Radius), R3(Anti-Pad Radius) and P(Pitch).

### A. Analysis of Mean($\mu$)

The regression equation for Mean of output is stated as below:

$$Mean(\mu) = -77.33 - 0.722R1 + 0.653R2 \\ + 3.33R3 - 0.742P \tag{4}$$

The differential-to-common mode S-parameters ($S_{dc}21$) is the smaller the better. In the Eq. (4), sensitivity coefficients for $\mu$ of R3 is the largest. So the R3 has the most significant effect on $\mu$; the second are P and R1; the R2 is small effect.

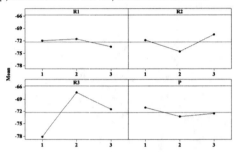

Fig. 4.  Main Effects Plot(Mean of output)

In Fig. 4, the R1(Antipad Radius) and P factors have horizontal slope; therefore, they are not as significant and there is shown the independent effect of each parameter on mean of output. Use the Fig. 4, we can optimize the differential-via' signal integrity to better. The factor of R1 use the level 3. R2 use the level 2; R3 use the level 1; P use the level 2.

### B. Analysis of Variance($\sigma^2$)

The regression equation for Variance of output is stated as below:

$$Variance(\sigma^2) = 15.93 + 2.65R1 + 7.35R2 \\ - 10.65R3 + 0.16P \tag{5}$$

The value of variance is small demonstrate the differential-via's have better robust SI in this the parameters' levels. So the R3 is also the most significant effect on $\mu$; the second are P and R1; the R2 is small effect.

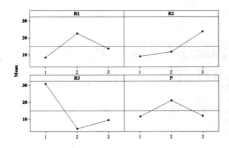

Fig. 5.  Main Effects Plot(Mean of Variance)

In the Fig. 5, there is shown the independent effect of each parameter on mean of variance. So we select R1 in level 1, R2 in level 1, R3 in level 2 and P use the level 1 to optimize the differential-via's robust output.

## V.  OPTIMIZATION

The most significant parameter/factor can be determined by the sensitivity coefficient of the regression equation. The absolute value of the variable's coefficient is larger; the factor is more effect in the output. Main effect plot diagrams as the graphical representations of the change in performance characteristics with variation in every level, giving a pictorial view of the variation of each factor and its effect on the performance, as each factor shifts from one level to another. So the main effect plot diagrams also can determine the optimum levels of design variables. A more extreme slope indicates a more significant effect on differential-via's signal integrity. There are shown the interaction effect. So the optimized design in this case is: R1=level1; R2=level2; R3=level3; P=level3. The optimized design parameter is in TABLE V. And the optimized differential-via's mold is shown in Fig. 6 and Fig. 7. The result of simulation in different dielectric material is shown in Fig. 8. We calculate the mean($\mu$) is -82.28dB and variance($\sigma^2$) is 29.17.

TABLE V.          OPTIMIZED DESIGN

| Parameter | Barrel Radius(R1) | Pad Radius(R2) | Anti-pad Radius(R3) | Pitch(P) |
|---|---|---|---|---|
| Optimized Design | 5mil | 10.5mil | 16mil | 42mil |

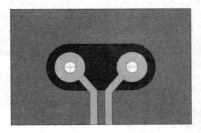

Fig. 6.  The Optimized Differential-via (Top)

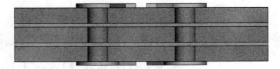

Fig. 7.  The Optimized Differential-via (Front)

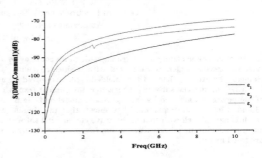

Fig. 8.  The Optimized Differential-to-common mode S-parameters ($S_{dc}21$dB)

## VI.  CONCLUSIONS

In this paper, we use the Taguchi Method and HFSS to study the differential-via's signal integrity. Use the result of simulation and statistical analysis, we can find:

- The factor of the anti-pad radius is the most effect on the output.

- To set a appropriate value of anti-pad radius, it optimize the differential-via's parameters so that there is small effect on SI in different dielectric material.

In this methodology, we use no large experiments to analysis and find the best optimized parameters. There is small effect on Signal Integrity in different dielectric constant. The design engineer, early in the program, can work with PCB suppliers to achieve a balance between differential-via performance. The methodology can be used to optimize any similar system.

REFERENCES

[1]  "Interconnect Signal Integrity Handbook", by Samtec, Inc. Aug. 2007.

[2]  Sammy G. Shina, Six Sigma for Electonics Design and Manufacturing, Professional Engineering Series, McGraw-Hill, 2002.

[3]  Genichi Taguchi, Introduction to quality Engineering, White Plain, NY: Uni Pub, 1986.

[4]  A. S. Hedayat, N. J. A. Sloane, and John Stufken, Orthogonal Arrays:Theory and Applications, Springer-Verlag New York Inc., New York, 1999.

[5]  Madhav Phadke, "Quality Engineering using Robust Design," Pearson Education, 2008.

[6]  Taguchi, G., Chowdhury S., Wu Y.: Taguchi's Quality Engineering Handbook. Wiley. 2005. ISBN 0-471-41334-8

[7]  E. G. A. Gaury and J. P. C. Kleijnen, "Risk Analysis of Robust System Design", in Proc. The 1998 Winter Simulation Conference, pp. 1533-1540.

# Modeling of the Cable-Induced Coupling into a Shielding Box Using BLT equation

Su-Fei Xiao, Zhen-Yi Niu, Jian-Feng Shi, Feng Liu
College of Electronic and Information Engineering, Nanjing University of
Aeronautics and Astronautics Nanjing 210016, China
xiaosufei1989@sina.cn

*Abstract*-The modeling of electromagnetic pulse coupling to a shielding device, which is connected with a shielded cable, is studied. The exciting fields of the shielded cable are solved by a full-wave commercial software. Thereafter, the current response of the device is simulated by the Agrawal's model and the BLT (Baum, Liu, Tesche) equation. It is shown that the simulated result is agreed well with that of full-wave commercial software, and the computational efficiency is improved over 80% by the proposed method.

## I. INTRODUCTION

With the development of the electronic technology and electromagnetic pulse (EMP) sources, it's important to analyze an EMP coupling to electronic devices and systems. A metal shielding box is widely used to protect the sensitive device from the electromagnetic interferences. However, there are mainly two inevitable coupling paths that the EMP interfere with the devices, apertures on the shielding box and the wires or shielded cables connect to the device. On the other hand, the presence of transmission lines through a shielding box increases computational difficulty as they provide additional electromagnetic coupling paths between outer and inner fields. Therefore, it is of great interest to develop a method to quickly predict the electromagnetic effects induced in a shielding box with various types of transmission lines.

Paletta et al. dealt with the application of electromagnetic field-to-transmission-line coupling models for large cable systems analysis [1]. The method uses a combination of a full-wave solver and transmission-line analysis. And an experiment has been performed on a prototype wiring installed in a Renault Laguna car to validate the efficiency of this methodology. The coupling of an incident electromagnetic wave to a device inside a shielding box penetrated by a wire or a shielded cable has received some attention in recent years. Lertsirimit et al. computed an electromagnetic wave coupling to a device on a printed circuit board inside a cavity from a wire penetrating a cavity aperture by a hybrid method. The exterior problem was analyzed by a full wave method, while the interior problem was analyzed by transmission-line theory [2]. Li et al. simulated an EMP induced interference current in circuits inside a shielding box by a wire penetrated through an aperture by the finite-difference time-domain (FDTD) method [3]. Hakan Bağci analyzed electromagnetic coupling into enclosures through coaxial cables by a fast hybrid time-domain method [4]. Sapuan et al. studied the shielding effectiveness and $S_{21}$ of a rectangular enclosure with apertures

and wire penetration experimentally and numerically by using the commercial software CST Microwave Studio [5]. Xie et al. analyzed an electromagnetic pulse coupling to a device from a wire penetrating a cavity aperture by applying the transient electromagnetic topology method [6]. The method uses a combination of the FDTD method and SPICE model.

There are mainly three models are used to describe the coupling of incident field to transmission lines, Taylor's model [7], Agrawal's model [8] and Rachidi's model [9]. Compare to Taylor's model and Rachidi's model, Agrawal's model has many advantages. First, the distributed voltage generators in Agrawal's model are directly equal to the incident electric field components tangent to the line. Second, the calculation of the BLT equation [10] is reduced because there are no equivalent current generators in the expression of source waves. Third, Agrawal's model requires less memory to store the data files of exciting fields . By using the Agrawal's model, only the incident electric field components tangent to the line and the electric fields along the terminals and the ground are needed. Therefore, the Agrawal's model is used to describe the coupling of incident field to the shielded cable in this paper.

The modeling of an electromagnetic pulse coupling to a shielding device is studied. The device is connected with a shielded cable that penetrates through an aperture on the shielding box. The method is based on the electromagnetic topology (EMT) technology [11]. The coupling of an external electromagnetic wave to the shielded cable is concerned while the influence of the shielded cable on external fields is neglected. Furthermore, the influence of the inner conductor on the shield layer of the shielded cable is neglected. The computation process of the method includes two steps: The exciting fields of the shielded cable are solved by a full-wave commercial software but the presence of the cables is at first neglected. Thereafter, the current response of the device is simulated by the Agrawal's model and the BLT equation (the key equation of EMT). The obtained result is compared with that solved by a full-wave commercial software. This paper is divided into four sections. In section 2, the computation model and the computation process of the proposed method is described. Results and discussion are provided in section 3 while conclude in section 4.

## II. COMPUTATION MODEL AND METHOD DESCRIPTION

978-7-5641-4279-7

### A. Computation Model

The schematic diagram of the system for computation is shown in Fig. 1. It consists of a shielded cable that penetrates through an aperture of a shielding box and then connects with a device inside the shielding box. The device is represented by a resistance $Z_2^{(i)}$ here. The length, the wide and the height of the shielding box is 41.0 cm, 10.5 cm and 27.3 cm respectively. And the back surface is 2 mm thick while others are 6 mm thick. The length of the shielded cable outside the shielding box and inside the shielding box is 2.57 m and 0.1 m respectively. The outer terminal loads of the shielded cable are both 100 Ω, and the loads between the inner conductor and the shield layer at the two terminations are both 50 Ω. The height of the shielded cable above the perfect ground plane is 0.2 m. The direction and polarization of the EMP is shown in Fig. 1

### B. Computation Process

The computation process includes two steps: a full-wave commercial software is used to compute the exciting fields of the shielded cable at first, thereafter the Agrawal's model and the BLT equation are used to compute the current response of the shield device. The method is based on the electromagnetic topology theory. The influence of the shielded cable on external fields is neglected. For the calculation of the exciting fields , the shielded cable is removed at first. And then compute the fields at the exact positions of the shielded cable. As the cable does not need to meshed, the computation of the external fields just spend a few time. The source terms of the BLT equation is obtained by the exciting fields using the Agrawal's model.

In this paper, Agrawal's model is used as the equivalent model to analysis the shielded cable. A full-wave commercial software is used to compute the exciting fields of the shielded cable. Therefore, the exciting fields at the exact positions of the shielded cable are saved into data files. The date files are used to compute the current response on the shield layer of the shielded cable by Agrawal's model. And then the applied sources on the inner conductor of the shielded cable is obtained by the transfer impedance $Z_t$ and transfer admittance $Y_t$. The current response of the device is computed by the BLT equation.

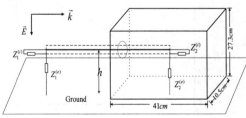

Figure 1 .Simple model of a shielded cable penetrates an aperture in a shielding box

There are two paths that the EMP coupling to the device as shown in Fig. 1, the shielded cable penetrating the shielding box and the aperture on the shielding box. In the computation model, the size of the aperture on the shielding box compare with the wavelength of the EMP is electrically small. Therefore the coupling effect through the aperture is neglected in this paper. As the heights of the outer part and the inner part of the shielded cable are different, the characteristic impedances of the two parts are different .The expression of the characteristic impedance $Z_c$ is

$$Z_c = \sqrt{\frac{L'}{C'}} \qquad (1)$$

where $L'$ is the inductance per unit length, and $C'$ is the capacitance per unit length. The expressions of them are

$$L' = \frac{\mu_0}{2\pi} \ln \frac{2h}{a} \qquad (2)$$

$$C' = \frac{2\pi\varepsilon_0}{\ln(\frac{2h}{a})} \qquad (3)$$

where $a$ is the radius of shield layer of the shielded cable, $h$ is the height of the shielded cable. The characteristic impedances is computed by these expressions. The characteristic impedance of the outer part is 329.55Ω, while the characteristic impedance of the inner part is 328.95Ω. As the characteristic impedance of the outer part is approximate to the characteristic impedance of the inner part, the shielded cable in Fig. 1 is regarded as a shielded cable over ground with the same height in this paper.

The shield layer of the shielded cable satisfies the good shielding approximation in this paper. The shielded cable is considered as two transmission line system, the external transmission line system which consists of the shield layer of the cable and the ground and the internal transmission line system which consists of the shield layer and the inner conductor of the cable [12]. The two transmission line systems are linked by the transfer impedance $Z_t$ and transfer admittance $Y_t$ of the shielded cable.

$$\begin{cases} V'_{si} = Z_t I_s \\ I'_{si} = -Y_t V_s \end{cases} \qquad (4)$$

where $I_s$ and $V_s$ are the current response and the voltage response of the shielded layer of the cable, they are computed by the exciting fields using Agrawal's model. $I'_{si}$ and $V'_{si}$ are the sources of the inner conductor of the shielded cable respectively.

The current response of the shielded layer is solved by

$$I_s(x) = \int_0^L G_I(x;x_s)V'_s(x_s)dx_s - G_I(x;0)V_1 + G_I(x;L)V_2 \qquad (5)$$

where $G_I(x;x_s)$ is a green function, it's represents the current response of voltage source per unit and connected with the characteristic impedance of the shield layer. The expression is

$$G_I(x;x_s) = \frac{e^{-\gamma L}}{2Z_c(1-\rho_1\rho_2e^{-2\gamma L})}(e^{\gamma x_<} - \rho_1 e^{-\gamma x_<}) \cdot$$
$$[e^{-\gamma(x_>-L)} - \rho_2 e^{\gamma(x_>-L)}]$$
(6)

where $x_<$ is the smaller between $x$ and $x_s$, $x_>$ is the bigger between $x$ and $x_s$. $V_s'$ is the distributed voltage sources on the shield layer of the shielded cable, and it is equal to tangential electric fields at the position of the shielded cable. $V_1$ and $V_2$ are the lumped voltage sources of the terminals of the shield layer in Agrawal's model. The expressions of them are

$$V_1 = -\int_0^h E_z^{inc}(0,z)dz$$
(7)

$$V_2 = -\int_0^h E_z^{inc}(L,z)dz$$
(8)

where $E_z^{inc}$ is the exciting fields along the terminals, it is obtained by the exciting fields. $h$ is the height of the shielded cable.

For the computation of transfer impedance $Z_t$, the diffusion and aperture penetration effects are taken into account in this paper. The expression of $Z_t$ is

$$Z_t' = Z_d' + j\omega L_a'$$
(9)

where

$$Z_d' = R_0 \frac{(1+j)d/\delta}{\sinh(1+j)d/\delta}$$
(10)

$R_0$ is the direct-current resistance of the shield layer. $\delta$ is the skin depth of the shield layer. $d$ is the diameter of the thin metal wires in the shield layer. $L_a'$ is the aperture leakage inductance.

The expression of transfer admittance $Y_t$ is

$$Y_t' = j\omega C S_s$$
(11)

where $C$ is the inner coax capacitance. $S_s$ is the electrostatic shield leakage parameter.

The internal sources of the inner conductor of the shielded cable $V_{si}'$ and $I_{si}'$ is calculated by transfer impedance $Z_t$ and transfer admittance $Y_t$ of the shielded cable, and then the BLT equation of the internal transmission line system is built.

$$\begin{bmatrix} I(0) \\ I(L) \end{bmatrix} = \frac{1}{Z_c}\begin{bmatrix} 1-\rho_1 & 0 \\ 0 & 1-\rho_2 \end{bmatrix}\begin{bmatrix} -\rho_1 & e^{\gamma L} \\ e^{\gamma L} & -\rho_2 \end{bmatrix}\begin{bmatrix} S_1 \\ S_2 \end{bmatrix}$$
(12)

where $\rho_1$ and $\rho_2$ are the reflection coefficient of the inner conductor. $\gamma$ is the propagation constant. $S_1$ and $S_2$ are the source terms of the equation. The expressions of them are

$$S_1 = \frac{1}{2}\int_0^L e^{\gamma^{(i)}x_s}[V_{si}'(x_s) + Z_c^i I_{si}'(x_s)]dx_s$$
(13)

$$S_2 = -\frac{1}{2}\int_0^L e^{\gamma^{(i)}(L-x_s)}[V_{si}'(x_s) + Z_c^i I_{si}'(x_s)]dx_s$$
(14)

where $Z_c^{(i)}$ is the characteristic impedance of the inner conductor.

Because the exciting fields obtained by the full-wave commercial software are in time-domain, all the results is post-processed in frequency-domain using fast Fourier transforms (FFT).The current response of the device is solved by BLT equation. As the BLT equation is a frequency-domain

equation, the results are in frequency-domain. The inverse fast Fourier transforms (IFFT) is needed to obtain the transient response of the device.

### III. RESULT AND ANALYSIS

The expression of the incident field waveform in the computation model is

$$E(t) = kE_0(e^{-\beta_1 t} - e^{-\beta_2 t})$$

where $E_0$=50 kV/m, $k$=1.3, $\beta_1$=4.0×10$^7$ s$^{-1}$, $\beta_2$=6.0×10$^8$ s$^{-1}$.The direction and polarization of the EMP are given in Fig. 1.The shielded cable for analyzing is a RG-58 cable. The internal characteristic impedance of the RG-58 cable is 50 Ω. The radius of the shield layer $a$=0.152 cm. The thickness of the shielded layer $\Delta$=0.127 mm. The direct-current resistance of the shield layer $R_0$=14.2 mΩ/m. The aperture leakage inductance $L_a$=1.0 nH/m, the electrostatic shield leakage parameter $S_s$=6.6×10$^7$ m/F.

The transfer impedance $Z_t$ is calculated by (9), and the amplitude of the shielded cable transfer impedance $Z_t$ is shown in Fig. 2.

After the exciting fields calculated by a full-wave commercial software, the current of the shield layer of the cable is computed by Agrawal's model. The current response on the shield layer of the cable is shown in Fig. 3.

To verify the accuracy of the proposed method, a full-wave commercial software is used to analyze the same problem in Fig. 1. The full-wave commercial software that used in this paper is a powerful and easy-to-use package for the analysis of conducted transmission, electromagnetic interference and electromagnetic susceptibility on complex cable structures. The current response on the load $Z_2^{(i)}$ in Fig. 1 obtained by the proposed method and full-wave commercial software are shown in Fig. 4. All the simulations are performed on a personal computer with the Intel(R) Pentium(R) Dual-Core CPU E5200 with 2.50 GHz and 2.0 GB RAM.

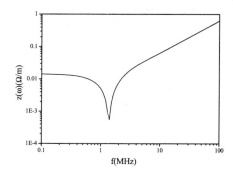

Figure 2. The amplitude of the shielded cable transfer impedance

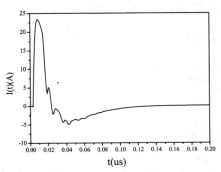

Figure 3. The current response of the shield layer of the cable

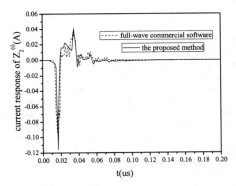

Figure 4. The current response of $Z_2^{(l)}$

From the results in Fig. 3 and Fig. 4, the shielding effectiveness of the shield layer is about 46 dB. The shield layer does a good job in protecting the internal electronics from the disturbance of the external EMP. In the proposed method, the cable does not need to be meshed in the process of compute the exciting fields . Furthermore, the transmission-line analysis just spent a few seconds. The calculation time of the proposed method is 17 minutes 36 seconds on a personal computer, including 17 minutes 30 seconds for the computation of the exciting fields and 6 seconds for the computation of the load current response, much less than 1 hour 37 minutes 29 seconds needed by the full-wave commercial software. As shown in Fig. 4, the result obtained by the proposed method and that obtained by the full-wave commercial software are in good agreement.

IV. CONCLUSION

The modeling of electromagnetic pulse coupling to a shielding device, which is connected with a shielded cable, is studied in this paper. The computation process of the method includes two steps: The exciting fields of the shielded cable are solved by a full-wave solver but the presence of the cables is at first neglected. Thereafter, the current response of the device is simulated by the Agrawal's model and the BLT equation. The results obtained by the proposed method and that solved by the full-wave commercial software verified the accuracy of the proposed method. These numerical results are helpful for further designing electromagnetic protection of the inner devices against the electromagnetic interference.

ACKNOWLEDGMENT

This research is support by the Aeronautical Science Foundation of China (20128052062) and A Project Funded by the Priority Academic Program Development of Jiangsu Higher Education Institutions.

REFERENCES

[1] Paletta, L., J. P. Parmantier, F. Issac, P. Dumas, and J. C. Alliot,"Susceptibility analysis of wiring in a complex system combining a 3-D solver and a transmission-line network simulation," *IEEE Trans. EMC*, Vol. 44, No. 2, 309-317, May 2002.

[2] Lertsirimit, C., R. Jackson, and D. R. Wilton, "An efficient hybrid method for calculating the EMC coupling to a device on a printed circuit board inside a cavity by a wire penetrating an aperture," *Electromagnetics*, Vol. 25, No. 7–8, 637–654, 2005.

[3] Li, X., J. Yu, Y. Li, Q. Wang, and Y. Zhang, "Simulation of the EMP coupling to circuits inside a shielding box by a wire penetrated with an aperture," *MAPE Proceedings*, 1345–1348, Hangzhou,China, 2007.

[4] Hakan Bağci, Ali E. Yılmaz, and Eric Michielssen, "A fast hybrid TDIE-FDTD-MNA scheme for analyzing cable-induced transient coupling into shielding enclosures," in *Proc. IEEE Int. Symp. EMC*, vol.3, pp.828-833,2005.

[5] Sapuan, S. Z. and M. Z. M. Jenu, "Shielding effectiveness of a rectangular enclosure with aperture and wire penetration," *ICMMT Proceedings*, 1–4, Guilin, China, 2007.

[6] H. Xie, J. Wang, D. Sun, and Y. Liu, "Analysis of EMP coupling to a device from a wire penetrating a cavity aperture using transient electromagnetic topology," *Journal of Electromagn. Waves and Appl.*, 2009,23:2313-2322.

[7] C. D. Taylor, R. S. Satterwhite, and C. W. Harisson, "The response of a terminated two-wires transmission-line excited by a nonuniform electromagnetic field," *IEEE Trans. Antennas Propagat.*, vol. 13, pp. 987–989, Nov. 1965.

[8] A. K. Agrawal, H. J. Price, and S. H. Gurbaxani, "Transient response of multiconductor transmission-lines excited by a nonuniform electromagnetic field," *IEEE Trans. Electromagn. Compat.*, vol. 22, pp. 119–129,May 1980.

[9] F. Rachidi, "Formulation of field to transmission-line coupling equations in terms of magnetic excitation field," *IEEE Trans. Electromagn. Compat.*, vol. 35, Aug. 1993.

[10] Baum, C. E., T. K. Liu, and F. M. Tesche, "On the analysis of general multiconductor transmission-line networks," *Interaction Note*, Vol. 350, 1–26, Nov. 1978.

[11] Baum C E. "How to think about EMP interaction," *Proc. 1974 Spring FULMEN Meeting*, Kirtland AFB, April 1974.

[12] Tesche F M, Ianoz M V, Karlsson T. *EMC Analysis Methods and Computational Models*. New York: Wiley, 1997.

# Magnetic Near-Field Mapping of Printed Circuit Board in Microwave Frequency Band

Shun-Yun Lin[1], Jian-Hua Chen[1] Ya-Ting Pan[1], and Tzung-Wern Chiou[2]

[1]Department of Electronics Engineering, Cheng Shiu University Kaohsiung city,Taiwan

[2]BWant Co., Ltd, Taiwan

*Abstract-*A near-field mapping system has been setup to scan the magnetic field above PCB circuits. A loop-type magnetic probe is used to measure the magnetic field up to 9 GHz. The probe is easily fabricated by printing technology. A set of quasi-periodic notches, which functions as a microstrip filter, is embedded between the feeding port and the sensor to introduce self-resonance suppression characteristics. In order to verify the performance of our new system, EM surface field measurements were conducted for a shorted-microstrip line and an open-microstrip. The comparison between the measured results and the reference data validate the good performance of our system.

## I. INTRODUCTION

With increasing complexity of high speed digital circuit board and rapid development of wireless devices built-in cell phone or wireless LAN, there are numerous design and fabrication issues. Especially, these circuitry modules sometimes cause electromagnetic interference in self equipment [1] or excessive electromagnetic radiation to generate electromagnetic interference (EMI) problems. With a map of electromagnetic field intensity on the circuit board [2-3], one can find the whole field distribution around device and find out the corresponding noise source.

In order to locate noise or to predict far-field emission level, a near-field measurement by magnetic field probe is a promising method. In ref. [4], an integrated RF magnetic field probe micro-fabricated using CMOS–silicon-on-insulator (SOI) technology obtains a further-miniature and high spatial-resolution function; A stripline magnetic near-field probe for high frequency band up to 10 GHz achieved an improved spatial resolution [5]; a set of orthogonal loops was proposed for a rapid E-, Hx- Hy- or circular H-fields measurement [6]; In ref. [7], a bond wired rectangular magnetic field probe, which effectively suppresses resonance behavior between probe and ground plane, was proposed to gauge much higher frequency region than conventional magnetic probe. According to requirement of electromagnetic sensor, broad bandwidth, high spatial resolution, and large isolation in sensed electrical and magnetic field are respect. On the other hand, by adding modified periodic loads to a microstrip line [8], the proposed integrated microstrip antennas not only retained good performance, but also eliminated harmonic resonances and spurious emission. Here, the geometric dimension and structure should be suitably chose to achieve desired performance.

In this study, firstly, we will present a description of field-mapping principle as well as the probe design and automatic measurement system. Secondly, a microstrip line with a short end and an open were characterized by our new system to verify its performance. The comparison between the measured results and the reference data from commercial microwave simulation software are presented to validate the performance of our system.

## II. MAGNETIC NEAR-FIELD MAPPING SYSTEM

### A. PRINTED MAGNETIC FIELD PROBE WITH ENHANCED PERFORMANCES

In this study, our previous design [9], a broadband magnetic field probe, is tested over the band 1~9 GHz. Figure 1 shows the computer-aided design (CAD) layout of the proposed probe. The planar circular loop with radius $R = 4$ mm and 1 mm width was printed on a fiberglass microwave substrate, which is with permittivity $\varepsilon_r = 4.4$ and thickness $h = 0.8$ mm. For real applications, the circular loop was connected to spectrum analyzer or receiver through a pair of microstrip line with 8.5 mm length and 1 mm width. Note that three pairs of notch were quasi-periodically embedded into the connecting portion. Two pairs of notch near to the feeding edge have the same dimension while the last is different from and larger than the former. It is well known that the microstrip with periodic loads acts as the band-reject filter at some frequency bands [8]. In this design, the embedded notches also introduce a filterable feature to suppress the self-resonance of the circular loop, moreover, improve the probe performances at much higher frequency. The geometry structure associated with the connector should be suitably chosen to achieve desired performance. After several simulations, we choose the notch width as 0.5, 0.5, and 1 mm and with the same length of 0.5 mm for the optimum characteristics. In addition to the loop radius, the location of the embedded notches is an important to the resonance suppression. Here, the space between the larger notch and the edge of the connecting portion was noted as $d = 3.5$ mm and used to represent the corresponding position of the notches.

The reflection coefficient $S_{11}$ of the proposed probe and the reference were presented. As shown in Figure 2, the first resonance is at about 3.5 GHz with 6 dB reflection for the reference probe. This self-resonance possibly interferes with the magnetic field penetrated through the circular loop. However, by the adding of the quasi-periodic notches, the $S_{11}$ of the proposed probe has improved on the

978-7-5641-4279-7

reference about 4 dB. Moreover, the operation bandwidth determined by $S_{11}$ less than 3 dB is up to 9 GHz.

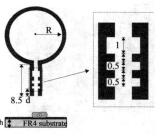

Figure 1 Configuration of the proposed magnetic probe. $R = 4$ mm, $d = 3.5$ mm, $h = 0.8$ mm.

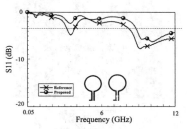

Figure 2 Measured $S_{11}$ against frequency in embedded notches.

In addition, the new probes we developed could provide advanced isolation between the desired $V_{emf}$ （caused by near magnetic field） and undesired $V_{emf}$ （caused by near electric field）. Figure 3 shows that the proposed probe has 10 dB isolation at least.

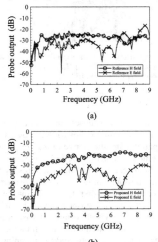

Figure 3 Measured probe output against frequency of (a) conventional magnetic probe, and (b) proposed magnetic probe.

## B. Automatic UWB Field Mapping System

According to the Faraday's law：

$$V_{emf} = -\int \frac{d\phi}{dt} = -\int \frac{d\vec{B}}{dt} \bullet d\vec{s};$$

The electromotive force $V_{emf}$ at the ends of loop probe is in proportion to the magnetic flux $\vec{B} \bullet \vec{S}$, which passes through the small loop above the device under test (DUT). In our study, the near field measurement system has been setup as shown in Figure 3. The loop-type probe functions as a magnetic field sensor, while a spectrum analyzer (SA) as a signal-detector. A fixture is designed to hold the sensor above the test planar circuit. The movement of fixture can be precisely adjusted to 0.1mm/step by a 3-D movable controller system. Moreover, a LabVIEW-based program is designed to automatically control the procedures of both mechanical movement and data collection from SA via a GPIB interface. With this measurement system, the broadband magnetic fields ($H_x(x,y)$ and $H_y(x,y)$) at a certain z-position could be automatically detected. Furthermore, the proposed probe is with more than 10 dB isolation between the desired and undesired $V_{emf}$, which is induced by the spurious emission. The accurate predictions of near magnetic field are promising.

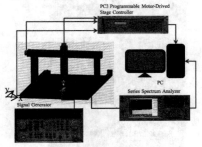

Figure 4 Configuration of measurement system

## III. EXPERIMENTAL RESULTS AND DISCUSSIONS

In order to validate the effectiveness of the proposed near-field mapping system, a directive fed 50-∧microsctrip line (MSL) with an open end and a short end were measured. The measurement results will be compared with simulation from the commercial microwave software Ansys HFSS.

The 2D magnetic field distribution with the proposed probe is studied. The probe examined an printed 50-$\Omega$ MSL with 50 mm length on a $50 \times 50$ mm$^2$ FR4 substrate. Both open-end MSL and short-end MSL were tested. Also, the simulative study from the commercial microwave software Ansys HFSS was performed. The printed circuit is scanned with 1m/step along with both x and y direction (shown in Figure 4). The probe directs at x- and y-axis to detect $H_x(x,y)$ and $H_y(x,y)$, respectively. Here, the MSL was 1 mm below the magnetic field sensor, and the signal fed to the MSL is 1800 MHz with the power of 0 dBm.

The 2D near magnetic fields of open-end MSL is studied in Figure 5. Also, the shot-end MSL is in Figure 6. Near the loading end and feeding end, the magnetic fields were distorted because of the non-ideal short/open/impedance matching condition. However, the experimental results still show that all the measurement results agree with the simulation results, which validated the magnetic field characterization ability of the new loop probe. In Figure 5(a), the measured results of $H_x$ show the characteristics of an open-circuited 50-$\Omega$ MSL. Whole field distribution is similar to half-wavelength, while the magnetic field was relative null near the open end ($Z_L=\infty$). In (b), the measured $H_y$ associates with a half-wavelength-distribution. Besides, to compare $H_x$ with $H_y$ (shown in (b)), the former has larger magnetic field and more concentrate on MSL.

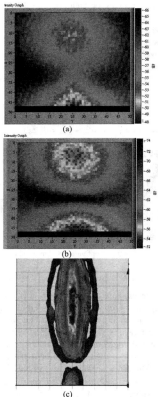

Figure 5 2D magnetic field distribution map of open-end MSL drawn with the proposed circular probe. (a) $H_x$, (b) $H_y$, and (c) simulated surface magnetic field from Ansys HFSS.

Figure 6 presents the 2D near magnetic fields of short-end MSL. In (a), the measured results of $H_x$ show the characteristics of an short-circuited 50-$\Omega$ MSL. Whole field distribution is similar to half-wavelength, while the magnetic field was relative large near the short end ($Z_L=0$). In (b), the measured $H_y$ associates with a half-wavelength-distribution.

Besides, to compare $H_x$ with $H_y$ (shown in (b)), the former has larger magnetic field and more concentrate on MSL.

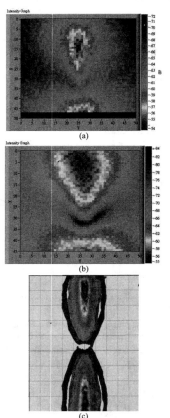

Figure 6 2D magnetic field distribution map of short-end MSL drawn with the proposed circular probe. (a) $H_x$, (b) $H_y$, and (c) simulated surface magnetic field from Ansys HFSS.

## IV. CONCLUSIONS

A measurement system is proposed for scanning the surface magnetic field of PCB circuits within wide frequency band up to 9 GHz. Compared with the conventional one, the novel magnetic probe embeds a set of periodic notches into the connection port. As a result, it not only suppresses the self-resonance but also spurious emissions. In this primary study, the proposed probe was applied to a 2D magnetic field mapping system. It scans the surface magnetic field of an open-end microstrip line and a short-end microstrip line. The agreement between the measurement and simulation validate the good characteristics of the probe and the mapping system. In the future, an advanced mapping system with fast detection will be developed for larger PCB circuits to solve the electromagnetic coupling problems for EMI/EMC etc.

ACKNOWLEDGMENT

This study was partly supported by Taiwan Science Council (NSC 101-2622-E-230 -006 -CC3).

REFERENCES

[1]. S. Aoki, F. Amemiya, A. Kitani, and N. Kuwabara, "Investigation of interferences between wireless LAN signal and disturbances from spread spectrum clocking", *IEEE Int. Symp. on EMC*, vol.2, pp.505-510, 2004.

[2]. J. Shi, M. A. Cracraft, J. Zhang, R. E. DuBroff, K. Slattery, and M. Yamaguchi, "Using near-field scanning to predict radiated fields," *IEEE Int. Symp. on EMC*, vol. 1, pp. 14–18, 2004.

[3]. C. P. Chen, K. Sugawara1, K. Li, H. Nihei, T. Anada, and C. Christopoulos, "Non-contacting Near-field Mapping of Planar Circuits in Microwave Frequency Band", *IEEE Int. Symp. on EMC*, pp.1-6, 2008.

[4]. M. Yamaguchi, S. Koya1, H. Torizuka1, S. Aoyama, and S. Kawahito, "Shielded-Loop-Type On chip Magnetic-Field Probe to Evaluate Radiated Emission From Thin-Film Noise Suppressor", *IEEE Trans. on magnetics*, vol. 43, pp. 2370-2372, 2007.

[5]. H. Funato and T. Suga, "Magnetic Near-field Probe for GHz band and Spatial Resolution Improvement Technique," *Int. Zurich Symp. on EMC*, pp. 284 – 287, 2006.

[6]. Tun Li; Yong Cheh Ho; Pommerenke, D., "Orthogonal loops probe design and characterization for near-field measurement," *IEEE Int. Symp. on EMC*, pp. 1-4, 2008.

[7]. J. M. Kim, W. T. Kim, and J. G. Yook, "Resonance-Suppressed Magnetic Field Probe," *IEEE Trans. on Microwave Theory and Tech.*, vol. 53, pp. 2693-2699, 2005.

[8]. [S. Y. Lin and Y. C. Chen "A Microstrip Filter with Adjustable Rejection Band," *IEEE Int. Symp. on Antennas Propagat.* pp. 594-597, 2005.

[9]. S. Y. Lin, S. K. Yen, W. S. Chen, and P. H. Cheng, "Printed Magnetic Field Probe with Enhanced Performances," 2009 Asia and Pacific Conference.